SPACE SCIENCE SERIES

Tom Gehrels, General Editor

Planets, Stars and Nebulae, Studied with Photopolarimetry
Tom Gehrels, editor, 1974, 1133 pages

Jupiter
Tom Gehrels, editor, 1976, 1254 pages

Planetary Satellites
Joseph A. Burns, editor, 1977, 598 pages

Protostars and Planets
Tom Gehrels, editor, 1978, 756 pages

Asteroids
Tom Gehrels, editor, 1979, 1181 pages

Comets
Laurel L. Wilkening, editor, 1982, 766 pages

Satellites of Jupiter
David Morrison, editor, 1982, 972 pages

Venus
D. M. Hunten, L. Colin, T. M. Donahue and V. I. Moroz, editors, 1983, 1143 pages

Saturn
Tom Gehrels and Mildred S. Matthews, editors, 1984, 968 pages

Planetary Rings
Richard Greenberg and Andre Brahic, editors, 1984, 784 pages

Protostars and Planets II
David C. Black and Mildred S. Matthews, editors, 1985, 1293 pages

Satellites
Joseph A. Burns and Mildred S. Matthews, editors, 1986, 1021 pages

The Galaxy and the Solar System
Roman Smoluchowski, John N. Bahcall and Mildred S. Matthews, editors, 1986,
485 pages

Meteorites and the Early Solar System
John F. Kerridge and Mildred S. Matthews, editors, 1988, 1269 pages

Mercury
Faith Vilas, Clark R. Chapman and Mildred S. Matthews, editors, 1988, 794 pages

Origin and Evolution of Planetary and Satellite Atmospheres
S. K. Atreya, J. B. Pollack and M. S. Matthews, editors, 1989, 881 pages

Asteroids II
Richard P. Binzel, Tom Gehrels and Mildred S. Matthews, editors, 1989, 1258 pages

Uranus
Jay T. Bergstralh, Ellis D. Miner and Mildred S. Matthews, editors, 1991,
1076 pages

The Sun in Time
C. P. Sonett, M. S. Giampapa, and M. S. Matthews, editors, 1991, 996 pages

HAZARDS
DUE TO COMETS
AND ASTEROIDS

HAZARDS
DUE TO COMETS
AND ASTEROIDS

Tom Gehrels

Editor

With the editorial assistance of
M. S. Matthews and A. M. Schumann

With 120 collaborating authors

THE UNIVERSITY OF ARIZONA PRESS
Tucson & London

About the cover:

Painting by William K. Hartmann. An Earth-approaching asteroid passes our planet at close range. Recent discoveries have increased the inventory of small Earth-approachers. Whether this one is on a collision course or will make a near miss is left to the viewer's imagination. The Sun's reflection off the ocean is seen in the middle of the Earth's disk.

About the back cover:

Hubble Image of Multiple comet impacts on Jupiter. Eight comet impact sites are visible in this image taken near the end of the week-long bombardment of comet Shoemaker-Levy 9. The dark features are made up of material "cooked" in the fireball, and then blasted high above Jupiter's cloud tops. Hubble's resolution, with the Planetary Camera, can resolve details in Jupiter's cloud tops as small as 200 miles (320 km) across. The picture is a combination of separate images taken through several color filters to create this "true color" rendition of Jupiter's multicolored clouds. The impact sites are located in Jupiter's southern hemisphere at a latitude of 44 degrees. The image was taken with the Wide Field Planetary Camera-2. Image courtesy of the Hubble Space Telescope Comet Team and NASA.

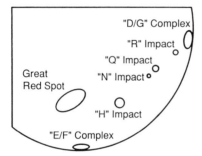

The University of Arizona Press
Copyright © 1994
The Arizona Board of Regents
All rights reserved

⊗ This book is printed on acid-free, archival-quality paper.
Manufactured in the United States of America

99   98   97   96   95   94   6   5   4   3   2   1

Library of Congress Cataloging-in-Publication Data

Hazards due to comets and asteroids / Tom Gehrels, editor ; with the
   editorial assistance of M. S. Matthews and A. M. Schumann ; with 120
   collaborating authors.
      p.   cm. — (Space science series)
   Includes bibliographical references and index.
   ISBN 0-8165-1505-0 (alk. paper)
      1. Comets.   2. Near-earth asteroids.   3. Astrogeology.
   I. Gehrels, Tom, 1925– .   II. Matthews, Mildred Shapley.
   III. Schumann, A. M.   IV. Series.
QB721.H29   1994   94-18759
363.3'49—dc20   CIP

British Cataloguing-in-Publication Data
A catalogue record for this book is available from the British Library.

CONTENTS

COLLABORATING AUTHORS

PREFACE

It is a special pleasure to present this book as the 24th in the Space Science Series. This volume documents the beginning of a new discipline, namely the detection and mitigation of hazardous comets and asteroids. It was a challenge to subject those topics to the Series' procedures for using the results of workshops, combining authors from various backgrounds, and refereeing and editing the material into textbook chapters.

The text presented here is based on at least four workshops: at San Juan Capistrano in April/May 1991, Los Alamos in January 1992, Tucson in January 1993 and Erice in May 1993. There was an even earlier meeting, in June 1981 at Aspen, Colorado. The amount of work that went into this enterprise, is estimated to be at least $1.6 M in terms of salaries and operations of the participants.

A. M. Schumann was coordinating editor of the book along with M. S. Matthews who incidently was awarded the 1993 Masursky Prize for the editing of various Space Science Series books. M. Guerrieri, S. J. Marinus, and C. K. Lobb assisted in various capacities.

The planet Jupiter provided the gravitational attractions for many of the asteroids and comets discussed in this book, including a rare but timely demonstration of impacts, at a safe distance from us.

We were able to successfully compile this volume thanks to the help of a total of 185 authors, advisers, referees and supporters as is detailed here, in the Table of Contents, and in the ACKNOWLEDGMENTS section. The production was supported by the Ballistic Missile Defense Organization, the Los Alamos National Laboratory, The University of Arizona, and The University of Arizona Press.

We thank you all for an interesting venture.

Tom Gehrels

PART I
Small Bodies

EARLY IMPACTS: EARTH EMERGENT FROM ITS COSMIC ENVIRONMENT

EUGENE H. LEVY
University of Arizona

Earth emerged from cosmic antecedents approximately four and a half billion years ago. The formation of Earth—as well as the other planets—was a result of accumulation of solid matter early in the history of our solar system, starting with the gentle accretion of small dust particles, and ultimately involving violent collisions among large objects—even planetary-scale bodies—moving at very high relative speeds. The subject of this volume is the last residual tail of planetary accretion: objects orbiting the Sun which occasionally collide with Earth. Only in the past few years has the reality and the magnitude of this hazard been appreciated. Scientists have an important role to play in helping society to understand and sensibly react—or not react—in the face of whatever dangers the continuing accumulation of our planet presents.

Our Milky Way Galaxy—one among at least a hundred billion similar galaxies in the observable universe—itself contains several hundred billion stars, along with an interstellar medium consisting of large assemblages of gas, dust, cosmic rays, and magnetic fields. Our universe is governed in large part by a struggle between the attractive, confining force of gravity, and the expansive influences of pressure, magnetic fields and rotation. The intricate interplay of these forces is believed to be responsible for virtually all of the large-scale cosmic architecture that we can observe in the sky. Modern science has given us a framework for understanding all terrestrial phenomena—including human existence—as an emergent consequence of natural physical laws playing against this cosmic context.

To human perception, the surrounding cosmos, and even Earth itself, seem nearly timeless, an unchanging stage on which the history of human affairs unfolds. However, astrophysical and geological evidence reveal the universe and Earth to be dynamic and evolving systems. The universe apparently took its present form between ten and twenty billion years ago. The matter of which we are made has come into existence only by virtue of having been cycled through generations of stars, wherein the nuclear processes that power starlight converted the primoridal hydrogen and helium to sequentially heavier and heavier atomic nuclei, ultimately producing the abundance of cosmic matter from which we and our planet formed.

Our own Sun has an expected lifetime—as a main-sequence star—of some ten billion years, after which it will undergo rapid changes until, exhausted of available nuclear fuel, it will die as a cooling white dwarf. More massive stars live starkly shorter, more brilliant, lives, going through entire life cycles in times that can, at the extremes, span only a few million years. Star formation is a continuing phenomenon; even today we can peer into nearby regions where large numbers of stars can be seen recently born or in the process of birth.

A star forms when the previously mentioned battle—between gravity and the forces of expansion—is won locally by gravity. In such cases, a fragment of interstellar gas and dust collapses under its own weight, falling inward to smaller and smaller size, and to higher and higher density. The interstellar gas rotates; it rotates by virtue of the slow rotation of the Galaxy, overall, as well as because of the "turbulence" induced by the colliding motions of the clouds. Conservation of angular momentum in a collapsing interstellar cloud induces the cloud to spin faster as it shrinks in size. At the same time, interstellar magnetic forces remove angular momentum, slowing the rotation. Eventually as the collapse accelerates, the magnetic forces can no longer keep up with the increasing rotation, and the resulting protostar is born with a high rate of rotation. Because of this fast rotation, centrifugal force prevents all of the matter from collapsing to a simple, compact sphere. Instead, a substantial fraction of the matter ends up in a disk spinning in the plane of the protostar's equator.

Approximately four and a half billion years ago, our own solar system formed, apparently in this very way. The planets of our solar system formed in the thin, extended disk of matter that orbited the Sun, approximately in the Sun's equatorial plane. The planets in our system seem to have formed as a result of the accumulation of the solid fraction of cosmic matter—that 2% of the material which could condense to solid form in the tenuous disk of matter orbiting our nascent Sun. This accumulation-of-solids-dominated planet formation proceeded as tiny dust particles—ice and rock—accumulated to larger and larger sizes. The particles stuck together through the combined actions of molecular surface forces and gravitational attraction, accumulating to larger and larger sizes, and ultimately forming the planets.

Thus, it seems that we owe our very existence to early collisional violence in our solar system. The direct evidence of this violence can easily be seen on the surfaces of those planets and moons which, lacking the obliterating effects of weather and geologic activity, still show us the characteristic pock-marked aspect left from the early "bombardment" in our solar system.

As might be expected, this planetary accumulation was not one hundred percent effective. Myriad small bodies—ranging from tiny pieces, to objects hundreds of kilometers across—remain in orbit about the Sun. These objects escaped both being incorporated into planets and being lost from the system. Asteroids and comets are best known among the populations of remaining small objects. Both asteroids and comets began to accumulate in the same manner as that material which ultimately ended up in the planets. However, in both cases, the gravitational influences of the already-forming planets intervened.

Asteroid orbits are largely confined to a torus in the space between the orbits of Mars and Jupiter. It is believed that the rapid formation of Jupiter interfered with the accumulation of asteroid-zone material to planetary dimensions. The strong Jovian gravity stirred the orbital motions of asteroidal bodies to the point that the relative collisional velocities became disruptive, rather than accretional. This halted the growth of a planet between Mars and Jupiter, and left the solar system with a large population of objects ranging in size from dust grains and pebbles to a few objects many hundreds of kilometers across. Because of continuing collisions and gravitational stirring, asteroidal orbits continue to evolve. Some asteroids, or asteroid-derived objects are ultimately put into orbits that cross the orbit of Earth, and many of these end up colliding with our planet.

Asteroids consist mostly of approximately that one-half percent of cosmic matter that condenses at moderate temperatures to form rock. Many asteroids are likely to be more or less homogeneous aggregations of such matter. However, it is clear that some large asteroids were heated to sufficiently high temperatures that they differentiated, much like the terrestrial planets. Dense, iron-nickel alloys sank to the interior, leaving behind mantles of igneous rock. Subsequent collisions disrupted some of these differentiated bodies, creating solid iron fragments from the deep interiors, as well as rock fragments from the surrounding iron-poor mantles. As a result, the asteroid-derived objects hitting Earth have compositions that can range from undifferentiated rock to nearly pure iron.

Comets formed in the more distant regions of the solar system, where not only rock, but also ice, persisted in solid form. Comets apparently consist of mixtures of ice, rock, and dust. Comet orbits were strongly disturbed by the gravity of the giant planets, as well as by galactic gravity, during the early years of the solar system. As a result some comets were ejected from the solar system, while others were scattered into the realm of the inner planets, ultimately colliding with the planets. Many comet orbits evolved to large distances from the Sun, perhaps to as far as 50,000 astronomical units. Today, as many as 10^{11} to 10^{12} comets remain, orbiting the solar system in the so-called Oort Cloud and Kuiper Belt. A variety of gravitational disturbances—including those from passing stars, interstellar clouds, and Galactic tides—perturb the motions of the distantly orbiting comets. From time to time, a comet is sent toward the center. Some penetrate into the realm of the planets; and a fraction of those become trapped into relatively short-period solar orbits.

In addition to these main asteroid and comet populations, it has become apparent that other comet-like and asteroid-like objects exist in our solar system. Fresh discoveries suggest the possible existence of a previously unknown population of near-Earth asteroids. Recently, several large, icy bodies—of which Chiron was the first—were discovered in the outer solar system. It is certain that many substantial objects remain to be discovered; and there may well be whole classes of objects yet to be identified.

These remnant objects, left over from the time of planet formation, are not fundamentally different from the original building blocks of the planets. In a sense, the accumulation of the planets is a process that continues today, as this remaining debris continues to fall on Earth and the other planets. These objects, and their collisions with Earth, are the subject of this book.

While the early bombardment in our solar system is obvious from even cursory examination of old planetary surfaces, less obvious is the extent to which planetary bombardment at all scales—from the trivial to the startling—remains a contemporary phenomenon in our system and on our planet. It is a commonplace knowledge that small objects—ranging in mass from milligrams to many tons—hit our planet frequently, usually burning up in the atmosphere, but occasionally with fragments reaching Earth's surface as meteorites. However, closer scrutiny reveals that significant impacts have occurred with attention-catching frequency. Indeed, in the United States, in northern Arizona, one of the most easily seen of the large terrestrial impact craters remains from an event that occurred just a few tens of thousands of years ago. The impact that has caught most people's imagination is the event that putatively caused the Cretaceous-Tertiary geological boundary, evidently involving global-scale disruption of the environment, and associated with massive biological extinctions, including the extinction of the dinosaurs.

But these are ancient history. We are part of a "now" generation, more interested in journalism than in the history of cosmogenesis. For that reason, it is worth noting that at least one significant impact has occurred in journalistic times; the impact at Tunguska in Siberia, near the beginning of this century. It is only an accident that this event is not more vividly etched on the public consciousness—an accident of geometry and time that placed the Tunguska object at an unpopulated location rather than at, say, Omaha, Nebraska. It is conceivable that other impact events have occurred within or near historical times. I have occasionally mused that one or two such events may be recorded vaguely in our cultural memory, perhaps meandering and evolving through generations of myth and legend possibly ending up as religious stories of miraculous events. We have no evidence one way or the other; but it is at least conceivable.

This book marks a significant step in the attempt to come to grips with such hazards. It is clear that the hazard is real. It is also clear that the hazard involves infrequent events that will occur with only a very small probability in any given interval of time. What takes the discussion of the impact hazard out of the realm of hand wringing and anxiety is the fact that we can at least contemplate the possibility of proactively instituting protective measures. That some such protective steps are technically feasible—at least at some level—is probably the easy part of the debate. More difficult is the recognition that many of the possible protection schemes incur their own large costs—both in financial and social measures; that many protection schemes entail their own risks; and that, in some cases, these risks, as well as the associated anxieties, may be large—as large, or larger, than the provoking hazards.

This is a difficult problem, having both technical dimensions and socio-political dimensions, and potentially interacting with other international concerns in complex, difficult to control ways. This problem has some similarities to the superpower nuclear arms race, from which our world may finally and tentatively be backing away. But here I want to emphasize the differences, especially insofar as those shape the essential role of the scientific community. In the arms race, the motivating dynamic was a political one. A dynamic in which scientists and engineers provided the technical tools, but, as a group, brought no special and unique wisdom to the table in making judgments about what to do. In the present case, the dynamic is different. The adversary is not another nation; the calculus is not one of political fears, anxieties, and motivations, for which we scientists have no special expertise. Rather the "adversary" is the physical world. In assessing *this* adversary, we scientists have special and unique expertise.

Because of this unique and essential expertise, I believe that we scientists have a crucial role to play, a critical responsibility to fulfill. It is very important that the informed community of scientists participates in defining the realistic extent of the hazard and in defining the sensible steps, if any, to take in response. It is also important for the scientific community to help weigh this hazard against the risks and dangers associated with conceivable protective measures. The scientific community can—I believe uniquely—gather the knowledge and develop the judgment necessary for informed decision-making and wisdom in this arena.

IMPACT DELIVERY OF VOLATILES AND ORGANIC MOLECULES TO EARTH

C. F. CHYBA
*National Security Council, White House Fellow**

T. C. OWEN
University of Hawaii at Manoa

and

W.-H. IP
Max-Planck-Institut für Aeronomie

Both models of solar system formation and the lunar cratering record indicate early Earth was subject to an intense bombardment of comets and asteroids that tapered off to its present comparatively low level by 3.5 Gyr ago. Geochemical estimates of the meteoritic component in the lunar crust, in the terrestrial mantle, and possibly in Mars-originating meteorites serve as consistency checks on these models, as do noble gas and isotopic abundances. It appears that Earth collected some $\sim 10^{22}$ kg of exogenous material subsequent to core formation ~ 4.4 Gyr ago. If carbonaceous asteroids or comets comprised a substantial fraction of this impacting mass, Earth could have collected most of its inventory of water and volatile elements as a late veneer from this source. Terrestrial impact erosion may have preferentially removed those volatiles that remained in the vapor phase. Dating of terrestrial fossils reveals that life had evolved on Earth by the end of the heavy bombardment. Delivery of exogenous organic molecules may have played an important role in the stocking of the terrestrial prebiotic organic inventory.

Microscopic fossils and fossil stromatolites (macroscopic structures formed by sediment-trapping microbes) both require life to have originated on Earth prior to 3.5 Gyr ago (Schopf and Walter 1983; Walter 1983; Schopf 1993). Controversial evidence for biologically mediated carbon isotope fractionation suggests that sophisticated (photosynthesizing) life may already have existed by 3.8 Gyr ago (Schidlowski 1988). The terrestrial origin of life must therefore have coincided with the last stages of the heavy bombardment of the inner solar system, during which those planetesimals "left over" from planetary accretion were largely swept up or scattered.

---

* The ideas contained in this chapter do not necessarily represent the views of the National Security Council or the United States Government.

Evidence for this bombardment, which extended for nearly the first billion years of Earth's history, comes from both theoretical studies of planetary formation (Wetherill 1977; Fernández and Ip 1983), and directly from the lunar cratering record (Basaltic Volcanism Study Project; hereafter BVSP 1981). For about its first 800 Myr, Earth's surface would have been subjected to frequent and devastating large impacts, possibly delaying the origin of life (Maher and Stevenson 1988; Oberbeck and Fogleman 1989a) or sterilizing the surface of the planet (Sleep et al. 1989). Apart from these most extreme cases, but depending on poorly known details of the impactor flux, early Earth could also have lost a contemporary atmosphere's worth of volatiles through atmospheric impact erosion (Melosh and Vickery 1989; Chyba 1991a).

Yet if the heavy bombardment devastated early Earth, it may also have had effects beneficial to the origin of life. "Impacts giveth and impacts taketh away," to take McKinnon's (1989) observation about comets only slightly out of context. There seems to be no question that the heavy bombardment had this Janus-like character. (Janus was the Roman solar deity who was the doorkeeper of heaven and patron of beginnings and ends [Bullfinch 1979]. As such, he had two faces, one for the rising, and one for the setting Sun.) At the same time that big impactors were devastating Earth's surface, some of them would have been delivering volatile elements and—although much less effectively—organic molecules useful for the origin of life.

It appears that the competition between impact delivery of new volatiles and impact erosion of those already present strongly favored the net accumulation of planetary oceans by the larger worlds of the inner solar system (Chyba 1990). In fact, this scenario is consistent with Earth receiving the bulk of its surface volatiles through impact delivery. Thus the heavy bombardment may ultimately have helped set the stage for the terrestrial origin of life, by delivering key biogenic elements (such as hydrogen, carbon, and nitrogen) to Earth's surface.

It is also possible that intact organic molecules, of direct relevance to the origins of life, were delivered to Earth in the heavy bombardment. The notion that exogenous organics may have played a role in the origin of life has its roots in the 1834 discovery by the Swedish chemist Baron Jöns Jacob Berzelius of organic molecules in the Alais meteorite (Berzelius 1834). The idea was made explicit in 1908 by Chamberlin and Chamberlin (1908), who proposed that impacts of organic-rich planetesimals may have been important to life's origins. In 1961, Oró suggested, on the basis of spectroscopic observations of carbon and nitrogen-containing radicals in cometary comae, that comets may have played a similar role (Oró 1961). This suggestion attracted renewed interest with the in situ discovery of organic-rich grains in the coma of comet Halley (Kissel and Krueger 1987a,b; Huebner and Boice 1992). The difficulty has always been in envisioning ways to deliver these organics intact to the surface of the Earth in significant abundance (Chyba 1991b).

Over the last two decades, a potential cometary source of prebiotic organics also attracted attention due to the emergence of a tentative consensus that the early terrestrial atmosphere was of an intermediate oxidation state, rich in carbon dioxide (CO_2) and molecular nitrogen (N_2), rather than a reducing one rich in methane (CH_4) and ammonia (NH_3)—though this debate is far from settled (see, e.g., Kasting et al. 1993). Such early CO_2-rich atmospheres are implied by "hot" accretion scenarios for Earth, in which core formation takes place quickly, leaving the upper mantle (and hence, outgassed carbon) in an oxidized state (Walker 1977,1986; Chang et al. 1983; Stevenson 1983; Harper and Jacobsen 1992). The short photodissociation lifetimes of CH_4 and NH_3 in model paleoatmospheres reinforce this conclusion (Kuhn and Atreya 1979; Levine and Augustsson 1985), though it is possible these lifetimes may have been extended to geologic time scales through the production of a high-altitude organic haze (Sagan and Chyba 1994). Syntheses of key prebiotic molecules would have been much more difficult in CO_2-rich atmospheres than in reducing ones (Chang et al. 1983; Schlesinger and Miller 1983a, b). An exogenous source of prebiotic organics on early Earth could provide an alternative way of stocking the prebiotic terrestrial organic inventory needed for the origin of life.

This chapter will assess the impact delivery of volatiles and intact organic molecules to the early Earth—the face of Janus for the rising Sun. It will not focus on Janus' other face, impact devastation. For completeness, however, we begin with a brief review of progress in the latter area.

I. IMPACT DEVASTATION ON EARLY EARTH

The largest lunar impact feature whose identification is secure is the South Pole-Aitken basin on the lunar farside, with a diameter $\sim$2200 km (Belton et al. 1992). These largest nearside basin, Mare Imbrium, has a diameter of 1160 km. The basins were excavated by impacting asteroids or comets with masses in the range $\sim10^{18}$ to 10^{19}kg ($\sim$50–150 km in radius). Because Earth's gravitational cross section at typical Earth-crossing asteroid velocities is some 24 times that of the Moon, Earth must have collected many more of the largest objects than did the Moon. Statistically, Earth should have been struck by about 17 objects larger than the biggest lunar impactor (Sleep et al. 1989, Chyba 1991a). We are therefore in the frustrating position of knowing that early Earth suffered many such enormously destructive events, but being required to rely on small-number statistical extrapolations from the Moon to estimate their frequency. Yet in the absence of a comprehensive geologic record of ancient Earth (Veizer 1983), we have no choice; knowledge of the early Earth, at least in this context, requires extrapolation from our knowledge of the Moon and nearby planets.

What does the lunar cratering record tell us of the environment of early Earth? Extrapolations of the kind just described imply that Earth was struck by $>4 \times 10^4$ objects as large or larger than the comet Halley nucleus (equivalent in volume to a sphere about 5 km in radius). Objects of this size are in the approximate size range necessary to cause impact erosion of the terrestrial atmosphere (Melosh and Vickery 1989), provided they have sufficiently high velocities (around 25 km s^{-1}). Nearly all asteroidal-type objects would have struck Earth at velocities below this, but something like half of short-period comet collisions would have been sufficiently fast to erode the atmosphere (Chyba 1990). The Earth could easily have lost a contemporary atmosphere's worth of atmospheric gases in this way, but this result is strongly dependent on still poorly constrained parameters, especially the exact size and velocity dependence of early impactors (Chyba 1991a).

More dramatically, it appears likely that prior to ~3.8 Gyr ago, Earth was struck several times by objects with impact energies sufficient to vaporize the entire terrestrial ocean. As shown by Sleep et al. (1989), such gargantuan impacts result in the formation of a 100-bar globe-encircling rock vapor atmosphere, which persists for several months before cooling sufficiently to condense out. During this time, the entire ocean is evaporated, leading to a runaway greenhouse effect that holds Earth's surface temperatures to ~2000 K for thousands of years. Such impacts would effectively heat-sterilize the Earth to a depth of some hundreds of meters. Whether this is equivalent to sterilizing the Earth depends on whether life had originated and had time to evolve into protected niches at depths sufficient to be insensitive to the surface heating (and otherwise not fatally influenced by altered surface conditions). It is of interest in this regard that there are extant terrestrial bacteria known to flourish at depths of several kilometers (Boston et al. 1992).

Were the origin of life still ongoing at the time of a surface-sterilizing giant impact, it seems likely that the result would have been to "reset the clock" for life's origins; there may therefore have been several "impact frustrations" of the origins of life (Maher and Stevenson 1988; Oberbeck and Fogleman 1989a). If this scenario is correct, terrestrial life today might be quite different if that final impactor had missed; "we" might then be the descendants of an earlier, perhaps very different, origin of life. These recent quantitative results support early post-Apollo suggestions that the origin of life on Earth would have had at most several hundred Myr between the time of the end of impact devastation and the oldest known fossil record (Sagan 1974).

In addition to the several most devastating ocean-evaporating impacts, Earth would sustain around five times more impacts sufficiently energetic to evaporate the ocean's 200-m-thick photic zone. In this kind of lesser catastrophe, adverse terrestrial surface conditions would persist for only several hundred years after impact (Sleep et al. 1989). The consequences of such an impact for obligate photosynthetic organisms would be grim. Statistically, it is likely that a photic-zone-destroying impact would occur subsequent to the final ocean-evaporating impact, so that the last global catastrophe life on early Earth would have surmounted was a destruction of the photic zone. This might seem to favor an origin of life in the deep oceans, possibly at hydrothermal vents (Corliss et al. 1981; Baross and Hoffman 1985; Waldrop 1990; Hennet et al. 1992). However, it has been argued that the highest-temperature vents would preferentially destroy, not synthesize, relevant prebiotic molecules

(Miller and Bada 1988,1989; Bada 1991), and there may be other difficulties with the deep-sea vent hypothesis as well (Towe 1991).

It has been suggested on the basis of molecular phylogeny studies (see, e.g., Woese 1987) that archaebacteria are the most "primitive" of extant organisms, and that the ancestral archaebacterium was a thermophile growing near the temperature of boiling water. This has also sometimes been taken as strong evidence for an origin of life at hydrothermal vents (see Waldrop 1991). Rather, it is consistent both with such an origin, or with a kind of post-origin "bottleneck" during which only those organisms that had evolved into deep sea niches survived. The hostile, impact-ridden environment of Earth's surface during the heavy bombardment is the obvious candidate for the creation of such a bottleneck.

II. THE TEMPERATURE-VOLATILES CONUNDRUM

From the point of view of the origin of terrestrial life, the distribution of volatiles in our solar system presents an apparent dilemma. The temperatures needed for liquid water, the *sine qua non* of life as we know it (Horowitz 1986), exist only close to the Sun in the inner solar system. Yet the biogenic elements are extremely depleted in the inner solar system, but comparatively abundant in the outer solar system—where, barring clever tidal or radiogenic heating schemes, temperatures are too low for liquid water to exist. McKay (1991) has depicted this dilemma graphically for the case of the element carbon. Figure 1 shows his compilation of the ratio of carbon atoms to total heavy atoms, normalized to solar abundance, for the objects of the solar system. Only beyond the middle of the asteroid belt does the carbon abundance approach solar, or cosmic, abundance.

The dilemma only appears to worsen when models of solar system formation are considered. Delsemme (1992) has reviewed progress in thermochemical equilibrium and chemical kinetics models of the solar accretion disk out of which the planets formed. Such models have consistently found it difficult to provide Earth with its known inventories of water, nitrogen, and carbon (see, e.g., Lewis 1974; Lewis et al. 1979; Cameron 1983; Prinn and Fegley 1989), due to temperatures around 1000 K in the region of terrestrial accretion. Delsemme (1992) has found similar temperatures by using the transition between volatile-poor S-type asteroids and comparatively volatile-rich C-type asteroids to peg the accretion disk's temperature to 450 ± 50 K at 2.6 AU, then extrapolating these temperatures inward to the region of the terrestrial planets. At these temperatures, over the range of likely pressures in the accretion disk, water, carbon, nitrogen, and other volatiles remain almost entirely in the gas phase, so are not present in the dust grains out of which the planet-forming planetesimals aggregate. These considerations suggest that Earth formed almost entirely devoid of the biogenic elements.

On the other hand, Wetherill (1990) has shown that the latter stages of the formation of the terrestrial planets were marked by collisions of large

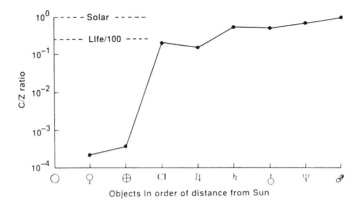

Figure 1.   Ratio of carbon (C) atoms to heavy (Z) atoms (all atoms heavier than He) throughout the solar system (figure from McKay 1991).

planetary embryos scattered across considerable heliocentric distances. Thus Earth during its accretion may have acquired some of its volatile inventory through this kind of radial mixing. Additionally, the latter stages of outer planet formation should have led to the scattering of extremely volatile-rich planetesimals ("comets") throughout the solar system (Fernández and Ip 1983; Shoemaker and Wolfe 1984; Delsemme 1992). Hartmann (1987,1990) has argued that spectral observations of solar system satellites, as well as the preponderance of CM clasts among the foreign fragments in polymict meteoritic breccias, provide evidence for an intense scattering of C-type asteroids during the first 10^8yr of solar system history. However, it is not at all clear that volatiles accreted by the Earth prior to 4.4 Gyr ago would have been retained. The terrestrial atmosphere present prior to the hypothesized Moon-forming impact might have been stripped by that event (Cameron 1986). Moreover, terrestrial water present before core formation at ~4.4 Gyr ago (~10^8yr after Earth's formation [Stevenson 1983,1990; Sun 1984; Swindle et al. 1986; Tilton 1988]) should have been efficiently destroyed by reacting with metallic iron according to $Fe + H_2O = FeO + H_2$; large quantities of H_2 produced in this way may have removed other degassed volatiles by hydrodynamic escape (Dreibus and Wänke 1987,1989). Some volatiles may also have been sequestered in the terrestrial core. It is possible that only those volatiles arriving at Earth subsequent to ~4.4 Gyr ago would ultimately have contributed to the extant terrestrial surface volatile inventory. We now examine different estimates of the magnitude of this putative late episode of volatile accretion.

III. DYNAMICAL MODELS OF OUTER PLANET FORMATION AND THE INFLUX OF COMETS INTO THE INNER SOLAR SYSTEM

The issue of cometary and asteroidal impacts with terrestrial planets is closely tied to the early history of outer planet formation. In the case of asteroids, the delivery of these objects from the main asteroid belt, located approximately between 2 and 4 AU, depends strongly on the resonant perturbation effect of Jupiter. As shown by Wisdom (1983) via numerical integrations, asteroids with orbital motion in near-commensurabilities with Jupiter could be perturbed into chaotic orbits with high eccentricities. As a result, these bodies would sooner or later become Earth- and/or Mars-crossing objects, e.g., the Apollo objects. About a hundred Earth-crossing asteroids have been observed so far; extrapolation to the full population suggests there are about 800 Apollo and Aten asteroids brighter than magnitude 17.7. Taking into account the comet population as well, Shoemaker et al. (1990) find that the production rate for terrestrial craters larger than 20 km in diameter is $(4.9\pm2.9)\times10^{-15}\mathrm{km}^{-2}\mathrm{yr}^{-1}$, which is in fine agreement with the terrestrial crater production rate derived from the geologic impact record over the past 120 Myr, $(5.4 \pm 1.7) \times 10^{-15}\mathrm{km}^{-2}\mathrm{yr}^{-1}$ (Grieve 1984; see the Chapter by Grieve and Shoemaker).

Estimates of the cometary contribution to the total contemporary impact flux vary considerably (Chyba 1987), but the situation is still more complicated for the early phase of collisional bombardment. For example, the population of the asteroid belt 4 Gyr ago might have been much greater than the present one, so the terrestrial asteroidal impact flux might have been enhanced by a large factor compared with today. Furthermore, it is unclear how the formation of proto-Jupiter might have influenced the growth of the asteroids. According to Safronov (1972), the accretion of small condensed icy bodies by Jupiter would have led to the continuous ejection of these objects to larger orbital distances; the loss of angular momentum should have then forced proto-Jupiter to move inward. Following this line of thinking, Torbett and Smoluchowski (1980) have made the interesting suggestion that the inward shift of Jupiter during its formative stage might also have led to the sweeping of the orbital resonant positions across the asteroid belt, dispersing large numbers of planetesimals. This scenario is far from certain, however, because outer planet formation appears to be closely coupled, and outer planet accretion and gravitational scattering of planetesimals must be treated in a mutually consistent manner.

A first attempt at this project was made by Fernández and Ip (1983,1984, 1987); Ip and Fernández (1988) by means of Monte Carlo simulation of the accretional process. These authors' starting point was to assume that both Jupiter and Saturn had already been formed in their present orbital positions, while proto-Uranus and proto-Neptune, with only a fraction of their current masses, were still in the early stage of accretion. In these studies, as a protoplanet grew to a certain critical mass, it would start to become a strong scatterer of its swarm of icy planetesimals, perturbing them into orbits of high eccentricity, or even escape from the solar system. (Volatile-rich icy planetesimals are here taken to be synonymous with "comets.") Because of the rapid dispersion of the planetesimal swarm in the planetary accretion zone, the gravitational capture process would be slowed down and eventually virtually terminated. Concomitantly with the gravitational accretion and scattering of the icy planetesimals by proto-Uranus and proto-Neptune, a large quantity of small icy bodies will be implanted in the outer skirt of the solar system (Kuiper 1951; Safronov 1972; Ip 1977; Fernández 1978,1985).

The results of the numerical simulations by Fernández and Ip (1983,1984, 1987) imply that Neptune was the principal contributor of icy planetesimals to the cometary reservoir, with a gravitational scattering time scale of hundreds of millions of years. Simultaneously, an appreciable amount of scattered mass, up to 10 to 100 Earth masses, could have reached the region of the terrestrial planets, after having been passed through Jupiter's orbit (where most of the objects would have been scattered outward or captured). Therefore, in association with the accretion of Uranus and Neptune, and the creation of the Oort cloud, the terrestrial hydrosphere could have been established as a result of this early bombardment of volatile-rich planetesimals from the outer solar system. However, it must be remarked that some published dynamical estimates of icy planetesimal mass scattered into the Earth (see, e.g., Delsemme 1992) differ by as much as an order of magnitude from these results.

What of the icy planetesimals scattered from Jupiter and Saturn? Delsemme (1992) points out that Earth would have accreted much more mass from these planets than from the Uranus and Neptune region. However, the time scales for scattering from the Jupiter and Saturn regions would have been much shorter, perhaps on the order of tens of Myr rather than hundreds of Myr (Ip 1977; Fernández 1985). The contribution to the terrestrial volatile inventory of this much greater, but much more short-lived reservoir, is difficult to assess for the reasons sketched at the end of Sec. II. What fraction of the Jovian and Saturnian planetesimals collided with Earth before the Moon-forming event? Before core formation? The answers to these questions remain unclear, turning on the relationship between several very uncertain time scales.

IV. IMPLICATIONS OF THE LUNAR CRATERING RECORD

Independently of planetary formation models, we may ask what the observed cratering records of the Moon and terrestrial planets imply about terrestrial accretion of volatiles and organics subsequent to 4.4 Gyr ago. Such an approach purports to minimize the model-dependence of the conclusions, by basing the calculations as much as possible on the available data (Chyba 1991a). In this and the following sections we pursue this approach, checking the results against important geochemical constraints. Extrapolations from the lunar cratering record appear to provide a natural explanation for the data available on siderophile abundances on the Earth, Moon, and possibly Mars. However, many uncertainties remain with this way of approaching the problem. We try to evaluate these throughout this discussion. The sources of error in these sorts of calculations have recently been summarized by Chyba (1993b).

A. Procrustean Fits to the Lunar Cratering Record

Attempts to estimate the impact environment of early Earth often begin with analytical fits to the lunar cratering record (Maher and Stevenson 1988; Oberbeck and Fogleman 1989a; Melosh and Vickery 1989; Chyba 1990). In practice, however, this procedure faces numerous difficulties—though these have often been disregarded in the literature. The oldest lunar province for which a radiometric date actually exists (the Apollo 16 and 17 uplands) is only 3.85 to 4.25 Gyr old; the ages of more heavily cratered provinces can at present only be estimated (BVSP 1981). Moreover, the entire interpretation of the heavy bombardment as representing exponentially decaying remnants of planetary formation is questioned by those favoring a lunar cataclysm (see, e.g., Ryder 1990; Dalrymple and Ryder 1993). Different choices for the decay rates fitted to lunar cratering data can lead to substantially different conclusions about terrestrial mass influx during the heavy bombardment (Chyba 1990). However, the more extreme of these choices can be excluded, as they are in contradiction with lunar and terrestrial geochemical data on meteoritic input (Chyba 1991a). Here we develop a model lunar bombardment history that is in good agreement with the geochemical constraints. Figure 2 shows cumulative lunar crater density as a function of surface age, from the Basaltic Volcanism Study Project (BVSP 1981) compilation.

Figure 2 also shows an analytical fit to these data. The data, for craters bigger than a diameter D, are well modeled by the equation

$$N(t, D) = \alpha[t + \beta(e^{t/\tau} - 1)](D/4 \text{ km})^{-1.8} \text{ km}^{-2} \qquad (1)$$

where $N(t, D)$ is the cumulative surface density of craters with diameters greater than D on a surface of age t (in Gyr), and α and β may be determined by two-dimensional χ^2 minimization (Melosh and Vickery 1989; Chyba 1990, 1991a). Equation (1) is just the mathematical statement of the observation that cratering has been roughly constant for the past 3.5 Gyr, while increasing exponentially into the past prior to that time. Because Fig. 2 is a cumulative crater plot, its data are fit by the integral (over time t) of this constant plus exponential, giving a sum of linear and exponential terms. The logarithmic ordinate leads to the term linear in t extrapolating to $-\infty$ as t approaches zero.

Choosing a heavy bombardment decay half-life of 100 Myr, or $t = 144$ Myr, is attractive because this is about the half-life found for the decrease of the primordial comet flux through the inner solar system in simulations of the formation of Uranus and Neptune (Fernández and Ip 1983; Shoemaker and Wolfe 1984). Such a choice might therefore at least correctly represent the cometary fraction of the heavy bombardment flux. Moreover, 100 Myr lies in the middle range of the values of t with which the lunar cratering record is reasonably consistent (Chyba 1990). In addition, an independent estimate of t by Oberbeck and Fogleman (1989b), which correlates absolute age estimates for lunar impact basins from a crustal viscosity model (Baldwin 1987a) with

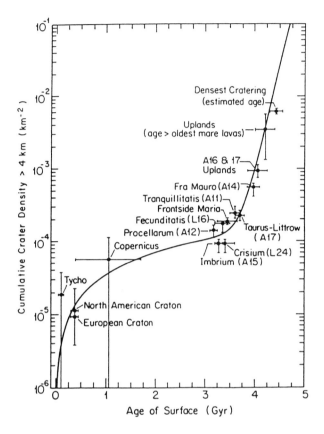

Figure 2. Cumulative lunar crater density as a function of surface age, with analytical fit (Eq. 1), using a 100 Myr half-life decay constant (figure from Chyba 1991a).

crater counts for these basins (Baldwin 1987b) yields $t = 150$ Myr. The choice $t = 144$ Myr in Eq. (1) gives $\alpha = 3.5 \times 10^{-5}$ and $b = 2.3 \times 10^{-11}$.

Fundamentally, choosing any single decay "constant" for the cratering flux is a procrustean exercise, as the impactor flux cannot actually have decayed at a constant rate, but must instead have made a transition from rapidly swept-up objects in Earth-like orbits, to objects from comparatively long-lived orbits (Wetherill 1977; Hartmann 1980; Grinspoon 1988). The fit described here therefore represents the impactor flux subsequent to the time after which those objects with short ($\sim$10 Myr) sweep-up time scales have yielded to more slowly ($\sim$100 Myr) decaying populations.

The initially excavated (or transient) crater diameter D_{tr} (in meters; MKS units are assumed throughout this discussion) is related to the mass m of an incident impactor by (Schmidt and Housen 1987)

$$m = \gamma v^{-1.67} D_{tr}^{3.80} \qquad (2)$$

where v is impactor velocity, and γ is a constant that depends on surface gravity, impactor and target densities, and impactor incidence angle θ (taken here to be the most probable impact angle, 45°):

$$\gamma = 0.31 g^{0.84} \rho^{-0.26} \rho_t^{1.26} (\sin 45° / \sin \theta)^{1.67}. \tag{3}$$

Here $\rho_t = 3000$ kg m^{-3} is the density of the lunar crust (Ryder and Wood 1977), $\rho = 2200$ kg m^{-3} is taken to be the density of a typical impacting asteroid, and $g = 1.67$ m s^{-2} is the gravitational acceleration at the lunar surface (Melosh 1989). These values give $\gamma = 1.6 \times 10^3$ kg s$^{-1.67}$.

The final crater diameter D_f is enlarged due to wall collapse over the initial (transient) crater diameter D_{tr}. Croft (1985) has suggested the relation

$$D_f \approx D_Q^{-0.18} D_{tr}^{1.18} \tag{4}$$

where D_Q is the diameter of the simple-to-complex crater transition. For the Moon, Croft takes $D_Q = 15$ km. Equation (4) is not applicable to diameters smaller than 15 km; formally it gives the impossible result $D_f < D_{tr}$ for such craters.

McKinnon et al. (1990), following McKinnon and Schenk (1985), argue that simple craters on the Moon near the simple-to-complex transition (which they define, on the basis of depth-diameter data, to be $D_C \approx 11$ km) are in fact ~ 15 to 20% wider than their original transient craters. McKinnon et al. (1990) put

$$D_f = k D_{tr}^{1.13} \tag{5}$$

with

$$k = \kappa D_c^{-0.13}. \tag{6}$$

Taking D_f to be 17.5% (the mean of 15% and 20%) larger than D_{tr} when $D_f = D_c$, combined with Eqs. (5) and (6), allows κ to be determined; one finds $\kappa = 1.2$, so that

$$D_f = 1.2 D_c^{-0.13} D_{tr}^{1.13} \tag{7}$$

with $D_c = 11$ km.

Equation (7) may be combined with Eq. (2) to relate impactor mass and velocity to final crater diameter D_f:

$$m = 0.54 \gamma v^{-1.67} D_c^{0.44} D_f^{3.36}. \tag{8}$$

D_f is, of course, what one observes on the Moon. Equation (1), with D interpreted as D_f, may then be combined with Eq. (8) to give the number of objects with mass $> m$ that have impacted a unit of area on the Moon as a function of time t:

$$n(> m, t) = \alpha[t + \beta(e^{t/\tau} - 1)][m/m(4 \text{ km})]^{-b} \text{ km}^{-2} \tag{9}$$

where $b = (1.8/3.36) = 0.54$, and $m(4$ km$)$ is given by Eq. (8) with $D_f = 4000$ m.

The total mass, $M(t)$, incident in objects with masses in the range m_{min} to m_{max} on a lunar surface of age t is:

$$M(t) = \int_{m_{max}}^{m_{min}} m[\partial n(> m, t)/\partial m]\mathrm{d}m \tag{10}$$

or

$$M(t) = \alpha[t + \beta(e^{t/\tau} - 1)][b/(1 - b)][m(4\text{ km})]^b m_{max}^{1-b}\text{km}^{-2} \tag{11}$$

where we have taken $m_{max} >> m_{min}$. Note that

$$M(t) \propto m_{max}^{0.46}[m(4\text{ km})]^{0.54} \propto v^{-1.67} D_{max}^{1.55} \tag{12}$$

so that $M(t)$ depends much more weakly upon D_{max}, the largest crater (basin) diameter used in the calculation, than one might have anticipated from the Schmidt-Housen scaling, Eq. (2). As discussed below, uncertainties in D_{max} and v results in final uncertainties in $M(t)$ of a factor of a few.

$M(t)$ from Eq. (11) may be gravitationally scaled to determine the total mass incident on Earth subsequent to a time t. Several parameters must first be determined; however, the mass m_{max} of the largest impactor (or, equivalently, D_{max}), and a "typical" impactor velocity v.

B. Comet and Asteroid Impact Velocities with the Earth and Moon

A common approach (BVSP 1981; Melosh and Vickery 1989) for determining "typical" impact velocities has been to use the rms impact velocity v_{rms} for the known Earth-crossing asteroids. This can at best provide a "snapshot" of asteroid-Earth collisions, as evolution of orbits due to planetary perturbations alter perihelia, which strongly influence collision probabilities and velocities. Note, however, that the calculations cited below do average over secular precession of asteroid perihelia, which for Earth-crossing asteroids occurs on time scales of 10^4 to 10^5yr (Shoemaker et al. 1979), short compared to typical asteroid lifetimes in the inner solar system. For the Earth-crossing asteroids these lifetimes are 10^7 to 10^8yr (Steel and Baggaley 1985). The implicit assumption of this approach, then, is that the current Earth-crossing asteroid swarm provides a "typical" distribution of Earth-crossing impactor velocities.

Using rms velocities in impact calculations skews "typical" velocities towards misleadingly high values, due to small numbers of high impact velocity objects (Chyba 1991a). Early work (Melosh and Vickery 1989; Sleep et al. 1989; Chyba 1990) used velocities calculated from the 20 Earth-crossing asteroids known in 1981 (BVSP 1981). But the orbits of 65 such objects were known by May 1989 (Olsson-Steel 1990); as discussed below, using these more extensive data yields typical impact velocities well below those found by using the BVSP (1981) data set (Chyba 1991a).

Figure 3a shows the percentage of asteroid-Earth collisions occurring at a given velocity, using Olsson-Steel's (1990) compilation of Aten and Apollo

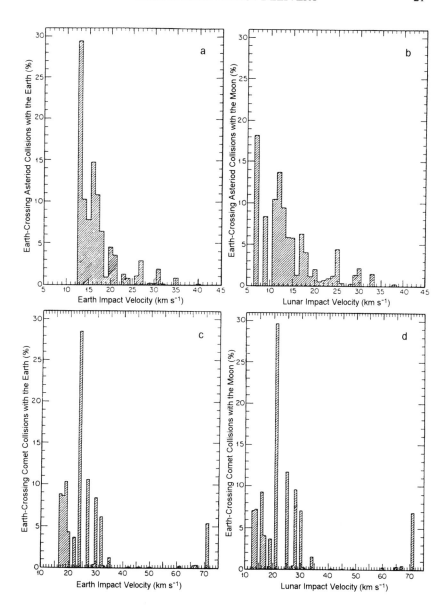

Figure 3. Percentage of Earth-crossing asteroid and short-period comet collisions with Earth and Moon as a function of impact velocity (figure from Chyba 1991a).

asteroids. Olsson-Steel's method for calculating individual asteroid collision probabilities and velocities (Kessler 1981; Steel and Baggaley 1985) results in minimum (v_{min}) and maximum (v_{max}) possible velocities at infinity (v_∞) for a given mean collision probability. Averaging these to find the values of v_∞ used here introduces an uncertainty of <0.5 km s^{-1}, small compared with other unknowns in the problem.

Given values of v_∞ and probabilities of terrestrial collisions for each Earth-crossing asteroid, it is simple to calculate the percentage of collisions with Earth as a function of impact velocity v, given by

$$v^2 = v_\infty{}^2 + v_{esc}{}^2 \tag{13}$$

where escape velocity $v_{esc} = 11.2$ km s^{-1} for Earth (see Fig. 3a). For terrestrial collisions, these data yield $v_{rms} = 18$ km s^{-1}, and an average velocity $v_{av} = 17$ km s^{-1}. However, about half the collisions occur at velocities below $v_{med} = 15$ km s^{-1}, which may therefore be taken as the "typical" value in the median sense. These values are well below those calculated from the smaller BVSP (1981) data set, for which $v_{rms} = 25$ km s^{-1}, $v_{av} = 20$ km s^{-1}, and $v_{med} = 17$ km s^{-1}.

Therefore, 15 km s^{-1} is the appropriate velocity to be used as a typical impact velocity for modeling contemporary collisions of large asteroids with the Earth. However, recent observations of small ($<$50 m diameter) Earth-crossing asteroids with the Spacewatch telescope reveal these objects to lie in unusually Earth-like orbits (Rabinowitz et al. 1993). Collision velocities are therefore lower; for these objects v_{med} is about 13 km s^{-1} (Chyba 1993a).

These calculations are easily extrapolated to the Moon, as shown in Fig. 3b. This is not merely a case of replacing v_{esc} for Earth in Eq. (13) with $v_{esc} = 2.4$ km s^{-1} for the Moon, because collision probabilities, which depend on the quantity $[1 + (v_{esc}/v_\infty)^2]$, also change. For lunar collisions, $v_{rms} = 16$ km s^{-1}, $v_{av} = 14$ km s^{-1}, and $v_{med} = 12$ km s^{-1}; this last value is, to within 2%, the median value of v_∞. The effects of the Moon's orbital velocity ($\sim$1 km s^{-1}) with respect to Earth may be neglected.

The Moon was probably much closer to Earth 4.5 Gyr ago than it is now (Ward 1975). Lunar orbital evolution time scales, however, suggest that the resulting enhanced gravitational focusing effects were unimportant subsequent to $\sim$4.4 Gyr ago (Chyba 1990). We will focus below on the impactor flux subsequent to this time, so may ignore this possible very early near-Earth stage of lunar history.

What of the Earth-crossing short-period (SP) comets? Models of the early cometary bombardment of the inner solar system (Fernández and Ip 1983; Shoemaker and Wolfe 1984) indicate that the flux of comets scattered directly from the Uranus-Neptune region (that is, following SP-like orbits) dominated by several orders of magnitude the flux of those (long-period) comets first scattered out to the Oort cloud, so it is reasonable to use contemporary SP comet orbits as analogs to those of the cometary component of the heavy bombardment. Figs. 3c and 3d show the results of using Olsson-Steel's (1987) compilation for 22 SP comets, following a procedure identical to that used above for asteroids. Obviously the statistics for SP comets are poor. The results are nearly identical to those found using Weissman's (1982) data. For SP comet collisions with the Earth, we find $v_{rms} = 30$ km s^{-1} and $v_{av} = 27$

km s^{-1}, whereas for lunar collisions, v_{rms} = 30 km s^{-1} and v_{av} = 26 km s^{-1}. Median collision velocities are more uncertain than for asteroids; roughly, v_{med} = 23 km s^{-1}. For cometary collisions with the Moon, v_{med} = 20 km s^{-1}, which is equal to v_∞ to within 1%.

Much of the "spike" at 24 km s^{-1} in Fig. 3a is due to comet P/Lexell, which was perturbed into an Earth-crossing orbit by Jupiter in 1770, and perturbed away again after only one orbit (Weissman 1982). Leaving Lexell out of these calculations increases the median cometary terrestrial impact velocity to about 25 km s^{-1}, a reminder of just how poor the cometary impact statistics remain.

We take the value v_{med} = 12 km s^{-1} for asteroid impacts on the Moon to be the appropriate value for v in Eqs. (8) and (11). Equation (8) then becomes:

$$m = 8.0 \times 10^{-3} D_f^{3.36} \qquad (14)$$

which gives m(4 km) = 1.0 × 10^{10}kg.

How sensitive are the results of Eq. (11) to different choices of v? From Eq. (12) we see that, were we to use the asteroidal v_{rms} instead of v_{med} in Eq. (11), $M(t)$ would decrease by a factor 1.6. If we used the cometary value of v_{med}, instead of the asteroidal one, $M(t)$ would decrease by a factor of 2.3. This range of values for v, then, introduces an error of a factor of a few into $M(t)$.

C. Maximum-Mass Lunar Impactors

Calculating $M(t)$ in Eq. (11) requires a choice of m_{max}. Chyba (1991a) chose m_{max} for the Moon to be the mass of the impactor that excavated the largest lunar basin, taken to be the ancient lunar farside basin South Pole-Aitken. However, there are problems with this choice. South Pole-Aitken was first tentatively identified photogeologically (see Wilhelms 1987); its existence was then clearly demonstrated by Galileo observations (Belton et al. 1992). Wilhelms (1987) suggests a pre-Nectarian age for South Pole-Aitken of 4.1 Gyr, and identifies (his Table 4.1) an average ring diameter of 2200 km. Other giant, highly-degraded lunar basins have been suggested, both on the basis of photogeological evidence (Wilhelms 1987), and subsurface mass concentrations (Campbell et al. 1969; O'Leary et al. 1969), but their existence remains controversial. In particular, Spudis et al. (1988) have called the reality of the proposed giant "Procellarum" basin into question by the claim that the large-scale, concentric structural pattern on the lunar nearside probably represents Imbrium rings (ranging up to 3200 km in diameter), rather than a "Procellarum" signature. If this argument is correct, one should perhaps worry that the 2200 km diameter South Pole-Aitken ring similarly represents a concentric feature other than the original transient crater diameter. Impacts larger than Imbrium would probably have excavated lunar mantle material, but no lunar mantle samples have ever been found (Sleep et al. 1989). However, Belton et al. (1992) have suggested, on the basis on Galileo multispectral imaging, that the South Pole-Aitken basin may contain lunar mantle material.

Combining Eqs. (8) and (11), and multiplying by the surface area of the Moon, gives an equation for the total mass incident on the Moon, subsequent to some time t, due to objects which excavated basins smaller than a final diameter D_f:

$$M(t) = 4.27 \times 10^7 [t + \beta e^{t/\tau}] D_f^{1.55}. \tag{15}$$

From Eq. (14) with D_f = 2200 km, the South Pole-Aitken object had a mass 1.6×10^{19}kg. Using South Pole-Aitken for D_f in Eq. (15) gives a total of 1.3×10^{20}kg of mass incident on the Moon subsequent to 4.4 Gyr ago (this choice of t will be discussed further below). In this case, the South Pole-Aitken object represented $\sim$14% of the mass incident upon the Moon subsequent to 4.4 Gyr ago. However, it has been shown (Wetherill 1975; Dones and Tremaine 1993) that for a cumulative power law distribution such as that of Eq. (9), viz.,

$$N(> m) \propto m^{-b} \tag{16}$$

the mass of the largest object in the distribution is related to the total mass of the distribution, M_{tot}, according to

$$m_{max} = \frac{2(1-b)}{(2-b)} M_{tot}. \tag{17}$$

With b = 0.54, Eq. (16) gives m_{max} = 0.63 M_{tot}. Thus there is an apparent inconsistency if one chooses the mass of the South Pole-Aitken object as m_{max} in Eq. (11). What mass m_{max} would be required in Eq. (11) for the ratio $m_{max}/M(t)$, with t = 4.4 Gyr, to be 63%? If m_{max} were some 19 times greater than the mass just calculated for the South Pole-Aitken object, or about 2.6×10^{20}kg, this condition would be satisfied.

It seems unwise to use this calculation to suggest an ancient impact of this magnitude (now presumably completely hidden by subsequent cratering) on the Moon. Where is the lunar mantle material, spread over the surface of the Moon, that such an impact would have excavated? It seems more likely that the vagaries of small-number statistics render hopeless these attempts to assess statistically the contribution of the few very largest impactors. Moreover, it is unclear that a simple power law distribution is valid up to the sizes of the largest impactors, although some authors have argued that this might have been the case (Dones and Tremaine 1993). Finally, the ratio $m_{max}/M(t)$ also depends on the time t at which $M(t)$ is evaluated. We argue below that t = 4.4 Gyr is the best choice, but a different choice would yield a different ratio for the mass of the South Pole-Aitken object to the total mass incident on the Moon subsequent to time t.

Is the number of observed lunar basins consistent with the power law of Eq. (15)? The answer is no, as illustrated in Fig. 4. The dotted line in Fig. 4 shows the total mass incident on the Moon integrated up to a given basin diameter, i.e., Eq. (15) as a function of D_f. The points connected by the solid line show cumulative masses for the recognized lunar basins (Wilhelms

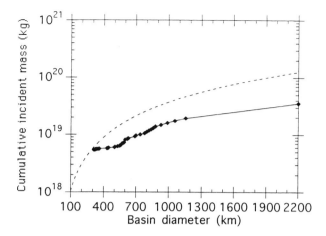

Figure 4.   Cumulative mass incident on the Moon during the heavy bombardment as a function of crater diameter. The dotted line assumes a simple power law (Eq. 15) that remains valid up the diameter of the South Pole Aitken basin (2200 km). The points show cumulative mass obtained by simply adding the masses (from Eq. 14) of the known lunar basins.

1987), excluding Procellarum. These masses are added to the mass given by Eq. (15) integrated up to 300 km, the crater diameter at which craters make the transition to multiring basins. (The power law is well obeyed up to this diameter [BVSP 1981]). Thus the rightmost point is the sum of the masses of all lunar impactors for the craters up to 300 km in diameter, plus the masses implied by all known lunar basins with radii up to the 2200 km of South Pole-Aitken. The next rightmost point is the same calculation, but stopping at 1160 km, for Imbrium.

Several conclusions are evident. First, the known basins are not consistent with an extrapolation of the power law, Eq. (15), up the to South Pole-Aitken diameter. Second, the discrepancy is nevertheless not that great; the cumulative mass implied by the power law extrapolation is 1.3×10^{20}kg; whereas the summation from individual basins yields 3.6×10^{19}kg, less than a factor of 4 smaller. Because it is likely that many lunar basins are no longer recognizable, the calculation using the simple summation may be viewed as a lower limit to the actual mass incident on the Moon during the heavy bombardment.

D. Extrapolation to the Earth

Our preceding treatment of the lunar impact record suggests that a net mass in the range $(0.4 \text{ to } 1.3) \times 10^{20}$kg was incident on the Moon subsequent to 4.4 Gyr ago. This is the appropriate mass to scale to Earth. It is not, however, the net mass actually accreted by the Moon since 4.4 Gyr ago; at least 50% of the objects striking the Moon during heavy bombarment were moving too fast for their mass to be accreted. Moreover, impact erosion of the Moon has also taken place (see Sec. VI below).

Scaling from the Moon to Earth requires accounting for Earth's larger gravitational cross section. A planet's gravitational cross section is $\sigma = \pi R_g^2$, where the gravitational radius R_g is given in terms of a planet's physical radius R and escape velocity v_{esc} by

$$R_g = R[1 + (v_{esc}/v_\infty)^2]^{1/2}. \tag{18}$$

With $v_\infty = 12$ km s^{-1}, the ratio of Earth's gravitational cross section to that of the Moon is 24. Multiplying $M(t = 4.4$ Gyr$)$ for the Moon by 24 gives (1.0 to 3.1) $\times$ 10^{21}kg accreted by Earth subsequent to 4.4 Gyr ago.

However, this scaling multiplication does not fully account for the role of Earth's larger gravitational cross section. It is likely that all the largest impactors incident on the Earth-Moon system were collected by the Earth. Quantifying this effect requires playing games with the statistics of small numbers, so should be treated with suspicion. Nevertheless, one such attempt would proceed as follows.

Following the approach of Sleep et al. (1989), use the statistical result that 24 out of 25 impactors incident on the Earth-Moon system should have been collected by Earth to write the probability P that the Moon was not hit by any of the largest n impactors as $P = (24/25)^n$. Then $P \approx 50\%$ provided $n \approx 17$. Thus if South Pole-Aitken is the largest lunar impactor subsequent to 4.4 Gyr ago, 17 larger objects should have hit Earth since that time. The question, of course, is how much larger. We may bravely employ Eq. (16) to calculate the mass distribution of these objects.

With the lunar value of $b = 0.54$ in Eq. (16), we find that if Earth collected 17 objects with masses $>1.6 \times 10^{19}$kg, Earth should have been hit by one object with mass $>3.0 \times 10^{21}$kg, the terrestrial maximum-mass impactor subsequent to 4.4 Gyr ago. How much more massive than 3.0×10^{21}kg was this object? Continuing on in our brave fashion, we may ask what mass must an impactor have had such that Earth would have collected 0.5 such objects. Equation (16) then gives

$$(17/0.5) = [(1.6 \times 10^{19} \text{ kg})/m]^{-0.54} \tag{19}$$

or $m = 1.1 \times 10^{22}$kg. The total mass M of all 17 impactors is, analogously, given by the sum

$$M = \sum_{i=1}^{17}(1.6 \times 10^{19} \text{ kg})[(i - 0.5)/17]^{-1.85} \tag{20}$$

or $M = 1.4 \times 10^{22}$kg. (Clearly there is enormous uncertainty in this result.) Summing the mass incident on Earth in these largest impactors with that previously found yields a net terrestrial accretion subsequent to 4.4 Gyr ago of (1.5 to 1.7) $\times$ 10^{22}kg. At the large-mass end of this range, the largest

terrestrial impactor contributes $\sim$65% of the total mass, in good agreement
with the requirements of Eq. (17).

Sleep et al. (1989) found that an impact energy of 2×10^{28} J is sufficient
to vaporize the entire terrestrial ocean, possibly sterilizing the Earth. For
a typical Earth impact velocity 15 km s^{-1}, this kinetic energy corresponds
to a mass 1.8×10^{20} kg. If 17 larger-than-South-Pole Aitken impactors are
taken to have hit Earth during the heavy bombardment, Eq. (20) shows that
Earth should have sustained about 5 Earth-sterilizing impacts between 4.4 and
3.8 Gyr ago.

The derivation of the maximum-mass terrestrial impactor found here
depends on the value of the exponent b in Eq. (16). While $b = 0.54$ has the
clear advantage of consistency with the ancient lunar cratering record, one
may ask how our results would differ for other plausible values. Values of b
for various classes of objects (comets, asteroid types) in the literature range
from about 0.5 to 1.0 (Chyba 1991a; Dones and Tremaine 1993). If we were
to adopt $b = 1.0$ in Eq. (17), the value differing the greatest from that used
here, the total mass collected by the Earth would be reduced by a factor of
about 4 compared with the values found above.

V. LUNAR, TERRESTRIAL, AND MARTIAN GEOCHEMICAL CON-STRAINTS

Given the uncertainties inherent in the kinds of lunar cratering calculations just
discussed, it is prudent to examine constraints on terrestrial volatile accretion
during the heavy bombardment provided by lunar, terrestrial, and Martian
geochemistry. Using Ir and Ni abundances, Sleep et al. (1989) estimate a
meteoritic component mixed into the upper half of the lunar crust (35 km) of
between 1% and 4%, with 2% their preferred value. This corresponds to a
meteoritic thickness of 700 m of material, or a total extra-lunar mass of about
8×10^{19} kg (0.2% of the current lunar mass), with an uncertainty of a factor
of 2. This mass would have been accreted subsequent to the solidification of
the lunar crust.

Sleep et al. (1989) cite samarium-neodymium (Sm-Nd) isotopic evidence that the upper lunar crust, ferroan anorthosite, solidified as early as
4.44±0.02 Gyr ago (Carlson and Lugmair 1988), whereas the age of KREEP
basalt (Carlson and Lugmair 1979) implies 4.36±0.06 Gyr ago for the solidification of the base of the crust. Therefore $t = 4.4$ Gyr seems to be the
appropriate choice in Eqs. (11) and (15). Because $t = 144$ Myr, an error in
this choice by 40 Myr will change $M(t)$ by less than 30%.

For Earth, one may use siderophile abundances in ultramafic xenoliths
(more-or-less unaltered solid mantle material brought to the surface by volcanic eruptions of basaltic magmas), as well as basalts, peridotites, and other
mantle-derived rocks, to estimate the meteoritic component mixed into the
Earth's mantle (Chou 1978; BVSP 1981; Sun 1984; Wänke et al. 1984;
Dreibus and Wänke 1987,1989; Newsom 1990). Among the xenoliths, ap-

parently unaltered (or nearly unaltered) mantle samples have been found only among the spinel-lherzolites, which represent upper mantle material from depths as great as 70 km (Wänke et al. 1984). The near-chondritic proportions of siderophile elements in these samples are evidently not the result of metal-silicate fractionation processes, as the partitioning of these elements upon fractionation is not the same (BVSP 1981). This conclusion is supported by detailed modeling of various core formation theories (Newsom 1990). It has therefore been suggested that the highly siderophile "noble metals," Ru, Rh, Pd, Re, Os, Ir, Pt, and Au, were accreted during the heavy bombardment, subsequent to terrestrial core formation (Chou 1978; BVSP 1981; Sun 1984; Newsom 1990). During core formation, virtually all noble metals in the mantle should have been incorporated into the core (Sun 1984; Newsom 1990). The nearly chondritic ratios of the noble metals excludes fractionated impactors (such as eucrites) as primary sources (Chou 1978).

Murthy (1991) has obtained an improved fit to terrestrial excess siderophile abundances by considering a high-temperature equilibrium core-mantle differentiation at 3000 to 3500 K. Although this approach remains controversial (see, e.g., Jones et al. 1992; O'Neill 1992), it yields mantle abundances that closely match the observed values for all but the noble metals (of which Murthy explicitly considers only three). However, it fails to reproduce the latter eight elements' CI relative abundance ratios without an appeal to additional mechanisms. Relative to CI abundances, the highly siderophile elements are present in the mantle at 0.01 to 0.1 times the other siderophiles (see Fig. 5). Therefore the heavy bombardment may deliver the highly siderophile elements in CI abundances, while making insignificant impact on the absolute or relative abundances of the remaining mantle siderophiles, i.e., those which Murthy's model is best able to explain. In any case, the abundances of highly siderophile elements in the mantle provide an upper limit on the post-core formation input from the heavy bombardment.

Noble metal abundances in upper mantle samples can be explained by ∼1% of CI carbonaceous chondrite input subsequent to core formation (Chou 1978; BVSP 1981; Sun 1984; Dreibus and Wänke 1989). How much total extraterrestrial mass this represents depends on whether one takes the post-core formation input to have been mixed throughout the entire, or only the upper, mantle. The upper mantle appears well mixed, as indicated by studies of isotope ratios and distributions of rare earth elements in mid-ocean ridge basalts; mixing in the upper mantle greatly reduces heterogeneities on a time scale of several hundred Myr (Turcotte and Kellogg 1986). Therefore it appears reasonable to take the CI chondrite abundances implied by available samples to represent typical upper mantle abundances. Assuming that the CI input is mixed only until this depth (Chou 1978), the CI abundance may be multiplied by the mass of the upper mantle to obtain the total mass accreted by Earth during the heavy bombardment.

The depth of Earth's upper mantle is given by the seismic discontinuity at 670 km (Stacey 1977; Turcotte and Kellogg 1986). Little earthquake activity

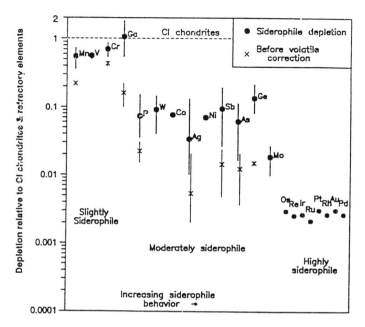

Figure 5.   The depletion of siderophile elements in the Earth's mantle relative to CI chondrite abundances. The elements are arranged roughly in order of increasing siderophile behavior. The eight highly siderophile "noble metals" are at the far right; uncertainties for these elements are not shown. Some elements are depleted due to both volatility and siderophility; the crosses indicate their actual depletions, while the circles show their corrected siderophile depletions (figure from Newsom and Sims 1991; copyright 1991 by the AAAS).

occurs below this depth; it is controversial whether subducted lithospheric slabs ever penetrate this layer (Frohlich and Grand 1990). However, even if a barrier to mantle convection exists at 670 km depth on contemporary Earth (Turcotte and Kellogg 1986), it remains possible that a hotter early Earth would have undergone whole mantle convection, and therefore mixed the extraterrestrial component throughout the entire mantle (Wänke et al. 1984).

The mass of the terrestrial upper mantle, calculated using a standard Earth interior model (Dziewonski et al. 1975; Stacey 1977), is about 1.1×10^{24}kg. Assuming this to be $\sim1\%$ CI chondritic then gives $\sim1.1 \times 10^{22}$kg as an estimate for the total mass accreted by Earth during the heavy bombardment, subsequent to core formation. This value is a lower limit, as it assumes only upper mantle mixing. An upper limit is given by taking mixing to occur throughout the entire mantle, which yields 4.0×10^{22}kg total accreted mass (taking the mass of the whole mantle to be 4.0×10^{24}kg [Stacey 1977]). Therefore the actual value is likely to lie in the range $(1 \text{ to } 4) \times 10^{22}$kg.

Wänke et al. (1984) and Dreibus and Wänke (1987,1989) have argued for whole-mantle mixing, and find a total CI component equivalent to 0.44% the mass of the Earth, or 2.6×10^{22}kg. (Given the mass of the Earth, $M_\oplus =$

5.98×10^{24} kg, these authors have effectively taken the CI component of the mantle to be 0.65%.) Newsom (1990) requires a late veneer chondritic component equivalent to 0.2% $M_\oplus$, or 1.2×10^{22} kg. Again, this value falls within the range, $(1 \text{ to } 4) \times 10^{22}$ kg, considered here.

Terrestrial core formation is thought to have taken place within $\sim 10^8$ yr of the formation of the Earth (Stevenson 1983,1990; Sun 1984). The Earth's age may be determined by iodine-plutonium-xenon (I-Pu-Xe) dating of the atmosphere and mantle, which suggests formation 75 to 100 Myr after t_0, the time of formation of primitive meteorites (Swindle et al. 1986). I-Pu-Xe dating of lunar samples suggests a lunar age of 63 ± 42 Myr after t_0, or between 80 Myr before and 30 Myr after terrestrial formation. (Therefore, if these dates as well as the impact-trigger hypothesis for the formation of the Moon are correct, the Moon-forming event occurred no more than 30 Myr subsequent to terrestrial accretion.) A period of 75 to 100 Myr for the formation of the Earth subsequent to meteorite formation agrees well with dynamical models for the formation of the terrestrial planets, which yield $\sim 10^8$ yr time scales for the accretion of the Earth from an initial population of planetesimals (Wetherill 1977; Greenberg 1989; see also Grinspoon 1988). Wetherill (1990) finds 99% accretion of the Earth after 77 Myr.

What is t_0? The oldest high-precision meteorite date comes from calcium-aluminum-rich inclusions of the CV chondrite Allende, which give a ^{207}Pb/-^{206}Pb model age of 4.559 ± 0.004 Gyr (Tilton 1988). This value implies that terrestrial formation was virtually complete by 4.49 to 4.46 Gyr ago. Earth's core would then have formed by ~ 4.4 Gyr ago. Therefore, it is appropriate to compare the extraterrestrial mass input implied by terrestrial geochemical data with that found by extrapolating from the lunar cratering record, using $t \approx 4.4$ Gyr in Eq. (11). The agreement between the two techniques is good. The value found via the lunar cratering extrapolation, $(1.5 \text{ to } 1.7) \times 10^{22}$ kg, lies within the range $(1 \text{ to } 4) \times 10^{22}$ kg implied by the geochemical data. The extrapolation is reasonably robust against a different choice for the time of terrestrial core formation; for example, changing the time of core formation by 100 Myr will change the result by only a factor of 2.

The important result is that the good agreement with the available lunar and terrestrial geochemical constraints suggests that Eq. (11) provides a reasonable, albeit procrustean, model for post-core formation terrestrial mass accretion during the heavy bombardment. Certain proposed fits (Maher and Stevenson 1988; Oberbeck and Foglemer 1989a, b) to other lunar cratering data sets (Wilhelms 1984) predict siderophile abundances 2 orders of magnitude in excess of those observed; such fits can probably be excluded as models for the early terrestrial impact environment.

The story developed here is also in reasonable agreement with geochemical data from Mars. An upper limit on the total amount of mass Mars may have accreted subsequent to core formation is placed by rhenium abundances in the SNC meteorites, whose parent body is almost certainly Mars. The low abundance of this noble metal puts an upper limit of ~ 0.1% on the chondritic contribution to the parent body mantle (Treiman et al. 1986). Taking Mars' core to comprise 18% of Mars' mass (BVSP 1981) of 6.4×10^{23}kg, this gives an upper limit of $\sim 5 \times 10^{20}$kg of chondritic material delivered to Mars subsequent to core formation, or about 1 to 2 km of material distributed over the Martian surface.

This limit is consistent with scaling from the lunar impact record. Estimated crater production rates (number of craters of a given diameter per unit area) for Mars relative to the Moon vary by factors of several around unity (BVSP 1981); here we take the two rates to be the same. Then to determine the relative mass incident on the two bodies, we must multiply by the ratio of their surface areas, 3.8, and also take into account how crater diameters scale with surface gravities and typical impact velocities. One finds that a crater of diameter D on Mars requires an impactor of mass ~ 4.3 times greater than on the Moon. Therefore a crater production rate for Mars equal to that of the Moon implies that ~ 16 times as much mass is incident on Mars as is incident on the Moon. Scaling to Mars, we find $\sim (6 \text{ to } 21) \times 10^{20}$kg exogenous material incident on Mars subsequent to ~ 4.4 Gyr ago. Not all of this material would have been retained in the impact. The simple lunar cratering scaling of Eq. (11) is therefore roughly consistent with Martian, lunar, and terrestrial geochemical constraints.

VI. IMPACT EROSION OF THE EARTH AND MOON

In Secs. IV and V, we used the lunar cratering record to estimate the total impactor mass incident on the Earth and Moon subsequent to ~ 4.4 Gyr ago, and compared this result to those implied by terrestrial and lunar geochemical data. However, such comparisons also require considering the erosion of the Earth and Moon by high-speed impacts; not all incident mass will be retained by the target world.

Impact erosion may occur both due to loss of high-velocity crater ejecta, and to expanding post-impact vapor plumes. Melosh and Vickery (1989) have presented a first-order treatment of the latter process. They find that large impacts may cause atmospheric erosion provided that the impactor strikes the planet at a velocity high enough for the vapor plume to form and expand at a speed greater than the escape velocity v_{esc}, and that the mass of the plume exceeds the air mass above the plane tangent to the point of impact. An impactor that satisfies these two criteria may be taken to be entirely lost as an escaping vapor plume from the target world. A mass of the target roughly equal to that of the impactor is simultaneously lost.

Melosh and Vickery (1989) find that the threshold impact velocity v_{min} for most of a vapor plume to exceed v_{esc} is given by

$$v_{min}^2 = 4(v_{esc}^2 + 2H_{vap}) \qquad (21)$$

where H_{vap} is the vaporization energy, 13 MJ kg^{-1} for silicates and 3 MJ kg^{-1} for ice (1 MJ kg^{-1} = 1 km^2 s^{-2}). For the Moon, v_{min} is a bit over 11 km s^{-1}

for silicate projectiles ("asteroids") and about 7 km s^{-1} for ice impactors ("comets"). As seen from Fig. 3d, virtually all vapor plumes resulting from cometary impacts on the Moon (for comets of any mass, as there is no atmosphere to be overcome) would be lost. In the case of asteroids, nearly 40% of lunar collisions occur at velocities below 11 km s^{-1}, and about 50% below 12 km s^{-1}. Recognizing that published treatments of impact erosion remain approximate, we take 50% as a rough estimate of the fraction of asteroidal collisions with the Moon that result in escaped vapor plumes.

For Earth, $v_{min} \approx 25$ km s^{-1} for asteroids, and about 23 km s^{-1} for comets. As seen in Figs. 3a and 3c, ~90% of Earth-colliding asteroids, and ~50% of Earth-colliding comets, impact Earth with velocities below their respective v_{min}. Therefore Earth retains nearly all asteroid-delivered volatiles, but only half of those brought in by comets; comets erode about as much terrestrial mass as they deliver.

Besides erosion by vapor plumes, we must also consider loss of target material in that fraction of crater ejecta propelled at velocities $> v_{esc}$. Quantitative treatments of this effect, like those of atmospheric erosion, remain at an early stage (Melosh 1989). Housen et al. (1983) have used scaling arguments to derive the functional form of an expression for the volume of crater ejecta with velocity greater than or equal to some velocity v_e. This may be combined with crater diameter-impactor mass scaling results (Schmidt 1980; Schmidt and Housen 1987) to determine how much ejecta is propelled at velocities $> v_{esc}$ for a projectile of a given mass and velocity incident upon a certain target world (Chyba 1991a), giving:

$$M_e(> v_{esc}) = 0.11(\rho/\rho_t)^{0.2}(v/v_{esc})^{1.2}m \qquad (22)$$

where $M_e(> v_{esc})$ is the mass of ejecta with velocity greater than v_{esc}, and ρ and ρ_t are the impactor and target densities, respectively. For the Moon, with an escape velocity 2.4 km s^{-1}, Eq. (22) leads to the conclusion that a net mass M of asteroids incident on the Moon will erode about a mass M of lunar material in the form of ejecta. The Moon will also lose mass from escaping impact vapor plumes. From the time of the heavy bombardment on, the Moon has experienced a net impact erosion from asteroid collisions.

Comets, with higher velocities, will tend to erode in high-speed ejecta slightly more lunar material than they deliver. Furthermore, virtually all cometary collisions will result in vapor plumes that expand away from the Moon at velocities greater than v_{esc}, carrying away impactor and target mass. A typical collision of a comet of mass m will add no mass to, and erode a mass ~$2m$ from the Moon.

Inserting terrestrial values into Eq. (22) gives $M(> v_{esc}) \approx 0.1m$, suggesting terrestrial erosion by high-speed ejecta is a small effect. This conclusion is reinforced by the likelihood that much of the ejecta escaping from Earth will be re-accreted from their resulting Earth-crossing heliocentric orbits. (In the lunar case, while some ejecta would be trapped in Earth orbit and re-accreted by the Moon, Eq. (22) shows that the majority would escape from geocentric orbit into a heliocentric one. Even when the Moon was as close as 10 Earth radii, Earth's escape velocity was only 3.5 km s^{-1} at the lunar orbit, scarcely bigger than lunar escape velocity. Hence, most of the material eroded from the Moon would eventually be collected by Earth, rather than re-accreted by the Moon.)

In Sec. IV, the lunar cratering record was used to show that $\sim 10^{20}$kg of material was incident upon the Moon in the heavy bombardment. At most half this mass was accreted, because $\sim 50\%$ of asteroids incident on the Moon are immediately lost in post-impact vapor plumes. (Virtually no cometary mass is retained.) Must the net amount of extra-lunar material added to the Moon be significantly further reduced due to impact erosion of the lunar surface?

Consider the result of a total mass M of asteroids incident on the Moon. Half this mass is accreted, while the remainder is lost. The latter fraction also results in $\sim 0.5M$ of lunar regolith lost in vapor plumes. Also, for an incident mass M, a regolith mass $\sim M$ is lost in high-speed ejecta. Therefore, the Moon experiences a net mass loss of $\sim M$. Subsequent to ~ 4.4 Gyr ago, taking $M \approx 10^{20}$kg, this represents ~ 1 km of material eroded from the lunar surface. This result does not, however, contradict the presence of $\sim 2\%$ extra-lunar siderophiles in the lunar crust. While a typical impactor will deliver 100% CI siderophile abundances, the resulting eroded lunar regolith material will be only $\sim 2\%$ CI material. Therefore the Moon will accrete a net mass $\sim (0.5M - 0.02M) \approx 0.5M$ of CI siderophile-abundance material-even as it experiences a net mass loss. That is, efficient regolith mixing (Sleep et al. [1989] take the extra-lunar material to be mixed through a megaregolith extending to a depth of 35 km) would guarantee that impact erosion will have little effect on extra-lunar abundances found in the crust.

VII. IMPACT DELIVERY OF VOLATILES AND TERRESTRIAL ABUNDANCES

If Earth collected (1 to 4) $\times 10^{22}$kg of material subsequent to core formation, what quantity of accreted volatiles is implied? Of course, this depends on the fraction of the heavy bombardment population composed of volatile-rich objects such as CI chondrites or comets. Given a net accreted terrestrial mass of 2×10^{22}kg (chosen to lie in the middle of the range allowed by the geochemical data), Table I (after Chyba 1991a) shows two candidate heavy bombardment compositions that yield an approximately terrestrial abundance of water. Implied abundances of certain other volatiles are also listed. If Earth's oceans were derived from such a heavy bombardment source, would this imply a terresrial abundance of any other element so far in excess of known terrestrial inventories as to rule the possibility out?

Column 4 of Table I considers the consequences if CI chondrites comprised 100% of the accreted mass. Wänke et al. (1984) have compiled a list of the abundances of some 56 elements in the terrestrial crust and mantle,

TABLE I

Estimated Terrestrial Volatile Inventories vs Accretion from Candidate Heavy Bombardment Fractions

Element or Compound	Terrestrial Inventory in Mantle and Crust (kg)	Terrestrial Surface Inventory (kg)[e]	100% CI Chondrite (H$_2$O Content 6%)[f]	25% Cometary (H$_2$O Content 43%)[g]
Carbon[a]	2.7×10^{20}	8.7×10^{19}	9.7×10^{20}	5.3×10^{20}
Nitrogen[b]	6.1×10^{19}	5.2×10^{18}	4.0×10^{19}	1.4×10^{20}
Sulfur[c]	1.4×10^{21}	2.1×10^{19}	1.5×10^{21}	8.8×10^{19}
Chlorine[c]	4.7×10^{19}	4.5×10^{19}	1.6×10^{19}	9.4×10^{17}
H$_2$O[d]	1.9×10^{21}	1.6×10^{21}	1.5×10^{21}	1.4×10^{21}

[a] C terrestrial abundance estimates in columns 2 and 3 from Turekian and Clark (1975) and Hayes et al. (1983), respectively.

[b] N terrestrial abundance estimates from Schidlowski et al. (1983).

[c] S and Cl terrestrial abundance estimates from Dreibus and Wänke (1989).

[d] H$_2$O terrestrial abundance estimates in columns 2 and 3 from Dreibus and Wänke (1989) and Turekian and Clark (1975), respectively. Terrestrial oceanic mass comprises 1.4×10^{21} kg H$_2$O (Walker 1977).

[e] Terrestrial surface abundances taken to include atmosphere, hydrosphere, sedimentary rocks, and crustal estimates.

[f] Chondritic abundances, exclusive of H$_2$O, from Dreibus and Wänke (1989). H$_2$O abundance from Chyba (1991a).

[g] Cometary abundances from Delsemme (1992); CI value assumes CI chondritic S/CI ratio. Only ~50% of cometary impacts occur at sufficiently low velocities for terrestrial accretion.

and normalized these to CI abundances. Given the resulting numerical values for noble metals, their list may be scanned for any elements present in the terrestrial mantle + crust in CI relative abundances below these values. Only these elements would be provided by a putative CI chondrite accretion in excess of the known terrestrial inventories.

These limits are exceeded for almost no elements. In the Wänke et al. (1984) compilation, only selenium (Se), sulfur (S), and carbon (C) would be delivered in quantities greater than the known terrestrial inventories for these elements if the heavy bombardment were 100% CI chondrite. A more recent estimate of the sulfur inventory on the Earth by Dreibus and Wänke (1989) substantially increases the estimate of S in the terrestrial mantle, and the apparent discrepancy is virtually removed. Se is identified by Wänke et al. (1984) as an element whose abundance, inferred from mantle samples, is suspect. Excluding this element, only C is delivered by a 100% CI chondrite heavy bombardment in excess of its estimated terrestrial abundance, and only by a factor of about 3. However, estimates of terrestrial mantle carbon abundances vary greatly from author to author. Dreibus and Wänke (1989) have said that discrepancies of factors of 4 between predicted and known C and N inventories are "not really disturbing, considering the poor knowledge on the mantle and even the crustal concentrations of these elements."

Column 3 of Table I gives estimates of terrestrial surface (atmosphere, hydrosphere, sedimentary rocks, and crust) volatile inventories. These estimates are, of course, on a firmer basis than the mantle estimates included in column 2; the former therefore provide lower limits for the terrestrial inventories.

In addition to C and N, Table I also gives abundances for water, sulfur and chlorine (Cl). The latter two elements are listed as some authors (Clark 1987; Dreibus and Wänke 1989) have suggested using these elements to constrain the chondritic or cometary fraction of the heavy bombardment. Table I shows that neither S nor Cl terrestrial inventories are smaller than those expected to be delivered by either a comet- or CI asteroid-delivered ocean of water. Finally, note that the deuterium to hydrogen ratio of comet Halley (0.6 to 4.8 $\times$ 10^{-4}) is consistent with the D/H ratio of Earth's oceans (1.6 $\times$ 10^{-4}; both are also consistent with meteoritic values, but are an order of magnitude higher than interstellar, Jovian, or Saturnian values (Eberhardt et al. 1987). Table I does not include noble gas abundances or isotopic data. These constraints will be discussed below in Sec. X.

Column 5 of Table I considers the possible role of a cometary fraction in the heavy bombardment. Comets could have supplied about a terrestrial ocean of water if they comprised ~25% of the heavy bombardment by mass.

Given the way the heavy bombardment impactor population's mass peaks in the largest objects, comets could have comprised 25% by mass of this population only if very large comets existed (see the Chapter by Bailey et al.). However, giant comets are known to exist; the trans-saturnian comet Chiron may have a diameter as large as 372 km (Sykes and Walker 1991) (as well as an orbit that may evolve into an Earth-crossing one [Hahn and Bailey 1990]) and the Great Comet of 1729 was a naked-eye object at a perihelion of about 4 AU (Kronk 1984). More recently, the discovery of the distant objects 1992 QB1 (Jewitt and Luu 1993) and 1993 FW (Luu and Jewitt 1993), both hundreds of kilometers in diameter, probably represent examples of giant Kuiper belt comets (Weissman 1993). A total of eight bodies beyond Neptune and Pluto, all between $\sim$100 and 250 km diameter, are now known (Weissman 1994).

VIII. IMPACT EROSION OF VOLATILES FROM THE EARTH

Melosh and Vickery (1989) have presented a simple analytical approximation to their more detailed numerical work (see, e.g., Vickery and Melosh 1990) addressing impact erosion of planetary atmospheres. Their treatment finds that an atmospheric cap of mass

$$m_{cap} = 2\pi P_o H R_\oplus / g \qquad (23)$$

above the plane tangent to the point of impact will be removed by impactors with masses above some threshold m^*. In Eq. (23), P_o is the terrestrial surface atmospheric pressure, H is the atmospheric scale height, 8.4 km for the contemporary Earth (Walker 1977), and g is terrestrial surface gravity. The threshold mass $m^* \approx m_{cap}$, which for the current 1 bar terrestrial atmosphere is 3.5×10^{15} kg.

Heavy bombardment should have resulted in substantial atmospheric erosion from the Earth. With m in Eq. (9) set equal to m^* from Eq. (23), scaling to the Earth gives $n(> m^*, 4.4 \,\text{Gyr}) = 1.3 \times 10^4$. For illustration, take comets to have represented a negligible fraction of the impactors, so that only $\sim$10% of the latter had sufficient velocity to cause erosion. Conservatively, then, Earth sustained $\sim$10³ atmosphere-eroding impacts subsequent to 4.4 Gyr ago. Again for illustration, take the mass of the early atmosphere to equal that of today's 5.3×10^{18} kg (Walker 1977). Then each eroding impact removed about 6.6×10^{-4} of the total atmospheric mass. Multiplying this result by 1.3×10^3 eroding impacts yields about an atmosphere of mass eroded by the heavy bombardment.

Treating the problem more quantitatively, the net effect of erosive impacts during the heavy bombardment may be calculated from integrals of the form

$$M(t) = \int_{m_{max}}^{m_*} Q(m)[\partial n(> m, t)/\partial m] dm. \qquad (24)$$

Erosion of condensed oceans appears to be an unimportant effect (Chyba 1990). Equation (24) gives the mass of atmosphere eroded, by setting $Q(m) = fm_{cap}$, where m_{cap} is from Eq. (23), and f is that fraction of the impactors that causes erosion. Performing the integral gives

$$M_{lost} = f\alpha[t + \beta(e^{t/\tau} - 1)]m_{cap}[m(4 \,\text{km})/m_*]^b \,\text{km}^{-2}. \qquad (25)$$

One concern about cometary volatile delivery scenarios is that a cometary source for Earth's water may imply an abundance of nitrogen in excess of known terrestrial inventories. The severity of this concern remains unclear in light of uncertainty in the size of the relevant terrestrial inventories (as well as the nitrogen abundance in comet Halley [Krankowsky 1991; Encrenaz et al. 1991]). For example, Turekian and Clark (1975) suggest 1.1×10^{19}kg of nitrogen resides in the Earth's atmosphere, crust, and upper mantle. Dreibus and Wänke (1989) give the terrestrial mantle N inventory as "?." Schidlowski et al. (1983) give as their mantle estimate 5.6×10^{19}kg N (used in Table I), but cite other estimates as high as 2.0×10^{20}kg. Here we simply demonstrate, using Eq. (25), that impact erosion could remove a substantial quantity of whatever "excess" N might have been delivered.

Consider the case of cometary delivery of the bulk of terrestrial volatiles (column 5, Table I). About half of the 25% of the collisions that were cometary, and about 10% of the remaining collisions, would have caused erosion of the atmosphere for masses $>m^*$. Hence $f \approx 0.2$ in Eq. (25). Consider a simple model in which half of all the N delivered by comets remains in the vapor phase in the early terrestrial atmosphere. Obviously, however, the exact quantity eroded depends on poorly known details of early Earth's atmosphere, in particular what fraction of N delivered to the early Earth remained as atmospheric N_2 (or ammonia, NH_3), and what fraction was incorporated into the sedimentary column.

To approximate cometary delivery and simultaneous atmospheric erosion, use Eq. (25), with m^* and m_{cap} scaled by Eq. (23) to an atmosphere with a mass equal to one in which half of the total nitrogen delivered, 7×10^{19}kg, is present in the atmosphere. Then some 3×10^{19}kg of atmospheric N_2 may be removed from Earth by impact erosion. This is about one-half of the "excess" cometary nitrogen predicted by comparing column 5 with column 2 of Table I. Considering the uncertainties of the various factors in Eq. (25), this result may be summarized by saying that atmospheric impact erosion on the early Earth could have removed a quantity of N_2 about equal to that which was delivered in excess. Given our ignorance of early terrestrial conditions, probably the best we can do here is to demonstrate the existence and magnitude of this effect.

IX. THE SOLAR SYSTEM CRATERING RECORD

Estimates of contemporary asteroidal and cometary cratering rates suggest that comets account for 10 to 30% of the recent production of terrestrial impact craters greater than 10 km in diameter (Shoemaker 1983), although the inclusion of "extinct" comets (Shoemaker and Wolfe 1986a) or possible comet showers (Shoemaker and Wolfe 1986b) could raise this fraction to above 50% (see the Chapter by Grieve and Shoemaker). How to extrapolate this result back to the early solar system is unclear.

Crater diameter-surface density distributions for the Moon, Mars, and Mercury have been taken to imply a common heliocentric impactor population for the heavy bombardment of the inner solar system (Strom 1987). But the disparities between the cratering records of the Moon and those of some outer planet satellites may argue that comets were not the major component of this population (Strom 1987). However, in the absence of absolute dating of outer satellite surfaces, this argument is not decisive. Moreover, Hartmann (1992) has found that a number of ancient lunar surfaces do not show the relative downturn in crater number at diameters below 30 km that has been a canonical part of the Moon-Mars-Mercury comparison. He suggests that this downturn in the diameter distribution is not the direct signature of the size distribution of the early impactor population, but rather is due to destruction of smaller craters by basaltic flooding. The correct interpretation of the observed cratering records throughout the solar system remains unclear, as do its implications for the cometary fraction of the heavy bombardment.

X. NOBLE GASES AND ISOTOPE RATIOS

If comets and asteroids carry volatiles in distinctly different proportions, or if some different fractionation of isotopes has occurred in the trapping of gases in these two types of objects, one could determine the relative contributions of comets and asteroids to the cratering record by studying the volatile inventories of Mars, Earth, and Venus. Unfortunately, such a project is inhibited by the many changes in the composition of planetary volatiles that have occurred during the lifetime of the solar system. Chemical reactions with planetary surfaces, escape from the upper atmospheres and interactions among the various atmospheric constituents all affect the composition of the volatile inventories we find today. On Earth, there have been the additional, major modifications brought about by the presence of living organisms (see Kasting 1993).

Under these circumstances, studies of the noble gases and their isotopes appear to offer the best hope. Here too, one must worry about changes in the original components, e.g., additions to some isotope abundances from radioactive decay of parent elements, and fractionation resulting from hydrodynamic escape (Pepin 1991). However, we are gradually accumulating enough information about noble gases in different reservoirs to feel optimistic about ultimately separating these various effects. Meanwhile, in this section, we shall simply indicate the potential of the noble gases for determining the relative importance of rocky and icy impactors (asteroids/meteoroids vs comets) in producing the observed cratering record by demonstrating the feasibility of a model for cometary delivery of the heavy noble gases (argon, krypton, and xenon).

The first clue that such a model may in fact be required is provided by the relative abundances of the noble gases in chondrites compared to the abundances found in the atmospheres of Earth and Mars (Fig. 6). Chondritic

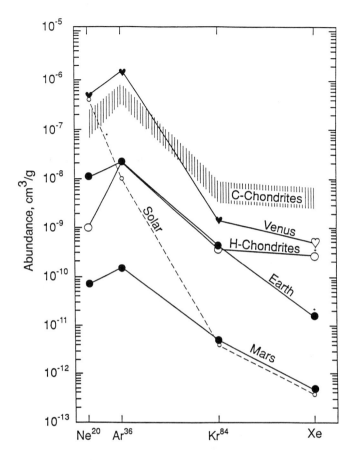

Figure 6.   A plot of the abundances of the noble gases in ordinary (H) chondrites,
carbondaceous (C) chondrites, and the atmospheres of Mars, Earth, and Venus. The
solar pattern of noble gas abundances, normalized to Martian ^{84}Kr, is shown for
comparison as a dashed line (figure from Owen et al. 1992).

meteorites contain too much xenon compared with the Mars-Earth pattern.
This problem has been known for many years, but the common solution
(burial of excess Xe in shales or ice) now appears to be untenable (Wacker
and Anders 1984). Thus the designation "planetary component" for the noble
gases in chondrites that exhibit the pattern shown in Fig. 6 has become a
misnomer. "Chondritic pattern" would be a better phrase.

This conclusion instantly places an upper limit on the amount of mass
the CI chondrites could have contributed to the Earth after the end of a
possible early period of substantial impact erosion of the terrestrial atmosphere
(see Sec. VIII). The typical xenon content of a CI is 5×10^{-9} cc g^{-1} for
^{132}Xe (Heymann 1971). The amount of ^{132}Xe in the Earth's atmosphere is
9×10^{16}cc, so if all of this is provided by carbonaceous chondrites, we obtain

$M_{Cl} \approx 2 \times 10^{22}$kg. This is certainly consistent with the range of (1 to 4) $\times$ 10^{22}kg considered here for impactor mass, but it leaves open the question of the source of the neon, argon, and krypton.

Owen et al. (1991,1992) have argued that laboratory experiments on the trapping of noble gases at low temperature (Laufer et al. 1987; Bar-Nun et al. 1988) have provided data implicating icy planetesimals (comets) as the carriers of these gases. The argument is illustrated in Fig. 7. If one makes the very simple assumption that both Mars and Earth obtain their noble gases from the same two reservoirs—a rocky inner reservoir consisting of gases that came into the planets with the rocks that comprise most of their masses and an external reservoir of gases supplied by icy impactors—we can draw a mixing line through the points representing Mars and the Earth on a three isotope plot (^{36}Ar/^{132}Xe vs ^{84}Kr/^{132}Xe) that also passes through a point for the noble gas abundances trapped by ice at a temperature of about 50 K. (This line appears curved in Fig. 7 because we use a logarithmic plot to accomodate the data.)

Unfortunately, there is an ambiguity in this interpretation, pointed out by Ozima and Wada (1993), viz., a straight line on the log-log plot of Fig. 7 also passes through these three points, and such a line would nicely represent equilibrium partitioning of noble gases between melt and gas. (Noble gas solubility in the melt is approximately proportional to $\exp(-Kr_i^2)$, where K is a numerical constant and r_i is the radius of an atom of noble gas i.) Thus one needs to examine additional evidence to determine whether it is possible to separate these two processes.

Owen et al. (1992) and Owen and Bar-Nun (1993) used data on noble gases in the Shergottite meteorites (the "S" in SNC) and terrestrial mantle basalts to buttress their case. Both of these types of rocks (from two different planets) exhibit noble gas abundances that fall along the proposed mixing line. At least some of the noble gas mixture in the Shergottites must have been emplaced by the shock that led to their expulsion from Mars (Bogard et al. 1986), while the trapping of gases in ice proceeds by adsorption in a microporous solid. It seems unlikely that such different processes should follow the same exponential law as partitioning between silicate melt and gas. There is some additional information that substantiates this conclusion. The isotopes of xenon in the different Shergottites show a distribution that correlates in an interesting way with the proposed mixing line. As (^{84}Kr/^{132}Xe) and (^{36}Ar/^{132}Xe) increase, so does ^{129}Xe/^{132}Xe. Furthermore, the meteorite EETA 79001, which lies closest to the Earth on the three isotope plot, also exhibits the closest match to Earth in the relative abundances of the other xenon isotopes. Equilibrium partitioning will only produce elemental, not isotopic fractionations. Owen (1992) and Owen et al. (1992) suggested that impact erosion of the early Martian atmosphere is required to enrich ^{129}Xe in the present atmosphere. Musselwhite et al. (1991) have pointed out that the greater water solubility of iodine compared to xenon would provide a means for sequestering the parent ^{129}I during the planet's early bombardment. Subsequent release of gas would enrich the atmosphere in ^{129}Xe. The differences

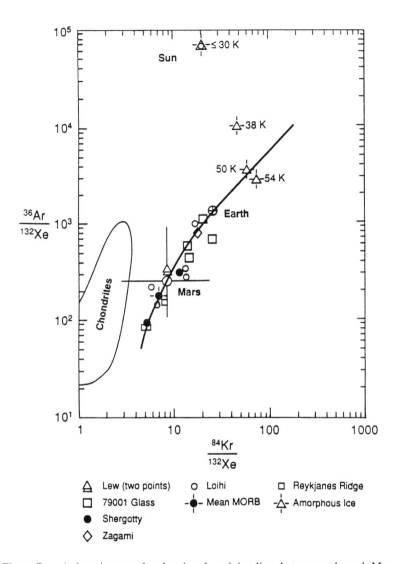

Figure 7.   A three-isotope plot showing the mixing line that passes through Mars, Earth, and the fractionated mixture of noble gases trapped in ice at ∼50 K. The positions of noble gas mixtures in all of the Shergottites for which such measurements have been made and in various submarine basalts are also indicated. Note that only an upper limit exists for Xe on Venus (Fig. 6), so the position of this planet on the plot is poorly defined (figure from Owen and Bar-Nun 1993).

in the ^{129}Xe content found in the SNCs today are then interpreted in terms of a difference in the amount of Martian atmospheric gas these meteorites have acquired.

Unlike the Earth, where mixing between the crust and mantle reduces isotopic differences in these two reservoirs on relatively short time scales,

Mars offers us a planet where such differences can survive for billions of years. One sees this in the SNC oxygen isotope systematics, where clear evidence for two distinct reservoirs of oxygen has been established by Karlsson et al. (1992). The oxygen in water in the SNCs is isotopically different from the oxygen in the host rocks, unlike the situation with water and rocks on Earth (Robert et al. 1992). (Note that this provides another line of evidence that supports the model developed in this chapter for volatile delivery by late-arriving impactors.)

It is clear that further studies of Mars, the SNC meteorites, and especially the comets are needed to resolve remaining ambiguities. Perhaps the most important effort is an attempt to determine noble gas abundances in comets. The model of Owen et al. (1992) and Owen and Bar-Nun (1993) implies that the primordial argon should retain a roughly cosmic abundance ratio with N_2, because both are trapped in ice with similar efficiency (Bar-Nun et al. 1988). Because about 70% of the nitrogen in the outer solar nebula was apparently in the form of N_2 (Owen 1994), this means a ratio N_2/Ar of about 3 to 10 for the gases trapped in the comet. This is a measurement that will soon be in the range of rocket-borne spectrometers (Stern et al. 1992; Feldman et al. 1991). Ultimately we will need to measure the xenon isotopes in comets, but this will require a deep space mission.

It is interesting to examine some of the implications of the cometary model for delivery of volatiles to Earth. Just considering the noble gases, and assuming that the efficiency of gas trapping by a comet forming at $T \approx 50$ K is the same as that measured for ice in the laboratory, a mass of $\sim 2.5 \times 10^{17}$ kg of ice is adequate to deliver all of the noble gases now in the atmosphere. This corresponds to a total cometary mass of $\sim 5 \times 10^{17}$ kg, well below the estimates given in Table I.

However, we have only delivered the correct amount of noble gases with this 50 K ice. The accompanying amounts of C,N, and H_2O are trivial compared with the Earth's inventory. Note that the ratio of $N_2/^{36}$Ar in the Earth's atmosphere is 2.5×10^4, compared with the value of 10 in the 50 K ice. To separate N from ^{36}Ar in order to achieve this high value, we need to invoke comets formed at much higher temperatures, in the Jupiter zone. For $T \gtrsim 100$ K, condensing ice will trap tiny amounts of argon and N_2, but comets formed at these temperatures will still carry large amounts of nitrogen because they will incorporate nitrogen compounds.

This composition is found in the list of abundant elements in Halley's comet (see, e.g., Krankowsky 1991; Encrenaz et al. 1991), where nitrogen is strikingly deficient relative to oxygen and carbon, which exhibit nearly solar abundances. (Note that this differs from the earlier conclusion of Anders and Grevesse [1989], who argued that C, N, and O in Halley were all present in solar ratios. It is consistent with the analysis of Geiss [1987,1988].) The reason for the apparent deficiency of nitrogen is probably different; N_2 will be rapidly lost from the active layers of such a comet, although it can be retained in the interior. In the $T \gtrsim 100$ K comets, however, N_2 (and Ar) would have been absent from the entire nucleus. If we simply use Halley's coma as an analogy for the composition of Jupiter-zone comets, the proportions of other elements derived from the Halley encounter provide an estimate of the total mass of $T \gtrsim 100$ K ice required to bring in the observed abundances of C,N, and H_2O. In the cometary model, a lower bound for this number corresponds to the mass of the Earth's oceans, because the original carbon and nitrogen abundances must have been depleted through impact erosion of the volatile compounds of these species. We then find for the mass of $T \gtrsim 100$ K ice, $M = 1.4 \times 10^{21}$kg, giving a total cometary mass of $\sim 2.8 \times 10^{21}$kg. This makes the cometary contribution to the final bombardment 15 to 30%, as was already suggested above (see Table I).

The important conclusion from this analysis—barring an explanation in terms of elemental fractionation due to preferential impact erosion of gas-phase volatiles—is that a mixture of comets formed at different places in the solar nebula is required to reproduce the volatile mixture observed on the Earth. Comets from the Uranus-Neptune region alone would deliver abundances of noble gases that would be several orders of magnitude too high. A second conclusion from this model is that the molecular nitrogen that dominates the Earth's atmosphere was delivered in the form of nitrogen compounds, not as N_2.

Neon provides another approach to the history of volatiles on these planets. This gas is not trapped in ice for temperatues >25 K (Bar-Nun et al. 1985) and therefore was presumably not brought to the planets in icy planetesimals. Owen et al. (1993) noted the close association of solar type neon with low values of $^{36}Ar/^{132}Xe$ and $^{84}Kr/^{132}Xe$ in submarine basalts, suggesting that both are manifestations of the internal rocky reservoir. In other words, the idea is that the neon was delivered to the planets in the rocky material that formed them. This appears in fact to be *required* by Hiyagon's (1994) demonstration that neon quickly diffuses out of rocks at relatively modest temperatures. Hence, there seems to be no way to subduct Ne-containing sediments to depths in the mantle from which they could be subsequently recirculated as submarine basalts.

We are then left with a dilemma. Solar neon has a value of $^{20}Ne/^{22}Ne = 13.7\pm0.3$ (Geiss 1973), whereas in atmospheric neon, $^{20}Ne/^{22}Ne = 9.8$. Evidently the original neon whose traces we now find in submarine basalts was fractionated by some process upon reaching the atmosphere (Craig and Lupton 1976). Something similar may have happened on Mars, where $^{20}Ne/^{22}Ne = 10.1\pm0.7$ (Wiens et al. 1986). On Venus, however, $^{20}Ne/^{22}Ne = 10.1\pm0.7$ (Istomin et al. 1982) and the amount of this gas, 3×10^{21}cc, is also high compared with the 7×10^{19}cc on Earth. Both Rayleigh diffusion and Jeans escape would reduce the abundance while fractionating the isotopes.

What caused the fractionation, and why did it only affect Ne and not C, N, or O? Perhaps the process producing the fractionation took place prior to the arrival of the bulk of the volatiles by cometary bombardment. Then the effect on CNO would have been overwhelmed by the new unfractionated material, whereas no new neon would have been included in the cometary volatiles. Why did the neon on Venus experience less fractionation than that on Mars and Earth? Whatever the reason, it is important to note that it is not possible to bring in the neon on Venus by CI chondrites, because the argon/krypton ratio on Venus is completely different from the "chondritic pattern" (Fig. 6). Owen et al. (1992) have argued that this difference results from the delivery of the Venus heavy noble gases by icy planetesimals formed at $T \lesssim 30$ K. Such objects are represented by Kuiper belt comets (Duncan et al. 1988; Jewitt and Luu 1993).

To summarize, it seems very clear that something other than chondrites brought the noble gases to the inner planets. Laboratory experiments on gas trapping in ice formed at low temperatures show that comets provide a credible alternative for argon, krypton and xenon. Barring elemental fractionation from impact erosion, comets formed at higher temperatures (in the Jupiter region) would be needed to supply the observed amount of H_2O, C, and N. (It is unclear that the time scales for comets scattered from the Jupiter region would be consistent with the roughly 100 Myr half-life seen in the lunar cratering record. However, if comets comprised only a fraction of the net impacting flux, this objection might not be important.) Planetary neon could have been delivered by the rocks comprising the bulk of the planets. If this interpretation is correct, there is no need to invoke CI chondrites as contributors to the Earth's volatile inventory. Much more work is needed before this model can be accepted, however, starting with a search for noble gases in comets and going on to better studies of the noble gases in the atmospheres of Mars and Venus.

XI. IMPACT DELIVERY OF PREBIOTIC ORGANICS

The heavy bombardment contributed to the early terrestrial prebiotic organic inventory in at least two ways. If the early Earth had a reducing (methane/ammonia-rich) atmosphere, impacts would have shock—synthesized organic molecules in that atmosphere with high efficiency (Bar-Nun et al. 1970; Barak and Bar-Nun 1975). In such an atmosphere, impact-driven shocks may have been the dominant source of atmospheric organic synthesis. But comets, carbonaceous chondrites, and interplanetary dust particles (IDPs) are rich in organic molecules, so may have contributed directly to terrestrial prebiotic inventories. In a reducing atmosphere, this exogenous contribution would have been swamped by *in situ* production, as well as by shock synthesis. However, if early Earth had an intermediate oxidation state (carbon dioxide-rich) atmosphere, direct exogenous organic delivery may have been the dominant source of organic material on early Earth. As this chapter is concerned with impact delivery of volatiles and organics, we will consider only the exogenous source here. The remaining sources have been summarized and quantitatively compared by Chyba and Sagan (1992), following the pioneering work by Miller and Urey (1959). Sources of uncertainty in these calculations have been catalogued and, to the extent possible, evaluated by Chyba (1993b).

A. Sources of Exogenous Organics

A variety of sources deliver exogenous organic molecules more-or-less intact to Earth today. These include (Anders 1989) those interplanetary dust particles small enough to be gently decelerated by the atmosphere, and meteorites large enough not to be completely ablated, but small enough to be substantially decelerated during their fall. Some impactors catastrophically explode during atmospheric passage, as occurred over Tunguska, Siberia, in 1908 (Chyba et al. 1993; Hills and Goda 1993). Catastrophic fragmentation of a CI carbonaceous chondrite took place over Revelstoke, Canada, in 1965; photomicrographs of recovered millimeter-sized fragments reveal unheated interiors (Folinsbee et al. 1967), within which organics should have survived. (An investigation of whether exogenous organics are in fact present in the Revelstoke fragments still remains to be performed.) Finally, the discovery of apparently extraterrestrial amino acids in Cretaceous-Tertiary (K/T) boundary sediments at Stevns Klint, Denmark, has been interpreted to suggest that a large fraction of cometary organics might in fact survive giant impacts (Zhao and Bada 1989), although both xenon measurements in K/T boundary sediments (Anders 1989) and hydrodynamic simulations of organic pyrolysis in impacts (Chyba et al. 1990) argue otherwise. Zahnle and Grinspoon (1990) have invoked the collection of cometary dust as an explanation; other explanations also seem possible (Chyba 1991b; Oberbeck and Aggarwal 1992), and are discussed in Sec. XI.G below.

All of these exogenous sources of organics should have been present on early Earth. Here we estimate their quantitative importance, giving our results as an exogenous organic mass flux through time, to allow the influx to be estimated for whatever epoch a particular model for the origin of life suggests is appropriate. There are numerous uncertainties in this approach, but these are no worse than many encountered in typical estimates of endogenous sources of prebiotic organics (see, e.g., Chyba and Sagan 1991,1992).

Organic mass flux for a particular source as a function of time will scale according to

$$\dot{m}(t) = \dot{m}[\dot{n}(t)/\dot{n}(o)] = \dot{m}(0)[1 + 1.6 \times 10^{-10} e^{t/\tau}] \qquad (26)$$

where $n(t)$ is given by differentiating Eq. (9) above with respect to time. In the following sections, we consider a variety of possible mechanisms for the delivery of intact exogenous organics to the prebiotic Earth. Each of these will turn out to be quantitatively unimportant compared to organic delivery by IDPs, so we will discuss only the latter source in detail. The less important sources have been discussed in detail by Chyba and Sagan (1992).

B. Asteroid and Comet Impacts

The model developed in Sec. IV for terrestrial mass accretion during the heavy bombardment is based on counts for lunar craters larger than 4 km in diameter. By Eq. (8), these correspond to impactors with radii greater than $\sim$100 m. What is the fate of organics in such bodies that impact Earth? Objects no greater than several hundred meters in diameter will typically catastrophically fragment ("airburst") while traversing the atmosphere (Passey and Melosh 1980; Melosh 1981; Chyba et al. 1993; Chyba 1993a; Chapter by Tagliaferri et al.); this case is treated below. Those objects that do not airburst are insufficiently decelerated ("aerobraked") by a 1 bar terrestrial atmosphere for their organics to survive the heat of impact (Chyba et al. 1990). Greenberg (see the Chapter by Greenberg and Remo) has suggested, on the basis of arguments that comets are very low density objects, that the physics of cometary impacts are fundamentally different from that of more dense objects, and that organics might thereby survive such collisions. This suggestion is yet to be thoroughly investigated. However, hydrodynamic impact simulations indicate (see, e.g., O'Keefe and Ahrens 1982) that the lower the density of the projectile, the more energy is partitioned into it upon impact.

Some early-Earth models suggest that Earth may have had a dense, $\sim$10 bar CO_2 atmosphere sometime prior to 3.8 Gyr ago (Walker 1986; Kasting and Ackerman 1986). In these atmospheres, comets with radii as large as $\sim$100 m could have been sufficiently aerobraked for most organics to have survived impact with the terrestrial ocean (Chyba et al. 1990)—if they had not first catastrophically disrupted in the atmosphere. Carbonaceous asteroids of similar size would not have been sufficiently aerobraked.

Following Delsemme's (1991) analysis of comet Halley data, we treat comets as 14% organic carbon by mass. If comets contributed 10% of the impacting mass in $\sim$100 m objects during the heavy bombardment, they would have contributed $\sim$10^{5}kg of organic carbon per year around the time of the origins of life $\sim$4 Gyr ago. It should be noted, however, that models of catastrophic fragmentation for small asteroids and comets in Earth's atmosphere (Passey and Melosh 1980; Melosh 1981; Chyba et al. 1993; Hills and Goda 1993; Chyba 1993a) strongly suggest that impactors in the 100-m-size range would airburst in the contemporary terrestrial atmosphere prior to reaching the surface, and *a fortiori* in an earlier, denser one.

C. Catastrophic Airbursts

An upper limit to organic delivery via airbursting asteroids and comets may be obtained by assuming that 100% of these objects' organics survive the airbursting of their host body. This assumption probably overestimates the contribution from this source by orders of magnitude; the Tunguska explosion had an energy of 10 to 20 megatons ($\sim$10^{24}Joules); it is difficult to imagine fragile organic molecules surviving this explosion. Nevertheless, the Revelstoke explosion of 20 kilotons did leave behind nearly pristine carbonaceous chondritic dust. In any case, even with these enormously favorable assumptions, the potential airburst source of organics on early Earth is an order of magnitude smaller than the contribution from IDPs.

D. Meteorites

Anders (1989) has estimated contemporary terrestrial accretion of intact organic matter from meteorites, finding about 8 kg of organic carbon per year. Despite uncertainties, this source is clearly of negligible importance.

E. Interplanetary Dust Particles (IDPs)

Anders (1989) has also estimated the amount of intact organic matter reaching the contemporary Earth in IDPs. Most of the mass in IDPs is in the $\sim 100\,\mu$m radius range, or, for a typical initial density (Love and Brownlee 1991) of 1 g cm^{-3}, in particles with masses $\sim 10^{-5}$g. The mass scaling in Eq. (11) implies most incident mass lies in the largest particles, but in fact this scaling breaks down for particles with masses below about 100 g, probably due to dust production from larger bodies (Kyte and Wasson 1986). The mass flux in particles below 100 g is observed (Hughes 1978; Grün et al. 1985) to increase with decreasing size until it peaks at $\sim 10^{-5}$g. This conclusion was recently confirmed by the examination of impact craters on the Long Duration Exposure Facility Satellite (Love and Brownlee 1993). This peak reaches a level 10^5 to 10^6 times higher than would be the case if the IDP mass spectrum did not deviate from power law scaling.

Anders (1989) takes IDPs to be $\sim 10\%$ organic carbon by mass, a mean content determined from 30 IDPs by X ray analysis (Schramm et al. 1989; see also Gibson 1992). He takes IDPs in the 10^{-12} to 10^{-6}g (~ 0.6–60 μm radius) range to be sufficiently gently decelerated during atmospheric entry to deliver their organics intact. Anders' 60 μm upper limit is in good agreement with results of both theoretical modeling (Love and Brownlee 1991) and direct examination of IDPs (Brownlee 1985). His lower limit is the size below which IDP organics would be destroyed by ultraviolet photolysis; this choice could vary greatly without significant quantitative effects on the final result.

It remains unclear whether the more fragile organic species would survive IDP atmospheric passage (Chyba 1993b), especially at the high velocities appropriate to cometary-derived particles (Love and Brownlee 1991). It is also unknown whether IDPs actually contain molecules of evident prebiotic interest, or whether the bulk of their organic material is in the (presumably) less interesting form of kerogen-like heteropolymers. Models predict cometary IDPs to be typically heated to around 900 K for several seconds during atmospheric entry (Flynn 1989), consistent with annealing temperatures for nuclear tracks in IDPs (Bradley et al. 1984). Atmosphere heating is a strong function of particle size and entry angle (Love and Brownlee 1991). A naive application of thermal decomposition rates from amino acids in solution (Chyba et al. 1990) suggests these molecules would survive around 800 K for only 1 s, but these rates are of dubious relevance to degradation of amino acids in a mineral matrix. (However, it is of interest that the alanine decarboxylation half-life seems to be the same whether performed in aqueous or solid phase [Bada 1991], suggesting the aqueous phase results may be more broadly applicable than has sometimes been suggested [Chyba 1991b].) Further laboratory work in this area is eagerly anticipated.

Earth is currently accreting (Kyte and Wasson 1986) $\sim 3 \times 10^6$kg yr^{-1} of 10^{-12} to 10^{-6}g IDPs, or $\sim 3 \times 10^5$kg yr^{-1} of intact organics (Anders 1989). How to scale this back in time through the heavy bombardment is unclear; would the mass influx peak at $\sim 10^{-5}$g have persisted? There is some suggestive, though hardly compelling, data relevant to this question.

The current terrestrial mass influx from IDPs found by Hughes (1978) from meteor observations is in good agreement with accretion rates inferred from terrestrial and lunar geochemical data. These data, summarized in Chyba and Sagan (1992), rely on a number of different sampling methods, and exhibit a remarkably constant net IDP mass flux over the past $\sim$3.6 Gyr, suggesting that the current IDP mass flux is at least not a recent anomaly. However, most of these data take the form of an integral over the IDP size distribution, so cannot prove that the current shape of the IDP flux curve has remained unchanging through time. In any case, they do not extend back prior to the end of the heavy bombardment.

We cannot say with confidence how to scale the IDP flux back beyond 3.6 Gyr ago. If we simply scale the IDP flux linearly with the lunar impact record, using Eq. (26), we find a flux of organic carbon due to IDPs 4 Gyr ago of about 10^8kg yr^{-1}. In an intermediate oxidation state atmosphere on early Earth, this could have been the quantitatively dominant source of prebiotic organics (Chyba and Sagan 1992).

F. Accretion of Interstellar Dust

Another source of exogenous organics on early Earth, independent of the heavy bombardment, would have been the terrestrial accretion of dust as during solar system passage through interstellar clouds. Greenberg (1981) has estimated that during its first 700 Myr, Earth should have passed through some 4 to 5 such clouds, accreting organic molecules during each 10^6yr passage at a rate 10^6 to 10^7kg yr^{-1}. Even during cloud passage, this source would have been 10 to 100 times smaller than that due to IDPs.

G. Lessons from the Cretaceous-Tertiary Boundary

Zhao and Bada (1989) have reported the discovery of high abundance of the presumably extraterrestrial amino acids α-amino-isobutyric acid (AIB) and isovaline at the Stevns Klint Cretaceous-Tertiary (K/T) boundary site. (AIB and isovaline are extremely rare in the biosphere, but common in meteorites; moreover, the K/T isovaline is racemic, and thought to have a racemization time scale much longer than 65 Myr [Bada 1991].) Yet, as discussed above, we would expect even the hardiest organics to be incinerated in the roughly 10^8 megaton explosion of the K/T collision. Yet the AIB/iridium ratio at Stevns/Klint is substantially higher than that found in the Murchison meteorite. More puzzling still, the amino acids were found tens of centimeters above and below the K/T boundary, but not in the boundary clay itself.

Zahnle and Grinspoon (1990) have taken this sedimentary record at face value, and suggested that Earth swept up the amino acids "gently and non-destructively" in cometary dust over a period of about 10^5 yr, analogous to the collection of organic-rich IDPs by Earth today, but in much greater concentration. The boundary clay itself would then record the impact of a fragment of the parent comet, an impact which amino acids did not survive (see also Steel 1992).

Other authors have championed the idea of amino acid production by shock synthesis in the K/T collision (Oberbeck and Aggarwal 1992). Simply taking the amino acid concentration at the boundary as that appropriate to a K/T-sized impact, it is possible to scale production (or delivery) to the impactor flux on early Earth. Such scaling has led Oberbeck and Aggarwal (1992) to conclude that amino acid production from cometary collisions on early Earth dominated that due to coronal discharge in a carbon dioxide atmosphere for the first 4 Gyr of Earth history. Understanding the global record of amino acids at the K/T boundary, and their delivery or production mechanism, is an important "ground truth" check on models of exogenous organic input on early Earth.

XII. IMPACT DELIVERY AND YOU

The apparent temporal coincidence between the termination of the heavy bombardment and the origin of life on Earth seems likely to have been more than just a coincidence. Late-accreting impactors, especially comets, probably played a major role in stocking an early, volatile-poor Earth with the critical "biogenic" elements. At the same time, the evolution of life may well have been impossible until the heavy bombardment largely subsided, by around 3.8 Gyr ago.

The delivery of intact organic molecules from extraterrestrial sources appears to be much more problematic. IDPs may well have been important, but the quantitative importance of this source turns on the nature of the early terrestrial atmosphere. In an early CO_2-rich atmosphere, IDPs may have been the dominant source of organics. Whether this is telling us that the origin of life on Earth was linked to exogenous sources not only at the elemental level, but also at the chemical level, remains unclear. The sphinx of evolution is the origin of life, and virtually all steps beyond the prebiotic synthesis of the first monomers remain unclear. Whatever role extraterrestrial organic molecules may have played, they at most provided an alternate mechanism for securing the one part of the puzzle that we can reproduce in the laboratory: the origin of organic monomers. Beyond this lies vast uncertainty.

REFERENCES

Anders, E. 1989. Pre-biotic organic matter from comets and asteroids. *Nature* 342:255–257.

Anders, E., and Grevesse, N. 1989. Abundances of the elements: Meteoritic and solar. *Geochim. Cosmochim. Acta* 53:197–214.

Anderson, D. L. 1993. ^{3}He from the mantle: Primordial signal or cosmic dust? *Science* 261:170–173.

Bada, J. L. 1991. Amino acid cosmogeochemistry. *Phil. Trans. Roy. Soc. Lond. B* 333:349–358.

Baldwin, R. B. 1987*a*. On the relative and absolute ages of seven lunar front face basins. I. From viscosity arguments. *Icarus* 71:1–18.

Baldwin, R. B. 1987*b*. On the relative and absolute ages of seven lunar front face basins. II. From crater counts. *Icarus* 71:19–29.

Barak, I., and Bar-Nun, A. 1975. The mechanisms of amino acid synthesis by high temperature shock-waves. *Origins of Life* 6:483–506.

Bar-Nun, A., Bar-Nun, N., Bauer, S. H., and Sagan, C. 1970. Shock synthesis of amino acids in simulated primitive environments. *Science* 168:470–473.

Bar-Nun, A., Herman, G., Laufer, D., and Rappoport, M. L. 1985. Trapping and release of gases by water ice and implications for icy bodies. *Icarus* 63:317–332.

Bar-Nun, A., Kleinfeld, I. and Kochavi, E. 1988. Trapping of gas mixtures by amorphous water ice. *Phys. Rev. B* 38:7749–7754.

Baross, J. A., and Hoffman, S. E. 1985. Submarine hydrothermal vents and associated gradient environments as sites for the origin and evolution of life. *Origins of Life* 15:327–345.

Belton, M. J. S., Head, J. W., Pieters, C. M., Greeley, R., McEwan, A. S., Neukum, G., Klaasen, K. P., Anger, C. D., Carr, M. H., Chapman, C. R., Davies, M. E., Fanale, F. P., Gierasch, P. J., Greenberg, R., Ingersoll, A. P., Johnson, T., Paczkowski, B., Pilcher, C. B., and Veverka, J. 1992. Lunar impact basins and crustal heterogeneity: New western limb and far side data from Galileo. *Science* 255:570–576.

Berzelius, J. J. 1834. Ueber meteorstein. *Ann. Phys. Chem.* 33:113.

Bogard, D. D., Hörz, F, and Johnson, P. H. 1986. Shock implanted noble gases: An experimental study with implications for the origin of Martian gases in Shergottite meteorites. *J. Geophys. Res.* 91:99–114.

Boston, P. J., Ivanov, M. V., and McKay, C. P. 1992. On the possibility of chemosynthetic ecosystems in subsurface habitats on Mars. *Icarus* 95:300–308.

Bradley, J. P., Brownlee, D. E. and Fraundorf, P. 1984. Discovery of nuclear tracks in interplanetary dust. *Science* 226:1432–1434.

Brownlee, D. E. 1985. Cosmic dust: Collection and research. *Ann. Rev. Earth Planet. Sci.* 13:147–173.

Bullfinch, T. 1979. *Bullfinch's Mythology* (New York: Crown Publishers).

BVSP (Basaltic Volcanism Study Project). 1981. *Basaltic Volcanism on the Terrestrial Planets* (New York: Pergamon Press).

Cameron, A. G. W. 1983. Origin of the atmospheres of the terrestrial planets. *Icarus* 56:195–201.

Cameron, A. G. W. 1986. The impact theory for the origin of the Moon. In *Origin of the Moon*, eds. W. K. Hartmann, R. J. Phillips and G. J. Taylor (Houston: Lunar and Planetary Inst.), pp. 609–616.

Campbell, M. J., O'Leary, B. T., and Sagan, C. 1969. Moon: Two new mascon basins. *Science* 164:1273–1275.

Carlson, R. W., and Lugmair, G. W. 1979. Sm-Nd constraints on early lunar differentiation and the evolution of KREEP. *Earth Planet. Sci. Lett.* 45:123–132.

Carlson, R. W., and Lugmair, G. W. 1988. The age of ferroan anorthosite 60025: Oldest crust on a young Moon? *Earth Planet. Sci. Lett.* 90:119–130.

Chamberlin, T. C., and Chamberlin, R. T. 1908. Early terrestrial conditions that may have favored organic synthesis. *Science* 28:897–911.

Chang, S., DesMarais, D., Mack, R., Miller, S. L., and Strathearn, G. E. 1983. Prebiotic organic syntheses and the origin of life. In *Earth's Earliest Biosphere*, ed. J. W. Schopf (Princeton, N. J.: Princeton Univ. Press), pp. 53–92.

Chou, C.-L. 1978. Fractionation of siderophile elements in the earth's upper mantle. *Proc. Lunar Planet. Sci. Conf.* 9:219–230.

Chyba, C. F. 1987. The cometary contribution to the oceans of primitive Earth. *Nature* 330:632–635.

Chyba, C. F. 1990. Impact delivery and erosion of planetary oceans in the early inner solar system. *Nature* 343:129–133.

Chyba, C. F. 1991*a*. Terrestrial mantle siderophiles and the lunar impact record. *Icarus* 92:217–233.

Chyba, C. F. 1991*b*. Extraterrestrial amino acids and terrestrial life. *Nature* 348:113–114.

Chyba, C. F. 1993*a*. Explosions of small Spacewatch objects in the Earth's atmosphere. *Nature* 363:701–703.

Chyba, C. F. 1993*b*. The violent environment of the origin of life: Progress and uncertainties. *Geochim. Cosmochim. Acta* 57:3351–3358.

Chyba, C. F., and Sagan, C. 1991. Electrical energy sources for organic synthesis on the early Earth. *Origins of Life* 21:3–17.

Chyba, C. F., and Sagan, C. 1992. Endogenous production, exogenous delivery and impact-shock synthesis of organic molecules: An inventory for the origins of life. *Nature* 355:125–132.

Chyba, C. F., Thomas, P. J., Brookshaw, L., and Sagan, C. 1990. Cometary delivery of organic molecules to the early Earth. *Science* 249:366–373.

Chyba, C. F., Thomas, P. J. and Zahnle, K. J. 1993. The 1908 Tunguska explosion: Atmospheric disruption of a stony asteroid. *Nature* 361:40–44.

Clark, B. C. 1987. Comets, volcanism, the salt-rich regolith, and cycling of volatiles on Mars. *Icarus* 71:250–256.

Corliss, J. B., Baross, J. A., and Hoffman, S. E. 1981. An hypothesis concerning the relationship between submarine hot springs and the origin of life on Earth. *Oceanologica Acta Suppl.* 4:59–69.

Craig, H., and Lupton, J. E. 1976. Primordial neon, helium, and hydrogen in ocean basalts. *Earth Planet. Sci. Lett.* 31:369–385.

Croft, S. K. 1985. The scaling of complex craters. *Proc. Lunar Planet. Sci. Conf.* 15, *J. Geophys. Res. Suppl.* 90:828–842.

Dalrymple, G. B., and Ryder, G. 1993. $^{40}Ar/^{39}Ar$ age spectra of Apollo 15 impact melt rocks by laser step-heating and their bearing on the history of lunar basin formation. *J. Geophys. Res.* 98:13085–13095.

Delsemme, A. H. 1991. Nature and history of the organic compounds in comets: An astrophysical view. In *Comets in the Post-Halley Era*, eds. R. L. Newburn, Jr., M. Neugebauer and J. Rahe (Dordrecht: Kluwer), pp. 377–428.

Delsemme, A. H. 1992. Cometary origin of carbon, nitrogen and water on the Earth. *Origins of Life* 21:279–298.

Dones, L., and Tremaine, S. 1993. Why does the Earth spin forward? *Science* 259:350–354.

Dreibus, G., and Wänke, H. 1987. Volatiles on Earth and Mars: A comparison. *Icarus* 71:225–240.

Dreibus, G., and Wänke, H. 1989. Supply and loss of volatile constituents during the accretion of terrestrial planets. In *Origin and Evolution of Planetary and Satellite*

Atmospheres, eds. S. K. Atreya, J. B. Pollack and M. S. Matthews (Tucson: Univ. of Arizona Press), pp. 268–288.

Duncan, M., Quinn, T., and Tremaine, S. 1988. The origin of short period comets. *Astrophys. J. Lett.* 328:69–73.

Dziewonski, A. M., Hales, A. L., and Lapwood, E. R. 1975. Parametrically simple Earth models consistent with geophysical data. *Phys. Earth Planet. Int.* 10:12–48.

Eberhardt, P., Dolder, U., Schulte, W., Krankowsky, D., Lämmerzahl, P., Berthelier, J. J., Woweries, J., Stubbemann, U., Hodges, R. R., Hoffman, J. H., and Illiano, J. M. 1987. The D/H ratio in water from comet P/Halley. *Astron. Astrophys.* 187:435–437.

Encrenaz, Th., Puget, J. L., and D'Hendecourt, L. 1991. Interstellar matter to comets: Elemental abundances in interstellar dust and in comet Halley. *Space Sci. Rev.* 56:83–92.

Feldman, P. D., Davidson, A. D., Blair, W. P., Bowers, C. W., Dixon, W. V., Durrance, S. T., Ferguson, H. C., Henry, R. C., Kimble, R. A., Kriss, G. A., Kruck, J. W., Long, K. S., Moos, W. H., Vancura, O., and Gull, T. R. 1991. HUT observations of Comet Levy (1990c). *Astrophys. J. Lett.* 379:37–40.

Fernández, J. A. 1978. Mass removed by the outer planets in the early early solar system. *Icarus* 34:173–181.

Fernández, J. A. 1985. The formation and dynamical survival of the comet cloud. In *Dynamics of Comets: Their Origin and Evolution*, eds. A. Carusi and G. B. Valsecchi (Dordrecht: D. Reidel), pp. 45–70.

Fernández, J. A., and Ip, W.-H. 1983. On the time evolution of the cometary influx in the region of the terrestrial planets. *Icarus* 54:377–387.

Fernández, J. A., and Ip, W.-H. 1984. Some dynamical aspects of the accretion of Uranus and Neptune: The exchange of angular momentum with planetesimals. *Icarus* 58:109–120.

Fernández, J. A., and Ip, W.-H. 1987. Time-dependent injection of Oort cloud comets into Earth-crossing orbits. *Icarus* 71:46.

Flynn, G. J. 1989. Atmospheric entry heating: A criterion to distinguish between asteroidal and cometary sources of interplanetary dust. *Icarus* 77:287–310.

Folinsbee, R. E., Douglas, J. A. V., and Maxwell, J. A. 1967. Revelstoke, a new type I carbonaceous chondrite. *Geochim. Cosmochim. Acta* 31:1625–1635.

Frohlich, C., and Grand, S. P. 1990. The fate of subducting slabs. *Nature* 347:333–334.

Geiss, J. 1973. Solar wind composition and implications about the history of the solar system. *Proc. 13th Intl. Cosmic Ray Conf.* 5:3375–3398.

Geiss, J. 1987. Composition measurements and the history of cometary matter. *Astron. Astrophys.* 187:859–866.

Geiss, J. 1988. Composition in Halley's comet: Clues to origin and history of cometary matter. *Rev. Modern Astron.* 1:1–27.

Gibson, E. K. 1992. Volatiles in interplanetary dust particles: A review. *J. Geophys. Res.* 97:3865–3875.

Greenberg, M. J. 1981. chemical evolution of interstellar dust—a source of prebiotic material? In *Comets and the Origin of Life*, ed. C. Ponnamperuma (Dordrecht: D. Rediel), pp. 111–127.

Greenberg, R. 1989. Planetary accretion. In *Origin and Evolution of Planetary and Satellite Atmospheres*, eds. S. K. Atreya, J. B. Pollack and M. S. Matthews (Tucson: Univ. of Arizona Press), pp. 137–164.

Grieve, R. A. F. 1984. The impact cratering rate in recent time. *Proc. Lunar Planet. Sci. Conf. 14, J. Geophys. Res. Suppl.* 89:403–408.

Grinspoon, D. H. 1988. Large Impact Events and Atmospheric Evolution on the Terrestrial Planets. Ph.D. Thesis, Univ. of Arizona.

Grün, E., Zook, H. A., Fechtig, H., and Giese, R. H. 1985. Collisional balance of the meteoritic complex. *Icarus* 62:244–272.

Hahn, G., and Bailey, M. E. 1990. Rapid dynamical evolution of giant comet Chiron. *Nature* 348:132–136.

Harper, C. L., and Jacobsen, S. B. 1992. Evidence from coupled [147]Sm-[143]Nd and [146]Sm-[142]Nd systematics from very early (4.5-Gyr) differentiation of the Earth's mantle. *Nature* 360:728–732.

Hartmann, W. K. 1980. Dropping stones in magma oceans: Effects of early lunar cratering. In *Proceedings of the Conference on the Lunar Highlands Crust*, (Houston: Lunar and Planetary Inst.), pp. 155–171.

Hartmann, W. K. 1987. A satellite-asteroid mystery and a possible early flux of scattered C-class asteroids. *Icarus* 71:57–68.

Hartmann, W. K. 1990. Additional evidence about an early intense flux of C asteroids and the origin of Phobos. *Icarus* 87:236–240.

Hartmann, W. K. 1992. The "Voyager paradigm" of lunar and planetary cratering: Further revisions from lunar studies. *Bull. Amer. Astron. Soc.* 24:1022 (abstract).

Hayes, J. M., Kaplan, J. R., and Wedeking, K. W. 1983. Precambrian organic geochemistry, preservation of the record. In *Earth's Earliest Biosphere*, ed. J. W. Schopf (Princeton, N. J.: Princeton Univ. Press), pp. 93–134.

Hennet, R. J.-C., Holm, N. G., and Engel, M. H. 1992. Abiotic synthesis of amino acids under hydrothermal conditions and the origin of life: A perpetual phenomenon? *Naturwiss.* 79:361–365.

Heymann, D. 1971. The inert gases. In *Handbook of Elemental Abundances in Meteorites*, ed. B. Bason (New York: Gordon and Breach), pp. 29–66.

Hills, J., and Goda, P. 1993. The fragmentation of small asteroids in the atmosphere. *Astron. J.* 105:1114–1144.

Hiyagon, H. 1994. Retention of solar helium and neon in IDPs in deep sea sediment. *Science* 263:1257–1259.

Horowitz, N. H. 1986. *To Utopia and Back: The Search for Life in the Solar System* (New York: W. H. Freeman).

Housen, K. R., Schmidt, R. M., and Holsapple, K. A. 1983. Crater ejecta scaling laws: Fundamental forms based on dimensional analysis. *J. Geophys. Res.* 88:2485–2499.

Huebner, W. F., and Boice, D. C. 1992. Comets as a possible source of prebiotic molecules. *Origins of Life* 21:299–316.

Hughes, D. W. 1978. Meteors. In *Cosmic Dust*, ed. J. A. M. McDonnell (New York: J. Wiley & Sons), pp. 123–185.

Ip, W.-H. 1977. On the early scattering processes of the outer planets. In *Comets-Asteroids-Meteorites: Interrelations, Evolution and Origin*, ed. A. H. Delsemme (Toledo, Oh.: Univ. of Toledo Press), pp. 485–490.

Ip, W.-H., and Fernández, J. A. 1988. Exchange of condensed matter among the outer and terrestrial protoplanets and the effect on surface impact and atmospheric accretion. *Icarus* 74:47–61.

Istomin, V. G., Grechnev, K. V., and Kochnev, J. V. A. 1982. Preliminary results of mass-spectrometric measurements on board the Venera 13 and 14 probes. *Pismo. Astron. Zh.* 8:391–398.

Jewitt, D., and Luu, J. 1993. Discovery of the candidate Kuiper belt object 1992 QB1. *Nature* 362:730–732.

Jones, J. H., Capobianco, C. J., and Drake, M. J. 1992. Siderophile elements and the Earth's formation. *Science* 257:1281–1282.

Karlsson, H. R., Clayton, R. N., Gibson, E. K., and Mayeda, T. K. 1992. Water in SNC meteorites: Evidence for a Martian hydrosphere. *Science* 255:1409–1411.

Kasting, J. F. 1993. Earth's early atmosphere. *Science* 259:920–925.

Kasting, J. F., and Ackerman, T. P. 1986. Climatic consequences of very high carbon dioxide levels in the Earth's early atmosphere. *Science* 234:1383–1385.

Kasting, J. F., Eggler, D. H. and Raeburn, S. P. 1993. Mantle redox evolution and the oxidation state of the Archean atmosphere. *J. Geology* 101:245–257.

Kessler, D. J. 1981. Derivation of the collision probability between orbiting objects: The lifetimes of Jupiter's outer moons. *Icarus* 48:39–48.

Kissel, J., and Krueger, F. R. 1987*a*. The organic component in dust from comet Halley as measured by the PUMA mass spectrometer on board Vega 1. *Nature* 326:755–760.

Kissel, J., and Krueger, F. R. 1987*b*. Organic dust in comet Halley. *Nature* 328:117.

Krankowsky, D. 1991. The composition of comets. In *Comets in the Post-Halley Era*, eds. R. L. Newburn, Jr., M. Neugebauer and J. Rahe (Dordrecht: Kluwer), pp. 855–878.

Kronk, G. W. 1984. *Comets: A Descriptive Catalogue* (Hillside, N. J.: Enslow).

Kuhn, W. R., and Atreya, S. K. 1979. Ammonia photolysis and the greenhouse effect in the primordial atmosphere of the Earth. *Icarus* 37:207–213.

Kuiper, G. P. 1951. On the origin of the solar system. In *Astrophysics*, ed. J. A. Hynek (New York: McGraw-Hill), pp. 357–424.

Kyte, F. T., and Wasson, J. T. 1986. Accretion rate of extraterrestrial matter: Iridium deposited 33 to 67 million years ago. *Science* 232:1225–1229.

Laufer, D., Kovchavi, E., and Bar-Nun, A. 1987. Structure and dynamics of amorphous water ice. *Phys. Rev. B* 36:9219–9227.

Levine, J. S., and Augustsson, T. R. 1985. The photochemistry of biogenic gases in the early and present atmosphere. *Origins of Life* 15:299–318.

Lewis, J. 1974. The temperature gradient in the solar nebula. *Science* 186:440–443.

Lewis, J., Barshay, S. S., and Noyes, B. 1979. Primoridal retention of carbon by the terrestrial planets. *Icarus* 37:190–206.

Love, S. G., and Brownlee, D. E. 1991. Heating and thermal transformation of micrometeoroids entering the Earth's atmosphere. *Icarus* 89:26–43.

Love, S. G., and Brownlee, D. E. 1993. A direct measurement of the terrestrial mass accretion rate of cosmic dust. *Science* 262:550–553.

Luu, J., and Jewitt, D. 1993. 1993 FW. *IAU Circ.* 5730.

Maher, K. A., and Stevenson D. J. 1988. Impact frustration of the origin of life. *Nature* 331:612–614.

McKay, C. P. 1991. Urey Prize lecture: Planetary evolution and the origin of life. *Icarus* 91:93–100.

McKinnon, W. B. 1989. Impacts giveth and impacts taketh away. *Nature* 338:465–466.

McKinnon, W. B., and Schenk, P. M. 1985. Ejecta blanket scaling on the Moon and mercury—And inferences for projectile populations. *Lunar Planet. Sci.* XVI:544–545 (abstract).

McKinnon, W. B., Chapman, C. R., and Housen, K. R. 1990. Cratering of the Uranian satellites. In *Uranus*, eds. J. T. Bergstralh, E. D. Miner and M. S. Matthews (Tucson: Univ. of Arizona Press), pp. 629–692.

Melosh, H. J. 1981. Atmospheric breakup of terrestrial impactors. In *Multi-Ring Basins*, eds. P. H. Schultz and R. B. Merril (New York: Pergamon Press), pp. 29–36.

Melosh, H. J. 1989. *Impact Cratering: A Geologic Process* (New York: Oxford Univ. Press).

Melosh, H. J., and Vickery, A. M. 1989. Impact erosion of the primordial atmosphere of Mars. *Nature* 338:487–489.

Miller, S. L., and Bada, J. L. 1988. Submarine hot springs and the origin of life. *Nature* 334:609–611.

Miller, S. L., and Bada, J. L. 1989. Origin of life. *Nature* 337:23.

Miller, S. L., and Urey, H. 1959. Organic compound synthesis on the primitive Earth. *Science* 130:245–251.

Murthy, V. R. 1991. Early differentiation of the Earth and the problem of mantle siderophile elements: A new approach. *Science* 253:303–306.

Musselwhite, D. S., Drake, M. J., and Swindle, T. D. 1991. Early outgassing of Mars supported by differential water solubility of iodine and xenon. *Nature* 352:697–699.

Newsom, H. E. 1990. Accretion and core formation in the Earth: Evidence from siderophile elements. In *Origin of the Earth*, eds. H. E. Newsom and J. H. Jones (New York: Oxford Univ. Press), pp. 273–288.

Newsom, H. E., and Sims, W. W. 1991. Core formation during early accretion of the Earth. *Science* 252:926–933.

Nier, A. O., and Schlutter, D. J. 1990. Helium and neon isotopes in stratospheric particles. *Meteoritics* 25:263–267.

Oberbeck, V. R., and Aggarwal, H. 1992. Comet impacts and chemical evolution on the bombarded Earth. *Origins of Life* 21:317–338.

Oberbeck, V. R., and Fogleman, G. 1989a. Impacts and the origin of life. *Nature* 339:434.

Oberbeck, V. R., and Fogleman, G. 1989b. Estimates of the maximum time required to originate life. *Origins of Life* 19:549–560.

O'Keefe, J. D., and Ahrens, T. J. 1982. The interaction of the Cretaceous/Tertiary bolide with the atmosphere, ocean, and solid Earth. In *Geological Implications of Impacts of Large Asteroids and Comets on the Earth*, eds. L. T. Silver and P. H. Schultz, Geological Soc. of America Special Paper 190 (Boulder: Geological Soc. of America), pp. 103–127.

O'Leary, B. T., Campbell, M. J., and Sagan, C. 1969. Lunar and planetary mass concentrations. *Science* 165:651–657.

Olsson-Steel, D. 1987. Collisions in the solar system—IV. Cometary impacts upon the planets. *Mon. Not. Roy. Astron. Soc.* 227:501–524.

Olsson-Steel, D. 1990. The asteroidal impact rate upon the terrestrial planets: An update. *Proc. Astron. Soc. Australia* 8:303–307.

O'Neill, H. 1992. Siderophile elements and the Earth's formation. *Science* 257:1282–1284.

Oró, J. 1961. Comets and the formation of biochemical compounds on the primitive Earth. *Nature* 190:389–390.

Owen, T. 1992. The composition and early history of the atmosphere of Mars. In *Mars*, eds. H. H. Kieffer, B. M. Jakosky, C. W. Snyder and M. S. Matthews (Tucson: Univ. of Arizona Press), pp. 818–834.

Owen, T. 1994. The search for other planets: Clues from the solar system. *Astrophys. Space Sci.*, in press.

Owen, T., and Bar-Nun, A. 1993. Noble gases in atmospheres. *Nature* 361:693–694.

Owen, T., Bar-Nun, A., and Kleinfeld, I. 1991. Noble gases in terrestrial planets: Evidence of cometary impacts? In *Comets in the Post-Halley Era*, eds. R. L. Newburn, Jr., M. Neugebauer and J. Rahe (Dordrecht: Kluwer), pp. 429–438.

Owen, T., Bar-Nun, A., and Kleinfeld, I. 1992. Possible cometary origin of heavy noble gases in the atmospheres of Venus, Earth and Mars. *Nature* 358:43–46.

Ozima, M., and Wada, N. 1993. Noble gases in atmospheres. *Nature* 361:693.

Passey, Q. R., and Melosh, H. J. 1980. Effects of atmospheric breakup on crater field formation. *Icarus* 42:211–233.

Pepin, R. O. 1991. On the origin and early evolution of terrestrial planet atmospheres and meteoritic volatiles. *Icarus* 92:2–79.

Prinn, R. G., and Fegley, B. 1989. Solar nebula chemistry: Origin of planetary, satellite and cometary volatiles. In *Origin and Evolution of Planetary and Satellite Atmospheres*, eds. S. K. Atreya, J. B. Pollack and M. S. Matthews (Tucson: Univ. of Arizona Press), pp. 78–136.

Rabinowitz, D. L., Gehrels, T., Scotti, J. V., McMillan, R. S., and Perry, M. L. 1993. The terrestrial asteroid belt: A new population of near-Earth asteroids. *Nature*, in press.

Robert, F., Rejou-Michel, A., and Javoy, M. 1992. Oxygen isotopic homogeneity of the Earth: New evidence. *Earth Planet. Sci. Lett.* 108:1–9.

Ryder, G. 1990. Lunar samples, lunar accretion and the early bombardment of the Moon. *Eos: Trans. AGU* 71:313, 322–323.

Ryder, G., and Wood, J. A. 1977. Serenitatis and Imbrium impact melts: Implications for large-scale layering in the lunar crust. *Proc. Lunar Sci. Conf.* 8:655–668.

Safronov, V. S. 1972. *Evolution of the Protoplanetary Cloud and Formation of the Earth and the Planets* (Moscow: Nauka), English trans. Israel Program for Scientific Translations (Jerusalem).

Sagan, C. 1974. The origin of life in a cosmic context. *Origins of Life* 5:497–505.

Sagan, C., and Chyba, C. 1994. The early faint sun "paradox": Organic shielding shielding of uv-labile greenhouse gases. *Icarus*, submitted.

Schidlowski, M. A. 1988. 3,800-million-year isotope record of life from carbon in sedimentary rocks. *Nature* 333:313–318.

Schidlowski, M. A., Hayes, J. M., and Kaplan, I. R. 1983. Isotopic inferences of ancient biochemistries: Carbon, sulfur, hydrogen, and nitrogen. In *Earth's Earliest Biosphere*, ed. J. W. Schopf (Princeton, N. J.: Princeton Univ. Press), pp. 149–186.

Schlesinger, G., and Miller, S. L. 1983*a*. Prebiotic synthesis in atmospheres containing CH_4, CO, and CO_2. I. Amino acids. *J. Molec. Evol.* 19:376–382.

Schlesinger, G., and Miller, S. L. 1983*b*. Prebiotic synthesis in atmospheres containing CH_4, CO, and CO_2. I. Amino acids. *J. Molec. Evol.* 19:383–390.

Schmidt, R. M. 1980. Meteor crater: Energy of formation—Implications of centrifuge scaling. *Proc. Lunar Planet. Sci. Conf.* 11:2099–2128.

Schmidt, R. M., and Housen, K. R. 1987. Some recent advances in the scaling of impact and explosion cratering. *Intl. J. Impact Eng.* 5:543–560.

Schopf, J. W. 1993. Microfossils of the early Archean apex chert: New evidence of the antiquity of life. *Science* 260:640–646.

Schopf, J. W., and Walter, M. R. 1983. Archean microfossils: New evidence of ancient microbes. In *Earth's Earliest Biosphere*, ed. J. W. Schopf (Princeton, N. J.: Princeton Univ. Press), pp. 214–239.

Schramm, L. S., Brownlee, D. E., and Wheelock, M. M. 1989. Major element composition of stratospheric micrometeorites. *Meteoritics* 24:99–112.

Shoemaker, E. M. 1983. Asteroid and comet bombardment of the Earth. *Ann. Rev. Earth Planet. Sci.* 11:461–494.

Shoemaker, E. M., and Wolfe, R.F. 1984. Evolution of the Uranus-Neptune planetesimal swarm. *Proc. Lunar Planet. Sci. Conf.* 15, *J. Geophys. Res. Suppl.* 2:780–781.

Shoemaker, E. M., and Wolfe, R. F. 1986*a*. Extinct Jupiter-family comets and cratering rates on the Galilean satellites. *Proc. Lunar Planet. Sci. Conf.* 17, *J. Geophys. Res. Suppl.* 2:799–780.

Shoemaker, E. M., and Wolfe, R. F. 1986*b*. Mass extinctions, crater ages and comet showers. In *The Galaxy and the Solar System*, eds. R. Smoluchowski, J. N. Bahcall and M. S. Matthews (Tucson: Univ. of Arizona Press), pp. 338–386.

Shoemaker, E. M., Williams, J. G., Helin, E. F., and Wolfe, R. F. 1979. Earth-crossing asteroids: Orbital classes, collision rates with Earth, and origin. In *Asteroids*, ed. T. Gehrels (Tucson: Univ. of Arizona Press), pp. 253–282.

Shoemaker, E. M., Wolfe, R. F., and Shoemaker, C. S. 1990. Asteroid and comet flux in the neighborhood of Earth. In *Global Catastrophes in Earth History*, eds. V. L. Sharpton and P. D. Ward, Geological Soc. of America Special Paper 247 (Boulder: Geological Soc. of America), pp. 155–170.

Sleep, N. H., Zahnle, K. J., Kasting, J. F., and Morowitz, H. J. 1989. Annihilation of ecosystems by large asteroid impacts on the early Earth. *Nature* 342:139–142.

Spudis, P. D., Hawke, B. R., and Lucey, P. G. 1988. Materials and formation of the Imbrium basin. *Proc. Lunar Planet. Sci. Conf.* 18, *J. Geophys. Res. Suppl.* 2:155–168.

Stacey, F. D. 1977. *Physics of the Earth* (New York: J. Wiley & Sons).

Steel, D. 1992. Cometary supply of terrestrial organics: Lessons from the K/T and the present epoch. *Origins of Life* 21:339–357.

Steel, D. I., and Baggaley, W. J. 1985. Collisions in the solar system—I. Impacts of the Apollo-Amor-Aten asteroids upon the terrestrial planets. *Mon. Not. Roy. Astron. Soc.* 212:817–836.

Stern, A. S., Green, J. C., Cash, W., and Cooke, T. A. 1992. Helium and argon abundance constraints and the thermal evolution of comet Austin (1989c1). *Icarus* 95:157–161.

Stevenson, D. J. 1983. The nature of the Earth prior to the oldest known rock record: The Hadean Earth. In *Earth's Earliest Biosphere: Its Origin and Evolution*, ed. J. W. Schopf (Princeton, N. J.: Princeton Univ. Press), pp. 32–40.

Stevenson, D. J. 1990. Fluid dynamics of core formation. In *Origin of the Earth*, eds. H. E. Newsom and J. H. Jones (New York: Oxford Univ. Press), pp. 231–249.

Strom, R. G. 1987. The solar system cratering record: Voyager 2 results at Uranus and implications for the origin of impacting objects. *Icarus* 70:517–535.

Sun, S.-S. 1984. Geochemical characteristics of Archean ultramafic and mafic volcanic rocks: Implications for mantle composition and evolution. In *Archean Geochemistry*, eds. A. Kröner, G. N. Hanson and A. M. Goodwin (Berlin: Springer-Verlag), pp. 25–46.

Swindle, T. D., Caffee, M. W., Hohenberg, C. M., and Taylor, S. R. 1986. I-Pu-Xe dating and the relative ages of the Earth and Moon. In *Origin of the Moon*, eds. W. K. Hartmann, R. J. Phillips and G. J. Taylor (Houston: Lunar and Planetary Inst.), pp. 331–357.

Sykes, M. V., and Walker, R. G. 1991. Constraints on the diameter and albedo of 2060 Chiron. *Nature* 251:777–780.

Tilton, G. R. 1988. Age of the solar system. In *Meteorites and the Early Solar System*, eds. J. F. Kerridge and M. S. Matthews (Tucson: Univ. of Arizona Press), pp. 259–275.

Torbett, M., and Smoluchowski, R. 1980. Sweeping of the Jovian resonances and the evolution of the asteroids. *Icarus* 44:722–729.

Towe, K. M. 1991. Hot little pond? *Science* 251:142.

Treiman, A. H., Drake, M. J., Janssens, M., Wolf, R., and Ebihara, M. 1986. *Geochim. Cosmochim. Acta* 50:1071–1091.

Turcotte, D. L., and Kellogg, L. H. 1986. Isotopic modelling of the evolution of the mantle and crust. *Rev. Geophys.* 24:311–328.

Turekian, K. K., and Clark, S. P. 1975. The non-homogenous accumulation model for terrestrial planet formation and the consequences for the atmosphere of Venus. *J. Atmos. Sci.* 32:1257–1261.

Veizer, J. 1983. Geologic evolution of the archean-early proterozoic Earth. In *Earth's Earliest Biosphere*, ed. J. W. Schopf (Princeton, N. J.: Princeton Univ. Press), pp. 240–259.

Vickery, A. M., and Melosh, H. J. 1990. Atmospheric erosion and impactor retention in large impacts, with application to mass extinctions. In *Global Catastrophes in*

Earth History, eds. V. L. Sharpton and P. D. Ward, Geological Soc. of America Special Paper 247 (Boulder: Geological Soc. of America), pp. 289–300.

Wacker, J. F., and Anders, E. 1984. Trapping of noble gases in ice and implications for the origin of the Earth's noble gases. *Geochim. Cosmochim. Acta* 48:2373–2380.

Waldrop, M. M. 1990. Goodbye to the warm little pond? *Science* 250:1078–1080.

Walker, J. C. G. 1977. *Evolution of the Atmosphere* (New York: Macmillan).

Walker, J. C. G. 1986. Carbon dioxide on the early Earth. *Origins of Life* 16:117–127.

Walter, M. R. 1983. Archean stromatolites: Evidence of the Earth's earliest benthos. In *Earth's Earliest Biosphere*, ed. J. W. Schopf (Princeton, N. J.: Princeton Univ. Press), pp. 240–259.

Wänke, H., Creibus, G., and Jagoutz, E. 1984. Mantle chemistry and accretion history of the Earth. In *Archean Geochemistry*, eds. A. Kröner, G. N. Hanson and A. M. Goodwin (Berlin: Springer-Verlag), pp. 1–24.

Ward, W. R. 1975. Past orientation of the lunar spin axis. *Science* 189:377–379.

Weissman, P. R. 1982. Terrestrial impact rates for long and short-period comets. In *Geological Implications of Impacts of Large Asteroids and Comets on the Earth*, eds. L. T. Silver and P. H. Schultz, Geological Soc. of America Special Paper 190 (Boulder: Geological Soc. of America), pp. 15–24.

Weissman, P. R. 1993. The discovery of 1992 QB1. *Eos: Trans AGU* 74:257, 262.

Weissman, P. R. 1994. Celestrial body-building. *Nature* 368:687–688.

Wetherill, G. W. 1975. Late heavy bombardment of the Moon and terrestrial planets. *Proc. Lunar Sci. Conf.* 6:1539–1561.

Wetherill, G. W. 1977. Evolution of the earth's planetesimal swarm subsequent to the formation of the earth and moon. *Proc. Lunar Sci. Conf.* 8:1–16.

Wetherill, G. W. 1990. Formation of the Earth. *Ann. Rev. Earth Planet. Sci.* 18:205–256.

Wiens, R. C., Becker, R. H., and Pepin, R. O. 1986. The case for a martian origin of the Shergottites, II. Trapped and indigenous gas components in EETA 79001 glass. *Earth Planet. Sci. Lett.* 77:149–158.

Wilhelms, D. E. 1984. Moon. In *The Geology of the Terrestrial Planets*, ed. M. H. Carr, NASA SP-469, pp. 107–205.

Wilhelms, D. E. 1987. *The Geologic History of the Moon*, U. S. Geological Survey Prof. Paper 1348.

Wisdom, J. 1983. Chaotic behavior and the origin of the 3/1 Kirkwood gap. *Icarus* 56:51–74.

Woese, C. R. 1987. Bacterial evolution. *Microbio. Rev.* 51:221–271.

Zahnle, K., and Grinspoon, D. 1990. Comet dust as a source of amino acids at the Cretaceous/Tertiary boundary. *Nature* 348:157–159.

Zhao, M., and Bada, J. L. 1989. Extraterrestrial amino acids in Cretaceous/Tertiary boundary sediments at Stevns Klint, Denmark. *Nature* 339:463–465.

THE IMPACT HAZARD

DAVID MORRISON
NASA Ames Research Center

CLARK R. CHAPMAN
Planetary Science Institute

and

PAUL SLOVIC
University of Oregon

This overview of the impact hazard characterizes the consequences of impacts as a function of meteoroid energy, assesses the probability of death from impacts, compares these risks with those of other natural disasters, and reports on preliminary studies of the public perception of impact risks. For impacts below 10 MT (equivalent TNT) energy there is virtually no risk because few meteoroids penetrate the atmosphere. Between 10 MT and the threshold for global catastrophe, impacts are a moderate source of risk, but substantially less so than more common natural disasters such as earthquakes, severe storms, or volcanic eruptions. The greatest hazard is associated with impacts at or a little above a threshold for global catastrophe, where we define a global catastrophe as one that leads to the death of >25% of the Earth's human population. Following the analysis of environmental effects of impacts by Toon et al. (see their Chapter) we estimate that this threshold lies between 10^5 and 10^6 MT, with a nominal value of 3×10^5 MT, corresponding to an average interval between events of about a third of one million years. Above this threshold the entire world population is at risk from impacts, which are the only known natural disasters capable of killing a substantial fraction of the population or, at still larger energies associated with mass extinction, of threatening the survival of the species. Simple arguments suggest that expenditure of up to several hundred million dollars per year might be appropriate in dealing with such disasters. As an extreme example of low-probability but high-consequence disasters, large impacts are without precedent in terms of public perception, but with increasing public awareness, demands may grow for action to deal with this newly identified hazard. Over the entire range of impact energies from the atmospheric cut-off to above the global threshold, the risk level increases with the energy of the impact. Therefore the most prudent strategies deal with detection and protection against the largest projectiles; any program to mitigate the impact hazard should begin with a comprehensive NEO census such as the proposed Spaceguard Survey.

I. INTRODUCTION

Recent widespread scientific and public recognition of a cosmic impact hazard results in part from an increased sensitivity to the historic role of impacts in planetary evolution. The past 25 yr have witnessed the discovery of numerous Earth-approaching comets and asteroids, while spacecraft images of planets and satellites show the craters that result from high-velocity impacts by these objects. However, the population of projectiles present today, few of which exceed diameters of about 10 km, are not capable of significant geological or geophysical modification to the Earth. It is only because the biosphere is sensitive to relatively small impact perturbations that objects a few kilometers in diameter pose a substantial hazard.

The effect of impacts upon the biosphere is most dramatically demonstrated by the discovery (Alvarez et al. 1980) that the K/T mass extinction resulted from the impact of one or more comets or asteroids with a mass (derived from the quantity of extraterrestrial material identified in the K/T boundary layer) of 10^{15} to 10^{16} kg. In this instance a relatively modest cratering event (produced by an object roughly the size of comet Halley) led to global collapse of ecosystems and the extinction of most terrestrial species, including the dinosaurs. It is clear that, far short of a mass extinction, an impact could lead to a lesser ecological catastrophe that might nevertheless kill large numbers of people and threaten the stability of society. Such a global catastrophe is qualitatively different from any other natural disaster and can be compared in its consequences only with the result of nuclear war. In addition, our planet is subject to much more frequent strikes by smaller meteoroids, of which the Tunguska explosion of 1908 is the best historical example (Krinov 1963).

Following the Alvarez discovery of the impact origin of the K/T extinction, NASA convened a workshop in 1981 in Snowmass, Colorado, to consider the impact hazard (Shoemaker 1982). Although the proceedings of that workshop were never published (see Chapman and Morrison 1989), conclusions reached there have influenced most subsequent work on this issue. In particular, the values for the terrestrial impact flux presented at that meeting and the discussion of a threshold for global environmental disaster do not differ greatly from those presented in this chapter.

A decade after the Snowmass meeting, NASA (at the request of the U. S. House of Representatives) asked a group of scientists (the Spaceguard Survey Working Group) to reconsider the magnitude of the hazard and to devise a strategy to greatly accelerate the discovery of potentially threatening asteroids. Much of the basic hazard discussion is derived from the work of that group (Morrison 1992) and our parallel publication of papers on the impact hazard (Morrison and Chapman 1992; Morrison 1993a, b; Chapman and Morrison 1994).

We begin this chapter with an overview of the physical and environmental consequences of hypervelocity impact and an estimate of the size-frequency distribution of the comet and asteroid impactors. We then calculate the risk from projectiles of different sizes and the cumulative risk from all impacts. These risks are compared with a variety of other natural and technological hazards in an effort to place this source of risk within a broader context of societal concerns. New information is also reported on public perception of the impact hazard and the ways this perception may influence public policy in dealing with this issue.

II. IMPACT FLUX

The flux of meteoroids striking the Earth is composed of near-Earth asteroids and short-period comets (collectively called near-Earth objects or NEOs), and of long-period comets. The asteroids and short-period comets have dynamical similarities; both reside in the inner solar system and generally impact the Earth with speeds of order 20 km s^{-1}. Physically, however, they span a wide range of properties, from metal (like the iron meteorites) through various types of rock (like the chondritic and achondritic meteorites) to the low-density, volatile-rich assemblages associated with the comets. Less is known about the rarer long-period comets, but they are probably also composed of low-density, volatile-rich material. Long-period comets strike with higher velocities, sometimes greater than 50 km s^{-1} (McFadden et al. 1989; Chapters by Chapman et al., by Shoemaker et al., and by Rahe et al.).

The implied range in projectile density spans an order of magnitude, and the range in physical properties (e.g., strength) is likely to be greater yet. At smaller sizes (diameters less than a few hundred meters), the physical properties are important in determining the fate of a meteoroid as it plunges into the Earth's atmosphere (Chyba et al. 1993; Hills and Goda 1993). For some purposes, however, the composition of the meteoroid is of little significance, because the threat from larger impacts is related to the kinetic energy of the objects, which is typically nearly 2 orders of magnitude greater than an equivalent mass of TNT. In this chapter, we adopt this kinetic energy (expressed in megatons (MT) of TNT, where 1 MT = 4.2×10^{15} Joules), rather than diameter or mass, as the most significant property of an impacting meteoroid.

By the end of 1992, 163 Earth-crossing asteroids had been cataloged (Chapter by Rabinowitz et al.); the largest is 1627 Ivar (diameter about 8 km, mass presumably about 10^{15} kg). Other still larger objects (up to 1036 Ganymed, with mass 2 orders of magnitude greater) are in Earth-approaching orbits but are not currently classed as Earth-crossing. The census is complete only for objects with diameters the size of Ivar or greater; for 1-km objects the completeness is less than 10% (Chapter by Bowell and Muinonen). From their large-amplitude variations in brightness, we know that many of these asteroids are irregular in shape (Chapters by Chapman et al. and by Ostro); at least two, 4769 Castalia (Ostro et al. 1990) and 4179 Toutatis (Ostro et al. 1993) appear to resemble contact binaries, consisting of two or more pieces in approximately a dumbbell configuration.

The integrated or average flux of meteoroids over the past 3 Gyr can be determined directly from crater densities observed on the lunar maria, which provide a convenient score-card for impacts in the Earth's vicinity. Current impact rates can also be derived, with less certainty, from the estimated population of NEOs and long-period comets together with calculations of their dynamical lifetimes. [For discussions of these flux rates see the following references: Shoemaker et al. (1979, 1990); Shoemaker and Wolfe (1982); Weissman (1982); Wetherill and Shoemaker (1982); Shoemaker (1983); Wetherill (1989); Weissman (1991); Ceplecha (1992); and Chapters by Rabinowitz et al. and by Shoemaker et al.] For this discussion we adopt the size-frequency distribution for impactors with energies between 1 KT (kiloton) and 10^8 MT published by Shoemaker (1983) and reproduced in the Spaceguard Survey Report (Morrison 1992). Recent observations of small (<10 MT energy) asteroids (Rabinowitz 1993; Rabinowitz et al. 1993) suggest a modest enhancement in the asteroid fluxes in this size range, but these objects (with the exception of rare irons) do not penetrate the atmosphere (Chyba 1993). Thus it is sufficient for our purposes to use the smooth curve (Fig. 1) based primarily on lunar crater data for estimations of impact hazard, which is in any case dominated by larger objects (>100 m diameter) where the recent asteroid observations are in good agreement with the lunar cratering curve.

The majority of impacts contributing to the integrated flux are near-Earth asteroids and short-period comets. In addition, occasional long-period comets impact our planet, although they probably amount to less than 25% of the asteroid flux at the same energy (Shoemaker and Wolfe 1982; Weissman 1982,1991; Chapter by Shoemaker et al.). There may also be short-term enhancements in the flux rate (e.g., comet showers) related to the breakup of comets or possibly asteroids (Clube and Napier 1990; Chapter by Steel et al.), but for purposes of this chapter it is adequate, given other uncertainties, to treat impacts as occurring randomly in time.

III. NATURE OF THE HAZARD

Based on the average flux of comets and asteroids striking the Earth, we can evaluate the danger posed by impacts of different magnitudes. Of particular interest are the threshold for penetration through the atmosphere, and the energy at which impacts begin to produce significant global environmental stress in addition to their direct blast damage.

We find the concept of energy thresholds to be useful for differentiating the qualitatively different effects of impacts, which span a range of 100 million (from 10 MT to 10^9 MT). The concept of a threshold does not necessarily imply a sharp transition from one scale of risk to another, however. The transition between local blast effects and global catastrophe, for example, may be quite gradual. It is also the case that the energy threshold at which an impact raises sufficient dust to influence the global climate will almost certainly depend on

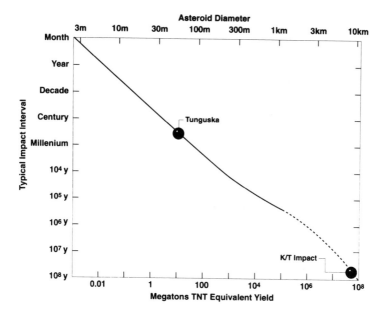

Figure 1. Cumulative energy-frequency curve for impacts on the Earth. The line is the "best estimate" from Shoemaker (1983) for the average interval between impacts equal to or greater than the indicated energy yield. Equivalent asteroid diameters are also shown, calculated assuming impact speed of 20 km s^{-1} and density of 3 g cm^{-3}. This is the same size distribution used in the Spaceguard Survey Report (Morrison 1992) and in our previous hazard discussion (Chapman and Morrison 1994).

the location of the impact. However, a threshold is useful for discussing the impact energies at which one class of physical effects gives way to another.

A. Penetration Through the Atmosphere

The atmosphere protects us from small impacts. The flux curve of Fig. 1 indicates that an impact with the energy of the Hiroshima nuclear bomb occurs roughly annually, while a 1 MT event is expected at least once per century. Obviously, however, such relatively common events have not been destroying cities or killing people. Even at megaton energies, most meteoroids break up and are consumed before they reach the lower atmosphere.

Meteoroids as large as a few tens of meters in diameter usually fail to penetrate into the lower atmosphere because they are subject to aerodynamic stresses that cause fragmentation and transverse dispersal at high altitude (Chyba et al. 1993; Hills and Goda 1993; Chyba 1993). The height of fragmentation depends primarily on the meteoroid's physical strength; only the strongest irons reach the surface in one piece. Loose cometary aggregates and carbonaceous meteorites fragment at altitudes above 30 km.

Although the altitude of initial fragmentation is nearly independent of projectile size, the history of the resulting fragments depends critically on size; smaller objects rapidly disperse while sufficiently large objects produce a dense cloud of fragments that continues to the ground as a coherent whole. If the object explosively disperses within about 20 km of the surface, the resulting airburst can be highly destructive. Numerical models of atmospheric fragmentation and dispersal show that rocky objects >50 m diameter (10 MT energy) and cometary objects >100 m (100 MT energy) penetrate deep enough to pose significant hazards (Chyba et al. 1993; Hills and Goda 1993; Chyba 1993).

If a large meteoroid penetrates to the troposphere or actually strikes the surface at velocities of tens of km s^{-1}, the sudden release of kinetic energy can result in damage analogous to that from nuclear bomb explosions of similar energies, but without associated radioactivity. The area of devastation scales approximately as the explosive yield to the 2/3 power and is somewhat greater for an airburst than for a groundburst explosion (Glasstone and Dolan 1977).

Tunguska provides a calibration, with a shock wave sufficient to topple trees over an area of nearly 10^5 hectares and a fireball that ignited fires over a smaller area near ground zero (Krinov 1966; Sekanina 1983; Pike 1991a, b; Chyba et al. 1993; Hills and Goda 1993). The yield of the Tunguska blast has been estimated at 10 to 20 MT from microbarograph measurements in Europe. If we assume that the radius of forest devastation would apply also to destruction of many buildings, the area of damage is given approximately by

$$A = 10^4 Y^{2/3} \tag{1}$$

where Y is the yield in MT and A is in hectares. Applying this formula we find that at 100 MT, the radius of destruction is 25 km and the fraction of the Earth's surface that is affected is 0.001%, while 10^5MT, the radius is 250 km and the area is about 0.1%. These are both examples of what we call local or regional events, rather than global catastrophes.

B. Tsunamis

The area of destruction is larger for impacts into oceans than for land impacts as a consequence of the great travel distances of impact-induced tsunamis. The radius of destruction for tsunamis scales approximately as the impact energy to the 1/2 power (Chapters by Hills et al. and Toon et al.), rather than as the 1/3 power for land impacts noted above. On the other hand, a tsunami results in large-scale destruction and loss of life only when it encounters a populated coastal region; a tsunami that dissipates before reaching a continental margin is a hazard only to oceanic islands.

From the expressions given by Hills et al., a 5-m height tsunami has a range of approximately 1000 km for a 1000 MT explosion; it is at roughly this magnitude that the tsunami from an oceanic impact has a substantial probability of reaching land and producing large-scale destruction. For yields greater than about 10^3MT, tsunamis associated with oceanic impacts contribute more to the hazard than the direct blast damage of impacts on land or in the continental margins.

C. Globally Catastrophic Impacts

At sufficiently great energies, an impact has global consequences. An obvious if extreme example is the K/T event 65 Myr ago. This impact released $>10^8$MT of energy and excavated a crater (Chicxulub in Mexico) at least 200 km in diameter (Sharpton and Grieve 1990; Hildebrand et al. 1991; Sharpton et al. 1992; Swisher et al. 1992; Chapter by Grieve and Shoemaker). Among the environmental consequences were devastating wildfires and changes in atmospheric and oceanic chemistry as well as a dramatic short-term perturbation in climate produced by some 10^{16}kg of submicrometer dust injected into the stratosphere (Alvarez et al. 1980; Toon et al. 1982; Wolbach et al. 1985; Covey et al. 1990; Gilmore et al. 1990; Sharpton and Grieve 1990; Chapter by Toon et al.).

The K/T impact darkened the entire planet for many months and precipitated a general destruction of terrestrial ecosystems. Fortunately, events of this magnitude are exceedingly rare. However, projectiles much smaller than the K/T impactor can still perturb the global climate by injecting dust into the stratosphere. An environmental shock that would leave most natural ecosystems intact can still severely curtail human agricultural production around the world. Few nations store one year's worth of food, so mass starvation could be expected (Harwell and Hutchinson 1989). Compounded by other possible effects of the impact, this agricultural disaster might result in collapse of global economic, social, and political structures. However, we do not know the degree of coupling of these effects, and it is very difficult to estimate the resilience of society to such massive environmental insults.

Chapman and Morrison (1994) define a globally catastrophic impact as one that *results in the deaths of more than a quarter of the world's population.* Such an event would affect the global climate in ways somewhat similar to those calculated for nuclear winter (Turco et al. 1983; Sagan and Turco 1990), leading to widespread loss of agricultural production and resulting in mass starvation. Although such a global agricultural catastrophe would be *worse* in loss of life and destruction of property than the effects of the great World Wars, it is *far smaller* than the K/T impact. A globally catastrophic impact might destabilize modern civilization, but it would not threaten extinction of the human species or produce a mass extinction of marine fauna that might be visible in the fossil record.

D. Threshold for a Globally Catastrophic Climate Perturbation

A drop of a few degrees C in surface temperature over many months is sufficient to reduce crop yields dramatically and precipitate large-scale starvation (Harwell and Hutchinson 1989; Covey et al. 1990; Turco et al. 1991). The energy threshold for a globally catastrophic impact is therefore determined by the explosive yield required to loft sufficient submicrometer dust into the stratosphere to lower the surface temperature by this amount.

The energy at which impacts lead to significant global climatic effects is considered in detail in the Chapter by Toon et al. They identify 10^5 MT as the energy for land impacts at which submicrometer dust yields an atmospheric opacity near unity, with climate effects probably similar to that of large volcanic eruptions such as Tambora (Stothers 1984); at this level dust can lower surface temperatures on at least a regional scale and lead to significant crop loss. Effects of the Tambora explosion on climate and agriculture are discussed in some detail by Sagan and Turco (1990, pp. 99–101). At ten times larger enegy (10^6 MT), for either land or ocean impacts, approximately 2×10^{-4} g cm^{-2} of submicrometer dust is injected into the atmosphere. The optical depth is about 10 with an atmospheric transmission of less than 1%, certainly sufficient to induce global freezing and large-scale crop loss. Toon et al. refer to the energy region between 10^5 and 10^6 MT as a "grey area between small effects and those that are obviously significant on the global scale." For a stony object striking at 20 km s^{-1}, 10^6 MT corresponds to diameter of about 2 km.

For purposes of this discussion, we adopt an energy of $10^{5.5\pm0.5}$ MT (or about 3×10^5 MT), in the middle of the "grey area" of Toon et al., as the nominal threshold for global catastrophe. For an asteroid with density 3 g cm^{-3} and an impact velocity of 20 km s^{-1}, the corresponding diameter is approximately 1.7 km. The average interval between impacts of this size or larger is 300,000 yr from Fig. 1.

It is important to recognize that the threshold for global catastrophe as we have defined it is uncertain by an order of magnitude in energy, and hence a factor of 2 in projectile diameter (from about 1 km to somewhat more than 2 km for a stony asteroid striking at 20 km s^{-1}). Our discussion attempts to be consistent with the chapter by Toon et al., but we cannot exclude more extreme values of the threshold. Chapman and Morrison (1994), who took a more conservative approach in their error analysis, considered a range from 1.5×10^4 MT (about 0.6 km diameter for an asteroid) to 10^7 MT (about 5 km diameter).

E. Misidentification of a Bolide

One additional source of hazard has frequently been discussed: the misidentification of a natural impact event with a nuclear attack (see, e.g., Shoemaker 1982 and Chapter by Tagliaferri et al.; see also "Erice Statement" quoted in Chapter by Morrison and Teller). During the periods of high international tension that characterized much of the cold war, it was feared that a natural event in the megaton range might be mistaken for a nuclear attack and thereby trigger a nuclear war, with catastrophic global consequences. A megaton event is expected once or twice per century, but most such explosions take place at high altitude and have no significant consequences at ground level.

There is no public record of such events having contributed to international tension during the cold war, and today surveillance satellites have the capability to detect natural high-altitude explosions over a wide range of energy (Reynolds 1993; Beatty 1994; Chapter by Tagliaferri et al.). However, such data are not available to many nations with emerging nuclear capability, and the proliferation of nuclear weapons causes continuing concern about the misidentification of such natural events.

IV. HAZARD ANALYSIS

We now address the scale of destruction expected for impacts and the numerical hazard associated with impacts of various magnitudes. By numerical hazard we mean the probability of death for an individual due to this event.

A. Impacts on Land

We begin with the destructive effects of airbursts or groundbursts on land or in shallow coastal areas. We identify the average zone of mortality associated with impacting objects larger than our threshold for penetration into the lower atmosphere (10 MT) with the area of blast devastation defined earlier from the Tunguska example. The average number of fatalities per impact, using the average world population density of 0.2 person per hectare of land, is:

$$N = 1 \times 10^3 Y^{2/3} \tag{2}$$

Thus an event with the energy of the Tunguska blast, occurring on land only once or twice per millenium, would on average cause about 14,000 deaths, although most such impacts can be expected to strike uninhabited parts of the planet, just as in 1908. Indeed, it is unlikely that a Tunguska-like impact would destroy even one city in the entire 10 millenium span of human history. (It is therefore not surprising that there is no historical record of such a catastrophe.)

B. Ocean Impacts

Approximately half the target area of the Earth consists of ocean basins, and impacts into deep water generate tsunamis of substantial size (Pike 1993; Chapter by Hills et al.). As the range of the tsunamis approaches the dimensions of the ocean basins, there is potential for destruction and loss of life far exceeding that associated with land impacts of comparable energy. The idea that the Mosaic Flood might have represented an impact-induced tsunami can be traced back as far as Laplace and even Halley (see the discussion of Sagan and Druyan [1985, pp. 276–279]).

To estimate the potential mortality from tsunamis, we use the expressions for wave height and depth of penetration into coastal plains given in the Chapter by Hills et al. As a tsunami approaches the shore, it experiences an increase in height of a factor of 10 to 40; thus a nominal 5-m ocean wave might break at the shore with a height of order 100 m. As shown by Hills et al., such a wave would penetrate a flat coastal plain to a distance inland of 22 km.

To calculate the casualties associated with such an event, we need to estimate the length of ocean margins that consist of flat coastal plains and the population density of such plains. For this order-of-magnitude estimate, we neglect the small-scale curvature of the ocean margins and adopt a length of the order 10^5 km. The associated area of coastal plains at risk from tsunamis, to a nominal width of 20 km, is 2×10^6 km^2 or 2×10^8 hectares. If the population density of coastal plains is 10 times the average for the land area of the Earth (2/hectare), the total population at risk from tsunamis is a few times 10^8, or roughly 10% of the world's population.

In the absence of modeling of the destruction of tsunamis on real coast-lines, two additional rough estimates can be made of the population at risk. Kopec (1971) calculates that 19% of the world's population would be inundated by a rise of 100 m in sea level, a figure that considerably overestimates the numbers that would be flooded by most tsunamis, which would be much less than 100 m high after penetrating a few kilometers inland. Another overestimate of the population fraction at risk can be obtained from the fact that 30% of the population in the world's 50 largest urban areas occupy coastal plains or harbors (Hoffman 1990). Although the estimates are crude, they all suggest that the fraction of the Earth's population living in areas subject to flooding by tsunamis is no more than 10%.

Such estimates of the population in coastal plains permit us to set a rough upper limit to the mortality that might be associated with impacts having energy in the gigaton range (10^3 to 10^5 MT). The geometry of the Earth's ocean basins suggests that any one impact is unlikely to affect more than 10% of the total coastline, hence to place at risk more than about 1% of the planet's population (that is, fewer than 10^8 people). This is the maximum level of mortality expected for an ocean impact; presumably the average impact would be less lethal, perhaps with the order of 10^7 deaths. In addition, current tsunami warning systems permit at least partial evacuation of coastal areas in the several hours that elapse before a typical deep-ocean tsunami reaches the shore. For comparison, the mortality associated with an average 10^4 MT land impact is of order 10^6. In a similar analysis, Pike (1993) concluded that the hazard from open ocean impacts is 4 to 8 times greater than for land impacts of similar energy. We conclude that the hazard associated with gigaton-energy impacts is dominated by ocean impacts, but the upper limit to the deaths associated with such events remains no more than 1% of the total population of the planet.

C. Globally Catastrophic Impacts

Above the threshold energy for global catastrophe (3×10^5 MT), the number of fatalities is (by definition) >1.5 billion. For example, for a land impact near 10^6 MT, the expected average mortality from the direct blast is (from the formula above) about 20 million, while indirect deaths are (by definition of the global threshold) at least 1.5 billion. This difference reflects the different areas affected: less than 1% of the Earth's surface for the direct blast, but the entire surface for the indirect effects.

When the threshold globally catastrophic impact is compared with a somewhat smaller impact in the deep ocean, the contrast is less. The maximum number of direct tsunami deaths from a 10^5MT ocean impact are of order 100 million, as compared with the postulated 1.5 billion indirect deaths from a global environmental disaster. However, there remain substantial uncertainties in the estimates for tsunamis of this size, which dwarf anything experienced directly, so all of these numbers must be used with great caution.

At still higher impact energies, other environmental as well as climatic effects come into play, leading to mass extinctions (Toon et al. 1982; Lewis et al. 1982; Zahnle 1990; Chapter by Toon et al.). However, because such events are extremely rare, they constitute a smaller numerical risk than do impacts near the global threshold.

D. Dependence of Risk on Impact Energy

In spite of substantial uncertainties in the expected consequences of impacts of various energies, we can still draw some interesting conclusions about relative risk. The nature of the hazard and the possible means of protecting ourselves vary greatly over the wide range in projectile size and energy from the atmospheric cutoff to the large mass extinction events.

We can illustrate our approach with a simplified estimate of the total casualties in the regime of local or regional impacts on land by using a single power-law fit to the data of Fig. 1. Our reference is an "average Tunguska" impact by a 60-m diameter projectile with an energy yield of 15 MT and a resulting mortality of 14,000; an equal or larger event occurs on land about once every 600 to 700 yr. Over the impact range of interest (from the 10 MT atmospheric cut-off up to the globally catastrophic case), the differential impact frequency can be represented approximately by

$$f(D) = 10D^{-2} \tag{3}$$

where the frequency is in yr^{-1} and the asteroid diameter D is in meters. Using the expression given previously for the average deaths as a function of impact yield, we find that the mortality is related to the asteroid diameter by the expression

$$N(D) = 4D^2 \tag{4}$$

where N is the number killed and D is in meters.

With these approximations, the total mortality is found by integrating from the minimum diameter for penetration through the atmosphere (50 m) up to the size of interest (D_{max}).

$$N = \int_{50}^{D_{max}} N(D)f(D) \, dD = 40 \ln (D_{max}/50). \tag{5}$$

TABLE I
Fatality Rates (Equivalent Annual Average Deaths) as a Function of Impact Energy

Type of Event	Energy (MT)	Diameter	World Deaths yr^{-1}
High atmosphere break-up	<10	<50 m	<1
Tunguska-like events	$10-2 \times 10^3$	50 m–300 m	55
Sub-global land impacts	$2 \times 10^3 - 5 \times 10^5$	300 m–2 km	30
Sub-global ocean impacts	$2 \times 10^3 - 5 \times 10^5$	300 m–2 km	300 (?)
Threshold global catastrophes	$10^5 - 10^6$	1 km–2 km	3000
Mass extinction events	$>10^7$	>4 km	<300

From this expression, the total mortality for impacts on land from the 50-m atmospheric cut-off to the global threshold at 1.7 km is about 140 deaths per year.

To obtain a more accurate assessment of the numerical risk as a function of projectile size or energy, we have integrated the cumulative damage over segments of the impactor size-frequency distribution, each segment represented by a best-fitting power law (Chapman and Morrison 1994; Chapter by Harris et al.). The results are summarized in Table I. If the effects of tsunamis from ocean impacts is added, the equivalent annual casualties may rise to several hundred. All of these calculated values are long-term averages, and they all have substantial uncertainties. However, it is possible to draw some semi-quantitative conclusions from this analysis, as follows.

The smaller, frequent events larger than the 10 MT atmospheric cut-off (what we may call Tunguska-class impacts) yield equivalent annual fatality rates of tens of deaths yr^{-1} for the current world population. In reality, of course, most of the casualties will be associated with rare events that strike in heavily populated areas, while the majority of these impacts produce practically no fatalities. The risks associated with Tunguska-class impacts, while not insignificant, represent mortality rates that are substantially less than those associated with many smaller and more frequent natural hazards, such as earthquakes, hurricanes, volcanic eruptions, floods, or mud slides. Thus the hazard from Tunguska-like impacts does not inspire special concern or justify heroic efforts either to predict such events or to attempt to avert them.

At the opposite, high-yield extreme, the K/T impact was vastly more devastating, but even though nearly everyone would be killed by such an event, they are so infrequent that the annual fatality rate is only about 60 yr^{-1} for the world's present population (6 billion people killed per 10^8 yr). Even with present limited knowledge, we can say that no asteroid exists in an Earth-crossing orbit that could cause a disaster of this magnitude. However, we cannot exclude the possibility of a large comet appearing at any time and dealing the Earth such a devastating blow—a blow that might lead to human extinction. This is the most extreme problem raised by this risk analysis—the possible extinction of humanity from a large comet at an annual probability level of $<10^{-8}$.

The greatest risks are associated with impacts near the threshold for global catastrophe; just above this threshold, we postulate that 1.5 billion people are killed. For the nominal threshold energy of 3×10^5 MT, such events occur roughly three times per million years, implying a fatality rate of a few thousand per year. The annual risk per individual is thus somewhat smaller than one in a million. For comparison, the fatality rate for all impacts below the threshold is a few hundred per year, yielding an annual risk of the order of 10^{-7}.

The total lifetime risk from both the globally catastrophic impact and the large tsunami is about 1:20,000 that an individual will die as the result of the impact of a comet or asteroid (Chapman and Morrison 1994). For

TABLE II

Summary of Impact Effects as a Function of Energy

Yield Y (MT)	Interval log T	NEO diameter		Crater D (km)	Consequences
<10					Upper atmosphere detonation of stones and comets; only irons (<3%) penetrate to surface.
10^1–10^2	3.0	75	m	1.5	Irons make craters (Meteor Crater); Stones produce airbursts (Tunguska). Land impacts destroy area the size of a city (Washington, Paris, Moscow).
10^2–10^3	3.6	160	m	3	Irons and stones produce groundbursts; comets produce airbursts. Land impacts destroy area size of large urban area (New York, Tokyo).
10^3–10^4	4.2	350	m	6	Impacts on land produce craters; ocean tsunamis becoming significant. Land impacts destroy area the size of a small state (Delaware, Estonia).
10^4–10^5	4.8	0.7	km	12	Tsunamis reach oceanic scales, exceed damage from land impacts. Land impacts destroy area the size of a moderate state (Virginia, Taiwan).
10^5–10^6	5.4	1.7	km	30	Land impacts raise enough dust to affect climate, freeze crops. Ocean impacts generate hemispheric scale tsunamis. Global destruction of ozone. Land impacts destroy area the size of a large state (California, France, Japan).

10^6–10^7	6.0	3 km	60	Both land and ocean impacts raise dust, change climate. Impact ejecta are global, triggering widespread fires. Land impacts destroy area the size of a large nation (Mexico, India).
10^7–10^8	6.6	7 km	125	Prolonged climate effects, global conflagration, probable mass extinction. Direct destruction approaches continental scale (Australia, Brazil, U. S.).
10^8–10^9	7.2	16 km	250	Large mass extinction (K/T).
$>10^9$				Threatens survival of all advanced forms of life.

our nominal range in threshold energies (10^5 to 10^6MT), the probability of death from impact ranges from about 1:10,000 to 1:40,000. For most of the Earth's population, this risk is small compared to many other causes of death, both natural and accidental, but it is not smaller (at least in the economically advanced nations) than the risk associated with other natural disasters. For an average American, for example, the risk of death from an impact may be greater than that from earthquake, flood, or severe storm, as is discussed in the next section.

Table II summarizes the effects of impacts as a function of energy from <10MT to $>10^9$MT. The most important conclusion from this analysis is that the hazard (the annual equivalent mortality) increases from small to large impacts across nearly this entire range of 8 orders of magnitude in energy. The power-law size-frequency distribution of the asteroid and comet population together with the physical consequences of the impacts results in a situation where the larger (and rarer) impacts are the most damaging, up to a few times the threshold for global catastrophe.

The nature of the hazard is further illustrated in Fig. 2 (adapted from Chapman and Morrison 1994), which displays schematically the dependence of risk on impact energy. The lower panel shows the average anticipated number of fatalities per event; the solid line is calculated using nominal values for land impacts, and the dotted line suggests the (highly uncertain) fatalities associated with deep-ocean impacts. A possible energy range from 10^5 to 10^6MT is shown for the transition from regional to global effects, at which point substantial fractions of the Earth's population will die in the aftermath of an impact. The upper panel illustrates the expected annual fatality rate as a function of impact energy.

E. Estimate of Economic Losses from Impacts

The primary concern about impacts is derived from their unique potential to kill a substantial fraction of the Earth's population and destabilize the global civilization. The hazard can also be analyzed in terms of its economic impact, however, and this approach may be useful in comparing it with other hazards. (Similar but more detailed economic arguments are given in the Chapter by Canavan.)

A simple estimate can be made of the economic cost of impacts in the regime between the atmospheric cut-off and the threshold for global catastrophe, if we assume that average economic loss, like average mortality, is proportional to the area of devastation. Only the multiplier is different. We assume that the total value of the world's developed property and economic infrastructure is $\$4 \times 10^{14}$ (Canavan 1993; Chapter by Canavan), to yield a total world value of approximately $\$1.5 \times 10^5$ per person.

For impacts with energies less than the global impact threshold, the approximate economic loss from impacts (in millions of dollars) is:

$$V = 6 \ln (D_{max}/50). \tag{6}$$

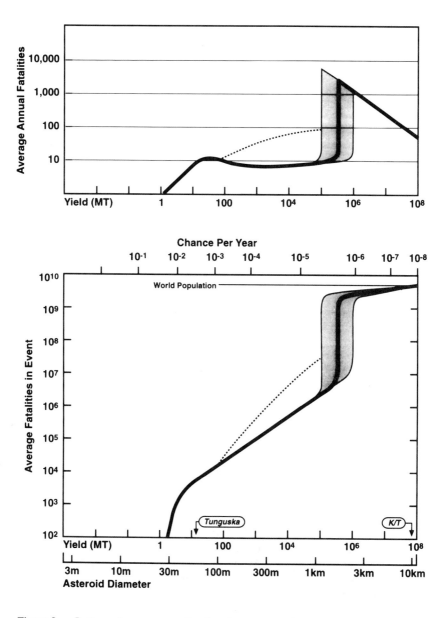

Figure 2. *Bottom*: Average mortality from impacts as a function of energy, for the current population of the Earth. The solid line from 10 to 10^5MT is for impacts on land or the continental shelves; the dashed line indicates enhanced mortality from tsunamis that result from deep ocean impacts. The shaded region shows the range of values associated with the choice of the threshold for global catastrophe (between 10^5 and 10^6MT). *Top*: Schematic representation of the average mortality (worldwide deaths per year) from impacts over the same range of energies, showing that the total hazard increases as the size of the impactor ranges from ordinary meteorites up to the global threshold at a nominal energy of 3×10^5MT. Scales for associated impact probabilites and asteroid diameters from Fig. 1 are also shown (figure is adapted from Chapman and Morrison 1994).

For the nominal global threshold of 3×10^5MT ($D = 1.7$ km), this value is about $20 million per year. If the effects of tsunamis are included, this number will rise to perhaps $100 million. This number (which is, we stress, very approximate) could be interpreted as a justifiable amount to spend in defending against impacts below the global threshold.

At and above the global threshold, the equivalent annual casualties are in the thousands. Application of a similar factor to convert from mortality to economic destruction then leads to a value for development of mitigation or defense systems (in advance of a specific threat) of several hundred million dollars per year. This simply says that most of the total economic "value" of the world—currently some 4×10^{14}—might be lost from a large impact once every million years or so, for an annual equivalent loss of hundreds of millions of dollars—a very crude way of assessing such a global catastrophe, but one that may be of some interest in considering the justification for possible mitigation schemes.

V. IMPACT HAZARD COMPARED WITH OTHER HAZARDS

In a rational world, society's response to the threat of impact by an asteroid or comet should be evaluated against other hazards that people face. In a typical year, nearly 1,000 people in the United States alone are killed as a result of being struck by a falling object. None of these objects, at least so far, has been a meteorite, comet, or asteroid. Some causes of mortality are much more common than falling objects; illness, war, pestilence, and famine head the list. Other potential causes of mortality have received prominent attention in the news media as well as government action even though nobody has actually died; reported deliberate tampering with imported grapes is one example from several years ago. In this section, we put the impact hazard into perspective by comparison with other causes of mortality, particularly those that are the closest analogs—natural disasters.

A. Current Risk Statistics

In the United States, motor vehicle accidents lead the list of hazards, followed by falls, poisoning by solids or liquids, drowning, fires and burns, suffocation, firearms, and poisoning by gas (National Safety Council 1989). Still other dangers are widely feared even though fewer than 100 people die per year in the U. S. (e.g., dog bites, lightning, poisonous snakes and spiders). All accidental deaths combined account for approximately 100,000 deaths yr^{-1} in the United States (half are motor-vehicle related), but they pale in comparison with cardiovascular and malignant diseases as a cause of mortality (U. S. Bureau of the Census 1991).

Table III illustrates some selected causes of mortality for the world population. The closest analogs for the impact hazard are other low-probability/-high-consequence disasters. These include natural disasters, which have accounted for only about 4% of the 150,000,000 catastrophic deaths in the world during the 20th century (Pike 1991*a*), and technological accidents. In contrast, the vast majority of catastrophic deaths are attributable to longer-acting famine, epidemics, and wars, which are less instructive as analogs to the impact hazard.

TABLE III
Ranked Selected Causes of Mortality (World)

1.	Major diseases (e.g., heart disease, cancer)
2.	War and genocide
3.	Epidemics
4.	Famine
5.	Other diseases (pulmonary, pneumonia, influenza)
6.	Major accidents (motor vehicle, falls)
7.	Suicide and homicide
8.	Cyclones and floods
9.	Lesser accidents (poisoning, fires/burns, suffocation)
10.	Earthquakes
11.	Rare accidents (falling object, railway accident, airline crash)
12.	Globally catastrophic impact
13.	Rare natural disasters (thunderstorms, volcanoes)
14.	Venomous bite or sting
15.	Tunguska-like locally catastrophic impact
16.	Airline hijacking aftermath
17.	Drinking water with EPA limit on TCE

Mortality rates due to natural disasters vary greatly from one part of the world to another. For example, the contemporaneous severe flooding of the summer of 1993 in the United States and India resulted in thousands of deaths in India but only a handful in the U. S. In the United States during the second half of the 20th century, natural disasters rank quite low as causes of mortality; for example, during the period 1966–1989, the annual death toll in the United States has averaged 142 from floods, 80 from tornados, and 27 from hurricanes (U. S. Bureau of the Census 1991), with earthquakes and tsunamis ranking even lower. Strict regulations and preventative measures have also reduced technological accidents (e.g., explosions, mining disasters, and building fires) to a very low level in the U. S. during recent years, despite some notable technological accidents during the first half of the twentieth century.

On a global scale, however, natural disasters are much more important causes of mortality than recent experience in the United States would suggest. In the developing world, natural disasters have been and remain a major cause of mortality. Eight natural disasters occurred between 1900 and 1985 that killed between 100,000 and 2,000,000 each (3 earthquakes, 3 floods, 1 cyclone, and 1 landslide, which occurred in China, Bangladesh, and Japan) (*World Map of Natural Hazards* 1988). This average 10-yr interval between huge natural disasters may be compared with the thousands of years between asteroid impacts large enough to cause comparable fatalities.

B. Qualitative Distinction of Large Impacts

While cosmic impacts do not compete with other natural disasters at smaller sizes, they can dominate the risk statistics at large sizes. In this way the impact hazard appears to be qualitatively different from other natural disasters. As we have already noted, the size distribution of asteroids and comets follows roughly a power law, and continues without bound to at least 10 km diameter (the approximate size of the largest known Earth-crossing asteroid) and probably to much larger sizes (comets are known, although not now in Earth-approaching orbits, that extend to over 100 km in size). This size distribution is what allows the finite possibility of an impact destroying civilization, or even killing everyone on the planet.

In contrast to the impact hazard, it appears that other natural hazards are bounded at magnitudes that do not lead to global catastrophe. The greatest fatalities from historic natural disasters have killed only tiny fractions of the world's population: earthquake (2 million), cyclone (300,000), landslide (100,000), tsunami (100,000), volcano (30,000 immediate deaths, 92,000 immediate plus secondary deaths), and avalanche (20,000) (Cornell 1982; Encyclopedia Britannica, 15th ed.). Of course, asteroid impact has killed nobody over the same period, so we must look at the fundamental nature of other natural disasters to see if they are bounded.

It is difficult to compare the magnitudes of different natural disasters because their effects on populations and infrastructure are so different. For example, the very largest earthquakes in history (Richter scale $\sim$9) have estimated seismic energies equivalent to a few hundred MT. Yet that energy is manifested in a way (shaking of the ground) that is very effective at killing people who live in poorly constructed homes or in hilly terrain subject to landslides. The Mt. St. Helens volcanic eruption has been estimated at $\sim$15 MT, similar to the Tunguska impact. However, the greatest damage from volcanoes is not from the direct blast effects but from either ground-hugging pyroclastic flows or (depending on the specific form of the eruption) from the indirect environmental effects of dust and aerosols ejected into the lower atmosphere and stratosphere. The same is true of the impact hazard; whether the killing is due to local blast damage, tsunamis, or starvation due to global climatic effects depends on the magnitude and location of the impact.

The frequency-magnitude distribution is poorly known for the largest natural disasters. For earthquakes with Richter magnitude >6, the cumulative frequency diminishes nearly a factor of 10 for each increase of 1 (corresponding to 32-fold increase in energy) in magnitude. This means that the very largest earthquakes are responsible for nearly all of the seismic energy released on our planet, a situation (like the impact hazard) that emphasizes the catastrophic potential of the biggest events. Earthquakes with a seismic moment exceeding 10^{31} dyne centimeters (roughly, Richter scale 9) may occur about once a century somewhere on Earth. Historical records during previous centuries are too imprecise to demonstrate an upper limit to earthquakes. However, unlike asteroidal or comet impact energies, there must be an upper bound to the energy of earthquakes, set by the geometry of seismic zones and the strength of crustal material (Chinnery and North 1975).

In general, it is not possible for unlimited strain or unlimited pressure to build up in the Earth's crust. Earthquakes and volcanic explosions are the mechanisms of relief. One must expect that there are upper bounds to the magnitude of volcanic explosions just as there must be upper bounds to earthquakes. The largest explosive volcanic event noted in the recent geologic record was the Toba, Sumatra, event 75,000 yr ago, with eruptive volumes at least a thousand times greater than Mt. St. Helens (Rampino et al. 1988; Sigurdsson 1990). There are great uncertainties about how effective such an explosion might be in producing a global climatic shock, but it is plausible that the greatest volcanic events could rival the effects of impacts by asteroids up to 1 km in diameter. However, the energies of volcanic explosions do not approach the magnitude of impacts by teraton-scale cosmic projectiles.

In human history, the greatest mortality from natural disasters has been from effects of extreme weather: windstorms and flooding. There certainly must be limits to the largest possible atmospheric storms. Tornados exemplify a dangerous meteorological phenomenon that is limited in scale to widths of <2 km, and other weather phenomena also have characteristic scales that are not exceeded. There also are limits to the magnitude of weather-related flooding. On time scales related to our personal experiences, the biggest floods generally do more damage than the cumulative effects of all smaller floods. Nevertheless, the natural scale of storms, the vertical structure of the atmosphere, and the moisture holding capacity of air all limit the quantity of rain that could theoretically fall within a given watershed.

Independent of the maximum energy or destructive power of different modes of natural disasters, they all (with the possible exception of explosive volcanism) differ from the globally catastrophic impact hazard in one important respect; they are localized. Even tsunamis, which can extend their reach around the world along ocean coastlines, cannot touch continental interiors. No matter how large the non-impact natural catastrophe, many nations would be unscathed by earthquakes, floods, or storms of the most exaggerated possible scale. Impacts are unique in producing global consequences at a scale that could threaten the entire world's population simultaneously.

C. Comparative Risk Summary

This discussion of comparative risk can be summarized by the schematic illustration in Fig. 3. For three representative classes of disasters and accidents, Fig. 3 presents the world fatality rate (deaths/yr) as a function of the number of deaths per catastrophic event. At one to a few deaths per event, automobile and other transportation accidents dominate the mortality. These are also the dominant forms of accidental death overall. Other human-caused accidents (chemical spills, mine explosions, Chernobyl) are of much smaller overall consequence, although individual events killing hundreds or thousands of people at once have great headline-producing potential. Natural disasters (especially floods, earthquakes, and cyclones) dominate the mortality for catastrophes that exceed about 100 fatalities/event. Tunguska-like impacts have an expected mortality rate hundreds of times less than those caused by these more familiar natural disasters.

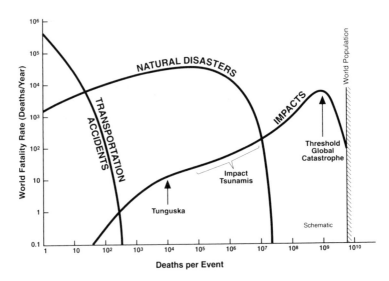

Figure 3.   Schematic indication of the fatality rate from accidents and disasters (worldwide deaths per year per factor of 10 in the abscissa) as a function of the number of deaths per event. The primary cause of accidental deaths is transportation accidents (including auto, train, and plane), which typically involve the deaths of fewer than 100 persons per incident. Large-scale natural disasters (floods, earthquakes, hurricanes, volcanic eruptions) cause deaths over a wide range of scales, up to rare but statistically important events that can kill millions. Only impacts, however, are capable of killing more than 100 million persons per event, and they dominate the hazard in the right-hand side of the figure.

As natural disasters approach their upper limits (at several million fatalities per event), cosmic impacts, occurring only once every 100,000 yr, begin to rival the mortality potential of natural disasters. Near the threshold for global catastrophe, impact has the unique potential for killing much of the world's population, and the probability of that happening (low) multiplied by the enormous mortality yields an average fatality rate that probably exceeds that of other natural disasters of any scale.

VI. PUBLIC PERCEPTIONS OF THE HAZARD

Whereas a great deal of study has been devoted to defining and characterizing the impact hazard, little effort has thus far been spent to understand how laypeople perceive this threat. Public perceptions of impact risks are important for several reasons. First, our general understanding of human response to risk will undoubtedly benefit from studies of people's response to this hazard, which is unique in its combination of very low probability and very great consequence. Second, public attitudes and perceptions influence government policies toward risk management. Third, understanding how the hazard is perceived is essential for effective education and communication efforts.

A. History of Risk-Perception Research

Social scientists have been studying risk perception extensively during the past 15 yr, examining the opinions that people express when they are asked to evaluate various natural and technological hazards. In these studies, researchers have sought to discover what people mean when they say that something is (or is not) "risky" and to determine what factors underlie these perceptions.

This research has demonstrated that the concept of risk means different things to different people. When technical experts judge risk, their responses correlate highly with expected annual fatalities or with probabilities of mortality, such as those discussed in the previous section of this chapter. However, the risk perceptions of laypeople are sensitive to factors other than fatalities, such as catastrophic potential, dread, controllability, and threat to future generations (Slovic 1987). As a result, public perceptions often differ considerably from judgments made by experts.

Research has compared public perceptions of risk and benefit from a great many activities, but most instructive for our purposes are the perceptions associated with radiation and chemical technologies. Nuclear power has a very high perceived risk and low perceived benefit, whereas diagnostic X rays have the opposite pattern. Similarly, nonmedical sources of exposure to chemicals suspected of being toxic (pesticides, food-additives, alcohol, cigarettes) are seen as very low benefit and high risk, while medical chemicals (prescription drugs, antibiotics, vaccines) are generally perceived as high benefit and low risk, despite the fact that they can be very toxic substances. The favorable attitudes and acceptance of risks from X rays and medicines demonstrate that perception of risk is conditioned by the context in which exposure occurs and the degree of trust in those responsible for managing the hazard; e.g., the medical profession vs industry.

B. Will the Public be Concerned About the Impact Hazard?

If a credible prediction of a catastrophic impact were made, the public would undoubtedly be quite frightened. Absent such a specific prediction, how might we expect people to respond to the statistical threat of impact, as it is reported in the news media? Will their level of concern be great enough to induce them to support expenditure of public funds to detect threatening asteroids or comets? In the absence of data on public perceptions regarding this specific hazard, we can find reasons from previous experience to predict both lack of concern and a high degree of concern. Reasons for expecting lack of concern and possible opposition to large expenditures are the following:

1. Natural hazards such as impacts tend to be less frightening than techno-logical hazards (Erickson 1990). People perceive nature as benign and react rather apathetically to the threat from natural hazards (Burton et al. 1978). Personal experience of a natural disaster is usually necessary to motivate action to reduce future risks.
2. Probabilities are typically more important than consequences in triggering protective actions (Slovic et al. 1977; Kunreuther et al. 1978); hence the impact probabilities may be too low and the risk apparently too remote in time to trigger concern, in spite of their high consequences.
3. People are often insensitive to very large losses of life. We will expend great effort to save an individual life, but in a context of impersonal numbers or statistics, the lives of individuals lose meaning. A threat that puts 100 people at risk is likely to be seen as quite serious, but we will probably respond identically to a hazard that threatens 2200 people and one that threatens 2300 (Kahneman and Tversky 1979).
4. People tend to prefer 100% insurance against a threat (Slovic et al. 1977; Kahneman and Tversky 1979). If impact defense systems cannot provide 100% protection, they may be undervalued.

On the other hand, there are also reasons to expect the public to be concerned enough about impact hazards to support action:

5. The risk is demonstrable (it happened to the dinosaurs) and is endorsed by credible scientists.
6. The potential consequences of large impacts are uniquely catastrophic and are qualitatively different from other natural hazards.
7. The probabilities of catastrophic impacts, while small, are not trivial. Considerable public funds are already being spent to deal with risks of even lower probability, such as death or injury from tornadoes or terrorist attacks.
8. Unless action is taken, the risk is unknown and uncontrollable. Lack of control, dread, and catastrophic potential are all qualities associated with high risk perception and strong desire for action to reduce risk (Slovic 1987).

In addition, we can expect increased public awareness of the hazard as the media report discoveries of new asteroids and comets and more frequent "near misses." The collision of comet Shoemaker-Levy 9 with Jupiter in July 1994 drew great public attention to the general impact issue (Chapman 1993). This awareness may lead to diverse reactions, ranging from incredulity that anyone could be concerned about such unlikely events to calls for public action to avert or reduce the risks.

C. Exploratory Research on Public Attitudes and Perceptions

Just as astronomers need observational data to determine the probability of asteroidal impact, social scientists need data to improve their understanding of risk perception and to forecast public attitudes toward impact detection and defense policies. Fortunately, data on public perceptions are relatively easy to acquire, by means of survey techniques. We now summarize the results from a two-part survey of attitudes and perceptions of the impact hazard (Slovic and Peterson 1993), carried out with a sample of 200 college students shortly after the *Newsweek* cover story on the impact hazard (November 1992). The participants were students at the University of Oregon with a median age of 20. This sample of students has been shown in previous studies to respond rather similarly to broader demographic samples of American adults. Before answering, each respondent was asked to read a seven-page briefing consisting of media articles on the impact threat. Despite the extensive recent media coverage of this topic, only about 25% of the respondents said they had heard about this hazard prior to participating in this study.

Part A of the survey asked people to rate 24 hazards on each of 11 scales. The hazards include cigarette smoking, motor vehicle accidents, AIDS, floods, earthquakes, nuclear power plant accidents, and an asteroid hitting the Earth. The scales include perception of risk to the American public, immediacy of risk, severity of consequences, ability of scientists to control the risk, threat to future generations, and potential for global catastrophe. Part B asked respondents to agree or disagree with a wide range of statements about the impact hazard dealing with such items as perceived risk, immediacy of the threat, support for establishing a tracking network, and attitudes toward development of a defense system.

The results of this survey showed that the impact risk ranks 14th out of 24 with regard to mean rating of risk to the American public. The impact risk was judged higher than risks from prescription drugs, medical X rays, bacteria in food, floods, and air travel, but lower than risks from earthquakes and hurricanes. Impact risks were rated as extreme with regard to being unknown to scientists and the public, distant in time (non-immediate), uncontrollable, and catastrophic. There was modest support for detection efforts but considerable opposition to use of weapons in space, even to deflect a threatening asteroid. The survey respondents indicated a strong preference for collecting more data on the risk before developing a defense system. Support for asteroid tracking and defense systems was greatest among those who tended to trust both the scientific community and the government, and it was lowest among those concerned about militarization of space and those who felt that the next major impact is likely to occur very far in the future.

It is of interest to look at the attitudes of the respondents toward the immediacy of the impact threat. Only 6% believed that a catastrophic impact would occur in the next 50 yr; 35% believed that one would occur more than 1000 yr from now, and 34% denied that at a threatening asteroid "could appear within the next 20 yr." When asked to interpret the statement that "scientists say that a civilization-threatening asteroid impact can be expected every 300,000 to 1,000,000 years," 56% felt that "we don't really have to worry about this threat in our own lifetimes," 38% agreed with the assertion that "no such asteroid will appear for thousands of years," and 33% agreed with the assertion that "this statement is not believable because no one can predict the future for hundreds of thousands of years."

While these results must be interpreted with caution because of the small and nonrepresentative sample used, they appear to demonstrate at least two interesting results. First is the perhaps surprising degree of credibility of the scientists who are expressing concern about this threat. Apparently the media have treated these activities in a positive light and have not interpreted these public statements as particularly ill-founded or self-serving. Second is the general problem of interpreting the immediacy of low-probability events. Considering the number of our scientific colleagues who have interpreted long average spacing between impacts to mean that this is a problem for future generations rather than the present, it should not surprise us that a sample of laypersons, most of whom are being exposed to this discussion for the first time, should be similarly confused.

VII. SPACEGUARD SURVEY

A number of options exist for dealing with the impact hazard, ranging from doing nothing (a reaction consistent with dismissal of the risk as negligible) to the development of elaborate planetary defense systems, such as the Spaceguard Survey envisioned in the science fiction novel *Rendezvous with Rama* (Clarke 1973) or the *ad hoc* defense against a fictitious impact from asteroid Icarus studied 25 yr ago at MIT (Kleiman 1968). Many of the chapters in this book are devoted to various aspects of possible defenses against asteroidal or cometary impacts. One of the unique aspects of this particular natural hazard is that it is possible to defend against incoming projectiles and, at least in principle, to avoid disaster entirely by deflecting any threatening object before it can strike the Earth.

Effective defense against asteroids and comets requires advanced knowledge of their orbits; the greater the warning time, the smaller the velocity change that must be imparted to render a threatening object harmless (Ahrens and Harris 1992; Chapter by Ahrens and Harris). Even if orbital deflection is impracticable, advanced warning would permit efforts to mitigate the damage of an impact, for example by evacuating population from the probable target area. The essential first step in any program to deal with the impact hazard is a system to discover and track Earth-crossing asteroids and comets.

One approach to such a survey requirement was developed by the NASA Spaceguard Survey Working Group (Morrison 1992) at the request of the U. S. Congress. The Spaceguard Survey discussed in that report is an international survey network of groundbased telescopes equipped with arrays of CCD detectors and sophisticated real-time data analysis systems (Gehrels 1991; Rabinowitz 1991). Such a network could increase the monthly discovery rate of Earth-crossing asteroids from a few to as many as a thousand (Chapter by Bowell and Muinonen). Implementation of this or a similar system would reduce the time scale required for a nearly complete census of large (>1 km diameter) Earth-crossers from several centuries (at the current discovery rate) to about 25 yr.

The proposed Spaceguard Survey is optimized for the detection of Earth-crossing asteroids with diameters >1 km, because these are the smallest asteroids that might threaten a global catastrophe (taking the range of uncertainty to be 10^5 to 10^6MT of energy). As shown by the hazard analysis presented in this chapter and its antecedents, the greatest risk is associated with objects above the global threshold. In addition, it is only these larger impacts that threaten society or, in extreme cases, the survival of the species. In practice, such a magnitude-limited survey would also discover many more smaller objects, but it would not attempt completeness for diameters <1 km within 25 yr. If the survey were continued, however, it would eventually catalog virtually all asteroids above the atmospheric cut-off (10 MT). The limitations of surveys of this sort are discussed in detail in the Chapter by Bowell and Muinonen.

The survey approach typically leads to discovery of an Earth-crossing asteroid many orbits in advance of any actual impact threat. Thus a warning of a decade or more would be provided, independent of the size of the asteroid (Ostro et al. 1991; Yeomans et al. 1992). Such a program could deal effectively with the threat of asteroidal collision. Only in the unlikely event that an object is predicted to strike the Earth is it necessary to take additional steps toward interception or deflection. However, no survey of this sort can provide similar long-term warning of cometary impact. Roughly 25% of the total impact risk is associated with previously unknown comets that descend into the inner solar system with little warning. Even the most elaborate telescopic system, either groundbased or orbital, cannot guarantee detection of such an incoming comet with more than about a year's warning (Morrison 1992; Chapter by Bowell and Muinonen). In practice, the great majority of the objects discovered by Spaceguard will be NEOs with diameter less than 1 km, which will be acquired at a rate of several hundred per month.

Search programs such as Spaceguard, which are based on current technology, represent efforts at risk reduction. They can provide advanced warning of most potentially catastrophic impacts, but not of all. We are unlikely to reduce the level of risk close to zero without the commitment of immense resources, a commitment that is surely inconsistent with the level of additional security that would be achieved.

VIII. CONCLUSIONS

While there is no longer any question that cosmic impacts have been important in the history of life on Earth, calculations of the contemporary hazard remain uncertain. While the average flux of incoming bodies is known to within a factor of 2 over a range of some 8 orders of magnitude in energy, there remain substantial uncertainties about possible "comet showers" or other non-random clustering of events. As a result of recent work, we now have a good understanding of the degree of protection against smaller impacts offered by the Earth's atmosphere, but our understanding of the physical consequences of impact becomes weaker as we move toward the largest (and most dangerous) cases. The greatest uncertainties, however, concern the ecological effects of large impacts (for example, on agricultural production), and the possible implications for human society.

In the analysis presented here we have adopted a nominal energy threshold of 3×10^5 MT (Chapter by Toon et al.) for transition to large-scale global effects, where a globally catastrophic impact is defined as one that leads to mass starvation and mortality of $>25\%$ of the world's human population. For purposes of calculation, we have carried uncertainties of a factor of 10 in this threshold energy. Our adopted range in threshold energies corresponds to average intervals between globally catastrophic events of approximately 0.2 to 1 Myr.

Even with the present uncertainties of the risk analysis, it is possible to derive several robust conclusions of a qualitative or semi-quantitative nature. We end this overview with a summary of these conclusions.

1. Impacting objects with energy of less than about 10 MT pose little danger; only rare metallic bodies in this size range can penetrate to the surface, and history demonstrates the rarity of significant mortality from iron meteorite falls. Only in the case of possible misidentification of a megaton-range atmospheric explosion with a nuclear attack should such impacts concern us.

2. Over a range of several orders of magnitude in energy, from the threshold for atmospheric penetration to the threshold for global catastrophe, the impact hazard to humans is smaller than the hazards associated with many other natural risks, such as those of earthquakes, floods, volcanic eruptions, and severe storms. Because of the concentration of human population near coastlines, the greatest danger is associated with tsunamis generated by large impacts in deep ocean.

3. Above some threshold energy (probably between 10^5 and 10^6 MT), impacts produce global environmental degradation sufficient to lead to massive crop failures and widespread mortality. Impacts just above this threshold dominate the hazard, with risks that are probably an order of magnitude higher than the cumulative effects of all smaller impacts. These impacts are also qualitatively different from other natural hazards, in that they have the potential to disrupt society as well as kill billions of people. However, such globally catastrophic impacts are rare, perhaps occurring only about three times per million years.

4. As an extreme example of low-probability/high-consequence disasters, large impacts are without precedent in terms of public perception and public policy. Issues associated with cosmic impacts and possible defenses against them are receiving public attention for the first time. With increasing public awareness, demands may grow for action to deal with the impact hazard.

5. Over the entire range of impact energies considered, the risk level increases with the energy (size) of the impact. Therefore the most prudent and cost-effective strategies deal first with the largest projectiles.

6. Any program to mitigate the impact hazard should begin with a comprehensive survey of Earth-crossing asteroids (such as the proposed Spaceguard Survey). Such a survey would provide decades of warning for asteroidal impacts. Better understanding of the numbers, orbital distributions, and physical properties of asteroids and comets are required in order to define an effective defense system.

Acknowledgments. We are grateful to our colleagues on the Spaceguard Survey Working Group and to others with whom we have discussed various aspects of the impact hazard over the past three years; we have benefitted greatly from these exchanges of views. We especially thank R. Binzel, E. Bowell, A. Harris, S. Ostro, E. M. Shoemaker, B. Toon, and K. Zahnle for their insightful comments and critiques of previous versions of this work, and G. Canavan, J. Hills, J. Pike, C. Sagan, and P. Weissman for sharing their results with us in advance of publication. This research was supported in part by grants from the NASA Planetary Astronomy Program.

REFERENCES

Ahrens, T. J., and Harris, A. W. 1992. Deflection and fragmentation of near-Earth asteroids. *Nature* 360:429–443.

Alvarez, L. W., Alvarez, W., Asaro, F. and Michel, H. V. 1980. Extraterrestrial cause for the Cretaceous-Tertiary extinction. *Science* 208:1095–1108.

Beatty, J. K. 1994. Secret impacts revealed. *Sky & Tel.* 87:26–27.

Burton, I., Kates, R. W., and White, G. F. 1978. *The Environment as Hazard* (Oxford: Oxford Univ. Press).

Canavan, G. 1993. The value of space defenses. In *Proceedings of the Near-Earth Object Interception Workshop*, eds. G. H. Canavan, J. C. Solem and J. D. G. Rather (Los Alamos: Los Alamos National Lab), pp. 261–274.

Ceplecha, Z. 1992. Earth influx of interplanetary bodies. *Astron. Astrophys.* 263:361–366.

Chapman, C. R. 1993. Comet on target for Jupiter. *Nature* 363:492–493.

Chapman, C. R., and Morrison, D. 1989. *Cosmic Catastrophes* (New York: Plenum Press).

Chapman, C. R., and Morrison, D. 1994. Impacts on the Earth by asteroids and comets: Assessing the hazard. *Nature* 367:33–40.

Chinnery, M. A., and North, R. G. 1975. The frequency of very large earthquakes. *Science* 190:1197–1198.

Chyba, C. F. 1993. Explosions of small Spacewatch asteroids in the Earth's atmosphere. *Nature* 363:701–703.

Chyba, C. F., Thomas, P. J., and Zahnle, K. J. 1993. The 1908 Tunguska explosion: Atmospheric disruption of a stony asteroid. *Nature* 361:40–44.

Clarke, A. C. 1973. *Rendezvous with Rama* (New York: Ballantine Books.

Clube, V., and Napier, B. 1990. *The Cosmic Winter* (Oxford: Blackwells).

Cornell, J. 1982. *The Great International Disaster Book*, 3rd ed. (New York: Scribners).

Covey, C., Ghan, S. J., Walton, J. J., and Weissman, P. R. 1990. Global environmental effects of impact-generated aerosols: Results from a general circulation model. In *Global Catastrophes in Earth History*, eds. V. L. Sharpton and P. D. Ward, Geological Soc. of America Special Paper 247 (Boulder: Geological Soc. of America), pp. 263–270.

Erickson, K. 1990. Toxic reckoning: business faces a new kind of fear. *Harvard Business Rev.* (Jan–Feb), pp. 118–126.

Gehrels, T. 1991. Scanning with charge coupled devices. *Space Sci. Rev.* 58:347–375.

Gilmore, I., Wolbach, W. S., and Anders, E. 1990. Early environmental effects of the terminal Cretaceous impact. In *Global Catastrophes in Earth History*, eds. V. L. Sharpton and P. D. Ward, Geological Soc. of America Special Paper 247 (Boulder: Geological Soc. of America), pp. 383–390.

Glasstone, S., and Dolan, P. J. 1977. *The Effects of Nuclear Weapons*, 3rd ed. (Washington, D. C.: U. S. Government Printing Office).

Harwell, M. A., and Hutchinson, T. C. 1989. *Environmental Consequences of Nuclear War II: Ecological and Agricultural Effects*, 2nd ed. (New York: Wiley and Sons).

Hildebrand, A. R., Penfield, G. T., Kring, D. A., Pilkington, M., Camargo, A. Z., Jacobsen, S. B., and Boynton, W. V. 1991. Chicxulub Crater: A possible Cretaceous/Tertiary boundary impact crater on the Yucatan Peninsula, Mexico. *Geology* 19:867–871.

Hills, J. G., and Goda, M. P. 1993. The fragmentation of small asteroids in the atmosphere. *Astron. J.* 105:1114–1144.

Hoffman, M. S., ed. 1990. *World Almanac* (New York: Pharos Books).

Kahneman, D., and Tversky, A. 1979. Prospect theory: An analysis of decision under risk. *Econometrica* 47:263–291.

Kleiman, L. A., ed. 1968. *Project Icarus* (Cambridge, Mass.: MIT Press).

Kopec, R. J. 1971. Global climate change and the impact of a maximum sea level on coastal settlement. *J. Geography* LXX(9):541–550.

Krinov, E. E. 1963. The Tunguska and Sikhote-Alin meteorites. In *The Moon, Meteorites, and Comets*, eds. B. M. Middlehurst and G. P. Kuiper (Chicago: Univ. of Chicago Press), pp. 208–234.

Krinov, E. E. 1966. *Giant Meteorites*, trans. J. Romankiewicz (Oxford: Pergamon Press).

Kunreuther, H. 1978. *Disaster Insurance Protection: Public Policy Lessons* (New York: Wiley).

Lewis, J. S., Watkins, G. H., Hartman, H., and Prinn, R. G. 1982. Chemical consequences of major impact events on Earth. In *Geological Implications of Impacts of Large Asteroids and Comets on the Earth*, eds. L. T. Silver and P. H. Schultz, Geological Soc. of America Special Paper 190 (Boulder: Geological Soc. of America), pp. 215–221.

McFadden, L., Tholen, D. J., and Veeder, G. J. 1989. Physical properties of Apollo, Aten, and Amor asteroids. In *Asteroids II*, eds. R. P. Binzel, T. Gehrels and M. S. Matthews (Tucson: Univ. of Arizona Press), pp. 442–467.

Morrison, D., ed. 1992. *The Spaceguard Survey: Report of the NASA International Near-Earth-Object Detection Workshop* (Pasadena: Jet Propulsion Laboratory).

Morrison, D. 1993*a*. An international program to protect the Earth from impact catastrophe: Initial steps. *Acta Astronautica* 30:11–16.

Morrison, D. 1993*b*. The impact hazard. In *Proceedings of the Near-Earth Object Interception Workshop*, eds. G. H. Canavan, J. C. Solem and J. D. G. Rather (Los Alamos: Los Alamos National Lab), pp. 49–61.

Morrison, D., and Chapman, C. R. 1992. Impact hazard and the international spaceguard survey. In *Observations and Physical Properties of Small Solar System Bodies*, eds. A. Brahic, J.-C. Gerard and J. Surdej (Liège: Université de Liège), pp. 223–229.

National Safety Council. 1989. *Accident Facts* (Chicago: National Safety Council).

Ostro, S. J., Chandler, J. F., Hine, A. A., Shapiro, I. I., Rosema, K. D., and Yeomans, D. K. 1990. Radar images of asteroid 1989 PB. *Science* 248:1523–1528.

Ostro, S. J., Campbell, D. B., Chandler, J. F., Shapiro, I. I., Hine, A. A., Velez, R., Jurgens, R. F., Rosema, K. D., Winkler, R., and Yeomans, D. K. 1991. Asteroid radar astrometry. *Astron. J.* 102:1490–1502.

Ostro, S. J., Jurgens, R. F., Rosema, K. D., Winkler, R., Howard, D., Rose, R., Slade, M. A., Yeomans, D. K., Campbell, D. B., Perillat, P., Chandler, J. F., Shapiro, I. I., Hudson, R. S., Palmer, P., and de Pater, I. 1993. Radar imaging of asteroid 4179 Toutatis. *Bull. Amer. Astron. Soc.* 25:1126 (abstract).

Pike, J. 1991*a*. The asteroid and comet impact hazard in the context of other natural and manmade disasters. Presented to the *International Conference on Near-Earth Asteroids*, San Juan Capistrano Research Inst., San Juan Capistrano, Ca., June 30–July 3.

Pike, J. 1991*b*. The sky is falling: The hazard of near-Earth asteroids. *Planetary Report* 11:16–19.

Pike, J. 1993. The Big Splash. Unpublished manuscript.

Rabinowitz, D. L. 1991. Detection of Earth-approaching asteroids in near real time. *Astron. J.* 101:1518–1529.

Rabinowitz, D. L. 1993. The size distribution of the Earth-approaching asteroids. *Astrophys. J.* 407:412–427.

Rabinowitz, D. L., Gehrels, T., Scotti, J. V., McMillan, R. S., Perry, M. L., Wisniewski, W., Larson, S. M., Howell, E. S., and Mueller, B. E. A. 1993. Evidence for a near-Earth asteroid belt. *Nature* 363:704–706.

Rampino, M. R., Self, S., and Stothers, R. B. 1988. Volcanic winters. *Ann. Rev. Earth Planet. Sci.* 16:73–99.

Reynolds, D. A. 1993. Fireball observation via satellite. In *Proceedings of the Near-Earth-Object Interception Workshop*, eds. G. H. Canavan, J. C. Solem and J. D. G. Rather (Los Alamos: Los Alamos National Lab), pp. 221–225.

Sagan, C. 1992. Bewteen Enemies. *Bull. Atomic Sci.* 48:24–26.

Sagan, C., and Druyan, A. 1985. *Comet* (New York: Random House).

Sagan, C., and Turco, R. 1990. *A Path Where No Man Thought: Nuclear Winter and the End of the Arms Race* (New York: Random House).

Sekanina, Z. 1983. The Tunguska event: No cometary signature in evidence. *Astron. J.* 88:1382–1414.

Sharpton, V. L., and Grieve, R. A. F. 1990. Meteorite impact, cryptoexplosion, and shock metamorphism: A perspective on the evidence at the K/T boundary. In *Global Catastrophes in Earth History*, eds. V. L. Sharpton and P. D. Ward, Geological Soc. of America Special Paper 247 (Boulder: Geological Soc. of

America), pp. 301–318.

Sharpton, V. L., Dalrymple, G. B., Marin, L. E., Ryder, G., Schuratz, B. C., and Urrutiafucugauchi, J. 1992. New links between the Chicxulub impact structure and the Cretaceous/Tertiary boundary. *Nature* 359:819–821.

Shoemaker, E. M., ed. 1982. Collision of Asteroids and Comets with the Earth: Physical and Human Consequences. Unpublished draft NASA Conference Report.

Shoemaker, E. M. 1983. Asteroid and comet bombardment of the Earth. *Ann. Rev. Earth Planet. Sci.* 11:461–494.

Shoemaker, E. M., Williams, J. G., Helin, E. F., and Wolfe, R. F. 1979. Earth-crossing asteroids: Orbital classes, collision rates with the Earth, and origin. In *Asteroids*, ed. T. Gehrels (Tucson: Univ. of Arizona Press), pp. 253–282.

Shoemaker, E. M., and Wolfe, R. F. 1982. Cratering time-scales for the galilean satellites. In *Satellites of Jupiter*, ed. D. Morrison (Tucson: Univ. of Arizona Press), pp. 277–339.

Shoemaker, E. M, Wolfe, R. F., and Shoemaker, C. S. 1990. Asteroid and comet flux in the neighborhood of the Earth. In *Global Catastrophes in Earth History*, eds. V. L. Sharpton and P. D. Ward, Geological Soc. of America Special Paper 247 (Boulder: Geological Soc. of America), pp. 155–170.

Sigurdsson, H. 1990. Assessment of the atmospheric impact of volcanic eruptions. In *Global Catastrophes in Earth History*, eds. V. L. Sharpton and P. D. Ward, Geological Soc. of America Special Paper 247 (Boulder: Geological Soc. of America), pp. 99–110.

Slovic, P. 1987. Perception of risk. *Science* 236:280–285.

Slovic, P., Fischoff, B., Lichtenstein, S., Corrigan, B., and Combs, B. 1977. Preference for insuring against probable small losses: Insurance implications. *J. Risk and Insurance* 44:237–258.

Slovic, P., and Peterson, K. 1993. Perceived Risk of Asteroid Impact. Unpublished manuscript.

Stothers, R. B. 1984. The great Tambora eruption of 1815 and its aftermath. *Science* 224:1191–1198.

Swisher, C. C., Grajaless-Nishimure, J. M., Montanari, A., Margolis, S. V., Claeys, P., Alvarez, A., Renne, P., Cedillopardo, E., Maurrasse, F. J. M. R., and Curtis, G. H. 1992. Coeval ^{40}Ar/^{39}Ar ages of 65 million years ago from Chicxulub crater melt rock and Cretaceous-Tertiary boundary tektites. *Science* 257:954–958.

Toon, O. B., Pollack, J. B., Ackerman, T. P., Turco, R. P., McKay, C. P., and Liu, M. S. 1982. Evolution of an impact-generated dust cloud and its effects on the atmosphere. In *Geological Implications of Impacts of Large Asteroids and Comets on the Earth*, eds. L. T. Silver and P. H. Schultz, Geological Soc. of America Special Paper 190 (Boulder: Geological Soc. of America), pp. 187–200.

Turco, R. P., Toon, O. B., Ackerman, T. P., Pollack, J. P., and Sagan, C. 1983. Nuclear winter: Global consequences of multiple nuclear explosions. *Science* 222:1283–1292.

Turco, R. P., Toon, O. B., Ackerman, T. P., Pollack, J. P., and Sagan, C. 1991. Nuclear winter: Physics and physical mechanisms. *Ann. Rev. Earth Planet. Sci.* 19:383–422.

U. S. Bureau of the Census. 1991. *Statistical Abstract of the United States 1991* (Washington, D. C.: U. S. Bureau of the Census).

Weissman, P. R. 1982. Terrestrial impact rates for long and short-period comets. In *Geological Implications of Impacts of Large Asteroids and Comets with the Earth*, eds. L. T. Silver and P. H. Schultz, Geological Soc. of America Special Paper 190 (Boulder: Geological Soc. of America), pp. 15–24.

Weissman, P. R. 1991. The cometary impactor flux at the Earth. In *Global Catastro-*

phes in Earth History, eds. V. L. Sharpton and P. D. Ward, Geological Soc. of America Special Paper 247 (Boulder: Geological Soc. of America), pp. 171–180.

Wetherill, G. W. 1989. Cratering of the terrestrial planets by Apollo objects. *Meteoritics* 24:15–22.

Wetherill, G. W., and Shoemaker, E. M. 1982. Collisions of astronomically observable bodies with the Earth. In *Geological Implications of Impacts of Large Asteroids and Comets with the Earth*, eds. L. T. Silver and P. H. Schultz, Geological Soc. of America Special Paper 190 (Boulder: Geological Soc. of America), pp. 1–14.

Wolbach, W. S., Lewis, R. S., and Anders, E. S. 1985. Cretaceous extinctions: Evidence for wildfires and search for meteoritic material. *Science* 230:167–170.

World Map of Natural Hazards 1988. (Munich: Münchener Rückversucherung-Gesellschaft).

Yeomans, D. K., Chodas, P. W., Keesey, M. S., Ostro, S. J., Chandler, J. F., and Shapiro, I. I. 1992. Asteroid and comet orbits using radar data. *Astron. J.* 103:303–317.

Zahnle, K. 1990. Atmospheric chemistry by large impacts. In *Global Catastrophes in Earth History*, eds. V. L. Sharpton and P. D. Ward, Geological Soc. of America Special Paper 247 (Boulder: Geological Soc. of America), pp. 271–288.

NEAR-EARTH OBJECT INTERCEPTION WORKSHOP

GREGORY H. CANAVAN and JOHNDALE C. SOLEM
Los Alamos National Laboratory

and

JOHN D. G. RATHER
NASA Headquarters

A NASA-sponsored workshop at Los Alamos evaluated the technologies for intercepting near-Earth objects (NEOs). It covered acquisition, tracking, and homing, inspection technologies; and the requirements for deflecting or destroying NEOs of various sizes and compositions, estimating which NEOs could be addressed by various technologies. This chapter estimates the requirements for distant deflection by mass drivers, kinetic energy impact, and nuclear explosions and for deflection on final approach by kinetic energy or nuclear explosives on various interceptors. For distant deflection, non-nuclear means could suffice, but for large NEOs detected on final approach, current boosters and nuclear explosives would be marginal. Larger payloads, more efficient explosives, and higher specific energy fuels are desirable.

I. INTRODUCTION

At irregular intervals, the Earth is struck by objects from space. The age of dinosaurs was apparently brought to an end by the impact of an asteroid of about 10 km diameter (Alvarez et al. 1980) Such impacts occur every few million years, while smaller objects strike more frequently, but do less damage. In 1908 a stony meteoroid of about 50 m in diameter exploded in the air above the Tunguska river in Siberia, devastating the countryside over thousands of square kilometers. (Krinov 1966; Sekanina 1983). Craters on the surface of the Earth and various moons suggest that such impacts occur every few hundred years and that NEOs of essentially all intermediate sizes impact at intermediate frequencies (Morrison 1990).

Recognizing the potential hazard of such impacts, in 1991 Congress mandated that the National Aeronautics and Space Administration (NASA) conduct two workshops on the detection and interception of NEOs. The Detection Workshop (see Morrison 1992; Chapter by Morrison et al.) defined the threat and defined a sensor net to detect NEOs a kilometer or larger that might impact the Earth in the next few decades at distances of a few AU. The Interception Workshop studied potential responses, examining the means and issues in deflecting or destroying NEOs on Earth-impacting trajectories.

[93]

In this chapter we discuss the presentations, issues, and conclusions of the Interception Workshop. It does not attempt to formulate priorities which are addressed in a separate report prepared by Rather et al. (1992). The usefulness of the two workshops was enhanced by considerable cross-fertilization and joint membership.

II. ASTRODYNAMICS OF INTERCEPTION

Detection and interception are interrelated. The farther away a threatening NEO is detected, the farther away it can be engaged and the easier it is to deflect or destroy. Table I taken from the Proceedings of the Working Group on the Astrodynamics of Interception gives a summary of likely intercepts and rough estimates of warning times, probabilities, interaction distances, and velocity increments for each. It also discusses the key issues associated with intercept times ranging from many orbits prior to impact to final approach, which establish the rough launch requirements for each type of intercept. For NEOs whose exact orbit could be established decades ahead of the projected impact, interaction could take place at a distance of several AU and would require a deflection of only centimeters per second. For such an object, it should be possible to send out missions to rendezvous with it at perihelion and apply enough impulse to shift the trajectory enough to just miss the Earth. A 2-km NEO could be deflected at perihelion by ejecting $\sim$40 tonnes of material. That is arguably feasible with nonnuclear means, although stressing. A single Russian Energia rocket could put about 100 tonnes low Earth orbit and accelerate about 30 tonnes to escape velocity. Higher specific impulses could in time significantly lower these mass requirements.

For a less certain orbit that only permits response time of about a year, the deflection velocity required increases by several orders of magnitude, and the masses and energies required for deflection increase with them. For very short-warning objects such as smaller-diameter asteroids or long-period comets, which are more difficult to detect, the interceptor would have to deflect or destroy the NEO during its final approach. In short-warning engagements, this whole detection range would not be available for deflection. A NEO approaching at a velocity $v = 30$ km s^{-1} detected at a range of $R = 1$ AU would permit a response time of at most $\sim R/v \sim 2$ months. An interceptor that flew out to it with the speed of chemical rockets of $V \simeq 3$ km s^{-1} would intercept it at a time $T = R/(V + v) \simeq R/v$ after detection at a range of about $R_i \simeq VT \simeq RV/(V + v) \simeq R/11 \simeq 0.1$ AU. The transverse velocity required would then increase to $\sim$13 m s^{-1} and the kinetic energy to about 250 KT. If the NEO was only detected at 0.1 AU and the intercept took place at about 0.01 AU, the interceptor would need an explosive energy of 25 MT. This illustrates the strong sensitivity of defenses to warning and interceptor performance in the terminal regime and demonstrates that for such engagements, the energies required are about at the limit of those that can now be generated. We consider this in more detail in Sec. IV. See the Chapter by Ahrens and Harris for a discussion of NEO threats and nuclear and non-nuclear responses. Along these lines, C. Phipps in the Workshop Summary (Rather et al. 1992) gives an independent basis for numerical estimates, which provides a road map for interception, deflection, and fracture in the form of an overall set of nomographs, into which point estimates fit in a consistent way.

III. ENERGY DELIVERY AND MATERIALS INTERACTION

Energy delivery and materials interaction determine how efficiently interceptor mass can be used by different concepts; hence, how large an NEO can be addressed by a given interceptor mass. For deflection several periods before impact, mass drivers, rockets, or Earth- or Moon-based lasers might suffice. For larger NEOs the energies required are much larger, and the most effective method of deflection is to blow off part of the NEO's surface and use reaction to produce the transverse velocity needed for deflection. The smallest NEOs might be negated by non-nuclear interceptors that used their kinetic energy to pulverize or deflect NEOs.

The Workshop Summary (Rather et al. 1992) argues that for NEOs 100 m or larger, nuclear explosives are mandatory for intercepts at ranges <1 AU. The most straightforward approach is to explode them at or near the NEO's surface. Detonations at an optimum sub-surface depth could be an order of magnitude more effective than surface detonations, but the apparatus needed for penetration could be so massive that it would cancel this advantage. Furthermore, subsurface explosions have a higher probability of fracturing the NEO into a few lethal chunks on nearly the original trajectory, which could produce an even less manageable threat. The probability of fracture is minimized by an enhanced radiation explosion at a distance of about half a radius above the surface of the NEO. Energy deposition by neutrons would heat and blow off a thin layer, whose recoil would deflect the NEO. While minimizing stresses and the need for knowledge about the NEO's interior, this approach would require about a factor of 100 more energy than a surface explosion.

Various authors in the Workshop Summary treat subjects that are subsequently expanded in this book. For instance, Solem and Snell (see their chapter) uses the increasing effectiveness of kinetic-energy deflection with interceptor velocity to effect deflection without nuclear explosives. In an unpublished report, Hyde et al. (1992) explore various ways of countering NEOs. It is available only in the form of briefing charts, which argue that NEOs of all diameters greater than 100 m are worth intercepting and that even smaller NEOs are worth addressing. However, the main challenge in intercepting small NEOs is detecting them.

In the Summary Report, Remo (see his chapter) uses analogs between meteorites and asteroids to classify NEOs in terms of mechanical strength and thermal properties, which reduce uncertainties in interception (see his chapter). In another paper, Remo shows that penetrator devices can place

TABLE I
Interception Case Definitions

Category	Time Warning	Action	Probability of Scenario for 1-km Objects		Typical Interaction Distance (AU)	Target ΔV (cm s⁻¹)	Object
			Now	Future			
1. Well-defined orbits Precursor missions are strongly advisable for detailed evaluation	Decades	Long-term missions	5% (95% objects presently unknown)	95%	2	1	ECAs only
2. More uncertain orbit Luxury of precursor mission may be absent	Years	Urgent response without much room for error	Unknown		2	10–100 (more error)	Newly-discovered ECAs
Intermediate warning time (but still urgent)							
Object motion is affected by nongravitational forces (cometary bodies)			Unknown			(less error)	Short-period comets

Target ΔV column uses units $\mathrm{cm\ s^{-1}}$.

3. Immediate threat	12 mos. to 1 mo.	Every available engineering measure. Continue to refine the orbit	95%	5%	0.1 (comet)	>1000 at 0.1 AU	Long-period comets
Best scenario: discovery at 10 AU			(5% of objects remain unknown after 20-yr search)			>100 at 1.0 AU	
Discovery initiates emergency					0.1–1 (ECA)	Impact velocities 10–40 km s^{-1}	Small, newly discovered ECAs
4. No warning	0–30 days	Evacuate impact areas	Unknown		0		Long-period comets and unknown ECAs

explosives within NEOs to optimize energy and momentum transfer and pulverization, and discusses penetrators' disadvantages in mass, particularly for conventional uranium alloy penetrator cores. In yet another paper, Hammerling and Remo give a broad technical analysis accessible to a wide audience, which treats repetitive and combined X-ray and neutron interactions, seeking synergistic interactions that could reduce energy requirements. It provides analytic treatments of optimal deposition and impulse for X-ray coupling from a 1 MT explosion above the NEO surface. Its treatment of neutron coupling is consistent with that of Solem. It concludes that both X-rays and neutrons should be used to reduce the total energies required for NEOs of moderate sizes. On another subject, Remo and Sforza discuss penetrators with high explosive or nuclear impulse generators for orbital adjustment; its results correspond roughly to Solem and others' point calculations.

Gertsch reviews experience with terrestrial blasting. It discusses the scaling of the "powder factor," the amount of energy and mass required to fragment a unit of rock, and the "geometry" that determines how energy is delivered in space and time, which have direct analogs in the nuclear yield and placement arguments in the papers above. It concludes that a large single point explosion is generally less efficient than an array of smaller explosions and concurs that uncontrolled fragmentation is the least desirable result and that the best defense is reasonably designed overkill. Most lessons from conventional explosive mining carry over directly to kinetic energy and nuclear explosive defenses.

IV. QUANTITATIVE ESTIMATES OF THE REQUIREMENTS FOR NEO DEFLECTION

This section provides simplified quantitative estimates of the requirements for deflection or destruction of NEOs and compares their scaling with the above predictions. It covers intercepts from distant deflection by mass drivers, kinetic energy impact, or nuclear explosions to large-angle deflection on final approach by kinetic energy or nuclear explosives on interceptors of various specific impulses. Detailed results are presented in a companion report (Canavan 1993). This paper estimates the NEO diameters that could be addressed with interceptors with the few tens of tonnes that could be put into deep space with current boosters and indicates how larger NEOs could be addressed by improved technology.

A. Distant Deflection

It is most efficient to deflect the NEO several orbits before impact. If the velocity of an initially circular orbit is changed by a parallel amount Δv, the deviation from its original orbit a time t later is (Ahrens 1992)

$$\delta \simeq 3\Delta vt. \tag{1}$$

If the velocity of a NEO with an initial eccentricity of 0.5 is given a parallel increment Δv at perihelion, the deviation at time t is $\delta \simeq \Delta vt$, but the smaller deviation of Eq. (1) is used for the conservative estimates below. It is desired to deflect the NEO enough to miss the Earth, which requires $\delta \simeq R_e$, the Earth radius. The velocity increment required at a time t prior to intercept is

$$\Delta v \simeq \frac{R_e}{3t} \qquad (2)$$

for $t \simeq 10$ yr, $\Delta V \simeq 6.4 \times 10^6$ m$(3 \times 10$ yr$\times 3 \times 10^7$ s yr$^{-1}) \simeq 0.7$ cm s^{-1}, which is quite small. For longer t, the velocity increment needed falls as $1/t$. If the increment is produced by the expulsion of an amount of mass Δm at a velocity $c \simeq 300$ m s^{-1}, the amount of mass required is

$$\Delta m \simeq \frac{m \Delta v}{c} \qquad (3)$$

where m is the mass of a NEO of density ρ and diameter D, which is

$$m = \frac{4}{3} \pi \rho \left(\frac{D}{2} \right)^3 \qquad (4)$$

which for $\rho \simeq 3 \times 10^3$ kg m^{-3} and $D = 100$ m gives $m \simeq 1.5 \times 10^9$ kg $= 1.5 \times 10^6$ tonne. This density corresponds to stony NEOs; denser metallic NEOs of the same diameter would present a roughly three-fold greater threat mass. Combining Eqs. (2) through (4) gives

$$\Delta m \simeq \frac{4 \pi \rho R_e}{9ct} \left(\frac{D}{2} \right)^3 \qquad (5)$$

as plotted in Fig. 1 as a function of t for $D = 0.1, 0.3, 1$, and 3 km. Δm is minimized by using a low c; hence, the $c = 0.3$ km s^{-1} assumed. For small diameters and $t = 256$ yr, $\Delta m \simeq 1$ tonne would be required for deflection. If only 1 yr was available for its deflection, about 300 tonnes would be required. At an intermediate time of 32 yr, the expulsion of about 10^4 tonnes would deflect a $D = 1$ km NEO by R_e, in accord with Ahrens' estimate (Ahrens 1992; see the Chapter by Ahrens and Harris). If only 4 yr were available, about 10^5 tonnes would be needed, which would amount to expelling the mass of about 10^4 interceptor payloads each year. For a 3 km NEO and only $t = 1$ yr for deflection, $\Delta m \simeq 10^7$ tonnes is indicated, which is large. The actual mass needed would be a factor of 2 to 3 larger, because times less than a decade stress the applicability of Eq. (2). If the mass was expelled by a mass driver of efficiency $\varepsilon \simeq 0.1$, the average electric power required would be about

$$P \simeq \frac{\Delta mc^2}{2t\varepsilon} \simeq \frac{4 \pi \rho R_e c}{18 t^2 \varepsilon} \left(\frac{D}{2} \right)^3 \qquad (6)$$

as shown in Fig. 2. P increases with c and D^3 and falls rapidly with t. Figure 2 assumes that power could be applied throughout. For maximum efficiency power would have to be concentrated earlier, nearer perigee, so the actual peak powers would be larger. For small D and long times, the powers are only a few W. For $D = 0.1$ km and $t = 8$ yr, the power would be about 100 W, which could be generated easily. For $t = 1$ yr, the power would increase to about 10 kW, which is at the limits of current capabilities. Current powers of 10 to 10,000 W could deflect NEOs with diameters of 1 to 3 km if a few hundred years were available. For a 3 km NEO and 1 yr, more than 100 MW would be required, which would require new technologies. The energy required to expel this mass is about

$$E \simeq \frac{\Delta mc^2}{2\varepsilon} \qquad (7)$$

shown in Fig. 3, which rescales Fig. 1 by a factor of $c^2/2\varepsilon$. The energies are shown in tonnes of high explosive equivalent. They are a few to a few tens of percent of the escape payloads needed to power mass drivers chemically. For 0.1 km and 1 yr, the energy would amount to about 30 tonnes, or 30 tonne$\times 4 \times 10^9$ Joule/tonne $\simeq 1.2 \times 10^{11}$ Joule. Current boosters can put at most 10 to 100 tonnes in low Earth orbit, so a payload of perhaps 30 tonnes to deep space is reasonable. If all of it was available for expulsion, and $t = 30$ yr was available, that could deflect a 0.3 km NEO. Deflecting a 3 km NEO in 10 yr would take 10^5 tonnes of high explosives, or an energy equivalent to a 0.1 megaton (MT) nuclear explosion. For deflection with a mass driver, it would be necessary to rendezvous with the NEO, which could reduce the useful payloads of current boosters to a few tonnes; this would probably not be adequate to support deflection.

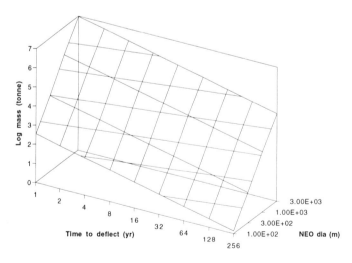

Figure 1.   Mass required to deflect NEOs of various sizes with a 300 m s^{-1} mass driver.

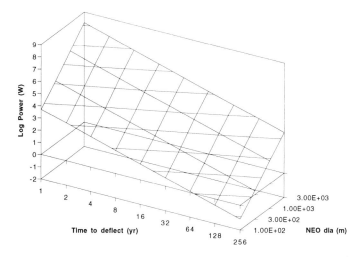

Figure 2.   Power required to deflect NEOs of various sizes with a 300 m s^{-1} mass driver.

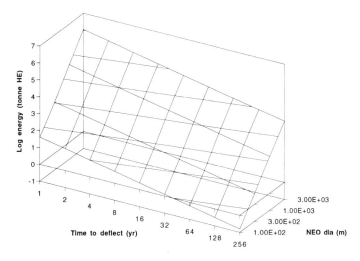

Figure 3.   Total energy required to deflect NEOs of various sizes with a 300 m s^{-1} mass driver.

For large NEOs, the energies become large enough that the energy source becomes a limitation. One million tonnes of fuel as payload to deep space would be impractical. This amount of energy could be provided by fission reactors for perhaps 10^3 tonnes of payload, although even that is large. It could also be provided by about 1 tonne of nuclear explosive, which could be delivered readily. Thus, for NEOs with $D{\simeq}0.1$ to 0.3 km and a few decades to deflect, non-nuclear energies and sources appear possible. For much larger NEOs and shorter times, the energies become large enough for nuclear explosives to offer advantages. Under those circumstances, the masses also become so large that higher volume concepts than mass drivers become desirable.

1. Kinetic Energy Deflection. For intermediate diameters and time scales, kinetic energy can be used to deflect NEOs. Using kinetic energy to expel mass at low velocities is both efficient and simple. If the final mass of the interceptor at impact is M_f, the velocity of the interceptor is V, and that of the NEO is v, the maximum kinetic energy of the impact is

$$E = M_f \frac{(V + v)^2}{2}. \tag{8}$$

Generally, $V < < v$, so Eq. (8) reduces to $E \simeq M_f V^2/2$, which can be achieved by maneuvering the interceptor in front of the NEO and letting it be overrun. About a fraction $\varepsilon \simeq 1/2$ of this energy could go into the kinetic energy of ejected material for a burst somewhat below the surface of the NEO (Solem 1992a). If so, the velocity and mass of the ejecta, v_e and M_e, would be given by

$$\varepsilon E \simeq \frac{M_e v_e^2}{2}. \tag{9}$$

The momentum associated with their expulsion is $M_e v_e$, so the NEO recoil velocity is

$$\Delta v = \frac{M_e v_e}{m} \tag{10}$$

which must equal that required by Eq. (2). Combining Eqs. (8) and (9) gives

$$M_e \simeq \varepsilon M_f \left(\frac{v}{v_e}\right)^2. \tag{11}$$

For near subsurface bursts, $v_e \simeq 100$ m s^{-1}; for optimally buried bursts, $v_e \simeq 10$ m s^{-1} (Solem 1992a). The latter would represent a significant saving, but achieving it would require knowledge of the NEO composition. Thus, the estimates below use $v_e \simeq 100$ m s^{-1}. For that value of v_e and $v \simeq 30$ km s^{-1}, kinetic energy gives a magnification of M_f to M_e of a factor of $(v/v_e)^2 \simeq (30$ km s$^{-1}/100$ m s$^{-1})^2 \simeq 10^5$, which brings the interceptor masses down into more reasonable ranges. Combining Eqs. (10) and (11) gives the interceptor mass required for a given NEO mass, which is

$$M_f \simeq \left(\frac{M_e}{\varepsilon}\right)\left(\frac{v_e}{v}\right)^2 = \left(\frac{m\Delta v}{v_e \varepsilon}\right)\left(\frac{v_e}{v}\right)^2 = \left(\frac{m}{\varepsilon}\right)\left(\frac{v_e \Delta v}{v^2}\right) \tag{12}$$

as shown in Fig. 4 for the ΔV from Eq. (2) and $v_e \simeq 100$ m s^{-1}. Interestingly, the mass required scales on m/v^2, so faster NEOs present less of a threat because of their higher specific energy. Masses range from a few kilograms to 10^5 tonnes, in rough accord with more detailed estimates. Figure 4 indicates that given 10 yr, a 100 kg interceptor could deflect a 0.3 km NEO. Ahrens (1992) estimates that a 200 kg interceptor would be required to deflect a 100 m NEO in that time. Equation (12) would scale Ahrens' result on mass to $\sim 33 \times 200$ kg $\simeq 5.4$ tonne. However, Ahrens uses the interceptor velocity rather than the NEO velocity in Eq. (12). The factor of 3 higher NEO velocity would reduce Ahrens' scaled mass to 540 kg, which is within factor of ~ 5 of Fig. 4. The remaining difference is presumably due to Ahrens and Harris' scaling of cratering from laboratory experiments, rather than the results of nuclear explosives, the basis of Eqs. (9) and (12).

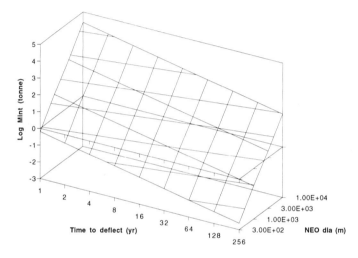

Figure 4.   Interceptor mass required to deflect NEOs of various sizes through collisions at 30 km s^{-1}.

Because all the interceptor has to do is maneuver in front of the NEO, perhaps a 30 tonne payload to escape velocity could support an interceptor final mass M_f of about 10 tonnes. If so, with $t \simeq 100$ yr, kinetic energy could deflect a $D = 3$ km NEO by the Earth's radius R_e. Even with $t = 2$ yr it could deflect a 1 km NEO, although the actual interceptor mass could be a factor of 2 to 3 higher, because a time this short presses the limits of validity of Eq. (2). If multiple or improved boosters could provide a 100 tonne interceptor, that could deflect a 3 km NEO at perigee about 10 yr before impact. Larger masses or shorter times would require larger masses than seem convenient with current, conventional boosters.

    2. *Nuclear Deflection at Perihelion.* Large nuclear explosions provide a specific energy $\phi \simeq 2$ MT/tonne $\simeq 8 \times 10^{12}$ J kg^{-1}, which corresponds to a velocity $\sqrt{\phi} \simeq 3,000$ km s^{-1}, about a factor of 100 greater than the ~ 30 km s^{-1} velocity of NEOs. If the explosion takes place below the surface of the NEO, the energy release and material expulsion is much the same as that by kinetic energy impact modeled above. The main difference is the replacement

of $v^2/2$ by ϕ in Eq. (12), which gives

$$M_{fnuc} \simeq \frac{m}{2\varepsilon} \frac{v_e \Delta v}{\phi} \qquad (13)$$

reducing the interceptor masses of Fig. 4 by

$$v^2/\phi \simeq (30 \text{ km s}^{-1}/3,000 \text{ km s}^{-1})^2 \simeq 10^{-4}$$

as shown by the auxiliary label. Such a reduction would make it possible to deflect the largest NEOs shown with a single current booster using a conventional nuclear explosive. For nuclear deflections at perihelion, only the NEO mass matters. These estimates are also in rough accord with more detailed calculations. Figure 4 indicates that deflecting a 10 km NEO in a decade would take about 1 tonne, or 2 MT. Ahrens estimates about 10 MT, which is again about a factor of 5 higher. Ahrens, however, literally treats a surface burst, rather than the slightly buried burst assumed above (Ahrens 1992). For a surface burst, nuclear test data indicate that v_e would increase about an order of magnitude (Solem 1992a), which would increase the mass requirements of Eq. (13) by a like amount. That would produce an energy of about 20 MT, which is within a factor of two of Ahrens' estimate. This agreement indirectly indicates the consistency of laboratory and nuclear experiments.

   3. *Standoff Nuclear Deflection.* Ahrens offers a simple estimate of the effect of standoff nuclear explosions, which have advantages due to more widespread deposition of energy (Ahrens 1992). He suggests that the explosion be placed a distance of about 40% of the NEO radius away from it. Then about 15% of the energy from the explosion is intercepted by a like fraction of the surface area of the NEO. If the energy deposition is large enough, it should spall off a shell of material of mass $M_e \simeq \rho D^2 L$. If a fraction ε of the energy E is in the form of X-rays and neutrons that are useful for heating and it is deposited in this shell, the energy per unit mass is $\varepsilon E/\rho D^2 L$. For a compression wave velocity in the NEO material of $c_p \simeq 2$ km s^{-1}, Ahrens argues that this heating gives an outward particle velocity of

$$v_p \simeq \frac{\varepsilon E}{\rho D^2 L c_p} \simeq \frac{\varepsilon E}{2mc_p} \frac{L}{D}. \qquad (14)$$

If all of the material in the shell breaks away, the momentum it carries is

$$M_e v_p \simeq \frac{\varepsilon E}{c_p} \qquad (15)$$

which by reaction equals the momentum transmitted to the NEO, $m\Delta v$. As $E = M_f \phi$, the final interceptor mass for standoff explosions is

$$M_{fso} \simeq \frac{c_p}{\varepsilon \phi} m\Delta v. \qquad (16)$$

Comparing this with Eq. (13) shows that the energy required for stand-off is greater than that for slightly subsurface bursts by a factor of about $M_{fso}/M_{fnuc} \simeq 2c_p/v_e \simeq 2 \times 2$ km s^{-1}/0.1 km s$^{-1} \simeq 40$. Figure 4 indicates that deflecting a 10 km NEO in a decade with a subsurface burst would take about 2 MT. With this factor of 40 penalty, a standoff explosion would require about 80 MT. Ahrens and Harris (1992) estimates 10 to 100 MT, which is quite close, although this agreement just indicates that the simplified treatment of Eqs. (14) to (16) retains the essential features of his original model.

This result is also in reasonable accord with the penalties estimated by other treatments, although the agreement is also likely to be partly fortuitous there. Solem (1992a) and Hyde et al. (1992) calculate penalties of factors of 30 to 50 for standoff explosions. However, they do so for later intercepts, and hence larger deflections and energies, for which the heating, $\varepsilon E/\rho D^2 L$, is larger than c_p. Thus, their results actually scale on the violent, few km s^{-1} blowoff velocity of the vaporized and partially ionized material ejected. Thus, their treatments essentially reduce to the use of $v_e \simeq 2$ to 3 km s^{-1} in Eq. (13), which produces about the same results as those from Eq. (16). Hammerling and Remo (1992) present a rough but detailed calculation of the interaction in this high energy density regime, which again leads to a similar estimate of the penalty for standoff.

B. Deflection on Final Approach

If it is necessary to intercept the NEO on final approach, the analysis changes in two ways. The first is that over shorter distances, the trajectory is relatively straight and the deflection is

$$\delta \simeq ut \qquad (17)$$

where t is the time between intercept and impact and u is the transverse velocity imparted to the NEO by hitting it on one side. The second is that the finite flyout velocity of the interceptor V must be taken into account. If the NEO of velocity v is detected at range R, the interceptor is launched immediately, and it quickly accelerates to V, then the intercept takes place at time

$$T_i = \frac{R}{V+v} \qquad (18)$$

at a range from the Earth of about

$$R_i = VT_i = \frac{RV}{V+v}. \qquad (19)$$

This reduces the distance over which the deflection can act, so the transverse velocity imparted must satisfy

$$\frac{u}{v} = \frac{R_e}{R_i} \qquad (20)$$

for the NEO to just miss the Earth. If the impact ejects a mass M_e at a velocity v_e, by conservation of momentum, $mu = M_e v_e$. by conservation of energy, as in Eq. (9), $E \simeq M_e v_e^2$, so that

$$u \simeq \frac{E}{mv_e}. \tag{21}$$

Thus, generating a deflection δ requires

$$\delta \simeq ut \simeq \frac{uR_i}{v} \simeq \frac{RV}{V+v}\frac{E}{mv_e v} \simeq \frac{RV}{V+v}\frac{M_f(V+v)^2}{2mv_e v} \tag{22}$$

where M_f is the final mass of the interceptor after acceleration to V, and Eq. (8) has been used. Introducing the rocket equation, $M_f \simeq M_i \exp(-V/c)$, where M_i is its initial mass, and rearranging gives

$$F = \frac{2\delta mv_e}{RM_i v} \simeq \frac{(V+v)V}{v^2}\exp\left(-\frac{V}{c}\right) \tag{23}$$

which can be maximized by the proper choice of V/v. The optimal V_{opt}/v is shown in Fig. 5 as a function of c/v. For $c << v$, which is the usual case for current rocket fuels, $V_{opt} \simeq c$. For advanced propulsion concepts with $c >> v$, $V_{opt} \simeq 2c$. Figure 6 shows F for this V_{opt} as a function of c/v. For $c << v$, $F \simeq c\exp(-1)/v$. For $v = 30$ km s^{-1}, typical chemical rocket fuels with $c \simeq 3$ km s^{-1} give $F \simeq 1/30$. For $c \simeq v$, $F \simeq 1$. For $c >> v$, $F \simeq (2c\exp(-1)/v)^2$, which increases rapidly to values of 10 to 100. Because the interceptor mass decreases with F, there is a great advantage in accelerating to very high velocities with high specific impulse fuels.

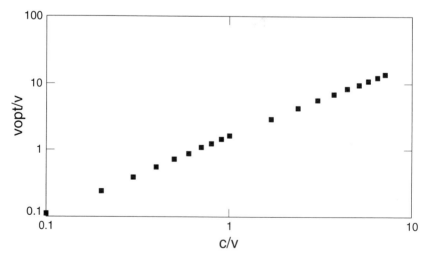

Figure 5.   Optimal velocity for kinectic-energy deflection on final approach as a function of interceptor specific efficiency.

1. Kinetic Energy Intercepts with Low Specific Impulse Fuels. For $c < < v$ and deflections just large enough for the NEO to miss the Earth, $\delta \simeq R_e$, the interceptor mass at escape velocity is

$$M_i \simeq \frac{2emR_e v_e}{Rc} \qquad (24)$$

which is shown in Fig. 7 as a function of detection range R for NEO diameters of 30, 100, 300, and 1,000 m. For the distant kinetic-energy intercepts of Eq. (12), the NEO mass enters in the combination m/v^2, so faster NEOs present less of a threat because of their high specific energy. For intercepts on final approach with low specific impulse fuels, only the NEO mass matters. Current boosters could give $M_i \simeq 10$ tonnes on an escape orbit, which would suffice to deflect a NEO with a diameter of about 50 m detected at 0.1 AU; 100 m detected at 0.8 AU; or 300 m detected at 26 AU. A 30 km s^{-1} NEO moves about 6 AU in a year. At that distance, a 10 tonne interceptor could deflect a 200 m diameter NEO. With development and integration of payloads in space, it might be possible to increase the available interceptor masses by an order of magnitude. If so, the NEO masses addressable at each range would increase by about a factor of 2. A 300 m NEO would take an interceptor mass of about 50 tonnes on final approach, which can be compared roughly to the value of ~1 tonne for deflection from Fig. 4. The factor of 50 difference between them can be roughly understood. The 1 tonne from Fig. 4 is the interceptor mass at impact; thus, its mass at escape would be about a factor of e larger, or about 3 tonnes. Moreover, Fig. 4 uses the three-fold larger deflection for a given time from Eq. (2) relative to that from Eq. (17), which is used in Fig. 7. For such short ranges that factor may not be appropriate, so the 3 tonnes should probably be increased to about 10 tonnes. The remaining difference comes from the different distances for deflection. In Fig. 4 the deflection is assumed to occur at the time indicated, which for 1 yr is about 6 AU. In Fig. 7, by Eq. (19) for $c < < v$ deflection occurs on final approach at only ~$RV/v \simeq R/10 \simeq 0.6$ AU, which requires a ten-fold larger deflection and mass. Thus, the two values correspond to within about a factor of 2.

This comparison indicates the value of much longer warning times. With ten-fold longer warning, the intercepts on final approach would reduce to the more distant ones, apart from the assumed factor of 3 greater deflection for intercepts several orbits ahead of impact. The lack of this warning results in a compound penalty of a factor of 3 for efficiency, and a factor of about 10 for flyout, for an overall penalty of about a factor of 30. Indeed, that follows directly from the masses from Eq. (12), multiplied by e to convert to an escape mass, to that from Eq. (24) for intercept on final approach. With $t = R/v$, the ratio of the masses for final and distant approaches is just a factor of ~$3v/c \simeq 3 \times 30$ km s^{-1}/3 km s$^{-1} \simeq 30$.

2. Kinetic Energy Intercepts with High Specific Impulse Fuels. For $c > > v$, if deflection is just large enough for the NEO to miss the Earth, $\delta \simeq R_e$,

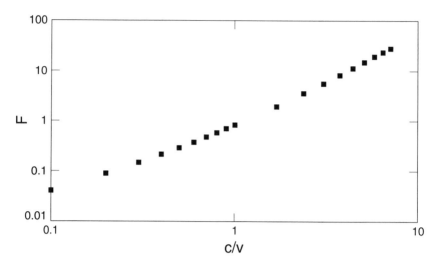

Figure 6. Optimal relative displacement for kinetic-energy deflection on final approach as a function of interceptor specific efficiency.

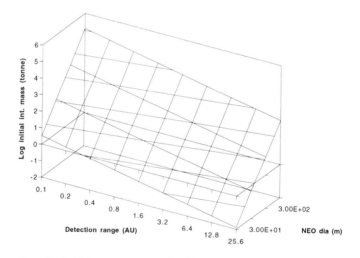

Figure 7. Optimal interceptor mass for kinetic-energy deflection as a function of NEO diameter and detection range.

the interceptor mass at escape velocity is

$$M_i \simeq \frac{e^2 m R_e v_e v}{2 R c^2} \qquad (25)$$

as shown in Fig. 8 for the $c = 100$ km s^{-1} that might be attained with nuclear explosive propulsion. For high specific energies, the NEO momentum again matters most. For a current escape mass of about 10 tonnes, a high specific

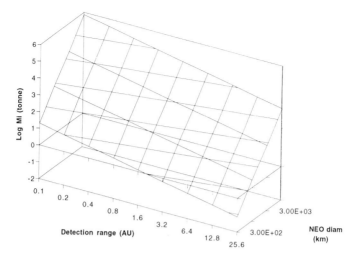

Figure 8.   Optimal interceptor mass for kinetic-energy deflection with high specific energy fuels as a function of NEO diameter and detection range.

impulse interceptor could deflect a 300 m NEO detected at 0.1 AU or a 1 km NEO detected at 6 AU (about 1 yr) away. That represents about a three-fold increase above the diameters that could be addressed with low specific impulse interceptors given those detection ranges. That increase is consistent with the ratios of Eqs. (24) and (25), which shows that the thousand-fold higher specific impulse of the latter reduces the interceptor mass by a factor of $100^2/0.1 \times 30 \simeq 3000$, which permits a given mass to deflect NEOs about a factor of $3000^{1/3} \simeq 14$ larger. This still does not address larger NEOs or shorter detection ranges. Note that M_i scales on mv, so it is not just the mass but the momentum of the NEO that matters. Faster comets would present more of a problem for limited interceptor mass. These results are consistent with other estimates. Solem calculates an interceptor escape mass of 2.5 tonnes for a 200 m diameter NEO detected at 1/100 AU (Solem 1992b); Fig. 8 gives 10 tonnes for a 300 m NEO detected at 0.1 AU, which scales to 10 tonnes $\times (\frac{1}{3})^3 \times 10 \simeq 4$ tonnes, in reasonable accord with Solem.

   3. *Nuclear Explosives on Final Approach.* For nuclear explosives, ϕ is so large compared to the kinetic energy of the interceptor and NEO that the energy delivered is $\sim M_f \phi$ and Eq. (23) can be simplified to

$$F_{nuc} = \frac{\delta m v v_e}{R M_i \phi} \simeq \frac{V}{V + v} \exp\left(-\frac{V}{c}\right). \qquad (26)$$

The resulting V_{opt}/v is shown in Fig. 9. For $c \ll v$, the usual case with chemical fuels, $V_{opt} \simeq c$. For the $c \gg v$ of advanced propulsion, V_{opt} saturates to roughly $2\sqrt{vc}$. The resulting F_{nuc} for this optimal V_{opt} is shown as a function of c/v in Fig. 10. For $c \ll v$, $F_{nuc} \simeq c \exp(-1)/v$. For current $c/v \simeq 0.1$, F_{nuc} is about 1/30. Increasing c by a factor of 100 would increase

F_{nuc} (and decrease the optimal interceptor mass) by about a factor of 10. For a deflection that would just miss the Earth ($\delta \simeq R_e$), for $c < < v$ and the optimal V, Eq. (26) reduces to

$$M_i \simeq \frac{emR_e v_e v^2}{Rc\phi} \tag{27}$$

as shown in Fig. 11, which essentially scales the mass for low specific impulse kinetic energy intercepts of Eq. (24) down by a factor of $v^2/2\phi \simeq 10^{-4}$; this brings most NEOs of concern within reach of practical interceptors. An interceptor of 10-tonne escape mass would deflect a 1 km NEO detected at 0.1 AU; a 3 km NEO detected at 2 AU; or a 10 km NEO detected at 26 AU. For comparison, Fig. 4 gives about 1/3 tonne for distant nuclear deflection of a 10 km NEO at 4 yr (about 24 AU). The ratio of masses for near and distant deflections is about 30 because of the compound penalties discussed above.

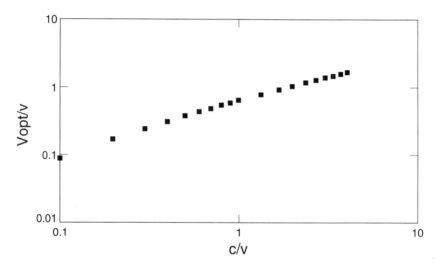

Figure 9.   Optimal velocity for nuclear deflection on final approach as a function of interceptor specific efficiency.

4. Fragmentation and Pulverization. Fragmentation and pulverization can either be viewed as design options or phenomena to be avoided. They occur when the mass ejected M_e approaches a fraction $f = 0.1$ to 0.2 of the total NEO mass. For distant kinetic energy deflections, Eq. (10) gives

$$\frac{M_e}{m} = \frac{\Delta v}{v_e}. \tag{28}$$

For distant intercepts a deflection time of 10 yr only gives a velocity of 0.7 cm s^{-1} and $v_e \simeq 100$ m s^{-1}, so the fraction of mass removed is about 10^{-4}, and no fragmentation is expected. For low-c kinetic energy deflections on final

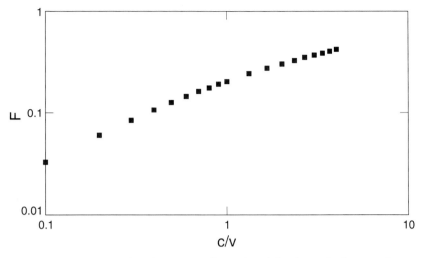

Figure 10.   Optimal relative displacement for nuclear deflection on final approach as a function of interceptor specific efficiency.

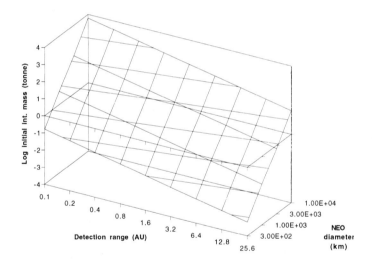

Figure 11.   Optimal interceptor mass for nuclear deflection as a function of NEO diameter and detection range.

approach, Eqs. (21) and (17) give

$$\frac{M_e}{m} \simeq \frac{v}{v_e}\frac{R_e}{R} \tag{29}$$

so that for $R<(v/v_e)R_e/f$, fragmentation is expected. Because $v/v_e \simeq 30/0.1 \simeq 300$, fragmentation would be expected for detections at $R<300 \times$

6400 km/0.1$\simeq$2 × 10^7 km$\simeq$0.1 AU. Because v_e is similar for subsurface nuclear explosions, their fracture criteria is about the same. Pulverization attempts to break the NEO down into parts too small to survive re-entry. That involves emplacing the explosives optimally, which produces lower v_e and hence a higher M_e/m. Whether or not that is useful depends on the size of NEOs that can penetrate and cause damage. It appears that stony NEOs down to a few tens of meters and metallic NEOs down to a few meters in diameter can survive re-entry, so pulverization would have to be very complete to be useful.

C. Intercept Capabilities of Fixed Boosters

This section summarizes the quantitative estimates given above of the masses required to deflect NEOs of various sizes and discusses their compatibility with current boosters. It starts with distant deflection by mass drivers, kinetic energy impact, or nuclear explosions and then proceeds to deflection on final approach by kinetic energy impactors or nuclear explosives carried by interceptors with various specific impulses. The estimates above give the masses required to deflect NEOs of varying diameters. This section inverts them to estimate the NEO diameters that could be addressed by interceptors of a given mass as a function of the time to detect or deflect. A useful benchmark on relevant masses is given by current booster capabilities, which could put at most 100 tonnes of payload into low Earth orbit and perhaps 20 to 30 tonnes into escape velocity.

Distant deflections are more efficient by about a factor of 3 in terms of displacement than intercepts on final approach, so they typically only involve velocity increments of centimeters per second. Distant intercepts can generally be accomplished with efficient trajectories and gravity assists, but needed maneuvers would still reduce interceptor masses at intercept to about 10 tonnes. Concepts requiring rendezvous would incur a further penalty in mass, reducing effective payloads to 1 to 3 tonnes. Assembly of interceptor payloads, rockets, and fuel in space might increase these payload masses by about a factor of 10. Improved rocket technology might increase them by a factor of 2 to 10. The impact of these potential increases is noted where appropriate below.

For distant intercepts of small NEOs, a wide range of concepts is possible, including efficient mass drivers. The masses ejected are very large, according to Fig. 1, but the source of that mass is the NEO itself, so they need not be a fundamental limit for long deflection times. The energies required, which range from tonnes to megatons of high energy equivalent according to Fig. 3, are also large, but provide a range of nonnuclear and nuclear source options. The limiting factor would appear to be the power available. Current sources could generate powers of perhaps tens of kW electrical for a few decades. Figure 2 indicates that such powers should be sufficient for ∼100 m diameter NEOs given a few years for deflection, or 3 km NEOs given a few centuries, as shown by the lowest curve on Fig. 12. According to Eq. (6), the NEO

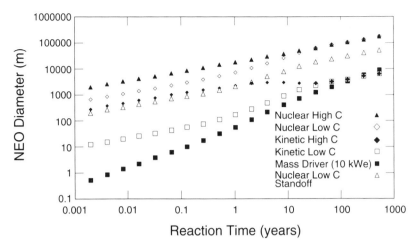

Figure 12.   NEO deflection capabilities for various deflection technologies and interceptor specific energies as a funciton of reaction time.

diameter that can be addressed with a given power scales as

$$D \propto \left(\frac{P\varepsilon}{c}\right)^{\frac{1}{3}} t^{\frac{2}{3}}. \tag{30}$$

D would be increased by a factor of 10 by a thousand-fold increase in P, but the greatest sensitivity is to deflection time t. Mass drivers are limited for short t, but have significant capabilities for long deflection times, which would appear to justify the development of compact and efficient sources of power for space. D is probably overestimated at 1 yr due to the violation of the range of validity of Eq. (2) and the neglect of the interceptor flyout time, which could become important there. These issues are addressed below in the discussion of how to join the interceptor mass requirements for distant and final approaches. This treatment also uses average power; higher peak powers would be required to concentrate thrust earlier near perihelion.

Kinetic energy deflection permits the NEO to use its own high specific kinetic energy to destroy itself. According to Fig. 4, the ~10 tonnes that might be maneuvered in front of it could deflect about a 1 km NEO in a few years or a 3 km NEO in a century (the second curve up on Fig. 12). Kinetic energy provides a useful interim capability at small t. According to Eqs. (12) and (2), the NEO diameter that can be addressed scales as

$$D \propto \left(\frac{M_f t}{\rho v_e}\right)^{\frac{1}{3}} v^{\frac{2}{3}} \tag{31}$$

which shows that the diameter negated by a given final interceptor mass increases with NEO velocity. A velocity a factor of 2 higher than the 30 km s^{-1} used in Fig. 12 would shift the kinetic energy curve up by about a factor of $2^{2/3} \simeq 1.6$, which is not insignificant. Three-fold more dense metallic NEOs would decrease D by a factor of $3^{1/3} \simeq 1.4$. Increasing M_f ten-fold would double D. The dependence on v_e is important. Penetrating to the optimal depth might increase v_e ten-fold from the 0.1 km s^{-1} used in Fig. 12, which would shift the curve for kinetic energy up by a factor of 2. If, however, achieving that optimal depth against imperfectly diagnosed NEOs required penetrator modifications that decreased M_f by a like amount, the two changes would cancel each other out.

As noted in the text, the masses shown on Fig. 4, which are based on nuclear tests, are about a factor of 5 lower than those estimated by Ahrens and Harris (1992; see also their Chapter) from laboratory data; this means that the diameters on Fig. 12 are uncertain by about a factor of $5^{1/3} \simeq 1.7$, which is small and not important. It occurs in an interesting region of potential non-nuclear intercepts. Its resolution would appear to be possible and worthwhile. This appears to be the largest discrepancy in the estimates presented at the NEO Interception Workshop.

Distant nuclear deflections essentially decrease all of the masses on Fig. 4 by a factor of about 10^4, as indicated in the auxiliary scale for nuclear explosives, which makes it possible to deflect NEOs with diameters >10 km by an Earth radius, given a few years. That scaling is shown by the top curve on Fig. 12, which is shifted up from that for kinetic energy by a factor of $(10^4)^{1/3} \simeq 20$. According to Eq. (13), the diameter addressable scales as

$$D \propto \left(\frac{M_{fnuc} \phi t}{v_e} \right)^{\frac{1}{3}} \tag{32}$$

showing that there are minor and approximately equal gains for increased payload, improved nuclear design, and greater deflection time. The dependence on v_e is shown explicitly to emphasize that Fig. 12 assumes the 0.1 km s^{-1} of a subsurface, but non-optimally buried, burst. If the burst was on the surface, v_e could increase an order of magnitude, which would decrease D by about a factor of 2. Conversely, an optimally buried burst could decrease v_e by an order of magnitude, which would increase D by about a factor of 2. If, however, achieving the optimal depth required a ten-fold reduction in the final interceptor mass, the changes would again offset one another.

Standoff explosions minimize sensitivity to NEO geometry and composition, but at a penalty that is generally estimated to be about a factor of 30 in interceptor energy and mass, or a factor of $30^{1/3} \simeq 3$ in D, as shown by the standoff curve in Fig. 12. There appears to be agreement on the size of the penalty for standoff, but this agreement appears somewhat fortuitous. More detailed work would appear justified, given the impact of standoff on an interesting region in diameter-deflection capability. Apart from these uncertainties, which largely have to do with the energy deposition assumed, the sensitivities for standoff explosions are the same as those for subsurface explosions shown in Eq. (32).

The efficiency of intercepts on final approach are reduced by about a factor of 3 due to the smaller deflections of Eq. (17) relative to those of Eq. (2) and another factor of 10 by the flyout time of interceptors with low specific impulse fuels. The overall reduction for kinetic and nuclear deflections is about a factor of 30. That gives a reduction of about a factor of $30^{1/3} \simeq 3$ in D, as shown by the curves on the left side of Fig. 12, which also follow from Eqs. (24) and (27) and Figs. 7 and 11. In plotting these curves, the detection distance R is taken to be $R \simeq vt$. The diameters possible with kinetic energy interceptors scale as

$$D \propto \left(\frac{M_i c v t}{v_e} \right)^{\frac{1}{3}} \tag{33}$$

which increases slowly with payload mass, fuel specific energy, and detection range, and falls slowly with v_e. Note, however, that this result corresponds to a region of some uncertainty in v_e.

The diameters possible with nuclear deflections scale as

$$D \propto \left(\frac{M_i c \phi v t}{v_e v^2} \right)^{\frac{1}{3}} \tag{34}$$

which also increases slowly with payload mass, fuel specific energy, and detection range, and falls slowly with v_e in a region of uncertainty in v_e. However, it also scales inversely with $v^{2/3}$, so that a two-fold higher velocity comet could decrease D by about a factor of 1.6. The diameters that can be addressed by interceptors of high specific energy and kinetic energy are also shown. In plotting them the detection range R is taken to be $R \simeq 2ct$, where $c = 100$ km s^{-1}, or $t = R/2c$, so that for these interceptors, a distance of $R = 1$ AU corresponds to a detection time of 1.5×10^8 km/2×100 km s$^{-1} \simeq 7.5 \times 10^5$ s $\simeq 0.025$ yr. With that amount of warning, Fig. 8 shows that a high c kinetic energy interceptor could deflect a ~ 1 km NEO. For very short warning times, their performance is intermediate between that of nuclear subsurface and standoff explosives. All of the curves are suspect below about 0.01 AU, or about 0.1 yr, where fragmentation should be significant according to Eq. (29). The curves for distant deflections and those for deflection on final approach are just joined smoothly over the interval from 1 to 10 yr, which corresponds roughly to the 6 to 60 AU interval in which the deflection should shift from the many-orbit value of Eq. (2) to the straight-line value of Eq. (17).

This prescription for joining the limiting results accounts for the difference of a factor of 3 in deflection efficiency, but leaves a further uncertainty unresolved. In this intermediate region the curves for addressable NEO diameters are actually double valued, depending on whether the time corresponds to detection or deflection; this amounts to a factor of 10 correction for interceptors of low specific impulse. The issue is illustrated by the recent concern over the comet Swift-Tuttle. Its possible impact time is over 100 yr, which is large compared to the time it would take a nuclear interceptor to maneuver into an efficient trajectory to intercept it at its line of ascending nodes about 5 AU (roughly 1 yr) away. If it were possible to deflect the comet now, the resulting 100 yr for deflection would allow current boosters and nuclear explosives to deflect even a ~60 km NEO with a subsurface burst or a 30 km NEO with a standoff burst. Unfortunately, it is too late to use all of this century for deflection, because Swift-Tuttle is now past perihelion, and receding from Earth at a very high velocity. Thus, the best that can be done is to intercept it at its line of ascending nodes, which is only about 5 AU or 1 yr away. With that reduced range, current boosters and nuclear explosives could deflect at most a ~5 km NEO, so that an increase in payload mass of about an order of magnitude would be needed to deflect the 10 km comet Swift-Tuttle with current technology. Fortunately, there is a century of warning, which should permit the development of advanced technology that could address that threat more efficiently, should the need arise.

Other undetected threats could be more stressing. If a 30 km s^{-1} NEO was detected today at 6 AU, that would only give about a 0.1 yr flyout time, in which case current boosters and nuclear explosives could only address about a 2 km NEO—even if they were orbiting in space and there was no delay in their launch. For such threats, either several order of magnitude larger boosters, much higher specific energy fuels and nuclear explosives, or other advanced technology would be required. The first sounds heroic, and the last is speculative, but the intermediate combination is interesting. From Fig. 10, accelerating the nuclear explosive with a high-c fuel would increase the effective deflection by about an order of magnitude, essentially by eliminating the interceptor flyout delay penalty, which would increase the addressable diameter by about a factor of 2 to about 5 km. In this limit D scales approximately as

$$D \propto \left(\frac{M_i \phi}{v_e v^2} \right)^{\frac{1}{3}} \tag{35}$$

so the other factor of 2 could be addressed by assembling larger payloads in space or using more efficient nuclear designs, which could significantly increase ϕ at large scale. It would also be possible to go for deeper penetration, which could reduce v_e by an order of magnitude for optimal burial—if that could be achieved without reducing the interceptor effective mass. All of these are possibilities, but it is clear that a 10 km NEO detected only a year (6 AU) away would stress current technologies to their limits.

D. Capabilities of Current Technologies

The conclusions of this section are summarized roughly in Fig. 12, which estimates the NEO diameters that could be addressed by interceptors of a given mass as a function of the time to deflect or detect the NEO. It indicates the NEO diameters that could be addressed with interceptors with the few tens of tonnes that could be put into deep space with current boosters. It covers intercepts ranging from distant deflection by mass drivers, kinetic energy impact, or nuclear explosions to deflection on final approach by kinetic energy or nuclear explosives on interceptors with various specific impulses. The text indicates how those curves could be altered by larger payloads, advanced technology, or varying NEO parameters. In general, the variations are less than factor of 2 shifts up or down. The lines are drawn by using the previous figures for the distant and final deflections, and connecting them over the interval from a few to a few tens of years.

Mass drivers and similar concepts are interesting for small NEOs and long times. With plausible increases in power and efficiency, they could compete with kinetic energy deflection of objects several hundred meters in diameter over time scales of decades. Distant kinetic energy impacts make somewhat better use of payload mass and NEO kinetic energy. They could address few hundred meter NEOs, given a few years to deflect, or few kilometer NEOs, given a few decades. There are, however, uncertainties in coupling efficiency, indicated by the factor of 4 to 5 different interceptor masses inferred from laboratory and nuclear experiments, and uncertainties in penetration and ejection efficiency for NEOs of uncertain composition.

Distant nuclear deflections decrease the masses required by a factor of about 10^4, which should make it possible to deflect NEOs with diameters over 10 km in a few years. There are only minor and approximately equal gains for increased payload, improved nuclear design, and greater deflection time. There are uncertainties in coupling efficiency and uncertainties in penetration and ejection efficiency. Standoff explosions minimize sensitivity to NEO geometry and composition, but at a penalty that is generally estimated to be about a factor of 30 in interceptor energy and mass, or about a factor of $\sim$3 in D. The agreement on the penalty for standoff appears to be somewhat fortuitous. More detailed treatments would appear to be justified, given the impacts of this penalty on an interesting region in the capability of diameter-deflection time. Standoff explosions also share the physical uncertainties in coupling of subsurface kinetic energy and nuclear explosions.

The efficiency of intercepts on final approach are reduced by about a factor of 3 due to the smaller deflections and another factor of 10 by the flyout time of interceptors with low specific impulses. The overall reduction is about a factor of 30, or a factor of $\sim$3 in D. The diameters that can be addressed with kinetic energy increase slowly with payload mass, fuel specific energy, and detection range, and fall slowly with v_e—in a region of significant uncertainty in v_e. Nuclear deflections share those sensitivities and add sensitivity to NEO velocity. Interceptors with high specific impulse and kinetic energy interceptors have performance roughly intermediate between that of nuclear subsurface and standoff explosives for short warning. All of these estimates are suspect below about 0.01 AU, or about 0.1 yr, which is where fragmentation should be significant.

Smoothly joining the curves for distant and final deflections over the interval from 1 to 10 yr, the 6 to 60 AU interval in which the deflection should shift from the many-orbit value to the straight-line value roughly accounts for the change in deflection efficiency, but it leaves unresolved the fact that the curves are actually double valued in this intermediate region. The interceptor masses required depend on whether the time indicated is available for deflection, or corresponds to detection—a factor of 10 correction for low specific impulse interceptors. With long detection times and prompt action, large NEOs can be deflected efficiently at long ranges. Even if the deflection must be delayed, early detection can still support the efficient deflection of NEOs several kilometers across on final approach. However, if large NEOs are not detected until final approach, current boosters and nuclear explosives are inadequate. Assembling larger payloads in space; using more efficient nuclear designs, and using much higher specific energy fuels could be required. Even then, it is clear that a 10 km NEO detected only about a year, or 6 AU away, would stress current technologies to their limits. Overall, it would appear that there is reasonably good agreement on estimates for the requirements for deflection of NEOs of various sizes by a wide range of technologies and that those technologies just cover all of the expected threats.

E. Sensitivities and Uncertainties

Uncertainties in the requirements for deflection or destruction by various means were discussed in the previous sections in the process of comparing the calculations by various groups, but it is useful to summarize the overall impact of those uncertainties in one place. That can conveniently be done in terms of the curves on Fig. 12, which summarize those calculations. The top curve could be labeled nuclear subsurface bursts, but actually it applies to bursts at depths sufficient for good but not optimal coupling. It assumes an ejecta velocity of about 100 m s^{-1}. Surface bursts could produce ejecta velocities an order of magnitude higher. Optimal penetration could produce ejecta velocities an order of magnitude smaller, much larger ejected masses, and hence much more impulse. That spread in velocities applies for both distant and near intercepts, and thus for all times indicated. By Eqs. (13) and (27), the NEO diameter that can be addressed by a given interceptor mass scales inversely with v_e, so the factor of 10 higher or lower v_e, for surface or optimal depth bursts would lead to a factor of $10^{1/3} \sim 2.2$ smaller or larger values of D, respectively. Thus, the line for subsurface bursts should really be viewed as a band of values with an overall spread of about a factor of 4, depending on the details of the emplacement of the explosive and penalties for the penetration mechanisms needed to reach those depths.

The next line down is for nuclear standoff explosions. While there is apparently good agreement on the penalty of about a factor of 40 in mass for standoff, and hence in the factor of $40^{1/3} \sim 3.4$ penalty in D, there is less agreement on the fundamentals of interaction and explosive efficiency. The most detailed calculations have been done by weapons laboratories in the high-fluence limit of intercepts on final approach; the most careful comparisons with laboratory results have been done by university researchers, through simulations of low-fluence, distant intercepts. It is perhaps fortuitous that their agreement is so close. These interaction issues are compounded by uncertainties in the efficiencies with which nuclear energy can be channeled into the various forms assumed for optimal coupling by different groups. Overall, the compound uncertainty appears to be less than an order of magnitude, which would result in a band of NEO diameters lying about a factor of 2 below the line shown. The next set of curves are for kinetic energy deflection. The large spread in diameters at low response times results from different assumptions as to the interceptor specific impulses that might result from unproved technologies. In addition to these propulsive efficiency uncertainties, there are additional uncertainties in the coupling efficiencies assumed, which are related to the uncertainties for nuclear explosives discussed above. The curves shown assume a v_e of 100 m s^{-1}. Lower energy impact or surface breakup could increase that by a factor of 10; better penetration could reduce it a like amount, leading to factor of $10^{1/3} \sim 2.2$ smaller or larger values of D, respectively. Thus, the line for kinetic energy impacts should also be viewed as a band of values with an overall spread of about a factor of 4, depending on the details of coupling to the NEO.

The dominant uncertainties in the bottom curve for mass drivers are exhibited explicitly in Eq. (30). In addition to uncertainties in the power that could be provided for extended periods and in the efficiency with which it could be converted into ejecta kinetic energy, there is about an order of magnitude uncertainty in the mass penalty that would be involved in soft landing the driver on the surface of the NEO, giving rise to another factor of about 2 uncertainty in the NEO diameters that could be addressed.

V. VEHICLES AND PAYLOADS

Interception involves two types of missions: precursors to measure NEO dimensions and composition, and intercepts to divert or destroy them. The former could involve relatively low-cost interceptors derived from current defensive technologies. For useful ranges, the latter would require much larger payload masses and higher velocities than those currently available. They could benefit from research on advanced upper stages, electric, and nuclear propulsion. The nuclear rocket discussed in the Workshop Summary by Solem illustrates their advantages.

The Workshop Summary (Rather et al. 1992) describes typical missions: precursors to examine NEOs, intercepts to divert them, and collisions to destroy large NEOs on short notice. The report summarizes the payload mass required for each, which increase rapidly with size. It also summarizes current capabilities for delivering such masses and energies to escape velocities. The discussion of DoD capabilities is stronger than that of NASA's capabilities due to the interests of the attendees at the Workshop. There would appear to be an adequate set of vehicles for flyby and rendezvous missions. There is also an adequate set of small payloads with relevant passive and active visible and infrared sensors for information gathering missions from U. S. strategic defense developmental programs, which are to be tested in the next one or two years.

In the Workshop Summary, Worden describes Clementine, a small satellite, whose main mission is to test the survivability and effectiveness of an applicable set of advanced missile defense vehicles and sensors in realistic operating space conditions. Clementine will test passive and active sensors that could be useful in characterizing NEO composition and configuration through a flyby of Geographos, one of the better-known Earth-crossing asteroids. Clementine is to approach within a few kilometers, from where its lidar could map Geographos' surface with resolutions of centimeters to meters. Its detailed spectral information should also support useful resource maps and materials measurements.

The options for diverting NEOs at long range appear feasible, but are marginal and undeveloped. There does not appear to be adequate payload for divert missions at closer range, other than nuclear diversion of small NEOs. The larger launchers of the former Soviet Union could improve the capability for longer range, higher-payoff divert and destroy missions, if they could be harnessed to international needs, as suggested by current trends. The launch vehicles and payloads needed for large NEOs that approach out of the Sun with little warning, as could be the case for long-period comets, appear well beyond current or planned capacity. Earlier warning with nonterrestrial sensors could reduce the intercept requirements to feasible levels.

VI. ACQUISITION, TRACKING, AND HOMING

NEOs must be deflected very precisely; thus, their initial trajectories and composition must be known precisely. NEOs discovered by visible sensors at ranges out to a few AU can be put into secure orbits by tracking observations with existing radars when they approach within about 0.1 AU of the Earth, as demonstrated at the Workshop by S. J. Ostro. Yeomans demonstrated that these orbits could then be extrapolated decades into the future with sufficient accuracy to assess possible hazards. Prono discussed advanced technologies derived from DoD-SDIO work on free-electron lasers that could extend those ranges to ~ 1 AU.

The Workshop Summary (Rather et al. 1992) describes the status of visible sensors, radars, and homing sensors. Visible cameras are quite adequate to re-acquire tracks established by the telescope survey system proposed by the Detection Workshop (Morrison 1992). Radars are a proven and valuable sensor for converting rough tracks into secure orbits that can be predicted forward with confidence. They can also measure key surface and geometric properties of NEOs. While the current radar network is adequate for current discovery rates, it would be swamped by the rate of discovery of the proposed telescope array. Radars in the Southern Hemisphere would be useful in speeding up search and securing orbit, as would brighter versions that could secure the orbits of NEOs detected visibly at longer ranges than the 0.1 AU of current radars. Homing sensors for interceptors are demanding but straight-forward derivatives of current defensive technologies. For a comet, however, obscuration of its nucleus by its coma may make detecting the appropriate target and precision impact more difficult.

In a Workshop paper by Canavan, scaling arguments are used to compare optical and radar search and track for NEO reacquisition. Passive optical search is superficially different, but is shown to scale similarly. Optical sensors are good for long-range search; modest telescopes of a few meter diameter appear suitable for reacquisition out to about 1 AU. Efficient visible detector arrays appear feasible and attractive; recent advantages in data processing could reduce false-track problems, increase speed, and automate searches. Radars are useful for track and characterization, although they have shorter ranges, because they are active, and hence have better metrics, which are needed for the precise trajectory information needed to secure the orbits of newly discovered objects. Existing defense radars could have some useful capabilities for near-Earth search and track. Bistatic geometries do not appear to offer advantages. Radar and optical sensors are largely complementary. Optical sensors are best for search at long range; radars are better for track and NEO characterization at shorter ranges.

In another Workshop Summary paper, Reynolds shows that existing defense visible and infrared warning satellites could have considerable value in augmenting searches for NEOs of relevant magnitudes. It reports the detection by satellite optical sensors of an intense flash of light over the Western Pacific Ocean on 1 October 1990. The sensors, though optimized for nuclear bursts, gave high-quality intensity-time data on a fireball of visual magnitude −23. Even better performance could be expected with more modern sensors and processing. It would be desirable to fuse defense data with civil NEO searches. Had this event been recorded a year later over the Mideast, it could have been misinterpreted as the use of a nuclear weapon in the Gulf War.

VII. ASSESSMENT OF FUTURE TECHNOLOGIES

The Workshop Summary (Rather et al. 1992) gives a wide-reaching, and somewhat controversial, view of the technologies that might be available for NEO interception at various future time periods. The array of technologies surveyed is encyclopedic, ranging from lasers to "brilliant mountains." The assessments of the risks associated with the different approaches were not without controversy. Many reviewers found the survey speculative, but the authors suspect that if anything, it is probably too conservative for the 50 to 100 yr time frame in which stressing threats are likely to develop. Scientists are generally too optimistic about what can be done in the next few years and too pessimistic about what can be done in the next few decades. Those who try to predict even to the end of the century are looked at askance by their peers. Anyone who speculates decades to centuries ahead tends to be dismissed as irresponsible. Recognizing that, the editors of the Workshop Summary have chosen to suspend judgment on the "Workshop Assessment of Future Technologies" and present it as written—not as a considered assessment, but as a collection of quantitative treatments of current concepts—they hope that a fuller discussion will thereby be stimulated in the next Workshop.

VIII. SUMMARY AND CONCLUSIONS

Following the mandate by the U. S. Congress, the Interception Workshop held at Los Alamos in January 1992 accomplished their assignment (see Rather et al. 1992). It evaluated the technical and policy issues involved in intercepting approaching NEOs. The Proceedings records the presentations and technical options reviewed, outlines the main points of agreement and disagreement, summarizes the status of the main remaining arguments, and outlines areas for further research. It addresses the argument raised by the Detection Workshop (see Morrison 1992) that NEOs pose a significant hazard to life and property that could disturb the Earth's climate on a global scale and the debate over the relative importance of small and very large NEOs.

Astronomers among the Workshop participants discussed results of and prospects for NEO searches, stressing threats from long-period comets, which are numerically less likely, but give less warning, and have very high kinetic energies. The damage from NEOs that break up in the atmosphere before hitting the ground was bounded for the first time, which made it possible to quantify the impact of smaller NEOs, and ultimately to reduce much of the confusion over the relative importance of very small and large NEOs. The Workshop made it clear that detection and interception are interrelated. The farther away a threatening NEO is detected, the farther away it can be engaged, and the easier it is to deflect or destroy. NEOs whose orbits are established decades or centuries ahead of impact could be addressed by impulses of centimeters per second at perihelion, which would shift the NEOs' trajectory just enough to miss the Earth. For less warning, more deflection would be required, ultimately requiring nuclear propulsion or explosion. For short-warning NEOs on collision courses, the interceptor would have to deflect or destroy the NEO on its final approach. The energies required are at the limit of those that can now be generated by man. It was agreed that the most effective method of deflection is to blow off part of the NEO's surface and use its reaction to obtain the transverse deflection of the NEO needed.

The advantages of emplaced, surface, and standoff explosions were discussed in a thorough and dispassionate manner. For intercept on final approach of NEOs 100 m or smaller, nuclear rockets and kinetic energy kill could suffice; for larger NEOs and shorter warning times, nuclear explosives appear essential. The largest, fastest NEOs are the easiest to detect, but intercepting them is a formidable task. Enough is understood about NEO materials and geometries to reduce uncertainties in interception to levels that could make tests profitable and safe. Precursor missions to measure the properties of NEOs could evolve from current defensive and civil technologies. There is an adequate set of lightweight passive and active sensors. Fast inspection and intercept could require larger payload masses and higher velocities than are currently available. The large launch capacities of the Energia could greatly improve the capability to inspect, divert, or destroy them.

As discussed in Sec. VI, NEOs must be deflected very precisely, which means that their initial trajectories and composition must be known precisely. NEOs discovered by visible sensors at ranges out to a few AU can be put into secure orbits by tracking observations with existing radars when they approach within about 0.1 AU of the Earth. These orbits could then be extrapolated decades into the future with sufficient accuracy to assess hazards. Homing sensors for the interceptors could be derived from current defensive technologies. Current radars are adequate for current discovery rates. Existing defense radars and optical and infrared sensors could have some useful capabilities for near-Earth search and track, for which their capabilities are complementary. The technologies that might be available at various future time periods is a controversial but stimulating subject. The array of technologies surveyed is broad. Many found the survey speculative; the authors suspect that it is probably too conservative for the 50 to 100 yr time frames on which the most stressing threats are likely to develop.

NEO impacts are infrequent, but potentially serious. The average loss rate has contributions from NEOs of all sizes. The relative importance of NEOs of different sizes was bounded for the first time at the Interception Workshop, although the issue was not settled. The discussion is carried close to its conclusion in this book. The Workshop received a reasonable amount of coverage in the media. The most publicity was generated by an editorial that reported the controversies raised in the Workshop, but most of that article, and the other coverage, gave thoughtful reviews of the role of NEOs in dinosaur extinction, frequency of impact, the damage they could produce, and the essentials of the proposed detection program.

There was significant, constructive critical reaction from participants, which is also summarized in the Proceedings (Rather et al. 1992). Comments by Morrison and Chapman were particularly helpful in clarifying the extent of agreement on the appropriate emphasis on NEOs of various sizes, readiness for test, the role of conventional and nuclear technologies, the relative priorities of detection and deflection, and of early and late intercept. This clarification paved the way for productive follow-up meetings on NEO Interception at the University of Arizona and Erice. In preparation for those meetings, it was useful to review the estimates of damage from small NEOs presented at the Workshop, obtain quantitative estimates of global effects, and complete the bounding calculations of economic loss. It would be useful to continue the exchange of information on NASA, DoD, and Russian launchers, payloads, and sensors, and solicit and integrate sensor inputs from multiple and national sources. It would also be appropriate to establish international cooperation on analysis, for which this book provides some of the data needed, and to begin discussions on means for implementing and controlling the defensive concepts for NEO impacts.

REFERENCES

Ahrens, T., and Harris, A. W. 1992. Deflection and fragmentation of near-Earth asteroids. In *Proceedings of the Near-Earth Object Interception Workshop*, eds. G. H. Canavan, J. C. Solem and J. D. G. Rather (Los Alamos: Los Alamos National Lab), pp. 89–110).

Alvarez, L. W., Alvarez, W., Asaro, F., and Michel, H. V. 1980. Extra-terrestrial cause for the Cretaceous-Tertiary extinction. *Science* 208:1095–1108.

Canavan, G. 1993. Estimates of NEO object deflection. Hazards Due to Comets and Asteroids, Jan. 4–9, Tucson, Ariz., Abstract book, p. 22.

Hammerling, P., and Remo, J. 1992. NEO interaction with X-ray and neutron radiation. *Proceedings of the Near-Earth Object Interception Workshop*, eds. G. H. Canavan, J. C. Solem and J. D. G. Rather (Los Alamos: Los Alamos National Lab), pp. 186–193.

Hyde, R. I., Colella, N., Ishikawa, M., Ledebuhr, A., Pan, Yu.-L., Pleasance, L., and Wood, L. 1992. Cosmic bombardment III: Ways and means of effectively intercepting the bomblets. In *Proceedings of the Near-Earth Object Interception Workshop*, eds. G. H. Canavan, J. C. Solem and J. D. G. Rather (Los Alamos: Los Alamos National Lab), pp. 155–162.

Krinov, E. 1966. *Giant Meteorites*, ed. M. Beynon; trans. J. Romankiewicz (Oxford: Oxford Univ. Press), pp. 125–252.

Morrison, D. 1990. Target Earth. *Sky & Telescope* 79:265–272.

Morrison, D., ed. 1992. *The Spaceguard Survey: Report of the NASA International Near-Earth Object Detection Workshop* (Washington, D. C.: NASA).

Rather, J. D. G., Rahe, J. H., and Canavan, G., eds. 1992. *Summary Report of the Near-Earth-Object Interception Workshop* (Washington, D. C.: NASA).

Sekanina, Z. 1983. The Tunguska Event: No cometary signature in evidence. *Astron. J.* 88:1382–1414.

Solem, J. 1992*a*. Interception of comets and asteroids on collision courses with Earth. In *Proceedings of the Near-Earth-Object Interception Workshop*, eds. G. H. Canavan, J. C. Solem and J. D. G. Rather (Los Alamos: Los Alamos National Lab), pp. 131–154.

Solem, J. 1992*b*. Nuclear explosive propelled interceptor for deflecting comets and asteroids on a potentially catastrophic collision courses with Earth. In *Proceedings of the Near-Earth-Object Interception Workshop*, eds. G. H. Canavan, J. C. Solem and J. D. G. Rather (Los Alamos: Los Alamos National Lab), pp. 121–130.

PART II
Searches, Orbit Determination, and Prediction

NEAR-EARTH OBJECTS: PRESENT SEARCH PROGRAMS

A. CARUSI
Istituto di Astrofisica Spaziale, Roma

T. GEHRELS
University of Arizona

E. F. HELIN
Jet Propulsion Laboratory

B. G. MARSDEN
Harvard-Smithsonian Center for Astrophysics

K. S. RUSSELL
Anglo-Australian Observatory

C. S. SHOEMAKER
Northern Arizona University

E. M. SHOEMAKER
U. S. Geological Survey

and

D. I. STEEL
Anglo-Australian Observatory

This chapter reviews the search programs that are being conducted at present, giving a description of methods and techniques being used in order to provide the reader with a complete overview, not only of the advancements that have been achieved, but also (implicitly) of the difficulties that may be faced in large scale searches. Only those programs which have a high degree of continuity are mentioned here. Sporadic researches contribute in a significant way to the total number of discoveries, but these searches are widely scattered so that it is not possible to mention them all. Follow-up programs are also discussed.

I. INTRODUCTION

The danger posed by near-Earth objects (NEOs) to mankind has been investigated for some years, and in the past two decades several books have been published and several meetings have been held on the subject. However the

international community has only recently been able to start coordinating research efforts in this field, and much is still to be done in such areas. In 1991/92
a NASA-sponsored committee, with international involvement, produced a
report (the Spaceguard Survey report) that recommended a strategy whereby
the majority of large impactors, producing global effects, could be detected
and tracked within the next 20 yr, and thus any impact by a short-period comet
or asteroid in the near-term (next 200 yr) predicted. A dedicated Working
Group of the International Astronomical Union (IAU) has been set up, mainly
devoted to provide an independent (of the Spaceguard report) assessment of
the hazard and discuss the steps that are feasible to tackle the problem, from
the astronomical search standpoint. Only an internationally coordinated effort
can reach the goal of discovering, following up and cataloging the majority,
if not the totality, of NEOs above some size limit. A future international
network, dedicated to this aim, will certainly build on the experience that has
been acquired by individual observers in recent years.

The fact of past and recurrent impacts of small solar system bodies
(asteroids and comets) on the Earth has received increased attention in the last
decade, following the publication in 1980 of the hypothesis that the faunal
mass extinctions at the Cretaceous-Tertiary (K-T) boundary were due to an
impact by one or more substantially sized asteroids or comets. This idea has
since received widespread support from a variety of investigations, including
the identification of platinum-group anomalies at various sites where the K-T
layer is exposed, enhanced amino acid content at the boundary, tektites and
micro-tektites at that level, and the discovery of at least two impact craters
(the largest in the Yucatan peninsula of Mexico) of the correct antiquity.

However, even before that controversial (at the time) hypothesis was put
forward, a number of observers had already started systematic searches for
near-Earth objects (NEOs), i.e., objects orbiting the Sun on Earth-approaching
paths that made them potentially dangerous to our planet.

The overwhelming majority of discoveries of NEOs have been made
with photographic instruments. Among 136 Earth-crossing asteroids discovered through May 1993, somewhat more than half were discovered through
dedicated programs and all but two of the remainder were discovered serendipitously from photographic exposures made for unrelated purposes. The bulk
of the discoveries have been made in the last 20 years and the rate of discovery
is rapidly increasing. This increase is due in part to improved technology but
principally to the dedicated search programs described in this chapter.

While a number of accidental discoveries have been made in the past
(e.g., obvious NEO trails being noticed on photographic plates exposed for
other purposes), the fraction of discoveries from this source will decline as
the dedicated programs increase their productivity, especially as the majority
of the larger (and thus brighter) objects are identified. The majority of the
discoveries made in dedicated searches have resulted from the programs run
by E. F. Helin and by E. M. and C. S. Shoemaker in the United States (see
Sec. II.A), which have been operating over the last two decades; most of

their work has been done using the 0.46-m Schmidt at Palomar Observatory, California.

Observing programs involving four large Schmidt telescopes (the Palomar 1.2-m Schmidt, the former CERGA [now OCA] Schmidt near Nice in France, the European Southern Observatory Schmidt in Chile, and the other now run jointly by the United Kingdom and Australia) have also contributed, but rather sporadically. In 1990 a program has been started on the U. K. Schmidt in Australia to scan rigorously all plates just after they are taken and this has been very successful (see Sec. II.B).

In 1981 a charge-coupled device (CCD) based NEO search instrument became operative, at the 0.91-m Steward Newtonian reflector at Kitt Peak. This was the Spacewatch program, directed by T. Gehrels. More recently the OCA Schmidt has had a CCD chip installed by A. Maury, with test observations for NEO searches now commencing. One can expect that this type of instrumentation will become more and more relevant in the future, as larger CCDs (as to the number of pixels) become operational and chip mosaicking is introduced. A 16-chip mosaic has already been tested on the Kiso Schmidt in Japan.

The intended scope of this chapter is to present an update on the search programs which are underway in 1993. For future projects, in particular the proposed Spaceguard Survey, the reader is referred to the Chapters by Morrison et al. and by Bowell and Muinonen. In Sec. II those surveys which are dedicated solely to NEO searches are reviewed. These are, to date, the major contributors to discoveries, although sporadic discoveries from other observers and observatories are not negligible. The present imbalance in sky coverage between the two hemispheres is clearly evident from the location of these survey programs.

In Sec. III photographic and CCD techniques are briefly reviewed and compared.

The relevance of follow-up observations is the subject of Sec. IV. It should be stressed that the amount of data likely to be produced in the ensuing years will render this point of major importance: a brief discussion on this topic, and on the need for international cooperation to achieve better results, is contained in Sec. V.

II. DEDICATED NEO SEARCH PROGRAMS

A. Northern Hemisphere

1. The Palomar Planet-Crossing Asteroid Survey (PCAS). December 1992 marked the completion of two decades of monthly observations of the Palomar Planet-crossing Asteroid Survey (PCAS), initiated and operated by E. F. Helin with the support and sponsorship of E. M. Shoemaker (Helin and Shoemaker 1977, 1979). Shoemaker was involved from 1973–1982, starting PACS in the latter year (see Sec. II.A.2 below), while Helin has been involved in PCAS from 1973 to the present. The primary aims of PCAS are the detection of asteroids in near-Earth space, and the determination through a systematic search strategy of the population and impact rates of these asteroids on the terrestrial planets.

Prior to the initiation of the formal PCAS program, Helin investigated the feasibility of the use of the 0.46-m Palomar Schmidt telescope coupled with various film emulsions. The wide angle (8.5° circular field of view) and short focal ratio ($f/2$) of the telescope together with the then-available films (such as Kodak spectroscopic 103a and IIa emulsions) produced a good system for the time. Pairs of exposures of 20 and 10 minutes of the same pre-selected star field were required to detect and confirm the elongated/trailed images of the asteroids. Caltech students assisted in this labor-intensive photographic operation. The discovery rate averaged one or two near-Earth asteroids (NEAs) per year and an occasional comet for 13,000 square degrees of sky surveyed *per annum* during its early stages.

In 1978 a custom-designed stereomicroscope was first tested by the PCAS team for scanning pairs of photographs of the same star field. The displacement of the asteroid image between films, against the stationary star background, renders a quasi-three-dimensional effect and causes the asteroid image to appear (to the brain) either above or below the plane of the film on the microscope stage. As a result of the use of the stereomicroscope and also shorter film exposures, the sky coverage and consequently the discovery rate were greatly increased. By 1981, after considerable testing and experimentation, pairs of films of 4 to 6 minute exposure, examined stereoscopically, had replaced the earlier technique. Magnitudes are calibrated with other stars on the field, based upon the selected areas used for comparison.

The system was further enhanced with the introduction of Kodak 4415 film, which has a panchromatic fine grain emulsion (see Sec. III.A.1 below). After hypersensitization, the films are exposed for six minutes to produce optimum results. This combination of hypered film and appropriate exposure time has extended the limiting magnitude, for non-trailed images, to approximately $V = 17.5$ or $B = 18.5$.

Helin with an observing team of two or three has continued the PCAS program through to the present. To maximize the use of observing time the telescope is kept in constant use, while other tasks involve dark-room developing and immediate review of the quality of the film, helping to set up the observer on preselected coordinates, and loading and reloading the film. As quickly as the films dry, they are labeled and scanned in order to identify unusual asteroids or comets as rapidly as possible. Discovery rates have steadily increased, as a result of the improvements indicated but also as a result of the growing experience and great dedication by everyone involved. Current annual sky coverage is 40,000 to 50,000 square degrees, which results in the discovery of about a dozen NEAs and several comets each year.

As of January 1993, PCAS has resulted in the discovery of over one-third of the known population of NEAs in addition to 14 comets and several thousand other asteroids. Among the NEAs discovered as a result of PCAS, the most remarkable objects are the Aten asteroids with semimajor axes of <1 AU and thus orbital periods of 1 yr. Such asteroids were unknown prior to the mid-1970s, when 1976 AA—now known as 2062 Aten—was identified (Helin and Shoemaker 1977). Since then over a dozen such asteroids have been found worldwide, half of them through the PCAS program. Another important contribution has been the detection of excellent spacecraft mission candidates, based on their orbital elements (small semimajor axis, low eccentricity and inclination). 4660 Nereus is one of about a half-dozen accessible, low ΔV objects discovered through PCAS which are under consideration for flyby and rendezvous missions (Lau and Hulkower 1987; McAdams 1991). PCAS observers recovered the Halley-type comet P/Brorsen-Metcalf, which had not been observed for over 70 yr, its predicted reappearance being overdue when PCAS efforts, based upon a published ephemeris, resulted in its recovery. 4015 1979 VA was discovered as an asteroidal object, but its orbital and physical characteristics caused speculation at the time of discovery that it might be a defunct comet. It was revealed in 1992 by E. Bowell that cometary images on Palomar Sky Survey plates taken in 1949 were identical with 1979 VA, and B. Marsden identified these with Comet Wilson-Harrington (1949 III). This photographic record seems to offer the first evidence of a comet evolving into an asteroid.

All of the PCAS discoveries are followed for as long as possible depending on sky placement and magnitude. As long as a new discovery is accessible and is sufficiently bright to be recorded on the telescope system, the PCAS team observe it astrometrically in order to achieve the longest possible observing arc and therefore the best-defined orbit. It is also reobserved at following oppositions to acquire the necessary three-apparition (or perhaps two-apparition) observations which lead to eventual official numbering, its orbit having been secured.

Plans are being made to upgrade from a photographic search to electronic detection. It is intended that an array of CCDs will be installed on an existing telescope to increase the discovery rate further. Fabrication of the camera will begin in October 1993, with the plans for the camera to be retrofitted to an existing groundbased electro-optical deep space surveillance (GEODSS) telescope in 1994. This will allow sky coverage equivalent to the present program, and will extend the limiting magnitude by two magnitudes. Both the Strategic Defense Initiative Organization and the U. S. Air Force Space Command have indicated their support of a CCD camera system and the installation of this system on the GEODSS instruments for the detection of NEOs. The GEODSS' primary mission, to monitor the sky for man-made objects, will naturally be unaffected by the insertion of a NEO search program because the former mainly utilizes dusk and dawn periods, the latter being conducted during darktime and directed towards opposition.

The International Near-Earth Asteroid Survey (INAS), an extension of PCAS, was established in the late 1970s to encourage and coordinate worldwide searches. A world-wide network of existing telescopes and interested astronomers are participants in this JPL-organized program (Helin and Dunbar 1984).

2. *The Palomar Asteroid and Comet Survey (PACS).* PACS was begun in 1983 as a long-term project to evaluate the populations and fluxes of Earth-crossing and other planet-crossing asteroids and comets. It is a collaborative project between the PACS observing team and E. Bowell of Lowell Observatory. Observations are carried out on the 0.46-m Schmidt telescope at Palomar Observatory by E. M. and C. S. Shoemaker, H. E. Holt and D. H. Levy, with the occasional assistance of students. The films are scanned chiefly by C. S. Shoemaker and Holt. Astrometry of the discovered objects has been carried out by C. S. Shoemaker and Northern Arizona University student assistants on the Mann Comparator of the U. S. Geological Survey and by students and other assistants on the PDS scanning microdensitometer at Lowell Observatory, under the supervision of Bowell.

For most of the duration of PACS, observing runs have been conducted during 11 or 12 lunations each year. In the first years of the survey the observations were carried out by E. M. and C. S. Shoemaker in only six to eight months of the year. Generally, seven nights have been requested for each lunation. Seven nights usually provide adequate contingency for weather and allow time for pairs of photographs to be taken of about 100 fields and for fields of special interest to be photographed on several nights. This coverage includes the most favorable part of the sky to scan for new asteroids and represents about 10 percent of the entire sky, after allowance is made for overlap of fields. Each field photographed with the 0.46-m Schmidt is circular and covers about 56 square degrees. A series of standard guide stars has been chosen such that the sky is covered almost continuously, with as little field overlap as is possible. The actual average overlap is about 40%; there are almost no gaps between the standard fields. Fields are photographed in alternate rows roughly following ecliptic parallels. Ideally, even-numbered rows are photographed one night, followed by odd numbered rows the next. The advantage of this strategy is that part of the overlap can be used to gain positions on additional nights for about 30% of the objects discovered. A disadvantage is that fast-moving objects may slip between the fields and be missed (the so-called "picket-fence" effect).

Of course, weather and other circumstances generally preclude carrying out the observations in a perfectly ideal sequence. On a night of average length, with continuously satisfactory observing conditions, about 50 films (representing 25 fields) can be exposed. More than 60 films have been exposed on long nights in the center of a dark run. Hence, four good nights out of seven generally suffice to meet the survey goals. This number is not always obtained, however, especially during the winter rainy season. The region photographed each lunation usually spans 5 to 6 hours in right ascension, roughly centered on opposition, and about 40 to 50 deg in declination. Special fields for follow-up astrometry may fall outside this region, particularly fields taken in comet searches. With average weather the total annual coverage of the sky is about

40,000 square degrees, with considerable overlap from month to month. Pairs of films generally are taken about 45 minutes apart to provide optimum parallax for stereopsis. Nominal exposures are six minutes with hypersensitized Kodak 4415 film. Somewhat longer exposures are used for follow-up astrometry of faint objects, recovery of special objects, and for special ecliptic fields. Film pairs are scanned for comets and fast-moving asteroids with a specially designed stereomicroscope as soon as is feasible during the observing run. In general, the scanning is not completed until return of the observing team from Palomar to Flagstaff, particularly when usable nights are clumped near the end of the observing run. In this case special arrangements must be made for follow-up of interesting objects discovered. When possible, follow-up observations are made with a CCD-equipped telescope by Bowell at Lowell Observatory.

International cooperation, coordinated through the Minor Planet Center, has been extremely important in securing essential follow-up astrometry of NEOs and comets. Objects discovered early in an observing run are followed to the end of the run with additional films. The use of seven nights gives a good opportunity for follow-up. Comets and asteroids that are especially fast moving are measured at Palomar or at Lowell Observatory (from films sent by overnight mail), or as soon as feasible after return to Flagstaff, in order to provide a basis for a preliminary orbit and ephemeris determined by the Minor Planet Center. Less critical asteroids are measured and reported before the beginning of the following observing run. The observing plan for each dark run includes a large block of contiguous standard search fields plus specific fields that are photographed on two or more nights for recovery or follow-up of previously discovered objects, including comets and NEAs found by other observers. Most of the follow-up fields are part of the contiguous search block; all fields photographed are suitable for search for new planet-crossing asteroids and comets.

Typically about 30 to 50 objects are followed each observing run for improved orbit determination. The observing goal for every asteroid discovered in the survey is to obtain a sufficiently precise orbit for numbering. The large majority of asteroids discovered in PACS remain unnumbered, although multiple-opposition orbits are available for about half of the planet crossers and high-inclination asteroids. Three members of the PACS team generally are present on an observing run. The three-person team is the minimum needed for efficient operations at the telescope, which include planning each night's observations, cutting and hypersensitizing the films, exposing and developing the films at the maximum rate possible, and scanning as many films as feasible during the the the run.

Since the inception of PACS in 1983, well over 10,000 unknown asteroids and about 100 comets have been detected in a total of about 7000 search fields photographed. Limitations of manpower have precluded measuring and reporting all the objects detected; hence the effort has been focused mainly on the planet-crossing bodies and on asteroids with unusual motion.

About 10% of these are now numbered; discovery of about 170 of the newly numbered objects are credited to the PACS team. Altogether, 43 Earth-approaching asteroids have been discovered, and about 30 others have been independently detected. Among the comets detected in PACS, 30 are credited as discoveries by members of the survey team; 14 of these are short-period comets. Many other comets were independently found close to the time that their discovery was reported by other observers. The frequency of discovery and independent detection of planet-crossing bodies as a function of the area of sky covered in the systematic PACS survey has provided a foundation for refining estimates of the population of Earth-crossing asteroids larger than about 1 km in diameter, their flux near the Earth, and the present cratering rate by asteroid impact on Earth (Shoemaker et al. 1990). These discoveries also provided the basis for calculating the asteroid impact and cratering rate on Venus and for estimating the mean age of the Venusian surface revealed in the Magellan radar images (Shoemaker et al. 1991). Holt's discovery of Apollo-type asteroid 4581 Asclepius (preliminary designation 1989 FC), which missed the Earth by about 650,000 km on 23 March 1989 and attracted much publicity, substantially increased public awareness of the hazard posed by asteroid impacts.

3. *The Spacewatch Program.* In 1981 the Spacewatch team began to develop new techniques for NEO searches by using CCDs in a novel way. The 0.91-m Newtonian $f/5$ reflector of the Steward Observatory at Kitt Peak was assigned to Gehrels and R. S. McMillan by P. A. Strittmatter, Director of that observatory. The dark time of each lunation is used for the CCD survey-ing, while one third of the month centered on full moon is used in a search for planets orbiting other stars, McMillan being the principal investigator (McMillan et al. 1994). The other team members of the two programs in 1993 include R. Jedicke, T. L. Moore, M. L. Perry, and J. V. Scotti. The 0.91-m reflector has become known as "the Spacewatch Telescope." The first CCD used, a 320×512 RCA chip, proved to be too small for the efficient discovery of NEAs, and 2048×2048 CCDs were not then available. However, the small CCD was used in a development program in order to explore the different techniques whereby such devices could be exploited in such programs, re-sulting in six distinct modes of CCD use being identified (Gehrels 1991), for searches for gamma-ray bursters, debris in geosynchronous orbits, satellites of asteroids, brown dwarfs, the hypothetical tenth major planet, cometesimals, various types of asteroids, and for comet recovery. From September 1989 until June 1992, a Tektronix 2048×2048 CCD chip was used covering a 38 arcmin wide field with a pixel size of $27 \ \mu\text{m}$, the largest ever made. This CCD, combined with the Spacewatch 0.91-m telescope, had a limiting mag-nitude of $V = 20.5$, at a 4σ detection level in $164.85/\cos(\text{declination})$ seconds exposure. Its quantum efficiency is reportedly 35%, compared to $\sim 70\%$ for the thinned 2048×2048 Tektronix CCD which was installed in August 1992. The latter CCD has a $24 \ \mu\text{m}$ pixel size and a limiting magnitude of 21.0 (4σ level, with a 146.53 sec integration for a scan at zero declination). The CCDs

are refrigerated to $-90°C$ using liquid nitrogen.

In a month under good conditions the Spacewatch team detects about 2000 new mainbelt asteroids and, on average, three NEOs. Only the latter are followed up with astrometric observations the night following the discovery, again a week later, in following months, and at following apparitions whenever possible. The Spacewatch discoveries range from the largest to the smallest NEOs. The largest are 5145 Pholus, 160 km in diameter (see the Chapter by Bailey et al.), and 1993 HA_2, 55 km, in the outer regions of the solar system. The smallest objects found are around 6 m in size. The flux of such small NEOs close by our planet, as detected using the Spacewatch Telescope, indicates an enhancement in their near-Earth flux as compared to a linear extrapolation of the larger (0.5 km plus) NEOs (Rabinowitz et al. 1993; Chapter by Rabinowitz et al.).

B. Southern Hemisphere

1. Anglo-Australian Near-Earth Asteroid Survey (AANEAS). The three searches described in Sec. II.A operate from the southwestern United States, and so cannot reach far southern declinations. Therefore a clear need arose for southern coverage, both to search for unknown objects at such declinations and also to allow follow-up of discoveries made by the U. S. teams as the NEOs move into the southern sky. Thus in 1990 a program was begun in Australia, called the Anglo-Australian Near-Earth Asteroid Survey, or AANEAS.

The *modus operandi* of AANEAS is distinct from the U. S. programs, and quite simple. The 1.2-m U. K. Schmidt Telescope (UKST, operated as part of the Anglo-Australian Observatory) is used for routine sky survey work, photographic plates and films with exposures of 20 to 180 minutes being taken so as to cover the whole southern sky both at different epochs and different wavelengths. Each plate covers an area $6.4° \times 6.4°$, or about 43 square degrees (which happens to be almost exactly one-thousandth of the whole sky). The UKST is operated for ~20 nights per lunation, with bright-of-moon being excluded; typically 6 to 10 plates or films are taken on a clear night, giving an overall coverage (after losses due to clouds, poor seeing, etc.) of around 40,000 square degrees *per annum*. In accord with the primary aims of the exposures, various broadband filters are used, for wavelengths from the ultraviolet to the infrared, whereas ideally for AANEAS the exposures would be unfiltered.

These plates and films are routinely scanned by eye using a binocular microscope, paying special attention to asteroid trails of anomalous length and/or orientation. To date the majority of this searching has been carried out by R. H. McNaught. Objects with angular speeds of $>0.5°$ per day are noted, along with those having high ecliptic latitude motion (because mainbelt asteroids have predominant motion in longitude). For more information on daily angular motions produced by different classes of objects, see the next to last paragraph of Sec. III.B. Trails found on plates/films from high ecliptic latitudes are also almost invariably due to objects of interest: high inclination objects, or bodies close to the Earth. The limiting *asteroidal* magnitude depends upon the angular speed of the asteroid in the sky plane, but is of the order of 19th stellar magnitude (3 to 4 magnitudes below that of the non-trailed objects) for red-filtered plates.

Any object of interest (plausible NEAs, Mars-crossers, Phocaeas, etc.) are then followed up on subsequent nights, most often using other telescopes on the same site, although at times weather conditions dictate that assistance is requested from observers at the Mount John Observatory in New Zealand, the Perth Observatory in Western Australia, or (for objects far enough north) the U. S. teams. Follow-up exposures with the UKST are only made if the object is too faint, and the ephemeris too uncertain, to allow detection otherwise, since UKST time is at a premium. The other two telescopes used are a 1 m reflector with 10 arcmin CCD, and a 0.5-m photographic Schmidt formally owned by the University of Uppsala, both of which are operated by the Australian National University as part of the Mount Stromlo and Siding Spring Observatories.

Steel and McNaught (1992) and Steel et al. (1992) have discussed many of the objects found (including six comets and 40 supernovae apart from the Earth-approaching asteroids which are the main aim of the AANEAS program).

The AANEAS program has led to the recovery of many other asteroids of interest (such as 1927 TC, 1980 WF, 6743 P-L, 1982 DB and 1989 PB), and the independent discovery or re-discovery of several others (the most recent example being the Apollo-type 1973 NA = 1992 OA, the asteroid with the highest known inclination at 68°). The program has also made over 50 "precovery" (identification of photographic detections of newly found asteroids in previous apparitions) observations of interesting objects and well over a dozen Aten-Apollo-Amors. It is believed that there are ~200 to 250 NEA tracks on UKST plates taken from 1973 to 1990 which await identification.

III. TECHNIQUES

A. Photographic Techniques

1. General Comments. Various techniques are used to detect and measure asteroids, but for NEOs the search process must be carried out promptly after the exposure in order to permit rapid follow-up. With angular motions of order 1° per day, and often directional ambiguity existing, clearly one can quickly lose an NEO using a telescope with a field of a few degrees unless follow-up observations are obtained within a night or two. While the human eye/brain image processing system is remarkable it remains a difficult process to detect asteroid images on photographic plates. The simplest way to enhance visibility is to use relatively long exposures because longer image trails are simpler to spot among the many diverse images present on the film. Stars are easily visually filtered out by an experienced scanner but some galaxies may also show as elongated images, although only a few possess the distinct length-to-width aspect ratios expected for fast moving asteroids.

While a long exposure does enhance asteroid trail visibility there are several drawbacks to such an approach. Clearly, with longer exposures fewer separate exposures can be taken, and the area of sky coverage is therefore reduced. Long exposures also produce photographs with darker backgrounds and are therefore more difficult to scan. They may be affected by moonlight, depending on the telescope f-ratio, and sky conditions. Nevertheless, the technique is useful where the long exposures are the by-product of another scientific program (as with the AANEAS group in Australia) or where specialized scanning equipment is unavailable.

A much improved method has been successfully pioneered by the two photographic groups PCAS and PACS (see Sec. II.A). These groups utilize stereo comparators to examine pairs of photographs of the same part of sky exposed a short time apart. Any moving object will be readily detected by eye because the image will appear to stand out above or below the plane of the film, all stationary objects appearing in that plane. It is a physiological effect, engraved in the human brain since childhood, that an object is seen to move with respect to its background, even with one eye. The apparent motion, transformed into a single image in the foreground, is brought into view by looking at two plates, by one eye each.

In stellar/galactic astronomy very faint images may be detected by using long sky-limited exposures; this is the point at which the background sky just starts to be detectable. However, this depends on the fact that the telescope can track each object perfectly during a long exposure. This is impossible for asteroids in general and the limiting magnitude is controlled by their apparent angular velocity on the sky. It follows that there is an optimum exposure beyond which no gain will be achieved, and the same applies to CCD observations. This exposure time can be estimated by assuming a range of angular velocities for the asteroids being sought and exposing for a time limited so as to keep the trailed asteroid image more or less within the seeing disc. This tends to be quite short, for fast moving objects. The gap in time between the two exposures is normally adjusted for operational convenience but will vary depending on the telescope's scale.

Once the object is detected, motion vectors are determined, either from the image movement between exposures or from the length of trail if only a single exposure is available. Selection of potential NEOs is carried out on the basis of this apparent angular velocity but is by no means a foolproof indicator because it is not unusual for such objects, when they are as far away as the main belt, to possess apparent motions indistinguishable from those of main-belt asteroids. If perfect identification were required, then it would be necessary to make sufficient astrometric observations of every object in order to compute approximate orbits. While this may be possible in principle, it is not an option given the present facilities and available personnel. Only those objects with anomalous motions are followed up in order to determine precise orbits. This identification technique is almost exactly equivalent to that used in the part that is called "trail detection" in the Spacewatch CCD search program (see Sec. III.B below), and is subject to the same limitations, although in the case of the CCD system the motion vector is more quickly and easily determined.

  2. Photographic Materials. A variety of materials has been used, but the most effective emulsions historically have been the IIIa-type emulsions coated on glass from Kodak, introduced twenty years ago, and more recently a panchromatic emulsion coated on a film base released in 1982, again from Kodak. The high-speed fine grain and good contrast characteristics of this new film (4415) make it the emulsion of choice for this work and it has been in use in all three photographic programs described in Sec. II for some years. The superiority of 4415 film has been detailed by Parker (1992).

  These photographic emulsions, as supplied, are very slow and must be treated in various ways to increase their sensitivity to faint images. This process can be rather labor intensive and involves soaking and/or baking in various dry gases. The very earliest emulsions had a quantum efficiency not greater than $\sim 1\%$, while the later IIIa emulsions are thought to reach 4 to 5% when optimally hypersensitized. The film in current use (i.e., 4415) may reach a quantum efficiency of $\sim 10\%$ under ideal conditions. This should be compared to values of 30%, easily obtained from thick CCD detectors, and values of 70 to 80% obtained for thinned CCDs. Clearly, photographic emulsions cannot compete with CCDs on the basis of speed.

  One point worth pointing out here is that, while absolute speed is essential for the discovery of the faintest asteroids, it is not necessary for subsequent observations where the motion vector of the asteroid is known. In this case photographic detectors can reach a similar depth to CCDs, although the required exposure will be greater. For example, in the case of the 1.2-m UKST, it is possible to reach a limiting magnitude of 22 to 23 with an exposure of 25 minutes. Even so, where an accurate orbit is available CCD observations are to be preferred for follow-up. This preference is due to: (i) the comparative ease with which they may be obtained; (ii) the higher astrometric quality derived using modern digital image processing techniques; and (iii) the availability of on-line star catalogs with field stars rapidly identified and their positions referenced.

  3. Sky Coverage. Photographic telescopes currently have a great advantage with regard to the area of sky which may be recorded in a single exposure. The 0.46-m Palomar Schmidt has a field of view covering 57 square degrees, and the 1.2-m UK Schmidt just over 43 square degrees. While CCDs have much higher quantum efficiencies, and the data obtained are available for digital analysis, they are limited to small fields of view, the physically largest CCD chip now available covering 4.9×4.9 cm compared to a 36×36 cm area for a UKST plate/film. Conversely, the technique of scanning at rates faster than the sidereal rotation rate, as described in Sec. III.B below, allows increased area coverage for CCDs. This is not at the cost of limiting magnitude for NEOs that move fast, and which would therefore produce a trail on longer exposures (see Table 2 of Rabinowitz 1991). A CCD telescope of large aperture can either scan over a wide field, or stare for faint follow-up. However, in the case of photographic programs there is also still scope for increasing the area of sky covered, in particular because many Schmidt telescopes are currently under-utilized.

B. CCD Techniques

There are basically two distinct CCD observational techniques, which may be termed "stare" and "scan." These can be understood quite simply by comparing them with photography with a camera; one can hold a camera still and make an exposure by staring at the scene, or one can have a film camera scan the scenery. The technique of staring with CCDs has been common at astronomical telescopes since about 1975. For NEO astrometry it has also been in usage for some years, for instance at the 1.5-m Wyeth telescope at the Oak Ridge Observatory, Massachusetts. Although this program was fully converted from photography to CCD only as recently as August 1989, this is the longest standing consistent CCD follow-up program. Before the Guide Star Catalog (GSC) of the Hubble Space Telescope was available, one would identify the small area of an exposure on the Palomar Atlas and use classical astrometric catalogs such as the AGK3, or that of the Smithsonian Astrophysical Observatory.

While it is a cumbersome procedure to transfer a small field (from the particular observation using a CCD) onto the large one (from the atlas or catalog), this procedure had been carried out with good results earlier in the history of long-focus photographic astrometry. With the availability of the GSC sufficient stars usually appear in the single frame of a stare exposure. A special application of this type of astrometry is by D. J. Tholen, who uses the precise readouts of telescopes at the Mauna Kea Observatory to find the coordinates of the center of each of his NEO exposures. Knowing the scale, it is merely a matter of counting pixels with respect to that center. As for the Spacewatch astrometry on comets and asteroids, such a process is carried out to a typical precision of about ± 0.5 arcsec; this is comparable to the precision of the GSC, which has some systematic errors depending on which plate (or part of a plate) derived from the 1.2-m Palomar or U. K. Schmidt telescopes was used to produce the catalog. The residuals $(O - C)$ of observations of this type with respect to the computed orbit are now generally on the order of ± 0.5 arcsec. Stare exposures are also used in photometry. Lightcurves are routinely obtained for instance by Wisniewski (Wisniewski and McMillan 1987).

Scanning with a CCD is more complicated with regard to software preparations; for example, the present Spacewatch observing method is based on about 8 man-years of computer programming. This represents a developmental phase, of course, and with this experience a new program could probably be implemented in a shorter time. For comparison with photographic techniques one can see that the labor comes up-front in the CCD operations, but once the computer programming has been done, the analysis can be in near-real time and in great detail and high precision (Rabinowitz 1991). The Spacewatch team was the first to accomplish routine scanning with CCDs in astronomy (Gehrels 1991,1994; Rabinowitz 1991; Scotti 1994). A new discipline became known ("Scannerscopy") in which the sky is surveyed with CCDs in a scanning mode, rather than by sequencing a series of stare exposures. This technique is efficient in that it saves the telescope time which would be used in stare mode in setting on new coordinates, and reading out the CCD—the next exposure awaiting the completion of the readout—which for scanning is done continuously during the successive exposure frames. "Flat fielding," to calibrate the nonuniform sensitivity of pixels, is rarely needed because all pixels in a column are sequentially exposed to the same object(s).

Up to the present, the Spacewatch scanning has been carried out simply by turning off the right ascension drive. While scanning at faster than the sidereal rate is more effective (see Table 2 of Rabinowitz 1991), this has not been possible with the first Spacewatch telescope for the following reasons: (i) the 0.91-m telescope was built in 1920, and it now has gears/bearings/preload problems that prevent smooth scanning on the sky by using its drives; (ii) because it is a Newtonian telescope, there are steel structures in the dome that would make rapid scanning hazardous because of possible impact and damage to equipment; (iii) the transfer rate of currently available CCDs is limited.

The motion on the sky must be precisely followed by slaving the charge transfer of the CCD to the drift rate of the image, while the CCD is read out continuously during the observing. The scan rate depends on the cosine of the declination δ; for the present 24 μm per pixel Spacewatch system the rate is $71557/\cos\delta$ microsec. This is based on the image scale which is derived from the astrometry, particularly from the solutions in the nonscan direction of declination; this scale is 1.07629 arcsec per pixel. The rotation of the Earth must be taken into account with respect to the fixed stars. The scan rate is therefore $1076290/1.002738 \times 15 \times \cos\delta$ microsec.

The end register reads the data into the computer which is then used to identify each image and store the x and y pixel coordinates. The scan is repeated two more times and the (x, y) coordinates are compared for each image. Nonmoving objects are ignored, and moving objects are reported with an approximate brightness, obtained from occasional calibration with a standard region, and with right ascension and declination coordinates obtained from comparison with a large number of stars in the scan; the coordinates of the calibration stars are taken from the GSC.

The reductions are made in near-real time; Rabinowitz (1991) described the procedure in detail, which it may be summarized as follows. The computer is a Solbourne Series 5/600 work station with three 62-bit processors and 32 megabytes of memory; an 8-bit gray-scale monitor with 1152×900

pixels is used for display. A Sun Unix operating system is used with a mouse-controlled, windowed interface. The data are recorded and archived on magnetic tape. During the first two scans the computer executes streak detection for objects that move sufficiently fast such that they display a recognizable trail during the pass of the image across the CCD frame, this lasting 146.53 sec on the equator (and changing with the cosine of the declination as noted above). This image inspection is also done by eye, with the observer visually examining the field as the CCD scan progresses. Such human interaction could be done by the software if the sensitivity were set at, say, 1σ, but then the number of reported false alarms (cosmic rays, etc.) would be very large. The human eye/brain system is an excellent discriminator of faint streaks. When the streak shows a brightness variation with rapid periodicity, up to say 20 sec, the object is discounted as being man-made (i.e., in geocentric orbits) and it is not followed any further. Asteroids are not known to spin at a faster rate than about one revolution every two hours, due to internal energy dissipation, so that fast spinners (as indicated by their lightcurves) are almost certainly man-made. Motion detection is reported by the software during the third scan, flagging each moving object on the screen (along with its position on the previous two scans) and reporting its rate of motion and position in right ascension and declination, and brightness.

This triple-scan technique avoids a large amount of ambiguity. However, at first sight it might be criticized as resulting in a loss of sky coverage and thus detection rate. In fact, it can be argued that the technique enhances the discovery rate. It is estimated that 0.7 mag is lost from the detection limit when only two exposures or scans are used. While the area coverage for this triple-scan operation decreases by a factor of 1.5 compared to a double scan (only), the volume of space within which objects of a set limiting absolute magnitude are detectable increases by a factor of about 2.7 as a result of this 0.7 mag gain, and the number of discoveries thereby increases by a factor of 1.8.

In the present operations about one-third of the objects originally reported during the scans are not real and must be eliminated by the observer; this fraction can be controlled by choosing the critical threshold and other sensitivity parameters. At the end of a night's observing, having flagged likely objects during the night, the computer analysis starts so as to produce more precise astrometry, using all GSC stars observed during each of the three scans, and to recall snapshots for each scan which the observer can inspect later during the day. Near opposition the motion of an object is a good indicator of its distance. A plot of these motions is given as Fig. 4 of Rabinowitz (1991). The typical daily motions are as follows: Jupiter Trojans $-0.16°$, Hilda asteroids $-0.18°$, and main-belt asteroids about $-0.20°$ to $-0.25°$. The daily latitudinal motion for all these may range up to $\pm0.13°$. The Hungaria asteroids have faster daily motions than the above: in longitude $-0.3°$ to $-0.4°$ or more, and in latitude $\pm0.5°$ or more. Objects progressively faster than the Hungarias are Mars-crossers, Amor, Apollo, and Aten asteroids. However, it is possible that objects very close to the Earth will demonstrate short streaks in the CCD frames because their motions come close to matching that of our planet, and indeed positive longitude motions are possible (and do occur) when the object's individual motion has overcome the reflex of the Earth when it is close by our planet. For instance, 1993 KA$_2$ was discovered from an eastward trail 200 arcsec long.

The expected rates of motion are used to define the interval between successive scans. For NEAs the optimum interval at the Spacewatch telescope is found to be about 10 minutes. Longer scans are, however, more economical, requiring less time for telescope setting, while the first frame of a scan is always a partial loss because of the ramping of the integration time (this is called the "ramp frame"). A compromise of 30 minutes has been used and is still used when the weather seems unreliable, but in order to also be able to identify more distant objects (such as 5145 Pholus, and 1993 HA$_2$), and yet not to affect deleteriously the detectability of fast-moving NEOs, an interval of 40 minutes is normally used.

IV. FOLLOW-UP OBSERVATIONS

A premise of the Spaceguard Survey (Morrison 1992) is that discovery and follow-up observations will be indistinguishable from each other. There will be automatic procedures for identifying and extracting all the observations of a particular object, and the resulting data will then automatically be used for the orbit determination.

However for the time being, and perhaps even for the foreseeable future, there is necessarily a distinction, a conscious decision being required to alert other observers to obtain further observations of a new discovery. It is therefore desirable that the discoverer provide enough information and in a timely enough fashion that the follow-up process has at least a fighting chance of success. At the very least the discoverer should be able to demonstrate that the object is real and moving; ideally, the information provided should be enough to obtain crude information about the object's orbit. In practice this minimum requirement applies only to amateur astronomers making visual discoveries of comets. Accurately measured positions should *always* be provided in the case of photographic or CCD discoveries. The photographic or CCD discoverer should also *plan to re-observe the object on a second night*. If the object is found on the last scheduled night of an observing run, and even if bad weather thwarts his/her plans, it is the responsibility of the discoverer to make alternate arrangements for securing the second-night data.

The positions on the two nights should then be communicated by electronic mail to the Minor Planet Center, where the staff can examine whether the object is already known, in the sense that an orbit determination is already available for it. Identification is straightforward in the case of numbered minor planets and other multiple-opposition minor planets, but a comparison is also made with orbits of other newly discovered objects, including Väisälä orbits, or other non-apsidal possibilities estimated from observations on only two nights (e.g., by the method discussed by Marsden [1991]). If there is no immediately recognized match the object is given a preliminary designation (e.g., 1983 TB), and a representative orbit for the object is incorporated into the data files. Searches are also made for identifications using short-arc orbits from other oppositions.

Balonek et al. (1992) have recently discussed the ease with which precisely simultaneous CCD observations of the same object can be coordinated at different sites and the distance directly determined by parallax. This has mainly been carried out for known objects and as a student exercise, but, with a baseline of 3000 km and telescopes of aperture 0.4 to 0.5 m, distances have been determined with an accuracy of 3% for objects as far away as 0.25 AU. Such coordination would be impractical for the first-night or second-night detection of a new discovery, but one can achieve the same effect by making observations at the discovery site on one or both nights several hours apart. The recent Palomar photographic discoveries 1993 BC_2 (= 1980 AA) and 1993 BW_2 (0.08 and 0.16 AU, respectively, from the Earth at discovery) both had better determinations of their orbits made due to such parallactic measurements. In each case three-position orbit determinations from observations on consecutive nights involved departures of some 40 arcsec from a great circle. Because the measuring engine on site at Palomar is incapable of consistently yielding positions on 0.46-m Schmidt films to better than 4 or 5 arcsec, the outcome was still rather satisfactory. Nevertheless, the fact that first-class measurements of the Palomar films are still not possible until the observers have returned to home base, sometimes as much as a week later, continues to be a ridiculous hindrance to the NEO enterprise (see Secs. II.A.1 and 2). Considering that CCD measures often nowadays show internal consistency to 0.2 arcsec and better, particularly with the use of a modern reference-star list like the PPM (Positions and Proper Motions) catalog, and with future editions of the GSC also being expected to show improvement, the use of diurnal parallax and observations on two nights has merit with regard to orbit determination for minor planets in the main belt (Marsden 1992).

Though not recommended as a general practice, it is sometimes possible (and very occasionally even necessary) to determine an orbit from observations on a single night. The most celebrated case is that of 1991 BA, followed by the Spacewatch team for some five hours, and computed to have passed closest to the Earth and into daylight after seven hours more (Scotti et al. 1991). Also observed for five hours, the Spacewatch discovery 1992 YD_3 was some 0.03 AU from the Earth; bad weather then intervened for several nights, and by the time conditions cleared the object had faded substantially and moved far to the south. The closest photographic analogy is perhaps 1991 JY, observed (and measured) at Palomar at the same time on two nights, but then observed in three groups spanning more than eight hours at Siding Spring, the data from the long austral winter night alone being adequate to show that the object was 0.11 AU from the Earth and in an Aten-type orbit.

Allowing, then, that the observations from the first two nights generally constitute the discovery/confirmation process, the follow-up stage is considered to begin at that juncture. Except for Spacewatch discoveries (because they are likely to be too faint for observation elsewhere), news of NEOs (whether minor planets or comets) is then relayed to other observers, almost exclusively in the *IAU Circulars*. Occasionally, the Central Bureau of Astronomical Telegrams makes specific requests of observers, but this is usually to support what is in the *IAU Circulars*. Of course all the relevant data, including subsequent updates, are given in the *Minor Planet Circulars* as soon as the monthly cycle permits. The principal professional observatories regularly and almost exclusively involved with follow-up (as opposed to discovery) are the Oak Ridge Observatory in Massachusetts (1.5-m reflector + CCD), the University of Victoria and Dominion Astrophysical Observatory (1.8-m reflector + CCD and a small Schmidt), and Mt. John University Observatory on New Zealand's south island (0.6-m reflector). At least for the brighter objects, there is also extensive involvement by amateur astronomers, most notably in Japan, but also in Italy and England; and with the increasing availability of convenient CCD systems, interest is growing in the United States. A particularly troublesome problem has been to find someone willing and able to obtain follow-up data for discoveries not announced until the bright-of-moon period is beginning, when most of the traditional observing programs have ceased operation for the month. It is encouraging that some of the CCD observers, particularly amateur astronomers (like J. Rogers in California, and A. Vagnozzi and V. S. Casulli in Italy), are now starting to fight bright moonlight to get the necessary observations, but the situation is still far from perfect.

Astrometric follow-up should not, of course, be restricted to the days immediately following discovery. Several new NEOs are followed into a second month, but there is a clear need for greater effort in securing the longest possible observed spans at the discovery apparitions (unless past observations can be found or radar detections are secured). If the spans are long enough, deliberate recovery attempts are appropriate on the next reasonably favorable occasion. Some monitoring is also needed at subsequent returns, particularly for short-period comets whose motions are likely to be affected by nongravitational forces (and then even if there have been radar observations in the past). It is interesting to note that a recent interval of little more than six months has brought no fewer than four accidental rediscoveries of NEOs that were none too well observed in the past. In addition to the aforementioned 1980 AA = 1993 BC_2, there were 1973 NA = 1992 OA, 1989 ML = 1992 WA (found quite far from the prediction because of an unexpectedly large error in the final 1989 observations; see also McAdams 1991) and 1991 CB_1 = 1993 BV_3 (near aphelion on each occasion and in 1993 almost unbelievably close to the prediction from the 1991 seven-day arc).

Follow-up physical observations, including radar, are also very desirable, although they tend to require more organization than does astrometry. With extensive planning, however, it is possible to obtain an extraordinary wealth of useful information, as exemplified by the late-1992 campaign on 4179 Toutatis when that object made the closest long-predicted approach of any object to the Earth of any celestial body other than the Moon.

V. CONCLUDING REMARKS

In this chapter a review has been presented of the search programs that are under way that mainly aim at discovering new NEOs. From what has been written one could have gained the impression that this subject is fairly well covered, and that little remains to be done.

This, however, is by no means the case. Three of the four search projects described here are based in the United States, the other being in Australia. This means that there is a ratio of 3/1 (judging only from the number of teams involved) in the observations made from the northern hemisphere with respect to the ones in the southern hemisphere. Equally well in terms of the places where follow-up observations only are obtained (as opposed to discovery efforts), as described in Sec. IV, it is clear that there are more observers in the northern rather than the southern hemisphere, and the former are far better equipped.

This imbalance may reflect in undesired selection effects in NEO discoveries. Moreover, only recently has a dedicated project been initiated in Europe, and is not yet operative (EUNEASO: EUropean Near-Earth Asteroid Search Observatory).

The most important point, however, is the modest emphasis given to follow-up observations. Up until now only a fraction of discovered NEOs has received sufficient attention to allow a fully reliable orbit to be computed. Some of the searches discussed in this chapter are in some ways not very suitable for this important part of the work (the follow-up); while it is appropriate that second-night and perhaps week-later observations be obtained with the same Schmidt telescope with which the discovery was made, as soon as a reasonable ephemeris is available (i.e., one allowing the NEO to be found within a 5 or 10 arcmin CCD image), it would be better if narrow-field telescopes fitted with CCDs took over the follow-up. An exception to this is the Spacewatch survey where in many cases the team on that program must itself accomplish follow-up of the NEOs that they find, due to the faintness of the objects ($V > 20$).

In the future we must look to a world-wide coordinated search program, which *must* include a coordinated and well-planned follow-up activity. Such a program would likely incorporate the present search and follow-up efforts, but with a major injection of new instruments and new teams (see the Chapter by Bowell and Muinonen). At present, communications are being handled by the Central Bureau of Astronomical Telegrams and the Minor Planet Center; however, a global discovery and follow-up network, whose data will be used in other steps of a long process, would require enhanced communication facilities and a volume of work (and number of workers) that is not feasible at present. Such a central nexus would also need directly to coordinate the efforts of the different teams, as opposed to the present situation where the search and follow-up teams work largely independently, but with informal arrangements for follow-up and collaboration.

Notwithstanding these difficulties, it is impressive how much our knowledge on NEOs has grown, and is likely to continue growing in the next years even if there are no new programs forthcoming. Thousands of asteroids are being discovered that cannot be followed-up under the present conditions. Most of these are mainbelt asteroids, but the knowledge of their frequency distribution is essential to an understanding of the dynamical and physical processes that are responsible for the replenishment of NEOs in the inner regions of the solar system. Further theoretical work is needed to understand these mechanisms, involving the activity of celestial mechanicians, experts in material properties, physicists and geologists. Coordination of the efforts in these fields is very desirable, and a linkage with the results of observations is essential.

REFERENCES

Balonek, T. J., Marschall, L. A., DuPuy, D. L. and Ratcliff, S. J. 1992. Transcontinental parallax measurements of asteroids 1991 TB_1 and 1992 JB. *Bull. Amer. Astron. Soc.* 24:1126 (abstract).

Gehrels, T. 1991. Scanning with charge-coupled devices. *Space Sci. Rev.* 58:347–375.

Helin, E. F., and Dunbar, R. S. 1984. International near-Earth asteroid search. *Lunar Planet. Sci.* XV:358 (abstract).

Helin, E. F., and Shoemaker, E. M. 1977. Discovery of 1976 AA. *Icarus* 31:415–419.

Helin, E. F., and Shoemaker, E. M. 1979. Palomar planet-crossing asteroid survey, 1973–1978. *Icarus* 40:321–328.

Lau, C. O., and Hulkower, N. D. 1987. Accessibility of near-Earth asteroids. *J. Guidance Control Dyn.* 10:225–232.

Marsden, B. G. 1991. The computation of orbits in indeterminate and uncertain cases. *Astron. J.* 102:1539–1552.

Marsden, B. G. 1992. Comments on search programs for near-Earth objects. In *Observations and Physical Properties of Small Solar System Bodies*, eds. A. Brahic, J.-C. Gerard and J. Surdej (Liège: Université de Liège), pp. 251–252.

McAdams, J. V. 1991. Mission Options for Rendezvous With the Most Accessible Near-Earth Asteroid–1989 ML. AIAA paper AAS 91-397.

McMillan, R. S., Moore, T. L., Perry, M. L., and Smith, P. H. 1994. Long, accurate time series measurements of radial-velocities of solar-type stars. *Astrophys. Space Sci.* 212:271–280.

Morrison, D., ed. 1992. *The Spaceguard Survey: Report of the NASA International Near-Earth-Object Detection Workshop* (Pasadena: Jet Propulsion Laboratory).

Parker, Q. A. 1992. Report on Kodak Tech-Pan 4415 Estar-Based Emulsion. Anglo-Australian Observatory Internal Document (Epping, N. S. W.: Anglo-Australian Observatory).

Rabinowitz, D. L. 1991. Detection of Earth-approaching asteroids in near real time. *Astron. J.* 101:1518–1559.

Rabinowitz, D. L., Gehrels, T., Scotti, J. V., McMillan, R. S., Perry, M. S., Wisniewski, W., Larson, S. M., Howell, E. S., and Mueller, B. E. A. 1993. Evidence for a near-Earth asteroid belt. *Nature* 363:704–706.

Scotti, J. V. 1994. Computer aided near-Earth object detection. In *Asteroids, Comets, Meteors 1993*, eds. A. Milani, M. DiMartino and A. Cellino (Dordrecht: Kluwer), pp. 17–30.

Scotti, J. V., Rabinowitz, D. L., and Marsden, B. G. 1991. Near miss of the Earth by a small asteroid. *Nature* 354:287–289.

Shoemaker, E. M., Wolfe, R. F., and Shoemaker, C. S. 1990. Asteroid and comet flux in the neighborhood of Earth. In *Global Catastrophes in Earth History*, eds. V. L. Sharpton and P. D. Ward, Geological Soc. of America Special Paper 247 (Boulder: Geological Soc. of America), pp. 155–170.

Shoemaker, E. M., Wolfe, R. F., and Shoemaker, C. S. 1991. Asteroid flux and impact cratering rate on Venus. *Lunar Planet. Sci.* 22:1253–1254 (abstract).

Steel, D., and McNaught, R. H. 1992. The Anglo-Australian near-Earth asteroid survey. *Australian J. Astron.* 4:42–48.

Steel, D., McNaught, R. H., and Asher, D. 1992. Did Icarus have a twin brother? *Minor Planet Bull.* 19:9–11.

Wisniewski, W. Z., and McMillan, R. S. 1987. Differential CCD photometry of faint asteroids in crowded star fields and non-photometric sky conditions. *Astron. J.* 93:1264–1267.

EARTH-CROSSING ASTEROIDS AND COMETS: GROUNDBASED SEARCH STRATEGIES

EDWARD BOWELL
Lowell Observatory

and

KARRI MUINONEN
Observatory, University of Helsinki

It is feasible to conduct a groundbased survey for near-Earth asteroids and comets that will lead to the discovery of a large fraction of those bodies down to subkilometer sizes. We describe survey strategies that will be effective in identifying near-Earth objects potentially hazardous to our planet, and that will provide advanced warning of the approach of hazardous long-period comets. Such surveys will also yield many discoveries of smaller bodies, some of which are potential hazards to a local or regional extent. A comprehensive survey requires monitoring a large volume of space to discover asteroids and comets whose orbits can bring them close to the Earth, and therefore mandates the frequent searching of large areas of the dark sky to a faint limiting magnitude. We here rework our findings for the proposed Spaceguard Survey, which comprises a worldwide network of 2- to 3-m-class wide-field telescopes, each provided with mosaic-CCD cameras and local near-real-time data analysis facilities. We also describe efforts to ramp up to the Spaceguard Survey using smaller telescopes.

I. INTRODUCTION

In this chapter, we present simulations of groundbased telescopic searches for near-Earth asteroids and near-Earth comets (collectively, near-Earth objects [NEOs]) that are based largely on material prepared for *The Spaceguard Survey* (Morrison et al. 1992). In that publication, the material appeared as Chapters 5 and 7 as parts of Chapters 3 and 6 and portions of three appendices. Here, we have reworked the Spaceguard Survey findings in light of more recent NEO discoveries, of work on the near-Earth asteroid population by Rabinowitz (1993) and Rabinowitz et al. (see their Chapter), and of our own maturing views of the NEO detection problem. We also report on some recent findings emerging from a new NEO search initiative at Lowell Observatory, particularly insofar as they indicate what might be achievable using a much more modest telescope than those recommended for the Spaceguard Survey.

A number of workers have examined the inversion problem of estimating the true NEO population from statistics of their discovery (for asteroids, see, e.g., Shoemaker et al. 1990; Benedix et al. 1992; Davis et al. 1993; Rabinowitz 1993; the Chapter by Rabinowitz et al.; for comets, Everhart 1967a, b; Shoemaker and Wolfe 1982; Fernández and Ip 1991; Weissman 1991; and the Chapter by Shoemaker et al.). However, we are aware only of our own efforts and those of Drummond et al. (1993) to use a model NEO population to predict where in the sky to search for NEOs, and thus how to design an efficient observational search program.

For purposes of this chapter, we define Earth-crossing asteroids (ECAs), as those whose orbits can intersect the capture cross section of the Earth as a result of long-range planetary perturbations. According to this definition, most Atens, most Apollos, and some Amors are ECAs, although some are protected from close Earth approach by resonance (see the Chapter by Rabinowitz et al. for more details). As of January 1994, there were about 180 known ECAs. We define Earth-crossing comets (ECCs) as those comets currently having perihelia less than the aphelion distance of the Earth (1.017 AU). They may be of any orbital period. Thus ECCs comprise a fraction of short-period Jupiter-family comets (orbital period $P <20$ yr), intermediate-period Halley-type comets ($20 \leq P \leq 200$ yr), and long-period comets (LPCs; $P >200$ yr).

Our goal in this chapter is to explore NEO search methods that favor the discovery of asteroids and comets that could cause global effects should they strike our planet, although we do extend our findings down to smaller bodies. We assume that the impact of asteroids and comets larger than 0.5 to 1 km in diameter could cause global rather than local effects. When considering possible NEO survey protocols, it is useful to refer to potentially hazardous NEOs. These are bodies whose orbits *apparently* come close enough to intersecting those of perturbing bodies (specifically, the terrestrial planets and Jupiter) that they could be gravitationally deflected onto a collision course with Earth on a time scale of, say, a century. We emphasize *apparently* to allow for uncertainty in orbital elements. Thus, at the moment of discovery, every asteroid and comet is a potentially hazardous NEO. In some cases, it will take an extensive series of observations to prove otherwise, though identification of the predominant mainbelt asteroids will generally take just a few days. We remark in more detail on potentially hazardous NEOs in Secs. V.D. and VII.B.

From the Spaceguard Report and from other considerations, we have devised the following precepts:

- Because it is clearly impossible to discover all hazardous NEOs, hazard reduction rather than hazard elimination should be the primary goal of an NEO search program. One must strive to detect as many NEOs as quickly and cost-effectively as possible.
- Although advances in detectors and computers are continuing apace, we take a conservative appraoch by restricting our study to current available technology.
- We treat only groundbased optical detection. It is clear that groundbased

radar detection will in the coming years be limited to the detection of nearby—and therefore mostly small—NEOs, and that spacebased detection systems are very much more costly, perhaps three orders of magnitude more so.

- NEOs larger than about 1 km in diameter pose the greatest threat. An NEO search should be optimized for their discovery.
- The incidence of cometary impacts on Earth appears to be less than that of asteroids. Accordingly, we tilt the design of a survey strategy toward discovery of ECAs.
- The search program should largely be automated. Discovery and most follow-up observations of NEOs can be taken care of by adopting an appropriate observing stragegy.
- NEOs are generally distinguishable from mainbelt asteroids by their motion, location, and/or morphology. However, all moving objects should be followed initially.
- NEO detection should be accomplished in near-real time.
- To reach reasonable detection completeness, an ECA search program must last at least 20 years. A perpetual search program is required for ECCs.
- The search strategy should evolve synergistically as knowledge of the NEO population accrues.

In Sec. II we describe how we have derived population models for ECAs and ECCs, and in Sec. III we use the population model for ECAs to get some idea of what the areal extent of an NEO survey needs to be. Starting, as a reference, with a whole-sky NEO survey, we draw general conclusions in Sec. IV about duration and completeness as a function of limiting magnitude, and then go on to allow for the effects of detection losses, to determine what search area and location it is useful to search, and to investigate NEO discovery completeness in realistic surveys. In Sec. V we build on the previous results to investigate the monthly discovery rate as a function of NEO diameter, how observational linkage and follow-up can be dealt with, and what criteria can be adduced for us to regard an NEO as potentially hazardous. Using the University of Arizona's Spacewatch Program as a model, we describe requirements for the Spaceguard Survey in Sec. VI: detector and telescope systems, magnitude limit and observing time, the use of CCD mosaics, the kind of sky-scanning regime, and computers and communications. In Sec. VII we outline one effort to ramp up to the Spaceguard Survey, and we sketch the components that may constitute an NEO survey. We end, in Sec. VIII, by commenting on areas of research we believe to be important for the future.

II. NEO POPULATION MODEL

A. Earth-Crossing Asteroids (ECAs)

Our ECA population model conforms to that of Rabinowitz et al. (see their Chapter). In particular, we have adopted their estimates of the joint distribution of orbital elements (semimajor axis, eccentricity, and inclination), and bias factors. We assume that the overall population of the smallest ECAs (diameter $D \lesssim 100$ m) is similar to that of the Spacewatch discoveries (i.e., is independent of their orbital elements and therefore whether or not they can approach the Earth very closely). We have not explicitly modeled the population of very small ($D \lesssim 50$ m, $H \gtrsim 25$ mag) ECAs that may constitute a distinct near-Earth belt (Rabinowitz et al. 1993) because the number of such objects currently known is too small. We express the cumulative frequency of ECAs by a series of power laws:

$$\log_{10} N = \begin{cases} 1.086(H - 13.2) & 15.0 > H \\ 0.391(H - 15.0) + 1.955 & 23.5 > H \geq 15.0 \\ 0.695(H - 23.5) + 5.279 & H \geq 23.5. \end{cases} \quad (1)$$

To transform from absolute magnitude to diameter requires knowledge of the albedo distribution. Luu and Jewitt (1989) found evidence that the ECA population comprises a mixture of S- and C-class asteroids partitioned in the ratio $n_S : n_C = 3 : 5 \ (\equiv v)$. As in Rabinowitz et al. (see their Chapter), we assume that $v = 1$. According to Tholen and Barucci (1989), the geometric albedos of the two broad taxonomic groups are $p_S \in [0.14, 0.17]$ and $p_C \in [0.04, 0.06]$. We adopt mean values: $\bar{p}_S = 0.155$ and $\bar{p}_C = 0.05$. We draw values of the mean slope parameter of the two albedo groups from Harris and Young (1988): $G_S = 0.23$ and $G_C = 0.09$. Then, from Bowell et al. (1989), asteroids of diameter 1 km have absolute magnitudes $H = 18.28$ mag (S) and $H = 18.90$ mag (C). If the partitioning ratio $v = 0.6$, we obtain $\bar{H} = 18.67$ mag, and for $v = 1$, we obtain $\bar{H} = 18.59$ mag. The concomitant model populations of ECAs larger than 1 km in diameter contain 1919 and 1538 members, respectively—a difference of 25%. Further, we assume v to be independent of an ECA's orbital elements. Our model population for $D > 1$ km is here 27% smaller than that used in the Spaceguard Report. Four effects contribute to the difference: (1) the modified cumulative frequency model (Eq. [1]); (2) an improved method of calculating the cumulative frequency of ECAs as a function of diameter (Muinonen et al. 1994a); (3) slightly different adopted values of p and G for S- and C-class asteroids; and (4) an improved joint element distribution. For $D \geq 10$ m, the ECA populations contain 2.6×10^8 (current model) and 1.5×10^8 (Spaceguard Report model) members. We emphasize that the difference between the ECA model used here and that of the Spaceguard Report is within the estimated errors at all diameters except for the relatively small number of objects larger than 2 km in diameter; for $D \geq 4$ km, for example, our model population now contains about 70 members, whereas the population used in the Spaceguard Report contained about 20 members. The current ECA population uncertainties are as follows; for $D \geq 1$ km, the estimated uncertainty is a factor of 2; and for $D \geq 10$ m, a factor of 4 or more.

B. Earth-Crossing Comets (ECCs)

It is certain that many dormant or extinct ECCs have, because of their asteroidal appearance, been termed ECAs and have been so modeled. Thus, we are largely concerned with *active* ECCs. The flux of ECCs has been studied by a number of workers. We start by considering LPCs. Fernández and Ip (1991) estimated that about three LPCs brighter than absolute magnitude $H_{10} = 10.5$ mag cross the Earth's orbit each year. From Weissman (1991), we estimate their nuclei to be larger than between 2.2 km (if the mean density $\bar{\rho} = 0.2$ g cm^{-3}) and 3.9 km ($\bar{\rho} = 1.2$ g cm^{-3}) in diameter. Following Shoemaker and Wolfe (1982), we assume that the annual flux of LPCs that reach heliocentric distances less than the aphelion distance of the Earth (1.017 AU) is given by

$$N(D) = N(1 \text{ km})D^{-1.97} \tag{2}$$

where D is in km and $N(1 \text{ km})$ is the flux of comets/year whose nuclei are larger than 1 km diameter. Thus, from Weissman, 14 yr^{-1} < $N(1 \text{ km})$ < 44 yr^{-1}.

However, a fundamental problem in assessing the flux of ECCs is the poorly known relationship between observed cometary brightness and nuclear diameter because almost all observed comets exhibit some degree of activity. Shoemaker and Wolfe (1982) examined the cumulative frequency of LPCs discovered before 1978, and deduced an annual flux of 230 Jupiter-crossing LPCs (by their definition, those having q < 5.527 AU) brighter than absolute blue magnitude $B(1,0) = 18$, at which magnitude they calculated a cometary nucleus to be 2.53 km in diameter. (These values imply that, at 5 AU heliocentric distance, a comet's observed brightness is, on average, 2.3 mag brighter than that of its bare nucleus and, further, that the nuclear geometric albedo averages $p = 0.07$.) Scaling to the Earth's aphelion distance (Shoemaker and Wolfe 1982), and using Eq. (2), we obtain $N(1 \text{ km}) = 184$/yr.

The dispersion of the aforementioned estimates of $N(1 \text{ km})$ makes it clear that the Earth-crossing LPC flux, as a function of diameter, is uncertain by a factor of several. We adopt $N(1 \text{ km}) = 100$/yr. Then, following Shoemaker and Wolfe (1982), we model the annual flux of Earth-crossing LPCs as a function of D (km) and perihelion distance q (AU) by

$$N(D) = \begin{cases} (125q - 25)\, D^{-1.97} & 0.5 \leq q \leq 1.017 \\ [97(q - 0.02)^2 + 32(q - 0.02)]\, D^{-1.97} & 0.02 \leq q < 0.5 \end{cases}$$

$$\tag{3}$$

Further, we assume that LPCs begin to exhibit comatic activity between heliocentric distances 10 AU$>r>$5 AU, and that the probability density for activity increases linearly with decreasing r to $r = 5$ AU, at which distance all LPCs are active. At smaller heliocentric distances we use

$$V = 14.53 - 5\log D + 5\log\Delta + 7.5\log r, \quad r \le 5 \text{ AU} \qquad (4)$$

where Δ is the geocentric distance, and the factor of 7.5 should better represent the r dependence of LPC brightness than the customarily used factor of 10 (D. W. E. Green, personal communication 1992).

We further assume that LPC orbital poles are isotropically distributed, which implies a uniform distribution of cos (inclination) in $[-1, +1]$ and of the longitude of the ascending node in $[0, 2\pi]$; and that the argument of perihelion and the mean anomaly are uniformly distributed in $[0, 2\pi]$. For computational convenience, we fixed the eccentricity at $e = 0.99$.

Shoemaker et al. (see their Chapter) have modeled the relative numbers of Earth-crossing Jupiter-family and Halley-type comets. They estimate that there are currently 40±0.3 dex [dex is a base-10 logarithmic measure of uncertainty; see Allen (1973)] active and roughly 800 inactive Jupiter-family comets, together with about 200 active and 3000 inactive Halley-type comets larger than 1 km in diameter. These numbers imply an Earth-orbit-crossing (i.e., comets having perihelia less than the Earth's aphelion distance) flux of about 5 active Jupiter-family comets and 2 active Halley-type comets per year, which together are roughly an order of magnitude less than the flux of LPCs and 2 orders of magnitude less than the flux of similar-size ECAs. Because the Jupiter family comets have much more asteroid-like orbits than the Halley-type comets they will automatically be discovered—active or inactive—during the course of a survey optimized for ECAs, and therefore need not be modeled separately. The Halley-type comets would be discovered in a survey optimized for the discovery of LPCs. According to our model ECA and ECC populations, the Earth-orbit-crossing flux of ECCs comprises between 20% (active) and 36% (active and inactive ECCs, including some already modeled as ECAs) of the total NEO flux. We stress that this estimate could be in error by a factor of 2 or more.

III. SPATIAL AND SKY-PLANE DISTRIBUTIONS OF NEOs

A. Conceptualization

Because ECCs are in all likelihood a lesser component of the total hazard, we concentrate on ECAs in this section. Without resorting to elaborate modeling, it is possible to put useful bounds on the nature and extent required of an NEO survey.

In Fig. 1, the positions of ECAs are plotted on planes in and perpendicular to the ecliptic. Assuming that the plots are representative of the spatial

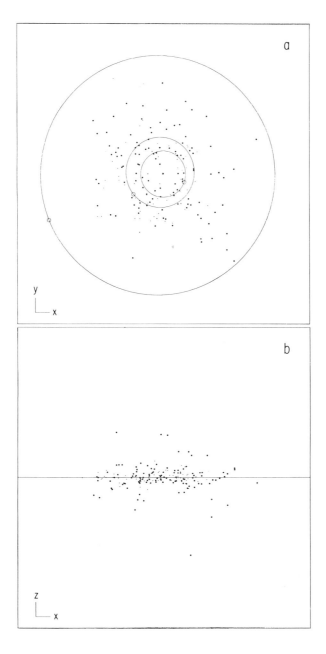

Figure 1.   Positions of 178 ECAs on 1 September 1993. Large dots are ECAs thought to be larger than 1 km in diameter; small dots are ECAs thought to be smaller than 1 km in diameter. (a) Ecliptic north polar projection of the inner solar system. The direction of the vernal equinox is to the right. Orbits and locations of Earth, Mars and Jupiter are also shown. (b) Ecliptic-plane projection of ECAs, on the same scale as (a), viewed from longitude 270°. The ecliptic is shown as the horizontal line.

distribution at other times, and ignoring the effects of discovery bias and the fact that some ECAs are misplotted because their orbits are imperfectly known, we can conclude:

1. About 15% are inside the Earth's orbit, and about 40% are inside Mars' orbit. Thus, assuming it to be difficult to observe within 90° of the Sun, a large proportion of ECAs are accessible to observation at any time.
2. The median geocentric distance of ECAs is 2.1 AU. At opposition at that distance, a 1-km ECA would appear to have $V = 19.5$ mag (S class) or 20.7 mag (C class).
3. ECAs outside the Earth's orbit that are thought to be larger than 1 km in diameter have median $V = 20.4$ mag. Thus a search to this magnitude limit that covered all the sky at solar elongations greater than 90° would be effective at detecting many ECAs.
4. Because of observational selection, small ECAs are concentrated toward the ecliptic plane.
5. ECAs are more widely dispersed in ecliptic latitude than mainbelt asteroids, though their median ecliptic latitude (about 6°) is quite similar.

Figure 2 is an ecliptic polar view of the solar system that illustrates the distances to which C-class asteroids of given diameter can be seen to limiting magnitude $V_{lim} = 22$ mag. Diameters are about a factor of two smaller for S-class asteroids, as they would be for an increase of 1.5 mag in V_{lim}. The plot is axisymmetric about the x axis and mirror symmetric about the y, z plane. The latter symmetry leads to the interesting consequence that ECAs are equally bright toward opposition and behind the Sun—though they cannot, of course, be observed in visible light at small solar elongations. Because of the sharp drop-off in brightness at large solar phase angles, C-class asteroids smaller than about 0.5 km in diameter are preferentially observable close to the Earth toward the antisun (an effect slightly enhanced by the opposition effect, which appears as a protuberance in each contour on the y axis). Asteroids larger than 0.5 km in diameter are observable in much larger volumes in all directions in the sky. We conclude:

1. The larger an ECA, the further away from Earth will it be visible at a given V_{lim}, and therefore the greater percentage of the time will it be discoverable.
2. At small diameters, the volume of space in which an ECA is visible is small, so a given small ECA is visible only for a relatively short time as it traverses the volume. Thus, to discover a given fraction of small ECAs, we expect that a survey to a given V_{lim} must be of longer duration than one for larger ECAs.
3. C-class ECAs of 1 km diameter are visible out to 2 AU geocentric distance—the median distance for the currently known ECAs. Therefore it ought to be possible to discover most ECAs larger than 1 km diameter in a search to $V_{lim} = 22$ mag within a reasonable time.

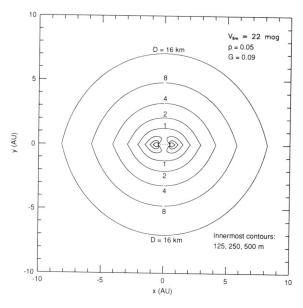

Figure 2.   Detectability of C-class asteroids ($p = 0.05$, $G = 0.09$) in the ecliptic plane as a function of diameter D, assuming $V_{lim} = 22$ mag. The effects of detection losses (see text) are not included. The origin of the coordinate system chosen is midway between the Earth and Sun, which are located at the ends of the dashed line.

B. Sky-Plane Distribution of ECAs

From Figs. 1 and 2, one expects a prevalence of small (faint) ECAs in the opposition and conjunction directions, and a concentration at low ecliptic latitudes. This is confirmed in Fig. 3, which shows the instantaneous number-density contours of ECAs larger than 0.5 km in diameter for limiting magnitudes $V_{lim} = 18, 20, 22$, and ∞ mag. At increasing V_{lim}, there is both a rapid increase in the number of discoverable ECAs and a greater preponderance near conjunction. At opposition, and ignoring detection losses (discussed below), about 145 deg^2 must be searched to $V_{lim} = 18$ mag to have a 50% chance of detecting an ECA; 37 deg^2 to $V_{lim} = 20$ mag; and 15 deg^2 to $V_{lim} = 22$ mag. Because, at $V_{lim} = 18$ mag, there are only a handful of ECAs (about 5) visible at any time, a survey aimed at discovering all ECAs larger than 1 km in diameter would last at least 1538 months ($\approx$25 yr) and actually much longer because of repeat detections.

IV. MODELING NEO SURVEYS

To quantify the above inferences and to understand what parameters are important for the design of an observational search program, more detailed modeling must be undertaken. To estimate the likely outcome of an NEO

Sky-Plane Distribution ($\log_{10}$ n(ECAs/sr)), $V_{\mathrm{lim}} = 18$ mag

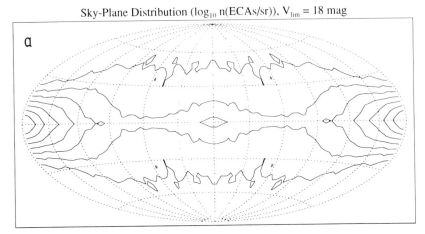

Sky-Plane Distribution ($\log_{10}$ n(ECAs/sr)), $V_{\mathrm{lim}} = 20$ mag

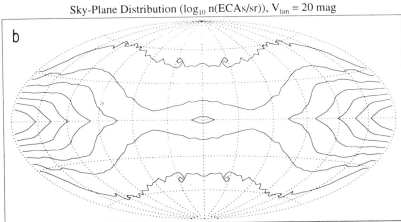

Figure 3. Modeled sky-plane number density (detection losses ignored) of ECAs larger than 0.5 km in diameter for three limiting magnitudes: (a) $V_{\mathrm{lim}} = 18$ mag, (b) $V_{\mathrm{lim}} = 20$ mag, (c) $V_{\mathrm{lim}} = 22$ mag, and (d) $V_{\mathrm{lim}} \to \infty$ mag. In these equal-area, Aitoff, projections opposition is at the center, and ecliptic longitudes are meridians. The Sun is at extreme left and right. Contours of the logarithm of the number density of ECAs per steradian are shown at an interval of 0.2 over the following ranges: (a) [0.65, 1.85], (b) [1.25, 2.45], (c) [1.65, 2.85], and (d) [2.1, 3.3].

search program and to devise a sound observing strategy, one must take a number of factors into account: limiting search magnitude, detection losses, search area and location, observing frequency, and survey duration. The survey simulations we describe below not only predict the percentage completeness of NEO discoveries as a function of diameter, but they also impose requirements on instrumentation and software, suggest some of the necessary capabilities of a global network of observing stations, and give pointers on follow-up and orbit-determination strategy.

Sky-Plane Distribution ($\log_{10}$ n(ECAs/sr)), V_{lim} = 22 mag

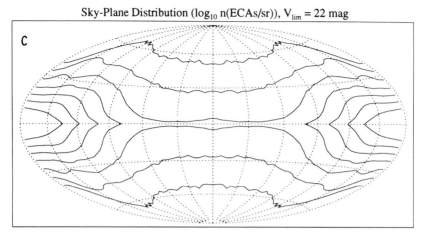

Sky-Plane Distribution ($\log_{10}$ n(ECAs/sr)), $V_{lim} \rightarrow \infty$

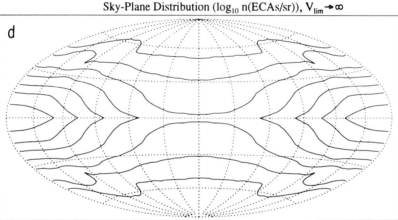

A. Whole-Sky ECA Survey

As a reference, consider the percentage completeness of ECAs discovered in a whole-sky survey as a function of diameter, limiting magnitude, and survey duration. Although a whole-sky survey is not achievable, modeling results give bounds on the more realistic surveys that we consider below. Figure 4 shows the results of simulations in which the effects of all the detection losses (Sec. IV.B) have been ignored, and in which the whole sky is searched once a month. At $V_{lim} = 18$ mag, about a magnitude fainter than the limit of the surveys at the 46-cm Palomar Schmidt telescope, even whole-sky surveys as long as 25 yr would not yield half of the largest ECAs ($D > 0.5$ km). At $V_{lim} = 20$ mag, somewhat inferior to the performance of the 0.91-m Spacewatch Telescope, about 80% of the ECAs larger than 1 km in diameter are accessible in only 3 yr. Searching the sky twice per month would make almost no difference after 25 yr, except at bright V_{lim}; for example, at $V_{lim} = 18$ mag, the percentage completeness of ECAs larger than 0.5 km in

diameter would increase from 42% to 47%, and of ECAs larger than 1 km in
diameter from 73% to 75%.

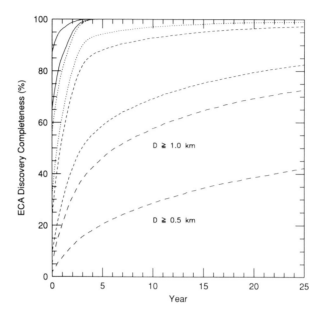

Figure 4.   Percentage discovery completeness of ECAs resulting from a whole-sky
survey. Four curve types pertain, from the bottom, to surveys having $V_{lim} = 18, 20$,
22, and 24 mag. For each V_{lim}, results for diameter thresholds $D \geq 0.5$ km (lower
curve) and $D \geq 1$ km (upper curve) are shown.

At fainter V_{lim} much greater completeness is attainable, and discovery
is characterized by a very rapid initial rate followed after some years by a
much slower approach toward completeness. Because one would expect the
detection rate of ECAs to be approximately constant from month to month,
recoveries must predominate more and more as time goes on. It is clear that
to discover, for example, 90% of ECAs larger than 1 km in diameter, a large
area of sky must be searched each month for a number of years to a magnitude
limit of $V_{lim} = 22$ mag or deeper. However, because surveys covering large
areas of sky each month must use scanning techniques, it is very difficult to
overcome the effects of sky brightness and readout noise at $V_{lim} \gtrsim 23$ mag.
Also, because of the rapid decline with time in the discovery rate of large
ECAs, there is no reason to undertake very long surveys for these bodies.
Prolonged surveys are mainly of value for adding to the completeness of the
discovery of small ECAs and for continued monitoring of ECCs. The results
of modeling a survey for ECCs are described in Sec. IV.C.

B. Detection Losses

In modeling more realistic surveys, one must allow for failure to detect NEOs because of observing procedures, equipment characteristics, and sky conditions. We have considered the following effects:

Trailing. Rabinowitz (1991) has given an expression, applicable to the Spacewatch Telescope, for computing the brightness/unit length of CCD camera images. Generalizing Rabinowitz' expression, if θ is the FWHM of a stellar image, ω is an NEO's angular motion, and t is the effective integration time, then the trailing loss ΔV_τ (in mag) is given by

$$\tau = \omega t / \theta$$

$$\Delta V_\tau = \begin{cases} 0 & \tau < 6/5 \\ 2.5 \ \log_{10}(5\tau/4 - 1/2) & 6/5 \leq \tau \leq 2 \\ 2.5 \ \log_{10} \tau & \tau > 2. \end{cases} \tag{5}$$

Note that τ is a dimensionless quantity. Experience using the Spacewatch camera has indicated a loss ΔV_τ of about 0.8 mag for asteroids moving at 0.5°/day—a motion not untypical of ECAs at opposition. For Spaceguard, ΔV_τ is considerably less, primarily due to shorter effective integrations (see Sec. VI.C). In our original calculations for the Spaceguard Report, we failed to make this distinction, with the consequence that, in simulated NEO surveys, the discovery completeness of ECAs is lower in the Spaceguard Report than here (the calculated discovery completeness of ECCs is largely unaffected). In other words, we now believe that ECAs of a given size are "easier" to discover than we did. Because of nonlinear response at faint light levels, photographic emulsions suffer greater trailing losses.

Confusion with Mainbelt Asteroids. ECAs are most readily identified by their unusual sky-plane motion. However, because some exhibit motion vectors similar to those of mainbelt asteroids, they cannot be unequivocally distinguished until their orbits or distances are known. Figure 5 shows daily motions of ECAs and mainbelt asteroids near opposition (the latter from Bowell et al. 1991) for $V_{\text{lim}} = 22$. At fainter V_{lim}, a greater fraction of ECAs are detectable at mainbelt asteroid distances. Thus their motions are somewhat more restricted, and confusion with mainbelt asteroids is slightly greater. At brighter V_{lim} the converse is true. Not plotted in Fig. 5 are Amors, many of which can be distinguished by prograde motion at opposition. We have found that, on the ecliptic and at $V_{\text{lim}} = 22$ mag, confusion with mainbelt asteroids increases monotonically from about 10% at opposition to 50% at 75° solar elongation. Such would be the loss during an "instantaneous" survey, in which no orbital information was available. Over the course of an extended survey, orbital information would allow ready distinction in most cases; additionally, almost all ECAs missed because of mainbelt asteroid confusion during one dark run would be discriminated at others. Moreover, most of the moving objects more than 40° from the ecliptic are NEOs.

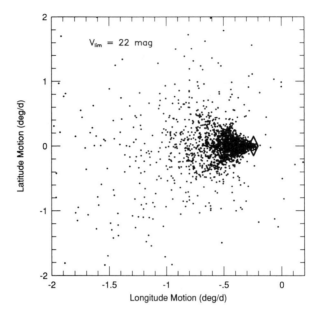

Figure 5.   Opposition daily motions in ecliptic longitude and latitude of ECAs
(points) and mainbelt asteroids (diamond-shaped region near $\dot{\lambda} = -0.25°$/day, and
$\dot{\beta} = 0°$/day). The motions of 2000 ECAs were modeled, allowing for detection
losses due to trailing as in a Spaceguard-like survey to $V_{\text{lim}} = 22$ mag. A few ECAs
(not plotted) moved as fast as 20°/day in longitude and 6°/day in latitude. Note the
unequal abscissa and ordinate scales.

Confusion with Stars and Galaxies. Because of the high surface density of
stars in the Milky Way, it will not be possible to survey at low galactic latitudes
b. At $V_{\text{lim}} = 22$ mag, we estimate the limit to be $|b| = 15°$. Accordingly, our
survey models account for this 30°-wide belt of avoidance. At high galactic
latitudes (e.g., *b*>40°), we estimate, using data from Allen (1973), that stellar
and galactic confusion is not a serious problem for $V_{\text{lim}} \lesssim 23$ mag (at which
magnitude star and galaxy counts are comparable). This is especially so when
multiple observations from one or more observing stations are made each
dark run. We have not attempted to model the effects of moonlight and the
zodiacal cloud on sky brightness.

"Picket-fence" losses, in which, when searching region A, an ECA is in
region B and vice versa, can be virtually eliminated by a suitable observing
strategy and by coordination of observations from different stations. Because
most ECAs retrograde through opposition (Fig. 5), scanning from east to west
is advantageous near the antisolar longitude.

Weather and equipment malfunction losses, being local, mandate a carefully chosen, global network of observing stations from which at least two
near-simultaneous observations of any part of the dark sky can be made. We
have allowed for them by making a conservative choice of the number of

hours of observing time in the search mode (see Sec. VI.D).

  Atmospheric extinction and seeing are also local effects. We have assumed that extinction losses for an unfiltered CCD detector amount to 0.12 mag/airmass and, in a multi-station survey program, can be neglected at solar elongation $E > 105°$. Thus, we adopt

$$\Delta V_e = \begin{cases} \dfrac{-0.12 \text{ mag}}{\cos(105° - E)} & 60° < E \leq 105° \\[2mm] 0 \text{ mag} & E > 105°. \end{cases} \tag{6}$$

  To model seeing losses, we assume an image has an FWHM of 1 arcsec at the zenith and that observations can be made near the zenith for $E > 105°$ (the Sun being a minimum of 15° below the horizon). After suitable scaling, we obtain

$$\Delta V_s = \begin{cases} -0.17 \cos^{-0.6}(105° - E) \text{ mag} & 60° < E \leq 105° \\[2mm] 0 \text{ mag} & E > 105°. \end{cases} \tag{7}$$

  The relative importance of the detection losses we have modeled is discussed in Sec. IV.D.

C. Search Area and Location

The reference case described in Sec. IV.A pertains to a hypothetical whole-sky survey for ECAs. We now turn to realistic search strategies, for both ECAs and ECCs, in which the region to be searched is limited to what is both practicable and necessary, and in which detection losses are allowed for. We ask: what area of sky is it necessary to search, and in what location, to discover a sample of NEOs that is reasonably complete to an acceptable diameter threshold?

  First, however, we must test our simulation method by comparing the predicted number of discoveries to the number actually found in a carefully controlled survey. For this purpose, by far the most suitable survey is that conducted by Spacewatch. Over a five-year interval, observational coverage has been a rather uniform 1500 deg^2/yr to a limiting magnitude of $V_{lim} = 20.5$ mag at the outset, and was improved to $V_{lim} = 20.8$ mag after emplacement of a better detector. We normalized the Spacewatch observations to a search window of $\pm 9.5°$ in ecliptic longitude and $\pm 5°$ in latitude, centered on opposition, we assumed that scanning took place at a steady rate eight times per year, and that $V_{lim} = 20.5$ mag was achieved for 25% of the observing area and 20.8 mag for the remainder. Table I gives the numbers of actual and predicted discoveries, the latter comprising three simulations having FWHM stellar image profiles of 2.0, 2.5 and 3.0 arcsec, which we believe encompass the range of seeing conditions actually encountered. Note that the increase in predicted faint discoveries in poorer seeing arises from fixing the detection SNR; broader seeing profiles imply more signal being spread into adjacent pixels and a reduction in trailing losses. Allowing for some tens of percent uncertainty in the simulations due to poor statistics, agreement is good. No doubt, further improvement could be effected by "fine tuning" the modeling to allow for effects such as a probabilistic V_{lim} and the fact that some of the Spacewatch discoveries were made by eye; but given the intrinsic uncertainties in model parameters, such fine-tuning cannot be warranted.

TABLE I
Simulations of Spacewatch Survey ECA Discoveries[a]

Absolute Magnitude	$H < 18$	$H < 19.5$	$H < 23$
Spacewatch (normalized)	23	36	60
Simulation: seeing = 2.0 arcsec	16	32	46
Simulation: seeing = 2.5 arcsec	17	35	60
Simulation: seeing = 3.0 arcsec	17	35	69

[a] Entries are the numbers of observed and predicted ECAs as a function of absolute magnitude H.

Next, we consider searching the maximum possible area of dark sky. It is practical to search a region extending to a solar elongation as small as 60°. In this and subsequent simulations we assume, for simplicity, that the integration time t and seeing θ are similar at all V_{lim}. (Actually, we have adopted $t = 14$ s and $\theta = 1$ arcsec, in accordance with operational values derived in Sec. VI.C.) Also, we assume that areal coverage is independent of V_{lim}. Together, these assumptions imply that larger telescopes would be used to attain fainter V_{lim}. Table II shows the calculated discovery completeness for a 25-yr survey for ECAs in which the entire dark sky is searched once per month. For $V_{lim} = 22$ mag, almost all ECAs larger than 1 km in diameter would be discovered. Potentially hazardous ECAs (described in more detail in Sec. V.D) are even more completely sampled because most of them would come within range of observation during a long survey. If it were possible to reach $V_{lim} = 24$ mag, we could virtually achieve total completeness of ECAs larger than 0.5 km in diameter.

TABLE II
Dark-Sky Survey Simulations for ECAs[a]

D	$V_{lim} = 22$ mag		$V_{lim} = 24$ mag	
50 m	11	21	41	57
0.1 km	39	52	79	87
0.5 km	95	97	99	99
1.0 km	99	98	100	100

[a] Entries are the percentage discovery completeness for the entire population and for potentially hazardous ECAs larger than a given diameter.

The ECCs reside most of the time in the outer solar system, and they can approach the inner solar system from any direction in space. Intermediate-

and long-period ECCs take about 16 months to travel from the distance of Saturn (9.5 AU from the Sun) to that of Jupiter (5.2 AU), and a little more than an additional year to reach perihelion. At any time, we estimate that about 500 ECCs are brighter than $V = 22$ mag. To model searches for ECCs, we introduce the concept of *warning time*: the interval between earliest possible detection (i.e., when the ECC brightens to V_{lim} as it approaches the inner solar system) and the first crossing of the Earth's aphelion distance from the Sun. Because there is equal probability that an ECC on a collision course with Earth would strike our planet before or after perihelion, our definition of warning time is somewhat conservative (most ECCs spend 2 to 3 months inside 1 AU). Marsden and Steel (see their Chapter) have considered the matter of LPC warning times in detail.

Because most ECCs are LPCs, assessing their discovery completeness requires that we model a perpetual survey. Table III shows the result of modeling such a perpetual monthly dark-sky survey for ECCs. Now, for $V_{lim} = 22$ mag and $D > 1$ km, the completeness with a short warning time of three months is 91%. For $V_{lim} = 24$ mag, we could achieve 96% discovery completeness. In contrast to ECAs, there is appreciable degradation of discovery completeness for ECCs arising from lack of observations at small solar elongations (where ECCs most often peak in brightness) and at low galactic latitudes.

For ECAs, Fig. 3 indicates that a search centered on opposition is optimum. We have simulated surveys that cover various areas of the sky and in which detection losses have been included. In particular, we computed the result of searching sky regions spanning various halfwidths in ecliptic longitude ($\Delta\lambda$) and latitude ($\Delta\beta$) (both in degrees). These are truncated lunes of area (in square degrees)

$$A = \frac{720°}{\pi} \Delta\lambda \sin \Delta\beta. \tag{8}$$

Figure 6 shows the results of simulating 25-yr surveys to $V_{lim} = 22$ mag and for ECAs larger than 0.5 km in diameter. Contours of percentage ECA discovery completeness show that to minimize the areal coverage needed to achieve a given discovery completeness, it is advantageous to search regions spanning a broader range of ecliptic latitude than ecliptic longitude. The same strategy holds for other V_{lim} and diameter thresholds. For plausible search areas (5000 to 10,000 deg^2/month, though not necessarily truncated lunes), one may anticipate about 80% discovery completeness at $V_{lim} = 22$ mag. However, coverage in both longitude and latitude must not be too small or some ECAs will pass through the search region undetected from one month to the next.

Even though it is thought that there are approximately equal numbers of S-class and C-class ECAs, we expect that more of S-class ECAs will be discovered during an NEO survey because of their higher surface albedo. For example, for a 25-yr survey of the region centered on opposition, 6000 deg^2 in extent, to $V_{lim} = 22$ mag and with a 0.5-km diameter threshold, about 77%

TABLE III

Dark-Sky Survey Simulations for ECCs[a]

$D >$	Warning Time	% ECCs Discovered	
(km)	(yr)	$V_{lim} = 22$ mag	$V_{lim} = 24$ mag
0.5	0.0	80	95
	0.25	74	93
	0.5	46	82
	1.0	6	29
1.0	0.0	93	97
	0.25	91	96
	0.5	75	92
	1.0	19	66
5.0	0.0	97	98
	0.25	97	97
	0.5	93	95
	1.0	76	88
	2.0	11	43
10.0	0.0	98	98
	0.5	94	96
	1.0	85	90
	2.0	30	84
	3.0	10	58

[a] Percentages for zero warning time correspond to overall discovery completenesses.

completeness might be achieved (Fig. 6). However, for S-class ECAs the discovery completeness would be 86%, whereas for C-class ECAs it would only be 69%. Thus inasmuch as S-class ECAs of a given diameter are likely to be more hazardous than C-class ECAs because of their greater density, the NEO survey would effectively lead to a more complete assessment of the total hazard.

Atens pose a special detection problem because some of them make very infrequent appearances that occur far from opposition. We calculate that only about 33% of the Atens larger than 0.5 km in diameter would be discovered in a 25-yr, 6000 deg^2/month survey having longitude coverage of $\pm 30°$ from opposition (although it should be recalled that only 16 Atens are known, so the bias-corrected estimate of their true number may be substantially in error). The discovery rate could be boosted to nearly 83% by searching the same area more broadly in longitude, and the overall ECA discovery rate would increase from 77% to 80% (the ECC discovery rate would increase, too). Another possible reason for doing so is because, at $V_{lim} = 22$ mag, the sky-plane density of ECAs increases on the ecliptic at solar elongations

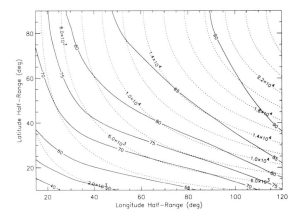

Figure 6.   Simulated 25-yr surveys centered on opposition and limited in ecliptic longitude and latitude. Dotted curves indicate monthly areal coverage in deg², and solid curves give the percentage discovery completeness of ECAs larger than 0.5 km in diameter.

smaller than about 120° (Fig. 3). Mitigating against searching at small solar elongations is the increased difficulty of observing large areas of sky near dusk and dawn, degraded seeing, and increased ECA-mainbelt asteroid confusion. We have not attempted to model the trade-offs among these conflicting effects.

D. Discovery Completeness

In what follows, it will be useful to consider a so-called standard search region of about 6000 deg², centered on opposition and extending ±30° in longitude and ±60° in latitude. This was the standard search region used in the Spaceguard Report. To increase the discovery completeness for a given search area, either the survey must be prolonged, the sky must be searched more frequently, V_{lim} must be increased, or detection losses must be reduced.

We first consider ECAs. As noted in Sec. IV.A, rapid decline with time in the discovery rate of ECAs at faint V_{lim} makes increasing the survey duration an ineffective strategy, particularly for large ECAs. For example, a whole-sky survey to $V_{lim} = 22$ mag and for $D \geq 0.5$ km could yield 97% completeness after 10 yr. After 20 yr, completeness would rise to 99% (Fig. 4). Observing a given region of the sky twice a month is likewise not very effective (see Sec. IV.A). Scanning the standard Spaceguard region twice a month to $V_{lim} = 22$ mag, the percentage completeness for $D > 0.5$ km would, after 25 yr, be only slightly increased (from 77% to 81%). For comparison, scanning twice the area once per month could lead to 84% completeness. However, as we will discuss in Sec. VI.C, it is really necessary to scan the NEO search region a number of times per month so as to derive orbits and allow linkage of observations. In addition, repeated monthly searching of a region centered on opposition would lead to a pronounced asymmetry in the distribution of ECAs discovered: because most faint ECAs retrograde through the opposition longitude (Fig. 5), they would be preferentially discovered in the morning sky after the first month's search. (Everhart [1967b] has remarked that a similar discovery pattern obtains for comets.) However, shifting the search region to the east would not be advantageous in the long term.

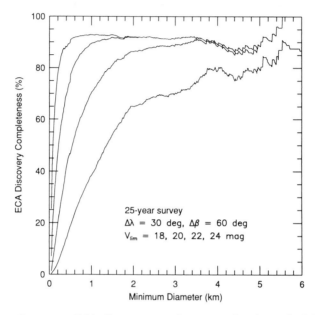

Figure 7.   Percentage ECA discovery completeness as functions of minimum diameter and V_{lim} for the standard Spaceguard search region (see Sec. IV.D). In the bias-corrected model population used, a number of large ECAs went undetected almost throughout the 25-yr duration of the simulated survey even at faint V_{lim}.

Figures 2 and 4 attest to the high value of mounting very deep surveys for ECAs, the key factor being the greatly increased volume of space in which ECAs of a given diameter can be detected. Figure 7 shows ECA discovery completeness as functions of V_{lim} and diameter threshold for the standard Spaceguard search region. At $V_{lim} = 20$ mag and for $D \geq 0.5$ km, one can expect a 25-yr survey to be only 47% complete, whereas at $V_{lim} = 22$ mag completeness rises to 77%. If the diameter threshold is relaxed to 1 km, completeness should increase to 70% and 89%, respectively. Table IV summarizes, for ECAs, the results of simulating a 25-yr survey of the standard Spaceguard region, and shows that, at $V_{lim} = 22$ mag, most of the ECAs larger than 0.5 km in diameter could be discovered.

Examination of the orbits of ECAs not discovered during simulated surveys reveals, as expected, that most of these bodies' orbits have large semimajor axes, high eccentricities, and/or inclinations such that either their dwell times near Earth are brief and infrequent or that they never come close to the Earth in their present orbits (of course, the latter class of ECAs pose no

TABLE IV
Standard Spaceguard Search Region Simulations for ECAs[a]

D	V_{lim} = 22 mag		V_{lim} = 24 mag	
50 m	7	21	25	34
0.1 km	25	32	55	63
0.5 km	77	79	91	88
1.0 km	89	87	93	90

[a] Percentage discovery completeness for the entire population and for potentially hazardous ECAs larger than a given diameter.

current hazard). Clearly, many such ECAs could be inactive Halley-type comets, though from our ECC population estimate in Sec. II.B, these bodies constitute perhaps 5% of the total NEO flux crossing the Earth's orbit.

For the standard Spaceguard search region, we find that mainbelt asteroid confusion is insignificant in surveys of all durations, except short surveys, to V_{lim} = 24 mag. Stellar confusion cannot be avoided, and amounts to 5% detection loss if the belt within 15° of the galactic plane is not searched. Because of the high inclination of the galactic plane with respect to the ecliptic, there is a slight advantage (2 or 3%) in searching regions broader in longitude rather than latitude. Trailing losses are very small. If necessary, they could be reduced by the implementation of streak-detection algorithms for fast-moving, mainly small ECAs (see Sec. V.A).

No survey can aspire to completeness in the discovery of ECCs, because new comets are constantly entering the inner solar system. Results for ECCs in a perpetual survey of the standard region to V_{lim} = 22 and 24 mag are given in Table V. The warning time is as defined in Sec. IV.C. We recall that it is equally likely that an ECC, if it is on a collision course, will strike the Earth on the outbound part of its orbit, increasing the warning time by a few weeks. The overall level of completeness, without regard to warning time, is 43% for $D > 1$ km, 61% for $D > 5$ km, and 65% for $D > 10$ km. Clearly, a survey designed for ECAs produces inferior results for ECCs, although the rate of discovery of these comets will be much greater than that achieved by current surveys, which rely on relatively small telescopes and sky-sweeping by amateur astronomers, and miss the great majority of the smaller LPCs.

V. SIMULATED SPACEGUARD SURVEY SCENARIOS

The simulations described above can be used to infer the nature of the observing activity during each monthly run of a major survey. In this section, we study the standard Spaceguard search region of 6000 deg²/month.

A. Discovery of Very Small ECAs

We have thus far not commented on very small ECAs discovered, although it is obvious that many tiny bodies, some just a few meters across, will be detected (see Tables II and IV). To estimate how many, 25-yr surveys of the

E. BOWELL AND K. MUINONEN

TABLE V
Perpetual Spaceguard Search Region Simulations for ECCs[a]

$D>$	Warning Time	% ECCs Discovered	
(km)	(yr)	V_{lim} = 22 mag	V_{lim} = 24 mag
0.5	0.0	29	48
	0.25	28	48
	0.5	19	40
	1.0	3	14
1.0	0.0	43	57
	0.25	42	57
	0.5	34	52
	1.0	9	29
5.0	0.0	61	67
	0.25	60	67
	0.5	56	64
	1.0	39	56
	2.0	7	27
10.0	0.0	65	72
	0.5	62	70
	1.0	51	65
	2.0	18	59
	3.0	7	45

[a] Percentages for zero warning time correspond to the overall discovery completenesses.

135,000-member model population of ECAs larger than 0.1 km in diameter were simulated. From Fig. 8, which shows the size-frequency distributions of ECA discoveries for various V_{lim}, it may be seen that many more ECAs smaller than the nominal 1-km diameter threshold would be discovered. For a survey to V_{lim} = 22 mag, one would expect about 270,000 discoveries, of which more than 85% are smaller than 0.1 km, 98% are smaller than 0.5 km, and 99.4% are smaller than 1 km in diameter. In other words, for every ECA larger than 1 km in diameter discovered in the standard survey, more than 150 smaller ones are likely to be discovered. We emphasize that the above estimates are based on a very uncertain extrapolation to ECAs smaller than those contained in our survey simulation.

B. Monthly Discovery Rate

What would be the discovery rate per dark run, assuming that the standard Spaceguard search region were scanned? Figure 9 indicates that, to V_{lim} = 22 mag, one can expect about 1000 ECA discoveries of all diameters during the first month. (We note, however, that the dashed curve is an extrapolation

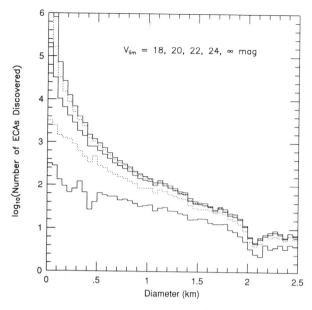

Figure 8.   Incremental size-frequency distributions of ECA discoveries for five V_{lim} (from bottom, 18, 20, 22, 24 mag, and ∞) resulting from simulated 25-yr surveys of the standard Spaceguard search region (see text).

to unmodeled ECAs smaller than 50 m in diameter, and could be in error by a factor of several.) This high initial discovery rate would tail off, perhaps by a factor of two, over the course of a 25-yr survey. The larger ECAs are preferentially discovered sooner, so whereas about 1 in 30 of the ECAs discovered will be larger than 1 km in diameter at the beginning of the survey, only about 1 in 4000 is expected after 25 yr. The corresponding ratios for 0.5-km-diameter ECAs are about 1 in 15 and 1 in 400, respectively. Thus, at a diameter threshold of 1 km, there may be little incentive to prolong a survey beyond about 15 yr, by which time about 85% ECA completeness can be achieved at $V_{lim} = 22$ mag. From Table IV and Eq. (3), we estimate that ECCs larger than 0.5 km in diameter will be discovered at a steady rate of about 10 per month.

C. Linkage and Follow-Up Observations

The majority of ECAs discovered will be visible only for short intervals because, being small, they must be close to the Earth to be detectable (Fig. 2). Indeed, during a 25-yr survey covering the standard Spaceguard region to $V_{lim} = 22$ mag, the geocentric distance of closest approach of ECAs larger than 0.5 km in diameter peaks inside 0.18 AU; for ECAs larger than 0.1 km in diameter, it peaks inside 0.13 AU. ECCs of nuclear diameter larger than 0.5 km have a corresponding maximum near 0.9 AU.

Linkage, that is the identification of images from one night to another, is

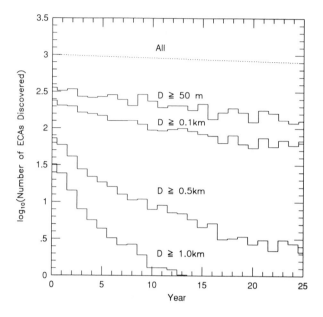

Figure 9.   Logarithm of the number of ECAs discovered *in the first month of each
year* of a simulated 25-yr survey of the standard Spaceguard region (see text) to
$V_{lim} = 22$ mag. Four histograms (top to bottom) are shown for all discoveries
(dashed) and for diameter thresholds of 0.1, 0.5, and 1 km. The dashed curve results
from a very uncertain extrapolation.

not usually thought of as a severe problem in photographic work, where the
number of images rarely exceeds a few hundred per plate (covering tens of
square degrees). However, at opposition on the ecliptic, and to $V_{lim} = 22$ mag,
we expect more than 100 asteroids/deg^2 to have angular motions similar to
those of mainbelt asteroids, together with a much smaller number of faster-
moving objects. Fortunately, the sky-plane motion vectors of most asteroids
and comets are preserved from night to night, especially near opposition,
and so can be used as a linkage criterion. A secondary criterion is apparent
magnitude, although this can change on a time scale of tens of minutes due
to rotational modulation of an irregular body. Even so, it becomes rapidly
more difficult to make accurate linkages as the number of nights' interval between observations of a given region increases. Thus, in photographic work,
a strategy often used is to make the first linkage a night or two after discovery,
and then to reobserve several nights later. Although this technique simplifies the linkage problem, it is not optimum for short-arc orbit determination
(Muinonen et al. 1994b). A complicating effect due to brightness variation
and changes in observing conditions is the intermittent detection of objects
near V_{lim}. Thus, although it will usually be easy to link fast-moving objects,
care will have to be taken for faint objects, especially those having motions
characteristic of mainbelt asteroids. Observing strategy will be driven, in

part, by the need to keep the interval between reobservation of a region short enough to engender few false linkages.

The success of recovery from one apparition to another depends critically on orbital accuracy (see, e.g., Muinonen and Bowell 1993*a*; Muinonen et al. 1994*b*). Because reasonably well-observed asteroids and comets tend, as time goes by, to drift away from their predicted positions in orbital longitude, they are usually sought by searching along the line of variation. Thus, in a large-scale NEO survey, the difficulty of recovering an NEO increases as the interval between its last observation and the first opportunity for recovery. Follow-up strategy during the discovery apparition can be influenced by the need to make a successful recovery one or more years hence. There is very rarely any difficulty in recovering an NEO already observed at two or more apparitions.

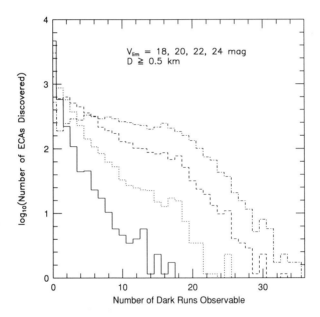

Figure 10.   Logarithm of the number of ECAs larger than 0.5 km in diameter discovered during a simulated 25-yr survey of the standard Spaceguard region (see text) as a function of the number of dark runs during which they are observable using scanning observations to $V_{lim} = 18$ (solid histogram), 20 (dotted), 22 (dashed) and 24 (dot-dash) mag. The leftmost bin, for zero dark runs, indicates the number of *undiscovered* ECAs: $\log n = 3.67$ ($V_{lim} = 18$ mag), 3.48 (20 mag), 3.11 (22 mag), and 2.72 (24 mag).

The number of dark runs (not necessarily consecutive) during which ECAs larger than 0.5 km in diameter can be detected in the standard Spaceguard search region is shown as a function of V_{lim} in Fig. 10. At $V_{lim} = 18$, 20, 22, and 24 mag, the percentages of ECAs detected in only one dark run are 54%, 32%, 13%, and 4%, respectively. The median numbers of dark runs in which ECAs are detectable are 1, 2, 4, and 9, respectively, although a few are reobservable more than 30 times at $V_{lim} = 22$ mag. At a diameter threshold of 1 km and for faint V_{lim}, the percentages of ECAs observed in only one dark run are typically a factor of 2 or 3 smaller, and the median numbers of dark runs are increased by about 50%. A similar simulation for ECCs produced less propitious results: For all V_{lim}, most ECCs are observable for a maximum of three dark runs. At $V_{lim} = 22$ mag, during 50 yr, one can expect a few ECCs to persist for more than 15 dark runs. These values imply that the achievable orbital accuracy for short- and intermediate-period comets will usually be good enough to ensure their recovery at a subsequent apparition.

We re-emphasize that the numbers, derived above, of ECAs detectable in only one dark run pertain to a repetitive scanning of the standard Spaceguard search region. If, in addition, follow-up observations were made out to, say, $\Delta\lambda = \Delta\beta = \pm 60°$ (from opposition) the percentage of ECAs larger than 0.5 km in diameter seen only once to $V_{lim} = 22$ mag would be reduced from 13% to 7%. Even greater safeguard against loss would be afforded by a follow-up strategy in which ECAs were reobserved as long as possible, using stare-mode integrations to $V_{lim} > 22$ mag, in any accessible region of the dark sky.

Because ECA losses after observation in one dark run can be reduced to a small fraction of the discoveries, it is possible that, for deep ECA surveys, follow-up can largely be ignored in favor of linkage of detections from one dark run or one apparition to another. In general, such linkage can be achieved unambiguously provided observations are not too sparse and false-positive detections too numerous. However, care must be taken not to lose the very ECAs that turn out to be hazardous to the Earth. Also, because of the large numbers of small ECAs that will be discovered, selection must be made, at least in part, on the basis of the diameter threshold selected for the survey. Both considerations call for a rapid, if only rough, estimate of the diameters of all ECAs discovered near the magnitude limit. To do this, the observed brightness can be combined with the distance gauged by means of diurnal parallax either from a single observing station or, better, from two or more well coordinated and optimally spaced stations. It will almost certainly be necessary to ignore follow-up of many of the ECAs smaller than the diameter threshold selected for the survey, though the observations should be archived, as should the mass of observations of mainbelt asteroids, in case future linkage becomes possible.

The parallactic method of distance determination is a powerful aid in orbit determination for objects observed over short orbital arcs. Not only can nearby asteroids and comets be distinguished from mainbelt asteroids and more distant comets, but preliminary orbits can often be established where solutions would otherwise be indeterminate. Indeed, it is often possible to determine useful orbital elements of very close NEOs from observations on one night only. Even more powerful is the augmentation of optical astrometry by radar time delay and Doppler frequency shift measurements. Ostro et al. (1991) used the ratio of a measurement to its uncertainty as a figure of merit M to compare optical and radar astrometry. They gave $M \sim 2 \times 10^5$ for optical astrometry, $1.4 \times 10^6 \leq M \leq 8.2 \times 10^7$ for delay astrometry, and $1.1 \times 10^3 \leq M \leq 1.3 \times 10^6$ for Doppler astrometry.

D. Potentially Hazardous NEOs

Not all NEOs pose a threat to the Earth. Many of them are in orbits that cannot, at present, bring them within a distance that need concern us. The potential threat of an ECA or an ECC can be gauged from the minimum distance of its orbit from that of the Earth and how that distance is likely to change on a time scale of, say, a century (it can be assumed that, at some time or another, many ECAs will be located near the minimum distance).

Consider the minimum orbital intersection distances (MOIDs) of known NEAs that have been observed at more than one apparition, and thus have well-determined orbits. MOIDs can be defined at a particular epoch or, because they vary with time due to planetary perturbations, over a specified interval. In Table VI we list MOIDs at epoch 1993 November 9, not only to the Earth, but also to the other terrestrial planets and Jupiter. In many cases, these planets have been responsible for significant perturbations, due to close approaches, during the 20th century. To compute MOIDs, we used the method developed by Sitarski (1968). We ignored asteroids, such as Trojans and Hildas, that are known to be protected from close planetary encounters by resonant motion. However, with the exception of the Mars Trojan (5261) Eureka, we did include other such asteroids whose orbits come close to those of the terrestrial planets, and have flagged them when they are known to be in resonance, because the orbits of newly discovered ECAs will not, in general, allow ready identification of resonant motion. Tabular entries give MOIDs in AU (whenever MOID<0.05 AU for the terrestrial planets and MOID<1.0 AU for Jupiter) at epoch 1993 November 9, and appended symbols give further indications of the behavior of MOIDs through the 20th century (see the notes to the table). Assuming our century to be typical, we remark as follows:

1. Of the 105 NEAs listed, 43 are known not to be ECAs by our definition, and therefore cannot approach the Earth closely. Although many of them can make close approaches to Mars, such approaches do not appear to be effective in producing close approaches to Earth within a century.
2. Three asteroids—(1915) Quetzalcoatl, (1981) Midas and (2201) Oljato— have orbits that are Earth-intersecting during the 20th century.
3. 12 asteroids have orbits that are planet-intersecting during the 20th century: 1 with Mercury, 1 with Venus, 3 with Earth, and 7 with Mars. However, in a few cases we may be ignorant of protective resonances.
4. From the list of Milani et al. (1989), we have identified 23 or 24 asteroids that are protected from planetary collision by resonance, some of them

TABLE VI
Minimum Orbital Intersection Distances (MOIDs) in AU,
at Epoch 1993 November 9[a]

Asteroid	Mercury	Venus	Earth	Mars	Jupiter
944 Hidalgo[b]	—	—	—	—	$0.2926i^{d,h}$
1009 Sirene[b]	—	—	—	0.0361^d	—
1011 Laodamia[b]	—	—	—	0.0333	—
1036 Ganymed[b]	—	—	g	0.0444	—
1139 Atami[b]	—	—	—	0.0291^f	—
1566 Icarus	—	—	0.0357^e	0.0489^f	—
1620 Geographos	—	—	0.0307^e	—	—
1685 Toro	—	g	f,g	f	—
1862 Apollo	—	$0.0016^{c,f}$	0.0252^f	0.0490^e	—
1863 Antinous	—	—	g	$0.0068^{e,i}$	g
1865 Cerberus	—	—	—	0.0476^e	—
1915 Quetzalcoatl	—	—	c,d	—	g
1921 Pala[b]	—	—	—	—	$0.9829^{d,e}$
1922 Zulu[b]	—	—	—	—	$0.6205^{d,f}$
1981 Midas	—	—	$0.0029^{c,f}$	—	—
2099 Öpik[b]	—	—	—	f	—
2101 Adonis	—	0.0135^f	0.0124^e	0.0070^e	—
2102 Tantalus	—	—	0.0441^e	—	—
2135 Aristaeus	—	$—^f$	g?	0.0036^c	—
2201 Oljato	—	0.0066^f	0.0008^c	0.0094^e	—
2212 Hephaistos	0.0075^c	—	g	—	—
2340 Hathor	0.0264^f	g	g	—	—
2368 Beltrovata[b]	—	—	g	g	—
2608 Seneca[b]	—	—	f	—	g
3040 Kozai[b]	—	—	g	0.0020^c	—
3122 Florence	—	—	g	g	—
3199 Nefertiti[b]	—	—	—	g	—
3200 Phaethon	—	0.0404^f	0.0221^e	—	—
3360 1981 VA	—	d	—	—	g
3361 Orpheus	—	—	g	0.0471^e	—
3362 Khufu	—	g	0.0139^e	—	—
3552 Don Quixote[b]	—	—	—	0.0342^d	$0.4488^{d,e,h}$
3554 Amun	—	0.0250	—	—	—
3671 Dionysus	—	—	g	—	—
3688 Navajo[b]	—	—	—	$0.0168^{d,e}$	$0.2355^{d,f}$
3757 1982 XB	—	—	g	—	g
3800 Karayusuf[b]	—	—	—	0.0120^f	—
3833 1971 SC[b]	—	—	—	0.0352	—
3838 Epona	—	—	—	0.0452^f	—
3988 1986 LA	—	—	—	0.0385^f	—
4015 Wilson-Harrington	—	—	0.0476^f	$0.0290^{d,f}$	—
4034 1986 PA	—	0.0332^e	0.0187^f	—	—
4179 Toutatis	—	—	0.0058^f	0.0424^f	—

Asteroid	Mercury	Venus	Earth	Mars	Jupiter
4183 Cuno	—	^{g}	^{g}	0.0180^e	—
4197 1982 TA	—	^{d,f}	—	$0.0421^{d,f}$	^{g}
4205 David Hughes^b	—	—	—	0.0145	—
4257 Ubasti	—	—	—	0.0226^e	—
4401 Aditi^b	—	—	^{g}	0.0197^d	—
4450 Pan	—	0.0186^f	0.0277^f	0.0170^d	—
4486 Mithra	—	0.0171	0.0460^f	0.0192^f	—
4503 Cleobulus^b	—	—	—	0.0071^f	^{d}
4581 Asclepius	—	0.0060	0.0300^f	—	—
4596 1981 QB	—	—	—	0.0298^f	—
4660 Nereus	—	—	0.0031^f	0.0382	^{g}
4769 Castalia	—	—	0.0199^f	—	—
4775 1927 TC^b	—	—	—	0.313	—
4947 Ninkasi^b	—	—	—	0.0109^e	—
4953 1990 MU	—	0.0455^f	0.0290^e	—	—
4957 1990 XJ^b	—	—	—	0.0488^e	—
5011 Ptah	—	^{g}	0.0257^e	0.0296^f	^{g}
5131 1990 BG	—	0.0023^e	—	0.0118^f	—
5143 Heracles	—	—	—	$0.0193^{d,f}$	—
5164 1984 WE$_i$^{b}	—	—	—	—	^{d,f,h}
5189 1990 UQ	—	—	0.0443^f	0.0106^f	—
5201 1983 XF^b	—	—	—	$0.0007^{c,d}$	^{g,h}
5335 Damocles^b	—	—	—	0.0495	^{h}
5370 Taranis	—	—	—	—	$0.3518^{d,e,h}$
5381 1991 JY	—	0.0189^f	—	—	—
5587 1990 SB^b	—	—	—	0.0398^e	—
5590 1990 VA	—	0.0035^f	—	—	—
5604 1992 FE	—	0.0058^f	0.0339^f	—	—
5620 1990 OA^b	—	—	—	0.0200^f	—
5621 1990 SG$_4$^{b}	—	—	—	0.0297	—
5626 1991 FE^b	—	—	—	0.0448^e	—
5641 1990 DJ^b	—	—	—	0.0041^e	—
5645 1990 SP	—	—	—	0.0052^f	—
5660 1974 MA	—	—	—	0.0225^f	—
5693 1993 EQ	—	0.0137^e	0.0052^f	—	—
5836 1993 MF	—	—	—	0.0157^c	—
1943 DF^b	—	—	—	0.0266	—
1982 TX^b	—	—	—	$0.0228^{c,d}$	—
1984 KB	—	0.0149^f	—	^{f}	—
1988 EG	—	0.0291^e	0.0245^e	0.0233^f	—
1988 XB^b	—	0.0338^e	0.0068^e	0.0179^f	—
1989 SL$_5$^{b}	—	—	—	$0.0047^{c,d}$	0.7766
1989 UQ	—	0.0036^e	0.0140^f	—	—
1989 WQ$_i$^{b}	—	—	—	0.0352^e	—
1990 KA^b	—	—	—	0.010^f	—
1991 CQ^b	—	—	—	$0.0099^{c,d}$	—

TABLE VI (cont.)

Asteroid	Mercury	Venus	Earth	Mars	Jupiter
1991 CB[i]	—	0.0433	—	—	—
1991 JW	—	—	0.0205	—	—
1991 PM$_5$[b]	—	—	—	0.0127[f]	—
1991 TB$_i$	—	—	—	0.0166[f]	—
1991 VK	—	—	0.0469[f]	—	—
1991 WA	—	0.0083[f]	—	—	—
1992 AE[b]	—	—	—	0.0427[e]	—
1992 SK	—	—	0.0464[e]	0.0326[f]	—
1992 TC[b]	—	—	—	0.0208[e]	—
1992 YC$_3$[b]	—	—	—	0.0103[e]	—
1993 BW$_3$	—	—	—	0.0122[f,j]	—
1993 HP$_i$	—	—	0.0049	—	—
1993 KH	—	—	0.0014[e]	—	—
1993 OV$_i$[b]	—	—	—	0.0271	—
1993 VW	—	—	—	0.0121[e]	—

[a] MOID<0.05 AU for Mercury, Venus, Earth and Mars; MOID<1.0 AU for Jupiter.

[b] NEA known not to be an ECA.

[c] MOID less than one planetary radius over one or more intervals during the 20th century.

[d] MOID varies by 0.02 AU or more during the 20th century.

[e] MOID on the whole monotonically decreases during the 20th century.

[f] MOID on the whole monotonically increases during the 20th century. Where a symbol is given but no numerical entry, MOID was less than one of the threshold distances given above at some time during the 20th century, but greater at the epoch 1993 November 9.

[g] Protected by resonance against close planetary approach according to Milani et al. (1989). Their calculations treated only planet-approaching asteroids discovered through November 1985.

[h] Inactive comets, according to Shoemaker and Shoemaker (1994). (4015) Wilson-Harrington exhibited cometary activity in 1949 (Bowell 1993).

[i] (1863) Antinous will have MOID less than one planetary radius in 2001.

[j] 1993 BW$_3$ had MOID less than one planetary radius for some years prior to 1897.

with more than one planet. However, Milani et al.'s calculations treated only planet-approaching asteroids discovered through November 1985 [numbered asteroids through (3330) and unnumbered asteroids through 1985 WA]. Of this sample, we calculate that 20 asteroids have MOIDs less than 0.05 AU from the Earth at some time during the 20th century; and of those, 9 or 10—about half—are known to be currently protected by resonances from collision with the Earth.

5. Few asteroids have MOIDs that change by more than 0.02 AU on a time scale of a century. All asteroids whose orbits approach that of Jupiter within 1 AU show such excursions, but only one Earth-approacher, (1915) Quetzalcoatl, does so.

From the above results, we derive the following (we believe conservative) definition: for the purposes of NEO surveys, we consider any asteroid having an inner-planet MOID less than 0.05 AU and/or Jupiter MOID less than 1.0 AU to be potentially hazardous on a time scale of a century unless it can be shown either that the body is protected from close Earth approach by resonance or will not actually approach the Earth closely. Likewise, any short-period comet exhibiting the same orbital characteristics is considered potentially hazardous, as is any intermediate- or long-period comet that appears to have an Earth MOID<0.05 AU.

Table II indicates that the completeness level for potentially hazardous ECAs is greater than that of the population as a whole. This is because, in a dark-sky survey, ECAs are preferentially observable when close to the Earth. However, for a 25-yr survey of the standard Spaceguard search region, only potentially hazardous ECAs smaller than a few tenths of a kilometer in diameter are so favored, presumably because larger ones tend to become visible outside the standard region. The discovery completeness of potentially hazardous ECCs is the same as that of the total population (Tables III and V).

E. Practical Considerations

It is inconceivable that a fully fledged network of observing stations will start operating simultaneously and at full efficiency. More likely, current photographic and CCD searches will be intensified in parallel with the development of new survey telescopes. There exists, therefore, an important opportunity to refine models of the NEO population and to test observing strategies. In particular, care should be taken to preserve the pointing histories of any systematic searches for NEOs so more reliable bias correction can be carried out as the known sample grows. When a full-up survey is in progress, it will be possible to refine the population model further. For example, if it is determined that Atens are more numerous than currently thought, an improved survey strategy could be designed to enhance their discovery. If physical observations diagnostic of albedo are made, our knowledge of the albedo and diameter distributions of ECAs could be greatly improved. Likewise, observations of ECCs, as they leave the inner solar system and revert to an inactive state, are much needed to improve our population model.

We have shown that potentially hazardous ECAs can be discovered at a sufficient rate that most of such larger members of the ECA population can be discovered and assessed within 25 yr. By prolonging the survey, the inventory of smaller ECAs can be brought to greater completeness. Indeed, we estimate that, using current technology to continue the Spaceguard Survey beyond 25 yr, we would stand a better-than-even chance, within about 150 yr, of discovering and identifying the ECA that might cause the next Tunguska-like event. In anticipation that huge strides in technological development will reduce this interval considerably, we can be almost certain that such an asteroid could be identified before Earth impact by means of a prolonged telescopic search.

The potential hazard of SPCs can be assessed in the same way as that of ECAs (allowing for nongravitational forces, where necessary). LPCs must be judged on a case-by-case basis by examining the evolution of the estimated MOID with Earth as the orbital arc is lengthened. Because ECCs enter the inner solar system at a near-constant rate, many of them for the first time, their potential for hazard to the Earth goes on forever. Thus, any survey of finite duration will be destined to ignore a significant fraction of the potential hazard posed to our planet. Only by continually monitoring the flux of ECCs in the Earth's neighborhood can we hope to achieve near-complete hazard assessment.

VI. THE SPACEGUARD SURVEY

In this section, we assess the instrumental requirements (telescopes, mosaics of CCD chips, computers, etc.) imposed by the observing strategy and follow-up research outlined above, and we comment on observational techniques and observing network operation. We concentrate on the requirements of a survey optimized for the discovery of ECAs, with the understanding that slightly different requirements are posed by a network optimized for an ECC search. To cover the requisite volume of search space, the survey must achieve a stellar limiting magnitude of at least $V_{lim} = 22$ mag, dictating telescopes of 2- to 3-m aperture equipped with CCD detectors. The most efficient use of such detectors is achieved if the pixel size is matched to the apparent stellar image size of about 1 arcsec FWHM, thus defining the effective focal length for the telescopes at about 5 m. According to the model explored in Sec. V, a plausible area of sky to be searched is about 6000 deg^2/month, centered on opposition, and extending to $\pm 30°$ in ecliptic longitude and $\pm 60°$ in ecliptic latitude. These considerations lead us to a requirement for multiple telescopes with moderately wide fields of view (at least $2°$) and mosaics of large-format CCD detectors. We develop these ideas in this section to derive a Spaceguard search program. This program is not unique (that is, an equivalent result could be obtained with other appropriate choices of telescope optics, focal-plane detectors, survey area, and locations), but it is representative of the type of international network that is vital if an effective NEO survey is to be undertaken.

A. Lessons from the Spacewatch Program

The Spacewatch Telescope, operated at the University of Arizona, has been described by Rabinowitz (1991) and Gehrels (1991). We remark on it here as it was until 1992. However, improvements to the CCD detector and detection software continue to be made. Spacewatch is the first telescope and digital detector system devised to carry out a semi-automated search for NEOs. As such, the lessons learned from its development and operation are invaluable when considering a future generation of scanning instruments. The Spacewatch system comprises a single 2048×2048-pixel CCD chip at the f/5

Newtonian focus of an equatorially mounted 0.9-m telescope. Each pixel subtends 1.21×1.21 arcsec^2. With the telescope drive turned off, the CCD camera scans the sky at the sidereal rate (such drift-scan mode of operation is often called time-delay integration), with an effective integration time of 165 s. In 1991, the Spacewatch Telescope achieved detection of celestial bodies to $V_{\text{lim}} = 20.5$ mag. One of the important demonstrations provided by the Spacewatch Telescope team is that image-recognition algorithms such as their Moving Object Detection Program are successful in making near-real-time discoveries of moving objects (asteroids and comets). False detections are almost eliminated by comparing images from three scans obtained one after the other. At present, the Spacewatch system makes detections by virtue of the signal present in individual pixels. With the incorporation of higher-speed computers, near-real-time comparison of individual pixels to measure actual image profiles would lead to a greater reduction in the most frequent sources of noise: cosmic ray hits and spurious electrical noise events.

In light of the successful performance of Spacewatch, we have dismissed a photographic survey. Even though sufficiently deep exposures and rapid areal coverage could be attained to fill the survey requirements using a small number of meter-class Schmidt telescopes (similar to the Oschin and U. K. Schmidts), there is no feasible way, either by visual inspection or digitization of the films, to identify and measure the images in step with the search. A photographic survey would fail for lack of adequate data reduction and follow-up. Future developments in electronics and data processing will further enhance the advantages of digital searches over the older analog methods using photography.

B. Detectors and Telescopes

The largest CCD chips readily available today contain 2048×2048 pixels, each about 25 μm on a side. Thus the chips are about 5×5 cm in size. For thinned chips, quantum efficiencies have attained a peak near 80%, and useful sensitivity is achievable from the near-ultraviolet to the near-infrared. To reach a stellar limiting magnitude of $V_{\text{lim}} = 22$ mag, we require the use of these CCDs at the focal plane of a telescope having an aperture of 2 m or larger, mostly operated when no bright moonlight is present.

In the coming decade, we envisage a trend toward smaller and more numerous CCD pixels covering about the same maximum chip area as at present. No great increase in spectral sensitivity is possible. At the telescope, the pixel scale must be matched to the image scale in good or adequate seeing conditions. In what follows, we assume a pixel scale of 1 arcsec/pixel (25 μm/arcsec or 40 arcsec/mm), which implies a telescope of 5.2-m focal length. For a telescope of 2-m aperture the focal ratio is f/2.6; for a 2.5-m, f/2.1; and for a 3-m, f/1.7.

A single 2048×2048 CCD chip is a very powerful data-gathering device, simultaneously detecting the signals from more than 4 million pixels, but it still falls short of the requirements for wide-field scanning imposed by an

NEO survey like the Spaceguard Survey. At the prime focus of a telescope of 5.2 m focal length, such a CCD chip covers a field of view on the sky about 0.5° on a side. However, we wish to scan an area at least 2° across (see below). Therefore, we require that several CCD chips be mosaicked in the focal plane.

Studies and planning are underway at the University of Arizona for an updated 1.8-m Spacewatch Telescope. The new telescope will be an excellent instrument to test and develop some of the instrumental and strategic considerations described here. Smaller new systems are being developed by groups at the Jet Propulsion Laboratory and at Lowell Observatory. We describe some of the work of the latter group in Sec. VII.A. From the viewpoint of the Spaceguard Survey, it is safe to assume that 2- to 3-m-class telescopes can be built having focal lengths near 5 m and usable fields of view between 2° and 3°. Refractive-optics field correction is probably required, and it appears advantageous to locate CCD mosaics at the prime focus of such instruments. Here, we indicate telescope functional requirements, but do not closely specify the size or design of survey telescopes.

C. Magnitude Limit and Observing Time

Exceptionally fine astronomical sites have more than 1000 hr yr^{-1} of clear, moonless observing conditions, during most of which good or adequate seeing prevails. More typically, 700 hr yr^{-1} of observing time is usable. We assume that the standard Spaceguard region of 6000 deg^2 is to be searched a number of times each month, and that initial NEO detection is made by two or three scans on the first night. Parallactic information, giving a useful estimate of an NEO's distance from the observer, is derived by four scans on a subsequent night, and an orbit is calculated from observations on a third night. Thus, nine or more scans of the search region are needed each month. In a given month, follow-up will be attempted for some of the NEOs that have moved out of the search region (mainly to the west). As a working value, we assume that 40 hr/month per telescope are available for searching.

The limiting stellar magnitude that can be observed by a telescope is a function of the ratio of the source brightness to that of the sky, the number of pixels n_p occupied by a stellar image, the pixel area p, the effective light-collecting area of the telescope A, the threshold SNR for detection S_n, and the integration time t. Rabinowitz (1991) has given suitable expressions, normalized to the performance of the Spacewatch Telescope which, at the time, achieved $V_{lim} = 20.5$ mag for $t = 165$ s, for $n_p = 9$, and at $S_n = 6$. Allowing for better performance of the Spaceguard system arising from improved detector quantum efficiency and improved image-recognition algorithms (see Sec. VII.A), we obtain

$$V_{lim} = 21.9 - 2.5 \log S_n \left(\frac{n_p p}{At}\right)^{1/2} \qquad (9)$$

where p is in $arcsec^2$, A is in m^2, and t is in seconds. For Spaceguard Survey telescopes, we assume that $A = 0.9\sqrt{\text{primary area}}$, $p = 1.0$ arcsec^2, and $n_p = 9$ (a conservative estimate in good seeing conditions). Then a single CCD should be able to achieve the survey requirement of $V_{\text{lim}} = 22$ mag with the combinations of telescope aperture and scan speed, in units of the sidereal rate, given in Table VII. As remarked in Sec. IV.B, detection losses due to trailing are much less for the Spaceguard Survey than for the Spacewatch Telescope. According to Table VII and Eq. (5), an NEO moving at $3°$/day would suffer a trailing loss of less than 0.1 mag when scanned by a 3-m telescope, whereas the same NEO detected by the Spacewatch system, which is constrained to drift scanning, would suffer a trailing loss of more than 2 mag.

TABLE VII
Spaceguard Survey Integration Time and Scan Rate as a Function of
Primary Mirror Diameter Required to
Achieve $V_{\text{lim}} = 22$ mag

Primary Diameter (m)	Integration Time (s)	Scan Rate ($\times$ sidereal)
2.0	21	6
2.5	14	10
3.0	10	14

D. CCD Mosaics and Number of Survey Telescopes

A single 2048×2048-pixel CCD chip, having an image scale of 1 arcsec/pixel, scans 0.14 deg^2/min at the sidereal rate. If 40 hr/month/telescope can be allotted to searching for NEOs over 6000 deg^2 to $V_{\text{lim}} = 22$ mag, and 10 scans per sky region are required for detection and rough orbital characterization of an NEO, then telescopes of the apertures considered above have the performance capabilities given in Table VIII. In computing values for the total number of CCD chips required in the worldwide network of Spaceguard Survey telescopes, we assumed that no two CCD chips together scan the same region of the sky, although there are particular advantages if two or more chips do sequentially scan a given region of the sky (see Sec. VII.A). These are minimum requirements for the telescopes; in practice, more scans may be needed for reliable automatic detection, and probably there will be some overlap of coverage among telescopes, even in a very carefully coordinated survey.

Searching to $\pm 60°$ ecliptic latitude implies sky coverage, over the course of a year, at almost all declinations. Thus telescopes must be located in both hemispheres. Usable fields of view of between $2°$ and $3°$ probably limit the number of CCD chips in a telescope's focal plane to about ten at the scales we have been considering. Most likely, four CCD chips/telescope can be accommodated in a linear array in the focal plane. Thus, it appears from Table VIIII that at most seven 2-m telescopes, five 2.5-m telescopes,

TABLE VIII
Spaceguard Survey Areal Coverage/Month/CCD and Total Number of
Chips as a Function of Primary Mirror Diameter Required to
Scan 6000 deg^2/month

Primary Diameter (m)	Area/Month/CCD (deg^2)	Total Number of CCDs Required
2.0	260	28
2.5	420	18
3.0	600	13

or four 3-m telescopes suffice to fulfill the search, follow-up, and physical observations requirements of the idealized 6000-deg^2 Spaceguard Survey. Should there remain spare observational capability, it could be used to enhance the detection rates of ECCs and Atens by scanning a few times per month outside the standard Spaceguard region. Additional space remaining in the focal plane could be filled by filtered CCD chips to provide colorimetry, which would give a first-order compositional characterization of some of the NEOs discovered while scanning.

To ensure that a single-point failure due to weather or other adverse factors does not hamper effective operation of the survey network, we conclude that three telescopes are required in each hemisphere. With fewer telescopes, orbital, and perhaps parallactic, information on NEOs would be sacrificed. The desirability of searching near the celestial poles calls for at least one telescope at moderate latitude in each hemisphere. Thus, the Spaceguard Survey requires a network of six or more telescopes, each 2- to 3-m in aperture, distributed in longitude and at various latitudes between, say, 20° and 40° north and south of the equator.

E. Scanning Regime

At high declinations, scanning along small circles of declination results in curvature in the plane of the CCD chip, with the result that star images do not trail along a single row of pixels. The problem can be avoided by scanning along a great circle. A good strategy would be to scan along great circles of which the ecliptic is a meridian, the pole being located on the ecliptic 90° from the Sun. Such scanning can be achieved using either equatorial or altazimuth telescope mounts, but it is probably more easily and cheaply accomplished using the latter. In either case, field rotation is required, as is currently routinely used at the Multiple Mirror Telescope in Arizona and at other installations. As an example, during the course of a scan 20° long at ecliptic latitude 10°, the CCD camera must be rotated 3.6° to compensate for field rotation.

At the proposed 1.8-m Spacewatch Telescope, it is planned to make three scans of each region of the sky (as is currently done at the 0.9-m Spacewatch Telescope). Each scan would cover 10° in 26 min, so the interval between the first and third scans is sufficiently long that objects moving as slowly as 1 arcmin/day could be detected. For the Spaceguard Survey, we envisage two or three quasi-longitudinal scans per sky region, about two hours apart. Thus, at a scan rate of 10 times sidereal rate, each scan could cover an entire strip of the 60°-wide search region, with a second search strip being interposed before the first was repeated. We assume that occasional false-positive detections will not survive scrutiny on the second night of observation, and thus will not significantly corrupt the detection data base.

F. Computer and Communications Requirements

Near-real-time detection of faint NEOs requires that prodigious amounts of data processing be accomplished at the telescope. If all detected sources (fixed and moving) are analyzed, the image-processing rate scales linearly with the number of sources recorded per second. The number of sources detected per second (the *object rate*), and therefore the computer requirements of the Spaceguard Survey, can be estimated from the performance of the Spacewatch Telescope. According to Rabinowitz (1991), the computer system in use at the Spacewatch Telescope can detect up to 10,000 sources in a 165-s effective integration. Thus its object rate is 60 s^{-1}. Scanning to $V_{\text{lim}} = 22$ mag requires detection of about 30,000 objects/deg^2 at moderate galactic latitudes (Allen 1973). At an image scale of 1 arcsec/pixel, using the scanning rates from Table VII, and allowing for a ten-fold increase in computer requirements to perform real-time image-profile analysis, we calculate the total Spaceguard Survey computer requirement to be 2000 to 3000 times that at the Spacewatch Telescope. Therefore, at each of six telescopes it would be 300 to 500 times that at Spacewatch. Such a requirement, although not easy to achieve, is possible using the newest generation of parallel processors.

In order of increasing demand on computer and storage resources, there are at least three levels of observational data storage that can be envisaged: (1) preservation of image-parameter or pixel data only for the moving sources detected; (2) preservation of image-parameter or pixel data for all sources detected (mostly stars); (3) storage of all pixel data. The first option is clearly undesirable, in part because data for very slow-moving NEOs mistaken as stars would be lost. The first two options have the disadvantage that there would be no way to search the data base, after the event, for sources whose brightnesses are close to V_{lim} and that may therefore not have been detected. The third option (the most attractive scientifically) may appear to result in serious problems of data storage and retrieval. However, we anticipate that, using fast, high-density Exabyte or similar technology now becoming available, the third option is tractable. As a practical matter, after-the-event searches should be minimized.

We estimate that about 1000 NEOs and 200,000 mainbelt asteroids of all sizes could be detected in the Spaceguard Survey each month—about two detections per second of observing time. Even though this detection rate exceeds the current worldwide rate by about 3 orders of magnitude, only moderate-speed data transmission is needed between observing sites and a central processing facility. Careful observational planning will be required to ensure efficient coverage of pre-programmed scan patterns, to avoid unintentional duplication of observations, to schedule the necessary parallactic and follow-up observations, and to optimize program changes resulting from shutdowns so as to maintain robustness of the survey. Successful operation of the Spaceguard Survey will also require the coordination and orbital computation capabilities of a modern central data clearinghouse.

VII. RAMPING UP TO THE SPACEGUARD SURVEY

A. The Lowell Observatory NEO Search

In this section, we describe a new initiative that has just begun at Lowell Observatory. The Lowell Observatory near-Earth object search (LONEOS) is an attempt to increase the NEO discovery rate using a mosaic-CCD camera on a small Schmidt telescope. LONEOS is in part designed to provide information that can eventually be exported to Spaceguard.

Our telescope is a 40-cm-aperture f/2.77 Schmidt with a 58-cm-diameter primary. We plan to replace the present corrector plate with a plate of about 58-cm aperture (making the telescope an f/1.91 instrument). A four-chip CCD camera is currently under construction. The unthinned 2048×2048 Loral Aerospace chips have 15 μm square pixels and peak quantum efficiency near 40%. The camera will resemble one of the two being used in a project to search for massive compact halo objects (Alcock et al. 1992). It will probably contain two pairs of chips in a square configuration, each chip having two readout amplifiers on the same side, thus allowing scan-mode operation. We estimate the maximum readout rate to be 0.5 Mpixel/s/chip, giving a minimum effective integration time of 34 s/chip. A stare-mode integration would cover an area of about 10 deg^2.

We compute the performance of the telescope/CCD camera combination using $A = 0.15$ m^2, $p = 7.8$ arcsec^2, $n_p = 9$, and $S_n = 3$ in Eq. (9). Each sky location is recorded on two chips during a single scan. As a consequence, we will be able to identify virtually all cosmic-ray hits, and to perform statistical tests on the reality of images. Indeed, our estimate of $S_n = 3$ is determined by tests we have recently carried out using extremely simple and fast source-detection algorithms in which the signal from the two separate images and in adjacent pairs of pixels is compared. (Actually, the tests indicate that we may be able to make secure detections down to $S_n \approx 2$, though this remains to be proved.) We obtain V_{lim} and scan rate (in units of the sidereal rate on the equator) as a function of integration time/chip as given in Table IX. No allowance has been made for NEO-image trailing; for a 100-s effective integration time, the image of an average NEO at opposition will exhibit a 0.7-pixel trail, which is fairly small compared to the image size. The extremely rapid scan rate required for the shortest effective integration time is noteworthy; it represents areal coverage at the rate of about 1000 deg^2/hr.

TABLE IX
Limiting Magnitude V_{lim} and Scan Rate ($\times$ sidereal) as Functions of
Effective Integration Time t for LONEOS

t (s)	V_{lim} (mag)	Scan Rate
34	19.3	23
100	19.9	8
316	20.5	2.4

Next, we estimate the sky coverage of LONEOS. As for the Spaceguard Survey, we assume a total observable time of 1000 hr/yr, of which 40 hr/month will be given over to scan-mode observing. The remaining time would be used for follow-up work, some of it in the stare mode (the actual operational mix of the two observing modes can be scheduled in response to discoveries). We further assume 90% observing efficiency (allowing for failures, down time, scan analysis, and telescope repointing), and that 9 scans/region/month will be made (3 scans/region on each of three nights). Thus a total of 4 hr of "new" sky scanning/month is anticipated which, given the model sky-plane distribution of ECAs in Fig. 3 and other considerations derived from our study of the Spaceguard Survey, might occupy an ecliptic longitude and latitude region of halfwidths ($\Delta\lambda$, $\Delta\beta$) as given in Table X. (We have ignored the fact that high southern latitudes cannot be scanned from Flagstaff. Obviously, the summer observing pattern would be biased toward northern ecliptic latitudes.)

TABLE X
Monthly Areal Coverage and Possible Semi-Extent in Ecliptic Longitude
$\Delta\lambda$ and Latitude $\Delta\beta$ as Functions of
Limiting Magnitude V_{lim} and Effective Integration Time t for LONEOS

t (s)	V_{lim} (mag)	Area (deg^2)	$\Delta\lambda$	$\Delta\beta$
34	19.3	4400	40°	29°
100	19.9	1300	20°	17°
316	20.5	400	20°	6°

Finally, we model the number of ECA discoveries per year as functions of V_{lim} and D. The calculation, pertaining to a 5-yr model survey, yields the average annual discovery rate given in Table XI. We have not modeled ECAs smaller than 0.1 km in diameter, so the total number of ECA discoveries is likely to be greater than indicated. The number of larger ECAs discovered is affected by the uncertainty in our model ECA population (Sec. II.A), but should not be in error by more than a factor of 2. At $V_{lim} = 19.3$ mag and $t = 34$ s, we estimate that 13 ECCs/yr larger than 1 km in diameter should be detected (some of them known, of course), one or two of which are larger than 5 km diameter. We conclude from these results:

TABLE XI
Annual Average Number of ECAs Discovered During the Course of
a Five-Year LONEOS Survey as a Function of
Limiting Magnitude V_{lim} and Diameter

| | V_{lim} (mag) | | |
Diameter (km)	19.3	19.9	20.5
0.1–0.5	196	91	18
0.5–1	70	48	15
1–2	45	33	12
>2	34	25	12
Total	345	197	57

1. Unlike the Spaceguard Survey, the total number of ECA discoveries does not rise with fainter V_{lim}. This is mainly because the search region shrinks in size with increasing V_{lim}, and partly due to the effects of trailing loss.

2. To maximize the discovery rate of the largest, most hazardous ECAs, it is essential to scan the largest possible area of sky (preferably near the ecliptic and near opposition) as fast as possible.

3. In the scan mode, using 34-s effective integration times, we should be able to discover about 80 ECAs/yr larger than 1 km in diameter. For comparison, the present worldwide discovery rate of such ECAs is about 20/yr, and the rate estimated for the first year of full-up operation of the Spaceguard Survey is about 400/yr (Fig. 9).

4. Because the areal sky coverage achievable is several times less than that necessary in a full-scale survey such as Spaceguard, the operation of two or more searches like LONEOS could lead to an almost proportionate increase in the discovery rate of NEOs. However, because of brighter V_{lim}, surveys using small telescopes cannot compete with the completeness of discovery achievable by the Spaceguard Survey.

B. Sketch of an NEO Survey Protocol

In Fig. 11, we summarize many of the aspects of an NEO survey in flowchart form. Figs. 11a and 11b illustrate the main pathway; Figs. 11c and 11d are, respectively, confined to the analysis of a single frame and to the analysis of sources (both fixed and moving) within the frame. The protocol is not unique, and indeed, is not in a mature state of development as presented here. In the following abbreviated description, Roman numerals are keyed to areas of Fig. 11.

I. There will probably be several moving-object catalogs; asteroids and comets likely to be found very close to their predicted positions; multiple-apparition asteroids and comets; single-apparition asteroids and comets for which two-body orbits have been determined; just-discovered and confirmed asteroids and comets, for which Väisälä-type or other short-term predictor orbits (perhaps several per object) have been included; unconfirmed moving sources. The various moving-source catalogs are referred to in other parts of the flowchart. In each case, ephemeris uncertainties must be computed, perhaps using the methods put forward by Muinonen and Bowell (1993*a*) and Muinonen et al. (1994*b*), so as to limit search and linkage problems.

II. We assume that scan- or stare-mode observations will be selected on the basis of predicted magnitudes and ephemeris uncertainty of target NEOs. Thus, much follow-up work can be accomplished by scanning, especially when the ephemeris uncertainty is large.

III. We have allowed for a variable V_{lim}, which requires selection of the scan rate.

IV. Because, at discovery, moving-source orbits can only be guessed at, all newly discovered asteroids and comets are potentially hazardous. They remain so until their orbits can be determined accurately enough to prove otherwise (see Sec. V.D.), and should therefore be given priority for follow-up.

V. The observing strategy should probably be updated frequently in response to the nature of discoveries being made. Much less frequently, an entirely fresh survey strategy, of the kind put forward in this chapter, can be implemented.

VI. We assume, for present purposes, that scans will be accumulated until a frame, similar to that from stare-mode integrations, has been acquired, and then analyzed as scanning proceeds uninterrupted.

VII. Cosmic-ray events will most likely be identified and eliminated by comparison of two or more frames of the same region, using failure to repeat from one frame to another and, perhaps, image moments, as criteria.

VIII. Astrometric mapping can be a very rapid process involving just a few catalog stars. Intrinsic field distortion can be pre-mapped using stare-mode integrations.

IX. We presume that data compression by a factor of at least a few can be accomplished before data storage. Other methods, such as those mentioned in Sec. VI.F, are possible.

X. It is desirable to develop a fixed-source catalog for use by those interested in non-NEO research.

XI. Working at the necessary low SNR, there will be many false-positive detections, most of which may be identified by correlation failure in repeat integrations on a given region. However, it will be useful to develop a statistical measure of source reality as part of the image-moment analysis (itself used to discriminate comet-like moving sources).

XII. Automatic trail and streak detection, although routine in fields such as Earth-satellite tracking, may not be implementable in the kind of near-real-time analysis we envisage for an automated NEO survey because of the already high demand on computer resources.

a   NEO SURVEY PROTOCOL

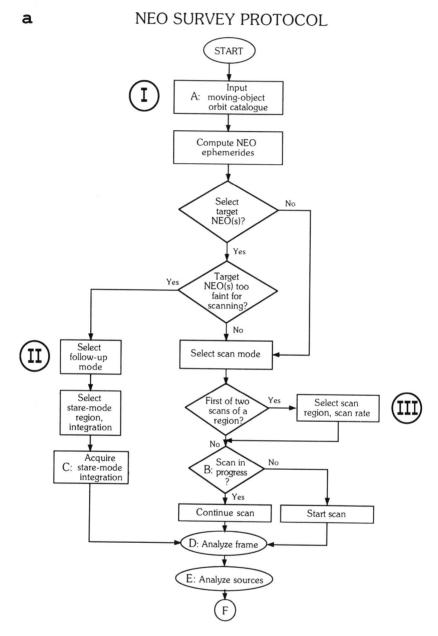

Figure 11 (a,b,c,d).   Flowchart showing principal components of an NEO survey protocol. Roman numerals are keyed to comments in Sec. VII.B.

VIII. FUTURE WORK

Much work is needed to bring feasibility studies, such as the one presented

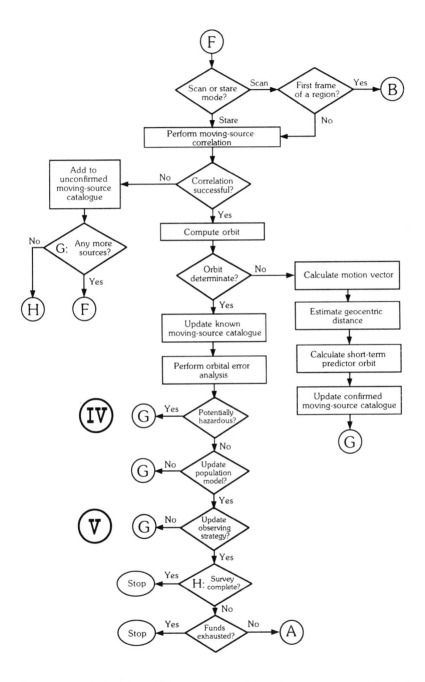

here, to practical reality. We comment on those elements we perceive to be most important.

C

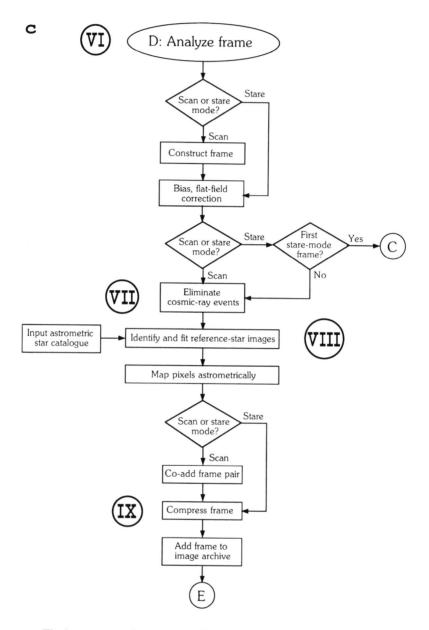

The key to mounting a successful groundbased NEO survey is to achieve a faint limiting magnitude using short integration times and to observe a large area of the dark sky, probably by scanning, a number of times each month. We have shown that, in a Spaceguard-like survey, wide-field, small-focal-ratio telescopes, of perhaps 3-m aperture, are required. Suitable optical and mechanical designs will be needed. Paving the focal surfaces of wide-field

d

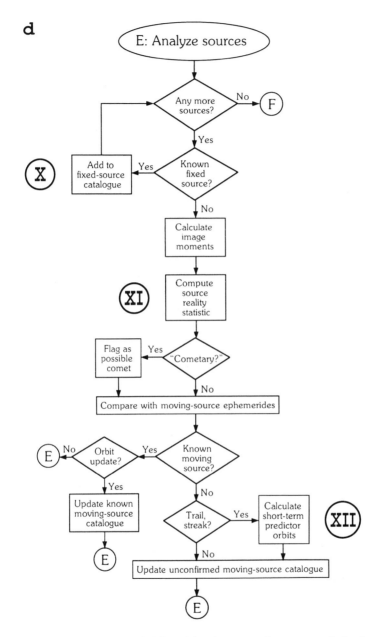

telescopes with mosaics of CCD chips is no routine matter; indeed, only recently have the first steps in this direction been taken. Study is required, not only of the engineering problems engendered by the use of CCD mosaics, but also of the disposition of CCD chips in the focal surface.

Near-real–time analysis of observational data acquired at a rate of MB/s/ chip places considerable stress on computer and data-storage capabilities. As an example of work now going on in this area, we cite the efforts of A. Maury and collaborators (personal communication 1993) to design NEO detection software using an associative string processor, which can attain a speed of about 1 GOPS. In the Lowell Observatory near-Earth object search, we will be using a Silicon Graphics computer, currently containing two processors (but shortly to contain four) that can simultaneously address common memory. Thus, we envisage pipeline data processing.

Attaining as deep a limiting magnitude as possible requires source-detection software that is capable of recognizing images at very low SNR. Working at SNR ≲ 3 implies the presence of a significant (perhaps large) number of false-positive source detections, the number rising as the threshold SNR decreases. Of course, false-positive detections will fail to correlate among successive scans of a given region unless they are extremely numerous. Thus there exists a trade-off between threshold SNR and number of scans. Optimizing that trade-off will require the development of software capable of making rapid and accurate source-detection tests.

In Sec. V.C, we have simplified the matter of making night-to-night linkages in regions of high asteroid surface density. Software will have to be devised both to minimize the number of false linkages and to allow for unusually rapid changes in motion. Work by Bowell et al. (1991), Marsden (1991), and Moynihan and Sykes (1993) can form the basis.

Estimating orbital accuracy, right from the moment of discovery, is of paramount importance both for linkage, follow-up, and recovery of NEOs, and for assessing whether an NEO is potentially hazardous. We have commented on the first matter in Sec. V.C. The degree of potential hazard can be framed in terms of probability of impact with the Earth, and first steps in developing the necessary methods have been taken by Chodas (1993), Muinonen and Bowell (1993a, b), and Yeomans and Chodas (see their Chapter). The problem becomes especially challenging—and sometimes intractable—when multiple close approaches are involved. It is clear that the combination of optical and radar astrometry should be attempted in critical cases.

Management of a full-scale NEO survey is a very complex undertaking. Although we envisage data acquisition and analysis largely as processes to be dealt with autonomously by operators at each observing station, overall control of the observing network will have to be maintained by a central processing facility. The facility will be responsible for scheduling observing patterns; establishing the mix among search, follow-up, and physical observations; changing the observing schedule in response to single-point failures and to the discovery of unusual NEOs; and computing orbits and continuously updating the fixed- and moving-source catalogs.

There is much room for improvement in the study of survey strategy. The known sample of ECAs is small enough that gross errors in modeling certain objects could have been incurred. For example, we know too little about the population of Atens, Halley-type comets (especially inactive ones), small NEOs of all kinds, and the possible near-Earth belt of ECAs. Moreover, our ideas about how best to model the NEO population and NEO surveys are bound to improve as CCD searches proliferate and the rate of discovery accelerates. The work we have presented here can easily be adapted to an assessment of the performance of spacebased survey protocols. One intriguing possibility, which eliminates the problems associated with detection at small solar elongations, is to construct a detector system in orbit around Venus. Ultimately, it appears that spacebased systems are the only way to achieve near-complete detection of ECCs.

Acknowledgments. It is a pleasure to thank L. H. Wasserman for energetic help in software development. E. M. Shoemaker has provided much scientific stimulus, as have a number of members of the team responsible for the Spaceguard Report. Most notably, workshop chairman D. Morrison's keen guidance helped keep our work in focus. We also thank summer students P. W. Tracadas and K. E. Daniels for helping scope out computer requirements and image-detection protocols for the Lowell Observatory Near-Earth Object Survey. Reviewers A. Carusi and D. L. Rabinowitz helped us clarify many points. Research funding was provided by NASA grants and by the Lowell Observatory endowment.

REFERENCES

Alcock, C., Axelrod, T. S., Bennett, D. P., Cook, K. H., Park, H.-S., Griest, K., Perlmutter, S., Stubbs, C. W., Freeman, K. C., Peterson, B. A., Quinn, P. J., and Rodgers, A. W. 1992. The search for massive compact halo objects with a (semi) robotic telescope. In *Robotic Telescopes in the 1990s*, ed. A. V. Filippenko (San Francisco: Astronomical Society of the Pacific), pp. 193–202.

Allen, C. W. 1973. *Astrophysical Quantities* (London: Athlone Press).

Benedix, G. K., McFadden, L. A., Morrow, E. M., and Fomenkova, M. N. 1992. Bias correction factors for near-Earth asteroids. In *Asteroids, Comets, Meteors 1991*, eds. A. W. Harris and E. Bowell (Houston: Lunar and Planetary Inst.), pp. 65–68.

Bowell, E. 1992. 1979 VA = Comet Wilson-Harrington (1949 III). *IAU Circ.* Nos. 5585 and 5586.

Bowell, E., Hapke, B., Domingue, D., Lumme, K., Peltoniemi, J., and Harris, A. W. 1989. Application of photometric models to asteroids. In *Asteroids II*, eds. R. P. Binzel, T. Gehrels and M. S. Matthews (Tucson: Univ. of Arizona Press), pp. 298–315.

Bowell, E., Skiff, B. A., Wasserman, L. H., and Russell, K. S. 1991. Orbital information from asteroid motion vectors. In *Proceedings of Asteroids, Meteors, Comets III*, eds. C.-I. Lagerkvist, H. Rickman, B. A. Lindblad and M. Lindgren (Uppsala: Uppsala University), pp. 19–24.

Chodas, P. W. 1993. Estimating the impact probability of a minor planet with the Earth. *Bull. Amer. Astron. Soc.* 25:1226 (abstract).

Davis, D. R., Friedlander, A. L., and Jones, T. D. 1993. Role of near-Earth asteroids in the Space Exploration Initiative. In *Resources of Near-Earth Space*, eds. J. Lewis, M. S. Matthews and M. L. Guerrieri (Tucson: Univ. of Arizona Press), pp. 449–472.

Drummond, J., Rabinowitz, D., and Hoffmann, M. 1993. On the search for near-Earth asteroids. In *Resources of Near-Earth Space*, eds. J. Lewis, M. S. Matthews and M. L. Guerrieri (Tucson: Univ. of Arizona Press), pp. 449–472.

Everhart, E. 1967*a*. Comet discoveries and observational selection. *Astron. J.* 72:716–726.

Everhart, E. 1967*b*. Intrinsic distributions of cometary perihelia and magnitudes. *Astron. J.* 72:1002–1011.

Fernández, J. A., and Ip, W.-H. 1991. Statistical and evolutionary aspects of cometary orbits. In *Comets in the Post-Halley Era*, eds. R. L. Newburn, Jr., M. Neugebauer and J. Rahe (Dordrecht: Kluwer), pp. 487–535.

Gehrels, T. 1991. Scanning with charge-coupled devices. *Space Sci. Rev.* 58:347–375.

Harris, A. W., and Young, J. W. 1988. Observations of asteroid phase relations. *Bull. Amer. Astron. Soc.* 20:865 (abstract).

Luu, J., and Jewitt, D. 1989. On the relative numbers of C types and S types among near-Earth asteroids. *Astron. J.* 98:1905–1911.

Marsden, B. G. 1991. The computation of orbits in indeterminate and uncertain cases. *Astron. J.* 102:1539–1552.

Milani, A., Carpino, M., Hahn, G., and Nobili, A. M. 1989. Dynamics of planet-crossing asteroids: Classes of orbit behavior (Project SPACEGUARD). *Icarus* 78:212–269.

Morrison, D., ed. 1992. *The Spaceguard Survey: Report of the NASA International Near-Earth-Object Detection Workshop* (Pasadena: Jet Propulsion Laboratory).

Moynihan, P., and Sykes, M. 1993. Asteroid motions. *Bull. Amer. Astron. Soc.* 25:1118 (abstract).

Muinonen, K., and Bowell, E. 1993*a*. Asteroid orbit determination using Bayesian probabilities. *Icarus* 104:255–279.

Muinonen, K., and Bowell, E. 1993*b*. Collision probability for Earth-crossing asteroids on stochastic orbits. *Bull. Amer. Astron. Soc.* 25:1116 (abstract).

Muinonen, K., Bowell, E., and Lumme, K. 1994*a*. Interrelating asteroid size, absolute magnitude, and geometric albedo distributions. *Astron. Astrophys.* 42:302.

Muinonen, K., Bowell, E., and Wasserman, L. 1994*b*. Orbital uncertainties of single-apparition asteroids. *Planet. Space Sci.*, 42:307.

Ostro, S. J., Campbell, D. B., Chandler, J. F., Shapiro, I. I., Hine, A. A., Velez, R., Jurgens, R. F., Rosema, K. D., Winkler, R., and Yeomans, D. K. 1991. Asteroid radar astrometry. *Astron. J.* 102:1490–1501.

Rabinowitz, D. L. 1991. Detection of Earth-approaching asteroids in near real time. *Astron. J.* 101:1518–1529.

Rabinowitz, D. L. 1993. The size distribution of the Earth-approaching asteroids. *Astrophys. J.* 407:412–427.

Rabinowitz, D. L., Gehrels, T., Scotti, J. V., McMillan, R. S., Perry, M. L., Wisniewski, W., Larson, S. M., Howell, E. S., and Mueller, B. E. A. 1993. Evidence for a near-Earth asteroid belt. *Nature* 363:704–706.

Shoemaker, E. M., and Wolfe, R. F. 1982. Cratering time scales for the Galilean satellites. In *Satellites of Jupiter*, ed. D. Morrison (Tucson: Univ. of Arizona Press), pp. 277–339.

Shoemaker, E. M., Wolfe, R. F., and Shoemaker, C. S. 1990. Asteroid and comet flux in the neighborhood of Earth. In *Global Catastrophes in Earth History*, eds. V. L. Sharpton and P. D. Ward, Geological Soc. of America Special Paper 247 (Boulder: Geological Soc. of America), pp. 155–170.

Sitarski, G. 1968. Approaches of the parabolic comets to the outer planets. *Acta*

Astron. 18:171–195.

Tholen, D. J., and Barucci, M. A. 1989. Asteroid taxonomy. In *Asteroids II*, eds. R. P. Binzel, T. Gehrels and M. S. Matthews (Tucson: Univ. of Arizona Press), pp. 298–315.

Weissman, P. R. 1991. Dynamical history of the Oort cloud. In *Comets in the Post-Halley Era*, eds. R. L. Newburn, Jr., M. Neugebauer and J. Rahe (Dordrecht: Kluwer), pp. 463–486.

DETECTION OF METEOROID IMPACTS BY OPTICAL SENSORS IN EARTH ORBIT

EDWARD TAGLIAFERRI
E. T. Space Systems

RICHARD SPALDING and CLIFF JACOBS
Sandia National Laboratories

SIMON P. WORDEN
Ballistic Missile Defense Organization

and

ADAM ERLICH
Comprehensive Technologies International

Between 1975 and 1992, spacebased infrared sensors operated by the U. S. Department of Defense (DoD) have detected 136 impacts of meteoroids worldwide. Three of these events were also detected by the Department of Energy (DoE) space-based visible wavelength sensors with very high time and intensity resolutions. All of the impacting objects detected appeared to have deposited the bulk of their kinetic energy at high altitudes while undergoing explosive disintegration. These detonations exhibit many of the characteristics of nuclear explosions, and therefore are of concern to agencies of the U. S. Government charged with detecting violations of nuclear proliferation treaties and monitoring development of nuclear weapons. Also of concern is that during times of international tension the misinterpretation of a meteoroid detonation as a nuclear explosion could result in an escalation of hostilities. Space-based sensor systems operated by the DoD and the DoE are described which are capable of detecting meteoroid impacts worldwide in the visible and infrared; the possibilities for future "dual use" sensors are also discussed.

I. INTRODUCTION

A. Background

Earth, in its orbit around the Sun, is continually encountering other bodies whose orbits cross the Earth's. They range in size from dust-like particles, the impacts of which into the atmosphere are too small and too numerous to count, to extremely infrequent but potentially catastrophic impacts of objects several kilometers in diameter. These larger objects include asteroids with Earth-crossing orbits (Apollos, Atens and Amors), and comets. Studies of the size of impactors vs the frequency of impacts have been made by examining impact craters on the surfaces of the Moon and Mars, and to a lesser extent, of the Earth itself. These studies show that, at least over geological time scales, the rate of impacts of objects is inversely proportional to their size. This is shown in Fig. 1, which presents the frequency of impact of objects as a function of the equivalent kinetic energy of the impactor (Shoemaker 1983). Note that here we include in the definition of "impact" objects that enter the Earth's atmosphere but do not survive to reach the surface; it turns out that most impacting objects are in this category. It should also be noted that there is no physical law which dictates such a frequency dependence, nor that rates should be constant over time. In fact, the probable mechanisms which cause most objects to end up in Earth-crossing orbits, such as asteroid-asteroid collisions or the close encounter of a comet or asteroid with a large planet, are likely to produce swarms of bodies resulting in rapid short-term fluctuations in the impact rate. There is also a considerable body of evidence that indicates that the orbits of some large comets are strewn with a trail of debris which the Earth encounters on a periodic basis; this is clearly evident in, for example, the meteor showers which occur regularly in the summer and early winter months in the northern hemisphere.

It should be pointed out that, because of the scanning nature of the infrared sensors and the manner in which the satellite data were collected, the true number of events was at least 10 times what we report here. The objects we observed exhibited energies of approximately 10 Kt equivalent down to something under one Kt; therefore, our observations indicate a much higher rate of impact of objects in this size range than indicated in Fig. 1. In this regard, they support the Spacewatch observations reported in the Chapter by Rabinowitz et al., which reports a much higher flux of objects in this size range in the near-Earth vicinity than would be expected on the basis of Fig. 1.

In this chapter we will discuss how satellite-based sensor systems can be used to validate such flux models, and perhaps even lead to determination of some of the characteristics of the impacting bodies. We will also discuss another attribute of impacting objects which is of military and political signif-icance. As can be seen from Fig. 1, objects with energies of up to 20 kilotons (KT) of TNT are predicted to impact the Earth once a year or so. This is the amount of energy normally associated with the detonation of nuclear weapons. In fact, the impact of such an object can easily be misinterpreted as a nuclear detonation, and such misinterpretations can lead to unfortunate repercussions.

B. Strategic Implications of Misinterpreted Meteoroid Impacts

On 1 October 1990, there was an explosion with an energy in excess of one kiloton of TNT that occurred 30 km above the Central Pacific Ocean (8°N, 142°E). This event, which is more fully described below, was observed by sensors on U. S. Department of Defense satellites. The information from these satellites was analyzed, and eventually, this event was determined to have been caused by what was probably a stony, 100 ton asteroid impacting

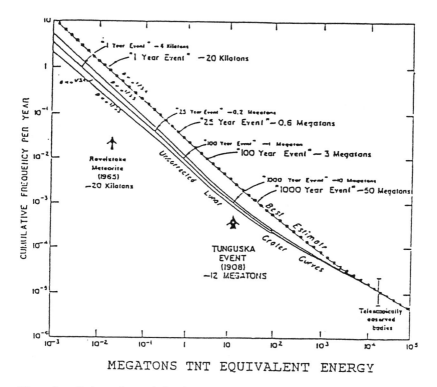

MEGATONS TNT EQUIVALENT ENERGY

Figure 1.   Estimated cumulative frequency distribution of kinetic energy of bodies colliding with the Earth.

the atmosphere (Jacobs and Spalding 1993). The data from this event were of a type which are not routinely processed by the operators of the satellite system. Flash data of this type is usually ignored if there is no crisis (Sandia National Laboratories 1992). In fact, according to an average calculated by Tagliaferri (1993), based upon DoD satellite information roughly eight asteroid detonations occur per year, but they are almost never actively detected nor processed.

The Central Pacific asteroid detonation was originally collected as a potential nuclear event, and it took several months, using the most sophisticated sensors and algorithms available, to determine the detonation's true source. This suggests that developing nations and potential combatants worldwide, with considerably less sophisticated equipment, might potentially misidentify one of these detonations as a nuclear attack and retaliate against the country's most likely aggressor. Therefore, the false identification of an asteroid detonation clearly could have global security implications.

C. Middle Eastern Scenario

The asteroid event of October 1990 is of particular significance because its date places it chronologically in the midst of the UN Desert Shield Operation.

If on October 1, 1990 the asteroid had detonated over Baghdad, the capital city of one of the belligerents in that conflict, the explosion would have had an equivalent luminosity to the Sun, and for several seconds, would have lit up an area hundreds of kilometers across. A clearly audible shock would have been heard on the ground and windows would have rattled throughout the area (this scenario assumes no inversion layer, which would have focused and enhanced ground effects). Iraq, in the midst of the war, might have assumed that one of its antagonists had launched a nuclear attack or a warning shot. It could conceivably have responded to the perceived attack by firing ballistic missiles (which it had in its arsenal at the time), at the suspected enemy.

Alternatively, in such a situation Israel could have interpreted the asteroid detonation as an Iraqi nuclear test and pre-emptively attacked Iraq's strikeforce (Tygiel 1991), just as they had pre-emptively attacked Arab forces during the 1968 Arab-Israeli War (Nagler 1990). Ultimately, the misidentification of the asteroid detonation might have resulted in large-scale warfare, which could have escalated to a global conflagration.

D. Discussion and Summary

During the height of the Cold War, the U. S. and U. S. S. R. were two highly stable protagonists with similar hair-trigger deterrent forces. Each expended considerable efforts to ensure that both man-made and natural occurrences were not misinterpreted as nuclear missile launches. For example, the North American Air Defense Command and its successor, the United States Space Command, have as a primary task the tracking of space debris in Earth orbit and the prediction of when and where orbital decay will result in re- entering space debris (spent rocket bodies, satellite fragments, etc.). The identification is conducted in order to differentiate the debris from an incoming missile. If the misidentification of space debris is worrisome, it seems that mistaking an asteroid detonation for a nuclear explosion would be even more threatening. Therefore, it seems reasonable and prudent that a similar effort on addressing asteroid impacts should be expended.

A comprehensive method for unambiguously and rapidly discriminating asteroid strikes from nuclear events is necessary. Moreover, a warning system with international legitimacy capable of distributing authoritative information to governments worldwide in near real-time is essential. Disseminating data which are considered credible by the international community could be a problem. Indeed, the urgent insistence by the United States that an asteroid event *is not* a nuclear explosion may actually serve to convince a regime that is not friendly with the U. S., that the explosion is a nuclear blast. Ultimately, international cooperation is essential for creating international legitimacy for the system. If several nations are involved in the system's deployment and operation, all nations could be more confident that the data they receive is credible.

Fortunately the framework for an internationally legitimate warning system is already being discussed. In June 1992, Presidents Bush and Yeltsin agreed in principle to cooperate on a system which was originally proposed by President Yeltsin to provide global protection against ballistic missile attack. At the heart of such a system is a distributed array of space-based sensors. The data from these sensors would be analyzed and potential attack warnings, assessments, and detailed target data would be disseminated to individual nations in near real-time. These nations could then use the sensor data in conjunction with indigenous anti-missile systems to protect their respective countries. The system that is envisioned would collect data similar to the information necessary to discriminate between asteroid impacts and nuclear detonations. Thus, only minor modifications would be necessary to provide the system with dual functionality. These minor additions would include algorithms capable of using discriminating features such as temperature, spectral characteristics, and time development of the fireball. At the same time, the data distribution network would have international legitimacy, thus permitting the timely dissemination of authoritative and credible information.

II. UTILITY OF SATELLITE-BASED OBSERVATIONS

Past efforts to improve the estimates of rate of impact include networks of human observers physically counting meteors, and wide-angle cameras (such as the Prairie Network) viewing the night skies. These efforts have helped to define various attributes of impacting objects, such as their velocities, trajectories, ablation rates, relative sizes, and altitudes of deepest penetration. However, because a network can cover only a small portion of the sky, can only operate at night, and is subject to the vagaries of weather, even the combined efforts of several such networks can observe only a small fraction of the events actually occurring. Add to this the fact that most of the Earth's surface is covered with water or is relatively inhospitable, and it is clear that such methods leave much to be desired, especially for observing the relatively infrequent larger events.

It is here that satellite-based sensors can make a significant contribution. A relatively small number of satellites in high altitude orbits (20,000 km or higher) can provide coverage of most of the Earth's surface. Because most impacting objects deposit their energy at relatively high altitudes above the ground, nearly all of the detectable signals are emitted above cloud tops, favorable to detection by satellite-based sensors. Further, satellite observations are not limited to night time; hence it is possible to have essentially continuous, day/night, all weather observations of the entire surface of the Earth.

In addition to aiding the understanding of impact rates, the detection and reporting in real time of meteoroid impacts, particularly larger ones, can also be of considerable importance in averting or minimizing emergencies that might otherwise result from their occurrence. For example, the impact of a sizeable object into an ocean can generate a tidal wave, or tsunami, which could inundate coastal areas thousands of miles away within hours and cause significant loss of life. Satellite monitoring may be the only means for reliably sensing and interpreting such an event. With appropriate reporting channels, it is conceivable that warnings could be broadcast early enough to allow evacuations to be initiated in threatened areas. As discussed above, the misinterpretation of a meteoroid impact as the detonation of a nuclear weapon could also have serious consequences. Finally, as will be explained more fully below, it may be prudent to divert aircraft away from an impact region to avoid encountering the debris cloud.

III. OBSERVABLES

A hypervelocity body entering the Earth's atmosphere produces a variety of remotely observable phenomena through conversion of its kinetic energy into other forms of energy. For example, high in the atmosphere the interaction of the impacting body with the air produces an ionized trail which reflects radio waves and is thus amenable to detection by radar. As the body penetrates deeper, air friction increases, heating its surface and causing material to be ablated from it at an ever-increasing rate. The heated body, ablated material, and the wake of heated air all emit infrared radiation which is detectable with infrared sensors. As the energy loss rate increases, the body surface and the impacted air are heated to incandescence, at which point they are detectable with visible light sensors (such as the human eye). As temperatures increase still further, ultraviolet radiation will be emitted which is detectable by ultraviolet sensors. Very sensitive ultraviolet sensors may also be able to observe the ionization decay emissions from the body's wake, because these emissions are expected to have line structures extending into the ultraviolet region.

With the possible exception of solid iron bodies, the larger bodies entering the atmosphere do not have the material strength to retain their structural integrity in the face of the extreme aerodynamic pressures involved in hypervelocity penetration to low altitudes. As a consequence, as pressure builds on the leading surfaces of the body, the stresses induced eventually exceed the cohesive strength of the body, and it breaks up. The breakup of the body rapidly increases its effective frontal area, abruptly increasing slowing. The result is literally an explosion, as the kinetic energy is quickly converted into heat, light, and shock. A fireball is produced which emits an intense flash of light, which, combined with the mechanical shock energy imparted to the air, dominates the energetics of the event. This fireball, which even for a relatively small body may be as intense as that produced by a nuclear explosion, can easily be detected and measured with satellite borne sensors.

IV. EXISTING SATELLITE SENSOR SYSTEMS

At any given time, there are literally dozens of satellites in orbit around the Earth with downward (Earthward) looking infrared, visible and ultraviolet

light sensors. In addition, there are at least a few satellites with radars, performing such tasks as terrain mapping, ice mapping, sea state determination,
and the tracking of ship movements. Virtually all of these sensors have the
sensitivity to detect the intense emissions or the trails produced by an entering
object. Why is it then, that so few reports have been generated reporting such
observations?

The primary reason is that these sensors are designed and operated for
other purposes and are thus tuned to best perform their intended mission, not
meteoroid detection. Given the sensitivity of the various sensors available
in wavelength areas of interest, trade-offs have been made between satellite
altitude, sensor instantaneous field of view and site revisit or scene scan time
requirements. For example, a multi-spectral imaging sensor flown on the
Landsat D satellites, called the Thematic Mapper, orbits the Earth at 700 km
altitude. Its visible sensor has an instantaneous field of view of 43×688 micro-
radians, corresponding to a spot size on the Earth's surface of approximately
30 m by 480 m. As the satellite moves along its track, the sensor scans the
spot back and forth across the track seven times a second, imaging a swath
185 km wide. With this scan geometry, it takes about 16 days to image the
entire Earth. However, the types of measurements that the Thematic Mapper
is tasked to carry out, such as determining crop yields or deforestation rates,
concern phenomena which do not change much in that time and hence this
revisit time is adequate for the intended mission. Detecting the fireball of an
impacting object is a different matter. During the fireball phase of a meteoroid
impact (approximately 1 s), the sensor will make 14 scans of 185 km each;
it will also have moved in orbit (along the track) approximately 7 km. The
fireball is not really a ball, but radiates from a volume shaped more like a
sausage, over a typical length of 10 km. Given a random orientation of the
track within the sensor field of view, the probability that the sensor will "see"
the impact of a meteoroid is approximately 1 in 120,000 (i.e., the ratio of
the area in which the meteoroid could impact and be detected to the area of
the Earth). However, even if the sensor did scan a fireball, the intensity of
the event would momentarily saturate the sensor, resulting in the observation
probably being discarded as anomalous data not useful to the imaging mission.
For much the same set of reasons, similar low probabilities will apply to most
of the other imaging systems, such as found on meteorological satellites,
ozone mappers, etc.

There are two space-borne sensor systems, however, which have demonstrated a useful combination of spectral response, sensitivity and field of view
to detect and make some measurements of impacting meteoroids. These
sensors are operated by the DoD for military missions. One of these is the
nuclear burst monitoring system, which operates in the visible region. The
other is a scanning infrared sensor. Both of these sensor systems are designed
to provide essentially worldwide surveillance for transient events. To accomplish their primary missions, these sensors are connected to large, high-speed,
real-time computer systems operating with sophisticated algorithms designed

to rapidly and automatically extract events of interest from the data stream being transmitted by the sensors to the ground. At present, these algorithms are tuned to filter out all but the events of significance to the mission operators. However, the infrared sensor has detected a number of meteoroid impacts, and fortunately, records of some of these have been kept. Although it is known that the visible wavelength sensor system has also observed fireballs associated with impacts, the records for this system are not as complete and as of the date of this writing, only three such detections have been preserved.

These two sensor systems will be discussed in the following sections. Although it will not be possible to disclose all of the details of the design and operation of these systems, it will be possible to show examples of the detections to date and to discuss some of the design considerations which would make future systems more effective.

V. TRANSIENT RADIOMETRY

Transient radiometry, as defined here, is a technique whereby relatively low intensity, short duration events or events with rapidly changing signatures are extracted from a much larger but slowly varying background. Application of this technique has permitted the design of sensor systems wherein a single sensor aboard a high altitude satellite viewing very nearly an entire hemisphere can routinely detect localized, short duration events such as lightning and meteoroid impacts. Because many events of military or political significance, such as launches of intercontinental ballistic missiles or nuclear weapon detonations fall in this category, sensors of this type are already deployed on enough DoD satellites to provide essentially 100% coverage of the Earth's surface 24 hours a day, seven days a week.

A. Visible Radiometry

The extraction of the rapidly changing signals of interest from the slowly changing background is performed by the simple expedient of using a high pass filter at the satellite. The filter effectively removes the extremely large DC component and greatly reduces the amplitude of the slowly varying background. By placing the sensor system on a satellite platform which is reasonably stable, the remaining background can be predicted well enough over the short times that events of interest are occurring that it can be subtracted during ground processing of the signals. The judicious application of thresholds can then permit the automatic extraction of the event.

The ability to do this depends on several attributes of the sensor, the satellite behavior and the behavior of natural phenomena. First the Sun is very well behaved over short time intervals in the visible region. Nearly all of the background light from the Earth is diffusely scattered sunlight. For a fully sunlit Earth disk with an average albedo of 0.3, the reflected power is approximately 10^{16} watts per steradian. Signals of interest from a meteoroid impact range from 10^9 watts per steradian upward. Hence, the solar intensity over short time scales needs to be stable enough to permit signals 10^7 times smaller than the maximum background to be extracted. If, over short time periods, solar intensity fluctuations of this order or larger were a common occurrence, this method of detecting transient phenomena would be severely limited.

Because of satellite orbital motion, sunlit portions of the Earth viewed by the sensor will be constantly changing, but for the high altitude satellites we are discussing here, these rates of change are slow, permitting reliable prediction and subtraction of the background. Weather-induced changes are also relatively slow, with the resultant variations in reflected sunlight being amenable to correction.

Probably the largest transient signals experienced by radiometers arise from sunglint. Glint occurs when the Sun-Earth-satellite geometry is such that the solar reflection point moves over particular regions of the Earth's surface. Reflections are strongest in regions of glassy smooth water surfaces, such as calm seas, slow moving rivers, and swampy areas. Considering the case of a satellite in geosynchronous equatorial orbit (approximately 36,000 km above the equator), the specular point (and therefore the glint region) moves from east to west across the sensors field of view at about half the Earth's rotation rate. This gives it an apparent velocity, when near the equator, of approximately 0.25 km s^{-1}. A minimum glint area diameter will be about 30 km; thus the specular intensity can go from zero to maximum in about two minutes. Because the glint intensity may be 10^6 to 10^7 times as strong as the impact signals of interest, detection of impacts using single detector element radiometry may be significantly hampered under such conditions.

In the absence of limitations to the detection and measurement of the fireball due to any of the above variations in background, photon noise associated with the background itself is a limiting factor. This manifests itself as shot noise in the sensor—in this case a silicon photovoltaic detector—which is proportional to the square root of the incident background power. It can be shown that the signal power limit for a high altitude observation is given by

$$P_{\text{noise rms}} = \sqrt{\frac{2e\Delta f}{RA_s} P_{bgd} D^2} \, (\text{w/sr}) \tag{1}$$

where $e = 1.6 \times 10^{-19}$ coulombs; Δf = signal bandwidth; D = satellite altitude; R = detector responsivity for solar spectrum background (0.25 amp/w); A_s = detector aperture area; and $P_{bgd} = 10^{16}$ w/sr for geostrationary orbit. This reduces to $P_{\text{noise}} \approx 4 \times 10^8 \sqrt{\frac{\Delta f}{A_s}}$ w/sr for a sensor in geostationary orbit. Figure 2 presents an example of the detector responsivity for a silicon detector.

Finally, the satellite on which the sensors reside needs to be well behaved so that satellite pointing drift and variations in rotation rate occur smoothly over relatively long time periods. This is necessary because, if the satellite were subject to random disturbances which caused the Earth to "jiggle" unpredictably in the field of view of the sensor over time periods comparable

Typical Si Responsivity

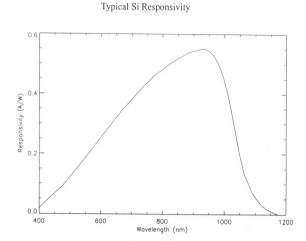

Figure 2.   Typical silicon detector responsivity (A/W) as a function of wavelength.

to the events we are trying to detect, the transient signals created would be difficult to separate from the event signals.

B. Optical Locators

Because it is militarily important to determine the location of any detected nuclear detonations, companion burst-locating sensors have been included in the payloads of some of the radiometer-equipped satellites. The locators must, by necessity, scan rapidly and therefore do not achieve the same level of sensitivity as the visible radiometers. Consequently, only larger events, such as the 15 April 1988 and the 1 October 1990 events discussed below, can be detected and located by the burst locators. For smaller events, one must rely on other systems to provide location information. For some fraction of these smaller events, the infrared sensors discussed below can provide the needed location information. When two or more burst locators detect a given event, it is also possible to derive the altitude of the fireball, and it is in principle possible to estimate both the body's velocity and track vector. For example, for the 1 October 1990 event the burst locators gave an estimated altitude of 30 km. Algorithms for unfolding velocity/track information have not yet been created, but there is no a priori reason why they cannot be developed.

C. Visible Fireball Lightcurves

In Figs. 3, 4 and 5, time-intensity plots of three different fireball events collected by satellite visible light transient radiometers are shown. Peak radiated powers of the three events were calculated to be 1.4×10^{12} watts per steradian for the 15 April 1988 event, 3.5×10^{11} watts per steradian for the 1 October 1990 event, and 6.5×10^{10} watts per steradian for the event of 4 October 1991. The peak visual magnitudes of these three events at the standard distance of 100 km are approximately -24.3, -22.8 and -21.0. In these estimates, the source (fireball) power was derived using the following formula:

$$P_{\text{source}} = [I/R] \times D^2 \qquad (2)$$

in watts per steradian, where I = signal current in amps per unit area of sensor aperture, R = measured responsivity of the silicon sensor to a 6000 K blackbody source (0.25 amps /watt) and D = satellite to event distance.

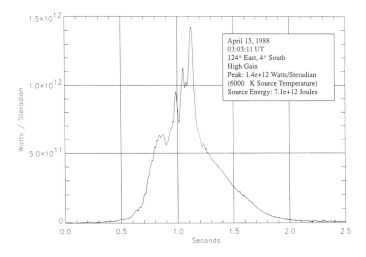

April 15, 1988
03:03:11 UT
124° East, 4° South
High Gain
Peak: 1.4e+12 Watts/Steradian
(6000 K Source Temperature)
Source Energy: 7.1e+12 Joules

Figure 3.   Visible lightcurve from the meteoroid impact of April 15, 1988. Presented is the visible light intensity (watts/steradian) vs time (seconds).

It is worth noting that estimates of peak radiated power using 6000 K as a source temperature are probably underestimates of the actual radiated power. For events with a source temperature of 6000 K, the response of silicon detectors (Fig. 2) is nearly optimum. It seems likely, however, that the fireball temperatures are really much hotter, and the silicon sensor response to these hotter radiations will be less. If this is true, the sources were considerably more intense than we are reporting here.

To convey something of the magnitude of these events, each would have had the same apparent brightness as the Sun when viewed by an observer 32, 19, and 8 km, respectively, from the events. Total radiated energies (again assuming a constant source temperature of 6000 K for the period of observation) were 7.1×10^{12}, 2.5×10^{12} and 5.5×10^{11} joules, or about the amount of energy radiated by atmospheric nuclear detonations of 5, 1.8 and 0.4 kilotons, respectively.

D. Infrared Radiometry

Existing infrared satellite sensors suited to detection of impact events operate differently than their visible light radiometer counterparts. Instead of staring

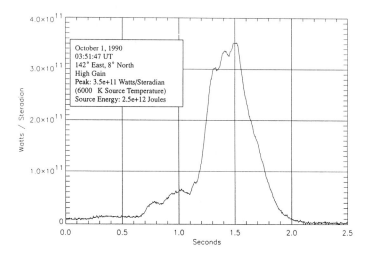

Figure 4.   Visible lightcurve from the meteoroid impact of October 1, 1990. Presented is the visible light intensity (watts/steradian) vs time (seconds).

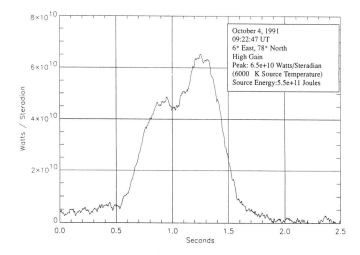

Figure 5.   Visible lightcurve from the meteoroid impact of October 4, 1991. Presented is the visible light intensity (watts/steradian) vs time (seconds).

at the entire Earth disk continuously, these sensors utilize arrays of detectors with relatively small instantaneous fields of view to scan repetitively the full Earth disk. Because the instantaneous fields of view of the detectors are much smaller, they are less affected by background. At the same time, the sensor collection apertures are significantly larger, so they can respond to much weaker sources than the visible wavelength radiometers. (In fact, infrared detections could be used by the operators of the visible light sensor to go back to their data records to retrieve what for that sensor are very low intensity level events, which otherwise would have been discarded as being too close to noise to be considered credible events.) However, this greater sensitivity does not necessarily translate into a better capability for detection and measurement of impact events. This is because the scan revisit intervals are designed for other purposes and are less than optimum for meteoroid impact detection.

As the visible radiometer waveforms in Figs. 3, 4 and 5 show, the duration of the explosive phase of even a fairly substantial impact event is quite short, lasting less than about two seconds overall. Although the event duration may appear somewhat longer when viewed in the infrared, it would still be necessary for a scanning sensor to revisit every second or so to have a high probability of scanning the fireball while it is brightest. For smaller events, the scan rate would have to be even higher. Current systems do not scan this rapidly; consequently, there is a good chance that a given sensor will miss the fireball phase of an event entirely. The picture is not all bleak, however, in that the probability that it will see some other portion of the impact event is much better. For example, prior to disintegration, the impacting body is undergoing rapid ablation, and there is therefore a fairly strong infrared emission taking place. After the flash, the heated mass of air and debris take some time to cool down, during which it is still detectable in the infrared. Finally, the trail and cloud of debris left in the atmosphere by the impact, when illuminated by sunlight, creates a distinct, localized source detectable in the infrared. We have observed the debris cloud from several events, and find they can persist for relatively long periods of time, particularly when they are at high altitudes. For example, we have one event in the data base where the debris cloud was tracked for two hours. Thus, the sudden appearance and subsequent persistence of a high altitude cloud are tell-tale indicators that a meteoroid impact has occurred. Of course, impacts which occur on the night side of the planet will not have a debris cloud illuminated by the Sun, and therefore will not produce a persistent infrared source; we also find that not all daytime detections have a persistent cloud associated with them.

From the above discussion we see that the slow scan speed has two consequences. First, the count of impacts is going to be low compared to the real number of events because some events will be missed entirely; second, the peak intensity recorded for the event cannot reliably be assigned to the time of peak energy release from the impactor. Hence, it in general will not be possible to determine the energy released by the impactor reliably using just the infrared data alone. However, the time and location of the event will be known to considerable precision.

Sensor noise is also an issue for infrared sensors. Typically, these sensors operate best (lowest noise, highest sensitivity) at low temperatures. With present-day technology, it is practical to maintain sensor temperatures of 200 K for time periods of up to several years on orbit using a single stage passive radiator. The sensitivity of infrared sensors is generally discussed in terms of the D^* of the sensor. This is given by the expression

$$D^* = (A_d \Delta f)^{\frac{1}{2}} \frac{V_s / V_N}{\Phi_{s,\lambda} \Delta \lambda} \text{cm Hz}^{\frac{1}{2}} \text{w}^{-1} \qquad (3)$$

where A_d = detector area; Δf = electronic bandwidth; V_s = detector signal voltage; V_n = detector noise voltage; and $\Phi_{s,\lambda}$ = incident signal power.

Curves of typical sensitivity of PbS detectors are presented in Fig. 6. Newer detectors utilizing HgCdTe are more sensitive than PbS, but require lower temperatures to operate effectively. This in turn requires more complex cooling systems, such as two stage passive coolers. However, the next generation of infrared satellite sensors will most likely be of this type.

Dealing with the infrared background is a significant factor when processing infrared sensor data. The primary source of background is reflections of sunlight. These are both from specular reflections, as discussed above for the visible sensors and infrared radiation scattered from cloud tops, snow fields, high sea states and abrupt discontinuities in terrain such as at land-sea interfaces. Because of the sensitivity of the infrared sensors, solar background is a more severe problem than for visible radiometers. For the infrared sensor, the area from which specular reflections occur can reach hundreds of km across. Fortunately, specular background is a geometric effect and is predictable. As a consequence, algorithms have been developed which effectively cope with background from that source. More troublesome are the random fluctuations in background, such as sunlight reflected from cloud tops which, except that they only occur over sunlit portions of the Earth disk, are unpredictable. Considerable effort has been spent in developing algorithms which operate in real time, and, surprisingly, cope rather successfully with this type of background as well.

E. Infrared Fireball Signatures

Since 1975 there have been 136 meteoroid impacts worldwide which were detected and recorded by the infrared sensors. As discussed above, the actual number of events which occurred during this time period was larger than this by some factor which, unfortunately, cannot be recovered. The nature of the operating environment of these sensors (as of this writing) is such that the data for events not related to the primary mission of the system have to be collected and recorded manually by system operators. Often, they are either involved in the primary mission or busy doing other tasks, and the events, although duly observed and reported by the sensor, are not recorded and therefore the data are lost. Also, as discussed above, the scan rate of the sensor is such that a significant fraction of the events will occur between scans, and therefore will be missed.

Nevertheless, these data represent a significant data base of observations, many of which have occurred over open ocean areas and therefore will not have been reported by the scientific community. In addition, the sensors can

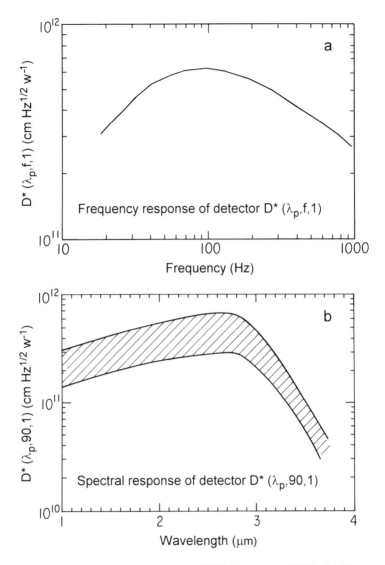

Figure 6.   (1) Typical sensitivity curves of PbS detectors at 193 K. (b) Frequency
response of detector D^* (figure after Wolf and Zissiz 1989).

just as effectively detect daytime events. In fact, as was discussed above and
will be amplified below, under some circumstances, daytime detections are
actually easier to see.

Figure 7 shows the geographic distribution of the events in the data base.
As would be expected, they are rather randomly distributed. In Fig. 8, the
number of events each month is shown.

Figures 9, 10 and 11 present the infrared intensity as a function of time

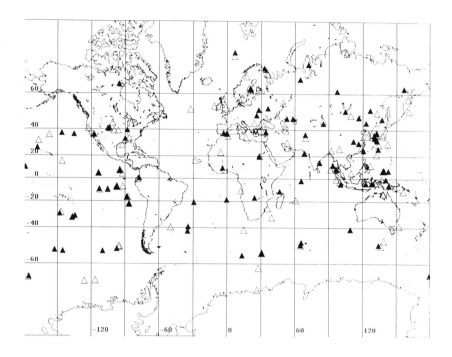

Figure 7.   Geographic distribution of meteoroid fireballs detected by infrared sensors.

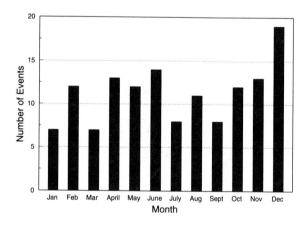

Figure 8.   Frequency of events per month between 1975 and 1992.

for three events selected from the data base. These three events were chosen because: (1) they are bright events; in each case the sensor must have scanned the fireball at close to the peak of the energy release of the event, because the detector elements sensing the event were all saturated; (2) they were detected by two satellites; this allowed a relatively accurate determination of the altitude of the explosive disintegration of the object; (3) they were all daytime events and show an attribute of daytime events (a persistent debris cloud detectable in the infrared) which is not seen in the nighttime events nor by the visible light sensor; and (4) the three events were also detected by the visible light sensors discussed above.

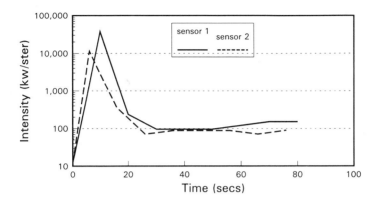

Figure 9.   Infrared time-intensity plot for event of April 15, 1988. Local time is 11:20; location 4° S, 124° E.

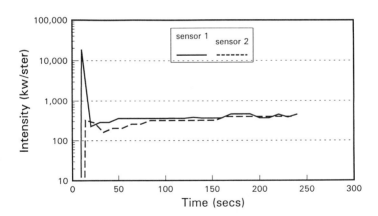

Figure 10.   Infrared time-intensity plot for event of October 1, 1990. Local time is 13:20; location 8° N, 142° E.

The first thing to note is that the burst of energy associated with the explosive disintegration of the impacting body is very short lived. We saw above from the visible lightcurves that the time of peak energy release was less than one second. Here, in the infrared data, we see that whereas one infrared sensor has a very high intensity, the second sensor, scanning just 3 or 4 seconds before or after the peak, registers intensities as much as 2 to 3

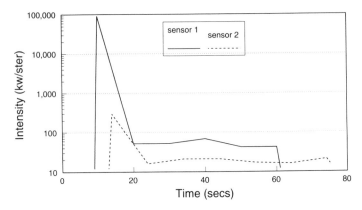

Figure 11.   Infrared time-intensity plot for event of October 4, 1991. Local time is 09:47; location 78° N, 6° E.

orders of magnitude down from the peak.

The second thing to note is the long tail due to sunglint off of the debris cloud. We have noted that in many cases, the debris cloud continues downward after the detonation of the main body of the meteoroid, and then slowly climbs again as the debris expands. This is shown in Fig. 12 for the event of 15 April 1988 (the data have been smoothed).

The persistence of the debris cloud after detonation raises some interesting possibilities; NASA, for example, has aircraft which are equipped to obtain dust samples (modified U-2 spy planes). Such samples would undoubtedly provide interesting clues as to the make up of the impacting object. On at least two occasions in the past, aircraft have successfully been employed to sample impact debris clouds. On 8 February 1969, an Air Force B57 collected debris particles from the Pueblito de Allende meteorite in Mexico, and on 4 January 1970, an Air Force plane collected debris samples from the Lost City, Oklahoma meteorite. These samples were distributed to and analyzed by several universities. It would appear that systematic sampling efforts would be an effective way to gather information on the makeup of impactors and may even permit determination of whether biological materials are introduced to the Earth through meteoroid impacts. Certainly, intercepting the debris before it has the possibility of being contaminated through contact with the ground or water should be important, and early notification of a large impact event is critical to obtaining good samples.

It is also possible that the debris cloud could pose a hazard to aircraft. As is evident in the April 1988 and October 1990 events, the amount of mass involved in larger impacts can be hundreds of tons and greater. That much material, dispersed into the atmosphere as ash-like or sand-like particles, is certain to present a hazard to aircraft for some period of time following the impact. One can expect the coarser grit to fall within hours. The finer, dust-

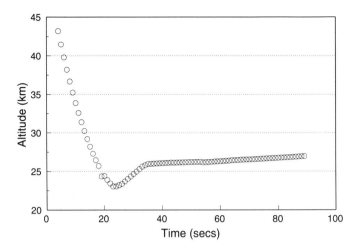

Figure 12.   Time-altitude history of debris cloud for event of April 15, 1988.

like debris though, can remain suspended, perhaps for days, before dissipating in density and altitude to where it no longer constitutes a hazard.

VI. FUTURE SATELLITE BASED SENSOR SYSTEMS

A. Planned Visible Light Systems

As of August 1993, an upgrade is being planned which will equip future global positioning system (GPS) satellites with visible light sensors capable of detecting and reporting meteoroid impacts. The GPS is a global navigation system developed by the DoD. It consists of a constellation of satellites in 12 hr orbits (20,200 km altitude) with an orbital configuration designed to provide visibility to at least 5 satellites at any time from any position on Earth. With appropriate equipment, users can determine their position in three dimensions in real time using GPS. The current GPS satellites also have instruments for detecting nuclear bursts. However, these instruments only report signals that meet a specific, rather narrow set of threshold criteria. Because the bulk of meteoroid impacts will generate signals that do not satisfy these requirements, they will routinely be discarded by the satellite. The planned GPS upgrade, called Block IIR, will have a new sensor system called the Global Verification and Location System (GVLS), from which radiometer background signals will be transmitted on a continuous basis. This will permit candidate transient signals to be extracted at ground stations in a manner similar to that used to obtain the waveforms shown above for the visible light curves in Figs.   3, 4 and 5.   Correlations of responses between satellites can then be used to achieve thresholds lower than are practical when using only a single sensor. This can be done, for example, by requiring that in order to be considered a candidate event, all sensor responses from the event must occur within a time window. Responses within the time window can then be correlated by peak intensity, pulse shape, and other signal attributes, with the time difference of arrival of the signals being used to locate the burst.

Funding for the GVLS is being provided by the DoE, with costs for integrating the sensors into the satellite payload being borne by the U. S. Air Force. Initial deployment of the Block IIR satellites is scheduled to begin in the 1996 time frame. Under the present DoD satellite launch philosophy, new satellites will be launched only as old satellites fail and need replacement. Hence, a complete constellation of Block IIR satellites may not exist until the year 2000.

There is in planning a follow-on to the Block IIR satellites, called Block IIF. These satellites are expected to have nuclear burst locator sensors as well. Discussions so far indicate sensors at least as capable as those on Block IIR, with a good probability that an imaging radiometer might be added to the Block IIR suite. Should this occur, better sensitivity and location information over that possible with the Block IIR sensors would naturally result.

One planned sensor system which undoubtedly will prove to have an extremely good capability for detecting and locating meteoroid impacts is the NASA Lightning Mapper. This instrument, developed by the Marshall Space Flight Center for the Geostationary Observational Environmental Satellites (GOES)(NEXT) weather satellite, is designed to monitor lightning activity over much of North America. Its 10-kilometer pixels and 2-millisecond scan interval should enable it to detect and locate impact events down to very small sizes (a few kilograms). Unfortunately, because of payload limitations which developed in the GOES program, the Lightning Mapper was dropped from the instrument complement for the scheduled launch. Re-instatement of the Lightning Mapper into some future GOES satellite is expected but not yet planned. The Lightning Mapper detects signals by performing scan-to-scan differencing on each pixel, accepting photons only within a 10 Å bandwidth centered on 7774 Å. Despite this narrow bandwidth and the fact that the differencing technique is optimized for short-pulse lightning and not for slower-developing impact signals, its sensitivity, and hence the number of impacts it may be expected to record, should be impressive.

There are no other new developments of which we are aware. However, with the current emphasis of "dual use" of DoD assets and the enormous capability of DoD satellite based sensors, with their attendant very sophisticated, dedicated data processing systems (most of which operate in real time), it is likely that increased use of these sensors for environmental monitoring purposes will be forthcoming. This will be especially important for the detection of transient events in remote locations such as meteoroid impacts, lightning, and volcanic eruptions.

B. Visible Light Sensor Systems Dedicated to Meteoroid Impact Detection

In all of the above discussions, the sensors involved were designed and are operated for purposes other than meteoroid impact detection. However, the design of sensor systems dedicated to meteoroid impact detection is not difficult. One relatively simple example of such a system is the placement of small, Earth-pointing video cameras on each of 3 or 4 geostationary TV relay satellites. The cameras need only weigh a few pounds, would have near-global monitoring capability, and use only a small fraction of the satellite's downlink capacity. Because of the inherent brightness of the fireballs associated with the impacts of larger objects, extracting the events and the location of the impact from the video signals on the ground should be a relatively trivial task. In fact, with standard cameras in use today, it would probably be necessary to reduce their sensitivity to better utilize their limited dynamic range.

Of course, with more investment more sophisticated sensors could be developed specially designed to detect a wider range of event sizes and to locate them more accurately. In addition, because most objects that we have observed to date appear to have detonated at altitudes of 30 km or so, the virtually unrestricted viewing from space permits accurate spectral measurements to be made over a wide range of wavelengths from the ultraviolet to the infrared. These would be useful in determining the temperatures of the fireballs, deriving information about the chemical makeup of the parent body, and perhaps thereby obtaining clues as to the type of object the parent body was.

C. Future Infrared Space Based Systems

There are plans to develop new space based infrared sensor systems for a variety of purposes. As with the presently deployed sensors, the new sensors will have the capability to detect impacting meteoroids. Present plans call for sensors with higher sensitivities and scan speeds, while maintaining the capability to monitor essentially the entire Earth continuously. Hopefully, the doctrine of "dual use" of DoD assets will apply to these new sensors as well and there will be opportunity afforded to allow some input to the design parameters of the sensors and to the design and operation of the software systems that will extract the desired information from the sensor data stream. The evolution of high speed computers now make it possible to perform environmental monitoring with the same system without any impact on the primary DoD missions.

Acknowledgments. The authors would like to thank The Aerospace Corporation and T. Stocker for their help in assembling the infrared event data base.

REFERENCES

Jacobs, C., and Spalding, R. 1993. Fireball observation by satellite-based Earth-monitoring optical sensors. Hazards Due to Comets and Asteroids, Jan. 4–9, Tucson, Ariz., Abstract book, p. 45.

Shoemaker, E. M. 1983. Asteroid and comet bombardment of the Earth. *Ann. Rev. Earth Planet. Sci.* 11:461–494.

Nagler, R. 1990. *Ballistic Missile Proliferation an Emerging Threat* (Arlington, Va.: System Planning Corp.), pp. 16 and 42.

Sandia National Laboratories. 1992. Proposal for the Detection of Asteroids Entering the Earth's Atmosphere Using Satellite Sensors. Unpublished.

Tagliaferri, E. 1993. Asteroid detection by space based sensors. Presented at Erice International Seminar on Planetary Emergencies, 17th Workshop: The Collision of an Asteroid or Comet with the Earth, April 28–May 4, Erice (Italy).

Tygiel, Y. 1991. Israel and the bomb. *The Nation* 252:191–192.

Wolf, W. L., and Zissiz, G. J., eds. 1985. *The Infrared Handbook*, rev. ed. (Ann Arbor, Mich.: Environmental Research Inst. of Michigan), pp. 11–51.

WARNING TIMES AND IMPACT PROBABILITIES
FOR LONG-PERIOD COMETS

B. G. MARSDEN
Harvard-Smithsonian Center for Astrophysics

and

D. I. STEEL
Anglo-Australian Observatory

The time taken by a long-period comet to move from its last opposition to possible Earth impact is considered as a function of the perihelion distance and inclination of the comet's orbit. Except for the possibility of an opposition of a comet in a low-inclination orbit immediately before impact, the lead time in the case of a pre-perihelion impact is found to range from 250 to 500 days. For a post-perihelion impact the time is generally shorter. In both cases there is a significant range of perihelion distances for which objects in orbits inclined steeply to the ecliptic would pass opposition near an ecliptic pole and thereby escape conventional searches, including the proposed Spaceguard Survey. This problem could be largely alleviated by extending the Spaceguard search region, especially in ecliptic longitude, but this would necessitate an increase in the number of search telescopes. A decision as to whether such an extension is worthwhile awaits a realistic assessment of the hazard posed by long-period comets as compared to short-period bodies; this assessment requires a much better knowledge of the masses of long-period comets. The mean impact probability per revolution for a theoretical distribution of long-period comets is 2 to 3×10^{-9}. Individual values are considerably larger for orbits that are close to the plane of the ecliptic or have perihelia near 1 AU, under which circumstances the impact probability may be 2 orders of magnitude higher. Because the distribution of inclinations i of long-period comets is expected to vary as $\sin i$, there are few such comets with $i < 10°$ or $i > 170°$. For a theoretical distribution that is uniform in perihelion distance q, the contribution to the Earth-impact hazard from comets with q near 1 AU is disproportionately high. The mean impact probability derived for the observed population of 411 long-period comets crossing the Earth's orbit is about 60% higher than the theoretical value, but reduces to near the latter if a handful of near-ecliptic, $q \approx 1$ AU comets is excluded.

I. INTRODUCTION

Long-period comets (LPCs) are defined to be those comets that do not completely orbit the Sun in less than 200 yr. The definition therefore includes comets that are not periodic at all and for which parabolic, and even hyperbolic, orbits have been computed. In particular, it includes comets that are making their first and only passages near the Sun from the Oort Cloud and

[221]

that are then perturbed entirely out of the solar system. The definition was adopted earlier in the present century for the entirely practical reason that it seemed unlikely that past observations of a comet having a period of more than 200 yr would be definitely identified. This reasoning is probably mis-guided: sooner or later, past records of such a comet will undoubtedly be found. Nevertheless, the definition is useful from the point of view of the possible hazard to the Earth. Two centuries seem to be as long as even the most farsighted individuals care to know of future disasters, as well as close to the limit for which any future asteroidal impacts can be computed with meaningful reliability.

With this 200-yr limit, the point about the LPCs is that their impacts on the Earth are completely unpredictable. The directions from which impacting LPCs come are also essentially random. It is true that the available cometary statistics of LPC orbits show some departures from randomness, but in view of the obvious extreme rarity of impacts such departures can be of no significance. Furthermore, although there are cases where LPCs share very similar orbits, the probability that two related LPCs will arrive at precisely the time of year necessary for a double impact during a 200-yr interval must be vanishingly small. The only well-populated set of related LPCs is the Kreutz sungrazing group, but the orientations of the Kreutz orbits are such that Earth impacts can not occur at present.

With regard to their potential hazard to mankind, not all LPCs are equal. Clearly, the brightness and the mass of a potential impactor are a significant consideration. In addition, the cometary orbit critically controls the prob-ability of a terrestrial collision, how long beforehand the comet might be discovered, and the velocity with which the impact would occur.

In the absence of demonstrated cases of Earth impact by an LPC, for an analysis of the above factors, it is sufficient to suppose that an LPC moves in a fixed parabolic orbit with the Sun at its focus, while the Earth is defined to move in a fixed circular orbit of radius 1 AU centered on the Sun and in the plane of the ecliptic. If the finite sizes of the Earth and the comet are ignored, impact can occur only if the comet is in the plane of the ecliptic when the comet is 1 AU from the Sun. Except in the unlikely event that the comet's orbit is entirely confined to that plane, there is some non-zero inclination i of the comet's orbit plane to the ecliptic, and at impact the comet must be crossing the ecliptic plane (i.e., have one of its nodes) at a distance of 1 AU from the Sun. By convention, when $0° \leq i < 90°$, the comet is said to be in a direct orbit, i.e., it travels about the Sun in the same general sense as the Earth; when $90° < i \leq 180°$ the orbit is termed retrograde, and the comet travels around the Sun in the opposite sense. The nodes are on the line (through the Sun) in which the two planes intersect, and they are measured in terms of the angle between this line and the line joining the positions of the Earth at the equinoxes. The longitude Ω of the ascending node is the angle in the ecliptic from the position of the Earth when autumn begins in the northern hemisphere to the node at which the comet is moving across

the ecliptic from south to north. The longitude of the descending node is
$\Omega\pm180°$, the value in the range $0°$ to $360°$ being specified. For a collision it is
also necessary, of course, for the longitude of the Earth (measured from the
same origin) to equal whichever of Ω or $\Omega\pm180°$ has the comet at 1 AU from
the Sun. The line of nodes also crosses the comet's orbit, the orientation of
which is measured in its orbit plane with respect to the line of apsides, which
connects the Sun and the comet's perihelion point, or the point in the orbit
that is closest to the Sun. The possibility of Earth impact obviously requires
that the comet's minimum distance from the Sun, its perihelion distance, is
$q<1$ AU (or <1.0167 AU, the Earth's aphelion distance, if the Earth's orbital
eccentricity is taken into account). It also restricts the angle between the
lines of nodes and apsides, specifically the argument of perihelion ω, which
is measured from the ascending node to the perihelion. For a given q, there is
then a fixed interval between the comet's relevant time of nodal passage and
its time T of perihelion passage. The position of the comet in its orbit at any
time is measured by the true anomaly, or the angle the object has traversed
from perihelion, and this is a function of q and the time difference from T.
Together with the orbital eccentricity e (equal to unity for a parabola), the
five quantities T, q, ω, Ω and i are taken to be the independent elements
that completely define the comet's orbit and allow the computation of the
comet's position at any time. If a comet really is periodic (as opposed to an
open-ended situation with the object perturbed out of the solar system in the
future), orbital precession may alter the nodal distances and rotate the lines
of apsides to make an impact possible on a future pass, but in terms of human
time scales we may consider LPCs as each having only one perihelion passage
on which a terrestrial impact may be possible.

II. DETECTION

The NASA Near-Earth Object Detection Workshop has proposed the establishment of the 'Spaceguard Survey' (Morrison 1992), a coordinated international network of six 2.5-m $f/2$ telescopes dedicated to the discovery,
confirmation and follow-up of asteroids and comets that could conceivably
be involved in devastating encounters with the Earth during the next couple
of centuries. In the Spaceguard report it was recommended that the searches
should be concentrated on the region of the sky centered on opposition (the
point in the sky directly opposite the Sun) and extending for a total of $60°$
in longitude along the ecliptic and for $\pm60°$ in ecliptic latitude. This Spaceguard search region (SSR) was selected to maximize the efficient discovery
of objects—both asteroids and comets—in direct orbits of low-to-moderate
inclination with revolution periods of a few years. With their short periods,
many of these objects get numerous opportunities to strike the Earth, and
the Spaceguard network could be used to monitor potential impactors on a
continuing basis. The telescopes would have a limiting magnitude of around
22, which is sufficient to follow most km-sized short-period objects out to

aphelion, where Kepler's second law ensures that they spend most of their time, at a sizable fraction of the distance to Jupiter.

In the Spaceguard report it was implied that this observing strategy would also be satisfactory for detecting potentially hazardous LPCs. Most estimates for the contribution of LPCs to the flux of Earth impactors put that value at 5 to 10% (e.g., see Shoemaker et al. 1990; Bailey 1991), although Marsden (1993) has suggested that the fraction may be as low as 2%, while Weissman (1990) finds that LPCs would typically impact the Earth at two to three times the speed of a low-inclination, short-period object, and thus that the energy of an impact from an LPC and a short-period object of comparable mass differ by an order of magnitude, Olsson-Steel (1987) concluded that the overall production rate of impact craters due to LPCs is 15 to 20 times smaller than that due to the other objects. He estimated that the impact rate of LPCs sufficient to produce craters 10 km across over the land area of the Earth is 1 per 7 Myr; for comparison Weissman's (1990) value would be 1 per 1.2 Myr while Marsden's (1993) would be closer to 1 per 20 Myr.

A. Theory

Given that $0° \leq v_0 < 180°$, there are for each node two (possibly coincident) true anomalies, $\pm v_0$, at which a collision could occur, provided that the comet's argument of perihelion $\omega = \mp v_0$ or $180° \mp v_0$, according as to whether the relevant node is the ascending or the descending one. As noted, v_0 is a function only of the perihelion distance q and takes the values $0°$, $60°$, $90°$, $120°$ and tends to $180°$ when $q = 1$, 3/4, 1/2, 1/4 and tends to 0 AU, respectively. The time interval between perihelion passage T and Earth impact is obviously zero when $q = 1$ AU, and if k is the usual Gaussian constant, approximately equal to the daily angular velocity of the Earth in its orbit, this time increases to a maximum of $2/(3k) \approx 38.8$ days when $q = 0.5$ AU and then decreases to $\sqrt{2}/(3k) \approx 27.4$ days as $q \to 0$. Decimal approximations to the time interval as a function of q are given in Table I.

TABLE I
Travel Time Between Perihelion
and Earth Orbit

q (AU)	t (days)	q (AU)	t (days)
0.0	27.4	0.5	38.8
0.1	31.2	0.6	38.1
0.2	34.3	0.7	36.0
0.3	36.7	0.8	31.9
0.4	38.2	0.9	24.3
		1.0	0.0

Meteor radiants are usually analyzed in terms of their elongations from the apex of the Earth's way, or the point in the ecliptic that is 90° west of the Sun and the geocentric radiant of an orbit having $q = 1$ AU, $i = 180°$. Because the SSR is centered on the opposition point, it will be convenient here to shift this convention by $90°$ and measure elongations from opposition. It then follows that, immediately before impact, an object with $q = 1$ AU will have a geocentric opposition elongation of $90°$, irrespective of the inclination. Otherwise, for a pre-perihelion impact the elongation is $\epsilon_0 < 90°$, while for a post-perihelion impact by an object in an orbit with the same q and i the elongation is $180° - \epsilon_0$. For a parabolic orbit ϵ_0 generally decreases from $90°$ at $q = 1$ AU, tending to $(\arctan 1/\sqrt{2}) \approx 35.264°$ as $q \rightarrow 0$ (again for all inclinations), but whenever $i < 90°$ there is a minimum of ϵ_0 somewhere in the range $0.5 \leq q < 1.0$ AU. The angle ϵ_0 has its absolute minimum of zero for $q = 0.5$ AU, $i = 0$, and other values of ϵ_0 for $q = 0.5$ AU are $(1/2 \arctan \sqrt{2})$ at $i = 30°$, $45°$ at $i = 60°$, $(\arctan \sqrt{2})$ at $i = 90°$, $60°$ at $i = 120°$, $(90° - 1/2 \arctan \sqrt{2})$ at $i = 150°$ and $(\arctan 2)$ at $i = 180°$. Decimal approximations to ϵ_0 for these and other values of q and i are given in Table II.

TABLE II

Angular Elongation, from Opposition for a Pre-Perihelion Impact, from Conjunction for a Post-Perihelion Impact[a]

$q \backslash i$	0	30	60	90	120	150	180
0.0	35.3	35.3	35.3	35.3	35.3	35.3	35.3
0.1	22.4	25.9	32.9	39.2	43.7	46.3	47.2
0.2	16.2	23.6	34.7	43.1	48.4	51.3	52.2
0.3	10.8	23.2	37.5	46.9	52.5	55.4	56.3
0.4	5.5	24.6	41.0	50.8	56.3	59.1	60.0
0.5	0.0	27.4	45.0	54.7	60.0	62.6	63.4
0.6	6.1	31.6	49.6	58.9	63.8	66.2	66.9
0.7	13.3	37.4	54.9	63.4	67.7	69.8	70.5
0.8	22.7	45.3	61.3	68.6	72.2	73.9	74.4
0.9	37.4	57.1	69.7	75.0	77.6	78.8	79.2
1.0	90.0	90.0	90.0	90.0	90.0	90.0	90.0

[a] Elongations ϵ_0 in degrees are shown as a function of i (deg) and q (AU).

Inverting one of the statements in the previous paragraph, we can say that, for a pre-perihelion catastrophe involving a parabolic orbit with $q = 0.5$ AU, $i = 0°$, the comet would be right in the center of the SSR as it impacts the pointlike Earth. The comet would have been inside the SSR for about one month. For small i the comet would be in the field at or shortly before impact over a wide range of q. As $q \rightarrow 1$ AU, opposition precedes a pre-perihelion impact by increasing amounts, up to more than 75 days when $q = 1$ AU (although if $i \gtrsim 35°$, the comet will then pass entirely beyond the latitude range of the SSR).

In any case, an interval of only 75 days between opposition and impact does not allow much opportunity for satisfactory orbit determination and the planning of ameliorative action. In view of the fact that oppositions occur very roughly at annual intervals, Fig. 1 shows that one can generally expect that from 250 to 500 days will elapse between the last "usable" opposition and a pre-perihelion impact. The 500-day interval corresponds to a direct orbit with $q \approx 1$ AU, and at such an opposition the comet's heliocentric distance $r_0 = 6.1$ AU; the opposition before that, some 875 days before impact (for $i = 0°$), puts the comet at $r_0 = 9.2$ AU. The 250-day interval corresponds to a retrograde orbit with $q \approx 1$ AU, and in this case $r_0 = 3.6$ AU; the intervals between oppositions are also shorter, the two previous oppositions occurring (for $i = 180°$) some 595 days ($r_0 = 6.9$ AU) and 950 days ($r_0 = 9.7$ AU) before impact.

Pre-Perihelion Impact

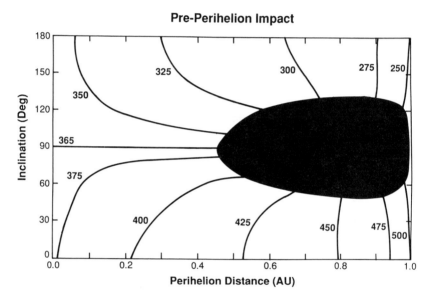

Figure 1.   The number of days by which the final opposition precedes a pre-perihelion impact, as a function of inclination and perihelion distance. Low-inclination cases of 75 days and less are excluded. The solid black region refers to cases where the latitude at opposition exceeds 60°.

It is important to note that when the comet's orbit plane is highly inclined to that of the Earth, the comet's latitude will place the comet beyond the northern or southern extremities of the SSR for at least some inclinations whenever $0.45 \lesssim q \lesssim 0.99$ AU. The unobservable zone corresponds to the solid black area in Fig. 1. This is particularly troublesome when one realizes that for LPCs arriving from a spherical source, the distribution of inclinations would be expected to follow a $\sin i$ distribution (Oort 1950; Porter 1963; Weissman 1982). Half of the cometary orbits would therefore have $60° < i < 120°$. In the known cometary population half of the LPCs with $q < 1$ AU have $52° \lesssim i \lesssim 128°$, and for this half the median $q \approx 0.61$ AU. Thus more than one out of four potential pre-perihelion LPC impactors fails to pass through the SSR. Although Fig. 1

shows the situation at the opposition preceding impact, the area of unobservability shrinks only slightly as one goes back to earlier oppositions and greater distances. The figure also shows that, for $q \lesssim 0.45$ AU, oppositions of objects in orbits of $i = 90°$ occur at precisely annual intervals before impact, and the same is true for all inclinations as $q \rightarrow 0$ (yielding $r_0 = 5.9$ AU, 9.1 AU, 11.9 AU, etc.).

Although post-perihelion hazards would never be discoverable immediately before impact, earlier Spaceguard detection of them is easier than for the pre-perihelion case, particularly for intrinsically fainter objects that would not be detectable at most of the heliocentric distances discussed in connection with pre-perihelion hazards; see Fig. 2. If 100 days from opposition to impact is considered to be sufficient warning, low-inclination comets with perihelia just inside the Earth's orbit could be handled; if not, it would be necessary to detect them at the previous opposition, somewhat more than 500 days before impact. The warning times for direct orbits increase as q decreases, again converging on an integral number of years as $q \rightarrow 0$. For post-perihelion impacts the unobservable cases for $i \approx 90°$ involve $0.08 \lesssim q \lesssim 0.48$ AU, and for larger q oppositions of comets with $i = 90°$ occur at odd multiples of half a year before impact. For retrograde orbits the minimum times from opposition to impact decrease with decreasing q from just under 250 days to 100 days at $q \approx 0.2$ AU. At that point it is appropriate to shift to the previous opposition, rather less than one year earlier. The opposition heliocentric distance shifts from 1.6 to 5.9 AU (for $i = 180°$) and renders visible comets in the range $130° \lesssim i \lesssim 150°$. For retrograde orbits with q much under 0.2 AU the more recent opposition disappears, and the minimum interval from opposition to impact again converges to one year.

B. Practice

If discovery 100 days before a prospective impact is the minimum acceptable leadtime, it is sobering to realize that only 80 comets, or less than 12% of the total set of known LPCs, have been discovered at a time interval $t > 100$ days before perihelion (which can be interpreted as representing the mean of a pre-perihelion and a post-perihelion impact). Furthermore, the 17 of these 80 comets that have $q < 1$ AU represent barely 4% of the known LPCs in this Earth-impacting perihelion-distance range.

Only five comets with $q < 1$ AU have been detected at $t > 160$ days (i.e., at heliocentric distance $r > 3.0$ AU) before perihelion: these are listed in the first section of Table III. It should be noted that the first detections of the three comets marked with asterisks were prediscovery observations identified some time after the discovery was actually made. Comet 1980 XV was discovered visually by an amateur astronomer at $t = 12$ days, $r = 0.5$ AU, magnitude $m = 6(!)$. The famous comets 1976 VI (West) and 1973 XII (Kohoutek) were found on later exposures made with the same telescopes responsible for the listed prediscovery observations, 1976 VI being found at $t = 154$ days, $r = 3.0$ AU, $m = 14–15$ (but not actually recognized until $t = 112$ days); and

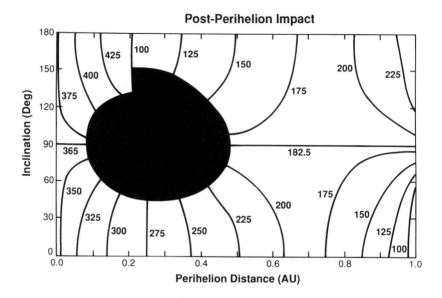

Figure 2.   The number of days by which the final opposition precedes a post-perihelion impact, as a function of inclination and perihelion distance. High-inclination cases of 75 days and less are excluded. The solid black region refers to cases where the latitude at opposition exceeds 60°.

1973 XII at $t = 296$ days, $r = 4.7$ AU, $m = 16$. The absolute magnitudes H_{10} were computed by taking the earliest observed magnitudes at their face value, and using the usual formula $H_{10} = m - 5 \log \Delta - 10 \log r$ (Δ being the comet's distance from the Earth in AU); the mean value for these five comets was 7.7. It is also important to note that, particularly in the cases of 1976 VI and 1973 XII, H_{10} values determined from estimates when the objects were nearer perihelion were up to 2 to 4 mag brighter. With this point in mind, these comets should be considered as slightly brighter than average, there being known LPCs that were up to 5 mag instrinsically brighter and perhaps down to 7 mag fainter.

Only seven LPCs have ever been detected at $t > 475$ days before perihelion. All of these comets had $q > 3$ AU and $r > 5.4$ AU, and they were generally exceptionally bright objects. They are listed in the second section of Table III. The mean value of H_{10} is 5.0, which is fainter than Halley's Comet but decidedly brighter than average. Again, comet 1984 XV involved a prediscovery detection; actual discovery was not made until after perihelion.

Success in following LPCs out to large distances after perihelion gives one rather limited insight as to the chance of detection before. Certainly, the careful use of a large telescope and accurate knowledge of an object's predicted position and brightness have allowed a handful of LPCs—exclusively again large-q objects—to be followed out to beyond $r = 10$ AU. The seven comets with $q < 1$ AU recorded for the longest time and to the greatest heliocentric

TABLE III

Observed Record Times and Distances

	H_{10}	m	t (days)	r (AU)
	$q < 1$ AU; first observation before perihelion			
1980 XV	9.4	16	164	3.1*
1976 VI	8.9	16–17	199	3.6*
1954 X	6.5	15	264	3.8
1992 XIX	7.7	16.5	292	4.2
1973 XII	5.8	16	333	5.2*
	$q > 3$ AU; first observation before perihelion			
1992h	8.3	19	493	5.5
1975 VIII	3.1	13	519	5.6
1957 VI	4.0	15.5	532	6.1
1905 IV	3.4	15.5	647	6.5
1977 IX	5.3	17–18	478	6.6
1982 I	5.7	16.5	760	7.3
1984 XV	5.3	19	946	8.6*
	$q < 1$ AU; last observation after perihelion			
1907 IV			297	4.4
1861 II			322	4.5
1948 I			355	4.9
1962 III			299	4.9
1957 V			342	5.0
1957 III			368	5.4
1990 XX			524	6.4

distances after perihelion were clearly exceptional objects, and they are listed in the last section of Table III. The more recent of these comets were followed down to magnitude ∼19 to 21, but they were all still active when last observed.

There is no real information on how H_{10} for an LPC can be reliably equated to a nuclear diameter. It is perhaps reasonable to say that a "typical" value of $H_{10} = 9$ (for a comet of $q < 1$ AU at "large" heliocentric distance on the way in to perihelion) corresponds to a nuclear diameter of a kilometer or two, large enough that an Earth impact would be globally disastrous. Except for the regions, shown in Figs. 1 and 2, where Spaceguard discovery would be geometrically impossible, it is necessary to detect comets at opposition at $r \approx 6$ AU and thus typically at $m \approx 20$. The Spacewatch team is discovering distant LPCs (e.g., 1992h) at $m \approx 19$; in view of this, Spaceguard might therefore be able to find something approaching three-quarters of the globally hazardous LPCs in time for reasonable warning. The point should be made that "new" comets, making their first approaches from the Oort Cloud, tend to be significantly brighter before perihelion than after perihelion at corresponding heliocentric distances (e.g., 1973 XII).

C. Extension of the SSR

Because of the blind spots shown in Figs. 1 and 2, it is useful to consider extending the range of the Spaceguard search. Increasing the range in ecliptic latitude is an obvious approach, but an increase in ecliptic longitude is also needed in order to give enough automatic coverage of discoveries to secure orbit determinations. Because of the need to observe beyond the celestial equatorial poles, extension of the latitude coverage would make it increasingly difficult to ensure the proper coordination of observing sites on the Earth's surface at all times of the year. Augmentation of the longitude search might therefore be considered without an increase in latitude. This would require a proportionate increase in the number of telescopes, by a factor of 3 to change the longitude span from $60°$ to $180°$, for example (except at high-latitude stations in the summertime), and it would have the effect of dramatically shrinking the blind spots for comets with $i \approx 90°$. It is near the extremes of the extended longitude ranges that these comets would be crossing the ecliptic. All comets would be detectable except for those with i within a few degrees of $90°$ and, in the limit, $0.78 \lesssim q \lesssim 0.89$ AU (pre-perihelion impact) and $0.21 \lesssim q \lesssim 0.34$ AU (post-perihelion impact). Because the comets would be crossing the ecliptic almost perpendicularly, it will still be necessary to cover the $120°$ in latitude in order to get coverage for orbit determinations and to avoid working too close to full Moon. The outlying latitude range could be as much as halved, however, if the longitude range were increased somewhat beyond $180°$. This kind of search mode has also been considered appropriate in connection with the discovery of Aten objects, which may have aphelion distances only slightly greater than 1 AU and which are therefore near opposition infrequently.

Despite this the blind spots would still not totally disappear. Near their centers the comets remain at almost identical distances from the Sun and the Earth. The isosceles triangle formed by comet, Sun and Earth changes its size and orientation, but not rapidly enough to allow observations of the comet from the Earth, even in the extended Spaceguard field (and with also an increase in latitude), in a time interval that is anywhere near the range of 100 to 500 days that we have been discussing. For a post-perihelion impact, prior detection requires the comet to be observable as much as 750 days earlier and at some 8 AU from the Sun. For a pre-perihelion impact, these values are extended to more than 850 days and 9 AU, although in such a case one should also be able to detect the comet within just a couple of weeks before impact.

III. IMPACT PROBABILITIES

One way of estimating the frequency of terrestrial impacts is to follow the orbits of observed comets and see how often close approaches to the Earth occur (see, e.g., Kresák 1978; Sekanina and Yeomans 1982); however, the data base available for such work is quite restricted, so a theoretical approach is required. This is not the case for near-Earth asteroids or short-period comets as a larger number of perihelion passages by such objects is available for study, making a realistic assessment of the impact rate possible (Milani et al. 1989).

An essential feature of the techniques used to calculate the probability of a hit by an ensemble of impactors is that ω and Ω are random. Öpik's (1951,1976) method has been extensively used, but it does have the drawback that one must assume that the impactor's orbit is a deep Earth-crosser (i.e., $q \lesssim 0.95$ AU). Because the impact probability rises steeply if the impactor has perihelion (or aphelion) near 1 AU, the use of Öpik's equations is questionable if a realistic estimate of the probability is to be gained (cf. Weissman 1982; Zimbelman 1984; Shoemaker et al. 1990). Alternative techniques, for pairs of arbitrary elliptical orbits that may be grazing, have been described by Wetherill (1967) and Kessler (1981). Here we use the latter approach (Steel and Baggaley 1985).

A. Theory

The impact probability for an LPC depends strongly upon q and i. Figure 3 shows the collision probabilities (P_c) and speeds (V_c) for parabolic orbits with $q = 0.1$, 0.5, 0.9 and 1.0 AU, and $i = 10°$ to $170°$ in $10°$ steps. Plots such as this have previously been presented by Steel (1992). The impact speed is the relative velocity between the Earth and the LPC added in quadrature with the terrestrial escape velocity (11.2 km s^{-1}). It is clear that there is a general increase in P_c with q, this being particularly marked as q approaches 1 AU. The reason for this is that the mean spatial density of an LPC increases around 1 AU if it has perihelion near that value, because it spends more time near the orbit of the Earth and a collision is more likely. This is discussed in more detail by Olsson-Steel (1987) and Steel (1992). Each of the loci also shows a characteristic dish-like shape, with lower and higher inclinations being associated with enhanced values of P_c; this is due to the reduction in the angle between the orbital planes that occurs in such cases, the comet spending more time at lower ecliptic latitudes, with a concomitant jump in P_c. However, the shapes are not symmetrical, a retrograde orbit having a higher P_c than the corresponding direct orbit; this results from the higher relative velocities that occur for retrograde orbits, the collision probability changing more or less linearly with the velocity, although the reduced collision cross section for high velocities (due to the lesser effect of gravitational focusing by the Earth) offsets this effect to a small degree.

For values of q ranging from 0.01 to 1.0 AU we considered LPCs with i changing in $1°$ jumps from $1°$ to $179°$. We calculated the value of P_c for each, weighting and averaging each by $\sin i$ to allow for the spherical distribution, and we applied a normalizing factor $\pi/2$, before calculating the mean of the ensemble (i.e., the inverse of $2/\pi$, which is the mean value of the weighting function $\sin i$). Although orbits with $i < 1°$ or $i > 179°$ have extremely high values of P_c, the $\sin i$ factor means that these are inconsequential with regard

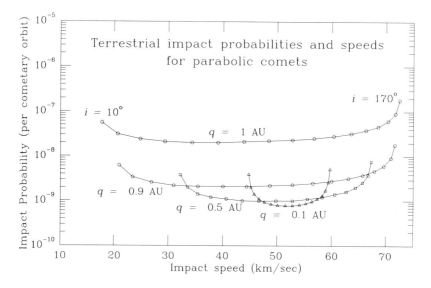

Figure 3.   Earth-impact probability plotted against impact speed for parabolic comets
with perihelion distances of 0.1, 0.5, 0.9, and 1.0 AU, with inclinations in each case
of 10° to 170° in 10° steps.

to the mean derived. We also calculated the mean impact speed, weighted
by $P_c \sin i$ in each case. The results are shown in Table IV. The final value
of q used was in fact 0.98 AU, which renders the maximal value of P_c for
any q (as with $q = 1$ AU the LPC and terrestrial orbits only half overlap, the
Earth's perihelion distance being 0.9833 AU). Previous values for the mean
impact probability have been evaluated at ~ 1.3 to 3.3×10^{-9} per cometary
orbit/per perihelion passage (Zimbelman 1984; Weissman 1982; Shoemaker
1984), and Bailey (1991) assumed $P_c = 2.5(\pm 1) \times 10^{-9}$. The "real" mean
impact probability depends upon the distribution of orbital elements, because
the enhanced value of P_c as $q \rightarrow 1$ AU (Fig. 3 and Table IV) leads to a
substantial contribution to the overall result from such orbits. Rather than
a uniform distribution in q (Weissman 1982; Table 4 of Zimbelman 1984),
an increasing flux with increasing q might be expected (Fernández 1981a;
Table V of Zimbelman 1984). Kresák and Pittich (1978) found that the best
match to observations was a distribution varying as $q^{0.5}$, which implies a
uniform space density independent of heliocentric distance; this would lead
to a significant enhancement in the effects of the $q \approx 1$ AU LPCs on the mean
impact probability. Contrary to the above, Everhart (1967a) found that the
discovered LPCs have a distribution in q that increased from a relative number
of 0.4 at $q = 0$ compared to unity at $q = 1$ AU, while Shoemaker and Wolfe
(1982) used their own observations to derive a distribution $n \simeq 500q' - 175$
where n is the number of LPCs with $q < q'$, valid in the range 0.5 to 1.1 AU.
Such a distribution would further weakly enhance (compared to an $n \sim q^{0.5}$

distribution) the effect of the $q \approx 1$ AU comets. Weissman (1985) investigated the theoretical distribution of q of new comets coming from the Oort cloud and found that the number might be expected to rise sharply through the planetary region. Straightforward averaging of the eleven values of P_c in Table IV (i.e., uniform q-distribution assumed) leads to an overall mean collision probability of 2.74×10^{-9} per orbit, but this is biased by the inclusion of the one very high P_c. Rather, it is necessary to perform a numerical integration for all values $0.005 < q < 1.0167$ AU, with a suitably small step size. Using 0.01 AU steps for $q < 0.9$ AU and 0.001 AU steps for $0.9 < q < 1.0167$ AU, we find that for a uniform distribution in q the mean impact probability is 2.21×10^{-9}. This is in agreement with the result of Weissman (1982), who used Öpik's equations for 250,000 random cometary orbits. If we assume a perihelion distribution that varies as $q^{0.5}$, then the added weight of the $q \approx 1$ AU orbits results in a mean P_c of 2.52×10^{-9}.

TABLE IV
Mean Terrestrial Impact Probablilties and Velocities for
Random Long-Period Comets Arriving from a Spherical Source

q (AU)	$\bar{P}c$ (per orbit)	$\bar{V}c$ (km s^{-1})
0.01	1.17×10^{-9}	52.8
0.1	1.22×10^{-9}	53.0
0.2	1.29×10^{-9}	53.3
0.3	1.38×10^{-9}	53.6
0.4	1.48×10^{-9}	53.9
0.5	1.61×10^{-9}	54.2
0.6	1.79×10^{-9}	54.5
0.7	2.06×10^{-9}	54.9
0.8	2.51×10^{-9}	55.2
0.9	3.55×10^{-9}	55.6
1.0	1.21×10^{-8}	56.0

In addition to the above, there is some evidence that LPCs are not uniformly distributed in $\sin i$, with a small retrograde excess occurring (Fernández 1981b; Matese et al. 1991; Fernández and Ip 1991). This may well be due to preferential discoveries of such objects, as suggested by Everhart (1967b), who also found that comets with i near 90° are underrepresented in the discovered sample; conversely, Everhart found that comets with q near 1 AU are more efficiently discovered, meaning that there may be a relative surplus of LPCs with high values of P_c in the sample. In view of the above it is fair to say that for a theoretical distribution of LPCs, with q- and i-distributions that are as yet uncertain in detail, the mean impact probability is in the range 2 to 3×10^{-9} per orbit, and the mean impact speed is 53 to 56 km s^{-1}.

B. Practice

Following Shoemaker (1984), we now consider the observed LPCs. Marsden and Williams (1992) have tabulated the orbits of all well-observed comets, and of the LPCs listed therein a total of 411 have perihelia that cross at least part of the Earth's current orbit (aphelion at 1.0167 AU), ranging from the Kreutz sungrazer 1981 XXI with $q = 0.0045$ AU (i.e., impacting the Sun) to comet 1900 II with $q = 1.0148$ AU. For each of these 411 the value of P_c and the extreme impact speeds V_{min} and V_{max} were calculated; these are slightly different, because the calculations here incorporate the effects of the Earth's orbital eccentricity, meaning that impacts at a range of heliocentric distances (0.9833–1.0167 AU) are possible. Here we define the characteristic impact speed V_c by averaging V_{min} and V_{max} (which differ little), incremented by the Earth's gravitational field. The values of P_c are plotted in Fig. 4 as a function of inclination, with different symbols being used for different values of q in order to show how P_c varies with that parameter (cf. Fig. 8 of Olsson-Steel 1987). The mean value of the impact probability ($\bar{P}_c$) and the standard deviation of the individual values are shown in Table V.

Inspection of Fig. 4 shows that there are a handful of LPCs that have very high P_c, and these tend to have q near 1 AU and/or orbital planes near the ecliptic (i near the minimal or maximal values). There are twelve LPCs with $P_c > 1 \times 10^{-8}$ per orbit, and six with $P_c > 2 \times 10^{-8}$. These six are as listed in Table VI. If the first three (those with $P_c > 5 \times 10^{-8}$) are excluded from the set of 411 LPCs, leaving 408, the value of $\bar{P}_c$ is reduced by ~30% and the standard deviation by a factor of 5 (Table V, second line). If the other three LPCs with $P_c > 2 \times 10^{-8}$ are also excluded, leaving a set of 405, $\bar{P}_c$ and its standard deviation suffer a further significant decrement (Table V, third line). The fifth entry in Table VI, comet 1743 I, is undoubtedly a short-period comet (Marsden and Williams 1992). Whilst none of the three sets of LPCs used in Table V could be said to be the best sample, it is clear that the few LPCs that have q near 1 AU and $i < 10°$ or (especially) $i > 170°$ have a disproportionate effect upon the calculation of $\bar{P}_c$.

TABLE V
Mean Terrestrial Impact Probabilities and Velocities for the
Observed Long-Period Comets

Data Set	$\bar{P}c \pm \sigma$ (per orbit)	$\bar{V}c$ (km s^{-1})
411 LPCs	$3.56 \pm 14.26 \times 10^{-9}$	58.2
408 LPCs	$2.53 \pm 2.78 \times 10^{-9}$	52.3
405 LPCs	$2.38 \pm 2.14 \times 10^{-9}$	52.0

Turning attention to the mean impact speed, we see that the five retrograde LPCs in Table VI produce $V_c > 70$ km s^{-1}, and thus with their high values of P_c bias the value of $\bar{V}_c$ upwards: in Table V, $\bar{V}_c = 58.2$ km s^{-1} if the entire set of 411 comets is used, reducing to 52.0 km s^{-1} if the six high-risk comets in Table VI are excluded. Individual impact speeds as a function of inclination are plotted in Fig. 5, again with different symbols for different

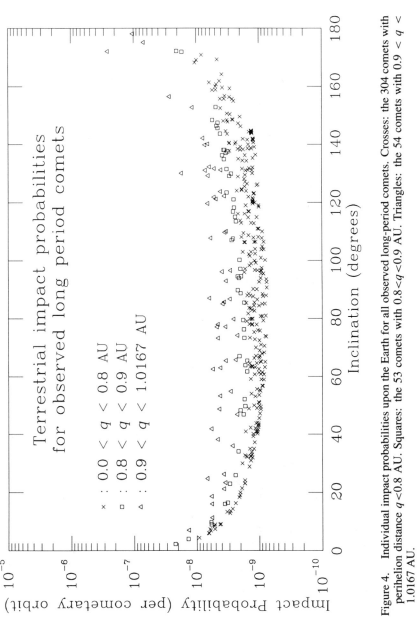

Figure 4.   Individual impact probabilities upon the Earth for all observed long-period comets. Crosses: the 304 comets with perihelion distance $q < 0.8$ AU. Squares: the 53 comets with $0.8 < q < 0.9$ AU. Triangles: the 54 comets with $0.9 < q <$ 1.0167 AU.

TABLE VI
Observed Long-Period Comets with Extremely High Impact Probabilities

Comet	q (AU)	e	i (deg)	P_c (per orbit)	V_c (km s^{-1})
1862 II	0.9813	1.0	172.11	2.60×10^{-7}	72.6
1864 II	0.9093	0.9964	178.13	1.03×10^{-7}	72.1
1759 III	0.9658	1.0	175.13	6.91×10^{-8}	72.6
1945 III	0.9981	1.0	156.51	2.68×10^{-8}	70.8
1743 I	0.8382	1.0	2.28	2.21×10^{-8}	22.2
1987 III	0.8696	0.9957	172.23	2.06×10^{-8}	71.6

q ranges, in order to illustrate the range of possible values. Deep Earth crossers show a modest increase in V_c as i increases, the trend being almost horizontal when $q \to 0$ (cf. the cluster of crosses at $140° < i < 144°$ and $V_c = 54$ to 55 km s^{-1} representing the Kreutz group of sungrazers). This contrasts with the shallow crossers, which have a more marked increasing trend in V_c with i. For the latter, encounter speeds (prior to enhancement by the Earth's attraction) may vary from $(\sqrt{2} - 1)$ times the Earth's orbital speed (i.e., ~ 12.3 km s^{-1}, rendering $V_c \simeq 16.6$ km s^{-1}) to $(\sqrt{2} + 1)$ times the same amount (i.e., ~ 71.9 km s^{-1}, rendering $V_c \simeq 72.8$ km s^{-1}, or 74.0 km s^{-1} if the Earth is at perihelion). The observed mean V_c is understandably a little larger than the value corresponding (for all q) to $i = 90°$ (i.e., $\sqrt{3}$ times the Earth's orbital speed, or ~ 51.6 km s^{-1}, rendering $V_c \approx 52.8$ km s^{-1}; this is also the value for all i as $q \to 0$).

IV. CONCLUDING REMARKS

It was shown in Sec. II that if surveys are restricted to the Spaceguard search region of ~ 6000 square degrees around opposition, as many as one-quarter of the potentially impacting LPCs could never be observed. Even for LPCs that do pass through this region, detection with reasonable warning requires the comets to be bright enough for discovery at up to 500 days before impact and at 6 AU from the Sun. The existing observational record contains very few cases where LPCs with perihelia inside the Earth's orbit have been observed at this distance, even after perihelion.

From the calculations in Sec. III, for both the theoretical distribution of LPC orbits and the observed distribution, it is clear that the terrestrial hazard is disproportionately affected by (i) those comets having perihelia near 1 AU and (ii) those comets with either very small or very large inclinations. For example, considering only LPCs with $q = 0.5$ AU, the mean probability of an impact for all inclinations is 3.43×10^{-9} per orbit, whereas taking only orbits with $i < 10°$ or $i > 170°$, the result is 1.67×10^{-8}, some five times higher. Such orbits are always close to the ecliptic and therefore within the field of the planned Spaceguard search region. Because of the $\sin i$ distribution there are few of these, but the hazard they pose is very substantial on a comet-by-comet

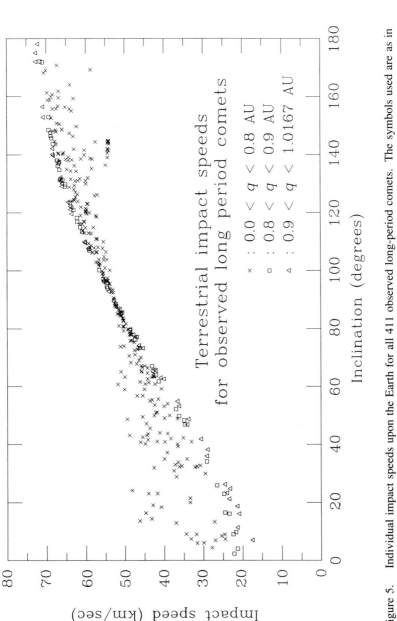

Figure 5.   Individual impact speeds upon the Earth for all 411 observed long-period comets.   The symbols used are as in Fig. 4.

basis. Spaceguard in its planned implementation might be expected to result, then, in the discovery of the few high-risk LPCs whilst missing a large fraction of the more numerous low-risk bodies with orbital planes highly inclined to the ecliptic. However, with regard to the overall risk the effects counteract, with the sin i variation in the number of LPCs cancelling the $1/\sin i$ variation in the collision probability: see p. 49 of Zimbelman (1984) for the Öpik implementation, Kessler (1981) or Steel and Baggaley (1985) for the better Kessler implementation.

While it only takes one impact from a random LPC to produce a global disaster, it follows that, purely from the point of view of the probability of an impact by a low-inclination vs a high-inclination object, it makes little difference as to which ecliptic latitudes the Spaceguard search telescopes are directed.

In order to discover the majority of the lower risk (on an individual basis) comets it would be necessary, as shown in Sec. II, to extend the Spaceguard search region in longitude from $\pm 30°$ centered on opposition to $\pm 90°$. This would increase the cost and complexity of the program considerably. A decision to do so, or even to recommend such a step, depends upon a realistic assessment of the long-period comet impact risk as compared to the hazard posed by short-period comets and asteroids. The main area of ignorance in this regard is in the masses and mass distribution of long-period comets.

REFERENCES

Bailey, M. E. 1991. Comet craters versus asteroid craters. *Adv. Space Res.* 11(6):43–60.

Everhart, E. 1967*a*. Intrinsic distributions of cometary perihelia and magnitudes. *Astron. J.* 72:1002–1011.

Everhart, E. 1967*b*. Comet discoveries and observational selection. *Astron. J.* 72:716–726.

Fernández, J. A. 1981*a*. New and evolved comets in the solar system. *Astron. Astrophys.* 96:26–35.

Fernández, J. A. 1981*b*. On the observed excess of retrograde orbits among the long-period comets. *Mon. Not. Roy. Astron. Soc.* 197:265–273.

Fernández, J. A., and Ip, W.-H. 1991. Statistical and evolutionary aspects of cometary orbits. In *Comets In The Post-Halley Era*, eds. R. L. Newburn, Jr., M. Neugebauer and J. Rahe (Dordrecht: Kluwer), pp. 487–535.

Kessler, D. J. 1981. Derivation of the collision probability between orbiting objects: The lifetimes of Jupiter's outer moons. *Icarus* 48:39–48.

Kresák, Ľ. 1978. Passages of comets and asteroids near the earth. *Bull. Astron. Inst. Czechoslovakia* 29:103–113.

Kresák, Ľ., and Pittich, E. M. 1978. The intrinsic number density of active long-period comets in the solar system. *Bull. Astron. Inst. Czechoslovakia* 29:299–309.

Marsden, B. G. 1993. To hit or not to hit. In *Proceedings of the Near-Earth Object Interception Workshop*, eds. G. H. Canavan, J. C. Solem and J. D. G. Rather (Los Alamos: Los Alamos National Lab), pp. 67–71.

Marsden, B. G., and Williams, G. V. 1992. *Catalogue of Cometary Orbits*, 7th ed. (Cambridge, Mass.: Minor Planet Center).

Matese, J. J., Whitman, P. G., and Whitmire, D. P. 1991. Gravitationally unbound comets move in predominantly retrograde orbits. *Nature* 352:506–508.

Milani, A., Carpino, M., Hahn, G., and Nobili, A. M. 1989. Dynamics of planet-crossing asteroids: classes of orbital behavior. *Icarus* 78:212–269.

Morrison, D., ed. 1992. *The Spaceguard Survey: Report of the NASA International Near-Earth-Object Detection Workshop* (Pasadena: Jet Propulsion Laboratory).

Olsson-Steel, D. 1987. Collisions in the solar system-IV. Cometary impacts upon the planets. *Mon. Not. Roy. Astron. Soc.* 227:501–524.

Oort, J. H. 1950. The structure of the cloud of comets surrounding the solar system, and a hypothesis concerning its origin. *Bull. Astron. Inst. Netherlands* 11:91–110.

Öpik, E. J. 1951. Collision probabilities with the planets and distribution of planetary matter. *Proc. Roy. Irish Acad.* 54A:165–199.

Öpik, E. J. 1976. *Interplanetary Encounters* (Amsterdam: Elsevier).

Porter, J. G. 1963. The statistics of cometary orbits. In *The Moon, Meteorites and Comets*, eds. B. M. Middlehurst and G. P. Kuiper (Chicago: Univ. of Chicago Press), pp. 550–572.

Sekanina, Z., and Yeomans, D. K. 1982. Close encounters and collisions of comets with the earth. *Astron. J.* 89:154-161.

Shoemaker, E. M. 1984. Large body impacts through geologic time. In *Patterns of Change in Earth Evolution*, eds. H. D. Holland and A. F. Trendall (Berlin: Springer-Verlag), pp. 15–40.

Shoemaker, E. M., and Wolfe, R. F. 1982. Cratering timescales for the Galilean satellites. In *The Satellites of Jupiter*, ed. D. Morrison (Tucson: Univ. of Arizona Press), pp. 277–339.

Shoemaker, E. M., Wolfe, R. F., and Shoemaker, C. S. 1990. Asteroid and comet flux in the neighborhood of Earth. In *Global Catastrophes in Earth History*, eds. V. L. Sharpton and P. D. Ward, Geological Soc. of America Special Paper 247 (Boulder: Geological Soc. of America), pp. 155–170.

Steel, D. 1992. Cometary supply of terrestrial organics: Lessons from the K/T and the present epoch. *Orig. Life Evol. Biosphere* 21:339–357.

Steel, D. I., and Baggaley, W. J. 1985. Collisions in the solar system-I. Impacts of the Apollo-Amor-Aten asteroids upon the terrestrial planets. *Mon. Not. Roy. Astron. Soc.* 212:817–836.

Weissman, P. R. 1982. Terrestrial impact rates for long and short-period comets. In *Geological Implications of Impacts of Large Asteroids and Comets on the Earth*, eds. L. T. Silver and P. H. Schultz, Geological Soc. of America Special Paper 190 (Boulder: Geological Soc. of America), pp. 15–24.

Weissman, P. R. 1985. Dynamical evolution of the Oort cloud. In *Dynamics of Comets: Their Origin and Evolution*, eds. A. Carusi and G. B. Valsecchi (Dordrecht: D. Reidel), pp. 87–96.

Weissman, P. R. 1990. The cometary impactor flux at the earth. In *Global Catastrophes in Earth History*, eds. V. L. Sharpton and P. D. Ward, Geological Soc. of America Special Paper 247 (Boulder: Geological Soc. of America), pp. 171–180.

Wetherill, G. W. 1967. Collisions in the asteroid belt. *J. Geophys. Res.* 72:2429–2444.

Zimbelman, J. R. 1984. Planetary impact probabilities for long-period comets. *Icarus* 57:48–54.

PREDICTING CLOSE APPROACHES OF ASTEROIDS AND COMETS TO EARTH

DONALD K. YEOMANS and PAUL W. CHODAS

Jet Propulsion Laboratory

The motions of all known Earth approaching asteroids and comets with reasonably secure orbits have been numerically integrated forward in time to A. D. 2200. Special care was taken to use the best available initial conditions including orbits based upon radar data. Each object was integrated forward with Earth and Moon perturbations treated separately, with general relativistic equations of motion and with perturbations by all planets at each integration step. For the active short-period comets whose motions are affected by the rocket-like effects of vaporizing ices, a nongravitational force model was employed. When a close approach to the Earth was sensed by the numerical integration software, an interpolation procedure was used to determine the time of the object's closest approach and the minimum separation distance at that time. For those objects making the closest approaches to Earth in the next two centuries, an error analysis was conducted to determine whether or not the object's error ellipsoid at the time of closest approach included the Earth's position (i.e., an Earth collision could not be ruled out). Although there are no obvious cases where a known near-Earth asteroid or comet will threaten the Earth in the next two centuries, there are a few objects that warrant special attention. The Aten type asteroid 2340 Hathor makes repeated close approaches to Earth and because most of its orbit lies within that of the Earth, it is often a difficult object to observe in a dark sky. For both asteroids and comets, there are generally dramatic increases in their position uncertainties following close planetary encounters. Because of their short observational data intervals, their unmodeled nongravitational effects, and the possibility of escaping early detection by approaching the Earth from the Sun's direction, long-period comets may present the largest unknown in assessing the long-term risk of Earth-approaching objects. Fortunately the frequency with which these objects approach the Earth is very small compared with the numerous approaches by the population of near-Earth objects with short periodic orbits. For the short-period comets, rocket-like outgassing effects and offsets between the observed center-of-light and the comet's true center-of-mass can introduce large uncertainties in their long-term orbital extrapolations. The uncertainty in the future motion of an active short-period comet is substantially larger than the motion of an asteroid with a comparable observational history. While asteroids dominate the list of close Earth approaches in the next two centuries, their motions are relatively predictable when compared to the active comets. For the rapidly growing population of known near-Earth asteroids and comets, efficient procedures are suggested for monitoring their long-term motions thus allowing early predictions of future close approaches to Earth.

I. INTRODUCTION

Collisions of asteroids and comets with the Earth is a topic so provocative and so prone to sensationalism that great care must be taken to assess the realistic hazards in the near future. The task of accurately predicting future close Earth approaches by known near-Earth objects is essential for studies of risk assessment. As the discovered population of near-Earth objects continues to grow, and the orbit determinations of previously known objects continues to improve, the motions of these objects should be routinely integrated forward for a few hundred years to investigate their orbital behavior. Note should be made of objects that can pass the Earth closely so they can be placed upon the "short list" of objects for which additional attention is required. In this fashion, possible Earth-threatening objects can be identified well in advance so that future astrometric observations can be scheduled. Once this short list of close Earth encounters has been compiled, error covariance analyses should be undertaken for each object to determine whether or not the object's position error ellipsoid includes the Earth's position at the time of the closest approach (i.e., an Earth–object collision cannot be ruled out). For the closest Earth approaches, impact probabilities should be computed in a realistic fashion. If useful risk assessments are to be conducted, future close Earth approach predictions for asteroids and comets must be accompanied by error analyses and impact probabilities.

The accurate prediction of asteroid and comet close approaches to Earth is also necessary for planning future groundbased and spacebased observation programs for these scientifically interesting objects. Given the very low probability of finding a truly threatening future encounter, this latter use of close approach predictions is, perhaps, of more immediate use.

In Sec. II, we briefly discuss the benefit of studying near-Earth objects when they are, in fact, near the Earth. Section III outlines the necessary steps for accurately monitoring the long-term numerical motions of near-Earth objects and presents the results of our numerical integrations to the year A. D. 2200 for all known near-Earth objects with reasonably well-known orbits. Section IV addresses the problems of trying to predict accurately the motions of some near-Earth objects and presents error analyses for those objects making the closest Earth approaches. Section V presents a summary and our main conclusions.

II. THE IMPORTANCE OF NEAR-EARTH APPROACHES BY ASTEROIDS AND COMETS

Because the signal-to-noise ratio for an observation of an asteroid or comet depends upon the inverse square of the topocentric distance, efforts have been made to conduct physical studies of these objects during close Earth approaches. This is especially true for radar observations because the signal-to-noise ratio is proportional to the inverse fourth power of topocentric distance. While the physical study of near-Earth objects during close Earth approaches is an obvious course to pursue, it is not as obvious how important astrometric observations are during these close Earth approaches. To a reasonable approximation, the power, or benefit, of optical astrometry improves linearly with the decreasing distance between the observer and the target object. Position measurements of an object that are accurate to 1 arcsec at a distance of 1 AU and 0.1 AU represent linear plane-of-sky errors of about 725 km and 72.5 km, respectively.

At close Earth approaches, radar astrometric observations can provide extremely powerful data for orbit improvement (Yeomans et al. 1987; Ostro et al. 1991; Yeomans et al. 1992). These radar Doppler and time delay measurements have far greater fractional precision than optical astrometric data but can only be taken during close approaches to Earth. The ability of radar data to reduce future ephemeris errors is most dramatic for newly discovered objects for which only short optical data intervals are available. Objects whose optical data intervals include several returns to opposition, have well-defined orbits even in the absence of radar data. The ideal data set for a near-Earth object includes the combination of optical astrometric data (plane-of-sky data) over long time intervals and precise radar measurements (line-of-sight data) during close approaches to Earth. The orbit that includes radar data as well as optical astrometric data can be more accurately extrapolated into the future than a similar orbit that is based upon only the optical data. As an example, an error analysis has been used to demonstrate the power of radar observations in reducing the ephemeris prediction error for minor planet 4179 Toutatis. As is evident from Figs. 1 and 2, Toutatis made a close Earth approach in December 1992 and will make an even closer Earth approach in September 2004. During the December 1992 Earth approach, 34 time delay (range) and 21 Doppler (range rate) observations were made during the interval from November 27 through December 18. Employing an error covariance analysis similar to that described by Yeomans et al. (1987), 1-sigma error ellipses were computed in the Toutatis orbit plane at the time of the Earth close approach on September 29, 2004. Figure 3 displays two error ellipses, the smaller one representing the expected 1-sigma position errors resulting from processing all optical observations (1934–1992) and the late 1992 radar data. To account for unmodeled error sources, the optical data were given noise values intentionally larger than the *rms* residuals from the orbit determination process. A data noise of 2 arcsec was used for the two observations in 1934 while the remaining optical data (1988–1992) were assigned noise values of 1.3 arcsec. The radar delay observations were given noise values of 15 microsec (1.5 km) and the Doppler observations were assigned values of 1 Hertz (1.8 cm s^{-1} for a frequency of 8510 MHz). Both these delay and Doppler noise values are very conservative. The larger error ellipse in Fig. 3 represents the 1-sigma position errors resulting from the processing of the optical data alone. Assuming no new astrometric observations are considered, the position errors for 4179 Toutatis during the 2004 close Earth approach will be nearly five times smaller as a result of the 1992 radar data.

Spacecraft mission planners have often taken advantage of close Earth passages to design low-cost flyby and rendezvous missions to comets and

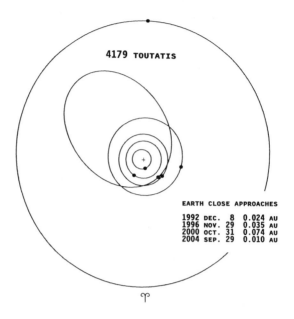

Figure 1.   An ecliptic plane projection of the orbit of asteroid 4179 Toutatis. The positions of the planets Mercury through Jupiter are denoted for the time of the close approach to Earth on December 8, 1992.

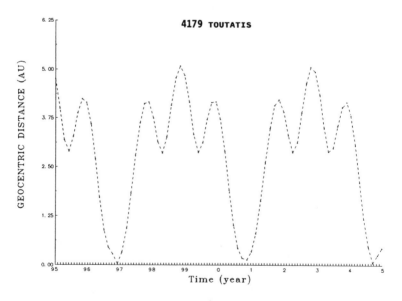

Figure 2.   Geocentric distance of asteroid 4179 Toutatis over the 1995 to 2004 time interval. Close Earth approaches occur in late 1996, 2000 and 2004.

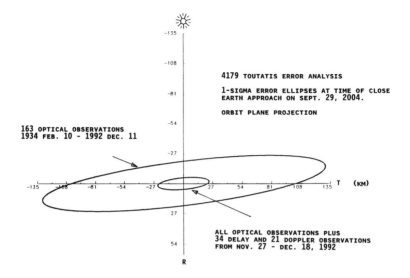

Figure 3.   Position error ellipse information for asteroid 4179 Toutatis at the time of the close approach to Earth on 29 September 2004. The ellipses represent the 1-sigma position errors in the orbit plane assuming the object's position is predicted using an orbit based only upon optical data through 11 December 1992 (larger ellipse) and assuming a prediction using optical data through 11 December 1992 and radar data through 18 December 1992 (smaller ellipse).

asteroids. A spacecraft encounter that takes place near Earth ensures a short communication distance, drives down the ephemeris uncertainties of the target object and allows excellent groundbased studies that nicely complement the *in situ* spacecraft observations. By selecting near-Earth object targets that have low orbital eccentricities, low inclinations and perihelia near 1 AU, transfer trajectories can be found that require only very modest energy requirements (McAdams 1991). As a crude rule of thumb, the easiest near-Earth objects to reach for rendezvous missions are those objects whose orbital characteristics are most like that of the Earth itself. An example of an asteroid flyby mission that will take advantage of a close Earth approach is the Deep Space Program Science Experiment (see the Chapter by Nozette et al.) that is scheduled to fly within 100 km of asteroid 1620 Geographos six days after the asteroid passes within 0.033 AU of the Earth in late August 1994 (Yeomans 1993).

III. LONG TERM PREDICTIONS OF NEAR-EARTH ENCOUNTERS

In an effort to establish the close Earth approaches of known objects in the next two centuries, orbits of near-Earth objects have been numerically integrated forward to the year 2200. Those asteroids whose perihelion distances are currently 1.3 AU or smaller and whose orbits are secure were selected for the integration process. Because comets often suffer large planetary perturbations, every periodic comet was included in our integrations, regardless

TABLE I

Earth Approaches of Comets and Asteroids to Within
100 Lunar Distances for the Interval 1994–2001[a]

Object		Date (TDB)		CA Dist (AU)
	1993 EA	1994 01	8.056	0.1543
3361	Orpheus	1994 03	2.455	.1497
4953	1990 MU	1994 05	30.441	.1418
5604	1992 FE	1994 06	16.740	.1565
2062	Aten	1994 06	20.360	.2514
1620	Geographos	1994 08	25.424	.0333
2100	Ra-Shalom	1994 10	12.945	.1549
	1989 VA	1994 11	15.168	.1820
2062	Aten	1995 01	12.072	.1268
2340	Hathor	1995 01	16.274	.1373
	1991 OA	1995 05	24.750	.1122
	1991 JX	1995 06	9.098	.0341
2062	Aten	1996 01	24.687	.2234
P/Honda-Mrkos-Pajdušáková		1996 02	4.587	.1702
5590	1990 VA	1996 03	31.361	.2253
2063	Bacchus	1996 03	31.670	.0678
1566	Icarus	1996 06	11.315	.1012
4953	1990 MU	1996 06	16.749	.2499
1685	Toro	1996 08	2.231	.2207
3103	1982 BB	1996 08	6.105	.1151
	1991 CS	1996 08	28.419	.0620
	1989 RS1	1996 09	16.095	.1958
	1989 UQ	1996 10	23.062	.1504
4947	Ninkasi	1996 10	23.190	.2131
4197	1982 TA	1996 10	25.639	.0846
3908	1980 PA	1996 10	27.860	.0613
4179	Toutatis	1996 11	29.955	.0354
	1991 VK	1997 01	10.695	.0749
	1989 UQ	1997 01	25.215	.2283
	1991 CS	1997 02	23.125	.2229
5590	1990 VA	1997 03	10.252	.2069
P/Encke		1997 07	4.840	.1901
3671	Dionysus	1997 07	6.890	.1144
	1988 XB	1997 07	8.698	.1080
4034	1986 PA	1997 08	28.607	.2061
2100	Ra-Shalom	1997 09	26.976	.1705
	1989 VA	1997 10	24.671	.2404
2340	Hathor	1997 12	21.839	0.1379

TABLE I (cont.)
Earth Approaches of Comets and Asteroids to Within
100 Lunar Distances for the Interval 1994–2001[a]

2102	Tantalus	1997 12	21.839	0.1379
3361	Orpheus	1998 02	12.772	0.1668
5590	1990 VA	1998 02	22.130	.2383
	1988 EG	1998 02	28.903	.0316
4183	Cuno	1998 06	9.683	.2079
	1987 OA	1998 08	19.187	.1019
	1991 RB	1998 09	18.475	.0401
1865	Cerberus	1998 11	24.747	.1634
	1989 UR	1998 11	28.689	.0800
	1992 SK	1999 03	26.265	.0560
1863	Antinous	1999 04	1.615	.1894
	1989 ML	1999 04	27.457	.2520
	1991 JX	1999 06	2.819	.0500
	1989 VA	1999 11	22.013	.1943
1685	Toro	2000 01	27.237	.2426
5604	1992 FE	2000 03	3.480	.2176
	1991 DB	2000 03	31.423	.1580
	1986 JK	2000 07	11.499	.1218
	1991 BB	2000 07	27.186	.1662
4486	Mithra	2000 08	14.365	.0466
4769	Castalia	2000 08	15.718	.2460
2100	Ra-Shalom	2000 09	6.039	.1896
	1991 CB1	2000 09	18.630	.2477
2340	Hathor	2000 10	25.249	.1970
4179	Toutatis	2000 10	31.188	.0739
4183	Cuno	2000 12	22.793	.1427
4688	1980 WF	2001 01	3.609	.1701
3362	Khufu	2001 01	3.687	.2174
4034	1986 PA	2001 04	3.047	.1465
3103	1982 BB	2001 08	6.314	.1161
	1987 QB	2001 08	16.760	.1631
	1991 FA	2001 12	14.947	.1923
	1990 SP	2001 12	27.921	.2298
3362	Khufu	2001 12	29.455	0.1596

[a] The entries, which are given in chronological order, include the object's name, date of closest approach and close approach distance (AU).

of its perihelion distances. For the long-term integrations, whose results are presented in Table II, initial orbits were considered secure if astrometric data existed for two or more apparitions. Some asteroids with single apparitions were also included if radar observations were available or if the optical data interval exceeded six months. For the short-term results displayed in Table I, these criteria were relaxed somewhat; this latter group consisted of 172 asteroids and 145 periodic comets.

A. Description of Numerical Integration to A. D. 2200

One of us (PWC) developed a special integration package whereby each object is sequentially integrated forward to a given time with orbital elements automatically output near the time of each perihelion passage. The most recent orbital information for each object was used to initialize these integrations. In addition, many near-Earth objects have orbits improved with radar data as well as the optical data (Yeomans et al. 1992) and when appropriate, these radar-based orbits were used to initialize the long-term integrations.

Once the integration of an object is underway, the step size is adjusted to maintain a local velocity error of less than 10^{-13} AU per day. The Jet Propulsion Laboratory Development ephemeris, DE200 (Standish 1990) was used throughout for the planetary perturbations that were computed at each time step. The outer planetary masses in DE200 were updated to include those values resulting from the Voyager spacecraft flybys. A special interpolation scheme is invoked each time the integrator senses a planetary close approach and if an object's orbit were perfectly known, close approach times would be accurately output to the one minute level. Depending upon the accuracy of an object's initial orbital elements, the error in the actual close approach time may be considerably larger.

General relativistic equations of motion were employed for all objects and the perturbations by the Earth and Moon were considered separately rather than treating their combined masses as being located at the barycenter. The general relativistic advancement in the line of apsides is an important consideration for objects whose eccentricity is large and whose semimajor axis is small. For each period, the advancement in the line of apsides is approximately $0.038/a(1-e^2)$ arcsec where e is the eccentricity of the object's orbit and a is the semimajor axis in AU. As an example, we note that for the four asteroids with the largest relativistic advances in their lines of apsides, 3200 Phaethon, 1566 Icarus, 2100 Ra-Shalom and 2340 Hathor, the perihelion advance in 200 yr amounts to 20.1, 19.9, 14.9 and 14.6 arcsec, respectively. To maintain consistency in the integration of asteroid and comet equations of motion, one should include general relativistic effects to account for the relativistic advancement in the lines of apsides, but more importantly the JPL planetary ephemerides most often used for asteroidal and cometary orbit determinations (DE118, DE200, etc.) were created using relativistic equations of motion. The use of the JPL planetary ephemerides and nonrelativistic equations of motion for a comet or asteroid will necessarily introduce an error in the asteroid's or comet's mean motion that is by no means negligible.

For objects making close Earth encounters, the Earth and Moon perturbations must be treated separately. In extreme cases, a satisfactory orbit cannot be computed without separating the Earth and Moon perturbations. For example, asteroid 1991 VG passed within 0.0031 AU of the Earth on Dec. 5.4, 1991 and within 0.0025 AU of the moon on Dec. 6.9. There are observational data on either side of this close approach. The orbital solution for 1991 VG was not successful until we abandoned the approximation of having the combined Earth and lunar masses located at the Earth-Moon barycenter.

Although we did not consider the perturbations of some of the larger asteroids (e.g., Ceres, Pallas and Vesta), these effects will be included in a future version of the JPL integration package. However, we note that the relative velocity difference between a perturbing asteroid and a near-Earth object will be relatively high so that these perturbations will rarely become important.

B. Tabular Information on Close Earth Approaches

Table I presents the asteroids and comets which makes close Earth approaches to within 100 lunar distances (0.257 AU) within the interval Jan. 1, 1993 through Jan. 1, 2001, and Table II presents the same information for those objects that will come within 10 lunar distances (0.0257 AU) during the interval 2001 to 2200. Table III lists the few cometary close Earth approaches (to within 50 lunar distances) over the interval 1993 to 2200. Table I is primarily for planning future astronomical observations while Tables II and III address the issue of risk from near-Earth objects in the next two centuries. In Tables II and III, in addition to the Earth close approach distances, the approximate minimum distance between the orbits of the object and the Earth are given in parentheses using a method described by Porter (1952). These latter distances are the closest that an object could be expected to approach the Earth if it were to arrive at just the right time. All near Earth objects known in mid-1993 were considered for inclusion in Tables I, II and III.

IV. ERROR ANALYSES AND IMPACT PROBABILITIES FOR CLOSE EARTH APPROACHES

A. Error Analyses for Closest Earth Approaches

For a few of those objects in Tables II and III which makes particularly close Earth approaches, an error covariance analysis was undertaken along the lines outlined by Yeomans et al. (1987) to estimate the ephemeris uncertainties at the time of the close Earth approach. These analyses were carried out for the two asteroids making the closest Earth approaches (2340 Hathor and 4660 Nereus) and for the periodic comet making the closest Earth approach in the next century (P/Finlay). Although there were several asteroids making closer Earth approaches, the 2060 Earth approach (to within 0.05 AU) of comet Finlay is the closest cometary approach in the coming century. It was included in the error analyses computations because the outgassing effects of

TABLE II

Predicted Earth Approaches of Comets and Asteroids to Within
10 Lunar Distances for the Interval 2001–2200[a]

Object		Date (TDB)		CA Dist (AU)
2340	Hathor	2086 10	21.670	0.0057(0.006)
2340	Hathor	2069 10	21.351	.0066(0.006)
2101	Adonis	2177 02	9.000	.0072(0.006)
4660	Nereus	2060 02	14.288	.0080(0.005)
4179	Toutatis	2004 09	29.568	.0104(0.006)
4581	Asclepius	2051 03	24.343	.0122(0.004)
4660	Nereus	2071 02	4.795	.0149(0.005)
	1991 OA	2070 07	13.675	.0149(0.003)
3361	Orpheus	2194 04	14.409	.0167(0.016)
4660	Nereus	2112 12	23.405	.0181(0.006)
4660	Nereus	2166 02	3.447	.0186(0.006)
4581	Asclepius	2133 03	25.160	.0187(0.005)
5011	Ptah	2170 03	23.262	.0191(0.019)
5011	Ptah	2193 03	18.580	.0193(0.019)
2101	Adonis	2102 07	10.004	.0195(0.009)
3362	Khufu	2169 08	22.539	.0197(0.011)
	1990 OS	2053 11	16.034	.0197(0.009)
3200	Phaethon	2093 12	14.453	.0198(0.013)
4179	Toutatis	2069 11	5.704	.0198(0.007)
	1990 OS	2195 11	10.553	.0208(0.007)
3362	Khufu	2045 08	22.060	.0209(0.014)
3361	Orpheus	2091 04	18.951	.0211(0.016)
3200	Phaethon	2189 12	13.642	.0215(0.003)
2101	Adonis	2143 07	10.945	.0222(0.007)
	1990 OS	2125 08	17.796	.0228(0.007)
4581	Asclepius	2183 03	21.643	.0230(0.006)
4953	1990 MU	2058 06	5.426	.0231(0.023)
2340	Hathor	2130 10	22.963	.0233(0.005)
	1988 EG	2110 02	28.810	.0236(0.023)
	1989 JA	2022 05	27.051	.0239(0.024)
	1988 EG	2041 02	27.799	.0241(0.024)
2340	Hathor	2045 10	21.341	.0242(0.006)
4769	Castalia	2046 08	26.817	.0250(0.022)
	1990 OS	2003 11	11.448	0.0250(0.010)

[a] The entries are given in order of their close approach distances. The quantities
in parentheses are the approximate minimum distances between the object's
orbit and that of the Earth.

active comets can introduce orbital position uncertainties far larger than for
asteroidal objects with comparable observation histories.

In an attempt to account for unmodeled error sources, observation noise values were purposely taken to be higher than the *rms* residual as determined from orbital computations. That is, despite the fact that modern astrometric observations of asteroids routinely achieve sub arc-second accuracy, we gave each observation a noise value of 1.3″ for the error analyses. As no simulated, future data were considered in these analyses, the position uncertainties quoted in Table IV represent the knowledge of the object's ephemeris given only the existing observations. With additional observations, the position uncertainties can be expected to shrink somewhat. In Table IV, the error estimates represent the uncertainty in the direction between the object and the Earth at the time of closest approach. For asteroids 2340 Hathor, 4660 Nereus, 4179 Toutatis and comet Finlay, the existing orbits were updated using recent observational data, general relativistic equations of motion and treating the Earth, Moon perturbations separately. These improved orbital elements were input into the long-term integrations used to generate the information displayed in Tables I through IV.

TABLE III
Cometary Approaches to Within 50 Lunar Distances of the Earth
(0.128 AU) for the Interval 1994–2200[a]

Object	Date (TDB)		CA Dist (AU)
Schwassmann-Wachmann 3	2006 05	10.688	0.0912(0.054)
Honda-Mrkos-Pajdušáková	2011 08	15.275	.0601(0.060)
Honda-Mrkos-Pajdušáková	2017 02	11.104	.0864(0.060)
Wirtanen	2018 12	18.464	.0846(0.013)
Finlay	2060 10	27.042	.0473(0.032)
Kowal 2	2060 12	10.766	.0928(0.063)
Schwassmann-Wachmann 3	2070 06	27.074	.1264(0.009)
Giacobini-Zinner	2112 10	8.355	.0469(0.030)
Honda-Mrkos-Pajdušáková	2130 02	18.324	.0756(0.065)
Tuttle	2130 12	23.972	.0890(0.089)
Halley	2134 05	8.006	.0881(0.074)
Tsuchinshan 1	2140 01	26.530	.1211(0.102)
Grigg-Skjellerup	2146 04	12.253	.0694(0.068)
Honda-Mrkos-Pajdušáková	2157 01	29.696	.1203(0.119)
Denning-Fujikawa	2190 11	20.720	0.0985(0.087)

[a] The entries, which are given in chronological order, include the comet's name, date of closest approach and close approach distance. The quantities in parentheses are the approximate minimum distances between the object's orbit and that of the Earth.

Because asteroid 4660 Nereus has a low inclination ($i = 1.4°$) and a perihelion distance just inside the Earth's orbit, it makes rather frequent Earth approaches. The closest of these is in February 2060 and at that time the component of the 3-sigma position uncertainty ellipse that lies along the asteroid-Earth line is about 21,000 km and hence well short of including the

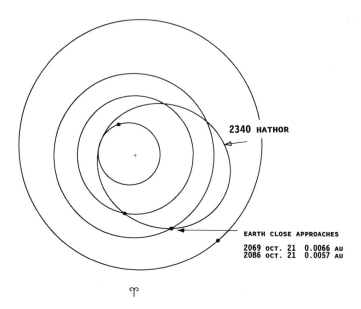

Figure 4.   An ecliptic plane projection of the orbit of asteroid 2340 Hathor. The positions of the planets Mercury through Jupiter are denoted for the time of the Earth close approach on 21 October 2069.

Earth. That is, an Earth collision is ruled out.

The Aten asteroid 2340 Hathor makes a close to Earth approach in October 2069 followed by an even closer approach 17 years later (see Figs. 4 and 5). As is evident from the information in Table IV and Fig. 6, the 1-sigma position uncertainties during the 2086 encounter are greatly increased as a result of the 2069 close Earth approach. Even so, the 3-sigma position uncertainty ellipse in 2086 does not include the Earth's position and a collision is again ruled out. From Fig. 4, it becomes evident why Aten-type asteroids like Hathor are so difficult to observe in a dark sky. Only 46 observations over the 1976 to 1983 interval were available for the orbit determination of 2340 Hathor.

Of the known short-period comets, comet Finlay will make the closest Earth approach in the 21st century (see Fig. 7). For the above error analyses for asteroids 4179 Toutatis, 4660 Nereus and 2340 Hathor, the observational data noise (1.3 arcsec) is the only assumed error source. For active comets like Finlay, there are also errors due to uncertain nongravitational effects and offsets between the comet's observed center of brightness and its true center of mass. The nongravitational effects are due to the rocket-like outgassing of the comet's nucleus and while these effects have been modeled (Marsden et al. 1973; Yeomans and Chodas 1989), there remain significant uncertainties in the behavior of these effects over long time intervals. For comet Finlay, the center of light was assumed to be offset toward the solar direction with a value

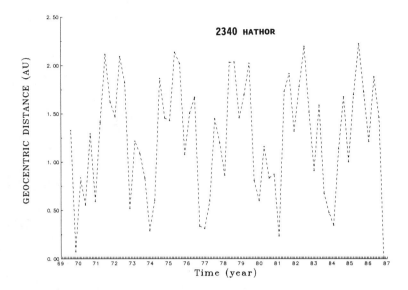

Figure 5. Geocentric distance of asteroid 2340 Hathor. The positions of the planets Mercury through Jupiter are denoted for the time of the close approach to Earth on 21 October 2069.

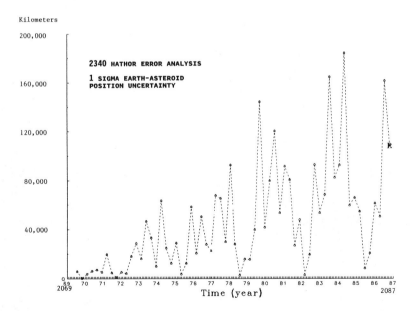

Figure 6. One-sigma Earth-asteroid position uncertainty estimates for asteroid 2340 Hathor over the 2069 to 2087 interval. Following the 2069 close approach to Earth, positions errors grow rapidly with time.

equal to 150 km at 1 AU and scaling with the inverse square of the heliocentric distance. We have assumed that the errors in the determination of the radial and transverse nongravitational parameters ($A1$, $A2$) are present but that we cannot solve for them; they are "consider" rather than "solve for" parameters. The $A1$ and $A2$ nongravitational parameters were considered to be 100% and 10% uncertain, respectively. At the time of the comet's close approach to Earth on October 27, 2060, the 3-sigma error ellipse axes on the plane of sky are 783,000 km and 3,420 km. These axes would be 156,000 km and 3,050 km without the errors introduced by the nongravitational parameters and the offset between the comet's center of light with respect to its center of mass. If comet Finlay were an inactive asteroid, its position uncertainties would be substantially less.

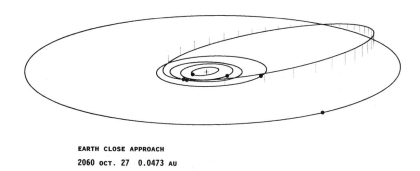

EARTH CLOSE APPROACH
2060 OCT. 27 0.0473 AU

Figure 7.   An orbital diagram for short-period comet P/Finlay. The positions of the planets Mercury through Jupiter are denoted for the time of the close approach to Earth on 27 October 2060.

B. Screening Potential Hazards As Near-Earth Objects Are Discovered

As each new asteroid or comet is discovered, an efficient screening process must be undertaken to establish whether or not the new object has the potential for approaching the Earth closely. With only a few astrometric observations to work with over a short time span, the initial orbits of each discovered object will be so uncertain that very little can be said about its potential as a future hazard. Initially, it would be sufficient to compute the minimum distance between the orbits of the Earth and the new object to determine how close the object can approach the Earth. Marsden (1992) pointed out that for most near-Earth objects, a simple two-body computation using current orbital elements can be used to identify which objects can closely approach the Earth's orbit in the future. However, planetary perturbations can be effective in altering the current orbital parameters of a near-Earth object, so that perturbed numerical integrations are required to identify actual close Earth approaches in the future.

TABLE IV

Earth Close Approach Circumstances and Associated Position
Uncertainties for the Three Closest to Earth Approaches of Asteroids
As Well as the Closest Cometary Approach in the 21st Century

Object	Date	Close Approach Distance	3-sigma Earth-object Position Uncertainty	
4660 Nereus	2060 Feb 14.3	0.0080 AU	21,000 km	= 0.00014 AU
2340 Hathor	2069 Oct 21.4	0.0066 AU	6,300 km	= 0.00004 AU
2340 Hathor	2086 Oct 21.7	0.0057 AU	335,000 km	= 0.0022 AU
P/Finlay	2060 Oct 27.0	0.0473 AU	412,800 km	= 0.0028 AU

If the initial, or a subsequent, preliminary orbit indicates that an object could pass within, say, 0.2 AU of the Earth's orbit, the object would be assigned to the "*A* list" of objects for which a long-term ephemeris integration will be required when a sufficient number of observations are available for a perturbed orbit computation. To generate the information given in Tables I, II, and III, we simply selected all periodic comets and those asteroids whose perihelion distances were less than 1.3 AU. Although a single planetary perturbation would not normally perturb an asteroid's perihelion distance by more than 0.05 AU, a series of future perturbations might be expected to move its perihelion distance by that amount. Because short-period comets can pass near Jupiter, their perihelion distances can evolve by more than this amount. In any case, with the increasing speed of modern-day computers, it would not be difficult to integrate routinely the motions of all near-Earth objects a few hundred years into the future and identify those objects making close approaches to Earth (see Sec. III). Those objects that pass close to the Earth in the future would then be assigned to the "*B* list" whereby a covariance error analysis would be conducted to determine how close the object's error ellipsoid approaches the Earth anytime in the near future. Those few objects whose future error position ellipsoids lie within, say, 10-sigma of the Earth's distance at a close approach would be singled out for an impact probability computation. This screening process could be achieved in a straightforward fashion using efficient software with little intervention by the user except, perhaps, in the case where one close Earth approach is followed by another or when poorly modeled nongravitational forces render the motion of an active comet particularly uncertain.

C. Predicting Impact Probabilities

The probability that a close Earth-approaching asteroid or comet will actually impact the Earth can be approximated via a procedure which examines the position error ellipsoid at the predicted time of closest approach to Earth. The error ellipsoid is a representation of the scale and orientation of a three-dimensional Gaussian probability density function. The probability that the object lies in a given region at a given time is simply the integral of the probability density function evaluated at that time over the volume of the region. If we take the region to be the figure of the Earth, this integral evaluation produces the probability that the object's position is within the Earth at the given time (here we ignore the dynamics associated with an impact). The result of this integral evaluation, however, is not the probability of impact, because it does not take into account the motion of the error ellipsoid past the Earth. The impact probability is the probability that the object's position will at any time lie within the figure of the Earth as it sweeps by the ellipsoid.

The element of time may be removed from the impact probability computation by projecting the error ellipsoid into the plane perpendicular to the velocity vector of the Earth relative to the object. We will refer to this plane as the "impact" plane. The error ellipsoid then becomes an error ellipse which represents the marginal probability density function describing the probability that the object will at some time pass through a given point on the impact plane. To first order, the figure of the Earth projects into a circle in this plane; the probability of Earth impact is computed by simply integrating the marginal probability density function over the area of this circle.

This problem of computing impact probabilities has been addressed in the past in the context of spacecraft studies motivated by requirements for planetary quarantine and avoidance of impact for spacecraft carrying nuclear materials. Efficient numerical techniques have already been developed for computing impact probabilities via the above procedure (see, e.g., Light 1965; Michel 1977). We are not aware of a previous application of this problem to natural bodies. To be sure, the above method for computing impact probabilities is only an approximation. It considers only linear variations about the predicted trajectory of the object, and so ignores differential perturbations. A more precise computation of impact probabilities could be obtained from a Monte Carlo approach, which would require a great deal more computation.

V. SUMMARY AND CONCLUSIONS

It is interesting to note that approximately one third of all near-Earth objects are discovered near their closest approach to Earth that they will experience in the next 200 years. This fact underscores the importance of rapidly following up new discoveries with observations (both passive and active) to study the object's physical characteristics and to refine its orbit. Often, the discovery apparition of a near-Earth object is the best opportunity to observe it for centuries to come.

Although there are no obvious cases where a known near-Earth asteroid or comet will threaten the Earth in the next two centuries, there are a few objects that warrant special attention. The Aten-type asteroid 2340 Hathor makes repeated close Earth approaches and because most of its orbit lies within that of the Earth, it is often a difficult object to observe. For both asteroids and comets, there is generally a dramatic increase in their position uncertainties following a close planetary encounter.

Long-period comets are often found on their first trip into the inner solar system. Because of their short observational data intervals, their unmodeled nongravitational effects and the possibility of their approaching the Earth from the direction of the Sun, long-period comets may present the largest unknown in assessing the long-term risk of Earth-approaching objects. Fortunately the number of close Earth approaches by these objects appears to be very small compared to the number of approaches by short periodic near-Earth objects. Even for the short-period comets, rocket-like outgassing effects and offsets between the observed center of light and the comet's true center of mass can sometimes introduce large uncertainties in the long-term extrapolation of cometary orbits. In addition, astrometric radar observations that might be expected to help refine its orbit are often difficult because the radar signal can bounce off a debris cloud of particles surrounding the nucleus rather than the nucleus itself. Astrometric radar observations exist only for two short-period comets, Encke and Grigg-Skjellerup. The uncertainties in the future motions of active short-period comets are substantially larger than those for an asteroid with a comparable observational history. While asteroids dominate the list of close Earth approaches in the next two centuries, their motions are relatively predictable when compared to the active comets.

Acknowledgments. The research described in this chapter was carried out by the Jet Propulsion Laboratory, California Institute of Technology, under contract with the National Aeronautics and Space Administration.

REFERENCES

Light, J. O. 1965. A Simplified Analysis Of The Launch Bias Necessary On Interplanetary Missions. JPL Tech. Memo 312-616.

Marsden, B. G. 1993. To hit or not to hit. In *Proceedings of the Near-Earth Object Interception Workshop*, eds. G. H. Canavan, J. C. Solem and J. D. G. Rather (Los Alamos: Los Alamos National Lab), pp. 67–71.

Marsden, B. G., Sekanina, Z., and Yeomans, D. K. 1973. Comets and nongravitational forces. V. *Astron. J.* 78:211–225.

McAdams, J. V. 1991. Mission Options for Rendezvous With the Most Accessible Near-Earth Asteroid–1989 ML. AIAA paper AAS 91-397.

Michel, J. R. 1977. A New Method For Accurately Calculating The Integral Of The Bivariate Gaussian Distribution Over An Offset Circle. JPL Tech. Memo 312, pp. 77–34.

Ostro, S. J., Campbell, D. B., Chandler, J. F., Shapiro, I. I., Hine, A. A., Velez, R., Jurgens, R. F., Rosema, K. D., Winkler, R., and Yeomans, D. K. 1991. Asteroid radar astrometry. *Astron. J.* 102:1490–1502.

Porter, J. G. 1952. *Comets And Meteor Streams* (New York: John Wiley and Sons), pp. 90–91.

Standish, E. M. 1990. The observational basis for JPL's DE 200, the planetary ephemerides of the Astronomical Almanac. *Astron. Astrophys.* 233:252–271.

Yeomans, D. K. 1993. Targeting An Asteroid: The DSPSE Encounter With Asteroid 1620 Geographos. AAS Paper 93-262.

Yeomans, D. K., Ostro, S. J., and Chodas, P. C. 1987. Radar astrometry of near-earth asteroids. *Astron. J.* 94:189–200.

Yeomans, D. K., and Chodas, P. C. 1989. An asymmetric outgassing model for cometary nongravitational accelerations. *Astron. J.* 98:1083–1093.

Yeomans, D. K., Chodas, P. C., Keesey, M. S., and Ostro, S. J. 1992. Asteroid and comet orbits using radar data. *Astron. J.* 103:303–317.

THE ROLE OF GROUNDBASED RADAR IN NEAR-EARTH OBJECT HAZARD IDENTIFICATION AND MITIGATION

STEVEN J. OSTRO
Jet Propulsion Laboratory

Groundbased radar is a key technique for the post-discovery reconnaissance of NEOs and is likely to play a central role in identification of possibly threatening objects during the foreseeable future. Delay-Doppler measurements are orthogonal to optical angle measurements and typically have a fractional precision between 10^{-5} and 10^{-9}, and consequently are invaluable for refining orbits and prediction ephemerides. The same measurements can provide two-dimensional images with resolution on the order of decameters. Imaging data sets with adequate coverage in subradar longitude/latitude can be used to determine the target's shape and spin vector. The active planetary radars use wavelengths that are sensitive to near-surface bulk density and structural scales larger than a few centimeters and, for comets, can penetrate optically opaque comas and reveal large-particle clouds. Upgrades of existing telescopes (especially Arecibo) will expand the range of groundbased radar and will optimize NEO imaging and astrometric capabilities. However, existing instruments are already oversubscribed, and observation of more than a small fraction of objects discovered in a Spaceguard-like survey will require radar telescopes dedicated to NEO reconnaissance.

I. INTRODUCTION

The goal of this chapter is to assess the potential role of groundbased radar in confrontation of the near-Earth-object (NEO) hazard now and in the future. The next section is a tutorial in current NEO radar techniques (Ostro 1993), with emphasis on measurements useful for orbit refinement and physical characterization. The state of NEO radar reconnaissance is described in Sec. III, which briefly summarizes pre-1993 observations and discusses expectations for NEO work after 1994, when telescope upgrades now underway should be finished. Sec. IV offers scenarios for radar involvement in NEO threat assessment during an era of a Spaceguard-like survey (Morrison 1992) and some speculations on possible developments during the next millennium.

Any discussion of the NEO hazard requires "boundary conditions" that define the domain of that discussion. In this chapter, the term "impact hazard" is meant to have a very broad connotation. In particular, I consider a *prediction* of an impact to be potentially hazardous, because it might provoke an economically or psychologically destructive societal response, even if the predicted collision were known to be insufficiently energetic to affect the global ecology. Moreover, under certain circumstances such a prediction could lead to development of a mitigation system whose existence would in itself introduce a significant risk to civilization (see, e.g., Sagan 1992; Sagan and Ostro 1994a, b). There will be a progression from dedicated search programs to identification of objects that might threaten collision within some time interval (e.g., the next century) to progressive refinement of each threat assessment. This underlying reality of the asteroid/comet hazard will eventually result in either classification of these objects as nonthreatening or decisions to take increasingly serious forms of action, beginning with spacecraft reconnaissance and proceeding to defensive operations, until we no longer believe that these objects might pose any danger. The overriding considerations throughout this entire process will be the state of our uncertainty about threatening objects and their trajectories, what can be done to reduce that uncertainty, and the cost of doing so. One can, in fact, view civilization's response to the NEO hazard as involving three realms of ignorance. First, and most fundamentally, although we know the gross character of the population and average collision rates, we have discovered only an insignificant fraction of the potential impactors. The Spaceguard survey is intended to be the initial step toward dispelling this kind of ignorance. The second kind of uncertainty concerns known objects' orbits and the circumstances of future close approaches. The third kind of uncertainty concerns the outcome of the impact of a specific object on a specific collision course, and hence the object's physical properties, including mass, dimensions, composition, internal structure, and multiplicity. If spacecraft inspection or defensive action are to be undertaken, then spin state, detailed surface properties, and the presence of accompanying swarms of macroscopic particles would be relevant as well. Groundbased radar is uniquely suited for cost-effective trajectory refinement and physical characterization, that is, for reducing both kinds of post-discovery uncertainty about potential NEO hazards. As will become clear in the text that follows, these two roles are inseparable in practice.

II. CURRENT NEO RADAR TECHNIQUES

The general stratagem of a radar observation is to transmit an intense, coherent signal with very well-known polarization state and time/frequency structure and then, by comparing those properties to the measured properties of the echo, deduce the properties of the target. The information content of an observation will depend on the echo strength, which must be at least several times greater than the *rms* fluctuation in the receiver's thermal noise. The signal-to-noise ratio (SNR) is proportional to factors describing the radar system and the target:

$$\text{SNR} \sim (\text{SYSTEM FACTOR})(\text{TARGET FACTOR})(\Delta t)^{1/2} \qquad (1)$$

where

$$\text{SYSTEM FACTOR} \sim P_{tx}A_{tx}A_{rcv}/\lambda^{3/2}T_{\text{sys}} \sim P_{tx}G_{tx}G_{rcv}\lambda^{5/2}/T_{\text{sys}} \qquad (2)$$

and

$$\text{TARGET FACTOR} \sim \hat{\sigma} D^{3/2} P^{1/2} / R^4. \tag{3}$$

Here Δt is integration time, P_{tx} is transmitted power, λ is wavelength, and T_{sys} is system temperature. A_{tx} and A_{rcv} are effective (i.e., illuminated) antenna apertures during transmit and receive, and are related to the corresponding antenna gains G_{tx} and G_{rcv} by $G/4\pi = A/\lambda^2$. In Eq. (3), the target properties are effective diameter D, spin period P, distance R, and radar albedo $\hat{\sigma}$, which is the ratio of radar cross section σ to projected area $\pi D^2/4$.

A. Telescopes

The two continuously active planetary radar telescopes are the Arecibo (λ=13 and 70 cm) and Goldstone (3.5 and 13 cm) instruments. For each, the shorter wavelength provides much greater sensitivity and is the exclusive choice for NEO work, and representative values of optimum system characteristics are $P_{tx} \sim$450 kW, gain $\sim 10^{7.1}$, and $T_{\text{sys}} \sim$25 K. The Arecibo 13-cm (S-band, 2380-MHz) system has for the past five years been twice as sensitive as the Goldstone 3.5-cm (X-band, 8510-MHz) system, but Goldstone can track targets continuously for much longer periods and has access to the whole sky north of $-40°$ declination. Two bistatic (two-station) experiments have been carried out at Goldstone, first with 1566 Icarus in 1968 and most recently with 4179 Toutatis in 1992. The latter used transmission from the 70-m antenna (DSS 14) and reception 22 km away with a new 34-m beam-waveguide antenna (DSS 13). Toutatis also was the target of a Russian-German bistatic experiment (the first non-U. S. asteroid radar observations), which used transmission from the Yevpatoria 70-m antenna in Crimea and reception at the Effelsberg 100-m antenna near Bonn, Germany (Zaytsev et al. 1993). Aperture-synthesis observations, employing 3.5-cm transmission from Goldstone and reception of echoes at the 27-antenna Very Large Array (VLA) in New Mexico, have been carried out for two NEOs, 1991 EE and 4179 Toutatis (de Pater et al. 1992,1994). That system can synthesize a beamwidths as small as 0.25 seconds of arc, vs 2 minutes of arc for single-dish observations. (The narrow beamwidths of planetary radars render them useless as NEO search instructments.)

The Goldstone 14-13 system and the Yevpatoria-Effelsberg 6-cm system are, respectively, about 50% and 30% as sensitive as the Goldstone monostatic (DSS-14) system. The Goldstone-VLA system is three times as sensitive as the Goldstone monostatic system, but only for targets with bandwidths no less than 381 Hz; for narrower echoes, including those from NEOs, the SNR falls off as the square root of the echo bandwidth.

The Arecibo telescope is being upgraded to increase its sensitivity by more than an order of magnitude by constructing a ground screen around the periphery of the dish, replacing high-frequency line feeds with a Gregorian configuration, doubling the transmitter power, and installing a fine-guidance pointing system. At Goldstone, installation of a new transmit-receive feed horn and a new data-acquisition system will help to optimize observations of close NEOs. Figure 1 shows the relative sensitivities of the primary planetary radar systems as a function of target declination.

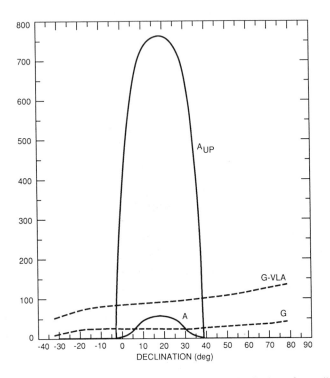

Figure 1.   Radar system sensitivities. The single-date SNR of echoes from a "typical" 1-km asteroid at a distance of 0.1 AU is plotted against declination for Goldstone (G), the Goldstone-VLA system (G-VLA), Arecibo (A), and the upgraded Arecibo (A_{UP}). The curve for the Goldstone-VLA system is an upper bound; see text.

The value of a radar observation increases in proportion to the echo strength. SNRs as large as 20 are usually adequate for detection and marginal resolution of the echoes. SNRs greater than 100 let one achieve enough resolution to be able to make simple statements about shape. With SNRs approaching 1000, the data permit detailed constraints on dimensions, and with SNRs at least as large as ~3000 one can make images that clearly show surface features. Crudely, one can expect the number of useful pixels in a data set to be of the same order as the SNR.

B. Disc-Integrated Measurements

In most modern radar observations, the transmission is circularly polarized and two parallel receiving channels are used to receive echoes in the same circular polarization as transmitted (the SC sense) and simultaneously in the opposite (OC) sense. The handedness, or helicity, of a circularly polarized wave is reversed on normal reflection from a plane mirror, so single backreflections from dielectric interfaces whose sizes and radii of curvature greatly exceed the wavelength yield echoes almost entirely in the OC polarization. SC echo power can arise from multiple scattering, from single backscattering from interfaces with wavelength-scale radii of curvature (e.g., rocks), or from subsurface refraction. Therefore the circular polarization ratio

$$\mu_C = \sigma_{SC}/\sigma_{OC} \qquad (4)$$

is a useful gauge of the target's near-surface, wavelength-scale complexity, or "roughness." When linear polarizations are used, it is convenient to define the linear polarization ratio

$$\mu_L = \sigma_{SL}/\sigma_{OL}. \qquad (5)$$

Both μ_L and μ_C would be zero for a perfectly smooth target. For all NEO radar measurements to date, $\mu_L < 1$ and $\mu_L < \mu_C$.

Widely used measures of radar reflectivity are the OC radar albedo

$$\hat{\sigma}_{OC} = \sigma_{OC}/A_{\text{proj}} \qquad (6)$$

where A_{proj} is the target's projected area, and the total power $(OC + SC = OL + SL)$ radar albedo $\hat{\sigma}_T$, which is four times the geometric albedo used in optical planetary astronomy. A smooth metallic sphere would have $\hat{\sigma}_{OC} = \hat{\sigma}_{SL} = 1$.

For solid-surfaced targets with low μ_C, the physical interpretation of the radar albedo is clear-cut, as the surface must be smooth at all scales within about an order of magnitude of the wavelength and the subsurface must lack structure at those scales down to several $1/e$ power absorption lengths L. Here we may interpret the radar albedo as the product $g\rho$, where ρ is the Fresnel power reflection coefficient at normal incidence and the backscatter gain g depends on target orientation, shape, and the distribution of surface slopes with respect to that shape. Large main-belt objects are expected to be covered with regoliths more than 15 m thick; in light of expectations about such objects' shapes and surface slope distributions, g is probably within a few tens of percent of unity, so $\hat{\sigma}_{OC}$ is a reasonable first approximation to ρ (Ostro et al. 1985). For the smaller, more irregularly shaped Earth-crossing asteroids, g might be a strong function of orientation, causing radar cross section to vary much more dramatically than A_{proj} as the object rotates. NEO albedos derived from observations with thorough rotation phase coverage might tend to "average out" variations in g, but possibly not enough to justify treating $\hat{\sigma}_{OC}$ as an approximation to ρ. Both ρ and L depend on interesting characteristics of the surface material, including bulk density, porosity, particle size distribution, and metal abundance (see, e.g., Ostro et al. 1991a and references therein).

If $\mu_C >> 0.1$, then physical interpretations are rarely unique, because models must consider not just the nature of the surface/space interface but also the regolith's structural and electrical properties, including the size distribution, spatial distribution, and scattering properties of subsurface rocks. One modeling complication is that multiply scattered radiation includes a diffusely scattered part as well as a "coherent backscatter" peak due to constructive interference between waves travelling on geometrically identical but time-reversed paths (Mishchenko 1992).

C. Time Delay and Doppler Frequency

In monostatic experiments, transmit/receive cycles, or runs, consist of transmission for a duration close to the signal's roundtrip time delay (until the first echoes are about to return), followed by reception of echoes for a similar duration. A bistatic configuration obviates transmit/receive cycling.

In continuous wave (cw) observations, one transmits an unmodulated, nearly monochromatic waveform and measures the distribution of echo power as a function of frequency. In ranging observations, modulation of the waveform permits measurement of the distribution of echo power in time delay as well. The echo time delay and Doppler frequency shift change continuously due to the relative motion of the target with respect to the radar. To avoid smearing of accumulated echoes in time and frequency, one tunes the receiver's front-end local oscillator according to an ephemeris based on an *a priori* orbit. Sometimes it is more convenient to take out the Doppler on the uplink, that is, to tune the transmitter continuously so the receiver sees a constant carrier frequency. In time-resolved experiments, one drifts the sampling time base according to the predicted rate of change of time delay to maintain constant registration of the samples with respect to the target's center of mass.

Time and frequency measurements have paramount importance in NEO radar astronomy, because the time-delay/Doppler-frequency distribution of echo power is the source of fine spatial resolution, and also because delay and Doppler are fundamental dynamical observables. In simple terms, the roundtrip time delay, τ, between transmission of a signal and reception of its echo is approximately $2R/c$, with c the speed of light and R the distance to the target. The time delay is 998 s for a target 1 AU from the radar, 2.6 s for the Moon, and typically between 10 s and 10 min for NEOs observed so far. The echo's Doppler frequency ν is approximately $2F_{tx}V_{rad}/c$, where F_{tx} is the transmitter carrier frequency and V_{rad} is the target's radial velocity; thus the magnitude of ν in Hertz is simply the radial velocity in half wavelengths per second.

Different parts of a rotating target have different velocities relative to the radar, so the echo will be dispersed in Doppler as well as in delay. The dispersion and the detailed functional form of the delay-Doppler distribution of echo power, $\sigma(\tau,\nu)$, depends on the target's size, shape, scattering characteristics, and orientation. For a sphere with diameter D and apparent rotation period P, echoes would have a delay depth $\Delta \tau_{TARGET} = D/c$ and a bandwidth $\Delta \nu_{TARGET} = (4\pi D \cos \delta)/\lambda P$, where δ is the angle between the radar line of sight and the target's equatorial plane. Radar experiments aim to constrain the target's properties by measuring $\sigma(\tau, \nu)$, perhaps with more than one combination of transmitted and received polarizations and perhaps as a function of time, i.e., as a function of the target's orientation and direction. Ideally, one would like to obtain $\sigma(\tau, \nu)$ with very fine resolution, sampling that function within cells whose dimensions, $\Delta \tau \times \Delta \nu$, are small compared to the echo dispersions (to achieve fine fractional resolution of the echoes) and as small as possible compared to the magnitudes of the echo's mean delay and Doppler (to permit refinement of the target's orbit).

D. Waveforms and Signal Processing

In cw experiments, complex voltage samples of the received signal are Fourier transformed and the resultant real and imaginary components are squared and summed to obtain an estimate of the power spectrum; the frequency resolution equals the reciprocal of the time series' length, i.e., of the coherence time. The number of fast Fourier transforms (FFTs) applied to data from a single transmit/receive cycle can range from one to tens of thousands. All NEOs are sufficiently narrowband for power spectra to be computed and accumulated in an array processor and recorded directly on magnetic tape, but often it is preferable to record voltages for post-real-time Fourier analysis, perhaps using FFTs of different lengths to obtain spectra at various frequency resolutions.

Delay resolution requires a modulated waveform. For example, with a coherent-pulsed-cw waveform, the transmitter's carrier-frequency oscillator operates continuously but power is radiated only during intervals that are one delay resolution cell long and occur at intervals called the pulse repetition period (PRP), which should exceed the target's delay depth to ensure that the echo will consist of successive, nonoverlapping range profiles. The reciprocal of the PRP is the maximum effective sampling rate at any given delay. Fourier transformation of N samples of the signal's complex voltage taken at the same position within each of N successive range profiles (i.e., the same delay relative to the delay of hypothetical echoes from the target's center of mass) yields the echo spectrum for the corresponding range cell on the target, with a frequency resolution $1/(N \times PRP)$. The bandwidth (1/PRP), and hence the recording rate and the spectral resolution achievable with a given FFT length, can be reduced by a factor of N_{coh} in real time by coherently summing N_{coh} successive, PRP-long time series of voltage samples.

Almost all NEO radar ranging has used a binary phase-coded cw waveform to simulate a coherent pulsed-cw waveform. The basic time interval Δt of the phase-coded waveform, called the baud, sets the delay resolution. Once every Δt seconds, the phase of the transmitted signal is either shifted by 180° or not, according to the value of the corresponding element in a binary code. Shift-register, "pseudo-random" binary codes, which are easy to generate and have very sharply peaked autocorrelation functions, are ubiquitous in radar astronomy. Shift-register code lengths equal $2M-1$, with M a whole number. In most delay/Doppler experiments the code is repeated continuously during the transmission, so the PRP is the product of the baud and the code length. The received signal is decoded by cross-correlating it with a replica of a single code cycle.

Current planetary radars work with waveforms that provide time resolution as fine as 0.1 μs. This limit, set by the 10-MHz modulation bandwidth of klystron amplifiers, corresponds to 15 m range resolution. Bounds on the frequency resolution $\Delta\nu$ are set primarily by the reciprocal of coherence times of recordable data sets; in monostatic experiments, $\Delta\nu \geq 1/\text{RTT}$.

E. Ephemerides and Delay-Doppler Astrometry

A singularly important aspect of NEO radar astronomy is the precision and reliability of time/frequency measurements that are made possible by high-speed data acquisition systems and stable, accurate clocks and frequency standards (see, e.g., Seidelmann et al. 1992). This fine measurement precision places stringent demands on the accuracy of NEO ephemerides, because the predictions of delay and Doppler [$\tau_{\text{eph}}(t)$ and $\nu_{\text{eph}}(t)$] must be accurate enough to prevent smearing of echoes, which would compromise the data's SNR and delay/Doppler resolution. For example, because

$$d\tau(t)/dt = -\nu(t)/F_{tx} \tag{7}$$

a Doppler prediction error of $\Delta\nu_{\text{eph}}$ will cause echoes integrated over one roundtrip time RTT to drift in time delay by $(-\Delta\nu_{\text{eph}}/F_{tx})\text{RTT}$. For new NEOs, errors in prediction ephemerides are very large and grow rapidly, because orbits must be estimated from optical astrometric data that span very short arcs. During initial radar observations of such an object at 0.04 AU (RTT~40 s) the delay uncertainty might be 0.4 s (~10 Earth radii) and the 2380-MHz Doppler uncertainty might be 1 kHz (or 17 μs of delay smear in a 40-s receive period). A delay measurement with a 40-μs baud could reduce the instantaneous delay uncertainty by 4 orders of magnitude, allowing one to generate more accurate delay-Doppler predictions, which would permit much more precise radar astrometry, and so on. This iterative, "bootstrapping" process has long characterized both the radar improvement of orbits and the refinement of the models and computational techniques that are the basis of the planetary ephemerides (see, e.g., Standish et al. 1992).

Current needs of NEO radar astronomy are served by the programs PEP770 at the Harvard-Smithsonian Center for Astrophysics and DE200 at the Jet Propulsion Laboratory. Prior to the advent of electronic networks, ephemerides were transported by magnetic tape. Nowadays, ephemerides are sent over the Internet and it is not uncommon for an NEO track on any given day to use an ephemeris based on all radar and optical astrometry reported through the previous day. In the near future, computation of updated ephemerides will be possible with on-site computers, permitting bootstrapping of new generations of ephemerides from delay-Doppler measurements as soon as they are made. With such a system, it should be possible to move from an initial detection to high-resolution imaging within one several-hour track. Most available NEO radar astrometry consists of an estimate of $\tau(t)$ or $\nu(t)$ for echoes received at the telescope's reference point at a specified Coordinated Universal Time (UTC) epoch t. For example, Arecibo observations are referenced to the center of curvature of the main reflector. Usually it is adequate to think of the offsets, $\tau_0(t)=\tau(t)-\tau_{eph}(t)$ and $\nu_0(t)=\nu(t)-\nu_{eph}(t)$, of the echoes from the ephemeris predictions as being constant over the pertinent measurement time scales, typically from one roundtrip time to a few hours. In practice, one measures τ_0 or ν_0 and reports $\tau(t)$ or $\nu(t)$ for a convenient epoch near the weighted mean time of the measurements.

Radar astrometry during the discovery apparition can ensure optical recovery of newly discovered NEOs, because delay-Doppler measurements have fine fractional precision and are orthogonal to optical, angular-position measurements (Yeomans et al. 1987). As discussed in the Chapter by Yeomans and Chodas, radar astrometry commonly improves upon the accuracy of optical-only ephemerides of newly discovered Earth-crossing asteroids by 1 to 3 orders of magnitude. Even for asteroids with very long astrometric histories and secure orbits, radar measurements can significantly shrink positional error ellipsoids for decades, with direct implications for the navigation of spacecraft to asteroids and predictions of extremely close approaches of asteroids to Earth.

F. Delay-Doppler Imaging

Radar can image NEOs if the echoes are strong enough. Continuous wave (cw) observations yield echo spectra, $\sigma(\nu)$, that can be thought of as one-dimensional images, or brightness scans across the target through a slit parallel to the asteroid's apparent (synodic) spin vector. The bandwidth of a target's instantaneous echo power spectrum is proportional to the breadth, measured normal to the line of sight, of the target's pole-on silhouette, and measurements of echo edge frequencies as functions of rotation phase can be used to estimate the shape (and the size in units of km/cosδ) of the convex envelope, or hull, of the silhouette as well as the frequency of hypothetical echoes from the asteroid's center of mass (Ostro et al. 1988).

Time-modulated waveforms yield a range profile $\sigma(\tau)$ and in most cases a delay-Doppler image $\sigma(\tau,\nu)$. Parallax effects and the curvature of the wave front at the target are negligible for groundbased radar observations, so contours of constant delay are intersections of the target's surface with planes perpendicular to the line of sight. Constant-Doppler planes are parallel to the line of sight and also parallel to the target's synodic spin vector. Thus one can imagine two orthogonal sets of parallel planes that cut the target into delay-Doppler cells like one dices a potato to make french fries. For a spherical target viewed equatorially, each bin of the dicer that contains the equator will define one surface cell, while each bin that cuts the target but does not contain the equator will define one surface cell in the northern hemisphere and one in the southern hemisphere; i.e., for most of the sphere there is a two-to-one mapping from surface coordinates to delay-Doppler coordinates, so a delay-Doppler image is north/south ambiguous. For an equatorial view, any pair of N/S ambiguous points have longitudes that are identical and latitudes that have the same magnitude but opposite signs, i.e., the two points are symmetrically located with respect to the target's equatorial plane. These points execute identical delay-Doppler trajectories as the target rotates. However, if the subradar latitude is nonzero, two points at any given longitude and opposite latitudes execute different delay-Doppler trajectories as a function of rotation phase, so inversion of delay-Doppler images taken at a variety of phases can overcome the N/S ambiguity. If the shape of the target is known *a priori*, one can solve for the global distribution of albedo (Hudson and Ostro 1990); this approach may be suitable for the largest main-belt asteroids. For NEOs, shape is the fundamental unknown, so it makes sense for an inversion to solve for the shape under the assumption of uniform scattering, at least at first (Hudson 1993). Such an inversion would also model the spin vector and the delay-Doppler trajectory of the target's center of mass, and hence would involve a rather complex parameter space. The estimation accuracy for the various parameters will depend on the geometrical leverage of the data (i.e., the subradar longitude/latitude coverage), the data's SNR and fractional delay-Doppler resolution, and of course the target's physical configuration. For example, images of a very flat object viewed off the equator would have no N/S ambiguities, whereas images of a highly nonconvex, multi-component target conceivably could have four-fold ambiguities. If an imaging data set has thorough subradar longitude coverage and also samples northern and southern middle latitudes, then the target's shape, center-of-mass location, and spin vector would be so well constrained that it would be possible to discern the existence of internal density variation.

Several other attributes of radar images deserve mention. First, whereas optical images are projections of brightness onto a plane normal to the line of sight, a delay-Doppler image is the projection of the target's radar brightness onto a plane containing the radar and parallel to the line of sight. Second, the term "radar image" usually refers to a measured distribution of echo power in delay, Doppler, and/or angular coordinates, while the term "radar map" usually refers to a display in target-centered coordinates of the residuals with respect to a model that parameterizes the target's average scattering properties. Third, radar images are time exposures, because the frequency resolution is the reciprocal of the duration of the time series that is coherently processed into a single power spectrum (one look), and the fractional self-noise in an incoherent sum of L looks is $L^{-1/2}$. Hence there is a trade-off between spatial resolution, self-noise, and motion-induced smearing. [This topic is addressed by Stacy (1993) in the context of high- resolution radar mapping of the Moon. All the topics in this section are discussed in detail by Ostro (1993 and references therein).]

III. THE CURRENT STATE OF NEO RADAR RECONNAISSANCE

A. Summary of Investigations Through 1992

Table I lists radar-detected comets and near-Earth asteroids. Papers reporting radar results for specific targets are cited in that table. Pre-1991 NEO radar astrometry is presented by Ostro et al. (1991b). Yeomans et al. (1992) outline techniques required to use delay-Doppler measurements in orbit estimations and give orbits for 34 targets based on the combined radar and optical data. As an example of recent radar astrometry, observations of 1991 JX with a 0.2 μs baud yielded a time-delay estimate whose fractional precision is 5×10^{-9} (Ostro et al. 1991c). Table II shows how radar astrometry during that object's discovery apparition yields a 300-fold improvement over optical-only prediction of that object's location decades after discovery. The most precise NEO radar measurements to date were obtained for Toutatis in 1992: a time-delay measurement with $\Delta\tau = 0.125$ μs and a fractional precision of 2×10^{-9}, and a Doppler measurement at $F_{tx} = 8510$ MHz with $\Delta v = 0.0083$ Hz (equivalent to $V_{rad} = 150$ μm s^{-1}) and a fractional precision of 2×10^{-8}. During the past decade, observations of newly discovered objects have revealed range-prediction errors from ~ 100 km to $\sim 100{,}000$ km (Fig. 2).

Radar cross sections and circular polarization ratios for most of the radar-detected Earth-crossing asteroids were reported by Ostro et al. (1991b). Figure 3 plots values of $\hat{\sigma}_{OC}$ vs μ_C for several Earth-crossing asteroids; there clearly is a great deal of diversity evident even in these objects' disc-integrated radar properties. In terms of imaging, the two most productive experiments so far were the observations of 4769 Castalia in 1989 and of 4179 Toutatis in 1992. The key Castalia images were taken in a 2.5-hr period, cover $\sim 220°$ of rotation phase at a nonzero subradar latitude, and consist of 64 frames, each of which was constructed from 26 looks and places a few dozen pixels on the target. The data are adequate to define the shape at a scale ~ 50 m, or $\sim 5\%$ of the object's maximum overall dimension, as well as the spin period and the subradar latitude (Hudson and Ostro 1994). The Toutatis imaging experiment spanned 2.5 weeks and $\sim 125°$ of geocentric direction, but no more than two synodic rotations. The most useful images place more than 1000 pixels on the target. However, because the resolution is so fine, rotational motion and ephemeris drift are evident over an interval containing enough looks to shrink the self-noise to a comfortable level. Hence, extraction of the full information contained in those images will rely on a model that parameterizes the asteroid's shape at very fine scales.

Both Castalia and Toutatis are strongly bifurcated objects. Very coarse-resolution images of 1627 Ivar and 1986 DA, as well as echo spectra for 2201 Oljato and 3908 (1980 PA), show clearly bimodal distributions of echo power. Thus some 20% of the radar-detected near-Earth asteroids show at least some indication of double-lobed structure; for another 20% the SNR was too low for useful spatial resolution. Melosh and Stansberry (1991) analyzed the occurrence of widely separated doublet craters on Earth and suggested

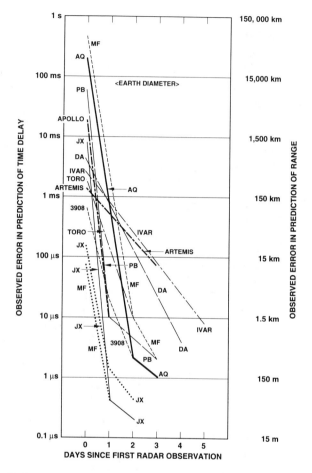

Figure 2. Radar reduction of instantaneous range-prediction error for selected asteroids vs days since the initial radar detection of the object (or, for 1685 Toro,
since the first detection during the 1988 apparition) at Goldstone (dotted curves) or
Arecibo (all other curves). All objects are Earth crossers except main-belt asteroid
105 Artemis. Two-letter abbreviations correspond to 1986 DA, 1989 PB, 1990 MF,
1991 AQ, and 1991 JX. For 1990 MF and 1991 JX, Arecibo astrometry was included
in calculation of ephemerides for Goldstone observations, which began eight days
and three days, respectively, after the last Arecibo observations.

that some 10% of the estimated ~2000 km-sized Earth-crossing asteroids
may be well separated binary asteroids. "Double" objects may therefore be
fairly common in the Earth-crossing asteroid population. Of course, it would
be highly desirable to know an object's gross physical configuration prior to
spaceborne reconnaissance.

B. Cometary Nuclei and Large-Particle Clouds

Because a coma is nearly transparent at radio wavelengths, radar is better

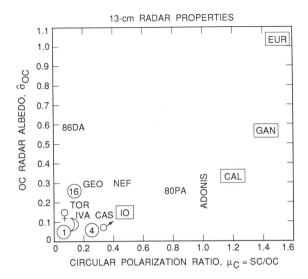

Figure 3. 13-cm radar properties for near-Earth asteroids 1986 DA, 3199 Nefertiti, 1620 Geographos, 1980 PA, 1685 Toro, 1627 Ivar, 4769 Castalia, and 2101 Adonis (whose albedo is uncertain), compared to those for other planetary targets. Symbols are used for the Moon, Venus, and Mars; the circled numbers denote main-belt asteroids 1 Ceres, 4 Vesta, and 16 Psyche; and rectangles identify the Galilean satellites Io, Europa, Ganymede, and Callisto.

equipped to inspect a cometary nucleus than are optical and infrared methods (see, e.g., Kamoun et al. 1982), and radar observations of several comets (Table I) have provided some useful constraints on nuclear dimensions, spin vectors, and surface morphologies. The most informative experiment to date, of IRAS-Araki-Alcock, which came within 0.03 AU of Earth in 1983 (Fig. 4), yielded echoes with a narrowband component from the nucleus as well as a much weaker broadband component attributed to large particles ejected mostly from the sunlit side of the nucleus (Harmon et al. 1989). The particles are probably each several centimeters in size and appear to be distributed within ∼1000 km of the nucleus, i.e., in the volume filled by particles ejected at several meters per second over a few days. Radar observations of comet Halley (Campbell et al. 1989) yielded echoes with a substantial broadband component but no component as narrowband as that expected from the nucleus, whose dimensions and spin vector were constrained by Giotto and Vega images. The echo's bandwidth and radar cross section suggest that it arises predominantly from coma particles with radii >2 cm. Hence at least two very different comets have been accompanied by swarms of large particles. There are obvious implications for spacecraft encounters with comets, including the extreme case of terminal interception.

TABLE I

Radar-Detected NEOs[a]

Year	Target	Reference (Site, λ, cm)
1968	1566 Icarus	1 (G, λ13); 2(H, λ3.8)
1972	1685 Toro	3 (G, λ13)
1975	433 Eros	4 (G, λ3.5, λ13); 5 (A, λ70)
1976	1580 Betulia	6 (A, λ13)
1980	1685 Toro[b]	7 (A, λ13)
	1862 Apollo	8 (A, λ13); 9 (G, λ3.5)
	comet Encke	10 (A, λ13)
1981	1915 Quetzalcoatl	8 (A, λ13)
	2100 Ra-Shalom	11 (A, λ13)
1982	comet Grigg-Skjellerup	12 (A, λ13)
1983	1620 Geographos	8 (A, λ13)
	comet IRAS-Araki-Alcock	13 (A, λ13); 14 (G, λ3.5)
	2201 Oljato	8 (A, λ13)
	comet Sugano-Saigusa-Fujikawa	15, 16 (A, λ13)
1984	2101 Adonis	8 (A, λ13)
	2100 Ra-Shalom[b]	8 (A, λ13)
1985	1627 Ivar	17 (A, λ13)
	1036 Ganymed	8 (A, λ13)
	comet Halley	18 (A, λ13)
	1866 Sisyphus	8 (A, λ13)
1986	1986 DA	19 (A, λ13)
	1986 JK	20 (G, λ3.5)
	3103 (1982 BB)	8 (A, λ13)
	3199 Nefertiti	8 (A, λ13)
1987	1981 Midas	8 (G, λ3.5)
	3757 (1982 XB)	8 (A, λ13)
1988	1685 Toro[c]	8 (A, λ13)
	3908 (1980 PA)	8 (A, λ13; G, λ3.5)
	433 Eros[b]	8 (A, λ13)
1989	4034 (1986 PA)	8 (A, λ13)
	1580 Betulia[b]	8 (G, λ3.5; A, λ13)
	1989 JA	8 (A, λ13; G, λ3.5)
	4769 Castalia	21 (A, λ13; G, λ3.5)
	1917 Cuyo	8 (A, λ13; G, λ3.5)
1990	1990 MF	8 (A, λ13; G, λ3.5)
	1990 OS	8 (G, λ3.5)
	4544 Xanthus	8 (A, λ13)
1991	1991 AQ	22 (A, λ13)
	1991 JX	22 (A, λ13; G, λ3.5)
	3103 (1982 BB)[b]	23 (G, λ3.5)
	1991 EE	23 (A, λ13); 24 (GV, λ3.5)

TABLE I (cont.)

1992	1981 Midas[b]	23 (G, λ3.5)
	5189 (1990 UQ)	23 (G, λ3.5)
	4179 Toutatis	25 (A, λ13; G, λ3.5);
		26 (YE, λ6); 27 (GV, λ3.5)

[a] Site abbreviations correspond to Goldstone, Haystack, Arecibo, Goldstone-VLA and Yevpatoria-Effelsberg. Observations are listed and cited chronologically. References: 1. Goldstein (1969a, b); 2. Pettengill et al. (1969); 3. Goldstein et al. (1973); 4. Jurgens and Goldstein (1976); 5. Campbell et al. (1976); 6. Pettengill et al. (1979); 7. Ostro et al. (1983); 8. Ostro et al. (1991b); 9. Goldstein et al. (1981); 10. Kamoun et al. (1982); 11. Ostro et al. (1984); 12. Kamoun (1983); 13. Harmon et al. (1989); 14. Goldstein et al. (1984); 15. Campbell et al. (1983); 16. Harmon, personal communication; 17. Ostro et al. (1990a); 18. Campbell et al. (1989); 19. Ostro et al. (1991a); 20. Ostro et al. (1989); 21. Ostro et al. (1990b); 22. Ostro et al. (1991c); 23. Ostro et al., unpublished research; 24. de Pater et al. (1992); 25. Ostro et al. (1993); 26. Zaytsev et al. (1993); 27. de Pater et al. (1994).
[b] Second apparition yielding radar detection.
[c] Third apparition yielding radar detection.

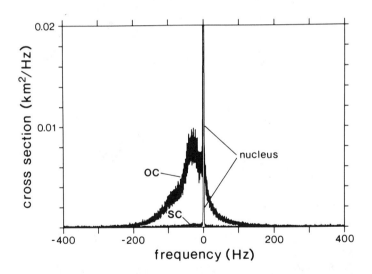

Figure 4.   Arecibo *OC* and *SC* echo spectra obtained for Comet IRAS-Araki-Alcock, truncated at 2% of the maximum *OC* amplitude. The narrowband echo from the nucleus is flanked by broadband echo from large particles in a cloud surrounding the nucleus. (Harmon et al. 1989.)

IV. SCENARIOS FOR RADAR RECONNAISSANCE OF POTENTIAL NEO HAZARDS

A. The Immediate Future

Arecibo and Goldstone have been the primary NEO radar observatories during the past decade. This situation is unlikely to change during the next decade, notwithstanding prospects for occasional G-VLA and Y-E bistatic observations of NEOs that make very close approaches with lead times of at least several months. Similarly, efforts to convert existing military radars to other applications may result in detection of some NEOs by those instruments, but it is unlikely that such endeavors will be very competitive with Arecibo/Goldstone experiments in the near future.

TABLE II
Effect of Delay-Doppler Measurements on Extrapolation of
1991 JX's Orbit from Discovery-Apparition Astrometry

Data Set	Positional Uncertainty[a]		
	Kilometers	Earth Radii	Lunar Distances
Optical	8,000,000	1260	21
Optical + Radar	25,000	4	0.07

[a] Uncertainties in the asteroid's April 2019 position predicted from 54 optical measurements from May 9 to July 3 are in the top row. The bottom row gives uncertainties for a prediction that includes 24 delay-Doppler measurements from June 5–15 (D. K. Yeomans, personal communication).

TABLE III
Range Limits (AU) for Radar Astrometry of New NEOs

Diameter	Arecibo[a]	Goldstone[a]
10 m	0.042	0.018
100 m	0.10	0.042
1 km	0.24	0.10
10 km	0.56	0.24

[a] These limits correspond to a single-date SNR of 20 and telescope sensitivities expected by 1995.

Table III lists the radar-astrometric range limits expected for Arecibo and Goldstone by 1995, when the hardware upgrades mentioned earlier (Sec. II.A) should be completed. Most of the optically discoverable Earth crossers traverse the joint Arecibo-Goldstone detectability window at least once every few decades. Arecibo, with nearly 40 times the sensitivity of Goldstone (Fig. 1), will see twice as far and cover three times as much volume as Goldstone, and hence will be the key instrument for NEO radar reconnaissance. Goldstone, with a solid angle window twice the size of Arecibo's and an hour-angle window at least several times wider than Arecibo's for any given target, will serve a complementary role, especially for newly discovered objects.

Discovery apparition geometry often is exceptionally favorable to radar reconnaissance (see the Chapter by Yeomans and Chodas). For this reason, and in view of the utility of radar data for orbit refinement and physical characterization, there is compelling motivation to do radar observations of newly discovered Earth crossers whenever possible. However, NEOs pass through each instrument's radar detectability window very rapidly and the timing of an object's discovery relative to its passage through the Arecibo and Goldstone windows is unpredictable.

Figure 5 plots SNRs that could have been attained by the upgraded (post-1994) Arecibo and Goldstone telescopes between June 1990 and June 1991 for Earth-crossing asteroids discovered during that period. As illustrated by that figure, it is common for Earth-approachers to be discovered after they have left one or both radar windows. One of Goldstone's jobs will be to follow-up discoveries that Arecibo cannot. Similarly, many new NEOs pass through both windows after discovery, but during intervals separated by days or weeks. In such situations, even if Arecibo observations are possible, Goldstone can extend the orbital coverage of the radar observations, thereby lengthening the astrometric arc and improving the orbit estimation. Furthermore, depending on the target's echo strength during the "first" instrument's observations and the quality of the optical astrometry used to make the initial radar ephemeris, measurements with high-resolution waveforms may not be possible before the target leaves the first instrument's window. In this case, the second instrument would inherit an improved delay-Doppler ephemeris that would permit quick progression to high-resolution waveforms, extension of the astrometric arc, and efficient determination of physical properties. Experiences with 1989 JA, 1989 PB, 1990 MF, 1991 JX, and Toutatis were like this.

Given the capabilities of existing radars, how much of the potential follow-up work on NEOs will actually be done? Minimal reconnaissance of a new NEO will require at least one block of time, probably at least two hours long, on one of a handful of possible dates, to be scheduled with extremely short notice (typically on the order of a few days to a few weeks). During 1989–1993, 11 new NEOs were observed at Arecibo and/or Goldstone under such circumstances, causing difficult scheduling adjustments. By 2000, the upgraded Arecibo might have nearly monthly opportunities to make thousand-pixel images of an Earth-crossing asteroid during a post-discovery apparition and weekly opportunities to do orbit-securing astrometry on a new object (Table IV). Imaging with enough coverage of subradar longitude and/or latitude to allow high-precision reconstruction of shape and spin vector probably would require at least one or two full tracks, i.e., much more than just an "astrometric" detection. It will not be easy for Arecibo to establish policy for dealing with an onslaught of target-of-opportunity situations. The observatory, a national center operated primarily for visitors engaged in passive radio astronomy and ionospheric physics, is oversubscribed and operates around the clock daily. (About 4% of the time has been used for planetary radar.) The pressure on the schedule after the upgrade will be severe. It seems prudent to assume that it may not be possible for Arecibo to look at more than a few tens of percent of the new NEOs that it could see. While some fraction of new NEO radar opportunities will be observable with Goldstone and perhaps the Y-E system, the pressure on DSN tracking allocations from flight projects is severe and is increasing, and the logistical impediments to short-notice Y-E observations are daunting.

TABLE IV
Intervals Between Earth-Crossing Asteroid Radar Opportunities[a]

Early 1990s

D (km)	SNR = 10,000 R (AU)	int'l	SNR = 1,000 R (AU)	int'l	SNR = 100 R (AU)	int'l	SNR = 20 R (AU)	int'l
3	0.078	45 yr	0.014	10 mon	0.2	4 mon	0.37	2 mon
1	0.052	2 yr	0.092	5 mon	0.16	7 wk	0.25	20 d
0.3	0.033	8 yr	0.059	2 yr	0.1	7 mon	0.16	3 mon
0.1	0.022	25 yr	0.038	8 yr	0.067	2 yr	0.10	1 yr

After Spaceguard

D (km)	SNR = 10,000		SNR = 1,000		SNR = 100		SNR = 20	
	R (AU)	int'l	R (AU)	int'l	R (AU)	int'l	R (AU)	int'l
3	0.078	10 yr	0.14	2 mon	0.24	3 wk	0.37	10 d
1	0.052	2 mon	0.092	2 wk	0.16	1 wk	0.25	2 d
0.3	0.033	1 mon	0.059	1 wk	0.1	2 d	0.16	1 d
0.1	0.022	1 mon	0.038	2 wk	0.067	3 d	0.10	1 d

[a] Mean intervals between Earth-crossing asteroid approaches close enough to yield four values of single-date SNRs with the upgraded Arecibo telescope (but not necessarily traversing Arecibo's declination window at closest approach). Nominal asteroid properties were assumed in calculation of distances (R) for each SNR and target diameter. The intervals were scaled from values taken by eye from Fig. 2 of Shoemaker et al. (1990).

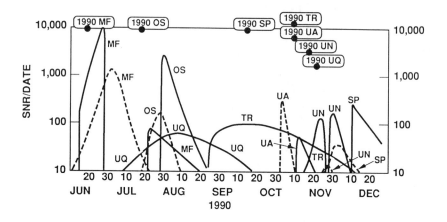

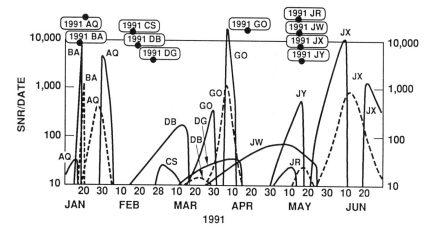

Figure 5. Radar investigations of new NEAs. Single-date SNRs during June 1990 to June 1991 for Earth-crossing asteroids discovered during that period, with upgraded (post-1994) capabilities assumed for Arecibo (solid curves) and Goldstone (dashed curves). The horizontal position of the dot on the border around each asteroid designation indicates the discovery date. Curves for seven other asteroids discovered during the period from June 1990 to June 1991, but not observable then, are not shown.

B. The Post-Spaceguard Era

If the Spaceguard survey comes to pass, the catalog of known Earth-crossing asteroids will swell to over 100 times its current size. Soon after the survey gets started, the frequency with which known Earth-crossing asteroids traverse the radar windows will dwarf that in the Fig. 5 simulation. Less than half way through the survey there will be several times as many catalogued Earth-crossing asteroids as there are numbered main-belt asteroids today. The initial orbits of newly discovered objects will be inaccurate, and the volume of follow-up work needed to secure orbits will rapidly become enormous. Reliable extrapolation of orbits will not be possible until the astrometric database matures, either through protracted optical follow-up over a long time span or with radar measurements over a very much shorter time span. The appeal of using radar for orbit refinement will grow in proportion to the number of predicted close approaches of new objects, because of the anxiety instilled by the uncertainties in the extrapolated orbits. Moreover, although sub-100-m objects are far below the threshold for global climatic effects, people may not find that qualification very comforting if estimates of the impact parameter for an object on an extremely close-approach trajectory have huge uncertainties. Of course, any predictions of very close approaches by large asteroids will be taken very seriously.

Consider the inevitable discovery of objects that threaten to come very close, e.g., within one lunar distance, keeping in mind the fact that the frequency of cislunar misses is ~3600 times higher than the collision frequency. For example, consider 10,000-megaton impacts (objects ~500 m in diameter), which are at the low end of the range of estimates for the local/global transition. Collisions with such objects are 63,000 yr apart on average, so cislunar misses are 18 yr apart. Spaceguard will discover ~6400 of those objects, or ~70% of the ~9200 in the population. One of those 6400 objects will make a cislunar miss every ~25 years on average. However, uncertainties in orbital extrapolations much more than a century into the future may not permit confident distinction of close-call trajectories from impact trajectories, at least until the optical astrometric time base is many decades long. Of course, this example could have considered the flux of objects within 10 lunar distances, or whatever; the important quantity is the impact parameter in units of its uncertainty.

Society will surely want to reduce uncertainties associated with predictions of extremely close approaches of NEOs, and radar is the most efficient groundbased technique for trajectory refinement. However, existing instrumentation will hardly be able to do discovery-apparition observations of the bulk of Earth-crossing asteroids found with Spaceguard. That task would require instrumentation dedicated to this work, at least as sensitive as the upgraded Arecibo, and able to see most of the sky. Two fully steerable radars, one each in the northern and southern hemispheres, designed specifically for NEO reconnaissance, would satisfy these requirements. (None of the active instruments was optimized for planetary radar astronomy.) With current technology, it should be possible to build a steerable telescope more sensitive than the upgraded Arecibo for as little as $ 100 M. For $ 1 B, it should be possible, even with 20th century technology, to build six globally distributed radar telescopes, each an order of magnitude more sensitive than the upgraded Arecibo. The radar and Spaceguard optical nets could be linked and highly automated, so radar could acquire new objects right after they were found. Radar sequences and ephemeris refinement could be run by an intelligent program that could progress from initial detection to imaging within minutes. Six telescopes could respond to ~100 new objects daily. In terms of the cost of information gained, each radar telescope would be considerably cheaper than a spacecraft flyby.

Once Spaceguard begins, there will be an outstanding imaging opportunity every month or so (Table IV). By then, the sophistication of inversion methods may allow delay-Doppler images to be piped in real time to software that will return a "running reconstruction" of the target's shape. In principle, an IAU telegram consisting of an animation file showing the three-dimensional model, properly oriented as a function of time, could be circulated globally within days of discovery.

C. The Next Millennium

At any time in the future, the role of groundbased radar in response to the NEO hazard will depend on the state of technology and the nature of civilization, neither of which can be confidently foreseen more than a few decades hence. If civilization endures through the next millennium, our descendents may decide to maintain a groundbased radar capability indefinitely as insurance against hazards from long-period comets (LPCs). Note that the risk of a civilization-ending impact during the next century is about the same as the risk of a civilization-ending LPC impact during the next millennium. Also note that whereas the warning time for an asteroid impact (in the post-Spaceguard era) is likely to be at least a century and hence more than adequate for mitigation at a comfortable pace, the warning time for an LPC impact would probably be less than one year. Moreover, LPC trajectory extrapolation will be hampered by obscuration of the nucleus and by uncertainties about nongravitational forces. Several cislunar misses by LPCs can be expected during the next millennium. The uncertainty in trajectory extrapolation after discovery of these objects could be terrifying, and any number of panic scenarios are possible. Needless to say, radar astrometry would be precious under such circumstances.

Acknowledgments. This research was conducted at the Jet Propulsion Laboratory, California Institute of Technology, under contract with the National Aeronautics and Space Administration.

REFERENCES

Campbell, D. B., Pettengill, G. H., and Shapiro, I. I. 1976. 70-cm radar observations of 433 Eros. *Icarus* 28:17–20.
Campbell, D. B., Harmon, J. K., Hine, A. A., Shapiro, I. I., Marsden, B. G., and Pettengill, G. H. 1983. Radar observations of comets IRAS-Araki-Alcock and

Sugano-Saigusa-Fujikawa. *Bull. Amer. Astron. Soc.* 15:800 (abstract).

Campbell, D. B., Harmon, J. K., and Shapiro, I. I. 1989. Radar observations of comet Halley. *Astrophys. J.* 338:1094–1105.

de Pater, I., Palmer, P., Snyder, L. E., Ostro, S. J., Yeomans, D. K., and Mitchell, D. L. 1992. Bistatic radar observations of asteroids 324 Bamberga, 7 Iris, and 1991 EE. *Bull. Amer. Astron. Soc.* 24:934 (abstract).

de Pater, I., Palmer, P., Mitchell, D. L., Ostro, S. J., Yeomans, D. K., and Snyder, L. E. 1994. Radar aperture synthesis observations of asteroids. In preparation.

Goldstein, R. M. 1969a. Radar observations of Icarus. *Icarus* 10:430–431.

Goldstein, R. M. 1969b. Radar observations of Icarus. *Science* 162:903–904.

Goldstein, R. M., Holdridge, D. B., and Lieske, J. H. 1973. Minor planets and related objects. XII. Radar observations of 1685 Toro. *Astron. J.* 78:508–509.

Goldstein, R. M., Jurgens, R. F., and Yeomans, D. K. 1981. Radar observations of Apollo. *Icarus* 48:59–61.

Goldstein, R. M., Jurgens, R. F., and Sekanina, Z. 1984. A radar study of comet IRAS-Araki-Alcock 1983d. *Astron. J.* 89:1745–1754.

Harmon, J. K., Campbell, D. B., Hine, A. A., Shapiro, I. I., and Marsden, B. G. 1989. Radar observations of comet IRAS-Araki-Alcock 1983d. *Astrophys. J.* 338:1071–1093.

Hudson, R. S. 1993. Three-dimensional reconstruction of asteroids from radar observations. *Remote Sensing Revs.* 8:195–203.

Hudson, R. S., and Ostro, S. J. 1990. Doppler radar imaging of spherical planetary surfaces. *J. Geophys. Res.* 95:10947–10963.

Hudson, R. S., and Ostro, S. J. 1994. Shape of asteroid 4769 Castalia (1989 PB) from inversion of radar images. *Science* 263:940–943.

Jurgens, R. F., and Goldstein, R. M. 1976. Radar observations at 3.5 and 12.6 cm wavelength of asteroid 433 Eros. *Icarus* 28:1–15.

Kamoun, P. G. 1983. Radar Observations of Cometary Nuclei. Ph.D. Thesis, Massachussetts Inst. of Technology.

Kamoun, P. G., Campbell, D. B., Ostro, S. J., Pettengill, G. H., and Shapiro, I. I. 1982. Comet Encke: Radar detection of nucleus. *Science* 216:293–296.

Melosh, H. J., and Stansberry, J. A. 1991. Doublet craters and the tidal disruption of binary asteroids. *Icarus* 94:171–179.

Mishchenko, M. I. 1992. Polarization characteristics of the coherent backscatter opposition effect. *Earth, Moon, Planets* 58:127–144.

Morrison, D., ed. 1992. *The Spaceguard Survey: Report of the NASA International Near-Earth-Object Detection Workshop* (Pasadena: Jet Propulsion Laboratory).

Ostro, S. J. 1993. Planetary radar astronomy. *Rev. Mod. Physics* 65:1235–1279.

Ostro, S. J., Campbell, D. B., and Shapiro, I. I. 1983. Radar observations of asteroid 1685 Toro. *Astron. J.* 88:565–576.

Ostro, S. J., Harris, A. W., Campbell, D. B., Shapiro, I. I., and Young, J. W. 1984. Radar and photoelectric observations of asteroid 2100 Ra-Shalom. *Icarus* 60:391–403.

Ostro, S. J., Campbell, D. B., and Shapiro, I. I. 1985. Mainbelt asteroids: Dual-polarization radar observations. *Science* 224:442–446.

Ostro, S. J., Connelly, R., and Belkora, L. 1988. Asteroid shapes from radar echo spectra: A new theoretical approach. *Icarus* 73:15–24.

Ostro, S. J., Yeomans, D. K., Chodas, P. W., Goldstein, R. M., Jurgens, R. F., and Thompson, T. W. 1989. Radar observations of asteroid 1986 JK. *Icarus* 78:382–394.

Ostro, S. J., Campbell, D. B., Hine, A. A., Shapiro, I. I., Chandler, J. F., Werner, C. L., and Rosema, K. D. 1990a. Radar images of asteroid 1627 Ivar. *Astron. J.* 99:2012–2018.

Ostro, S. J., Chandler, J. F., Hine, A. A., Shapiro, I. I., Rosema, K. D., and Yeomans,

D. K. 1990*b*. Radar images of asteroid 1989 PB. *Science* 248:1523–1528.

Ostro, S. J., Campbell, D. B., Chandler, J. F., Hine, A. A., Hudson, R. S., Rosema, K. D., and Shapiro, I. I. 1991*a*. Asteroid 1986 DA: Radar evidence for a metallic composition. *Science* 252:1399–1404.

Ostro, S. J., Campbell, D. B., Chandler, J. F., Shapiro, I. I., Hine, A. A., Velez, R., Jurgens, R. F., Rosema, K. D., Winkler, R., and Yeomans, D. K. 1991*b*. Asteroid radar astrometry. *Astron. J.* 102:1490–1502.

Ostro, S. J., Harmon, J. K., Hine, A. A., Perillat, P., Campbell, D. B., Chandler, J. F., Shapiro, I. I., Jurgens, R. F., and Yeomans, D. K. 1991*c. Bull. Amer. Astron. Soc.* 23:1144 (abstract).

Ostro, S. J., Jurgens, R. F., Rosema, K. D., Winkler, R., Howard, D., Rose, R., Slade, M. A., Yeomans, D. K., Campbell, D. B., Perillat, P., Chandler, J. F., Shapiro, I.I., Hudson, R. S., Palmer, P., and de Pater, I. 1993. Radar imaging of asteroid 4179 Toutatis. *Bull. Amer. Astron. Soc.* 25:1126 (abstract).

Pettengill, G. H., Shapiro, I. I., Ash, M. E., Ingalls, R. P., Rainville, L. P., Smith, W. B., and Stone, M. L. 1969. Radar observations of Icarus. *Icarus* 10:432–435.

Pettengill, G. H., Ostro, S. J., Shapiro, I. I., Marsden, B. G., and Campbell, D. B. 1979. Radar observations of asteroid 1580 Betulia. *Icarus* 40:350–354.

Sagan, C. 1992. Between Enemies. *Bull. Atomic Soc.* 48:24–26.

Sagan, C., and Ostro, S. J. 1994*a*. Dangers of asteroid deflection. *Nature* 368:501.

Sagan, C., and Ostro, S. J. 1994*b*. Long-range consequences of interplanetary collisions. *Issues Sci. Tech.* X(4):67–72.

Seidelmann, P. K., Guinot, B., and Doggett, L. E. 1992. Time. In *Explanatory Supplement To The Astronomical Almanac*, ed. P. K. Seidelmann (Mill Valley, Ca.: University Science Books), pp. 39–93.

Shoemaker, E. M., Wolfe, R. F., and Shoemaker, C. S. 1990. Asteroid and comet flux in the neighborhood of Earth. In *Global Catastrophes in Earth History*, eds. V. L. Sharpton and P. D. Ward, Geological Soc. of America Special Paper 247 (Boulder: Geological Soc. of America), pp. 155–170.

Standish, E. M., Newhall, X. X., Williams, J. G., and Yeomans, D. K. 1992. Orbital ephemerides of the sun, moon, and planets. In *Explanatory Supplement to the Astronomical Almanac*, ed. P. K. Seidelmann (Mill Valley, Ca.: University Science Books), pp. 279–323.

Stacy, N. J. S. 1983. High-Resolution Synthetic Aperture Radar Observations of the Moon. Ph.D. Thesis, Cornell University.

Yeomans, D. K., Ostro, S. J., and Chodas, P. W. 1987. Radar astrometry of near-Earth asteroids. *Astron. J.* 94:189–200.

Yeomans, D. K., Chodas, P. W., Keesey, M. S., Ostro, S. J., Chandler, J. F., and Shapiro, I. I. 1992. Asteroid and comet orbits using radar data. *Astron. J.* 103:303–317.

Zaytsev, A. L., Sokolsky, A. G., Wielebinski, R., Vyshlov, A. S., Grishmanovsky, V. A., Altenhoff, W. J., Rzhiga, O. N., Shor, V. A., Koluka, Yu. F., Shubin, V. A., Krivtsov, A. P., Zaytseva, O. S., Margorin, O. K., and Nabatov, A. S. 1993. 6-cm radar observation of (4179) Toutatis. In *Asteroids, Comets, Meteors 1993*, LPI Contrib. No. 810, p. 325 (abstract).

PART III
NEO Populations and Impact Flux

THE POPULATION OF EARTH-CROSSING ASTEROIDS

DAVID RABINOWITZ
Carnegie Institution of Washington

EDWARD BOWELL and EUGENE SHOEMAKER
Lowell Observatory

and

KARRI MUINONEN
Observatory, University of Helsinki

We define Earth-crossing asteroids (ECAs) as those whose orbits can intersect the capture cross section of the Earth as a result of long-range perturbations by the planets. About 50% of the asteroids currently classified as Amors (having perihelia $1.017 < q < 1.3$ AU) are ECAs; we term them Earth-crossing Amors. In addition, a small number of asteroids currently classified as Apollos (having semimajor axes $a > 1.0$ AU and $0 < q < 1.017$ AU) are not ECAs. Owing to secular decreases of eccentricity during an advance of the argument of perihelion, their orbits do not intersect Earth's. As of early August 1993, about 180 ECAs had been discovered, of which somewhat less than 10% are Atens (having $a < 1.0$ AU and aphelia $Q > 0.983$ AU), two-thirds are Apollos, and one quarter are Earth-crossing Amors. Almost half of the ECAs have received permanent catalog numbers, implying that their orbits are well known, while moderately reliable orbits are in hand for most of the rest. Almost 20 ECAs are considered lost, meaning that they will most likely only be reobserved serendipitously. ECA discovery completeness declines with decreasing diameter, and hence increasing absolute magnitude H. Discovery is thought to be complete to $H = 13.2$ mag, about 35% complete to $H = 15.0$ mag, 15% complete to $H = 16.0$ mag, and 7% complete to $H = 17.7$ mag. Our main purpose in this chapter is to estimate the size of the ECA population as a function of diameter. To do this we have assumed that about half the ECAs are low-albedo, mainly C-class objects, and that the remainder are moderate-albedo, mainly S-class objects. We have computed the observational biases in the known sample of ECAs as a function of the orbital elements a, eccentricity e, and inclination i, and this has allowed us to construct a model of the true population that we can test against the observed discoveries of ECAs during searches at the Palomar 46-cm Schmidt telescope and at the University of Arizona Spacewatch telescope. Our model indicates that there are about 20 ECAs larger than 5 km diameter, about 1500 larger than 1 km diameter, and 135,000 larger than 100 m diameter. At smaller sizes (e.g., diameters ~ 10 m), it appears that the number of ECAs that can approach the Earth closely is enhanced relative to the general population and includes a relatively large fraction of low-eccentricity objects having $q \sim 1.0$ AU. Using our model ECA population, we have computed collisional probabilities, and have compared them with the cratering histories of the Earth and Moon. Finally, we have examined the sample of known ECAs to determine which of them are potentially hazardous, by which we mean ECAs having orbits that apparently bring them within about 0.05 AU of Earth, but which at the same time have been poorly enough observed that their orbital longitudes are quite unknown on time scales of a century or more. We tabulate the orbits of the known ECAs together with the uncertainties of their orbital elements.

I. INTRODUCTION

In this chapter, we follow the definition of Earth-crossing asteroids (ECAs) given by Shoemaker et al. (1979,1990). ECAs are those asteroids whose orbits can intersect the capture cross section of the Earth as a result of long-range perturbations by the planets. In this case, "long-range" refers to intervals of tens of thousands of years; it is certain that greater numbers of objects, some of them at present in the main belt, some of them at present displaying cometary activity, can become ECAs over longer time scales. Thus the population of ECAs comprises Aten asteroids (semimajor axis $a < 1.0$ AU and aphelion distance $Q > 0.983$ AU—the perihelion distance of the Earth), most Apollo asteroids ($a \geq 1.0$ AU and perihelion distance $q \leq 1.017$ AU—the aphelion distance of the Earth), and those Amor asteroids ($a > 1.0$ AU and $1.017 < q \leq 1.3$ AU) whose orbits can intersect Earth's capture cross section as a result of long-range perturbations. Another class of ECA, whose orbits lie at present entirely within Earth's orbit, doubtless exist, although no such objects are currently known.

The population of Earth-crossing, periodic comets (ECCs) is dealt with in the Chapter by Shoemaker et al. The distinction between ECAs and ECCs is normally made on the basis of morphology: of the presence or absence of cometary activity. A typical 1 km diameter ECC nucleus, when near the Earth, is surrounded by an atmosphere of gas and dust many orders or magnitude larger in volume, and is generally completely obscured from view. However, it has been known for many years that the visible activity of ECCs persists for only a small fraction of their dynamical lifetimes, and that a few can decay relatively quickly into objects that are indistinguishable from ECAs. Consequently, the population of objects commonly called ECAs consists both of bodies that originated as mainbelt asteroids and as comets (short- and long-period). Binzel et al. (1992) have estimated that as many as 40% of ECAs originated as comets. In this chapter, we classify ECAs solely on a morphological basis, without regard to origin.

As of August 1993, about 180 ECAs had been discovered, of which about 15 are Atens, about 125 are Apollos, and about 40 are Earth-crossing Amors. Most have diameters in the range 0.1 to 10.0 km, smaller objects not being detectable until the recent advent of computer-aided inspection of charged-coupled device (CCD) images (Gehrels 1991; Rabinowitz 1991). With these advances and the optimization of traditional photographic techniques (Helin and Dunbar 1990), the discovery rate of ECAs has been such as to double the known population in about 5 yr. Sixty-eight ECAs have received permanent catalog numbers, implying that their orbits are accurately established and that their locations in space can, for the most part, be well predicted for a few centuries hence. Moderately reliable orbits are in hand for about 70 others. Most of the remainder are considered lost, meaning that further observation of them will occur only through serendipitous re-discovery.

As the number of discovered ECAs has grown, estimates of their total number above a given size threshold have been revised and reformulated. Lower bounds for the total number larger than ~1 km have been estimated from the frequency of chance rediscoveries (Whipple 1967; Wetherill 1976). Direct estimates of the total population have been determined from the number of ECAs discovered and volume of space searched in systematic telescopic surveys. Given assumed orbit, size, and albedo distributions for the ECA population, the discovery fraction of the survey can be calculated, and hence the size of the ECA population determined (Öpik 1963; Helin and Shoemaker 1979; Shoemaker et al. 1979; Shoemaker et al. 1990). The relative number of Atens, Apollos, and Amors in the ECA population can also be determined from survey results, but only after the effects of observational bias on the observed ratios have been corrected. Detailed models of the discovery circumstances in telescope searches for ECAs have been used to determine the unbiased orbit and size distributions (Benedix et al. 1992; Rabinowitz 1993). The impact rates of ECAs with the Earth have been determined by direct observation of the near-Earth flux (Kresák 1978; Rabinowitz 1993) or by calculating the mean collision probabilities with the Earth of the known orbits and multiplying by the population size (Shoemaker et al. 1979,1990). These results are consistent with the terrestrial cratering record over the last 120 Myr, and with the lunar cratering record over the last 3.3 Gyr.

In this chapter we describe the size distribution of the ECA population as a series of power laws relating the cumulative number to diameter D. This is determined in telescopic surveys from the distribution of values that are observed for absolute magnitude H. The diameter distribution is then derived by assuming a distribution of ECA albedos, which are known to span the same range as mainbelt asteroids (McFadden et al. 1989). Because the two most common taxonomic types in the main belt (C and S) also dominate the ECAs, the distribution of ECA diameters can be calculated assuming a fixed proportion for these two types and weighting the albedos accordingly. In this chapter, we assume equal numbers of each type for a given ECA diameter. Half have a C-type albedo (0.05) and half have a S-type albedo (0.155). We also describe the orbital distribution of the ECAs, debiased for observational selection, by integral probability distributions in the orbital elements a, e (eccentricity), and i (inclination).

It is believed that all ECAs brighter than about $H = 13.5$ mag, corresponding to a diameter of 12 km for C-class asteroids and 6 km for S-class asteroids, have been discovered (Shoemaker et al. 1990). About 35% of ECAs brighter than $H = 15$ mag (diameters of 6 and 3 km, respectively) have been discovered. At $H = 16$ (4 and 2 km), the estimated completeness is only 15%, while at $H = 17.7$ mag (2 and 1 km), it is only about 7%. The largest ECAs are (1627) Ivar and (1580) Betulia, both of them Earth-crossing Amors, and each with a diameter of about 8 km. The smallest ECA yet discovered is probably 1993 KA2, about 4 to 8 m across, which passed within 0.0010 AU (less than half the distance to the Moon) in May 1993. Recent results by Rabinowitz (1993,1994; Rabinowitz et al. 1993) indicate that small ECAs (5 m<D<50 m) comprise a special population whose orbits are concentrated near the Earth, and whose numbers may be enhanced compared to the general ECA population. At H = 29, the relative number is 40 times greater (to within a factor of 2 uncertainty) than a power law extrapoation of the magnitude frequency of km-sized ECAs.

II. DEFINITION OF EARTH-CROSSING ASTEROIDS

The definition of ECAs quoted above has been given and described in detail by Shoemaker et al. (1979). It is a broader definition than that previously adopted by most researchers: namely, an asteroid whose orbit currently overlaps that of the Earth. Shoemaker et al.'s definition had to await the development of suitable analytical techniques (Williams 1969) or practicable methods of numerical integration to solve the secular variations of asteroid orbits.

Asteroids whose orbits overlap that of the Earth undergo intersections chiefly because of precession of the major axis. In a full cycle of precession, the heliocentric distance to each point of intersection of the asteroid's orbit on the orbit plane of the Earth oscillates between the perihelion and aphelion distances. Intersection occurs when either the ascending or descending node of the asteroid's orbit coincides with the Earth's orbit. If the orbits of the asteroid and the Earth overlap continuously, four crossings take place during one rotation of the line of apsides. Precession periods of ECAs range from about 4,000 to 100,000 yr; the fastest processors cross as frequently as once per thousand years.

However, there are more complicated patterns of precession. Secular variation of the eccentricities and, in some cases, of the semimajor axes of Earth-crossing Amors leads to part-time overlap with the Earth's orbit. Conversely, some Apollo asteroids lose overlap and become Amors part of the time. Apollos revolving on certain unusual orbits can lose overlap with Earth before intersection occurs because of rapid changes in eccentricity during precession. This happens in the case of the largest Apollo, (1866) Sisyphus. Thus Sisyphus cannot, at present, impact the Earth, and has not been considered in our analysis.

Table I lists the known ECAs as of May 1991. For those ECCs and ECAs with well-defined orbits, Table II describes close approaches during the interval July 1994 to December 2019. The appendix at the end of this chapter lists the nonrelativisitic, least-squares orbital elements for each of the ECAs listed in Table I and additional asteroids with q<1.017 AU discovered June 1991 to July 1993 (Muinonen and Bowell 1992; Muinonen and Bowell 1993). Also listed is the leak metric that measures overall orbital quality.

These orbital elements were determined from optical observations assuming Gaussian observational noise.

III. POPULATION ESTIMATES

Efforts to describe the size and orbit distribution of the ECA population have always been limited by the incompleteness of the known sample and biases within the known sample towards those ECAs that are most easily observed. Nearly all searches for ECAs have been conducted with groundbased telescopes which can only detect objects within a finite volume of space. Those objects with orbits that bring them most often into the search volume, and keep them there the longest, dominate the observed population. Because the extent of the search volume depends on diameter and albedo (the brightest objects can be seen the farthest away), the observed population is biased towards the largest objects and those with the highest albedos. Any attempt to estimate the size of the complete population of ECAs and to describe the orbit and size distribution must take into account the extent of the discovery volume and these observational biases.

The most accurate estimate for the total number of ECAs with $H \leq 17.7$ (larger than ~ 1 km) is based on the discoveries made with the 46-cm Schmidt telescope at Palomar Observatory (Shoemaker et al. 1990), where most of the known ECAs larger than ~ 1 km have been discovered. The result is based on: (1) an estimate of the size and shape of the detection volume given the limiting magnitude and number of fields photographed to make these discoveries, the position in the sky of the exposed fields relative to the Sun, and the angular rate and solar phase angle of the ECAs at discovery; (2) an estimate of the fractional dwell time of the ECA population within the detection volume given their observed distribution of orbits; and (3) the assumption that ECAs larger than 1 km have the same size distribution as faint mainbelt asteroids and the distribution revealed by young craters on the moon. The accuracy of the estimate depends on correctly assessing the threshold magnitudes of detection, on estimating the distribution of photometric functions for the asteroids residing in the detection volume, and on the reliability of the inferred magnitude distribution.

The assumption that the ECAs larger than ~ 1 km share the same size distribution as the main-belt asteroids has been confirmed independently by analysis of the ECAs detected with the 0.92 m Spacewatch telescope of the University of Arizona (Rabinowitz 1993). Employing a computer to search images of the sky scanned with a CCD, the Spacewatch telescope has extended the observed sizes for ECAs from ~ 100 m down to ~ 5 m. These observations allow the incremental flux of ECAs in the vicinity of the Earth to be derived directly from the incremental number of Earth approachers observed as a function of H. A simple method is to evaluate the search volume as a function of H. Assuming only that the observed Earth approachers have velocities near the Earth that are characteristic of the known population, and

TABLE I

Earth-Crossing Asteroids[a]

Aten Asteroids

Object		H	Approx. diameter (km)	Depth (AU)	q (AU)	a (AU)	e	i (deg)	P_s (Gyr^{-1})	P_o (Gyr^{-1})	V_i (km s^{-1})
Provisional Designation	No. & name										
1989 VA	—	17.00	1	—	0.292	0.729	(0.60)	(29)	—	(4)	(22)
1978 RA	2100 Ra-Shalom	16.05	2.4	0.388	0.445	0.832	0.465	13.1	6.3	6.7	17.9
1954 XA (L)	—	18.9	0.5	0.203	0.475	0.777	0.389	5.04	34.0	30	14.5
1984 QA	3362 Khufu	18.10	0.7	0.559	0.481	0.990	0.514	8.37	5.3	6.2	19.8
1976 UA	2340 Hathor	20.26	(0.2)	0.356	0.486	0.844	0.424	6.27	14.0	14	16.3
1986 TO	3753	14.4	3	—	0.499	0.998	(0.50)	(22)	—	(3)	(22)
1991 JY	5381	16.50	2	0.204	0.572	0.940	0.391	44.3	2.9	2.4	26.7
1989 UQ	—	19.0	0.5	0.240	0.645	0.915	0.295	1.95	42	43	14.1
1990 VA	5590	19.5	0.4	0.346	0.662	0.985	0.328	13.7	5.4	5.7	16.5
1986 EB	3554 Amun	15.82	2.0	0.299	0.730	0.974	0.251	21.6	5.4	5.0	17.4
1976 AA	2062 Aten	16.80	0.9	0.223	0.743	0.966	0.231	18.0	7.1	6.5	16.0

Apollo Asteroids

Object		H	Approx. diameter (km)	Depth (AU)	q (AU)	a (AU)	e	i (deg)	P_s (Gyr^{-1})	P_o (Gyr^{-1})	V_i (km s^{-1})
Provisional Designation	No. & name										
1983 TB	3200 Phaethon	14.60	6.9	—	0.140	1.271	(0.89)	(22)	—	(1.4)	(35)
1949 MA	1566 Icarus	16.40	0.9	0.844	0.205	1.078	0.810	18.0	1.8	2.2	30.6
1978 SB	2212 Hephaistos	13.87	5	0.929	0.240	2.163	0.889	10.0	0.44	1.2	34.6
1990 UO (L)	—	20.5	0.2	0.771	0.276	1.234	0.776	24.1	0.9	1.4	30.5
1990 SM	—	16.5	1	0.727	0.395	2.157	0.817	11.3	0.4	1.1	30.0
1974 MA	5660	14.0	5	—	0.973	1.757	0.446	53.4	0.9	2.5	32.0
							0.772	32.8			
5025 P-L (L)	—	15.9	2	—	0.420	(4.2)	(0.90)	(6.2)	—	(~1)	(32)
1984 KB	—	15.5	1.4	0.760	0.429	2.221	0.807	3.37	1.1	3.2	28.8
1986 WA	3838 Epona	15.4	3	—	0.452	1.505	(0.70)	(29)	—	(1)	(29)
1991 AM	5828	16.5	1	0.613	0.454	1.695	0.732	19.7	0.5	1.0	28.0
1991 AQ	—	17.5	1	—	0.500	2.159	(0.77)	(3.2)	—	(4)	(27)
1947 XC	2201 Oljato	15.25	1.4	0.657	0.511	2.174	0.765	1.33	2.3	6.1	26.4
1936 CA	2101 Adonis	18.70	1	0.620	0.512	1.875	0.727	2.03	2.8	6.2	25.4
1971 UA	1865 Cerberus	16.84	1.0	0.504	0.526	1.080	0.513	14.9	2.5	3.1	20.9
1971 FA	1864 Daedalus	14.85	(3.1)	0.491	0.526	1.461	0.640	15.9	1.0	1.6	26.0
1982 TA	4197	14.5	1.8	—	0.529	2.300	(0.77)	(12)	—	(1)	(27)
1990 BG	5131	14.0	5	0.409	0.544	1.486	0.634	26.3	0.6	1.1	26.3
1985 PA	3752 Camillo	15.5	2	—	0.551	1.414	(0.61)	(32)	—	(1)	(27)
1987 SY	4450 Pan	17.1	1	0.522	0.555	1.442	0.615	1.85	6.4	10.5	22.2
1979 XB (L)	—	19.0	0.5	(0.56)	0.566	2.264	(0.75)	(10)	(0.5)	(1.2)	(25)
1989 PB	4769 Castalia	16.9	1.5	0.478	0.568	1.063	0.466	9.68	4.2	5.1	18.9

TABLE I (cont.)

Apollo Asteroids

Object		H	Approx. diameter (km)	Depth (AU)	q (AU)	a (AU)	e	i (deg)	P_s (Gyr^{-1})	P_o (Gyr^{-1})	V_i (km s^{-1})
Provisional Designation	No. & name										
1990 MU	4953	14.3	3	0.487	0.569	1.622	0.649	26.4	0.6	0.9	26.5
1991 CB1	—	18.0	1	0.484	0.580	1.686	0.656	9.36	1.2	2.1	23.4
1987 KF	4341 Poseidon	15.6	3	0.512	0.593	1.836	0.677	5.87	1.5	2.9	23.3
1986 PA	4034	18.1	1	0.438	0.606	1.060	0.428	10.0	4.5	5.3	18.0
1937 UB (L)	—	17.0	1	0.459	0.624	1.639	0.619	5.64	2.2	3.7	21.7
1989 QF	—	18.0	1	0.418	0.639	1.155	0.447	5.27	6.4	8.1	17.9
1987 OA	—	18.5	1	0.414	0.641	1.490	0.570	11.0	1.6	2.4	21.1
1932 HA	1862 Apollo	16.25	1.4	0.423	0.647	1.471	0.560	6.13	2.8	4.3	20.3
1988 EG	—	19.0	1	0.399	0.664	1.270	0.477	2.71	8.8	12	18.3
1990 TG1	—	15.0	3	—	3:1 commensurability						
1989 UR	—	18.0	1	0.362	0.675	1.080	0.375	11.5	4.4	5.1	17.1
1989 FC	4581 Asclepius	20.5	0.2	0.371	0.679	1.023	0.336	4.41	13.3	15.1	15.4
1959 LM	4183 Cuno	14.5	4	0.425	0.699	1.981	0.647	7.42	1.3	2.2	21.3
1991 GO (L)	—	19.0	0.5	0.380	0.706	1.930	0.634	9.55	1.1	1.9	21.2
1991 BA (L)	—	28.5	0.006	—	0.713	2.161	(0.67)	(2.2)	—	(5)	(21)
1977 HB	2063 Bacchus	16.4	1	0.330	0.719	1.078	0.333	8.99	6.5	7.2	15.8
1989 DA	—	18.0	1	0.602	0.728	2.166	0.664	5.08	1.5	2.8	20.8
1973 EA	1981 Midas	15.0	1	—	0.735	1.776	0.586	5.5	3.8	0.7	30.7
							0.652	41.3			
1983 VA	—	16.5	2	0.485	0.737	2.615	0.718	6.81	0.7	1.6	21.6
1990 HA	—	17.0	1	0.599	0.747	2.567	0.709	3.94	1.1	2.7	21.1
1990 UA	—	19.50	0.4	0.373	0.750	1.721	0.564	0.81	8.3	12.9	18.8
1989 FB	4544 Xanthus	7.1	1	0.250	0.761	1.042	0.270	13.3	6.6	6.4	15.5
1948 OA	1685 Toro	14.23	5.2	—	0.769	1.368	0.438	9.15	(4)	(4.2)	17.2
1988 XB	—	17.5	1	0.263	0.779	1.467	0.469	3.98	6.2	7.9	17.0

TABLE I (cont.)
Apollo Asteroids

Object			Approx. diameter (km)	Depth (AU)	q (AU)	a (AU)	e	i (deg)	P_s (Gyr⁻¹)	P_o (Gyr⁻¹)	V_i (km s⁻¹)
Provisional Designation	No. & name	H									
1950 DA (L)	—	15.8	2	0.266	0.783	1.683	0.535	10.7	1.9	2.4	18.7
1978 CA	—	17.8	1.9	0.293	0.784	1.125	0.303	25.5	3.6	3.0	19.5
1988 TA	—	21.0	0.2	0.262	0.789	1.541	0.488	3.42	6.3	8.2	17.1
1990 UQ	5189	17.5	1	0.262	0.789	1.551	0.491	3.76	5.8	7.5	17.2
1988 VP4	5731	15.8	3	0.328	0.790	2.263	0.651	10.3	1.0	1.5	20.1
1973 NA	5496	15.5	3	—	0.803	2.447	0.672	67.9	0.5	0.4	40.5
							0.891	51.8			
1990 UN (L)	—	23.5	0.06	0.363	0.808	1.709	0.527	3.42	5.3	7.0	17.4
1951 RA	1620 Geographos	15.60	2	0.210	0.808	1.245	0.351	14.2	3.8	3.9	16.7
1987 SB	4486 Mithra	15.4	3	0.595	0.813	2.202	0.631	1.06	5.5	8.8	18.5
1977 HA	2135 Aristaeus	17.94	1	—	0.816	1.600	(0.49)	(23)	(2.0)	(1.5)	(21)
6344 P-L (L)	—	21.9	0.1	0.207	0.822	2.619	0.686	3.71	—	1.6	19.1
1982 HR	3361 Orpheus	19.03	0.8	—	0.822	1.209	(0.32)	(2.7)	—	(21)	(14)
1990 SP	5645	17.0	1	—	0.827	1.355	(0.39)	(13)	—	(4)	(17)
6743 P-L	5011 Ptah	17.0	1	0.224	0.837	1.635	0.488	6.88	3.8	4.5	16.7
1991 EE	—	17.5	1	—	0.838	2.266	(0.63)	(10)	—	(2)	(19)
1991 BB	—	16.0	2	—	0.866	1.186	(0.27)	(39)	—	(3)	(24)
1991 DG	—	18.5	0.6	0.175	0.870	1.427	0.390	11.0	4.4	4.4	15.8
1990 OS	—	20.0	0.3	0.251	0.878	1.672	0.475	0.86	17.2	19.3	15.5
1991 BN (L)	—	20.0	0.3	0.168	0.885	1.443	0.387	7.58	6.4	6.5	15.0
1990 SS	—	19	0.6	—	0.886	1.703	(0.48)	(19)	—	(2)	(19)

TABLE I (cont.)
Apollo Asteroids

Provisional Designation	Object	No. & name	H	Approx. diameter (km)	Depth (AU)	q (AU)	a (AU)	e	i (deg)	P_s (Gyr⁻¹)	P_o (Gyr⁻¹)	V_i (km s⁻¹)
1975 YA	2102 Tantalus		15.3	2	—	0.888	1.290	0.312	62.5	2.5	1.5	34.8
								0.744	49.1			
1989 JA		—	16.5	2	0.158	0.893	1.769	0.495	14.8	2.8	2.3	17.5
1986 JK		—	19.0	1	—		5:2 commensurability			—	—	—
1983 LC (L)		—	19.0	0.5	0.334	0.904	2.629	0.656	1.54	5.8	7.6	16.8
1991 JW		—	19.5	0.5	0.104	0.913	1.038	0.120	8.34	30.0	28.9	12.6
1991 CS		—	17.5	1	—	0.932	1.123	(0.17)	(37)	—	(5)	(22)
1989 AZ		—	19.5	0.4	0.138	0.935	1.649	0.433	9.92	5.9	5.0	15.1
1987 QA	4257 Ubasti		15.8	3	—		—	—	—	—	—	—
1990 MF		—	18.7	0.6	0.163	0.949	1.747	0.457	3.02	16.6	15.2	14.1
1948 EA	1863 Antinous		15.54	1.8	0.076	0.951	2.260	0.579	23.1	—	1.3	19.9
1979 VA	4015 Wilson-Harrington		15.99	5	0.074	0.955	2.645	0.639	5.03	—	0.84	15.5
1981 VA	3360		16.20	1.8	0.346	0.958	2.462	0.611	38.6	0.7	1.6	26.6
								0.751	20.8			
1982 DB	4660 Nereus		18.30	1	0.092	0.969	1.490	0.350	4.87	22.5	20.3	13.1
1982 BB	3103		15.38	1.5	0.054	0.974	1.407	0.308	22.0	—	3.9	17.3
1989 UP		—	20.7	0.2	—	0.988	1.864	(0.47)	(3.9)	—	(13)	(14)
1989 AC	4179 Toutatis		14.0	5	—		3:1 commensurability			—	—	—
1976 WA	2329 Orthos		14.9	3	0.295	0.990	2.404	0.588	32.6	1.8	2.2	23.3
								0.702	16.9			
1991 JX		—	18.5	0.6	—		3:1 commensurability			—	—	—
1984 KD	3671 Dionysus		16.3	1.0	—	1.011	2.198	(0.54)	(14)	—	(1.5)	(16)
1989 VB		—	20.0	0.3	—	1.026	1.865	(0.45)	(1.6)	—	(13)	(13)

Amor Asteroids

Provisional Designation	No. & name	H	Approx. diameter (km)	Depth (AU)	q (AU)	a (AU)	e	i (deg)	P_s (Gyr^{-1})	P_o (Gyr^{-1})	V_i (km s^{-1})
1982 DV	3288 Seleucus	15.0	2.8	0.319	0.709	2.032	0.651	5.46	—	1.4	21.0
1980 WF	4688	18.6	0.6	0.299	0.730	2.231	0.673	5.38	—	1.1	20.9
1988 SM	—	18.0	1	—	—	—	—	—	—	—	—
1918 DB	887 Alinda	13.76	4.2	—	3:1 commensurability						
1978 DA	2608 Seneca	17.52	0.9	—	3:1 commensurability						
1991 FA	—	17.5	1	0.262	0.766	2.164	0.646	4.51	—	1.7	19.8
1991 FB	—	18.5	0.5	0.218	0.810	2.356	0.656	10.0	—	0.84	19.7
1991 FE	5626	14.9	4	0.116	0.912	2.451	0.628	3.54	—	1.6	16.4
1980 PA	3908	17.4	1	0.090	0.938	1.926	0.513	2.80	—	4.0	14.7
1987 WC	—	19.5	0.4	—	—	—	—	—	—	—	—
1991 DB	—	18.5	0.6	0.079	0.949	1.720	0.448	14.2	—	3.1	16.0
1950 KA	1580 Betulia	14.52	7.4	(0.27)	0.957	2.196	0.564	48.5	0.50	1.7	30.6
							0.792	26.4			
1981 QB	4596	16.0	2	0.064	0.963	2.240	0.570	28.5	—	0.75	21.8
1932 EA1	1221 Amor	17.7	1	0.061	0.966	1.921	0.497	12.0	—	1.5	15.4
1990 VB	—	16.0	2	0.058	0.970	2.444	0.603	18.5	—	0.34	18.1
1960 UA	2061 Anza	16.56	(2.7)	0.050	0.978	2.263	0.568	3.41	—	1.3	14.2
1968 AA	1917 Cuyo	13.9	3	0.051	0.978	2.149	0.545	19.3	—	1.3	17.8
1982 XB	3757	18.95	0.5	0.046	0.981	1.838	0.466	2.55	—	5.1	13.4
1929 SH	1627 Ivar	13.20	8.1	0.042	0.986	1.863	0.471	7.98	—	1.8	14.0
1973 EC	1943 Anteros	15.75	1.8	0.041	0.987	1.431	0.310	9.86	—	3.5	13.4

TABLE I (cont.)
Amor Asteroids

Provisional Designation	No. & name	H	Approx. diameter (km)	Depth (AU)	q (AU)	a (AU)	e	i (deg)	P_s (Gyr^{-1})	P_o (Gyr^{-1})	V_i (km s^{-1})
1983 RD	3551	16.75	0.9	—	—	—	—	—	—	—	—
1953 EA	1915 Quetzalcoatl	18.97	0.3			3:1 commensurability					
1987 SF3	—	19.0	0.5	0.039	0.989	2.252	0.561	3.56	—	1.1	13.9
1991 JR (L)	—	22.5	0.1	0.038	0.991	1.407	0.296	10.3	—	5.7	13.4
1981 ET3	3122 Florence	14.2	4	0.036	0.992	1.768	0.439	20.0	—	2.1	17.0
1987 QB	—	19.0	0.5	0.021	1.006	2.795	0.640	2.34	—	0.26	14.7
1985 WA	—	19.0	0.5			5:2 commensurability					
1972 RA	2202 Pele	16.8	2.0	0.018	1.010	2.291	0.559	8.70	—	0.10	14.8
1980 AA	5797	19.5	(0.5)	0.009	1.020	1.892	0.461	3.55	—	0.30	13.2
1986 LA	3988	18.3	1	−0.002	1.026	1.545	0.336	13.4	—	—	—
1986 DA	—	16.0	2.3			5:2 commensurability					
1986 RA	5370 Taranis	16.0	4			2:1 commensurability					

[a] For each object whose orbit can evolve to intersect that of the Earth, the following information is given: the absolute magnitude H, the approximate diameter, the depth interior to the Earth's orbit to which the asteroid can evolve, the orbital perihelion distance q, semimajor axis a, eccentricity e, and inclination i. The inclination is referred to the invariable plane of the solar system and each of these orbital elements represents a mean value at the time of Earth orbital crossing. P_s is the estimated probability of collision with the Earth (number per Gyr) using the equations of Shoemaker et al. (1979), while P_o is the same probability calculated from the equation of Öpik (1951). V_i gives the approximate impact speed for an Earth collision. For some objects, it is not possible to compute mean orbital elements at the time of crossing and for these objects, osculating elements (given in parentheses) have been used as rough approximations. Where two sets of elements are given, there are two different conditions of crossing. For those objects whose motions are commensurate with Jupiter, no orbital elements are given because they have chaotic (unpredictable) motions over the long term. Those objects whose orbits are not sufficiently secure are probably lost and may be found only by a serendipitous search. These objects have an "L" following their names. This table is based upon the work of E. M. Shoemaker and is current through May 1991.

that their probability of detection is constant everywhere inside the volume of space searched to find them, the incremental flux at a given H is proportional to the number observed divided by the search volume. No model of the orbits of the ECAs nor their individual discovery biases is required by this method.

The results of this analytic method are confirmed by a Monte Carlo program that models the discovery biases of a telescopic search (Rabinowitz 1993). Assuming a distribution of ECA orbits consistent with the known sample of discovered objects and an arbitrary size distribution, the program models the selection effects of the search. It determines how many ECAs should be detected as a function of size and which orbits will be preferentially observed. When the detection limits (limiting visual magnitude, total area and position of searched fields, and criteria for recognition) of the Spacewatch telescope are assumed, matching the simulated results to the actual observations fixes the true size distribution. Bowell and Muinonen used a similar program to design search strategies for a future survey (Morrison 1992, Ch. 5). Given the size-distribution determined by Rabinowitz (1993), their program predicts discoveries consistent with the Spacewatch observations.

The search model developed by Rabinowitz (1993) has also been used to debias the orbital distribution of the known ECAs larger than ~ 1 km by modeling the search conducted at Palomar with the 46 cm Schmidt. In this case, the size distribution determined by the Spacewatch discoveries is assumed along with orbits uniformly distributed in a, e and i. Given the detection limits of the 46 cm Schmidt, the orbit distribution of simulated discoveries then represents the bias in the Palomar search as a function of a, e and i. Correcting the observed population for these biases yields the debiased distribution.

Figure 1 shows the estimated number of ECAs, N, larger than a given diameter D for the complete population (Muinonen et al. 1994). It is based on the population estimate for ECAs larger than ~ 1 km determined by Shoemaker et al. (1990) and the size-distribution curve determined by Rabinowitz (1993). There are about 20 larger than 5 km, 1500 larger than 1 km, and 135,000 larger than 100 m. The uncertainty in N is a factor 2 for $D < 2$ km. Sections of this curve can be described mathematically as $N = kD^{b}$ where k is a constant and b is the power-law exponent. For $D > 3.5$ km, $b = -5.4$; for $0.070 < D < 3.5$ km, $b = -2.0$, and for $0.010 < D < 0.070$ km, $b = -3.5$. At small sizes (5 to 10 m) the number of ECAs is enhanced by a factor of 40 (to within an uncertainty of a factor of 2) compared to the size distribution of ECAs larger than 50 m (Rabinowitz 1994). Below 5 m , however, the slope of the distribution must decrease (b is less negative) in order to match the fluxes observed for bright meteors (Ceplecha 1988).

Figures 2 and 3 show the results of the simulation performed by Rabinowitz (1993) in order to determine the observational bias in the orbits of the known ECAs larger than ~ 1 km. The search conducted with 46 cm Schmidt at Palomar Observatory was simulated as described above. The orbital elements of the input population were randomly chosen so that they uniformly

TABLE II

ECA Close Approaches to Earth (<0.1 AU) During the Interval
July 1994 Through December 2019 [a]

Designation		Approach Distance (AU)	Date of Close Approach (year mon day)			Solar Elong. (deg)	V (mag)
1620	Geographos	0.0333	1994	8	25.42	121	12.0
	1991 JX	0.0323	1995	6	11.85	108	15.2
	1993 QA	0.0684	1996	2	6.92	142	14.1
2063	Bacchus	0.0678	1996	3	31.67	109	13.0
4197	1982 TA	0.0846	1996	10	25.64	98	13.4
3908	1980 PA	0.0613	1996	10	27.86	146	14.4
4179	Toutatis	0.0354	1996	11	29.95	39	13.4
	1991 VK	0.0749	1997	1	10.69	6	32.3
	1988 EG	0.0317	1998	2	28.90	91	14.3
	1992 SK	0.0560	1999	3	26.27	44	17.2
	1991 JX	0.0342	1999	6	10.46	116	13.5
4486	Mithre	0.0466	2000	8	14.36	86	12.0
4179	Toutatis	0.0739	2000	10	31.19	7	27.9
	1991 VK	0.0718	2002	1	16.50	9	28.2
4660	Nereus	0.0290	2002	1	22.51	18	22.1
5604	1992 FE	0.0768	2002	6	22.26	92	14.4
	1991 JX	0.0313	2003	6	1.42	115	13.3
4179	Toutatis	0.0104	2004	9	29.57	62	8.7
	1988 XB	0.0729	2004	11	21.96	74	15.4
	1993 VW	0.0862	2005	4	24.90	109	13.6
	1992 UY4	0.0402	2005	8	8.42	131	12.4
1862	Apollo	0.0752	2005	11	6.80	80	14.1
	1991 VK	0.0679	2007	1	21.51	15	23.4
1862	Apollo	0.0714	2007	5	8.64	90	13.7
2340	Hathor	0.0600	2007	10	22.24	70	18.1
4450	Pan	0.0408	2008	2	19.93	96	13.0
4179	Toutatis	0.0502	2008	11	9.52	12	22.2
	1993 KH	0.0992	2008	11	22.33	89	17.0
	1991 JW	0.0813	2009	5	23.96	143	15.6
	1990 SS	0.0994	2011	3	17.37	116	16.2
	1991 VK	0.0650	2012	1	26.00	22	20.8
4179	Toutatis	0.0463	2012	12	12.28	119	9.4
	1984 KB	0.0790	2013	11	11.54	87	12.6
2340	Hathor	0.0482	2014	10	21.89	120	15.8
1566	Icarus	0.0539	2015	6	16.65	100	13.7
	1991 VK	0.0647	2017	1	25.54	22	20.7
5604	1992 FE	0.0336	2017	2	24.42	80	13.2
	1984 KB	0.0985	2017	5	27.63	82	13.3
	1991 VG	0.0568	2017	8	7.36	154	23.9

3122	Florence	0.0472 2017	9	1.50	152	8.9
	1989 UP	0.0471 2017	11	4.24	123	15.9
3361	Orpheus	0.0607 2017	11	25.69	80	16.3
3200	Phaethon	0.0689 2017	12	16.96	111	11.1
	1991 VG	0.0472 2018	2	11.91	92	25.1
1981	Midas	0.0896 2018	3	21.89	98	12.5
	1992 YD3	0.0406 2018	12	27.39	124	21.5

[a] Only asteroids whose orbital arcs exceed 100 days (as of January 1994) are included.

covered the ranges $0.0 < a < 6.0$ AU, $0 < e < 1$, and $0° < i < 90°$. Values of H were randomly chosen so that they were consistent with the size distribution described by Fig. 1. The a, e values for the simulated discoveries are shown by Fig. 2, where the concentration of points is directly proportional to the observational bias. Most conspicuous is the bias towards those orbits with $0.1 < e < 0.3$, $1.0 < a < 1.3$ AU and q near the Earth's orbit (as shown by the curve labeled $q = 1.0$ AU). Such objects have the smallest radial velocities near the Earth and therefore the highest dwell times within their detection volumes. For orbits with higher eccentricities ($e > 0.3$), there is also a bias towards orbits with Q near, or just outside the Earth's orbit (as shown by the curve labeled $Q = 1$ AU). Again, these objects have large dwell times near the Earth because their radial velocities are low. This bias towards large dwell times is also seen in Fig. 3, which shows the number of simulated objects versus i. The bias is towards those orbits with the smallest i, and therefore the lowest transverse velocities near the Earth.

Tables III and IV show incremental distributions, $P(a, e)$ and $P(i)$, that represent the normalized probability that an ECA larger than ~ 1 km chosen at random will have semimajor axis and eccentricity in the range 0 to a, 0 to e and inclination in the range $0°$ to i, respectively. These result from the debiasing procedure performed by Rabinowitz (1993), using the bias functions shown by Figs. 2 and 3 to correct the observed orbits. $P(a, e)$ and $P(i)$ are weighted less strongly towards Atens and other ECAs with low eccentricities than the observed distribution, which is consistent with the weighting functions used by Shoemaker et al. (1990). Based on the results of Rabinowitz et al. (1993), it is believed that these distributions do not apply for ECAs smaller than ~ 50 m because they are not weighted strongly enough towards orbits with perihelia near 1 AU and low eccentricities. The number of house-sized ECAs so far discovered is enough to show that their orbits are peculiar, and may form a belt concentrated near the Earth's orbit, but the number discovered is not yet enough to characterize this distribution (Rabinowitz 1994).

IV. COMPARISON WITH CRATERING RECORDS

Cratering records on Earth and the Moon have been used to check the population estimates, size and orbit distributions described in the preceding section.

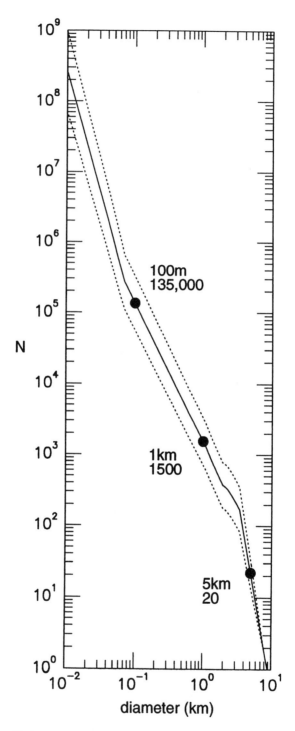

Figure 1.   Estimated number of Earth-crossing asteroids *N* larger than a given diameter.

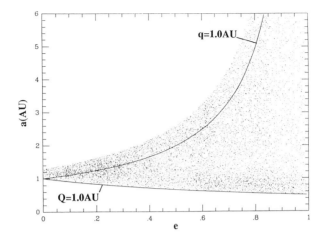

Figure 2.   Semimajor axis a versus eccentricity e for a simulated population of Earth approachers (asteroids with $q < 1.3$ AU) showing the observational bias of the observed population larger than $\sim$1 km.

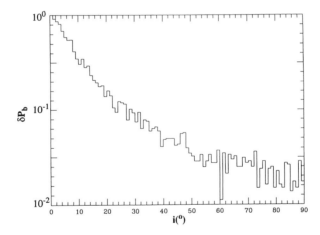

Figure 3.   The number of Earth approachers N versus inclination i for the population of asteroids shown by Fig. 2.

The terrestrial cratering record reveals the impact rate of bodies that have left craters larger than 10 km within the last $\sim$100 Myr, whereas the lunar record reveals the impact rate and size distribution of the impacting flux within the last 3.3 Gyr. In order to test our descriptions of the ECA population, however, we must either start with the observed flux in the vicinity of Earth (see, e.g., Rabinowitz 1993) or translate population number at a given size to impact rate at a given size (Shoemaker et al. 1990). The latter requires a computation of collision probability for each of the known ECAs and an evaluation of the

TABLE III
A Debiased Integral Distribution for Earth-Approacher Orbits

a (AU)	e	$P(a, e)$	a (AU)	e	$P(a, e)$
0.7	0.35	0.00746	1.7	0.75	0.35661
0.7	0.55	0.01457	1.9	0.35	0.37541
0.9	0.15	0.02223	1.9	0.45	0.41676
0.9	0.25	0.03681	1.9	0.55	0.42302
0.9	0.45	0.04755	1.9	0.65	0.44545
0.9	0.55	0.05168	1.9	0.75	0.45607
1.1	0.25	0.05665	2.1	0.35	0.45759
1.1	0.35	0.06679	2.1	0.45	0.53012
1.1	0.45	0.08233	2.1	0.55	0.54851
1.1	0.85	0.08886	2.1	0.65	0.55751
1.3	0.15	0.09482	2.1	0.75	0.59054
1.3	0.25	0.09954	2.1	0.85	0.61035
1.3	0.35	0.11176	2.3	0.45	0.62155
1.3	0.45	0.11963	2.3	0.55	0.66118
1.3	0.75	0.13128	2.3	0.65	0.69666
1.3	0.85	0.14367	2.3	0.75	0.74239
1.5	0.15	0.14534	2.5	0.55	0.78862
1.5	0.25	0.15194	2.5	0.65	0.83170
1.5	0.35	0.16420	2.5	0.75	0.85413
1.5	0.45	0.17802	2.7	0.55	0.89059
1.5	0.55	0.21472	2.7	0.65	0.92408
1.5	0.65	0.22333	2.7	0.75	0.93791
1.5	0.75	0.23625	2.9	0.55	0.94092
1.7	0.25	0.24087	2.9	0.65	0.95867
1.7	0.35	0.25529	3.3	0.65	0.96577
1.7	0.45	0.29909	3.7	0.65	0.97016
1.7	0.55	0.32385	4.3	0.75	0.98142
1.7	0.65	0.34560	4.3	0.85	1.00000

mean impact probability for the debiased population. We must then translate impact rate by ECAs of a given diameter to an equivalent cratering rate at a given crater diameter. This last step requires that we (1) assume albedos and bulk densities in order to determine impactor masses; (2) use the distribution of ECA orbits to determine impact velocities and energies at Earth or the Moon, taking into account gravitational acceleration; and (3) predict crater diameters as a function of impacting energy using scaling laws. Large uncertainties are introduced because we must assume values for the unknown albedos and densities of ECAs and use impact scaling laws that have not been tested in the energy range of asteroid impacts. For collisions by ECAs smaller than ∼50 m , further uncertainty is introduced because these objects have an unusual distribution of orbits yet to be characterized quantitatively

TABLE IV
A Debiased Integral Distribution for
Earth-Approacher Inclinations

i	$P(i)$
2°.5	0.05109
7.5	0.13270
12.5	0.22984
17.5	0.36625
22.5	0.50655
27.5	0.63445
32.5	0.67948
37.5	0.79288
42.5	0.84097
52.5	0.87800
57.5	0.92014
62.5	0.95748
67.5	1.00000

(see preceding discussion).

Even when the cratering rates from ECAs are correctly determined, however, the contribution from comets must be included before a comparison can be made with the observed terrestrial and lunar craters. In the Chapter by Shoemaker et al., these authors estimate that comets account for about half of the terrestrial impact craters larger than 20 km, and may be responsible for the majority of the largest impact craters on Earth. The uncertainties in their estimates are high owing to the inability of astronomers to observe most cometary nuclei directly and to the uncertainty of the size distribution for comets. For impacts by objects smaller than $\sim$1 km, the contribution by active comets is small compared to ECAs and is not reflected in the cratering record. The inactive nuclei of comets may be a major contribution (Wetherill 1991), but they are already accounted for as part of the ECA population (see discussion in Sec. I).

In the face of these difficulties, it is therefore remarkable that impact rates have been derived which match the rate determined from lunar and terrestrial craters to within a factor of 2 (Shoemaker et al. 1979, 1980; Shoemaker 1983; Steel 1994). For example, Shoemaker et al. (1990) estimate that the total production of terrestrial craters larger than 20 km is $(4.9\pm2.9)\times10^{-15}$ km^{-2}yr^{-1} while Grieve (1984) estimates $(5.4\pm2.7)\times10^{-15}$ km^{-2}yr^{-1} from the geologic record of impact for the last 120 Myr. Shoemaker et al. (1990) also estimate a value for the number of lunar craters larger than 30 km that is about twice that observed by Wilhelms (1987) in the 3.3 Gyr record on the Moon. Although the estimate agrees with the lunar record given the uncertainties, it may be that the mean cratering rate has increased in late geologic time. A possible cause is a recent (within $\sim$100 Myr) increase in the flux of comets. Wetherill (1989) has independently derived cratering rates on Earth using a Monte Carlo simulation of the processes that bring asteroids from the main belt to Earth-crossing. His results also match the observed record to within the uncertainties of the estimation, and allow a ~30% contribution from active comets to the cratering record.

Shoemaker (1983) has also shown that the size distribution of young lunar craters is consistent with the size distribution of the ECAs shown by Fig. 1 for ECAs larger than ~50 m . For impacts by smaller ECAs, however, there are not enough lunar craters to account for the observed ECA flux. The cause for this discrepancy may be that ECAs smaller than 50 m have a distribution of orbits different from the orbits of larger ECAs. Recent discoveries with the Spacewatch telescope show that these smallest ECAs have perihelia concentrated near 1 AU and relatively low eccentricities (Rabinowitz et al. 1993), so that their impact velocities with the Moon are lower than previously expected. Correcting for the lower impact velocities brings crater results into agreement with Fig. 1 (Rabinowitz 1993). There are other explanations, however. It may be that small ECAs have unusually high albedos or low densities compared to larger ones. Taking either of these differences into account would lower the slope of the observed size distribution and lower the estimated flux. Also, there is disagreement about how to interpret the size distribution of small craters on the Moon. Neukum et al. (1975) determine a size-frequency for lunar craters smaller than ~1 km that is consistent with Fig. 1 for ECAs smaller than ~50 m. Recent images of (951) Gaspra returned by the Galileo satellite also show a steep size distribution for small craters (Belton et al. 1992). Gaspra being a mainbelt asteroid, the crater record reveals the size-distribution of other mainbelt asteroids, not ECAs.

V. CONCLUSIONS

We have described the size and orbit distributions of the Earth-crossing aster-oids. Although the observed population is incomplete and biased by obser-vational selection, the total number larger than ~1 km is known to within a factor of 2. Integral distributions describing the unbiased distribution of their orbits have been determined by modeling the selection effects of telescope surveys. The number versus size is determined from telescope obervations down to ~5 m , and it is in agreement with the cratering record on Earth and the Moon down to asteroid diameters of ~50 m . Below that range, the cratering record might be brought into agreement by taking into account the relatively high fraction of orbits with perihelia near 1.0 AU and low eccentric-ities among the small ECAs. Otherwise, ECAs smaller than ~50 m may have higher albedos or lower densities than larger ECAs. Active comets are not a significant contribution to the terrestrial cratering record for crater diameters smaller than 10 km.

Acknowledgments.   We thank C. Chapman, D. Steel and A. Harris for useful comments. This work was partly funded by grants from the National

Aeronautics and Space Administration and the Air Force Office of Scientific Research.

REFERENCES

Belton, M. J. S., Veverka, J., Thomas, P., Helfenstein, P., Simonelli, D., Chapman, C., Davies, M. E., Greely, R., Greenberg, R., Head, J., Murchie, S., Klaasen, K., Johnson, T. V., McEwen, A., Morrison, D., Neukum, G., Fanale, F., Anger, C., Carr, M., and Pilcher, C. 1992. Galileo Encounter with 951 Gaspra: First pictures of an asteroid. *Science* 257:1647–1652.

Benedix, G. K., McFadden, L. A., Morrow, E. M., and Fomenkova, M. N. 1992. Bias correction factors for near-Earth asteroids. In *Asteroids, Comets, Meteors 1991*, eds. A. W. Harris and E. Bowell (Houston: Lunar and Planetary Inst.), pp. 65–68.

Binzel, R. P., Shui, Xu, Bus, S. J., and Bowell, E. 1992. Origins of near-Earth asteroids. *Science* 257:779–782.

Ceplecha, Z. 1988. Earth's influx of different populations of sporadic meteoroids from photographic and television data. *Bull. Astron. Inst. Czechoslovakia* 39:221–236.

Gehrels, T.G . 1991. Scanning with charge-coupled devices. *Space Sci. Rev.* 58:347–375.

Grieve, R. A. F. 1984. The impact cratering rate in recent time. *Proc. Lunar Planet. Sci. Conf.* 14, *J. Geophys. Res. Suppl.* 89:B403–B408.

Helin, E. F., and Dunbar, R. S. 1990. Search techniques for near-Earth asteroids. *Vistas in Astron.* 33:21–37.

Helin, E. F., and Shoemaker, E. M. 1979. The Palomar planet-crossing asteroid survey, 1973–1978. *Icarus* 40:321–328.

Kresák, Ľ. 1978. The mass distribution and sources of interplanetary boulders. *Bull. Astron. Inst. Czechoslovakia* 29:135–149.

McFadden, L. A., Tholen, D. J., and Veeder, G. J. 1989. Physical properties of Aten, Apollo, and Amor asteroids. In *Asteroids II*, eds. R. P. Binzel, T. Gehrels and M. S. Matthews (Tucson: Univ. of Arizona Press), pp. 442–467.

Morrison, D., ed. 1992. *The Spaceguard Survey: Report of the NASA International Near-Earth-Object Detection Workshop* (Pasadena: Jet Propulsion Lab).

Muinonen, K., and Bowell, E. 1992. Orbital uncertainties for Earth-crossing asteroids. *Bull. Amer. Astron. Soc.* 24:942.

Muinonen, K., and Bowell, E. 1993. Asteroid orbit determination using Bayesian probabilities. *Icarus* 104:255–279.

Muinonen, K., Bowell, E., and Lumme, K. 1994. Interrelating asteroid size, albedo, and magnitude distributions. *Astron. Astrophys.*, in press.

Neukum, G., Konig, B., and Fechtig, H. 1975. Cratering in the earth-moon system: Consequences for age determination by crater counting. In *Proc. Lunar Sci. Conf.* 6:2597–2620.

Öpik, E. J. 1951. Collision probabilities with the planets and distribution of interplanetary matter. *Proc. Roy. Irish Acad.* 54A:165–199.

Öpik, E. J. 1963. The stray bodies in the solar system. Part 1. Survival of cometary nuclei and the asteroids. *Adv. Astron. Astrophys.* 2:219–261.

Rabinowitz, D. L. 1991. Detection of Earth-approaching asteroids in near real-time. *Astron. J.* 101:1518–1529.

Rabinowitz, D. L. 1993. The size distribution of the Earth-approaching asteroids. *Astrophys. J.* 407:412–427.

Rabinowitz, D. L. 1994. The size and shape of the near-Earth asteroid belt. *Icarus,* in press.

Rabinowitz, D. L., Gehrels, T., Scotti, J. V., McMillan, R. S., Perry, M. L., Wisniewski, W., Larson, S. M., Howell, E. S., and Mueller, B. E. A. 1993. Evidence for a near-Earth asteroid belt. *Nature* 363:704–706.

Shoemaker, E. M. 1983. Asteroid and comet bombardment of the Earth. *Ann. Rev. Earth Planet. Sci.* 11:461–494.

Shoemaker, E. M., Williams, J. G., Helin, E. F., and Wolfe, R. F. 1979. Earth-crossing asteroids: orbital classes, collision rates with Earth, and origin. In *Asteroids,* ed. T. Gehrels (Tucson: Univ. of Arizona Press), pp. 253–282.

Shoemaker, E. M., Wolfe, R. F., and Shoemaker, C. S. 1990. Asteroid and comet flux in the neighborhood of Earth. In *Global Catastrophes in Earth History,* eds. V. L. Sharpton and P. D. Ward, Geological Soc. of America Special Paper 247 (Boulder: Geological Soc. of America), pp. 155–170.

Steel, D. 1994. Collisions in the solar system—VI. An updated terrestrial impact rate for Apollo asteroids. *Mon. Not. Roy. Astron. Soc.,* submitted.

Wetherill, G. W. 1976. Where do the meteorites come from ? A re-evaluation of Earth-crossing Apollo objects as sources of chondritic meteorites. *Geochim. Cosmochim. Acta* 40:1297–1318.

Wetherill, G. W. 1989. Cratering of the terrestrial planets by Apollo objects. *Meteoritics* 24:15–22.

Wetherill, G. W. 1991. End products of cometary evolution: Cometary origin of Earth-crossing bodies of asteroidal appearance. In *Comets in the Post-Halley Era,* vol. 1, eds. R. L. Newburn, Jr., M. Neugebauer and J. Rahe. (Dordrecht: Kluwer), pp. 537–556.

Whipple, F. L. 1967. *Proc. Roy. Soc. London* 296A:304.

Wilhelms, D. E. 1987. *The Geological History of the Moon,* (Washington, D. C.: U. S. Government Printing Office).

Williams, J. G. 1969. Secular Perturbations In The Solar System. Ph.D. Thesis, Univ. of California, Los Angeles.

Orbital Elements for the Known Earth-Crossing Asteroids[a]

ECA		M_o (deg)	ω (deg)	Ω (deg)	i (deg)	e (deg)	a (AU)	c	t_o (ET)
887	Alinda	296.085093(20)	349.737638(35)	110.888391(30)	9.2487365(96)	0.558256530(58)	2.492898071(16)	6.47	1993 01 13
1221	Amor	293.135929(28)	26.281619(70)	171.538548(25)	11.892460(15)	0.435535116(66)	1.9194478628(30)	6.09	1993 01 13
1566	Icarus	25.613096(34)	31.222221(21)	88.1585672(86)	22.881611(39)	0.82673956(17)	1.0780400787(25)	5.90	1993 01 13
1580	Betulia	47.155842(16)	159.276848(51)	62.3881602(66)	52.118137(54)	0.490227733(95)	2.1945640132(44)	5.89	1993 01 13
1620	Geographos	344.000840(12)	276.6127278(76)	337.3921650(45)	13.3261534(68)	0.335388107(33)	1.244726293(14)	6.72	1993 01 13
1627	Ivar	341.707780(12)	167.458269(37)	133.286695(31)	8.4394710(96)	0.396423601(75)	1.8633179022(18)	6.43	1993 01 13
1685	Toro	245.040588(13)	126.853997(80)	274.469713(74)	9.375434(17)	0.43602263(13)	1.3670742168(17)	6.29	1993 01 13
1862	Apollo	322.416937(13)	285.644988(63)	35.933729(62)	6.356364(28)	0.559929970(64)	1.4710823240(45)	6.22	1993 01 13
1863	Antinous	48.608878(19)	266.861114(75)	347.577929(64)	18.423520(53)	0.60585073(14)	2.2605567764(75)	5.86	1993 01 13
1864	Daedalus	39.90170(40)	325.45251(11)	6.793224(68)	22.180885(55)	0.61488798(47)	1.460921525(95)	5.17	1993 01 13
1865	Cerberus	222.377247(44)	325.113430(58)	213.067013(46)	16.088161(27)	0.46701472(27)	1.0802302971(47)	6.03	1993 01 13
1915	Quetzalcoatl	342.334235(36)	347.919463(54)	163.059590(38)	20.457752(48)	0.57438789(14)	2.536290624(38)	5.95	1993 01 13
1917	Cuyo	9.068634(12)	194.143380(50)	188.519110(33)	23.984480(24)	0.505275000(91)	2.149446157(17)	6.05	1993 01 13
1943	Anteros	186.331984(46)	338.129677(74)	246.481228(72)	8.702008(16)	0.255924733(93)	1.4303296301(57)	6.20	1993 01 13
1981	Midas	108.732227(10)	267.679032(24)	357.188402(17)	39.844564(73)	0.64983417(11)	1.7761477164(96)	5.82	1993 01 13
2061	Anza	170.243959(50)	156.12072(14)	207.88532(13)	3.747665(29)	0.53664721(16)	2.267551639(15)	5.97	1993 01 13
2062	Aten	189.29382(29)	147.94617(16)	108.698318(27)	18.93209(10)	0.1825649(12)	0.9665654851(92)	5.40	1993 01 13
2063	Bacchus	107.90499(35)	55.05685(14)	33.34292(13)	9.424193(71)	0.34935117(93)	1.077586392(53)	5.25	1993 01 13
2100	Ra-Shalom	145.425812(93)	355.938668(52)	170.962455(28)	15.755829(31)	0.43649244(25)	0.8320573269(76)	5.79	1993 01 13
2101	Adonis	95.797853(26)	41.68021(72)	351.27598(69)	1.364835(23)	0.76392302(10)	1.875501093(14)	6.05	1993 01 13
2102	Tantalus	195.13444(13)	61.627931(39)	94.409384(15)	64.00812(11)	0.29861868(13)	1.290091717(26)	5.53	1993 01 13
2135	Aristaeus	311.37497(41)	290.65076(11)	191.45078(17)	23.037784(88)	0.50316336(63)	1.59991618(14)	5.27	1993 01 13
2201	Oljato	8.667953(38)	95.92833(19)	76.91195(19)	2.516057(58)	0.71102113(18)	2.175263636(41)	5.80	1993 01 13
2202	Pele	299.247319(55)	217.198419(92)	170.414755(46)	8.783179(15)	0.513030109(95)	2.289344116(31)	5.97	1993 01 13
2212	Hephaistos	195.197032(68)	208.36756(12)	28.441184(94)	11.781832(31)	0.83336948(46)	2.16817113(13)	5.60	1993 01 13
2329	Orthos	140.824486(39)	145.80433(13)	169.54562(12)	24.422407(43)	0.65904164(16)	2.401320875(35)	5.83	1993 01 13

APPENDIX (cont.)

2340	Hathor	66.34434(26)	39.74105(55)	211.706679(44)	5.849522(31)	0.4495616(24)	0.843960675(12)	5.46	1993 01 13
2608	Seneca	277.59350(21)	33.87872(14)	169.618128(20)	15.337711(51)	0.5816585(23)	2.49039487(30)	5.34	1993 01 13
3103	82BB	278.640886(46)	253.758901(36)	129.933873(34)	20.936326(26)	0.35475692(22)	1.405974001(16)	6.10	1993 01 13
3122	Florence	180.318382(64)	27.495637(91)	336.267179(88)	22.185276(43)	0.42252003(30)	1.769086496(63)	5.76	1993 01 13
3200	Phaethon	193.10096(22)	321.792986(36)	265.611442(38)	22.091168(40)	0.89011526(17)	1.271320507(91)	5.37	1993 01 13
3288	Seleucus	256.03653(11)	349.175687(79)	218.854019(84)	5.929287(12)	0.457457944(98)	2.032403301(96)	5.87	1993 01 13
3360	81VA	339.574872(52)	60.588250(65)	245.523616(61)	21.734685(64)	0.74272787(15)	2.465041271(93)	5.75	1993 01 13
3361	Orpheus	70.508761(25)	301.55222(19)	189.86669(19)	2.681725(15)	0.32261776(13)	1.209243877(43)	6.28	1993 01 13
3362	Khufu	264.43127(41)	54.85782(23)	152.67548(11)	9.91384(11)	0.4686348(38)	0.98945769(14)	5.16	1993 01 13
3551	83RD	26.793908(25)	193.003432(42)	174.043865(22)	9.516344(12)	0.48798230 9(54)	2.091774486(18)	6.23	1993 01 13
3554	Amun	207.83375(28)	359.33179(12)	358.717593(27)	23.364514(65)	0.28050823(68)	0.973737453(80)	5.30	1993 01 13
3671	Dionysus	222.82313(21)	203.671981(43)	82.458111(11)	13.625497(23)	0.543047692(97)	2.19446793(31)	5.47	1993 01 13
3752	Camillo	45.762368(99)	312.232745(56)	148.034243(17)	55.547066(46)	0.3024312 7(24)	1.413667118(44)	5.84	1993 01 13
3753	86TO (83UH)	356.64764(17)	43.62524(38)	126.40343(49)	19.81181(10)	0.5148529(21)	0.997733720(23)	5.19	1993 01 13
3757	82XB	13.885488(20)	16.903394(80)	75.120097(54)	3.8750238(94)	0.446181515(43)	1.8354109190(48)	6.15	1993 01 13
3838	Epona	152.33845(24)	49.426982(52)	235.775873(84)	29.282093(45)	0.70169014(91)	1.50468034(30)	5.24	1993 01 13
3908	80PA	211.1011492(89)	125.52308(24)	262.07274(23)	2.1690993(51)	0.458715350(26)	1.9244562485(47)	6.37	1993 01 13
3988	86LA	120.96531(14)	86.59274(10)	230.037287(64)	10.767334(20)	0.31677765(13)	1.54453223(11)	5.63	1993 01 13
4015	Wilson-Harrington	33.1315146(81)	90.86533(34)	271.06243(33)	2.784993(21)	0.62272309(13)	2.6409237855(79)	5.97	1993 01 13
4034	86PA	219.5602(25)	296.44233(17)	158.14660(12)	11.166507(80)	0.4440594(45)	1.05994990(86)	4.36	1993 01 13
4179	Toutatis	15.0534843(75)	276.2767(12)	126.4864(12)	0.466728(30)	0.63985498(13)	2.505463873(65)	5.99	1993 01 13
4183	Cuno	39.09947(16)	235.11778(23)	295.96175(19)	6.76761 6(51)	0.63730519(63)	1.980499982(22)	5.59	1993 01 13
4197	82TA	315.833458(19)	119.256517(72)	10.199856(47)	12.213583(52)	0.77280043(26)	2.298159429(32)	5.78	1993 01 13
4257	Ubasti	152.192194(58)	278.887901(50)	169.373276(30)	40.705338(69)	0.46852649(20)	1.647001645(84)	5.76	1993 01 13
4341	Poseidon	125.977072(38)	15.49842(23)	108.24790(24)	11.863381(58)	0.67884927(34)	1.835354932(34)	5.72	1993 01 13
4450	Pan	70.15107(27)	291.30416(21)	312.29255(21)	5.522791(17)	0.58644721(48)	1.44185139(26)	5.39	1993 01 13
4486	Mithra	256.86818(26)	168.50397(68)	82.53386(64)	3.040080(23)	0.66223668(67)	2.19852277(11)	5.53	1993 01 13
4544	Xanthus	44.85104(53)	333.61060(12)	24.131978(64)	14.148281(98)	0.2503425(20)	1.04209237(42)	4.93	1993 01 13

4581	Asclepius	312.20641(55)	254.98701(50)	180.56235(40)	4.906306(87)	0.3570731(59)	1.022770831(82)	4.87	1993 01 13
4596	81QB	118.732079(62)	248.242605(76)	154.576242(32)	37.129614(87)	0.51791526(14)	2.240482043(48)	5.73	1993 01 13
4660	Nereus	17.28930(54)	157.7883(10)	314.8744(12)	1.425253(26)	0.36029561(38)	1.48965586(11)	5.40	1993 01 13
4688	80WF	220.58298(38)	212.80764(21)	241.80058(15)	6.408467(33)	0.51445287(14)	2.23322611(11)	5.67	1993 01 13
4769	Castalia	335.32031(81)	121.15831(12)	325.811475(20)	8.893799(38)	0.483128(24)	1.06314282(62)	4.91	1993 01 13
4953	90MU	145.92697(15)	77.39006(12)	78.12571(14)	24.420744(59)	0.65714242(44)	1.62117773(15)	5.55	1993 01 13
5011	Ptah	107.80990(60)	105.37052(23)	11.08845(17)	7.404580(66)	0.4999637(67)	1.63486225(24)	4.88	1993 01 13
5131	90BG	167.621764(40)	135.682740(65)	110.541234(51)	36.371868(56)	0.56988761(35)	1.486430977(97)	5.73	1993 01 13
5143	Heracles	121.025876(68)	226.32347(20)	310.80515(20)	9.193678(31)	0.77147303(21)	1.834174013(17)	5.84	1993 01 13
5189	90UQ	100.44531(10)	159.4467(13)	135.4459(12)	3.576927(37)	0.47786150(70)	1.55089481(83)	5.33	1993 01 13
	37UB	335.19(22)	90.7351(52)	35.98575(56)	6.226(25)	0.6241(24)	1.6410(80)	2.32	1937 10 27
	50DA	336.91(19)	223.6063(67)	357.5033(45)	12.184(48)	0.5036(28)	1.6881(82)	2.29	1950 02 25
	54XA	195.5(3.7)	48(14)	201(17)	5.3(3.3)	0.42(18)	0.748(67)	0.59	1954 12 06
5660	73NA	12.704150(66)	118.23833(36)	101.122854(27)	68.01377(17)	0.6384104(13)	2.433904602(47)	5.12	1992 06 27
	74MA	330.0(1	126.91(42)	302.77(11)	37.60(23)	0.7587(42)	1.748(36)	1.83	1974 07 02
	78CA	306.951(47)	102.085(37)	161.3920(10)	26.095(20)	0.21474(16)	1.12462(30)	3.35	1978 04 02
	79XB	346.39(33)	75.473(37)	86.0376(89)	24.97(12)	0.7173(45)	2.292(33)	1.34	1979 12 12
5797	80AA	20.022(16)	167.85718(70)	299.36546(17)	4.1851(21)	0.44343(28)	1.8914(10)	3.37	1980 03 02
	83LC	349.38(44)	184.65(23)	159.83(26)	1.513(12)	0.7107(83)	2.644(68)	2.10	1983 06 30
	83VA	3.47202(59)	11.70576(86)	77.54248(86)	16.24767(70)	0.692280(42)	2.61146(33)	4.34	1984 01 01
	84KB	31.8322(90)	334.8635(47)	171.2687(48)	4.62985(25)	0.762403(51)	2.22204(40)	3.78	1984 07 19
	85WA	6.18007(59)	350.89177(14)	43.85094(12)	9.74948(23)	0.601509(24)	2.84478(18)	4.55	1985 12 01
	86DA	15.13113(85)	126.79960(26)	65.14720(19)	4.298517(69)	0.585429(14)	2.81206(10)	4.69	1986 06 19
	86JK	357.29762(16)	232.56977(13)	62.775968(90)	2.141851(36)	0.679706(11)	2.801634(95)	4.90	1986 06 19
5370	Taranis	341.636140(54)	161.157414(99)	177.907190(25)	18.998069(28)	0.63156610(13)	3.34895537(26)	5.79	1992 06 27
	87OA	319.370(29)	235.2876(17)	180.4680(56)	9.0211(29)	0.59551(26)	1.49651(58)	3.21	1987 07 31
	87QB	19.0785(39)	156.08277(28)	153.69931(24)	3.45594(18)	0.593337(53)	2.80400(39)	4.05	1987 11 01

APPENDIX (cont.)

Designation								Date
87SF3		21.7696(76)	133.5902(13)	187.84228(27)	3.31122(46)	0.535070(98)	2.25292(51)	3.79	1987 11 01
87WC		54.1626(74)	308.0872(25)	51.9912(11)	15.8456(28)	0.233998(76)	1.36219(11)	3.77	1987 12 15
88EG		267.4958(13)	241.44074(36)	183.00283(16)	3.479892(54)	0.491876(57)	1.2691037(25)	4.72	1990 06 30
88SM		29.613(20)	312.8560(83)	1.1348(23)	10.9465(39)	0.34395(21)	1.66482(62)	3.45	1988 10 15
88TA		58.6925(43)	104.36848(16)	195.38968(11)	2.53897(11)	0.478559(28)	1.541216(66)	4.10	1988 12 05
5731	88VP4	337.69334(11)	215.49546(33)	282.86496(33)	11.661459(59)	0.6528101(14)	2.2631435(79)	5.51	1988 12 05
88XB		33.3308(26)	279.80792(44)	73.70020(29)	3.12072(16)	0.481578(29)	1.467366(60)	4.27	1988 12 05
89AZ		34.270(11)	111.47245(56)	295.89015(63)	11.7600(26)	0.46801(12)	1.64545(33)	3.64	1989 01 31
89DA		7.7923(18)	138.64985(43)	349.814868(95)	6.44711(54)	0.543629(61)	2.16201(29)	4.15	1989 03 01
89JA		353.60609(19)	231.63124(13)	61.72790(10)	15.24906(25)	0.484380(93)	1.770410(29)	4.95	1989 06 23
89QF		81.782(16)	239.4799(13)	344.89459(59)	3.9416(12)	0.41298(11)	1.151466(94)	3.54	1989 09 15
89UP		13.45528(49)	17.298312(69)	53.44187(11)	3.866002(84)	0.473009(14)	1.863872(49)	4.78	1989 12 31
89UQ		232.1180(18)	14.85848(89)	178.7273(11)	1.27392(17)	0.265243(31)	0.9156443(71)	4.36	1989 11 15
89UR		282.084(32)	289.246(19)	234.679(11)	10.340(15)	0.35642(40)	1.08038(18)	3.08	1989 11 15
89VA		184.6736(78)	2.7922(12)	225.66844(95)	28.753(37)	0.59449(27)	0.728630(38)	3.10	1989 11 15
89VB		10.75(11)	329.605(13)	38.946(14)	2.139(11)	0.4611(34)	1.866(12)	2.37	1989 11 05
90HA		353.9614(15)	307.6842(14)	185.2834(10)	3.88460(30)	0.693003(70)	2.57289(52)	4.14	1990 01 31
90MF		354.76736(38)	113.47611(77)	210.91587(82)	1.858280(83)	0.456169(21)	1.747106(64)	4.63	1990 07 28
90OS		337.098(73)	19.586(23)	348.402(20)	1.1100(17)	0.4617(11)	1.6777(30)	2.79	1990 07 28
90SM		20.17(14)	105.399(51)	138.306(30)	11.4914(72)	0.7648(10)	2.0693(84)	2.64	1990 09 30
5645	90SP	10.086695(47)	47.991277(66)	45.930277(70)	13.516990(87)	0.3873124(26)	1.3548553(41)	5.51	1990 12 31
90SS		345.49799(15)	115.733142(36)	0.149993(46)	19.39799(16)	0.4749172(51)	1.702397(15)	5.20	1990 12 31
90TG1		66.84(83)	33.35(11)	205.106(41)	8.95(13)	0.6874(62)	2.472(16)	1.72	1990 10 15
90UA		23.17(56)	203.52(34)	103.14(35)	0.947(12)	0.5406(74)	1.683(23)	1.90	1990 10 15
90UN		337.689(83)	97.0019(47)	8.4689(77)	3.6887(90)	0.5287(12)	1.7120(37)	2.67	1990 10 30
90UO		293.4(2.3)	333.06(41)	205.775(17)	30.0(2.3)	0.763(17)	1.241(25)	1.15	1990 10 30
5590	90VA	189.71267(59)	34.40862(23)	216.478095(70)	14.18744(18)	0.2792351(33)	0.98535260(14)	5.10	1990 12 31
90VB		22.6674(44)	101.1041(24)	255.46294(82)	14.52850(52)	0.527058(47)	2.44388(27)	4.12	1990 12 31

ID	Label								Date
5828	91AM	331.463(47)	152.598(32)	125.603(12)	30.0217(80)	0.69532(23)	1.6948(17)	3.18	1991 01 31
91AQ		342.60(13)	239.602(19)	342.816(26)	3.2044(36)	0.7718(13)	2.1812(97)	2.41	1991 01 15
91BA		347.46(50)	70.749(47)	118.8809(25)	1.961(20)	0.6823(86)	2.243(53)	1.89	1991 01 18.3
91BB		259.31008(17)	322.812435(73)	295.063807(39)	38.483014(94)	0.2724714(10)	1.1864474(11)	5.51	1991 01 31
91BN		96.85(13)	80.31(11)	269.355(57)	3.443(11)	0.3978(15)	1.4428(11)	2.46	1991 01 31
91CS		74.994(27)	249.270(26)	156.9398(14)	37.1084(60)	0.164655(16)	1.12301(11)	3.84	1991 02 15
91CB1		236.6(3.0)	346.9(2.3)	317.89(48)	15.9(2.3)	0.623(45)	1.6856(55)	1.08	1991 02 15
91DB		31.8643(21)	50.94955(13)	158.487177(64)	11.43269(56)	0.401844(26)	1.71644(76)	4.34	1991 06 30
91DG		22.01564(22)	63.06418(18)	180.34532(11)	11.15078(14)	0.362822(41)	1.4273560(75)	5.25	1991 06 30
91EE		10.04627(17)	115.043861(54)	169.168606(25)	9.757321(59)	0.6240386(44)	2.245274(24)	5.21	1991 09 01
91FA		51.843(15)	91.4860(35)	339.7705(23)	3.08213(39)	0.44611(14)	1.97831(19)	3.50	1991 03 20
91FB		345.059(23)	218.3073(10)	19.1140(21)	9.1979(43)	0.56267(41)	2.3706(23)	3.21	1991 03 15
5626	91FE	114.77497(33)	231.04954(28)	173.59183(22)	3.856164(31)	0.4549400(45)	2.1942163(15)	5.52	1991 06 30
91GO		21.47(52)	88.533(24)	25.0379(15)	9.511(86)	0.6513(58)	1.909(27)	1.89	1991 04 15
91JR		341.002(33)	207.0020(25)	60.2242(17)	10.115(20)	0.25990(72)	1.4036(14)	2.98	1991 05 15
91JW		98.22372(40)	301.71765(38)	54.11385(17)	8.72337(27)	0.1183708(44)	1.0380085(16)	5.14	1991 12 31
91JX		1.09255(12)	64.40165(25)	212.97142(20)	2.31059(11)	0.599656(35)	2.52123(22)	4.43	1991 06 30
5381	91JY	160.43200(24)	37.43901(11)	58.5885288(89)	48.973348(79)	0.2959019(12)	0.9749312(23)	5.49	1991 06 30
91LH		110.31(27)	203.74(10)	281.056(82)	52.87(38)	0.73164(49)	1.3549(21)	2.07	1991 06 15
91OA		14.0616(21)	317.20262(18)	306.67352(15)	5.51685(28)	0.586936(40)	2.50891(25)	4.25	1991 08 15
91RB		328.08(12)	68.708(14)	359.599(12)	19.580(41)	0.4846(12)	1.4524(27)	2.64	1991 09 15
5786	91RC	126.400(96)	8.2448(67)	161.3891(58)	23.169(44)	0.82614(39)	1.08101(33)	2.68	1991 09 15
91TT		333.52(14)	217.62(10)	192.5213(14)	14.847(75)	0.1619(12)	1.1955(18)	5.03	1991 10 07
91TU		339.75(27)	220.51(28)	193.5512(13)	7.6368(91)	0.3290(28)	1.4106(62)	0.94	1991 10 07
91TB1		349.56499(13)	103.56935(13)	6.281171(54)	23.46713(41)	0.3521176(53)	1.453680(11)	1.34	1991 12 31
91TB2		340.9(4.4)	195.7(3.0)	297.0(4.2)	8.63(61)	0.834(36)	2.38(34)	2.70	1991 10 05
91TF3		18.9(1.9)	303.240(44)	6.7231(71)	14.07(55)	0.532(29)	2.05(13)	1.56	1991 10 07

APPENDIX (cont.)

		M_o	ω	Ω	i	e	a	c	t_o
91VA		22.233(51)	313.2923(49)	37.65671(80)	6.549(14)	0.35338(94)	1.4311(19)	2.85	1991 11 01
91VE		96.950(76)	193.3648(62)	62.165(16)	7.203(15)	0.66369(61)	0.89051(18)	2.75	1991 11 15
91VG		303.73025(14)	261.13631(67)	213.93190(42)	0.378706(13)	0.0671028(37)	1.0396299(29)	5.12	1991 11 15
91VH		55.6777(38)	206.9721(34)	139.5111(14)	13.91831(86)	0.143921(14)	1.136582(25)	4.40	1991 11 30
91VK		356.947462(17)	173.19679(31)	295.16993(32)	5.416902(17)	0.506407102(85)	1.844281074(32)	6.11	1991 12 31
91WA		37.150(22)	241.769(11)	66.7841(18)	39.6611(64)	0.64262(10)	1.57596(54)	3.51	1991 12 10
91XA		11.64(49)	308.432(58)	77.248(11)	5.202(87)	0.562(12)	2.234(61)	1.80	1991 12 05
92BC		324.56(31)	76.943(63)	123.555(13)	14.38(12)	0.3532(32)	1.4234(68)	2.22	1992 01 31
92BF		214.071(42)	336.2782(31)	315.7083(19)	7.255(23)	0.27078(69)	0.90785(13)	2.93	1992 01 31
92CC1		101.4619(21)	21.85820(88)	349.327148(58)	36.89909(15)	0.3748638(65)	1.39159537(36)	4.63	1992 01 31
92DU		21.24(18)	121.88(18)	338.0063(22)	25.179(80)	0.17665(94)	1.1625(16)	2.73	1992 01 31
5604	92FE	210.5381(70)	82.2400(44)	312.2378(37)	4.79225(22)	0.4053930(91)	0.927107(22)	3.96	1992 06 27
92HE		1.708709(39)	262.59955(26)	27.32064(16)	37.36947(46)	0.571763(12)	2.240892(67)	4.80	1992 06 27
92HF		338.914(39)	128.0152(23)	213.6390(42)	13.244(25)	0.56069(65)	1.3884(11)	2.95	1992 06 27
92JB		56.2509(31)	306.72028(43)	218.541131(64)	16.06544(63)	0.360065(21)	1.556790(52)	4.27	1992 06 27
92JD		72.738(92)	285.747(90)	222.700226(65)	13.568(10)	0.032146(28)	1.034285(72)	3.71	1992 05 05
92LC		20.769(24)	89.601(14)	62.0122(25)	17.8470(19)	0.70560(17)	2.5188(18)	3.48	1992 06 27
92LR		348.70586(80)	67.50223(35)	233.42644(33)	2.02485(10)	0.408829(27)	1.830223(89)	4.49	1992 06 27
92NA		338.866(28)	7.829(13)	349.6429(96)	9.7723(29)	0.55166(35)	2.3925(20)	3.23	1992 06 27

[a] This table lists the mean anomaly (M_o), argument of perihelion (ω), longitude of ascending node (Ω), inclination (i), eccentricity (e), and the semimajor axis (a) at the epoch t_o (equinox J2000.0) for the Earth-crossing asteroids listed in Table I and additional asteroids with $q < 1.017$ AU discovered June 1991–July 1993. In parentheses, we give the 1-σ uncertainties in units of the last significant figure. We also tabulate the orbital quality, $c = -\log_{10}[\theta(\text{rad})]$, where θ is the leak metric at epoch t_o.

THE FLUX OF PERIODIC COMETS NEAR EARTH

EUGENE M. SHOEMAKER
Lowell Observatory

PAUL R. WEISSMAN
Jet Propulsion Laboratory

and

CAROLYN S. SHOEMAKER
Lowell Observatory

A total of 26 active Earth-crossing periodic comets has been discovered; half are in the Jupiter family (J-f) and half in the Halley family (H-f). Because of observational and physical selection effects, it is inappropriate to use the raw statistics for total discovered H-f and J-f comets to estimate the flux of Earth-crossing comets. The vast majority of periodic comets may be dormant or extinct, in the sense that they do not or cannot exhibit a detectable coma. The extinct comets are much more difficult to detect than active comets; only 15 candidate extinct J-f and H-f comets have been recognized, among which only 2 are Earth-crossing. The current number of active Earth-crossing J-f comet nuclei ≥ 1 km diameter is estimated to be about $40 \times / \div 2$, and the number of extinct Earth-crossing J-f comet nuclei ≥ 1 km diameter is very roughly estimated at 780. To the same level of completeness of discovery of Earth-crossing J-f comets, there may be about 10 to 20 times as many active Earth-crossing H-f comets as have been discovered, i.e., about 140 to 270. The ratio of extinct to active H-f comets may be about 10 to 20. Altogether, active and extinct periodic comets may account for about 20% of the production of terrestrial impact craters larger than 20 km diameter. The time required for discovery of 95% of the extinct Earth-crossing H-f comets larger than 1 km diameter with the proposed Spaceguard Survey may be of the order of several millennia.

I. INTRODUCTION

Accurate assessment of the near-Earth flux of comet nuclei presents an especially difficult challenge. In this chapter we treat the flux of periodic comets (period ≤ 200 yr); the long-period comets (period > 200 yr) are discussed in the chapter by Marsden and Steel. Our conclusion, based on very limited statistics, is that impact of periodic comets may account for approximately 20% of the terrestrial craters larger than 20 km diameter, a fraction roughly comparable with that produced by the long-period comets.

[313]

There are strong observational selection effects in the discovery of active comets. Additionally, periodic comets passing through the inner planetary region may be dormant or "extinct" during much of their dynamical evolution. The extinction of activity is thought to be due to the development of an insulating layer of nonvolatile material that accumulates on the surface of a periodic comet after many passages near the Sun. Here we will refer to objects revolving on orbits similar to those of known active periodic comets, but which have exhibited no detected cometary activity, as extinct comets. As used in this way, the term extinct comet is based solely on the orbit of the object and on available observations of its appearance. Higher-resolution images or a longer record of observations might reveal some apparently extinct objects to be weakly active, and a sudden decrease in perihelion distance might return an extinct comet to an active state. Extinct comets are much more difficult to detect than active ones of similar nucleus size. Because they are point-like or stellar in appearance they are reported as asteroids by the Minor Planet Center at Cambridge, Massachusetts, and are given asteroid designations.

As will be seen below, extinct comets passing near the Earth are likely an order of magnitude more numerous than active periodic comets. But only a small number of extinct comets with orbits similar to those of active periodic comets have been identified, and, of those, only two or three are Earth-crossing. Hence there is hardly any statistical basis for evaluating the flux of extinct Earth-crossing comet nuclei directly from observation of these bodies. We must utilize the more abundant but still meager discoveries of the active Earth-crossing periodic comets and try to understand the selection effects that govern the discovery of active and extinct comets, if we are to assess the population and flux of comet nuclei in the inner solar system and to evaluate the rate of collision of these bodies with Earth. Understanding observational selection effects and the process of cometary extinction is also vital to the identification of the source regions for periodic comets.

In addition to extinct comets with orbits similar to those of active comets, a substantial number of extinct comet nuclei probably are represented among the ordinary Earth-crossing asteroids (Wetherill 1991). These objects cannot be readily separated on the basis of orbital parameters from the remainder of the Earth-crossing asteroids, which are discussed fully in the chapters by Rabinowitz et al. and by Bowell and Muinonen. With one exception, they will not be discussed further in this chapter.

II. OBSERVATIONS OF ACTIVE EARTH-CROSSING PERIODIC COMETS

A total of 26 active Earth-crossing periodic comets has been discovered (Table I); 13 are in the Jupiter family (period ≤ 20 yr) and 13 in the Halley family (20 yr<period ≤ 200 yr). Here we define Earth-crossing to mean that the current osculating perihelion distance q is less than the Earth's current aphelion distance of 1.017 AU. P/Giacobini-Zinner currently has a q of 1.034 AU,

slightly beyond Earth's aphelion distance but was, in fact, Earth-crossing when discovered in 1900 (q = 0.93 AU). The orbit of P/Giacobini-Zinner has slowly evolved to a larger perihelion distance, owing to Jupiter perturbations. Conversely, P/Tuttle, which was not Earth-crossing when discovered in 1790, has recently evolved to a perihelion distance of 0.998 AU. The Earth-crossing comets constitute 9% of the discovered active comets of the Jupiter family (J-f), and 62% of the discovered comets of the Halley family (H-f).

Some of the active Earth-crossing comets have been observed on only one apparition, whereas others have been seen repeatedly. The greatest number of observed returns is for P/Encke (54 out of 62 returns since its discovery in 1786). However, the longest period that any comet has been under observation is P/Halley, which has been seen on 30 consecutive apparitions since 240 B. C.

Of the three single apparition J-f comets, two (P/Helfenzrieder and P/Blanpain) were never recovered and are considered lost, and one (P/Lexell) is no longer a short-period comet (see below). Four of the single apparition H-f comets (P/Levy, P/Bradfield 2, P/Mellish, and P/Wilk) have not yet passed perihelion again since their discovery. Only two of the J-f comets that have been seen on more than one apparition are lost: P/Brorsen and P/Biela (see below); none of the multiple apparition H-f comets are lost. The failure to recover short-period comets could have a number of explanations, including poor original orbit determinations (usually only applicable to single apparition comets), poor observing geometry at return apparitions, nucleus disintegration, or nucleus evolution to a dormant state that makes them more difficult to detect and/or recognize as comets.

Several of the comets in Table I are illustrative of important cometary phenomena. P/Lexell was discovered when it approached to within 0.015 AU (2.2×10^6 km) of the Earth in 1770, the closest known cometary encounter with Earth on record. Integration of the orbit backward in time showed that the comet had been thrown into its Earth-crossing orbit by a close encounter with Jupiter in 1767, at a jovicentric distance of 0.020 AU. Prior to the Jupiter encounter, P/Lexell was in a low eccentricity orbit with a perihelion distance of 3 AU, and had never been observed. The new orbit had a period of 5.6 yr, which was close to half of Jupiter's period, leading to a second, even closer Jupiter encounter in 1779 at 0.0015 AU (about 3 Jupiter radii, or half the distance of Io's orbit). The comet was thrown into a long-period orbit with perihelion just beyond Jupiter, aphelion at about 80 AU, and a period of over 260 yr. Thus, P/Lexell is no longer an immediate threat to the Earth. The large changes in its orbit are an example of the chaotic dynamics of short-period comets.

P/Swift-Tuttle, the parent of the Perseid meteor shower, was discovered in 1862 and predicted to return in 1980 (Marsden 1973). However, some of the 1862 observations used in the orbit determination appear to have been in error. An alternative orbit solution (Marsden 1973), based on the possible identification of Swift-Tuttle with a comet observed briefly in 1737, led to a predicted return in 1992, which turned out to be correct. This subsequently led

TABLE I
Earth-Crossing Periodic Comets[a]

	Discovery	q	e	i	P	H	CP	CP	u
		AU		deg	yr		10^9 orb	10^9 yr	km s^{-1}
Jupiter-Family Comets									
Encke	V	0.331	0.855	11.9	3.28	9.8	3.32	.956	29.9
Grigg-Skjellerup	V	0.995	0.664	21.1	5.10	12.5	7.65	1.50	15.3
Machholz	V	0.126	0.958	60.1	5.24	13.0	.874	.168	43.6
Honda-Mrkos-Padjdusakova	V	0.541	0.822	4.2	5.30	13.5	9.47	1.79	24.4
Schwassmann-Wachmann 3	Ph	0.936	0.694	11.4	5.35	12.0	7.64	1.43	13.2
Brorsen	V	0.590	0.810	29.4	5.46	8.6	1.63	.298	27.6
Hartley 2	Ph	0.953	0.719	9.3	6.26	10.5	10.7	1.72	12.4
Biela	V	0.861	0.756	12.5	6.62	7.8	4.78	.722	16.5
Denning-Fujikawa	V	0.780	0.820	8.7	9.01	15.0	5.54	.614	18.9
Tuttle	V	0.998	0.824	54.7	13.73	8.0	4.52	.334	33.3
Helfenzrieder	U	0.406	0.848	7.9	4.35	6.8	4.96	1.13	28.1
Blanpain	V	0.892	0.699	9.1	5.10	8.5	7.38	1.45	14.1
Lexell	V	0.674	0.786	1.6	5.60	7.7	26.8	4.79	20.5
Means		0.699	0.789	18.6	6.19	10.3	7.32	1.30	22.9
Weighted mean									19.9

Row groupings (left margin): Multiple Apparitions (Encke through Tuttle); One Apparition (Helfenzrieder, Blanpain, Lexell).

Halley-Family Comets

	Discovery method	q	e	i	P	H	CP (per 10^9 orbits)	CP (per 10^9 yr)	u
Multiple Apparitions									
Crommelin	V	0.735	0.919	29.1	27.4	9.3	1.81	.0661	26.6
Tempel-Tuttle	V	0.982	0.904	162.7	32.9	8.5	37.9	1.16	69.9
Brorsen-Metcalf	V	0.479	0.972	19.3	70.5	7.8	2.06	.0292	31.0
Pons-Brooks	V	0.774	0.955	74.2	70.9	5.1	1.47	.0206	44.7
Halley	U	0.587	0.967	162.2	76.0	5.5	4.89	.0651	66.4
Swift-Tuttle	U	0.963	0.960	113.6	135	4.0	4.77	.0348	59.9
Herschel-Rigollet	V	0.748	0.974	64.2	155	7.8	1.36	.0088	41.0
One Apparition									
Levy	V	0.983	0.929	19.2	51.3	8.0	28.0	.544	16.9
Pons-Gambart	V	0.807	0.946	136.5	57.5	7.0	3.13	.0542	64.8
de Vico	V	0.664	0.963	85.1	76.3	7.2	1.26	.0166	49.2
Bradfield 2	V	0.420	0.978	83.1	81.9	11.0	.958	.0114	49.0
Mellish	V	0.190	0.993	32.7	145	7.3	1.26	.0089	41.0
Wilk	V	0.619	0.981	26.6	187	10.4	1.69	.0091	29.3
Means		0.698	0.957	77.6	88.6	7.6	6.97	.156	45.4
Weighted mean									52.3

[a] Discovery method: U = unaided eye; V = visual search with telescope; Ph = photographic. Orbital elements are from Marsden and Williams (1993); q = perihelion distance, e = eccentricity, i = inclination, P = period, CP = collision probability with Earth per 10^9 orbits or 10^9 yr, and u = asymptotic encounter velocity. H is absolute total magnitude. Values of H for recently observed comets are listed in annual issues of the *Comet Handbook* (published in the *International Comet Quarterly*) and unpublished estimates by D. W. E. Green. For comets which have not been observed since 1985, the values of H are the mean of values of H by Vsekhsvyatskii (1964).

to identifications, first suggested by Hasegawa (1979), of previous apparitions of P/Swift-Tuttle in ancient Chinese records in 69 B. C. and 188 A. D. (Yau et al. 1994). Initial integration of the orbit forward in time, after the comet was recovered in 1992, resulted in a prediction of a possible very close approach to Earth in 2126 (Marsden 1992). After identification of the earlier apparitions, however, it became clear that the comet will miss Earth by a healthy margin, 0.153 AU. The history of evolving solutions for the orbit of P/Swift-Tuttle is illustrative of the difficulty of accurately determining periodic comet orbits and predicting returns, until at least three returns have been reliably identified (see the Chapter by Yeomans and Chodas).

P/Biela, which was observed on several perihelion passages, was seen to have split in 1846, returned as a double comet in 1852, and was never observed again. The nucleus appears to have disrupted spontaneously, though this cannot be absolutely confirmed. Thus, the comet likely does not pose a threat to the Earth anymore. The comet is the parent body of the Andromedid meteor stream. Such disruption events are fairly common for long-period comets; about 10% of dynamically new long-period comets and 4% of returning long-period comets break up without an obvious cause (Weissman 1980). Disruption events are less common for short-period comets, about 1% of which break up per perihelion passage. Comets have also been known to break up when passing within the Roche limit of the Sun (the Kreutz family of Sun-grazing comets), and of Jupiter (P/Brooks 2 in 1886, and P/Shoemaker-Levy 9 in 1992).

III. SELECTION EFFECTS IN THE DISCOVERY OF PERIODIC COMETS

One notable feature of the statistics given in Table I is that the same number of active H-f and J-f Earth-crossing comets have been discovered. This is in marked contrast to the ratio of total number of known active H-f and J-f comets (21:151). The explanation for this difference lies, at least in part, in strong selection effects in the discovery of H-f comets. No active H-f comets have been discovered with $q > 1.574$, and only 38% have $q > 1.0$ AU. In contrast, more than 90% of J-f comets have $q > 1.0$ AU and many are now being found photographically with $q > 2.0$ AU. The low rate of discovery of H-f comets, especially with $q > 1.0$ AU, can be attributed, in part, to two selection effects: (1) the much longer average period of H-f comets, which results in many fewer perihelion passages and fewer opportunities for discovery for a given period of observations than for J-f comets; and (2) the broader distribution of inclination of H-f comets, which results in a bias against discovery by photographic observations that are concentrated along the ecliptic and near opposition.

The large majority of the active Earth-crossing Jupiter-family comets and nearly all the Earth-crossing Halley-family comets have been discovered visually with small to modest-sized telescopes or with the unaided eye (Table I), generally at small to moderate solar elongations. This is the region where techniques of visual search, which cover a broad range of ecliptic latitude, tend to minimize bias against discovery of high-inclination comets. However, comets must be relatively bright (generally visual magnitude 11 or 11.5 or brighter) to be discovered with the telescopes commonly used for visual search. The success of the visual discoverers has depended on the rapid brightening of active comets as they approach the Sun. Further, the visual detection of a comet rests on the recognition of its coma. Hence the methods of the visual observer essentially preclude the discovery of extinct comets, which lack comae and are rarely brighter than magnitude 15.

A check on the completeness of the currently known, active Earth-approaching comets is provided by analysis of dust trails found in IRAS All-Sky maps by Sykes et al. (1986). These dust trails are in the orbits of short-period comets and are the precursors of cometary meteoroid streams. Sykes and Walker (1992) showed that 8 of 17 identified trails were associated with known short-peroid comets, and that all of the identified comets were near perihelion at the time of the IRAS observations. This led them to infer that all short-period comets have such dust trails, but that they were only observable (by IRAS) when they were relatively close to the Sun. No trails were associated with any of the known asteroids, or with any long-period comets.

The existence of 9 additional dust trails that do not have any known comets associated with them suggests that there are still a significant number of undiscovered short-period comets in the inner solar system. Although the IRAS data set does contain some observational selection effects, it is a relatively unbiased all-sky survey. Based on the statistics of Sykes and Walker (1992), the total number of short-period comets with perihelia $\lesssim 1.8$ AU is roughly twice the known number. We would expect that completeness is somewhat greater for the Earth-crossing comets because of their smaller perihelion distances. However, this still implies on the order of 15 to 25 active J-f Earth-crossing comets that are bright enough to be readily detected.

Most photographic discoveries of comets are made at large solar elongation, commonly near the ecliptic, and all extinct comets have been discovered in photographic surveys that are mostly confined to a region near opposition. It is a striking fact that only 2 out of the 26 known active Earth-crossing comets have been discovered photographically, although a few were found independently by photographic methods after visual discovery. Photographic searches can detect much fainter comets than visual searches, but they simply have not covered as broad an area of the sky.

There is a very large difference in the inclination distribution between active H-f and J-f Earth-crossing comets. H-f comets exhibit a broad range in inclination, including several retrograde objects such as P/Halley and P/Swift-Tuttle, whereas only two high-inclination J-f comets have been found, P/Machholz and P/Tuttle, both of which were discovered by visual search. One explanation that has been advanced for this difference is that the J-f comets are derived from a source region, the Kuiper belt, that is different from the presumed Oort cloud source for the long-period and H-f comets (Duncan et al. 1988; Quinn et al. 1990). This explanation might be unsatisfactory, however. Observations of radar meteors and meteoroid streams suggest that extinct high-inclination and retrograde J-f comets might be present on short-period Earth-crossing orbits (Olsson-Steel 1988). The inferred short-period high-inclination and retrograde Jupiter-family meteor parent bodies apparently have been rendered almost but not quite inactive, so that they have escaped discovery but are still active enough to liberate meteoroids. Alternatively, the parent bodies may be dormant, or have totally dissipated, and all that is being observed is the remnant meteoroid streams; such streams have lifetimes of 10^4 to 10^5yr, even without replenishment. Statistics of short-period meteor orbits suggest that the distribution of inclination for extinct J-f comets may resemble that of H-f comets.

IV. COLLISION PROBABILITIES

Orbits of the discovered comets can be used to gain some insight into the rate of collision of the periodic comets with Earth. It should be clearly understood, however, that, so far as we know, none of the discovered comets can collide in the next few centuries, and it is unlikely that any of the known periodic comets will ever strike Earth. They are much more likely to be ejected from the solar system or even to hit Jupiter than to hit a terrestrial planet.

A variety of methods have been used for statistically estimating the probability of collision of two bodies revolving on overlapping orbits. The basic approach was first outlined by Öpik (1951) and later refined by Wetherill (1967), Öpik (1976), Shoemaker et al. (1979), and Kessler (1981). These methods depend on the assumption that the orbits of both bodies are relatively stable and that they intersect occasionally as a result chiefly of precession of the apsides or secular perturbabion of the eccentricity of one or both orbits. As we have seen, however, the orbits of periodic comets, particularly J-f comets are highly chaotic, primarily because of relatively frequent encounters of the comets with Jupiter. Hence the probability of collision with Earth of any specific periodic comet cannot be reliably estimated from its present orbital elements by the methods of Öpik or by the other related approaches.

We can adopt the point of view, on the other hand, that the orbital elements a, e and i of the discovered periodic comets, are a more or less representative sample of element sets, over time, of the comets of a given family and that the arguments of perihelion, over time, may be approximately distributed at random. To the extent that these assumptions are valid, the mean probability of collison obtained from the observed sample of orbits by appropriate refinement of Öpik's method may be taken as an estimate of the mean for the population of comets in a given family.

A version of Öpik's method suggested by Kessler (1981) was used to calculate collision probabilities shown in Table I. The Earth's orbit was taken as elliptical, with its present eccentricity, and probabilities were integrated around the orbit. This procedure does not treat very low inclination (<1 deg) cases correctly and also tends to overestimate collision probabilities for comets with perihelion distances close to Earth's perihelion distance. Collision probabilities shown in Table I for the orbital elements corresponding to P/Temple-Tuttle and P/Levy, for example, are likely somewhat high. The probabilites listed agree well with estimates obtained by Olsson-Steel (1987) using Kessler's more complete method. By far the greatest uncertainty stems not from the approximations employed but from the fact that the observed sample of orbits is small and a few cases of high probability, particularly for H-f comets, strongly influence the statistical means. More stable estimates of the mean collision probability for each family could be obtained by integrating the motion of each comet to generate a much larger sample of orbital elements.

The estimated mean collision probability with Earth is $1.3 \times 10^9 \text{yr}^{-1}$ for the Jupiter family and $1.6 \times 10^{-10} \text{yr}^{-1}$ for the Halley family. The lower mean collision probability for the H-f comets is due almost entirely to their longer orbital periods. The mean asymptotic encounter velocities (velocities relative to the Earth prior to acceleration by Earth's gravity) are 22.9 km s^{-1} for J-f comets and 45.4 km s^{-1} for H-f comets. However, the most probable encounter velocities, found by weighting the individual velocities by the probability of impact, are 19.9 km s^{-1} and 52.3 km s^{-1} for J-f and H-f comets, respectively. For the J-f comets, the highest collision probabilities correlate, in general, with the lower inclinations, and thus the lower encounter velocities. But for the H-f comets, impact probabilities rise again for the retrograde comets, which have the highest encounter velocities. Note that the velocity at which the comet would strike the Earth's surface (in the absence of atmospheric retardation), is given by the root-sum-square of the encounter velocity and the Earth escape velocity, 11.2 km s^{-1}.

Using Schmidt-Holsapple scaling for impacts (Melosh 1989), a J-f comet with a mass of 1.5×10^{15} g would be required to create a 20-km diameter crater on the Earth, assuming a nucleus density of 1.0 g cm^{-3}, a target rock density of 3.5 g cm^{-3}, a crater slumping factor of 1.6, and the most probable encounter velocity found above. The corresponding mass for an H-f comet is 2.4×10^{14} g, lower because of the higher impact velocity of the H-f comets. Other crater scaling laws (see, e.g., Gault et al. 1975; Shoemaker and Wolfe 1982) give somewhat larger or smaller crater diameters for a given impactor size and velocity; the differences reflect the uncertainties in extrapolating such estimates from laboratory and nuclear experiments.

V. POPULATION OF EARTH-CROSSING JUPITER-FAMILY COMET NUCLEI

The present understanding of the activity history of J-f comets and actual discoveries of candidate "extinct" comets leads us to expect that the activity of most J-f comets has been quenched by repeated prior passages substantially nearer the Sun than their current q (cf., Rickman et al. 1991; Rickman 1994). To date, 14 candidate extinct J-f comets have been discovered (Table II); two of them (including 5025 P-L, whose orbit is very uncertain) are Earth-crossing.

The criterion adopted in this chapter for recognition of extinct comets among objects of asteroidal appearance is based on the Tisserand invariant with respect to Jupiter's orbit T_j

$$T_j = 1/A + 2[A(1 - e^2)]^{1/2} \cos i \tag{1}$$

where A is the semimajor axis of the orbit in units of Jupiter's semimajor axis. The Tisserand invariant provides a fairly reliable test of whether a body can make a sufficiently close approach to Jupiter that its orbit can be strongly modified during this approach. Objects orbiting under the perturbing control of Jupiter have Tisserand invariants less than 3.00. With one exception, all active comets have $T_j < 3.00$; the exception is P/Encke, the comet with the shortest known period, which revolves on a relatively stable orbit inside the orbit of Jupiter and is safe from strong Jupiter perturbation. (An asteroidal object (4015) Wilson-Harrington, which briefly displayed cometary activity in 1949 and also revolves on a stable orbit, is discussed below.) Some asteroids with $T_j \approx 3.00$ are protected from encounters with Jupiter by resonances or other dynamical behavior; for all other asteroids with stable orbits, T_j is greater than 3.00.

A number of objects listed in Table II have osculating aphelion distances Q inside the orbit of Jupiter. So also do a number of active J-f comets. Even when the orbits do not overlap Jupiter's, approaches to Jupiter within about 0.5 AU result in strong perturbations and chaotic orbital evolution. Orbits of this type have evolved from larger orbits that did overlap Jupiter's orbit at an earlier time, and the objects can return gradually or suddenly to overlapping orbits. For convenience, we call this chaotic orbital behavior Jupiter-crossing. Two fairly high inclination ($i = 44°$ and $38°$) objects in Table II have current aphelion distances as low as 3.95 and 3.98 AU. However, relatively high i orbits of this type undergo large oscillations in i and e and corresponding large variations in Q as a result of secular (long-range) perturbations by Jupiter. T_j for these objects is well within the range for Jupiter-crossing.

Shoemaker et al. (1986) estimated the total population of extinct J-f comets brighter than $B(1,0) = 18$ at 1460 ± 650, from a set of candidates recognized at that time, including six of the objects listed in Table II. This estimate was based on the assumptions that (944) Hidalgo is the brightest extinct comet with $q \leq 2.0$ AU and that the magnitude-frequency distribution and the frequency distribution of q are similar in form to the distributions observed for active J-f comets. This derived population of extinct J-f comet candidates is about 9 times higher than the population of active J-f comet nuclei brighter than $B(1,0) = 18$ estimated by Shoemaker and Wolfe (1982). As noted above, all of the extinct comets have been discovered photographically,

TABLE II[a]
Extinct Periodic Comets[a]

	q AU	Q AU	i deg	$1/t_i$	P yr	T_j	$B(1,0)$
Extinct Jupiter-Family Comets							
5025 P-L	0.439	7.96	6.2	1.0	8.61	2.03	16.9
1992 LC[b]	0.737	4.30	17.8	—	4.00	3.00	16.0
1982 YA	1.123	6.29	34.6	23.8	7.14	2.40	17.5
(3562) Don Quixote	1.208	7.25	30.8	13.6	8.70	2.31	14.5
(5370) 1986 RA	1.232	5.46	19.0	7.9	6.12	2.73	17.0
1991 XB	1.250	4.69	16.1	6.3	5.12	2.93	19.0
1983 XF	1.457	4.79	4.2	1.0	5.51	2.97	16.0
1992 AB	1.468	5.11	40.8	9.4	5.96	2.51	15.0
1991 YA	1.532	3.95	44.3	9.2	4.54	2.81	15.5
1991 BY2	1.548	3.98	37.9	7.9	4.48	2.92	14.0
1984 BC	1.559	5.32	22.4	4.8	6.38	2.77	17.0
(5164) 1984 WE1	1.835	5.49	19.8	3.3	7.01	2.79	14.0
1988 JB1[b]	1.879	4.39	20.1	3.2	5.56	2.99	15.0
(944) Hidalgo	1.999	9.68	42.4	6.1	14.11	2.07	11.6
Means	1.376	5.62	25.5	7.5	6.66	2.66	15.6
Weighted mean			31.7				
Extinct Halley-Family Comet							
(5335) Damocles	1.579	22.16	61.9	40.9	1.15	14.5	

[a] Orbital elements and constants listed are as follows: q = perihelion distance, Q = aphelion distance, i = inclination, $1/t_i$ = bias factor (see text), P = period, T_j = Tisserand invariant with respect to Jupiter's orbit, $B(1,0)$ = absolute B magnitude calculated on the basis of a linear phase function that neglects the opposition affect. All the values of $B(1,0)$ listed except for (944) Hidalgo are derived from magnitudes estimated from photographic plates and films. Values of T_j less than 3.00 are characteristic of Jupiter-family comet orbits.

[b] 1992 LC and 1988 JB1 have values of T_j close to 3.00; it is uncertain whether they are Jupiter-crossing but they are likely to be extinct comets.

relatively near opposition. When observational selection against discovery of high inclination objects in the photographic surveys is taken into account, the true ratio of extinct to active comets probably is substantially higher than 9.

The mean i of discovered extinct J-f comets is about 1.4 times that of the known, active, Earth-crossing J-f comets; the true mean i of extinct J-f comets probably is higher that that observed, owing to observational bias against discovery of high-inclination objects. The bias corrected mean i can be estimated as follows. Most of the candidate extinct comets have been discovered close to perihelion. On the assumption that the arguments of perihelion ω of J-f comets are randomly distributed, the probability t_i that the perihelion, as seen from the Sun, lies at or below ecliptic latitude $|\ell|$ is given by

$$t_i = \frac{\sin^{-1}(\sin|\ell|/\sin i)}{90°}, \quad i > |\ell| \tag{2}$$

$|\ell|$ can be related to the range of eclipti latitude$\pm\alpha$ photographed in asteroid surveys approximately by

$$|\ell| \approx \sin^{-1}\left[\frac{q-1}{q}\sin|\alpha|\right]. \tag{3}$$

Average $|\alpha|$ for a wide variety of photographic surveys may be taken to be about 20°. While ω is not randomly distributed (Shoemaker and Wolfe 1982), the effect of failure of this assumption on the analysis presented here is small. The bias against discovery of high inclination (i.e., $i>20°$) objects is given by $1/t_i$, and is listed in Table II for an adopted value of $|\alpha| = 20°$. When the observed inclinations are weighted by the bias factors $1/t_i$ the weighted mean i is found to be 31.7°, about 6° higher than the mean of the observed inclinations. The mean bias factor is 7.5. This high value (or its reciprocal) illustrates one of the reasons why so few extinct comets have been discovered.

Levison and Duncan (1994) have carried out numerical integrations of the orbits of all of the known short-period comets (neglecting nongravitational forces). They find that the mean inclination of J-f comets with $q<2.5$ AU rises from the present value for active comets of $\sim17.8°$ to a steady-state value of $\sim27°$, in about a 1.8×10^4 yr period. Because the physical lifetimes of short-period comets are estimated to be only $\sim2 \times 10^3$ to 10^4 yr (Rickman 1994; also, see below), virtually all of the comets would be expected to become extinct before the mean evolved to these higher inclinations.

The bias factor listed in Table II for discovery of an object with the observed q and i of Hidalgo is 6.1. But Hidalgo, unlike the other extinct comets listed in Table II, is detectable rather far from perihelion. Most low inclination mainbelt asteroids that reach apparent magnitude $B = 17$ at opposition have now been discovered. At this magnitude, Hidalgo is detectable at a heliocentric distance of 4.0 AU at opposition. At a distance of 4.0 AU, extinct comets with $q = 2.0$ AU, $Q = 5.6$ AU (mean Q of Table II) and the absolute magnitude of Hidalgo would be detectable about 25% of the time. Further, if $i = 31.7°$ (weighted mean, Table II) and $\alpha = 20°$, detectability near opposition would occur within the average latitude limits of asteroid surveys about 25% of the time for a random distribution of ω. Thus the mean frequency of discovery opportunities is about $0.25 \times 0.25 = 0.06$ per opposition. Hidalgo was discovered in 1920; there would have been a mean of about 60 oppositions for objects with $q = 2.0$ AU, $Q = 5.6$ AU ($a = 3.8$ AU) since that time. Full coverage of the $\pm 20°$ latitude band (or an equivalent area) to a threshold of B mag 17 has not been approached, however, until the last half dozen years. This is the reason, for example, that the discovery of the next extinct comet was not reported until 1982; 5 of the 14 have been found in the last 3 yr. We estimate that mean annual coverage of the $\pm 20°$ band over the last 73 yr has been no better than about 15%. Hence the estimated mean probability of discovery of an extinct comet with $q = 2.0$ AU, $Q = 5.6$ AU, $i = 31.7°$ and the absolute magnitude of Hidalgo is about $0.15 \times 60 \times 0.06 = 0.5$. We conclude that there is about a 50% chance per object that an extinct comet as bright as Hidalgo (with $q = 2.0$) would have been discovered. Given that one object this bright has been found, the most probable population with $q \leq 2.0$ AU is two. If so, the estimated ratio of extinct to active J-f comets is 2×9, or roughly about 18.

Additionally, one can roughly estimate the ratio of active to extinct comets by comparing the active physical lifetimes vs the dynamical lifetimes of these objects. The latter figure is known fairly well as a result of modeling planetary perturbations (Wetherill 1975; Levison and Duncan 1994). Typical J-f comets have mean dynamical lifetimes of 10^4 to 10^6 yr, or about 2×10^3 to 2×10^4 orbits. Levison and Duncan found a median dynamical lifetime for all known periodic comets of 4.5×10^5 yr. Physical lifetimes of periodic comets are more difficult to ascertain, but for example, comets Encke, Halley and Swift-Tuttle are known to have made many tens of returns, and Halley is estimated to have an active lifetime in the inner solar system between 400 and 2000 returns (Jones et al. 1989; Hajduk 1987; Weissman 1987). Further, Halley shows little evidence of becoming extinct in the near future. Thus, the dynamical lifetimes of short-period comets may be anywhere between 1 and 50 times that of their physical lifetimes as active objects. This is not a very definitive limit but is in keeping with the estimate of 18 above. A similar calculation by Levison and Duncan (1994) finds an extinct: active ratio between 5 and 20, based on the same range of physical lifetimes assumed here.

Numerous researchers have addressed the question of whether comets do in fact evolve to dormant, asteroidal-appearing objects (see, e.g., Degewij and Tedesco 1982; Weissman et al. 1989). The recent discovery that asteroid 4015 was previously discovered as a (weakly) active short-period comet, Wilson-Harrington, in 1949 (Bowell and Marsden 1992), has provided unequivocal evidence that this happens. Asteroid (4015) Wilson-Harrington, which has an osculating q of 0.996 AU, is definitely Earth-crossing (Shoemaker et al. 1990); because it has a relatively stable orbit and is unable to make close approaches to Jupiter it is not included here as a Jupiter-family object (although we do include P/Enke, which also has a stable orbit, in the statistics for active J-f comets). There also are cases like comet Biela, where the nucleus of the comet may have spontaneously and completely disintegrated, leaving no detectable remnant except for the resulting meteoroid stream. At present, there is no good way of estimating what fraction of comets go to each of these end-states.

It should be borne in mind that the objects we have classified here as extinct comets, with the exception of Hidalgo, have not been closely scrutinized for possible cometary activity. They have only been reported as asteroidal in appearance on the basis of photographic observations, generally with plates or films taken with fairly short focal length, wide-field telescopes. Some of these objects may be proven to be detectably active with high resolution images. A case in point is P/Shoemaker-Levy 2, which was classified on discovery as asteroid 1990 UL3. This object was strictly asteroidal in appearance on films taken with the Palomar 0.46-m Schmidt telescope, but was found to exhibit a coma and tail on CCD images taken by S. Larson and D. Levy with a 1.5 m telescope. Further, it is entirely possible and even likely that some or perhaps most apparently dormant or extinct comets will become detectably active if or when their perihelion distances are appreciably decreased by a future encounter with Jupiter. Thus our estimate of the population of extinct comets merely refers to those objects which would not be reported as active with present techniques of discovery. From the standpoint of the hazard of their impact with Earth, their classification is of minor importance.

To check whether a population of two extinct J-f comets at $B(1,0) \leq 10.6$ and $q \leq 2.0$ AU is consistent with the discoveries of fainter objects, we assume, as before, that the magnitude distribution for the extinct comets has the form found for nuclear magnitudes of active J-f comets (Shoemaker and Wolfe 1982),

$$\Sigma N \propto e^{0.91H}. \tag{4}$$

Here, H may represent either $B(1,0)$ or absolute visual magnitude. Cumulative frequency ΣN, is listed in Table III, as a function of $B(1,0)$ for discovered extinct J-f comets up to $B(1,0) = 16.25$. An approximate estimate for the population of extinct J-f comets is obtained by multiplying the frequency of discovered objects by 7.5, the mean bias factor found from Table II. Cumulative frequencies calculated from Eq. (4) on the basis that $\Sigma N = 2$ at $B(1,0) = 11.6$ are shown in the column headed $\Sigma N_{predicted}$. It may be seen that the estimate of the population obtained from multiplying the frequency by the mean bias factor agrees fairly well with $\Sigma N_{predicted}$ up to $B(1,0) = 14.25$. For extinct comets fainter than $B(1,0) = 15.25$, the bias corrected frequency drops well below $\Sigma N_{predicted}$, as expected; even where these objects are not far from perihelion, their apparent magnitudes at opposition generally are fainter than the $B = 17$ threshold of likely detection. When extrapolated to $B(1,0) = 18$, the predicted number of extinct J-f comets with $q \leq 2.0$ AU is about 680, and the total population of all extinct J-f comets with $q \leq 5.53$ AU is estimated at about 3000. On the basis of the fit to $\Sigma N = 4$, the statistical uncertainty of the predicted population is about ±50% (1σ). Greater uncertainty probably resides in the form of the magnitude distribution when the population is extrapolated to $B(1,0) = 18$. Also, the reported values of $B(1,0)$ must be treated with caution; except in the case of Hidalgo, they are all derived from a limited number of apparent magnitudes estimated from photographs.

TABLE III

Comparison of Cumulative Frequency of Discovered Extinct Comets ($q \leq 2.0$ AU) with Bias Corrected Frequency and Frequency Predicted from the Assumed Form of the Magnitude-Frequency Distribution[a]

$B(1,0)$	$\Sigma N_{discovered}$	$\Sigma N_{corrected}$	$\Sigma N_{predicted}$
11.6	1	2	2
14.25	3	23	22
14.75	4	30	35
15.25	6	45	55
15.75	7	53	87
16.25	9	68	138

[a] Notes: $\Sigma N_{discovered}$ = cumulative frequency of discovered objects up to $B(1,0)$ = 16.75, listed in Table II. $\Sigma N_{corrected} = 7.5 \times \Sigma N_{discovered}$, except for frequency at $B(1,0) = 11.6$ (absolute B magnitude for Hidalgo). The estimated bias correction at $B(1,0) = 11.6$ is explained in text. $\Sigma N_{predicted}$ is the cumulative frequency calculated from proportionality (Eq. 4) by adopting $\Sigma N = 2$ at $B(1,0) = 11.6$.

To determine the number of extinct J-f comets that are Earth-crossing, we utilize the distribution of q of active J-f comets found by Shoemaker and Wolfe (1982). When observational selection effects and completeness of discovery have been accounted for, 18% of active J-f comets with $q \leq 2.0$ AU are estimated to be Earth-crossing. The estimated number of Earth-crossing extinct J-f comets to $B(1,0) = 18$, then, is 18%×680≈120. The typical diameter of a $B(1,0)$ comet nucleus was calculated by Shoemaker and Wolfe (1982) to be 2.53 km for an assumed geometric albedo in the B band of 0.03. Extrapolated to 1-km diameter, the cumulative number of Earth-crossing extinct J-f comets is about 780.

If 120 extinct comets brighter than $B(1,0) = 18$ or about 780≥1 km diameter are Earth crossing, it is pertinent to ask how many should have been discovered. Assuming the color ratio $B - V$ is 0.7 and that the magnitude distribution is given by Eq. (4), 230 of the extinct J-f Earth-crossers should have absolute V magnitudes brighter than 18. About 6% of ordinary Earth-crossing asteroids brighter than $V(1,0) = 18$ are thought to have been discovered through 1989 (Shoemaker et al. 1990). In order to make a comparison, we must correct for the mean period of J-f comets, which is about twice as great as that of Earth-crossing asteroids, and for the difference in observational bias against discovery of high-inclination objects. Observed mean i of extinct J-f comets is about 1.7 times higher than the mean for ordinary

Earth-crossing asteroids; the bias against discovery of high-inclination extinct J-f comets may be about two to three times stronger than for the ordinary Earth crossers. Combining the effects of the differences in mean period and mean i, we estimate that about 1% of the extinct J-f Earth crossers brighter than $V(1,0) = 18$ should have been detected. This expectation may have been fulfilled by the discovery of 5025 P-L and 1992 LC. As the pace of discovery quickens in the future, many more extinct J-f Earth crossers almost certainly will be found.

VI. RATE OF IMPACT OF JUPITER-FAMILY COMETS WITH EARTH

From the distribution of nuclear magnitudes of active J-f comets (Shoemaker and Wolfe 1982) and the fraction of active J-f comets that are Earth crossing, the number of active Earth-crossing J-f comets brighter than absolute nuclear B magnitude 18 can be estimated at about 30, with an uncertainty range of about a factor of 2. After correcting for unresolved coma (average correction of 2.3 mag, as estimated by Shoemaker and Wolfe [1982]) and adopting a mean nucleus albedo of 0.03, the current number of active Earth-crossing J-f comet nuclei larger than 1 km diameter is estimated to be about 40. This is only 2% of the number of Earth-crossing asteroids larger than 1 km in diameter estimated by Shoemaker et al. (1990). Moreover, the mean collision probability of $1.3 \times 10^{-9} \mathrm{yr}^{-1}$ for J-f comets is about three times lower than the mean collision probability of Earth-crossing asteroids. For these reasons, the active J-f comets were neglected by Shoemaker et al. (1990) in calculating cratering rates on Earth. The estimated impact rate of 1-km diameter and larger active J-f comet nuclei is $40 \times 1.3 \times 10^{-9} \mathrm{yr}^{-1} = 0.5 \times 10^{-7} \mathrm{yr}^{-1}$.

Our perspective on the hazard of comet impact is somewhat different in the case of extinct Earth-crossing J-f comets, which are estimated to be $\sim$40% as numerous as the ordinary Earth-crossing asteroids. The mean collision probability of the extinct J-f Earth crossers is likely to be somewhat lower than that of the active J-f comets, owing to the higher mean inclination of the extinct comets. The bias corrected mean i of extinct J-f comets ($31.7°$) is 70% higher than mean i of active Earth-crossing J-f comets ($18.6°$). Collision probabilities are directly proportional to the encounter velocities and inversely proportional to $\sin i$. Assuming the same mean q and e for extinct and active J-f Earth crossers, the ratio of their mean encounter velocities would be 1.16. As a rough guess, the mean collision probability of extinct J-f Earth crossers may be $1.16 \times (\sin 18.6°/\sin 31.7°) \times 1.3 \times 10^{-9} \mathrm{yr}^{-1} = 0.9 \times 10^{-9} \mathrm{yr}^{-1}$, or about 4.5 times lower than the mean collision probability of ordinary Earth-crossing asteroids. The estimated impact rate of extinct J-f comets larger than 1-km diameter is $780 \times 0.9 \times 10^{-9} \mathrm{yr}^{-1} = 0.7 \times 10^{-6} \mathrm{yr}^{-1}$, or about 9% of the impact rate of ordinary Earth-crossing asteroids.

The weighted mean rms impact velocity of active J-f Earth crossers is 23.6 km s^{-1}. Because of their expected higher mean inclination, the mean impact velocity of the extinct J-f comets should be about 1.13 times higher or about 26.7 km s^{-1}. This is 1.5 times the weighted *rms* impact velocity of ordinary Earth-crossing asteroids. The impact energy is proportional to the square of the velocity, which may be about 2.2 times higher for extinct J-f comets, and to the density of the impacting bodies. The density of comet nuclei was estimated by Shoemaker et al. (1990) to be about 1.2 g cm^{-3} or about 1.6 to 2.2 times lower than the density derived for fragmented stony asteroids by Shoemaker et al. (1979). The densities of both the comet nuclei and the Earth-crossing asteroids are very uncertain, although there is perhaps a wider diversity of opinion about the density of comet nuclei (see the Chapter by Rahe et al.). (In this connection, it should be borne in mind that some unknown but probably significant fraction of ordinary Earth-crossing asteroids are also extinct comets.) To first order, the mean impact energy per unit volume may be about the same for extinct J-f comets and ordinary Earth-crossing asteroids. Thus, as impact crater diameter can be scaled as a power function of energy, extinct J-f comets may produce about 9% as many impact craters as ordinary Earth-crossing asteroids at crater diameters larger than about 20 km. Comet nuclei smaller than about 0.5 to 1 km in diameter probably are disrupted in Earth's atmosphere (cf., Chyba et al. 1993; Melosh 1981) and do not produce craters recognizable in the geologic record, although they certainly could produce devastating effects at the Earth's surface.

VII. POPULATION OF EARTH-CROSSING HALLEY-FAMILY COMET NUCLEI

The number of active, Earth-crossing Halley-family comets is likely much greater than the number discovered. As shown by Shoemaker and Wolfe (1982), the completeness of discovery of comets is proportional to the frequency of opportunity of discovery or to the number of perihelion passages over a given interval of time. The mean period of Earth-crossing H-f comets, 88.6 (± 14.3, 1σ) yr is about 14.3 ($+6.2$, -3.6) times longer than the mean period of Earth-crossing J-f comets, 6.16 (± 0.76) yr. Thus, among the Earth crossers, discovery of H-f comets should be about 11 to 21 times less complete than discovery of J-f comets. This difference is reflected in the average absolute total magnitudes (H) of the two classes of comets (Table I). Mean H for the H-f comets (7.6$\pm$0.57), is 2.7$\pm$1.33 magnitudes (i.e., about 3.5 to 40 times) brighter than the mean for J-f comets (10.3$\pm$0.76). On the basis of the magnitude distribution given by Eq. (4), the difference in mean magnitude is equivalent to a difference in cumulative frequency by the factor $e^{0.91(2.7\pm1.13)} = 11.7 \times / \div 3.3$. Within the statistical errors, the factor 11.7 $\times$ / $\div$ 3.3 is consistent with the range of 11 to 21 obtained by comparing frequencies of perihelion passages. To the same level of completeness of discovery (by magnitude) of Earth-crossing J-f comets, there probably are roughly 10 to 20 times as many active Earth-crossing H-f comets as have been discovered. Interestingly, this yields a ratio of H-f to J-f comets among the active Earth crossers that is similar to the ratio predicted by Quinn et al. (1990) for periodic comets derived from the Oort cloud (when account is taken of dynamical lifetimes).

Data on the sizes of the nuclei are so limited that no definitive conclusions can be drawn as to whether the discovered H-f comets are intrinsically larger, on average, than the discovered J-f comets. The one well-measured active J-f comet, P/Halley, appears to have nucleus dimensions (mean radius $\sim$5.5 km) that are typical of the best determined J-f comets. However, most of the J-f comets for which direct estimates of the size are available, are likely larger than the average or typical active J-f Earth crossers. Until the physical and dynamical evolution of these objects is better understood, and their source reservoirs (Oort cloud or Kuiper belt) identified, estimates of their populations will remain highly uncertain.

Taking our estimate of completeness of discovery, the number of active Earth-crossing H-f comets corresponding to the 13 known active J-f comets is about 140 to 270, of which 13 have been discovered (Table I). This number implies that, on average, about 2 to 3 of the active Earth-crossing H-f comets pass perihelion each year. Typically they are 2.7±1.3 magnitudes fainter than the comets discovered by visual observers. About half should be detectable with dedicated photographic surveys of the region now scanned by visual observers.

There are strong reasons for expecting that active H-f comets, like the J-f comets, are accompanied by a large number of extinct H-f comets. Presumably H-f and J-f comets are both relatively mature populations that include both physically young and old members. The members of each family probably have made comparable numbers of perihelion passages close to the Sun. Two dynamical effects lead to orbital evolution of periodic comets through episodes of low q, which is thought to lead to loss of activity or extinction of comets after q has again increased. The first effect, which has long been recognized, is due chiefly to close approaches to Jupiter that cause comparatively large, chaotic changes in the orbit of a comet, in particular in q. For all periodic comets, q tends to jump back and forth between higher and lower distances. Comet Lexell, discussed earlier, is a good example of such an evolution. The frequency of these jumps is typically once every few tens of revolutions in the case of J-f comets (Carusi et al. 1985). New J-f comets sometimes are discovered during the first passage through perihelion after a jump to lower q, when the comet rather suddenly becomes bright enough to be detected.

A second dynamical effect arises from long range, secular perturbations that drive a slow oscillation of q as the orbit precesses. The change in q for high-inclination orbits can be dramatic; comets with i near 90° during one phase of precession of the apsides become Sun-grazers (Bailey et al. 1992). A half cycle of oscillation typically takes about 500 revolutions. Many H-f comets, because of their broad distribution of orbital inclination, may experience large oscillation of q due to secular perturbations. Jitter in q due to planetary encounters is superimposed on these oscillations. Only those comets with q near or below a previously experienced minimum are likely to show activity that is detectable by the usual methods of visual or photographic search.

The ratio of extinct to active H-f comets cannot be determined from observations at the present time, as only one candidate extinct H-f comet, (5335) Damocles, has been discovered so far. This object, with a probable diameter near 20 km, may be larger than any of the known active H-f comet nuclei, but probably is smaller than the largest extinct H-f comet. The much larger relatively inactive comet nuclei (2060) Chiron and (5145) Pholus do not cross the orbit of Jupiter and therefore are not classified here as H-f comets, although they provide direct evidence about the population of Saturn-family comets and about the existence of very large periodic comet nuclei.

Thus, we are forced to fall back on rather weak theoretical arguments to estimate the number of extinct, Earth-crossing H-f comets. Probably nearly all J-f comets, in the course of their dynamical evolution to short period, have passed through a stage of intermediate period orbits corresponding to Halley-family orbits. (Here we define Halley family strictly on the basis of orbital period, without regard to the distribution of i or q.) For those H-f comets that are truly the precursors of J-f comets, the steady state number of active plus extinct H-f comets should exceed the number of J-f comets by a factor equal to the ratio of the dynamical lifetimes of the two families. This factor may be roughly equal to the ratio of the mean periods of the two families. If we use the ratio of mean periods of discovered Earth-crossing H-f and J-f comets to estimate the total number of Earth-crossing H-f comets (active and extinct), the estimated ratio of extinct to active H-f comets is between 10 and 20 to 1. This is only a very crude estimate, however, as it is also necessary to understand the evolution to low-q of both H-f and J-f comets. The low-q, high-i H-f comets may well be derived chiefly from a different source region than the J-f comets, and they tend to be dynamically more stable than the J-f comets. Also, the arguments in the previous section regarding the ratio of the physical and dynamical lifetimes for short-period comets, and the possibility of evolution to total disintegration rather than physical dormancy, would apply here.

Assuming that the ratio of extinct to active H-f comets is as high as 10:1 to 20:1, the contribution of extinct H-f comets to the near-Earth periodic comet flux should be very roughly comparable with that of extinct J-f comets. While the total population of H-f comets may be 10 to 20 times higher than that of J-f comets, the mean frequency with which individual H-f comets cross Earth's orbit is 10 to 20 times lower than the mean for individual J-f comets. Although the Earth-crossing H-f and J-f comets have similar probabilities of collision with Earth per orbit, the H-f comets have roughly twice the impact velocity of the J-f comets (Table I). Together, H-f and J-f periodic comets may account, very roughly for about 20% of the production of terrestrial impact craters larger than 20 km diameter, a fraction about equal to the estimated contribution of long-period comets (cf., Shoemaker et al. 1990; Weissman 1990). Because of the very small number of ordinary Earth-crossing asteroids larger than 5 km diameter (Shoemaker et al. 1990), periodic and long-period comets combined may be responsible for the majority of the largest impact craters on Earth, including those which may be associated with mass extinction events.

If some substantial fraction of J-f comets has evolved dynamically from H-f comets, it is still necessary to explain why discovered active J-f comets have predominantly low-inclination orbits, whereas active H-f comets are found with a broad range of inclination. An important dynamical distinction between J-f comets and H-f comets is that most J-f comets have aphelia in the neighborhood of Jupiter, resulting in low mean encounter velocities with Jupiter. Encounters with Jupiter, therefore, tend to produce much larger jumps in q for J-f comets than for H-f comets. While the high-inclination J-f comets have higher encounter velocities relative to Jupiter than the low-inclination J-f comets, they are nevertheless subject to larger jumps in q, on average, than high-inclination H-f comets. The combined effect of periodic decreases of q due to secular perturbations and intermittent decreases in q due to Jupiter encounters may have extinguished all but a very small fraction of high-inclination J-f comets. A detailed study of the dynamical evolution of cometary perihelia and the physical response of comet nuclei is required in order to arrive at a full understanding of the activity histories of J-f and H-f comets. The low number of active high-inclination J-f comets might reflect a history of repeated evolution of q to small solar distances that has been more prolonged, in terms of number of revolutions, than in the dynamical history of H-f comets. Alternatively, if the J-f comets come predominantly from the Kuiper belt, as suggested by Duncan et al. (1988), then there likely are few very high-inclination extinct J-f comets.

VIII. THE CHALLENGE OF DISCOVERING THE EARTH-CROS-SING PERIODIC COMETS

If the populations of active and extinct J-f and H-f comets are as high as suggested here, they constitute a significant fraction of the total impact hazard. Discovery of these objects, on the other hand, particularly of the extinct comets, will be much more challenging than the discovery of ordinary (very short period) Earth-crossing asteroids. Extinct H-f comets will be particularly difficult to find, because of their long periods, broad distribution of orbital inclinations, and possibly because of low albedos. Even with an intensive search program such as the proposed Spaceguard Survey (Morrison 1992; Chapter by Morrison et al.), discovery of the Earth-crossing extinct H-f comet nuclei larger than 1 km diameter to 95% completeness will take vastly longer than the 25 yr estimated to achieve comparable completeness for ordinary Earth-crossing asteroids. The mean period of H-f comets is about 30 times higher than that of ordinary Earth-crossing asteroids; the broad distribution inclinations requires scanning 4 to 5 times as much area of the sky for equivalent efficiency of detection; and the mean magnitude of extinct H-f comets of a given size may be more than a magnitude fainter than the mean magnitude of ordinary Earth crossers, owing to a combination of lower albedo and higher phase angles at detection. The time required for discovery of 95% of the extinct H-f comets larger than 1 km diameter with the proposed Spaceguard system would be measured in millennia rather than in decades.

Acknowledgments. We thank H. Levison and M. Duncan for stimulating discussions and for sharing results of their work in advance of publication. D. Rabinowitz and D. Steel provided very helpful criticism of an early draft of this chapter. The Shoemakers thank Palomar Observatory for granting extensive observing time on the 18-inch Schmidt telescope over the past decade, enabling them to participate in the discovery of active and extinct periodic comets. Part of our research on this topic has been supported by NASA contracts. Weissman's work was supported in part by the Planetary Geology and Geophysics Program and was performed in part at the Jet Propulsion Laboratory under contract with NASA.

Note Added in Proof

Since the submission of this manuscript, two more extinct comets were discovered in the spring of 1994, 1994 EQ3 (q = 2.98 AU, Q = 6.45 AU, T_j = 2.86) and 1994 JC (q = 1.65 AU, Q = 5.11 AU, T_j = 2.73). 1994 EQ3, at H = 11.3, is within one magnitude as bright as Hidalgo; 1994 JC is much fainter. After the discovery of 1994 EQ3 by K. Endate and K. Watanabe in Japan, we were able, with the help of E. Bowell, to identify images of this object on films that we had taken with the Palomar 46-cm Schmidt telescope over three months in 1984. This, in turn, led to identifications of observations made in 1965, 1956, 1954, and even as far back as 1937. Thus the orbit of this ~30 km diameter Jupiter-crosser is now well established. The ongoing discovery of extinct comets, particularly large, bright objects like 1994 EQ3, illustrates just how incomplete our knowledge is of these important bodies.

REFERENCES

Bailey, M. E., Chambers, J. E., and Hahn, G. 1992. Origin of sungrazers: A frequent cometary end-state. *Astron. Astrophys.* 257:315–322.
Bowell, E., and Marsden, B. G. 1992. (4015) 1979 VA. *IAU Circ.* 5585.
Carusi, A., Kresák, Ľ., Perozzi, E., and Valsecchi, G. B. 1985. *Long-Term Evolution of Short-Period Comets* (Bristol: Adam Hilger).
Chyba, C. F., Thomas, P. J., and Zahnle, K. J. 1993, The 1908 Tunguska explosion: Atmospheric disruption of a stony asteroid. *Nature* 361:40–45.
Dejewij, J., and Tedesco, E. F. 1982. Do comets evolve into asteroids? Evidence from physical studies. In *Comets*, ed. L. L. Wilkening (Tucson: Univ. of Arizona

Press), pp. 665–695.

Duncan, M., Quinn, T., and Tremaine, S. 1988. The origin of short-period comets. *Astrophys. J. Lett.* 328:69–73.

Gault, D. E., Guest, J. E. Murray, G. J., Dzurisin, D., and Malin, M. C. 1975. Some comparisons of impact craters on Mercury and the Moon. *J. Geophys. Res.* 80:2444–2460.

Hajduk, A. 1987. Meteoroids from comet Halley: The comet's mass production and age. *Astron. Astrophys.* 187:925–927.

Jones, J., McIntosh, B. A., and Hawkes, R. L. 1989. The age of the Orionid meteor shower. *Mon. Not. Roy. Astron. Soc.* 200:281–291.

Kessler, D. J. 1981. Derivation of the collision probability between orbiting objects: The lifetimes of Jupiter's outer moons. *Icarus* 48:39–48.

Levison, H. F., and Duncan, M. J. 1994. The long-term dynamical behavior of short-period comets. *Icarus*, 108:118–136.

Marsden, B. G. 1973. The next return of the comet of the Perseid meteors. *Astron. J.* 78:654–662.

Marsden, B. G. 1992. Periodic comet Swift-Tuttle. *IAU Cir.* 5636.

Marsden, B. G., and Williams, G. V. 1993. *Catalogue of Cometary Orbits 1993* (Cambridge, Mass.: Smithsonian Astrophysical Observatory).

Melosh, H. J. 1981 Atmospheric breakup of terrestrial impactors. In *Multi-Ring Basins*, eds. P. H. Schultz and R. B. Merrill (New York: Pergamon Press), pp. 29–35.

Melosh, H. J. 1989. *Impact Cratering: A Geologic Process* (New York: Oxford Univ. Press).

Morrison, D., ed. 1992. *The Spaceguard Survey: Report of the NASA International Near-Earth-Object Detection Workshop* (Pasadena: Jet Propulsion Laboratory).

Olsson-Steel, D. I. 1987. Collisions in the solar system—IV. Cometary impacts upon the planets. *Mon. Not. Roy. Astron. Soc.* 227:501–524.

Olsson-Steel, D. I. 1988 Meteoroid streams and the zodiacal dust cloud. In *Catastrophes and Evolution: Astronomical Foundations*, ed. S. V. M. Clube (Cambridge: Cambridge Univ. Press), pp. 169–193.

Öpik, E. J. 1951. Collision probabilities with the planets and the distribution of interplanetary matter. 54A:165–199.

Öpik, E. J. 1976. *Interplanetary Encounters* (Amsterdam: Elsevier).

Quinn, T., Tremaine, S., and Duncan, M. 1990. Planetary perturbations and the origin of short-period comets. *Astrophys. J.* 355:667–679.

Rickman, H. 1994. Physico-dynamical evolution of aging comets. In *Interrelations Entre la Physique et la Dynamique des Petits Corps du Système Solaire*, in press.

Rickman, H., Kamel, L., Froeschlé, C., and Festou, M. C.. 1991. Nongravitational effects and the aging of periodic comets. *Astron. J.* 102:1446–1463.

Shoemaker, E. M., and Wolfe, R. F. 1982. Cratering time scales for the Galilean satellites of Jupiter. In *Satellites of Jupiter*, ed. D. Morrison (Tucson: Univ. of Arizona Press), pp. 277–339.

Shoemaker, E. M., Williams, J. G., Helin, E. F., and Wolfe, R. F. 1979. Earth-crossing asteroids: Orbital classes, collision rates with Earth, and origin. In *Asteroids*, ed. T. Gehrels (Tucson: Univ. of Arizona Press), pp. 253–282.

Shoemaker, E. M., Wolfe, R. F., and Shoemaker, C. S. 1986. Extinct Jupiter-family comets and cratering rates on the Galilean satellites. *Lunar Planet. Sci.* XVII:799–800 (abstract).

Shoemaker, E. M., Wolfe, R. F., and Shoemaker, C. S. 1990. Asteroid and comet flux in the neighborhood of the Earth. In *Global Catastrophes in Earth History*, eds. V. L. Sharpton and P. D. Ward, Geological Soc. of America Special Paper 247 (Boulder: Geological Soc. of America), pp. 155–170.

Sykes, M. V., and Walker, R. G. 1992. Cometary dust trails, I: Survey. *Icarus* 95:180–210.

Sykes, M. V., Lebofsky, L. A., Hunten, D. M., and Low, F. 1986. The discovery of dust trails in the orbits of periodic comets. *Science* 232:1115–1117.

Vsekhsvyatskii, S. K. 1964. *Physical Characteristics of Comets* (Washington, D. C.: NASA/NSF/Israel Program for Scientific Translation).

Weissman, P. R. 1980. Physical loss of long-period comets. *Astron. Astrophys.* 85:191–196.

Weissman, P. R. 1987. How typical is Halley's comet? In *Diversity and Similarity of Comets*, eds. E. J. Rolfe and B. Battrick, ESA SP-278 (Noordwijk: European Space Agency), pp. 31–36.

Weissman, P. R. 1990. The cometary impactor flux at the Earth. In *Global Catastrophes in Earth History*, eds. V. L. Sharpton and P. D. Ward, Geological Soc. of America Special Paper 247 (Boulder: Geological Soc. of America), pp. 171–180.

Weissman, P. R., A'Hearn, M. F., McFadden, L. A., and Rickman, H. 1989. Evolution of comets into asteroids. In *Asteroids II*, eds. R. P. Binzel, T. Gehrels and M. S. Matthews (Tucson: Univ. of Arizona Press), pp. 880–920.

Wetherill, G. W. 1967. Collisions in the asteroid belt. *J. Geophys. Res.* 72:2429–2444.

Wetherill, G. W. 1975. Late heavy bombardment of the moon and terrestrial planets. *Proc. Lunar Sci. Conf.* 6:1539–1561.

Wetherill, G. W. 1991. End products of cometary evolution: Cometary origin of Earth-crossing bodies of asteroidal appearance. In *Comets in the Post-Halley Era*, vol. 1, eds. R. L. Newburn, Jr., M. Neugebauer and J. Rahe (Dordrecht: Kluwer), pp. 537–556.

Yau, K. D. C., Yeomans, D. K., and Weissman, P. R. 1994. The past and future motion of comet P/Swift-Tuttle. *Mon. Not. Roy. Astron. Soc.* 266:305–316.

COLLISIONAL LIFETIMES AND IMPACT STATISTICS OF NEAR-EARTH ASTEROIDS

WILLIAM F. BOTTKE, JR., MICHAEL C. NOLAN,
RICHARD GREENBERG and ROBERT A. KOLVOORD
University of Arizona

Impact probabilities and impact velocity distributions are computed for collisions among populations of orbiting bodies using a formalism that takes into account the geometry of intersecting Keplerian orbits. We consider impacts of near-Earth asteroids (NEAs) by various populations, including other NEAs, mainbelt asteroids, and terrestrial planets. Contrary to current meteorite delivery models, a typical NEA is twice as likely as a mainbelt asteroid to be disrupted by impact with a mainbelt body. Lifetimes of NEAs against collision with terrestrial planets are similar to results of Monte Carlo studies. Rates of impact on terrestrial planets, as a function of NEA size, and the corresponding impact velocity distributions are important for consideration of impact hazards.

I. NEAR-EARTH ASTEROIDS: POPULATION AND EVOLUTION

The near-Earth asteroids (NEAs) are fragments of mainbelt asteroids on Earth-crossing or near-Earth crossing orbits that can potentially impact the Earth. Impact from the largest bodies of the NEA population could conceivably produce catastrophic effects on a global scale. For this reason, we need to better understand the size, number, composition, and collisional and dynamical evolution of the NEA population.

The Earth and other terrestrial planets currently reside in a large population of comets and asteroids referred to as near-Earth objects or asteroids (NEOs or NEAs, respectively). These objects have been classified into three groups based on their orbits: Amors, Apollos, and Atens, where Amors are defined as having a perihelion distance between 1.017 AU and 1.3 AU, Apollos have semimajor axes equal to or larger than 1 AU and a perihelion distance less than or equal to 1.017 AU, and Atens have semimajor axes smaller than 1 AU and a aphelion distance greater than 0.983 AU (Shoemaker et al. 1979). The total number of bodies in these groups larger than 1 km has been estimated to be at least 2000 (for further detail, see the Chapter by Rabinowitz et al.).

The near-Earth asteroids can end their existence in one of three ways: (a) collision with a planet, (b) collision with another asteroid which produces fragmentation, or (c) dynamical ejection from the solar system through a close approach with Jupiter. The time scale for all these events (10–100 Myr) is substantially shorter than the age of the solar system (4.6 Gyr), so it is unlikely that many (if any) NEAs remain from the primordial population (Wetherill 1988). In fact, the NEA population would be quickly depleted by collisional and dynamical processes if constant replenishment from one or more sources did not maintain its current size.

There is some evidence that the size of the NEA population is in steady state and has been in steady state for at least 3 Gyr. Shoemaker et al. (1979) inferred this result from the lunar cratering record, which indicates that crater production has been relatively constant for the last 3 Gyr, and from Earth's cratering record, where the surviving crater population agrees within a factor of 2 with the lunar cratering record (for more detail, see the Chapter by Neukum and Ivanov). Lunar and terrestrial craters are both too numerous to be represented solely by comet impacts (Shoemaker et al. 1979; Grieve and Dence 1979a, b).

A known source for much of the NEA population is the main asteroid belt. Collisions between mainbelt asteroids can produce cratering or catastrophic disruption, both of which provide ejecta which can escape the parent body. If the debris is ejected with the appropriate trajectory and velocity, it may reach a resonance zone, from which chaotic orbital evolution may transport that material to the terrestrial planet region (Greenberg and Nolan 1989). Asteroids in these resonances have their eccentricities increased on a short timescale, e.g., an asteroid in the 3:1 resonance may become Earth-crossing in ~1 Myr (Wisdom 1983). As an asteroid's eccentricity increases, it may experience a close encounter with either Mars or Earth, which can modify the asteroid's semimajor axis enough to remove it from resonance.

The collisional evolution of these mainbelt fragments continues as they orbitally evolve in the terrestrial planet region. These new NEAs frequently strike other asteroids, either other NEA or mainbelt bodies, again yielding cratering or catastrophic disruption. Weak or small NEAs are destroyed relatively quickly. Larger or stronger bodies can survive much longer, because their collisional fragmentation requires larger, rarer projectiles. The meteorites that survive to reach Earth are merely the remnants of a long succession of collisional events, which Wetherill has referred to as a "collisional hierarchy" (Wetherill 1988).

The irradiation record of meteorites is one indicator of collisional evolution among NEAs. It dates how long the surface of the meteorite parent body had been exposed to galactic cosmic rays. Some meteorites show complex irradiation histories, evidence of multiple collisional events (and surface burial) over time. Ordinary chondrites, one of the most common meteorites, are not composed of strong material, and therefore wear away more easily than some other asteroids. H chondrite meteorites display a peak cosmic-ray exposure age of only 6 to 7 Myr, with most less than 20 Myr (Caffee et al. 1988). This age implies that most chondritic material must be ground to sub-meteorite size on such time scales. However, iron meteorites are much stronger than ordinary chondrites, allowing them to survive long enough to reach cosmic-ray exposure ages as high as 1 Gyr (Caffee et al. 1988).

A second indicator of collisional and dynamical evolution among NEAs is "orbital-maturity," defined by an impacting NEA's encounter trajectory with Earth. Once an asteroid enters the terrestrial planet regime, its orbital evolution is controlled by planetary close approaches and distant perturbations. When NEAs first leave resonance they often have perihelia near the Earth. Such asteroids are classified as "orbitally-immature." This immaturity is reflected in a tendency to fall during the PM hours on Earth because, with perihelia near 1 AU, these asteroids are moving tangentially to the Earth's orbit and at higher heliocentric velocity. Hence, they overtake the Earth from its trailing side, where local times are PM. Because weak bodies cannot survive long enough to reach orbital maturity, ordinary chondrite meteorites tend to fall in the PM.

Asteroids which do not first collide with the Earth can be perturbed into orbits with perihelia significantly less than 1 AU. Such orbits do not produce an excess of PM falls among meteorites, and are classified as orbitally mature. Iron and stony-iron meteorites, with long cosmic-ray exposure ages, are usually sufficiently strong to avoid catastrophic fragmentation and achieve orbital maturity before reaching Earth; weaker chondrite and achondrite meteorites must originate as collisional ejecta from a large orbitally mature parent body to achieve orbital maturity. The bodies that do not collide with the Earth can be further perturbed into Venus-crossing orbits. Some of these bodies do collide with Venus, but others encounter Venus and Earth multiple times, allowing them to move throughout the terrestrial planet-crossing portion of (a, e) space. Eventually, asteroids experiencing repeated close approaches with the terrestrial planets may dynamically evolve into Jupiter-crossing orbits. Typical Jupiter-crossing asteroids are quickly ejected from the solar system ($\sim$1 Myr).

Thus, the meteorite record suggests the following about the collisional and dynamical processes affecting NEAs: (a) small mainbelt asteroids reaching resonances may not survive long enough to impact a terrestrial planet; (b) only large or strong asteroids (e.g., irons) will survive long enough to become orbitally mature; and (c) meteorites showing orbital maturity but short cosmic-ray exposure ages must have been "protected" by surface material on orbitally mature parent bodies before reaching Earth. For more detailed reviews of meteorite delivery to Earth, see Greenberg and Nolan (1989,1993).

II. CATASTROPHIC DISRUPTION RATES FOR NEAR-EARTH AS-TEROIDS

Studies of the collisional histories of asteroids have generally invoked approximations of impact velocities and collision frequencies. Recently, as knowledge of asteroid statistics and properties has grown, and with increasingly sophisticated understanding of the processes involved, more precise representations of impact velocity distributions and of collision probabilities are required. An examination of terrestrial impact hazards is a good example of an application that requires these improved calculations.

Most calculations assume that a single velocity can characterize the impacts involving a population of asteroids, and that particle-in-a-box approximations, based on gas dynamics, can give approximate collision frequencies. Wetherill (1967) compared the particle-in-a-box method with an improved calculation method based on earlier work by Öpik (1951). He showed that the particle-in-a-box method was a reasonably accurate approximation for most applications, although errors could be a factor of two or more.

Using these methods, Wetherill (Wetherill and Williams 1968; Wetherill 1976) calculated the amount of collisional debris produced by the fragmentation of Apollo asteroids and found it agreed with the estimated meteorite flux at Earth. Wetherill also estimated the collisional lifetime of a typical Apollo object and compared its value to dynamical lifetimes found using Monte Carlo calculations. He found that most larger Apollo asteroids (>1 km), if they do not hit a terrestrial planet, evolve dynamically into Jupiter-crossing orbits and are dispersed before a catastrophic disruption event can occur.

Using a formulation by Bottke et al. (1994), we can now compute accurate velocity probability distributions and impact rates for collisions among pairs of bodies in any orbiting population. Bottke et al.'s algorithm is based on the formulation of impact probabilities by Greenberg (1982; see also Bottke and Greenberg 1993), which further improved the approach of Wetherill (1967) and Öpik (1951) by introducing a way to avoid singularities in the integration. The method of Bottke et al. allows construction of complete velocity probability distributions as well as impact probabilities.

Velocity distributions have also been published recently by Namiki and Binzel (1991) and by Farinella and Davis (1992). As noted by Bottke et al. (1994), these results actually represent histograms of the average impact velocity for each pair of bodies in a population, which are significantly different from the actual velocity distributions.

The algorithm of Bottke et al. can be applied to collisions (a) among asteroids in a given population, (b) between asteroids in two different populations, or (c) between a given target planet and a population of bodies crossing its orbit. Calculation (c) is directly applicable to determination of the planetary impact hazard. However, results for impacts among asteroids [(a) and (b)] are equally important. For example, they determine the lifetimes of near-Earth asteroids, which along with the statistics of craters on the Earth and Moon help us to compute the number of NEAs and the flux of new bodies needed to maintain the population. Assuming (as is often done; see, e.g., Wetherill 1979) that the NEA population is in a steady state, the lifetimes determine the flux of new bodies (from the main belt or from comets) needed to replenish the population.

In this chapter we show the collision rates and relative-velocity distributions of NEAs among themselves, with the main belt, and with the terrestrial planets.

III. IMPACT VELOCITIES AMONG ASTEROIDS

A. Velocity Distributions Among Mainbelt Asteroids

First, as a reference, we consider the probability distribution for collision velocities among mainbelt asteroids (Bottke et al. 1994). Here we assume that the population of 682 asteroids with diameters larger than 50 km (Farinella and Davis 1992) is representative of the orbital distribution of all bodies in the main belt, as discovery of these asteroids is nearly complete (Cellino et al. 1991). Furthermore, we assume that this population is sufficiently representative of a distribution of phases of secular perturbations, so that we can use current osculating orbital elements. Secular perturbations do not significantly affect collision probabilities and velocities for mainbelt asteroids. This result has been verified by comparing the results obtained with both osculating elements and proper elements (Farinella and Davis 1992) and by direct numerical integration of mainbelt asteroid orbits (Davis et al. 1992). Because the selected set of orbits is assumed to be representative of orbits independent of size, the velocity distributions do not take into account the size of the asteroids. Throughout this chapter probabilities are given on a "per collision cross section" basis (actually "per cross section radius squared" in accordance with Wetherill's [1967] definition of intrinsic collision probability).

In our calculations we determine the velocity probability distribution for each of the pairs of asteroids in the population (there are $682 \times 681/2$ such pairs) and then we sum those distributions to obtain the results shown in Fig. 1. As described by Bottke et al. (1994), our procedure yields the total probability distribution and is therefore more complete than calculations by Farinella and Davis (1992) and by Namiki and Binzel (1991) who obtained average impact velocities for each pair and then presented histograms of those averages as a representation of the velocity distribution of the entire population.

Each of the distributions for an individual pair has spikes and irregularities, which are real and physically meaningful as discussed by Bottke et al. (1994), but in Fig. 1 the sum over many pairs smoothes out the distribution. (Other cases with fewer samples, e.g., Fig. 2, are in fact spiky.) The distribution in Fig. 1 has a median value of 5.025 km s^{-1}, close to the canonical value of 5 km s^{-1} used in most asteroid collisional modeling. However, the distribution is quite broad and a single average value is not a good representation. The mean value is 5.292 km s^{-1}, the *rms* value is 5.769 km s^{-1}, and there is a significant probability of impacts at greater than 10 km s^{-1}. Note that the area under this curve, divided by the number of pairs in the population, is equal to the mean intrinsic collision probability (defined by Wetherill 1967), in this case 2.86×10^{-18} km^{-2} yr^{-1}.

B. NEAs Hitting Mainbelt Asteroids

Next we consider the velocity distribution for impacts between 224 known NEA orbits and the 682 mainbelt orbits described above. The 224 NEA orbits are those of the known Apollos and Amors as of January 1993 (calculated

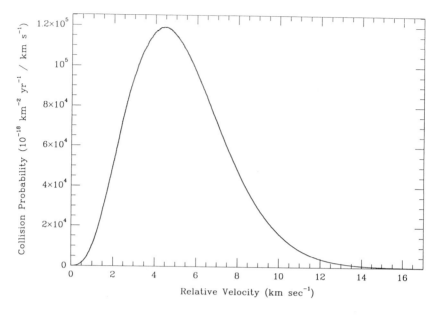

Figure 1. Probability distribution for impact velocities among mainbelt asteroids. The orbits of the 682 mainbelt bodies with diameters larger than 50 km are assumed to be representative of all mainbelt orbits. The distribution smoothes out the spikiness of individual pairs, due to the large number of pairs that are added together. The area under the curve, divided by the total number of possible colliding pairs, is the intrinsic collision probability for an average pair of intersecting asteroid orbits, 2.86×10^{-18} km^{-2} yr^{-1}. The median value of 5.025 km s^{-1} is close to the canonical value of 5 km s^{-1} used in most asteroid collisional modeling. However, the distribution is quite wide and a single average value is not a good representation: the mean value is 5.292 km s^{-1}, the rms value is 5.769 km s^{-1}.

by the Minor Planet Center). Only 182 NEAs contribute; the others do not reach the main belt. The tallest curve (curve 3) in Fig. 2 represents the sum over all possible pairs consisting of one NEA and one mainbelt asteroid. This curve retains some spikes from the distributions of individual pairs. We emphasize that these irregularities are real, not computational noise, although they would be smoothed if we had a larger known sample of NEA orbits. The mean, median, and rms values from this distribution are 10.238, 9.755, and 10.882 km s^{-1}, respectively, all about twice as large as the mainbelt values in the previous section. The intrinsic collision probability (the mean over the total number of pairs) is 2.18×10^{-18} km^{-2} yr^{-1}, comparable to that among mainbelt asteroids (previous section).

We also show separate curves for impacts between Amors and the main belt (curve 1) and for impacts between Apollos and the main belt (curve 2). Note that of the 114 Apollos, 92 reach the main belt, so 22 have zero probability of collision. Similarly, of the 96 Amors, 6 do not contribute to the impact statistics. None of the Atens are included in the calculations represented in

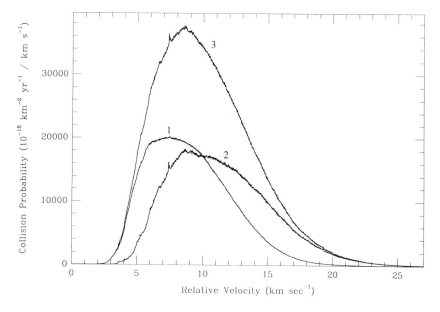

Figure 2. Probability distribution for impact velocities between mainbelt asteroids and NEAs. Curves 1 and 2 include only Amors and Apollos, respectively, while curve 3 is for the combined population (the sum of curves 1 and 2). The mean, median, and *rms* values from curve 3 are 10.238, 9.755,and 10.882 km s^{-1}, respectively. The mean intrinsic collision probability (from the area under curve 3) is 2.18 × 10^{-18} km^{-2} yr^{-1}.

Fig. 1, because as a population they do not have a significant overlap with the main belt. Impact velocities are significantly higher for Apollos than for Amors.

C. NEAs Hitting other NEAs

In Fig. 3 we show the velocity probability distribution for all possible pairs of NEAs hitting other NEAs (curve 4). The mean, median, and *rms* values from this distribution are 18.539, 17.355, and 20.606 km s^{-1}, respectively, all much higher than velocities for impacts involving mainbelt bodies (previous sections). These high collision velocities (relative to those in the main belt) result from the large eccentricities present in the NEA population. The intrinsic collision probability (the mean over the total number of pairs) is 15.34 × 10^{-18} km^{-2} yr^{-1}, much higher than for impacts with the main belt. These high collision probabilities are produced through a combination of (a) high collision velocities, and (b) a smaller available volume. We also show the separate distributions for Amors, for Apollos, and for Atens impacting all other NEAs (curves 1, 2, and 3, respectively).

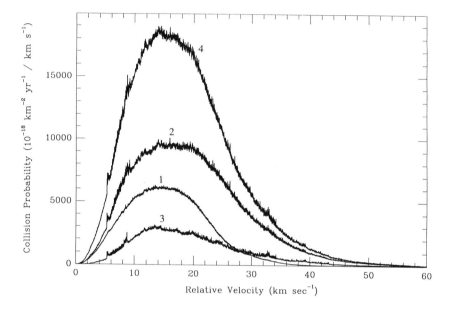

Figure 3. Impact-velocity probability distributions for populations of known NEAs impacting all other NEAs. Curves 1, 2 and 3 are respectively for all known NEAs impacting Amors, Apollos, and Atens. Curve 4 is the sum of the other three, representing the distribution for impacts among all NEAs. The mean, median, and *rms* values from this distribution are 18.539, 17.355, and 20.606 km s^{-1}, respectively. The mean intrinsic collision probability is 15.34 $\times$ 10^{-18} km^{-2} yr^{-1}.

IV. FREQUENCY OF CATASTROPHIC FRAGMENTATION

The information on collision probabilities and velocity distributions described in Sec. II, combined with physical models of the outcomes of impacts, allows us to make improved calculations of the frequency of collisional events that can catastrophically fragment a given target asteroid.

Consider, for example, a hypothetical average NEA whose velocity distribution with respect to the main belt is identical to curve 3 in Fig. 2. Suppose this NEA has a given size ("target diameter"). Further, assume that the main-belt size-frequency distribution follows the distribution suggested by Belton et al. (1992), specifically that it is extrapolated as a power-law (incremental index -2.95) from the small asteroids of the Palomar-Leiden survey (van Houten et al. 1970) and then steepens to an incremental power-law index of -3.5 for asteroids smaller than 175 m in diameter:

$$dn = 2.7 \times 10^{12} D^{-2.95} dD \text{ for } D > 175 \text{ m} \qquad (1)$$

$$dn = 4.7 \times 10^{13} D^{-3.50} dD \text{ for } D < 175 \text{ m}. \qquad (2)$$

We also adopt the strength-scaling laws of Housen and Holsapple (1990) which give the criteria (projectile size and impact velocity) for catastrophic disruption of the target. With all this information, we can calculate the frequency (events per year) of all impacts into the target NEA that result in catastrophic disruption.

The results for this example are shown in Fig. 4a. For a steeper assumed mainbelt power-law distribution (index −4.0 for the smaller asteroids, as suggested by the Gaspra studies of Greenberg et al. 1993), we obtain the steeper curve. We do similar calculations for a target asteroid in the main belt (again, a hypothetical asteroid with the same velocity distribution (Fig. 1) as the total population). These results are also shown in Fig. 4a. The lowest curve in Fig. 4a shows the rates for impacts of NEAs (assuming they follow the "Spaceguard" size-distribution for Earth crossers reported by Morrison (1992), and shown here as Fig. 11 below) into an NEA target, assuming the velocity distribution of Fig. 3, curve 4. The short disruption lifetimes for smaller asteroids seems reasonable, given that typical chondrites show orbital immaturity among falls and short cosmic-ray exposure ages (see Sec. I for details).

The implications of Fig. 4a are significant:

1. Compared with rates of disruption of mainbelt asteroids by other mainbelt asteroids, we find that NEAs are disrupted twice as often by encounters with mainbelt asteroids. This result may seem surprising, given that the NEAs spend relatively little time in the main belt during their orbital periods. However, their relatively high impact velocities (compare Fig. 2 with Fig. 1) mean that numerous, much smaller mainbelt bodies can destroy NEAs, which explains the higher frequency of catastrophic events. Note that this factor of two result is insensitive to changes in the size distribution assumed for the main belt, unless it is very different from a power-law distribution for small bodies.

   The disruption rates we find are within an order of magnitude of values estimated by Wetherill (1988). Our result is more precise than the estimate by Greenberg and Chapman (1983) that NEAs and mainbelt asteroids are disrupted at roughly equal rates by mainbelt asteroids. It also tilts the balance in the opposite direction from that suggested by Wetherill (1988), who estimated that NEAs would be disrupted slightly less often by mainbelt collisions.

2. The lower curve in Fig. 4a shows that average NEAs are disrupted 1 to 2 orders of magnitude less frequently once their aphelia become collisionally decoupled from the main belt. At this point in their evolution, lifetimes of such NEAs (e.g., Atens) become dominated by collisions with terrestrial planets instead of catastrophic fragmentation by smaller projectiles. These longer lifetimes may help explain the population of small, Spacewatch-discovered bodies with low eccentricities near the Earth. It also emphasizes why some meteorite types show evidence of orbital maturity and long cosmic-ray exposure ages (see Sec. I for details).

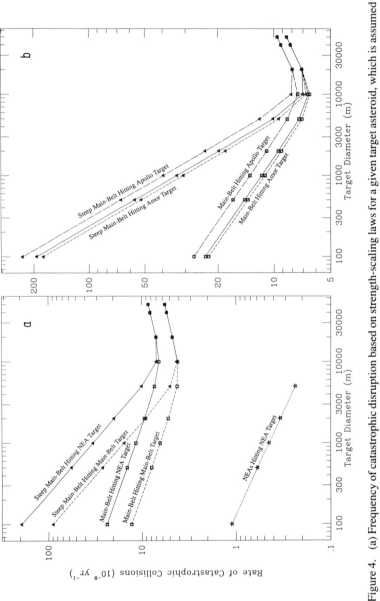

Figure 4. (a) Frequency of catastrophic disruption based on strength-scaling laws for a given target asteroid, which is assumed to be an average member of its orbital class, where average means it has the velocity distribution (from Fig. 1, 2 or 3) for impacts with another population. Each curve is labeled with the corresponding target and impacting population. For the main belt, the size distribution of bodies with $D < 175$ m has an incremental power-law index of -3.5, or -4.0 in the "steep" case. (b) Frequency of catastrophic disruption based on strength-scaling laws for a target asteroid that is either an average Apollo, average Amor, or average NEA (Apollos plus Amors, solid curve), impacted by small mainbelt objects following either the -3.5 or -4.0 (steep) power-law.

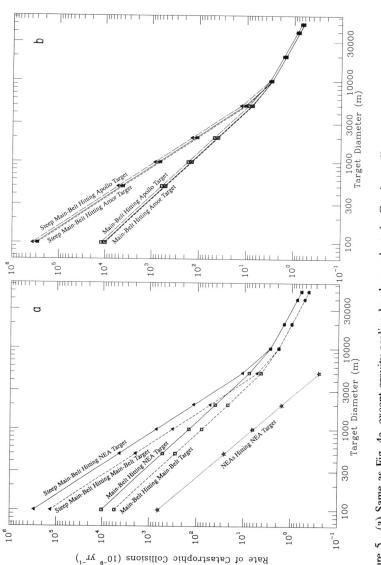

Figure 5. (a) Same as Fig. 4a, except gravity scaling has been adopted. Gravity scaling curves in Figs. 5 are probably appropriate for targets over 10 km, with a transition between the two scaling laws occurring from 300 m to 10 km. The major implications are unchanged from Fig. 4a. (b) Same as Fig. 4b, except gravity scaling has been adopted.

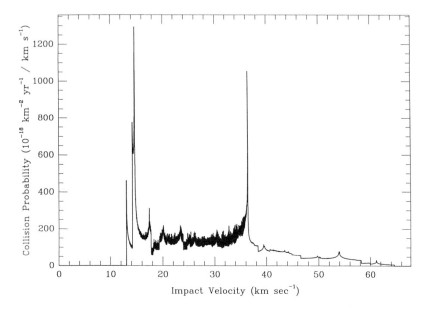

Figure 6.   Impact probability distribution for impacts with Mercury by the 16 (of 224) NEAs that are Mercury crossers. The impact velocity takes into account the gravity of Mercury. The spikes in the distribution represent the effects of individual asteroids in this small sample. Mean, median, and rms velocities are 30.2, 29.5, and 32.5 km s^{-1}, respectively. Mean intrinsic collision probability is 328.3×10^{-18}km^{-2}yr^{-1}.

In Fig. 4b we show the disruption rates for average Amor and Apollo targets. For reference, the solid curves in Fig. 4b, for an average NEA target, are the same curves as the solid ones in Fig. 4a. These results show that Apollo asteroids are disrupted more frequently by the main belt than Amors are, primarily due to the high collision velocities. Also, the rate of catastrophic collisions for NEAs is highly dependent on the size distribution of smaller bodies ($D < 500$ m) in the main belt, but not on whether they are Apollos or Amors. The similarity between results for Apollos and Amors indicates that these results are probably not very sensitive to observational biases in NEA statistics.

The criteria for disruption used in Figs. 4a and b are based on the strength-scaling laws of Housen and Holsapple (1990). This rate turns up for asteroids > 10 km because the size-frequency distributions have more projectiles. However, note that strength scaling becomes inappropriate at some unknown size, probably between 1 and 10 km. Nolan et al. (1992) and Greenberg et al. (1993) describe hydrocode modeling that shows that small asteroids may resist catastrophic disruption more readily than suggested by strength scaling, surviving impacts that create craters almost as large as the target body. Also, they found that crater sizes on asteroids greater than about 10 km in diameter are closer to the values given by gravity-scaling formulations, even though gravity scaling

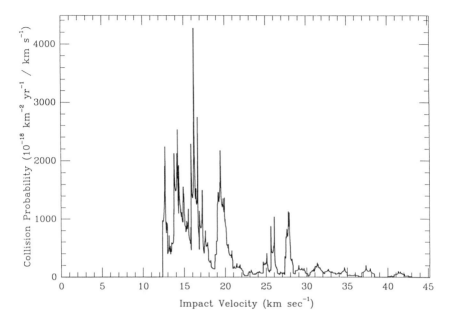

Figure 7. Impact probability distribution for impacts with Venus by the 57 NEAs that are Venus crossers. The impact velocity takes into account the gravity of Venus. Mean, median, and *rms* velocities are 19.2, 16.9, and 20.2 km s^{-1}, respectively. Mean intrinsic collision probability is 180.2×10^{-18}km^{-2}yr^{-1}.

had not been originally thought to be applicable to such small target bodies. For example, in the case of 951 Gaspra, application of gravity scaling rather that strength scaling nearly doubles the asteroid's average lifetime against catastrophic disruption (1 Gyr vs 500 Myr, respectively) (Greenberg et al. 1993). If we adopt the gravity-scaling laws of Schmidt-Holsapple (Melosh 1989; see also Bottke et al. 1994) instead of strength scaling, Figs. 4a and b would be replaced by Figs. 5a and b, respectively. The values in Figs. 4 are probably relevant for targets smaller than a few hundred meters, and those in Figs. 5 are relevant for those greater than 10 km, with a gradual transition over the mid ranges. The lifetimes of mainbelt asteroids are discussed further by Bottke et al. (1994).

V. IMPACTS WITH TERRESTRIAL PLANETS

The probability distributions for impact velocities of planet crossers hitting each of the terrestrial planets are shown in Figs. 6, 7, 8, and 9. The orbits used for these calculations are the set of all known planet-crossing NEAs. The velocities shown in these plots have been augmented by the gravitational acceleration of the respective planets, so they represent the actual velocity on impact. Mean, median, and *rms* impact velocity for each distribution, as well as the average intrinsic collision probability (specifically the area under the velocity distribution curve divided by the number of planet crossers), are given in the caption of each figure.

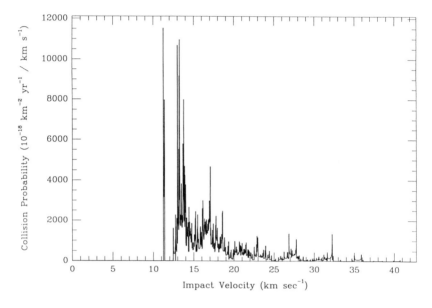

Figure 8.   Impact probability distribution for impacts with Earth by the 128 NEAs that are Earth crossers. The impact velocity takes into account the gravity of the Earth. Mean, median, and *rms* velocities are 17.2, 16.1, and 17.9 km s^{-1}, respectively. Mean intrinsic collision probability is 100.4×10^{-18} km^{-2} yr^{-1}. The double spike near 11.2 km s^{-1} (equal to the Earth's escape velocity) is due to the single asteroid, 1991 VG, which has an orbit nearly identical to the Earth's.

In general, the numbers of separate distributions that are summed to produce these results are relatively small (equal to the number of known planet crossers), so they are quite spiky. This effect is minimal for Mars, with 205 crossers, but is dominant for Venus and Mercury, with few crossers. Recall that these spikes are not computational noise, but represent the actual irregular distributions of velocity probabilities for specific orbits. In addition, particular NEAs with small inclinations or nearly tangential orbits may contribute disproportionately to both the spikiness of the overall distributions as well as the collision probabilities for each asteroid class (Milani et al. 1990). However, it is not clear that removing these anomalous asteroids from our calculation will increase our accuracy significantly, because the known population of NEAs may or may not be a biased sample of the total NEA population. Future asteroid discoveries should allow better statistics and results.

    We can use these results to estimate the lifetimes of planet crossers. For example, for Earth crossers we multiply the mean intrinsic collision probability (units: 10^{-18} km^{-2} yr^{-1}) by the square of the gravitational cross-sectional radius (not by the cross-sectional area, because the π is already

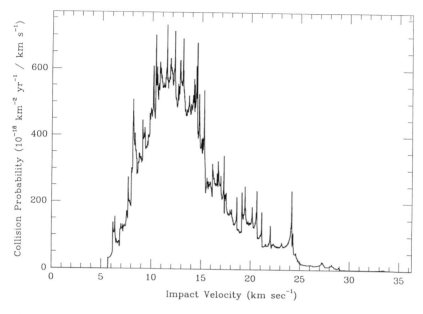

Figure 9. Impact probability distribution for impacts with Mars by the 205 NEAs that are Mars crossers. The impact velocity takes into account the gravity of Mars. Mean, median, and *rms* velocities are 13.6, 12.8, and 14.2 km s^{-1}, respectively. Mean intrinsic collision probability is 25.2 $\times$ 10^{-18} km^{-2} yr^{-1}. One asteroid has contributed some small probability of an impact at >33 km s^{-1}.

included in the definition of the intrinsic collision probability) to obtain the collision rate. The lifetime of a typical Earth crosser is then the reciprocal of that rate or about ~100 Myr. Results for the other planets, including for subsets of the crossing populations, are shown in Fig. 10.

In order to convert these results to the frequency of impacts of bodies of a given size on each terrestrial planet, we need to know how many such impactors there are. If we assume, for example, that all classes of planet crossers have size distributions proportional to the Earth-crossing population given by the Spaceguard Survey (Morrison 1992) as shown in Fig. 11, we obtain the frequency of impacts on terrestrial planets as a function of impactor size, shown in Fig. 12. We find agreement with earlier Monte-Carlo numerical experiments (see, e.g., Wetherill 1979), in that planet crossers are much more likely to hit the Earth or Venus than the other planets.

Shoemaker et al. (1979,1990) also obtained results for the frequency of Earth impacts from Earth-crossing asteroids. Shoemaker et al. (1990) estimated the mean collisional lifetime for Earth-crossing NEAs, Atens, and Apollos to collide with Earth to be 238, 93, and 244 Myr, respectively. Our results (given in Fig. 10, are 134, 67, and 153 Myr, respectively) differ with Shoemaker et al. for at least a couple of reasons: (a) the NEA population we use in our calculation is more complete than that of Shoemaker (1990),

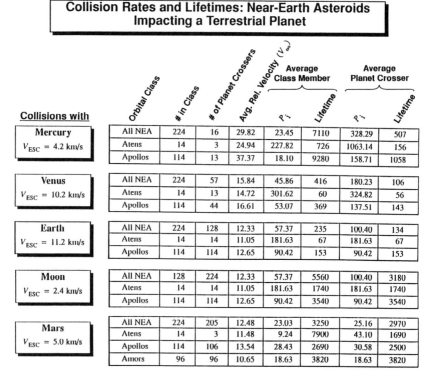

Figure 10. Mean intrinsic collision probabilities (units: 10^{-18}km^{-2}yr^{-1}) for various groups of small bodies whose members may impact various target planets. The average relative velocities (units km s^{-1}) shown do not include gravitation. Lifetimes are the mean duration before impacting the target planet (units: Myr).

e.g., we include the Spacewatch asteroids which tend to have large collision probabilities (see Chapter by Rabinowitz et al.), and (b) his calculations use Öpik's method, modified to include the secular variations of orbital elements due to Earth; our calculations do not account for secular effects.

Shoemaker and Wolfe (1987) calculated the collision rate for Apollo and Aten asteroids striking Venus and the Earth-crossing asteroids striking the Moon. For Venus, they found the mean collisional lifetime of the 22 then-known Venus crossers to be 127 Myr, similar to the value of 106 Myr in our Fig. 10 for 57 Venus crossers. For the Moon, they found the mean collisional lifetime of the 51 then-known Earth and Moon crossers to be 4150 Myr, comparable to the value 3180 Myr for 128 NEAs, as shown in Fig. 10. As in the case of the Earth, the inclusion of recent Spacewatch asteroids lowers the total mean collisional lifetime of the population.

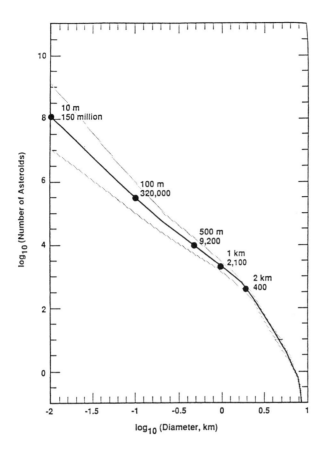

Figure 11.   Size distribution for Earth crossers from the Spaceguard survey (Morrison 1992).

VI. CONCLUSIONS

The results given in this chapter for probability distributions for asteroidal impact velocities may be applicable to problems regarding impact hazards in a variety of ways. The impact rates in Fig. 12 are certainly directly applicable. Other results, such as the frequency of disruption of NEAs, will also be useful components of evaluations of the potential hazard.

These results have implications for other aspects of planetary science. For example, impact records on terrestrial bodies are often used as benchmarks for estimating populations of small bodies elsewhere in the solar system. The information about impact velocity distributions may affect estimates of the size distribution of very small planet crossers, and the comparative rates of collisional disruption of mainbelt asteroids relative to NEAs will help provide an appropriate accounting as the targeted precision of comparative impact rate studies improves.

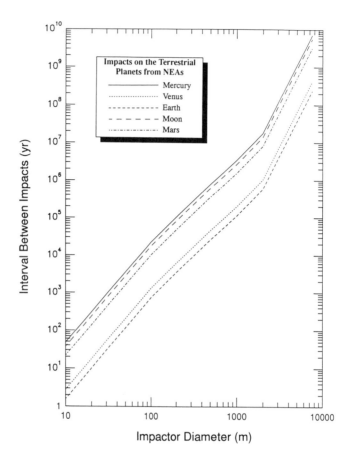

Figure 12. Mean interval between impacts (i.e., 1/frequency) on each of the terrestrial planets by NEAs larger than any given impactor size, assuming the NEA size distribution is the same as in Fig. 11. Planet crossers are much more likely to hit the Earth or Venus than the other planets.

Estimates of the mainbelt contribution to NEA populations may also be improved by the use of the results in this chapter. For example, Wetherill (1979,1985,1988) treated the NEAs as a steady-state population. Removal from the NEA population was found to be largely by ejection to Jupiter crossing or impact with the Earth and Venus, at rates consistent with our results in Figs. 6 through 9. The source of NEAs was assumed to be broken pieces of mainbelt asteroids that had been disrupted by impacts with other mainbelt asteroids and then evolved toward the terrestrial planets. To the extent that disruption of mainbelt asteroids was inadequate to maintain the steady-state population of NEAs, Wetherill could estimate the fraction of NEAs that must have come from other sources. This fraction varied in different papers by

Wetherill as understanding of asteroid transport processes improved. In order to estimate the disruption rates for mainbelt asteroids, Wetherill (1985,1988) first estimated the disruption rate for NEAs from cosmic-ray exposure ages of meteorites and other observational evidence. He then assumed that frequency of such fragmentation for a mainbelt asteroid would be comparable or even somewhat greater. Here (Sec. III) we have shown that it would more likely be a factor of 2 less. Modifying Wetherill's calculations accordingly would yield a requirement for greater cometary component supplying NEAs, or, alternatively, additional mainbelt transport mechanisms, such as additional resonance routes (Froeschlé and Greenberg 1989; Scholl et al. 1989).

The results reported here will thus be applicable in a variety of fields. They do, however, remain subject to modification as additional statistics on continually discovered NEAs become available and as models of the criteria for catastrophic disruption become increasingly sophisticated. Such information may then provide improved input for our algorithms for impact velocity distributions and disruption frequencies.

Acknowledgments. We thank D. Tholen, D. R. Davis, and P. Farinella for providing us with the orbital elements necessary for the calculations presented in this chapter. We also thank P. Farinella and an anonymous reviewer for their careful and constructive reviews. This work was supported by a grant from NASA's Planetary Geology and Geophysics Program.

REFERENCES

Belton, M. J. S., Veverka, J., Thomas, P., Helfenstein, P., Simonelli, D., Chapman, C., Davies, M. E., Greely, R., Greenberg, R., Head, J., Murchie, S., Klaasen, K., Johnson, T. V., McEwen, A., Morrison, D., Neukum, G., Fanale, F., Anger, C., Carr, M., and Pilcher, C. 1992. Galileo encounter with 951 Gaspra: First pictures of an asteroid. *Science* 257:1647–1652.

Bottke, W. F., and Greenberg, R. 1993. Asteroidal collision probabilities. *Geophys. Res. Lett.* 20:879–881.

Bottke, W. F., Nolan, M. C., and Greenberg, R. 1994. Velocity distributions among colliding asteroids. *Icarus* 107:255–268.

Caffee, M. W., Goswami, J. N., Hohenberg, C. M., Marti, K., and Reedy, R. C. 1988. Irradiation records in meteorites. In *Meteorites and the Early Solar System*, eds. J. F. Kerridge and M. S. Matthews (Tucson: Univ. of Arizona Press), pp. 205–245.

Cellino, A., Zappalá, V., and Farinella, P. 1991. The size distribution of main-belt asteroids from IRAS data. *Mon. Not. Roy. Astron. Soc.* 253:561–574.

Davis, D. R., Farinella, P., and Carpino, M. 1992. Asteroid collisional frequencies: Variations in space and time. *Lunar Planet. Sci.* XXIII:283–284 (abstract).

Farinella, P., and Davis, D. R. 1992. Collision rates and impact velocities in the main asteroid belt. *Icarus* 97:111–123.

Froeschlé, Cl., and Greenberg, R. 1989. Mean motion resonances. In *Asteroids II*, eds. R. P. Binzel, T. Gehrels and M. S. Matthews (Tucson: Univ. of Arizona Press), pp. 827–844.

Greenberg, R. 1982. Orbital interactions: A new geometrical formalism. *Astron. J.* 87:184–195.

Greenberg, R., and Chapman, C. R. 1983. Asteroids and meteorites: Parent bodies and delivered samples. *Icarus* 55:455–481.

Greenberg, R., and Nolan, M. C. 1989. Delivery of asteroids and meteorites to the inner solar system. In *Asteroids II*, eds. R. P. Binzel, T. Gehrels and M. S. Matthews (Tucson: Univ. of Arizona Press), pp. 778–804.

Greenberg, R., and Nolan, M. C. 1993. Dynamical relationships of near-Earth asteroids to main-belt asteroids. In *Resources of Near-Earth Space*, eds. J. Lewis, M. S. Matthews and M. L. Guerrieri (Tucson: Univ. of Arizona Press), pp. 473–492.

Greenberg, R., Nolan, M. C., Bottke, W. F., and Kolvoord, R. 1993. Collisional history of Gaspra. *Icarus* 107:84–97.

Grieve, R. A. F., and Dence, M. R. 1979*a*. The terrestrial cratering record. I. Current status of observations. *Icarus* 38:212–229.

Grieve, R. A. F., and Dence, M. R. 1979*b*. The terrestrial cratering record. II. The crater production rate. *Icarus* 38:230–242.

Housen, K. R., and Holsapple, K. A. 1990. On the fragmentation of asteroids and planetary satellites. *Icarus* 84:226–253.

Melosh, H. J. 1989. *Impact Cratering: A Geologic Process* (New York: Oxford Univ. Press), pp. 112–125.

Milani, A., Carpino, M., and Marzari, F. 1990. Statistics of close approaches between asteroids and planets: Project SPACEGUARD. *Icarus* 88:292–335.

Morrison, D., ed. 1992. *The Spaceguard Survey: Report of the NASA Near-Earth Object Detection Workshop* (Pasadena: Jet Propulsion Laboratory).

Namiki, N., and Binzel, R. 1991. 951 Gaspra: A pre-Galileo estimate of its surface evolution. *Geophys. Res. Lett.* 18:1155–1158.

Nolan, M. C., Asphaug, E., and Greenberg, R. 1992. Numerical simulation of impacts on small asteroids. *Bull. Amer. Astron. Soc.* 24:959–960.

Öpik, E. J. 1951. Collisional probabilities with the planets and the distribution of interplanetary matter. *Proc. Roy. Astron Acad.* A54:165–199.

Scholl, H., Froeschlé, Ch., Kinoshita, H., Yoshikawa, M., and Williams, J. G. 1989. Secular resonances. In *Asteroids II*, eds. R. P. Binzel, T. Gehrels and M. S. Matthews (Tucson: Univ. of Arizona Press), pp. 845–861.

Shoemaker, E. M., and Wolfe, R. F. 1987. Crater production on Venus and Earth by asteroid and comet impact. *Lunar Planet. Sci.* XVIII:487–489 (abstract).

Shoemaker, E. M., Williams, J. G., Helin, E. F., and Wolfe, R. F. 1979. Earth-crossing asteroids: Orbital classes, collision rates with Earth, and origin. In *Asteroids*, ed. T. Gehrels (Tucson: Univ. of Arizona Press), pp. 253–282.

Shoemaker, E. M., Wolfe, R. F., and Shoemaker, C. S. 1990. Asteroid and comet flux in the neighborhood of Earth. In *Global Catastrophes in Earth History*, eds. V. L. Sharpton and P. D. Ward, Geological Soc. of America Special Paper 247 (Boulder: Geological Soc. of America), pp. 155–170.

Wetherill, G. W. 1967. Collisions in the asteroid belt. *J. Geophys. Res.* 72:2429–2444.

Wetherill, G. W. 1976. Where do meteorites come from? A re-evaluation of the Earth-crossing Apollo objects as sources of chondritic meteorites. *Geochim. Cosmochim. Acta* 40:1297–1317.

Wetherill, G. W. 1979. Steady-state population of Apollo-Amor objects. *Icarus* 37:96–112.

Wetherill, G. W. 1985. Asteroidal source of ordinary chondrites. *Meteoritics* 20:1–22.

Wetherill, G. W. 1988. Where do the Apollo objects come from? *Icarus* 76:1–18.

Wetherill, G. W. 1989. Cratering of the terrestrial planets by Apollo objects. *Meteoritics* 24:15–22.

Wetherill, G. W., and Williams, J. G. 1968. Evaluation of the Apollo asteroids as sources of stone meteorites. *J. Geophys. Res.* 73:635–648.

Wisdom, J. 1983. Chaotic behavior and the origin of the 3/1 Kirkwood gap. *Icarus* 56:51–74.

van Houten, C. J., van Houten-Groeneveld, I., Herget, P., and Gehrels, T. 1970. The Palomar-Leiden survey of faint minor planets. *Astron. Astrophys. Suppl.* 2:339–448.

CRATER SIZE DISTRIBUTIONS AND IMPACT PROBABILITIES ON EARTH FROM LUNAR, TERRESTRIAL-PLANET, AND ASTEROID CRATERING DATA

G. NEUKUM
DLR, Institute for Planetary Exploration, Berlin

and

B. A. IVANOV
Institute for Dynamics of the Geosphere, Moscow

The terrestrial planets, the Earth's Moon, and the asteroids Gaspra and Ida show striking similarities in their production crater size distribution characteristics. The investigation of the crater populations on Gaspra and Ida yield information on the crater size distribution in the source region of the bodies largely responsible for cratering the planets of the inner solar system. Comparison of these data especially with the lunar impact record, which is our most reliable data base, confirms the complex shape of the distribution curve (standard distribution), and lends strong support to the idea of a common population of bodies impacting the inner planets and largely stemming from the asteroid belt. The steepening of the production crater size distribution at sizes $D \leq 1$ km is confirmed to be due to the characteristics of the size distribution of the primary impactor production and not to an admixture of objects from secondary cratering processes. The impact hazard for craters, or objects, is assessed for the Earth and the other terrestrial planets relative to the lunar case through application of the lunar production size-frequency distribution which is well known for the crater diameter range 10 m $< D < 1000$ km. Current or past impact rates can be calculated for any size of crater or projectile and are given for specific crater sizes and respective projectile sizes as well as for projectile energies. The impact hazard for projectiles, e.g., capable of forming 1-km craters on solid surfaces of the Earth (if the atmosphere were absent) is assessed for the present and results in a production rate of one crater every 1600 yr for the surface area of the whole Earth, or every 6000 yr for the area of the continents; for 100-km craters, e.g., the corresponding number is one event every 27 Myr for the whole Earth.

I. INTRODUCTION

The cratering record on the terrestrial planets, Earth's Moon, and the asteroids Gaspra and Ida provides information on the size distribution, time scales and the origin of the impacting objects. Unlike the Earth, where the shielding effect of the atmosphere prevents the small bodies from reaching the ground or where substantial erosional and depositional effects by the geologically very active environment erase craters smaller than 10 km in diameter in a few tens or hundreds of millions of years, the crater populations on the atmosphereless terrestrial planets and the asteroids have not changed too much through time after formation of the planetary surface units which function as "impact counters" (Öpik 1960), i.e., yield an image of the mass distribution of the impactors. Especially from lunar data, the cratering record for the Earth-Moon system and hence for the Earth directly after application of the appropriate scaling relationships can be inferred with a high degree of reliability (uncertainty less than a factor of 2 in production rate) for craters in the size range $D \leq 1$ km not accessible directly on Earth. In this way, a good measure of the impact hazard on the Earth for projectiles potentially capable of forming craters in this size range on Earth can be obtained for the past history and the present time. Even in the large-crater size range ($D \geq 10$ km), the Moon is a more reliable source of information than the Earth where only the record on the old shields (cratons) can be assessed with sufficient reliability (factor of 2 in impact rate) for the period of the last ≈ 400 Myr. The recent results from the Galileo mission which has been able to provide close-up images of the asteroids Gaspra and Ida during its passage through the asteroid belt make it possible to compare the cratering record on the asteroidal surfaces with that on the Earth's Moon and the other terrestrial planets; thus, for the first time in the history of planetary exploration, we are able to analyze the impactor distribution, or their image, the crater size distribution, in the source region of the bodies largely responsible for cratering the inner solar system planets. It is the purpose of this chapter, to review the wealth of cratering data which have been derived primarily during the Apollo and early post-Apollo times for the Moon, to make appropriate comparison with the data subsequently derived from the investigations of the other terrestrial planets by a variety of authors and to combine those data with the latest results for Gaspra and Ida (see the Chapter by Cheng et al.). This is done in a novel way by rigorously comparing the crater size distributions on the different planetary objects in an analytical way, taking into account all major parameters which affect the impact process scaling from one planet to another. The resulting data are related to the lunar data which are our best data base for studying the cratering record of the Earth-Moon system. In this way, the characteristics of the inner solar system impactor family are assessed and impact probabilities on Earth are derived to a high degree of accuracy.

II. THE METHOD OF DETERMINING CRATER SIZE-FREQUENCY DISTRIBUTIONS ON PLANETARY SURFACES AND DERIVING CRATERING RATES

The method to achieve relative and absolute age determination from crater statistics or vice versa, if the age of the surface unit is known, to achieve the derivation of impact rates from measured crater frequency data, was developed for the Moon by different groups concentrating on different aspects (see, e.g., Öpik 1960; Shoemaker 1962; Baldwin 1964,1971; Hartmann 1965,1966; Soderblom 1970; Neukum 1971,1981,1982,1983; Neukum and Wise 1976). In order to accomplish one's task for a specific surface unit, it is necessary to verify by careful examination of the imagery and precise measurement that (1) a production frequency can be measured; (2) the structure to be dated is really one homogeneous geologic unit (no or negligible resurfacing in terms of destruction of craters of the population measured); (3) only superimposed craters and no relic or ghost craters from an underlying older unit are measured; (4) secondary craters and volcanic craters are eliminated; (5) the area of measurement is determined accurately; and (6) the size of craters is measured with highest possible precision since the exponential size-frequency distribution dependence amplifies small errors in diameter as large errors in crater density per unit area (i.e., crater frequency).

The measurement of production size-frequency distributions and derivation of cratering rates is only possible if the crater population has not yet reached a state of equilibrium. The central issue for our purposes is not whether a surface is saturated, but whether the observed crater size-frequency distribution is essentially the same as, or close to, the production size-frequency distribution. A production size-frequency distribution is one that has retained its original form, and, therefore, reflects the projectile size-frequency distribution. In the absence of extensive erosion and deposition, a lightly cratered surface preserves the crater production population. Even on surfaces where craters have been destroyed by the cratering process, the production size-frequency distribution can still be preserved (Neukum and Dietzel 1971; Chapman and McKinnon 1986; Woronow 1977,1978). If one is dealing with a production population, then the projectile size-frequency distribution can be recovered by providing the correct values to parameters in an appropriate crater scaling law.

There are essentially two ways a production size-frequency distribution can be changed. One is by geologic (endogenic) processes such as erosion and deposition that destroy pre-existing craters, and the other is by the cratering process itself. Depositional and erosional processes destroy small craters more easily than large craters, resulting in a preferential loss of small craters relative to large ones. This results in an increase in the population index at smaller crater sizes, i.e., the slope index (crater frequency vs crater diameter on a log-log plot) becomes less negative. At an extreme, all craters smaller than some threshold diameters may be destroyed.

The relationship between a crater production size-frequency distribution and age is as follows: The crater population on one planet approximately represents the mass-velocity distribution of the impactors responsible for the cratering record. Under the assumption of a mean impact velocity, a meteorite population (i.e., the impactors) of the mass distribution $n(m, t)$ in the mass interval $(m, m + dm)$ causes a crater size-frequency distribution $n(D, t)$ in the crater diameter interval $(D, D + dD)$ for a specific exposure time t. The function $n(D, t)$ is termed differential distribution (number per unit area per diameter at time t). The relationship between differential distribution and differential cratering rate (number per unit area per diameter per time at time t) is given by:

$$n(D, t) = \int_0^t \varphi(D, t') \, dt' \qquad (1)$$

where t is the exposure time with respect to the age of the crater population ($t > 0$). The cumulative crater frequency $N(D, t)$ (number per unit area of all craters with diameters equal to or larger than D and which were formed during exposure time t) is given by

$$N(D, t) = \int_D^\infty n(D, t) dD' = \int_D^\infty \int_0^t \varphi(D', t') dD' dt'. \qquad (2)$$

Note that this is in the continuous approximation; in reaiity, it is the sum of discrete numbers. The cumulative cratering rate $\Phi(D, t)$ is given by

$$N(D, t) = \int_0^t \Phi(D, t') dt' \quad \text{or} \quad \Phi(D, t) = \frac{\partial N(D, t)}{\partial t}. \qquad (3)$$

The function $\varphi(D, t)$ can be separated into a function $g(D, t)$ that reflects the underlying crater size distribution and into a function $f(t)$ that reflects the general functional dependence of cratering rate on time. Therefore one can write

$$\varphi(D, t) = g(D, t) \cdot f(t) \qquad (4)$$

or

$$\Phi(D, t) = \int_D^\infty g(D', t) f(t) dD' = G(D, t) f(t). \qquad (5)$$

If the size distribution is not directly dependent on time (i.e., the diameter distribution is the same over the whole exposure time), then the function $\varphi(D, t)$ can be separated in

$$\varphi(D, t) = g(D) \cdot f(t) \qquad (6)$$

and

$$n(D, t) = g(D) \int_0^t f(t') dt' \qquad (7)$$

or $n(D, t) = g(D) \cdot F(t)$ where $F(t) = \int_0^t f(t')dt'$. The cumulative crater frequency then is

$$N(D, t) = \int_D^\infty \int_0^t g(D')dD'f(t')dt' = G(D)F(t) \qquad (8)$$

which is a production distribution.

The production crater frequencies $n(D, t)$ or $N(D, t)$ show the same value, respectively, for areas exposed for an equal length of time on one planet for the same crater diameter or diameter interval regardless of location on the planet, provided the impactor flux is isotropic and target material composition have no influence. In this case, we find the simple relationship

$$N_1(D, t_1)/N_2(D, t_2) = F(t_1)/F(t_2). \qquad (9)$$

This means that the ratio of the crater production population frequencies on two areas (1 and 2) is directly proportional to the ratio of their functional dependence on time. From this relationship a relative age sequence of geologic units can be determined by direct comparison of their superimposed impact crater frequencies; i.e., the frequency of superimposed craters (per unit area) for a specific diameter or a specific diameter interval is a direct measure of relative age, termed relative crater retention age. Vice versa, if the age of the surface units is known in combination with functional dependence of N on time, it is then possible to determine impact rates as a function of time.

For direct comparison of the size-frequency distributions of two or more units, it is necessary to measure in the same crater size range or, if that is not possible, to compare crater frequencies through application of the functional dependence of the production distribution (called standard or calibration distribution by Neukum and Wise [1976]) and to convert the measured frequencies in this way to the values which they would show at one and the same diameter if they could be measured there.

Two types of crater size-frequency distribution plots are used in this chapter: (1) the cumulative plot, and (2) the relative or "R" plot. In the cumulative size-distribution, the log cumulative crater frequency is plotted against the log crater diameter and is in the form $N \propto D^\alpha$ where N is the cumulative crater frequency, D is the crater diameter and α is the population or slope index. The "R" plot displays information on the differential size distribution, and is the ratio of the observed distribution to the function $dN \propto D^{-3}dD$. Because most large crater populations have slope or population indices within the range of ± 1 of the function D^{-3}, they plot as nonsloping or moderately sloping lines on these log/log plots. On an "R" plot, a horizontal line has a differential -3 slope index; one sloping down to the left at an angle of $45°$ has a differential -2 slope index and one sloping down to the right at 45° has a differential -4 slope index. The vertical position of the curve is a measure of crater density; the higher the curve, the greater the crater density per unit area. For a given single-sloped cumulative slope index, the equivalent differential slope index is decreased by 1, e.g., a cumulative -1 is a differential -2, etc.

III. PRODUCTION CRATER SIZE-FREQUENCY DISTRIBUTIONS OF THE EARTH-MOON SYSTEM

An accurate analysis of the crater populations of the Moon on the basis of Lunar Orbiter and Apollo data, in combination with an assessment of the cratering record on terrestrial cratons, forms the basis for the determination of the time dependence of the impact rate in the Earth-Moon system and for comparison with the cratering record on the other terrestrial planets and the asteroids Gaspra and Ida. Comprehensive compilations of the results of different authors are found in Hartmann et al. (1981), in Neukum (1983) and in Strom and Neukum (1988). These results will be reviewed in comparison with new results for the asteroids Ida and Gaspra in the Chapter by Cheng et al.

A. Analysis of the Production Crater Size-Frequency Distribution of the Earth's Moon

Areas of the Moon of very different ages show crater populations in the state of production covering a wide range in ages. The maria with ages between approximately 3 and 3.8 Gyr, the lunar highlands older than 3.8 Gyr, and some young impact craters allow the analysis of the crater frequencies over an age period of ≈ 4 Gyr to almost the present. As discussed in Sec. II, the determination of the production size-frequency distribution and of the impact rate, as a function of time, can be solved empirically through the measurement of the crater frequencies $N(D, t) = G(D) \cdot F(t)$ or $n(D, t) = g(D) \cdot F(t)$, respectively, which should be determined over as large a range in D and t as possible.

B. The Lunar Standard or Calibration Distribution

A detailed analysis of the function $G(D)$ and $N(D, t)$, respectively, for various, but always fixed ages t has been performed by a number of authors. Shoemaker et al. (1970a) found $N \propto D^{-2.9}$ for $D > 3$ km for mare areas of the Moon of different ages.

Hartmann and Wood (1971) give $N \propto D^{-2}$ for craters with $D > 1$ km, and Baldwin (1971) and Hartmann et al. (1981) find $N \propto D^{-1.8}$ for about the same range. In both papers no time dependence of the diameter distribution was found, e.g., $N(D, t) = G(D) \cdot F(t)$.

Chapman and Haefner (1967) argued for the first time, that their measurements seemed to indicate no simple power law for the crater size distribution, of the form $N \propto D^\alpha, \alpha = $ const, but instead that the exponent α seemed to show a dependence on crater size, $\alpha = \alpha(D)$. Also the measurements cited above (Shoemaker et al., Hartmann and Wood, and Baldwin) could be satisfactorily explained with such a power law dependence.

A detailed analysis of the distribution law has been performed by Neukum et al. (1975), followed up by Neukum and König (1976) and König (1977). Crater populations from lunar sites of various ages, but on geologically homogeneous areas with undisturbed crater distributions have been analyzed over a wide range in diameters.

The areas of measurement and their crater populations have been selected in such a way that the crater frequencies measured in the different areas do overlap for the smallest and largest diameters, respectively. In this way, the ratio $N_i/N_k = C_{ik}$ can be calculated for different populations, or, in other words, the frequencies can be normalized by "shifting" the plots in $\log N$-direction. Figure 1 shows a smooth curve as the result from a normalization of the frequencies upon each other. From this finding we can draw the following conclusions:

1. The distributions in the diameter range 20 m $\leq D \leq 1$ km from areas of ages between 0.1 and 3 Gyr coincide, and the whole data set can be combined into a single size-frequency distribution. The very fact, that it is easily possible to normalize individual distributions successively into a smooth function, indicates that there are no significant variations with time in the crater size distribution dependence. A similar conclusion can be drawn for the underlying size and velocity distribution of the crater-producing impactors.

2. For diameters larger than about 1 km, the distributions coincide for areas in the age range from 3 to more than 4 Gyr. This means that the crater size distribution dependence in the time interval from ≈ 4 Gyr to about 1 Gyr has not undergone any significant changes. Correspondingly, this is also valid for the size and velocity distribution of the crater producing impactors. The lower time limit of about 1 Gyr in this interpretation comes from the fact that the crater populations have been built up over the whole time scale (of more than 3 Gyr) and any changes within the last 1 Gyr would affect the measurements only to a minor extent.

3. The production crater size distribution of the lunar surface cannot be determined directly for certain size and age ranges, either because there are no statistically significant data available, or the distribution is in a state of equilibrium ("saturation"), due to a superposition of impacts. Such conditions occur, e.g., for $D < 300$ m and ages of more than 3.5 Gyr or $D < 1$ km and $t > 4$ Gyr.

4. The lunar crater populations can be described with a single size distribution function over the range in which measurements can be made. This means one can use *one* standard function to compare the crater frequencies with reference to the same diameter, even in the case where the distribution law might be different for ranges which are not accessible.

5. The distribution function does not follow a single power law, but its

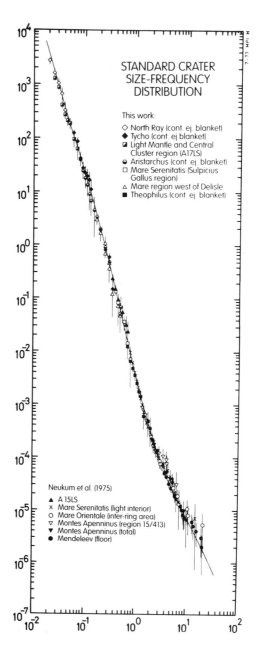

Figure 1. Normalized production distribution of the Moon with an approximation through a 7 th order polynomial in log *D* (figure from König 1977).

structure in the diameter range 20 m$\leq D \leq$20 km is so complex, that log$N(D)$ can only be represented satisfactorily with a polynomial of order $\geq$7 in log$D$ (see, e.g., Neukum et al. 1975; König 1977). The fit in Fig. 1 is done with a 7th order polynomial.

6. It is quite apparent from the refined measurements and interpretations presented here, that the former power law interpretations of the distributions can now be seen as approximate descriptions over limited size ranges of the real distribution which is more closely represented by the polynomial of the standard distribution.

The diameter range D>5 km, which is statistically somewhat poorly represented, has been studied more thoroughly by Neukum (1983), in particular for arriving at a description of the distribution of the populations in the range above 20 km crater diameter. These populations can be measured on very old areas of the lunar highlands. In addition, it is possible to quantify the distribution of Copernican-Eratosthenian craters (post-mare distribution) with $D \geq$20 km over the whole front-side of the Moon (Wilhelms 1979). In this way the distribution in the range $D \geq$20 km can be measured over a large age interval ($\approx$4.4 to 1 Gyr).

The connection of the frequencies for $D \leq$20 km with those for $D \geq$20 km for the determination of the distribution over the whole accessible diameter range, from less than 100 m to more than 100 km, can be achieved in several ways:

1. Direct combination of the frequencies for $D \leq$20 km and $D \geq$20 km in areas where data from both diameter ranges are available, which, through their similar ages, relates the distribution of the small craters to that of the large ones, directly. Examples are the populations on Imbrium Ejecta and Orientale Ejecta (cf., Fig. 2).

2. Determination of frequencies of small craters $D \leq$20 km and large craters $D \geq$20 km on different areas, whose ages are known. The connection of the frequencies for the determination of the distribution over the whole diameter range can be done through normalization of the individual frequencies onto the time integral of the impact rate $F(t)$, which of course has to be known. In particular for ages t<3 Gyr, all measurements indicate, on the average, a constant impact rate. Therefore, crater populations in areas which fall in that age span can be easily compared with each other, because the frequencies correlate linearly with time ($F(t) \propto t$).

An analysis of the crater populations as described above has been performed in several cases. An example of method 2. can be seen in Fig. 3.

All frequencies fall, after normalization on $F(t) \propto t$, on a smooth curve which is the lunar standard distribution. It therefore appears that the distribution determined here is a good approximation for the lunar crater production size distribution for the last 3 Gyr, over the investigated range of diameters. The consistency of the representation indicates that the assumption of a con-

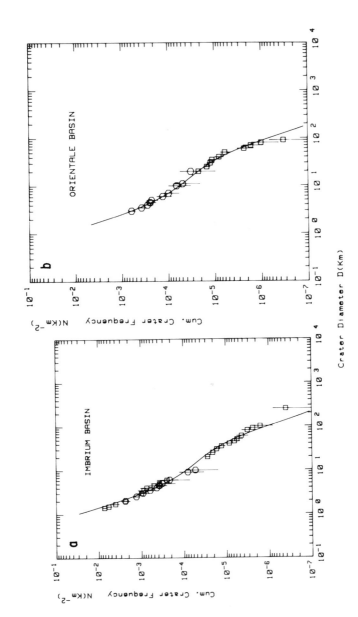

Figure 2. Frequencies measured for small craters ($D<20$ km) and large craters ($D>20$ km) separately on areas of Imbrium- and Orientale-ejecta, respectively (data from Neukum et al. 1975 and Wilhelms 1979).

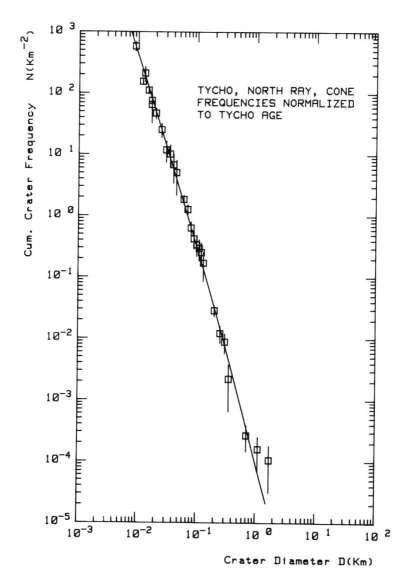

TYCHO, NORTH RAY, CONE
FREQUENCIES NORMALIZED
TO TYCHO AGE

Crater Diameter D(Km)

Figure 3. Crater frequencies for various younger lunar areas, normalized on the measured frequency of craters superimposed on crater Tycho, under the assumption of a constant impact rate ($F(t) \propto t$). The frequencies coincide with the lunar standard distribution (fitted curve).

stant impact rate in the last 100 Myr is correct to a high degree of probability.

A synthesis of the data from Neukum et al. (1975), Neukum and König (1976) and König (1977) and the most recent data for $D \geq 20$ km allows the determination of the size distribution in the diameter range 10 m $\leq D \leq 300$ km (Neukum 1983). An approximation of the normalized logarithmic cumulative frequencies with an 11th degree polynomial in $\log D$ fits the complex structure of the distribution with sufficient accuracy. The polynomial has the form

$$\log N = a_0 + a_1 \log D + a_2 (\log D)^2 + \ldots + a_{11} (\log D)^{11}. \qquad (10)$$

Its coefficients are: $a_1 = -3.6269$; $a_2 = 0.4366$; $a_3 = 0.7935$; $a_4 = 0.0865$; $a_5 = -0.2649$; $a_6 = -0.0664$; $a_7 = 0.0379$; $a_8 = 0.0106$; $a_9 = -0.0022$; $a_{10} = -5.180 \times 10^{-4}$; $a_{11} = 3.970 \times 10^{-5}$.

The formula yields a cumulative crater frequency N per km^2, if the diameter D is given in km. The coefficients are valid for the range $10\,\mathrm{m} \leq D \leq 300$ km, but *not* outside this range. On average, the quality of the fit in the range $10\,\mathrm{m} \leq D \leq 300$ km is better than 50% (standard deviation). The distribution function agrees to within 50%, over the specified range of validities, with previously determined expressions (Neukum et al. 1975; König 1977).

In order to derive the general mathematical expression for the cumulative distribution function for a specific area, one has to add a term to the polynomial formula, which refers to the age t_0 of the area. $\log N = a_0 + a_1 \log D + a_2(\log D)^2 + \ldots + a_{11}(\log D)^{11} + \log F(t_0)$, where $F(t_0)$ is the time integral of the impact rate to which the area has been exposed since its formation. For the numerical approximation we put $a_0 + \log F(t_0) = -2.5340, N(D = 1$ km$) = 2.92 \times 10^{-3}$.

The standard distribution allows us now to compare crater frequencies of populations in different diameter ranges, and to determine the relative ages of the geologic formations on which the crater frequencies have been measured, or, if the age of the geologic formation is known, to derive impact rates. It is useful, to refer to the same reference diameter, normally $D = 1$ km or $D = 10$ km. The distribution function is displayed in Fig. 4 in comparison with distribution laws from older references (Shoemaker et al. 1970a; Baldwin 1971; Hartmann and Wood 1971), which used a constant distribution index α in the form $N \propto D^{\alpha}$. Such simple power laws are only valid within a very limited range of diameters, and their application can lead to substantial errors, as can be seen in Fig. 4.

The lunar production crater size distribution function (termed standard or calibration distribution) is a very complex function with values for the distribution index between -1 and -4. This can be seen particularly well from the representation of the relative frequencies in Fig. 5. One can clearly see the deviation from the postulated distribution with $\alpha = -2$ (horizontal line) in the range $D > 5$ km. In the range $D < 1$ km the mean value of α is around -3, which means the frequency rises steeply towards smaller craters. These peculiarities are discussed below.

The distribution of craters in the diameter range $D \geq 20$ km on the oldest parts of the lunar crust has been an issue for scientific debate for years. A series of authors (see, e.g., Hartmann and Wood 1971; Shoemaker 1970; Hartmann 1972) consider these populations as being in a state of "saturation," i.e., in equilibrium with destructive processes, predominantly those of direct impact

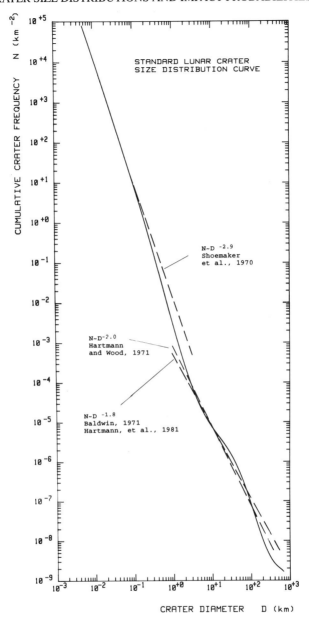

Figure 4. Comparison of the standard crater size distribution function as derived by Neukum (1983) with other distributions published in the literature. These distributions have been normalized in an appropriate way and have been appropriately displayed for their stated range of validity. The reader should be aware that there is some ambiguity in the literature about the range of validity of these distributions at the small-crater size end. It was well known at the times of the early work that the distributions turned up below ≈2 km crater size. This, however, was largely interpreted as a contribution of secondary craters to the population of primary ones showing the flatter distribution characteristics.

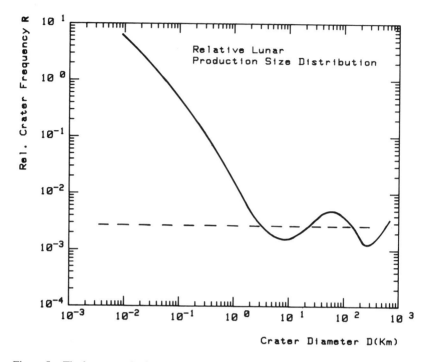

Figure 5. The lunar standard crater size distribution (solid curve) in relative frequency representation. The dashed horizontal line, R = const., corresponds to a cumulative distribution $N \propto D^{-2}$, and a differential distribution $n \propto D^{-3}$, respectively.

superposition. This seemed to be consistent with a distribution $N \propto D^{-2}$, which had been derived approximately, and which would have been expected to arise from such conditions, under certain asssumptions. A number of authors realized deviations from the D^{-2} distribution. Hartmann (1984) derived an empirical fit of a -1.83 power law to what he identified as a saturation equilibrium curve found on the Moon and on several other bodies. Another group's interpretation was that at least the larger craters ($D \geq 30$ to 50 km) were not so densely packed, that a significant number of craters of this size would have been erased through superposition (Neukum 1971; Neukum and Wise 1976; Baldwin 1971; Woronov 1977; Strom 1977). A solution to this problem is now possible. The measurements show clearly, that the distributions on the oldest areas of the Moon, for $D \geq 20$ km, do not follow a D^{-2} equilibrium law. What we are essentially dealing with is a production distribution. This statement is confirmed by results from an analysis of younger highland and mare populations. Figure 6 presents the frequencies for pre-Nectarian, Nectarian and Eratosthenian-Copernican populations in comparison to highland crater frequencies. The distributions are identical, within their error bars, which means the distributions have been

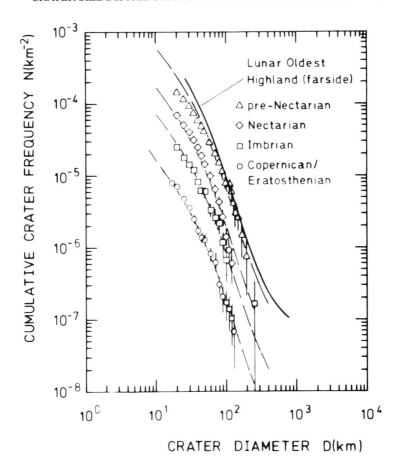

Figure 6. Size distributions as measured for lunar craters with diameters $D \geq 20$ km. The data are in good agreement with the lunar standard distribution (solid and dashed curves). This confirms the stability of the distribution over more than 4 Gyr (data from Wilhelms 1979).

stable over the whole time interval.

Strom (1977) argued that the size distributions of the post-mare craters are different from those of the pre-mare craters. This finding was not supported by Hartmann (1984) who did not find any such difference. The argument has been discussed again by Strom and Neukum (1988). The distribution of the Eratosthenian-Copernican (i.e., post-mare craters) is shown in Fig. 7 in the way of a relative plot.

The scatter in the data is considerable, but in our opinion there is no contradiction with the conclusion that it is the same as for older populations as shown in Fig. 8. Unfortunately, relative crater frequency plots for the post-mare craters in this size range ($D \geq 10$ km) do not give unambiguous information on the distribution behavior. The cumulative distribution for the

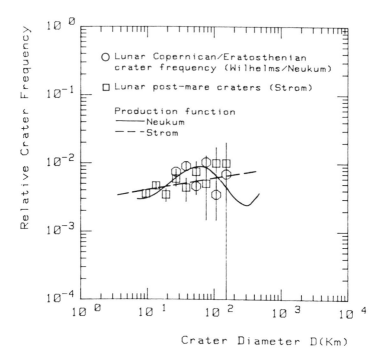

Figure 7. Comparison of data sets for lunar post-mare and Eratosthenian plus Copernican craters. Also given are functions as advocated by Strom and Neukum (1988), respectively.

Copernican-Eratosthenian craters as shown in Fig. 6, however, which gives the coarser distribution characteristics, is very suggestive of a distribution which has not changed through time, i.e., is identical to the pre-mare size distributions (Imbrian, Nectarian, Pre-Nectarian).

C. Primary vs Secondary Impact Craters at Sizes $D \leq 1$ km

A series of authors (cf., Shoemaker 1965; Brinkmann 1966; Soderblom et al. 1974) have interpreted the steep rise in the distributions around $D = 1$ km toward smaller crater sizes as a contribution of secondary craters. Shoemaker termed these secondary craters "background secondaries," because there were no primary sources discernible. The idea of the process is, that large craters ($D \geq 10$ km) will produce ejecta, which fly over large distances around the Moon on ballistic trajectories and produce impact craters in the range $D \leq 1$ km, statistically distributed over the lunar surface, and which resemble the lunarwide distribution of primary impacts.

Soderblom et al. (1974) have used these secondary crater arguments in order to explain the distributions on Mars for $D < 1$ km, that show, similar to the lunar ones, a steep rise in frequency. Their arguments will be examined in the following.

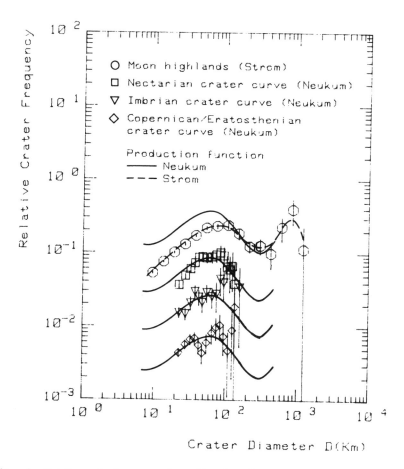

Crater Diameter D(Km)

Figure 8. Relative crater frequency plots of data from the lunar highlands, Nectarian-aged, Imbrian-aged, and Eratosthenian-aged surfaces (from Strom and Neukum 1988). The highland data (Strom) are given in $\sqrt{2}$ intervals. The other data (Neukum) are given in much finer binning in order to bring out the structure of the distributions over the more limited size range better. These data have been smoothed by a floating average procedure to take out some of the statistical noise. Production functions are given in comparison. There is a slight disagreement of the authors in the interpretation of the data; Strom believes the lunar highland data totally reflect the production function for Imbrian to pre-Nectarian (highlands) ages, whereas Neukum believes that the production function as given here is not totally reflected in the highlands data but that the highlands distribution suffered some loss of craters (factor of 2 to 3) at size <50 to 100km from cratering and noncratering processes.

Under the assumption of a simple exponential function $N \propto D^{\alpha}$ for the primary distribution, Soderblom et al. derive a relationship between D_c, the diameter, at which the frequencies of the primary and secondary craters are equal, and $D_{\max}$, the diameter of the largest primary crater that contributes to the secondary crater population:

$$D_c = D_{\max} \left[\frac{\alpha}{(\beta - \alpha)k^\beta} \right]^{1/(\alpha-\beta)} \tag{11}$$

where $\beta = -3.5$ is the distribution index of the secondary crater distribution:

$$N_{\sec} = (D_{\sec}/kD)^\beta. \tag{12}$$

$N_{\sec}$ is the cumulative frequency of the secondary craters produced by a primary crater of diameter D, and $D_{\sec}$ is the diameter of the secondary crater and k is a constant. The value of $\beta = -3.5$ originates from the observed distribution of lunar secondary craters in strewn fields of primary craters and from investigations of experimental nuclear explosion craters (Shoemaker 1965).

Soderblom et al. (1974) conclude from their investigations, that $D_c \approx 1$ km corresponds to a value which is compatible with contributions from primary craters of a size of $D \approx 50$ km (which are the largest craters these authors have measured), and conclude further that the steeply rising diameter distribution for Martian craters of diameter $D \leq 1$ km also is a contribution of secondary craters.

This interpretation neglects the fact that $D_{\max}$ is a function of the age and an accidental coincidence for one test area and one age does not imply a generally valid conclusion.

A correct treatment of the problem for the lunar case gives for a simple exponential size distribution law of the form $N \propto D^\alpha$, under consideration of the time dependence of the crater frequency $N \propto D^\alpha \cdot F(t)$, for $N(D_{\max}) = 1$, (the largest crater with diameter $D_{\max}$, which contributes to the secondary crater population)

$$D_{\max} \propto F(t)^{-1/\alpha}. \tag{13}$$

For two areas of the Moon with different age, t_1 and t_2, one gets from $D_c \approx D_{\max}$:

$$D_c(t_2)/D_c(t_1) = F(t_2)^{-1/\alpha}/F(t_1)^{-1/\alpha} \tag{14}$$

Taking a distribution of primary craters with $\alpha \approx -2$, as observed in the range $D \geq 10$ km, and $t_1 = 3.2$ Gyr, and $t_1 = 3.9$ Gyr, one obtains

$$D_c(t_2)/D_c(t_1) \approx \sqrt{10}. \tag{15}$$

One should therefore, as shown in Fig. 9, find a correponding shift of the value of D_c with the age of the population. But this is not observed. This fact is a strong argument against a secondary crater distribution in the range $D \leq 1$ km, and advocates clearly for the interpretation that we are dealing with a primary crater distribution over the whole range.

A supporting argument for the interpretation as a primary crater distribution is provided by comparing the younger ($t \leq 100$ Myr) crater populations

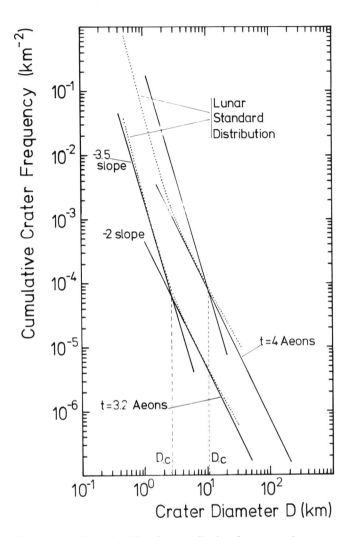

Figure 9. Theoretical relationship of a contribution from secondary craters to the
primary crater population, taken as $N \propto D^{-2}$. The value of D_c should show a shift
in relation with time, which is not observed in the lunar crater populations (standard
distribution, dotted curve) (1 Aeon = 1×10^9yr).

of the Moon in the range $D > 10$ m with measurements on lunar rocks in the
range $D < 1$ cm. In this case one has to normalize onto the time integral $F(t)$
and assume, as inferred from the observations, that the impact rate has been
constant over the time interval considered, e.g., $F(t) \propto t$. In Fig. 10 these
data are compared. Those in the range $D < 1$ cm originate from microcrater
measurements on lunar rocks (Neukum 1971; Hörz et al. 1974; Fechtig et
al. 1975) and are in agreement with *in situ* measurements on interplanetary
dust by spacecraft detectors (Grün 1981). The steep distribution observed

in the range $D>10$ m is only compatible with the measurement in the range $D<1$ cm. An extrapolation with $N \propto D^{-2}$ from large craters with $D \geq 1$ km would lead to a discrepancy of several orders of magnitude.

Our finding that the steepening of the distribution toward smaller crater sizes at $D \approx 1$ km is due to primary impacts and not to secondary craters is much supported by some recent results: the characteristics of the crater size distributions on the asteroids Gaspra (Chapman et al. 1993) and Ida (Neukum 1993, unpublished data) (cf., the respective discussion further down in this chapter) and the recent direct astronomical observation of the population of impactors responsible for those craters in the Earth-Moon system (Chapter by Rabinowitz et al.).

IV. THE PRODUCTION CRATER SIZE DISTRIBUTIONS ON THE TERRESTRIAL PLANETS

For an assessment of the NEO size distribution and flux, it is quite useful to evaluate the relationship of this population of objects with the crater populations found on the other terrestrial planets. Such discussions will help us understand in a more general way what kind of underlying solar system impactor population is responsible for the cratering record of the Earth-Moon system and what the hazard of impactors on Earth has been in the past and today, and will be in the future.

A. Mars

Like for the Moon there is no agreement in literature concerning the size distribution of craters on Mars. The production crater size distribution (standard distribution) of Mars in the range of 1 km$<D<20$ km has been determined by Neukum and Wise (1976). More detailed analyses have been performed by Neukum et al. (1978), Neukum and Hiller (1981), and Neukum (1983). The distribution demonstrates a qualitatively similar characteristic like the lunar distribution with a steepening at $D \leq 1$ km. In the literature there exists a number of reference distributions following a simple exponential distribution $N \propto D^{\alpha}$ (see, e.g., Hartmann [1973], with $\alpha = -2$). These distributions are valid only in a small range of crater diameters and lead to discrepancies in the interpretation of measured data when used outside this range.

A more detailed analysis performed by Neukum (1983) shows that, unfortunately, there is no uniform production crater size distribution for Mars, and that in fact different water contents (or other compositional effects of the Martian upper crustal rocks) have produced somewhat discrepant distributions on different geologic formations on Mars. These effects predominantly influence craters with $D>5$ km (see, e.g., Carr 1981). The deviations of the distributions for the areas on Mars which have been investigated so far, lead to an uncertainty of about a factor 2 in the cumulative crater frequency N at a given diameter. The average standard distribution for Mars is shown in Fig. 11 (from Neukum 1983), where it is compared with the lunar standard

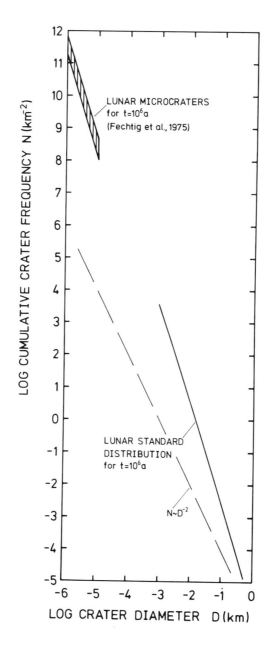

Figure 10.   Age-normalized ($t = 10^6$yr), cumulative distributions in the diameter range D <1 cm (lunar microcraters) and D>10 m. The solid curve is the lunar crater production distribution (standard distribution) for an age of 10^6yr (from Neukum 1983).

distribution. It is evident that the differences in the size dependence between the distributions are so large, that for a comparison of the crater frequencies of Mars with those on the Moon, it is necessary to clearly specify the diameter range. Otherwise errors up to a factor of 5 or more can occur. Unfortunately, this fact has been largely ignored.

The mathematical expression for the average crater size distribution for Mars (Neukum 1983) is a polynomial in $\log D$ of the form

$$\log N = a_0 + a_1 \log D + a_2 (\log D)^2 + \ldots + a_{11} (\log D)^{11} \qquad (16)$$

with its coefficients are: $a_1 = -2.9076$; $a_2 = 1.1870$; $a_3 = 0.3842$; $a_4 = -0.3955$; $a_5 = -0.1652$; $a_6 = 5.8655 \times 10^{-2}$; $a_7 = 2.6348 \times 10^{-2}$; $a_8 = -3.6585 \times 10^{-2}$; $a_9 = -1.6354 \times 10^{-3}$; $a_{10} = 7.3875 \times 10^{-5}$; $a_{11} = 2.8421 \times 10^{-5}$. a_0 is variable, and contains the age dependence of the distribution.

The polynomial is valid in the range $100 \text{ m} < D < 300 \text{ km}$, but *not* outside these limits. Polynomials of lower order approximate the rather complicated size distribution only insufficiently. The expression given above agrees to within 50% in N with the results in Neukum and Wise (1976) and Neukum and Hiller (1981).

B. Mercury

Measurements of craters with $D > 20$ km and of the large basins ($D > 200$ km) in the most densely cratered areas of Mercury yield the cumulative crater size-frequency distribution of Fig. 12. (Neukum 1981,1983). This distribution is compared with that of the Moon, presented as relative frequencies (Fig. 13). There is a striking similarity in the characteristics of the distributions on both planets, also with similar changes of the distribution index, in particular the bending-over characteristics at D around 100 km (Mercury) and around 60 km (Moon).

The great similarities of the distributions for the Moon and Mercury indicate a corresponding similarity in the mass-velocity distribution of the crater-producing impactors. In fact the distributions are identical within their measurement errors, if we assume $D_{(Mercury)} = 1.6 \cdot D_{(Moon)}$ for the same crater-producing impactor mass.

C. Venus

With the recent imaging radar data from the Magellan mission, it is now possible to analyze the large-size crater populations on Venus. As shown in Fig. 27 below, the distribution characteristics above $D \approx 20$ km crater diameter are very similar to the lunar one. Below 20 km crater diameter, the Venus atmosphere prevents the build-up of a crater production size-frequency distribution as an image of the impactor velocity-mass distribution. A detailed comparison of the Venus data with the lunar ones will be made in Sec. VII of this chapter.

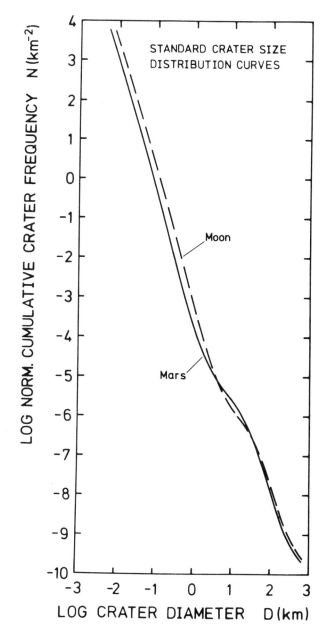

Figure 11. Comparison between the average standard crater size distributions for Mars and the Moon (from Neukum 1983).

D. Asteroids Gaspra and Ida

With the recent flybys of the Galileo mission and the images acquired by the

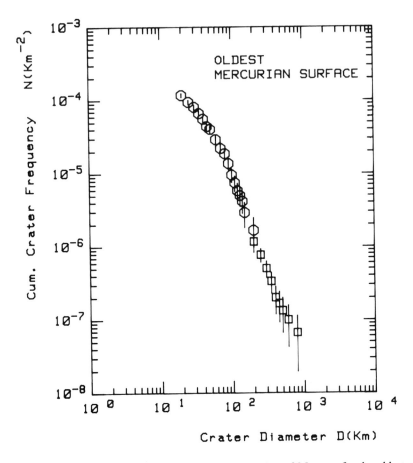

Figure 12. Cumulative crater size-frequency distribution of Mercury for the oldest
crater populations (from Neukum 1983).

Solid State Imaging System (Fig. 30 below) it is now possible to characterize
the production crater size-frequency distributions in the asteroid belt in the
diameter range ≈100 m to 10 km (Chapman et al. 1993; Belton et al. 1994).
The distributions are very lunar-like, showing all the features of the characteristic steepening at ≤1 km crater diameter, and the flattening between 1 km
and 10 km crater size.

   These findings are an excellent confirmation of the views long held by
one of the authors (G. N.) that the steep characteristics of the crater size distributions on the Moon and on Mars and Mercury for crater sizes $D \leq 1$ km
(distribution index about -3 to -3.5 cumulative) are due to the size distribution characteristics of the crater-forming impactors stemming largely from
the asteroid belt and clearly rule out the view by others who have interpreted
the steep distributions as secondary craters. We would also argue from these

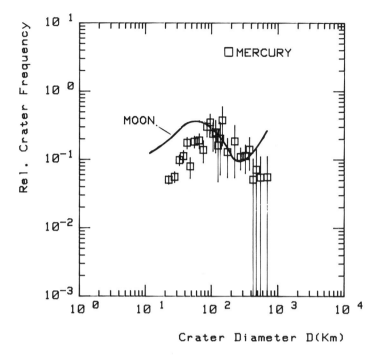

Figure 13.   Relative crater size-frequency distribution of Mercury of the oldest populations in comparison with the relative lunar standard distribution (oldest population)
(from Strom and Neukum 1988).

results that the contribution of comets to the impact record in the inner solar
system and especially in the Earth-Moon system cannot be a high percentage,
very probably at the 10 to 20% level or below, unless by chance the size
distribution of comets mimics the asteroidal one.

A detailed comparison of the Gaspra and Ida data with the lunar standard
distribution is done in Sec. VII of this chapter by applying appropriate scaling
laws in a rigorous treatment.

V. THE IMPACT CHRONOLOGY OF THE EARTH-MOON SYSTEM

The lunar landers of the Apollo missions, which returned rock samples to
be radiometrically dated, have made it possible to correlate the ages of the
rocks sampled at the landing sites with the crater frequencies derived from
the investigation of image data of the landing sites. The investigation of
the terrestrial cratering record on the old shields of the Earth yields data for
the past ≈400 Myr. The correlation of crater frequency with age data leads
to an empirical determination of the impact chronology, i. e., the functional
dependence of the accumulated crater frequency on age or exposure time.

Two scientific goals can be achieved through application of this functional dependence:

1. Absolute age determination for any region of the Moon from measurements of crater frequencies on images, and
2. Derivation of the rate of impacts on the Moon and on the Earth as a function of time.

A variety of techniques have been applied by various authors in order to correlate crater frequencies with radiometric ages (Shoemaker et al. 1970*a*, *b*; Neukum 1971,1977; Baldwin 1971; Hartmann 1972; Soderblom and Boyce 1972; Soderblom et al. 1974; Neukum et al. 1975). A comparison between impact chronologies of the different authors has been made by Neukum and Wise (1976), and Neukum (1981). Figure 14 gives a summary of the results. The interpretations of the various authors all coincide within a factor of 2 to 3.

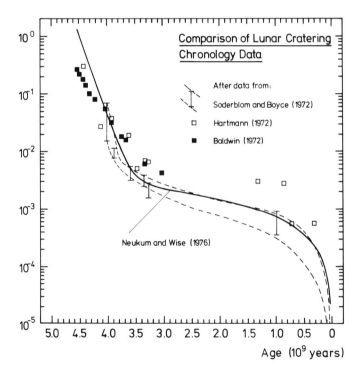

Figure 14.   Comparison of lunar cratering chronology data (Neukum 1981).

Neukum (1983) has re-assessed the data by partly going back to the original crater frequency and age data and making appropriate correlations where new information was available. The results are summarized in the following. Figure 15 is a view of the Moon's near side including the landing sites of the American Apollo missions and the Soviet Luna missions. A more detailed description of the landing sites and the nature of collected lunar samples is given in Taylor (1975). Some crater populations, the rock age characteristics of the landing sites, and the relevant structures which have been used to determine the impact chronology, are described briefly (cf., also Basaltic Volcanism Study Project 1981; Neukum 1983). The measurements of the crater frequencies and the radiometric age data which have been applied to determine the impact chronology are summarized in Neukum (1983).

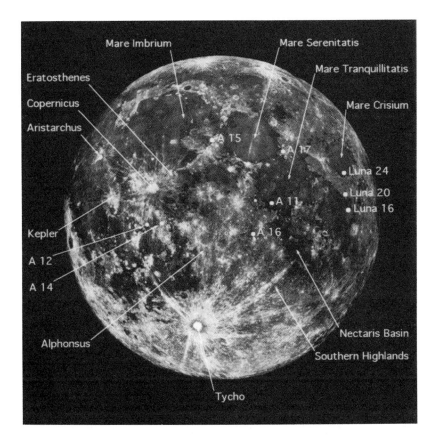

Figure 15. Lunar nearside showing the locations of Apollo and Luna landing sites, and some prominent features.

In the literature the most common interpretation of highland rock ages suggests that large impacts have dominated over local events. The age distribution reflects the putting back of the radiometric clock by the formation of the large lunar ring basins. Due to the cumulated age frequency near 4 Gyr, Tera et al. (1974) concluded that the impact rate must have been extremely high during this period. They called this unusual high bombardment "terminal lunar cataclysm." Later analysis by various authors (see Neukum 1983, and references therein) demonstrate that there are obviously different groups of ages present in the data. However, the interpretation of age distributions and correlation with various impact basins is still controversial. Generally, two opinions exist (also see Basaltic Volcanism Study Project 1981):

(a) Formation of nearly all of the large impact basins of the Moon's near side within less than 100 Myr (advocated by supporters of the "terminal lunar cataclysm" theory).
(b) Continuous decrease of the impact rate and formation of the large basins between 4.4 and 3.8 Gyr ago.

The possibility of peaks of impact rate at certain times in the bombardment history of the Earth-Moon system remains controversial. We present arguments in favor of the theory that the impact rate (including the large basins) has undergone a steep but continuous decrease between 4.4 and 3.8 Gyr ago and that the "terminal lunar cataclysm" is neither supported by the radiometric age data nor by the geologic observations. The age data are interpreted in such a way, that the peak of the age distribution nearly reflects the putting back of the radiometric clock due to a large impact event. This is controversial to a number of other authors' thinking (see Jessberger 1981; Basaltic Volcanism Study Project 1981) that the age of an event is given by the youngest radiometric age obtained from totally molten inclusions in the rock. We do not follow this interpretation because local impacts ensuing the basin forming events may have contributed to impact metamorphism and could also have lowered the radiometric age. We interpret the youngest melting processes to be at least partly local effects.

A more detailed investigation of the Apollo 16 rocks reveals two age frequencies (Maurer et al. 1978). We interpret the peak at 4.1 Gyr to be the age of the Nectaris ejecta at the Apollo 16 landing site. A similar interpretation is given by Wetherhill (1981).

The correlation between the ages of highland rocks and crater frequencies is based upon the ages and crater frequencies of the lunar landing sites. However in some cases genetic relations must be taken into account, for example the crater frequencies measured on Nectaris material which has been deposited far away (i.e., on the basin rim). The corresponding age is extracted from Apollo rock ages with respect to the genetic relations (e.g., age of the Nectaris basin obtained from Apollo 16 data).

The crater frequencies of crater North Ray (Apollo 16 landing site) and Cone (Apollo 14 landing site) have been measured on their ejecta blankets (König 1977; Moore et al. 1980). The ages obtained are exposure ages of the ejecta material and show values of 50 and 24 Myr, respectively (Drozd et al. 1974).

A summary of lunar crater frequencies and assigned ages for mare and highland regions for the determination of the lunar cratering chronology is given in Table I. In addition we list the production frequency of terrestrial phanerozoic craters (Northern America and Northern Europe), recalculated to lunar impact conditions, and the assigned age of the craton (data from Grieve and Dence [1979] reinterpreted).

The lunar impact chronology is empirically derived from the correlation between crater frequencies and radiometric ages. The correlation is taken from Neukum (1983) and is nearly identical with the correlation of Neukum (1977) and Neukum and Wise (1976).

The mathematical expression of the cratering chronology curve for lunar impact conditions is

$$N(1) = 5.44 \times 10^{-14}[\exp(6.93 \times t) - 1] + 8.38 \times 10^{-4}t. \qquad (17)$$

$N(1)$ is the cumulative crater frequency per km^2 at $D = 1$ km; t is the age in units of 10^9yr.

Using this correlation displayed in Fig. 16 (solid line), it is possible to obtain absolute ages for any surface of the whole Moon by measurements of crater frequencies from imaging data.

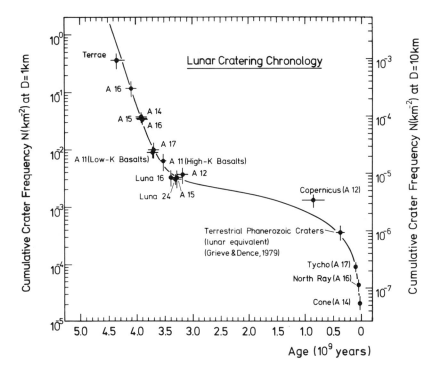

Figure 16.   Lunar cratering chronology (Neukum 1983).

Figure 16 gives the dependence of N on t for $D = 14$ km and $D = 10$ km. The general dependence on crater diameter is defined by the lunar

TABLE I

Crater Frequencies of Lunar Structures and Terrestrial Phanerozoic Craters with Assigned Absolute Ages[a]

Lunar or Terrestrial Strucutre	$N(D=1)$ (km^{-2})	$N(D=10)$ (km^{-2})	Age Gyr
Highland (Terrae)	$(3.6\pm1.1)\times10^{-1}$	9.2×10^{-4}	4.35 ± 0.10
Nectaris Basin (A16)	$(1.2\pm0.4)\times10^{-1}$	3.1×10^{-4}	4.10 ± 0.10
Serenitatis Basin (A17)			3.98 ± 0.05
Descartes Formation (A16)	$(3.4\pm0.7)\times10^{-2}$	8.7×10^{-5}	3.90 ± 0.10
Imbrium Basin (A15)	$(3.5\pm0.5)\times10^{-2}$	8.9×10^{-5}	3.91 ± 0.10
Fra Mauro Formation (A14)	$(3.7\pm0.7)\times10^{-2}$	2.6×10^{-4}	3.91 ± 0.10
Taurus Littrow Mare (A17)	$(1.0\pm0.3)\times10^{-2}$	9.4×10^{-5}	3.70 ± 0.10
Mare Tranquillitatis (A11)	$(9.0\pm1.8)\times10^{-3}$	2.3×10^{-5}	3.72 ± 0.10
Mare Tranquillitatis (A11)	$(6.4\pm2.0)\times10^{-3}$	1.6×10^{-5}	3.53 ± 0.05
Mare Imbrium (A15)	$(3.2\pm1.1)\times10^{-3}$	8.2×10^{-6}	3.28 ± 0.10
Oceanus Procellarum (A12)	$(3.6\pm1.1)\times10^{-3}$	9.2×10^{-6}	3.18 ± 0.10
Mare Fecunditatis (Luna 16)	$(3.3\pm1.0)\times10^{-3}$	8.4×10^{-6}	3.40 ± 0.04
Mare Crisium (Luna 24)	$(3.0\pm0.6)\times10^{-3}$	7.6×10^{-6}	3.30 ± 0.10
Copernicus (A12)	$(1.3\pm0.3)\times10^{-3}$	3.3×10^{-6}	0.85 ± 0.20
Tycho (A17)	$(9.0\pm1.8)\times10^{-5}$	2.3×10^{-7}	0.109 ± 0.004
North Ray (A16)	$(4.4\pm1.1)\times10^{-5}$	1.1×10^{-7}	0.0500 ± 0.0014
Cone (A14)	$(2.1\pm0.5)\times10^{-5}$	5.3×10^{-8}	0.0260 ± 0.0008
Phanerozoic Crater (Northern American and Europe, lunar equivalent)	$(3.6\pm1.1)\times10^{-4}$	9.2×10^{-6}	0.375 ± 0.075

[a] Data from Neukum 1983.

production crater size frequency distribution (the lunar standard distribution). Any equivalent frequencies for any crater diameter can be derived from this empirical relationship in converting the values for 1 km or 10 km to other diameters through application of the lunar standard size distribution curve.

The cumulative impact rate Φ is given by the time derivative of the impact chronology relationship, $\partial N / \partial t = \Phi$ for $D = 1$ km:

$$\Phi(1) = 3.77 \times 10^{-13} \exp(6.93 \times t) + 8.38 \times 10^{-4} \ (\text{km}^{-2} 10^{-9} \ \text{yr}^{-1}) \ (18)$$

with t inserted in units of 10^9 yr.

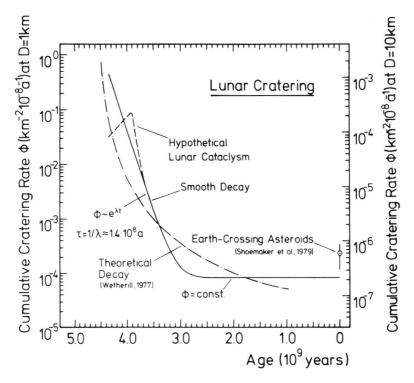

Figure 17.   Time dependence of the lunar cratering rate (Neukum 1983).

This functional dependence is shown in Fig. 17 in units of $\text{km}^{-2} 10^{-8} \text{yr}^{-1}$ in comparison with other interpretations (e.g., "hypothetical terminal cataclysm"). There is an exponential decrease of the impact rate between 4.5 and 3.5 Gyr ("smooth decay") with a time constant $\tau = 1/\lambda = 1.44 \times 10^8$ yr and a half-life value $T_{1/2} = 1.0 \times 10^8$ yr, respectively. For $t < 3$ Gyr, a constant value of the cumulative impact rate of $\Phi = 8.38 \times 10^{-5} \ (\text{km}^{-2} 10^{-8} \text{yr}^{-1})$ results. For $t < 1$ Gyr, the constancy of the impact rate is rather well defined in this region. In former investigations (Hartmann 1972; Neukum et al. 1975;

Neukum and König 1976; Shoemaker et al. 1979) data for $t < 1$ Gyr seemed to indicate a higher impact rate than the average over the last 3 Gyr.

Particularly the terrestrial data (phanerozoic craters) have been systematically too high to fit with the impact rate which has been constant over the last 3 Gyr. After recalculating the terrestrial frequencies extracted from investigations of Grieve and Dence (1979) to lunar impact conditions, Neukum (1983) found them to fit well with the lunar data as given in Fig. 16. Only the value for Copernicus appears somewhat too high. The value of the crater frequency for Copernicus is assured several times by measurements but it remains possible that the interpretation of the age of Copernicus, based on Apollo 12 data, is wrong.

The adjustment of the terrestrial phanerozoic data to lunar impact conditions by Neukum (1983), however, has been done by applying a simple scaling law where gravity scaling and post-impact modification of crater walls have not been considered. Application of rigorous scaling of lunar to terrestrial impact conditions, as done below in Sec. VII.B, again leads to a discrepancy of a factor of 1.5 to 2.0 between the terrestrial and the lunar data in terms of derived impact rates for recent times, the terrestrial data lying systematically higher by that amount.

VI. INFLUENCE OF CRATERING MECHANICS ON SIZE DISTRIBUTIONS

In order to compare crater size distributions and cratering rates on different planets in a rigorous way, we need to take into account many parameters such as gravity, atmospheric characteristics, crustal strength, density and structure as well as differences in the projectile flux, size-frequency distribution and impact velocity spectrum of the impacting bodies. As we do not have at hand all the information listed above, we can only make estimates based on certain assumptions.

In the following subsections we explore the influence of cratering mechanics on the size-frequency distributions for different planets.

A. General Approach

According to experimental and theoretical results for a given planetary body, the frequency distributions of crater diameters may be divided into several specific ranges, where different target parameters control the relationship between projectile mass and velocity and the final crater dimensions. For smaller craters one may suppose a *strength* regime of cratering, in which the final crater dimensions are governed by projectile parameters and the target material strength. Larger craters are formed in a *gravity* regime, in which the energy needed to uplift target material in a planetary gravity field is larger than the energy dissipated due to plastic work. The largest craters experience *structural modifications* after a transient cavity formation; modification processes give rise to craters with central peaks, multi-ring basins and other forms of *complex craters* and *basins*. The boundary diameters of the above listed regimes of cratering vary from planet to planet, and the ratio, for example, of crater rim diameter to projectile diameter is not a constant even for the same projectile velocities.

The goal of this section is to present some model comparisons of crater size-frequency distribution curves on various terrestrial planets. We start by outlining the role of cratering mechanics.

B. Scaling Relations: Vertical Impact

During the last 20 yr much progress has been made in our understanding of the cratering scaling relations. Reviews of the main ideas of scaling are given by Melosh (1989), Schmidt and Housen (1987) and Holsapple (1987).

Here we use the thoroughly elaborated approach of Schmidt and Housen (1987), which is partially supported by their experimental data from high-G centrifuge modeling. The approach is based on the concept of a *coupling parameter* which is a combination of projectile parameters and target density. The coupling parameter allows a comparison of the cratering efficiency to be made for various impact conditions. To introduce the coupling parameter here we shall try to avoid the highly formalized π-theorem (presented by Schmidt and Housen (1987), which uses scaling relations for dimensional parameters. Let us use the coupling parameter in the form:

$$CP = a(\delta/\rho)^\nu v^\mu f(\vartheta) \tag{19}$$

where a is projectile diameter, δ is projectile density, ρ is target density, v is impact velocity, ν and μ are experimentally derived exponents, $f(\vartheta)$ is a function of oblique impact efficiency which depends on the angle of impact ϑ.

For *strength craters* Schmidt and Housen (1987) use the scaling relation in the form:

$$\frac{\rho V}{m} = A_1 \left(\frac{\rho}{\delta}\right)^{1-3\nu} \left(\frac{Y}{\rho v^2}\right)^{-3\mu/2} \tag{20}$$

where m is the projectile mass, V is crater volume, Y is an effective target material strength, A_1 is a coefficient of proportionality, the numerical value of which depends on the form of the presentation.

Here we rewrite this equation collecting all the parameters of a projectile into a single value of a coupling parameter (Eq. 19):

$$\frac{V}{CP^3} = A_V \left(\frac{Y}{\rho}\right)^{-3\mu/2} . \tag{21}$$

Assuming the geometric similarity of all craters, the crater radius R is proportional to the cube root of volume, $V^{1/3}$. For such similar craters Eq. (20) may be rewritten in the form:

$$\frac{R}{CP} = A_2 \left(\frac{Y}{\rho}\right)^{-\mu/2} \tag{22}$$

where CP is a coupling parameter (Eq. 19).

For *gravity craters* the scaling relationship by Schmidt and Housen is:

$$\frac{\rho V}{m} = B_1 \left(\frac{\rho}{\delta}\right)^{\frac{2+\mu-6\nu}{2+\mu}} \left(\frac{ga}{v^2}\right)^{\frac{-3\mu}{2+\mu}} \tag{23}$$

where g is the gravitational acceleration and B_1 is a numerical constant.

Making the same rearrangement as before, we derive:

$$\frac{V}{CP^3} = B_V \left(Vg^3\right)^{-\mu/2}. \tag{24}$$

For geometrically similar craters where R is proportional to $V^{1/3}$, Eq. (20) may be rewritten as:

$$\frac{R}{CP} = B_2(Rg)^{-\mu/2}. \tag{25}$$

Equations (22) and (25) have the same functional form; all projectile related parameters are collected in a single coupling parameter and the right side of these equations depends only on the target parameters.

For given effective strengths of target and planetary gravity, we may define a crater radius R_{sg} for which the ratio R/CP will be the same for strength and gravity presentations:

$$\frac{R_{sg}}{CP} = A_2 \left(\frac{Y}{\rho}\right)^{-\mu/2} = B_2 \left(R_{sg}g\right)^{-\mu/2}. \tag{26}$$

For a given Y and g, Eq. (26) defines an effective boundary between the gravity and strength regimes of cratering. For crater radii around R_{sg}, strength and gravity have comparable influence on the determination of final cratering dimensions, so one needs to combine Eqs. (22) and (23) to make the strength-gravity transition in a smooth and more realistic way. The simplest approach is to write a generalized expression, using the exponent $-\mu/2$:

$$\frac{R}{CP} = \frac{B_2}{(A_3Y/\rho + Rg)^{\mu/2}} \tag{27}$$

where A_3 and B_2 are "evolved" constants from Eqs. (22) and (25). A similar approach has been proposed earlier by Holsapple and Schmidt (1979). The strength member A_3Y/ρ may be rewritten in the form:

$$R_{sg}g = A_3Y/\rho.$$

With this definition Eq. (27) may be rewritten thus:

$$\frac{R}{CP} = \frac{B_2}{[g(R_{sg} + R)]^{\mu/2}} \tag{28}$$

for crater diameters:

$$\frac{D}{CP} = \frac{B_D}{[g(D_{sg} + D)]^{\mu/2}} \qquad (29)$$

and for crater volumes:

$$\frac{V^{1/3}}{CP} = \frac{B_V}{[g(V_{sg}^{1/3} + V^{1/3})]^{\mu/2}}. \qquad (30)$$

Figure 18 illustrates the geometrical meanings of Eqs. (27–30), showing the smooth transition from the strength to the gravity regime of cratering. The influence of gravity begins to be noticeable at crater diameters larger than $0.25D_{sg}$; the influence of strength may be important up to $D \approx 4D_{sg}$. The transition from simple to complex craters, also shown in Fig. 18, is discussed later.

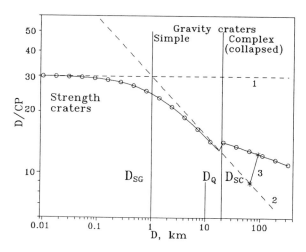

Figure 18. The ratio of crater diameter to the coupling parameter (in arbitrary units) as a function of crater diameter. Regions of different regimes of cratering are designated. Dashed lines correspond to the pure strength (1) and gravity (2) scaling. Thin vertical lines show the strength/gravity effective boundary diameter D_{sg} (the diameter of intersection of strength and gravity scaling laws 1 and 2), and the diameter of simple/complex crater transition. Line 3 with asterisks shows the "trajectory" of a crater diameter increasing due to gravity collapse of a transient cavity. The diameter of intersection of gravity and modified scaling relationships, D_Q, may be smaller than D_{sc}. The scale of D corresponds approximately to the cratering on the Moon.

Equations (29) and (30) may be tested using experimental data obtained from explosion and impact cratering. Figures 19a,b show the data by Holsapple and Schmidt (1979) in comparison with the data on surface and shallow burst explosion cratering. Here the cube root of the explosion energy is used as an analog of a coupling parameter.

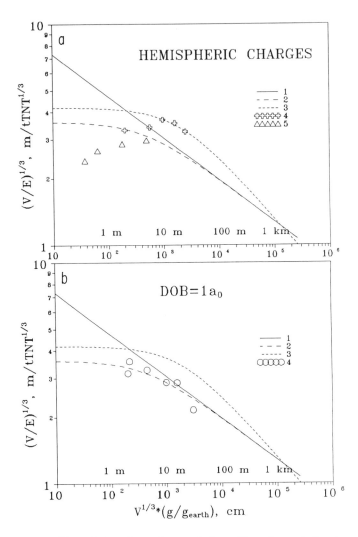

Figure 19. Scaling relations for surface explosions. The cube root of crater volume $V^{1/3}$ is used as a linear scale to measure cratering efficiency. The cube root of energy, measured in tons of TNT is used here as an explosion analog of a coupling parameter. (a) Surface burst. Solid and dashed lines are approximations to the centrifuge explosion data of Holsapple and Schmidt (1979) in the form of Eq. (30): 1 - dry sand; 2 = alluvium; 3 = oil-based clay. The large scale explosion data from Vortman (1968) are shown for comparison: 4 = craters at Watching Hill Site; 5 = craters at Drowing Ford Site. No correction for different coupling for field and laboratory explosions has been done. (b) Craters from high explosion and nuclear bursts in alluvium at the depth of burst equal to one TNT charge radius a_0 Lines are the same as in (a). Large scale explosion data are shifted down to fit the laboratory curve for alluvium (2). The scaled volume definitely decreases with the scale of event reflecting a transition from strength to gravity regime of cratering.

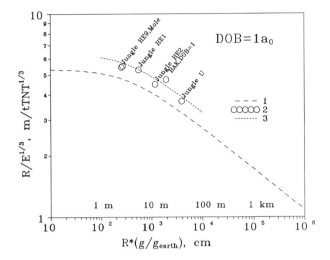

Figure 20. Scaling of crater radii for the same experiments as in (a) (points 2). Experimental point "BAK" is an average of two centrifuge experiments with shallow bursts (Schmidt 1978). The dashed line 1 is the approximation to the data of Holsapple and Schmidt (1979) for contact bursts in alluvium. The dotted line 3 is the same relationship shifted up to take into account the higher energy coupling of shallow buried explosions. As in the previous figure, the data illustrate a transition from the strength to gravity regime of cratering in the vicinity of crater radii of 5 m (diameter of 10 m).

Presentation of data in the form Eqs. (27–30) for a given impact or explosion source allows us to choose a proper value of relative efficiency, fitting data to the same D/CP (or $D/E^{1/3}$, or $(V/E)^{1/3}$) values simply by "moving" data points up and down on the logarithmic plot. The strength-to-gravity transition in this case depends only on target strength and gravity. In Fig. 20 data for shallow buried explosions in alluvium (2) are approximated with the curve (1) derived from centrifuge data of Holsapple and Schmidt (1979), and shifted up (3) to take into account the higher energy "coupling" efficiency for shallow bursts.

Numerical values of the strength-gravity boundary diameter D_{sg} are still poorly known. Data from centrifuge experiments by Holsapple and Schmidt (1979) presented in the form of Eq. (29) give $D_{sg} = 6\pm2$ m for the alluvium model and $D_{sg} = 100\pm50$ m for the oil-based clay model. Cratering data for shallow buried high energy explosions in alluvium give $D_{sg} = 10\pm5$ m at DOB (depth of burst) equal to one charge radius (Fig. 20). High energy explosions at the surface of clay soils (Fig. 19a) produced craters, the scaled volumes of which increased with the size of the event. The strength-gravity transition here may be roughly estimated to be in the diameter range of 50 to 100 m.

Although data for hard rock are very limited, they indicate a decrease of effective strength as the size of events increases (Schmidt 1980). A value of D_{sg} of 100±50 m seems to be a reasonable first approximation on Earth. Numerical simulation by Bryan et al. (1978) showed that gravity played an important role during the formation of Meteor Crater, Arizona (D = 1.2 km). So the value of D_{sg} seems to be well below 1 km.

Housen et al. (1983) have shown that all simple gravity craters have geometrically similar rim profiles and maximum rim heights h linearly proportional to crater diameters. (The same conclusion was published in Russian earlier by Ivanov [1979].) Based on the linear relationship $h(D)$ one may conclude that the value of D_{sg} may be as low as 300 m on the Moon (lowest diameter used by Pike [1977]) and 600 m on Mars (lowest diameter used by Pike and Davis [1984]). The same range of the lunar D_{sg} value (about 300 m) has been estimated by Moore et al. (1974) from the radial extent of continuous ejecta (see also discussion in Croft [1985]). Scaled to the terrestrial gravity (Eq. 31, below), the above mentioned values correspond to D_{sg} values of 50 m which seem to be too small for rocky surfaces. Taking into account the gradual character of the strength-to-gravity transition, and a wide scattering of rim height measurements both from the Moon and Mars, it is hard to say how accurate an estimate based on linear $h(D)$ relations is. At the same time dynamic weakening of rock during a crater formation may lead to relatively small *effective* values of strength in comparison to the strength of single rock fragments.

The effect of intense fracturing due to meteoroid bombardment (on the Moon) and near-surface weathering (on Mars) may result in small apparent values of target strength, which gradually increase with depth. If so, our model with a single value of effective strength would be too simple.

The upper limit of the effective strength may be deduced from the model of the crater collapse. Melosh (1989) estimates the apparent strength (cohesion) of planetary crusts at around 3 MPa (30 bar) for all terrestrial planets. Note that some collapse models treat the strength not as a target material constant but as a parameter dependent on the strain rate. The model used here is the simplest end-member model, which needs to be elaborated much more thoroughly in the future. Here we use the lunar value of $D_{sg} \geq 300$ m in gravity scaling to other planets (Eq. 31).

C. Scaling for Modified (Collapsed) Craters

Above some critical diameter D_{sc} impact craters on all planetary bodies have so-called complex morphology (central peaks, concentric rings, etc.). Formation of complex craters seems to be connected with a modification of a transient cavity in the gravity field of a planet. The gravity collapse of a transient cavity makes the diameter of a final modified crater larger in comparison with a hypothetical simple crater which would be created by the same impact in the absence of the gravity collapse. While mechanical models of crater modification are not yet well established (see, e.g., Melosh 1989), geologic and morphologic data allow us to estimate, more or less reliably, the hypothetical equivalent simple crater diameter D_{s_eq}:

$$D_{s\_eq} = D_Q^{(1-\omega)} D^\omega \qquad (31)$$

where D_Q is a coefficient of linear dimension and D is the rim diameter of the final (observed) crater. Croft (1985) gives the value of the exponent summarizing a number of geologic data as:

$$\omega = 0.85 \pm 0.04. \qquad (32)$$

The recent model by Holsapple (1993) proposes a value of ω about 0.9, not far from the upper limit given by Croft (1985).

Numerically the value of D_Q is a boundary diameter of intersection of various relationships for simple and complex craters. It may not coincide with a critical diameter D_{sc} which separates the ranges of observed simple and complex crater morphology. Typically, the value of D_Q is 1.5 to 2 times smaller than D_{sc}.

In order to calculate the scaling relationship for modified craters one needs to replace the value of D in Eq. (24), for a simple crater, with the value of D_{s_eq}, defined with Eq. (31).

A general review of the simple-to-complex transition has been published by Pike (1980). Most of these data reflect boundary diameters between ranges of different morphology styles of impact craters. It is hard to say what is the best value of D_Q that should be used in the Eq. (31) in order to estimate the crater diameter increase due to modification on a given planetary body. As a first approximation for terrestrial planets one may use the inverse proportionality of D_{sc} and D_Q to gravity g (Pike 1980).

Values used in this chapter are collected in Table II. For Venus the simple-to-complex crater transition is not well defined: all craters larger than approximately 10 to 12 km are "normal" complex craters, and smaller craters are formed with fragments of projectiles destroyed in the dense Venusian atmosphere (Schaber et al. 1992).

VII.   COMPARISON OF SIZE-FREQUENCY CURVES AND CRA-TERING RATES—INTERPLANETARY CORRELATIONS

Using the scaling of impact craters as described above we can estimate diameters of impact craters formed by the same projectile on different planets. As we choose the lunar crater size-frequency standard distribution as a basis for interplanetary comparison, let us derive the ratio of crater diameters on the Moon D_m and on a given planet D_x given the same projectile diameter.

The scheme of the comparison is as follows:

Step 1. Choose the crater diameter D_x for a specific planet x;

TABLE II
Boundary Crater Diameters for Simple-to-Complex Transition[a]

Planet	Gravity (m s^{-1})	D_Q (km)	Parameter Related to Crater Rim Diameter	D_{sc} (km)
Moon	1.62	10	Depth	19
		13	Interior volume[b]	
		20	Rim height	
Mercury	3.63	9	Depth	16
		16	Rim height	
Mars	3.74	3.1	Depth	6
		4.1	Rim height[c]	
Earth	9.81		Crystalline target	4[d]
		2[e]	True depth	
			Sedimentary target	2.5[d]
		<1[e]	True depth	

[a] Most data from Pike (1980).
[b] Croft (1978).
[c] Pike and Davis (1984).
[d] Largest simple crater (Brent crater; Grieve et al. 1981).
[e] Intersection of true depth/diameter curves (Grieve et al. 1981).

Step 2. If $D_x > D_{sc\_x}$ (modified crater), calculate the diameter of an equivalent simple craterD_{s_eq} (Eq. 31);

Step 3. Using the scaling law for simple craters (Eq. 24), calculate the CP/B_D ratio;

Step 4. Calculate the coupling parameter ratio on the Moon and on planet x using the impact velocities and projectile/target density ratios. To simplify the problem in this chapter, we use the same ratio of target/projectile densities for all terrestrial distributions; for each planet (still poorly defined) we use one ("average" or "typical") value of impact velocity for every planet. These values correspond to averaged impact values of planet-crossing asteroids from the main belt ($e = 0.6$) derived earlier by Horedt and Neukum (1984). The same value will be used below to compare gravity enhancement of cross sections of different planets (see Table III below). Parameters adopted here are close to values previously used by Hartmann (1977);

Step 5. Calculate the simple crater diameter on the Moon formed with the same projectile (but with a proper impact velocity);

Step 6. If the diameter of a simple lunar crater is greater than D_{sc_Moon}, calculate the complex crater diameter (Eq. 31).

Figures 21a–d show the ratios of D/D_{Moon} for Mercury, Mars, Earth and

TABLE III

Impact Conditions for Terrestrial Planets[a]

Object	Impact Velocity V_i (km s^{-1})	Escape Velocity V_{esc} (km s^{-1})	Approach Velocity u_∞ (km s^{-1})	Gravity Focusing[b] f_p	f_p/f_{Moon}
Moon	14.1	2.38	13.9	1.03	1.0
Mercury	23.6	4.2	23.2	1.03	1.0
Venus	19.3	10.3	16.32	1.4	1.36
Earth	17.8	11.2	13.83	1.66	1.61
Mars	12.4	5.0	11.35	1.19	1.16

[a] Horedt and Neukum (1984).

[b] The ratio of the gravity cross section to the geometric one for a given approach velocity u_∞.

Venus (for the last two planets without correction for the atmosphere) as a function of D. In order to compare size-frequency distributions we need to make one more step:

Step 7. Using the production function (Eq. 10); cf., Fig. 22, calculate the number of craters with diameters equal to and larger than D_{Moon}, N_{DMoon}.

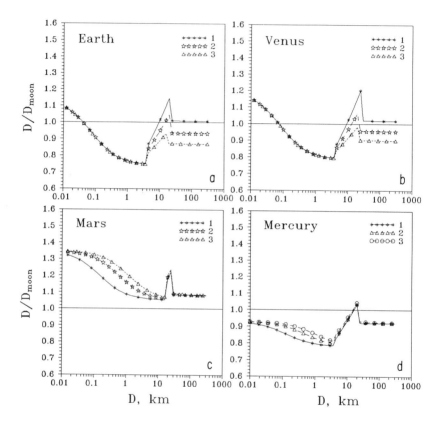

Figure 21. Comparison of crater diameters on various planets for projectiles of the same diameter. Changes due to surface gravity and different impact velocity are taken into account in the presented model. For Earth and Venus the practically useful range of diameters lies above $D = 1$ km. (a) Earth; $D_{sg\_Moon} = .3$ km; values of ω: $1 = 0.81$; $2 = 0.85$; and $3 = 0.89$. (b) Venus; $D_{sg\_Moon} = .3$ km; values of ω: $1 = 0.81$; $2 = 0.85$; $3 = 0.89$. (c) Mercury; $\omega = .85$; values of D_{sg_Moon} are: $1 = 0.3$ km; $2 = 1.2$ km; $3 = 2.4$ km. (d) Mars; $\omega = .85$; values of D_{sg_Moon} are: $1 = 0.3$ km; $2 = 1.2$ km; $3 = 2.4$ km.

As the "input" crater on planet x is formed with the same projectile population, the cumulative number of craters on the planet x will be equal to:

$$N > D = \left(\frac{F_x}{F_m}\right)\left(\frac{T_x}{T_m}\right) N_{DMoon} \tag{33}$$

where F_x/F_m is the projectile flux ratio and T_x/T_m is the relative age of the investigated area. Traditionally some estimates of the F_x/F_m ratio are used to evaluate a crater retention age T_x.

The projectile flux ratio depends on the projectile flux "at infinity" (beyond the gravity field of the target planet) and upon the gravitational focusing effect which increases as the projectile velocity "at infinity" becomes small in comparison with the escape velocity of a target planet. In this chapter we compare size-frequency curves for:

$$\left(\frac{F_x}{F_m}\right)\left(\frac{T_x}{T_m}\right) = 1. \tag{34}$$

In this way we may estimate the geometry of the crater production curve but cannot draw any conclusions about relative crater retention ages of planetary surfaces. The astronomically derived F_x/F_m ratio is a very important and interesting issue for comparative planetology but is beyond the scope of this discussion.

Some data for terrestrial planets, and two asteroids (Gaspra and Ida), are summarized below. In each case the lunar crater production size-frequency distribution curve (standard distribution; Neukum 1983) is used as the basis for interplanetary comparisons.

The result of this comparison is the *relative crater production rate*, namely the ratio of production rates of craters with the *same diameter D* on a test planet and on the Moon, assuming the same projectile flux at infinity. In the approach adopted here, the relative crater production rate takes into account (within the limits of the models used) all particulars of the crater formation process, average impact velocity difference, and gravity focusing. The reader need only choose an appropriate value of the projectile flux at infinity to compare absolute values of crater production rate. An improvement of our model, as a future task, can be the incorporation of an impact velocity spectrum which leads to some modification of the gravity focusing effect (slow projectiles will have larger collisional probability cross sections than fast projectiles). The model results presented here, however, are considered good approximations to the more general problem.

A. The Moon

We use the generalization of lunar data in the form of the cumulative crater production size-frequency distribution, approximated by a polynomial of 11th degree as treated in Sec. III:

$$logN = a_0 + \sum_{k=1}^{11} a_k (logD)^k \tag{35}$$

Figure 22 demonstrates the crater production size-frequency distribution curve (1) in comparison with the estimated distribution of hypothetical strength craters, which would be formed with the same projectiles in the absence of gravity. Diameters of strength craters are linearly proportional to the projectile diameter (for a given impact velocity and an angle of impact), so the size distribution of hypothetical strength craters describes approximately the size-frequency distribution of projectiles.

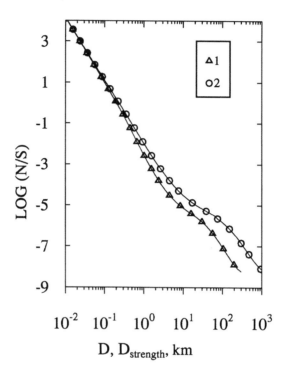

Figure 22. The calibration crater cumulative production curve (1) for the Moon. The calculated hypothetical strength crater diameter, $D_{strength}$, for every lunar crater (2) becomes larger as gravity decreases crater dimensions. As $D_{strength}$ is proportional to the projectile dimension (for a given velocity and angle of impact) this plot gives an indication of the projectile size-frequency distribution. The strength/gravity transitional diameter for this plot is 300 m. The vertical scale uses arbitrary units.

B. Earth

The relative cratering rate for the Earth as a function of crater diameter is shown in Fig. 23 for probable variations of the exponent in Croft's (1985) modification model $\omega = 0.085\pm0.15$. To an accuracy of 15% the terrestrial relative cratering rate is the same as on the Moon for craters larger than 10 km, where higher terrestrial gravity is compensated for a higher level of transient crater modification and a larger impact velocity on the Earth. For $D<10$ km the terrestrial relative cratering rate decreases down to 0.5 at $D = 1$ km, whereas on the Moon and on Earth impacts form simple (non-modified)

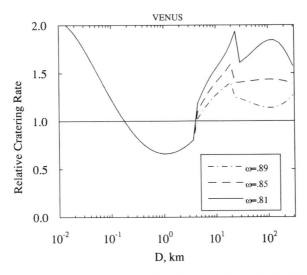

Figure 23.   The relative cratering rate on Earth in comparison with the Moon (number of craters larger than a particular diameter for equal crater retention age). Model dependence for different exponent w in the collapse Eq. (31) are shown.

gravity craters.

The fitting of the lunar crater size-frequency curve to the data for terrestrial cratons is illustrated in Fig. 24. Only one correction to the list of terrestrial craters by Grieve and Shoemaker (see their Chapter) has been done; according to recent research, the original rim diameter of Puchezh-Katynki crater in Russia is estimated to be 40 km, not 80 km as assumed previously (Pevzner et al. 1992). The age of the structure is about 180 Myr, and not 220 Myr (V. L. Masaitis, personal communication, 1993).

The model crater retention age, based upon the lunar chronology, for terrestrial cratons lies in the range 700 ± 100 Myr (the uncertainty of 100 Myr is related only to variations of parameters in the crater collapse model and does not include the accuracy of the lunar and terrestrial dating itself). The age is approximately twice as large as the age of the crater accumulation at North American and European cratons, estimated to be about 375 Myr (Grieve and Dence 1979). This is a long-known problem (see, e.g., the Chapter by Grieve and Shoemaker). Neukum (1983) found coinciding values for the lunar and terrestrial data after applying a less rigorous scaling law for reducing the terrestrial data to lunar conditions (cf., Sec. V). The rigorous treatment, as shown here, yields the discrepancy again. Possible explanations for this discrepancy are: (i) low accuracy of the crater retention age. Standard deviations are typically of the order of $\pm50\%$ and error bars for lunar and terrestrial cratering rates overlap; (ii) there may have been approximately a two-fold increase (peak) in the crater-forming projectile flux in recent times, in comparison with the average flux over the last 3 Gyr; Here we discuss

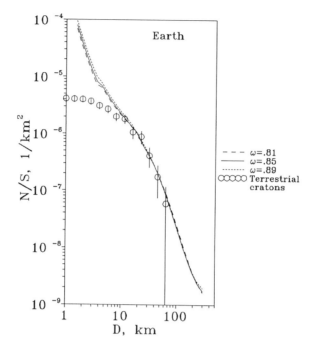

Figure 24. Cumulative size-frequency distribution for craters on terrestrial cratons approximated with the model recalculation lunar curve. Note that the model curve deviates from a simple power law, showing that craters as small as 10-km may be used to estimate the terrestrial distribution.

briefly the second possibility using the Moon/Earth comparison of crater size-frequency distributions.

Previously Grieve and Dence (1979) and Grieve (1984) ascribed the deficit of craters with $D < 20$ km, compared to the $D^{-1.8}$ cumulative distribution, to the erosion of small craters. The crater production size-frequency curve, believed to be the proper curve and recalculated for terrestrial conditions, deviates from the $D^{-1.8}$ power law, thus, in the accuracy limits of the model, the relative deficit of craters is observed only for $D < 10$ km.

Figure 25 illustrates the dependence of the cumulative number of craters older than a given age at terrestrial cratons for diameters $D \geq 20$ km. Grieve (1984) proposed that the lower-appearing accumulation rate for $T > 120$ Myr may be due to the erosion of small craters. If one supposes that a peak in the cratering rate occurred at about 100 Myr ago, then the data could be approximated as a sum of a constant "background" cratering rate and an exponentially decaying "burst" component. An average constant cratering rate, though, would be compatible with the data within their error limits as shown in Fig. 25 by the dashed line fitting the data.

Independent of the lunar cratering rate estimates, we can use the recalculated lunar cratering curve to extrapolate the size-frequency data for terrestrial

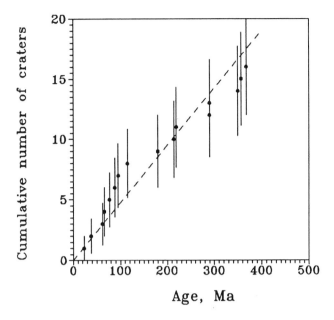

Figure 25.   Cumulative number of craters with $D \geq 20$ km on terrestrial cratons vs age.

cratons to the diameter $D = 1$ km. This gives an areal density of cratering events of 4×10^{-4}km^{-2} for the last 400 Myr. For the whole area of the Earth's surface of 5×10^{8}km^{2}, the estimated value suggests that about 200,000 bodies struck the Earth during this time. The energy release for every event was between 5 Mt TNT (2×10^{16}J) for 1-km craters. On average, 50×50 km^{2} area "cells" on Earth have been hit at least once by an asteroidal or cometary projectile which would have formed a crater ≥ 1 km diameter if the effect of the atmosphere were neglected. The average time between subsequent 1-km crater formation events is about 1600 yr for the surface of the whole Earth; for continents the time span would be about 6000 yr.

C. Venus

Figure 26 shows the cratering rate on Venus relative to that on the Moon for equal fluxes at infinity and Fig. 27 displays the fitting of the lunar cumulative curve (recalculated for Venus impact conditions) to the Magellan cratering data (Schaber et al. 1992). The well-known shielding effect of the Venusian atmosphere leads to a relative decrease in the crater number below $D = 30$ km. This is slightly smaller than the previous estimate by Ivanov (1990), who used a simple power law size-frequency distribution as an input in modeling the shielding effect. As the shielding effect depends upon the angle of impact (more oblique projectiles travel longer paths through the atmosphere), the correct model needs to take into account the angular distribution of impactors. So the fitting below 30 km cannot be done in the simplified model ("one impact velocity, one angle of impact") used here.

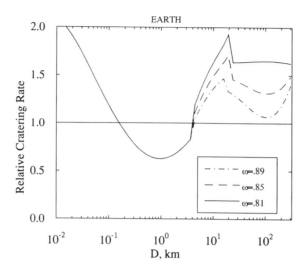

Figure 26.   Relative cratering rate on Venus. See explanation in caption of Fig. 24.

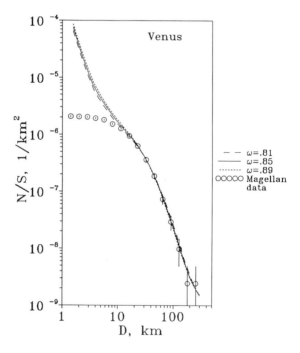

Figure 27.   Approximation to the Venusian cumulative crater size distribution (Magellan data) using the recalculated lunar curve. A deficit of Venusian craters smaller than 20 to 30 km reflects the shielding effect of the Venusian atmosphere.

The model age for Venus is 650±100 Myr (the uncertainty being due to cratering model parameter variations) assuming the same projectile flux "at infinity." We have here the same problem as for the Earth of possible time variations of the cratering rate. In addition, we need to correct the projectile flux in the vicinity of Venus. The current point of view is that the correction factor is probably not far from unity. While one cannot use direct age calculations for Venus, the fact that the crater areal density on Venus and on terrestrial cratons is approximately the same (Ivanov and Basilevsky 1987), suggests that the average age of the Venusian surface is about 0.5 Gyr. More detailed discussions of the Venusian chronology have been published by Schaber et al. (1992) and Phillips et al. (1992).

D. Mercury and Mars

Figures 28 and 29 illustrate relative cratering rates on Mercury and Mars. The main difference between these two planets is the much higher impact velocities for Mercury than for Mars (Table III). Small strength craters are twice as abundant on Mercury as on the Moon, while on Mars their number is small compared with the areal density of complex craters. The similarity of abundance of complex craters on the planets is connected mostly with the fact that both the impact velocity and the simple/complex crater transition diameter depend upon on the same parameter, namely the gravity of the planet.

E. Gaspra and Ida

The images of the asteroids Gaspra and Ida (Fig. 30) taken during the recent Galileo flybys show that their crater population is dominated by fresh craters. The small gravity fields of the asteroids allow us to assume that all these craters have been formed as strength ones. So the fitting of the lunar standard crater size curve to the Gaspra and Ida data depends mainly on the supposed strength/gravity transition on the Moon. The fitted curves for several values of D_{sg} are shown in Figs. 31a,b. Gaspra's data confirm that the steepness of the lunar standard crater size distribution is an effect of the primary impactor size-frequency distribution, and *not* an effect of secondary crater admixture at small sizes to a flat distribution with a supposed population index of -2 cumulative at large sizes. The data for Ida show a flattening of the distribution for sizes $D > 1$ km. At smaller sizes, the Ida crater population is in a state of cratering equilibrium, i.e., shows a -2 distribution which lies slightly lower in density than the equivalent equilibrium distribution expected for densely cratered areas of the Earth's Moon.

VIII. CONCLUSIONS

1. The complex shape of the lunar production crater size-frequency distribution curve (standard distribution; Neukum [1983]) cannot be made compatible with simple power-law distributions for the underlying impactor distribution

TABLE IV

Impact Accumulation Rates and Mean Time Intervals on Earth for Impacts of Specific Energies and Projectile or Crater Diameter Sizes

D^a (km)	L^b (m)	E_{kin}^c (J)	E_{kin}^c (Mt TNT)	N_{400}^d (km^{-2})	T_{mean}^e
0.01	1.60×10^{-1}	1.11×10^9	2.65×10^{-7}	7.659×10^3	1 day
0.1	2.10×10	2.20×10^{12}	5.26×10^{-4}	3.22×10^0	88 days
1	3.50×10^1	1.08×10^{16}	2.58×10^0	4.83×10^{-4}	1.60×10^3 yr
5	1.92×10^2	1.76×10^{18}	4.21×10^2	7.10×10^{-6}	1.10×10^5 yr
10	4.21×10^2	1.85×10^{19}	4.43×10^3	3.04×10^{-6}	2.58×10^5 yr
20	9.22×10^2	1.95×10^{20}	4.66×10^4	1.49×10^{-6}	5.25×10^5 yr
50	2.601×10^3	4.38×10^{21}	1.05×10^6	1.75×10^{-7}	4.48×10^6 yr
100	5.701×10^3	4.61×10^{22}	1.10×10^7	2.94×10^{-8}	2.67×10^7 yr
200	12.494×10^3	4.85×10^{23}	1.16×10^8	5.15×10^{-9}	1.52×10^8 yr
300	19.772×10^3	1.92×10^{24}	4.60×10^8	2.53×10^{-9}	3.10×10^8 yr

[a] D = crater rim diameter on an "atmosphereless" Earth.
[b] L = projectile diameter for $V_{imp} = 17.8$ km s^{-1} based on a scaling law according to Schmidt and Housen (1987).
[c] E_{kin} = projectile kinetic energy; 1 Mt TNT = 4.2×10^{15} J.
[d] N_{400} = frequency of craters which accumulate on an "atmosphereless" Earth over a time interval of 400 Myr.
[e] T_{mean} = mean interval between events with energies $\geq E_{kin}$.

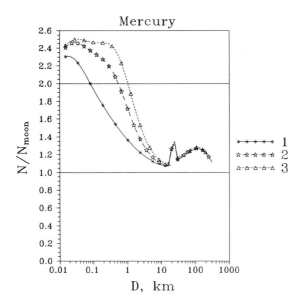

Figure 28. Relative cratering rate on Mercury. The effect of various values of D_{sg} are shown. Curves correspond to the lunar D_{sg} values: 1 = 300 m; 2 = 1.2 km; and 3 = 2.4 km. The larger number of small craters on Mercury reflects higher impact velocities (same projectiles form larger craters). Note that complex craters are only 20% more numerous than on the Moon for the same projectile flux.

taking into account the known scaling law for impact cratering. The exception is the range 300 m $<D<$ 4 km, where the value of the strength/gravity transitional diameter of 300 m gives a projectile size-frequency distribution very close to a simple power law with an exponent -3 (for the cumulative distribution). Other changes in the slope of the calibration curve are connected with peculiarities of the projectile size-frequency distribution and cannot be attributed to any cratering mechanics effect at different sizes.

2. The cratering rate for a specific planet relative to the Moon for the same flux at infinity in the diameter range 100 m to 1000 km for a given crater diameter may be changed by a factor of 2 due to effects of cratering mechanics and scaling laws. In the regime of large craters, where craters are complex (modified) both on a test planet and on the Moon, the cratering rate relative to the Moon is a constant close to unity to within 10 to 15%.

3. The apparent discrepancy of cratering rates on the Moon (averaged for 3.2 Gyr) and the Earth (for the last 0.5 Gyr) cannot be attributed to the effects of cratering mechanics alone. Reasons may be fluctuations in cratering rate or possible slight changes in the size distribution dependence with time or, more likely, a slight error in the shape of the standard distribution as determined for the Moon.

4. The lunar calibration distribution is a useful tool for approximating

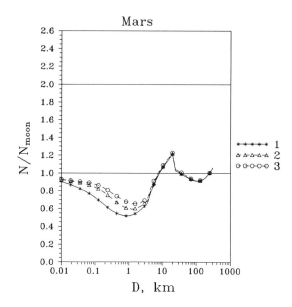

Figure 29. Relative cratering rate on Mars. Designations as in the previous figure for Mercury. Due to smaller impact velocities at Mars there may be a relative depletion of craters with diameters of the order of 1-km in comparison with complex craters. The situation is the inverse of that for Mercury, where craters with diameters of 0.1 to 1 km may be twice as numerous than on the Moon for the same level of complex crater areal density.

the size-frequency distributions of craters on other terrestrial planets provided the appropriate scaling corrections are made. This fact is strongly suggestive of one and the same family of impactors responsible for the cratering record on the terrestrial planets and the members of the asteroid belt. This, in turn suggests that the majority of these impactors are derived from the asteroid belt and that the contribution of cometary bodies is probably small. The steep increase in the distribution at sizes $D \leq 1$ km is confirmed to be due to primary craters and gives support to the idea that the impact hazard of smaller bodies on Earth is much larger than thought before.

5. The erosional limit of terrestrial craters may be twice as large (10 to 12 m) as was previously supposed (20 to 22 km) by comparison with a simple power law of a size-frequency distribution.

6. The shielding effect of the Venusian atmosphere may begin to be important at smaller crater diameters than was previously supposed by comparison with a simple power law of size-frequency distribution.

7. Through application of the lunar standard distribution for terrestrial conditions, current impact probabilities and recent impact rates or accumulation of craters of any size on an "atmosphereless" Earth for the last 400 Myr have been calculated for certain discrete diameters as shown in Table IV. For craters with $D \geq 1$ km, e.g., the resulting number is about 200,000 impact

Figure 30.   Images of the asteroids Gaspra (long axis dimension ≈17 km) and Ida (long axis dimension ≈55 km) as taken by the Solid State Imaging Experiment of the Galileo mission during flyby.

events for the surface of the whole Earth. Every area of the Earth's surface with average dimensions 50×50 km^2 has been hit by an impact event of these scales at some time in the past. The average time between subsequent 1-km crater formation events in recent time is about 1600 yr for the surface of the whole Earth and about 6000 yr for the continents. For 100-km craters, the corresponding number is one event every 27 Myr for the surface of the whole Earth.

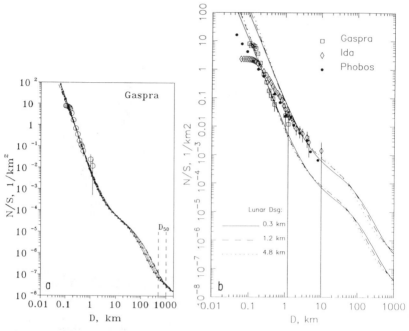

Figure 31. Cumulative crater size-frequency distributions measured on the asteroids Gaspra and Ida in comparison with the crater frequency data for the Martian Moon Phobos and the lunar standard crater size distribution curve modelled to Gaspra and Ida impact conditions for a set of gravity-to-strength crater transitions according to the cratering mechanics relations (D_{sg} values) as described in Sec. VI.

REFERENCES

Baldwin, R. B. 1964. Lunar crater counts. *Astron. J.* 69:377–392.

Baldwin, R. B. 1971. On the history of lunar impact cratering: The absolute time scale and the origin of planetesimals. *Icarus* 14:36–52.

Basaltic Volcanism Study Project. 1981. *Basaltic Volcanism on the Terrestrial Planets* (New York: Pergamon Press).

Belton, M. J. S., Chapman, C. R., Veverka, J., Klaasen, K. P., Harch, A., Greeley, R., Greenberg, R., Head, J. W., III, McEwen, A., Morrison, D., Thomas, P. C., Davies, M. E., Carr, M. H., Neukum, G., Fanale, F. P., Davis, D., Anger, C., Gierasch, P. J., Ingersoll, A. P., and Pilcher, C. B. 1994. First images of 243 Ida. *Science*, submitted.

Brinkmann, R. T. 1966. Lunar crater distribution from Ranger VI photographs. *J. Geophys. Res.* 71:340–342.

Bryan, J. B., Burton, D. E., Cunningham, M. E., and Lettis, L. A. 1978. A two-dimensional computer simulation of hypervelocity impact cratering: Some preliminary results for Meteor Crater, Arizona. *Proc. Lunar Planet. Sci. Conf.* 9:3931–3964.

Carr, M. H. 1981. *The Surface of Mars* (New Haven, Conn.: Yale Univ. Press).

Chapman, C. R., and Haefner, R. R. 1967. A critique of methods for analysis of the diameter-frequency relation for craters with special application to the Moon. *J. Geophys. Res.* 72:549–557.

Chapman, C. R., and McKinnon, W. B. 1986. Cratering on planetary satellites. In *Satellites*, eds. J. A. Burns and M. S. Matthews (Tucson: Univ. of Arizona Press), pp. 492–580.

Chapman, C. R., Veverka, H., Belton, M. J. S., Neukum, G., and Morrison, D. 1993. Cratering on Gaspra. *Icarus*, submitted.

Croft, S. K. 1978. Lunar crater volumes: Interpretation by models of impact cratering and upper crustal structure. *Proc. Lunar Planet. Sci. Conf.* 9:3711–3733.

Croft, S. K. 1985. The scaling of complex craters. *Proc. Lunar Planet. Sci. Conf.* 15, *J. Geophys. Res. Suppl.* 90:828–842.

Drozd, R. K., Hohenberg, C. M., Morgan, C. J., and Ralston, C. E. 1974. Cosmic-ray exposure history at the Apollo 16 and other lunar sites: Lunar surface dynamics. *Geochim. Cosmochim. Acta* 38:1625–1642.

Fechtig, H., Gentner, W., Hartung, J. B., Nagel, K., Neukum, G., Schneider, E., and Storzer, D. 1975. Microcraters on lunar samples. In *Proceedings of the Soviet-American Conference on Cosmochemistry of the Moon and the Planets*, June 4–8 (1974), Moscow, NASASP-370 (1977), pp. 453–472.

Grieve, R. A. F. 1984. The impact cratering rate in recent time. *Proc. Lunar Planet. Sci. Conf.* 14, *J. Geophys. Res. Suppl.* 89:B403–B408.

Grieve, R. A. F., and Dence, M. R. 1979. The terrestrial cratering record II. The crater production rate. *Icarus* 38:230–242.

Grieve, R. A. F., Robertson, P. B., and Dence, M. R. 1981. Constraints on the formation of ring impact structures, based on terrestrial data. Multi-Ring Basins. *Proc. Lunar Planet. Sci. Conf.* 12:37–57.

Grün, E. 1981. Physikalische und chemische Eigenschaften des interplanetaren Staubes—Messungen des Mikrometeoritenexperimentes auf Helios. *Habilitation Dissertation for Faculty Membership*, Univ. of Heidelberg.

Hartmann, W. K. 1965. Terrestrial and lunar flux of large meteorites in the last two billion years. *Icarus* 4:157–165.

Hartmann, W. K. 1966. Early lunar cratering. *Icarus* 5:406–418.

Hartmann, W. K. 1972. Paleocratering of the Moon. Review of post-Apollo data. *Astrophys. Space Sci.* 17:48–64.

Hartmann, W. K. 1973. Martian cratering 4. Mariner 9 initial analysis of cratering chronology. *J. Geophys. Res.* 78:4096–4116.

Hartmann, W. K. 1977. Relative crater production rates on planets. *Icarus* 31:260–276.

Hartmann, W. K. 1984. Does crater "saturation equilibrium" occur in the solar system? *Icarus* 60:56–74.

Hartmann, W. K., and Wood, C. A. 1971. Moon: Origin and evolution of multi-ring basins. *The Moon* 3:3–78.

Hartmann, W. K., Strom, R. G., Weidenschilling, S. J., Balsius, K. R., Woronow, A., Dence, M. R., Grieve, R. A. F., Diaz, J., Chapman, C. R., Shoemaker, E. M., and Jones, K. L. 1981. Chronology of planetary volcanism by comparative studies of planetary cratering. In *Basaltic Volcanism on the Terrestrial Planets* (New York: Pergamon Press), pp. 1049–1128.

Holsapple, K. A. 1987. The scaling of impact phenomena. *Intl. J. Impact Eng.* 5:343–355.

Holsapple, K. A. 1993. The size of complex craters. *Lunar Planet. Sci. Conf.* XXIV:665–666 (abstract).

Holsapple, K. A., and Schmidt, R. M. 1979. A material-strength model for apparent crater volume. *Proc. Lunar Planet. Sci. Conf.* 10:2757–2777.

Hörz, F., Brownlee, D. E., Fechtig, H., Hartung, J. B ., Morrison, D. A., Neukum,

G., Schneider, E., Vedder, J. F., and Gault, D. E. 1974. Lunar microcraters: Implications for the micrometeoroid complex. *Planet. Space Sci.* 23:151–172.

Horedt, G. P., and Neukum, G. 1984. Planetocentric versus heliocentric impacts in the Jovian and Saturnian satellite system. *J. Geophys. Res.* 12:10405–10410.

Housen, K. R., Schmidt, R. M., and Holsapple, K. A. 1983. Crater ejecta scaling laws: Fundamental forms based on dimensional analysis. *J. Geophys. Res.* 88:2485–2499.

Ivanov, B. A. 1979. Simple model of cratering. *Meteoritika* 38:68–85 (in Russian).

Ivanov, B. A. 1990. Venusian impact craters on Magellan images: View from Venera 15/16. *Earth Moon and Planets* 50/51:159–174.

Ivanov, B. A., and Basilevsky, A. T. 1987. A comparison of crater retention ages on the Earth and Venus. *Solar System Res.* 21:84–89.

Jessberger, E. K. 1981. Die Ar-Ar-Altersbestimmungstechnik: Prinzipien, Entwicklungen und Anwendungen in der Kosmochronologie. *Habilitation Dissertation for Faculty Membership*, Univ. of Heidelberg.

König, B. 1977. Untersuchung von prim ren und sekundären Einschlagsstrukturen auf dem Mond und Laborexperimente zum Studium des Auswurfs von Sekundärteilchen. Ph.D. Thesis, Univ. of Heidelberg.

Maurer, P., Eberhardt, P., Geiss, J., Grögler, N., Stettler, A., Brown, G. M., Peckett, A., and Krähenbühl, U. 1978. Pre-Imbrian craters and basins: Ages, compositions and excavation depths of Apollo 16 breccias. *Geochim. Cosmochim. Acta* 42:1687–1720.

Melosh, H. J. 1989. *Impact Cratering: A Geologic Process* (New York: Oxford Univ. Press).

Moore, H. J., Boyce, J. M., and Hahn, D. A. 1980. Small impact craters in the lunar regolith Q their morphologies, relative ages and rates of formation. *Moon and Planets* 23:231–252.

Moore, H. J., Hodges, C. A., and Scott, D. H. 1974. Multi-ringed basins illustrated by Orientale and associated features. *Proc. Lunar Sci. Conf.* 5:71–100.

Neukum, G. 1971. Untersuchungen über Einschlagskrater auf dem Mond. Ph.D. Thesis, Univ. of Heidelberg.

Neukum, G. 1977. Lunar cratering. *Phil. Trans. Roy. Soc. London A* 285:267–272.

Neukum, G. 1981. Surface history of the terrestrial-type planets. In *Proceedings of the Alpbach Summer School*, ed. W. R. Burke, ESA SP-164 (Noordwijd: European Space Agency), pp. 129–137.

Neukum, G. 1982. Ancient cratering records of the terrestrial-type planets. *Lunar Planet. Sci. Conf.* XXIII:588–589 (abstract).

Neukum, G. 1983. Meteoritenbombardement und Datierung planetarer Oberflächen. Habilitation Dissertation for Faculty Membership, Ludwig-Maximilians-University of Munich.

Neukum, G., and Dietzel, H. 1971. On the development of the crater population on the Moon with time under meteoroid and solar wind bombardment. *Earth Planet. Sci. Lett.* 12:59–66.

Neukum, G., and Hiller, K. 1981. Martian ages. *J. Geophys. Res.* 86:3097–3121.

Neukum, G., and König, B. 1976. Dating of individual lunar craters. *Proc. Lunar Sci. Conf.* 7:2867–2881.

Neukum, G., and Wise, D. U. 1976. Mars: A standard crater curve and possible new time scale. *Science* 194:1381–1387.

Neukum, G., Hiller, K., Henkel, J., and Bodechtel, J. 1978. Mars chronology. In *Reports of Planetary Geology Program 1977–1978*, NASA TM-79729, pp. 172–174.

Neukum, G., König, B., and Arkani-Hamed, J. 1975. A study of lunar impact crater size- distributions. *The Moon* 12:201–229.

Öpik, E. 1960. The lunar surface as an impact counter. *Mon. Not. Roy. Astron. Soc.* 120:404–411.

Pevzner, L. A., Kirjakov, A. F., Vorontsov, A. K., Masaitis, V. L., Mashchak, V. S., and Ivanov, B. A. 1992. Vorotilovskaya drillhole: First deep drilling in the central uplift of large terrestrial impact crater. *Lunar Planet. Sci. Conf.* XXIII:1063–1064 (abstract).

Phillips, R. J., Raubertas, R. F., Arvidson, R. E., Sarkar, I. C., Herik, R. R., Izrnberg, N., and Grimm, R. E. 1992. Impact craters and Venus resurfacing history. *J. Geophys. Res.* 97:15923–15948.

Pike, R. J. 1977. Size-dependence in the shape of fresh impact craters on the Moon. In *Impact and Explosion Cratering*, eds. D. J. Roddy, R. O. Pepin and R. B. Merrill (New York: Pergamon Press), pp. 489–509.

Pike, R. J. 1980. Control of crater morphology by gravity and target type: Mars, Earth, Moon. *Proc. Lunar Planet. Sci. Conf.* 11:2159–2189.

Pike, R. J., and Davis, P. A. 1984. Toward a topographic model of Martian craters from photoclinometry. *Lunar Planet. Sci. Conf.* XV:645–646 (abstract).

Schaber, G. G., Strom, R. G., Moore, H. J., Soderblom, L. A., Kirk, R. L., Chadwick, D. J., Dawson, D. D., Gaddis, L. R., Boyce, J. M., and Russel, J. 1992. Geology and distribution of impact craters on Venus: What are they telling us? *J. Geophys. Res.* 97:13257–13301.

Schmidt, R. M. 1978. Centrifuge simulation of the JOHNIE BOY 500 ton cratering event. *Proc. Lunar Planet. Sci. Conf.* 9:3877–3889.

Schmidt, R. M. 1980. Meteor Crater: Energy of formation implications from centrifuge scaling. *Proc. Lunar Planet. Sci. Conf.* 11:2099–2128.

Schmidt, R. M., and Housen, K. R. 1987. Some recent advances in the scaling of impact and explosion cratering. *Intl. J. Impact Eng.* 5:543–560.

Shoemaker, E. M. 1962. Interpretation of lunar craters. In *Physics and Astronomy of the Moon*, ed. Z. Kopal (New York: Academic Press), pp. 283–359.

Shoemaker, E. M. 1965. Preliminary Analysis of the Fine Structure of Mare Cognitum. JPL-TR-32-700 (Pasadena: Jet Propulsion Lab).

Shoemaker, E. M. 1970. Origin of fragmental debris on the lunar surface and the history of bombardment of the Moon. Oral presentation, I Seminario de Geologia Lunar, May, Univ. of Barcelona (Rev. Jan. 1971).

Shoemaker, E. M., Hait, M. H., Swann, G. A., Schleicher, D. L., Dahlem, D. H., Schaber, G. G., and Sutton, R. L. 1970*a*. Lunar regolith at Tranquillity Base. *Science* 167:452.

Shoemaker, E. M., Batson, R. M., Bean, A. L., Conrad, C., Jr., Dahlem, D. H., Goddard, E. N., Hart, M. H., Larson, K. B., Schaber, G. G., Schleicher, D. L., Sutton, R. L., Swann, G. A., and Waters, A. C. 1970*b*. Preliminary geologic investigation of the Apollo 12 landing site, Part A. In *Geology of the Apollo 12 landing site, Apollo 12 Preliminary Science Report*, NASA SP-235.

Shoemaker, E. M., Williams, J. G., Helin, E. F., and Wolfe, R. F. 1979. Earth-crossing asteroids: orbital classes, collision rates with Earth, and origin. In *Asteroids*, ed. T. Gehrels (Tucson: Univ. of Arizona Press), pp. 253–282.

Soderblom, L. A. 1970. A model for small impact erosion applied to the lunar surface. *J. Geophys. Res.* 75:2655–2661.

Soderblom, L. A., and Boyce, J. M. 1972. Relative ages of some near-side and far-side terra plains based on Apollo 16 metric photography. In *Apollo 16 Preliminary Science Report*, NASA SP-315, pp. 29-3P–29-6.

Soderblom, L. A., Condit, C. D., West, R. A., Herman, B. M., and Kreidler, T. J. 1974. Martian planetwide crater distributions: Implications for geologic history and surface processes. *Icarus* 22:239–263.

Strom, R. G. 1977. Origin and relative age of lunar and mercurian inter-crater plains.

Phys. Earth Planet. Interiors 15:156–172.

Strom, R. G., and Neukum, G. 1988. The cratering record on Mercury and the origin of impacting objects. In *Mercury*, eds. F Vilas, C. R. Chapman and M. S. Matthews (Tucson: Univ. of Arizona Pess), pp. 336–373.

Taylor, S. R. 1975. *Lunar Science: A Post-Apollo View* (New York: Pergamon Press).

Tera, F., Papanastassiou, D. A., and Wasserburg, G. J. 1974. Isotopic evidence for a terminal lunar cataclysm. *Earth Planet. Sci. Lett.* 22:1–21.

Vortman, L. J. 1968. Craters from surface explosions and scaling laws. *J. Geophys. Res.* 73:4621–4636.

Wetherhill, G. W. 1981. Nature and origin of basin-forming projectiles. Multi-ring basins. *Proc. Lunar Planet. Sci. Conf.* 12:1–18.

Wilhelms, G. W. 1979. Relative ages of lunar basins. In *Reports of Planetary Geology Program 1978–1979*, NASA TM-80339, pp. 135–137.

Woronow, A. 1977. Crater saturation and equilibrium: A Monte Carlo simulation. *J. Geophys. Res.* 82:2447–2456.

Woronow, A. 1978. A general cratering history model and its implication for lunar highlands. *Icarus* 34:76–88.

THE RECORD OF PAST IMPACTS ON EARTH

RICHARD A. F. GRIEVE
Geological Survey of Canada

and

EUGENE M. SHOEMAKER
United States Geological Survey

There are currently ~140 known hypervelocity impact craters on Earth. They range up to over 200 km in diameter (*D*) and have ages from recent to PreCambrian. The highly active terrestrial geologic environment has, however, served to blur the record of impact. Few terrestrial impact craters are preserved in their original form, most having been eroded and some even have been tectonized. In addition, ~30% of the known craters are buried by post-impact sediments and knowledge of their character is limited to drill-hole and geophysical data. The terrestrial cratering record has also some intrinsic biases. The majority (~60%) of the known craters are <200 Myr old. There is a marked deficiency of smaller craters <20 km in diameter. The known craters are spatially biased towards relatively well-studied cratonic areas of the Earth. In analyses of the record, as a whole, these biases must be taken into account. Nevertheless, an average cratering rate estimate of $5.6\pm2.8\times10^{-15}\mathrm{km}^{-2}\mathrm{yr}^{-1}$ for $D\gtrsim20$ km can be derived from the known record. This is comparable with rate estimates based on astronomical observations of Earth-crossing bodies and a factor of two higher than the rate estimate from post-mare cratering on the Moon. This can be interpreted as a recent increase in the cratering rate due to a higher proportion of cometary impacts, but the uncertainties involved are considerable and this cannot be considered categorical. Similarly, evidence for periodicities due to cometary showers in the impact record are equivocal and cannot be found consistently, because of the generally large uncertainties (for the purposes of time-series analyses) in estimates of individual crater ages. The evidence for impacting body compositions indicates that a variety of impactor types formed terrestrial craters and can be interpreted as being compatible with most of the largest craters being formed by cometary bodies. The less than categorical answers to questions such as periodicity and variations in impacting body types from interrogating the present knowledge base indicate the need for additional, systematic and detailed study of the Earth's past record of impact cratering.

I. INTRODUCTION

The concept that the Earth is subjected to the impact of interplanetary bodies is not new. It failed, however, to take root until recently. The first studies in the early 1900s of the now famous Meteor or Barringer Crater, Arizona by D. M. Barringer and colleagues and the 1908 Tunguska event by L. A. Kulik and colleagues, which felled 2000 km^2 of Siberian forest, raised more controversy than acceptance of the occurrence of terrestrial impact events. There was, however, a slow increase in the number of recognized small terrestrial impact craters with associated meteorite fragments till the late 1960s, when shock metamorphic effects were accepted as reliable criteria for assigning an impact origin to terrestrial structures. This resulted in a major increase in the number of recognized impact craters, particularly large craters. The results of the planetary exploration programs of the 1970s demonstrated that impact is a common and important geologic process throughout the solar system. These led to a more general acceptance of terrestrial impact structures but there was little overall appreciation of the forces involved and the potential of impact to affect the evolutionary history of the Earth. Impact was regarded largely as a planetary process, with the Earth somehow not fully regarded as a planet. A major change in attitude occurred, however, following discoveries of chemical and physical evidence of impact at the Cretaceous/Tertiary (K/T) boundary in the early 1980s. While originally hotly debated, there has been increasing consensus that, at least in this case, large-scale impact can result in sufficient global environment damage to produce mass extinctions in the biosphere. Those topics are also discussed in Part VI of this book.

It is legitimate to ask why it took so long for the scientific community to recognize that impact is a process that affects the Earth (Marvin 1990). There are probably many reasons; some cultural, some related to the nature of the terrestrial impact record itself. For example, the impact record of the Earth is the least obvious of any of the terrestrial planets. The highly active geologic environment of the Earth serves to remove and obscure impact craters. Those that are preserved are often highly modified by erosional processes. Thus, the record of impact on Earth is harder to read than that of the other terrestrial planets. It also contains biases. As a result, care must be exercised in its interpretation. This contribution summarizes the known record of impact on Earth and outlines its inherent biases resulting from terrestrial geologic activity. It also considers the contribution the record can make to addressing questions regarding the impact rate, its variability with geologic time and the nature of the impacting bodies.

II. THE TERRESTRIAL IMPACT RECORD

Approximately 140 terrestrial impact craters are currently known. The principal facts of these craters are listed in Table I. The present discovery rate of new terrestrial impact craters is $\sim$3 to 5 per year. All the structures listed in Table I have evidence of an impact origin through the documented occurrence of meteoritic material and/or shock metamorphic features. To various degrees they also have a number of other aspects in common, such as form, structure, geophysical characteristics, etc. There are a number of known terrestrial structures that have some of these aspects but lack documented shock metamorphic features. Although some of these are more than likely impact in origin, they are not included in Table I for consistency. Events associated with such phenomena as the 1908 Tunguska explosion, the late Pliocene meteorite debris found over ~300,000 km² of the South Pacific (Kyte and Brownlee 1985; Kyte et al. 1988), the late Eocene impact wave deposits on the northeast coast of the U. S. A. (Poag et al. 1992), which may be related to the N. American tektite strewn field, and others are not included in Table I, as they have no confirmed associated crater.

A. Spatial Distribution

The locations of known terrestrial craters are shown in Fig. 1. Virtually all are on land. The exceptions are: Montagnais (Table I), which occurs on the continental shelf off Nova Scotia, Canada; Chicxulub (Table I), which extends into the Gulf of Mexico from the Yucatan Peninsula, Mexico, and Ust-Kara (Table I), which extends into the Kara Sea, Russia. The Ust-Kara structure is poorly known and it has been suggested that the impact lithologies believed associated with it are, in fact, related to its twin structure Kara (Table I), making Kara a much larger structure with a diameter of ~120 km (Nazarov et al. 1992). A number of structures currently on land were formed under water in epicontinental seas or on continental margins. No impact structures are known from the world's ocean basins. Some sites have been suggested and oceanic craters undoubtedly exist. The present level of knowledge of the ocean floors, however, is insufficient to confirm an impact origin. Ocean floor spreading and subduction also plays a role in the obliteration of oceanic impact craters.

The spatial distribution of known craters is not random. There are concentrations in N. America, Australia, and Europe through to the eastern part of the Commonwealth of Independent States (Fig. 1). This can be attributed to the fact that these are cratonic areas, either exposed PreCambrian Shield or platform sediments overlying Shield, where there have been programs to identify and study impact craters. As the geologic knowledge of other major cratonic areas, such as in S. America and Africa increases, so will undoubtedly the number of known craters in these areas. Very few known craters occur outside cratonic areas, which, with their relatively low levels of tectonic and erosional activity, are the most suitable surfaces for the acquisition and preservation of craters in the terrestrial geologic environment. A few craters have been heavily tectonized, e.g., Beaverhead, U. S. A. and Sudbury, Canada (Table I), and/or occur in mountainous areas, e.g., Garnos, Norway and Kara-Kuhl, Tajikistan (Table I). In the latter cases, the craters were formed after the tectonic event that led to the formation of the mountain belts.

Not all known terrestrial impact craters are exposed at the surface. Many contain post-impact sediments and approximately 30% are completely buried by cover rocks. The latter were generally discovered through geophysical anomalies, which are associated with impact craters (Pilkington and Grieve 1992) and subsequently explored through drilling.

TABLE I

Principal Facts of Known Terrestrial Impact Craters

Crater Name	Location	Latitude	Longitude	Diameter (km)	Age (Myr)	Age (s.d.)
Acraman	South Australia, Australia	S32 1	E135 27	90	>570	—
Ames	Oklahoma, U.S.A.	N36 15	W98 10	16	470	30
Amguid	Algeria	N26 5	E4 23	0.45	<0	—
Aouelloul	Mauritania	N20 15	W12 41	0.39	3.1	0.3
Araguainha Dome	Brazil	S16 46	W52 59	40	249	19
Avak	Alaska, U.S.A.	N71 15	W156 38	12	100	5
Azuara	Spain	N41 10	W0 55	30	<130	—
B.P. Structure	Libya	N25 19	E24 20	2.8	<120	—
Barringer	Arizona, U.S.A.	N35 2	W111 1	1.186	0.049	0.003
Beaverhead	Montana, U.S.A.	N44 36	W113 0	60	~600	—
Bee Bluff	Texas, U.S.A.	N29 2	W99 51	2.4	<40	—
Beyenchime-Salaatin	Russia	N71 50	E123 30	8	<65	—
Bigach	Kazakhstan	N48 3	E82 0	7	6	3
Boltysh	Ukraine	N48 45	E32 10	24	88	3
Bosumtwi	Ghana	N6 32	W1 25	10.5	1.03	0.02
Boxhole	Northern Territory, Australia	S22 37	E135 12	0.17	0.03	—
Brent	Ontario, Canada	N46 5	W78 29	3.8	450	30
Campo del Cielo	Argentina	S27 38	W61 42	0.05	<0.004	—
Carswell	Saskatchewan, Canada	N58 27	W109 30	39	115	10
Charlevoix	Quebec, Canada	N47 32	W70 18	54	357	15
Chicxulub	Yucatan, Mexico	N21 20	W89 30	180	64.98	0.05
Chiyli	Kazakhstan	N49 10	E57 51	5.5	46	7

Clearwater Lake East	Quebec, Canada	N56 5	W74 7	22	290	20
Clearwater Lake West	Quebec, Canada	N56 13	W74 30	32	290	20
Connolly Basin	Western Australia, Australia	S23 32	E124 45	9	<60	—
Crooked Creek	Missouri, U.S.A.	N37 50	W91 23	7	320	80
Dalgaranga	Western Australia, Australia	S27 45	E117 5	0.021	0.027	—
Decaturville	Missouri, U.S.A.	N37 54	W92 43	6	<300	—
Deep Bay	Saskatchewan, Canada	N56 24	W102 59	13	100	50
Dellen	Sweden	N61 55	E16 39	15	89	2.7
Des Plaines	Illinois, U.S.A.	N42 3	W87 52	8	<280	—
Dobele	Latvia	N56 35	E23 15	4.5	300	35
Eagle Butte	Alberta, Canada	N49 42	W110 35	19	<65	—
El'gygytgyn	Russia	N67 30	E172 5	18	3.5	0.5
Flynn Creek	Tennessee, U.S.A.	N36 17	W85 40	3.55	360	20
Garnos	Norway	N60 39	E9 0	5	500	10
Glasford	Illinois, U.S.A.	N40 36	W89 47	4	<430	—
Glover Bluff	Wisconsin, U.S.A.	N43 58	W89 32	3	<500	—
Goat Paddock	Western Australia, Australia	S18 20	E126 40	5.1	<50	—
Gosses Bluff	Northern Territory, Australia	S23 50	E132 19	22	142.5	0.5
Gow Lake	Saskatchewan, Canada	N56 27	W104 29	5	<250	—
Gusev	Russia	N48 21	E40 14	3.5	65	—
Haughton	Northwest Territories, Canada	N75 22	W89 41	24	23.4	1
Haviland	Kansas, U.S.A.	N37 35	W99 10	0.015	<0.001	—

TABLE I (cont.)

Crater Name	Location	Latitude	Longitude	Diameter (km)	Age (Myr)	Age (s.d.)
Henbury	Northern Territory, Australia	S24 35	E133 9	0.157	<0.005	—
Holleford	Ontario, Canada	N44 28	W76 38	2.35	550	100
Ile Rouleau	Quebec, Canada	N50 41	W73 53	4	<300	—
Ilumetsa	Estonia	N57 58	E25 25	0.08	>0.002	—
Ilyinets	Ukraine	N49 6	E29 12	4.5	395	5
Janisjärvi	Russia	N61 58	E30 55	14	698	22
Kaalijärvi	Estonia	N58 24	E22 40	0.11	0.004	0.001
Kaluga	Russia	N54 30	E36 15	15	380	10
Kamesnk	Russia	N48 20	E40 15	25	65	2
Kara	Russia	N69 5	E64 18	65	73	3
Kara-Kul	Tajikistan	N39 1	E73 27	52	<25	—
Kardla	Estonia	N57 0	E22 42	4	455	—
Karla	Russia	N54 54	E48 0	12	10	—
Kelly West	Northern Territory, Australia	S19 56	E133 57	10	>550	—
Kentland	Indiana, U.S.A.	N40 45	W87 24	13	<300	—
Kursk	Russia	N51 40	E36 0	5.5	250	80
Lac Couture	Quebec, Canada	N60 8	W75 20	8	430	25
Lac La Moinerie	Quebec, Canada	N57 26	W66 37	8	400	50
Lappajärvi	Finland	N63 9	E23 42	23	77.3	0.4
Lawn hill	Queensland, Australia	S18 40	E138 39	18	>515	—
Liverpool	Northern Territory, Australia	S12 24	E134 3	1.6	150	70
Lockne	Sweden	N63 0	E14 48	7	540	10

Name	Location	Latitude	Longitude			
Logancha	Russia	N65 30	E95 48	20	25	20
Logoisk	Belarus	N54 12	E27 48	17	40	5
Lonar	India	N19 59	E76 31	1.83	0.052	0.006
Macha	Russia	N59 59	E118 0	0.3	<0.007	—
Manicouagan	Quebec, Canada	N51 23	W68 42	100	214	1
Manson	Iowa, U.S.A.	N42 35	W94 31	35	65.7	1
Marquez	Texas, U.S.A.	N31 17	W96 18	22	58	2
Middlesboro	Kentucky, U.S.A.	N36 37	W83 44	6	<300	—
Mien	Sweden	N56 25	E14 52	9	121	2.3
Misarai	Lithuania	N54 0	E23 54	5	395	145
Mishina Gora	Russia	N58 40	E28 0	4	<360	—
Mistastin	Newfoundland & Labrador, Canada	N55 53	W63 18	28	38	4
Montagnais	Nova Scotia, Canada	N42 53	W64 13	45	50.5	0.76
Monturaqui	Chile	S23 56	W68 17	0.46	1	—
Morasko	Poland	N52 29	E16 54	0.1	0.01	—
New Quebec	Quebec, Canada	N61 17	W73 40	3.44	1.4	0.1
Nicholson Lake	Northwest Territories, Canada	N62 40	W102 41	12.5	<400	—
Oasis	Libya	N24 35	E24 24	11.5	<120	—
Obolon'	Ukraine	N49 30	E32 55	15	215	25
Odessa	Texas, U.S.A.	N31 45	W102 29	0.168	<0.05	—
Quarkziz	Algeria	N29 0	W7 33	3.5	<70	—
Piccaninny	Western Australia, Australia	S17 32	E128 25	7	<360	—
Pilot Lake	Northwest Territories, Canada	N60 17	W111 1	6	445	2

TABLE I (cont.)

Crater Name	Location	Latitude	Longitude	Diameter (km)	Age (Myr)	Age (s.d.)
Popigai	Russia	N71 30	E111 0	100	35	5
Presqu'ile	Quebec, Canada	N49 43	W78 48	12	<500	—
Pretoria Salt Pan	South Africa	S25 24	E28 5	1.13	0.2	—
Puchezh-Katunki	Russia	N57 6	E43 35	80	220	10
Ragozinka	Russia	N58 18	E62 0	9	55	5
Red Wing	North Dakota, U.S.A.	N47 36	W103 33	9	200	25
Riachao Ring	Brazil	S7 43	W46 39	4.5	<200	—
Ries	Germany	N48 53	E10 37	24	15.1	1
Rio Cuarto	Argentina	S30 52	W64 14	4.5	<0.1	—
Rochechouart	France	N45 50	E0 56	23	186	8
Roter Kamm	Namibia	S27 46	E16 18	2.5	3.7	0.3
Rotmistrovka	Ukraine	N49 0	E32 0	2.7	140	20
Sääksjärvi	Finland	N61 23	E22 25	5	560	12
Saint Martin	Manitoba, Canada	N51 47	W98 32	40	219.5	32
Serpent Mound	Ohio, U.S.A.	N39 2	W83 24	8	<320	—
Serra da Cangalha	Brazil	S8 5	W46 52	12	<300	—
Shunak	Kazakhstan	N47 12	E72 42	3.1	12	5
Sierra Madera	Texas, U.S.A.	N30 36	W102 55	13	<100	—
Sikhote Alin	Russia	N46 7	E134 40	0.027	0	—
Siljan	Sweden	N61 2	E14 52	55	368	1.1
Slate Islands	Ontario, Canada	N48 40	W87 0	30	<350	—
Sobolev	Russia	N46 18	E138 52	0.053	<0.001	—
Soderfjärden	Finland	N63 2	E21 35	6	550	—

Spider	Western Australia, Australia	S16 44	E126 5	13	>570	—
Steen River	Alberta, Canada	N59 31	W117 37	25	95	7
Steinheim	Germany	N48 40	E10 4	3.8	14.8	0.7
Strangways	Northern Territory, Australia	S15 12	E133 35	25	<470	—
Sudbury	Ontario, Canada	N46 36	W81 11	200	1850	3
Tabun-Khara-Obo	Mongolia	N44 6	E109 36	1.3	3	—
Talemzane	Algeria	N33 19	E4 2	1.75	<3	—
Teague	Western Australia, Australia	S25 52	E120 53	30	1685	5
Tenoumer	Mauritania	N22 55	W10 24	1.9	2.5	0.5
Ternovka	Ukraine	N48 1	E33 5	12	280	10
Tin Bider	Algeria	N27 36	E5 7	6	<70	—
Tookoonooka	Queensland, Australia	S27 0	E143 0	55	128	5
Tvaren	Sweden	N58 46	E17 25	2	455	—
Upheaval Dome	Utah, U.S.A.	N38 26	W109 54	10	<65	—
Ust-Kara	Russia	N69 18	E65 18	25	73	3
Vargeao Dome	Brazil	S26 50	W52 7	12	<70	—
Veevers	Western Australia, Australia	S22 58	E125 22	0.08	<1	—
Vepriaj	Latvia	N55 6	E24 36	8	160	30
Vredefort	South Africa	S27 0	E27 30	140	1970	100
Wabar	Saudi Arabia	N21 30	E50 28	0.097	0.006	0.002
Wanapitei Lake	Ontario, Canada	N46 45	W80 45	7.5	37	2
Wells Creek	Tennessee, U.S.A.	N36 23	W87 40	12	200	100
West Hawk Lake	Manitoba, Canada	N49 46	W95 11	2.44	100	50
Wolfe Creek	Western Australia, Australia	S19 18	E127 46	0.875	<0.3	—
Zapadnaya	Ukraine	N49 44	E29 0.18	4	115	10
Zeleny Gai	Ukraine	N48 42	E32 54	2.5	120	20
Zhamanshin	Kazakhstan	N48 24	E60 58	13.5	0.9	0.1

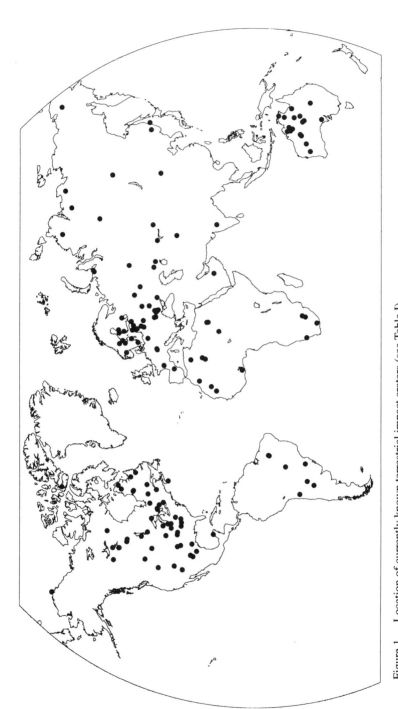

Figure 1. Location of currently known terrestrial impact craters (see Table I).

B. Temporal Distribution

Unlike craters on other planetary bodies, it is possible to determine the time of impact for individual terrestrial craters. Approximately 40% of the known craters have isotopic ages, determined from analyses of such isotopically reset lithologies as impact melt rocks. Generally, the isotopic ages are K-Ar or more recently ^{40}Ar–^{39}Ar ages. Fine-grained, often clast-rich, impact melt rocks are not particularly easy to date isotopically, because of inherited Ar from the clasts. Care must be taken and the ^{40}Ar-^{39}Ar methodology is preferable over K-Ar (Bottomley et al. 1989). In some cases, simple K-Ar ages can be in considerable error. For example, a single K-Ar analysis on an impact melt sample from Mistastin, Canada gave an age of 202±25 Myr (Currie 1971). Others gave 38±6 and 36±4 Myr, which correspond to the ^{40}Ar-^{39}Ar plateau age of 38±4 Myr from several samples (Mak et al. 1976). In only a few cases are impact melt lithologies of sufficient grain size to permit such techniques as Rb-Sr mineral isochrons (Deutsch et al. 1992; Reimold et al. 1990). Precise U-Pb ages have been obtained from the analysis of shocked zircons (Krogh et al. 1984,1992) and new zircons crystallized from impact melt rocks (Hodych and Dunning 1992).

The remainder of the known craters have biostratigraphic or stratigraphic ages. In some cases, the biostratigraphic ages, on such units as crater-filling sediments, are as precise as isotopic ages. In other cases, stratigraphic ages are only maximum estimates of age, the age being listed only as less than the age of the target rocks (Table I). In such cases, the degree of erosion can be used to further constrain the age. For example, the age of the Slate Islands, Canada is based on its similar erosional level to Charlevoix, Canada (Table I; Halls and Grieve 1976). In this case, both are similar sized structures, 54 km and 30 km, respectively, and occur in areas of broadly similar geologic history. Erosional rates, however, can vary considerably over relatively short distances, particularly in areas that have been glaciated. In addition, some craters have been buried, preserved and only recently exhumed. For example, the relatively small, but old, Brent, Canada, Janisjärvi, Russia and Sääksjärvi, Finland craters (Table I) probably owe their preservation to burial by cover sediments soon after formation.

Crater age estimates, therefore, are a mixture of accuracies and precisions. Caution must be exercised when using these ages to calculate parameters such as cratering rate estimates and as input into time-series analyses for searches for periodicities and links to other geologic process (see, e.g., Rampino and Stothers 1984b; Stothers and Rampino 1990). Some broad trends, however, are clear. As with the spatial distribution, the temporal distribution of known terrestrial craters is not random (Fig. 2). It is highly biased towards younger ages, with over 60% being younger than 200 Myr. This is not a reflection of a cratering rate increase but is a function of erosion and sedimentation. As terrestrial impact craters are surface features in a highly active geologic environment, they can be removed or buried from observation relatively

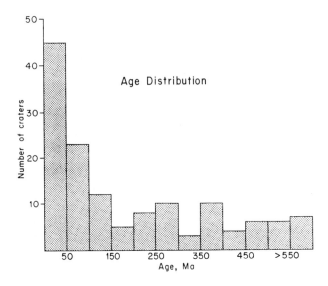

Figure 2.   Histogram of ages of known terrestrial impact craters (see Table I). Note bias towards young ages.

rapidly. The rate at which this occurs varies with the geologic history of the area. For example, it has been estimated that craters with $D \lesssim 20$ km can be effectively removed and not recognized in exposed Shield areas that have been glaciated in >120 Myr (Grieve 1984). Conversely, the interior of Australia, which has had a remarkably stable geologic history has a relatively high number Proterozoic-aged impact structures (5 or 6, or $\sim 30\%$ of the known craters in Australia (Table I)) and the Russian platform has a relatively high number of impact structures of Mesozoic age (Table I), because of post-impact burial by cover rocks.

C. Size Distribution

Terrestrial impact craters range up in size to over 200 km in diameter (Table I). Many of the listed diameters, however, are only estimates, with erosion having removed the original topography. For structures with minimal geological and geophysical information, there is considerable uncertainty in these estimates. In some cases, erosion has removed essentially all the topographic expression of the original crater and what remains is a geological anomaly, with a roughly circular shape. In a few cases, the original negative topographic expression of the crater has been replaced by positive topography. For example, Gosses Bluff, Northern Territory, Australia (Table I) is currently expressed topographically as a 5 km diameter annular topographically high ring of erosionally resistant sandstones. Other geological and geophysical data suggest the original diameter of the crater was 22 km (Milton et al. 1972) and what remains is the erosional remanent of the interior of a central uplift.

There is an overall bias in the number of terrestrial impact craters of a

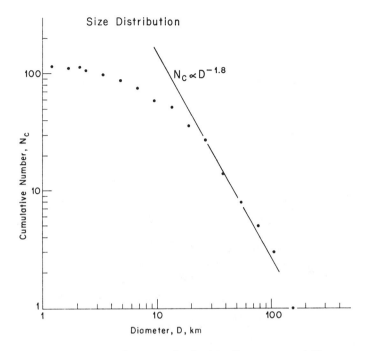

Figure 3.   Logarithmic size-frequency distribution of known terrestrial impact craters with diameter $D>1$ km. Note deficit of craters with $D<20$ km from $N_c \propto D^{-1.8}$ distribution.

particular size. At large diameters, the cumulative size-frequency distribution of terrestrial impact craters is similar to that on other planetary bodies (Fig. 3). At diameters below ~ 20 km, however, the distribution falls off and there is an increasing deficit of craters at smaller diameters (Fig. 3). This type of deficit is generally taken to indicate the effects of resurfacing processes (Basaltic Volcanism Study Project 1981). In the terrestrial case, erosion and burial are highly effective at removing or obscuring the topographic and geologic signature of smaller craters. This shape to the size-frequency distribution of terrestrial craters appears to be an inherent property of the record, as it has persisted as more craters have been added to the known list over the years (Grieve 1991). It also results in limitations on cratering rate calculations from terrestrial crater counts. For example, in the lunar case, it is recommended that craters $\lesssim 4$ km in diameter not be counted, so as not to include large secondary craters within the primary crater population (Basaltic Volcanism Study Project 1981). In the terrestrial case, the recommended cut off is craters $\lesssim 20$ km, so as to avoid the portion of the size-frequency distribution that is generally deficient in craters (Grieve 1984).

III. CRATER MORPHOLOGY

Planetary craters are recognized essentially by their morphology. Due to the active geologic environment, terrestrial impact craters are recognized not only by their morphology but also by their geologic structure. In the most highly eroded examples, terrestrial impact craters no longer have an obvious crater form, i.e., basically a circular depression, and are recognized solely by their geologic structure, which represents the subsurface root of the original crater form.

Figure 4.   Oblique aerial photograph of 1.2 km diameter Barringer or Meteor Crater, Arizona, U. S. A. This relatively young crater is a classic example of a well-preserved simple crater.

Relatively uneroded terrestrial craters display the basic progression from simple to complex forms with increasing crater diameter observed on other terrestrial planets. Simple craters have the form of a bowl-shaped depression with a structurally upraised rim. The rim area is overlain by ejecta deposits and the crater floor represents the top of a sub-surface breccia lens. The canonical example of a terrestrial simple crater is the Barringer or Meteor Crater, Arizona, U. S. A. (Fig. 4). Barringer has a partially preserved exterior ejecta deposit, because of its young age of ~50,000 yr (Sutton 1985). Most simple craters, however, are considerably more degraded than Barringer and, in some cases, the rim area has been completely removed by erosion and the interior filled with post-impact sediments (Fig. 5).

Simple craters occur up to diameters of ~4 km on Earth. Above 4 km, terrestrial craters generally have a complex form. As with craters on other

Figure 5.   Vertical aerial photograph of originally 3.8 km diameter Brent crater, Ontario, Canada. This simple crater is 450 Myr old and has been eroded and filled with post-impact sediments. It no longer has an obvious crater form due to terrestrial geologic activity (compare with Fig. 4.).

planetary bodies, however, there appears to be some overlap between simple and complex forms near the transition diameter. Some of this can be ascribed to differences in target rock properties, with complex craters occurring in sedimentary targets at diameters greater than 2 km. Complex crater forms are characterized by structurally complex and faulted rim areas, a flat annular trough and uplifted topographically high central structures (Fig. 6). Studies at terrestrial craters indicated that the central structures contain rocks uplifted from deeper levels (see, e.g., Milton et al. 1972). Various lines of evidence indicate that complex craters are a highly modified crater form with respect to simple craters. Although some details are not well understood, the basic principles of crater mechanics in the formation of simple and complex craters have been established (see, e.g., Melosh 1989).

Terrestrial complex craters also show the secondary forms observed on other planetary bodies, such as central peak craters, peak-ring craters and ring basins. It is not known if there are examples of terrestrial multi-ring basins.

Figure 6.   Oblique aerial photograph of the 3.8 km complex crater Steinheim, Germany. This relatively well-preserved crater has an obvious central peak, flat floor, containing the village of Steinheim and structural rim. The crater, however, has been eroded and the floor partially filled with sediments.

The largest known terrestrial craters are: Chicxulub (180 km), Sudbury (200 km) and Vredefort (140 km) (Table I). Chicxulub is buried by ~1 km of platform sediments and is very poorly known (Hildebrand et al. 1991) and may be as large as 300 km. Sudbury is eroded and highly tectonized (Fig. 7) (Milkereit et al. 1992), but appears to have had an interior ring (Stöffler et al. 1989; Grieve et al. 1991b). Vredefort (Fig. 8) is too highly eroded, with essentially only the crater floor preserved (Dietz 1961) to give any indication of its original morphology. It too may have been originally as large as 300 km. The lack of definitive evidence for multi-ring impact structures on Earth because of the terrestrial environment serves to illustrate the caution that must be applied when appraising the form of terrestrial impact structures. Almost without exception, exposed terrestrial impact structures have been modified by erosion. Some buried structures, e.g., Chicxulub, which formed in an area of continuous post-impact sedimentation, presumably have preserved their original morphology. They are, however, poorly known, because they can only be reconstructed from spot information, such as from drill holes, and extrapolations from geophysics (Pilkington and Grieve 1992).

Care must be exercised when comparing the modified morphologic elements of craters within the terrestrial record and, in particular, when comparing terrestrial and planetary craters (Pike 1985). Original morphologic elements can be enhanced, modified or removed by the various types of erosional processes on Earth. For example, Haughton (Table I) has a topographic ring (Fig. 9), which has led to its description as a peak-ring crater (Robertson and Sweeney 1983). More recent analyses indicate that these rings are not reflected in the geology or geophysical data (Grieve 1988). They are erosional artifacts rather than primary features. In addition, similar-sized terrestrial craters can have different apparent morphologies, depending on the depth of erosion and the degree to which target lithologies have contributed to differential erosion.

Morphometric relations are affected by erosion to varying degrees. For example, the correlation between the physical height of the central uplift and crater diameter for terrestrial craters is poor, in contrast to the good correlation in lunar data (Pike 1977). Other morphometric parameters are better defined. For example, there is a more obvious correlation between the diameter of the central uplift and crater diameter for terrestrial craters. This is because the diameter of the central uplift, in addition to being defined topographically, can be defined geologically by mapping the occurrence of uplifted target rocks, even when the topographic expression has been completely removed.

The intent of the discussion above has been to outline the basic character of the terrestrial impact record and to emphasize that it differs from that on other planets in the degree to which it is complicated by terrestrial geologic processes. Data compilations, such as in Table I, should not be used without understanding the underlying character of the data. Unfortunately, this is not always the case.

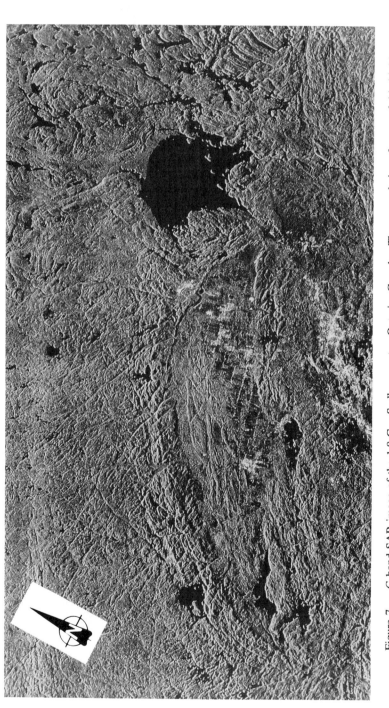

Figure 7. C-band SAR image of the 1.8 Gyr Sudbury crater, Ontario, Canada. The most obvious feature of this highly eroded and tectonized impact crater is the elliptical outline of the Sudbury Igneous Complex and the interior Sudbury Basin. The present shape, long axis (~60 km), of the Igneous Complex is due to post-impact Penokean deformation. The Igneous Complex is interpreted as the eroded remnant of the interior impact melt of an original impact crater ~200 km in diameter. The form of Sudbury may have been originally that of a multi-ring basin. The younger, 37 Myr, 7.5 km diameter Wanapitei impact crater is superimposed upon the Sudbury structure in the middle right. Wanapitei appears as a partially circular lake and shows obvious concentric fracturing in the surrounding target rocks.

Figure 8.   Vertical Space Shuttle photograph of the 1.97 Gyr, 140 km diameter
Vredefort crater, South Africa. This ancient crater has been eroded to below the
original crater floor and appears as a circular geological anomaly. It is partially
covered to the south and east by younger volcanic units.

IV. CRATERING RATE

Due to the inherent biases of the terrestrial impact record, calculations of the
terrestrial cratering rate from the known record must be limited to relatively
young (Fig. 2), large (Fig. 3) craters. This results in the rate estimate being
based on relatively small numbers of craters and, thus, having large uncertainties. For example, Shoemaker (1977) calculated a terrestrial cratering
rate of $2.2\pm1.1\times10^{-14}$ km^{-2} yr^{-1} for craters with diameter, $D\gtrsim10$ km. This
was based on the occurrence of 4 craters in an area of 0.7×10^{6} km^2 in the
Mississippi lowlands of the United States. Grieve and Dence (1979) derived
an estimate of $3.5\pm1.3\times10^{-15}$ km^2 yr^{-1} for craters with $D\gtrsim20$ km, based
on 15 craters from the cratonic areas, exposed Shield and platform cover,
of North America and eastern Europe-western Russia. Both of these studies

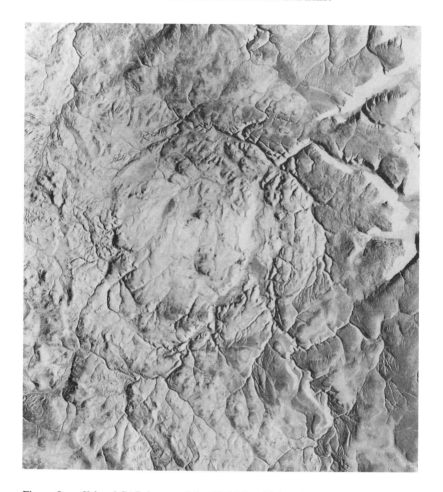

Figure 9. X-band SAR image of the 23.4 Myr, 24 km diameter Haughton crater, Northwest Territories, Canada. This crater appears to have a 12 km diameter interior ring, which is more obvious than the outer rim area. This interior ring is enhanced by differential erosion and, although it must be ultimately controlled by original crater morphology, does not correspond directly to the structural interior ring seen in lunar peak ring basins.

are limited to crater counts in areas where there has been thorough searches for craters and it was believed the majority of large impact structures had been recognized. When the Shoemaker (1977) rate estimate is extended to $D \gtrsim 20$ km through $N \propto D^{-1.8}$ (Fig. 3), it corresponds to $6.3 \pm 3.2 \times 10^{-15}$ km^{-2} yr^{-1}, which is almost twice as high as the estimate of Grieve and Dence (1979).

This inconsistency, and the debate as to whether the cratering rate in the Earth-Moon system had remained constant over the last ~ 3.0 Gyr (Neukum et al. 1975; Shoemaker et al. 1979; Wetherill and Shoemaker 1982; Young

1977), prompted Grieve (1984) to re-examine the earlier rate estimate from the N. American and European-Russian cratonic areas. From an examination of the degree of erosion at the specific craters and consideration that the depth to which a crater is recognizable in the crust increases at a slower rate than crater diameter, Grieve (1984) concluded that the original listing of craters considered by Grieve and Dence (1979) consisted of two populations; a younger group ($\lesssim$120 Myr) in which erosion was not a factor in their recognition and an older group ($\gtrsim$120 Myr) in which craters $\gtrsim$20 km may have been removed from the counted sample by erosion. From a reanalysis of the original data, Grieve (1984) revised the estimate of the cratering rate upwards to $5.5\pm2.7\times10^{-15}$ km^{-2} yr^{-1} for $D \gtrsim$20 km and craters $\lesssim$120 Myr in age. This rate estimate is more comparable to that of Shoemaker (1977).

We have repeated part of this re-analysis to determine if there have been any changes based on recent additional knowledge of the terrestrial cratering record. The craters considered and their age estimates are listed in Table II. Table II shows some minor differences in the original estimates of ages, compared to the previous listing (Table II in Grieve 1984). It has also one additional crater, Lappajärvi (Table II). Lappajrvi is not a recent discovery but represents a case where there has been a revision upwards in size. Based on recent drilling (Pipping 1991) and geophysics (L. Pesonen 1992, personnal communication), Lappajärvi's estimated original diameter is now 23 km not 17 km, as previously considered. The fact that no new large craters have been reported in these areas since 1979 or, in the case of the area considered by Shoemaker (1977), since 1977, suggests that the assumptions regarding the completeness of search of these areas is reasonable. The two distributions noted by Grieve (1984) are evident in the cumulative number of craters with time (Fig. 10), with the younger group having a steeper cumulative number with time slope. The detailed arguments of why and how the older group is incomplete due to erosional processes can be found in Grieve (1984). If only the younger group is used to calculate the recent terrestrial cratering rate, the estimated rate is $5.6\pm2.8\times10^{-15}$ km^{-2} yr^{-1} for $D \gtrsim$20 km and $\lesssim$120 Myr. This is essentially the same as the earlier estimate. The uncertainties attached to all these estimates are large, $\pm$50%. These reflect concerns regarding completeness of search and small number statistics.

As noted previously (Grieve 1984; Shoemaker et al. 1979,1990), estimates of the terrestrial cratering rate are approximately a factor of two higher than the post-mare cratering rate on the Moon (Fig. 11). In constructing Fig. 11, an attempt was made to account for the different impact conditions on the Moon. For example, assuming the impact of bodies of equivalent composition, approach velocity and size, the relative dimensions of the resultant lunar and terrestrial transient cavities can be calculated from scaling relations, based on dimensional analyses of experimental craters (Schmidt and Housen

TABLE II

Impact Craters with $D \gtrsim 20$ km on N. American and European Cratons

Crater	Diameter[a] (km)	Age[b] (Myr)	Source[c]
Haughton	24	23±1	Jessberger 1988
Mistastin	28	38±4	Mak et al. 1976
Kamensk	25	65±2	Movshovich and Milyavskii 1986
Manson	35	66±1	Kunk et al. 1989
Lappajärvi	23	77±1	Jessberger and Reimold 1980
Boltysh	24	88±3	Boiko et al. 1985
Steen River	25	95±7	Carrigy 1968
Carswell	39	115±10	Bottomley et al. 1989
Manicouagan	100	214±1	Hodych and Dunning 1992
Saint Martin	40	219±32	Reimold et al. 1990
Puchezh-Katunki	80	220±10	Fel'dman et al. 1984
Clearwater East	22	290±20	Reimold et al. 1981
Clearwater West	32	290±20	Reimold et al. 1981
Slate Islands	30	<350	Halls and Grieve 1976
Charlevoix	54	357±15	Rondot 1971
Siljan	55	368±1	Bottomley et al. 1989
Sudbury	200	1850±3	Krogh et al. 1984

[a] Estimated original diameter.
[b] All ages are based on isootpic data, except Kamensk, which is based on biostratigraphy and Slate Islands, which is based on comparative erosional level.
[c] Source for age estimate.

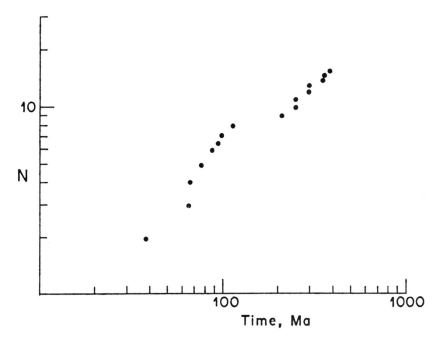

Figure 10.   Cumulative number of craters with $D > 20$ km with age on the N. American and European cratons. Note the two apparent distributions of younger and older craters, suggesting that some craters > 120 Myr in age may have been removed from the record by erosion.

1987), such as:

$$D_{tc} = 1.16 \left[\frac{\rho_p}{\rho_t} \right]^{1/3} D_p^{0.78} V_i^{0.44} g^{-0.22} \qquad (1)$$

where D_{tc} is transient cavity diameter, ρ_p and ρ_t are density of projectile and target, respectively, D_p is projectile diameter, V_i is impact velocity, g is planetary gravity, and units are cgs (Grieve and Cintala 1992). The effect of the lower lunar gravity in increasing the size of the transient cavity, relative to an equivalent terrestrial impact, is partly offset by the lower impact velocity and higher density of the mare target rocks on the Moon. Mare impacts are only considered, as we are concerned with the post-mare cratering rate.

More importantly, the lower lunar gravity results in relatively less enlargement of the transient cavity during cavity collapse and modification to form the final complex craterform. The relative dimensions of the final craters can be derived from the empirical relation:

$$D_{tc} = D_t^{0.15} D^{0.85} \qquad (2)$$

where D_t is the transition diameter between simple and complex craterforms (Croft 1985). In the case of craters on the lunar mare, this is taken to

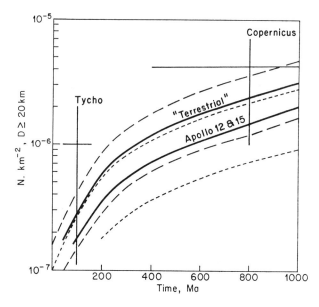

Figure 11.   Semi-logarithmic plot of cumulative number of lunar craters with $D \gtrsim 20$ km with time, based on the terrestrial cratering rate averaged over the last 120 Myr recalculated for lunar conditions (see text) and "observed" post-mare lunar cratering rate averaged over the last ~3.2 Gyr. Note only small overlap in uncertainties of the two rate estimates, the higher rate based on the terrestrial data and the better fit of this rate to the Tycho and Copernicus data.

be ~20 km, compared to 4 km for terrestrial craters. The net result of these differences is that a large model complex lunar crater is slightly smaller (20%) than its terrestrial counterpart formed by an impacting body of equivalent composition and size. An additional factor is the differences in the normalized gravitational radius:

$$R_g/R = \sqrt{1 + V_{esc}^2/V_i^2} \qquad (3)$$

where R_g and R are gravitational and physical radius, respectively, and V_{esc} is escape velocity of the planetary body. This factor varies with impact velocity, decreasing in importance with increasing impact velocity. Thus, asteroidal, as opposed to cometary, impacts are towards the maximum end of this correction factor. For asteroidal bodies with an average approach velocity of 14 km s^{-1}, the terrestrial impact flux is ~ 1.25 times higher than the lunar flux. Thus, for asteroidal impacts and assuming $N \propto D^{-1.8}$, the model impact cratering rate on the Moon would be $3.0 \pm 1.5 \times 10^{-15}$ km^{-2} yr^{-1} for $D \gtrsim 20$ km (Fig. 11). This is a minimum model estimate, as a percentage of the impacts will be due to higher velocity comets. In these cases, the effect of the smaller gravitational radius of the Moon will be minimized. The observed post-mare cratering rate on the Moon, based on crater counts at the Apollo 12 and 15 sites, is $3.3 \pm 1.7 \times 10^{-14}$ km^{-2} yr^{-1} for $D \geq 4$ km (Basaltic Volcanism Study

Project 1981). Assuming $N \propto D^{-1.8}$, this translates to an "observed" rate of $1.8 \pm 0.9 \times 10^{-15}$ km^{-2} yr^{-1} for $D \geq 20$ km (Fig. 11).

With all the observational and model uncertainites, it is probably unjustified to state categorically that the current cratering rate, based on terrestrial data, is higher than that integrated over the last ~ 3.2 Gyr, based on lunar data. There is, however, some supporting evidence. For example, the crater counting data for Tycho and Copernicus, if interpolated to $D \gtrsim 20$ km, are consistent with the present rate extrapolated back to 1 Gyr before present (Fig. 11). Assuming the reverse, that is, a constant rate defined by the Apollo 12 and 15, the Tycho and Copernicus data points are too high (Fig. 8.4.1, Basaltic Volcanism Study Project 1981), although they are within the uncertainty limits. As noted previously, Australia has a better preserved sample of Proterozoic-aged impact craters than anywhere else on Earth. From these data, Shoemaker and Shoemaker (1990) have estimated a Proterozoic terrestrial cratering rate of $4.5 \pm 2.0 \times 10^{-15}$ km^{-2} yr^{-1} for $D \gtrsim 20$ km, which is more similar to the present rate than the one calculated from the lunar data integrated over ~ 3.2 Gyr. While the ages of the Tycho and Copernicus events are not well constrained and the Australian estimate is based on relatively few craters, (although the cumulative-size frequency of the Australia craters suggests that most of the large craters have been identified [Shoemaker and Shoemaker 1990]), they are all consistent with a higher rate than suggested by the lunar data.

On the other hand, the cratering rate for the Apollo 12 site is $\sim 50\%$ higher than that for the Apollo 15 site for $D \gtrsim 4$ km (Table 8.4.1, Basaltic Volcanism Study Project 1981). The post-mare cratering rate based on only the Apollo 12 site is $2.4 \pm 1.2 \times 10^{-15}$ km^{-2} yr^{-1} for $D \gtrsim 20$ km, not significantly different from the model rate calculated from the terrestrial data. In addition, it is difficult to imagine a mechanism to increase the cratering rate with time, beyond the increasing involvement of cometary impacts. Shoemaker et al. (1990) have suggested that a decrease in the amplitude of the oscillation of the Sun normal to the galactic plane has led to an increase in the flux of long-period comets by possibly a factor of 2. This requires, however, that more than half the larger terrestrial craters be the result of cometary impacts, which is somewhat inconsistent with the similarity between the estimated cratering rates based on terrestrial craters and that based on astronomical observations of current Earth-crossing bodies (Shoemaker et al. 1990). In addition, in a recent study of the cometary flux, Weissman (1990) concludes that long- and short-period comets and random cometary showers account for only up to $\sim 25\%$ of the terrestrial impact flux.

In summary, we are equivocal as to whether there has been an increase in the terrestrial cratering rate since ~ 3.2 Gyr. It is a matter of interpretation of data with considerable attached uncertainties, plus the manipulation of these data using a number of relationships and assumptions, which have their own degrees of uncertainty. One of us (E. M. S.) favors, but does not champion strongly, an increase due to increased cometary impacts. The other (R. A. F. G.) is less convinced because of the inherent uncertainties. An improvement in the quality of the data is required, either through the discovery and dating of a larger number of terrestrial craters or the direct age dating of a number of individual lunar craters.

V. PERIODICITY OF IMPACTS

Considerable interest in the potential for impact to disrupt the biological balance on Earth (see, e.g., Silver and Schultz 1982; Sharpton and Ward 1990) followed the initial reports of evidence for the involvement of large-scale impact at the Cretaceous-Tertiary (K/T) boundary and the suggestion that the related mass extinctions were the result of a major impact event (Alvarez et al. 1980; Ganapathy 1980). When Raup and Sepkoski (1984) reported evidence for a periodicity to the marine extinction record over the past 250 Myr, a number of works followed claiming a similar or equivalent periodicity in the terrestrial cratering record (see, e.g., Alvarez and Muller 1984; Davis et al. 1984; Rampino and Stothers 1984a; Whitmire and Jackson 1984). They suggested that this was the result of periodic cometary showers and linked these showers to extinction events. In some cases, these showers were linked to a variety of other global geological phenomena (Rampino and Stothers 1984b). These claims were based on time-series analysis of subsets of the terrestrial cratering record from the then most current published listing in Grieve (1982). There were a variety of selection criteria but they generally conformed to large ($\gtrsim$5 km), young ($\lesssim$250 Myr) and "well-dated" ($\pm$20 Myr) craters. The book edited by Smoluchowski et al. (1986) was mostly dedicated to these issues.

Grieve et al. (1985) argued against these conclusions, noting that the biases in the terrestrial impact record were not considered and that the periods were defined more by the sampling criteria than their reality. Periodicity in the cratering record, however, figured in several contributions in Smoluchowski et al. (1986). These drew upon the earlier studies or were a re-analysis of the data in Grieve (1982). Shoemaker and Wolfe (1986) analyzed their own data compilation of terrestrial craters with ages $\lesssim$250 Myr and with $\lesssim$20 Myr uncertainty and diameters $\gtrsim$5 km. They concluded that the record may consist of a mix of random (asteroidal and cometary) and periodic (cometary) impacts. Grieve et al. (1988) argued that if the uncertainties in crater ages are taken into account, a consistent period cannot statistically be detected and that periodicities in the cratering record are questionable to weak. They also argued that, to have confidence in the reality of any period, the age uncertainties attached to individual ages had to be <10% of the period in question, which was in this case ~30 Myr. This is not a general property of the terrestrial cratering record (Table I). Heisler and Tremaine (1989) reached a similar conclusion based on different statistical arguments. Weissman (1990), using an updated listing of craters (Grieve 1987), also found no evidence for periodic cometary showers and challenged the proposed mechanisms for producing periodic cometary showers. He did not, however, rule out random cometary showers (Weissman 1990).

Despite these arguments, periodic cometary showers, as defined by time-series analysis of the terrestrial cratering record, are still featured (Yabushita 1992) and suggested as a causative agent for various geologic phenomena on Earth (see, e.g., Stothers and Rampino 1990). The original listing that was used to define periodic cometary showers by most workers was Grieve (1982), which contained 103 known terrestrial impact craters. The most current list contains 139 known craters (Table I). In view of these additional data, as well as refinements to age estimates, we have undertaken a time-series analysis of part of the record. In doing this, we do not dismiss the size (Fig. 3) and temporal (Fig. 2) biases in the terrestrial record. Nor do we exclude the problem of the relatively high age uncertainties in the record, which if taken into account, essentially preclude defining any consistent period (Grieve et al. 1988; Heisler and Tremaine 1989). As in previous analyses, consideration is limited to craters with $D \gtrsim 5$ km and ages $\lesssim 250$ Myr with $\lesssim 20$ Myr uncertainty (Table III).

For comparison with previous work (see, e.g., Grieve et al. 1985,1988), we have continued to use an adaptation of the method described by Broadbent (1955,1956) to search for a constant time interval (period) between known impact events. Briefly, the procedure is to examine the hypothesis that a series of crater ages, y_i, may be expressed as:

$$y_i = A + r_i t \, (i = 1, 2 \ldots) \tag{4}$$

where A (phase) and t (period) are constants, and R_i is zero or a positive integer. A similar model for periodicity was used by Rampino and Stothers (1984a, b). As a measure of the goodness-of-fit of the data to a period, we use the variable Q, which is an *rms* measure of q_i, the departure from an integer of the individual observed ages divided by the period in question such that:

$$Q = \left\{ 1/n \sum_{i=1}^{n} (q_i)^2 \right\}^{1/2}. \tag{5}$$

For perfectly periodic data, Q is zero. For random data, Q is an approximately normally distributed variable with a mean of 0.29 and a standard deviation (s.d.) of $0.13/n^{1/2}$. Perfectly periodic ages combined with random ages yield Q values between 0.29 and zero, depending on the relative proportions of periodic and random data. We consider Q values less than 0.29–3 s.d. as indicating the signal of a statistically significant period. This corresponds to less than 1 chance in 100 that the detection of a specific period is the result of a fortuitous combination of random data. A similar level of significance is quoted for periods determined from previous analyses of subsets of the observed cratering record (Alvarez and Muller 1984; Rampino and Stothers 1984a). A 50:50 mixture of periodic and random ages, which is similar in

TABLE III

Impact Craters with $D \gtrsim 5$ km and Ages ≤ 250 Myr with ≤ 20 Myr Uncertainty

Crater	Diameter (km)	Age (Myr)	Method	Source
Zhamanshin	13.5	0.90±0.10	Ar-Ar	Deino and Becker 1990
Bosumtwi	10.5	1.03±0.02	K-Ar	Gentner et al. 1967
El'gygytgyn	18	3.5±0.5	K-Ar	Gurov and Gurova 1980
Bigach	7	6±3	Strat.[a]	Kiselev and Korotushenko 1986
Karla	12	10±10	Strat.	Masaitis et al. 1980
Ries	24	15.1±1.0	Ar-Ar	Stauchacher et al. 1982
Haughton	24	23.4±1.0	Ar-Ar	Jessberger 1988
Logancha	20	25±20	Strat.	Fel'dman et al. 1985
Popigai	100	35±5	Ar-Ar	Bottomley and York 1989
Wanapitei	7.5	37±2	Ar-Ar	Bottomley et al. 1979
Mistastin	28	38±4	Ar-Ar	Mak et al. 1976
Logoisk	17	40±5	Strat.	Masaitis et al. 1980
Montagnais	45	50.5±0.8	Ar-Ar	Bottomley and York 1988
Ragozinka	9	55±5	Strat	Vishnevsky and Lagutenko 1986
Marquez	22	58±2	Strat.	Sharpton and Gibson 1990
Chicxulub	180	64.98±0.05	Ar-Ar	Swisher et al. 1992

Kamensk	20	65±2	Strat.	Movshovich and Milyaskii 1986
Manson	35	65.7±1.0	Ar-Ar	Kunk et al. 1989
Kara	65	73±3	Ar-Ar	Koeberl et al. 1990
Ust-Kara	25	73±3		Twin structure to Kara
Lappajärvi	23	77.3±3.0	Ar-Ar	Jessberger and Reimold 1980
Boltysh	24	88±3	K-Ar	Boiko et al. 1985
Dellen	15	89.0±2.7	Rb-Sr	Deutsch et al. 1992
Steen River	25	95±7	K-Ar	Carrigy 1968
Avak	12	100±5	Strat.	Kirschner et al. 1992
Carswell	39	115±10	Ar-Ar	Bottomley et al. 1989
Mien	9	121.0±2.3	Ar-Ar	Bottomley et al. 1989
Tookoonooka	55	128±5	Strat.	Gorter et al. 1989
Gosses Bluff	22	142.5±0.5	Ar-Ar	Milton and Sutter 1987
Rochechouart	23	186±8	Rb-Sr	Reimold and Oskierski 1987
Manicouagan	100	214±1	U-Pb	Hodych and Dunning 1992
Puchezh-Katunki	80	220±10	Strat.	Fel'dman et al. 1984
Araguainha	40	249±19	Rb-Sr	Engelhardt et al. 1992

[a] Biostratigraphy.

proportion to that suggested by Shoemaker and Wolfe (1986), still yields a clear periodic signal (Grieve et al. 1988). The results on running the algorithm on the data in Table III produced no detectable period at the level of Q less than 0.29–3 s.d. The run with the lowest Q value for any period-phase combination is illustrated in Fig. 12, which indicates a weak signal at ~30 Myr, for a phase of +4 Myr. This period is similar to some previous claims (see, e.g., Alvarez and Muller 1984; Shoemaker and Wolfe 1986) but the signal is even weaker than that in previous analyses (e.g., compare Fig. 12 and Figs. 1 and 4 in Grieve et al. 1985 and 1988, respectively).

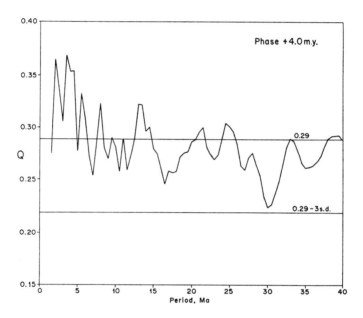

Figure 12.   Plot of statistic Q (see text for definition) against period for crater age data in Table III showing the lowest Q value for all runs with phases between 0 and 20 Myr. Lowest Q is ~30 Myr but fails to meet the criterion of being less than 3 s.d. from the mean for Q (see text for details and discussion).

If the terrestrial cratering record has a periodic signal, it will not be perfect. This is due, in part, to orbital dynamics, which will smear out over several million years the time of impact of an injection of cometary bodies (Hut 1984). More importantly, real age data always have an attached uncertainty, due to experimental error, variability of isotopic ages from different samples and, in some cases, uncertainties in the absolute age of biostratigraphic indicators. To simulate age uncertainties, the algorithm determining Q for each possible phase and period was run 100 times with noise added to each age. The noise was added in such a manner that after 100 cycles it conformed to a normal distribution around each age, with a standard deviation equal to the uncertainty attached to that age estimate in Table III. Using this procedure, the 30 Myr signal was detected in 3% of the runs at the 0.29–3 s.d. level. This level of detection, however, is no higher than a number of other "periods" detected when the age estimates were allowed to randomly fluctuate within their attached uncertainties. The faint signal at 30 Myr is driven largely by two age groupings of craters: four craters with ages between 35 and 40 Myr and three craters with ages around 65 Myr (Table III). In neither case, however, is either group exactly in phase due to the statistical contribution of other craters to defining the lowest Q value.

The ages of some of the craters in Table III differ from those used previously in Grieve et al. (1988), because of more recent isotopic data. For example, Manicouagan is now estimated at 214±1 Myr compared to 212±2 Myr and Dellen is now estimated at 89.0±2.7 Myr compared to 109.6±1 Myr. There are also small differences with some of the ages used in Shoemaker and Wolfe (1986). This illustrates a feature of our current knowledge of the past record of impact on Earth that bears on the question of periodicity. Namely, not only is the precision of some age estimates for craters poor but so is the accuracy. Debating which data set of age estimates is more accurate would resolve little at this time, considering the effects of age uncertainties in detecting a consistent period.

In summary, we find no compelling evidence for the occurrence of a period in the cratering record due to cometary showers. Statistical indications of a period are not strong and, when age uncertainties are considered, period detection may be fortuitous. The question of periodicity in the cratering record is still highly debatable. A campaign to obtain additional age estimates is required. They must also be both accurate and precise ages.

VI. IMPACTING BODY COMPOSITIONS

Small terrestrial impact craters are generally the result of the impact of iron bodies. This is due to the effective removal of weaker stony and icy bodies by atmospheric crushing, dispersal and retardation (Melosh 1981). These small craters are Recent in age and it is often possible to recover pieces of the impacting body from material that spalled off the body prior to impact. At larger diameters, the impacting bodies have undiminished velocities and the impacting body ceases to exist as a physical entity on impact. Shock pressures and corresponding post-shock temperatures are sufficient to melt and vaporize the body (Ahrens and O'Keefe 1977). The one exception is the <100,000 year old Rio Cuarto cratering field (Table I), where a piece of chondritic meteorite has been recovered (Bunch and Schultz 1992). The largest crater of this 10 crater field is 4.5 km in its maximum dimension and it is believed that the preservation of a piece of the impacting body is due to the oblique nature of the impact event, which resulted in decapitation of the impacting body (Schultz and Lianza 1992).

It is possible, however, to get some indication of projectile composition at large terrestrial impact craters. Although the impacting body is destroyed, some of the vaporized impactor material can be admixed with the impact lithologies; particularly, this is the case in impact melt rocks. Thus, through analyses for siderophile, and other trace elements, which are rare in the terrestrial crust and relatively more abundant in certain meteorites, it is possible to detect a meteoritic signal. It would appear, therefore, that the terrestrial record of impact has the potential to discriminate impactor types.

In practice, however, a number of factors militate against the easy application of this procedure. Meteoritic contamination in, for example, impact melt rocks is generally at the 1% level or less, with abundances of siderophile elements at the ppb level. Corrections must also be made for the terrestrial contribution to the siderophile, or other trace element, inventory in impact lithologies. Complications associated with this endogenic correction vary in proportion to the complexity of the lithologies in the target area. Fractionation of siderophile elements can occur, either directly through the vapor and melt phases produced in the impact event (Attrep et al. 1991; Evans et al. 1992; Mittlefehldt et al. 1992) or indirectly through terrestrial processes, such as weathering and alteration (Janssens et al. 1977; Lambert 1982). These factors place limitations on the identification of impacting body composition. An excellent summary of the status of impacting body identification studies and the criteria for assigning a compositional estimate is given in Palme (1982). In some cases, however, careful studies have resulted in the identification of impacting body type, even, in a few cases, down to the level of known meteorite classes. In other cases, the identification of impacting body is more controversial and, in yet others, no meteorite signal has been detected.

Given these caveats, we have assembled from the literature estimates of impacting body compositions at large terrestrial craters (Table IV), in order to examine whether or not there are variations. Such a variation might be expected if the largest terrestrial impact craters were preferentially the result of the impact of cometary, as opposed to asteroidal, bodies (Shoemaker et al. 1990). The sample of impacting body types is dominated by chondritic bodies (Table IV). This is not unexpected, given the even greater dominance of chondrites as a percentage of meteorite falls (Dodd 1981). Table IV, however, may be somewhat misleading. For example, the reliability of impacting body identification in many of the craters in the Commonwealth of Independent States is fairly low, as they are generally based on relatively few elements and elemental ratios. Also, the geochemical signature of chondritic impacting bodies is the most easily identifiable, through their relatively high abundances of both siderophile elements and Cr. Identification of non-chondritic impacting bodies, e.g., at Mistastin (Table IV), can be more of a negative result that the body was not a chondrite rather than the firm indication that it was, in this case, an iron body. Similar arguments apply to achondrites, which as differentiated bodies have compositions most similar to terrestrial rocks.

The largest known terrestrial craters appear, however, to have been

TABLE IV

Impacting Body Types at Large Terrestrial Craters

Crater	Diameter (km)	Body Type	Source
New Quebec	3.4	Chondrite	Grieve et al. 1991
Brent	3.8	Chondrite	Palme et al. 1981
Gow	4	Iron?	Wolf et al. 1980
Rio Cuarto	4.5[a]	Chondrite	Bunch and Schultz 1992
Ilyinets	4.5	Iron	Val'ter 1975, unpublished data
Sääksjärvi	5	Chondrite	Palme et al. 1980
Wanapitei	7.5	Chondrite	Wolf et al. 1980
La Moinerie	8	[b]	Unpublished data
Mien	9	Stone?	Palme et al. 1980
Bosumtwi	10.5	Iron	Jones et al. 1981
Ternovka	12	Chondrite	Val'ter 1988
Nicholson Lake	12.5	Achondrite	Wolf et al. 1980
Zhamanshin	13.5	Chondrite?	Masaitis et al. 1980
Dellen	15	Stone?	Unpublished data
Obolon	15	Iron	Val'ter and Ryabenko 1977
Lappajärvi	17	Chondrite	Gobel et al. 1980
El'gygytgyn	18	Achondrite	Val'ter et al. 1982
Clearwater East	22	Chondrite	Palme et al. 1979
Rochechouart	23	Chondrite?	Wolf et al. 1980
		Iron?	Janssens et al. 1977
Ries	24	Achondrite?	Morgan et al. 1979
		Chondrite?	El Goresy and Chao 1976
Boltysh	25	Chondrite	Val'ter et al. 1987
Mistastin	28	Iron?	Wolf et al. 1980
		Achondrite?	Palme et al. 1978a, b
Clearwater West	32	[b]	Palme et al. 1978a, b
Kara	65	Chondrite	Masaitis et al. 1980
Manicouagan	100	[b]	Palme et al. 1978a, b
Popigai	100	Chondrite	Masaitis and Raikhlin 1985
Chicxulub	180	Chondrite	K/T boundary deposits; Swisher et al. 1992

[a] Longest dimension of largest crater.
[b] Search for meteoritic signature unsuccessful, due either to a sampling problem or to an achondritic impacting body.

formed by chondritic bodies, with the exception of Manicouagan (Table IV). While this does not confirm the suggestion of Shoemaker et al. (1990) that large craters are generally the result of cometary impacts, it is consistent with it. This assumes that cometary bodies have chondritic abundances (Wetherill 1974). Table IV also assumes that the chondritic meteoritic signature at the K/T boundary is the result of the Chicxulub crater (Hildebrand et al. 1991).

Recent analytical data from Chicxulub support its K/T age (Swisher et al. 1992) and indicate elevated Ir values in the impact melt rock (Sharpton et al. 1992). It is not known at present, however, if this Ir anomaly is complemented by anomalies in other siderophile elements and that they have chondritic relative abundances. In the case of Manicouagan, there is no apparent meteoritic signature in the melt rocks (Palme et al. 1978a, b,1981). This may reflect a sampling problem, with the inhomogeneous distribution of meteoritic material within this voluminous, originally $\sim$1000 km^3, impact melt sheet. Only four samples of melt have been analysed to date for siderophile elements. They were, however, widely separated (Palme et al. 1981). There is a small Cr enrichment, which may indicate an achondritic impacting body, but the situation is complicated by the occurrence of relatively Cr-rich mafic gneisses in the target rocks (Palme et al. 1981).

In summary, the evidence for the composition of impacting bodies supplied by the terrestrial impact record indicates a variety of impacting body types (Table IV). It is generally also compatible with the suggestion that the largest terrestrial craters are due to cometary bodies. It must be remembered, however, that the identification of impacting body type at some craters is tentative and, in some cases, the quality and quantity of the analytical data, on which the interpretation is based, is relatively low. As with the question of periodicity, analysis of the current knowledge-base results in a somewhat equivocal interpretation and what is required is a systematic analytical campaign to upgrade the available data base.

VII OTHER RECORDS OF IMPACTS

In addition to the occurrence of actual impact craters, the terrestrial stratigraphic column records a small number of other impact events. The evidence is rare, probably reflecting the extent of detailed examination of the stratigraphic record. The evidence generally takes the form of tektites or layers of microspherules (Table V), either microtektites or microkrystites, the latter being microtektite-like bodies but with crystallites (Glass and Burns 1988). Some of the known tektite and microtektite strewnfields have been related to specific impact craters, others have not (Table V).

Recent work on the North American microtektite strewnfield has demonstrated that it is in reality two (Glass and Burns 1987) or perhaps three (Keller et al. 1987) separate fields, which are closely associated in time. They occurred within 1 Myr of each (Keller et al. 1987) at about 35 Myr (Bottomley and York 1988; Glass et al. 1985). There is, however, some disagreement over the absolute age and they may be $\sim$ Myr younger (Montanari 1990). The occurrence of up to as many as three impact events in a short time period, as recorded by these microtektite strewnfields, could be interpreted as the result of a cometary shower. Although originally referred to as the North American strewnfield, the now defined two or three layers are spatially discrete but overlapping and together form an approximately equatorial band over two-thirds

TABLE V
Other Evidence of Terrestrial Impact Events

Evidence	Age (Myr)	Associated Crater	Source[a]
Australasian tekites and microtekites	0.71±0.10	—	Genter et al. 1970
Darwin glass	0.74±0.04	—	Genter et al. 1969
Ivory Coast tektites and microtektites	1.09±0.20	Bosumtwi	Genter et al. 1970
Urengoite tektites	>1	—	Masaitis et al. 1989
Meteoritic debris, S. Pacific	~2	—	Kyte et al. 1988
Moldavite tektites	14.7±0.7	Ries	Genter et al. 1967
Libyan desert glass	28.5±1.3	—	Genter et al. 1969
N. American tektites, microtektites, and mi-crokrystites[b]	34–35	—	Keller et al. 1987
Impact wave deposits, U. S. east coast[c]	~35	—	Poag et al. 1992
Haitian microtektites	65.01±0.08	Chicxulub	Swisher et al. 1992
S. China microtektites	~365	L. Taihu[d]	Wang 1992
Belgian microtektites	~365	Charlevoix or Siljan	Claeys et al. 1992

[a] Source of age estimate.
[b] May contain two or three layers, which may represent discrete events. See text for discussion.
[c] May be related to N. American tektites.
[d] May be large (>70 km) crater in S. China (He et al. 1991).

of the Earth (see, e.g., Fig. 1 in Keller et al. 1987). Some of the craters, associated with other known strewnfields are not particularly large (Table V) and the average cratering rate suggests that ∼3 impacts capable of producing craters with $D \gtrsim 20$ km are expected somewhere on Earth every 1 Myr. One could make the case, therefore, that there is probably nothing unusual about these occurrences. They are only unusual in the sense that they are known and studied.

Attention was first drawn to this section of the stratigraphic record by the occurrence of the (single) North American strewnfield. Interest was further heightened by the implications of the K/T impact and extinction event, as the age of the strewnfield is close to the Eocene-Oligocene boundary (see, e.g., Glass 1982). It was this increased attention that resulted in the new discoveries. We have little doubt that if the rest of the stratigraphic record was subjected to the same level of expert scrutiny for impact-related materials that there would be additional discoveries. This is already occurring. For example, microspherules, believed to be microtektites, have been reported recently in Upper Devonian strata in China (Wang 1992) and Belgium (Claeys et al. 1992). They have similar ages but have been linked to two different impact craters (Table V).

VIII. THE IMPACT HAZARD

Although the geologic community was slow to recognize and study terrestrial impact craters, it is now obvious that large impact events have occurred on the Earth throughout geologic time. The hypothesis that major terrestrial impacts could lead to mass extinctions is also not new but it was not until Alvarez et al. (1980) and Ganapathy (1980) presented evidence for enhanced siderophile element abundances in K/T boundary clays and interpreted them in terms of an impact event that there was evidence for the hypothesis. Since 1980, the chemical and physical evidence for a major impact event at the K/T boundary has continued to grow until there is currently a compelling case. From estimates of the average terrestrial cratering rate, K/T-sized impact events occur on time scales of ∼100 Myr. At present, however, the K/T is the only mass extinction event with a firm link due to an impact event. Others have been suggested (see, e.g., Bice et al. 1992; McLaren and Goodfellow 1990) but they lack the considerable observational data of the K/T boundary. No doubt, as the stratigraphic and impact records are better known, stronger cases will be able to be made for linking other impact events to events in the biosphere.

The original killing mechanism, suggested for the K/T boundary, was the cessation of photosynthesis due to a global dust cloud (Alvarez et al. 1980). Modeling of the effects of loading the atmosphere with 10^{18} to 10^{19} g of dust are in general agreement with this scenario (Toon et al. 1982) but are heavily dependent on initial parameters, such as dust particle size. It is apparent, however, from these models that dust-loadings orders of magnitude less produce similar effects. This raises the question of why there are no obvious global mass extinction events associated with known large craters in the 100 km diameter range, which produce sufficient global dust according to the models to block photosynthesis (Grieve and Sharpton 1986). It may well be that there were other factors in the K/T impact that resulted in even more catastrophic global effects.

One recent suggestion is that the target rocks had an effect. The 180 km Chicxulub crater (Table I) is of the appropriate size, age and location (Hildebrand et al. 1991; Swisher et al. 1992) to be a prime candidate for being produced in the K/T event. Other impact craters, however, have also been suggested as products of the K/T event. Nevertheless, the Chicxulub target contains both anhydrite and carbonates, which would volatilize on impact (Brett 1992; O'Keefe and Ahrens 1989; Sigurdsson et al. 1991). As sulphur aerosols, derived from the anhydrite, are more effective in reducing light passage than dust (Rampino and Self 1982) and remain in the atmosphere longer (Cadle et al. 1976), they would enhance the global darkness effect. They would also return to the Earth as sulphuric acid, estimated at 600 billion tons (Brett 1992), and, if the return was rapid, this would have an additional serious effects on the biosphere. As high sulphur-bearing rocks are not a common rock-type, their involvement in the K/T event may have contributed to its "uniqueness," in terms of the potential of large impact events to degrade seriously the global environment and produce a mass extinction event.

Other impacts have presumably affected the biosphere, although they did not produce global mass extinctions. Modeling of the K/T event suggests that as little as 10^{16} g of dust is sufficient to reduce photosynthesis to 10^{-3} of normal for at least one growing season (Gerstl and Zardecki 1982). This corresponds to the impact of an asteroidal body in the kilometer size-range, capable of producing a crater in the 15 to 20 km size-range, according to the scaling relations used earlier. Impacts of this size are sufficient to produce atmospheric blow-out about the impact site (Melosh 1989) and, thus, have the capability of distributing globally some of their high-speed, early time ejecta. As noted earlier, based on the estimate of terrestrial crater rate, a few (2–4) such events occur on the Earth every million years. While such events are not likely to have a major effect on the biosphere, their near equivalence in effect to a nuclear winter suggests that they would have disastrous global results for human civilization.

IX CONCLUSIONS

Due to the high level of terrestrial geologic activity, the record of terrestrial impact as evidenced by impact craters is far from complete. There are, however, sufficient known craters with well established ages to determine an average cratering rate, which is applicable to the present-time. This rate is higher than that estimated from the post-mare lunar record but both rate estimates have overlapping uncertainty estimates. It is not clear, therefore, if this rate increase is real, due, possibly, to an increase in cometary impacts. When the knowledge base on terrestrial craters is interrogated for the purposes of determining whether there is an element of periodicity or whether the formation of large craters is due largely to cometary impacts, the results are equivocal. The fault lies, partly, with the relatively small number of craters that have the required data and, partly, with the quality of such data. For example, age uncertainties, in terms of both accuracy and precision, make the search for periodicities little more than a mathematical exercise. In some cases, the required information is in the terrestrial impact record but is not presently available.

The terrestrial impact record is currently the major source of ground truth regarding large-scale impact events. It is a potentially powerful source of information on many aspects of impact phenomena. The detailed study of terrestrial craters by the geologic community is a relatively recent endeavour and has tended to concentrate on observations that can constrain the nature of cratering mechanics and the chemical and physical changes produced in the target rocks by hypervelocity impact. These processes are now fairly well understood in terms of first principles. Given the catastrophic nature of the impact process, and its inevitability, the terrestrial impact record ought to be also studied in sufficient detail to answer some of the outstanding questions concerning the character of the flux of impacting bodies. Clearly, this is currently not the case. While several terrestrial impact craters are recognized each year, the discovery of new craters cannot be the ultimate goal. What is required are further detailed studies beyond the discovery publication. As some of the questions to be addressed require the derivation of relatively sophisticated analytical data (e.g., ages, relative siderophile abundances), systematic campaigns of study are required.

Acknowledgments. Reviews by B. Ivanov and an anonymous reviewer are gratefully acknowledged, as is the assistance of M. Ford and J. Smith in producing the final manuscript. Part of this work was completed while R. A. F. Grieve was an Alexander von Humboldt-Stiftung Forschungspreisträger at the Institut für Planetologie, Münster, receipt of which is also gratefully acknowledged.

REFERENCES

Ahrens, T. J., and O'Keefe, J. D. 1977. Equations of state and impact-induced shock-wave attenuation on the moon. In *Impact and Explosion Cratering*, eds. D. J. Roddy, R. O. Pepin and R. B. Merrill (New York: Pergamon Press), pp. 639–656.

Alvarez, L. W., Alvarez, W., Asaro, F., and Michel, H. V. 1980. Extraterrestrial cause for the Cretaceous-Tertiary extinction. *Science* 208:1095–1108.

Alvarez, W., and Muller, R. A. 1984. Evidence from crater ages for periodic impact on the Earth. *Nature* 308:718–720.

Attrep, M., Orth, C. J., Quintana, L. R., Shoemaker, C. S., Shoemaker, E. M., and Taylor, S. R. 1991. Chemical fractionation of siderophile elements in impactites from Australian meteorite craters/ *Lunar Planet. Sci.* XXII:39–40.

Basaltic Volcanism Study Project. 1981. *Basaltic Volcanism on the Terrestrial Planets* (New York: Pergamon Press).

Bice, D. M., Newton, C. R., McCauley, S., Reiners, P. W., and McRoberts, C. A. 1992. Shocked quartz at the Triassic-Jurassic Boundary in Italy. *Science* 225:443–446.

Boiko, A. K., Val'ter, A. A., and Vishnyak, M. M. 1985. On the age of the Boltysh depression. *Geologicheskii zhurnal* 45:86–90 (in Russian).

Bottomley, R. J., and York, D. 1988. Age measurement of the submarine Montagnais impact crater. *Geophys. Res. Lett.* 15:1409–1412.

Bottomley, R. J., and York, D. 1989. The dating of impact melt rocks using the ^{40}Ar-^{39}Ar method. *Lunar Planet. Sci.* XX:101–102 (abstract).

Bottomley, R. J., York, D., and Grieve, R. A. F. 1979. Possible source craters for the North American tektites—a geochronological investigation. *Eos: Trans. AGU* 60:309 (abstract).

Bottomley, R. J., York, D., and Grieve, R. A. F. 1989. $^{40}Argon$-$^{39}Argon$ dating of impact craters. *Proc. Lunar Planet. Sci. Conf.* 20:421–431.

Brett, R. 1992. The Cretaceous-Tertiary extinction: A lethal mechanism involving anhydrite target rocks. *Geochim. Cosmochim. Acta* 56:3603–3606.

Broadbent, R. S. 1955. Quantum hypotheses. *Biometrika* 42:45–57.

Broadbent, R. S. 1956. Examination of a quantum hypothesis based on a single set of data. *Biometrika* 43:32–44.

Bunch, T. E., and Schultz, P. H. 1992. A study of the Rio Cuarto loess impatites and chondritic impactor. *Lunar Planet. Sci.* XXIII:179–180 (abstract).

Cadle, R. D., Kiang, C. S., and Louis, J.-F. 1976. The global scale dispersion of the eruption clouds from major volcanic eruptions. *J. Geophys. Res.* 81:3125–3132.

Carrigy, M. A. 1968. Evidence of shock metamorphism in rocks from the Steen River structure, Alberta. In *Shock Metamorphism of Natural Materials*, eds. B. M. French and N. M. Short (Baltimore: Mono Book Corp.), pp. 367–378.

Claeys, P., Casier, J.-G., and Margolis, S. V. 1992. Microtektites and mass extinctions: Evidence for a late Devonian asteroid impact. *Science* 257:1102–1104.

Croft, S. K. 1985. The scaling of complex craters. *Proc. Lunar Planet. Sci. Conf.* 15:828–842.

Currie, K. L. 1971. Geology of the resurgent cryptoexplosion crater at Mistastin Lake, Labrador. *Canada Geol. Surv. Bull.* 207:1–62.

Davis, M., Hut, P., and Muller, R. A. 1984. Extinction of species by periodic comet showers. *Nature* 308:715–717.

Deino, A. L., and Becker, T. A. 1990. Laser-fusion $^{40}Ar/^{39}Ar$ ages of acid Zhaman-shinite. *Lunar Planet. Sci.* XXI:271–272 (abstract).

Deutsch, A., Buhl, D., and Langenhorst, F. 1992. On the significance of crater ages—new ages for Dellen (Sweden) and Araguainha (Brazil). *Tectonophysics*, 216:205–218.

Dietz, R. S. 1961. Vredefort ring structure: Meteorite impact scar? *J. Geol.* 69:499–516.

Dodd, R. T. 1981. *Meteorites: A Petrologic-Chemical Synthesis* (New York: Cambridge Univ. Press).

El Goresy, A., and Chao, E. C. T. 1976. Evidence of the impacting body of the Ries crater—the discovery of Fe-Cr-Ni veinlets below the crater bottom. *Earth Planet.*

Sci. Lett. 31:330–340.

Engelhardt, W. V., Matthai, S. K., and Walzebuck, J. 1992. Araguainha impact crater, Brazil. I. The interior part of the uplift. *Meteoritics* 27:442–457.

Evans, N. J., Goodfellow, W. D., Gregoire, D. C., and Veizer, J. 1992. Ruthenium/iridium ratios in the Cretaceous-Tertiary Boundary clay: Implications for global dispersal and fractionation within the ejecta cloud. *International Conf. on Large Meteorite Impacts and Planet Evolution* LPI Contrib. No. 790 (Houston: Lunar and Planetary Inst.), p. 25 (abstract).

Fel'dman, V. I., Suzonovoa, L. V., and Nosova, A. A. 1984. The geological structure and petrography of impactites of the Puchezh-Katunki astrobleme (Volga region). *Byulleten' Moskovoskogo Obshchestva Ispytatelei Prirody* 59:53–64 (in Russian).

Fel'dman, V. I., Mironov, Yu. V., Melikhov, B. A., Ivanov, B. A., and Basilevsky, A. T. 1985. Astroblemes on trap rock: Structural features and differences from impact structures on other targets. *Meteoritika* 44:139–145 (in Russian).

Ganapathy, R. 1980. A major meteorite impact on the Earth 65 million years ago: Evidence from the Cretaceous-Tertiary boundary clay. *Science* 209:921–923.

Gentner, W., Kleinmann, B., and Wagner, G. A. 1967. New K-Ar- and fission track ages of impact glasses and tektites. *Earth Planet. Sci. Lett.* 2:83–86.

Gentner, W., Storzer, D., and Wagner, G. A. 1969. New fission track ages of tektites and related glasses. *Geochim. Cosmochim. Acta* 33:1075–1081.

Gentner, W., Glass, B. P., Storzer, D., and Wagner, G. A. 1970. Fission track ages and ages of deposition of deep-sea microtektites. *Science* 168:359–361.

Gerstl, S. A. W., and Zardecki, A. 1982. Reduction of photosynthetically active radiation under extreme stratospheric aerosol loads. In *Geological Implications of Impacts of Large Asteroids and Comets on the Earth*, eds. L. T. Silver and P. H. Schultz, Geological Soc. of America Special Paper 190 (Boulder: Geological Soc. of America), pp. 201–210.

Glass, B. P. 1982. Possible correlation between tektite events and climatic changes? In *Geological Implications of Impacts of Large Asteroids and Comets on the Earth*, eds. L. T. Silver and P. H. Schultz, Geological Soc. of America Special Paper 190 (Boulder: Geological Soc. of America), pp. 251–256.

Glass, B. P., and Burns, C. A. 1987. Late Eocene crystal-bearing spherules: Two layers or one? *Meteoritics* 22:265–279.

Glass, B. P., and Burns, C. A. 1988. Microkrystites: A new term for impact-produced glassy spherules containing primary crystallites. *Proc. Lunar Planet. Sci. Conf.* 18:455–458.

Glass, B. P., Burns, C. A., Crosbie, J. R., and DuBois, D. L. 1985. Late Eocene North American microtektites and clinopyroxene-bearing spherules. *Proc. Lunar Planet. Sci. Conf.* 16, *J. Geophys. Res. Suppl.* 90:D175–D196.

Gobel, F., Reimold, U., Baddenhausen, H., and Palme, H. 1980. The projectile of the Lappajärvi crater. *Zeit Naturforsch* 35a:197–203.

Gorter, J. D., Gostin, V. A., and Plummer, P. S. 1989. The enigmatic sub-surface Tookoonooka complex in south-west Queensland: Its impact origin and implications for hydrocarbon accumulations. In *The Cooper and Eromanga Basins, Australia*, ed. B. J. O'Neil (Adelaide: Society of Petroleum Engineers), pp. 441–456.

Grieve, R. A. F. 1982. The record of impact on Earth: Implications for a major Cretaceous/Tertiary impact event. In *Geological Implications of Impacts of Large Asteroids and Comets on the Earth*, eds. L. T. Silver and P. H. Schultz, Geological Soc. of America Special Paper 190 (Boulder: Geological Soc. of America), pp. 25–37.

Grieve, R. A. F. 1984. The impact cratering rate in recent time. *Proc. Lunar Planet. Sci. Conf.* 14, *J. Geophys. Res. Suppl.* 89:B403–B408.

Grieve, R. A. F. 1987. Terrestrial impact structures. *Ann. Rev. Earth Planet. Sci.* 15:245–270.

Grieve, R. A. F. 1988. The Haughton impact structure: Summary and synthesis of the results of the HISS project. *Meteoritics* 23:249–254.

Grieve, R. A. F. 1991. Terrestrial impact: The record in the rocks. *Meteoritics* 26:175–194.

Grieve, R. A. F., and Cintala, M. J. 1992. An analysis of differential impact melt-crater scaling and implications for the terrestrial impact record. *Meteoritics* 27:526–538.

Grieve, R. A. F., and Dence, M. R. 1979. The terrestrial cratering record II. The crater production rate. *Icarus* 38:230–242.

Grieve, R. A. F., and Sharpton, V. L. 1986. The K/T impact event: Some implications from the evidence. *Lunar Planet. Sci.* XVII:289–290 (abstract).

Grieve, R. A. F., Sharpton, V. L., Goodacre, A. K., and Garvin, J. B. 1985. A perspective on the evidence for periodic cometary impacts on Earth. *Earth Planet. Sci. Lett.* 76:1–9.

Grieve, R. A. F., Sharpton, V. L., Rupert, J. D., and Goodacre, A. K. 1988. Detecting a periodic signal in the terrestrial cratering record. *Proc. Lunar Planet. Sci. Conf.* 18:375–382.

Grieve, R. A. F., Bottomley, R. B., Bouchard, M. A., Robertson, P. B., Orth, C. J., and Attrep, M. 1991*a*. Impact melt rocks from New Quebec Crater, Quebec, Canada. *Meteoritics* 26:31–39.

Grieve, R. A. F., Stöffler, D., and Deutsch, A. 1991*b*. The Sudbury Structure: Controversial or misunderstood? *J. Geophys. Res.* 96:22753–22764.

Gurov, E. P., and Gurova, E. P. 1980. Shock metamorphosed rocks of the El'gygytgyn meteorite crater in the Chukchi National Okrug. *Meteoritika* 39:102–109 (in Russian).

Halls, H. C., and Grieve, R. A. F. 1976. The Slate Islands: A probable complex meteorite impact structure in Lake Superior. *Canadian J. Earth Sci.* 13:1301–1309.

He, Y., Xu, D., Lu, D., Shen, Z., Lin, C., and Shi, L. 1991. Preliminary study on the origin of Taihu Lake: Inference from shock deformation features in quartz. *Chinese Sci. Bull.* 36:847–851.

Heisler, J., and Tremaine, S. 1989. How dating uncertainties affect the detection of periodicity in extinctions and craters. *Icarus* 77:213–219.

Hildebrand, A. R., Penfield, G. T., Kring, D. A., Pilkington, M., Camargo, A. Z., Jacobsen, S. B., and Boynton, W. V. 1991. Chicxulub Crater: A possible Cretaceous/Tertiary boundary impact crater on the Yucatan Peninsula, Mexico. *Geology* 19:867–871.

Hodych, J. P., and Dunning, G. R. 1992. Did the Manicouagan impact trigger end-of-Triassic mass extinction? *Geology* 20:51–54.

Hut, P. 1984. How stable is an astronomical clock which can trigger mass extinction on earth? *Nature* 311:638–641.

Janssens, M. J., Hertogen, J., Takahashi, H., Anders, E., and Lambert, P. 1977. Rochechouart meteorite crater: Identification of projectile. *J. Geophys. Res.* 82:750–758.

Jessberger, E. K. 1988. [40]Ar-[39]Ar dating of the Haughton impact structure. *Meteoritics* 23:233–234.

Jessberger, E. K., and Reimold, W. U. 1980. A late Cretaceous [40]Ar-[39]Ar age for the Lappajärvi impact crater, Finland. *J. Geophys.* 48:57–59.

Jones, W. B., Bacon, M., and Hastings, D. A. 1981. The Lake Bosumtwi impact crater, Ghana. *Geol. Soc. of America Bull.* 92:342–349.

Keller, G., D'Hondt, S. L., Orth, C. J., Gilmore, J. S., Oliver, P. O., Shoemaker, E. M., and Molina, E. 1987. Late Eocene impact microspherules: Stratigraphy, age and geochemistry. *Meteoritics* 22:25–60.

Kirschner, C. E., Grantz, A., and Mullen, M. W. 1992. Impact origin of the Avak structure, Arctic Alaska, and genesis of the Barrow gas fields. *American Assoc. Petro. Geol. Bull.* 76:651–679.

Kiselev, N. P., and Korotushenko, Yu. G. 1986. The Bigach astrobleme in eastern Kazakhstan. *Meteoritika* 45:119–121 (in Russian).

Koeberl, C., Sharpton, V. L., Harrison, T. M., Sandwell, D., Murali, A. V., and Burke, K. 1990. The Kara/Ust-Kara twin impact structure; A large-scale event in the Late Cretaceous. In *Global Catastrophes in Earth History*, eds. V. L. Sharpton and P. D. Ward, Geological Soc. of America Special Paper 247 (Boulder: Geological Soc. of America), pp. 223–238.

Krogh, T. E., Davis, D. W., and Corfu, F. 1984. Precise U-Pb zircon and baddeleyite ages for the Sudbury area. In *The Geology and Ore Deposits of the Sudbury Structure*, eds. E. G. Pye, A. J. Naldrett and P. E. Giblin (Toronto: Ministry of Natural Resources), pp. 431–446.

Krogh, T. E., Kamo, S. L., Bohor, B. F., and Satterly, J. 1992. U-Pb isotropic results for single shocked and polycrystalline zircons record 550–65.5-Ma ages for a K-T target site and 2700–1850-Ma ages for the Sudbury impact event. *International Conf. on Large Meteorite Impacts and Planet Evolution*, LPI Contrib. No. 790 (Houston: Lunar and Planetary Inst.), pp. 44–45 (abstract).

Kunk, M. J., Izett, G. A., Haugerud, R. A., and Sutter, J. F. 1989. [40]Ar-[39]Ar dating of the Manson impact structure: A Cretaceous-Tertiary boundary crater candidate. *Science* 244:1565–1568.

Kyte, F. T., and Brownlee, D. E. 1985. Unmelted meteoritic debris in the late Pliocene iridium anomaly: Evidence for the ocean impact of a nonchondritic asteroid. *Geochim. Cosmochim. Acta* 49:1095–1108.

Kyte, F. T., Zhou, Z., and Wasson, J. T. 1988. New evidence on the size and possible effects of a late Pliocene oceanic asteroid impact. *Science* 241:63–65.

Lambert, P. 1982. Anomalies within the system: Rochechouart target rock meteorite. In *Geological Implications of Impacts of Large Asteroids and Comets on the Earth*, eds. L. T. Silver and P. H. Schultz, Geological Soc. of America Special Paper 190 (Boulder: Geological Soc. of America), pp. 57–68.

Mak, E. K., York, D., Grieve, R. A. F., and Dence, M. R. 1976. The age of the Mistastin Lake crater, Labrador, Canada. *Earth Planet. Sci. Lett.* 31:345–357.

Marvin, U. B. 1990. Impact and its revolutionary implications for geology. In *Global Catastrophes in Earth History*, eds. V. L. Sharpton and P. D. Ward, Geological Soc. of America Special Paper 247 (Boulder: Geological Soc. of America), pp. 147–154.

Masaitis, V. L., amd Raikhlin, A. J. 1985. The Popigai crater formed by the impact of an ordinary chondrite. *Doklady Akademii Nauk SSSR* 286:1476–1478 (in Russian).

Masaitis, V. L., Danilin, A. I., Mashchak, M. S., Raikhlin, A. I., Selivanovskaya, T. V., and Shadenkov, E. M. 1980. *The Geology of Astroblemes* (Leningrad: Nedra), (in Russian.)

Masaitis, V. L., Ivanov, M. A., Ezersky, V. A., Kozlov, V. S., and Reshetnyak, N. B. 1989. The West Siberian dispersion area of tektite-like glasses. *Doklady Akademii Nauk SSSR* 304:1419–1423 (in Russian).

McLaren, D. J., and Goodfellow, W. D. 1990. Geological and biological consequences of giant impacts. *Ann. Rev. Earth Planet. Sci.* 18:123–171.

Melosh, H. J. 1981. Atmospheric breakup of terrestrial impactors. In *Multi-Ring Basins*, eds. P. H. Schultz and R. B. Merrill (New York: Pergamon Press), pp. 29-35.

Melosh, H. J. 1989. *Impact Cratering: A Geologic Process* (New York: Oxford Univ. Press).

Milkereit, B., Green, A., and Sudbury Working Group. 1992. Deep geometry of the Sudbury structure from seismic reflection profiling. *Geology* 20:807–811.

Milton, D. J., and Sutter, J. F. 1987. Revised age for the Gosses Bluff impact structure, Northern Territory, Australia, based on $^{40}Ar/^{39}Ar$ dating. *Meteoritics* 22:281–289.

Milton, D. J., Barlow, B. C., Brett, R., Brown, A. R., Glikson, A. V., Manwaring, E. A., Moss, F. J., Sedmik, E. C. E., Canson, J., and Young, G. A. 1972. Gosses Bluff impact structure, Australia. *Science* 175:1199–1207.

Mittlefehldt, D. W., See, T. H., and Horz, F. 1992. Dissemination and fractionation of projectile materials in the impact melts from Wabar Crater, Saudi Arabia. *Meteoritics* 27:361–370.

Montanari, A. 1990. Geochronology of the terminal Eocene impacts: An update. In *Global Catastrophes in Earth History*, eds. V. L. Sharpton and P. D. Ward, Geological Soc. of America Special Paper 247 (Boulder: Geological Soc. of America), pp. 607–616.

Morgan, J. W., Janssens, M. J., Hertogen, J., Gros, J., and Takahashi, H. 1979. Ries impact crater, southern Germany: Search for meteoritic material. *Geochim. Cosmochim. Acta* 43:803–815.

Movshovich, E. V., and Milyavskii, A. E. 1986. New data on the formation conditions and age of the Kamensk and Gusev astroblemes. *Meteoritika* 45:112–118 (in Russian).

Nazarov, M. A., Badjukov, D. D., and Alekseev, A. S. 1992. The Kara structure as a possible K/T impact site. *Lunar Planet. Sci.* XXIII:969–970 (abstract).

Neukum, G., Konig, B., Fechtig, H., and Stozer, D. 1975. Cratering in the Earth-Moon systems: Consequences of age determinations by crater counting. *Proc. Lunar Sci. Conf.* 6:2597–2620.

O'Keefe, J. D., and Ahrens, T. J. 1989. Impact production of CO_2 by the Cretaceous-Tertiary extinction bolide and the resultant heating of the Earth. *Nature* 338:247–249.

Palme, H. 1982. Identification of projectiles of large terrestrial impact craters and some implications for the interpretation of Ir-rich Cretaceous/Tertiary boundary layers. In *Geological Implications of Impacts of Large Asteroids and Comets on the Earth*, eds. L. T. Silver and P. H. Schultz, Geological Soc. of America Special Paper 190 (Boulder: Geological Soc. of America), pp. 223–233.

Palme, H., Goebel, E., and Grieve, R. A. F. 1979. The distribution of volatile and siderophile elements in the impact melt of East Clearwater (Quebec). *Proc. Lunar Planet. Sci. Conf.* 10:2465–2492.

Palme, H., Grieve, R. A. F., and Wolf, R. 1981. Identification of the projectile at Brent crater, and further considerations of projectile types at terrestrial craters. *Geochim. Cosmochim. Acta* 45:2417–2424.

Palme, H., Janssens, M. J., Takahashi, H., Anders, E., and Hertogen, J. 1978*a*. Meteoritic material at five large impact craters. *Geochim. Cosmochim. Acta* 42:313–323.

Palme, H., Rainer, W., and Grieve, R. A. F. 1978*b*. New data on meteoritic material at terrestrial impact craters. *Lunar Planet. Sci.* IX:856–858 (abstract).

Palme, H., Rammensee, W., and Reimold, U. 1980. The meteoritic component of impact melts from European impact craters. *Lunar Planet. Sci.* XI:848–850 (abstract).

Pike, R. J. 1977. Size-dependence in the shape of fresh impact craters on the moon. In *Impact and Explosion Cratering*, eds. D. J. Roddy, R. O. Pepin and R. B. Merrill (New York: Pergamon Press), pp. 489–509.

Pike, R. J. 1985. Some morphologic systematics of complex impact structures. *Meteoritics* 20:49–68.

Pilkington, M., and Grieve, R. A. F. 1992. The geophysical signature of terrestrial impact craters. *Rev. Geophys.* 30:161–181.

Pipping, F. 1991. Lappajärvi Impact Crater: Drilling Continued. Geological Survey of Finland Special Paper 12, pp. 33–36.

Poag, C. W., Powars, D. S., Poppe, L. J., Mixon, R. B., Edwards, L. E., Folger, D. W., and Bruce, S. 1992. Deep Sea Drilling Project Site 612 bolide event: New evidence of a late Eocene impact-wave deposit and a possible impact site, U. S. east coast. *Geology* 20:771–774.

Rampino, M. R., and Self, S. 1982. Historic eruptions of Tambore (1815), Kratatau (1883), and Agung (1963), their atmospheric aerosols, and climate impact. *J. Quart. Res.* 18:127–143.

Rampino, M. R., and Stothers, R. B. 1984*a*. Terrestrial mass extinction, cometary impacts and the Sun's motion perpendicular to the galactic plane. *Nature* 308:709–712.

Rampino, M. R., and Stothers, R. B. 1984*b*. Geological rhythms and cometary impacts. *Science* 226:1427–1431.

Raup, D. M., and Sepkoski, J. J. 1984. Periodicity of extinctions in the geologic past. *Proc. National Acad. Sci.* 81:801–805.

Reimold, W. U., and Oskierski, W. 1987. The Rb-Sr age of Rochechouart impact structure, France, and geochemical constraints on impact melt-target rock-meteorite compositions. In *Research in Terrestrial Impact Structures*, ed. J. Pohl (Weisbad, Germany: Fried Vieweg and Sohn), pp. 94–114.

Reimold, W. U., Grieve, R. A. F., and Palme, H. 1981. Rb-Sr dating of the impact melt from East Clearwater, Quebec. *Contrib. Min. Petrol.* 76:73–76.

Reimold, W. U., Barr, J. M., Grieve, R. A. F., and Durrheim, R. J. 1990. Geochemistry of the melt and country rocks of the Lake St. Martin impact structure, Manitoba, Canada. *Geochim. Cosmochim. Acta* 54:2093–2111.

Robertson, P. B., and Sweeney, J. F. 1983. Haughton impact structure: Structural and morphological aspects. *Canadian J. Earth Sci.* 20:1134–1151.

Rondot, J. 1971. Impactite of the Charlevoix structure, Quebec, Canada. *J. Geophys. Res.* 76:5414–5423.

Schmidt, R. M., and Housen, K. R. 1987. Some recent advances in the scaling of impact and explosion cratering. *Intl. J. Impact Eng.* 5:543–560.

Schultz, P. H., and Lianza, R. E. 1992. Recent grazing impacts on the Earth recorded in the Rio Cuarto crater field, Argentina. *Nature* 355:234–237.

Sharpton, V. L., and Gibson, J. W. 1990. The Marquez dome impact structure, Leon County, Texas. *Lunar Planet. Sci.* XXI:1136–1137 (abstract).

Sharpton, V. L., and Ward, P. D., eds. 1990. *Global Catastrophes in Earth History*, Geological Soc. of America Special Paper 247 (Boulder: Geological Soc. of America).

Sharpton, V. L., Dalrymple, G. B., Marin, L. E., Ryder, G., Schuraytz, B. C., and Urrutia-Fucugauchi, J. 1992. New links between the Chicxulub impact structure and the Cretaceous/Tertiary boundary. *Nature* 359:819–821.

Shoemaker, E. M. 1977. Astronomically observable crater-forming projectiles. In *Impact and Explosion Cratering*, eds. D. J. Roddy, R. O. Pepin and R. B. Merrill (New York: Pergamon Press), pp. 617–628.

Shoemaker, E. M., and Shoemaker, C. S. 1990. Proterozoic impact record of Australia. In *International Workshop on Meteorite Impact on the Early Earth*, LPI Cont. No. 746 (Houston: Lunar and Planetary Inst.), pp. 47–48 (abstract).

Shoemaker, E. M., and Wolfe, R. F. 1986. Mass extinctions, crater ages and comet showers. In *The Galaxy and the Solar System*, eds. R. Smoluchowski, J. N.

Bahcall and M. S. Matthews (Tucson: Univ. of Arizona Press), pp. 338–386.

Shoemaker, E. M., Williams, J. G., Helin, E. F., and Wolfe, R. F. 1979. Earth-crossing asteroids: Orbital classes, collision rates with Earth, and origin. In *Asteroids*, ed. T. Gehrels (Tucson: Univ. of Arizona Press), pp. 253–282.

Shoemaker, E. M., Wolfe, R. F., and Shoemaker, C. S. 1990. Asteroid and comet flux in the neighborhood of Earth. In *Global Catastrophes in Earth History*, eds. V. L. Sharpton and P. D. Ward, Geological Soc. of America Special Paper 247 (Boulder: Geological Soc. of America), pp. 150–170.

Sigurdsson, H., Bonte, Ph., Turpin, L., Chaussidon, M., Metrich, N., Steinberg, M., Pradel, Ph., and D'Hondt, S. 1991. Geochemical constraints on source region of Cretaceous/Tertiary impact glasses. *Nature* 353:839–842.

Silver, L. T., and Schultz, P. H., eds. 1982. *Geological Implications of Impacts of Large Asteroids and Comets on the Earth*, Geological Society of America Special Paper 190 (Boulder: Geological Soc. of America).

Smoluchowski, R., Bahcall, J. N. and Mathews, M., eds. 1986. *The Galaxy and the Solar System* (Tucson: Univ. of Arizona Press).

Stauchacher, T., Jessberger, E. K., Dominik, B., Kirsten, T., and Schaeffer, O. A. 1982. ^{40}Ar-^{39}Ar ages of rocks and glasses from the Nordlinger Ries crater and the temperature history of impact breccias. *J. Geophys.* 51:1–11.

Stöffler, D., Avermann, M., Bischoff, L., Brockmeyer, P., Deutsch, A., Dressler, B. O., Lakomy, R., and Muller-Mohr, V. 1989. Sudbury, Canada: Remnant of the only multi-ring (?) impact basin on Earth. *Meteoritics* 24:328 (abstract).

Stothers, R. B., and Rampino, M. R. 1990. Periodicity in flood basalts, mass extinctions and impacts; a statistical view and a model. In *Global Catastrophes in Earth History*, eds. V. L. Sharpton and P. D. Ward, Geological Soc. of America Special Paper 247 (Boulder: Geological Soc. of America), pp. 9–18.

Sutton, S. R. 1985. Thermoluminescence measurements on shock-metamorphosed sandstone and dolomite from Meteor Crater, Arizona. 2. Thermoluminescence age of Meteor Crater. *J. Geophys. Res.* 90:3690–3700.

Swisher, C. C., Grajales-Nishimura, J. M., Montanari, A., Margolis, S. V., Claeys, P., Alvarez, W., Renne, P., Esteban, C.-P. Maurrasse, F., Curtis, G. H., Smit, J., and McWilliams, M. D. 1992. Coeval ^{40}Ar/^{39}Ar ages of 65.0 million years ago from Chicxulub crater melt rock and Cretaceous-Tertiary boundary tektites. *Science* 257:954–958.

Toon, O. B., Pollack, J. B., Ackerman, T. P., Turco, R. P., McKay, C. P., and Liu, M. S. 1982. Evolution of an impact-generated dust cloud and its effects on the atmosphere. In *Geological Implications of Impacts of Large Asteroids and Comets on the Earth*, eds. L. T. Silver and P. H. Schultz, Geological Soc. of America Special Paper 190 (Boulder: Geological Soc. of America), pp. 187–200.

Val'ter, A. A. 1975. Interpretation of the Il'inetsy structure in the Vinnitsa Oblast, Ukraine, as an astrobleme. *Doklady Akademii Nauk SSSR* 224:1377-1380 (abstract).

Val'ter, A. A. 1988. Geochemical features of the meteoritic material in the impactites of the Terny astrobleme. In *Veshchestvo i proiskhozhdenie meteoritov*, pp. 85–92 (in Russian).

Val'ter, A. A., and Ryabenko, V. A. 1977. Impact craters of the Ukrainian shield. *Izd-vo Naukova Dumka, Kiev*, pp. 1–154 (in Russian).

Val'ter, A. A., Barchuk, I. F., Bulkin, V. S., Ogorodnik, A. F., and Kotishevskaya, E. Yu. 1982. The El'gygytgyn meteorite: Probable composition. *Pis'ma v Astronomicheskii Zhurnal* 8:115–120.

Val'ter, A. A., Kolesov, G. M., Fel'dman, V. I., and Kapustkina, I. G. 1987. Contamination of the Boltysh astrobleme impactites with meteoritic matter. *Doklady*

462 R. A. F. GRIEVE AND E. M. SHOEMAKER

Akademii Nauk SSSR 295:164–167 (in Russian).

Vishnevsky, S. A., and Lagutenko, V. N. 1986. The Ragozinka astrobleme: An Eocene crater in the central Urals. *Doklady Akademii Nauk SSSR* 14:1–42 (in Russian).

Wang, K. 1992. Glassy microspherules (microtektites) from an Upper Devonian limestone. *Science* 256:1547–1550.

Weissman, P. R. 1990. The cometary impactor flux at the earth. In *Global Catastrophes in Earth History*, eds. V. L. Sharpton and P. D. Ward, Geological Soc. of America Special Paper 247 (Boulder: Geological Soc. of America), pp. 171–180.

Wetherill, G. W. 1974. Solar system sources of meteorites and large meteoroids. *Rev. Earth Planet. Sci.* 2:303–331.

Wetherill, G. W., and Shoemaker, E. M. 1982. Collision of astronomically observable bodies with the Earth. In *Geological Implications of Impacts of Large Asteroids and Comets on the Earth*, eds. L. T. Silver and P. H. Schultz, Geological Soc. of America Special Paper 190 (Boulder: Geological Soc. of America), pp. 1–13.

Whitmire, D. F., and Jackson, A. A. 1984. Are periodic mass extinctions driven by a distant solar companion? *Nature* 308:713–715.

Wolf, R., Woodrow, A. B., and Grieve, R. A. F. 1980. Meteoritic material at four Canadian impact craters. *Geochim. Cosmochim. Acta* 44:1015–1022.

Yabushita, S. 1992. Periodicity and decay of craters over the past 600 Myr. *Earth, Moon, and Planets* 58:57–63.

Young, R. A. 1977. The lunar impact flux, radiometric age correlation, and dating of specific lunar features. *Proc. Lunar Sci. Conf.* 8:3457–3473.

ARE IMPACTS CORRELATED IN TIME?

D. I. STEEL
Anglo-Australian Observatory and The University of Adelaide

D. J. ASHER
Anglo-Australian Observatory

W. M. NAPIER
Royal Observatory, Edinburgh

and

S. V. M. CLUBE
University of Oxford

We show on the basis of the orbital distribution of known Earth-crossing objects that there is at least one meteoroid stream (the Taurid Complex) which also contains large objects of asteroidal as well as cometary appearance. The sizes of these objects range from ~1 mm to several kilometers. This implies that the stream will produce episodes of major atmospheric detonations on Earth with important consequences for civilization on time scales ~10 to 10^4 yr; that is, impacts are not random in time, but correlated. Present search and mitigation plans (e.g., the proposed Spaceguard survey) are predicated on stochastic catastrophism time scales of ~10^5 to 10^8 yr and therefore neglect the most likely immediate hazard to mankind.

I. INTRODUCTION

The time-averaged (~10^9 yr) space density of Earth-crossers expressed as a power law distribution function of mass and extrapolated from the more conspicuous lunar craters to lower input energies (Shoemaker 1983) leads to a rate of sub-asteroidal inputs to Earth in the kiloton to megaton range (radii ~1–10 m) which is some 2 orders of magnitude less than the contemporary flux whether this be directly (Ceplecha 1992; Chapter by Tagliaferri et al.) or indirectly (see Chapter by Rabinowitz et al.) observed. A contemporary excess of this kind is most plausibly understood as an episodic process involving frequently occurring fragmentation events associated with defunct comets of normal size (Kresák 1978a; Clube 1987), the characteristic upper bound in the size of the fragments produced in these events being ~50 to 100 m. The *prima facie* evidence from Chinese records is of fragmentations occurring most often in the Taurid stream (Astapovič and Terent'jeva 1968; Hasegawa 1992) in accordance with the seasonally enhanced sporadic flux of meteoroids intercepted by the Earth (Štohl 1986). Furthermore the extrapolated population at lower input energies, being $\sim$1% of the total flux, is reassuringly consistent with the meteoritic influx to the Earth (Hughes 1978) if, as expected, the Earth-crosser population at higher masses comprises a substantial fraction of meteoritic bodies from the asteroid belt (e.g., 50%; Wetherill 1988). Given that meteoroid erosion rather than active comet emission is the prime source of zodiacal dust (Grün et al. 1985) and noting that the aphelia of larger meteoroids observed as fireballs are suggestive of a predominantly Taurid stream source (Dohnanyi 1978), the cometary asteroidal component of the Earth-crosser population first suggested by Öpik (1963) is expected to include a substantial Taurid component (Clube and Napier 1984) as has now been independently confirmed (Asher et al. 1993). The apparent presence of a substantial concentration of material at the heart of the Taurid complex (Asher and Clube 1993) is suggestive of inputs to Earth which are closely correlated in time, the subject we explore here.

II. METEOR SHOWERS

We begin with the evidence from meteors and move towards the central tenet of coherent catastrophism (Steel 1991), that as one considers larger particles, of sizes $\sim$10 to 100 m but also including some km-sized bodies, cometary disintegrations lead to constrained clusters of such objects which will have repetitive intersections with the Earth when (i) the node is near 1 AU; and (ii) the cluster passes its node when the Earth is nearby. This would mean that impacts by cluster objects would occur at certain times of year, every few years (depending upon the relationship between the cluster's orbital period and that of the Earth), but only when precession had brought a node to 1 AU, and so on time scales of a few kyr. It seems from the evidence (Sec. I) that there is at least one such cluster of material currently existing, which over the past 20 kyr has produced episodes of atmospheric detonations with significant consequences for the terrestrial environment, and for mankind. This time scale is fairly well established; Whipple's (1940) study of Jovian perturbations on Taurid orbits yielded an age of $\sim$14 kyr for the stream, a time scale of this order being confirmed by subsequent research (Whipple and Hamid 1952; Babadzhanov et al. 1990; Steel et al. 1991).

Anyone watching a meteor shower sees abundant evidence that the influx of extraterrestrial material to our planet is not random in time. A larger fraction of bright meteors (typical sizes 1 cm and larger) appear to be members of showers than the fraction of fainter meteors detected using powerful radars (typical sizes 1 mm and smaller). The general trend is thus for the larger meteoroids to be in streams and the smaller ones to be more dispersed in space. We discuss the reasons for this, and various other details of the points summarized in this chapter, in Asher et al. (1994).

Most meteor showers are observed annually, as the Earth passes the solar longitude at which one of the stream nodes occurs, with roughly equal activity from one year to the next. That is, the meteoroids are distributed quite uniformly along the stream by the effects of radiation pressure and differential ejection velocities without significant concentrations (see, e.g., McIntosh 1991; Steel 1994), so that the number passing perihelion (or a node) at any time is constant. Thus the small particles are spread evenly around the orbit within $\sim 10^2$ to 10^3 yr, although there will still be concentrations near the nucleus of an active parent comet, particularly for the larger particles. Therefore when the Earth passes close by an active parent a meteor storm may be observed, for example every 33 yr in the case of the Leonids because the parent comet (P/Tempel–Tuttle) has a period close to this, and a node very near 1 AU, so that the Earth passes close by the nucleus on a cyclic basis (Yeomans 1981; McIntosh 1991; Kresák 1993).

However, while many meteor showers are observed annually, with broadly consistent activity, this situation does not continue indefinitely because orbital precession takes the node of the stream away from a heliocentric distance of 1 AU on a typical time scale of centuries, followed by a lengthy hiatus before their orbital evolution again allows Earth-intercept. This also results in different branches of a stream producing showers of differing activities at different times (McIntosh 1991; Steel et al. 1991; Babadzhanov and Obrubov 1992). This cyclic, recurrent, activity of meteor showers, meteor storms, and (we suggest) impacts by much larger bodies in complexes, is an essential feature of the orbital evolution under planetary perturbations of such structures.

It is well known that comets often break asunder producing large fragments (see, e.g., Bailey et al. 1990; Yeomans 1991); consider the case of P/Shoemaker–Levy 9. Such physical disintegrations may occur due to passage through the Roche radius of a planet, or in interplanetary space due to thermal stresses or possibly meteoroid impacts (Fernández 1990; Babadzhanov et al. 1991). While many cometary splits result in only minor fragments being lost from a major remnant core, this is not always the case, as evidenced by the Kreutz group of sungrazing comets (Marsden 1989) and several other well-documented splits (Sekanina 1982; Kresák 1991). The size of the largest particles/bodies that may be lost from the nucleus of a comet under normal-type decay, so as to form a meteoroid stream as described by McIntosh (1991), as opposed to catastrophic splitting into a number of similar-sized bodies with much smaller debris, is limited by the gas outflow from the nucleus (Whipple 1951; but see Hajduk 1991). For very large comets (above tens of km), which hold most of the cometary mass in the solar system, it is difficult to explain the release of bodies of mass above $\sim 10^6$ g unless some other form of ejection is occurring. Apart from cataclysmic disruption, the possibilities are differential erosion amongst natural heterogeneities of the appropriate size in the bulk of cometary material and/or the rocket effect as light volatiles evaporate from newly exposed large fragments. Smaller comets can produce such fragments in normal-type decay, but with a limited mass supply. It therefore appears that the disruption of large cometary nuclei is the major source of 10^6 to 10^{12} g ($\sim$1–100 m) bodies. A cometary origin for objects of such sizes as observed by the Spacewatch telescope has been discussed by Rabinowitz (1993). The terrestrial effects of incoming 10 to 100 m objects are of most interest here, although the environmental/climatic perturbations caused by rather smaller objects may also be significant over more extended periods. The role of giant comet break-up in perturbing the terrestrial environment is discussed in the Chapter by Bailey et al.

After such cometary disruptions it is to be expected that many particles, ranging from mm and smaller to $\sim$1 km sizes, will have been released, all in basically similar orbits. This is exactly what is being observed at the time of writing in the case of P/Shoemaker–Levy 9, although in a joviocentric rather than heliocentric orbit. The bulk of the larger bodies will remain in orbits of very similar (a, e) such that the secular perturbations are very similar for each, whereas the smaller particles will spread around the original orbit by the action of differential ejection velocities and radiation pressure, becoming more dispersed in the other orbital elements as their initial variations in (a, e) lead to strong differential perturbations (cf., Steel et al. 1991). Further, if the larger ($\sim$10–100 m) fragments happen to inhabit a mean-motion resonance with Jupiter, as is often observed to occur both in reality and in numerical integrations, then it is to be expected that a cluster of such objects will orbit the Sun grouped not only in the orbital elements $(a, e, i, \omega$ and $\Omega)$, but also in the mean anomaly, M. Such clusters, with associated smaller particles, may be evidenced by the cometary trails described by Sykes and Walker (1992) which Kresák (1993) has interpreted as being the source of meteor storms; cf., Clube and Asher (1990). The presence of these trails with mean anomalies distinct from the apparently-associated comet in each case (and in particular P/Encke) is *prima facie* observational evidence for the scenario that we have described above in a series of logical steps based upon well-known phenomena.

The question that we address, then, is what we would observe if an event like the break-up of P/Shoemaker–Levy 9 were to have occurred in the case of a giant comet deflected from a Jupiter-family orbit in the recent ($<10^5$ yr) past into sub-Jovian space; the likelihood of such an event is detailed in the Chapter by Bailey et al. The answer is that we would expect to find a substantial number of large bodies forming a broad but not fully-relaxed stream, with a wider dispersed complex of smaller meteoroids. We believe that the Taurid meteoroid stream indeed fits this picture.

Taking first the smaller bodies, observed as meteors in the atmosphere, it was demonstrated long ago that the Taurid meteor showers are linked to P/Encke, are exceptional, and are of great scientific interest (Whipple 1940; Whipple and Hamid 1952). Whipple (1967) suggested that P/Encke and its associated stream could be the major source of the zodiacal dust cloud. While most meteor showers last for about a week (e.g., the Perseids and Halleyid showers have large orbits but are retrograde and so affected little by the giant planets; at the other end of the orbit size scale the Geminids have a small aphelion distance and so do not approach Jupiter closely; others have complex orbital evolution under the control of Jupiter, such as the Quadrantid/Comet Machholz complex (Babadzhanov and Obrubov 1992; Steel 1994), the Taurids have aphelia about 0.5 to 1.0 AU within Jupiter, causing spreading about the stream and thus diffuse showers lasting for some months. Thus Štohl (1986) showed that about 50% of the observed "sporadic" background meteors are not sporadic at all, but instead consist of a broad stream of meteoroidal material identifiable with the Taurids.

The large bodies, in our visualization of giant comets being differentiated objects as well as having heterogeneous compositions, containing sub-units of both predominantly refractory (i.e., asteroidal) and volatile (i.e., cometary) material, may appear either asteroidal or cometary in a telescope. In the next section we show through a study of the orbital elements of the discovered Earth-crossing asteroids and comets that the Taurid stream does indeed contain objects ranging in size from ~ 10 m up to several kilometers.

III. THE TAURID COMPLEX

A. The Taurid Meteor Showers and Other Terrestrial/Lunar Phenomena

The Taurid Complex (hereafter TC) of interplanetary objects is known to contain several subordinate nighttime (optical and radar) and daytime (radar) meteor showers, P/Encke, and many Apollo-type asteroids (Clube and Napier 1984; Štohl and Porubčan 1990; Steel et al. 1991; Asher et al. 1993). In the next subsection we present updated evidence for the inclusion of these asteroids into the TC.

The TC also appears to contain other large objects which in the present epoch have orbits intersected by the Earth in the last few days of June each year; one was the $\sim 10^{11}$ g object which entered the atmosphere over the Tunguska region of Siberia in 1908 (Kresák 1978b). There is strong evidence that the TC contains many boulder-sized ($\lesssim 10^6$ g) objects; for example clusters of lunar impacts were seismologically detected in late June 1975 from a direction consistent with the β-Taurid radiant (Oberst and Nakamura 1991). Clube and Asher (1990) pointed out that the IRAS-detected trails associated with several short-period comets (including P/Encke), and thought to be comprised largely of mm-sized particles and larger (Sykes and Walker 1992), may contain a large population of boulder-sized bodies. Passage through such a trail, as must occur from time to time, would result in many coherent impacts upon the Earth (cf., Kresák 1993). The 1975 lunar events, with the terrestrial ionosphere being disturbed at the same time (Kaufmann et al. 1989), presumably by otherwise-unobserved large meteoroid ablation effects, are but an example of the passage of the Earth–Moon system through the periphery of such a complex, assuming the latter is spatially confined and replenished by a trail, such as would arise in the case of a resonant source (cf., Asher and Clube 1993).

Apart from the dynamical arguments based upon the dispersion of the orbits, the recognition that the broad Taurid meteoroid stream is a dominant fraction of the near-ecliptic sporadic meteoroid population (Štohl 1986) additionally demonstrates the recent origin of the TC, because the meteoroid collisional lifetime for such particles is of order 100 kyr (Grün et al. 1985; Olsson-Steel 1986). Thus the TC-derived zodiacal dust cloud may have been of rather higher density over the past 20 kyr implying that the collisional lifetime of the meteoroids was reduced, i.e., the collision rate would vary as $\sim S^2$ rather than S, where S is the spatial density, if the TC were by itself significantly boosting the zodiacal cloud, as suggested by Whipple (1967). Because the Poynting-Robertson lifetime for the smaller particles is of the order of 10 kyr, the decay products are quickly lost; in fact for dust grains with small perihelion distances like the TC, the Poynting-Robertson effect leads to rapid orbit circularization well within 1 AU. At sizable masses, up to $\sim 10^8$g, the sporadic fireball population shows a tendency to have semimajor axes and aphelia typical of the TC (cf., Dohnanyi 1978).

To summarize, a variety of lines of evidence indicate that the Taurid meteoroid stream must be of recent ($\lesssim 20$ kyr) origin. While the physical limitations on the lifetimes of small meteoroids as described above do not apply to macroscopic bodies, any such bodies found to be within the stream must have been produced within such a time scale because otherwise differential perturbations would have led to their orbital dispersal (Asher et al. 1993). In the next section we show that there are indeed many large objects in the TC.

B. The Taurid Asteroids

We have claimed previously that the TC not only includes P/Encke and many meteoroids/meteor showers but many less readily observed larger bodies as well (Clube and Napier 1984; Olsson-Steel 1987; Steel 1992). The existence of km-sized asteroids in the TC is apparently doubted by some workers in the field of minor solar system bodies. This doubt has never been substantiated in the scientific literature so far as we are aware but to convince skeptics, we clearly require indisputable evidence. The data base of orbits of Earth-crossing asteroids, with the known number increasing over the past decade to more than 150, now allows a convincing statistical demonstration.

Meteoroid streams make themselves apparent through showers, which imply similar elements $(a, e, i, \omega$ and $\Omega)$. The condition that the orbit intersect that of the Earth to enable a meteor to be observed applies a restriction on the values of ω which is not applied to asteroidal observations (except that small asteroids must pass close by the Earth to be observable). As the Taurid meteor showers are widely spread in Ω, the nighttime showers persisting from October through to January and even February (Štohl and Porubčan 1990), a criterion using Ω is inappropriate. (Note that this spread in Ω is due to $\tilde{\omega}$ being conserved to first order while Ω and ω change much more rapidly under secular perturbations; the January-February TC meteors are produced by a different ω-precession cycle from those of October-November.) We therefore follow the example of studies in which interrelationships between multiple showers and parent objects are indicated by looking for similarities in a, e, i and $\tilde{\omega}$ (see, e.g., Cook 1973), because the longitude of perihelion $\tilde{\omega}$ varies little over time scales of $\sim 10^4$yr. In order to select asteroids from the list of known orbits we define a reduced D-criterion (c.f., Drummond 1981):

$$D^2 = (\frac{a_1 - a_2}{3})^2 + (e_1 - e_2)^2 + (2 \sin \frac{i_1 - i_2}{2})^2 \qquad (1)$$

where the relative scaling of a and e is broadly in line with that produced by reasonable dynamical models. The reference orbit, based upon observed meteors, is given by $a_1 = 2.1$ AU, $e_1 = 0.82$ and $i_1 = 4°$ (see, e.g., Steel et al. 1991). We impose longitude selection after applying the D-criterion. We use narrower bounds than did Štohl and Porubčan, as we are interested here in a critical selection of Apollos which are in the core of the TC.

Because the inclination varies by a factor of a few over $\sim 10^3$yr time scales, we use Brouwer's (1947) secular perturbation theory to adjust i to be the minimum value that ever occurs, before applying the D-criterion. This is not necessary in the case of meteoroids because i is constrained to be low by the fact that the particle orbit must intersect the Earth's orbit for a meteor to be produced (see, e.g., Steel et al. 1991). We have also allowed for the corresponding variations in eccentricity e, even though these are much less significant with regard to the D-criterion than the i-variations.

Table I (similar to Table 1 of Asher et al. 1993) but updated to include the most recent data—all NEA's through 1993 November—shows the 25 asteroids with the smallest resulting values of D. The elements listed there are current osculating values, not adjusted for secular perturbations. Appended to the list is P/Encke, the only active comet known in the TC (though there is some evidence for cometary activity associated with 2201 Oljato). The orbit used for 5025 P–L is due to E. Bowell, whereas the orbit used for that asteroid by Asher et al. (1993) was from the original computations in the Palomar-Leiden survey; 5025 P–L was only observed over a short arc and so has an uncertain orbit.

Next we consider whether these asteroids, selected only on the basis of similarity of orbital size, shape and inclination (a, e, i) to the TC, turn out to be aligned with the TC on the basis of their longitudes of perihelion $\tilde{\omega}$ ($= \Omega + \omega$). We assume that $\tilde{\omega}$ will be uniformly distributed in the absence of a reason (i.e., one or more common progenitors) for groupings. It is also notable that the high TC value of the eccentricity ($e_1 = 0.82$) is more representative of cometary rather than typical Apollo-asteroidal orbits, and the perihelion distance ($q_1 = 0.38$ AU) is smaller than usual among the discovered Apollo orbits; that is, orbits selected using our D-criterion are noteworthy in themselves as being atypical of the bulk of Apollo asteroid orbits.

A reasonable range of $\tilde{\omega}$ to take as corresponding to the TC, based on the core of the Taurid meteoroid stream/meteor showers, is $140° \pm 40°$. Of the first 5, 10, 15, 20 and 25 asteroids in Table I, respectively, 2, 6, 8, 12 and 13 are within this range. Individually, these longitude alignments would happen by chance with probabilities 0.31, 0.011, 0.0083, 0.00029 and 0.0011 if $\tilde{\omega}$ were randomly oriented.

The significance of these probabilities is discussed in more detail by Asher et al. (1994), but it must be clear to any reader that at a very high confidence level we have demonstrated that of the discovered Earth-crossing asteroids (mainly km-sized) there is a subset aligned with the Taurid meteoroids. In order to portray the situation more vividly we plot the values of D and $\tilde{\omega}$ in Fig. 1. An inspection of Table I shows that there is an increasing number of non-Taurid asteroids as the limit on D is increased. We can justify independently a cut-off at $D \simeq 0.20$ (see Asher et al. 1994), but here we will merely note that this is the limit normally used in studies of meteoroid stream/parent body interrelationships (Drummond 1981). We also note that Štohl and Porubčan (1990) identify the Taurid showers continuing into early February, implying a larger limit on $\tilde{\omega}$ than the 180° used above, and adopting limits of $100° < \tilde{\omega} < 190°$ encompasses the $\sim$10 m asteroid 1991 BA which has $\tilde{\omega} \simeq 190°$, having been observed in mid-January (of course at its node); in fact Štohl and Porubčan (1992) independently associated 1991 BA with the Southern ρ-Geminids, an outlying stream in the broad TC. Figure 1 then clearly shows the relationship between these asteroids and the TC meteors; 13 of the 19 asteroids plotted appear to be members of the TC (with perhaps one or two having such values of $\tilde{\omega}$ by chance), representing $\sim$10% of the entire discovered Apollo population.

Table I and Fig. 1 demonstrate another remarkable alignment. Apart from the 13 TC asteroids, of the other six, five have $\tilde{\omega}$ in the range 222° to 251°. Obviously this is unlikely to have occurred by chance; Asher et al. (1994) discuss the formal probability of such a chance alignment. This cluster we term the Hephaistos group for the best-known (and largest) member. We identify this group and the TC group as significant discoveries in the context of asteroid streams (see also Obrubov 1991).

The similarity in (a, e, i) of the TC and Hephaistos groups is plausibly due to an ancient splitting of a common progenitor, but the difference of $\sim$100° in $\tilde{\omega}$ may argue against this. Clube and Napier (1984) pointed out that short-period comets perturbed by Jupiter into relatively stable Earth-crossing orbits will preferentially have new orbital elements $a \simeq 2.28$ AU, $e \simeq 0.84$, and very low inclinations, remarkably close to the values of P/Encke and Hephaistos. It is possible therefore that the two groups are the remnants of independent large comets, corresponding to the existence of a prominent "entry corridor" into the sub-Jovian region (Napier 1984). The numerical experiments of Wetherill (1991) seem to have confirmed this suggestion. He investigated the decoupling of comets from Jupiter, searching in his simulations for particles reaching $Q < 4.35$ AU. Wetherill's results (his Figs. 4 and 5) show a preponderance of final orbits with $a = 2.2$ to 2.6 AU and $0.7 < e < 0.8$ (i.e., similar to the TC asteroids, P/Encke, and the other bodies in Table I).

We might expect the proportion of asteroids found to be members of

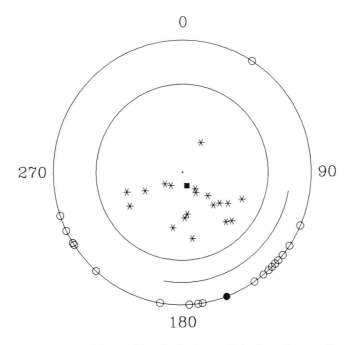

Figure 1. The Taurid asteroids. In the inner circle the polar coordinate is the longitude of perihelion $\tilde{\omega}$ going from 0° to 360° in a clockwise sense, while the radial coordinate is the value of D from Table I. The stars represent the 19 asteroids with $D \lesssim 0.2$, while the solid square is for P/Encke. This is clearly a nonrandom distribution. On the outer circle are plotted (as open symbols) the values of $\tilde{\omega}$ for the 19 asteroids, with P/Encke being the solid symbol. If these were not associated in any way, then one would expect these to be randomly distributed in $\tilde{\omega}$. The arc between the two circles running from $\tilde{\omega} = 100°$ to 190° represents the orientation of the Taurid meteoroids observed as showers on the Earth. Thirteen of the 19 asteroids appear to be associated with these smaller particles, showing that the stream consists of macroscopic objects as well as meteoroids. The other cluster at $\tilde{\omega} = 222°$ to 251° (the Hephaistos group) is discussed in the text.

the TC to be even higher among newly discovered sub-km objects if our hypothesis that the TC is particularly rich in smaller asteroids, due to its having an origin in the disintegration of a giant comet, is correct. In this context we note that the two smallest telescopically observed asteroids, which comprise the two closest observed misses of the Earth, are both members of the TC (1991 BA and 1993 KA$_2$).

Finally, we note that an alternative approach to the orbital analysis, which avoids the subjectivity inherent in the application of the D-criterion but which yields results virtually identical to the above, has also been described (Napier 1993). In this approach, an hypothesis is set up; *significant numbers of asteroids are co-orbiting with P/Encke*, and tested against the null hypothesis, *the numbers of any such co-orbiters do not exceed chance expectancy*. The

TABLE I
Asteroids with Orbital Elements (a, e, i) Similar to
the Taurid Complex Meteoroids[a]

Aligned with Taurids	Others	a (AU)	q (AU)	e	i deg	D	$\tilde{\omega}$ deg
	1991 AQ	2.16	0.50	0.77	3	0.05	222
	2212 Hephaistos	2.17	0.36	0.83	12	0.06	237
1993 KA$_2$		2.23	0.50	0.77	3	0.06	141
1984 KB		2.22	0.52	0.76	5	0.07	146
	2101 Adonis	1.88	0.44	0.76	1	0.10	33
1991 TB$_2$		2.40	0.39	0.84	9	0.10	133
2201 Oljato		2.18	0.63	0.71	3	0.12	173
	1990 SM	2.16	0.49	0.78	12	0.12	244
5143 Hercales		1.83	0.42	0.77	9	0.13	177
5025 P-L		2.14	0.65	0.70	3	0.13	136
4197 1982 TA		2.30	0.52	0.77	12	0.14	129
1991 BA		2.24	0.71	0.68	2	0.16	190
4341 Poseidon		1.84	0.59	0.68	12	0.16	124
	4486 Mithra	2.20	0.74	0.66	3	0.17	251
	1990 TG$_1$	2.48	0.76	0.69	9	0.18	238
1988 VP$_4$		2.26	0.78	0.65	12	0.19	138
1991 GO		1.96	0.66	0.66	10	0.19	114
4183 Cuno		1.98	0.72	0.64	7	0.19	171
1990 HA		2.58	0.79	0.69	4	0.20	133
	1983 VA	2.61	0.81	0.69	16	0.21	89
	1983 LC	2.63	0.76	0.71	2	0.21	344
	1991 EE	2.25	0.84	0.62	10	0.22	283
	4179 Toutatis	2.51	0.90	0.64	0	0.23	43
	4344 P-L	2.62	0.94	0.64	5	0.25	58
1932 UB Hermes		1.65	0.62	0.62	6	0.25	127
P/Encke		2.21	0.33	0.85	12	0.04	161

[a] D is the difference criterion defined in the text, determined using the orbital elements of the observed Taurid meteor showers. The longitude of perihelion $\tilde{\omega}$ is clearly nonrandom (as vividly shown in Fig. 1), indicating that the Taurids continue to km-sizes.

hypothesis may be regarded as *a priori* because it was formulated largely from physical rather than statistical considerations (Clube and Napier 1982), at a time when the properties of the asteroidal Earth-crossing population were barely known.

To test the hypothesis, the 118 Earth-crossers known to March, 1992 were taken and the distribution of osculating orbital elements used to define a probability distribution in the reduced phase-space $(a, e, i, \tilde{\omega})$. The test then consists of comparing the actual numbers of asteroids in the phase space around P/Encke's elements with the numbers expected by chance extraction from this distribution. Beginning with a small element around the comet (in the phase space) and expanding it, one finds that the 4-space encompasses five asteroids when ~0.2 are expected by chance, six when ~0.35 are expected, and so on, until the signal of co-orbiters is lost in the noise of chance associations. In this way it is found that, at a confidence level (each) of ≳99.9%, six asteroids are associated with P/Encke, while a further five are associated with the comet at a confidence level of ≳99%. Observational selection effects are inadequate to lead to the preferential discovery of objects with orbital elements so closely similar to those of P/Encke.

IV. CONCLUSIONS

A commonly held view is that the predominant danger posed to mankind by extraterrestrial objects incoming at hypersonic speeds is due to the occasional arrival of km-plus asteroids and comets, these cataclysmic events occurring randomly in time (see Morrison 1992, and various chapters in this book). This we term *stochastic catastrophism*, which says that it is only these large impactors, above the threshold causing global effects, that are significant to mankind as a whole (as opposed to smaller objects causing only local damage). Such mega-impacts occur on time scales of $\sim 10^5$ yr, with evidence for past such events coming from observations such as the populations and orbits of these objects in space, large impact craters on the Earth, Moon, and other terrestrial planets, mass faunal extinctions in the paleontological record, and so on.

In this chapter we promulgate a contrasting idea (Clube and Napier 1990; Clube 1994*a*), which is supported by astronomical and other physical observations in the present era, and indeed the historical record (such as Chinese and Japanese fireball records; Hasegawa 1992). Under this hypothesis— *coherent catastrophism*—the global hazard on shorter time scales ($< 10^5$ yr) is dominated by smaller objects in the size range ~50 to 300 m which arrive not randomly in time but rather in epochs of high activity. We have discussed elsewhere the mass distribution which results in this size range being identified as the regime of maximum hazard (Asher et al. 1994), and we have also detailed the dynamical arguments which lead to the definition of the epochs of high influx. These epochs, lasting a few centuries every few millennia, are the result of precession, the orbit(s) of the complex(es) of such bodies being brought around to have a node near 1 AU on the latter time scale. In those epochs impacts occur at the same times of year for extended periods, though with the annual impact rate varying because of the concentration of many large meteoroids in mean anomaly in the complex's orbit by their recent low-velocity ejection and/or by a mean-motion resonance with Jupiter.

The dominant contemporary complex is that associated with the four Taurid meteor showers and Comet P/Encke, having been produced in the hierarchical disintegration of a giant comet which arrived in the inner solar

D. I. STEEL ET AL.

system $\sim 2 \times 10^4$yr ago (cf., the Chapter by Bailey et al.). The objects which we believe dominate the extraterrestrial threat to mankind have been largely unobserved until now, although the operation of the Spacewatch telescope has led to the recognition that the population of ~ 10 to 100 m asteroids is 2 orders of magnitude higher than previously believed (but see Clube and Napier 1986). We note that the two smallest Spacewatch objects (1991 BA and 1993 KA$_2$) are apparently members of the Taurid Complex. We predict that future Spacewatch discoveries (this being the only search program currently capable of detecting these smaller objects) will demonstrate a preponderance of objects in Taurid-type orbits; should the *Spaceguard Survey* (Morrison 1992) go ahead then our hypothesis will be verifiable quite rapidly. Already when we consider all known (mostly >0.5 km) Earth-crossing asteroids we find that at a very high confidence level there are two alignments of high-e, low-q, low-i orbits, one with the orientation of the Taurid meteors and P/Encke, and the other finding its archetype in 2212 Hephaistos; these two groups may or may not be genetically linked. The Taurid Complex asteroid group contains $\sim 10\%$ of the known Earth-crossers, indicating the order of their contribution to the impact rate by ~ 1 km objects; the fractional contribution is much higher for 50 to 300 m impactors according to our model (Asher et al. 1994).

Coherent catastrophism, then, says that our major problem is not the once-per-10^5yr large impact with global effects, but rather the multiple-Tunguskas (and larger) which occur in clusters every $\sim 10^3$ years at such times as the orbital evolution of the Taurid Complex (and other similar structures) results in temporary (century or two) nodal intersections with the orbit of the Earth. While global catastrophes of the scale of a km-plus impact do not occur on these shorter time scales, nevertheless such persistent storms of smaller impactors produce deleterious effects which would lead to a sharp decline of civilization as we know it, as they have done in historical times (Clube 1994a, b).

Acknowledgments. We are grateful to A. Carusi for a thoughtful review of our original submission. The work of D. I. S. has been supported by the Australian Research Council, and D. J. A. by the U. K. Science and Engineering Research Council. We dedicate this publication to the memory of Jan Štohl, one of the leading experts on the Taurid Complex, who will be sadly missed.

REFERENCES

Asher, D. J., and Clube, S. V. M. 1993. An extraterrestrial influence during the current glacial-interglacial. *Quart. J. Roy. Astron. Soc.* 34:481–511.

Asher, D. J., Clube, S. V. M., and Steel, D. I. 1993. Asteroids in the Taurid Complex. *Mon. Not. Roy. Astron. Soc.* 264:93–105.

Asher, D. J., Clube, S. V. M., Napier, W. M. and Steel, D. I. 1994. Coherent catastrophism. *Vistas Astron.* 38:1–27.

Astapovič, I. S., and Terent'jeva, A. K. 1968. Fireball radiants of 1–15th centuries. In *Physics and Dynamics of Meteors*, eds. Ĺ. Kresák and P. M. Millman (Dordrecht: D. Reidel), pp. 308–319.

Babadzhanov, P. B., and Obrubov, Yu. V. 1992. Evolution of short-period meteoroid streams. *Cel. Mech. Dyn. Astron.* 54:111–127.

Babadzhanov, P. B., Obrubov, Yu. V., and Makhmudov, N. 1990. Meteor streams of Comet Encke. *Solar System Res.* 24:12–19.

Babadzhanov, P. B., Wu, Z., Williams, I. P. and Hughes, D. W. 1991. The Leonids, Comet Biela and Biela's associated meteoroid stream. *Mon. Not. Roy. Astron. Soc.* 253:69–74.

Bailey, M. E., Clube, S. V. M. and Napier, W. M. 1990. *The Origin of Comets* (Oxford: Pergamon Press).

Brouwer, D. 1947. Secular variations of the elements of Encke's Comet. *Astron. J.* 52:190–198.

Ceplecha, Z. 1992. Influx of interplanetary bodies onto Earth. *Astron. Astrophys.* 263:361–366.

Clube, S. V. M. 1987. The origin of dust in the Solar System. *Phil. Trans. Roy. Soc.* A323:421–436.

Clube, S. V. M. 1994*a*. Hazards from space: Comets in history and science. In *How Science Works In a Crisis: The Mass Extinction Debate*, ed. W. Glen (Stanford: Stanford Univ. Press), pp. 152–169.

Clube, S. V. M. 1994*b*. The global incidence of meteoric airbursts (fireballs) recorded in China and the associated eschatological response in Europe, 0–1900 A. D. *Quart. J. Roy. Astron. Soc.*, submitted.

Clube, S. V. M., and Asher, D. J. 1990. The evolution of Proto-Encke: Dust bands, close encounters and climatic modulations. In *Asteroids, Comets, Meteors III*, eds. C.-I. Lagerkvist, H. Rickman, B. A. Lindblad and M. Lindgren (Uppsala: Univ. of Uppsala Press), pp. 275–280.

Clube, S. V. M., and Napier, W. M. 1982. Spiral arms, comets and terrestrial catastrophism. *Quart. J. Roy. Astron. Soc.* 23:45–66.

Clube, S. V. M., and Napier, W. M. 1984. The microstructure of terrestrial catastrophism. *Mon. Not. Roy. Astron. Soc.* 211:953–968.

Clube, S. V. M., and Napier, W. M. 1986. Mankind's future: An astronomical view. *Interdisc. Sci. Rev.* 11:236–247.

Clube, S. V. M., and Napier, W. M. 1990. *The Cosmic Winter* (Oxford: Basil Blackwells).

Cook, A. F. 1973. A working list of meteor streams. In *Evolutionary and Physical Properties of Meteoroids*, eds. C. L. Hemenway, P. M. Millman and A. F. Cook, NASA SP-319, pp. 183–191.

Dohnanyi, J. S. 1978. Particle dynamics. In *Cosmic Dust*, ed. J. A. M. McDonnell (New York: J. Wiley), pp. 527–605.

Drummond, J. D. 1981. A test of comet and meteor shower associations. *Icarus* 45:545–553.

Fernández, J. A. 1990. Collisions of comets with meteoroids. In *Asteroids, Comets,*

476 D. I. STEEL ET AL.

Meteors III, eds. C.-I. Lagerkvist, H. Rickman, B. A. Lindblad and M. Lindgren (Uppsala: Univ. of Uppsala Press), pp. 309–312.

Grün, E., Zook, H. A., Fechtig, H. and Giese, R. H. 1985. Collisional balance of the meteoritic complex. *Icarus* 62:244–272.

Hajduk, A. 1991. Evolution of cometary debris: Physical aspects. In *Comets in the Post-Halley Era*, eds. R. L. Newburn, Jr., M. Neugebauer and J. Rahe (Dordrecht: Kluwer), pp. 592–606.

Hasegawa, I. 1992. Historical variation in the meteor flux as found in Chinese and Japanese chronicles. *Cel. Mech. Dyn. Astron.* 54:129–142.

Hughes, D. W. 1978. Meteors. In *Cosmic Dust*, ed. J. A. M. McDonnell (New York: J. Wiley), pp. 123–185.

Kaufmann, P., Kuntz, V. L. R., Paes Leme, N.M., Piazza, L. R., Vilas Boas, J. W. S., Brecher, K., and Crouchley, J. 1989. Effects of the large June 1975 meteoroid storm on Earth's ionosphere. *Science* 246:787–790.

Kresák, Ľ. 1978a. The comet and asteroid population of the Earth's environment. *Bull. Astron. Inst. Czechoslovakia* 29:114–125.

Kresák, Ľ. 1978b. The Tunguska object: A fragment of comet Encke? *Bull. Astron. Inst. Czechoslovakia* 29:129–134.

Kresák, Ľ. 1991. Evidence for physical aging of periodic comets. In *Comets in the Post-Halley Era*, eds. R. L. Newburn, Jr., M. Neugebauer and J. Rahe (Dordrecht: Kluwer), pp. 607–628.

Kresák, Ľ 1993. Cometary dust trails and meteor storms. *Astron. Astrophys.* 279:646–660.

Marsden, B. G. 1989. The sungrazing comet group. II. *Astron. J.* 98:2306–2321.

McIntosh, B. A. 1991. Debris from comets: The evolution of meteor streams. In *Comets in the Post-Halley Era*, eds. R. L. Newburn, Jr., M. Neugebauer and J. Rahe (Dordrecht: Kluwer), pp. 557–591.

Morrison, D., ed. 1992. *The Spaceguard Survey: Report of the NASA International Near-Earth-Object Detection Workshop* (Pasadena: Jet Propulsion Laboratory).

Napier, W. M. 1984. The orbital evolution of short-period comets. In *Asteroids, Comets, Meteors*, eds. C.-I. Lagerkvist and H. Rickman (Uppsala: Univ. of Uppsala Press), pp. 391–395.

Napier, W. M. 1993. Earth-crossing asteroid groups. In *Meteoroids and Their Parent Bodies*, eds. J. Štohl and I. P. Williams (Bratislava: Astron. Inst. of the Slovak Acad. of Sci.), pp. 123–126.

Oberst, J., and Nakamura, Y. 1991. A search for clustering among the meteoroid impacts detected by the Apollo lunar seismic network. *Icarus* 91:315–325.

Obrubov, Yu. V. 1991. Complexes of minor solar system bodies. *Soviet Astron.* 35:531–537 (English trans.).

Olsson-Steel, D. 1986. The origin of the sporadic meteoroid component. *Mon. Not. Roy. Astron. Soc.* 219:47–73.

Olsson-Steel, D. 1987. Asteroid 5025 P–L, comet 1967 II Rudnicki, and the Taurid meteoroid complex. *The Observatory* 107:157–160.

Öpik, E. J. 1963. The stray bodies in the solar system. Part I. Survival of cometary nuclei and the asteroids. *Adv. Astron. Astrophys.* 2:219–262.

Rabinowitz, D. L. 1993. The size distribution of Earth-approaching asteroids. *Astrophys. J.* 407:412–427.

Sekanina, Z. 1982. The problem of split comets in review. In *Comets*, ed. L. Wilkening (Tucson: Univ. of Arizona Press), pp. 251–287.

Shoemaker, E. M. 1983. Asteroid and comet bombardment of the Earth. *Ann. Rev. Astron. Astrophys.* 11:461–494.

Steel, D. 1991. Our asteroid-pelted planet. *Nature* 354:265–267.

Steel, D. 1992. Additions to the Taurid Complex. *The Observatory* 112:120–122.

Steel, D. I. 1994. Meteoroid streams. In *Asteroids, Comets, Meteors 1993*, eds. A. Milani, M. Di Martino and A. Cellini (Dordrecht: Kluwer), in press.

Steel, D. I., Asher, D. J., and Clube, S. V. M. 1991. The structure and evolution of the Taurid Complex. *Mon. Not. Roy. Astron. Soc.* 251:632–648.

Štohl, J. 1986. The distribution of sporadic meteor radiants and orbits. In *Asteroids, Comets, Meteors II*, eds. C.-I. Lagerkvist, B. A. Lindblad, H. Lundstedt and H. Rickman (Uppsala: Univ. of Uppsala Press), pp. 565–574.

Štohl, J., and Porubčan, V. 1990. Structure of the Taurid Meteor Complex. In *Asteroids, Comets, Meteors III*, eds. C.-I. Lagerkvist, H. Rickman, B. A. Lindblad and M. Lindgren (Uppsala: Univ. of Uppsala Press), pp. 571–574.

Štohl, J., and Porubčan, V. 1992. Dynamical aspects of the Taurid meteor complex. In *Chaos, Resonance and Collective Dynamical Phenomena In The Solar System*, ed. S. Ferraz-Mello (Dordrecht: Kluwer), pp. 315–324.

Sykes, M. V., and Walker, R. G. 1992. Cometary dust trails. I. Survey. *Icarus* 95:180–210.

Wetherill, G. W. 1988. Where do the Apollo objects come from? *Icarus* 76:1–18.

Wetherill, G. W. 1991. End products of cometary evolution: cometary origin of Earth-crossing bodies of asteroidal appearance. In *Comets in the Post-Halley Era*, eds. R. L. Newburn, Jr., M. Neugebauer and J. Rahe (Dordrecht: Kluwer), pp. 537–556.

Whipple, F. L. 1940. Photographic meteor studies. III. The Taurid shower. *Proc. Amer. Phil. Soc.* 83:711–745.

Whipple, F. L. 1951. A comet model. II. Physical relations for comets and meteors. *Astrophys. J.* 113:464–474.

Whipple, F. L. 1967. On maintaining the meteoritic complex. In *The Zodiacal Light and the Interplanetary Medium*, ed. J. L. Weinberg, NASA SP-150, pp. 409–426.

Whipple, F. L., and Hamid, S. E. 1952. On the origin of the Taurid meteor streams. *Helwan Obs. Bull.* 41:1–30.

Yeomans, D. K. 1981. Comet Tempel-Tuttle and the Leonid meteors. *Icarus* 47:492–499.

Yeomans, D. K. 1991. *Comets: A Chronological History of Observations, Science, Myth, and Folklore* (New York: J. Wiley).

HAZARDS DUE TO GIANT COMETS: CLIMATE AND SHORT-TERM CATASTROPHISM

M. E. BAILEY
Liverpool John Moores University

S. V. M. CLUBE
University of Oxford

G. HAHN
DLR, Institut für Planetenerkundung, Berlin

W. M. NAPIER
Royal Observatory Edinburgh

and

G. B. VALSECCHI
IAS-Planetologia, Roma

Orbital integrations of the outer solar system bodies (2060) Chiron and (5145) Pholus reveal that they evolve into Jupiter-family cometary orbits on time scales 10^4 yr to 10^6 yr. The half-life for Chiron to become Jupiter-crossing is 1 to 2×10^5 yr, that for Pholus roughly 10 times longer. Assuming a cumulative size distribution proportional to d^{-2}, giant comets with diameters $d \gtrsim 100$ km enter the Jupiter family at intervals on the order of $(2 \text{ to } 20) \times 10^4$. Perturbations cause some to evolve on to sub-Jovian orbits similar to those of P/Encke or asteroids in the Taurid Complex while others pass close to the Sun or Jupiter. Long-term integrations of the asteroid (5335) Damocles, in a high-inclination Halley-type cometary orbit, and of Halley-type comets generally, also demonstrate dynamical evolution on short time scales in the range 10^4 to 10^6 yr. Such bodies evolve from parabolic orbits with perihelia q within the orbit of Jupiter to sungrazing orbits similar to those of the Kreutz group or the future orbit of P/Machholz. These results show that giant comets are inserted on to short-period Earth-crossing orbits on time scales 2 orders of magnitude shorter than that (10^7–10^8 yr) for dynamical evolution of Earth-crossing asteroids from the main belt. The hazard due to comets and asteroids therefore encompasses not only the collision of large bodies with the Earth and the frequency and duration of the small-q state, but also effects due to passage of the Earth through streams of debris from recently fragmented giant comets. We highlight the most active Earth-crossing asteroid-meteoroid stream, namely the Taurids, and the effects of debris from bodies in high-inclination sungrazing orbits exemplified by the Kreutz group. The Taurid Complex represents the dominant hazard on time scales of most immediate concern to civilization ($\lesssim 10^4$ yr). Both low-inclination and high-inclination cometary populations illustrate the importance of giant comets as sources of material in Earth-colliding orbits. Giant comets wholly or partially originate from a variable near-parabolic flux, so their long-term terrestrial effects will be time dependent and predominantly cometary over meteoritic. The existence of large comets in the inner solar system leads to prolonged periods of stratospheric dusting, producing optical depths in the range ~0.1 to 1.0 on a variety of time scales. Climatic consequences of this influx are reviewed in the light of palaeoclimatic evidence and known threshold, mode-switching and global feedback mechanisms. Cometary debris, ranging in size from $\lesssim 1$ μm to $\gtrsim 1$ km, plays a crucial role in biological evolution (e.g., the K/T boundary event) and is the source of a complex civilization hazard. Search strategies should look beyond the mere enumeration of current near-Earth objects.

I. INTRODUCTION

Recent years have seen rapid growth in our knowledge of the dynamics and physical makeup of the so called small bodies—comets and asteroids—in the solar system. Imaging of Halley's comet demonstrated the remarkably low albedo of this object, and immediately led to an increase in the perceived size of a typical cometary nucleus from 2 to 3 km to a figure closer to 5 to 10 km, with a corresponding order-of-magnitude increase in the assumed cometary mass (cf., masses of P/Halley quoted by Hughes [1983,1987a]). Observations of P/Halley also established the idea that cometary nuclei contain a significant fraction of nonvolatile or (carbonaceous) chondritic material: icy dirtballs (Keller 1989; Sykes 1993) rather than the canonical dirty snowball (Whipple 1950).

This picture is consistent with observational evidence that cometary mass loss involves ejection of material with a net dust-to-gas mass ratio greater than unity (Green et al. 1987; Curdt and Keller 1990; McDonnell et al. 1991,1993), and in a form where the total mass is dominated by millimeter-sized particles or larger. Further insight into the mass loss mechanism was provided by the infrared astronomical satellite IRAS, which discovered streams of dust associated with both short-period comets and unseen, presumably asteroidal bodies, extending both in front of and behind the presumed cometary source (Sykes et al. 1986; Sykes and Walker 1992). These observations, and those of the cometary trail generated by comet P/Shoemaker-Levy 9, expected to bombard Jupiter during late July 1994 (Carusi et al. 1993; Sekanina 1993; Yeomans 1993), suggest that the disintegration process giving rise to trails is generally hierarchical in character. Dust trails are a common structure in the inner solar system, and their presence provides an explanation for occasional intense meteoroid showers—so-called meteor storms—caused by passage of the Earth through their dense cores (Kresák 1980a,1993; Clube 1987). Taken together these results provide new insight into both comets and the distribution of material they produce in the inner solar system.

This chapter considers the way the perceived hazard is affected due to cometary debris. Taking the long view the hazard is essentially unbounded, dominated by impacts of asteroids and cometary nuclei and by large-scale accretion of dust; in the short run the most likely events are impacts by bodies less than a kilometer in size, causing devastation on a local or regional scale and intermittent stratospheric dusting. Throughout this chapter, we emphasize the importance of giant comets, those with diameters $d \gtrsim 100$ km, in determining the detailed space and mass distributions of potential impactors near the Earth.

In fact, giant comets are the main carriers of mass to the inner solar system, the principal orbital paths through which they arrive being represented by Chiron and the Kreutz Group progenitor, respectively. These orbital regimes are respectively of low ($\lesssim 20°$) and high ($\gtrsim 20°$) inclination and originate from the isotropic long-period cometary flux as a result of planetary deflections. The low-inclination regime may also be fed by evolution from trans-Neptunian reservoirs, such as the Kuiper belt, an intermediate-period cometary flux (possibly enhanced by planetary deflections from an intermittently increased long-period flux), or an inner core of the Oort cloud (presumably of primordial origin). However, while the relative contribution of these additional sources (if they exist) is not known, it is clear that low-inclination orbits feed bodies into the short-period Jupiter family whereas high-inclination orbits feed bodies into the Halley family. The chaotic nature of these orbital transitions allows comets with perihelion distances far beyond the Earth to be transferred into Earth-crossing orbits on astronomically short time scales of $\lesssim 10^6$ yr.

These time scales are substantially shorter than those obtained for the dynamical lifetime of bodies in sub-Jovian phase space (10^7–10^8 yr), so any census of potential Earth-crossers must include bodies of cometary origin which are not presently Earth-crossing. Moreover, because giant comets dominate the cometary mass function and arrive in Earth-crossing orbits at intervals on the order of 10^5 yr, undergoing disruption and eventual dispersion as dust close to their orbital planes on time scales of 10^5 to 10^6 yr, compared to 10^4 to 10^5 yr for comets a few km in size, it is to be expected that the total flux of material incident on the Earth per 10^5 yr will be substantially greater than that brought in by a typical 1 km asteroid impacting the Earth during such an interval.

The apportionment of this material among smaller bodies depends on the disruption sequence and orbital spread. However, material arriving in near-Earth space with diameters $\lesssim 10$ m appears to be largely cometary in origin and only a few percent meteoritic (i.e., originating in the main asteroid belt), and it is therefore to be expected that the incident population will include many larger sub-km meteoroids, in accordance with observations of fireballs (Ceplecha 1992; Chapter by Tagliaferri et al.) and the latest Spacewatch findings (Rabinowitz 1993a; Rabinowitz et al. 1993). Thus, while occasional collisions of the Earth with kilometer-sized asteroids and cometary fragments impart major shocks to the terrestrial system at intervals $\sim 10^5$ to 10^6 yr (Grieve 1991; Morrison 1992; Chapter by Grieve and Shoemaker), dust and co-orbiting streams of meteoroids and small bodies with diameters in the range 10 to 300 m should also be counted as exerting a significant influence on shorter time scales ($\sim 10^2$ to 10^4 yr; Clube 1987,1994a). These astronomical findings have fundamentally changed our perception of the Earth's near-space environment and provide a unique opportunity, for example through deflection or destruction of a potential impactor (Bailey 1993; Canavan et al. 1992), for mankind actively to intervene in cosmic processes. Such processes, up to now, have dominated the course of biological evolution on the Earth and provided the celestial backdrop against which regression in civilization and periods of major social change have occurred in historical times.

II. GIANT COMETS

The idea that the observed short-period comet system might be dominated by the fragmentation of a single large progenitor goes back more than a century (Alexander 1850,1851; Bredichin 1889; Bobrovnikoff 1931). Recent work by Clube and Napier (1984,1986a, b,1990) has brought the theory up to date and into the frame of modern discussions including solar system dynamics, long-term terrestrial evolution and the nature of the celestial hazard to civilization. Empirically, it has been known for some time that giant comets may dominate the mass function (Clube and Napier 1984), while there is a generally presumed tendency among Oort cloud members towards increased resilience and reduced luminosity as a result of splitting and physical evolution during close perihelion passage (Oort 1950; Oort and Schmidt 1951; Clube 1992).

These results accord naturally with the short-period comet excess (Rickman 1990,1992) and the presence of a large cometary asteroid, Chiron, which has a high probability of having formerly been much closer to the Sun (Hahn and Bailey 1990). On dynamical grounds Chiron is most likely to be a significantly evolved object, although the *prima facie* evidence also indicates that bodies this size may be differentiated (see, e.g., Yabushita 1993), a factor which should be borne in mind when discussing their origin in the solar nebula, protoplanetary disk or elsewhere. Irrespective of detailed considerations about their origin, however, the idea that disintegrating giant comets may play a significant role in terrestrial evolution, notably the K/T extinction (Napier and Clube 1979; Clube and Napier 1984), has proved attractive and has now been followed up by a number of groups (Clube and Napier 1986a,1990; Zahnle and Grinspoon 1990; Steel 1991; Hut et al. 1991a, b; Shoemaker 1991; Shoemaker and Izett 1992a, b; Shoemaker 1993). The question arises, therefore, what is their evolution and how do they affect the inner solar system? In this section we review the evidence for massive comets and then consider recent results on their long-term dynamical evolution in Sec. III. The following Secs. IV and V discuss implications of these results for bodies in Earth-crossing orbits and for the K/T extinction event, emphasizing the resulting short-term and long-term terrestrial effects. Our conclusions are summarized in Sec. VI.

A. Luminosity Distribution

The main direct argument for the existence of giant comets comes from the observed long-period cometary flux as a function of total visual absolute magnitude H_{10}. The detailed investigation by Everhart (1967), who modeled a range of observational selection effects in order to obtain the intrinsic brightness distribution, produced a double power-law distribution with a break close to $H_{10} = 5.4$. In terms of the long-period flux ($P > 200$ yr) brighter than absolute magnitude H_{10}, Everhart's cumulative brightness distribution normalized to one long-period comet brighter than $H_{10} = 7$ passing perihelion per year with perihelion distance <1 AU may be written (Bailey 1991)

$$F_{LP}(\leq H_{10}) = \begin{cases} 1.61 \times 10^{-2} \times 10^{0.27 H_{10}} - 0.25 & \text{for } H_{10} > 5.4 \\ 1.40 \times 10^{-4} \times 10^{0.59 H_{10}} & \text{for } H_{10} < 5.4 \end{cases} \quad (1)$$

The mean interval between passages of a zero-magnitude comet within the Earth's orbit is thus on the order of 7000 yr. However, by restricting his analysis to a relatively small sample which did not include the historically brightest comets, it is possible that the logarithmic slope at the bright end, $a_1 = 0.59$, may be too high. An alternative approach, adopted for example by Hughes and Daniels (1982), Donnison (1986) and Hughes (1987b), is to accept the observed brightness distribution more or less at face value. This leads to a weaker dependence of the long-period flux on total visual absolute magnitude, namely

$$F_{LP}(\leq H_{10}) = 8 \times 10^{-3} \times 10^{0.30 H_{10}} \quad (2)$$

where we have again normalized the equation so that $F_{LP}(\leq 7) = 1$ comet AU^{-1} yr^{-1}.

A third approach, taken by Shoemaker and Wolfe (1982), is to count observed long-period comets as a function of absolute blue nuclear magnitude B. Their discussion implies

$$F_{LP}(\leq B) = 237 \times \exp[0.91(B - 18)] \quad (3)$$

which reduces to

$$F_{LP}(\leq H_{10}) = 4.2 \times 10^{-2} \times 10^{0.32 H_{10}} \quad (4)$$

when the mean relationship between B and H_{10} derived by Bailey et al. (1992a), namely $B \simeq 8.5 + 0.8 H_{10}$, is used to intercompare the two figures. This suggests that the slope of the total magnitude distribution is indeed close to 0.3, although the predicted flux from Eq. (4) at $H_{10} = 7$ is 7 times larger than that from Eq. (2). Because the detailed arguments of Shoemaker and Wolfe (1982) suggested that they had possibly overestimated the flux by about this amount, in what follows we assume that a law of the general form of Eq. (2), with an accuracy of about a factor of 2, describes the present long-period comet flux versus total absolute magnitude. These figures imply that a relatively massive long-period comet with total absolute magnitude $H_{10} \lesssim 0$ passes within the Earth's orbit about once every 125 yr.

The process of converting these results to nuclear diameter d, and hence mass, for example using a formula of the form

$$d = A \times 10^{-\alpha H_{10}} \text{ km} \qquad (5)$$

involves large uncertainties (Bailey 1990a). Hughes (1985,1987a,1990) has adopted values for A in the range 70 to 140 and $\alpha = 0.2$ (luminosity proportional to surface area), while Bailey (1990a,1991) has suggested $A \simeq 188$ and $\alpha = 0.16$ (i.e., between $\alpha = 0.2$ and $\alpha = 0.13$, the latter corresponding to luminosity proportional to mass). On this basis we would expect a zero magnitude comet to have a diameter on the order of 150 ± 30 km, while the long-period flux versus nuclear diameter relation would then have a logarithmic slope a_1/α in the range -1.5 to -2. These figures necessarily involve extrapolation, because a giant comet has not been observed in the inner solar system in modern times, but the evidence is consistent with a long-period flux at the bright end given approximately by

$$F_{\text{LP}}(\geq d) \approx 1 \times (d/5 \text{ km})^{-2} \text{ comets AU}^{-1} \text{ yr}^{-1}. \qquad (6)$$

For comparison, the independent analysis of Shoemaker et al. (1990) and Shoemaker and Shoemaker (1990) leads to a formula of the same general form, but with a numerical coefficient of 9 instead of 1. We thus estimate that a giant long-period comet with diameter > 100 km crosses the Earth's orbit about once every 400 yr, while a body of similar size passes within the orbit of Jupiter about once in a human lifetime. A further comparison may be made with the recently disrupted comet P/Shoemaker-Levy 9 (1993e), which had a visual magnitude close to 13 at the distance of Jupiter, suggesting $H_{10} \simeq 3$ and an original diameter in the range between 30 and 60 km. However, we note that recent work has indicated a much smaller diameter for the original nucleus of this body (Scotti and Melosh 1993; Sekanina 1993; Waddington 1993), suggesting that the mean diameter-magnitude relation should probably not be used for a body as unusual as this split comet (see Fig. 9 below). Of course, the disagreement may also indicate the presence of systematic errors in the calibration of cometary masses using total magnitudes, involving questions of methodology and issues such as coma contamination (Jewitt 1991). The problem of determining cometary masses at the bright end, and of understanding the detailed correlation with total absolute magnitude remains an important source of uncertainty in determining the cometary flux through the inner solar system as a function of size. However, although we expect that the numbers given in Eq. (2) are uncertain by at least a factor of 2, the general conclusion that giant comets frequently visit the inner solar system cannot easily be avoided, and is consistent with observations of comet Sarabat in 1729 which became a naked-eye object despite having a perihelion beyond 4 AU. The comet was observed for a period of almost 6 months, and had a total absolute magnitude close to -3.

B. Further Evidence

The second direct argument for the existence of giant comets comes from observations of the Kreutz sungrazing group, the result of hierarchical fragmentation of a giant comet that was first observed around 370 BC (Marsden 1989). The size of the progenitor is difficult to assess, but Öpik (1966) estimated an initial diameter of 114 km.

Observations also demonstrate the existence of at least two families of outer solar system bodies, with diameters, adopting conventional low albedos, in the range between 100 and 300 km: (2060) Chiron and (5145) Pholus, and (among others, see Table II below) 1992 QB$_1$ and 1993 FW (Jewitt and Luu 1993). The orbits of the first two of these objects resemble those achieved by Jupiter-family comets during different phases of their evolution, and there is a significant probability that both Chiron and Pholus will become part of the Jupiter family within the next 0.1 to 1.0 Myr (Hahn and Bailey 1990; Bailey et al. 1992b). The dynamics of the others remain uncertain, definitive orbits not yet being established, but if such bodies become Neptune crossing on a time scale comparable with the age of the solar system, the outer belt of large objects could be an important source of comets for the Jupiter family.

Because the inferred giant-comet injection rate from these sources, $\approx$ 10 Myr^{-1}, is comparable to the rate of impacts of kilometer-size asteroids on Earth, the significance of giant comets in discussions of the long-term extraterrestrial hazard is not in question. If the cumulative diameter distribution is quadratic or steeper, as indicated by Eq. (6), the discovery of a few very large bodies, presumably comets, suggests the presence of a great many comets of ordinary size moving in similar kinds of orbit. Only one of these smaller bodies has yet been discovered (Rabinowitz 1993b), namely 1993 HA$_2$ with a diameter on the order of 80 km; but it seems likely that many more will be found.

Other suggestions of the presence of giant comets are less direct. Dones (1991) has reassessed the evidence that Saturn's rings are relatively young, concluding that the process of angular momentum transport and erosion by micrometeoroids probably sets an upper limit to the ring lifetime on the order of 10^8 yr. Assuming a cometary flux similar to that inferred from the existence and dynamical lifetime of Chiron, and a cumulative diameter distribution proportional to d^{-2}, it is possible to understand the existence of massive rings about Saturn even if their lifetime is relatively short; planetary rings are produced by the breakup of large comets passing within the Roche limit of the planet. Theories of the formation of the outer planets also point in the same direction; Stern (1991,1992) showed that the existence of the Pluto-Charon binary implies, on a probabilistic argument, that there must once have been

$\sim 10^3$ bodies of comparable size in the outer protoplanetary disc. Similarly, to explain the tilts of the rotation axes of Uranus and Neptune some 50 Earth-size bodies have to be invoked in order for the necessary collisions to appear reasonably likely. With a cumulative diameter distribution proportional to d^{-2}, an enormous number of Chiron-size bodies must once have been present, many of which would have been scattered (like ordinary comets) into the outer Oort cloud or its inner core to form part of the present long-period influx.

We conclude that there are strong arguments for the existence of large bodies on cometary orbits. They enter the inner solar system as short-period comets at mean intervals on the order of 10^5 yr by two routes, both directly and indirectly from the long-period flux, and appear as either Halley-type or Jupiter-family comets, respectively, evidence for the latter being provided in particular by observed Chiron-like bodies. The low-inclination family may, of course, also be augmented by bodies from outer solar system reservoirs such as the Kuiper belt, the trans-Neptunian cometary disk or the inner Oort cloud; these, if present, would produce a significant flux of low-inclination Chiron-like objects dynamically indistinguishable from those captured from the parabolic flux by Jupiter or Saturn.

III. DYNAMICAL EVOLUTION

A. Jupiter Family

Jupiter family short-period comets are usually defined to be those with periods <20 yr, low inclinations (usually ≲30°), and Tisserand parameters with respect to Jupiter close to 3. The origin of the Jupiter family is still unresolved (Everhart 1972,1977; Rickman 1990; Bailey 1992), the observed bodies being too numerous (by at least a factor of 10) to be produced by steady-state capture from the observed low-inclination near-parabolic flux. This suggests capture from another source region (Bailey 1983,1990b), abandonment of the steady-state hypothesis (for example, by postulating a recent enhancement of the near-parabolic flux; Clube and Napier 1986a), or, in accordance with the presumed evolutionary behavior of long-period comets, the presence of disintegrating large members (Clube 1992). A number of possibilities exist for the hidden source: a dense inner core of the Oort cloud (Bailey 1986; Stagg and Bailey 1989), an extended trans-Neptunian cometary disk (Fernández and Ip 1983) or a flattened Kuiper belt (Fernández 1980; Duncan et al. 1988). Arguments in favor of the latter have gained strength from the discovery of objects such as 1992 QB$_1$ and 1993 FW, and it is widely assumed that the Kuiper belt (if it exists) occupies a region similar in spatial extent to the volume swept out by these bodies, possibly extending to several hundred AU or more (Torbett 1989; Torbett and Smoluchowski 1990; Quinn et al. 1990; Duncan and Quinn 1993) and merging into the extended cometary disk and inner core of the Oort cloud predicted by the planetesimal theory (Fernández 1985; Duncan et al. 1987; see Bailey et al. 1990 for a review). However, the zones between neighboring outer planets cannot be the source of a significant present-day flux of periodic comets as suggested by Duncan et al. (1988), because the dynamical lifetime of bodies formed in these regions is too short (Gladman and Duncan 1990).

Of course, the degree to which one accepts such arguments for a primordial comet belt depends on the extent to which a flattened system might also be created through dynamical evolution of comets from another source, for example, ejection of bodies from the Jupiter family, itself possibly enhanced by surges in the long-period flux. Long-term investigations into the motion of bodies close to the 5/2 resonance with Jupiter show at least one case of an asteroid scattered into a Chiron-like cometary orbit (Farinella et al. 1993a; cf., Hahn et al. 1991), while spikes in the lunar and terrestrial cratering records (Baldwin 1985; Lindsay and Srnka 1975; Goswami and Lal 1978; Schultz 1987) are suggestive of occasional comet showers acting throughout the lifetime of the solar system. Whether bodies in Chiron-like orbits predominantly originate by capture from the near-parabolic flux, ejection from sub-Jovian orbits, or inward drift from the outer solar system remains to be seen; any further discoveries of such bodies and of their numbers versus heliocentric distance will provide scope for discrimination between the several possibilities.

Irrespective of arguments about their origin, however, once potential Jupiter-family comets are in orbits that allow close encounters with one or other of the major planets, they undergo a process of multi-stage capture involving passage of the body from the control of one planet to the next by successive close encounters, until they finally reach orbits controlled by Jupiter. This, of course, is one of the characteristic features of Chiron-like orbits, and the result provides strong support for the conclusion that large bodies the size of Chiron are indeed occasionally transferred into orbits identical to those of observed Jupiter family comets.

B. Centaurs

Asteroid (2060) Chiron, discovered by Kowal in 1977, was the first discovered member of the so-called Centaur group (Kresák 1980b), a family of massive bodies (presumably giant comet nuclei) moving on low-inclination orbits of moderate eccentricity with perihelia close to the orbit of Saturn and periods in the approximate range 50 to 150 yr. Chiron is observed to be outgassing, with both a coma and tail (Hartmann et al. 1990; West 1991; Larson and Marcialis 1992), although the second Centaur, (5145) Pholus, despite having an extremely red spectrum, suggestive of a cometary link (Fink et al. 1992; Mueller et al. 1992), is currently inactive. The orbital parameters and approximate diameters of the three known bodies classified as Centaurs are shown in Table I. The large size of these bodies suggests (assuming a frequency distribution proportional to d^{-2} or steeper) that there must be many smaller bodies still to be discovered.

Early dynamical studies (Kowal et al. 1979; Oikawa and Everhart 1979; Scholl 1979) demonstrated extreme sensitivity of Chiron's long-term evolution to uncertainties in the assumed initial conditions, illustrating the chaotic

TABLE I

Orbital Elements and Diameters of Chiron, Pholus and 1993 HA$_2$

Name	a AU	e	q AU	T	i (deg)	d (km)	Source
2060 Chiron	13.7488	0.38482	8.4580	1996 Feb 9.3	6.93	150–300	EMP 1993
5145 Pholus	20.4801	0.57585	8.6866	1992 Jul 22.6	24.68	130–180	MPC 19850
1993 HA$_2$	24.8661	0.52344	11.8502	1992 Jun 16.5	15.57	50–100	MPC 22410

nature of the orbit after a few close encounters with Saturn. These studies highlighted three important results. First, a significant tendency, due to the asymmetric frequency distribution of energy changes in close encounters with Saturn (Oikawa and Everhart 1979; Carusi et al. 1990), for both the future and past orbit to be one of shorter period and smaller perihelion distance, meaning that Chiron has a high probability to evolve towards the dynamical control of Jupiter. Secondly, that the motion is dominated, on time scales of order 10^3 to 10^4 yr, by low-order mean-motion resonances with Saturn; and thirdly, that the half-life for Chiron to be ejected from the solar system is on the order of 1 to 2 Myr, showing that this particular giant comet has probably been in its present orbit for much less than the age of the solar system.

More recent investigations have calculated the long-term evolution of a much larger number of test particles with initial orbital elements similar to those of Chiron and Pholus, for time scales in the range 0.1 to 1.0 Myr (Hahn and Bailey 1990; Bailey et al. 1992b,1993; Asher and Steel 1992,1993; Nakamura and Yoshikawa 1993). Detailed results for 1993 HA$_2$ are not yet available, although one would expect a somewhat longer dynamical lifetime for this body than for either Chiron or Pholus, mainly because it is not in a Saturn-crossing orbit. Calculations of the long-term dynamical evolution of Chiron and Pholus have generally confirmed the main conclusions mentioned above, although there are detailed differences between the dynamical evolution of the two bodies primarily due to the higher inclination of Pholus and its longer orbital period. Following Öpik's theory of close encounters (Carusi et al. 1990), which is applicable to this case as 75 to 80% of significant encounters involving low-eccentricity, low-inclination orbits are found to be planet-crossing (Manara and Valsecchi 1991), these orbital factors lead to a relatively small asymmetry in the frequency distribution of orbital energy changes of Pholus in close encounters with Saturn as compared with Chiron (Bailey et al. 1993). In fact, the asymmetry in the largest energy perturbations is determined by the value of the angle θ between the direction of motion of the planet and the planetocentric velocity of the small body at the time of the encounter (Valsecchi 1992). If this angle, given by $\cos \theta = U_y/U$, where $U_y = \sqrt{A(1 - e^2)} \cos i - 1$ is the component of velocity of the comet in the direction of motion of the planet and $U = [3 - 1/A - 2\sqrt{A(1 - e^2)} \cos i]^{1/2}$ is the relative speed of the comet with respect to the planet, is <90° negative energy perturbations predominate, leading to shorter orbital periods, and the small body tends to be handed inwards. Here $A = a/a_S$, and a and a_S are the semimajor axes of the orbits of Pholus and Saturn, respectively. For encounters with Saturn, the controlling planet for both Chiron and Pholus, the encounter angle is 71.8° and 81.4°, respectively, implying a more random evolution for Pholus and hence a reduced probability per unit time for transfer to the control of Jupiter or to a short-period Earth-crossing orbit.

The main results for Chiron and Pholus may thus be summarized as follows. First, the development of chaos, and hence unpredictability in the evolution of the orbit, occurs soon after the first close encounter with Saturn. In the case of Chiron this is some 2000 yr in the past and 12,000 yr in the future (Hahn and Bailey 1990), and for Pholus on the order of 4000 yr in both the past and future (Bailey et al. 1993; Nakamura and Yoshikawa 1993). Both Chiron and Pholus have a finite probability of having been Jupiter-family comets as recently as $\sim 10^4$ yr ago, and it is therefore reasonable, for example, to assume that giant comets may have been present in the inner solar system at some time during the last glacial maximum circa 20,000 yr ago, suggesting a possible causal connection between cometary dust and ice ages (see Sec. V.C). Secondly, the half-life for transfer to the control of Jupiter is on the order of 0.15 Myr and 1.6 Myr for Chiron and Pholus, respectively; and thirdly, the half-lives for ejection of Chiron and Pholus from the solar system are about 1.4 and 2.1 Myr. The probability that at some time during its dynamical evolution Chiron has been or will become a Jupiter-family short-period comet is 85 to 90%, while the corresponding probability for Pholus is in the range between 40 and 50%. These results are based on the integration of 83 Chiron-like orbits for ± 0.1 Myr (Hahn and Bailey 1990) and more than 81 Pholus-like orbits for ± 1.0 Myr (Bailey et al. 1992*b*,1993). Figures 1 through 4 illustrate these aspects of the long-term evolution of four test particles with initial orbits similar to those of the observed bodies Chiron and Pholus.

It is of interest to determine the probability that a body the size of Chiron may collide with the Earth. Giant comets with $d \gtrsim 100$ km enter the Jupiter family at a rate on the order of 5 to 50 Myr^{-1}, and about 20% of these enter short-period Earth-crossing orbits for individual periods on the order of 10^3 yr (cf., Rickman et al. 1992). Because the collision probability per short-period comet is about 10^{-9} yr^{-1} (Olsson-Steel 1987), the giant-comet collision probability with the Earth is on the order of $(1–10) \times 10^{-12}$ yr^{-1}, leading to a mean interval between impacts in the range 10^{11} to 10^{12} yr—a comfortably long time. Having regard to the massive energy input of such mantle-piercing impacts, however, it is conceivable that one or more events of this kind occurred up to and including the late bombardment phase (Sleep et al. 1989) with consequences that remain relevant to the state of the mantle today. This is not, however, to underrate the importance of lesser impacts, which produce craters on Earth with diameters greater than D at a rate on the order of $3 \times 10^{-6} (D/20 \text{ km})^{-2}$ (Grieve 1987,1991; Bailey 1991), nor the indirect effects of giant-comet fragmentation, which occasionally inject large quantities of cometary debris into the inner solar system on astronomically short time scales.

C. Halley-Types

Halley-type comets represent the short-period tail of the observed near-parabolic flux. Their inclinations span the whole range between 0 and 180° (though there is a slight preference for direct orbits), and although their perihelion distances are usually $\lesssim 1$ AU, recent investigations into their long-term dynamical evolution (MEB, in preparation) are suggestive of initial orbits

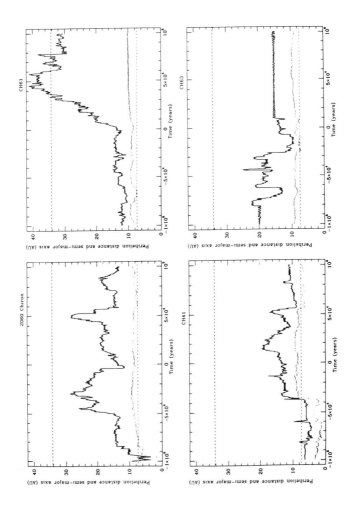

Figure 1. Evolution of the semimajor axis (solid line) and perihelion distance (dotted line) for four variational orbits similar to those of Chiron, illustrating the range of different behaviors observed over time scales ±0.1 Myr. the horizontal dashed lines correspond to semimajor axes of comets with periods 20 and 200 yr, respectively (figure after Hahn and Bailey 1990).

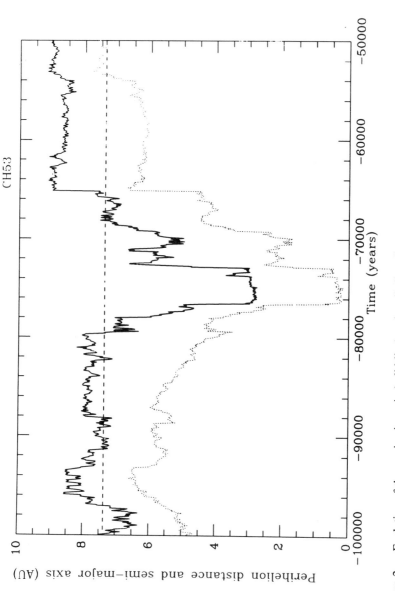

Figure 2. Evolution of the semimajor axis (solid line) and perihelion distance (dotted line) during the first 50 Kyr of the Chiron-like test particle CH53. The horizontal dashed line corresponds to a period $P = 20$ yr. This test particle evolved into a Jupiter-family short-period orbit with aphelion close to Jupiter's perihelion ~75 Kyr ago for a total time on the order of 4 Kyr. The total dust ejected during this time scale by a comet the size of Chiron is estimated to lie in the range 10^{17} to 10^{18} kg, 4 to 40 times the present mass of dust in the zodiacal cloud (figure after Hahn and Bailey 1990).

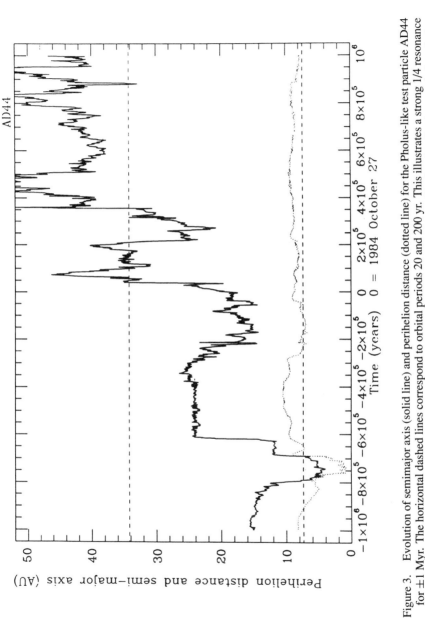

Figure 3. Evolution of semimajor axis (solid line) and perihelion distance (dotted line) for the Pholus-like test particle AD44 for ±1 Myr. The horizontal dashed lines correspond to orbital periods 20 and 200 yr. This illustrates a strong 1/4 resonance with Saturn lasting ~200 Kyr centered around $t = -500$ Kyr, and an extended phase of Jupiter-family short-period evolution lasting almost 100 Kyr. During this time the body spend nearly 25 Kyr in orbits with perihelion distance <2 AU, becoming Earth-crossing on several occasions (figure after Bailey et al. 1993).

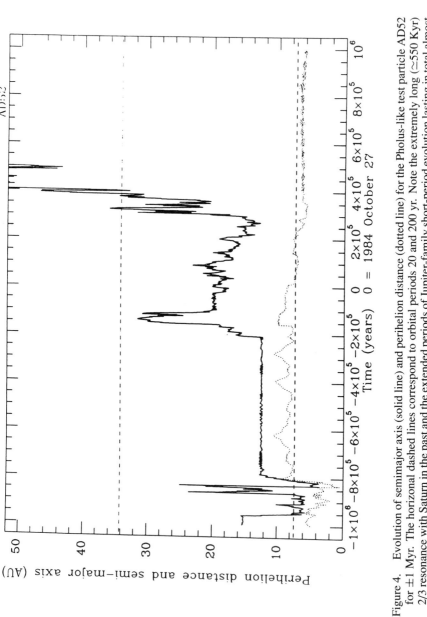

Figure 4.   Evolution of semimajor axis (solid line) and perihelion distance (dotted line) for the Pholus-like test particle AD52 for ±1 Myr. The horizonal dashed lines correspond to orbital periods 20 and 200 yr. Note the extremely long (≃550 Kyr) 2/3 resonance with Saturn in the past and the extended periods of Jupiter-family short-period evolution lasting in total almost 100 Kyr. This body twice became Earth-crossing (not resolved on this diagram), for individual times each lasting ~10³yr (figure after Bailey et al. 1993).

with perihelion distances up to 6 AU or more. Halley-types are conventionally defined as comets with orbital periods in the range 20 to 200 yr and
values of the Tisserand parameter T_J with respect to Jupiter less than 2.0.
This may be expressed in terms of the relative velocity U of the comet in a
close encounter with Jupiter by $T_J = 3 - U^2$, where U is defined as before
in terms of the relative inclination i of the two orbits, the eccentricity e and
dimensionless semimajor axis $A = a/a_J$ (where a_J is the radius of Jupiter's
orbit) by $U^2 = 3 - 1/A - 2\sqrt{A(1 - e^2)} \cos i$.

It is important to note that both the upper and lower limits on the periods
defining Halley-type comets are conventional and have no underlying dynamical significance; a short-period high-inclination orbit like that of P/Machholz
(with $P = 5.25$ yr) should probably be counted as a Halley type. Numerical
integrations of Halley-type comets demonstrate that when ejection occurs it
usually takes place during an episode of nodal crossing with either Jupiter or
Saturn, suggesting that it is probably the reverse process which initially leads
to the capture of Halley-type comets from the near-parabolic flux.

Dynamical studies of the evolution of Halley-type comets illustrate three
principal results. First, between periods of nodal crossings with the major planets a graph of the evolution of semimajor axis versus time shows
prominent mean-motion resonances with Jupiter (Dvorak and Kribbel 1990),
typically of the form $1/n$, $2/n$ or higher, where n in a small integer ranging
up to of order 10. This feature of observed Halley-type orbits has been remarked upon many times (Carusi et al. 1987a, b), but long-term integrations
demonstrate that the phenomenon is of general importance. Mean-motion
resonances typically last 10^4 to 10^5 yr, corresponding to the mean interval
between periods of close encounters with Jupiter and Saturn, but occasionally,
due to protection from close encounters at the nodal crossing, a particularly
strong resonance may persist for much longer. Mean-motion resonances give
Halley-type orbits great dynamical stability, a result which could be important in assessing the short-term hazard due to individual bodies. For example,
comet P/Swift-Tuttle, in the 1/11 resonance with Jupiter, seems certain to
undergo repeated close passages to the Earth during the coming millennia
(Chambers 1993; Yau et al. 1994).

Secondly, whether or not the orbit is in resonance with Jupiter, Halley-
type comets experience strong secular perturbations which lead to large systematic changes in perihelion distance and inclination, and correlated variations in both these orbital parameters (Kozai 1962,1979; Carusi et al. 1988,
Bailey and Hahn 1992; Bailey et al. 1992c). Such perturbations may be amplified by a resonance, but the main effect can be explained as the response of the
comet to the smoothed-out gravitational field of Jupiter in its orbit. To a first
approximation the square of the comet's orbital angular momentum perpendicular to the plane of the ecliptic, namely $K = a(1 - e^2) \cos^2 i$, is a constant
of the motion, implying that when i is near 90° for a given perihelion distance,
the constant is small and secular evolution to a small perihelion distance becomes dynamically possible. This mechanism (cf., Quinn et al. 1990) causes

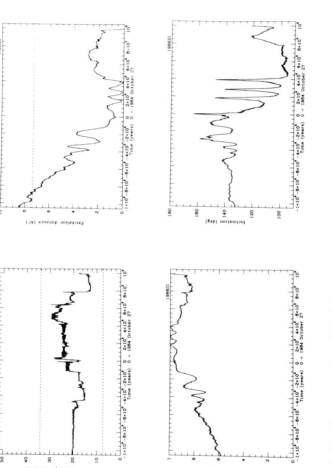

Figure 5. Evolution of a test particle with initial elements identical to those of P/Halley (1986 III). The horizontal dashed lines on the semimajor axis plot correspond to orbital periods of 20 and 200 yr. Note the prominent mean-motion resonances characteristic of these types of orbits, the systematic secular change in perihelion distance at roughly constant semimajor axis and inclination, and the coupled oscillations of perihelion distance and inclination, the maxima in i correlating with successive minima in q. The body first evolves into an orbit of very small perihelion distance about 2×10^5 yr in the future.

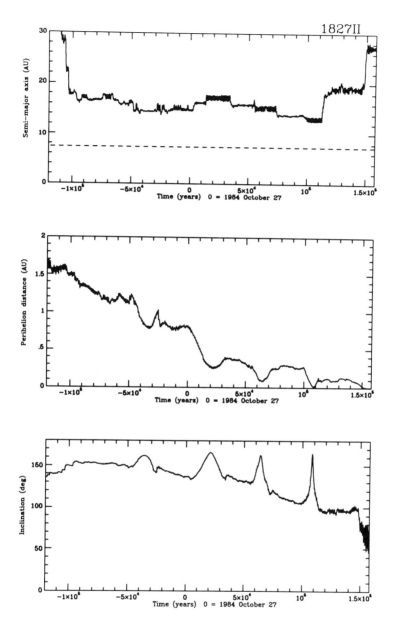

Figure 6.   Evolution of a test particle with initial elements identical to those of P/Pons-Gambart (1827 II), showing coupled swings in perihelion distance and inclination within 200 Kyr of the present and evolution to a probable sungrazing end-state (figure after Asher et al. 1993a).

Halley-type comets with inclinations in the approximate range $90° \pm 15°$ to evolve rapidly on to almost sungrazing orbits (Bailey et al. 1992c), while

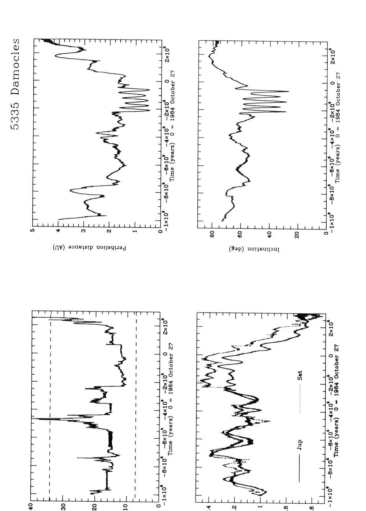

Figure 7.   Evolution of a test particle with initial elements identical to those of the Halley-type asterid (5335) Damocles = 1991 DA from −1.0 Myr to 0.3 Myr. This illustrates a sequence of prominent Jovian resonances centered on t = −900, −600, −400 and −300 Kyr (the 2/11, 1/5, 1/5 and 1/6, respectively), the large secular changes in perihelion distance and inclination over this time scale, and the coupled secular oscillations in q and i that brought the object into an Earth-crossing orbit at various times between −230 and −50 Kyr (figure after Asher et al. 1993a).

similar results apply to Halley-type orbits of all inclinations on a time scale
on the order of 1 Myr. Figures 5 through 8 show the evolution in semimajor
axis and perihelion distance for four bodies in such orbits: P/Halley, P/Pons-
Gombart, (5335) Damocles = 1991 DA, and P/Machholz, each integrated for
various lengths of time in model solar systems including the mutual interactions of the 4 major planets Jupiter, Saturn, Uranus and Neptune. The plots of
semimajor axis demonstrate the importance of resonant behavior, while those
showing the perihelion distance and inclination illustrate the large systematic
changes that affect chaotic Halley-type orbits, and the prominent coupled oscillations of inclination and perihelion distance that frequently drive the latter
to Earth-crossing values <1 AU and occasionally to almost sungrazing.

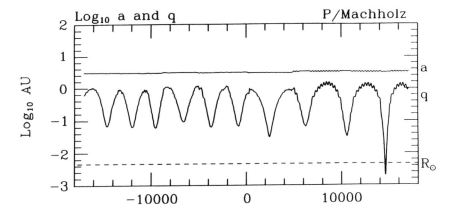

Figure 8.   Evolution of semimajor axis a and perihelion distance q for a test particle
with initial elements identical to those of P/Machholz in a model solar system
including the mutual perturbations of the four outer planets (Jupiter, Saturn, Uranus
and Neptune). Time is measured in years A. D. The evolution of this orbit culminates
in a collision with the Sun about 10 Kyr in the future (figure after Bailey et al. 1992c).

D. Disruption and Splitting

A possible model for the origin of Jupiter-family comets is that they predominantly arise through the fragmentation of larger bodies undergoing evolution
in orbits of small perihelion distance in the inner solar system. Rickman
(1990) has emphasized that the problem of explaining the number of comets
in the Jupiter family might in principle be overcome by successive splitting
of ordinary-size cometary nuclei, although the general similarity of the orbits
of bodies in the Taurid Complex (see Sec. IV.B; also see the Chapter by Steel
et al.) appears to favor a model involving disintegration of a single massive
progenitor with an overall dynamical age for the system on the order of 10^4 yr
(Porubčan and Štohl 1987; Steel et al. 1991, and their Chapter; Asher et al.
1993b, c).

Roughly 10% of the observed Jupiter family are known or suspected to have split, sometimes leaving comets that have persisted as active bodies for at least several revolutions. In most cases the reason for splitting is not well understood, a notable example being P/Van Biesbroeck and P/Neujmin 3 which arose from a common parent around 1840, the close encounter with Jupiter in 1850 (to within 0.1 AU) causing the bodies to diverge in their orbits to yield two independent comets (Carusi et al. 1985,1987c). However, a mechanism applicable to several examples is tidal disruption of the cometary body during close passage to either the Sun or Jupiter, the encounter taking the comet to within a few radii of the central object. Encounters within Jupiter's Roche lobe are known to cause splitting into many pieces, for example comet P/Brooks 2 in 1886, for which the largest fragment continues to survive as an active object, and the more recent example of P/Shoemaker-Levy 9. The latter demonstrates survival of about 20 fragments for at least a year, although most if not all appear destined for destruction in July 1994 by direct physical collision with Jupiter (Carusi et al. 1993; Sekanina 1993).

Apart from these examples, several other Jupiter-family comets are known to have passed within 3 planetary radii of Jupiter, notably P/Lexell in 1779 and P/Gehrels 3 in 1970 (Carusi et al. 1985). Neither provides direct evidence for splitting, the first because P/Lexell was never subsequently seen since 1770, the second because P/Gehrels 3, discovered in 1975, was found only when short-lived fragments, if any, had already disappeared. Nevertheless, such observations allow us to estimate that at least one comet passes through the Roche lobe of Jupiter per century, a figure which is presumably a lower limit because both the comets of this century, P/Gehrels 3 and P/Shoemaker-Levy 9 (see Fig. 9), have exceptionally high values of the Tisserand parameter, slightly above 3, and hence large perihelion distances making them less likely to be discovered. The same comment applies to a lesser extent to P/Brooks 2, because following the encounter with Jupiter it had a Tisserand parameter $T_J \simeq 2.9$ and a perihelion distance $q \simeq 2$ AU. Observations of chains of craters on the Jovian satellites Ganymede and Callisto are also consistent with frequent cometary disruption events during close passages to Jupiter, indicating a mean recurrence interval on the order of 10^2 yr (Melosh and Schenk 1993).

If giant comets with $d \gtrsim 100$ km represent $\sim 0.1\%$ of the Jupiter-family population (expected for a d^{-2} cumulative size distribution), then we would expect such comets to be disrupted by Jupiter at a mean rate on the order of 10 Myr^{-1}, each event providing conditions for a major enhancement in the Jupiter-family population. Apart from the Taurid Complex progenitor, the most obvious prototype for such an event is the Kreutz sungrazing group. This appears to have been formed through hierarchical fragmentation of a massive progenitor and its remnants during successive exceptionally close passages to the Sun (Marsden 1967,1989). The Kreutz group demonstrates that fragmentation of giant comets may produce many comets of ordinary size that persist for several revolutions.

Figure 9.   Image of the recently fragmented comet P/Shoemaker-Levy 9, taken by J. Luu at Stanford University and D. Jewitt at University of Hawaii on 1993 March 27 with the University of Hawaii 2.2 m telescope (figure courtesy of D. Jewitt).

It is also interesting to note that Roche-lobe splitting is closely related to the long-term dynamical evolution; it either involves comets of high Tisserand parameter with the potential to undergo frequent temporary satellite captures with Jupiter, or high-inclination Halley-types with the potential to become sungrazers. Furthermore, some comets of high Tisserand parameter (e.g., P/Encke), which move entirely in sub-Jovian orbits, may be driven by secular perturbations into orbits that become sungrazing (Levison and Duncan 1993). Thus, an enhanced Jupiter family may also be produced by the disintegration of a Jupiter-family giant comet while in a sub-Jovian orbit such as that of P/Encke, disintegration occurring as a result of evolution on to a sungrazing or Jupiter-grazing orbit, or simply by collisions with mainbelt asteroids (see, e.g., Fernández 1981,1990), or some other process, causing the fragments to re-enter the Jovian sphere of influence and produce the observed Jupiter-family. Although such a proposal might seem unduly elaborate, an ongoing process of this nature involving substantially asteroidal remnants of the original Taurid progenitor is considered to be a dominant influence on the present terrestrial environment (Asher and Clube 1993). A hierarchical fragmentation picture along one or other of these lines is supported by observations of both the Taurid Complex and the Kreutz group, and is amenable to several dynamical tests (Pittich and Rickman 1994). Cometary breakup is a natural process, and represents a direct way by which small bodies (meteoroids, asteroids and active cometary fragments) might be injected into Earth-crossing orbits with potentially catastrophic consequences, having implications both for the origin of comets (Bailey et al. 1990; Bailey 1992; Clube 1992) and the extraterrestrial hazard.

IV. EARTH-CROSSING ASTEROIDS

The population of Earth-crossing asteroids, which we loosely define as bodies with short-period orbits decoupled from close encounters with Jupiter and perihelion distances $\lesssim 1$ AU, represents a very diverse collection of objects. Some originate in the main asteroid belt, for example, as a result of collisions leading to injection into chaotic regions of phase space associated with the 3/1 and 5/2 mean-motion resonances with Jupiter, which in turn pump up the orbital eccentricities so they eventually become Mars-crossing, subject to stronger perturbations by Jupiter and finally Earth-crossing. Others, for example (2212) Hephaistos, (4015) Wilson-Harrington and comet P/Encke, presumably originate in the Jupiter-family or Halley-type short-period comet systems (cf., Table II), becoming decoupled from strong Jovian perturbations by processes involving secular resonances, close encounters with the terrestrial planets, and nongravitational forces associated with cometary outgassing, splitting or collisions with other bodies. Mantling or devolatilization of a cometary nucleus while in such an orbit seems likely to produce a low-activity object essentially indistinguishable from the more primitive types of asteroid. A significant fraction of near-Earth asteroids (the Amors) are not currently Earth-crossing, but instead move in orbits in which secular perturbations drive the perihelia down to $q \lesssim 1$ AU, while recent observations (Rabinowitz et al. 1993) indicate the existence of a new population of small bodies with semimajor axes and orbits close to that of the Earth. The Earth-crossing asteroid population thus contains bodies in a variety of orbits from a variety of sources: asteroids from the main belt and a significant number of known comets, suspected comets and cometary fragments.

Partly as a result of this diversity, and partly due to the complexity of the physical and dynamical evolution, the proportion of bodies arising from each source is poorly known. In the past, despite arguments such as those of Öpik (1963), there has been a general tendency to assume that the majority of Earth-crossing asteroids come from the asteroid belt (see review in Bailey et al. 1990), but it is now recognized that cometary sources are at least as important as asteroids in the main belt (see, e.g., Wetherill 1988). Here we emphasize the role played by giant comets, the possible ways by which comets might in principle be scattered into sub-Jovian Encke-type orbits, and the implications for short-term variability of the impact hazard.

A. Orbital Transitions

Long-term integrations of bodies in Chiron-like orbits together with an assumed cumulative diameter distribution proportional to d^{-2} lead to the prediction that giant comets with $d \gtrsim 100$ km are injected into the Jupiter family at mean intervals on the order of 2 to 20×10^4 yr. Some pass exceptionally close to the Sun or Jupiter and undergo tidal disruption, while others evolve on to Earth-crossing orbits of small perihelion distance similar to that of the Chiron-like test particle CH53 illustrated in Fig. 2. During these Jupiter-family phases of evolution, which individually last for periods on the order of 10^4 yr, the physical or purely dynamical evolution referred to above may cause the giant comet to be transferred on to an Earth-crossing orbit of potentially much longer lifetime (10^7–10^8 yr) typical of that estimated for most Earth-crossing asteroids and comets such as P/Encke (Wetherill 1975,1991). Because outgassing appears unlikely to produce sufficiently large nongravitational forces to affect the motion of a very massive comet, we conclude that tidal disruption, close encounters to Jupiter or a terrestrial planet and effects due to secular perturbations by Jupiter and Saturn, represent the most likely routes by which giant comets arrive on orbits similar to those of observed Earth-crossing asteroids. The probability of such orbital transitions cannot yet be quantified, but indications of their possible efficiency may be obtained from the investigation of Earth-crossing asteroid orbits undertaken by Milani et al. (1989).

These authors, in what they described as Project Spaceguard, calculated the dynamical evolution of 89 Earth-crossing asteroids for ± 0.1 Myr. They found that a surprisingly large fraction (5–10%) evolved on this time scale into typical cometary orbits, including examples of Jupiter-family comets, long-period comets and even cases of capture or ejection. These results show that a significant fraction of observed Earth-crossing asteroids are dynamically connected to a cometary source, their orbital types and the pattern of transitions between the different orbital classes (notably the disguised comets, Oljatos and Alindas) delineating the main dynamical paths followed by such bodies. The small number of objects so far investigated means that the frequency of orbital transitions is poorly known, but the work was notable in demonstrating a much closer connection between comets and Earth-crossing asteroids than had previously been assumed.

For example, despite the fact that no direct transition of an individual cometary orbit into a decoupled Encke-type orbit was found in any of the Spaceguard integrations, the effects of very close encounters to the terrestrial planets (illustrated in Fig. 5a of Milani et al.) was clearly demonstrated. The occurrence of significant jumps in semimajor axis during the 0.2 Myr period covered by the integrations suggests that this process affects a few orbits on a much shorter time scale than estimated from Monte-Carlo studies (cf., Wetherill 1975). Nongravitational forces (not considered in the Spaceguard integrations) resulting from ordinary cometary activity and tidal breakup or fragmentation in close encounters will tend to enhance these effects, leading to significant orbital evolution on time scales $\lesssim 1$ Myr.

These results were based on objects discovered up to 1986. Many more discoveries of apparently asteroidal objects moving on cometary orbits have since been reported, bringing the present sample up to more than 30 (see Table II). Long-term integrations of some of these orbits, notably those lying close to the 5/2 mean-motion resonance with Jupiter (Hahn et al. 1991; Farinella et al. 1993a), often reveal evolution on to typical cometary trajectories, while (944) Hidalgo also belongs to the cometary class (M. Dahlgren 1993, personal communication) and is a potential sungrazer (Nakamura and Yoshikawa 1993). Evidence supporting the existence of dormant or low-activity comets residing in typical Earth-crossing asteroid orbits has been reviewed by Rickman (1985), Hahn and Rickman (1985) and Weissman et al. (1989); recent arguments include objects such as (3200) Phaethon, the assumed parent body of the Geminid meteor stream (Williams and Wu 1993), (2201) Oljato, with possible cometary properties inferred from spectroscopic and other observations (see, e.g., McFadden et al. 1993), and (4015) Wilson-Harrington, recently identified as identical to comet Wilson-Harrington 1949 III (Bowell 1992). Such bodies are similar to extreme low-activity comets such as P/Arend-Rigaux and P/Neujmin 1, which during some apparitions exhibit little or no outgassing. They show that a number of bodies classified on physical grounds as asteroids may, in fact, be dormant comets, and have a chaotic long-term dynamical evolution.

Table II contains all known asteroids in cometary orbits with aphelia $Q \gtrsim 4.4$ AU, including outer solar system objects such as the Centaurs and the first four Kuiper-belt candidates. Four bodies, namely (1373) Cincinnati, (1921) Pala, (1922) Zulu and (3688) Navajo, have been omitted from this list because they are associated with the apparently stable 2/1 mean-motion

TABLE II

Asteroids in Cometary Orbits[a]

Number	Name or Initial Designation	a AU	e	i deg	q AU	Q AU	H mag	d km
	1984 QY$_1$	2.97	0.917	15.5	0.25	5.70	14.0	11.2
	1991 TB$_2$	2.40	0.836	8.7	0.39	4.40	17.0	2.8
	5025 P-L	4.20	0.895	6.2	0.44	7.96	17.9	1.9
	1983 LC	2.63	0.709	1.5	0.77	4.50	21.0	0.4
	1983 VA	2.61	0.692	16.2	0.80	4.42	18.5	1.4
	1986 JK	2.80	0.680	2.1	0.90	4.71	19.0	1.1
	1985 WA	2.84	0.604	9.8	1.12	4.56	18.5	1.4
	1982 YA	3.71	0.697	34.6	1.12	6.29	18.5	1.4
	1987 QB	2.80	0.594	3.5	1.14	4.47	19.0	1.1
5324	1987 SL	2.96	0.615	19.5	1.14	4.78	14.1	10.7
	1986 DA	2.81	0.585	4.3	1.17	4.46	16.0	4.5
3552	Don Quixote	4.24	0.713	30.8	1.22	7.26	13.0	17.8
5370	1986 RA	3.35	0.632	19.0	1.23	5.47	15.6	5.4
	1991 XB	2.97	0.580	16.1	1.25	4.69	18.1	1.7
	1992 UB	3.07	0.583	15.9	1.28	4.85	16.0	4.5
	1989 SL$_5$	2.99	0.537	23.7	1.39	4.60	16.0	4.5
	1992 EB$_1$	3.38	0.571	21.6	1.45	5.31	16.5	3.5
5201	1983 XF	3.13	0.533	4.1	1.46	4.79	14.8	7.8
	1992 AB	3.28	0.553	40.8	1.47	5.10	14.0	11.2
	1992 RN$_1$	3.18	0.512	6.0	1.55	4.81	15.0	7.1
	1984 BC	3.44	0.547	22.4	1.56	5.32	16.0	4.5
5335	Damocles	11.88	0.867	61.9	1.58	22.18	13.3	15.5
	1992 WW$_2$	3.07	0.448	15.7	1.69	4.44	12.5	22.4
	1992 XA	3.47	0.477	24.8	1.82	5.12	17.0	2.8
5164	1984 WE$_1$	3.66	0.502	19.8	1.82	5.49	13.0	17.8
944	Hidalgo	5.84	0.657	42.4	2.00	9.67	10.8	49.0
2060	Chiron	13.75	0.385	6.9	8.46	19.04	6.6	338.7
5145	Pholus	20.48	0.576	24.7	8.69	32.27	7.3	245.4
	1993 HA$_2$	24.87	0.523	15.6	11.85	37.88	9.5	89.1
	1992 QB$_1$	43.83	0.088	2.2	39.99	47.67	7.5	223.8
	1993 FW	43.91	0.041	7.7	42.13	45.69	7.0	281.7
	1993 RO	32.3	0.0	2.5	32.3	32.3	8.0	177.8
	1993 RP	35.4	0.0	2.8	35.4	35.4	9.5	89.1
	1993 SB	33.1	0.0	2.3	33.1	33.1	8.0	177.8
	1993 SC	34.5	0.0	5.6	34.5	34.5	7.0	281.7

[a] Orbits with aphelia $Q \geq 4.4$ AU, sorted by increasing perihelion distance q, and divided into Apollo-Amors, Mars-crossers, Halley-types, Hidalgo-types, Chirons and Centaurs, and outer solar system bodies with q beyond Neptune. The diameters d have been included for illustrative purposes, calculated assuming a constant geometric albedo of 0.037 typical of C-type asteroids. The orbits of the outermost bodies are particularly uncertain.

resonance with Jupiter (Nobili 1989), but it would be interesting to know
(cf., Varvoglis 1993) whether such orbits really are stable for time scales
comparable with the age of the solar system. We emphasize that with the
exception of (2060) Chiron and (4015) Wilson-Harrington (with $q = 0.996$ AU
and $Q = 4.29$ AU) none of these apparently cometary asteroids has ever shown
direct evidence of cometary activity such as a coma or tail, and their supposed
cometary origin is based solely on orbital dynamics, which resembles the
chaotic motion of active comets. In summary, we expect that a significant
number of near-Earth asteroids, probably at least 50% (cf., Wetherill 1988),
will eventually prove to have a cometary rather than a main-belt asteroidal
source.

B. Taurids: Examples of the Oljato Class?

Of particular interest is the fact that a significant fraction of observed Earth-
crossing asteroids appears to be associated with the Taurid Complex (Asher
and Clube 1993). Asher et al. (1993b, c) have identified 15 such asteroids
using a modified D-criterion, and show that they fall into three separate
dynamical groups with different orbital orientations, suggesting (cf., Napier
1993) possible formation during distinct episodes of fragmentation of a former parent body. Depending on how one defines membership of the Taurid
Complex, some 20 asteroids are known to be in such orbits, the latest addition
being 1993 KA_2, a small body of diameter just a few meters which passed
exceptionally close to the Earth ($\sim$150,000 km) on 1993 May 20.

Of the identified Taurid Complex members, 6 (or 7 if the Alinda object
1983 LC is included) were studied in the Spaceguard integrations. Of these,
4 (or 5) were classified as Oljatos, and 1, namely (2212) Hephaistos, made a
transition from the comet class to the Oljato. A relatively high percentage of
the Taurid Complex asteroids thus shows dynamical characteristics typical of
cometary orbits. Because the main feature of the Oljato class is a tendency for
chaotic dynamical evolution, then if this is interpreted as cometary behavior,
the cometary link for at least some of the Taurid Complex asteroids is secure.
We also note that the orbits of Taurid Complex members frequently have
perihelia close to or within the orbit of Venus, suggesting that encounters with
both the Earth and Venus are important in the medium term, especially in
view of the known importance of nearly tangent planetary encounters in the
long-term dynamical evolution of comets.

It is worth emphasizing that although the Spaceguard integrations showed
only (2212) Hephaistos as evolving towards a small perihelion distance ($q \lesssim$
0.1 AU), a number of Earth-crossing asteroids, including several bodies in the
Taurid Complex, are in orbits that become sungrazing on surprisingly short
time scales (10^5–10^6 yr; Yoshikawa 1992; Levison and Duncan 1993). This is
illustrated by the case of P/Encke, which becomes sungrazing within $\sim$$10^5$ yr
of the present with aphelion well within 1 AU of Jupiter's long-term minimum perihelion distance; but whether this implies that the general dynamical
path for comets to become Earth-crossing asteroids involves a single entry

corridor to sub-Jovian space, for example, through distant encounters with Jupiter (Napier 1983) or through pumping of the eccentricity during secular resonances, is not yet known. It is nevertheless significant that a number of secular resonances cross the region in (a, e, i)-space in which many Earth-crossing asteroids reside, and an important issue deserving further study is the role of the ν_6 secular resonance, in which the frequency of revolution of the longitude of perihelion of the comet or asteroid matches that of the longitude of perihelion of Saturn. At low inclinations this resonance affects semimajor axes in the range 2.0 to 2.2 AU (Wetherill 1979; Scholl 1987; Froeschlé and Froeschlé 1992; Farinella et al. 1993b), with effects that are relatively insensitive to the eccentricity, even for large values of the latter. Objects in the resonance can undergo large oscillations of eccentricity at nearly constant semimajor axis, and the main question to be addressed is whether these long-term variations, leading to Jupiter-tangent and even Jupiter-crossing orbits, are an efficient mechanism that links bodies in the Taurid Complex to typical Jupiter-family comets. To summarize this discussion, it is now clear that there are several possible routes by which giant comets could in principle be inserted on to relatively stable Earth-crossing orbits in which they might remain for at least their physical lifetime (10^5–10^6 yr).

C. Nodal Intersections: Short-Term Catastrophes

We refer the reader to the Spaceguard integrations (Milani et al. 1989) for further details of the dynamical classification and orbital behavior of Earth-crossing asteroids over these time scales, and to Milani et al. (1990) for an analysis of the rate of close approaches to the terrestrial planets. Here we emphasize two important aspects of the impact hazard posed by such bodies. First, for an individual orbit close encounters do not occur at random, but are bunched together around the times of nodal crossings. This means that periods in which many close approaches occur, possibly lasting several hundred years, will usually be followed by a much longer interval in which only shallow or no close approaches occur. Depending on the precise orbit, the interval between nodal crossings may vary from a few thousand years up to about 10^5 yr. Also, depending on the geometry of the orbit in relation to that of the planet, a particular orbit, such as a low-inclination nearly tangent orbit, may yield an exceptionally large number of close approaches. This was illustrated in the Spaceguard integrations by (4660) Nereus = 1982 DB and the spurious Aten-like orbit 1954 XA that together contributed almost 30% of all the close encounters registered for the 89 objects analyzed over ±0.1 Myr by Milani et al. (1990).

We demonstrate the effects of these short-term variations in the encounter rate by considering the high-inclination Apollo asteroid (5496) 1973 NA = 1992 OA, integrated for ±0.1 Myr in a 6-planet solar system model (Earth,. . ., Neptune). Figures 10 and 11 show that in contrast to small changes in semimajor axis produced by close encounters with the terrestrial planets, distant perturbations by Jupiter and Saturn drive large secular changes in the orbit which primarily affect the perihelion distance and inclination. Furthermore, despite having an aphelion distance as high as 4.7 AU, this particular orbit avoids close encounters with Jupiter due to libration of the argument of perihelion (Milani et al. 1989; Kozai 1980). So far as the potential hazard represented by such a body is concerned, for example were it to represent a clump of giant-comet meteoroidal debris, the important result to be emphasized is the nonrandom frequency distribution of close encounters. A similar result, highlighting the bunching of observed close encounters due to individual orbits each with a fixed geometry and phase relative to the Earth, was obtained by Hahn (1991), who calculated the distribution of close approaches of Earth-crossing asteroids for ±100 yr from the present. Figure 11 shows the same general phenomenon, but on a variety of time scales.

The asteroid (5496) 1973 NA = 1992 OA thus experiences roughly 10 close encounters within several hundred years, followed by a gap lasting some thousands of years. The pattern is repeated, but modulated on a 10^4 yr time scale by the secular oscillations in perihelion distance and inclination. Clearly, the exact pattern depends on the precise orbit and its phase relative to the Earth, but the result highlights the short-term variations in the impact hazard on the various time scales of most immediate concern to civilization; dynamical considerations alone impose apparent periodicities on time scales ranging from 10^2 to 10^4 yr. Behavior of this kind in the case of the Taurid Complex progenitor has already been explored by Asher (1991) and Asher and Clube (1993), and in general, because the dominant hazard at a given time will be that associated with debris originating from the most recently disintegrated Earth-crossing giant comet, strategies to map the predicted streams of small bodies in the inner solar system should be given high priority. The phenomenon of short-term catastrophism, sometimes thought to have no obvious physical explanation (Yau et al. 1993), is a predictable consequence of an impactor flux dominated by a relatively small number of streams. It is also demonstrated by (i) the intervals ranging from decades to centuries between years of intense meteor storms, for example, those associated with the Leonid and Perseid showers caused by comets P/Tempel-Tuttle and P/Swift-Tuttle, respectively (Sekanina 1974); (ii) historical variations in the fireball flux during the past two millennia (Hasegawa 1992; Clube 1994a); and (iii) the apparently greater rate of deaths and damage by meteorite impact in the past compared to the present (Yau et al. 1993).

In summary, dynamical investigations show that giant comets are frequently injected into short-period orbits in the inner solar system. The orbits are generally chaotic and a significant fraction of Chirons, the source orbits for the Jupiter family, and of high-inclination Halley-types, including associated devolatilized or inert asteroidal remnants, not only remain to be discovered but also may evolve into orbits similar to those of observed Earth-crossing asteroids. Strategies to assess the hazard due to kilometer-size Earth-crossers should consider how to identify a greater proportion of these largely unknown source orbits for the Earth-crossing asteroid population, both high-inclination

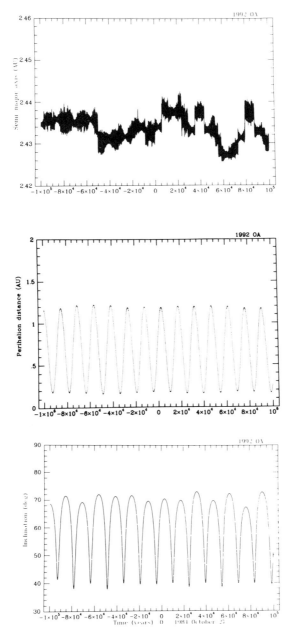

Figure 10.   Evolution of a test particle with initial elements identical to those
of the high-inclination Apollo asteroid (5496) 1973NA = 1992OA, integrated for
±0.1 Myr in a model solar system including the mutual perturbations of the six
planets (Earth+Moon, Mars, Jupiter, Saturn, Uranus and Neptune).

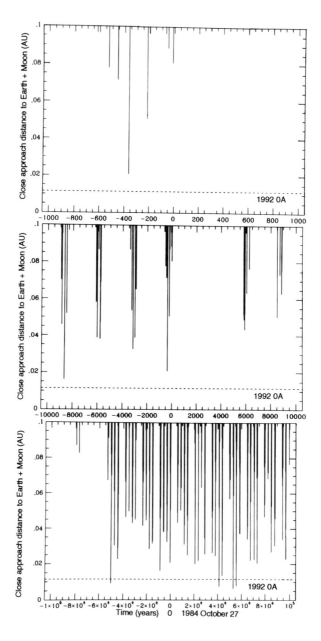

Figure 11.   Close appraches of (5496) 1973NA = 1992OA to the Earth over time-scales ranging from $\pm 10^3$ to $\pm 10^5$ yr.

types and those with $q \gtrsim 1$ AU, and consider how to discover the small bodies currently moving in streams (cf., Drummond 1991), especially those in the Taurid Complex. This system comprises a significant fraction of known Earth-crossing asteroids, and contains particles in Earth-crossing orbits known to collide with the Earth during early November and late June each year (see the Chapter by Steel et al.; Clube 1994*b*).

V. TERRESTRIAL EFFECTS

Whipple (1967), and Kresák (1980*a*) had previously considered the possibility of much larger than average giant comets, but Hoyle and Wickramasinghe (1978) and Clube and Napier (1984,1986*a*,1990) were among the first specifically to highlight the important role in terrestrial evolution played by such bodies. The latter argued that the mass influx to the inner planetary system is dominated by giant comets in short-period orbits, and specifically that the disintegration products of such bodies would arrive in bombardment episodes controlled by the Galactic field (Napier and Clube 1979; Clube and Napier 1982,1994). According to this thesis, giant comets are the primary cause of mass extinctions, as a result of derivative material deposited in Earth-crossing orbits, and primarily originate from the Oort cloud. The main purpose of these ideas (Bailey et al. 1990) was to explain how stochasticity and periodicity could naturally arise as aspects of terrestrial catastrophism both in the long term (i.e., the galacto-terrestrial relationship of Holmes [1927]) and in the short term (i.e., the historical record); they were formulated to meet complexities of the terrestrial record to which more narrowly focused modeling of the kind promoted by Alvarez et al. (1980) was not well suited. The latter authors advocated the "one-off" impact of a single main-belt asteroid as the cause of the K/T extinction, and set aside the previously favored cometary hypothesis (Öpik 1958; Urey 1973; Napier and Clube 1979; Hsü 1980). Nevertheless, despite difficulties with the concept of a single random impact (Hallam 1987,1989; Hut et al. 1987; Cisowski 1990), subsequent investigators have tended to follow the simpler astronomical model, with the result that from about 1980 to 1990 the stray asteroid hypothesis became for many authors the generally accepted paradigm (see, e.g., Alvarez 1983).

A. Cometary Catastrophism

Developments during the 1980s, in which the giant comet picture was widely ignored, have been reviewed by Bailey et al. (1990) and Steel (1991); see also Hoyle (1993). Since then the situation has changed and there has been a substantial swing back to physical models guided by a more complete astronomical picture. There have now been several claims that a giant comet explanation of the K/T extinction is preferable (Zahnle and Grinspoon 1990; Hut et al. 1991*a, b*; Shoemaker 1991,1993; Shoemaker and Izett 1992*a, b*; Bhandari et al. 1993), and it is clear that despite the potential for divergence amongst those considering the fundamental implications of such a picture, several groups are now working on broadly the same model.

According to the view developed in Secs. III and IV, debris from the most recent large comet captured into the inner solar system is currently observed as a broad meteoroid stream known as the Taurid Complex (Steel et al. 1991; Steel 1992). This comprises comets, small asteroids, boulders and dust, together totalling roughly 10% of known Earth-crossing asteroids and all but a few percent of the known meteoroidal sources of dust (Asher et al. 1993b, c; Asher and Clube 1993). The enhanced stratospheric extinction due to accretion of dust and larger bodies from such a stream provides an explanation for recent episodes of climatic deterioration known as little ice ages, while major enhancements in the stratospheric dust load provide an *a priori* astronomical mechanism capable of triggering major glaciations. These have occurred throughout the past 2 Myr, a time scale which is interestingly close to that of a comet shower (see, e.g., Fernández 1992) or the expected dynamical lifetime of either a Halley-type giant comet or a Jupiter-family Chiron.

In addition, observations suggest that the near-parabolic flux injects at least one giant comet within the orbit of Jupiter per century. Adopting an inclination-averaged probability on the order of 2×10^{-3} for capture into an observable Halley-type orbit with $P < 200$ yr (Quinn et al. 1990), giant Halley-type comets should be produced at a rate on the order of 20 per Myr. In general, one would expect the giant comet to survive for at least its dynamical lifetime, $\sim 10^6$ yr, although mantling, devolatilization or dynamical evolution to a sungrazing orbit may occur much faster than this depending on the initial orbit.

These results show that there are now two significant sources of giant comets in the inner solar system: Halley-type and Chiron-like intermediate-period orbits, respectively. Whereas Chirons typically return to orbits with perihelia close to Saturn within $\sim 10^4$ yr, possibly to return ~ 5 to 10 times at mean intervals on the order of 10^5 yr, Halley-type comets persist in the inner solar system for a time comparable to the ejection time scale $\sim 10^6$ yr. Given that giant comets are likely to be copious sources of dust and interplanetary debris throughout their short-period phases, Halley-types could in principle produce prolonged episodes of deterioration in the Earth's near-space environment, similar to that caused by a giant comet which happened to be captured into a sub-Jovian Encke-type orbit. Furthermore, although Halley types might become mantled or devolatilized, they have a significant probability to be tidally disrupted while on a sungrazing orbit. Halley-type giant comets thus represent a significant source of streams of high-velocity material in Earth-crossing orbits. Collisions between the Earth and large bodies within such streams would be particularly damaging compared to collisions of similar-sized bodies in Encke-type orbits, owing to their much higher relative velocity. Collisions by such bodies with the terrestrial planets may also be favored by the fact that the sungrazing phase correlates with an extremum of the inclination ($\cos^2 i \simeq 1$), leading to breakup and the formation of a massive meteoroid stream occurring preferentially when the inclination is either very small or very large, when the impact probability per revolution has a maximum (see the Chapter by Marsden and Steel).

The different evolution displayed by Halley-type and Jupiter-family short-period comets therefore suggests that evidence for exceptionally long-lived episodes of terrestrial bombardment in the geologic record might be associated with the evolution and decay of Halley-type giant comets, particularly if fragmentation of the body in a short-period sungrazing orbit were to coincide with intersection of the stream of debris with the Earth's orbit. The complex dynamical evolution of such material, susceptible both to close encounters with Jupiter and to strong secular perturbations, is well illustrated by the meteoroid streams associated with P/Machholz (Froeschlé and Scholl 1986; Babadzhanov and Obrubov 1992; Froeschlé et al. 1993). A dense stream with properties similar to that of the Kreutz sungrazing group, containing meteoroids and cometary debris ranging in size from Pygmy sungrazers ($d \sim 100$ m, comparable to the Tunguska projectile) up to km-sized cometary nuclei, illustrates how such a model might produce long-lived environmental stress on the Earth, and demonstrates the importance of understanding climatological and other effects associated with intermittent stratospheric dusting lasting ~ 10 to 10^3 yr. It seems probable therefore that the most extreme effects of impacts, the great mass extinctions detected at roughly 30 Myr intervals in the geologic record, and correlated at the K/T boundary, for example, with the Chicxulub crater (Grieve 1993), may each be understood as due to fragmentation of a giant comet in a short-period orbit that produced a massive debris stream intersecting the orbit of the Earth.

B. K/T Extinctions

The predicted effects of giant comets (see, e.g., Clube and Napier 1984,1986a) included (i) multiple impacts on the Earth at extinction horizons, including swarms of small bodies; (ii) associated concentrations of extraterrestrial material too high to be explicable by impacts; (iii) prolonged climatic deterioration and ocean regressions; (iv) a complex depositional history; and (v) prompt effects associated with large impacts. A number of geologic discoveries made subsequently to the above astronomical predictions are more easily understood as the consequences of giant-comet breakup, rather than of a stray impact, and may even be associated with the time scales involved in cometary evolution.

First, depositional and palynological evidence of a double impact layer in the western USA has been presented by a number of workers (Fastovsky et al. 1989; Wolfe 1991; Shoemaker and Izett 1992a, b). Estimates of the interval range from a few months up to $\sim 10^2$ yr, which are orders of magnitude shorter than those expected for a comet shower, but consistent with accretion from a swarm of cometary material within the debris stream of a short-period Earth-crossing giant comet. At least one large crater (Chicxulub, at 64.98 ± 0.06 Myr) appears to be contemporaneous with the K/T horizon (at 64.3 ± 1.2 Myr for the nonmarine boundary; Baadsgaard et al. 1988), but others have yet to be securely identified (the Manson crater, a leading candidate, has recently been redated at 73.8 ± 0.3 Myr; cf., Kunk et al. 1989; Izett et al. 1993). Less direct evidence for multiple impact includes the global combustion of the biomass, which is difficult to achieve with even a 10 km impact and which seems to call for a storm of Tunguskas; and even the presence of isotopic, mineralogical and other signatures which appear to require both continental and oceanic impact sites (Cisowski 1990; Hildebrand and Boynton 1990; Robin et al. 1993). No meteoritic refractory minerals or primordial noble gases have been found in K/T carbonaceous separates. This implies total vaporization of the bolide, which is unlikely (Wolbach et al. 1985), and is more consistent with the impact of one or more extremely weak impactors, and possibly again with a Tunguska storm.

Secondly, there is evidence for prolonged meteoroidal dust deposition at the K/T boundary associated with giant-comet disintegration. Amino acids of probable extraterrestrial origin appear to have been laid down over a $\sim 10^5$ yr period around the boundary (Zhao and Bada 1989), the interval before the boundary also being associated with pronounced cooling ($\Delta T \sim 4$ to $5°C$) in Antarctic surface waters (Stott and Kennett 1988). Amino acids would have been destroyed in the $\sim 10^5$ K fireball of an impactor (Zahnle and Grinspoon 1990; Clube and Napier 1990), but their presence and the observed cooling are consistent with injection of material in the form of Brownlee particles from at least one very large comet. Such dusting is also indicated by the concentrations of probable extraterrestrial material at K/T basal layers worldwide: these have a mean value $\sim 12\%$, as against $\sim 10^{-2}$ to 10^{-3} for most impact ejecta (Kyte et al. 1985). At one site (Woodside Creek, New Zealand), the basal clay has an extraterrestrial component $\sim 100\%$ (Schmitz 1988). Chemical profiles at the Meghalaya K/T boundary section (Bhandari et al. 1993) also provide strong evidence for cometary fragmentation over the period of amino-acid formation, while the differing Os/Ir ratio at the major extinction peak (~ 0.37, as opposed to ~ 1.50 in the broad Ir enrichment zone), is indicative of a reduced Os value at this boundary, possibly due to fractionation within the impactor. Multiple iridium peaks, corresponding to deposition over several 10^5 yr, have been reported at a number of K/T sections (Michel et al. 1985; Crocket et al. 1988).

Third, there is evidence of trauma acting on even more extended time scales (Officer et al. 1987; Cisowski 1990). A steady decline in diversity of major marine fossil groups such as the rudists and inoceramid bivalves, correlated with a rapid increase in the isotopic ratio $^{87}Sr/^{86}Sr$ of ocean water (Nelson et al. 1991), itself suggestive of increased continental weathering due to acid rain produced by bolide impacts (Macdougal 1988), appears to have been under way for a few Myr before the K/T extinction horizon, these groups disappearing about 1 Myr before the mass extinctions. This characteristic time scale is consistent with both the active lifetime of a comet in a Halley-type orbit (Figs. 5–7) and with the capture and dynamical evolution of giant comets from the parabolic flux following an Oort cloud disturbance.

On longer time scales, the only currently viable astronomical explanation for the apparent periodicity of many terrestrial phenomena (see, e.g., Rampino and Stothers 1984,1988; Sepkoski and Raup 1986; Rampino and Caldeira 1992,1993) is periodic disturbance of the Oort comet cloud. This calls for episodes of comet bombardment triggered by gravitational tides primarily caused by the galactic disk and massive concentrations of matter therein such as spiral arms and molecular clouds (Clube and Napier 1994; Matese et al. 1994; see Bailey et al. [1990] for earlier references).

Clearly one should not draw too many general conclusions from a single extinction boundary (cf., Holser et al. 1989; Hodych and Dunning 1992), but it does appear that although a massive impact did occur at the K/T boundary (Hildebrand and Boynton 1991; Sharpton et al. 1993) the stray asteroid theory cannot account for the full range of the geologic evidence. The potential for extinguishing life on a massive scale thus primarily belongs to occasional giant comets. It remains to assess the hazard they pose on time scales of human rather than geologic concern.

C. Dust Input From A Giant Comet

Dust in the inner planetary system derives both from the main asteroid belt (Sykes and Greenberg 1986; Flynn 1989) and from comets (Whipple 1967) although the relative contribution from each remains uncertain over all mass intervals (Leinert and Grün 1990). Evidence that a large short-period comet made a major contribution to the current meteoroidal dust population ($\sim$0.1 mm–10 cm) comes from the fact that many sporadic meteoroids, the chief constituent of the meteoroid population, belong to the so-called Štohl streams (Štohl 1983); these seem to form a bridge between the Taurid Complex and the zodiacal cloud (see the Chapter by Steel et al.). Additional evidence of the role played by a giant Taurid progenitor includes (i) the exceptional Encke trail which passes inside 1 AU (Sykes and Walker 1992); (ii) the apparent concentration of Taurid debris in the 7/2 mean-motion resonance with Jupiter (Asher and Clube 1993); (iii) the high frequency of exceptional asteroid fragmentations in the Taurid stream indicated by the best available historical airburst record (Hasegawa 1992; Clube 1994a, b); and (iv) the predominance among climatic and climate-proxy data of solar-terrestrial frequencies consistent with a dominant dust generating source in the 7/2 resonance (Asher and Clube 1993).

Table III lists the power-law index α of the differential diameter distribution, $n(d) \propto d^{-\alpha}$, of cometary grains from various sources. The 'Štohl meteoroids' refer to sporadic and stream meteors examined by Štohl (1987), whose size distribution peaks at around a millimeter, while the dust size distributions in the tails of P/Encke and P/d'Arrest have been derived by Fulle (1990). The power-law index for the dust in the coma of P/Schwassmann-Wachmann 1 is somewhat shallower than these (3.3 ± 0.3 according to Fulle [1992]) but still in the same sense, while much smaller particles are ejected during outbursts (Cochran et al. 1982). These distributions have generally been derived down to about $10\,\mu$m, and are such that progressively smaller particles make a declining contribution to the overall mass but an increasing contribution to the total cross section. The impact detectors aboard the Giotto spacecraft (McDonnell et al. 1987) revealed that at close encounter particles in the diameter range 1 to $10\,\mu$m dominated the geometric area of the dust (cf., Evlanov et al. 1991).

TABLE III

Size Distribution Indices of Cometary Grains[a]

Particles	α
Štohl meteoroids	3.7
P/Encke tail	3.6
P/d'Arrest tail	3.8
P/Schwassmann-Wachmann 1 coma	3.3
P/Halley inner coma	3.5

[a] See text for details.

The overall mass loss rate of a large comet thrown into a short-period orbit may be estimated following the procedure described by Kresák and Kresáková (1987). These authors adopted a mass loss calibration based on the 1986 apparition of comet Halley, which produced $\sim 3 \times 10^3$kg s^{-1} of dust particles $\lesssim 10\,\mu$m in radius (McDonnell et al. 1986,1987). However, IRAS observations (Sykes 1993) and modeling of the dust tail of P/Schwassmann-Wachmann 1 (Fulle 1992) suggest that the total mass loss rate due to particles up to centimeter dimensions may be up to 2 orders of magnitude more than that due to $\lesssim 10\,\mu$m grains. Then assuming an r^{-4} law for the mass loss rate $\dot{M}$ as a function of heliocentric distance r, a comet 200 km in diameter thrown into an Encke-like orbit (cf., Fig. 2) would have $\dot{M} \sim 10^{15}$kg yr^{-1} and a characteristic lifetime $M/\dot{M} \sim 3 \times 10^3$yr, leaving aside dormant or inactive phases (Kresák 1986). Thus, during each century of the giant comet's active lifetime, the mass of dust released is expected to be an order of magnitude greater than that of the present zodiacal cloud, currently about 2.5×10^{16}kg for particles of mass $\lesssim 1$ g (Whipple 1967).

Once in the zodiacal cloud, the particles undergo rapid collisional evolution. The lifetime τ_c of particles in the size range 50 to $500\,\mu$m against disruptive collisions at 1 AU is currently 10^4 to 10^5yr (Grün et al. 1985). With a zodiacal cloud 2 orders of magnitude more massive than the present one, $\tau_c \simeq 10^2$ to 10^3 yr for these larger particles, with even shorter lifetimes for smaller ones. Thus, collisional disintegration over a wide range of sizes takes place in less than the active lifetime of the comet, providing a copious source of micron- and submicron-sized particles in the zodiacal cloud. The Poynting-Robertson lifetime of a micron-sized particle is $\sim 10^3$ yr. Ten-micron radius particles ejected from Encke-like orbits spiral into the Sun in $\sim 2 \times 10^4$yr, with mean inclinations of order $10°$ to $15°$ (Jackson and Zook 1992), while particles smaller than $\sim 0.1\,\mu$m are lost through radiation pressure. It appears, therefore, that the injection of an exceptionally large comet (such as Sarabat or Chiron) into an Encke-like orbit would enhance the mass of the zodiacal cloud in the few-micron range by at least 2 orders of magnitude for the few millennia of its active lifetime.

If the zodiacal cloud during such a phase is modeled as a disk of mass 5×10^{18}kg, radius 1 AU and thickness 0.2 AU, then the mass of dust swept up by the Earth annually is $\sim 7.5 \times 10^8$ tons (allowing for a factor ~ 2 due to gravitational focusing), or more than 4 orders of magnitude greater than the current influx of microparticles (Love and Brownlee 1993). Currently the distribution of particles with respect to heliocentric distance r is proportional to $r^{-1.3}$ in the range $0.3 \lesssim r \lesssim 1$ AU, but this is for a somewhat evolved system; uncertainty in the detailed distribution affects the calculation at the factor-of-two level.

A particle of radius 3 μm and density 1 g cm^{-3} has a settling time through the atmosphere of about a year (Kasten 1968). Assuming a mean extinction cross section for particles of radius a on the order of πa^2, the stratospheric optical depth produced by material from such a zodiacal cloud is $\tau \sim 0.5$, larger if the particles are smaller, nonspherical or fluffy. Allowing for modeling and other uncertainties, giant comet disintegration will result in prolonged stratospheric dusting for 10^3 to 10^4 yr, the dust veil producing optical depths in the approximate range 0.05 to 3.0, with generally a comparable further extinction arising from interplanetary dust between the Earth and the Sun (cf., Yabushita and Allen 1989). The optical properties of the cometary grains are a function of size, composition and roughness of the surface, the effects of which have been modeled by Lamy and Perrin (1983) who find that the extinction and scattering increase strongly with surface roughness. Brownlee particles, having very irregular surfaces, are expected to be strongly scattering, implying that much sunlight incident on the Earth would be scattered back into space.

The climatic role of refractory submicron-sized particles in the wake of a disintegrating Chiron is uncertain. They are detected in cometary comae (Mukai 1985) and ejected from the solar system by radiation pressure, but they nevertheless enter the atmosphere as comminution products of extremely fragile cometary meteoroids (Fechtig 1982; Clube 1987) and are observed, for example, as fundamental constituents of micrometeorites (Maurette et al. 1993). The settling time of a 0.3 μm particle through the stratosphere is an order of magnitude longer than that of a 3 μm one (Kasten 1968), hence a relatively small admixture of tiny grains could have a disproportionate effect. Hoyle (1984) and Hoyle and Wickramasinghe (1978,1990) have noted that the scattering angle of a ~ 0.1 μm particle is about 90°, so if the stratosphere were to be appreciably dusted by very small particles, backscattering of sunlight incident on the Earth would again be expected, leading to further cooling.

It should be emphasized that in these calculations we are dealing with a giant comet whose inactive or dormant phases are not important (cf., Clube and Napier 1984), a situation which probably best applies during the early stages of giant comet evolution in sub-Jovian space and which presumably plays a significant role in the production of ice ages (see below). For a more realistic case, allowing the possibility of heterogeneous composition with some degree of differentiation (see, e.g., Whipple 1992), we would envisage an early active phase (say $\sim 10^4$ yr) when such calculations are likely to be relevant and a subsequent less active phase (possibly $\gtrsim 10^4$ yr) during which short-lived enhancements of the zodiacal dust may occur as a result of temporary activity and overproduction of dust. In general, it is to be expected that comparable terrestrial effects will then only arise as a result of passages through the giant comet trail (i.e., the threshold for an ice age may not be reached). Another possibility, of course, is that variable production of zodiacal dust may occur as the comet encounters its own debris (e.g., in a resonance; cf., Asher and Clube 1993). Leaving these arguments aside, whatever the precise optical properties of the grains and the manner in which they are injected at the top of the Earth's atmosphere, the resultant extinction is quite adequate to produce both long-term and short-term effects on the global climate and hence on the biosphere. The ice-albedo instability models of Budyko (1969), for example, predict that even a 2 to 5% sustained decline in insolation could yield a runaway glaciation.

D. Climatic Consequences

For dust concentrations with optical depths $\tau \sim 1$, glacial conditions would be expected for the duration of the dusting. Global cooling is inevitable, but the process is complex and involves a variety of nonlinear feedback loops and threshold effects. Reliable modeling requires a holistic treatment, which may never in fact be realistic because of the complexity of the Earth system (Broecker 1987). Here we outline some of the phenomena likely to be involved in the cooling process.

The first effect of cooling of the atmosphere is a lowering of its water vapor content. Because water vapor is a major greenhouse gas, the cooling is enhanced by this process, by a factor ~ 3. However, a cooler atmosphere may be cloudier, resulting in a system which may tend to trap surface heat depending on the clouds of different types which form as the atmosphere cools. Cirrus clouds have high albedo, reflecting more short-wave radiation and so adding to the effects of dust. It is thus likely that the hydrological cycle will play an important but uncertain role in cloud formation mechanisms, with a further uncertainty in the optical properties of the clouds. Even without the likely effects of the ocean, current models may not correctly represent cloud feedbacks; a comparison by Cess et al. (1989) of 14 general circulation models showed cloud-climate feedback ranging from modest negative to strong positive.

The hydrosphere may also enter the equations through the thermohaline circulation created by differential salinity of the North Atlantic relative to the North Pacific (Broecker 1987). This difference is caused by the higher evaporation rate of North Atlantic waters relative to the North Pacific. About 30% of solar heat incident on the Atlantic is transferred to Europe, and cessation or deflection of the Gulf Stream would have a catastrophic effect on European climate. In fact, small changes in the Atlantic thermohaline circulation can have prolonged effects on the whole climate system (Dickson et al. 1988). The converse also applies; climatic cooling, by reducing differential evaporation rates between the northern Atlantic and Pacific oceans, would tend to shut down the thermohaline conveyor belt, so initiating a harsh climatic regime throughout Europe, similar to that currently experienced in northern Scandinavia.

It is unlikely that these climatic effects would be confined to Europe, however. Large-scale variations in snowfall over Eurasia in spring affect the Asian summer monsoon, and are strongly connected to the atmosphere over North America (Barnett et al. 1988). Thus, in addition to the atmosphere/hydrosphere feedbacks, continental processes need to be considered. Finally, the cloud-climate feedbacks may also be coupled to the cryosphere. Simple models by Budyko and Sellers have shown that a decrease in incident sunlight by only a few percent could cause polar ice to spread over the entire planet, by virtue of an ice-albedo instability. Atmospheric ice crystals may then maintain the glacial state, and an oceanic impact or similar perturbation might be required to dislodge the system (Hoyle 1981,1984), implying that glaciations may come to an abrupt end.

These underlying conditions pertain whether or not additional astronomical processes have to be invoked to explain the long-term climatic record of the Earth. The question to be answered by modeling, then, is whether the inherent instabilities in the global climate system are capable in themselves of reproducing the record (in a statistical sense initially) or whether an improved fit would be obtained by invoking further likely astronomical processes, for example (i) Milankovitch modulation; (ii) *indirect* dust injection, for example through asteroid impacts elevating terrestrial dust into the stratosphere; and (iii) *direct* dust injection, for example through comminution of giant comet debris above the stratosphere. Although the stratospheric optical depths obtained for the second and third of these may be comparable to one another, the expected effects of a cosmic winter induced by comet dusting (i.e., through comminution of giant comet debris; Clube and Napier 1990) should not be confused with those expected from volcanic emissions or the nuclear winter scenario (Turco et al. 1983) which in the astronomical context considers effects due to impact-induced elevation of terrestrial dust coupled with smoke due to secondary fires. What distinguishes them in addition to particle size is duration: in the case of giant comet disintegration, crucial time constants involved in the climatic/hydrological system such as the cooling time of the oceans or the growth time of ice caps may be equaled or exceeded. In the case of a volcanic eruption or a single massive impact, the sky is cleared in a matter of months. The climatic response is likely to be qualitatively different in the two cases, especially as many climate models reveal that discontinuous transitions from ice-free to glaciated states may require the application of stress for $\gtrsim 10^3$ yr (Crowley and North 1988,1991).

Severe coolings of the Earth by as much as 5°C, sudden in onset and lasting several hundred years, are a further well-established part of the palaeoclimatic record which may have an origin in relatively brief, intense dustings

(Clube and Napier 1990). A 1% decline in the solar constant yields $\sim 1.1°C$ of global cooling (Wickramasinghe et al. 1989a, b), comparable to that which pertained during the Little Ice Age (Crowley and North 1991). With modern agricultural practices and dependence on green revolution crops, a global cooling of this order at the present day would have catastrophic results (Dotto 1986).

VI. CONCLUSIONS

Recent astronomical results suggest that giant comets with diameters $d \gtrsim$ 100 km, observed both in the form of intermediate-period Chirons and in the long-period flux, may be captured into short-period Jupiter-family or Halley-type orbits at a rate on the order of 5 to 50 Myr^{-1}, some evolving into stable orbits similar to those of P/Encke or P/Machholz. Large bodies in such orbits represent a significant environmental hazard due to direct impacts of km and sub-km sized fragments and accretion of dust, the latter released in large quantities during evolution of the giant comet in an orbit of small perihelion distance. The dust alone is sufficient to trigger global cooling and a possible ice age; multiple Tunguskas (Kyte et al. 1980) could prove equally damaging to life.

In this chapter we have emphasized the importance of giant comets both for the interpretation of the canonical K/T mass extinction and the present distribution of solid material in the inner solar system. The evidence of the Taurid Complex, comprising comets, small asteroids, boulders and dust, suggests that we are now living in the post-breakup phase of the most recent giant comet to be inserted into a short-period Encke-type Earth-crossing orbit. The same evidence indicates that this Complex remains a potential multiple Tunguska hazard on short time scales (e.g., during nodal intersections at ~ 2000 yr intervals) and that any discussion of future impacts should, as a matter of course, include a discussion of the historical record as well (Clube 1994b). A correct assessment of the celestial hazard must allow for the fact that much cometary debris circulates in streams, leading to significant enhancements of the accretion rate involving small bodies for periods lasting from years to decades. During such phases the enhanced stratospheric dust load could be sufficient to trigger episodes of global cooling similar to that observed during historical periods known as little ice ages; impacts by meteorites and Tunguska-sized bodies associated with such streams represent an additional hazard. The simultaneity of the Little Ice Age and a corresponding significant enhancement of the Taurid meteoroid flux (Asher and Clube 1993) provides suggestive evidence of the most frequent hazard at present faced by civilization.

Astronomical surveys intended to quantify the immediate risk of a rare globally devastating encounter with a km-sized asteroid within the next hundred years or so should consider that bodies currently in orbits of large perihelion distance may be inserted on to Earth-crossing orbits within time scales of this order and that many sub-km sized Earth-crossing bodies also exist. The latter, chiefly fragmentation and disintegration products of comets, are predicted to move in streams that encounter the Earth at intervals on the order of 10 to 10^3 yr, the shorter end of the range relating to possible commensurabilities in the encounter rate depending on the stream's initial orbital period and spread. Such streams provide a hazard to civilization that is extremely variable as a function of time. A correct assessment of the near-term risk thus requires surveys capable not only of discovering (a) the current population of km-sized Earth-crossing asteroids and bodies of similar size in high-inclination orbits, but also (b) comets and asteroids in orbits of large perihelion distance, which may become Earth-crossing within the next $\sim 10^2$ yr, and (c) streams of small bodies predicted to exist in the inner solar system, which may become Earth-approaching within the same time scale. An enhanced Spaceguard program (Chapter by Morrison et al.) could identify these potential hazards and provide effective warning of enhanced accretion events, necessary if strategies to alleviate the risk are to be implemented. Such a program presents a crucial challenge for the future.

Acknowledgments. It is a pleasure to thank L. Dones and an anonymous referee for detailed comments on a preliminary version of this chapter. MEB thanks the Royal Society for partial support.

REFERENCES

Alexander, S. 1850. On the classification and special points of resemblance of certain of the periodic comets; and the probability of a common origin in the case of some of them. *Astron. J.* 1:147–150.

Alexander, S. 1851. On the similarity of arrangement of the asteroids and the comets of short period, and the possibility of their common origin. *Astron. J.* 1:181–184.

Alvarez, L. W. 1983. Experimental evidence that an asteroid impact led to the extinction of many species 65 million years ago. *Proc. Natl. Acad. Sci. USA* 80:627–642.

Alvarez, L. W., Alvarez, W., Asaro, F., and Michel, H. V. 1980. Extraterrestrial cause for the Cretaceous-Tertiary extinction. *Science* 208:1095–1108.

Asher, D. J. 1991. The Taurid Meteoroid Complex. Thesis, Univ. of Oxford.

Asher, D. J., and Clube, S. V. M. 1993. An extraterrestrial influence during the current glacial-interglacial. *Quart. J. Roy. Astron. Soc.* 34:481–511.

Asher, D. J., and Steel, D. I. 1992. Future orbital evolution of giant comet/asteroid (5145) Pholus = 1992 AD. In *Observations and Physical Properties of Small Solar System Bodies*, eds. A. Brahic, J.-C. Gerard and J. Surdej (Liège: Université de Liège), pp. 263–266.

Asher, D. J., and Steel, D. I. 1993. Orbital evolution of the large outer solar system object 5145 Pholus. *Mon. Not. Roy. Astron. Soc.* 263:179–190.

Asher, D. J., Bailey, M. E., Hahn, G., and Steel, D. I. 1993*a*. Asteroid 5335 Damocles and its implications for cometary dynamics. *Mon. Not. Roy. Astron. Soc.*, in press.

Asher, D. J., Clube, S. V. M., and Steel, D. I. 1993*b*. The Taurid Complex asteroids. In *Meteoroids and their Parent Bodies*, eds. J. Štohl and I. P. Williams (Bratislava: Slovak Acad. of Science), pp. 93–96.

Asher, D. J., Clube, S. V. M., and Steel, D. I. 1993*c*. Asteroids in the Taurid Complex. *Mon. Not. Roy. Astron. Soc.* 264:93–105.

Baadsgaard, H., Lerbekmo, J. F., and McDougall, I. 1988. A radiometric age for the Cretaceous-Tertiary boundary based upon K-Ar, Rb-Sr, and U-Pb ages of bentonites from Alberta, Saskatchewan, and Montana. *Canadian J. Earth Sci.* 25:1088–1097.

Babadzhanov, P. B., and Obrubov, Yu. V. 1992. Evolution of short-period meteoroid streams. *Cel. Mech. Dyn. Astron.* 54:111–127.

Bailey, M. E. 1983. Is there a dense primordial cloud of comets just beyond Pluto? In *Asteroids, Comets, Meteors*, eds. C.-I. Lagerkvist and H. Rickman (Uppsala: Uppsala University), pp. 383–386.

Bailey, M. E. 1986. The near-parabolic flux and the origin of short-period comets. *Nature* 324:350–352.

Bailey, M. E. 1990*a*. Cometary masses. In *Baryonic Dark Matter* eds. D. Lynden-Bell and G. Gilmore (Dordrecht: Kluwer), pp. 7–35.

Bailey, M. E. 1990*b*. Short-period comets: Probes of the inner core. In *Asteroids, Comets, Meteors III*, eds. C.-I. Lagerkvist, H. Rickman, B. A. Lindblad and M. Lindgren (Uppsala: Uppsala University), pp. 221–230.

Bailey, M. E. 1991. Comet craters versus asteroid craters. *Adv. Space Res.* 11:(6)43–60.

Bailey, M. E. 1992. Origin of short-period comets. *Cel. Mech. and Dyn. Astron.* 54:49–61.

Bailey, M. E. 1993. Applied astronomy saves the world! *Physics World* 6:(2)22–23.

Bailey, M. E., and Hahn, G. 1992. Orbital evolution of 1991 DA: Implications for near-Earth asteroids. In *Periodic Comets*, eds. J. A. Fernández and H. Rickman (Montevideo: Universidad de la República), pp. 13–24.

Bailey, M. E., Clube, S. V. M., and Napier, W. M. 1990. *The Origin of Comets* (Oxford: Pergamon Press).

Bailey, M. E., Chambers, J. E., and Hahn, G. 1992*a*. Detection of comet nuclei at large heliocentric distances. *Mon. Not. Roy. Astron. Soc.* 254:581–588.

Bailey, M. E., Chambers, J. E., Hahn, G., Scotti, J. V., and Tancredi, G. 1992*b*. Transfer probabilities between Jupiter and Saturn-family orbits: Application to 1992 AD = 5145. In *Observations and Physical Properties of Small Solar System Bodies*, eds. A. Brahic, J.-C. Gerard and J. Surdej (Liège: Université de Liège), pp. 285–287.

Bailey, M. E., Chambers, J. E., and Hahn, G. 1992*c*. Origin of sungrazing comets: A frequent cometary end-state. *Astron. Astrophys.* 257:315–322.

Bailey, M. E., Chambers, J. E., Hahn, G., Scotti, J. V., and Tancredi, G. 1993. Chaotic orbital evolution of 5145 Pholus. In preparation.

Baldwin, R. B. 1985. Relative and absolute ages of individual craters and the rate of infalls on the Moon in the post-Imbrium period. *Icarus* 61:63–91.

Barnett, T. P., Dumenill, L., Schlese, U., and Roeckner, E. 1988. The effect of Eurasian snow cover on global climate. *Science* 239:504–507.

Bhandari, N., Gupta, M., Pandey, J., and Shukla, P. N. 1993. Chemical Profiles in K/T Boundary Section of Meghalaya, India: Cometary, Asteroidal or Volcanic. Preprint.

Bobrovnikoff, N. T. 1931. The origin of asteroids. *Publ. Astron. Soc. Pacific* 43:324–333.

Bowell, E. 1992. (4015) 1979 VA = Comet Wilson-Harrington 1949 III. *IAU Circ.* No. 5585.

Bredichin, T. 1889. Sur l'origine de comètes périodiques. *Ann. Obs. Moscou (Ser. 2)* 2:1–17.

Broecker, W. S. 1987. Unpleasant surprises in the greenhouse? *Nature* 328:123–126.

Budyko, M. I. 1969. The effect of solar radiation variations on the climate of the Earth. *Tellus* 21:611–619.

Canavan, G. H., Solem, J. C., and Rather, J. D. G., eds. 1992. *Proceedings of the Near-Earth-Object Interception Workshop* (Los Alamos: Los Alamos National Lab).

Carusi, A., Kresák, Ľ., Perozzi, E., and Valsecchi, G. B. 1985. First results of the integration of motion of short-period comets over 800 years. In *Dynamics of Comets: Their Origin and Evolution*, eds. A. Carusi and G. B. Valsecchi (Dordrecht: D. Reidel), pp. 319–340.

Carusi, A., Kresák, Ľ., Perozzi, E., and Valsecchi, G. B. 1987a. High-order librations of Halley-type comets. *Astron. Astrophys.* 187:899–905.

Carusi, A., Perozzi, E., and Valsecchi, G. B. 1987b. Perturbation computations and numerical modelling experiments. In *The Evolution of the Small Bodies of the Solar System*, eds. M. Fulchignoni and Ľ. Kresák (Amsterdam: North-Holland), pp. 191–201.

Carusi, A., Kresák, Ľ., Perozzi, E., and Valsecchi, G. B. 1987c. Long-term resonances and orbital evolutions of Halley-type comets. In *Interplanetary Matter*, vol. 2, eds. Z. Ceplecha and P. Pecina (Prague: Czechoslovak Acad. Sci.), 29–32.

Carusi, A., Kresák, Ľ., Perozzi, E., and Valsecchi, G. B. 1988. On the past orbital history of Comet P/Halley. *Cel. Mech.* 43:319–322.

Carusi, A., Valsecchi, G. B., and Greenberg, R. 1990. Planetary close encounters: Geometry of approach and post-encounter orbital parameters. *Cel. Mech. Dyn. Astron.* 49:111–131.

Carusi, A., Marsden, B. G., and Valsecchi, G. B. 1993. The Probable collision of P/Shoemaker-Levy 9 (1993e) with Jupiter in 1994. Poster paper presented at Asteroids, Comets, Meteors 1993, IAU Symp. 160, June 14–18, Belgirate (Italy).

Ceplecha, Z. 1992. Influx of interplanetary bodies onto Earth. *Astron. Astrophys.* 263:361–366.

Cess, R. D., Potter, G. L., Blanchet, J. P., Boer, G. J., Ghan, S. J., Kiehl, J. T., Le Truet, H., Li, Z.-X., Liang, X.-Z., Mitchell, J. F. B., Morcrette, J.-J., Randall, D. A., Riches, M. R., Roeckner, E., Schlese, U., Slingo, A., Taylor, K. E., Washington, W. M., Wetherald, R. T., and Yagai, I. 1989. Interpretation of cloud-feedback as produced by 14 atmospheric general circulation models. *Science* 245:513–516.

Chambers, J. E. 1993. The long-term threat posed by Comet Swift-Tuttle. *Icarus*, submitted.

Chapman, C. R. 1993. Comet on target for Jupiter. *Nature* 363:492–493.

Cisowski, S. M. 1990. A critical review of the case for, and against, extraterrestrial impact at the K/T boundary. *Surveys Geophys.* 11:55–131.

Clube, S. V. M. 1987. The origin of dust in the solar system. *Phil. Trans. Roy. Soc. London* A323:421–436.

Clube, S. V. M. 1992. The fundamental role of giant comets in Earth history. *Cel. Mech. Dyn. Astron.* 54:179–194.

Clube, S. V. M. 1994a. Hazards from space: Comets in history and science. In *How Science Works in a Crisis: The Mass Extinction Debate*, ed. W. Glen (Stanford: Stanford Univ. Press), pp. 152–169.

Clube, S. V. M. 1994b. The global incidence of meteoric airbursts (fireballs) recorded in China and the associated eschatological response in Europe, 0–1900 A. D. *Quart. J. Roy. Astron. Soc.*, submitted.

Clube, S. V. M., and Napier, W. M. 1982. Spiral arms, comets and terrestrial catastrophism. *Quart. J. Roy. Astron. Soc.* 23:45–66.

Clube, S. V. M., and Napier, W. M. 1984. The microstructure of terrestrial catastrophism. *Mon. Not. Roy. Astron. Soc.* 211:953–968.

Clube, S. V. M., and Napier, W. M. 1986a. Giant comets and the galaxy: Implications of the terrestrial record. In *The Galaxy and the Solar System*, eds. R. Smoluchowski, J. N. Bahcall and M. S. Matthews (Tucson: Univ. of Arizona Press), pp. 260–285.

Clube, S. V. M., and Napier, W. M. 1986b. Mankind's future: An astronomical view. Comets, ice ages and catastrophes. *Interdisciplinary Sci. Rev.* 11:236–247.

Clube, S. V. M., and Napier, W. M. 1990. *The Cosmic Winter* (Oxford: Basil Blackwell).

Clube, S. V. M., and Napier, W. M. 1994. Galactic periodicities in the terrestrial record. *Mon. Not. Roy. Astron. Soc.*, submitted.

Cochran, A. L., Cochran, W. D., and Barker, E. S. 1982. Spectrophotometry of Comet Schwassmann-Wachmann 1. II. Its color and CO^+ emission. *Astrophys. J.* 254:816–822.

Crocket, J. H., Officer, C. B., Wezel, F. C., and Johnson, G. D. 1988. Distribution of noble metals across the Cretaceous/Tertiary boundary at Gubbio, Italy: Iridium variation as a constraint on the duration and nature of Cretaceous/Tertiary boundary events. *Geology* 16:77–80.

Crowley, T. J., and North, G. R. 1988. Abrupt climatic change and extinction events in Earth history. *Science* 240:996–1002.

Crowley, T. J., and North, G. R., 1991. *Paleoclimatology* (New York: Oxford University Press).

Curdt, W., and Keller, H. U. 1990. Large dust particles along the Giotto trajectory. *Icarus* 86:305–313.

Dickson, R. R., Meincke, J., Malmberg, S.-A., and Lee, A. J. 1988. The "Great Salinity Anomaly" in the Northern North Atlantic 1968–1982. *Prog. Oceanography* 20:103–151.

Dones, L. 1991. A recent cometary origin for Saturn's rings? *Icarus* 92:194–203.

Donnison, J. R. 1986. The distribution of cometary magnitudes. *Astron. Astrophys.* 167:359–363.

Dotto, L. 1986. *Planet Earth in Jeopardy* (Chichester: Wiley).

Drummond, J. 1991. Earth-approaching asteroid streams. *Icarus* 89:14–25.

Duncan, M. J., and Quinn, T. 1993. The long-term dynamical evolution of the solar system. *Ann. Rev. Astron. Astrophys.* 31:265–295.

Duncan, M. J., Quinn, T., and Tremaine, S. 1987. The formation and extent of the solar system comet cloud. *Astron. J.* 94:1330–1338.

Duncan, M. J., Quinn, T., and Tremaine, S. 1988. The origin of short-period comets. *Astrophys. J. Lett.* 328:69–73.

Dvorak, R., and Kribbel, J. 1990. Dynamics of Halley-type comets for 1 million years. *Astron. Astrophys.* 227:264–270.

Everhart, E. 1967. Intrinsic distributions of cometary perihelia and magnitudes. *Astron. J.* 72:1002–1011.

Everhart, E. 1972. The origin of short-period comets. *Astrophys. Lett.* 10:131–135.

Everhart, E. 1977. Evolution of comet orbits as perturbed by Uranus and Neptune. In *Comets, Asteroids, Meteorites: Interrelations, Evolution and Origins*, ed. A. H. Delsemme (Dordrecht: D. Reidel), pp. 99–104.

Evlanov, E. I., Prilutskii, O. F., and Fomenka, M. N. 1991. Mass of comet Halley dust particles from results of the Puma experiment. *Kosm. Issled.* 29:641–646 (in Russian).

Farinella, P., Froeschlé, C., and Gonczi, R. 1993a. Meteorite delivery. In *Asteroids,*

Comets, Meteors 1993, IAU Symp. 160, eds. A. Milani, M. DiMartini and A. Cellino, (Dordrecht: Kluwer).

Farinella, P., Gonczi, R., Froeschlé, Ch., and Froeschlé, C. 1993*b*. The injection of asteroid fragments into resonances. *Icarus* 101:174–187.

Fastovsky, D. E., McSweeney, K., and Norton, L. D. 1989. Pedogenic development at the Cretaceous-Tertiary boundary, Garfield County, Montana. *J. Sedim. Petrol.* 59:758–767.

Fechtig, H. 1982. Cometary dust in the solar system. In *Comets*, ed. L. Wilkening (Tucson: Univ. of Arizona Press), pp. 370–382.

Fernández, J. A. 1980. On the existence of a comet belt beyond Neptune. *Mon. Not. Roy. Astron. Soc.* 192:481–491.

Fernández, J. A. 1981. The role of collisions with interplanetary particles in the physical evolution of comets. *Moon and Planets* 25:507–519.

Fernández, J. A. 1985. The formation and survival of the Oort cloud. In *Dynamics of Comets: Their Origin and Evolution*, eds. A. Carusi and G. B. Valsecchi (Dordrecht: D. Reidel), pp. 45–70.

Fernández, J. A. 1990. Collisions of comets with meteoroids. In *Asteroids, Comets, Meteors III*, eds. C.-I. Lagerkvist, H. Rickman, B. A. Lindblad and M. Lindgren (Uppsala: Uppsala University), pp. 309–312.

Fernández, J. A. 1992. Comet showers. In *Chaos, Resonance and Collective Dynamical Phenomena in the Solar System*, ed. S. Ferraz-Mello (Dordrecht: Kluwer), pp. 239–254.

Fernández, J. A., and Ip, W.-H. 1983. On the time evolution of the planetary influx in the region of the terrestrial planets. *Icarus* 54:377–387.

Fink, U., Hoffmann, M., Grundy, W., Hicks, M., and Sears, W. 1992. The steep red spectrum of 1992 AD: an unusual asteroid covered with organic material? *Icarus* 97:145–149.

Flynn, G. J. 1989. Atmospheric entry heating: A criterion to distinguish between asteroidal and cometary sources of interplanetary dust. *Icarus* 77:287–310.

Froeschlé, Ch., and Froeschlé, C. 1992. Collective resonance phenomena on small bodies in the solar system. *Cel. Mech. Dyn. Astron.* 54:71–89.

Froeschlé, C., and Scholl, H. 1986. Numerical investigations on a possible gravitational breaking of the Quadrantid meteor stream. In *Asteroids, Comets, Meteors II*, eds. C.-I. Lagerkvist, B. A. Lindblad, H. Lundstedt and H. Rickman (Uppsala: Uppsala University), pp. 555–558.

Froeschlé, C., Gonczi, R., and Rickman, H. 1993. New results on the connection between P/Machholz and the Quadrantid meteor streams: Poynting-Robertson drag and chaotic motion. In *Meteoroids and Their Parent Bodies*, eds. J. Štohl and I. P. Williams (Bratislava: Slovak Acad. of Science), pp. 169–172.

Fulle, M. 1990. Meteoroids from short period comets. *Astron. Astrophys.* 230:220–226.

Fulle, M. 1992. Dust from short-period comet P/Schwassmann-Wachmann 1 and replenishment of the interplanetary dust cloud. *Nature* 359:42–44.

Gladman, B., and Duncan, M. 1990. On the fates of minor bodies in the outer solar system. *Astron. J.* 100:1680–1693.

Goswami, J. N., and Lal, D. 1978. Temporal changes in the flux of meteorites in the recent past. *Moon and Planets* 18:371–382.

Green, S. F., McDonnell, J. A. M., Perry, C. H., Nappo, S., and Zarnecki, J. C. 1987. P/Halley dust coma: Grains or rocks? In *Symposium on the Diversity and Similarity of Comets*, eds. E. J. Rolfe and B. Battrick, ESA SP-278 (Noordwijk: European Space Agency), pp. 379–384.

Grieve, R. A. F. 1987. Terrestrial impact structures. *Ann. Rev. Earth Planet. Sci.* 15:245–270.

Grieve, R. A. F. 1991. Terrestrial impact: The record in the rocks. *Meteoritics* 26:175–194.

Grieve, R. A. F. 1993. When will enough be enough? *Nature* 363:670–671.

Grün, E., Zook, H. A., Fechtig, H., and Giese, R. H. 1985. Collisional balance of the meteoritic complex. *Icarus* 62:244–272.

Hahn, G. 1991. Close encounters of near-Earth asteroids during 1900–2100. *Adv. Space Res.* 11:(6)29–41.

Hahn, G., and Bailey, M. E. 1990. Rapid dynamical evolution of giant comet Chiron. *Nature* 348:132–136.

Hahn, G., and Rickman, H. 1985. Asteroids in cometary orbits. *Icarus* 61:417–442.

Hahn, G., Lagerkvist, C.-I., Lindgren, M., and Dahlgren, M. 1991. Orbital evolution studies of asteroids near the 5:2 mean motion resonance with Jupiter. *Astron. Astrophys.* 246:603–618.

Hallam, A. 1987. End-Cretaceous mass extinction event: Argument for terrestrial causation. *Science* 238:1237–1242.

Hallam, A. 1989. Catastrophism in geology. In *Catastrophes and Evolution: Astronomical Foundations*, ed. S. V. M. Clube (Cambridge: Cambridge Univ. Press), pp. 25–55.

Hartmann, W. K., Tholen, D. J., Meech, K. J., and Cruikshank, D. P. 1990. 2060 Chiron: Colorimetry and cometary behavior. *Icarus* 83:1–15.

Hasegawa, I. 1992. Historical variations in the meteor flux as found in Chinese and Japanese chronicles. *Cel. Mech. Dyn. Astron.* 54:129–142.

Hildebrand, A. R., and Boynton, W. V. 1990. Proximal Cretaceous-Tertiary boundary impact deposits in the Caribbean. *Science* 248:843–847.

Hildebrand, A. R., and Boynton, W. V. 1991. Cretaceous-Tertiary ground zero. *Natural History* (June), pp. 47–53.

Hodych, J. P., and Dunning, G. R. 1992. Did the Manicouagan impact trigger end-of-Triassic mass extinction? *Geology* 20:51–54.

Holmes, A. 1927. *The Age of the Earth—An Introduction to Geological Ideas* (London: Benn).

Holser, W. T., Schönlaub, H.-P., Attrep, M., Boeckelmann, K., Klein, P., Magaritz, M., Orth, C. J., Fenninger, A., Jenny, C., Kralik, M., Mauritsch, H., Pak, E., Schramm, J.-M., Stattegger, K., and Schmöller, R. 1989. A unique geochemical record at the Permian/Triassic boundary. *Nature* 337:39–44.

Hoyle, F. 1981. *Ice* (London: Hutchinson).

Hoyle, F. 1984. On the causes of ice-ages. *Earth, Moon, and Planets* 31:229–248.

Hoyle, F. 1993. *The Origin of the Universe and the Origin of Religion* (Rhode Island: Moyer Bell).

Hoyle, F., and Wickramasinghe, N. C. 1978. Comets, ice ages and ecological catastrophes. *Astrophys. Space Sci.* 53:523–526.

Hoyle, F., and Wickramasinghe, N. C. 1990. Backscattering of Sunlight by Ice Grains in the Mesosphere. Weston preprint no. 9.

Hsü, K. J. 1980. Terrestrial catastrophe caused by cometary impact at the end of the Cretaceous. *Nature* 285:201–203.

Hughes, D. W. 1983. Temporal variations of the absolute magnitude of Halley's comet. *Mon. Not. Roy. Astron. Soc.* 204:1291–1295.

Hughes, D. W. 1985. The size, mass loss and age of Halley's comet. *Mon. Not. Roy. Astron. Soc.* 213:103–109.

Hughes, D. W. 1987a. The history of Halley's comet. *Phil. Trans. Roy. Soc. London* A323:349–367.

Hughes, D. W. 1987b. On the distribution of cometary magnitudes. *Mon. Not. Roy. Astron. Soc.* 226:309–316.

Hughes, D. W. 1990. Cometary absolute magnitudes, their significance and distribution. In *Asteroids, Comets, Meteors III*, eds. C.-I. Lagerkvist, H. Rickman, B. A. Lindblad and M. Lindgren (Uppsala: Uppsala University), pp. 327–342.

Hughes, D. W., and Daniels, P. A. 1982. Temporal variations in the cometary mass distribution. *Mon. Not. Roy. Astron. Soc.* 198:573–582.

Hut, P., Alvarez, W., Elder, W. P., Hansen, T., Kauffman, E. G., Keller, G., Shoemaker, E. M., and Weissman, P. R. 1987. Comet showers as a cause of mass extinctions. *Nature* 329:118–126.

Hut, P., Shoemaker, E. M., Alvarez, W., and Montanari, A. 1991*a*. Astronomical mechanisms and geologic evidence for multiple impacts on Earth. *Proc. Lunar Planet. Sci. Conf.* 22:603–604.

Hut, P., Shoemaker, E. M., Alvarez, W., and Montanari, A. 1991*b*. Multiple impacts at Cretaceous-Tertiary boundary time: Break-up of a single comet? International Conference on Near-Earth Asteroids, San Juan Capistrano Research Inst., San Juan Capistrano, Ca., June 30–July 3, Abstract book, p. 16.

Izett, G. A., Cobban, W. A., Obradovich, J. D., and Kunk, M. J. 1993. The Manson impact structure: ^{40}Ar/^{39}Ar age and its distal impact ejecta in the Pierre Shale in Southeastern South Dakota. *Science* 262:729–732.

Jackson, A. A., and Zook, H. A. 1992. Orbital evolution of dust particles from comets and asteroids. *Icarus* 97:70–84.

Jewitt, D. 1991. Cometary photometry. In *Comets in the Post-Halley Era*, vol. 1, eds. J. R. Newburn, Jr., M. Neugebauer and J. Rahe (Dordrecht: Kluwer), pp. 19–65.

Jewitt, D., and Luu, J. 1993. Discovery of the candidate Kuiper belt object 1992 QB$_1$. *Nature* 362:730–732.

Kasten, F. 1968. Falling speeds of aerosol particles. *J. Appl. Meteorology* 7:944–947.

Keller, H. U. 1989. Comets—dirty snowballs or icy dirtballs? In *Physics and Mechanics of Cometary Materials*, eds. J. Hunt and T. D. Guyenne, ESA SP-302 (Noordwijk: European Space Agency), pp. 39–45.

Kowal, C. T., Liller, W., and Marsden, B. G. 1979. The discovery and orbit of (2060) Chiron. In *Dynamics of the Solar System*, ed. R. L. Duncombe (Dordrecht: D. Reidel), pp. 245–250.

Kozai, Y. 1962. Secular perturbations of asteroids with high inclination and eccentricity. *Astron. J.* 67:591–598.

Kozai, Y. 1979. Secular perturbations of asteroids and comets. In *Dynamics of the Solar System* ed. R. L. Duncombe (Dordrecht: D. Reidel), pp. 231–237.

Kozai, Y. 1980. Asteroids with large secular orbital variations. *Icarus* 41:89–95.

Kresák, Ľ. 1980*a*. Sources of interplanetary dust. In *Solid Particles in the Solar System*, eds. I. Halliday and B. A. McIntosh (Dordrecht: D. Reidel), pp. 211–222.

Kresák, Ľ. 1980*b*. Dynamics, interrelations and evolution of the systems of asteroids and comets. *Moon and Planets* 22:83–98.

Kresák, Ľ. 1986. On the aging process of periodic comets. In *20th ESLAB Symposium on the Exploration of Halley's Comet*, vol. II, eds. B. Battrick, E. J. Rolfe and R. Reinhard, ESA SP-250 (Noordwijk: European Space Agency), pp. 433–438.

Kresák, Ľ. 1993. Meteor storms. In *Meteoroids and Their Parent Bodies*, eds. J. Štohl and I. P. Williams (Bratislava: Slovak Acad. of Science), pp. 147–156.

Kresák, Ľ., and Kresáková, M. 1987. The contribution of periodic comets to the zodiacal cloud. In *Interplanetary Matter*, vol. 2, eds. Z. Ceplecha and P. Pecina (Prague: Czechoslovak Acad. of Science), 265–271.

Kunk, M. J., Izett, G. A., Haugerud, R. A., and Sutter, J. F. 1989. ^{40}Ar-^{39}Ar dating of the Manson impact structure: a Cretaceous-Tertiary boundary crater candidate. *Science* 244:1565–1568.

Kyte, F. T., Zhou, Z., and Wasson, J. T. 1980. Siderophile-enriched sediments from the Cretaceous-Tertiary boundary. *Nature* 288:651–656.

Kyte, F. T., Zhou, Z., and Wasson, J. T. 1985. Siderophile interelement variations in the Cretaceous-Tertiary boundary sediments from Carvaca, Spain. *Earth Planet. Sci. Lett.* 73:183–195.

Lamy, P. L., and Perrin, J. M. 1983. Optical properties of rough cometary grains. In *Asteroids, Comets, Meteors* eds. C.-I. Lagerkvist and H. Rickman (Uppsala: Uppsala University), pp. 273–277.

Larson, S., and Marcialis, R. 1992. 2060 Chiron. *IAU Circ.* No. 5669.

Leinert, C., and Grün, E. 1990. Interplanetary dust. In *Physics of the inner heliosphere. 1 Large-scale Phenomena*, eds. R. Schwenn and E. Marsch (Berlin: Springer-Verlag), pp. 207–275.

Levison, H. F., and Duncan, M. J. 1993. *The SwRI Catalog of Short-Period Comet Orbit Integrations* (San Antonio: Southwest Research Inst.).

Lindsay, J. F., and Srnka, L. J. 1975. Galactic dust lanes and lunar soil. *Nature* 257:776–778.

Love, S. G., and Brownlee, D. E. 1993. A direct measurement of the terrestrial mass accretion rate of cosmic dust. *Science* 262:550–553.

Macdougal, J. D. 1988. Seawater strontium isotopes, acid rain, and the Cretaceous-Tertiary boundary. *Science* 239:485–487.

Manara, A., and Valsecchi, G. B. 1991. Dynamics of comets in the outer planetary region. I. A numerical experiment. *Astron. Astrophys.* 249:269–276.

Marsden, B. G. 1967. The sungrazing comet group. *Astron. J.* 72:1170–1183.

Marsden, B. G. 1989. The sungrazing comet group. II. *Astron. J.* 98:2306–2321.

Matese, J. J., Whitman, P. G., Innanen, K. A., and Valtonen, M. J. 1994. Modulation of the Oort cloud comet flux by the adiabatically changing galactic tides. *Icarus*, in press.

Maurette, M., Kurat, G., Perreau, M., and Engrand, C. 1993. Microanalysis of Cap-Prudhomme Antarctic micrometeorites. *Microbeam Analysis* 2:239–251.

McDonnell, J. A. M., Alexander, W. M., Burton, W. M., Bussoletti, E., Clark, D. H., Grard, R. J. L., Grün, E., Hanner, M. S., Hughes, D. W., Igenbergs, E., Kuczera, H., Lindblad, B. A., Mandeville, J.-C., Minafra, A., Schwehm, G. H., Sekanina, Z., Wallis, M. K., Zarnecki, J. C., Chakaveh, S. C., Evans, G. C., Evans, S. T., Firth, J. G., Littler, A. N., Massonne, L., Olearczyk, R. E., Pankiewicz, G. S., Stevenson, T. J., and Turner, R. F. 1986. Dust density and mass distribution near comet Halley from Giotto observations. *Nature* 321:338–341.

McDonnell, J. A. M., Alexander, W. M., Burton, W. M., Bussoletti, E., Evans, G. C., Evans, S. T., Firth, J. G., Grard, R. J. L., Green, S. F., Grün, E., Hanner, M. S., Hughes, D. W., Igenbergs, E., Kissel, J., Kuczera, H., Lindblad, B. A., Langevin, Y., Mandeville, J.-C., Nappo, S., Pankiewicz, G. S. A., Perry, C. H., Schwehm, G. H., Sekanina, Z., Stevenson, T. J., Turner, R. F., Weishaupt, U., Wallis, M. K., and Zarnecki, J. C. 1987. The dust distribution within the inner coma of comet P/Halley (1992i): Encounter by Giotto's impact detectors. *Astron. Astrophys.* 187:719–741.

McDonnell, J. A. M., Lamy, P. L., and Pankiewicz, G. S. 1991. Physical properties of cometary dust. In *Comets in the Post-Halley Era*, vol. 2, eds. J. R. Newburn, Jr., M. Neugebauer and J. Rahe (Dordrecht: Kluwer), pp. 1043–1073.

McDonnell, J. A. M., McBride, N., Beard, R., Bussoletti, E., Colangeli, L., Eberhardt, P., Firth, J. G., Grard, R., Green, S. F., Greenberg, J. M., Grün, E., Hughes, D. W., Keller, H. U., Kissel, J., Lindblad, B. A., Mandeville, J.-C., Perry, C. H., Rembor, K., Rickman, H., Schwehm, G. H., Turner, R. F., Wallis, M. K., and Zarnecki, J. C. 1993. Dust particle impacts during the Giotto encounter with comet Grigg-Skjellerup. *Nature* 362:732–734.

McFadden, L. A., Cochran, A. L., Barker, E. S., Cruikshank, D. P., and Hartmann, W. K. 1993. The enigmatic object 2201 Oljato: Is it an asteroid or an evolved comet? *J. Geophys. Res.* 98:3031–3041.

Melosh, H. A., and Schenk, P. 1993. Split comets and the origin of crater chains on Ganymede and Callisto. *Nature* 365:731–733.

Michel, H. V., Alvarez, W., and Alvarez, L. W. 1985. Elemental profile of iridium and other elements near the Cretaceous/Tertiary boundary in hole 577B. *DSDP Initial Reports* 86:533–538.

Milani, A., Carpino, M., Hahn, G., and Nobili, A. M. 1989. Project SPACEGUARD: Dynamics of planet-crossing asteroids. Classes of orbital behavior. *Icarus* 78:212–269.

Milani, A., Carpino, M., and Marzari, F. 1990. Statistics of close approaches between asteroids and planets: Project SPACEGUARD. *Icarus* 88:292–335.

Morrison, D., ed. 1992. *The Spaceguard Survey: Report of the NASA International Near-Earth-Object Detection Workshop* (Pasadena: Jet Propulsion Laboratory).

Mueller, B. E. A., Tholen, D. J., Hartmann, W. K., and Cruikshank, D. P. 1992. Extraordinary colors of asteroidal object (5145) 1992 AD. *Icarus* 97:150–154.

Mukai, T. 1985. Small grains from comets. *Astron. Astrophys.* 153:213–217.

Nakamura, T., and Yoshikawa, M. 1993. Orbital evolution of giant comet-like objects. *Cel. Mech. Dyn. Astron.* 57:113–121.

Napier, W. M. 1983. The orbital evolution of short-period comets. In *Asteroids, Comets, Meteors*, eds. C.-I. Lagerkvist and H. Rickman (Uppsala: Uppsala University), pp. 391–395.

Napier, W. M. 1993. Earth-crossing asteroid groups. In *Meteoroids and Their Parent Bodies*, eds. J. Štohl and I. P. Williams (Bratislava: Slovak Acad. of Science), pp. 123–126.

Napier, W. M., and Clube, S. V. M. 1979. A theory of terrestrial catastrophism. *Nature* 282:455–459.

Nelson, B. K., MacLeod, G. K., and Ward, P. D. 1991. Rapid change in strontium isotopic composition of sea water before the Cretaceous-Tertiary boundary. *Nature* 351:644–647.

Nobili, A. M. 1989. Dynamics of the outer asteroid belt. In *Asteroids II*, eds. R. P. Binzel, T. Gehrels and M. S. Matthews (Tucson: Univ. of Arizona Press), pp. 862–879.

Officer, C. B., Hallam, A., Drake, C. L., and Devine, J. D. 1987. Late Cretaceous and paroxysmal Cretaceous/Tertiary extinctions. *Nature* 326:143–149.

Oikawa, S., and Everhart, E. 1979. Past and future orbit of 1977 UB, object Chiron. *Astron. J.* 84:134–139.

Olsson-Steel, D. I. 1987. Collisions in the solar system–IV. Cometary impacts upon the planets. *Mon. Not. Roy. Astron. Soc.* 227:501–524.

Oort, J. H. 1950. The structure of the cloud of comets surrounding the solar system and a hypothesis concerning its origin. *Bull. Astron. Inst. Netherlands* 11:91–110.

Oort, J. H., and Schmidt, M. 1951. Differences between new and old comets. *Bull. Astron. Inst. Netherlands* 11:259–269.

Öpik, E. J. 1958. On the catastrophic effects of collisions with celestial bodies. *Irish Astron. J.* 5:34–36.

Öpik, E. J. 1963. The stray bodies in the solar system. Part 1. Survival of cometary nuclei and the asteroids. *Adv. Astron. Astrophys.* 2:219–262.

Öpik, E. J. 1966. Sun-grazing comets and tidal disruption. *Irish Astron. J.* 7:141–161.

Pittich, E. M., and Rickman, H. 1994. Cometary splitting—a source for the Jupiter family? *Astron. Astrophys.* 281:579–587.

Porubčan, V., and Štohl, J. 1987. On orbits and associations of meteor streams with comets P/Halley and P/Encke. In *Symposium on the Diversity and Similarity of Comets*, eds. E. J. Rolfe and B. Battrick ESA SP-278 (Noordwijk: European Space Agency), pp. 435–440.

Quinn, T., Tremaine, S., and Duncan, M. 1990. Planetary perturbations and the origin of short-period comets. *Astrophys. J.* 355:667–679.

Rabinowitz, D. L. 1993a. The size distribution of the Earth-approaching asteroids. *Astrophys. J.* 407:412–427.

Rabinowitz, D. L. 1993b. 1993 HA2. *IAU Circ.* No. 5789.

Rabinowitz, D. L., Gehrels, T., Scotti, J. V., McMillan, R. S., Perry, M. L., Wisniewski, W., Larson, S. M., Howell, E. S., and Mueller, B. E. A. 1993. Evidence for a near-Earth asteroid belt. *Nature* 363:704–706.

Rampino, M. R., and Caldeira, K. 1992. Episodes of terrestrial geologic activity during the past 260 million years: A quantitative approach. *Cel. Mech. Dyn. Astron.* 54:143–159.

Rampino, M. R., and Caldeira, K. 1993. Major episodes of geologic change: correlations, time structure and possible causes. *Earth Planet. Sci. Lett.* 114:215–227.

Rampino, M. R., and Stothers, R. B. 1984. Geological rhythms and cometary impacts. *Science* 236:1427–1431.

Rampino, M. R., and Stothers, R.B. 1988. Flood basalt volcanism during the past 250 million years. *Science* 241:663–668.

Rickman, H. 1985. Interrelations between comets and asteroids. In *Dynamics of Comets: Their Origin and Evolution*, eds. A. Carusi and G. B. Valsecchi (Dordrecht: D. Reidel), pp. 149–172.

Rickman, H. 1990. Origin and evolution of the Jupiter family. In *Nordic-Baltic Astronomy Meeting*, eds. C.-I. Lagerkvist, D. Kiselman and M. Lindgren (Uppsala: Uppsala University), pp. 257–273.

Rickman, H. 1992. Structure and evolution of the Jupiter family. *Cel. Mech. Dyn. Astron.* 54:63–69.

Rickman, H., Bailey, M. E., Hahn, G., and Tancredi, G. 1992. Monte Carlo simulations of Jupiter family evolution. In *Periodic Comets*, eds. J. A. Fernández and H. Rickman (Montevideo: Universidad de la República), pp. 55–64.

Robin, E., Froget, L., Jéhanno, C., and Rocchia, R. 1993. Evidence for a K/T impact event in the Pacific Ocean. *Nature* 363:615–617.

Schmitz, B. 1988. Origin of microlayering in worldwide distributed Ir-rich marine Cretaceous/Tertiary boundary clays. *Geology* 16:1068–1072.

Scholl, H. 1979. History and evolution of Chiron's orbit. *Icarus* 40:345–349.

Scholl, H. 1987. Dynamics of asteroids. In *The Evolution of the Small Bodies of the Solar System*, eds. M. Fulchignoni and Ľ. Kresák (Amsterdam: North-Holland), pp. 53–78.

Schultz, P. H. 1987. Possible non-random impact fluxes on the moon in recent time (<100 Myr). *Eos: Trans. AGU* 68:344.

Scotti, J. V., and Melosh, H. J. 1993. Estimate of the size of comet Shoemaker-Levy 9 from a tidal breakup model. *Nature* 365:733–735.

Sekanina, Z. 1974. Meteoric storms and formation of meteor streams. In *Asteroids, Comets, Meteoric Matter*, eds. C. Cristescu, W. J. Klepczynski and B. Milet (Romania: Editura Academiei Republicii), pp. 239–267.

Sekanina, Z. 1993. Disintegration phenomena expected during collision of Comet Shoemaker-Levy 9 with Jupiter. *Science* 262:382–387.

Sepkoski, J. J., and Raup, D. M. 1986. Periodicities in marine extinction events. In *Dynamics of Extinctions*, ed. D. K. Elliot (New York: Wiley-Interscience), pp. 3–36.

Sharpton, V. L., Burke, K., Caamargo-Zanoguera, A., Hall, S. A., Lee, D. S., Marin, S. E., Suárez-Reynoso, G., Quezada-Muñeton, J. M., Spudis, P. D., and Urrutia-Fucugauchi, J. 1993. Chicxulub multiring impact basin: Size and other characteristics derived from gravity analysis. *Science* 261:1564–1567.

Shoemaker, E. M. 1991. Geological and Astronomical Evidence for Comet Impact and Comet Showers During the Past 100 Million Years. LPI Contrib. No. 765 (Houston: Lunar and Planetary Inst.).

Shoemaker, E. M. 1993. The impact history of the Earth and terrestrial planets. Presented at Erice International Seminar on Planetary Emergencies, 17th Workshop: The Collision of an Asteroid or Comet with the Earth, April 28–May 4, Erice (Italy).

Shoemaker, E. M., and Izett, G. A. 1992a. Stratigraphic evidence from Western North America for multiple impacts at the K/T boundary. *Lunar Planet. Sci.* XXIII:1293–1294 (abstract).

Shoemaker, E. M., and Izett, G. A. 1992b. Did the "Big One" pack a one-two punch? *Sky & Tel.* (July), p. 8.

Shoemaker, E. M., and Shoemaker, C. S. 1990. The collision of solid bodies. In *The New Solar System*, 3rd. ed., eds. J. K. Beatty and A. Chaikin (Cambridge, Mass.: Sky Publishing Co.), pp. 259–274.

Shoemaker, E. M., and Wolfe, R. F. 1982. Cratering timescales for the Galilean satellites. In *Satellites of Jupiter*, ed. D. Morrison (Tucson: Univ. of Arizona Press), pp. 277–339.

Shoemaker, E. M., Wolfe, R. F., and Shoemaker, C. S. 1990. Asteroid and comet flux in the neighborhood of Earth. In *Global Catstrophes in Earth History*, eds. V. L. Sharpton and P. D. Ward, Geological Soc. of America Special Paper 247 (Boulder: Geological Soc. of America), pp. 155–170.

Sleep, N. H., Zahnle, K. J., Kasting, J. M., and Morowitz, H. J. 1989. Annihilation of ecosystems by large asteroid impacts on the early Earth. *Nature* 342:139–142.

Stagg, C. R., and Bailey, M. E. 1989. Stochastic capture of short-period comets. *Mon. Not. Roy. Astron. Soc.* 241:507–541.

Steel, D. I. 1991. Our asteroid-pelted planet. *Nature* 354:265–267.

Steel, D. I. 1992. Additions to the Taurid Complex. *Observatory* 112:120–122.

Steel, D. I., Asher, D. J., and Clube, S. V. M. 1991. The structure and evolution of the Taurid Complex. *Mon. Not. Roy. Astron. Soc.* 251:632–648.

Stern, A. 1991. On the number of planets in the outer Solar System: evidence of a substantial population of 1000-km bodies. *Icarus* 90:271–281.

Stern, A. 1992. Where has Pluto's family gone? *Astronomy* 20:(9)40–47.

Štohl, J. 1983. On the distribution of sporadic meteor orbits. In *Asteroids, Comets, Meteors*, eds. C.-I. Lagerkvist and H. Rickman (Uppsala: Uppsala University), pp. 419–424.

Štohl, J. 1987. Meteor contribution by short-period comets. *Astron. Astrophys.* 187:933–934.

Stott, L. D., and Kennett, J. P. 1988. Cretaceous-Tertiary boundary in the Antarctic: Climate cooling precedes biotic crisis. In *Global Catastrophes in Earth History Abstracts*, Oct. 20–23, Snowbird, Utah, LPI Contrib. No. 673 (Houston: Lunar and Planet. Inst.), p. 184–185.

Sykes, M. V. 1993. Great balls of mire. *Nature* 362:696–697.

Sykes, M. V., and Greenberg, R. 1986. The formation and origin of the IRAS zodiacal dust bands as a consequence of single collisions between asteroids. *Icarus* 65:51–69.

Sykes, M. V., and Walker, R. G. 1992. Cometary dust trails. I. Survey. *Icarus* 95:180–210.

Sykes, M. V., Hunten, D. M., and Low, F. J. 1986. Preliminary analysis of cometary dust trails. *Adv. Space Res.* 6:(7)67–78.

Torbett, M. V. 1989. Chaotic motion in a comet disk beyond Neptune: The delivery of short-period comets. *Astron. J.* 98:1477–1481.

Torbett, M. V., and Smoluchowski, R. 1990. Chaotic motion in a primordial comet disk beyond Neptune and comet influx to the solar system. *Nature* 345:49–51.

Turco, R. P., Toon, O. B., Ackerman, T. P., Pollack, J. B., and Sagan, C. 1983. Nuclear winter: global consequences of multiple nuclear explosions. *Science* 222:1283–1300.

Urey, H. C. 1973. Cometary collisions and geological periods. *Nature* 242:32–33.

Valsecchi, G. B. 1992. Close encounters, planetary masses and the evolution of cometary orbits. In *Periodic Comets*, eds. J. A. Fernández and H. Rickman (Montevideo: Universidad de la República), pp. 81–96.

Varvoglis, H. 1993. Large orbital eccentricities and close encounters at the 2:1 resonance of a dynamical system modelling asteroidal motion. *Astron. Astrophys.* 275:301–308.

Waddington, G. 1993. Simulation of the nuclear train of Comet Shoemaker-Levy 9. *The Astronomer* 30:135–138.

Weissman, P. R., A'Hearn, M. F., McFadden, L. A., and Rickman, H. 1989. Evolution of comets into asteroids. In *Asteroids II*, eds. R. P. Binzel, T. Gehrels and M. S. Matthews (Tucson: Univ. of Arizona Press), pp. 880–920.

West, R. M. 1991. A photometric study of (2060) Chiron and its coma. *Astron. Astrophys.* 241:635–645.

Wetherill, G. W. 1975. Late heavy bombardment of the moon and terrestrial planets. *Proc. Lunar Sci. Conf.* 6:1539–1561.

Wetherill, G. W. 1979. Steady-state populations of Apollo-Amor objects. *Icarus* 37:96–112.

Wetherill, G. W. 1988. Where do the Apollo asteroids come from? *Icarus* 76:1–18.

Wetherill, G. W. 1991. End products of cometary evolution: Cometary origin of Earth-crossing bodies of asteroidal appearance. In *Comets in the Post-Halley Era*, vol. 1, eds. J. R. Newburn, Jr., M. Neugebauer and J. Rahe (Dordrecht: Kluwer), pp. 537–556.

Whipple, F. L. 1950. A comet model. I. The acceleration of Comet Encke. *Astrophys. J.* 111:375–394.

Whipple, F. L. 1967. On maintaining the meteoritic complex. In *The Zodiacal Light and the Interplanetary Medium*, ed. J. Weinberg, NASA SP-150, pp. 409–424.

Whipple, F. L. 1992. The activities of comets related to their aging and origin. *Cel. Mech. Dyn. Astron.* 54:1–11.

Wickramasinghe, N. C., Hoyle, F., and Rabilizirov, R. 1989a. Greenhouse dust. *Nature* 341:28.

Wickramasinghe, N. C., Hoyle, F., and Rabilizirov, R. 1989b. Extraterrestrial particles and the greenhouse effect. *Earth, Moon, and Planets* 46:297–300.

Williams, I. P., and Wu, Z. 1993. The Geminid meteor stream and asteroid 3200 Phaethon. *Mon. Not. Roy. Astron. Soc.* 262:231–248.

Wolbach, W. S., Lewis, R. S., and Anders, E. 1985. Cretaceous extinctions: Evidence for wildfires and search for meteoritic material. *Science* 230:167–170.

Wolfe, J. A. 1991. Palaeobotanical evidence for a June 'impact winter' at the Cretaceous/Tertiary boundary. *Nature* 352:420–423.

Yabushita, S. 1993. Thermal evolution of cometary nuclei by radioactive heating and possible formation of organic chemicals. *Mon. Not. Roy. Astron. Soc.* 260:819–825.

Yabushita, S., and Allen, A. J. 1989. On the effect of accreted interstellar matter on the terrestrial environment. *Mon. Not. Roy. Astron. Soc.* 238:1465–1478.

Yau, K., Weissman, P., and Yeomans, D. 1993. Human casualties and structural damage due to meteorite falls found in East Asian histories. Hazards Due to Comets and Asteroids, Jan. 9–12, Tucson, Ariz., Abstract book, p. 86.

Yau, K., Yeomans, D., and Weissman, P. 1994. The past and future motion of Comet P/Swift-Tuttle. *Mon. Not. Roy. Astron. Soc.* 266:305–316.

Yeomans, D. 1993. Comet (1993e) Shoemaker-Levy 9 and death to the Jovian. Poster paper presented at Asteroids, Comets, Meteors 1993, IAU Symp. 160, June 14–18, Belgirate (Italy).

Yoshikawa, M. 1992. Numerical investigations of motions of resonant asteroids in the three-dimensional space. *Cel. Mech. Dyn. Astron.* 54:287–290.

Zahnle, K., and Grinspoon, D. 1990. Comet dust as a source of amino acids at the Cretaceous/Tertiary boundary. *Nature* 348:157–160.

Zhao, M., and Bada, J. L. 1989. Extraterrestrial amino acids in Cretaceous/Tertiary boundary sediments at Stevns Klint, Denmark. *Nature* 339:463–465.

PART IV
Physical Properties

PHYSICAL PROPERTIES OF NEAR-EARTH ASTEROIDS: IMPLICATIONS FOR THE HAZARD ISSUE

CLARK R. CHAPMAN
Planetary Science Institute

ALAN W. HARRIS
Jet Propulsion Laboratory

and

RICHARD BINZEL
Massachusettes Institute of Technology

Physical properties of NEAs are less well characterized than for large mainbelt asteroids. Their properties are of great interest for the scientific understanding of small bodies and for eventual resource utilization. Here we consider their properties in terms of the impact hazard, for which they are relevant in two ways: (a) implications for hazardous impacts, such as different modes of break-up during atmospheric penetration; and (b) implications for potential mitigation techniques. NEAs are believed to be derived from both mainbelt asteroids and comets, and thus they may have physical and compositional properties ranging from loosely aggregated cometary ices to solid metal. Generally, observed NEA colors and inferred compositions appear similar to those of mainbelt asteroids. NEA spins range from rapid (2.3 hr) to practically non-spinning (many days). They are generally small bodies, with minimal gravity, hence are expected to lack deep regoliths. A few, and possibly many, NEAs are compound bodies rather than monolithic, cohesive objects. The diversity of physical properties of NEAs leads to various behaviors during atmospheric entry for smaller objects of different types. The diversity requires that mitigation involve either (a) precursor missions to identify the nature of an object to be deflected or destroyed; (b) technologies that are largely independent of the nature of the object; or (c) highly flexible technologies, capable of immediately assessing physical characteristics and dealing with a wide range of possibilities.

I. INTRODUCTION

The physical properties of Aten, Apollo, and Amor asteroids have been reviewed by McFadden et al. (1989). The purpose of the present review is more limited. Although we present a tabulation (see Table I) of the most up-to-date physical parameters for near-Earth asteroids (NEAs) available as of 1994, our discussion is restricted to those physical and compositional parameters that are relevant to the subject of this book, the impact hazard. There are two chief hazard-related reasons for interest in the physical nature of NEAs. First, the potential effects of an impact can vary depending on the specific nature of the impactor. In many cases, the physical nature is not important; to first order, consequences are dictated by the kinetic energy of the impactor. However, for a range of smaller-sized objects, the composition of the impactor and its effective body strength determines whether it explodes harmlessly in the upper atmosphere, penetrates deeply enough to cause a dangerous airburst, or actually reaches the ground. The second, and more important, reason for considering physical properties of NEAs relates to proposed methods of mitigation. Any mitigation technology to deflect or destroy an impactor involves some type of physical interaction with the object. As we discuss below, the inferred range of physical properties of NEAs is very great and profoundly influences what methods can be successfully and safely used to prevent a dangerous impact from happening.

NEAs are not so well characterized as large mainbelt asteroids because they are very small and, despite their near-Earth status, they are usually about as far away from Earth as other asteroids. They are generally very faint, and many of our remote-sensing techniques cannot be applied with precision to such objects. On occasion, NEAs (especially those capable of being a hazard to Earth) come much closer, but then they are available for only a short period of time (hours to weeks). The vagaries of weather and telescope scheduling procedures interfere with attempts to characterize these objects. Thus the sample of physically characterized NEAs is small, and the quality of the data is generally lower than for mainbelt asteroids of similar magnitudes. Although occasional observing campaigns can be mounted to obtain comprehensive data on one object, we must await dedicated programs, preferably using dedicated telescope(s) and instrumentation—or at least routine access to a properly equipped telescope—to markedly improve on our statistical sampling of the physical properties of NEAs. Coordination with NEA discovery programs would be particularly important, as most NEAs are discovered during highly favorable passages of short duration, and we usually have to wait many years for a follow-up opportunity.

An exception to the difficulty of observing NEAs is use of radar, which is greatly facilitated by the occasional proximity of these objects compared with the always much more distant mainbelt asteroids. This has been a particularly potent method for sampling the properties of some NEAs and has led to some of our most robust views about their nature (including actual delay-doppler maps of shapes, configurations, and even surface features for a few of them).

NEAs are believed to be derived from both the main asteroid belt and from evolved short-period comets (Wetherill 1988). Those from the main asteroid belt are generally small, multi-generation collisional fragments from larger mainbelt objects, which have been dynamically converted into Earth-approaching orbits by chaotic processes associated with the 3:1 Kirkwood gap and other resonances. Those derived from the cometary population are dormant or dead comets, which differ from their still-active cousins only in the degree to which—through relatively long-term residence in the inner solar system (thousands of years or longer)—they have lost accessible volatiles and thus no longer appear to be cometary. However, just as asteroid-derived NEAs may be expected to resemble their siblings in the asteroid belt, inactive comets may be expected to be generally cometary in body character (whatever "cometary" means). Of course, we know rather little about what either mainbelt asteroids or comets are like, so our inferences about the physical traits of NEAs are quite uncertain.

Table I gives provisional values of observable properties for Earth-approaching asteroids (Apollos, Atens, and those Amors and other objects that could, theoretically, intersect Earth's orbit; nearly all have perihelion distance $q < 1.3$ AU), derived from McFadden et al. (1989) as updated from the literature and unpublished files (as noted) by one of us (AWH). We address the properties of the NEAs in three categories: inferred mineralogy, surface properties, and geophysical properties and configurations.

II. MINERALOGY OF NEAs

Gaffey et al. (1993) have reviewed asteroid reflectance spectroscopy, the observational technique responsible for most of our information about asteroid compositions and mineralogy. The crudest but most broadly applied type of spectroscopy is colorimetry, where the reflectance spectrum of an object is characterized by photometric measurements in several (typically 3 to 8) broad-band wavelength ranges in the visible and near-infrared. Asteroid taxonomy is derived from statistical analysis of such data. Observations of reflectance spectra over a broader range of wavelengths and/or with higher spectral resolution have been made only for a smaller number of brighter objects. Such data refine the distinctions between NEAs and help to provide real constraints on implied mineralogy. Other techniques, chiefly thermal infrared (radiometric) measurements, provide geometric albedos (and sizes) of NEAs. Additional techniques provide ancillary information about composition. The most important is radar, which is particularly sensitive to the metal content of the near-surface layers of an object—a trait that is ambiguous in spectral reflectance data alone. For instance, Ostro et al. (1987; see also the Chapter by Ostro) have claimed unambiguous proof of the metallic nature for at least one NEA.

Our sampling of NEAs for physical observations is strongly biased. The biases are analogous to, but quantitatively different from, those that affect the statistics of mainbelt asteroid taxonomy (Gradie et al. 1989). Proposed corrections for bias among NEAs (Luu and Jewitt 1989) suggest that the NEA population has approximately the same proportion of S type and C type (or C-like) asteroids as in the main belt. Because McFadden et al. (1989) conclude that nearly all taxonomic types found in the main belt (especially the inner belt, near the resonance escape hatches) are found among the NEAs, and that

TABLE I[a]
Physical Properties of NEAs

Asteriod	a	e	i	H	Albedo	Diam.	Cl	Period	Amplitude	Radar
433 Eros	1.46	.223	10.8	11.16	0.12	22.	S	5.270	0.05–1.5	2
887 Alinda	2.49	.559	9.3	13.76	0.31	4.2	S	73.97	0.35	
944 Hidalgo	5.84	.658	42.4	10.77	d	38.	D	10.064	0.6	
1036 Ganymed	2.67	.537	26.5	9.45	0.19	39.	S	10.31	0.12–0.45	1
1221 Amor	1.92	.435	11.9	17.70	m	1.0			0.1*	
1566 Icarus	1.08	.827	22.9	16.40	0.51	1.0	S	2.273	0.05–0.22	1
1580 Betulia	2.19	.490	52.1	14.52	0.08	5.8	C	6.130	0.21–0.50	2
1620 Geographos	1.24	.335	13.3	15.60	0.25	2.0	S	5.223	1.10–2.03	1
1627 Ivar	1.86	.397	8.4	13.20	0.14	8.1	S	4.797	0.25–0.60	1
1685 Toro	1.37	.436	9.4	14.23	0.31	3.4	S	10.196	0.6–0.8	3
1862 Apollo	1.47	.560	6.3	16.23	0.25	1.5	Q	3.065	0.15–0.60	1
1863 Antinous	2.26	.606	18.4	15.54	0.24	2.1	S	4.02	0.12	
1864 Daedalus	1.46	.615	22.2	14.85	m	3.7	SQ	8.57	0.85	
1865 Cerberus	1.08	.467	16.1	16.84	0.22	1.2	S	6.800	1.48–2.0	
1866 Sisyphus	1.89	.539	41.1	13.00	0.16	8.2	S	2.48	0.12	1
1915 Quetzalcoatl	2.54	.574	20.5	18.97	0.21	0.5	S	4.9	0.26	1
1916 Boreas	2.27	.451	12.9	14.93	m	3.5	S			
1917 Cuyo	2.15	.505	24.0	13.90	m	5.7		2.697	0.43	1
1943 Anteros	1.43	.256	8.7	15.75	0.17	2.3	S	long	0.04	
1951 Lick	1.39	.062	39.1	14.70	m	3.9	A	4.417	0.28*	
1980 Tezcatlipoca	1.71	.365	26.8	13.92	0.25	4.3	S	7.853	0.74*	

1981 Midas	1.78	.650	39.8	15.00	m	3.4	S	5.218	0.85*	1
2061 Anza	2.26	.537	3.7	16.56	d	2.6	TCG:	11.50	0.3	
2062 Aten	.97	.183	18.9	16.80	0.26	1.1	S		0.34	2
2100 Ra-Shalom	.83	.437	15.8	16.05	0.06	3.4	C	19.79		1
2101 Adonis	1.87	.764	1.4	18.70	m	0.6			>0.1	1
2201 Oljato	2.18	.711	2.5	15.55	0.54	1.4	S?	24.?	0.05*	
2212 Hephaistos	2.17	.833	11.8	13.87	m	5.7	SG			
2340 Hathor	.84	.450	5.9	20.26	m	0.3	CSU		0.84	
2368 Beltrovata	2.11	.414	5.2	15.21	0.27	2.3	SQ	5.9	0.5	
2608 Seneca	2.49	.581	15.3	17.52	0.21	0.9	S	8.	>1.0	
3102 1981 QA	2.15	.449	8.4	16.70	m	1.6	QRS	147.8	0.72–0.9	2
3103 1982 BB	1.41	.355	20.9	15.21	0.64	1.5	E	5.709	0.25*	
3122 1981 ET3	1.77	.423	22.2	14.20	m	4.9		5 or 10	0.12	
3199 Nefertiti	1.57	.284	33.0	14.84	0.42	2.2	S	2.82	0.12*	1
3200 Phaethon	1.27	.890	22.1	14.60	0.09	6.9	F	4.0	1.0	
3288 Seleucus	2.03	.457	5.9	15.00	0.22	2.8	S	75.		
3360 1981 VA	2.47	.743	21.7	16.20	0.17	1.8			0.32*	
3361 Orpheus	1.21	.323	2.7	19.03	1	0.3	V	3.58		
3362 Khufu	.99	.469	9.9	18.10	0.21	0.7			0.11–0.15	
3551 1983 RD	2.09	.488	9.5	16.81	0.37	0.9	V	4.930	>0.41	
3552 Don Quixote	4.24	.714	30.8	13.00	0.03	19.0	D	7.7	0.16*	
3554 Amun	.97	.280	23.4	15.82	0.20	2.0	M	2.53	0.26	
3671 Dionysius	2.19	.543	13.6	16.30	m	1.9		2.4		

TABLE I[a] (cont.)
Physical Properties of NEAs

Asteriod	a	e	i	H	Albedo	Diam.	Cl	Period	Amplitude	Radar
3691 1982 FT	1.77	.284	20.4	14.50	m	4.3		days?	*	1
3757 1982 XB	1.84	.446	3.9	18.95	0.18	0.5	S	9.012	0.20	1
3908 1980 PA	1.93	.458	2.2	17.30	0.23	1.0	V	4.426	0.25–0.46	
3988 1986 LA	1.54	.317	10.8	18.30	m	0.7		8.	0.2†	
4015 1979 VA	2.64	.623	2.8	15.99	d	3.4	CF	3.556	0.06	
4034 1986 PA	1.06	.444	11.2	18.10	m	0.8				1
4055 Magellan	1.82	.327	23.2	14.50	0.23	3.4	V	7.5	0.5‡	
4179 Toutatis	2.51	.641	.5	14.00	m	5.4	S	200.	1.	
4197 1982 TA	2.30	.773	12.2	15.40	0.37	1.8				
4544 Xanthus	1.04	.250	14.1	17.10	m	1.3				1
4688 1980 WF	2.23	.515	6.4	18.60	0.18	0.6	SQ			
4769 Castalia	1.06	.483	8.9	16.90	m	1.4		4.088	1.0	
4954 Eric	2.00	.449	17.5	12.50	m	10.8	S	>11.	0.6*	1
5332 1990 DA	2.16	.456	25.4	14.90	m	3.6	S	5.816	0.36	
5370 1986 RA	3.35	.631	19.0	15.90	d	3.6	C		0.02	
1977 VA	1.86	.394	3.0	19.40	m	0.4	XC			
1978 CA	1.12	.215	26.1	16.90	0.09	1.9	D?	3.756	0.8	
1980 AA	1.89	.444	4.2	19.40	m	0.4	S	2.697	0.12	
1984 BC	3.44	.457	22.4	16.00	d	3.4	D			
1984 KB	2.22	.765	4.9	16.60	0.21	1.4	S			
1986 DA	2.82	.582	4.3	15.90	0.15	2.3	M	3.58	0.32	1
1986 JK	2.80	.680	2.1	18.90	d	0.9	C		0.05	1
1988 TA	1.54	.479	2.5	20.90	d	0.4	C			
1989 DA	2.16	.544	6.4	17.90	m	0.9		7.867	0.12*	

	a	e	i	H	Albedo	Diam.	Cl	Rotation period	Amplitude	
1989 JA	1.77	.484	15.2	16.40	m	1.8			*	1
1989 UP	1.86	.473	3.9	20.60	m	0.3		6.983	1.2*	
1989 VA	.73	.595	28.8	16.90	m	1.4			>0.15*	1
1989 VB	1.86	.461	2.1	19.90	m	0.4		>24.	>0.3*	
1990 HA	2.57	.692	3.9	16.90	m	1.4		8.51	0.06*	
1990 KA	2.20	.433	7.6	15.90	m	2.3		6.	>0.4	
1990 MF	1.75	.456	1.9	18.50	m	0.7				1
1990 OS	1.67	.459	1.1	19.90	m	0.4				1
1990 SA	1.96	.430	37.5	16.90	m	1.4	S		*	
1990 TR	2.14	.437	7.9	14.50	m	4.3	S	6.25	0.19	
1990 UA	1.72	.552	1.0	19.40	m	0.4		long	0.08	
1990 UP	1.33	.169	28.1	20.50	m	0.3		>20.	>0.07*	
1991 AQ	2.16	.769	3.2	17.50	m	1.1			*	1
1991 EE	2.25	.624	9.8	18.50	m	0.7		3.00	0.12*	1
1991 JX	2.53	.599	2.3	18.50	m	0.7				1
1992 AC	2.10	.421	16.1	13.60	m	6.3			<0.02*	

* An asterisk in the amplitude column indicates that the photometry, including color information, is from the unpublished files of the late Wieslaw Wisniewski, one of the most productive and dedicated observers in recent years.

† Unpublished data from E. Bowell.

‡ Unpublished data from A. Harris.

a = semimajor axis; e = eccentricity; i = inclination; H = absolute magnitude, H-G system; Albedo = visual albedo. When not measured but inferred from taxonomic class, listed as d for "dark" (0.06), m for "medium" (0.15), or l for "light" (0.4). Diam. is measured diameters (km) from radiometry, or for inferred albedos, computed from H and above albedo values. Cl designates taxonomic classification. Rotation period is given in hours. Lightcurve amplitude is given in magnitudes, or observed range of amplitudes. The number of apparitions for which successful radar observations have been obtained is given in the last column.

the albedo distributions are similar, it is reasonable to conclude that the NEAs have roughly the same compositional distribution as the mainbelt asteroids.

There are caveats to this conclusion. For example, there is a suggestion that there may be a higher fraction of Qs (a rare type with spectra like ordinary chondrites) among the NEAs than in the main belt; although 1862 Apollo is the only confirmed Q among 55 classified NEAs (see Table I), Q is an allowed type for 4 others, while only a single Q-like asteroid (3628 Boňěmcová) has been identified in the much larger population of classified mainbelt asteroids (Binzel et al. 1993). The similarity between NEA and mainbelt populations also raises the question about where the dead comets are, because such comets are not expected to be a major component of the main belt. The answer may be that dormant comets are very low albedo objects, that are poorly sampled among NEAs so far (Hartmann et al. 1987).

The mainbelt asteroid population, and thus the NEA population, is dominated by materials that we find among collected meteorites, but it almost surely contains weaker materials that cannot usually penetrate the atmosphere. The former materials include rocks of various strengths, stony-metal assemblages, and nickel-iron alloy. The signature of metal in reflectance spectroscopy is not definitive (most M-types are thought to be metallic), but the inference of metallic composition has been strengthened in the cases of several M-type objects by strong radar backscatter. 1986 DA is an M-type NEA with a strong radar backscatter indicative of metallic composition (Ostro et al. 1987). It is possible that some or all of the common S-type asteroids are stony-irons, so a significant minority of NEAs may have the strength of stony-irons and be much stronger than ordinary rocks. How they actually respond to external forces may depend on their ductility or brittleness, which depends in turn on ambient temperature and nickel content.

A substantial fraction of asteroidal materials (probably of the C or C-like taxonomic classes of carbonaceous composition) are expected to be too weak to penetrate our atmosphere and become part of meteorite collections. There is a general belief, although unproven, that most cometary materials would also be too weak to penetrate the atmosphere intact and survive as meteorites. Thus, as far as we know, we do not have any samples that could be considered to be good analogs of cometary material; we can only speculate about the physical nature of dead comets. Presumably they are like active comets, except that their mantles are thicker. The oldest dead comets might be totally depleted of icy volatiles (see the Chapter by Rahe et al.).

III. SURFACE PROPERTIES

Little is known about the surface properties of NEAs. Asteroids in general are expected to have soil-like regoliths on their surfaces, but less mature, (i.e., less reworked, less altered) than for the well known case of the Moon. The ability of an asteroid to develop and maintain a regolith is expected to diminish with size. Objects smaller than ~10 km in diameter are increasingly likely to have thin regoliths, or none at all, unless the object contains part of an ancient regolith developed on a larger, precursor body. Indeed, solar radiation pressure should "blow away" any micron-sized regolith component on a body less than a few km in diameter, so even if small asteroids retain some regolith, it should become coarser with decreasing size of the body. Radiometric data for several NEAs reduced using the "standard thermal model" (which invokes regolith) yield results that are inconsistent with other data sets, which seems to imply that a number of NEAs have thin or nonexistent regoliths (Lebofsky et al. 1979).

IV. GEOPHYSICAL PROPERTIES AND CONFIGURATIONS

The chief data for assessing the spins and shapes of the largest sample of NEAs are photometric lightcurves. Although some objects have been studied in great detail, most of the results are less reliable than comparable lists for mainbelt asteroids, due to the short durations that most NEAs are available for observation. Table I lists NEA spin periods, ranging from 2.27 hr for 1566 Icarus up to ~200 hr for 4179 Toutatis. The period for Toutatis remains poorly known as of this writing; it may be tumbling in a state of non-principle axis rotation (Harris 1994), thus, its period may be inherently indeterminate. An asteroid's spin could presumably complicate a mitigation mission. For this reason, it is well to note that observed NEAs typically spin once every 4 to 8 hr, with some spinning much more rapidly (but never so rapidly that they are in danger of coming apart) and others much more slowly. In fact, there is a small but real, and as yet unexplained, excess of very slowly rotating NEAs compared with a Maxwellian distribution. Because the excess exists for small mainbelt asteroids as well as NEAs (Binzel et al. 1992a, b), the explanation is probably not NEA-specific (e.g., that slowly rotating ones are extinct comets). Indeed, the similarity in spin characteristics between the main belt and the NEAs may indicate a rather small cometary component (<25 to 40%) in the NEA population, because cometary spins are statistically different on average (Binzel et al. 1992a, b).

Lightcurve amplitudes indicate that many NEAs are markedly non-spherical. Several, like 1566 Icarus, have low-amplitude lightcurves, indicating that they have nearly circular equatorial profiles and may even be close to spherical. But lightcurve amplitudes of several tenths of a magnitude are common, and amplitudes in excess of one magnitude have been observed. Perhaps it should go without saying, but we will say it anyway: no mitigation scenario should invoke the assumption of a quasi-spherical asteroid. Many NEAs are irregularly shaped bodies, spinning once every few hours.

The best information about NEA shapes, by far, comes from radar. There have been tantalizing hints in some early radar data for two-lobed shapes of a few objects, but the first clear-cut case was for 4679 Castalia (Ostro et al. 1990), that was found to have roughly a "contact binary" shape. Far more detailed radar results were achieved for 4179 Toutatis in late 1992 (Ostro et al. 1993; Chapter by Ostro). Toutatis is revealed to be a very irregularly shaped object, also roughly "contact binary" in shape. It is too soon to be certain, but it appears that a significant fraction of NEAs may be compound bodies, with two or several large pieces loosely resting on each other due to their weak mutual gravity. Indeed, the "rubble pile" concept that had been applied to larger asteroids may also apply to smaller asteroids as well. How such small, compound objects can be formed so readily is not yet understood, but there is independent evidence (e.g., the existence of doublet craters on the Earth, Moon, and Mars) that is consistent with a significant percentage of such objects being binary.

Comets have often been seen to split. Often but not always the splitting has resulted from passage within the Roche limit of a planet (e.g., the break-up of P/comet Shoemaker-Levy 9 [1993e]) or close approach to the Sun. Probably the tensile strength of many comets is not large, and the same could be true of the remnant, dormant population among the NEAs. Therefore, both the cometary component of the NEAs and the fraction of rocky objects that are compound in configuration warn us that we should not regard an approaching hazardous object as necessarily a monolithic, single object.

V. IMPLICATIONS FOR HAZARD RELATED TO ATMOSPHERIC ENTRY

If one of the larger of the known NEAs (>1 km diameter) were to strike the Earth, its physical properties would make essentially no difference. To such an impactor, the Earth's atmosphere would be thin, and the object would do damage to the Earth in proportion to the kinetic energy it carries when it explodes at the Earth's surface.

The numerous objects smaller than several hundred meters in diameter may break up in the atmosphere, if they are made of loosely aggregated, weak, low density materials, i.e., "cometary." Even rocky objects break up in low-altitude airbursts if they are smaller than ~70 m, like the ~50 m object that caused the Tunguska explosion. Strong iron meteorites of all sizes can penetrate the Earth's atmosphere practically unscathed. For more details on how the physical properties of the impactor affect its passage through the atmosphere (see, e.g., Chyba et al. 1993; Hills et al. 1993; Chapters by Chyba et al. and by Tagliaferri et al.). The implications of this chapter are simply that there is a wide variety of materials among the NEAs, ranging from loose, weak aggregates to solid objects made of metal.

The possibility that a significant proportion of NEAs may be compound bodies raises the possibility of multiple impacts. As we have learned from the example of comet Shoemaker-Levy 9, it is possible for such bodies to be tidally disrupted by passage through a Roche zone and then have the disrupted pieces collide with the planet. This scenario has been considered for the Earth by Melosh and Stansberry (1991) and is under further development. Passage within the Earth's Roche limit is several times more likely than direct impact, and the question then concerns the ultimate fate of such objects. There has been controversy over the existence of "asteroid streams" (Drummond 1991), and tidal disruption of NEAs may be one explanation for such observations.

VI. IMPLICATIONS FOR MITIGATION

One of the most dangerous possible outcomes of an attempt to deflect or destroy an approaching asteroid is the unintended production of several large pieces, which might then do more damage than the single, intact body. For example, a swarm of objects distributed around the Earth could create widespread firestorms (both from the direct bolide and from hot ejecta) that could be more severe than from a single impact. If a threatening asteroid is to be destroyed, the method must be sufficient to pulverize an object (or compound object) made of the strongest possible material, which is strong metal. If it is to be deflected, it must be guaranteed that the forces applied actually move the object intact, without inducing stresses that would cause the body to split (unless the mitigation scenario additionally considers dealing with all such pieces that might result).

If the asteroid to be deflected is a "rubble pile" of unconsolidated pieces, then the largest single impulsive ΔV which can be applied is of the order of the surface escape velocity of the asteroid, or ~ 1 m s^{-1} per km diameter of the body (see the Chapter by Ahrens and Harris). This is because an impulse (e.g., from a surface or stand-off nuclear blast) is unevenly distributed throughout the body, and the various pieces will acquire a dispersion of velocities of the same order as the average ΔV; thus if that impulse exceeds the escape velocity, the pieces will become unbound from one another. For initially solid, rocky objects larger than 100 m, an impulse great enough to produce a ΔV of ~ 0.1 m s^{-1} would also lead to widespread fracturing of the body, so the above limitation would apply to that case, as well.

NEAs are inherently heterogeneous bodies of great diversity. Any engineering application will have to involve detailed study of the individual object in question to ascertain its particular physical and compositional characteristics. Likewise, experimentation on destroying or deflecting a benign asteroid may be of little value; the specific characteristics of the threatening object are likely to be different. Extrapolation from existing groundbased data and current speculations would be most unwise. To date, there has been no "ground truth" obtained for any NEA, even though there is the Clementine mission to Geographos underway. Even the Galileo fly-bys of the mainbelt asteroids Gaspra and Ida were at considerable distances and involved only quick-look pictures and remote-sensing data (Chapman 1994). There is much about our current concepts of the make-up of asteroids and dead comets that could be far from the truth. It would be desirable to build up a data base concerning the properties of diverse NEAs so that, if a threatening object is found, we will have a larger and more reliable context in which to assess its characteristics.

It is beyond the scope of this chapter to develop particular approaches to mitigation that take our poor knowledge of physical properties into account. But we may briefly offer the following generic recommendations: (1) better characterization of NEAs as a population is highly desirable. In addition, precursor missions to assess the properties of the specific hazardous body are essential, or else such assessment capabilities must be built into the deflection mission itself. Such assessment should concentrate on issues such as body strength (e.g., coherent body or rubble pile), body shape and dynamics, and composition. (2) Mitigation procedures must be designed to be as independent of the specific shapes, compositions, heterogeneity, etc., of the body as possible. (3) A controlled deflection scheme such as stand-off explosions (Ahrens and Harris 1992) or low-thrust propulsion would help minimize the stresses that might unintentionally disrupt the body into a more dangerous swarm of unpredictable character instead of moving it.

Acknowledgments. This work has been supported by NASA grants and contracts, at the Planetary Science Institute a division of Science Applications International Corporation; the Jet Propulsion Laboratory, California Institute of Technology and at the Massachusetts Institute of Technology.

REFERENCES

Ahrens, T. J., and Harris, A. W. 1992. Deflection and fragmentation of near-Earth asteroids. *Nature* 360:429–433.
Binzel, R. P., Shui, X., Bus, S. J., and Bowell, E. 1992a. Small main-belt asteroid lightcurve study. *Icarus* 99:225–237.
Binzel, R. P., Xu, S., Bus, S. J., and Bowell, E. 1992b. Origins for the near-Earth asteroids. *Science* 257:779–782.
Binzel, R. P., Xu, S., Bus, S. J., Skrutskie, M. F., Meyer, M., Knezek, P., and Barker, E. S. 1993. The asteroid-meteorite connection: The discovery of a main belt ordinary chondrite asteroid. *Meteoritics* 28:324 (abstract).
Chapman, C. R. 1994. The Galileo encounters with Gaspra and Ida. *Planet. Space Sci.*, in press.
Chyba, C. F., Thomas, P. J., and Zahnle, K. J. 1993. The 1908 Tunguska explosion: Atmospheric disruption of a stony asteroid. *Nature* 361:40–44.
Drummond, J. D. 1991. Earth-approaching asteroid streams. *Icarus* 89:14–25.
Gaffey, M. J., Burbine, T. H., and Binzel, R. P. 1993. Asteroid spectroscopy: Progress and perspectives. *Meteoritics* 28:161–187.
Gradie, J. C., Chapman, C. R., and Tedesco, E. F. 1989. Distribution of taxonomic classes and compositional structure of the asteroid belt. In *Asteroids II*, eds. R. P. Binzel, T. Gehrels and M. S. Matthews (Tucson: Univ. of Arizona Press), pp. 316–335.
Harris, A. W. 1994. Tumbling asteroids. *Icarus* 107:209–211.

Hartmann, W. K., Tholen, D. J., and Cruikshank, D. P. 1987. The relationship of active comets, "extinct" comets, and dark asteroids. *Icarus* 69:33–50.

Hills, J. G., and Goda, M. P. 1993. The fragmentation of small asteroids in the atmosphere. *Astron. J.* 105:1114–1144.

Lebofsky, L. A., Lebofsky, M. J., and Rieke, G. H. 1979. Radiometry and surface properties of Apollo, Amor, and Aten asteroids. *Astron. J.* 84:885–888.

Luu, J., and Jewitt, D. 1989. On the relative numbers of C types and S types among near-Earth asteroids. *Astron. J.* 98:1905–1911.

McFadden, L. A., Tholen, D. J., and Veeder, G. J. 1989. Physical properties of Aten, Apollo, and Amor asteroids. In *Asteroids II*, eds. R. P. Binzel, T. Gehrels and M. S. Matthews (Tucson: Univ. of Arizona Press), pp. 442–467.

Melosh, H. J., and Stansberry, J. A. 1991. Doublet craters and the tidal disruption of binary asteroids. *Icarus* 94:171–179.

Ostro, S. J., Campbell, D. B., Hine, A., and Shapiro, I. I. 1987. Radar echoes from asteroid 1986 DA indicate a metallic composition. *Bull. Amer. Astron. Soc.* 19:840 (abstract).

Ostro, S. J., Chandler, J. F., Hine, A. A., Rosema, K. D., Shapiro, I. I., and Yeomans, D. K. 1990. Radar images of asteroid 1989 PB. *Science* 248:1523–1528.

Wetherill, G. W. 1988. Where do the Apollo objects come from? *Icarus* 76:1–18.

CLASSIFYING AND MODELING NEO MATERIAL PROPERTIES AND INTERACTIONS

JOHN L. REMO
Quantametrics Inc.

Asteroid material properties are interpreted in terms of analogs with known meteorites and interactions with kinetic and nuclear energy to establish a classification of near-Earth objects based on mechanical strength, thermal properties and atomic composition. Relationships between meteorites and asteroids are inferred from recovered meteorites, their fall trajectories, morphology, physical, chemical, and isotopic data and from asteroid spectra. Experimental techniques for determining the equation of state and mechanical properties for high (shock) and lower strain rates are outlined. Preliminary data on meteorite abundances, mechanical and thermal properties, shock wave effects, magnetic remanence and nuclear absorption cross sections are presented. Materials science and dynamic modeling approaches to assist in characterizing kinetic and nuclear interactions with NEOs are suggested. Recommendations for a NEO material classification scheme and an interception interactions matrix are outlined.

I. INTRODUCTION

This chapter is an extension of earlier work (Remo 1993*a*) describing analogs between material properties of meteorites and asteroids in order to establish a classification scheme for near-Earth objects (NEOs) in terms of material properties thereby reducing uncertainty in planning strategy for orbit modifying interception. In attempting to achieve this goal, experimental and dynamic modeling methods are described which can assist in generating data. Data bases are initiated here, which must be considerably extended to provide the necessary materials properties to support interaction models for an exploratory or orbit modifying mission. Astronomical, physical, chemical, and spectral evidence has long been used to establish relationships between asteroids and known meteorites. Based on data included in tables outlining mechanical and thermal properties, X-ray, gamma-ray, and neutron absorption coefficients, shock wave effects, and magnetic properties of meteorites, a properties classification scheme for NEO materials is presented to assist in NEO hazard mitigation strategies. It is not our purpose to engage in a discussion on the origin and evolution of either meteorites or asteroids. The literature on these subjects is extensive and still not complete. Our goal is to provide an empirical methodology anticipating the range of NEO properties.

Original objectives of the earlier work remain. First, was to outline associations between meteorite material properties and asteroid observational spectra in order to establish a classification of NEO materials based on anticipated dynamic responses of meteorite mechanical and thermal properties to various kinetic and radiation energy interactions. Second, was to outline an experimental mechanical methodology associated with variable strain rates, initiate a data base on some available laboratory measurements and encourage additional experimental work taking into account internal (parent body) pressures and temperatures of materials in space. These objectives are only a first step which are limited in detail and, in most cases, incomplete. However, additional experimentation should generate interest or controversy that will provide support and motivation for continued research leading to a better understanding of the microstructural properties of meteorite materials and their evolution. These material properties can then be compared, through modeling, to the behavior of meteorites and their (assumed) parent bodies on a macrostructural (astrogeological) level.

II. PROPERTIES OF METEORITES AND ASTEROIDS

A. Compositions and Relative Mass Abundances of Meteorites

A meteorite is a recovered object that has survived entry and transit through the atmosphere and may generally (but not always) be regarded as a fragment originating from the collision of asteroids and may therefore by considered representative of material properties of asteroids. This association between meteorites and asteroids establishes a basis for the interpretation of asteroid properties in terms of meteorite properties. However, many NEOs may not have material properties that are meteorite-like. For instance, many NEOs may be more comet nuclei-like (see the Chapter by Greenberg and Remo). Nonetheless, this chapter will concentrate on analyzing the properties of meteorites as surrogates for at least a fraction of the NEOs (presumably NEAs) because this material is available for laboratory study.

Table I gives the proportion of meteorites based on (observed) falls and finds. Proportions of meteorite types among finds are different, being heavily weighted to the more distinctive FeNi and stony irons. The (mass) proportions of FeNi and stony-iron meteorite (observed) falls are actually very small with respect to the total number of meteorites collected. Many regard fall statistics (as opposed to finds) a more accurate representation of the proportions of meteoritic material orbiting in the vicinity of the Earth and by implication the proportional amount of NEA material. There are other factors to consider. While meteorites recovered on the Earth have been found in all sizes, the irons tend to be much larger than the surviving stones. The largest meteorite discovered, the FeNi Hoba, had an (estimated) original impacting mass of about 75,000 kg. Other massive iron meteorites include the Cape York irons which have a collective mass of about 58,000 kg. The largest FeNi meteorite found in the U. S. is the Willamette, weighing 12,700 kg. There are several

TABLE I

Major Meteorite Composition Types and Abundances of
Well-Classified Falls and Finds[a]

	Falls	Finds	
Stones	95.7%	52.3%	
Chondrites[b]	87.3%	51.3%	
Achondrites[c]	8.3%	1.0%	
Stony Irons	1.2%	5.4%	
Pallasites[d]	0.3%	3.8%	
Mesosiderites[e]	0.9%	1.6%	
Irons (FeNi)	3.1%	42.3%	
Hexahedrites (H):			<6 wt % Ni; kamacite, body-centered-cubic (bcc) structure; no taenite, face-centered-cubic (fcc).
Octahedrites (O):			6 to 7 wt % Ni; the most abundant FeNi, contains kamacite and taenite in the Widmanstatten pattern formed by an extremely slow cooling rate.
Ataxites (D):			16 to 35 wt % Ni; Ni rich alloy composed of taenite with trace quantities of kamacite, yielding a microscopic Widmanstatten pattern.

a The relative (number) abundance of well-classified meteorite types as observed from falls and finds is given in percent. Total weight among finds is very different than the observed fall values, being heavily weighted toward the larger irons. Less distinctive among stones, achondrites have lower "find" values. Micrometeorites or space dust material found in the Earth's atmosphere represent the bulk of extraterrestrial material acquired daily ($\sim 10^2$ to 10^3 tons). Antarctic meteorites are not included in this table.

b Agglomeration of early solar system materials that underwent little if any chemical change since their formation. Composed either of tiny silicate grains of olivine and pyroxene along with minor sulfides, oxides, feldspathoids, and clay minerals or graphite mixed with magnetite. Most chondrites are speckled with FeNi kamacite and taenite.

c Silicate rich, nonvolatile element distribution, extremely diverse igneous types of rock the product of partial melting and re-crystallization. Compositions range from almost mono-mineralic olivine and pyroxene to those of terrestrial and lunar basalts.

d Stony-iron composed almost exclusively of magnesium olivine crystal grains enclosed in a continuous matrix of FeNi.

e Agglomeration of re-crystallized mechanical mixtures in roughly equal proportions of metal and silicates often having brecciated mixtures.

FeNi and two stony-iron meteorites weighing more than 4000 kg. In contrast, the largest known stony meteorite fell as a shower of fragments in Jilin, China, with a collective mass of 1750 kg.

Selection processes may favor survival of the stronger FeNi meteorites (Table I) through atmospheric hypersonic entry within the terrestrial atmosphere, high-speed impact and the resultant cratering. In addition, resistance to weathering of some FeNi is significantly better than stones. The superior structural strength and higher density of FeNi component of meteorite parent bodies also allows them to remain as larger, intact pieces in the collisional space environment; fragments of stony meteorite parent bodies undergo greater fragmentation. Stony meteorites, besides weathering more easily, may enter the Earth's atmosphere as smaller pieces with less chance to survive entry or impact. They may be discovered preferentially in special circumstances such as exist in the Antarctic. Although there are reasons to presume the meteorite observed fall proportions are appropriate for the NEA populations, it is certainly not conclusive. Problems of long- and short-term orbital evolution depending on models of meteorite parent bodies must be considered when extrapolating to asteroid compositions from limited fall and find data (Shoemaker 1983). If a greater proportion of the larger NEA fragments tend to be composed of FeNi, it is a serious materials factor for NEO interaction strategies.

B. Evidence for the Origin of Meteorites from Asteroids

Evidence implying asteroids as meteorite parent bodies includes:

1. Orbits of five observed and recovered ordinary chondrites (Pribram, Lost City, Farmington, Innisfree, and Dhajala) had highly elliptical orbits with aphelia in or near the asteroid belt.
2. The common fusion crusts and other surface features indicate high speed passage through the Earth's atmosphere and shock metamorphism indicates hypervelocity impact.
3. Meteorite craters are indicative of hypervelocity impact. Found in the vicinity of some (recent) craters are shocked and melted (iron) meteorite fragments. High pressure polymorphs of target material are also found within the crater. Meteorites bear no resemblance to the terrain on which they are found.
4. Chemical, petrologic, isotopic, and cosmic ray (exposure) data differ from corresponding terrestrial samples.
5. Direct, but imprecise, evidence associating asteroids and meteorites can be interpreted from a combination of (albedo) spectral reflectance measurements in the visible and infrared (Chapman 1976; Chapman and Gaffey 1979; Fanale et al. 1992; Chapter by Chapman et al.).

C. Asteroid Composition Types

Both mainbelt asteroids as well as NEAs have been classified photometrically using reflectance spectra in the visible through infrared and in a more limited manner by radar reflectance measurements, thereby measuring some important mineralogical characteristics that set boundaries for plausible analogs of meteorite/asteroid materials for the most populous classes (McFadden et al. 1989; Gaffey et al. 1989; Lipshutz et al. 1989; Binzel et al. 1991). Additional detailed astrodynamic data on asteroid size, spin rate, orientation, and surface properties is provided by radar reflection (Ostro et al. 1991; Ostro 1993; see also his chapter). A comparison of asteroid composition types, meteorite mineralogy and mechanical properties is given in Table II. It has been suggested that from the population of classified NEAs, the (major types) S:C:M ratio is approximately 7:3:1, while among the hundreds of classified mainbelt asteroids the ratio is approximately 7:5:1 (Tholen 1984). In Table II the most abundant types of asteroid composition are categorized with possible meteorite analogs that will have generally similar mechanical, thermal, and atomic properties. These permit classification of meteorite-like NEOs into three structural classes. Based on the inferred surface mineralogy (composition) and the possible meteorite analogs the composition types B, C, F, and G as well as D and P are likely to be mechanically (structurally) weak and can be placed in the same category. Similarly, one may assume that E and S have compositions similar to chondrites and achondrites and are somewhat stronger mechanically than the B, C, D, F, G, and P types. Continuing, we consider the M type asteroids to be predominantly metallic and structurally resemble metallic meteorites which are mechanically the strongest class, as indicated from Table III and other work (Johnson and Remo 1974; Remo and Johnson 1974,1975; Johnson et al. 1979; Remo 1992a).

Although NEAs can be divided into three structural classes with a relatively limited range of (meteoritic) materials, there is an extensive range of mechanical and thermal properties on the micro and macroscopic scales caused by inhomogeneities originating from circumstances of origin and evolution. Most data from which the mineralogical properties of asteroids is inferred comes from telescope observations of asteroids which are generally faint and must be observed within a narrow spectral window. Figures 1, 2, and 3 compare reflectance spectra of minerals, asteroids and meteorites. Data in Tables III, IV, V, and VI which describe the mechanical and thermal properties of meteorites are taken from ideal individual samples with relatively homogeneous microstructure (Krinov 1960; Öpik 1958; Wasson 1974).

Observational factors relating NEAs to meteorite and mainbelt asteroid analogs include:

1. Mainbelt asteroids may replenish the NEA population. Although mainbelt asteroids in the size range of NEAs ($\sim$5 km) are difficult to characterize due to their greater distances, their spin rates and shape distributions appear to be similar (Binzel et al. 1992).

2. Superficial modification from physical processes may have substantially modified the reflective surface to the degree that reflected optical data is not definitive for classification.

TABLE II

Most Abundant Near-Earth Asteroid Composition Types, Possible Meteorite Analogs and Mechanical Strength

Asteroid Type	Spectral Properties	Inferred Surface Mineralogy
I		
B, C, F & G (primitive & metamorphic)	Low albedo, ~2 to 7% Neutral, slight blue & strong UV absorption.	Hydrated silicates & carbon/organics/opaques.
Possible meteorite analogs: Carbonaceous chondrites. CI1 to CM2 as well as assemblages produced by aqueous alteration and/or metamorphism of CI/CM precursor materials; low crushing strength		
II		
D & P (primitive)	Low albedo, ~2 to 7%; red spectral reflectivity.	Carbon/organic rich silicates
Possible meteorite analogs: Carbonaceous chondrites; organic rich, primitive, cosmic dust grains; CI1 to CM2 plus organics; low crushing strength.		
III		
E & R (igneous)	E: high albedo, >23% appears slightly red R: moderate albedo, very red with strong IR absorption due to pyroxene.	Enstatite and/or other iron free silicates
Possible meteorite analogs:: (E) Enstatite chondrite and enstatite achondrite; (R) pyroxene-olivine achondrite; moderate structural strength.		

IV
M
(igneous)

Neutral spectral reflectivity; moderate albedo; high radar reflectivity.

Metal (FeNi) with possible traces of silicate.

Possible meteorite analogs: FeNi metal with possible silicate inclusions; enstatite, FeNi metal, or a combination of both enstatite chondrites derived from differentiated parent bodies; very strong structurally within metallic phases.

V
S, V & subclasses
(igneous and/or metamorphic)

Moderately high albedo, ~7 to 23% red; absorption band at 0.9 to 1.0 and near 2 microns; broad absorption band in blue & UV; wide spectral variety.

Olivine, pyroxene & FeNi

Possible meteorite analogs: (S) Ordinary chondrites and/or stony irons; possible parent body of chondrites. Only extreme metal poor and olivine poor members of the S group have spectra that approach ordinary chondrites; moderate structural strength if chondritic; strong if stony-iron. (V) basaltic achondrites; eucrites.

TABLE III

Mechanical Properties of FeNi Meteorites

1. Density	Iron Nickel	7.29 to 7.88 g cm^{-3}
	Mesosiderites	5.20 to 6.20 g cm^{-3}
	Pallasites	4.74 g cm^{-3}
2. Hardness	Brinell, correlated with Ni content—Taenite	90 to 660
	Kamacite	90 to 380
	Brinell, hard inclusions	
	(cohenite, troilite, schreibersite and chromite)	950
	Vickers	200 to 350
3. Tensile strength	FeNi	0.58 to 1.8 × 10^{10} d cm^{-2}
	Hard mineral inclusions	3 × 10^{10} d cm^{-2}
4. Young's modulus (tensile)		2.0 × 10^{12} d cm^{-2}
5. Compressive strength	FeNi	1.1 to 3.4 × 10^{10} d cm^{-2}
	Hard mineral inclusions	6.3 × 10^{10} d cm^{-2}
6. Compressibility	Bulk modulus (all sides compress ion)	1.67 × 10^{12} d cm^{-1}
7. Surface tension		1200 d cm^{-1}
8. Coefficient of viscosity	Molten meteoritic iron at	0.026 to 0.019 poises (d s cm^{-2})
	effective temperature of vaporization	0.02
		0.01
9. Lattice parameters	Kamacite, bcc cubic	2.876 × 10^{-8} cm
	Taenite, fcc cubic	3.622 × 10^{-8} cm

TABLE IV

Thermal Properties of FeNi Meteorites

1. Melting point (with 10% Ni)	1770 K
2. Boiling point	3508 K
3. Average specific heat	
(a) Solid (0 to 1500 C)	6.91×10^6 erg/g-deg
(b) Liquid	6.66×10^6 erg/g-deg
(c) Gaseous	$C_p = 3.72 \times 10^6$ erg/g-deg
	$C_v = 2.23 \times 10^6$ erg/g-deg
4. Latent heat of vaporization	3.72 ev/atom = 6.40×10^{10} erg/g
5. Latent heat of fusion	2.69×10^9 erg/g
6. Thermal conductivity (deg/cm)	4×10^6 erg/cm^2-s
7. Vapor pressure	
(a) Iron	log p = 10.607 to 16120/T
(b) Nickel	log p = 10.725 to 16120/T
8. Lattice energy	4.18 ev per atom or 96,000 cal/mole

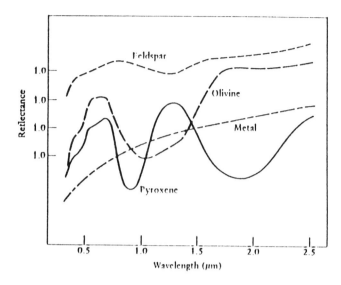

Figure 1.   Individual absorption bands of solar reflectance spectra for feldspar,
olivine and pyroxene minerals and iron-nickel metal are shown as a function of
wavelength in microns. As a rule, linear spectra are indicative of (FeNi) metal
components while absorption spectra are indicative of silicate minerals. Meteoritic
and presumably related asteroid materials composed of a mixture of these materials
is (ideally) likely to possess composite spectra of these curves. It should be noted
that the reflectance spectra is only indicative of the surface material composition
over an optical (superficial) layer.

3. Effects of coherent backscatter at small phase angles (Hapke et al. 1993)
   and shadow hiding (D. Harris 1961) can effect the interpretation of photometric data.
4. Effects of large scale (>1 cm) inhomogeneities within given microstructural phases may produce a diverse surface. Totally different mineral
   phases in physical and structural contact could result from collisions.
   The results of such masking could create uncertainty in interpreting the
   material properties of NEOs, e.g., Gaspra (Belton et al. 1992).
5. Even if the homogeneous microstructure composition is ideally homogeneous, there may have been thermal, radiation, or impact interactions
   throughout the history of the asteroid which altered its structural integrity.
   Effects of small-scale inhomogeneities (<1 cm) in the microstructure (inclusions or grain and phase boundaries) can collectively present large- or
   small-scale discontinuities to a (propagating) shock wave.
6. NEO objects may not be contiguous and therefore lack overall structural
   integrity. NEO objects may also be composed of two, for example,
   4179 Toutatis (Ostro 1991,1993), three, or a many-body aggregate of
   relatively large components in the range of hundreds of m to tens of km.
7. Some NEOs may be derived from extinct or dormant comet nuclei, implying a density of ~0.2 to 1.5 g cm^{-3} and may not be easily identified.

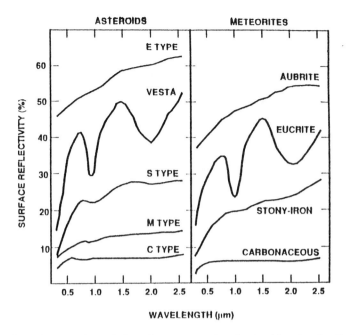

Figure 2.   A comparison of the reflectance spectra of asteroid and meteorite types which may be interpreted as associating some meteorite and asteroid types.

It is also important to realize that measurements in Tables III through VI were carried out in terrestrial laboratories, at room temperature and atmospheric pressure. At low temperatures (<200 K) the mechanical and thermal properties of materials are likely to change substantially.

Models for asteroid internal structure may include zones of metamorphic grades producing different chondrite petrologic types within which stony-iron regions enclose a central FeNi core where differentiation occurs yielding FeNi materials with variable % Ni. Variants of this basic model include isolated metallic pockets surrounded by metamorphic grades of silicates. Some bodies may be undifferentiated (no metallic core) while others may be composed of non-metamorphosed primitive solar material or volatile-depleted comet cores. However, even from an ideal set of models for NEO parent bodies, a unique sequence of asteroid fragmentation and meteorite populations cannot be determined. Because the mechanical and thermal properties of asteroids have not been measured, it is not possible to predict NEA composition. Ultimately, we must assume that meteorites are close analogs of asteroid fragments.

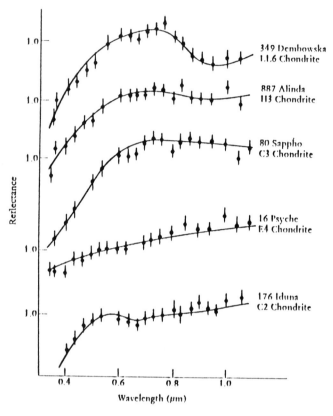

Figure 3.   A comparison of reflectance spectra between selected asteroids (error bars) and some chondrite classes (curves). Because the absolute albedo depends on the particle size on the reflecting layer (surface scattering) in addition to the molecular composition of the layer, the spectral reflectance is re-normalized. In particular, the albedo in (c) is assigned a value of 1.0 at the wavelength of 0.56 microns. It is, therefore, the relative shapes of the spectral curves which provide the basis for comparison of the spectra.

III. EXPERIMENTAL MODELING OF NEO DYNAMIC PROPERTIES

A. NEO Deflection and Debris Considerations

If a given amount of kinetic energy is used to alter the NEO velocity vector, proper planning utilizing the available data of material properties is necessary to assure that the kinetic energy is coupled into the NEO orbital adjustment without seriously affecting structural integrity. Therefore, a goal of NEO materials research must be to develop a capability for predicting reaction of these NEO materials in hypervelocity (>2–6 km s^{-1}) impacts. The kinetic energy must provide the desired orbital adjustment via momentum transfer while minimizing the probability of this momentum being transferred to a

TABLE V

Mechanical Properties of Stony Meteorites

1. Density:	Range 2.38 to 3.84 g cm^{-3}
2. Mean atomic weight:	23
3. Porosity	micro = 2.7 to 3.59% macro = 3.0 to 12.4%
4. Surface tension	at fusion = 400 d/cm at volatilization = 360 d/cm
5. Compressive strength:	0.062 to 3.7 kb
6. Bulk modulus of terrestrial stone analogs:	(appears to be unrelated to petrologic type) Pressure = 1 km cm^{-2}–10^4 km cm^{-2}
Olivine diabase	7.4–8.3×10^{11} d/cm^2
Peridotite dunite	10.6–12.6×10^{11} d/cm^2
7. Young's modulus of terrestrial stone analogs:	
Olivine diabase	10.01–10.9×10^{11} d/cm^2
Peridotite dunite	14.7–17.4×10^{11} d/cm^2
8. Viscosity:	1.3×10^4 poises at 1400 C
9. Seismic velocity:	Ordinary chondrites (a) Longitudinal–2050 to 4200 m s^{-1} (b) Transverse–600 to 1220 m s^{-1}
10. Electrical resistivity:	10^4 to 10^8 ohm cm; average value = 10^5 ohm cm
11. Reflectivity (total albedo):	0.25 to 0.65; if surface is gouged, scratched, or generally pitted (in space values are lower)

TABLE VI

Thermal Properties of Stony Meteorites

1. Average melting point: 1350 to 1800 K		
	(a) Magnesia olivine	1890 K
	(b) Iron olivine	1100 K
	(c) Magnesia pyroxene	1554 K
	(d) Ferrosilicate	1100 K
2. Boiling or dissociation point:		2960 K
Mean lattice energy per vapor molecule:		3.5 Ev
3. Evaporation:		$n = 3.08 \times 10^{-4} p\, T^{-1/2}$ g cm^{-2} s; p = d/cm^2
4. Average specific heat (fayalite)	Solid, 0 to 1200°C	8.95×10^6 erg/g-deg
	Liquid	1.1×10^7 erg/g-deg
5. (a) Latent heat of fusion (fayalite):		2.65×10^9 erg g^{-1}
(b) Latent heat of vaporization (average for meteoritic stone)	from mean lattice energy	6.05×10^{10} erg g^{-1}
6. Thermal conductivity (at 50°C):		3.6 to 5.8×10^{-3} cal/(s-deg-cm)

diverging spray of dangerously large secondary particles.

It is vital that attempts to eliminate the NEO threat should not produce fragments >100 m diameter on separate orbits from the initial NEO. Such fragments would be difficult to track and interdict on short notice. Dispersal of such fragments can obscure input to sensors and interfere with sequential interactions. Good coupling must therefore be achieved which will optimally transfer momentum to the NEO.

Specific research objectives that will assist in mitigating uncertainties associated with kinetic interactions at impact include:

1. Making analytic estimates of the total ejecta debris from a range of kinetic interactions with a given NEO material;
2. Calculating the anticipated size distribution of statistical fragments from both uniform and heterogeneous NEO material;
3. Calculating distributions of ejecta momenta and trajectory data sufficient to predict details of the impulse caused by the debris sprayed both on down-stream structures and in the vicinity of the target;
4. Experimentally evaluating effects of low (<250 K) temperatures on impact properties and its effect on fragmentation;
5. Experimentally assessing the effects of inclusions, grain boundaries and inhomogeneities on material strength and effects on fragment size distributions;
6. Experimental studies of momentum transfer to NEO-like bodies;
7. Experimental studies of crater structure, impact velocity and fragmentation spectra as a function of impactor energy and material properties.

Problems identified above which may affect the accuracy of momentum transfer estimates are primarily related to the equation of state and material properties. Surface morphology issues, such as whether the surface is comprised of loosely bound aggregates or surface regoliths, or is monolithic, may also be important in interdiction planning.

B. Kinetic Hypervelocity Impact, Shock Waves, and NEO Material Properties

Hypervelocity impact is a kinetic energy interaction that generally produces states of high compression and expansion. The response of materials to hypervelocity impact spans a wide range of material behavior from high impact pressures and temperatures where nonequilibrium effects prevail in the region of direct impact to the relatively lower pressures where mechanical properties dominate. The study of the response of materials to dynamic loading is based on continuum mechanics. Sophisticated experimental methods have been developed for both on- and off-Hugoniot measurements, and useful equations of state have been produced for many classes of materials which can be used over a wide range of densities and temperatures.

When a projectile impacts a target or when high radiative energy densities are hydrodynamically coupled into materials, the highest pressures are reached by a shock wave. As the shock wave propagates through the target material it is usually attenuated, recording progressively lower pressures. During a single impact event, depending on the energy, a wide range of states can be produced. These states include changes in electronic and chemical structure, phase transitions, melting, plastic flow, and simple elastic behavior. When release waves appear, as when compression waves reflect from a free surface, an expansion follows the shock compression. Materials may undergo tensile failure (spall) where release waves interact, and fragment. For smaller NEOs this may cause material dispersal. When a high-speed projectile impacts a shallow target, perforation of the target often results, depending on the relative material strength and thickness, and an opening is formed. Target and projectile debris may be ejected from this opening and may also be backscattered. This debris generally consists of solid fragments with smaller amounts of melted and vaporized material.

For NEO impact problems there is need for analysis of the projectile and target response and subsequent motion of the target debris. An additional concern is the destructive potential of the debris when interacting with secondary objects such as a subsequent interaction system.

As an example of the dependence of system behavior on input parameters, consider the case of a penetrating emplacement of a nuclear or conventional explosive within the NEO. The penetration depth is a critical parameter for predicting such aspects of system behavior as ejecta velocity and mass distributions. This depth is a function of impact velocity and projectile (explosive delivery system) and target material properties of the NEOs in turn. A debris cloud is produced; the problem is characterizing it. In the case of very high velocity impacts, sufficient energy is available to vaporize some material of the target upon release. The presence of a significant vapor phase will markedly increase the divergence of an expanding ejecta cloud; momentum and energy will be spread over a large region, and there will be a decrease both in the destructive potential (if the NEO is to be destroyed) and in the momentum transfer (if the NEO is to undergo a velocity change). In the case of lower-velocity impacts, comparatively less vapor phase will be produced.

Empirical evidence indicates that NEO material properties vary widely among materials of interest. Such properties include both the intrinsic material properties associated with the mineralogy and chemistry and the gross physical properties such as inhomogeneities and body dynamics of the overall assemblage.

The material properties of the NEO may vary from location to location as well. This can make an important difference in the blow-off fraction and momentum coupling for situations of a nuclear thermal high specific impulse interception vehicle arriving at different locations on the NEO.

Impact experiments must be carried out with a variety of large metal (alloy) projectiles at the (low) temperatures expected in the NEO interdiction zones on several solid targets corresponding to the NEO material classifications. The impact velocity should be in the range of 2 to 6 km s^{-1}, higher where possible. The suite of experiments should be capable of providing a data base for verifying theoretical and computational predictions of primary target failure, momentum coupling mechanisms, and the characteristics of secondary debris. The data base should be of use for studies involving homogeneous and nonhomogeneous materials. Such experiments can serve as the basis for establishing a theoretical framework for predicting fragmentation within the (target) multi-temperature and multi-phase environments associated with a high-velocity impact event. Such fragmentation theory can be coupled with a three-dimensional Eulerian hydrocode to predict the characteristics of large-scale NEO interactions.

The knowledge of the material properties required to model NEO kinetic interactions with hypervelocity projectiles and the associated shock wave effects must include an understanding of the equation of state, dynamics of the body, and the geological and mechanical structure. Knowledge of these properties will require an organized effort combining laboratory experimentation with computer modeling.

A major problem with experiments on the Earth on the analog materials of the NEO is the atmospheric gas and water contained within the sample. The low temperatures of space are more easily reproduced, and hence are not as much of a problem. The adsorption problem is much less critical for FeNi samples which, except for nodules and faults, are apparently impervious to such gases. As this adsorption will affect the outcome of experiments, measurements must ultimately be carried out which will accurately simulate the anticipated regions of NEO interaction. Perhaps theory can assist our interpretations.

During the acquisition and targeting phase of the NEO orbital adjustment, an extensive reconnaissance effort involving remote sensing and contact probes will provide data that will be processed to yield a reliable interpretation of the NEO material, the gross structure of the NEO, and its mass. The successful interpretation of this data and of the NEO classification will set up the optimal interaction mechanism for modification of the NEO's orbit.

The following sections describe three experimental techniques appropriate for determining the equation of state and mechanical properties of some meteorite materials.

C. Charpy Impact Specimens: Dynamical Property Measurements of High Strain Rates

A direct method to probe the dynamic strength of materials is the Charpy impact technique which utilizes a pendulum hammer raised to a horizontal position to impact and (ideally) break the specimen at the bottom of its swing, losing some of its kinetic energy in the process. The energy absorbed by the specimen determines the height to which the hammer goes after it has passed through the bottom of its swing and has broken the specimen. The height recorded is directly interpreted in terms of the energy absorbed by the specimen.

A Charpy specimen is a prism of square cross section 5.4 cm in length and 1 cm in width and depth, with a transverse "V" notch at the center of one of its long faces. The Charpy test is routinely used to evaluate engineering materials such as steels, which usually contain imperfections and inclusions and are not unlike some meteorites. The "V" notch in the Charpy specimen provides a source of stress concentration that is generally more effective than sources produced by inclusions, therefore allowing the effect of the "V" notch to dominate the fracture process. This is not true for the case of unnotched tensile specimens, and so the results of the Charpy test are less likely to be influenced by the inclusions than are the results of the tensile tests. However, if the size of the inclusions is comparable to the size of the Charpy specimen the test is no longer valid, because it is now the mechanical properties of the inclusion that are being measured. Therefore, the Charpy test will measure characteristics of the material, provided that the inclusions are relatively small.

Experiments on meteoritic irons using the Charpy test have yielded results (Remo and Johnson 1975) which confirmed the ductile to brittle transition predicted to occur in the vicinity of 200 K for octahedrites (Johnson and Remo 1974). Specifically, one of two equivalent specimens tested at 195 K was found to be fully brittle and absorbed only 6.9×10^4 dyne cm of energy during fracture. The other specimen tested at 300 K was partially ductile, absorbing 2.8×10^5 dyne cm. Additional extensive experimental data supporting this result, and its extension to different classes of FeNi meteorites with varying Ni content, has not yet been published (see Remo 1993b). In another Charpy "V" notch impact experiment at 77 K and 195 K on specimens prepared from a high nickel ataxite, it was found that the specimen tested at 77 K was fully brittle while the one tested at 195 K was fully ductile. An interpretation of this result is that high nickel material is ductile at the ambient temperature of an asteroid, which is taken to be about 150 K, and does not fracture in a brittle manner. This confirms a hypothesis put forward by Remo and Johnson (1974) according to which this failure of high nickel material to fragment may explain the larger average size of recovered high nickel meteorites. Recent shock experiments (Furnish et al. 1993) on Henbury indicate, as is expected at this much higher loading rate that the ductile-brittle transition temperature did not exist at ~200 K. However, this transition is expected to be present at a much higher temperature due to the very high strain rate constitutive response in the uniaxial shock environment. Fracture events depend critically on relaxation effects for yield/failure surface interactions which are stress state dependent.

While the Charpy impact work provided insight into the temperature and composition (% Ni) dependence of the meteorite material strength, it is only the first step in characterizing the overall response of meteorites, as NEO analogs, to impacts and energy densities which may be required for orbital management.

D. Dynamical Property Measurements of Ultrahigh Strain-Rate

The most widely used technique for measuring materials properties under high strain rates is the planar impact experiment, which may involve gun, explosive or electrical launchers. Other methods may include shock driving by in-contact explosives or laser bursts. Let us concentrate on the widely-used planar impact methods utilizing gun launchers.

Material properties measurements under ultrahigh strain-rate loading and unloading have been greatly facilitated by time-resolved interferometry technique (see, e.g., Chhabildas 1987; Davison and Graham 1979) for measuring shock wave structures. The most widely used such method is the VISAR, or Velocity Interferometer System for Any Reflector (Barker and Hollenbach 1972). In addition to Hugoniot data, it has become possible to obtain information about yield strengths, shock viscosity, release trajectories, multiwave structures, spall properties and the strength of materials in the shocked state. Techniques have been developed to determine the dynamic material properties at strain rates of 10^5 to 10^{10} s^{-1} at stresses ranging from less than 1 GPa to about 250 GPa (see, e.g., Chhabildas et al. 1990). In particular, rate-dependent effects and release hysteresis can be characterized (Furnish et al. 1992). These sets of data are crucial in evaluating strain-rate-dependent viscoplastic or viscoelastic material models.

For the present area of study, samples are likely to contain heterogeneities (mm or larger scale). Such imperfections constrain experimental design slightly. The most important constraints include the need for buffers between the sample and the diagnostic and the need to choose carefully samples on the basis of non destructive testing to avoid results strongly influenced by a particular imperfection.

Other important experimental considerations for natural samples may include a need to preserve volatiles content (especially water), friability of a specimen rendering difficult its fabrication into a usable gas-gun sample, and whether the sample can withstand the kilogravity to megagravity environment of a gun launch without damage.

The most generally usable configuration for gas-gun testing places the sample in the target, and is shown in Fig. 4. It is especially appropriate for measuring loading wave profiles, Hugoniot states and strength properties . If a window material can be chosen which is an approximate shock impedance match for the sample (such as Z-cut sapphire for iron, or lithium fluoride for aluminum or granite), a continuous release path may be measured; otherwise the pressure and particle velocity of a single point on the release (or reshock) of the sample may be determined. If the shock impedance of the window is chosen as much lower than that of the sample, spall properties of the sample may be measured as well as Hugoniot properties.

Analysis of the velocity profile from such a test consists of determining the precursor and Hugoniot states from the transit time across the sample (hence velocity of the observed waves), then extracting available release or spall

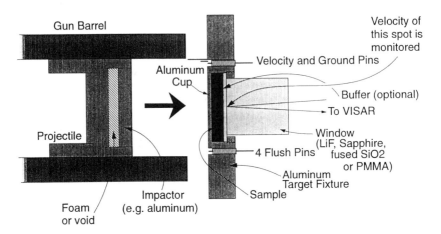

Figure 4.   Forward-ballistic impact geometry.

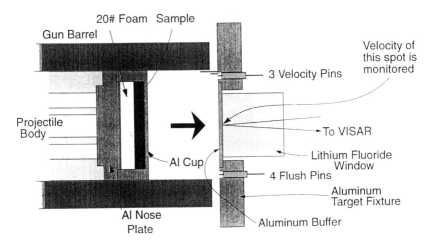

Figure 5.   Reverse-ballistic impact geometry.

information (Furnish 1993). If the window is a fairly close impedance match to the sample and the waveforms entering and leaving the sample are known (easiest if no buffer is used), Lagrangian integration of the wave velocities yields a table relating stress, strain, time, shock velocity and wave velocity (Furnish et al. 1992). For many materials, the strain rate during loading varies approximately as the fourth power of the Hugoniot stress (Swegle and Grady 1985). If buffers are used, wavecode modeling of the experiment to match the observed waveform may yield the pressure-volume path, although this procedure is somewhat more laborious. If the window is a poor impedance match for the sample, the average amplitude of the waveform "plateau" may

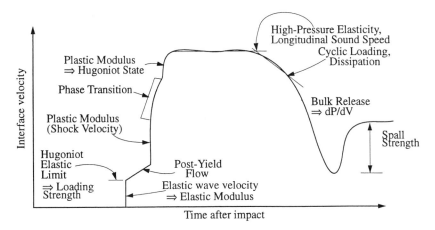

Figure 6.   Wave profile and physical property correspondence for the transmitted wave.

be used with the Hugoniot of the window to calculate a single partial release or reshock state of the sample. Finally, if the sample has spalled, the amplitude of the pull-back signal may yield the spall strength of the sample (Chhabildas et al. 1990).

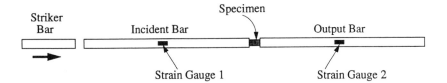

Figure 7.   Schematic of the Split Hopkinson pressure bar.

For conditions where a window matching the shock impedance of the sample cannot be found, Hugoniot and continuous release paths may be measured using the geometry shown in Fig. 5 (Furnish 1993). This configuration, called "reverse-ballistic," has been used extensively for measuring Hugoniot and release properties of rocks and grouts. It may be used with water-saturated samples (as may the forward-ballistic configuration). It does not give any information about loading characteristics, such as precursors and material strength, and in fact may give misleading results if incorrect assumptions about these quantities are made. Under certain conditions, however, a family of such tests can be used to determine precursor conditions. It is normally most useful if the dynamic strength of the material is small compared to the Hugoniot stress.

Within these constraints, gas gun tests are able to produce a wide range of data of materials properties for materials undergoing ultrahigh strain-rate deformation. Impact velocities may range from essentially zero to over 7 km s^{-1}

with minor variations in the configurations discussed here, depending on the
launcher system used. Under certain conditions gram-sized samples may be
launched to 12 km s^{-1}. Hence these properties can be evaluated over a wide
range of stress (density) states.

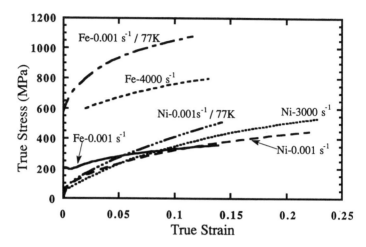

Figure 8.   Stress-strain behavior of iron and nickel under various temperature and
strain rate conditions.

In summary, the waveform of Fig. 6 illustrates representative properties
which can be obtained for metallic or stony NEA material (represented in
Fig. 6 for a forward-ballistic test).

E. Constitutive Response of NEO Materials

Accurate modeling of high energy interactions with NEO material requires
understanding of the total loading impulse imparted into the object and its
effect on material behavior. Following the shock wave produced by the
detonation of explosives or by high velocity impact, significant inertial deformation occurs within the loaded material as well as various fracture and
fragmentation processes due to multiwave interactions from nonplanar loading and effects of shock release. Development of physically based material
models contribute to the description of this complex loading history; therefore it requires an understanding of the influence of the initial shock-loading
impulse on the subsequent dynamic deformation and fracture. The dynamic
material response driven by inertia from the shock is controlled by how the
shock prestraining has altered the material microstructure and its effect on the
stress-strain behavior. The dynamic fracture (spallation) behavior of NEO
materials is similarly dependent on the starting microstructure, magnitude of
the shock prestrain, and duration of the shock process.

Detailed description of the total deformation response of the NEO materials therefore requires: (1) characterization of the starting microstructure of

the NEO materials using metallographic and electron microscopic techniques; (2) influence of shock loading on the structure/property behavior of the NEO materials; (3) influence of strain rate and temperature on the mechanical response of NEO materials; and (4) dynamic fracture (spallation) behavior of NEO materials.

The relationship of stress versus strain for a material (termed the constitutive response) is measured utilizing a variety of experimental techniques. Because mechanical properties such as strength and ductility can vary as a function of temperature, strain rate, and strain, it is often necessary to determine these properties under conditions that closely match the expected deformation rates in service. In the instance of NEO modeling, the constitutive data needed encompasses a wide range of strain rates (10 to 3 s^{-1} to shock levels) and temperatures (the cryogenic temperatures representative of deep space versus the rapidly elevated temperatures of high-impulse loading interaction or re-entry conditions).

Measurement of the mechanical properties of materials encompasses a very broad field depending on the property of interest, stress state, temperature, loading rate, environment, etc., and as such will not be reviewed in this section. Those interested in the diversity and details of mechanical testing should consult volume 8 of the *ASTM Metals Handbook* (1985) which covers mechanical testing. For the purposes of providing input to those interested in modeling NEO materials, the constitutive and fracture properties, particularly at high strain rates, are of most immediate concern for use in large-scale hydrocode calculations. The constitutive response of metallic and nonmetallic materials, as a function of strain rate and temperature, are most often measured in compression to avoid inertial and wave propagation effects.

Assessment of these properties over a range of strain rates and temperatures comprises the necessary data to benchmark most material models aimed at providing algorithms to quantify the constitutive behavior of materials. Past experience has graphically illustrated the danger of extrapolating these models outside the range over which parametric data were measured. At low strain rates, ~10 to 4 or 10 to 3 s^{-1}, which are termed quasi-static levels, the stress-strain behavior is normally probed using standard screw-driven testing equipment employing right-regular cylindrical samples. Similar sample configurations can also be utilized at strain rates up to nominally 100 s^{-1} with the addition of hydraulic servo-controlled test machines. With increasing strain rate additional load and strain measurement complexities are imposed. The highest possible strain rates in a uniaxial compression test under uniform deformation conditions is achieved using the Split-Hopkinson pressure bar (SHPB). Using this device the constitutive response of metallic samples can be probed over the strain rate range of ~200 to 10^4 s^{-1}. Combination of the quasi-static testing mentioned previously with SHPB data will allow assessment of the stress-strain response of a material over ~7 orders of magnitude in strain rate. This range represents sufficient data to formulate materials models capable of grasping most of the crucial physics of the deformation response of the material in question with the exception of the shock region discussed separately. A schematic of the SHPB is shown in Fig. 7.

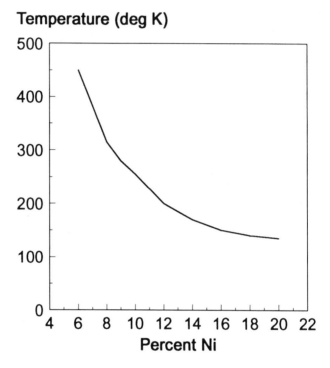

Figure 9.   The ductile-brittle transition temperature as a function of weight % Ni is plotted based on Charpy impact tests on hexahedrites, octahedrites, and ataxites. This indicates that both composition as well as temperature determine the material strength of meteoritic irons.

The SHPB consists of two elastic pressure bars that have a sample sandwiched between them as described by Follansbee (1985).

Analysis and characterization of the constitutive response of FeNi, chondrite and achondrite meteorite analogs to NEAs will necessarily need to be specific to the type of meteorite to be modeled as there exist substantial differences between metallic types compared to the stony types. Due to the fractured nature of the stony types surviving Earth impact, the ability to conduct detailed constitutive studies on these materials representative of NEA materials is significantly impaired, although obtainable.

To understand the constitutive response of metallic meteorite materials, such as a Fe-Ni metallic type, it is instructive to examine the relations governing the deformation behavior of metals and alloys which have been developed over the past 50 yr. The flow stress of a metal or alloys can be expressed in terms of the following components:

$$\sigma = \sigma_p + \sigma_d + \sigma_a \tag{1}$$

where σ_p is the lattice friction term or Peierls stress, σ_d is the stress due to dislocation-dislocation interactions, and σ_a is the athermal stress is due to long-range obstacles such as grain boundaries, second-phase particle, etc.

The differences in the strengths of various materials can be examined using this formulation. The Peierls component σ_p represents the force to move a dislocation along a slip plane of a crystal by overcoming the interatomic forces across the plane by a series of movements determined by the periodic stress field of the lattice. The yield stress of body-centered cubic (BCC) metals, such as Fe and Ta, are markedly temperature dependent, particularly at low temperatures, in contrast to face-centered cubic (FCC) metals such as Ni or Cu due to this force. Overcoming the lattice friction stress energy wells is aided by thermal activation as a dislocation moves from one equilibrium position to the next in the slip plane. Accordingly this stress is dependent on the temperature and strain rate of the deformation. In polycrystalline metals, moving dislocations encounter other dislocations from other slip planes, giving rise to a σ_d term. The stress necessary to pass forest dislocations is dependent on the dislocation density, which in turn is dependent on strain, strain rate, and temperature. Lastly, the athermal σ_a component arises from long range barriers to dislocation motion, such as grain or phase boundaries and second phase particles, and can be considered as a fixed value for a given material condition.

Data on the response of metallic materials can be formulated in a series of equations which attempt to model each of the components described above. An illustration of the type of differences observed in the two predominant metals in some meteorites, i.e., iron and nickel, as a function of temperature for a given strain rate (Charpy) is shown in Fig. 8.

Consistent with the increased σ_p term for polycrystalline unalloyed Fe compared to Ni, decreasing temperature or increasing strain rate is observed to have a drastic effect on the constitutive response of Fe. Both high-rate deformation and testing at cryogenic temperatures are observed to increase drastically the yield strength of Fe. The constitutive response, in particular the yield strength when in an annealed state, of Ni is conversely seen to be virtually invariant as a function of strain rate and temperature over the range measured. Alloys of Fe-Ni, such as in meteorites, depending on whether they are Fe- or Ni-rich, may exhibit a stress-strain response as a function of temperature and strain rate which lies somewhere between the extremes exhibited by pure Ni or pure Fe or some combination of both. The addition of nonmetallic inclusions or a major volume fraction of a stony component will significantly alter the stress-strain response and our ability to model and predict their response. This figure illustrates the need for characterizing a range of NEO materials to

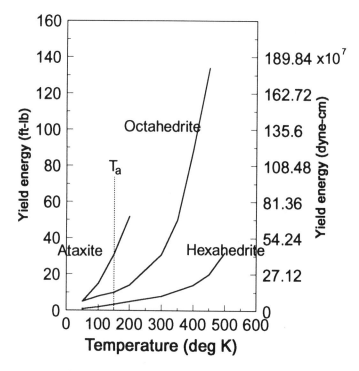

Figure 10.   The yield energies under Charpy impact are shown as a function of temperature for hexahedrites, octahedrites, and ataxites.

generate a data base from which accurate deformation models and material modeling algorithms can be developed. Accurate models of this type are crucial to the development of physically based constitutive models to be used in predictive hydrocodes intended for modeling NEO events.

IV. FeNi MATERIAL PROPERTIES

A. Iron (FeNi) Asteroid/Meteorite Analogs

Predominantly iron meteorites (M type analogs) are sometimes relatively easy to detect by radar due to their higher reflectivity. Because of their structure and density, they are the most lethal for a given size. Observations indicate that there are currently at least two M type NEAs (1986 DA and 1986 EB) which may have a mineralogy identifiable with iron meteorites (Ostro et al. 1991; Tedesco and Gradie 1987). Additional M type asteroids are thought to exist in the main belt. These asteroids, either partially or completely composed of iron-nickel, are generally thought to be fragments of large cores and/or localized metal reservoirs of differentiated parent asteroids. Such hard metallic pockets will initially be protected from fragmentation by impact absorbing silicate layers. Their survival, after the outer layers were (presumably) stripped away by collision, would demonstrate an ability to withstand a high velocity impact with other asteroid bodies. Their survivability and lethality is based upon three physical properties: (1) relatively high mechanical strength, especially when Ni enriched (taenite phase); (2) high density (7.8 g cm^{-3}); (3) hardness which is an impediment to external penetration, coupling, and possible mitigation efforts.

B. Mechanical Properties of FeNi NEAs

The presumed mechanical strength of M type asteroids can be an important asset or a formidable problem for NEO orbit-modifying interactions. One of the assets of such robust materials is their ability to absorb the large amount of energy (impulsive forces) necessary to undergo the required change in momentum to adjust their orbit. Another advantage is the relatively simple modeling afforded by an ideal iron-nickel surface which is not complicated by several (modeling) variables associated with a variety of mineralogical components as well as the presence of extensive regolith and/or breccia. Disadvantages in dealing with M type asteroids include resistance to penetrator devices, presence of a network of inclusions that may introduce faults which can weaken structural properties, and large-scale discontinuities such as stony iron mixtures or large mineral inclusions (Johnson et al. 1993). Also, little data exist for larger-scale hypersonic impact and explosive effects in meteoritic metals, although we are currently carrying out impact tests which include release and reloading over a stress range of 2 to 20 GPa, triaxial quasi-static tests, ultrasonic attenuation, sound velocity measurement, and variable strain rate experiments on metallic meteorites (Furnish et al. 1993). Material properties and design methods to optimize the effectiveness of penetrator devices for M type asteroids or planetary surfaces are discussed elsewhere by Remo (1993). The high density of M type asteroids indicates a large inertial mass to size ratio requiring a precisely targeted payload delivery system. Some mechanical properties of FeNi meteorites are listed in Table III. The data are based on a limited number of select samples which do not take into account inhomogeneities and other variations in the material properties. In addition, the actual asteroid analogs to these meteorite properties may be composed of internal layers or regions of silicate material mixed with the iron nickel phases. Such a structure will present gross mechanical discontinuities to a penetrating projectile or high explosive. Ideally, the velocity vector of the asteroid should be changed with a minimum amount of work done on the NEO structure.

An example of the dependence of the gross mechanical property on microstructure, or in particular the weight % Ni, is the ductile-brittle transition temperature as shown in Fig. 9. The yield strength for an impact depends on the Ni content. In Fig. 10, the impact yield energy for three iron meteorite classes, octahedrite, hexahedrite, and ataxite are plotted as a function of

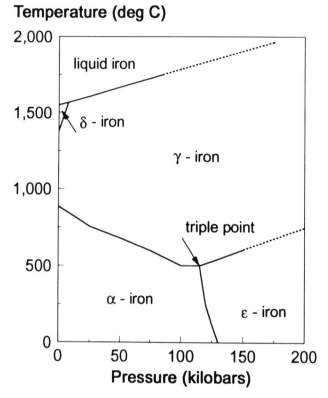

Figure 11.   This phase diagram of pure iron shows the forms of stable iron as a
function of pressure and temperature. The α and δ forms are body centered cubic,
the γ is face centered cubic, and the ϵ is the hexagonal close packed structure.

temperature. Other factors such as effects of micro-inclusions as a function
of temperature, strain rate, and shock impact on the ductile-brittle transition
are also important and are currently being studied. As the FeNi system is well
understood and the meteorites are homogeneous, phase diagrams (Figs. 11 and
12) are generally applicable and facilitate an understanding of the evolution
of the microstructure.

C. Thermal Properties of FeNi Meteorites

Thermal measurements on FeNi meteorites in the literature have generally
been carried out at room temperature and at atmospheric pressure. This is
especially significant for melting and boiling points as well as for the determination of latent heat of fusion and vaporization. An advantage of FeNi
is that because they do not absorb water vapor, electrical and thermal conductivity measurements are probably reliable. Much can be learned about
thermal properties and the history of FeNi meteorites from a study of both the

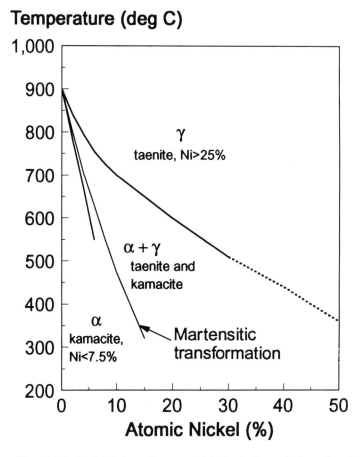

Figure 12.   This is the FeNi phase diagram which also indicates the boundary for the martensitic under non-equilibrium conditions.

metallic phases and mineral inclusions. Thermodynamic properties of FeNi meteorites are given in Table IV. Modification of NEO orbits should couple minimal energy into thermal modes of melting and vaporization. However, alteration of the asteroid trajectory is likely to require large amounts of energy imparted in a very short time. Examples of such heat generating methods include high-speed impact by penetrators with or without high energy or nuclear devices, external, surface, and buried nuclear explosives with associated X-ray, gamma-ray, and neutron photo-ablation and vaporization. Laser and focused solar radiation will also generate large amounts of heat.

Little effort has been directed towards understanding meteorite thermal properties in terms of the technology of materials science, or mining engineering. Characterization of impulsive loading from hypervelocity impact and associated shock, radiative scattering, and thermal cycling effects over a

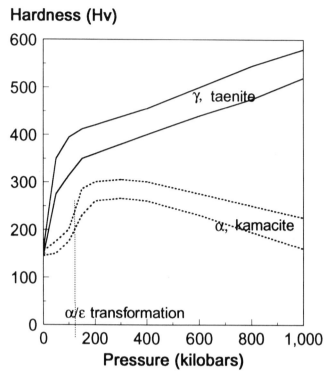

Hardness (Hv)

Figure 13.   The Vickers hardness is plotted against the pressure, in kilobars, for artificially shocked samples of taenite and kamacite from the Odessa meteorite. Taenite is initially steeply shocked by the shock pressure and then continues, at a lower rate, to increase its hardness at higher shock intensities. On the other hand, the maximum hardness occurs for taenite at about 200 Kb and then slowly decreases with increasing shock pressure. Relaxation heat from the higher shock pressures anneals, recrystallizes and transforms the body centered cubic kamacite to the α_2 structure as opposed to the face centered cubic taenite which can continue to acquire hardness.

broad temperature and pressure range must be added to the data base. Such data can assist in validating computer codes that may be used to simulate response of complex materials to arbitrary stress loading and to determine strain levels required for a material to fail either locally or catastrophically.

D. Shock Processes in FeNi Meteorites

Most of the FeNi have been subjected to shock pressures greater than 130 kbar. Shock induced effects in FeNi meteorites can be divided into four categories (Buchwald 1975):

1. Shock induced plastic deformation: 10 to 130 kbar; kamacite undergoes simple twinning, plastic deformation and micro-cracking. At low rates of

deformation the metallic matrix is cold worked and included minerals are distorted and brecciated. Vickers hardness is about 325. Taenite hardness (peak at 475 to 560) is generally higher than co-existing kamacite (see Fig. 13). Troilite inclusions undergo simple twinning.

2. Shock induced solid state transformation; >123 kbar; kamacite undergoes a diffusionless transformation to hexagonal close packed ϵ iron of high density with a microhardness of 300 to 350.

3. Shock induced melting; reflection and attenuation of the shock wave at grain and phase boundaries and external surfaces induce localized heating and selective melting of silicates, troilite, cohenite, schreibersite, daubreelite, chromite, etc. FeNi phases rarely exhibit bulk heating as a result of shock interactions.

4. Shock induced thermal annealing and orbital entry heating; thermal heating above 750°C converts kamacite to taenite and upon cooling into a distorted (martensitic) α_2 phase under nonequilibrium conditions (see Fig. 12). α_2 is in heat affected rim zones. In taenite, heating above 700°C redistributes carbon and upon rapid cooling produces bainitic-martensitic structures.

E. Suggested Research for the Material Properties of Metallic (M) Asteroids)

To address issues associated with NEO orbit modification, a systematic and detailed study should be carried out on FeNi and related stony iron meteorite analogs of M type asteroids. To generate critical new data for predictive models on impulsive interactions with asteroids and comets, a detailed understanding of mechanical, thermal, electric, and magnetic properties in a space environment is required. The properties in particular that need to be understood are:

1. Low and high strain rate impact properties;
2. Shock wave propagation and microstructural effects;
3. Effects of inclusions, dislocations, and grain boundaries on plastic structure and fracture characteristics;
4. Radiation (X-ray, gamma-ray and neutron) absorption cross sections, scattering and related thermal and mechanical effects;
5. Electrical, magnetic, mechanical and thermal properties in a simulated space environment;
6. Pulsed and CW laser opacities and associated induced plasmas over a variety of wavelength;

Accurate values are also required of thermodynamic properties that will assist in modeling NEO interactions with X-rays, gamma-rays and neutron flux absorption cross sections as well as with laser induced plasma effects. In addition, many iron meteorite properties listed in Tables III and IV should be re-measured with emphasis on stony iron mixtures as well as on the effects of inclusions and imperfections within the FeNi matrix.

V. STONY ASTEROIDS

A. Stony Asteroids

Physical properties of metallic meteorites are more homogeneous than stony meteorites. The origin of FeNi meteorites is regarded to have occurred when liquid metal solidified yielding steel-like alloys kamacite and taenite which have been extensively studied by metallurgists. While kamacite and taenite differ from steel, they still retain many of the same properties and resemble a manufactured material in uniformity. Stony asteroids on the other hand do not possess such a uniform structure which easily allows their characterization in terms of an analogous industrial material. Stony meteorites are mineral assemblages which generally lack a homogeneous structure; inclusions, and grain boundaries are the rule. For these reasons it will be difficult to characterize their response to impact. Among the many minerals abundantly found in ordinary chondrites are small stony crystal spherules or droplets called chondrules. Enstatite chondrites have high abundances of the mineral enstatite ($MgSiO_3$). Carbonaceous chondrites contain the highest proportion of volatile elements while ordinary chondrites and enstatite chondrites are depleted in volatiles.

Stony asteroids can be divided into three groups with significantly different physical and chemical properties affecting their thermal and mechanical properties:

1. Primitive (C, D, and P types) dominate the outer part of the asteroid belt;
2. Metamorphic (F, G, B, and T) peak in the central region of the asteroid belt;
3. Igneous (S and E) are common in the inner part of the belt. The metallic (M) asteroids may also be considered in this group, although its materials properties are very different from those of the S and E;

Table II categorizes asteroid types based on their visual infrared spectra and relationships to known meteorite types (analogs).

Tables V and VI outline mechanical and thermal properties of stony meteorites. To a much greater degree than for FeNi meteorites, values for the stony meteorites present a limited sampling of ideal specimens and do not take into account inhomogeneities, structural, or chemical variations within a given meteorite class. Experiments to determine accurately the chemical and physical properties in a space environment of stony meteorites should be undertaken. Additional research on stony meteorites should follow along the lines suggested for the metallic meteorites in terms of understanding strain, compression, and shock failures as they relate to inclusions and inhomogeneities.

B. Shock Processes in Stony Meteorites

Shock processes are important in the evolution of meteorites because this mechanism is regarded to have disrupted the asteroids. Shock waves will also

TABLE VII

Petrological Shock Indicators in Meteorites[a]

Silicate Systems		Minimum Shock Pressures (Kb)
Pyroxene	Majorite	135
Plagioclase	Maskelynite	300
Plagioclase, pyroxene	Glass	450
Olivine	Ringwoodite	450
FeNi Systems		
Neumann lamellae in kamacite		10
Acicular kamacite		130
Recrystallization of kamacite		750
Graphite	Cubic diamond	130
Graphite	Hexagonal diamond	700

[a] From Wasson 1974.

play a major role in NEO orbital interactions because large amounts of energy are rapidly delivered (high specific momentum). It is possible that over 90% of the ordinary chondrites contain shocked olivine crystals. An extensive review of shock processes in terrestrial, lunar, and meteoritic materials is given by Stoffler (1972). The effect of shock waves over a range of pressures is outlined below for the (chondritic) meteoritic mineral olivine:

Pressure

0–50 kbar: Fracture occurs in directions different from those produced at low strain rates.

150–225 kbar: Fracturing continues with plastic deformation and slight disorientation of adjacent crystal domains.

1225–350 kbar: Fracturing continues with the initiation of mosaicism.

350–450 kbar: Fracturing and mosaic extinction continue, accompanied by microbrecciation or granulation along margins of olivine grains.

450–590 kbar: Olivine grains begin to recrystallize into aggregates of smaller, strain-free grains, polycrystallinity.

Similar changes occur in low Ca pyroxenes but at higher pressures. However, it is difficult to generalize the above pressure response to other minerals.

VI. MAGNETIC PROPERTIES OF METEORITE PHASES

Magnetism has been found in most if not all classes of meteorites. Magnetic properties of asteroids are important for studying the origin and evolution of the solar system because certain types of remanent magnetism may be an indicator of an early solar T-Tauri stage. The discovery of what appears to be an anomalous (very high) remanent magnetic field associated with 951 Gaspra (Kivelson 1993, personal communication) highlights the importance of using

TABLE VIIIa

Magnetic Phases in Meteorites

Material	Composition	Magnetism Type	$T_C/T_N \,^\circ C$	M_S emu g^{-1}
Kamacite	FeNi	Ferromagnetic	~750	90–140[a]
Taenite	FeNi	Paramagnetic or		
		Ferromagnetic	0–600[a]	180–220[a]
Pure iron	Fe	Ferromagnetic	770	~218
Pure nickel	Ni	Ferromagnetic	358	~54.4
Magnetite	Fe_3O_4	Ferromagnetic	580	92
Trevorite	$NiFe_2O_4$	Ferromagnetic	585	51
Pyrrhotite	$Fe_{0.87-1.00}S$	Ferromagnetic	<320	<19.5–62
Troilite	FeS	Antiferromagnetic	320	
Schreibersite	$(FeNi)_3P$	Ferromagnetic	<420	
Cohenite	$(FeNi)_3C$	Ferromagnetic	<215	
Fayalite	Fe_2SiO_4	Antiferromagnetic	−147	
Ferrosilite	$FeSiO_3$	Antiferromagnetic	−233	

[a] Magnetic properties depend on the Ni content.

magnetic properties as a possible diagnostic of NEO composition. Magnetic properties of meteorites are determined from the magnetic properties of individual phases and their interactions within the meteorite which is primarily the vector sum of magnetization induced by an external field and the remanent magnetization which is independent of the field. Remanent magnetization includes thermo-, isothermal-, viscous-, and chemical-remanent magnetization. For meteorites and asteroids impact (shock heating) and solar wind induced effects must also be considered. There are many problems in the interpretation of the thermomagnetic and other remanent curves for (iron bearing) phases of meteorites because Curie temperatures can be ambiguous. Additional ambiguity can arise if diffusion takes place across and within the grain boundaries during the heating cycle. Field orientation, electromagnetic boundary conditions, chemical composition and inclusions also affect remanent magnetism. Data, primarily from Wasson (1974), on the magnetic properties of individual phases and the remanent magnetism are listed in Tables VIIIa and b, for Curie, T_c and Neel T_N, temperatures (susceptibility is maximum and below which antiferromagnet spins are ordered in an antiparallel arrangement with net zero moment), and the maximum induced remanence, M_S.

TABLE VIIIb
Meteorite Remanent Magnetism

Meteorite Type	Magnetic Remanence (emu g^{-1})
FeNi, pallasites and siderites	0.01 to 0.3
Metal bearing chondrites	10^{-4} to 0.06
Achondrites and chondrites with <0.5% metal	5×10^{-5} to 2×10^{-3}

VII. EFFECTS OF NUCLEAR RADIATION ON ASTEROID MATERIAL

Stand-off nuclear explosives (SNE) may mitigate the NEO threat by transferring momentum via induced blow-off directly by debris kinetic energy, by ablation and related effects from X-rays, gamma-rays and high-energy neutrons. A simple model (Hammerling and Remo 1993) can estimate the impulse from a SNE and the velocity change imparted by the various nuclear products to the different NEO material compositions. A perceived advantage of SNE is that it appears to be the mechanism least sensitive to current uncertainties in NEO composition, shape and structure. This hypothesis must be tested by detailed computer modeling. Also, as the SNE radiates approximately half the surface, it is less likely to impart a substantial additional (rotation) angular momentum. Results from this simple stand-off momentum transfer model (Table IX) illustrate the dependence on asteroid material properties (atomic composition) on the velocity change, ΔV. Detailed computations will require the application of multi-group radiation, neutron transport, and a detailed equation of state coupled to a full hydrodynamic code. The data in Table IX indicates the velocity change, ΔV, imparted to the NEO by the SNE nuclear radiation components. The velocity change, ΔV, will depend on the absorption cross section, μ, and $\mu^{-1/2}$, which depends, in turn, on the opacity of the NEO target (atomic) material properties. Table IX, for example, shows composites generated from atomic abundances for three types of meteorite.

TABLE IX
Approximate Absorption Coefficients

cm^2 g$-$1	NVF Chondrites	VF Chondrites	FeNi
μ_{X-ray}	35	70	180
$\mu_{neutron}$	0.05	0.05	0.028
$\mu_{gamma-ray}$	0.05	0.05	0.05

The total SNE induced velocity change will be approximately the sum of the induced velocity change from the momentum transfer due to each radiation component of absorption and the debris

$$\Delta V \approx \Delta V_{X-ray} + \Delta V_{gamma-ray} + \Delta V_{neutron} + \Delta V_{debris}. \qquad (2)$$

Also, for example, a SNE may have a yield partitioned into the fractions:

$$\eta_{X-ray} \approx 0.60, \eta_{neutron} \approx 0.20, \eta_{debris} \approx 0.15, \text{ and } \eta_{rest} \approx 0.05. \qquad (3)$$

Each component in the above example contributes to the total ΔV, angular momentum, thermal heating and vaporization, and structural alteration, depending on the design of the SNE. Specific nuclear devices can be utilized or developed that will optimize momentum coupling to particular NEO compositions by yielding a partition of nuclear radiation that will have a high absorption cross section.

Based on the relative atomic abundance (composition) for each of three meteorite types selected, Table IX describes the corresponding composite absorption cross sections. It is clear, for instance that the metal-rich FeNi will have a much greater X-ray absorption cross section than the chondritic (stony) meteorites. Also, although there are differences between meteorite types in the neutron absorption cross section, these are not as extreme as in the case of the X-ray absorptions. When the absorption cross sections for meteorites/asteroids differ significantly, the NE will therefore impart substantially different velocity changes, depending on the particular NEO interaction composition and the design of the NE. X-rays are very abundant and do not deeply penetrate, transferring their momentum to the NEO over a short distance while traveling near the speed of light. Their interaction time is relatively short. However, neutrons penetrate deeper and travel at only a fraction of the speed of light. Their interaction time is relatively long.

It is also important to understand that the NE radiation may change the nature of the ablating surface. For instance, volatile components may be depleted after the first few interactions, thereby changing the atomic composition of the outer layer and consequently changing the absorption coefficient which determines the ΔV. To provide an example, we have chosen to compute cross sections for ordinary chondrites with both non volatile fractionated and volatile fractionated compositions. For X-ray absorption, there is a marked increase in the absorption cross section for the volatile fractionated chondrites; because the volatiles are vaporized away, the NEO surface may become metal and Si enriched. Such factors must be very carefully considered if one uses NE to interact with a NEO. The composite ΔV is very sensitive to both the NEO materials and the NE design. In particular, the X-rays absorption is very sensitive to details of the atomic compositions and radiation spectrum and should be averaged over a Planck distribution and detailed computations must be carried out to determine reliable opacities.

As described in the simple example above, the nuclear radiation absorption cross section properties of meteorites, which are regarded to be material analogs to asteroids, suggests that there is a considerable amount of work that must be carried out to have a reliable understanding of the interaction of NEOs with NE. This is especially true in the case comet-like (as opposed to FeNi NEOs) which are presumed to have a very large proportion of volatile components. Before planning nuclear interactions with a NEO, a reliable determination of its chemical composition should be carried out. A very important research goal for NEO hazard mitigation is the generation of realistic NEO nuclear interaction models over a broad range of meteorite/asteroid types (chemical compositions), sizes, and shapes for nuclear devices of varying yield, energy partition, and detonation distance. This implies that to carry out effectively a NEO nuclear interaction mission, there must be a substantial NEO reconnaissance program.

VIII. NEO MATERIALS INTERACTION

A. Properties Data Needed for NEO Orbit Interaction

The required material properties will depend on the method of interaction chosen for NEO orbital modification. Because one must also anticipate a broad range of NEO properties, there are many materials characteristics which must be well understood. If we divide the interactions into three major methods of energy delivery (nuclear, high explosive, and kinetic energy), and four other less technologically developed methods (thrusters, mass drivers, lasers, and solar sails), the required material properties for each type of interaction can be outlined.

Table X is an attempt to provide just such an overview of the materials properties required for several different types of NEO orbit modifying interactions. This tabulation is not evaluating, comparing, or recommending any of the energy delivery methods. The purpose is primarily to highlight the type of materials properties data required to carry out the orbit modifying interdiction. For each of these energy delivery methods, comment is offered as to whether the method will be effective for a remote (R) or terminal (T)

TABLE X
Materials Data Necessary for NEO Orbital Interaction

Energy Delivery	Interdiction	Required Materials Data
Nuclear (4×10^6 MJ/kg at ~ 0.1c-c)		
Below NEO surface	R & T	D, EOS, G and M
At NEO surface	R & T	A, D, EOS, G, M and Th
Above NEO surface (stand-off NE)	R & T?	A, D and Th
High explosive (6 MJ/kg at 6–9 km s^{-1}, detonation velocity)		
Below surface	R & T	D, EOS, G and M
Kinetic energy (50 MJ/kg) at 10 km s^{-1} (relative velocity with target)		
Deflection	R & T	D, EOS?, G and M
Pulverization (small, friable NEO)	R & (T?)	D, EOS, G and M
Other		
Attached thruster	R	CR, D, M and VFR
Mass driver	R	CR, D, G, M and VFR
Laser	R	A, D and Th
Solar sail	R	D and M

  A = atomic composition; X-ray, gamma-ray and neutron absorption cross sections and/or optical opacity.
  C = speed of light.
CR = cutting rate.
  D = dynamic; center of mass, inertial moment and stress tensor.
EOS = equation of state.
  G = geology and mineral composition.
  M = mechanical structure; surface and interior.
Th = thermodynamic properties; conductivity, melting and vaporization energy, etc.
VFR = vibrational fracture rate.
  R = remote; momentum transferred over a long period of time, long time scales to react.
  T = terminal; momentum rapidly transferred at high energy density, short reaction time scale.

interdiction. Also, these interdiction methods can provide a diagnostic of the NEO, depending on the sensors available.

The energy methods listed in Table X do not address several issues, among which are those associated with some particular characteristics of NEOs which may provide unique opportunities for deflection or breakup. For example, if a comet-like NEO is composed of loosely agglomerated components of ice, dust, and organics, then it may be possible to impart a large angular momentum either internally by inducing or enhancing asymmetric jetting or by externally introducing a momentum vector such as by impact or radiation. If the increased angular momentum is sufficiently large, it might be able to disrupt loosely bound segments which may be more easily mitigated or even avoided. Bearing in mind the difficult nature of characterizing NEOs, about which so little is known, the next section will attempt to categorize NEOs into four groups in terms of their material properties.

B. Recommended NEO Materials Categorization

Mechanical, thermal, magnetic and atomic properties of NEO materials can divided into three meteorite/asteroid related groups and one comet related group:

NEO material 0: NEOs identified as similar to extinct comet nuclei are expected to have a very low density 0.2 to 1.5 g cm^{-3}, poor mechanical strength, variable refractory organic (C, H, N, and O), ice, some Fe, Mg, and silicate compositions in, for example, ice and organic refractory mantles, and silicate cores. There can be many variations especially for dormant or extinct comets having experienced selective molecular and atomic losses thereby affecting overall nuclear radiation absorption. Thermal properties strongly depend on temperature; if composed of amorphous ice, the conductivity can be extremely low. Structural integrity widely varies depending on the structure of individual comets. The magnetic field is likely to be limited to (minor) metallic components as listed in Table VIIIa.

NEO material 1: NEOs identified as composed of materials similar to the structurally weakest (friable) meteorites, and resemble (primitive) asteroid classes; B, C, D, F, G, P, and T which contain Si, O, Fe, Mg and S, have low density and conductivity, high melting point, and low albedo with substantially higher X-ray absorption than NEO material 0. Volatiles can be driven off by radiation, thereby increasing X-ray absorption. Structural integrity will depend on porosity and faults and will therefore tend to have low crushing strengths. Possible magnetic field depends on the meteorite type listed in Table VIIIb, which is based on phases in Table VIIIa.

NEO material 2: NEOs identified as composed of metamorphic and igneous materials corresponding to asteroid classes E, R, S, and V with similar thermal properties and richer in Si, Fe, and Mg than NEO material 1. This group is structurally stronger than the NEO material 1 group with less porosity and a higher crushing strength with higher density, conductivity, melting point, and albedo. Volatiles can be driven off, resulting in an additional increase in X-ray absorption. Possible magnetic field is similar to NEO material 1, except that the NEO materials 2 are igneous materials and this is likely to effect the remanent magnetism.

NEO material 3: NEOs identified as resembling metallic (FeNi) meteorites and corresponding to M type asteroids. This group, resembling stainless steel in many ways, has very high density and virtually no porosity (except for nodules) thereby possessing the strongest mechanical structure, and most uniform thermal characteristics. Both electrical and thermal conductivity are extremely high with the capability of possessing a very large permanent (remanent) ferromagnetic field (see Tables VIIIa and VIIIb). When Ni rich, these are ductile at low (asteroid) temperatures while the Ni poor (<7.5 wt%) are brittle at low temperatures. No volatile components are present except for S- or C-rich nodules which are a minor component and tend to have very high X-ray absorption and relatively low neutron absorption. Fortunately, the relative abundance of this strong material is thought to be low.

The above classification is not ideal, but is a simple starting point. If a NEO is classified into one of the above categories, it does not necessarily mean that it is homogeneous, because it might represent a differentiated fragment or fused collision fragments from two or more different type of asteroid-like components, appearing as a macro-heterogeneous object falling into two or more above groups. In applying the identification of NEO material, it should be borne in mind that the inferred meteorite/asteroid composition types based on the visual-infrared and radar reflectance spectra of asteroid surfaces and interpreted through asteroid modeling cannot be substituted for data obtained from asteroid sampling missions.

Binzel and Xu (1993) have recently presented evidence that the spectral properties of the asteroid Vega, Fig. 3, is a source of basaltic achondrite (eucrite, diogenite, and howardite) meteorites. On of the particularly significant aspects about this discovery is that 20 small (diameter <10 km) mainbelt asteroids were measured and 12 had orbits similar to Vesta. It is likely that these orbitally associated fragments are materially related which further implies that the surface spectra of Vega is also indicative of major (interior) portions of the asteroid Vega's overall composition. Even more recently a discovery of a mainbelt asteroid, (3628) Boznemcova, shows a spectral match with ordinary chondritic material (Binzel et al. 1993). The substantial evidence for the association of basaltic achondrites with Vega and the association of any ordinary chondrite with (3682) Boznemcova is very important to the meteorite/asteroid association in general (Gaffey 1993), and to the goal of using meteorites as asteroid analogs in particular (Remo 1993).

The NEO materials 1, 2, and 3 groups are reminiscent of the original C (carbonaceous), S (siliceous), and M (metal) spectral classes used to account for the original asteroid types associated with carbonaceous chondrites, ordinary chondrites, and irons (Chapman et al. 1975) and provided an excellent start for the classification of asteroids in general. However, much spectroscopic and radar analysis has been done in the past twenty years and revised taxonomies of a more detailed nature are warranted in terms of explaining the evolution of asteroid groups. The NEO materials categorization presented in this chapter is concerned with establishing material properties of NEOs in terms of response to kinetic, HE, and NE interactions. Indeed, the NEO categorization would not have been possible without the previous extensive spectroscopic work.

The seriously planned and safe NEO asteroid mitigation mission must take into account interactions with objects from each of the four NEO materials groups described in Table X. Table XI, below, generalizes the characteristics of nuclear interaction, volatile depletion, and structural integrity in a very approximate manner. This tabulation is merely a first step, basically only serving as an outline for further work. Detailed modeling based on observational and experimental data will extend both the range and detail of these listed characteristics. A third parameter can be added to represent NEO mass or size, which will then establish the approximate magnitude of the energy requirement for orbital interactions and velocity changes.

The NEO interactions in Table XI describes trends in material properties based on some major nuclear, mechanical, and thermal interaction characteristics to roughly outline changes going from NEO materials 0 to 3. Laser beam power propagation and material interactions can be extremely complex, and cannot be understood within such a simple framework. Beam degradation is a major problem for coherent energy delivery through the Earth's atmosphere.

The object of NEO materials research is to fill in Tables X and XI in detail as well as to extend Table XI into a third dimension to account for NEO mass or size to understand the type of interaction that can be effectively carried out to establish the desired NEO orbital changes. However, data on the response of meteorite material in conditions such as temperature, (space) vacuum or hydrogen enrichment must be generated to simulate the interactions outlined in Table X. Until such data is available, much of the nuclear, kinetic, and laser interaction modeling will be unreliable.

IX. CONCLUSIONS

This chapter has a bias toward FeNi materials which are not regarded to constitute the bulk of recovered meteorites, and are generally regarded to be even much less representative of the NEO population. However, because FeNi meteorites are more homogeneous than stony meteorites and the FeNi mechanical and thermal properties closely resemble stainless steel, a vast array of testing methodologies and reference standards exist to characterize responses over an extended range of temperatures and pressures. Therefore, the experimental analysis of FeNi meteorites can serve as a guide for testing to be carried out on the less homogeneous stony meteorites which is where the future work should be concentrated. However, the natures of comets appear to be so distinct in their structural, mechanical, and thermal properties, (i.e.,

TABLE XI
Simplified NEO Interactions Matrix Trends

	NEO Materials 0	NEO Materials 1	NEO Materials 2	NEO Materials 3
Nuclear				
Neutron absorption	← moderate →		← weaker per unit mass →	
X-ray absorption	← weak →		← very strong →	
Nuclear interaction	← stand-off, surf., & subsurf. →			← stand-off →
Volatile depletion	← significant →		← moderate →	← marginal →
Structural integrity	← very low →	← low →	← moderate →	← strong →
Laser absorption	wavelength must be carefully selected to optimize for surface absorption. Plasma cloud effects can be minimized with pulsing, but will require extensive experimental research and computational modeling.			

ice, volatiles, and dust) from FeNi meteorites that many of the testing methods used may not be applicable. Perhaps innovative techniques developed from the analysis of stony meteorites can find applications in describing the material properties of comets.

In regard to meteorites as representative of asteroid material, it may simply be assumed that based on inferences from limited evidence, meteorites are likely to be fragments of asteroids. Other interpretations may be possible, but the association between asteroids and meteorites, although tenuous for some mineral assemblages such as the ordinary chondrites, is at this time a reasonable working hypothesis. With regard to comets, it is unfortunate that its material cannot be tested in a laboratory; there do not appear to be any fragments of this volatile material available. This underscores the need for a comet probe/penetrator mission.

Some of the mechanical, thermal, magnetic, and atomic properties have been outlined and underscores the fact that, in terms of the knowledge of material properties necessary for modifying an NEO orbit, very little is known. This is true whether the NEO interaction involves nuclear or conventional explosives, kinetic impact, laser ablation, solar energy or any other plausible long- or short-term method. Therefore, before substantial serious thought or planning can be given to interdiction, detailed and systematic meteorite experimental work coupled with a commitment to asteroid reconnaissance and sampling must be carried out in order to appropriately characterize responses to NEO materials at the high-energy (interaction) levels and low ambient (space) temperatures that are likely to be associated with orbit modifications. This research should be coordinated with enhanced acquisition, tracking, and homing technologies with an emphasis on innovative telescope-based, CCD-chip-based optical detection systems interfaced with computer image (parallel) processing. Without enhanced data, based on NEO observation in conjunction with both macro- and microstructural analysis of meteorites, NEO interception or orbit modification cannot be effectively planned and may have dangerous consequences if attempted prematurely. This is especially true in the case of the utilization of nuclear devices.

Acknowledgments. The material in Secs. III.D and E was provided by M. Furnish and G. T. Gray, respectively. I wish to thank C. Chapman for helpful suggestions. I also wish to thank R. Warasila for critical comments. Acknowledgement is made to J. D. G. Rather for encouragement. This work was supported by Quantametrics internal research funding.

REFERENCES

Metals Handbook, 9th ed. 1985. (Metals Park, Ohio: ASTM).

Barker, L. M., and Hollenbach, R. E. 1972. Laser interferometer for measuring high velocities of any reflecting surface. *J. Appl. Phys.* 43:4669–4675.

Belton, M. J. S., Veverka, J., Thomas, P., Helfenstein, P., Simonelli, D., Chapman, C., Davies, M. E., Greely, R., Greenberg, R., Head, J., Murchie, S., Klaasen, K., Johnson, T. V., McEwen, A., Morrison, D., Neukum, G., Fanale, F., Anger, C., Carr, M., and Pilcher, C. 1992. Galileo Encounter with 951 Gaspra: First pictures of an asteroid. *Science* 257:1647–1652.

Binzel, R. P., and Xu, S. 1993. Chips off of asteroid 4 Vesta: Evidence for the parent body of basaltic achondrite meteorites. *Science* 260:186–191.

Binzel, R. P., Barruci, M. A., and Fulchignoni, M. 1991. The origin of asteroids. *Sci. Amer.* 265:88–94.

Binzel, R. P., Xu, S., Shelte, J. B., and Bowell, E. 1992. Origins for the near Earth asteroids. *Science* 257:779–782.

Binzel, R. P., Xu, S., Schelte, J., Bus, J., Skrutskie, M. F., Meyer, M. R., Knezek, P., and Barker, E. S. 1993. Discovery of a main-belt asteroid resembling ordinary chondrite meteorites. *Science* 262:1541–1543.

Buchwald, V. 1975. *Handbook of Iron Meteorites* (Berkely: Univ. of California Press).

Chhabildas, L. C. 1987. Survey of diagnostic tools used in hypervelocity impact studies. *Intl. J. Impact Eng.* 5:205–220.

Chhabildas, L. C., Barker, L. M., Asay, J. R., and Trucano, T. G. 1990. Relationship of fragment size to normalized spall strength for materials. *Intl. J. Impact Eng.* 10:107–124.

Chapman, C. R. 1976. Asteroids as parent bodies: The astronomical perspective. *Geochim. Cosmochim. Acta* 40:701–719.

Chapman, C. R., and Gaffey, M. J. 1979. Reflectance spectra for 277 asteroids. In *Asteroids*, ed. T. Gehrels (Tucson: Univ. of Arizona Press), pp. 655–687.

Chapman, C. R., Morrison, D., and Zellner, B. 1975. Surface properties of asteroids: A synthesis of polarimetry and spectrophotometry. *Icarus* 25:104–130.

Davison, L., and Graham, R. A. 1979. Shock compression in solids. *Physics Reports* 55:255–379.

Fanale, F., Clark, B. E., and Bell, J. F. 1992. A spectral analysis of ordinary chondrites, S-type asteroids, and their component minerals: Genetic implications. *J. Geophys. Res.* 97:20863–20873.

Follansbee, P. S. 1985. The Hopkinson Bar. In *Mechanical Testing*, vol. 8, *Metals Handbook*, 9th ed. (Metals Park, Ohio: ASTM), pp. 198–203.

Furnish, M. D. 1993. Recent advances in methods for measuring the dynamic response of geological materials to 100 GPa. *Intl. J. Impact Eng.*, in press.

Furnish, M. D., and Chhabildas, L. C. 1992. Dynamic material properties of refractory materials. In *High Strain Rate Behavior of Refractory Materials and Alloys*, eds. R. Asfahani, E. Chen and A. Crowson (Warrendale, Penn.: The Minerals, Metals, and Materials Society), pp. 229–240.

Furnish, M. D., Chhabildas, L. C., Steinberg, D. J., and Gray, G. T., III. 1992. Dynamic behavior of fully dense molybdenum. In *Shock Compression of Condensed Matter 1991*, eds. S. C. Schmidt, R. B. Dick, J. W. Forbes and D. G. Tasker (New York: Elsevier Sci. Pub. Co.), pp. 419–422.

Furnish, M., Gray, G. T., and Remo, J. 1993. Dynamical behavior of octahedrite from the Henbury meteorite. AIRAPT/Am. Phys. Soc. Conf., June 28–July 2, Colorado Springs, Colo.

Gaffey, M. J. 1993. Forging an asteroid-meteorite link. *Science* 260:167–168.

Gaffey, M. J., Bell, J. F., and Cruikshank, D. P. 1989. Reflectance spectroscopy and asteroid surface mineralogy. In *Asteroids II*, eds. R. P. Binzel, T. Gehrels and M. S. Mathews (Tucson: Univ. of Arizona Press), pp. 98–127.

Goldstein, J., and Ogilivie, R. E. 1965. The growth of the Widmanstatten pattern in metallic meteorites. *Geochim. Cosmochim. Acta* 29:893–920.

Hammerling, P., and Remo, J. L. 1992. NEO interaction with X-ray and neutron radiation. In *Proceedings of the NEO Interception Workshop*, eds. G. H. Canavan, J. C. Solem and J. D. G. Rather (Los Alamos: Los Alamos National Lab), p. 186–193.

Hapke, B. W., Nelson, R. M., and Smythe, W. D. 1993. The opposition effect of the moon: The contribution of coherent backscatter. *Science* 260:509–511.

Harris, D. 1961. Photometry and colorimetry of planets and satellites. In *Planets and Satellites*, eds. G. Kuiper and B. Middlehurst (Chicago: Univ. of Chicago Press), pp. 272–342.

Johnson, A. A., and Remo, J. L. 1974. A new interpretation of the mechanical properties of the Gibeon meteorite. *J. Geophys. Res.* 79:1142–1146.

Johnson, A. A., Remo, J. L., and Davis, R. B. 1979. The low temperature impact properties of the meteorite Hoba. *J. Geophys. Res.* 84:1683–1688.

Johnson, A. A., Weeks, S., and Remo, J. L. 1993. A microstructural study of the hexahedrite coahuila. *J. Intl. Meteor. Soc.*, submitted.

Kaufman, L., and Cohen, M. C. 1956. The martensitic transformation in the iron-nickel system and the structure of metallic meteorites. *Acta Metallurgica* 9:941–944.

Kivelson, M. G., Bargatze, L. F., Khurana, K. K., Southwood, J. J., Walker, P. J., and Coleman, P. J., Jr. 1993. Magnetic signatures near Galileo's closest approach to Gaspra, *Science*, 261:331–334.

Krinov, E. L. 1960. *Principle of Meteoritics* (New York: Pergamon Press).

Lipshutz, M. E., Gaffey, M. J., and Pellas, P. 1989. Meteorite parent bodies: size, number, and relation to present day asteroids. In *Asteroids II*, eds. R. P. Binzel, T. Gehrels and M. S. Matthews (Tucson: Univ. of Arizona Press), pp. 740–777.

McFadden, L. A., Tholen, D. J., and Veeder, G. J. 1989. Physical properties of Aten, Apollo, and Amor asteroids. In *Asteroids II*, eds. R. P. Binzel, T. Gehrels and M. S. Matthews (Tucson: Univ. of Arizona Press), pp. 442–467.

Öpik, E. J. 1958. *Physics of Meteorite Flight In The Atmosphere* (New York: Interscience).

Ostro, S. J. 1989. Radar observation of asteroids. In *Asteroids II*, eds. R. P. Binzel, T. Gehrels and M. S. Matthews (Tucson: Univ. of Arizona Press), pp. 192–212.

Ostro, S. J., Campbell, D. B., Chandler. J. F., Hine A. A., Hudson, R. S., Rosema, K. D. and Shapiro, T. J. 1991. Asteroid 1986 DA: Radar evidence for a metallic composition. *Science* 252:1399–1404.

Remo, J. L. 1993a. Asteroid/meteorite analogs and material properties. In *Proceedings of the NEO Interception Workshop*, eds. G. H. Canavan, J. C. Solem and J. D. G. Rather (Los Alamos: Los Alamos National Lab), pp. 163–175.

Remo, J. L. 1993b. Penetrator device applications and NEO materials properties. In *Proceedings of the NEO Interception Workshop*, eds. G. H. Canavan, J. C. Solem and J. D. G. Rather (Los Alamos: Los Alamos National Lab), pp. 182–185.

Remo, J. L., and Johnson, A. A. 1974. The ductile-brittle transition in meteoritic irons. *Meteoritics* 9:209–213.

Remo, J. L., and Johnson, A. A. 1975. A preliminary study of the ductile-brittle transition under impact conditions in material from a octahedrite. *J. Geophys. Res.* 80:3744–3748.

Shoemaker, G. 1983. Asteroid and comet bombardment of the earth. *Ann. Rev. Earth Planet. Sci.* 11:461–494.

Takahashi, T., and Basset, C. 1964. High pressure polymorphs of iron. *Science*

145:483–486.

Tedesco, E. F., and Gradie, J. 1987. Discovery of M class objects among the near-Earth ateroid population. *Astron. J.* 93:738–746.

Tholen, D. J. 1984. Asteroid Taxonomy from Cluster Analysis of Photometry. Ph.D. Thesis, Univ. of Arizona.

Stoffler, D. 1972. Deformation and transformation rock-forming minerals by natural and experimental shock processes I: Behavior of minerals under shock compression. *Fortschr. Mineral.* 516:256–289.

Swegle, J. W., and Grady, D. E. 1985. Shock viscosity and prediction of shock wave rise times. *J. Appl. Phys.* 58:692–701.

Wasson, J. T. 1974. *Meteoritics* (New York: Springer-Verlag).

Zukas, E. G. 1969. Metallurgical results from shock-loaded iron alloys applied to a meteorite. *J. Geophys. Res.* 74:1993–2001.

PROPERTIES OF COMETARY NUCLEI

J. RAHE
NASA Headquarters

V. VANYSEK
Charles University, Prague, Czech Republic

and

P. R. WEISSMAN
Jet Propulsion Laboratory

Active long- and short-period comets contribute about 20 to 30% of the major impact events on the Earth. Cometary nuclei are irregular bodies, typically a few to ten kilometers in diameter, with masses in the range 10^{15} to 10^{18} g. The nuclei are composed of an intimate mixture of volatile ices, mostly water ice, and hydrocarbon and silicate grains. The composition is the closest to solar composition of any known bodies in the solar system. The nuclei appear to be weakly bonded agglomerations of smaller icy planetesimals, and material strengths estimated from observed tidal disruption events are fairly low, typically 10^2 to $10^4 \mathrm{N} \ \mathrm{m}^{-2}$. Density estimates range between 0.2 and 1.2 g cm^{-3} but are very poorly determined, if at all; a probable best guess is ~ 1.0 g cm^{-3}. As comets age they develop nonvolatile crusts on their surfaces which may eventually render them inactive, similar in appearance to carbonaceous asteroids. However, dormant comets may continue to show sporadic activity and outbursts for some time before they become truly extinct. Alternatively, some cometary nuclei may disintegrate entirely to meteoroid streams. The source of the long-period comets is the Oort cloud, a vast spherical cloud of $\sim 10^{13}$ comets surrounding the solar system and extending to interstellar distances. The likely source of the short-period comets is the Kuiper belt, a ring of perhaps 10^8 to 10^{10} remnant icy planetesimals beyond the orbit of Neptune, though some short-period comets may also have originated as long-period comets from the Oort cloud. Rendezvous spacecraft missions to comets are required to characterize these potential impactors properly.

I. INTRODUCTION

Active long- and short-period comets account for ~ 20 to 30% of the major impact events on the Earth, those where the crater diameter is > 10 km (Shoemaker et al. 1990; Weissman 1990a). Virtually the entire mass of a comet is concentrated in its solid, relatively small nucleus which has a typical diameter of several to ten kilometers. If such a body, with an estimated average mass of 10^{15} to 10^{18} g would strike the Earth with a velocity ranging between 16 and 72 km s^{-1}, the energy released in the impact would be in the range 10^{27} to 10^{31} erg, equivalent to 10^5 to 10^9 megatons of explosives. The consequences of such a catastrophic impact on the Earth's global environment could be extremely serious and have long-lasting effects.

Our direct knowledge of cometary nuclei is relatively poor. When active, they are unresolvable using Earth-based telescopes, buried deep within the bright cometary comae. When inactive (and it is uncertain if cometary nuclei are ever really inactive while they are within the planetary system), they are often too distant and faint for many diagnostic techniques used in asteroid studies to be applied to them. The only spacecraft missions to comets so far have been the ultra-fast flybys of comet Halley in 1986 by the Giotto, Vega, and Suisei spacecraft, and the follow-on Giotto flyby of comet Grigg-Skjellerup in 1992 (the ISEE 3 spacecraft passed through the tail of comet Giacobini-Zinner in 1985, but made no direct measurements of the cometary nucleus). Although these missions provided a wealth of new information, they left many questions about the nature of cometary nuclei and cometary processes unanswered. Those unanswered questions require future cometary exploration missions which will rendezvous with a cometary nucleus and follow it through its orbit, watching the onset and decline of cometary activity, and studying the nucleus in detail at close range.

As part of this introduction to comets, it is useful to provide the reader with definitions as to what we mean by the terms "comet," "active comet," "dormant comet," etc., and to explain the differences between comets and asteroids. Unfortunately, no formal definitions exist. What follows are general definitions which should be acceptable to most researchers in this field.

Comets are generally distinguished from asteroids as being icy rather than rocky bodies. Comets are believed to have formed in the outer part of the planetary system, beyond about 5 AU, where water ice and other more volatile ices can condense. Asteroids are generally thought of as having formed interior to 5 AU, where ices could not condense, or if they did condense, were driven off by subsequent heating events. In reality, there is no sharp boundary between the zones of asteroidal and cometary formation, and there is a spectrum of compositions seen in the asteroid belt from differentiated, highly refractory objects (e.g., iron and basaltic achondrite asteroids) to undifferentiated, volatile-rich objects (e.g., the carbonaceous asteroids). These latter objects often contain considerable water and some of it may be in the form of ices; there is spectroscopic evidence for frosts on the surfaces of some carbonaceous asteroids. However, comets are typically defined as objects with a substantial fraction of their composition, $\sim$25 to 50, made up of water ice.

The observational test of whether an object is an asteroid or a comet is generally taken to be the ability to produce a visible coma of outflowing gas and dust. An object producing such a coma is usually termed an "active comet." Note that this is a subjective definition because the ability to observe the coma is a function of both the observing geometry (e.g., distance from the

Earth, solar elongation) and the capabilities of the instrument used to make the observation.

Because cometary activity is generally a result of solar heating, the activity of comets will vary as they move around their eccentric orbits. Many researchers regard a comet as "active" even though it may be near aphelion in its orbit, where solar heating is too low to sublimate a substantial enough amount of ices to produce a visible coma. Other researchers refer to such a state as "dormant," implying that the comet will become active only when it again approaches close enough to the Sun. A comet may be perturbed into an orbit where it never gets warm enough to produce a visible coma and thus is dormant throughout its orbit. Many observers refuse to regard such objects as comets because they never produce visible coma, and classify them as asteroids, or more formally, minor planets. Other researchers insist that these are comets because of their icy composition, and that they would display the normal range of cometary phenomena if they were again brought close to the Sun.

These conflicting, subjective definitions, one based on observed behavior and the other on detected or inferred composition, often lead to considerable confusion and conflict among researchers in this field. We will take the view that a comet is defined by its bulk composition and that an icy object in orbit around the Sun (not an icy satellite of one of the giant planets) is a comet. In addition, we will consider active comets to be ones that display cometary activity at some point in their current orbit. The term dormant comet will be used to designate a comet that is either in an orbit that does permit it to generate a substantial coma, or has physically evolved (see discussion of cometary surface crusts below) so that it does not show a coma in its present orbit.

The term "extinct comet" is also used to refer to a comet that has physically evolved so that it no longer can produce a visible coma anywhere in its orbit. We will use "dormant" to signify comets that might be reactivated if they were perturbed to a smaller perihelion distance or if their surface crusts could be removed or broken up, and "extinct" to signify comets that cannot be reactivated by any means. Note that both dormant and extinct comets will appear asteroidal (i.e., star-like) and may be misclassified as asteroids. This has actually happened on many occasions.

These definitions are not perfect. However, they provide a framework for the discussion that will follow. Two other terms that require definition are long- and short-period comets. In this case, a formal definition does exist. A long-period comet has an orbital period greater than 200 yr, whereas a short-period comet has a period less than 200 yr. The dividing line at 200 yr is arbitrary and reflects the ability at some time in the past of orbit computations to recognize returns of previously observed comets.

In the following, we review the current state of knowledge of cometary nuclei with emphasis on relevant parameters for the hazards problem, both in defining the estimated hazard and in formulating impact mitigation technologies. We also provide some background on the nature of cometary nuclei and cometary processes for those not familiar with comets, and references to more comprehensive reviews in those areas. Finally, a brief description of cometary dynamics is given, along with suggestions for further reading on that topic. Detailed estimates of cometary impact probabilities and impact rates are left to the chapters on that subject for long-period comets (Chapter by Marsden and Steel) and short-period comets (Chapter by Shoemaker et al.). Some consequences of impacts peculiar to comets will be briefly discussed.

II. DIMENSIONS AND ALBEDOS OF COMETARY NUCLEI

The only resolved images of a cometary nucleus are those obtained in 1986 of the comet Halley nucleus by the Giotto spacecraft multicolor camera (Fig. 1; Keller 1987) and those, with somewhat lower resolution, obtained by the Vega 1 and 2 spacecraft cameras (Fig. 2; Sagdeev et al. 1986a). The nucleus appears as a highly irregular ellipsoid with dimensions of $\sim 15 \times 8 \times 7$ km. It is one of the darkest bodies in the solar system with a measured surface albedo of 0.035 to 0.045. Active areas on the Halley nucleus (the sources of the visible jets) were confined to discrete areas comprising about 10% of the surface area visible in Fig. 1, or perhaps 20 to 30% of the total nucleus surface area. Irregular surface topography with typical scale lengths of hundreds of meters was visible, but difficult to interpret because of the modest resolution of the Giotto and Vega images. The nucleus occupies a volume of ~ 365 km^3 (Szegö 1991). Because of the high velocity of the flybys, the mass of the Halley nucleus could not be measured; it has been estimated indirectly (see below).

Early estimates of the dimensions of cometary nuclei were based on observations of comets at moderate solar distances (i.e., 2 to 4 AU), where they often displayed a "stellar" appearance and were assumed to be inactive (Roemer 1965; Roemer and Lloyd 1966; Roemer et al. 1966). Shoemaker and Wolfe (1982) used such observations to estimate a size distribution for cometary nuclei and compared their results with that predicted from the observed crater size distributions on Ganymede and Callisto, which they showed were dominated by cometary impacts. Shoemaker and Wolfe found that the estimated size distribution from the distant comet observations predicted far too many craters. To correct this discrepancy, they suggested that most of the brightness of the distant nuclei was due to unresolved comae, and that the nuclei themselves contributed only $\sim 12\%$ of the light, on average. This conclusion was later confirmed by deep CCD exposures which showed that many distant comets did indeed have substantial comae (see Jewitt 1991, and references therein). Given this fact, plus the high variability of cometary activity at large solar distances (Meech 1991), it is now clear that such distant observations alone cannot provide definitive radius measurements.

Past estimates of cometary albedos were fairly high (see, e.g., Delsemme and Rud 1973), which led to relatively small estimates for the dimensions of

Figure 1.   Composite image of the nucleus of comet Halley taken by the Giotto spacecraft on 14 March 1986. The nucleus is silhouetted against the bright dust coma. The view is from a phse angle of ~113°; the Sun is at the left. Bright dust jets emanate from discrete active areas on the nucleus, while other parts of the surface appear to be inactive. The resolution in the image vaires from 800 m/line-pair at lower right to ~80 m/line-pair in the upper left.

cometary nuclei, also based in part on observations of what were assumed to be bare cometary nuclei at large heliocentric distances. This in turn led to fairly low estimates for nucleus masses, making them relatively unimportant among potential terrestrial impactors. However, the spacecraft images of comet Halley's nucleus, as well as more recent photometric and radiometric measurements of other nuclei (see below) indicate that the albedo of most cometary nuclei appears to range from 0.02 to 0.1, with a typical value of about 0.04. Thus, the nuclei are much larger than previously thought, and mass estimates have increased accordingly.

Application of modern radiometric techniques for determining asteroid dimensions and albedos have now been successfully applied to cometary nu-

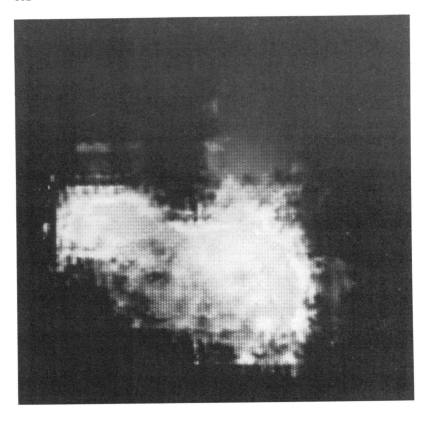

Figure 2.   Image of the Halley nucleus taken by the Vega 2 spacecraft on 9 March
1986 at a range of ~8000 km, at near-zero phase. The "peanut" shape of the nucleus
and the bright southern jet are clearly visible.

clei, with the observations being sensitive enough to discriminate the nuclei
from the dust comae, and/or to insure that bare nuclei are indeed being measured. A listing of radii, albedos, and other relevant parameters for eight
cometary nuclei is given in Table I. The columns in Table I are: the mean radius in km, the ratio between the major and minor axes of the nucleus (derived
from rotation lightcurves), the visual albedo, comments on the nucleus color,
the rotation period in hours, and the measured gas production rate and the
distance from the Sun at which it was measured. All of the listed comets have
short-period orbits, with the exception of comet IRAS-Araki-Alcock (IRAS-
Araki-Alcock was a long-period comet with a period of $\sim 10^3$ yr, which likely
means that it is dynamically old and has made many perihelion passages close
to the Sun). The dimensions of the nuclei can usually only be determined
with an accuracy of about ± 0.5 km; the estimate for Chiron is only an upper
limit, though a radius of ~ 140 km and albedo of 0.04 seems likely (Campins
et al. 1992). The elongated, prolate shape of Halley is seen to be typical for
other comets as well.

TABLE I
Observed Cometary Nuclei[a]

Comet	R_{eff}^{b} (km)	Axial Ratio	p_v	Color	P_{rot} (hr)	$Q(r)$ (s^{-1})	r (AU)
Halley	5.5[c]	2.0	0.04	Neutral, red	53, 89, 177 ?	6×10^{29}	0.8
Arend-Rigaux	5.2	1.9[d]	0.03	Neutral, red	13.6	2×10^{26}	1.58
Neujmin 1	10.4	1.6[d]	0.02	Very red	12.7	2×10^{26}	1.68
Schwassmann-Wachmann 1	15.4, 8.6 ?[d]	2.6 ?[d]	0.04, 0.13 ?	Very red[d]	14.0, 32.3 ?[d]		
Tempel 2	5.6	1.9	0.02	Very red	8.9	2×10^{27}	1.71
Encke	<2.2, 1.?	1.8[d]			15.1[d]	6×10^{28}	0.76
IRAS-Araki-Alcock	4.0 ?				48–72 ?	2×10^{28}	1.03
Chiron	<186.	Small	>0.027	Neutral	5.92		

[a] From Weissman and Campins (1993); references to original data sources are given in that paper.

[b] $R_{eff} = \sqrt{ab}$ where a and b are the projected semi-axes at maximum light.

[c] Actual Halley nucleus dimensions: $15 \times 8 \times 7$ km.

[d] New values from Meech et al. (1993) and Jewitt (1991).

The size distribution of cometary nuclei is not well known (see next section) and much smaller cometary nuclei as well as much larger ones may exist, e.g., the Great Comet of 1729 that could be seen with the naked eye at 4 AU from the Earth and Sun. The largest known cometary nucleus may be 2060 Chiron, an outer solar system "asteroid" (a = 13.7 AU, q = 8.5 AU) with geometric albedo >0.027 and corresponding radius <186 km (Sykes and Walker 1991). Despite its original classification as an asteroid, Chiron has been observed to display sporadic outbursts and a cometary coma (Meech and Belton 1989,1990), even near its aphelion of 18.9 AU (Bus et al. 1991a), as well as cometary CN emission (Bus et al. 1991b).

The objects listed in Table I are not an unbiased sample of cometary nuclei. They are all relatively large objects which were inactive enough (with the exception of comet Halley) for their nuclei to be directly observed, or at least discriminated from their dust comae. Comets Arend-Rigaux and Neujmin 1 are generally regarded as near-extinct and often appear stellar if they are not particularly close to Earth. Chiron was classified as an asteroid for over a decade before its sporadic cometary behavior was first noted in 1988–89. All of these objects have passed perihelion in their present orbits many times, and thus are far from pristine Oort cloud comets. However, they are still typical of many of the comets that pose a hazard to the Earth.

III. DENSITIES AND MASSES

The ratio of mass to volume yields the average density of cometary nuclei. In practice, the density is very difficult to determine because both mass and volume are so uncertain. Attempts to date to measure the masses of cometary nuclei have generally been based on comparing the nucleus activity with estimates of the nongravitational forces determined from the comet's motion. This is a tenuous method and involves many unknown parameters, particularly with regard to the momentum exchange between the nucleus and the evolving gases. For comet Halley, estimates of the bulk density based on modeling of nongravitational forces range from 0.2 to 1.2 g cm^{-3} (Rickman 1986,1989; Sagdeev et al. 1988; Peale 1989), with error bars covering an even wider range of values. The uncertainties involved in estimating nongravitational forces appear to make this method too unreliable.

An alternative, indirect method for estimating density is to consider the low-temperature, low velocity accretion of ice and dust grains in the solar nebula, yielding values ∼0.3 to 0.5 g cm^{-3} (Greenberg and Hage 1990). However, such estimates are based solely on theory and ignore subsequent thermal and physical processing of cometary materials, which will tend to compact the ice-dust mix. Also, other researchers using similar methods but different assumptions have found somewhat higher densities, ∼1.2 g cm^{-3} (Shoemaker and Wolfe 1982). Boss (1994) found an upper limit of ∼2 g cm^{-3} based on e disruption of comet Shoemaker-Levy 9 when it passed within the Roche limit of Jupiter in 1992.

Densities determined for somewhat larger though still modest sized bodies in the outer solar system; Pluto, Charon and Triton are all $\sim$2 g cm^{-3} (Beletic et al. 1989; Tyler et al. 1989). The relatively high dust-to-gas estimates derived from IRAS observations of cometary dust trails by Sykes and Walker (1992), and the high dust-to-gas ratio of 2.0 found for comet Halley from Giotto measurements (McDonnell et al. 1991), tend to support the possibility that these higher densities may also be present in comets. Also, measurements of the density of collected interplanetary dust particles (IDPs), believed to be derived from cometary nuclei, yield values of $\sim$0.7 to 1.2 g cm^{-3} (Fraundorf et al. 1982), despite the fact that the volatile ices have been lost from these particles. In all probability, the best guess for the density of a typical cometary nucleus is $\sim$1.0 to 1.2 g cm^{-3}.

Mass distributions for comets have been estimated indirectly from the distribution of the intrinsic (or "absolute") magnitudes, $H_{10}{}^{a}$, of the observed comets. In theory, the gas-producing ability of a cometary nucleus, and hence its coma brightness, should be proportional to the sublimating surface area, which leads to the relationship: $H_{10} \propto 2.5 \log A$, where A is the projected area of the nucleus. Assuming a roughly spherical nucleus, its mass will vary as $A^{1.5}$. Thus, $\log m$ (where m is the nucleus mass) should vary as $0.6 H_{10}$. This simple-minded model ignores such factors as the fraction of the nucleus surface that is active, the dust-to-gas ratio, the heliocentric distance at which the comet is observed, etc. Thus, any suggested mass distributions employing this method should be viewed with some skepticism.

Empirical estimates by researchers have found slopes somewhat shallower than 0.6 for the mass distribution. One suggested distribution by Weissman (1990a) is shown in Fig. 3. This distribution is based on the distribution of absolute magnitudes (with coma) after correction for observational selection effects, as found by Everhart (1967). For comets brighter than $H_{10} = 11$, Weissman found the mass-magnitude relationship

$$\log m = 20.0 - 0.4 H_{10} \tag{1}$$

where m is the nucleus mass in grams and a density of 0.6 g cm^{-3} is assumed. The magnitude $H_{10} = 11$ corresponds to a radius of 1.2 km and a mass of 4×10^{15}g. Using the magnitude distribution in Fig. 3, the average cometary nucleus, brighter than $H_{10} = 11$, has a mass of 3.8×10^{16}g.

Bailey and Stagg (1988) derived a similar mass-magnitude relationship

$$\log m = 19.9 - 0.5 H_{10} \tag{2}$$

though with a steeper slope such that mass decreases more sharply with increasing absolute magnitude. Fernandez and Ip (1991) combined Eq. 2 with

---

[a] The absolute magnitude H_{10} is defined as the magnitude the active comet sould have at 1 AU from the Sun and 1 AU from the Earth, assuming that the comet's brightness varied as r^{-4} with heliocentric distance, and r^{-2} with geocentric distance.

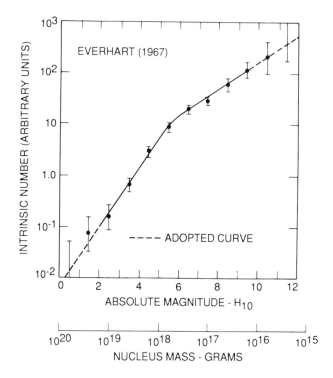

Figure 3.   Relative distribution of absolute magnitudes H_{10} for long-period comets after correction for observational selection effects, as found by Everhart (1967). The scale of nucleus masses at the bottom of the figure was derived with Eq. (1) (Weissman 1990a).

Everhart's brightness distribution to give the cumulative number distribution relationship, $N(m) \propto m^{-0.58}$ for masses $< 10^{17}$g and $N(m) \propto m^{-1.16}$ for $m > 10^{17}$g.

IV. NUCLEUS STRUCTURE

The surface morphology and the internal structure of cometary nuclei are unknown. A variety of models have been suggested: Whipple (1950,1951); Donn et al. (1985); Weissman (1986); Gombosi and Houpis (1986); see also reviews by Donn (1991) and Rickman (1991). Several of these models are illustrated in Fig. 4. The current consensus is that the typical nucleus can be described as a weakly bonded, fractal assemblage of smaller icy-conglomerate planetesimals of diameter $\sim$1 km, possibly "welded" into a single body by thermal processing and sintering. Each icy planetesimal fragment is likely a porous and fragile body composed of fine-grained refractory material, intimately mixed with organic grains and volatile ices. Dynamical scattering of planetesimals by the outer planets may have resulted in mixing and assemblage of planetesimals formed in different temperature regimes; observational evidence for such heterogeneity has been found for comet Halley (Mumma et al. 1993a). In some cases (see, e.g., Fink 1992; Schleicher et al. 1993) observed compositional differences in comets are so great as to suggest formation in distinctly different regions of the solar nebula (i.e., the Jupiter-Saturn zone rather than the Uranus-Neptune zone).

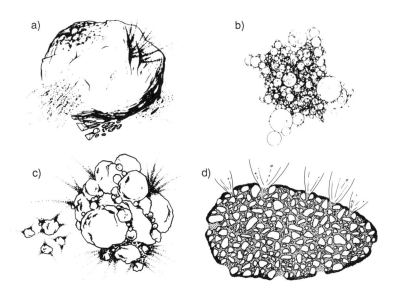

Figure 4.   Four suggested models for the structure of cometary nuclei: (a) the icy conglomerate model (Whipple 1950; drawing from Weissman and Kieffer 1981); (b) the fractal model (Donn et al. 1985); (c) the primordial rubble pile (Weissman 1986); and (d) the icy-glue model (Gombosi and Houpis 1986). All but (d) were suggested prior to the Halley spacecraft encounters in 1986.

Based on the images of comet Halley as well as observations of other comets, a tri-axial, irregular ellipsoid may be an acceptable model for a cometary nucleus. Several attempts have been made to explain the non-sphericity (Daniels and Hughes 1981; Jewitt and Meech 1988; see also Donn 1991), which seems to be a common property of cometary nuclei. These studies were based mainly on random-walk schemes for the aggregation processes. Although some results indeed lead to irregularities and deviation from sphericity, a tendency toward tri-axial shapes was obtained only by the numerical experiments by Jewitt and Meech (1988).

Mass-loss may also change the shape of cometary nuclei. At the time of the Giotto flyby, comet Halley was losing gas and dust at a rate of $\sim 3 \times 10^7$ to 10^8 g s^{-1} (see Hughes 1991). The total estimated water loss per orbital revolution is $\sim 3 \times 10^{14}$ g (Feldman et al. 1987). Adding in other volatiles and assuming a dust-to-gas ratio of 2 raises the total mass loss to $\sim 1.1 \times 10^{15}$ g, equivalent to the loss of a surface layer of average thickness ~ 3.0 meters with density 1.0 g cm^{-3}. Because activity appears to come from discrete areas on the nucleus surface, an average erosion rate of ~ 10 to 15 meters per revolution could be expected from those active areas. Therefore, one might expect that mass loss could lead to some irregular evolution of surface features, though possibly not to gross changes in the shape of the overall nucleus. Note that these estimates are for comet Halley only, and will vary considerably with other cometary nuclei as a function of their size, dynamical age, perihelion distance, active surface fraction, etc.

V. ROTATION OF COMETARY NUCLEI

Although there are about 60 reported determinations of rotational periods of cometary nuclei (see, e.g., Sekanina 1981; Whipple 1982), only a few appear to be reliable, and even they do not necessarily describe the true rotational state. The determination of spin state, i.e., rotation period and orientation of the spin axis, is based on the search for periodicities in time-series of some continuously varying observed property of the comet, mainly a variation in its brightness.

The difficulties one encounters when attempting to determine the spin parameters of cometary nuclei can be demonstrated with the case of comet Halley. Although early published results indicated that the nucleus rotated with a period of 2.2 days around the shortest axis (Sekanina and Larson 1984; Kaneda et al. 1986; Sagdeev et al. 1986b), i.e., in the "ground" state, strong evidence later emerged for a period of 7.4 days (Millis and Schleicher 1986; Stewart 1987). More detailed examination of the data suggested that the nucleus of comet Halley rotates as an asymmetric top (Belton et al. 1992), in which it is assumed that the nucleus rotates around both the long and short axes. In the most likely mode, the long axis executes a precessional motion around the space-fixed total rotational angular momentum vector with a period near 3.7 days, while performing a "nodding" with a period of about 7.3 days (Samarasinha and A'Hearn 1991).

The available data on rotation for seven periodic comets, including comet Halley, indicate that the cometary nuclei rotate in an excited energy state (Belton 1991). The characteristic time t for an oblate rotator (and to first approximation, for other shapes) with frequency ω to relax to a state of principal axis rotation (i.e., the time necessary for damping of the wobble motion) is given by (Burns and Safronov 1973)

$$t = \mu q Q \omega^3 / \rho R^2 \qquad (3)$$

where μ is the rigidity or shear modulus of the cometary nucleus, ρ is the density, Q is a dimensionless measure of internal energy dissipation per rotation cycle, R is the "mean" radius and $q = 38/5$ for a nearly homogeneous body. Peale and Lissauer (1989) suggest $t = 10^6 Q$ yr for cometary nuclei, assuming the shear modulus for ice, with $Q \leq 10^2$ and even $Q \leq 1$, while Burns and Safronov suggest typical values for Q for asteroids of 10^2 to 10^3. The rigidity of the cometary material is most likely much lower than that for asteroids and the shear modulus could be $\sim 10^7 \text{N m}^{-2}$, which is about 0.01 times that of solid water ice. Even with $Q = 1$ the relaxation time in the rotational characteristics of cometary nuclei could be $t = 10^6$yr, comparable to typical dynamical lifetimes of short-period comets, and likely much longer than their physical lifetime as active objects. The excitation into higher rotational states for cometary nuclei is most probably a result of nongravitational forces from jetting on the nucleus surface. Because the damping time scale is long compared to the orbital period, the excitation of the rotational state is probably a cumulative process.

Another question is the orientation of the spin vector. The original spin characteristics may be modified and/or randomized by nongravitational forces and mass loss. As already noted, cometary nuclei can lose substantial mass during their active phases, on the order of 0.01 to 0.1% per perihelion passage for active comets like Halley or the brighter long-period comets. This process is not expected to lead to dramatic changes in the spin (Peale and Lissauer 1989), but may be responsible for a secular, systematic drain of the angular momentum resulting in "spin-down" of the nucleus. This may account for the apparently longer mean rotation periods for cometary nuclei as compared with comparably sized asteroids.

VI. MATERIAL STRENGTHS

Little is known about the strengths of cometary materials. Observational evidence, i.e., splitting of comets, as well as theoretical considerations suggest that the cometary nuclei are poorly consolidated bodies. Statistics show that about 10% of "dynamically new" comets (those coming in from the Oort cloud for the first time) randomly disrupt during their first perihelion passage. Similar random disruption events are observed for $\sim 4\%$ of long-period comets making subsequent returns, and $\sim 1\%$ of short-period comets (Weissman 1980; see also Sekanina 1982). The disruption events show no obvious correlation with time relative to perihelion, perihelion distance, orbital inclination, or the ecliptic plane. Presumably, the events are associated with thermal stresses generated by the heating the nuclei receive as they approach and recede from the Sun. The disruption events could be regarded as a selection process, whereby comets which are likely to split, do so rather rapidly, while others are more stable and survive many perihelion passages (Weissman 1980; Sekanina 1982). Comet Halley appears to belong to the latter group.

A second class of observed disruption events are those caused by passage through the Roche limit of the Sun or a planet. This has been seen for many members of the Kreutz group of Sun-grazing comets (Marsden 1967,1989) and for two short-period comets which passed close to Jupiter: P/Brooks 2 (in 1886) and comet Shoemaker-Levy 9 (in 1992; see Shoemaker et al. 1993). Although all members of the Kreutz group pass within a solar radius of the Sun's photosphere, and some have been observed to impact the Sun, not all of them are observed to disrupt during their perihelion passage. In the cases of the two short-period comets, both were discovered as multiple nuclei, after their close passages to Jupiter.

There are two different concepts for analytical studies of tidal break-up: "tidal disruption" and "tidal failure." Boss et al. (1991) defines "tidal disruption" as a process whereby a body is tidally separated into two or more pieces which subsequently move on individual orbits. This concept can be applied to the case of inviscid bodies which are held together by gravity and not by internal forces. The "tidal failure" concept is based on a comparison of tidal stresses to material strengths of solid bodies (Aggarwal and Oberbeck 1974; Dobrovolskis 1990; Sridhar and Tremaine 1992). The latter concept is the better for obtaining an understanding of material strengths in cometary nuclei. A similar comparison can be made between centrifugal forces and material strengths if the size and rotation of the nucleus is known.

Sekanina (1982) determined tensile strengths of $\sim 10^2$ to 10^4 N m^{-2} from an analysis of the tidally split Sun-grazing comets 1882 II and 1965 VIII. For a Sun-grazing cometary nucleus with radius = 5 km, density = 0.5 g cm^{-3}, and perihelion = 0.005 AU, the minimal required value of tensile strength to survive perihelion passage is about 3×10^3 N m^{-2}. Sekanina and Yeomans (1985) determined similar values for P/Brooks 2 which passed within the Roche limit of Jupiter in 1886. Sekanina (1993) estimated a tensile strength for comet Shoemaker-Levy 9, which split during a pass within the Roche limit of Jupiter in July 1992, of 1.5×10^2 N m^{-2}, assuming a density of 0.5 g cm^{-3}. However, that estimate was based on an assumed perijove distance of 1.62 R_J, and subsequent orbit solutions have given a value closer to 1.3 R_J. Scaling to the newer perijove distance and to our best guess density of 1.0 g cm^{-3} raises the tensile strength estimate to $\sim 6 \times 10^2$ N m^{-2}. Whipple (1982) derived an upper limit on tensile strength of about 10^4 N m^{-2} from cometary spin and size statistics. Note that all of these values are highly uncertain because of the lack of detailed knowledge of the density and dimensions of the cometary nuclei.

These values are all very low as compared with values for common materials. For example, the "breaking strain" for rocky materials is $\sim 4 \times 10^6$ N m^{-2}, and $\sim 2 \times 10^6$ N m^{-2} for solid water ice. Thus, the available evidence points to very low material strengths for cometary nuclei. Note, however, that these low values apply to the bulk tensile strength of the cometary nuclei, the strength with which the individual fragments are welded together. Internal strengths for the fragments themselves may be much higher.

VII. SURFACES PROCESSES

The source of cometary activity is the sublimation of volatile ices on the nucleus surface (Whipple 1950,1951). The evolving gases, mostly water, carry with them micron-sized grains of silicate dust and organics, forming the comet's extended atmosphere, or coma. Outflowing dust and gases in the coma interact both physically and chemically. Ions formed in the coma are accelerated by Lorentz forces in the solar magnetic field to form the bright Type I plasma tails, while dust grains are blown back by solar radiation pressure to form the broader, more curving Type II tails.

The surface temperature of an icy-conglomerate mix exposed to sunlight is found by balancing the incoming solar radiation with outgoing thermal radiation, heat conducted into the cometary interior, and energy used in sublimation (Watson et al. 1963; Weissman and Kieffer 1981)

$$S_o r^{-2}(1 - A) \cos i = \epsilon \sigma T^4 - K \delta T / \delta Z|_{z=0} - L \delta m / \delta t \qquad (4)$$

where S_o is the solar constant, r is the heliocentric distance of the comet, A is the surface bond albedo, i is the local solar zenith angle, ϵ is the emissivity, σ is the Stefan-Boltzmann constant, T is the temperature, K is the conductivity, $\delta T / \delta z$ is the temperature gradient evaluated at the surface, L is the heat of sublimation, and $\delta m / \delta t$ is the mass loss (sublimation) rate. At large solar distances both the conduction and sublimation terms are small and the surface temperature acts similarly to an inactive asteroid. At about 5.8 AU, water ice located at the sub-solar point on the nucleus can begin to sublimate but the total production rate remains low because of the very small surface area involved. At about 3 AU, water ice sublimation begins to become significant for the entire nucleus, at surface temperatures $\sim$160 to 175 K. By 1.5 AU, sublimation typically dominates the energy outflow, buffering the nucleus surface temperaures at $\sim$200 to 220 K. Other more volatile ices, such as CO and HCN, can begin to diffuse out of the ice-dust mix at large solar distances, and will continue to be liberated as the overlying ice-dust layers are sublimated away, and the solar heating wave penetrates to greater depths within the nucleus.

Few comets actually match this highly idealized physical picture, though their gross behavior usually follows it somewhat. The situation is complicated, in part, by the existence of lag deposits of large nonvolatile grains which develop on the nucleus surfaces, insulating the ices beneath them (Brin and Mendis 1979; Fanale and Salvail 1984). Estimates of the thickness of these lag deposits, or crusts, range from a few centimeters to meters. If the crusts become sufficiently thick and insulating, they can essentially "turn-off" the cometary ices beneath them. Alternatively, build-up of gas pressure from sublimating ices beneath the crusts may cause them to rupture, resulting in sudden visible outbursts. Sublimation may also lead to development of unusual surface morphologies, as have been seen on terrestrial glaciers and icefields.

It had been predicted that the surface of the nucleus of comet Halley would be free of crust because of the high activity that comet reaches at its perihelion of only 0.587 AU. Thus, it was somewhat of a surprise to discover that ~70 to 80% of the nucleus was covered by an apparently inert crust. Subsequent studies of other short-period comets (A'Hearn 1988; Weissman et al. 1989) showed that the fraction of active surface area appears to decline with cometary age, reaching less than 1% for some of the older short-period comets such as P/Arend-Rigaux and P/Neujmin 1. This leads to the interesting possibility that comets might evolve to completely inactive, dormant objects that would be asteroidal in appearance (see below).

Although comets are the most pristine bodies in the solar system and have essentially been in "cold storage" in the Oort cloud and Kuiper belt over most of their lifetimes, there are a number of physical processes which may have modified them in various ways. These include: irradiation by galactic cosmic rays and solar wind protons (Moore et al. 1983; Johnson et al. 1987), heating by supernovae and passing stars (Stern and Shull 1988), competing erosion and accretion by interstellar dust grains (Stern 1986,1990), and conversion from amorphous to crystalline ice (Smoluchowski 1981; Prialnik and Bar-Nun 1988). These, and several other possible modifying processes are depicted in Fig. 5. All of these processes are fairly modest when compared with typical planetary processes such as giant impacts or differentiation and core formation. The low degree of processing is evidenced by the high volatile content of comets.

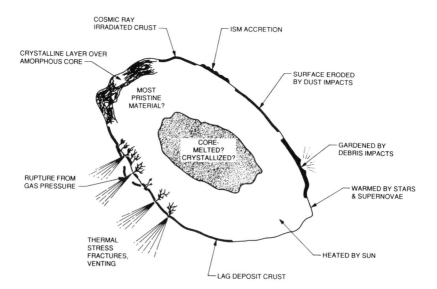

Figure 5.   Suggested physical processes which may alter the surfaces or interiors of cometary nuclei over their lifetimes, including during storage in the Oort cloud and during passage close to the Sun (figure from McSween and Weissman 1989).

Radiation processing is particularly interesting because it leads to development of both nonvolatile and highly volatile materials in near-surface layers; the latter may possibly be free radicals, trapped in the ice-dust matrix. The processed surface layer is expected to be on the order of one to several meters thick. The nonvolatile materials may form a fairly inert surface crust which would need to be removed, at least in part, for the comets to show any activity at all. At the same time, the inter-mixed volatile materials may provide an energy source at large solar distances as they are slowly warmed, and result in cracking and removal of the radiation-processed crust. That removal may be aided by the amorphous-to-crystalline ice conversion which is exothermic, and occurs for the first time as a comet approaches 5 AU from the Sun. New comets from the Oort cloud are anomalously bright at large solar distances, ∼5 AU, on their first perihelion passage and these processes may explain that distant activity. Useful reviews on nucleus thermal and physical processing are provided by Rickman (1991) and Weissman (1990*b*).

VIII. DORMANCY OF COMETARY NUCLEI

In some well-documented cases, periodic comets with well-determined orbits could not be recovered although they were in a relatively favorable position for Earth-based observers. In other cases, comets were recovered but appeared asteroidal, with no visible coma. This suggests that some cometary nuclei could be inactive for long periods of time, making their recovery difficult. However, such reports must be examined carefully because of the varying quality of observations and recovery searches. For example, Millis et al. (1988) reported that P/Arend-Rigaux was stellar in appearance for three consecutive orbital periods, but also noted that all three apparitions were quite far from Earth, and thus not the best observing geometry. P/Tempel 1 has been suggested as having been dormant for more than 10 returns (Kresák 1991), but those returns correspond to a time when the orbit of the comet was perturbed to a substantially larger perihelion distance, where it would have been less active and thus more difficult to observe.

Studies of the observed returns of comets Halley and Swift-Tuttle over more than 2000 yr each (Stephenson et al. 1985; Yau et al. 1994) have shown that naked-eye discoveries only occurred when the comets attained a brightness of magnitude $V = 3.4$ to 4.0, considerably brighter than the limit of $V = 6.0$ normally assumed for naked-eye observations of stars. This result is obvious when one considers the extended, diffuse nature of cometary comae. Thus, even though some comets may have achieved naked-eye brightness in the past, i.e., brighter than $V = 6.0$, it is quite reasonable that they could have been missed. In addition, statistics compiled by Kresák (1991) show that the fraction of "lost" comets has steadily decreased with time, suggesting that past failed recoveries may have been more a matter of poor instrumentation or poor orbit determination.

Nevertheless, some clear cases of lost comets do exist. P/Denning-Fujikawa passed close to the Earth in 1829 and should have been discovered by naked-eye observers, but was not. The same seems to be true for P/Brorsen in 1835 and possibly P/Finlay in 1827 (Kresák 1987). Comet P/Tuttle-Giacobini-Kresák was independently rediscovered on two separate occasions.

Dormant periods among short-period comets are likely related to the development of surface crusts of nonvolatile materials (see previous section). Kresák (1991) estimates an active lifetime for short-period comets of 200 to 500 returns. On the other hand, Weissman (1979,1982a) suggested that the lack of observed long-period comets with perihelia <0.4 AU and semimajor axes <400 AU may be due to the development of surface crusts, which would have to form in about 20 returns. Note that these long-period comets have much smaller perihelia than the short-period comets referred to by Kresák, and thus the two estimates are not necessarily in conflict. Variations in cometary activity from one perihelion passage to the next may also be due to precession of nucleus spin vectors, causing different active areas to be exposed to solar insolation, as suggested for comet Encke by Sekanina (1988).

It is almost certain that the physical lifetime of short-period comets as active objects is shorter than their dynamical lifetimes in the planetary region. Periods of dormancy interspersed with short periods of reactivation and possibly outbursts may be the final stages of a nucleus' active life. However, whether the final state of a nucleus' evolution is to an extinct, asteroidal object or to total disintegration into a meteoroid stream, is still an unanswered question.

IX. EXTINCT COMETS AMONG THE NEAR-EARTH ASTEROIDS

The origin of the near-Earth asteroids has been the subject of an ongoing debate among solar system scientists for several decades now. Because they are in planet-crossing orbits, these objects have dynamical lifetimes of only $\sim 3 \times 10^7$ yr (Wetherill 1975). Thus, the population of Earth-crossing asteroids must be continuously replenished. The origin arguments largely split along discipline lines. In general, observers believed that the near-Earth asteroids were derived from the main asteroid belt between Mars and Jupiter, because they appeared similar spectroscopically to mainbelt objects. On the other hand, dynamicists favored a cometary source because of the lack of known dynamical mechanisms to move sufficient numbers of asteroids from the main belt to near-Earth orbits.

However, in the past 20 years two dynamical mechanisms for delivery of mainbelt asteroids have been recognized which could account for at least half of the estimated population of near-Earth asteroids (Wetherill 1988). The two mechanisms are chaotic motion at mean-motion resonances with Jupiter (semimajor axes where the asteroid's orbital period is a small integer multiple of Jupiter's period, i.e., 3:1, 5:2; Wisdom 1987), and secular resonances where the effect of mutual perturbations of the giant planets can force high eccentricities among the mainbelt asteroids (Williams and Faulkner 1981). At the same time, observational evidence has continued to mount of anomalous Earth-crossing asteroids whose characteristics and behavior (i.e., albedo, shape, spin rate, outgassing, radar reflectivity, dynamical motion, etc; see Weissman et al. 1989) might be explained if the objects were indeed extinct cometary nuclei. The recognition that $\sim$70% of the surface of comet Halley was covered by an apparently inert crust was important in lending credibility to the idea that comets could evolve into dormant, asteroidal-appearing objects.

Various tests of cometary versus asteroidal origin have been put forward, usually based on statistical differences between the size, shape, rotation periods, spectral properties, meteor stream associations, and orbital dynamics of their parent populations (see Weissman et al. 1989). However, none of these tests are definitive, and many near-Earth objects display contradictory combinations of physical and dynamical attributes. The one agreed-upon test of a cometary origin is the ability to generate an appreciable cometary coma around the nucleus. A number of independent observations do exist of what may well be sporadic activity from some near-Earth asteroids.

The possible evolution of short-period comets to inert-appearing, near-Earth asteroids has been reviewed extensively by Öpik (1963), Wetherill (1971,1991), Degewij and Tedesco (1982), Kresák (1985), Fernandez (1988), and Weissman et al. (1989). The reader is referred to those papers and references therein for a more complete discussion of the problem. The reviews by Degewij and Tedesco and by Weissman et al. are particularly recommended as they arrive at rather opposite conclusions as to the likelihood of extinct or dormant comets among the near-Earth asteroids, and thus provide good examples of the two sides in the debate.

A list of comet-like asteroids and a detailed discussion of the characteristics that make each object a member of the list is given in Weissman et al. (1989). Probable cometary candidates identified among the numbered asteroids include: 944 Hidalgo, 2060 Chiron, 2101 Adonis, 2201 Oljato, 2212 Hephaistos, 3200 Phaethon and 3552 Don Quixote. Possible candidates include: 1566 Icarus, 1580 Betulia, 1620 Geographos, 1685 Toro, 1862 Apollo, 1866 Sisyphus, 1917 Cuyo, 1981 Midas and 2062 Aten.

As noted earlier, Chiron is indeed cometary as demonstrated by its sporadic activity around its orbit. Another example of a comet that apparently evolved to an asteroidal object is asteroid 4015 (1979 VA) which was independently discovered and catalogued as comet Wilson-Harrington in 1949, when it briefly showed coma activity (Bowell and Marsden 1992). Future observational surveys are likely to discover additional transitional objects like 4015 Wilson-Harrington, and to provide additional evidence for objects of cometary origin among the near-Earth asteroids.

TABLE II

Spectral Identifications in Comets

	Spectral Range
Coma	
CN, ^{13}CN, C_2, $^{12}C^{13}C$, C_3, CH, NH, NH_2, OH [OI], Na, Ca, Cu, Cr, Mn, Fe, Ni, K, Co, SH CO^+, CH^+, CO_2^+, N_2^+, OH^+, Ca^+, H_2O^+	Visibile
H, C, O, S, OH, CO, CS, CO_2^+, CO^+, CN^+, C^+, S_2	Ultraviolet
H_2O, H_2CO, CO_2, CO, CH_3OH, (OCS)^a "C-H" feature near 3.4 μm; complex organic compounds, "silicate" emission feature near 10 μm	Infrared
OH, HCN, H_2CO, H_2CS, H_2S, OCS, H_2O (CH_3CN, NH_3)^b	Radio
Ions of H, N, O, C, S_2, H_3O, NH_4, NH_3, NH_2, CH_5, CH_4, CH_3, C_3H_3, C_3H_4, C_3H, CS_2, CS, H_3S, $(H_2CO)_n$, H_2O, CO, N_2, C_2H_4, many molecules, probably hydrocarbons, with mass up to 105 amu. "CHON" grains (dust grains with hydrocarbon mantles or composed entirely of light elements, i.e., H, C, N, O molecules)	*In situ* measurements by mass spectrometers during spacecraft flybys of comet Halley
Plasma tail	
Ions of CO, CH, CO_2, N_2, OH, H_2O, CN, OH, NH_4, SH	Visible and ultraviolet

^{a} Tentative identification.
^{b} Identification claimed in the past, but not confirmed in comet Halley.

X. CHEMICAL COMPOSITION

The present knowledge of the chemical composition of cometary nuclei is inferred from measurements of neutral and ionized gases in the coma and tail, and from dust grains. These data are obtained primarily through ground-based spectroscopy at ultraviolet, visible, infrared, and radio wavelengths, and by *in situ* measurements by the spacecraft that encountered comet Halley in 1986. Earth-orbiting instruments and rocket-borne payloads have also provided valuable data.

Prior to the mid-1980s, most gaseous species accessible to groundbased observers were photo-dissociation products of parent molecules that subli-mated from the cometary nucleus. However, improvements in infrared and radio instrumentation, as well as *in situ* measurements by flyby spacecraft at Halley, led to direct measurements of the parent molecules. The abundances of parent molecules in comet Halley and other comets were recently critically reviewed by Mumma et al. (1993*a*). The observed molecules, radicals, and ions in comets are summarized in Tables II and III.

The dominant molecule in the volatile component is H_2O, representing $\sim 80\%$ of the total abundance of volatiles. Other species clearly present in comet Halley at more than 1% relative to water are CO, CO_2, H_2CO and CH_3OH. Volatile abundances vary from comet to comet: CO ranges from 1 to 30% of the water abundance (A'Hearn and Festou 1990); CH_3OH ranges from 1 to 7% (Mumma et al. 1993*b*).

In situ mass-spectroscopy of dust grains in comet Halley revealed the presence of two distinct compositional classes of particles: refractory sili-cates which had been expected, and organic grains, termed "CHON" because they contained only the light elements C, H, O, and N. Larger grains with more complex compositions were apparently composed of assemblages of these two particle types. The compositional signature of the cometary grains closely matches that of anhydrous olivine IDPs (interplanetary dust particles), providing support for the belief that these recovered grains are cometary. An example of a suspected cometary IDP is shown in Fig. 6. The IDP is a botryoidal ("cluster-of-grapes") assemblage of submicron silicate and or-ganic grains. The inter-grain spaces were presumably formerly filled with cometary ices. Organic grains apparently also act as a source of volatiles in the cometary coma. This was demonstrated by jet structures visible in the emission lines of CN in Halley (A'Hearn et al. 1986), and by an increase in the relative abundance of CO farther from the cometary nucleus (Eberhardt et al. 1987), indicating the existence of an extended source, i.e., volatile grains in the coma.

Estimates of the dust-to-gas ratio in comets have ranged from 0.5 to 2, with current best estimates for Halley now tending toward the latter, higher value (McDonnell et al. 1991). Analysis of IRAS dust trails, i.e., large (~ 0.01 to 1 cm in radius) cometary grains in the orbits of short-period comets, have suggested even higher dust-to-gas ratios for many known comets (Sykes and Walker 1992). Note that these ratios refer to measurements made in cometary comae and/or dust trails, and may not reflect the true dust-to-gas ratio in cometary nuclei.

Atomic abundances of elements obtained from mass-spectroscopy of the dust in the coma of comet Halley, normalized to magnesium, are summarized in Table IV. These abundances are in excellent agreement with solar com-position (with the exception of H and He), demonstrating that comets have undergone very little physical processing, and that they are indeed the best obtainable source of original solar nebula material.

Bulk ratios of the stable isotopes of the major elements, including C,

TABLE III
Relative Abundances of Probable Prent Molecules in Cometary Comae[a]

Molecule	Relative Abundance	Comments
Comet Halley		
H_2O	100	Remote and *in situ* detections[b]
CO	~7	Direct (native) source[c]
	~8	Distributed source[c]
H_2CO	0–5	Variable[d]
CO_2	3	Infrared (Vega 1 IKS)
CH_4	<0.2–1.2	Groundbased infrared[e]
	0–2	Giotto IMS[f]
NH_3	0.1–0.3	Variable[g]; based on NH_2
	1–2	Giotto NMS[h]
HCN	0.1	Variable; groundbased radio
	<0.02	Giotto IMS
N_2	~0.02	Groundbased N_2^+ emission
SO_2	<0.002	Ultraviolet (IUE)
H_2S	—	Giotto IMS[i]
CH_3OH	~1	Giotto NMS and IMS[j]
Other Comets		
CO	20	West (1976 VI)
	2	Bradfield (1979 X)
	1–3	Austin (1989c1)
CH_4	<0.2	Levy (1990c)[k]
	1.5–4.5	Wilson (1987 VII)[l]
CH_3OH	1–5	Variable[m]
H_2CO	0.1–0.04	If a parent species[n]
HCN	0.03–0.2	Several comets[o]
H_2S	0.2	Austin (1989c1); Levy (1990c)[p]
S_2	0.025	IRAS-Araki-Alcock (1983 VII)[q]

[a] Table from Mumma et al. (1993a).
[b] Water was detected directly by infrared spectroscopy (Mumma et al. 1986; Combes et al. 1986) and by mass spectroscopy (Krankowsky et al. 1986).
[c] CO was detected directly at ultraviolet wavelengths (Feldman et al. 1987; Woods et al. 1986,1987) and in neutral mass spectra (Eberhardt et al. 1987). A tentative detection at infrared wavelengths (Combes et al. 1988) provided production rates in agreement with the native source.
[d] The H_2CO abundance is variable, relative to water. The largest value found for comet Halley was 4.5%±0.5%, measured by both IKS and Giotto NMS (also IMS), but at other times the production rates were 10 times smaller (Mumma and Reuter 1989). The values retrieved for comets Austin and Levy were much smaller than the values found in comet Halley.
[e] Retrieved from a single spectral line of CH_4. The range reflects the uncertainty in rotational temperature for cometary CH_4 (50–200 K). The extrapolation from a single line to the ensemble production rate is therefore highly uncertain. See Kawara et al. (1988).
[f] The production rate retrieved from the neutral mass spectra on Giotto is highly model dependent, and could be zero in comet Halley (Allen et al. 1987; Boice et al. 1990). Brooke et al. (1991) recently found CH_4<0.2% in comet Levy (1990c).

TABLE III (cont.)

[g] Assuming that NH_2 is produced solely from NH_3 (Magee-Sauer et al. 1989; Mumma et al. 1990; Krasnopolsky and Tkachuk 1991; Wyckoff et al. 1991).

[h] Allen et al. (1987). Boice et al. (1990) re-analyzed the NMS spectra, retrieving 1%, while Ip et al. (1990) retrieved 0.5% from the Giotto IMS data.

[i] Marconi et al. (1990). An incorrect lifetime was used in deriving the abundance of H_2S (see Crovisier et al. 1991).

[j] See Geiss et al. (1991) and Eberhardt et al. (1991).

[k] Based on groundbased infrared spectroscopy (Brooke et al. 1991).

[l] Based on airborne infrared spectroscopy (Larson et al. 1989).

[m] The value in comet Levy was ∼1%, but in comet Austin it was about 5% (Bockelée-Morvan et al. 1991; Bockelée-Morvan, personal communication; Hoban et al. 1991).

[n] 0.1% in comet Austin (1989c1) and 0.04% in comet Levy (1990c). The production rate would be about ten-fold larger if formaldehyde were a daughter product (Colom et al. 1992).

[o] In P/Brorsen-Metcalf (1989 X), Austin (1989c1), Levy (1990c) (Bockelée-Morvan et al. 1990).

[p] Crovisier et al. (1991).

[q] Kim et al. (1990).

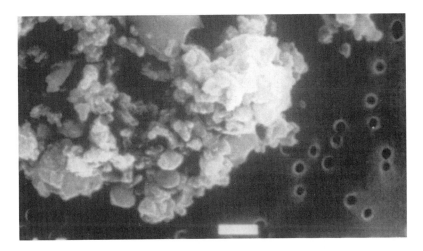

Figure 6.   Interplanetary dust particle (IDP) recovered by high flying U-2 aircraft. The suspected cometary dust particle is an assemblage of sub-micron silicate and hydrocarbon grains. Inter-grain spaces were likely formerly filled by cometary ices. A $1.0\,\mu m$ scale (10^{-6}m) is at lower right. IDPs are also referred to as Brownlee particles because of the pioneering work of D. Brownlee in collecting and studying them.

TABLE IV
Average Atomic Abundances of Elements in Halley Dust Grains[a]

Element	Halley		Solar System	CI-Chondrites
	Dust	Dust and Ice		
H[b]	2025	4062	2,600,000	492
C	814	1,010	940	70.5
N	42	95	291	5.6
O	890	2,040	2,216	712
Na	10	10	5.34	5.34
Mg	100	100	100	100
Al	6.8	6.8	7.91	7.91
Si	185	185	93.1	93.1
S	72	72	46.9	47.9
K	0.2	0.2	0.35	0.35
Ca	6.3	6.3	5.69	5.69
Ti	0.4	0.4	0.223	0.223
Cr	0.9	0.9	1.26	1.26
Mn	0.5	0.5	0.89	0.89
Fe	52	52	83.8	83.8
Co	0.3	0.3	0.21	0.21
Ni	4.1	4.1	4.59	4.59

[a] Table from Jessberger and Kissel (1991). For comparison, the new solar system abundances and the CI-chondrite composition are also given (Anders and Grevesse 1989). The solar photospheric abundances of the listed elements are practically indistinguishable from the solar system abundances, with the exception of Fe, which has a photospheric abundance ratio of 123 (Anders and Grevesse 1989).

[b] From *short* spectra only ("short spectra" refers to one of the internal instrument modes of the PIA and PUMA mass spectrometers onboard the Giotto and Vega spacecraft, respectively).

O and S, in the gas phase in cometary comae appear to be in agreement with other measured isotopic ratios in the solar system; these are shown in Table V. The deuterium/hydrogen ratio in cometary ices matches the range of D/H in Uranus, Neptune, and Titan, as well as meteorites and terrestrial ocean water. These are all bodies which may have received their volatiles in the form of condensed solids (i.e., ices, hydrates). In contrast, the D/H ratio in comets is about an order of magnitude more than in Jupiter and Saturn, which likely received the bulk of their volatiles in the form of nebula gases. One explanation of such an enrichment is ion-molecule reactions at low temperatures in a dense gas phase environment in the pre-solar nebula.

Earth-based spectroscopic measurements of the carbon isotope ratio $^{12}C/^{13}C$ found for CN in the coma of comet Halley gave values of 89 ± 17 (Jaworski and Tatum 1991) and 95 ± 12 (Kleine et al. 1994), both consistent with the terrestrial value. Mass spectrometry of individual Halley dust

TABLE V

Isotopic Ratios in Comets and Other Reservoirs[a]

Species	Solar System	Local ISM[b]	Comets
D/H	2×10^{-5}	1.5×10^{-5}	0.6×10^{-4} to 4.8×10^{-4c}
D/H		1×10^{-4} to 1×10^{-2d}	
$^{12}C/^{13}C$	89	43 ± 4^e	$95\pm12^{f,j}$
$^{12}C/^{13}/C$		65 ± 20^g	70 to 130^h
$^{12}C/^{13}C$		12 to 110^d	10 to 1000^i
$^{14}N/^{15}N$	270	$\approx400^g$	$>200^j$
$^{16}O/^{18}O$	490	≈400	<450
$^{24}Mg/^{25}Mg$	7.8		variable^i
$^{25}Mg/^{26}Mg$	0.9		$<2^i$
$^{32}S/^{34}S$	22.6		22^i
$^{56}Fe/^{54}Fe$	15.8		15^i

[a] Table adapted from Vanysek 1991. See Vanysek (1991) for discussion and references for above data.
[b] ISM = interstellar matter.
[c] Data for P/Halley only.
[d] Range of observed values in dense ISM.
[e] From visual spectra.
[f] From groundbased observation of CN bands in Halley.
[g] From radio astronomical data.
[h] From groundbased observations of C_2 spectra.
[i] From *in situ* mass spectrometry of Halley dust.
[j] New values from Kleine et al. (1994).

grains showed variations in $^{12}C/^{13}C$ between 10 and 1000, likely reflecting the nonequilibrium chemistry in the pre-solar nebula and in the dense cloud core out of which the solar system formed. However, the bulk of the isotopic evidence suggests that comets formed out of the same compositional reservoir and at the same time as the rest of the planetary system.

XI. COMETARY DYNAMICS

Cometary orbits are classified as either long- or short-period, depending on whether their orbital periods are greater than or less than 200 yr, respectively. The long-period orbits are randomly oriented on the celestial sphere, whereas the short-period comets are generally confined to direct orbits with inclinations less than $\sim35°$. In recent years the short-period orbits have been additionally subdivided into two groups: the Jupiter-family comets with periods less than 20 yr, virtually all of which are in low inclination orbits, and the Halley-family comets with periods, $20<P<200$ yr, which tend to include high inclination comets as well. Long-period orbital periods range up to $\sim10^7$yr. The designation "P/" in front of a comet's name is often used to denote a short-period comet. It is only in the last several decades that comets have been recognized to be primordial members of the solar system, and not captured from interstellar space. Approximately one-third of all long-period comets observed passing through the planetary system are on weakly hyperbolic orbits. However, integration of the orbits backward in time to points outside the planetary region, and conversion from a heliocentric to a barycentric coordinate system, showed that those comets in fact had highly eccentric but still gravitationally bound orbits. Planetary perturbations, primarily by Jupiter, scatter the long-period comets in orbital energy, either ejecting them on hyperbolic orbits or capturing them to more tightly bound ellipses.

The successful explanation of the observed energy distribution of long-period comet orbits was provided by Oort (1950) who proposed that the planetary system was surrounded by a distant spherical cloud of comets extending roughly halfway to the nearest stars. Random passing stars and galactic tidal forces perturb the comet cloud and provide the flux of long-period comets into the planetary region. Current estimates for the population of the dynamically active outer Oort cloud are $\sim 10^{12}$ comets (Weissman 1991). There is also expected to be a massive inner comet cloud with 5 to 10 times the population of the outer cloud, but in orbits that are not easily perturbed except by very close stellar passages or encounters with giant molecular clouds in the Galaxy (Duncan et al. 1987). The source of the Oort cloud is presumed to be icy planetesimals ejected by the growing proto-planets in the outer solar system, in particular Uranus and Neptune.

An additional suggested cometary reservoir is the Kuiper belt, a disk of remnant icy planetesimals beyond Neptune, proposed by Kuiper (1951). Because of their long orbital periods and the expected decreasing density of material in the solar nebula accretion disk beyond Neptune, this material never accreted into a large planetary body. Duncan et al. (1988) showed that this material was the likely source of the short-period comets, in particular the low inclination Jupiter-family comets. The more randomly inclined Halley-family comets may be long-period comets from the Oort cloud which have been random-walked in energy down to their low semimajor axes by planetary perturbations, or Kuiper belt comets which have been scattered to higher inclination orbits. The first two members of the Kuiper belt, 1992 QB_1 and 1993 FW, were discovered in 1992–93 (Jewitt and Luu 1992; Luu and Jewitt 1993; Weissman 1993).

Returns of known short-period comets and the possibility of impacts on the Earth can be predicted with high accuracy, though not quite as well as for asteroids because of nongravitational forces on the comets resulting from jetting of volatiles from the nucleus surfaces. Twenty-six short-period comets have been discovered in Earth-crossing orbits, though some of them are currently lost (possibly disrupted) or no longer Earth-crossing (see the Chapter by Shoemaker et al.). Returns of long-period comets can, in general, not be predicted; they appear randomly in time.

The most recent Catalogue of Cometary Orbits (Marsden and Williams 1993) lists 1392 cometary apparitions, of which 681 are long-period comets. The remaining 711 are appearances by 174 short-period comets, 107 of them on two or more returns. The most observed comet is P/Encke with 56 returns, and the longest observed comet is P/Halley, seen on every return since 240 B. C. Recent useful reviews on the Oort cloud and cometary dynamics are provided by Weissman (1991) and Fernandez and Ip (1991), and on short-period comets by Weissman and Campins (1993). An excellent recent review on cometary origin is that by Mumma et al. (1993a).

XII. CONSEQUENCES OF COMETARY IMPACTS

To first order, the impact of a comet on the Earth will be similar to an asteroid impact, in that it will deposit a large quantity of kinetic energy at some location on the surface. However, it is worth considering some subtle differences in the nature of the impacts as a result of the typically higher velocities and lower densities and material strengths of cometary impactors. Some minor differences resulting from the different composition of comets are also possible.

A comet approaching the Earth on an impact trajectory will be more likely to disrupt tidally as it crosses the Roche limit, because of both its lower density and lower material strengths than for rocky or iron asteroids. However, there is insufficient time for the fragments to disperse prior to impact. The geocentric distance d_E at which a comet of density equal to 1 g cm^{-3} (and zero strength) will begin to disrupt is given by

$$d_E = 2(\rho_E/\rho_c)^{1/3} R_E = 3.53 R_E \qquad (5)$$

where ρ_E and ρ_c are the density of the Earth and the comet, respectively, and R_E is the radius of the Earth. A short-period comet traveling at a mean impact velocity of 29 km s^{-1} (Weissman 1982b) will take 556 sec, or 9.3 min to travel from the Roche limit to the Earth's surface, assuming a normal impact. If the fragments were to separate at 1 m s^{-1}, they would strike at most only 0.6 km apart (if the velocity impulse was tangential), still likely well within the mutual craters they would each form, assuming impactors more than a few hundred meters in size. Impact times and separation distances will be smaller for the faster long-period comets, but would be greater for oblique impacts. However, only near-grazing impactors would likely have sufficient time to allow substantial crater separation.

Because of their low material strengths and typically higher velocities, comets will tend to break up in the atmosphere at higher altitudes due to aerodynamics stresses, than for comparably sized asteroids. For example, Chyba et al. (1993) estimate that a cometary impactor with the 15 megaton energy of the 1908 Tunguska explosion would have deposited its energy at an altitude of ~23 km for a short-period comet, and ~29 km for a (higher velocity) long-period comet, as compared with ~9 km for a typical stony asteroid. The minimum size cometary impactor which could survive intact to strike the surface of the Earth is likely a few hundred meters in diameter (P. Thomas, personal communication).

Passage of a hyper-velocity impactor through the Earth's atmosphere has been suggested as a possible source of atmospheric pollution as a result of the frictional heating of the air along the impactor's path (Lewis et al. 1982). Pollutants would include NO_2 and HNO_3 and might result in global smog and/or acid rain. Because of their higher velocities and propensity for fragmentation, comets will likely produce more such pollutants than comparably sized asteroids entering at lower velocities.

Contamination of the Earth's biosphere by toxic cometary hydrocarbons, in particular HCN, was suggested as a possible cause of extinctions associated with impacts (Hsü 1980). However, the bulk of the cometary material will be vaporized in the impact fireball and few cometary molecules would survive the impact. The vaporized material would, of course, provide a rich volatile reservoir that would form new compounds as the fireball cools. However, the fraction of vaporized cometary material is still likely small as compared with the vaporized target rock and/or ocean, and thus likely a minor contributor to any subsequent atmospheric pollution.

The estimated diameter in meters of a crater formed by an impactor is given by Melosh (1989) from Schmidt-Holsapple scaling as

$$D = 1.8\rho_i^{0.11}\rho_t^{-0.33}g^{-0.22}L^{0.13}W^{0.22} \tag{6}$$

where ρ_i and ρ_t are the densities of the impactor and the target, respectively, g is the acceleration of gravity, L is the diameter of the impactor, and W is the impact energy (all in m.k.s. units). If one takes $W = 0.5m_iv_i^2 = 0.667\pi\rho_i(L/2)^3v_i^2$, where m_i and v_i are the mass and velocity of the impactor, respectively, then

$$D = 1.34\rho_i^{0.33}\rho_t^{-0.33}g^{-0.22}L^{0.79}v_i^{0.44} \tag{7}$$

For impacts of equal energy W the lower density of the comet will result in a smaller equivalent crater than for a rocky or iron asteroid, even though the comet may be somewhat larger in diameter. Note, however, that the dependence on impactor density goes only as the cube root of density, so the effect will not be large. Cometary impactors might also be expected to create shallower craters because the low material strength of the comets will lead them to deposit most of their kinetic energy relatively close to the target surface.

XIII. DISCUSSION

Active long- and short-period comets provide between 20 and 30% of the potential major impactors on the Earth (Shoemaker et al. 1990; Weissman 1990a). The uncertainty in the cratering estimates reflect the current poor state of knowledge of the cometary mass distribution, as well as the lack of detailed knowledge of the flux of both long- and short-period comets.

Additional uncertainty comes from the possible existence of large numbers of extinct cometary nuclei. Comets may represent a particularly difficult threat to deal with because of the short warning times which might occur for long-period comets on an impacting trajectory.

Further research into the nature of cometary nuclei is required to provide the knowledge necessary for understanding their potential threat and for developing technologies for deflection or destruction of hazardous nuclei. These should include groundbased studies such as the proposed Spaceguard Survey (Morrison 1992), as well as follow-up physical studies using Earth-based and Earth-orbiting telescopes. However, the small size of the cometary nuclei demands that they can only be adequately explored with spacecraft missions which can observe the nuclei at close range and directly sample cometary materials. In particular, rendezvous missions are required which can collect materials at low velocities such that they are not vaporized or otherwise altered, and which can observe the nucleus as it evolves from its dormant state at large heliocentric distances, through its active phases around perihelion, and back to dormancy as it recedes from the Sun. By necessity, spacecraft missions can only be planned to short-period comets whose returns can be accurately predicted. However, the possibility of a fast fly-through of a new long-period comet is an intriguing idea and one worthy of some additional study.

One mission that would have gone a great way towards providing the necessary information was the Comet Rendezvous Asteroid Flyby mission (Weissman and Neugebauer 1992), which was canceled in 1992 for budgetary reasons. CRAF's complement of remote sensing and *in situ* dust and gas sampling instruments would have provided a quantum leap forward in our knowledge of cometary nuclei, as well as providing an accurate bulk density measurement and other relevant observations of a cometary nucleus. The demise of that mission was a tremendous loss for this area of research.

The European Space Agency is currently planning a comet rendezvous mission, called Rosetta (Schwehm and Langevin 1991), which is similar to CRAF in many ways, and would provide much of the same information on the nucleus structure, morphology, density, and composition. Those data are vital to further studies of the impact hazard problem. Although earlier Rosetta designs emphasized a sample-return mission, the current mission plan is for *in situ* measurements of the coma and nucleus of a typical short-period comet from a close orbit. The payload may also include a surface lander package. Thus, Rosetta should be given strong support by the space science community and by those interested in the impact hazard problem. Current plans are to launch Rosetta in 2003 with rendezvous occurring in 2010.

Other missions under consideration include several small, Discovery class missions being studied by NASA. Most of these missions involve flybys, often multiple flybys, of cometary targets. As such, these missions can only provide a modest amount of information on individual nuclei and will lack the detailed measurements possible with Rosetta. In particular, nucleus density will be very difficult to determine with flyby missions because of the low density of the comets, the high speed of the flybys, the poor determination of the nucleus volume (half hidden in shadow during the flyby), and the need to stay some modest distance from the nuclei for spacecraft safety as well as remote sensing needs. In addition, the limited payloads of the Discovery class missions are likely to leave many important questions unanswered. Multiple flyby missions can have some value in addressing the question of cometary diversity.

One Discovery class mission with substantially higher potential is a planned low-cost rendezvous mission called Cometary Coma Chemical Composition, also known as C4. Because this mission will perform a rendezvous, it can determine the nucleus mass and density to high accuracy, can collect and analyze materials over a wide range of heliocentric distances, and can observe the rise and fall of cometary activity. In addition to dust and gas sampling mass spectrometers, the mission would include a simple CCD camera to be used for both science and spacecraft navigation.

Comets are the least well-characterized of the Earth approaching population of potential impactors. This is an unacceptable situation from the point of view of the impact hazard, both in terms of defining the hazard and in designing potential mitigation architectures. Future groundbased studies and the proposed Spaceguard Survey will add greatly to our knowledge of comets and their population statistics. However, only rendezvous spacecraft missions have the capability to determine the necessary physical parameters of concern to the impact hazard. It is our fervent hope that such missions will finally become a reality within the next decade.

Acknowledgments. The authors thank D. Yeomans, B. Marsden, S. Wyckoff and M. A'Hearn for useful reviews of an earlier draft of this chapter. This work was supported in part by the NASA Planetary Geology and Geophysics Program, and was performed in part at the Jet Propulsion Laboratory under contract with the National Aeronautics and Space Administration.

Note Added in Proof

Asphaug and Benz (1994) have modeled the breakup of comet Shoemaker-Levy 9 as a primordial rubble pile of hundreds or thousands of smaller cometesimals. They find that following the passage within the Roche limit of Jupiter in 1992, the swarms of particles reaccreted under their own mutual gravity, with the number of final fragments being a function of the density of the cometesimals. To obtain the 21 observed fragments of comet Shoemaker-Levy 9, they find the cometesimals should have had densities of 0.4 to 0.9 g cm^{-3}, or somewhat higher, 1.0 to 1.3 g cm^{-3}, if the comet was a prograde rotator. Densities lower than 0.4 g cm^{-3} and retrograde rotations seem to be clearly ruled out.

The subsequent impact of the Shoemaker-Levy 9 fragments into Jupiter in July 1994 provided additional evidence in support of the rubble pile model for cometary nuclei. In particular, fragments appeared not to penetrate as deeply into the atmosphere as predicted, suggesting that they possibley entered the atmosphere as already partially dispersed rubble swarms. Several smaller fragments appear to have produced bright meteor displays but very little else in the way of visual phenomena. Only a few of the largest fragments appear to have penetrated to the level of the ammonia clouds in the Jovian atmosphere. More detailed analyses of the Shoemaker-Levy 9 impact observations are certain to add considerable knowledge to the characterization of cometary nuclei.

REFERENCES

Aggarwal, H. R., and Oberbeck, V. R. 1974. Roche limit of a solid body. *Astrophys. J.* 191:577–588.

A'Hearn, M. F. 1988. Observations of cometary nuclei. *Ann. Rev. Earth Planet. Sci.* 16:272–293.

A'Hearn, M. F., and Festou, M. C. 1990. The neutral coma. In *Physics and Chemistry of Comets*, ed. W. F. Huebner (Berlin: Springer-Verlag), pp. 66–112.

A'Hearn, M. F., Hoban, S., Birch, P. V., Bowers, C., and Klinglesmith, D. A., III. 1986. Cyanogen jets in comet Halley. *Nature* 324:649–651.

Allen, M., Delitsky, M., Huntress, W., Yung, Y., Ip, W.-H., Schwenn, R., Rosenbauer, H., Shelley, E., Balsiger, H., and Geiss, J. 1987. Evidence for methane and ammonia in the coma of comet Halley. *Astron. Astrophys.* 187:505–512.

Anders, E., and Grevesse, E. 1989. Abundances of the elements: Meteoritic and solar. *Geochim. Cosmochim. Acta* 53:197–214.

Asphaug, E., and Benz, W. 1994. Density of comet Shoemaker/Levy 9 deduced by modelling breakup of the parent 'rubble pile'. *Nature* 370:120–124.

Bailey, M. A., and Stagg, C. R. 1988. Cratering constraints on the inner Oort cloud and implications for cometary origins. *Mon. Not. Roy. Astron. Soc.* 235:1–32.

Beletic, J. W., Goody, R. M., and Tholen, D. J. 1989. Orbital elements of Charon from speckle interferometry. *Icarus* 79:38–46.

Belton, M. J. S. 1991. Characterization of the rotation of cometary nuclei. In *Comets in the Post-Halley Era*, eds. R. L. Newburn, Jr., M. Neugebauer and J. Rahe (Kluwer: Dordrecht), pp. 691–721.

Belton, M. J. S., Julian, W. H., Anderson, A. J., and Mueller, B. E. A. 1992. The spin and homogeneity of comet Halley's nucleus. *Icarus* 93:183–193.

Bockelée-Morvan, D., Crovisier, J., Colom, D., Despois, D., and Paubert, G. 1990. Observations of parent molecules in comets P/Brorsen-Metcalf, Austin (1989c1) and Levy (1990c) at millimetre wavelengths: HCN, H_2S, H_2CO, and CH_3OH. In *Formation of Stars and Planets*, ed. B. Battrick, ESA SP-315 (Noordwijk: European Space Agency), pp. 143–148.

Bockelée-Morvan, D., Colom, P., Crovisier, J., Despois, D., and Paubert, D. 1991. Microwave detection of hydrogen sulphide and methanol in comet Austin (1989c1). *Nature* 350:318–320.

Boice, D. C., Huebner, W. F., Sablik, M. J., and Konno, I. 1990. Distributed coma sources and the CH_4/CO ratio in comet Halley. *Geophys. Res. Lett.* 16:1813–1816.

Boss, A. P. 1994. Tidal disruption of periodic comet Shoemaker-Levy 9 and a constraint on its mean density. *Icarus*, in press.

Boss, A. P., Cameron, A. G. W., and Benz, W. 1991. Tidal disruption of inviscid planetesimals. *Icarus* 92:165–178.

Bowell, E., and Marsden, B. G. 1992. 4015 (1979 VA) = Comet Wilson-Harrington (1949 III). *IAU Circ.* 5585.

Brin, G. D., and Mendis, D. A. 1979. Dust release and mantle development in comets. *Astrophys. J.* 229:402–408.

Brooke, T. Y., Tokunaga, A. T., Weaver, H. A., Chin, G., and Geballe, T. R. 1991. A sensitive upper limit on the methane abundance in comet Levy (1990c). *Astrophys. J. Lett.* 372:113–116.

Burns, J. A., and Safronov, V. 1973. Asteroid nutation angles. *Mon. Not. Roy. Astron. Soc.* 165:403–411.

Bus, S. J., Stern, S. A., and A'Hearn, M. F. 1991a. Chiron: Evidence for historic cometary activity. In *Asteroids, Comets and Meteors 1991* (Houston: Lunar and Planet Inst.), p. 34 (abstract).

Bus, S. J., A'Hearn, M. F., and Schleicher, D. G. 1991b. Detection of CN emission from (2060) Chiron. *Science* 251:774–777.

Campins, H., Jewitt, D., and Telesco, C. 1992. 2060 Chiron. *IAU Circ.* 5457.

Chyba, C. F., Thomas, P. J., and Zahnle, K. J. 1993. The 1908 Tunguska explosion: Atmospheric disruption of a stony asteroid. *Nature* 361:41–44.

Colom, P., Crovisier, J., Bockelée-Morvan, D., Despois, D., and Paubert, G. 1992. Formaldehyde in comets: I. Microwave observations of P/Brorsen-Metcalf (1989 X), Austin (1989c1), and Levy (1990c). *Astron. Astrophys.* 264:270–281.

Combes, M., Moroz, V. I., Crifo, J. F., Bibring, J. P., Coron, N., Crovisier, J., Encrenaz, T., Sanko, N., Grigoriev, A., Bockelée-Morvan, D., Gispert, R., Emerich, C., Lamarre, J. M., Rocard, F., Krasnopolsky, V., and Owen, T. 1986. Detection of parent molecules in comet Halley from the IKS-Vega experiment. In *Exploration of Halley's Comet*, vol. 1, eds. B. Battrick, E. J. Rolfe and R. Reinhard, ESA SP-250 (Noordwijk: European Space Agency), pp. 353–358.

Combes, M., Moroz, V. I., Crovisier, J., Encrenaz, T., Bibring, J. P., Grigoriev, A. V., Sanko, N. F., Coron, N., Crifo, J. F., Gispert, R., Bockelée-Morvan, D., Nikolsky, Yu. V., Krasnopolsky, V. A., Owen, T., Emerich, C., Lamarre, J. M., and Rocard, F. 1988. The 2.5–12 μm spectrum of comet Halley from the IKS-VEGA experiment. *Icarus* 76:404–436.

Crovisier, J., Despois, D., Bockelée-Morvan, D., Colom, P., and Paubert, G. 1991. Microwave observations of hydrogen sulfide and searches for other sulphur compounds in comets Austin (1989c1) and Levy (1990c). *Icarus* 93:246–258.

Daniels, P. A., and Hughes, D. W. 1981. The accretion of cosmic dust: A computer experiment. *Mon. Not. Roy. Astron. Soc.* 195:1001–1009.

Degewij, J., and Tedesco, E. F. 1982. Do comets evolve into asteroids? Evidence from physical studies. In *Comets*, ed. L. L. Wilkening (Tucson: Univ. of Arizona Press), pp. 665–695.

Delsemme, A. H., and Rud, D. A. 1973. Albedos and cross-sections for the nuclei of comets 1969 IX, 1970 II, and 1971 II. *Astron. Astrophys.* 28:1–6.

Dobrovolskis, A. R. 1990. Tidal disruption of solid bodies. *Icarus* 88:24–38.

Donn, B. 1991. The accumulation and structure of comets. In *Comets in the Post-Halley Era*, eds. R. L. Newburn, Jr., M. Neugebauer and J. Rahe (Dordrecht: Kluwer), pp. 335–359.

Donn, B., Daniels, P. A., and Hughes, D. W. 1985. On the structure of the cometary nucleus. *Bull. Amer. Astron. Soc.* 17:520 (abstract).

Duncan, M., Quinn, T., and Tremaine, S. 1987. The formation and extent of the solar system comet cloud. *Astron. J.* 94:1330–1338.

Duncan, M., Quinn, T., and Tremaine, S. 1988. The origin of short-period comets. *Astrophys. J. Lett.* 328:69–73.

Eberhardt, P., Krankowsky, D., Schulte, W., Dolder, U., Lämmerzahl, P., Berthelier, J. J., Woweries, J., Stubbeman, U., Hodges, R. R., Hoffman, J. H., and Illiano, J. M. 1987. The CO and N_2 abundance in comet P/Halley. *Astron. Astrophys.* 187:481–484.

Eberhardt, P., Meier, R., Krankowsky, D., and Hodges, R. R. 1991. Methanol abundances in comet P/Halley from in-situ measurements. *Bull. Amer. Astron. Soc.* 23:1161 (abstract).

Everhart, E. 1967. Intrinsic distributions of cometary perihelia and magnitudes. *Astron. J.* 72:1002–1011.

Fanale, F. P., and Salvail, J. R. 1984. An idealized short-period comet model: Surface insolation, H_2O flux, dust flux, and mantle evolution. *Icarus* 60:476–511.

Feldman, P. D., Festou, M. C., A'Hearn, M. F., Arpigny, C., Butterworth, P. S., Cosmovici, C. B., Danks, A. C., Gilmozzi, R., Jackson, W. M., McFadden, L. A., Patriarchi, P., Schleicher, D. G., Tozzi, G. P., Wallis, M. K., Weaver, H. A., and Woods, T. N. 1987. IUE observations of comet P/Halley: Evolution of the ultraviolet spectrum between September 1985 and July 1986. *Astron. Astrophys.* 187:325–328.

Fernandez, J. A. 1988. End-states of short-period comets and their role in maintaining the zodiacal dust cloud. *Moon & Planets* 41:155–161.

Fernandez, J. A., and Ip, W.-H. 1991. Statistical and evolutionary aspects of cometary orbits. In *Comets in the Post-Halley Era*, eds. R. L. Newburn, Jr., M. Neugebauer and J. Rahe (Dordrecht: Kluwer), pp. 487–535.

Fink, U. 1992. Comet Yanaka (1988r): A new class of carbon-poor comet. *Science* 257:1926–1929.

Fraundorf, P., Hintz, C., Lowry, O., McKeegan, K. D., and Sandford, S. A. 1982. Determination of the mass, surface density, and volume density of individual interplanetary dust particles. *Lunar Planet. Sci.* XIII:225–226 (abstract).

Geiss, J., Altwegg, K., Anders, E., Balsiger, H., Ip, W.-H., Meier, A., Neugebauer, M., Rosenbauer, H., and Shelley, E. G. 1991. Interpretation of the ion mass spectra in the mass per charge range 25–35 amu/e− obtained in the inner coma of Halley's comet by the HIS-sensor of the Giotto IMS experiment. *Astron. Astrophys.* 247:226–234.

Gombosi, T. I., and Houpis, H. L. F. 1986. An icy-glue model of cometary nuclei. *Nature* 324:43–46.

Greenberg, J. M., and Hage, J. I. 1990. From interstellar dust to comets: A unification of observational constraints. *Astrophys. J.* 361:260–274.

Hoban, S., Mumma, M., Reuter, D. C., DiSanti, M., Joyce, R. R., and Storrs, A. 1991. A tentative identification of methanol as the progenitor of the 3.52 μm emission feature in several comets. *Icarus* 93:122–134.

Hsü, K. J. 1980. Terrestrial catastrophe caused by cometary impact at the end of the Cretaceous. *Nature* 285:201–203.

Hughes, D. W. 1991. Possible mechanisms for cometary outbursts. In *Comets in the Post-Halley Era*, eds. R. L. Newburn, Jr., M. Neugebauer and J. Rahe (Dordrecht: Kluwer), pp. 825–851.

Ip, W.-H., Rosenbauer, H., Schwenn, R., Balsiger, H., Geiss, J., Meier, A., Goldstein, B. E., A. J., Lazarus, Shelley, E., and Kettmann, G. 1990. Giotto IMS measurements of the production rate of hydrogen cyanide in the coma of comet Halley. *Ann. Geophys.* 8:319–325.

Jaworski, W. A., and Tatum, J. B. 1991. Analysis of the Swings effect and Greenstein effect in comet Halley. *Astrophys. J.* 377:306–317.

Jessberger, E. K., and Kissel, J. 1991. Chemical properties of cometary dust and a note on carbon isotopes. In *Comets in the Post-Halley Era*, eds. R. L. Newburn, Jr., M. Neugebauer, and J. Rahe (Dordrecht: Kluwer), pp. 1075–1092.

Jessberger, E. K., Christoforidis, A., and Kissel, J. 1988. Aspects of the major element composition of Halley's dust. *Nature* 332:691–695.

Jewitt, D. 1991. Cometary photometry. In *Comets in the Post-Halley Era*, eds. R. L. Newburn, Jr., M. Neugebauer and J. Rahe (Dordrecht: Kluwer), pp. 19–65.

Jewitt, D., and Luu, J. 1992. 1992 QB$_1$. *IAU Circ.* 5611.

Jewitt, D. C., and Meech, K. J. 1988. Optical properties of cometary nuclei and a preliminary comparisons with asteroids. *Astrophys. J.* 328:974–986.

Johnson, R. E., Cooper, J. F., Lanzerotti, L. J., and Strazzula, G. 1987. Radiation formation of a non-volatile comet crust. *Astron. Astrophys.* 187:889–892.

Kaneda, E., Hirao, K., Takagi, M., Ashihara, O., Itoh, T., and Shimizu, M. 1986. Strong breathing of the hydrogen coma of comet Halley. *Nature* 320:140–141.

Kawara, K., Gregory, B., Yamamoto, T., and Shibai, H. 1988. Infrared spectroscopic observation of methane in comet P/Halley. *Astron. Astrophys.* 207:174–181.

Keller, H. U. 1987. The nucleus of comet Halley. In *Diversity and Similarity of Comets*, eds. E. J. Rolfe and B. Battrick, ESA SP-278 (Noordwijk: European Space Agency), pp. 447–454.

Kim, S. J., A'Hearn, M. F., and Larson, S. M. 1990. Multi-cycle fluorescence: Application to S$_2$ in comet IRAS-Araki-Alcock 1983d. *Icarus* 87:440–451.

Kleine, M., Wyckoff, S., Wehinger, P., and Peterson, B. 1994. The carbon isotope abundance ratio in comet Halley. *Astrophys. J.*, submitted.

Krankowsky, D., Lämmerzahl, P., Herrwerth, I., Woweries, J., Eberhardt, P., Dolder, U. Herrmann, U., Schulte, W., Berthelier, J., Illiano, J. J., Hodges, R. R., and Hoffman, J. H. 1986. In situ gas and ion composition measurements at comet Halley. *Nature* 321:326–329.

Krasnopolsky, V. A., and Tkachuk, A. Yu. 1991. TKS-Vega experiment: NH and NH$_2$ bands in comet Halley. *Astron. J.* 101:1915–1919.

Kresák, L. 1985. The aging and lifetimes of comets. In *Dynamics of Comets: Their Origin and Evolution*, eds. A. Carusi and G. B. Valsecchi (Dordrecht: D. Reidel), pp. 279–302.

Kresák, L. 1987. Dormant phases in the aging of periodic comets. *Astron. Astrophys.* 187:906–908.

Kresák, L. 1991. Evidence for physical aging of periodic comets. In *Comets in the Post-Halley Era*, eds. R. L. Newburn, Jr., M. Neugebauer and J. Rahe (Dordrecht: Kluwer), pp. 607–628.

Kuiper, G. P. 1951. On the origin of the solar system. In *Astrophysics*, ed. J. A. Hynek (New York: McGraw Hill) pp. 357–424.

Larson, H. P., Weaver, H. A., Mumma, M. J., and Drapatz, S. 1989. Airborne infrared spectroscopy of comet Wilson (1986l) and comparisons with comet Halley. *Astrophys. J.* 338:1106–1114.

Lewis, J. S., Watkins, G. H., Hartman, H., and Prinn, R. G. 1982. Chemical consequences of major impact events on Earth. In *Geological Implications of Impacts of Large Asteroids and Comets on the Earth*, eds. L. T. Silver and P. H. Schultz, Geological Soc. of America Special Paper 190 (Boulder: Geological Soc. of America), pp. 215–221.

Luu, J., and Jewitt, J. 1993. 1993 FW. *IAU Circ.* 5730.

Magee-Sauer, K., Scherb, F., Roesler, F. L., Harlander, J., and Lutz, B. L. 1989. Fabry-Perot observations of the NH$_2$ emission from comet Halley. *Icarus* 82:50–60.

Marconi, M. L., Mendis, D. A., Korth, A., Pin, R. P., Mitchell, D. L., and Reme, H. 1990. The identification of H$_3$S$^+$ with the ion of mass per charge (m/q) 35 observed in the coma of comet Halley. *Astrophys. J. Lett.* 352:17–20.

Marsden, B. G. 1967. The Sun-grazing comet group. *Astron. J.* 72:1170–1183.

Marsden, B. G. 1989. The Sun-grazing comet group. II. *Astron. J.* 98:2306–2321.

Marsden, B. G., and Williams, G. V. 1993. *Catalogue of Cometary Orbits*, 8th ed. (Cambridge, Mass.: Smithsonian Astrophysical Observatory).

McDonnell, J. A. M., Lamy, P. L., and Pankiewicz, G. S. 1991. Physical properties of cometary dust. In *Comets in the Post-Halley Era*, eds. R. L. Newburn, Jr., M. Neugebauer and J. Rahe (Dordrecht: Kluwer), pp. 1043–1073.

McSween, H. Y., and Weissman, P. R. 1989. Cosmochemical implications of the physical processing of cometary nuclei. *Geochim. Cosmochim. Acta* 53:3263–3271.

Meech, K. J. 1991. Physical aging in comets. In *Comets in the Post-Halley Era*, eds. R. L. Newburn, Jr., M. Neugebauer and J. Rahe (Dordrecht: Kluwer), pp. 629–669.

Meech, K. J., and Belton, M. J. S. 1989. 2060 Chiron. *IAU Circ.* 4770.

Meech, K. J., and Belton, M. J. S. 1990. The atmosphere of 2060 Chiron. *Astron. J.* 100:1323–1338.

Meech, K. J., Belton, M. J. S., Mueller, B. E. A., Dicksion, M. W., and Li, H. R. 1993. Nucleus properties of P/Schwassmann-Wachmann 1. *Astron. J.* 106:1222–1236.

Melosh, J. 1989. *Impact Cratering: A Geologic Process* (New York: Oxford Univ. Press).

Millis, R. L., and Schleicher, D. G. 1986. Rotational period of comet Halley. *Nature* 324:646–649.

Millis, R. L., A'Hearn, M. F., and Campins, H. 1988. An investigation of the nucleus and coma of P/Arend-Rigaux. *Astrophys. J.* 324:1194–1209.

Moore, M. H., Donn, B., Khanna, R., and A'Hearn, M. 1983. Studies of proton-irradiated ice mixtures. *Icarus* 54:388–392.

Morrison, D., ed. 1992. *The Spaceguard Survey: Report of the NASA International Near-Earth Object Detection Workshop* (Pasadena: Jet Propulsion Laboratory).

Mumma, M. J., and Reuter, D. C. 1989. On the identification of formaldehyde in Halley's comet. *Astrophys. J.* 344:940–948.

Mumma, M. J., Weaver, H. A., Larson, H. P., Davis, D. S., and Williams, M. 1986. Detection of water vapor in Halley's comet. *Science* 232:1523–1528.

Mumma, M. J., Reuter, D., and Magee-Sauer, K. 1990. Heterogeneity of the nucleus of comet Halley. *Bull. Amer. Astron. Soc.* 22:1088 (abstract).

Mumma, M. J., Stern, S. A., and Weissman, P. R. 1993a. Comets and the origin of the solar system. In *Protostars and Planets III*, eds. E. H. Levy and J. L. Lunine (Tucson: Univ. of Arizona Press), pp. 1177–1252.

Mumma, M. J., Hoban, S., Reuter, D. C., and DiSanti, M. 1993b. Methanol in recent comets: Evidence for two distinct cometary populations. *Asteroids, Comets, Meteors 1993*, IAU Symp. 160, June 14–18, Belgirate (Italy), p. 227 (abstract).

Oort, J. H. 1950. The structure of the cloud of comets surrounding the solar system and a hypothesis concerning its origin. *Bull. Astron. Inst. Netherlands* 11:91–110.

Öpik, E. J. 1963. The stray bodies in the solar system. Part I. Survival of cometary nuclei and the asteroids. *Adv. Astron. Astrophys.* 2:219–262.

Peale, S. J. 1989. On the density of Halley's comet. *Icarus* 82:36–49.

Peale, S. J., and Lissauer, J. J. 1989. Rotation of Halley's comet. *Icarus* 79:396–430.

Prialnik, D., and Bar-Nun, A. 1988. The formation of a permanent dust mantle and its effect on cometary activity. *Icarus* 74:272–283.

Rickman, H. 1986. Masses and densities of comets Halley and Kopff. In *The Comet Nucleus Sample Return Mission*, ed. O. Meliter, ESA SP-249 (Noorwijk: European Space Agency), pp. 195–205.

Rickman, H. 1989. The nucleus of comet Halley: Surface, structure, mean density, gas and dust production. *Adv. Space Res.* 9(3):59–71.

Rickman, H. 1991. The thermal history and structure of cometary nuclei. In *Comets in the Post-Halley Era*, eds. R. L. Newburn, Jr., M. Neugebauer and J. Rahe (Dordrecht: Kluwer), pp. 733–759.

Roemer, E. 1965. Observations of comets and minor planets. *Astron. J.* 70:397–402.

Roemer, E., and Lloyd, R. E. 1966. Observations of comets, minor planets, and satellites. *Astron. J.* 71:443–457.

Roemer, E., Thomas, M., and Lloyd, R. E. 1966. Observations of comets, minor planets, and Jupiter VIII. *Astron. J.* 71:591–601.

Sagdeev, R. Z., Avanesov, G. A., Shamis, V. A., Szegö, K., Merenyi, E., Smith, B. A., Ziman, Ya. L., Krasikov, V. L., Tarnopolsky, V. L., and Kuzim, A. A. 1986*a*. TV experiment in Vega mission: Image processing technique and some results. In *Exploration of Halley's Comet*, vol. 2, eds. B. Battrick, E. J. Rolfe and R. Reinhard, ESA SP-250 (Noordwijk: European Space Agency), pp. 295–305.

Sagdeev, R. Z., Krasikov, V. A., Szegö, K., Toth, I., Smith, B., Larson, S., Merenyi, E., Shamis, V. A., and Tarnopolsky, V. L. 1986*b*. Rotation and spin axis of comet Halley. In *Exploration of Halley's Comet*, vol. 2, eds. B. Battrick, E. J. Rolfe and R. Reinhard, ESA SP-250 (Noordwijk: European Space Agency), pp. 335–338.

Sagdeev, R. Z., Elyasberg, P. E., and Moroz, V. I. 1988. Is the nucleus of comet Halley a low density body? *Nature* 308:240–242.

Samarasinha, N. H., and A'Hearn, M. F. 1991. Observational and dynamical constraints on the rotation of comet P/Halley. *Icarus* 93:194–225.

Schleicher, D. G., Bus, S. J., and Osip, D. J. 1993. The anomalous molecular abundances of comet P/Wolf-Harrington. *Icarus* 104:157–166.

Schwehm, G. H., and Langevin, Y. 1991. In *Rosetta: A Comet Nucleus Sample Return Mission*, ESA SP-1125 (Noordwijk: European Space Agency), p. 189.

Sekanina, Z. 1981. Rotation and precession of cometary nuclei. *Ann. Rev. Earth Planet. Sci.* 9:113–145.

Sekanina, Z. 1982. The problem of split comets in review. In *Comets*, ed. L. L. Wilkening (Tucson: Univ. of Arizona Press), pp. 251–287.

Sekanina, Z. 1988. Outgassing asymmetry of periodic comet Encke, II. *Astron. J.* 96:1455–1475.

Sekanina, Z. 1993. Disintegration phenomena expected during collision of comet Shoemaker-Levy 9 with Jupiter. *Science* 262:382–387.

Sekanina, Z., and Larson, S. M. 1984. Coma morphology and dust-emission pattern of periodic comet Halley. II. Nucleus spin vector and modeling of major dust features in 1910. *Astron. J.* 89:1408–1425.

Sekanina, Z., and Yeomans, D. K. 1985. Orbital motion, nucleus precession and splitting of periodic comet Brooks 2. *Astron. J.* 90:2335–2352.

Shoemaker, E. M., and Wolfe, R. F. 1982. Cratering time scales for the Galilean satellites. In *Satellites of Jupiter*, ed. D. Morrison (Tucson: Univ. of Arizona Press), pp. 277–339.

Shoemaker, E. M., Wolfe, R. F., and Shoemaker, C. S. 1990. Asteroid and comet flux in the neighborhood of Earth. In *Global Catastrophes in Earth History*, eds. V. L. Sharpton and P. D. Ward, Geological Soc. of America Special Paper 247 (Boulder: Geological Soc. of America), pp. 155–170.

Shoemaker, E. M., Levy, D., and Shoemaker, C. S. 1993. Comet Shoemaker-Levy 1993e. *IAU Circ.* 5725.

Smoluchowski, R. 1981. Amorphous ice and the behavior of cometary nuclei. *Astrophys. J. Lett.* 244:31–34.

Sridhar, S., and Tremaine, S. 1992. Tidal disruption of viscous bodies. *Icarus* 95:89–99.

Stephenson, F. R., Yau, K. K. C., and Hunger, H. 1985. Records of Halley's comet on

Babylonian tablets. *Nature* 314:587–592.

Stern, S. A. 1986. The effects of mechanical interaction between the interstellar medium and comets. *Icarus* 68:276–283.

Stern, S. A. 1990. ISM induced erosion and gas dynamical drag in the Oort cloud. *Icarus* 84:447–466.

Stern, S. A., and Shull, J. M. 1988. The thermal evolution of comets in the Oort cloud by passing stars and stochastic supernovae. *Nature* 332:407–411.

Stewart, A. I. F. 1987. Pioneer Venus measurements of H, O, and C production in comet Halley near perihelion. *Astron. Astrophys.* 187:369–374.

Sykes, M. V., and Walker, R. G. 1991. Constraints on the diameter and albedo of 2060 Chiron. *Science* 251:777–780.

Sykes, M. V., and Walker, R. G. 1992. Cometary dust trails, I: Survey. *Icarus* 95:180–210.

Szegö, K. 1991. P/Halley, the model comet in view of the imaging experiment aboard the VEGA spacecraft. In *Comets in the Post-Halley Era*, eds. R. L. Newburn, Jr., M. Neugebauer and J. Rahe (Dordrecht: Kluwer), pp. 713–732.

Tyler, G. L., Sweetnam, D. N., Anderson, J. D., Borutzki, S. E., Campbell, J. K., Eshelman, V. R., Gresh, D. L., Gurrola, E. M., Hinson, D. P., Kawashima, N., Kursinski, E. R., Levy, G. S., Lindal, G. F., Lyons, J. R., Marouf, E. A., Rosen, P. A., Simpson, R. A., and Wood, G. E. 1989. Voyager radio science observations of Neptune and Triton. *Science* 246:1466–1473.

Vanysek, V. 1991. Isotopes in comets. In *Comets in the Post-Halley Era*, eds. R. L. Newburn, Jr., M. Neugebauer and J. Rahe (Dordrecht: Kluwer), pp. 879–896.

Watson, K., Murray, B. C., and Brown, H. 1963. The stability of volatiles in the solar system. *Icarus* 1:317–327.

Weissman, P. R. 1979. Physical and dynamical evolution of long-period comets. In *Dynamics of the Solar System*, ed. R. L. Duncombe (Dordrecht: D. Reidel), pp. 277–282.

Weissman, P. R. 1980. Physical loss of long-period comets. *Astron. Astrophys.* 85:191–196.

Weissman, P. R. 1982a. Dynamical history of the Oort cloud. In *Comets*, ed. L. L. Wilkening (Tucson: Univ. of Arizona Press), pp. 637–658.

Weissman, P. R. 1982b. Terrestrial impact rates for long and short-period comets. In *Geological Implications of Impacts of Large Asteroids and Comets on the Earth*, eds. L. T. Silver and P. H. Schultz, Geological Soc. of America Special Paper 190 (Boulder: Geological Soc. of America), pp. 15–24.

Weissman, P. R. 1986. Are cometary nuclei primordial rubble piles? *Nature* 320:242–244.

Weissman, P. R. 1990a. The cometary impactor flux at the Earth. In *Global Catastrophes in Earth History*, eds. V. L. Sharpton and P. D. Ward, Geological Soc. of America Special Paper 247 (Boulder: Geological Soc. of America), pp. 171–180.

Weissman, P. R. 1990b. Physical processing of cometary nuclei since their formation. In *Comet Halley: Investigations, Results, Interpretations*, vol. 2, ed. J. Mason (Chichester: Ellis Horwood), pp. 241–257.

Weissman, P. R. 1991. Dynamical history of the Oort cloud. In *Comets in the Post-Halley Era*, eds. R. L. Newburn, Jr., M. Neugebauer and J. Rahe (Dordrecht: Kluwer), pp. 436–486.

Weissman, P. R. 1993. The discovery of 1992 QB$_1$. *Eos: Trans. AGU* 74:257–263.

Weissman, P. R., and Campins, H. 1993. Short-period comets. In *Resources of Near-Earth Space*, eds. J. S. Lewis, M. S. Matthews and M. L. Guerrieri (Tucson: Univ. of Arizona Press), pp. 569–617.

Weissman, P. R., and Kieffer, H. H. 1981. Thermal modeling of cometary nuclei. *Icarus* 47:302–311.

Weissman, P. R., and Neugebauer, M. 1992. The comet rendezvous asteroid flyby mission: A status report. In *Asteroids, Comets, Meteors 1991*, eds. A. W. Harris and E. Bowell (Houston: Lunar and Planetary Inst.), pp. 629–632.

Weissman, P. R., A'Hearn, M. F., McFadden, L. A., and Rickman, H. 1989. Evolution of comets into asteroids. In *Asteroids II*, eds. R. P. Binzel, T. Gehrels and M. S. Matthews (Tucson: Univ. of Arizona Press), pp. 880–920.

Wetherill, G. W. 1971. Cometary versus asteroidal origin of chondritic meteorites. In *Physical Studies of Minor Planets*, ed. T. Gehrels, NASA SP-267, pp. 447–460.

Wetherill, G. W. 1975. Late heavy bombardment of the moon and terrestrial planets. In *Proc. Lunar Sci. Conf.* 6:1539–1561.

Wetherill, G. W. 1988. Where do the Apollo objects come from? *Icarus* 76:1–18.

Wetherill, G. W. 1991. End products of cometary evolution: Cometary origin of Earth-crossing bodies of asteroidal appearance. In *Comets in the Post-Halley Era*, eds. R. L. Newburn, Jr., J. Rahe and M. Neugebauer (Dordrecht: Kluwer), pp. 537–556.

Whipple, F. L. 1950. A comet model I. The acceleration of comet Encke. *Astrophys. J.* 112:375–394.

Whipple, F. L. 1951. A comet model II. Physical relations for comets and meteors. *Astrophys. J.* 113:464–474.

Whipple, F. L. 1982. The rotation of comet nuclei. In *Comets*, ed. L. L. Wilkening (Tucson: Univ. of Arizona Press), pp. 227–250.

Williams, J. G., and Faulkner, J. 1981. The positions of secular resonance surfaces. *Icarus* 46:390–399.

Wisdom, J. 1987. Chaotic dynamics in the solar system. *Icarus* 72:241–275.

Woods, T. N., Feldman, P. D., Dymond, K. F., and Sahnow, D. J. 1986. Rocket ultraviolet spectroscopy of comet Halley and abundance of carbon monoxide and carbon. *Nature* 324:436–438.

Woods, T. N., Feldman, P. D., and Dymond, K. F. 1987. The atomic carbon distribution in the coma of comet P/Halley. *Astron. Astrophys.* 187:380–384.

Wyckoff, S., Tegler, S. C., and Engel, L. 1991. Ammonia abundances in four comets. *Astrophys. J.* 368:279–286.

Yau, K., Yeomans, D., and Weissman, P. 1994. The past and future motion of comet P/Swift-Tuttle. *Mon. Not. Roy. Astron. Soc.*, 266:305–316.

A CURRENT WORKING MODEL OF A COMET NUCLEUS AND IMPLICATIONS FOR NEO INTERACTIONS

J. MAYO GREENBERG
Huygens Laboratory, University of Leiden

and

JOHN L. REMO
Quantametrics Inc.

The chemical and physical properties of the interior and surface of a comet nucleus are derived from the aggregated interstellar dust model. Such properties can be used to model kinetic, radiative, and other hazard mitigating interactions. These interaction mechanisms depend critically on the relative atomic compositions and morphological stuctures of the interior and exterior of the comet which are shown to be different.

I. INTRODUCTION

Large (>1 km) long period comets may constitute about 5 to 10% of the NEO threat and, on average, traverse cis-lunar space once a century producing catastrophic terrestrial impacts imparting 10^{21} J every few thousand years. Impacts of smaller asteroids imparting 10^8 to 10^{20} J may occur every few centuries (see the Chapter by Rabinowitz et al.). To mitigate such threats one must be able to model interactions with nuclear, kinetic, and other orbit modifying agents (see the Chapters by Marsden and Steel and Yeomans and Chodas).

Those features of a NEO comet nucleus that must be understood in order to model physical processes produced by possible mitigating procedures include: (1) chemical composition; (2) overall (macro) morphological structure such as density and strength; and (3) material state and micro properties such as thermal conductivity, amorphous, and crystalline properties. While there are many difficulties interpreting NEO material properties in general, cometary NEOs present some unique problems partly because there are no sizeable fragments available for analyses. On the other hand, near-Earth asteroids can be associated with meteorites of different classes (Gaffey et al. 1993; Chapter by Remo) and these are large enough to derive representative mineralogical compositions as well as morphological structure and strength. Meteorites which are handled in the laboratory may be used to infer the properties of their parent body asteroids. On the other hand, no such large pieces of a comet are known to be available. The closest we may come is with some of the examples from interplanetary particles collected in the atmosphere. However, we have gained confidence in being able to assign basic properties to comets as NEOs because of the remarkable success of the interstellar dust model of comet nuclei in predicting the hitherto unexpected properties of comets revealed by the Vega and Giotto space probes.

TABLE I*

Comparison of Comet and C1 Chondrite Compositions[a]

Element	Averaged Solar Abundances	Comets	C1 Chondrites
H	31×10^3	15	1.5
C[b]	12	3	0.7
N	3	>0.1	0.05
O	21	21	7.5
Si	1–1.35	1.0	1.0
Metals		1	1

* Adapted from Delsemme 1982.
[a] C1 chondrites appear to be similar to comets that have lost most of their volatile components. In particular, H is depleted in C1 chondrites by a factor of 10 with respect to comets, while C and O are depleted in chondrites by factors of 4 and 3, respectively. Si and metals (nonvolatiles) are about the same for comets and meteorites.
[b] This carbon abundance was derived from the volatiles in the comet coma. The "missing" carbon (relative to solar system abundances) is contained in the organic refractory component of the dust (Greenberg 1982,1983; Kissel and Krueger 1987).

In this chapter we outline a model of a comet nucleus which, based on observational and theoretical studies, is a suitable chemical-physical model for which physical interactions with kinetic and radiation forces may be reasonably predicted. With the amount of remote data and with the evidence deduced from the Giotto and Vega 1, 2 *in situ* measurements a reconstruction of the comet nucleus may be derived using the interstellar dust in its final protosolar cloud phase as a basic starting point. The distinction between comet and (C1 chondrite analog) asteroid formation and evolution (Greenberg 1991) is explicitly indicated in Table I where the implication is that the major difference in the atomic composition between comets and C1 chondrite asteroids is depletion of the volatile elements in the latter.

II. INTERSTELLAR DUST MODEL

We consider interstellar dust grains as the basic building blocks of comets. The relative amount and kinds of volatiles of interstellar dust will be related to the volatiles of comets. Along with the refractory components of the dust they may be used to derive the chemical and physical properties of comets and comet dust, which are quite close to the observed properties. We summarize some examples of observed interstellar dust mantle molecules with particular relevance to the comet volatile composition. Figure 1 shows a schematic of dust grains as seen in molecular clouds, such as may be expected in the protosolar cloud. This dust model is representative of large grains as seen in low-density clouds after condensibles have accreted as icy mantles, and small carbonaceous particles became inclusions in the mantles at the late stage of cloud contraction before the formation of the protosolar nebulae (Greenberg and Mendoza-Gómez 1993).

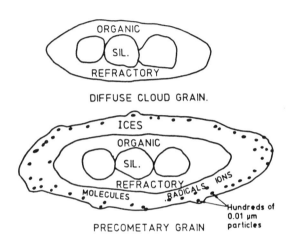

Figure 1. Interstellar grains as core-mantle structures.

Relatively refractory organic molecules are an abundant constituent of the interstellar dust. As shown by Greenberg (1983), the "carbon depletion mystery" posed by Delsemme (1982) is completely explained by the abundant carbon in the organic refractory component of interstellar dust.

Table II gives a summary of the molecular species identified or inferred in each component of the average large presolar interstellar grain pictured in Fig. 1. Estimation of relative abundances are to be taken as indicative but not conclusive because they must be based on an uncertain extrapolation of our knowledge of dust/gas chemistry in collapsing clouds.

III. THE MORPHOLOGICAL STRUCTURE OF COMET NUCLEI

It is generally believed that comets formed in the outer regions of the solar nebula. A major question is whether the original composition is entirely that of the interstellar dust in the cloud from which the solar system was born or whether some of the dust volatiles evaporated and underwent the subsequent chemistry in the solar nebula before becoming incorporated in comets. While this may be argued for some time the general trend is to believe that all—or

TABLE II

Molecules Directly Observed in Interstellar Grains and/or
Strongly Inferred from Laboratory Spectra and Theories
of Grain Mantle Evolution

	Molecule		Comment[a]
$(1)^b$	H_2O	O	M2
(0.1)	CO	O	M2
(0.3)	H_2S	I	M2
(0.01)	NH_3	O	M2
(0.03)	H_2CO	O	M2
(?)	$(H_2CO)_n$	I	M2
(0.01)	OCN^-	O	M2
(0.01)	NH_4	O	M2
(0.05)	CH_3OH	O	M2
(0.01)	OCS	O	M2
(0.05)	CO_2	O	M2
(0.01)	CH_4	O	M2
(?)	S_2	I	M2
	Complex organic	O	M1
	"Silicate"	O	C
	"Carbonaceous"	(O,I)	B
	PAHs	(O,I)	B

[a] O = observed; M1 = inner mantle; M2 = outer mantle; B = small; I = inferred;
C = core.
[b] Estimation of relative abundances in interstellar dust icy mantles.

almost all—the volatiles of the interstellar dust are brought forward into the comet (Greenberg 1986; Mumma et al. 1993). An immediate implication of this is that comet nuclei are basically aggregates of the protosolar interstellar dust grains. Then the only remaining question is what are these grains? On the basis of chemical modeling of interstellar clouds (van Dishoeck et al. 1993) and from observations of grains and grain mantles in the infrared and from observations of interstellar polarization, a putative model of the interstellar dust is that it consists of at least three and possibly four components of various sizes and compositions (Greenberg 1984; Chlewicki 1985). The major fraction of the dust (by volume) resides in elongated particles whose thickness is of the order of several tenths of microns consisting of a core of silicate ($Mg_{a1}Fe_{a2}Si_{a3}O_{a4}$) material, an inner mantle of predominantly carbon (carbonaceous) material ($C_{b1},H_{b2},O_{b3},N_{b4}$) and an outer mantle of frozen volatile molecules dominated by H_2O ice. Additional components of large carbon-rich molecules (e.g., PAHs) and small carbonized organics with sizes <0.01 μm make up the rest of the particles (Allamandola et al. 1989; Jenniskens et al. 1993). Furthermore, all the condensible elements are bound in the dust in the last stage of protosolar cloud contraction. There is an initially wide particle size distribution but the aggregation process is dominated by the larger (sub micron) particles whose size distribution consists of core-mantle particles with variable mantle thickness. The aggregation of these particles is here represented in first approximation by an aggregation of average uniformly sized interstellar particles.

As the putative preservation of the interstellar dust volatiles implies gentle aggregation, the elongation alone prevents the formation of a compact body. Something like a bird's nest structure of randomly aligned sticks is easily conjectured (Greenberg 1986; Greenberg and Gustafson 1981) and is explicitly modeled in Fig. 2a.

It is beyond the scope of this chapter to go through all the arguments concerning the kind of packing of this microstructure and what this kind of dust aggregation leads to; reference is made to several sources of information from which deductions of the ultimate mean comet nucleus density may be inferred. Classically, the density of comet nuclei has been derived from their splitting which results from gravitational gradients near massive objects like Jupiter and the Sun. From observation of the splitting of a comet in close orbit to Jupiter it has been suggested that the nucleus is fragile with a mean density of ≤ 0.30 g cm^{-3} (Sekanina and Yeomans 1985). Similar deductions have been made from the effects of nongravitational forces induced by the momentum of evaporating volatiles (Rickman 1991). While interplanetary particles may be descended from comet debris they have been modified by their thousands of years (on average) sojourn in the interterplanetary medium (Brownlee 1978). Using them directly as building blocks for comet nucleus material would lead to a comet nucleus density of the order of 1 g cm^{-3} (if they were fully packed) and furthermore their organic content would be too low by a factor of at least 5. It is to be noted here that using cosmic abundances of the elements as a constraint on the chemical composition of the interstellar dust leads to an absolute maximum density of about 1.54 cm^{-3} for a fully compact nucleus (Greenberg and Hage 1990). Finally, using material which is most resembling the comet nucleus—namely the comet dust—it has been possible to apply the most complete set of observational constraints to derive a comet nucleus density of $\rho_c \approx 0.3$ g cm^{-3} with an upper limit of $\rho_c \approx 0.6$ g cm^{-3} (Greenberg and Hage 1990). A reinvestigation of the comet dust porosity required to provide a source of distributed molecules in the coma of comet Halley places an even stricter constraint on comet densities, leading to values $\rho_c \leq 0.25$ g cm^{-3} (Greenberg and Shalabiea 1993). Combining all the information to date and weighing it toward the most quantitative and direct observations suggests that comets have densities not higher than $\rho_c = 0.3$ g cm^{-3}. A simplified model of a piece of a comet as pictured in Fig. 2a represents an aggregate of 100 average size (large) interstellar grains with a structural porosity of $P = 0.8$. The density derived for the model shown in Fig. 2a is 0.28 g cm^{-3}.

This implies that comets are at least 80% empty space or have a porosity of 0.8 and this will have a major effect for nuclear, kinetic, or solar energy orbit modifying interactions. For example, the primary components of a stand-off

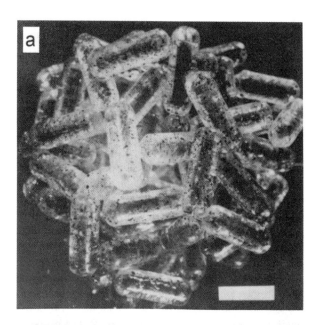

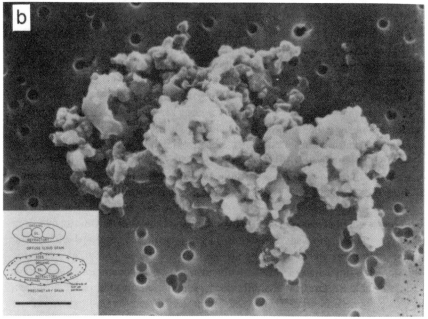

Figure 2.   (a) Model of an aggregate of 100 average interstellar dust grains representing a piece of comet is depicted. Each particle as represented corresponds to a protosolar particle on which all condensibles have accreted. The mean mantle thickness corresponds to a size distribution of thicknesses starting from zero. The packing factor of the particles is about 0.2 (80% empty space) and leads to a mean mass density of 0.28 g cm^{-3} and an aggregate diameter of 5 μm (the bar is 1 μm). (b) Highly porous chondritic interplanetary dust particle (IDP). Note that the "bird's nest" particle (a), the IDP (b) and average core-mantle particle (inset) are equally scaled to 1 μm.

nuclear explosion are X-rays and neutrons. Both the low comet density as well as their porosity will reduce the absorption cross section. In particular, the photoelastic cross section is dependent on the atomic number raised to the fourth power, and the total cross section for high-energy neutrons depends on the atomic mass number (see the Chapter by Shafer et al.).

It is interesting and important to note that porosities much higher than this, such as $P > 0.95$, which characterizes comet material from which all volatiles have been evaporated, can still be reasonably solid—though fragile—structures. An example is provided by silica glasses as pictured in Fig. 3 which consist of $\sim 0.1 \ \mu$m silica particles in a matrix with packing factor 0.2. ($P \sim 0.8$). Even lower packing factors can lead to reasonable rigidity where the fluffy silica "glass," (such as aerogels proposed for impact and capture of interplanetary particles) has a porosity of $P > 0.995$, i.e., 99.5% empty space. Probably, this type of rigidity is made possible by means of linked particles which are multiply connected via interlocking loops. In any case, rigidity and mechanical strength is not strictly defined by density so long as one is dealing with small gravitational forces. What happens when shocks occur in such structures is a subject for future study. Preliminary computational modeling (Chapter by Shafer et al.) of momentum coupling efficiencies for ice and for 43% porous (typical) snow indicate dramatically lower values for snow due to dissipation effects driven by irreversible porosity compaction at low stress levels. At 25 GPa most of the void porosity for the 43% porous snow is removed. Therefore, it is likely that the effect on the (even higher) porosity of comets on momentum coupling is very likely to be more important than the solid state phase transition that will occur at higher pressures; coupling efficiency at low pressure is inversely proportional to the porosity.

It may be important to note not only the mean morphological structure of the dust aggregates but the morphological structure of the individual dust components. In the mean, each dust grain is surrounded by a mantle of amorphous ices whose physical properties may be dramatically different from crystalline ices. In fact, Kouchi et al. (1992) have shown that the thermal conductivity of the ices resembling the primordial interstellar dust mantles could well be as little as 10^{-7} that of crystalline ice. We note that the speed of sound in such a material would be, accordingly, exceedingly low—perhaps only tens of centimeters per second so that the transfer of energy within each grain is strongly limited to within the mantle structures. One should note, however, that the degree of amorphicity of the water ice will depend on the degree of previous heating by solar radiation to which the comet has been exposed. Crystallinity in a Jupiter family comet could extend to hundreds of meters below the surface (Tancredi et al. 1993). But in a new comet this depth of penetration is likely to be negligible (Harayuma et al. 1993).

IV. ATOMIC ABUNDANCES IN COMETS

Based on the interstellar dust model we break down the chemical components

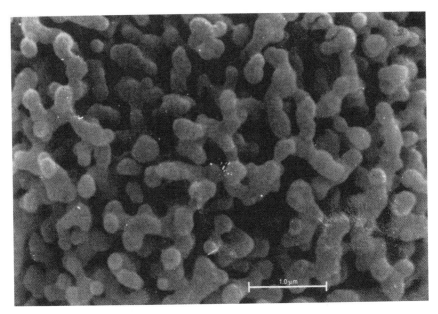

Figure 3.   A silica glass sample with mass porosity $P = 0.8$ with the mean silica
grains approximately equal in size to interstellar grains. This sample is a rigid
structure even though the voids occupy 80% of the total volume.

TABLE III

Mass Fraction Distribution of the Basic Chemical
Components of a Mean Aggregate of Interstellar Dust:
A Comet Nucleus Interior

Component	Mass Fraction
Silicate	0.20
Small carbonaceous particles and large (PAHs)	0.06
Organic refractories	0.19
H_2O	0.40
CO	0.03
CO_2	0.02
Other (CH_3OH, H_2CO, etc.)	0.10

by mass as shown in Table III.

The relative abundances of the atomic constituents in the interior are
assumed to be as if all condensible atoms are contained in the nucleus. The
H is under-abundant because it is only contained in combinations with O, C,
N in the organic refractory and in the volatiles. In the organic refractory we
assume, based on what has been measured in the laboratory, that there is about
1.5 H for every C. In the volatiles, which are dominated by oxygen, one may
assume about 2H for each O. Note that if one uses 40% of the nucleus mass

as H_2O this leads to a fractional abundance of ≥ 0.75 for O and for each there are 2 H.

TABLE IV

Estimated Atomic Fractional Abundances as Derived Using Organic Refractory Composites[a]

	(1) Nucleus	(2) New Comet Crust	(3) Periodic Comet Crust
H	0.527	0.37	0.32
C	0.151	0.37	0.32
N	0.031	0.07	0.01
O	0.250	0.1	0.16
Mg	0.0144	0.03	0.07
Si	0.0136	0.03	0.07
Fe	0.0122	0.03	0.06

[a] Greenberg and Mendoza-Gómez (1993).

Table IV presents estimated atomic fractional abundances for: (1) the comet nucleus using solar abundances as given by Cameron (1982); (2) the crust of a new comet based on the assumption that not 19% but 80% of the mass of the outer crust is organic; and (3) the crust of a periodic comet.

These fractions for the nucleus are close to what is observed in comet Halley with the exception of nitrogen which is notably under-abundant but this is not critical for the working model presented here.

No one currently knows the thickness of the crust of comets. It may be anything from centimeters to meters. Such a crust is defined as being devoid of volatiles, leaving only the carbonaceous and silicate materials. The crust on a new comet would be different from that on a periodic comet. It would contain cosmic-ray-processed ices in addition to the interstellar organics and such material could be perhaps 1 m deep. Its conjectured presence (Strazzula and Johnson 1991) appears to have been confirmed by recent observations (Hanner et al. 1992; Fink 1992) of dynamically new comets as interpreted by Greenberg et al. (1993). The porosity of the cosmic-ray irradiated crust of a new comet is probably greater even than the porosity of the nucleus.

For the crust of a periodic comet one may to a first approximation simply eliminate the volatile components and assume that also the very small carbonaceous particles and PAHs have been carried off with the evaporating ices. This leaves only the silicate cores with their organic refractory mantles. The relative abundances in such a melange of residual matter is given in the last column of Table IV. Note the relative abundances of the rockies relative to the organics is about twice as large in the "crust" of a periodic comet as in the "crust" of a new comet. The periodic comet crust may approach the properties of the "observed" comet dust which implies a porosity of $P = 0.975$ and a density $\rho_{pcc} = 0.05$ g cm^{-3}.

From Tables IV and V it is easy to understand that the arrangement of atoms on the surface of and within the comet will determine the absorption

TABLE V

X-Ray and Neutron Absorption Cross Sections for the Most Abundant Elements in Comets

Element	X-rays (cm^2/g)			Neutron (barn $= 10^{24}$ cm^2/g)	
	5	10	25 Kev	5	14 Mev
H	0.084	0.006	—		
C	18.9	1.91	~0.09		
N	30.4	3.4	~0.17		
O	49	5.8	0.4	1.0	1.5
Mg	158	20.4	1.3	2.0	1.65
Si	244	33	8.4	2.5	1.85
Fe	138	176	44.5	3.7	2.5

and associated specific momentum transfer over a given distance and time. Each layer will have a set of absorption coefficients for X-rays and neutrons which will determine the effective momentum transfer to that layer from each radiation component. The debris cloud will primarily effect only the comet surface over which it will transfer momentum by simple collision.

V. DEPLETION OF VOLATILES AND PLASMA CLOUDS

The cross sections listed in Table V are for an ideal target which has not been depleted of volatile components; for sequential nuclear energies, the cross section would also change sequentially. Another factor that must be taken into account are plasma clouds generated by the interaction of preceding nuclear energies radiation and debris which ionize comet target material and generate additional debris especially from sublimating surface regions. This evolution of comet material may be approximated by the values in Table IV only for certain types of irradiation.

The interaction of the radiative flux with comet plasma can present formidable problems of X-ray momentum coupling to a volatile target. The amount of plasma producing volatile and nonvolatile material liberated from the comet by the nuclear radiation will depend on the irradiation flux, the type of material, and its location on or within the comet. It is clear from Table V that the X-ray absorption coefficients are generally much greater than the neutron absorptions. This implies that if volatiles are depleted at the surface, enhanced X-ray absorption (if not scattered by the plasma cloud) will occur near the comet surface over extremely short distances, while the neutrons will still tend to penetrate further and couple over a longer distance as well as a longer period of time. Enhancing elements such as Mg, Si, or Fe on the comet surface by depletion of volatiles will be very effective in absorbing X-rays. However, it must be determined through detailed computations what effects depleted volatiles and enhanced plasma scattering will cause on (especially X-ray) radiation coupling.

Therefore, because chemical (atomic), morphological, and material properties will strongly influence the interaction dynamics of comets with nuclear as well as kinetic or solar energy, we must be able to provide as accurate a model as possible based on current observational and theoretical studies.

REFERENCES

Allamandola, L. J., Tielens, A. G. G. M., and Barker, J. R. 1989. Interstellar polycyclic aromatic hydrocarbons: The infrared emission bands, the excitation/emission mechanism, and the astrophysical implications. *Astrophys. J. Suppl.* 71:733–775.

Brownlee, D. 1978. Microparticle studies by sample techniques. *Cosmic Dust*, ed. J. A. M. McDonnell (New York: J. Wiley), pp. 295–336.

Cameron, A. G. W. 1982. Elemental and nuclidic abundances in the solar system. In *Essays in Nuclear Astrophysics* eds. C. A. Barnes, D. D. Clayton and D. N. Schramm (Cambridge: Cambridge Univ. Press), pp. 23–43.

Chlewicki, G. 1985. Observational Constraints on Multimodal Interstellar Grain Populations. Thesis, Univ. of Leiden.

Delsemme, A. H. 1982. Chemical composition of cometary nuclei. In *Comets*, ed. L. L. Wilkening (Tucson: Univ. of Arizona Press), pp. 85–130.

Fink, U. 1992. Comet Yanaka (1988r): A new class of carbon-poor comets. *Science* 257:1926–1929.

Gaffey, M. J., Burbine, T. H., and Binzel, R. P. 1993. Asteroid spectroscopy: Progress and perspectives. *Meteoritics* 28:161–187.

Greenberg, J. M. 1982. What are comets made of?—A model based on interstellar dust. In *Comets*, ed. L. L. Wilkening (Tucson: Univ. of Arizona Press), pp. 131–163.

Greenberg, J. M. 1983. Laboratory dust experiments—tracing the composition of cometary dust. In *Cometary Exploration*, ed. T. I. Gombosi (Budapest: Hungarian Academy of Sciences), pp. 23–54.

Greenberg, J. M. 1984. The structure and evolution of interstellar grains. *Sci. Amer.* 250:124–135.

Greenberg, J. M. 1986. Evidence for the pristine nature of comet Halley. In *The Comet Nucleus Sample Return Mission*, ed. O. Meliter, ESA SP-249 (Noordwijk: European Space Agency), pp. 47–55.

Greenberg, J. M., and Gustafson, B. 1981. A comet fragment model for zodiacal light particles. *Astron. Astrophys.* 93:35–42.

Greenberg, J. M., and Hage, J. I. 1990. From interstellar dust to comets: A unification of observational constraints. *Astrophys. J.* 361:260–274.

Greenberg, J. M., and Mendoza-Gómez, C. X. 1993. Interstellar dust evolution: A reservoir of prebiotic molecules. In *The Chemistry of Life's Origins*, eds. J. M. Greenberg, C. X. Mendoza-Gómez and V. Pirronello (Dordrecht: Kluwer), pp. 1–33.

Greenberg, J. M., and Shalabiea, O. 1993. Comets as a reflection of interstellar medium chemistry. In *Comets, Asteroids, and Meteorites 1993*, IAU Symp. 160, eds. A. Milani, M. DiMartini and A. Cellino (Dordrecht: Kluwer), pp. 327–342.

Greenberg, J. M., Singh, P. D., and de Almeida, A. A. 1993. What is new about the new comet Yanaka (1988n)? *Astrophys. J.* 414:L45–L48.

Hanner, M. S., Russell, R. W., Lynch, D. K., and Brooke, T. Y. 1993. Infrared spectroscopy and photometry of comet Austin 1990 V. *Icarus* 101:64–70.

Jenniskens, P., Baratta, G. A., Kouchi, A., de Groot, M. S., Greenberg, J. M. and Strazzula, G. 1993. Carbon dust formation on interstellar grains. *Astron. Astrophys.* 273:583–600.

Kissel, J., and Krueger, F. R. 1987. The organic component in dust from comet Halley as measured by the PUMA mass spectrometer onboard Vega 1. *Nature* 326:755–760.

Kouchi, A., Greenberg, J. I., Yamamoto, T., and Mukai, T. 1992. Extremely low thermal conductivity of amorphous ice: Relevance to comet evolution. *Astrophys. J. Lett.* 388:73–76.

Mumma, M. J., Weissman, P. R., and Stern, S. A. 1993. Comets and the origin of the solar system: Reading the Rosetta Stone. In *Protostars and Planets III*, eds. E. H. Levy and J. I. Lunine (Tucson: Univ. of Arizona Press), pp. 1177–1252.

Rickman, H. 1991. The thermal history and structure of cometary nuclei. In *Comets in the Post-Halley Era*, vol. 2, eds. R. L. Newburn, Jr., M. Neugebauer and J. Rahe (Dordrecht: Kluwer), pp. 733–760.

Sekanina, Z., and Yeomans, D. K. 1985. Orbital motion, nucleus precession, and splitting of periodic Comet Brooks 2. *Astron. J.* 90:2335–2352.

Strazzulla, G., and Johnson, R. E. 1991. Irradiation effects on comets and cometary debris. In *Comets in the Post-Halley Era*, vol. 1, eds. R. L. Newburn, Jr., M. Neugebauer and J. Rahe (Dordrecht: Kluwer), pp. 243–275.

Tancredi, G., Rickman, H., Greenberg, J. M. 1993. Thermochemistry of cometary nuclei. In *Evolution of Jupiter Family Comets*, Thesis, Uppsala University.

van Dishoeck, E. F., Blake, G. A., Draine, B. T., Lunine, J. I., 1993. The chemical evolution of protostellar and protoplanetary matter. In *Protostars and Planets III*, eds. E. H. Levy and J. I. Lunine (Tucson: Univ. of Arizona Press), pp. 163–241.

PART V
Space Exploration

MISSIONS TO NEAR-EARTH OBJECTS

ANDREW F. CHENG
Applied Physics Laboratory

J. VEVERKA
Cornell University

C. PILCHER
NASA Headquarters

and

ROBERT W. FARQUHAR
Applied Physics Laboratory

Key physical and chemical properties of near-Earth objects (NEOs) can be determined only by dedicated space missions. While Earth-based remote sensing can obtain unique and invaluable information on global properties of NEOs and NEO populations, particularly in view of the number and diversity of NEOs, spacecraft missions to NEOs are needed to determine their detailed surface morphology, their composition, density and internal structure, to name just a few examples. Flyby missions to one or more NEOs, rendezvous missions to orbit around them, and sample return missions have all been proposed for the NASA Discovery Program of small planetary missions. The near-Earth asteroid rendezvous (NEAR) mission is the first launch in the Discovery Program, scheduled for a 1996 launch to a flyby of asteroid 2968 Iliya followed by a rendezvous with 433 Eros. The Clementine mission for the Ballistic Missile Defense Organization is scheduled for a 1994 launch to orbit the Moon and then to perform a flyby of asteroid 1620 Geographos. We will summarize the scientific objectives and technical implementation of these and other proposed missions to NEOs.

I. INTRODUCTION

Two classes of solar system objects can appear on orbits that pass close to Earth's orbit; these are asteroids and comets.

The distinction between asteroids and comets is made primarily on the basis of their appearance in a telescope; if there is evidence for outgassing activity or tail formation, the object is called a comet, but otherwise the object is denoted an asteroid. Comets have significant volatile content, whereas asteroids have a more refractory, rocky and/or metallic, composition. Comets are also typically found on higher energy, higher inclination orbits around the Sun. However, there are no widely agreed upon quantitative criteria to distinguish comets from asteroids, and at a sufficiently low level of activity a comet can easily be identified as an asteroid. Indeed the asteroid 4015, previously known as 1979 VA, was recently shown to be the same object as comet Wilson-Harrington.

Because the distinction between asteroids and comets is at least partly a matter of convention, it is useful to define simply the class of near-Earth objects (NEOs) containing both comets and asteroids. For convenience, we define NEOs to be objects other than planets or their satellites with orbital perihelion distances <1.3 AU. In more familiar terms, the so-called Apollo asteroids, those with semimajor axis $a \geq 1$ AU and perihelion $q \leq 1.017$ AU, are included; these asteroids cross the radius of Earth's orbit. The Aten asteroids ($a < 1$ AU and aphelion >0.983 AU) are also included and are also Earth-crossers. The Amor asteroids ($1.017 < q < 1.3$ AU) are included but lie outside of Earth's orbit. In addition, some of the short- and the long-period comets (classified according to whether the period exceeds 20 yr) are members of the NEO population. The NEO definition involves no criterion for minimum aphelion, but as a practical matter there are no known examples of small bodies with aphelion $<<1$ AU (i.e., such that the orbit would lie entirely within that of Venus).

The usefulness of defining NEOs simply by $q < 1.3$ AU is that such objects are in orbits that, when subjected to gravitational perturbations from Earth and other planets, can evolve into orbits that impact Earth (Shoemaker et al. 1990; Weissman 1990). Approximately 230 NEOs have well determined orbits as of 1993. The known population is believed to be complete to an absolute magnitude of about 13.5, with the degree of completeness decreasing rapidly for fainter objects. It is clear that there are basically two sets of scientific requirements essential to understanding the influence of NEO impacts on the past evolution of Earth's atmosphere and biosphere, as well as to quantifying and understanding the future collision hazard. The first is to survey the NEO populations, to increase the completeness of the survey at the faint end, and to determine the distributions in orbital parameters and in spectral type. The second is to determine the physical nature and the composition for the various types of NEOs. In addition, the study of NEOs is of great scientific importance in its own right.

Dedicated space missions to NEOs will be essential to obtaining a better knowledge of their fundamental nature. We will now outline the scientific rationale for exploration of NEOs, provide an overview of the current status of NEO science, and summarize results from the recent Galileo flyby of the main-belt asteroid 951 Gaspra. Although this chapter will focus on missions to NEOs, we will present some Gaspra results as these represent our only detailed information to date on any specific asteroid. In Sec. II we will outline the various mission options for exploration of NEOs and the expected science returns, technical issues, advantages and disadvantages of each. Section III will present the near-Earth asteroid rendezvous (NEAR) mission which is currently planned as the first launch in the NASA Discovery Program of small planetary missions. For a presentation of the Clementine mission, also known as the Deep Space Program Science Experiment, we refer the reader to the Chapter by Nozette et al. The Clementine mission is supported by the Ballistic Missile Defense Organization (BMDO). We conclude with a brief look at future exploration of NEOs.

A. Rationale for Exploration of NEOs

We now outline the rationale for systematic exploration of NEOs, both in the specific context of collision hazards and in the larger context of solar system exploration. The NEOs are a numerous, extremely diverse class of objects and are certainly interesting if only for their proximity to Earth, which has important consequences. NEOs are the primary source population of large objects that impact Earth and that have influenced the evolution of the atmosphere and of life on Earth. As Earth's nearest and most accessible neighbors aside from the Moon, the NEOs are also particularly well suited as targets for small missions, because NEO missions place relatively low demands on spacecraft propulsion, power, and communication systems. Finally, NEOs are a potentially important source of raw materials for the future exploration and utilization of space.

In the larger context of solar system exploration, the NEOs are of scientific interest for several reasons. One is that they preserve clues to early solar system processes and to conditions during the formation and early evolution of the planets. Records of these early epochs can be found in asteroids, comet nuclei, and meteorites, but these are objects that have undergone varying degrees of processing (e.g., thermal, collisional, chemical) and that therefore preserve records of different processes, different regions, or different epochs of the early solar system. We have access to the meteoritic record but cannot relate it to the history or the physical nature of any specific solar system object, and we cannot correlate it to the lunar record. We do not yet have detailed access to the records in asteroids or in comet nuclei; we lack detailed knowledge of the composition and physical nature of both asteroids and comet nuclei, and we do not understand clearly how the asteroid and the comet populations are related to each other or to the meteorites.

More specifically, the NEOs are the source of most of the meteorites (small Earth impactors) and include both asteroids and comets. Some of the objects denoted near-Earth asteroids (NEAs) may actually be extinct or inactive comet nuclei. The NEAs themselves are extremely diverse for an additional reason: some appear to be differentiated bodies. These may be fragments of terrestrial protoplanets that did not survive intact to the present epoch. Other NEAs appear to be much more primitive bodies that may be related to primitive chondritic meteorites. A continuing controversy centers on whether the S-type NEAs, the most common spectral type among the known NEA population, are the source of ordinary chondrites, the most common type of meteorites. The fundamental nature of cometary members of the NEO population is very uncertain, but it is at least possible that these comet nuclei have preserved refractory and volatile materials from the early solar nebula or even from the interstellar medium.

It is important to bear in mind that NEOs are efficiently removed, over times of 10 to 100 Myr, from the region of the solar system in which we presently find them, as a result of the orbital evolution mentioned above leading to collisions or close gravitational encounters with the Sun or planets. Hence the NEOs are no longer located near where they formed, although they do preserve clues to the nature of processes in the early solar system. Cometary nuclei may originate within a distant reservoir of objects very loosely bound to the Sun, called the Oort cloud, or from a closer reservoir called the Kuiper disk. NEAs may originate from the main belt of asteroids and may be derived from mainbelt asteroids via collisions and chaotic gravitational interactions. This idea is supported by spectral similarities between NEAs and main belt asteroids; most spectral types found in the main belt are represented among NEAs, although at least one type (Q) of NEA is not known to exist in the main belt (McFadden et al. 1989). In addition, some NEAs are likely to be extinct or dormant comets.

Aside from this class of science issues, there is another important set of problems that can be addressed by detailed study of NEOs. Namely, the NEOs are objects of a size range that has never been systematically explored; the largest known NEOs are about the size of the Martian moons Phobos and Deimos, while typical known NEAs are km-sized objects. Fainter objects, of course, are less likely to be discovered, so the known population becomes progressively less complete at lower luminosity (corresponding to smaller size on the average). Surface processes on NEAs, such as cratering, regolith formation and transport, are likely to be different from those on significantly larger bodies.

B. Lessons Learned from the Galileo Flyby of 951 Gaspra

On October 29, 1991 the Galileo spacecraft made the first ever flyby of an asteroid, 951 Gaspra, which is an irregularly shaped S-asteroid with a mean diameter of about 12 km (Belton et al. 1992a; see also the Chapter by Neukum and Ivanov). While Gaspra is larger than most NEAs, it is comparable in size to the largest ones, such as Eros. The Galileo experience provides valuable lessons as to what can and cannot be learned during asteroid flybys and reaffirms the need for the detailed studies that only orbital missions can provide (NEAR Report 1986; Veverka et al. 1991, personal communication). The Galileo flyby provided important new information on the shape, surface morphology, and collisional history of Gaspra, but did not resolve fundamental issues concerning the asteroid composition.

Flying within 1600 km of Gaspra, Galileo obtained spectral data using the Near Infrared Mapping Spectrometer (NIMS), possibly detected Gaspra's magnetic signature (Kivelson et al. 1993) and returned 57 images of the asteroid. The images were obtained on approach using the Galileo camera

(SSI), a 1.5 m focal length telescope (Belton et al. 1992*b*). The highest resolution image was obtained some 10 minutes before closest approach from a range of 5300 km and had a resolution of 54 m/px. Images were taken for almost the full rotation period of the asteroid (~7 hr). Some of the earliest views obtained from a range of 200,000 km (2 km/px) are already useful in defining the gross shape of the asteroid. The best color coverage (0.41, 0.56, 0.89, 0.99 μm) was obtained 36 minutes before closest approach from a range of 16000 km (160 m/px) at a phase angle of 39°.

The major results of the encounter include the confirmation that Gaspra is irregularly shaped and that cratering has played a major role in its evolution. Small craters (diameter $d < 1$ km) are very abundant and there are no features having the morphology of classical craters with $d > 3$ km. Carr et al. (1993) separate craters on Gaspra into two populations based on morphology and size. Type 1 craters include well-defined, bowl-shaped craters ranging in diameter up to 1.4 km. The population of these craters follows a power law with a cumulative slope of about -3.1. This steep slope is similar to that observed for comparable craters on the Moon and Mars and suggests that the population is a production function (see also Belton et al. 1992*a*; Chapman et al. 1993).

No prominent geologic boundaries are evident, either in terms of differences in the morphology of small features or in the distribution of craters; nor are there evident large scale albedo or color patterns. Those that exist are subtle and probably reflect recent impact and regolith processes rather than significant lateral heterogeneities within Gaspra (Belton et al. 1992*a*; Carr et al. 1993; Helfenstein et al. 1993).

The basic characteristics of the asteroid derived from Galileo data agree with those inferred from ground-based observations. The rotation pole, the rotation period and the sense of rotation all agree well with previous Earth-based results (Davies et al. 1993). The asteroid's mean radius of 6.1 km is in excellent agreement with determinations by Tholen et al. (1993). Gaspra's shape as derived by Thomas et al. (1992) shows it to be an irregular object with principal diameters $18.2 \times 10.5 \times 8.9$ km. Qualitatively the shape is somewhat cone-shaped with a more pointed end near longitude zero and a fat end diametrically opposite. The axial ratios found by Thomas et al. (1992) are in close agreement with the predictions of Magnusson et al. (1992) based on lightcurve analysis. The geometric albedo derived by Helfenstein et al. (1993) of 0.23 ± 0.06 is similar to that found by Tholen et al. (1993) from groundbased radiometry.

During the Galileo flyby it was possible to image the asteroid's northern hemisphere and areas south of the equator to about 40 to 50°S. Thus there are good data on about 80% of the asteroid's surface. Thomas et al. estimate that the surface area of the asteroid, used to determine the surface density of craters, is known to $\pm 10\%$ (525 ± 50 km^2) and that the volume, needed to determine the mean density, is known to about $\pm 20\%$ (954 ± 200 km^3). The Galileo flyby did not yield a mass estimate. For an asteroid as small as Gaspra, a mass estimate would have required miss distances of several hundred km which were precluded for safety of the spacecraft.

Gaspra's shape has several remarkable properties. Based on limb irregularities, Gaspra is one of the most irregular objects studied to date (Thomas et al. 1992). The most interesting features of Gaspra's shape are several planar areas, one of which is a 6 km×6 km area flat to within 200 m. The planarity of this facet and the fact that the normal to this area is coplanar with the normals to the other two large flat areas is intuitively consistent with the derivation of present Gaspra from a larger body by catastrophic collision.

Also found on Gaspra are several large depressions of uncertain origin, some as large as 9 km wide and 1 to 2 km deep, that may be related to past cratering events. Greenberg et al. (1993) have suggested that at least some of the large depressions correspond to crater cavities, their unusual morphology resulting from their unusually large size compared to Gaspra's diameter. On the other hand, Chapman et al. (1993) argued that they cannot be craters produced by the current population of impactors, but are more likely spallation scars.

Another apparent manifestation of severe impacts preserved on Gaspra are grooves, long linear depressions with more or less pitted segments, clearly visible in the highest resolution image; the larger ones are also discernible in the 160 m/px sequence. According to Veverka et al. (1992), typical grooves on Gaspra are several km long, 200 to 300 m wide and probably 10 to 20 m deep. At least two sets are visible: the most prominent set roughly parallels a direction offset about 15° from the long axis of Gaspra. The grooves are similar, but less well developed than those on Phobos (Thomas et al. 1979). Two aspects of the grooves are important. First, the pitted morphology of some segments can be interpreted as evidence of a regolith (Horstmann and Melosh 1989). Second, the Gaspra-wide pattern of grooves suggests a global structural coherence to Gaspra. The most likely origin for the grooves is that they are fractures (possibly modified by downslope motion of regolith) that results from one or more severe impacts, a scenario similar to that advanced by Thomas et al. (1979) for the origin of grooves on Phobos. Unlike the Phobos situation, however, it has not proved possible to associate the grooves on Gaspra with any particular impact structure still preserved on the asteroid.

The Galileo images illustrate dramatically that at least one small asteroid has experienced intense collisional evolution. Gaspra's irregular shape, the occurrence of grooves and planar facets, as well as the observed crater size distribution, all attest to the asteroid's violent collisional history. Whether typical asteroids of Gaspra's size are coherent structures or piles of rubble remains a much debated issue. The Galileo images provide evidence of significant structural continuity across the asteroid. Thomas et al. (1992) show that three of the flat facets as well as one of the two prominent groove sets follow a body-wide structure parallel to a direction about 15° from the asteroid's long axis. This congruence suggests that present Gaspra is a single body derived by collisions from a larger precursor. Various estimates have been derived of how often bodies the size of Gaspra at 2.3 AU suffer catastrophic collisions, and of how long the craters visible on Gaspra's surface have taken to accumulate. Estimates of average disruption times range from some 500 to 600 Myr (Chapman et al. 1993) to about 1 Gyr (Greenberg et al. 1993) while estimates of the cratering age of the current surface vary from 20 to 300 Myr (Carr et al. 1993; Chapman et al. 1993; Greenberg et al. 1993). Although these estimates are uncertain, it is clear that collisions have been important in the evolution of Gaspra, and by extension, other asteroids. We should expect most small asteroids today to be collisional fragments.

Indications are that Gaspra is a fairly homogeneous body. As small pieces of larger asteroids, many NEAs should also be fairly homogeneous (we exclude here those NEAs derived from cometary precursors). Larger asteroids among the NEAs may be more interesting in that they have a higher probability of retaining evidence of some heterogeneities of their larger parents.

It has not proven possible so far to derive quantitative estimates of the relative mineralogical makeup of Gaspra's surface from the spectral data obtained by NIMS (and by groundbased observers) or from the SSI color data. While it is clear that both olivine and pyroxene are present (Granahan et al. 1993) the relative amount of nickel-iron metal in the surface layers cannot be determined reliably. Such a determination would be one way of distinguishing whether S-asteroids are akin to stony-irons (more metal) rather than to ordinary chondrites (less). Another way of possibly determining the difference would be from a measurement of average density (higher density corresponding to the stony-iron hypothesis); unfortunately, Galileo did not obtain a mass measurement and the average density of Gaspra remains unknown. Granahan et al. did suggest variations in the relative abundances of pyroxene and olivine across Gaspra that may eventually provide means of demonstrating that Gaspra is derived from a body much more differentiated than seems appropriate for an ordinary chondrite.

Another outstanding issue concerns the amount of regolith on Gaspra. Available remote sensing of asteroids has been interpreted to suggest that larger asteroids tend to have surface textures different from smaller ones, with smaller bodies being rougher and having less or no regolith cover. Model calculations are dependent on the assumed surface mechanical properties. Calculations predict that an object of Gaspra's size will retain either negligible or modest amounts of regolith if the surface is "hard" or "soft," respectively (Belton et al. 1992a; Veverka et al. 1986). The Galileo data include evidence that some regolith is present on Gaspra. For example, Belton et al. (1992a) noted that the subtle color and albedo variations observed are best explained if at least a thin regolith is present. Specifically most of the color variations seen involve the prominence of the 1 μm olivine band. The color variations tend to be associated with small, fresh-appearing craters on ridges, a situation that led Belton et al. (1992a) and Carr et al. (1993) to suggest that regolith with a stronger 1 μm signature is being exposed by craters along ridges and gradually migrates downslope, where it mixes with older material. Additional evidence for a regolith comes from the morphology of grooves. The pitted appearance of many groove segments is best explained if a moderate amount of regolith (10–20 m) is present (Veverka et al. 1993). Estimates of the degree of regolith cover could be refined if thermal inertia measurements or ultrahigh resolution images (1–2 m/px) were available.

It is interesting that the three small bodies for which high resolution images are available—Phobos, Deimos, and Gaspra—look remarkably different in terms of the amount of regolith covering their surfaces. On Deimos, for example, a body similar in mean radius to Gaspra, many craters have been filled in by loose surface materials and there is ubiquitous evidence of significant downslope movement of regolith in the form of streamers of higher albedo material (Thomas and Veverka 1980). Neither of these is observed on Gaspra. The retention of regolith on small bodies remains a poorly understood phenomenon. Galileo observations of Ida will provide additional data for testing models. However, observations of NEAs well under a km in diameter will be of fundamental importance in that almost all models indicate that such small bodies should not retain any regolith, a suggestion that appears consistent with existing radar observations of NEAs (Ostro 1989).

II. NEO MISSION OPTIONS

Options for scientific study of NEOs include not only deep space missions but also Earth-based investigations, both groundbased and orbital. However, given the number and the diversity of the NEOs, it is obvious that deep space missions can be sent to only a small fraction of the NEO population and that specific information on the vast majority of individual NEOs will be available only from Earth-based remote sensing. Searches for NEOs and studies of NEO population statistics, such as distributions in spectral type and orbital parameters, will be areas in which Earth-based studies will continue to play a dominant role. Earth-based studies of individual objects will also continue to be important for unusual or interesting objects that are not planned as targets for upcoming missions; to prepare for upcoming missions by refining knowledge of positions, shape, rotation, and spectra for planned targets; and to make use of techniques such as radar reflectance that early low-cost space missions will likely be unable to accomodate.

Deep space missions are conventionally classified as follows: flybys, orbiters (rendezvous), landers, and sample return missions. Of course, a given mission can combine more than one of these types. The Galileo mission, for example, is a Jupiter Orbiter/Probe that will execute two flybys of mainbelt asteroids. The Clementine mission will orbit the Moon and then perform a flyby of an NEA, while the NEAR mission will perform a main-belt asteroid flyby and then orbit an NEA. However, it has been possible to combine mission types only in certain cases, e.g., a major outer planet mission like Galileo that must pass through the main asteroid belt and that has sufficient resources to enable asteroid flybys at a small fraction of the mission cost. In addition, one or more NEO flybys can often be included in rendezvous missions to targets that are sufficiently accessible and that have sufficiently numerous launch opportunities, like the Moon or NEAs.

The chief advantage of NEO flybys is their relatively low cost, which enables in principle the return of some detailed information on a relatively large number of targets, particularly because missions can sometimes be designed for multiple flybys or to combine flybys with a rendezvous. Increasing the number of targets that can be examined by spacecraft is especially important for studying objects that are extremely numerous and diverse, like NEOs. Another advantage of NEO flybys is that they provide opportunities to obtain flight experience with new space technology. The Clementine mission under the auspices of the Ballistic Missile Defense Organization (see the Chapter by Nozette et al.) exemplifies the type of mission whose rationale combines technology development with science return.

The chief disadvantage of flybys is the short time available for detailed scientific investigation. The time spent within twice the closest approach distance is about $3.5b/v$, where b is the closest approach distance and v is the relative speed. For typical $b \sim 1000$ km and $v \sim 15$ km s^{-1}, this time is ~ 4 min and is much less than the rotation period of the NEO, so that only a portion of the surface (the illuminated portion) can be imaged at maximum resolution. The time within a given distance scales linearly with the distance, up to times of a few weeks. The characteristic observation time for an instrument is set by a combination of instrument and trajectory parameters. For important classes of composition measurement like X-ray and gamma-ray spectroscopy, the time spent close enough to the target to obtain useful measurements is typically less than the integration times required, so that flybys are limited in compositional data return. Among the types of compositional data that can be obtained in flybys are: ultraviolet/visible/near-infrared reflectance spectroscopy to study mineralogy and volatiles; and mass spectrometry of volatiles and grains in a comet coma.

With radio tracking during a flyby, the mass of the target can be determined if it is large enough and b is small enough. The density (which constrains the composition) is difficult to determine accurately in a flyby because the volume, as well as the mass, needs to be measured. With a high resolution imager, it is possible to make useful shape determinations of the full illuminated surface over an entire asteroid rotation period, from well beyond the distance b. However, depending on the location of the rotation pole, a significant portion of the surface may not rotate into sunlight and a volume determination may not be possible. For the Galileo asteroid flybys, the closest approach distances were chosen large enough to avoid any possibility of collision with debris near the target, hence no mass determinations were possible. For the Giotto Halley encounter, the closest approach distance was so small that nongravitational interactions with cometary gas and dust prevented a mass determination.

Rendezvous missions are more demanding and hence more costly than flybys, because of the need to match the heliocentric orbit of the target precisely. A rendezvous mission, however, has enormous scientific advantages stemming from the possibility of close observations over extended time periods, as discussed by the Discovery Program Science Working Group (Veverka et al. 1991, personal communication) for the NEAR mission. The NEAR mission may also include an additional asteroid flyby. During the rendezvous phase, NEAR will obtain global imaging data at high resolution, combined with global mineralogic mapping, global elemental composition, and determinations of density, spin state, topography, internal structure, and intrinsic magnetic field (assuming that an altimeter and a magnetometer, in addition to imaging and geochemical experiments, can be accomodated within the mission cost cap).

For the NEAR mission, it is anticipated that the scope, sensitivity and accuracy of the compositional measurements will enable the target asteroid's composition to be related to those of known meteorite classes. Specifically, if the target is an S-type asteroid, for example, the goal would be to determine whether this example of S-type material is like ordinary chondritic material or a more evolved material like that in stony iron meteorites. Of course, it is possible that the S-type asteroids are themselves a diverse class with distinct compositions, and furthermore that a single asteroid may, when mapped at high spatial and spectral resolution, show evidence of distinct compositional units and/or structural inhomogeneity due to accretion or differentiation in a parent body.

Any such results would be a great advance in our understanding of the inner solar system and its history. Nevertheless, the compositional data from a rendezvous mission, particularly from a cost-constrained mission like NEAR, will not be sufficiently quantitative and detailed to permit identification of a meteoritic parent body, meaning that even if the target asteroid were a source of a particular meteorite or specific meteorite class, such a relationship could not be established with reasonable confidence. Such a result can be obtained only by return of a bulk sample from an NEO for analysis in terrestrial laboratories (Swindle et al. 1991).

A mission to return samples to Earth from multiple sites on a near-Earth asteroid would have profound implications for our understanding of processes and conditions in the early solar system, going far beyond the advances to be expected from the NEAR mission. Sample return from a near-Earth asteroid would elevate asteroid science—as well as meteorite science if a definite link can be established—into a wholly new regime (Swindle et al. 1991). Detailed petrological, chemical, age, and isotopic analyses would firmly establish the formation conditions, formation age, and gross dynamical history for a target body of known identity. This detailed history and compositional analysis of an NEA would allow, for the first time, the lunar chronology to be related to that for a known body outside the Earth-Moon system and would moreover allow the asteroid's petrology, elemental abundances, and isotope ratios to be compared with those for terrestrial, lunar, and meteoritic materials. Of course the NEA sample return may lead to clear links with more than one specific meteorite or meteorite types if there is significant asteroidal heterogeneity, or alternatively an asteroid sample may not lead to any such linkages if it represents a new class of planetary material without analog in the present meteorite collection. Any of these outcomes would be fundamentally important. Likewise, sample return from a comet nucleus would be enormously important as reviewed recently by Schwehm and Langevin (1991). Other recent reviews of sample return science have been given by Drake et al. (1987), Ryder et al. (1989), and Gooding et al. (1989).

Many samples of NEAs are believed to have been "returned" to Earth already in the form of meteorites, and the issue has been raised as to why more samples need to be taken *in situ*. However, a similar situation exists for the Moon and Mars, because samples of both bodies are believed to be already represented in the meteorite collection. Nevertheless it is still extremely important scientifically to obtain more samples *in situ* and to study their geologic context. The significance of "geologic context" for a small body like an asteroid differs from the case of planetary bodies, since it is not even known if asteroids can consist of more than one geologic unit. Confirmation that an NEA is inhomogeneous, whether because of differentiation or because it is an aggregate of different precursor bodies, would be a major discovery. The possibility of such a "geologic context" motivates the return of multiple samples from the target NEA. Nevertheless, the asteroid may be homogeneous and its mere identity may be the most important aspect of "geologic context"; that is, the most important result may be that a particular type of meteoritic material can be established to originate on a particular NEA. At present there are two large data bases on primitive bodies, the astronomical data base and the meteoritic one, whose relationships are by and large unclear. If firm relationships can be established, the value of both data sets would be greatly enhanced. Samples from several different types of asteroids will be needed to understand these relationships fully.

Moreover, it is likely that asteroid surfaces contain much material that is simply not represented in the meteorite collection, because it is not strong enough and/or refractory enough to survive high speed re-entry. Any volatile or semi-volatile material, and much regolith material, would not be expected to survive except perhaps in the form of interplanetary dust particles. Asteroid sample return can be expected to shed new light on regolith formation processes and the nature of space weathering, and it will provide laboratory ground truth for spectroscopic remote sensing, extending the power of groundbased observations.

A fourth mission option for investigation of NEOs is to deploy a landed instrument package to make *in situ* measurements on the surface of the body. Such a mission generally involves a rendezvous prior to the instrument deployment, as does the sample return mission prior to sampling. A rendezvous phase for these missions is desirable to map the surface and the gravity field, to identify landing/sampling site(s), and to obtain their geological contexts. Safe deployment of a landed package on an NEO is also greatly facilitated from rendezvous orbit. Both the landed package and the sample return characterize only a single site (for each landed package or for each returned sample), whereas a rendezvous orbital survey obtains global data. Multiple samples, or multiple landed packages, would allow sampling possible heterogeneity of an NEO. The ultimate accuracy and sensitivity of compositional analyses from a landed package can be expected to exceed that for a rendezvous mission, given sufficient resources for the landed instruments. However, the science return from any conceivable landed package cannot begin to compare with that from a sample return (Drake et al. 1987; Gooding et al. 1989). To begin with, much larger, more capable and more up-to-date analytical instrumentation can be used in terrestrial laboratories compared to what can be landed on an NEO. Moreover, many critical analyses require extensive, sophisticated chemical and/or physical preparation procedures that are impractical in deep space. Finally, as the Viking experience makes clear, we have only limited ability to anticipate just which preparations and analyses will be needed *in situ* to achieve our scientific objectives.

A landed instrument package must meet certain technical requirements. The landed package must be able to transmit data, either to Earth or to the orbiter; both of these will in general be moving in the sky (because of asteroid rotation) and neither will necessarily be in view at any given time. Moreover, if a landed package must operate more than a few hours in order to make the required measurements and return data, it will typically have to survive eclipse on the nightside of the NEO.

The deployment of a landed instrument package is comparable in technical difficulty to obtaining a "touch-and-go" sample (where the sample is obtained rapidly by a pyrotechnic device so that contact with the NEO surface lasts only a fraction of a second, and there is no long duration landing). Landed packages may represent an attractive option to make certain important measurements *in situ*, such as composition and physical state of ices and/or clathrates on the cometary nucleus, in view of the expected difficulty of returning such material to Earth and preserving it in its pristine state (or close to it). The return of pristine volatile samples from a comet nucleus is much more technically difficult and expensive than is return of rock samples from a near-Earth asteroid. This difference stems mainly from three factors: (1) the inaccessibility of comet orbits, which greatly increases propulsion requirements for both the trips to and from the comet; (2) the resulting re-entry conditions, which are much more demanding than for NEA sample return; and (3) the cometary science requirement of cryogenic return of volatile samples, which requires new technology development not only for re-entry but also for later handling, curation, and analysis (Schwehm and Langevin 1991). In view of these technical difficulties for volatile sample return, landed packages are an attractive option for *in situ* study of cometary volatiles. In contrast, returned samples from an NEA can be handled like lunar samples using existing protocols, techniques, and facilities. The relative ease and merits of NEA sample return, as well as possible benefits to space exploration, have been discussed previously by Davis et al. (1990).

It is clear that the trade-offs among various mission types are complicated and depend at least on the nature of the target in question (asteroid or comet) and the specific science objectives, as well as cost and technical issues. It is likely that Earth-based observations and each of the various types of deep space missions will have some role in the future exploration of NEOs.

III. THE NEAR MISSION

The near-Earth asteroid rendezvous (NEAR) is the first launch in the NASA Discovery Program of small planetary missions costing no more than $150 million each (launch plus 30 days, in FY 1992 dollars). The Discovery Program is modeled after the Explorer program for astrophysics and space physics, and it has received an FY 1994 new start. It is intended to enable rapid response to emerging scientific opportunities; to increase the breadth of activities in solar system exploration; to enhance timeliness of new information return, to provide increased access to space; to expand industrial, academic and public involvement in solar system exploration missions; and to facilitate cooperative ventures with other space agencies. The NEAR mission will be managed for NASA by the Johns Hopkins University Applied Physics Laboratory (APL), which will also fabricate the spacecraft and provide a facility instrument payload.

The Discovery NEAR mission will emphasize focused science return. NEAR will be the first rendezvous with an asteroid and promises to answer important, fundamental questions concerning the nature and origin of NEAs. NEAR could clarify relationships between asteroids, comets, and meteorites and may reveal clues to the nature of the planetesimals from which terrestrial planets formed. Results from NEAR should greatly enhance the interpretation of remote sensing data acquired on other asteroids, whether obtained in Earth-based observations or in spacecraft flybys (e.g., Galileo, Clementine).

A. Science Objectives

The primary scientific goals of the NEAR mission are to characterize an asteroid's physical and geological properties and to infer its elemental and mineralogical composition. NEAR will determine the density, shape, and spin state of the asteroid, reveal aspects of its interior structure, and characterize its surface morphology. NEAR will furthermore make spatially resolved geochemical measurements and identify heterogeneity of geologic units if present. Table I lists some scientific questions addressed by the Discovery NEAR mission, and Table II lists the facility instrument payload.

The asteroid's shape, volume, and spin state will be inferred from visible imaging of the entire surface. Such high resolution global imaging generally requires rendezvous orbit over an extended period of months. When

TABLE I

Science Questions[a]

Is the asteroid related to a known meteorite type or types?

Is the body chemically primitive or differentiated?

What were the dominant geologic processes in the asteroid's evolution?

Is there evidence of endogenic activity?

Is there evidence that the asteroid is a fragment of a larger body?

Could the asteroid be a rubble pile?

Is the asteroid possibly related to comets?

What is the characteristic morphology and texture of the surface, and how do they compare with those observed on larger bodies?

Is there any evidence of a magnetic field?

Is there any circum-asteroid gas or dust?

How does the asteroid interact with the solar wind?

---

[a] All questions but the last are addressed at least in part by Discovery NEAR mission.

combined with spacecraft tracking data, from which the asteroid mass can be determined, the density can be derived so as to constrain the asteroid's bulk composition. Detailed mapping of the asteroid gravity field may also reveal interior inhomogeneities. The shape and surface morphology, as determined from imaging, will help distinguish whether the asteroid is a collisional fragment or an agglomeration of smaller bodies. The spin state will also constrain the asteroid's strength and internal structure while providing an important clue as to its possible collisional history.

Elemental composition data will be obtained by gamma-ray and X-ray spectroscopy. These data will enable a vitally important chemical comparison between the asteroid and meteorites and will establish whether the asteroid is a primitive or differentiated body. The surface mineralogy will also be investigated with a visible/near-infrared spectral mapper, which would map the entire surface at moderate spatial resolution (less than that of the imager). The mineralogic mapper will study mafic minerals and search for other species. The elemental and mineralogic measurements may establish a link between the asteroid and a meteorite type (e.g., ordinary chondrite, mesosiderite). These geochemical measurements and imaging data may alternatively provide evidence for past or present cometary activity. This might be also be inferred from surface features like vents, crusts, or collapsed structures or even from detection of dust near the target.

B. Instrumentation

Table II summarizes the facility instrument payload for NEAR. The imager is a CCD camera covering 0.4 to 1 μm. The X-ray/gamma-ray spectrometer has two sensor systems. The X-ray spectrometer uses three gas proportional counters with balanced filters to separate Mg, Al, and Si lines, while Ca and Fe lines can be resolved directly. These gas proportional counters are collimated to a 5° field of view. An X-ray solar monitor is included to measure directly the solar excitation. The gamma ray sensor uses a NaI scintillator with a BGO active shield to remove background from the spacecraft and cosmic rays. It will measure several elements, including Fe, Si, K, and possibly H and Th. The infrared spectrograph covers approximately the range 0.8 to 2.6 μm and disperses the spectrum from a single spot on the asteroid onto line array detectors. A one-dimensional scan mirror is provided, so the spectrograph field of view can be boresighted with the imager or scanned more than 90° away.

TABLE II
NEAR Facility Instruments[a]

Visible Imager	95×161 μr resolution
	$2.25° \times 3°$ field of view
	8-position filter wheel
X/γ-ray Spectrometer	Al, Mg, Si, Fe, Ca, S,
	H, Th, K
Near-Ir Spectrograph	$\sim$0.8–2.6 μm spectral range,
	spectral resolution 2 to 3%
Magnetometer	Sensitivity 5 nT
	(local field)
Laser Altimeter[a]	Range 50 km,
	resolution 6 m
Radio Science[a]	Two-way Doppler to 0.1 mm s^{-1}

[a] Engineering subsystems.

In addition to the basic imaging and geochemical experiments described above, the APL design for the NEAR spacecraft includes a body-mounted magnetometer to search for any intrinsic asteroidal magnetic field, with a sensitivity of 5 nT local field. This instrument could provide valuable information on the nature and origin of the asteroid. A laser ranger is also included as an engineering subsystem for close-in navigation; it uses a diode pumped, solid state laser transmitter. This could provide valuable information on asteroid shape and topography.

C. Mission Design

The NEAR mission is currently scheduled for a February 1996 launch to a flyby of asteroid 2968 Iliya followed by a rendezvous with asteroid 433 Eros in late December 1998. The mission profile is summarized in Fig. 1. The asteroids Ganymede and Eros are the two largest of the known NEAs and are both larger than Phobos and Deimos, the moons of Mars. Eros, discovered in 1898, was the first known Mars-crosser and approached within 0.15 AU of Earth in 1975. At that time it became the target of a dedicated ground observing campaign (Zellner 1976) and extensive radar observations (see Ostro et al. 1990). Eros is an S-type asteroid and is covered by a regolith. Its rotation period is 5.27 hr, and its approximate dimensions are $36 \times 15 \times 13$ km, so it is highly nonspherical and much larger than typical NEAs. In this respect Eros is comparable to typical mainbelt asteroids.

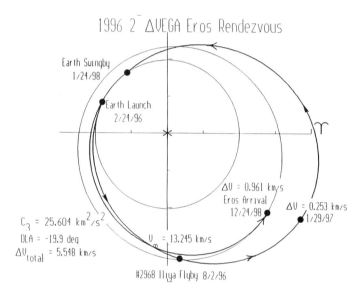

Figure 1.   Mission profile for the near-Earth asteroid rendezvous (NEAR). The quantity C3 is a conventional measure of launch energy relative to that of low-Earth orbit, the DLA is the direction of launch asymptote, and the total Δv is a measure of the total deterministic propulsion capability needed to perform the mission.

The trajectory to Eros is a two-year ΔVEGA with the Earth swingby in January 1998. The launch opportunities for this type of mission to Eros occur only about once every seven years. The spacecraft will spend one year in rendezvous at Eros at a minimum orbit radius of 50 km or less.

D. Spacecraft and Rendezvous Operations

The NEAR spacecraft will be launched by the Delta II-7925, whose launch capability for the Eros mission is 805 kg. The NEAR spacecraft uses a dual mode propulsion system with post-launch ΔV capability of $\sim$1.4 km s^{-1}. The NEAR spacecraft design will emphasize simplicity and robustness. The spacecraft is solar powered and uses an X-band communications system. The spacecraft has fixed solar panels, a fixed high gain antenna system, fixed instruments, and a passive thermal design. A solid state recorder with 0.5 Gbit capacity is provided. The spacecraft is three-axis stabilized, and attitude determination uses Sun sensors, an inertial reference unit, and a star camera. The telecommunications system, the attitude processors and the command and data handling system processors are fully redundant.

The nominal rendezvous orbit for NEAR will be such that the high gain antenna is always pointed to Earth, the instruments are pointed to the asteroid, and the solar panels are within about 30° of full illumination. The nominal orbit plane is near the terminator and is desirable for imaging science. The nominal orbit altitude of 50 km or less allows the geochemical experiments to measure the asteroid composition over the course of the rendezvous. Enough data, in the form of images and spectra, can be taken within a few minutes to fill the spacecraft memory and to require several hours for transmission to Earth.

IV. INTERNATIONAL MISSIONS

Thus far we have discussed only U. S. missions, and we turn now to a brief outline of approved and proposed international missions to explore NEOs. A comet rendezvous mission, the current version of ROSETTA (Schwehm and Langevin 1991), was selected as the third cornerstone of the European Space Agency (ESA) program. The Japanese Institute of Space and Astronautical Science (ISAS) is studying a near-Earth asteroid rendezvous mission. The German Center of Applied Space Technology and Microgravity (ZARM), University of Bremen, has proposed a program of NEO flyby missions called Imaging of Near-Earth Objects (INEO).

A. ROSETTA

The ROSETTA mission was originally conceived as a comet nucleus sample return mission (Schwehm and Langevin 1991) that would have obtained a bulk (kg-sized) volatile sample and preserved its temperature and stratigraphy for Earth return and intensive analyses in terrestrial laboratories. With the cancellation of the NASA comet rendezvous asteroid flyby (CRAF) mission and the likelihood that NASA will not be able to commit major resources to a comet nucleus sample return in the immediate future, the ROSETTA mission was redefined as a comet rendezvous mission, with a surface landed package for *in situ* science. ROSETTA as a comet rendezvous has been selected as the third cornerstone of the ESA program. The ROSETTA spacecraft will be solar powered and three axis-stabilized. Possible targets include comet Schwassmann-Wachmann 3, Wirtanen, and Finlay. The Wirtanen mission, for example, launches in 2003, with a Mars flyby, two Earth flybys, and two asteroid flybys leading to comet rendezvous in 2011. Rendezvous orbital science would commence in 2012, once the comet is close enough to the Sun, and continue through perihelion in 2013. Landed *in situ* science experiments are under study. Possible NASA contributions would include a launch vehicle, use of the Deep Space Network, and a portion of the science payload.

B. ISAS NEAR

A near-Earth asteroid rendezvous mission is under study by ISAS. This Japanese NEAR mission would have two principal objectives, mapping of the surface and hovering within one foot of the asteroid surface for extended periods (possibly months). The hovering phase would make use of a target plate to be dropped on the asteroid surface and is a technology demonstration.

C. Imaging of Near Earth Objects (INEO)

The INEO program consists of four exploratory flybys of NEOs with small spin-stabilized spacecraft (Iglseder et al. 1992). The first INEO mission would perform high resolution imaging and broadband color mapping, using the flight spare of the Halley Multicolor Camera flown on Giotto, and it would investigate the dust environment with existing dust counters. Subsequent INEO missions would fly different imaging systems, a magnetometer, and infrared spectroscopic instruments. Numerous mission opportunities have been identified for the period 1995 to 2001 including single, double and triple flybys of asteroids and comets. The spacecraft can be launched as piggyback payloads on large vehicles like the Ariane IV or as stand-alone payloads on smaller vehicles. The spacecraft mass is about 225 kg. With a piggyback launch on ARIANE for a launch cost of $5 million, the cost of one mission will be about $20 million.

V. OUTLOOK

Since the time of the previous review (Veverka et al. 1989), some promising developments have occurred. We can hope that a systematic scientific exploration of NEOs is now underway with Clementine and NEAR, and that these two missions will be implemented successfully and without delay. We have made a case that even after the NEAR mission, the scientific exploration of NEOs will have barely begun. Given the large number of NEOs, their extreme diversity, and the fundamental scientific importance of questions about their nature and origin, additional missions to NEOs are needed. Further systematic scientific exploration of NEOs is a natural component of the Discovery Program.

Looking further ahead, we can envision new technology developments playing a role in enabling future comprehensive scientific exploration of NEOs. Ultra-lightweight sensors and spacecraft components may allow small, inexpensive spacecraft to meet focused science objectives. A large number of missions may then be possible at a reasonable cost, allowing us to explore more fully the diversity of NEOs. In addition, advanced propulsion technology such as solar electric propulsion may well prove useful.

Note Added in Proof

The Clementine mission successfully mapped the Moon but did not achieve a flyby of Geographos. The Galileo flyby of 243 Ida made the startling discovery of a satellite of Ida. The NEAR mission, as of late 1994, will not be targeted to a flyby of 2968 Iliya, although the trajectory remains very similar to that shown in Fig. 1.

REFERENCES

Belton, M. J. S., Veverka, J., Thomas, P., Helfenstein, P., Simonelli, D., Chapman, C., Davies, M. E., Greeley, R., Greenberg, R., Head, J., Murchie, S., Klaasen, K., Johnson, T. V., McEwen, A., Morrison, D., Neukum, G., Fanale, F., Anger, C., Carr, M., and Pilcher, C. 1992a. Galileo encounter with 951 Gaspra: First pictures of an asteroid. *Science* 257:1647–1652.

Belton, M. J. S., Klaasen, K. P., Clary, M. C., Anderson, J. L., Anger, C. D., Carr, M. H., Chapman, C. R., Davies, M. E., Greeley, R., Anderson, D., Bolef, L. K., Townsend, T. E., Greenberg, R., Head, J. W., Neukum, G., Pilcher, C. B., Veverka, J., Gierasch, P. J., Fanale, F. P., Ingersoll, A. P., Masursky, H., Morrison, D., and Pollack, J. B. 1992b. The Galileo solid state imaging experiment. *Space Sci. Rev.* 60:413–455.

Carr, M. H., Kirk, R. L., McEwen, A., Veverka, J., Thomas, P., Head J. W., and Murchie, S. 1994. The Geology of Gaspra. *Icarus* 107:61–71.

Chapman, C. R., Veverka, H., Belton, M. J. S., Neukum, G., and Morrison, D. 1993. Cratering on Gaspra. *Icarus*, submitted.

Davies, M. E., Colvin, T. R., Belton, M. J. S., Veverka, J., and Thomas, P. C. 1994. The direction of the north pole and control network of asteroid 951 Gaspra. *Icarus* 107:18–22.

Davis, D. R., Hartmann, W. K., Friedlander, A., Collins, J., Niehoff, J., and Jones, T. 1990. The Role of Near-Earth Asteroids in the Space Exploration Initiative. SAIC Study No. 1-120-232-S28.

Drake, M., Boynton, W., and Blanchard, D. 1987. The case for planetary sample return missions 1: Origin of the solar system. *Eos: Trans. AGU* 68:105–109.

Gooding, J., Carr, M. H., and McKay, C. 1989. The case for planetary sample return 2: History of Mars. *Eos: Trans. AGU* 70:745–755.

Granahan, J. C., Fanale, F. P., Robinson, M., Carlson, R. W., Kamp, L. W., Klaasen, K. P., Belton, M., Cook, D., Edwards, K., McEwen, A. S., Soderblom, A. L., Carcich, B. T., Helfenstein, P., Simonelli, D., Thomas, P. C., and Veverka, J. 1993. Galileo's multispectral synergistic view of 951 Gaspra. *Eos: Trans. AGU* 74:197.

Greenberg, R., Nolan, M. C., Bottke, W. F., Kolvoord, R. A., and Veverka, J. 1994. Collisional History of Gaspra. *Icarus* 107:84–97.

Helfenstein, P., Veverka, J., Thomas, P. C., Simonelli, D. P., Lee, P., Klassen, K., Johnson, T. V., Breneman, H., Head, J. W., Murchie, S., Fanale, F., Robinson, M., Clark, B., Granahan, J., Garbeil, H., McEwen, A. S., Kirk, R. L., Davies, M. E., Neukum, G., Mottola, S., Wagner, R., Belton, M. J. S., Chapman, C. R., and Pilcher, C. 1994.Galileo photometry of asteroid 951 Gaspra. *Icarus* 107:37–60.

Horstmann, K. C., and Melosh, J. H. 1989. Drainage pits in cohensionless materials: Implications for the surface of Phobos. *J. Geophys. Res.* 94:12433–12441.

Iglseder, H., et al. 1992. INEO: Imaging of near Earth objects. In *Proceedings of XXIX COSPAR Plenary Meeting*, Aug. 28–Sept. 5, 1992, Washington, D. C., p. 607.

Kivelson, M., Bargatze, L., Khurana, K. J., Southwood, D., Walker, R., and Coleman, P. 1993. Magnetic signatures near Galileo's closest approach to Gaspra. *Science* 261:331–334.

Magnusson, P., Barucci, M. A., Binzel, R. P., Blanco, C., DiMartino, M., Goldader, J. P., Gonano-Beurer, M., Harris, A. W., Michalowski, T., Mottola, S., Tholen, D. J., and Wisniewski, W. Z. 1992. Asteroid 951 Gaspra: Pre-Galileo physical model. *Icarus* 97:124–129.

McFadden, L., Tholen, D., and Veeder, G. 1989. Physical properties of Aten, Apollo and Amor asteroids. In *Asteroids II*, eds. R. P. Binzel, T. Gehrels and M. S. Matthews (Tucson: Univ. of Arizona Press), pp. 442–467.

NEAR. 1986. *Report of the NEAR Science Working Group*, JPL Pub. 86-71 (Pasadena: Jet Propulsion Laboratory).

Ostro, S. J. 1989. Radar observations of asteroids. In *Asteroids II*, eds. R. P. Binzel, T. Gehrels and M. S. Matthews (Tucson: Univ. of Arizona Press), pp. 192–212.

Ostro, S. J., Rosema, K., and Jurgens, R. 1990. The shape of Eros. *Icarus* 84:334–351.

Ryder, G., Spudis, P., and Taylor, G. J. 1989. The case for planetary sample return missions 3: Origin and evolution of the Moon and its environment. *Eos: Trans. AGU* 70:1495–1509.

Schwehm, G., and Langevin, Y., eds. 1991. *ROSETTA: Comet Nucleus Sample Return*, ESA SP-1125 (Noordwijk: European Space Agency).

Shoemaker, E. M., Wolfe, R. F., and Shoemaker, C. S. 1990. Asteroid and comet flux in the neighborhood of Earth. In *Global Catastrophes in Earth History*, eds. V. L. Sharpton and P. D. Ward, Geological Soc. of America Special Paper 247 (Boulder: Geological Soc. of America), pp. 155–170.

Swindle, T., Lewis, J., and McFadden, L. 1991. The case for planetary sample return missions 4: Near Earth asteroids and the history of planetary formation. *Eos: Trans. AGU* 72:473–480.

Tholen, D. J., Goldader, J. D., Herbst, T. M., Spencer, J. R., Golisch, W. F., Griep, D. M., and Kaminski, C. 1993 Radiometry, photometry, and astrometry of 951 Gaspra in 1991. *Icarus*, submitted.

Thomas, P. C., and Veverka, J. 1980. Downslope movement of material on Deimos. *Icarus* 42:234–250.

Thomas, P. C., Veverka, J., Duxbury, T., and Bloom, A. 1979. The grooves on Phobos: Distribution, morphology, and possible origin. *J. Geophys. Res.* 84:8457–8477.

Thomas, P. C., Belton, M. J. S., and the Galileo SSI Team. 1992. Surface processes on Gaspra. *Eos: Trans. AGU* 73:334.

Veverka, J., Thomas, P., Johnson, T. V., Matson, D., and Housen, K. 1986. The physical characteristics of satellite surfaces. In *Satellites*, eds. J. Burns and M. S. Matthews (Tucson: Univ. of Arizona Press), pp. 342–402.

Veverka, J., Langevin, Y., Farquhar, R., and Fulchignoni, M. 1989. Spacecraft exploration of asteroids; the 1988 perspective. In *Asteroids II*, eds. R. P. Binzel, T. Gehrels and M. S. Matthews. (Tucson: Univ. of Arizona Press), pp. 970–996.

Veverka, J., Belton, M., Chapman, C., and the Galileo Imaging Team. 1992. Gaspra: Overview of Galileo imaging results. *Eos: Trans. AGU* 73:335.

Weissman, P. R. 1990. The cometary impactor flux at the earth. In *Global Catastrophes in Earth History*, eds. V. L. Sharpton and P. D. Ward, Geological Soc. of America Special Paper 247 (Boulder: Geological Soc. of America), pp. 171–180.

Zellner, B. 1976. Physical properties of asteroid 433 Eros. *Icarus* 28:149–153.

DoD TECHNOLOGIES AND MISSIONS OF RELEVANCE TO ASTEROID AND COMET EXPLORATION

S. NOZETTE
Ballistic Missile Defense Organization

L. PLEASANCE
Lawrence Livermore National Laboratory

D. BARNHART
USAF Phillips Laboratory

and

D. DUNHAM
Applied Physics Laboratory

During the past decade the Department of Defense (DoD), and its Ballistic Missile Defense Organization (formally the Strategic Defense Initiative) have invested heavily in space technology, focused on the development of lighter and more cost effective components and systems. With the end of the Cold War many of these technologies can be made available to the civilian space community. The exploration of asteroids and comets offers a fruitful venue for application of these technologies. To further these efforts in dual use application, NASA and BMDO have collaborated on the Clementine missions, the first of which was launched 25 January 1994, to flight qualify these technologies, by mapping the Moon and by performing a flyby of asteroid 1620 Geographos in the Fall of 1994. A flyby of 1983 RD in October 1995 is currently under study. These missions will flight test most of the defense technologies of potential utility in the study of asteroids and comets. These relevant technologies, current flight test programs, and their potential modification and application to asteroid and comet exploration are examined in this chapter.

I. INTRODUCTION

The Ballistic Missile Defense Organization (BMDO; formally the Strategic Defense Initiative Organization) of the U. S. Department of Defense (DoD) has been developing strategic defense technologies over the last decade. These technologies have been incorporated into components and systems which are much lighter in weight and consume much less power than previous generations of spaceflight components. There is significant benefit in the use

of these technologies for civilian space science research. The DoD investment has been much greater than the concurrent NASA investment in space
technology during the decade of the 1980s (Farmer 1992; Augustine 1990).
However, the applications must be selected carefully to achieve these benefits.

The technologies of specific interest to address the study of asteroids and
comets in near-Earth space include large and small launch vehicles with rapid
response times, advanced on-orbit guidance and control systems and software,
advanced telecommunications systems, lightweight propulsion and maneuvering systems, advanced sensors and reconnaissance components, lightweight
structural materials, and more efficient space power systems. Lower cost
management and operations of space programs has been demonstrated by
DoD as well.

This chapter examines areas of specific technology as well as ongoing
flight test programs and future mission concepts which can utilize the available
technologies and flight test results. Small body studies are potentially the
most interesting due to the synergies of requirements in both the scientific and
defense applications. These include the requirements for low mass (<200 kg)
and low cost ($<\$50$ M) mass producible spacecraft with multispectral sensors
and high capability attitude control and propulsion systems. Given the desire
to study many different small bodies, a highly capable low cost, maneuverable,
mass produced spacecraft which can be launched on any number of potential
launchers could revolutionize the exploration and characterization of these
objects. The authors of the Chapter by Cheng et al. recognize this potential.

II. DoD/BMDO TECHNOLOGY DEVELOPMENT

The advanced development efforts pursued by the DoD over the past 10
years have focused on several technology areas such as: launch systems,
autonomous defensive and offensive weapons, and component and subsystem level miniaturized hardware. Access to space is still considered a large
cost driver, thus programmatic decisions drove technology investments towards smaller and lighter weight devices. Thus creating smaller and lighter,
lower cost devices with improved performance has been a high priority.
Lightweight, mass produced technology developed for tactical (e.g., cruise
missiles, and "smart" munitions) applications also has been transitioned into
space application. These tactical systems are typically an order of magnitude
less costly than an equivalent mass of space hardware, yet perform in highly
stressing environments (e.g., being shot out of a cannon, or launched by rocket
from an aircraft, ship, or armored vehicle). Improvements in thermal and radiation tolerance have recently been made allowing these technologies to be
used in the space environment.

BMDO concentrated on a set of these core technologies to develop and
demonstrate lightweight, small, very high performance space vehicles. These
were directly developed for defensive surveillance and interceptor systems.
Specific areas of interest and work over the past have been in lightweight radia-

tion hardened avionics, sensors, power, attitude control system and propulsion development, and ultimately smaller integrated devices and spacecraft. These are explored in the sections below.

A. Avionics, Data Processing, and Storage

Avionics include inertial measurement units based on ring laser gyro and interferometric fiber optic gyro, integrated with accelerometers, and controlled by Application Specific Integrated Circuits. These units have achieved masses on the order of 600 g, with drift rates of 0.1 to $1°hr^{-1}$. Also developed are computing systems based on 32 bit Reduced Instruction Set Computing, and 0.8 to 1.2 micron complimentary metal oxide semiconductor technology, as well as advanced concepts in SiC, and C (diamond) semiconductor devices. These latter materials offer the potential for very high radiation tolerance and very dense packaging due to their unique crystal structure. Processing systems already developed can approach 40 million instructions per second in a radiation hardened (100 krad (Si)) package. These processing systems and their associated support devices have masses on the order of 1 kg and represent the computing power of modern high performance workstations, or small supercomputers (e.g., Cray 1). Parallel architectures with multiple processors allow even more powerful systems. Hybrid circuit board construction will also allow substantial additional reductions in mass.

Supporting the processing systems are static and dynamic random access memory (DRAM) and hybrid circuits. Radiation hardened solid state memories based on 4 and 16 megabit DRAM have been constructed in the 2 to 8 Gbit range with mass on the order of 3 kg, with throughputs of between 20 and 40 Mbits/sec and bit error rates of less than 1 part in 10^9, and radiation tolerance from 50 to 100 krad (Si). Experimental efforts in fiberoptic data bus and harness and space qualified optical and magnetic disc drives are also underway. These processing systems can use supporting software derived from the extensive scientific and commercial base in languages such as C, supported by operating systems such as UNIX. Effective C to Ada interfaces have also been developed and flight tested, greatly enhancing the potential software development options.

B. Telemetry Systems

Communication technology based on Microwave Monolithic Integrated Circuits can support deep space communications into the Ka band (30 Ghz). Patch and phased array antennas also offer link and crosslink potential for probes and subprobes with electronic steering (no moving parts). Lightweight solid state laser and optical systems also offer the potential for laser communications over interplanetary distances.

C. Power Systems

Substantial efforts have gone into space power system development for non-nuclear applications. Batteries such as common pressure vessel nickel hydrogen and lithium ion offer the demonstrated energy density of up to 220 w-hr/kg, a factor of 10 better than current spacecraft batteries. Solar arrays based on gallium arsenide (GaAs), deposited on thin (3 mil) substrates of germanium have been produced and are undergoing flight testing. When mounted on composite panels these systems can produce greater than 300 w m^{-2}and on the order of 230 w/kg at one astronomical unit solar distance. Technologies under development such as multijunction concentrating GaAs cells, amorphous silicon, and InP and CuInSe2 could provide up to 1 kw/kg in the next decade. This is accomplished by depositing these materials at fairly low temperature, under 600°C, on a very thin substrate of kapton and metal film. When combined with improved power distribution techniques and the lower power requirements of lightweight components, solar powered exploration of the main asteroid belt becomes feasible.

D. Sensors

Sensor systems have been developed and tested from the extreme ultraviolet (0.08 micron) to the long wave infrared (9–15 micron), and in between. Active radar, lidar and neutral particle beam sensors are also required for many applications. Emphasis has been placed on detection, acquisition, and tracking of both hot and cold targets against space and Earth backgrounds. These sensors could be integrated into orbital platforms and kinetic kill vehicles (KKVs). The KKV was created to be used as a nonexplosive device for ballistic missile defense, thus spurring related technology development in miniaturization of aerospace hardware. The KKVs are small, 3 axis spacecraft systems that have the autonomous guidance, navigation, and control functions necessary for close-approach body maneuvering. They may be deployed by ground launched missile, manned or unmanned aircraft, sea launched missiles, or as a space based system. The need to proliferate these sensors required research efforts aimed at production at the lowest weight, power and unit costs, as discussed previously. The infrared requirements for sensitivity and long life drove research into lightweight active cooling based on Stirling cycle mechanical refrigerators. For short duration cooling requirements, expendable Joule Thompson refrigerators can cool detectors to less than 20 K for several minutes. Infrared detectors have been actively cooled to 65 K, using detector materials such as, InSb, PtSi, and HgCdTe, with up to 512 by 512 pixels array size. These focal plane arrays were also required to be mass producible. In the ultraviolet and visible spectrum most effort has been given to commercially derived charge coupled devices (CCDs) with coatings providing ultraviolet induced fluorescence, to allow detection in the near ultraviolet (0.25 micron). The far ultraviolet detection has focused on microchannel plates and novel detectors based on SiN, SiC, or C (diamond). Active imaging and ranging sensors based on diode pumped Nd: YAG lasers at 1.06 micron, with 0.2 to 1 J/pulse and 10 nanosecond pulse duration and compact imaging MIMIC based synthetic aperture radar have also been demonstrated. The current state of the art will allow a package of high resolution sensors with spectral coverage from

0.25 to 10 micron, with either hyperspectral resolution based on wedge filter, grating, or interferometric dispersion, or discrete filter wheels, with lidar, and having instantaneous fields of view of less than 40 microradian, for less than 8 kg and 50 w power consumption. Also developed for star tracker applications are wide field of view optical systems and data processing for viewing large areas of sky (60 degrees square) for moving targets. These systems may be adapted to ground based telescopes such as Schmidt cameras to allow wide area scanning using CCDs and supporting data processing hardware. Large format CCDs (4096×4096) have also been developed which could be applied to this groundbased application.

Research in directed energy technologies has produced entirely new sensor capability. Folded, deployable adaptive optics, and neutral particle beam (NPB) technologies offer powerful tools to examine planetary objects. By folding the optics and using an active secondary to correct for aberration and vibration, a meter scale optical system may be collapsed into compact launch packages. The NPB utilizes compact accelerators and particle sources and detectors. These NPB systems have achieved substantial reductions in weight, cost, and complexity, as demonstrated by the Beam Aboard Rocket (Farmer 1991) experiment flown in 1989. Compact NPB technology has been advanced substantially by the former Soviet Union. Employed as discrimination devices the NPB projects a 2 to 10 Mev/nucleon beam of protons (neutral H) or deuterons (neutral D) over distances of tens of kilometers, in the presence of magnetic fields. These particles impact on surfaces exposed to vacuum and cause X-ray, gamma ray, and neutron emission from the surface. These may be measured by sensors on the spacecraft to determine precise surface composition. These technologies, when combined with the lightweight power and electric components described above, could be integrated into a multi-asteroid sampler which could encounter many bodies in the main asteroid belt. Useful NPB systems can be developed with power requirements on the order of 50 Kwe for under $300 M including launch on former Soviet boosters (e.g., Proton).

E. Propulsion

The KKV application required the development of lightweight high ΔV (6 km s^{-1}) propulsion systems. The requirements are for axial thrust for achieving high axial velocity, and divert propulsion for end game maneuvering. Both solid and liquid systems have been developed for these applications. Substantial improvements in specific impulse, to about 350 s for storable propellants have been demonstrated by use of fluorinated oxidizers (ClF5). Improvements in mass fraction have also been made by application of lightweight materials in tank construction. A novel approach to improvement in small propulsion system mass fraction has been the development of miniature (100 g) high-pressure (1000 psi) pumps based on hydrazine decomposition. These systems allow mass fractions of greater than 9 to 1, as opposed to between 2 and 3 to 1 for pressure fed propulsion systems, by eliminating the need for thickwalled high-pressure propellant and pressurant tanks.

Electric propulsion technology based on Xenon Hall effect thrusters has been demonstrated up to 1 Kw/thruster by the former Soviet Union. Acquisition of this technology for use in DoD spacecraft has provided a solar electric propulsion capability for small spacecraft. When combined with lightweight components and power systems such electric propulsion capability can open new options for high ΔV exploration of small bodies (e.g., comets). These systems are undergoing flight qualification at the present time and could provide major contribution to small body exploration by expanding the range of characteristic velocities available for a given launch vehicle.

F. Structures

The use of advanced materials such as graphite epoxy, ceramic metal matrix composites, and fiber woven and wrapped structures has provided substantial (up to 40%) weight reductions in spacecraft structures. Structure composition is usually a combination of lightweight metal and composite wound carbon filament for reduced mass and increased strength. The structure on lighter vehicles also serves a dual purpose to house major components as well as provide propulsion lines and linkages. In addition, improved manufacturing techniques and industrialization of composite material production has made these materials very cost competitive with standard aluminum and titanium. High temperature materials such as HfC have also been developed for use in high performance propulsion systems (rocket nozzles and combustion chambers).

G. Spacecraft Design and Testing

Currently DoD is undertaking a series of test flight experiments to flight qualify many of the above mentioned technologies. The most relevant for purposes of scientific space application is the Clementine program. It was designed to flight qualify a range of DoD developed technology in a mission of interest to NASA and the science community. Project Clementine is an ABM Treaty compliant science experiment using an integrated DoD developed sensor payload to do a polar lunar orbit and a flyby of asteroid Geographos in late August 1994, during the closest approach of this asteroid to Earth. An extended mission to 1983 RD in October 1995 is also under study. The payload will consist of the technology being developed by the BMDO Technology Directorate (ultraviolet/visible sensors, infrared sensors, computer, startracker, propulsion, attitude control system, and software) with enhanced features (solar panels/batteries, materials, memory, more powerful lidar, and kick stages). Follow on missions are being planned.

The mission objective of the experiment will be long life testing of the technology in a realistic, stressing, space environment. This can be accomplished with a trajectory design that involves a two month lunar orbit mapping phase followed by a flyby of the asteroid 1620 Geographos in August 1994. The natural space objects provide excellent targets for the sensors, at realistic closing velocities, time scales, and ranges not obtainable in Earth orbit without the use of a costly target spacecraft. The trajectory designed for Clementine includes the following phases: low-Earth parking orbit; lunar transfer phase; lunar mapping phase; Earth orbit with lunar swingby; and interplanetary trajectory to intercept Geographos. The spacecraft is being built at the U. S. Naval Research Laboratory in Washington, D. C., and will use sensors, attitude control systems and software supplied by the Lawrence Livermore National Laboratory. Several other organizations are involved, especially NASA with communications support through the Jet Propulsion Laboratory's Deep Space Network, and operations support from both the Goddard Space Flight Center and JPL.

The near-Earth asteroid flyby also provides a realistic test of sensors and autonomous navigation which would be required under actual mission scenarios, following a long and degrading exposure to radiation, over a long time period to allow accurate measurement of this degradation. The launch occurred on 25 January 1995 on schedule. The main instrumentation on Clementine consists of four cameras, one with a laser-ranging system. The cameras include an ultraviolet/visual camera, a longwave infrared camera, the laser-ranger (lidar)/high-resolution camera, and a near-infrared camera. The spacecraft will also have two star trackers, to be used mainly for attitude determination, but they will also serve as wide-field cameras for various scientific and operational purposes. The spacecraft consists of an octagonal cylinder about 2 m high. A 110-pound thruster for ΔV maneuvers is on one end of the cylinder and a high-gain fixed dish antenna is on the other end. Two arrays of rotatable solar panels protrude from opposite sides; by rolling the spacecraft and rotating the panels, full solar illumination of the panels can be achieved. The sensors are all located on one side 90° away from the solar panels. Clementine has 12 small attitude control jets that are used to orient the spacecraft to point the cameras to desired targets. During the Geographos flyby, Clementine will be rotated using the attitude jets to keep the cameras pointed at the asteroid. A flyby distance of 100 km has tentatively been selected that should allow the spacecraft to keep the cameras on Geographos.

Demonstration of autonomous navigation, including autonomous orbit determination, is a major goal of the Clementine missions. Demonstrations will be conducted at times in Earth orbit, in lunar orbit, and during the heliocentric cruise to Geographos. Due to the relatively high flyby speed, the on-board computers and attitude control system will be programmed to perform the rotation and pointing sequence needed to optimally image Geographos. The parameters of the flyby can be determined accurately enough from on-board approach imaging to perform the flyby imaging only about two minutes before the flyby. Preliminary results from studies by the Naval Research Laboratory, Lawrence Livermore National Laboratory, Alamos and Jet Propulsion Laboratory indicate that the flyby terminal navigation and pointing is feasible. The three organizations are developing an integrated plan to perform this critical phase that will result in the first *in-situ* observations of a near-Earth asteroid. Clementine imaging sensors can produce high quality science data in the following areas.

Lunar surface composition. Clementine is using two six-position filter wheels on the ultraviolet/visible and infrared imagers to create a twelve-channel imaging spectrometer. Orbital parameters can be adjusted to produce global maps of lunar surface color provinces; with filter selection, these maps can be interpreted in terms of surface mineralogy by using mineral absorption features in the near-IR and selected chemistry (e.g., empirically for Ti content from ultraviolet/visible channels). Given the planned orbital parameters surface resolution of 100 to 200 m/pixel over the entire Moon will be obtained. High resolution imaging with four color visible spectral observation is planned to provide <30 m/pixel resolution over selected areas of the lunar surface using the lidar receiver. The scale of surface resolution will be limited by the data compression used and the downlink data rate. The use of a long-wave infrared camera with 1 to 200 m/pixel resolution in the broad (9–10.5) micron band will allow limited measurements of thermal properties of the lunar regolith.

Lunar terrain. Near side gravity is being derived from tracking data. With slight modification of the firing rate (1–8 Hz), the lidar can serve as a laser altimeter, mapping the global figure of the Moon simultaneously with acquisition of imaging and spectral data through a filter wheel. For passive lidar receiver imaging four filters are available for multi-spectral imaging, with a clear filter for panchromatic imaging.

Monochrome frame images taken at the same point on the Moon from two different spacecraft positions can provide a data base for global geodetic control. Lidar measurements are currently planned for a latitude band of 30°N to 30°S, at 1 to 8 per sec. This will provide a 200 m diameter spot sampled at 40 m height resolution. Altimetry will be coupled with near simultaneous high resolution imaging to locate the altitude spot in the frame. Higher pulse rates (up to 40 s^{-1}) are feasible on an intermittent basis, allowing for more detailed coverage of selected areas.

Near-Earth asteroid observation. The fast (10.8 km s^{-1}) flyby of Geographos will provide data on surface composition, figure, gravity, rotation rate, and morphology in the same manner as the lunar data. Several hundred images will be obtained to 5 m or greater resolution. In addition, the close proximity of Geographos to Earth (8×10^6 km) is a unique opportunity to both observe Geographos by groundbased sensors and close up with Clementine, providing unique ground truth science obtainable in no other way. This will allow direct relationship between groundbased observation and close up observation to aid in calibration of other groundbased asteroid data. In order to provide more target opportunities a groundbased search will be conducted from Lowell Observatory, the University of Arizona and the U. S. Air Force site on Maui Hawaii, also utilizing advanced sensors and software from DoD sponsored research. A flyby in October 1995 to asteroid 1983 RD, a U-type object, is currently being considered.

III. SOLAR SYSTEM EXPLORATION APPLICATIONS OF DoD TECHNOLOGY

The successful completion of the Clementine I mission will accomplish three primary NASA Objectives: (1) a low spectral resolution (12-channel) global lunar mineral map; (2) a lunar altimetry data set from the lidar; and (3) a close look at a near-Earth asteroid (Shoemaker and Nozette 1993; COMPLEX 1992). In addition, Clementine has the ability to produce a set of monochrome image frames that, coupled with the laser altimetry, could provide a geodetic control network for the Moon, an essential data set for the lunar digital terrain model (cartographic data base). With these data in hand, alternative and/or advanced sensors could be flown on future NASA science missions. Given the need for extensive global lunar composition information, global surface chemistry data collection would still be necessary, utilizing X-ray, gamma ray, and related sensors. Given Clementine experience, a new thermal emission spectrometer could also be developed, producing data in a wholly new spectral region (5 to 50 microns), one in which unique mineralogical information is obtainable. Alternatively, a second-generation imaging spectrometer could be flown, thus producing high spatial (up to 20 m/pixel) and spectral (256 channels) resolution data to follow up on and characterize Clementine discoveries in greater detail. Such a system could be developed as a modification of the Clementine sensor suite, by combining the narrow and wide field optics with a selectable optical path and dispersive system, either a wedge filter or a Fourier spectrometer, to allow pixel-coregistered multi-spectral imaging. The altimetry data collected by Clementine might make a dedicated lunar terrain laser data set redundant, but this conclusion depends strongly on both the mode of lidar operation on Clementine and on the amount of time it is able to spend in lunar orbit. However, even if we do obtain a partial global figure from Clementine, an altimeter on a future lunar terrain mapper would still be desirable to build up the altimetric data base for the Moon to a significantly higher density. Each laser altimeter data point provides an absolute elevation on the Moon, a feature which improves both the topographic and geodetic fidelity of the global cartographic data base. Depending on how successful a Clementine geodesy experiment is (e.g., can enough overlapping frame images be obtained for the global control net), it may still be necessary to fly a separate geodetic framing camera on a future NASA lunar missions. This can easily be developed from the existing Clementine system with modifications.

For example, additional thermal control may be added to the lidar allowing a greater sustained firing rate. Lack of mechanical filter wheels combined with turning mirrors would allow shorter exposures of to reduce image smear to 1 to 2 pixels at 100 km altitude. The spacecraft attitude control system, based on startracker and lightweight momentum wheels is expected to allow low smear when combined with turning mirrors, achieving a pointing accuracy of less than 150 microradian.

Modification of the Clementine sensor suite and lunar orbital parameters of a Clementine spacecraft, (e.g., 100 km circular), would provide 20 to 40 m/pixel multi-spectral imaging combined with 1 to 2 m high resolution imaging, with a 40 m lidar spot size. The major limitation then becomes data handling, and transmission. Current Clementine link limitations of 128 kbs with 2 Gbit solid state recorder would not allow extended coverage without modifications. By extending the time in lunar orbit and enhancing the data storage and downlink an order of magnitude improvement may be possible. For example, current developmental efforts will result in a solid state recorder with a factor of 3 increase in memory for the same weight (4 kg) in 18 months. Moving to X-band from S-band should allow a factor of 3 increase in data rate, Ka band a factor of 10. A dedicated lunar mission of this type would fulfill many if not all requirements of the Lunar Scout mission for a total estimated cost of $70 to $100 M including launch on a commercial expendable launch vehicle. Commercial launch vehicles such as Orbital Science Corporation's "Taurus," and the proposed Lockheed LLV have projected launch costs of under $30 M. Refurbishment of decommissioned ICBM's has projected costs of $4 to $30 M, ranging from Minuteman 3, MX, or Titan IIG (T IIG) vehicles. Additional costs to use a refurbished T IIG with a subsatellite for lunar far side gravity determination would still allow a total mission cost under $100 M. These costs are based on actual costs incurred for Clementine and other DoD programs. Bulk buys for additional flights could further reduce these costs. The demonstration on Clementine of autonomous spacecraft operation is also expected to further reduce ground support costs.

As configured, the existing Clementine spacecraft could rendezvous with a selected set of near-Earth asteroids, provided spacecraft lifetimes of greater than one year are proven. The basic Clementine spacecraft component suite proposed could also be dispatched to Mars (Phobos/Diemos), Mercury (with thermal modifications), mainbelt asteroids (with enhanced PV materials or concentrator cells), or outer planets (with RTGs). Lander/rover technology could incorporate high mass fraction propulsion technology from the BMDO interceptor program. A lander based on KKV technology could deliver 50 kg to the surface of Ceres or Vesta with limited mobility, multi-spectral imaging, and communications. Space qualified batteries of 220 w-hr/kg (depending on charge cycle) would allow significant operation in a small lander/rover. These could be emplaced on small bodies for operation over weeks/months depending on power requirements. Several such landers could be dispatched to promising objects for under $100 M, including launch costs based on one T IIG or commercial expendable launch vehicles (e.g., Taurus).

There are several approaches to utilization of DoD developments described above in an asteroid/comet encounter. The first is to approach the body on a path to observe it at some distance; the second is to travel to a point in space to approach, intercept its path, and encounter the body either as kinetic impact or as "soft" rendezvous for further scientific study. Interception of trailing or leading debris or gas field would also fall under the second approach. The latter concept could use the KKVs as autonomous and independent spacecraft, not dependent upon the planetary spacecraft. Alternatively, one could use an architecture that relies upon remote telemetry from the KKV to the mother spacecraft at some time to send data back to Earth. Effectively, the main spacecraft will become a relay station back to Earth for a KKV/asteroid encounter.

The KKV(s) would be "dropped-off" by the mother spacecraft on a predetermined course for a single or multiple body mission. At some predetermined time in the spacecraft's orbit, the mother spacecraft would continue on to another asteroid encounter. Applications that directly exploit the KKV's attributes are Earth-crossing asteroid sensing and impact missions, lunar and planetary low orbit fly-by sensing, and multiple asteroid fly-by, rendezvous, and sample return. KKV characteristics can be examined and modified as they relate to the asteroidal bodies motion and requirements for subsequent maneuvering. Planned Clementine II and III missions are investigating the potential of deploying KKVs as impact and soft landing probes to a set of near Earth objects to study cratering, and composition of the ejecta. The ability to sample the structure and composition of objects using the "hard" impact could allow much more information to be gathered on many near small bodies which are dynamically unsuitable for low energy rendezvous.

The application of former Soviet NPB and other directed energy technologies could produce an "asteroid explorer" combining optical, infrared, and active and passive X-ray, gamma ray, and neutron sensors, to determine surface elemental composition. Such a vehicle could be propelled by solar electric power and rendezvous with a number of mainbelt asteroids. Small KKV based sample return vehicles could be carried along and dispatch samples back to Earth for study. This capability could provide the only cost effective means to fulfilling the recommendations with regard to the exploration of the small solar system bodies (COMPLEX 1980; U. S. Congress, Office of Technology Assessment 1992).

IV. SUMMARY

The completion of a lunar orbital and asteroid mission phase of Clementine will provide valuable information relevant to NASA strategic needs, and begin the process of transferral of DoD developed technology to NASA and the civilian scientific community. By obtaining early data suitable for reconnaissance needs, the data from Clementine could permit the flight of lower cost and/or more scientifically capable instruments on future NASA science missions. Clementine data is contributing to several essential lunar and asteroid data bases, which subsequent missions can build upon and expand. Clementine has opened a new phase of potentially much lower cost unmanned exploration by initiating the transfer of low mass, high performance DoD derived technology to NASA and the civilian scientific community. Clementine provides a demonstration of how DoD technology may be used by NASA in a collaborative and relatively low cost way, thus minimizing the risk to NASA of developing and incorporating new technology into future programs.

To fully incorporate DoD technology, joint efforts with NASA should begin with: mission design, sensor modification and calibration, computing system modification, propulsion, power systems, and operations. These modifications will be necessary to fully meet NASA as opposed to DoD requirements. DoD satellite technology and spacecraft can also be applied directly to other civilian applications such as Earth observation systems, global resource monitoring, ocean mapping, and telecommunications. The only way the civilian community will reap the benefits of DoD technology is through joint efforts which share expertise as well as hardware.

REFERENCES

Augustine, N. 1990. Report of the Advisory Committee on the Future of the U. S. Space Program. (Washington, D. C.: Government Printing Office).

Farmer, D. 1992. Using Today's Strategic Defense Initiative (SDI) Technologies to Accomplish Tomorrow's Low Cost Space Missions. IAF-92-0752 (Washington, D. C.: International Astronautical Federation).

COMPLEX (Committee on Planetary and Lunar Exploration). 1980. Strategy for the Exploration of Primitive Solar System Bodies—Asteroids, Comets, and Meteoroids: 1980–1990 (Washington, D. C.: National Academy of Sciences).

COMPLEX (Committee on Planetary and Lunar Exploration. 1992. Scientific Assessment of the Strategic Defense Initiative Organization's Integrated Sensor Experiment (Clementine). (Washington, D. C.: National Academy of Sciences).

Shoemaker, E., and Nozette, S. 1993. Clementine: An inexpensive mission to the Moon and Geographos. *Lunar Planet. Sci.* XXIV:1299–1300 (abstract).

U. S. Congress, Office of Technology Assessment. 1992. NASA's Office of Space Science and Applications: Process, Priorities, and Goals. (Washington, D. C.: U. S. Government Printing Office).

HUMAN EXPLORATION OF NEAR-EARTH ASTEROIDS

T. D. JONES
NASA/Johnson Space Center

D. B. EPPLER
Science Applications International Corporation

D. R. DAVIS
Planetary Science Institute

A. L. FRIEDLANDER and J. McADAMS
Science Applications International Corporation

and

S. KRIKALEV
Russian Space Agency

The advent of space infrastructures supporting expeditions to the Moon and Mars will bring dozens of near-Earth objects (NEO) within reach of human explorers. The growing discovery rate of NEOs has already yielded many attractive mission profiles with multi-week stay times and durations under a year. Such flights are within the current experience humans have to microgravity exposure, and well within the duration of planned Mars expeditions. Beyond broadening the base of deep-space experience necessary for Mars flights, NEO missions will open up a new "planetary" surface to detailed scientific scrutiny, offer in-depth space resource assessment, and provide early experience in NEO surface operations useful in any future interception scenarios. Early asteroid missions by humans will strengthen the integrity of any future lunar/Mars program (LMP) planning. Using hardware developed for human missions to Mars, NEO missions offer a major increase in the scientific return of the LMP program for a marginal increase in the cost of a lunar/Mars capability. Analagous to the rehearsal of the Apollo 8 and 10 lunar landing missions, NEO flights can test Mars mission hardware relatively close to Earth, and include a robust abort capability. They can increase dramatically an exploration into the planning of access to space resources, and offer ideal profiles for tests of nuclear propulsion flight. Finally, NEO missions will maintain the programmatic momentum of a Mars exploration program during the interval between the return to the Moon and the first Mars expedition. Asteroid field exploration presents obstacles, but none are insurmountable in light of lunar and orbital EVA experience. A major operational challenge will be extended periods of field work (probably weeks in length) in a microgravity environment. Exploratory EVAs will use a combination of manned maneuvering units (MMU) for controlled flight over the asteroid, and a platform-mounted workstation with remote manipulator system (RMS).

The MMU has the mobility to conduct close reconnaissance and survey work, while the RMS platform frees an astronaut from flying and navigation tasks during intensive work at sites of high interest. Surface operations will require an anchor and tether system, and a strategy for NEO-unique control for equipment and collected samples. Humans enroute to an NEO will face a suite of deep-space hazards. Life support, microgravity exposure, health maintenance, and long-term operations issues can be resolved on space stations, both Mir and subsequent international facilities. Strategies to minimize radiation exposure, a particularly daunting hazard, must be developed for any deep-space voyage. On the plus side, the relatively short flight times and robust abort-to-Earth capability of NEO missions represent a logical extension of human spacefaring capabilities beyond the Moon. As the groundbased NEO search accelerates, and robotic precursor flights take shape, the asteroid and Mars exploration communities must work actively and cooperatively to open an early window to NEO exploration. These objects, potentially dangerous to Earth, can instead become a rich focus for scientific and resource exploration.

I. NEAR-EARTH OBJECTS WITHIN OUR GRASP

Human exploration of near-Earth objects (NEOs) will become possible with the development of space infrastructures supporting human expeditions to Mars. NEOs (they include both asteroids and comets) present a small, yet real risk of collision with the Earth. Yet these objects are also accessible, attractive scientifically, and offer a varied assortment of resources. Their overall attractiveness as exploration targets, coupled with planned growth in human space-faring capability, presents us with a glittering opportunity. Sending humans to asteroids will boost both the scientific and programmatic integrity of any Mars program. We can, and should, undertake NEO missions under the umbrella of the larger human exploration effort.

NEOs are a population of small bodies, ranging from 40 km to less than 1 km in diameter, moving on orbits approaching (or crossing) that of the Earth (McFadden et al. 1989). A formal definition we adopt here is that used by Helin and Shoemaker (1979): a near-Earth object is an asteroid or comet with a perihelion less than 1.3 AU. The first NEO, 433 Eros, was discovered in 1898 by G. Witt of Berlin. Discoveries continued at a slow pace for the next few decades—only 14 were known by 1950, and just 28 by 1970. However, the discovery rate has increased significantly in the last decade due to the dedicated efforts of search programs led by Helin (Helin and Dunbar 1990), by Shoemaker (Shoemaker et al. 1990), and, more recently, by the Spacewatch program at the University of Arizona (Gehrels 1991); the programs are described in the Chapter by Carusi et al. As of mid-1993, there are more than 204 known NEOs, with 1992's total of 26 making that year a banner one for their discovery. Of these 204 asteroids, only a small fraction have been studied in sufficient detail to derive properties other than their orbits and very approximate sizes.

The orbits of near-Earth asteroids are quite diverse, with a wide range of inclination (i) and eccentricity (e). Most NEOs have aphelia that lie within the region of the main asteroid belt, between Mars and Jupiter. Individual NEOs are transient bodies on the geologic (and astronomical) time scale: because their orbits bring them close to Earth, Mars, and Venus, they will occasionally have their orbits significantly perturbed by, or even collide with, one of these planets. Calculations have shown that on the time scale of several tens of Myr, the population would be significantly reduced by such planetary encounters. This time scale is short compared with the 4.6 Gyr age of the solar system, so clearly, the NEO population is not a remnant from the formation of the solar system. Instead, they must be continually resupplied from several sources. NEO orbits point to the main asteroid belt as the principal resupply source. Short-period comets offer another source for NEOs: evidence suggests that these comets become inactive, i.e., lose their ability to blow off dust and gas, after repeated orbits about the Sun (Weissman et al. 1989). What remains is probably an extinct comet core: volatile-depleted (at least in its outer layers), very dark, with an orbit similar to those observed for some NEOs. The best estimates (with a large degree of uncertainty) is that about 60% of NEOs originate in the asteroid belt, while some 40% are derived from extinct comets (Hartmann et al. 1987; Weissman et al. 1989). Asteroid-derived NEOs are widely viewed as fragments of rock and/or metal parent bodies, disrupted by impact. NEOs originating in the main belt are more likely to have low eccentricity, low inclination orbits, and be more accessible to piloted or robotic spacecraft.

Earth-crossing asteroids are a subset of the more general class of near-Earth objects that we study in this chapter. The known population of such asteroids is only a sample of the much larger true population. The Earth-crossing population includes objects that have a perihelion distance between 1.02 and ≈1.08 AU, and either presently cross Earth's orbit or will evolve into Earth-crossing orbits at some future time. Shoemaker et al. (1990) have estimated the size distribution of this true Earth-crossing asteroid population, taking into account the number discovered through 1989, and the fraction of the asteroid-populated volume of space searched in the discovery process. About 60% of the known population of NEOs are Earth-crossing asteroids. Assuming this percentage is characteristic of the total NEO population, Shoemaker et al. estimate there are 1700±800 Earth-approaching asteroids larger than ~1 km in diameter.

The number of NEOs increases dramatically at smaller sizes, a result of well-understood power-law size relations among asteroid fragments. For example, if we consider objects larger than 0.5 km in diameter, the estimated number rises sharply to 5700±2600 objects. Clearly, there are a large number of asteroids yet to be discovered. We have found less than 10% of the estimated 1700 bodies larger than 1 km diameter (Davis et al. 1993). These estimates of the NEO population's total size are generally supported by the statistical rate at which these objects are rediscovered (Shoemaker et al. 1990; see the Chapter by Rabinowitz et al.).

The scientific attractiveness of NEOs stems from the origin of many as samples of mainbelt asteroids, which represent, in turn, the left-over debris from the truncated process of planet formation in the region between Mars and Jupiter. Groundbased telescopic studies of asteroid spectra have identified three or four main compositional classes, and half a dozen or more minor ones (Tholen and Barucci 1989). The NEOs are a "grab-bag" mixture of different asteroid classes from the main belt, with perhaps some new varieties not yet identified in the main belt. The most common objects in the outer part of the main belt are the black C class, thought to be similar to the carbonaceous chondrite meteorites. Dominating the inner portion of the main belt is the S class, apparently composed of common, rock-forming silicate minerals, along with metal (Gradie et al. 1989). The true nature of the S class fuels one of the most volatile controversies in asteroid studies today. Are these S-class bodies the sources of the ordinary chondrite meteorites—primitive, never-melted stones that are the most common type of meteorite? Alternatively, are they collisionally exposed metal-rich cores of bodies that partially melted in the past? Or does the S class encompass objects of both origins? NEOs also preserve evidence of the history of the early Sun, the original composition of inner solar system planetesimals, and the volatile inventories present during formation of the terrestrial planets. Missions to NEOs could address these and other fundamental questions, leading to an explosion in knowledge about the early history of the solar system. Additional scientific rationale for missions to NEOs is given in the recent study by the Discovery Science Working Group on the Near-Earth Asteroid Rendezvous mission (Discovery 1991; NEAR 1993).

Detailed knowledge of NEO composition will open the way for recovery of potential NEO resources, yet another reason for our heightened interest in these objects. Water is very likely to be present in abundance on a significant fraction of them, especially the C-class objects and extinct comet nuclei (Jones et al. 1990; Weissman and Campins 1993). Indeed, these objects are likely to be the only *significant* sources of water between the oceans of Earth and the frozen deposits on Mars (polar caps, permafrost, and water of hydration). The great advantage of asteroidal water is its location outside the trap of a deep gravity well. This advantage is especially relevant to Mars exploration, given the most recent evidence from spectroscopic studies by Bell et al. (1993) and the Phobos-2 mission (Langevin et al. 1991) that the surface soils of the Martian satellites Phobos and Deimos are severely dehydrated. The water derived in the mainbelt NEOs is at least present in the form of hydrated minerals; Jones et al. (1990) found that about two-thirds of the mainbelt C asteroids show spectral evidence for water of hydration. In addition, water may exist as ice in the interior of asteroids or inactive comets (Nelson et al. 1993; Weissman and Campins 1993). While water is the most precious NEO resource, these objects can also provide bulk mass (for shielding large space structures in Earth orbit), without the expense of lifting it from a deep planetary gravity well. Detailed discussion of the resource potential inherent in NEOs can be found in Lewis and Hutson (1993; see also Nichols 1993).

The NEO population has long played an important role in shaping the surfaces of the terrestrial planets. On Earth, NEOs and comets have likely played a hand in altering the evolution of Earth's biosphere. Current research into the effects of the K/T impact provides an extreme example of the potential effects of asteroid collisions. The exploration of NEOs by human crews promises to characterize NEO physical properties in sufficient detail to support the development of practical interception and diversion strategies.

This chapter examines the rationale for piloted missions to NEOs, particularly within the context of a larger effort to return humans to the Moon, and extend our reach to Mars. We discuss the challenges facing astronauts performing field investigations on the surface of an asteroid, and survey some of the very real technical challenges which must be overcome before piloted NEO missions become possible. Finally, we will list some steps that can be taken now to make asteroids exploration by humans an early reality.

II. RATIONALE FOR PILOTED ASTEROID MISSIONS

The major advantage of early human flights to NEOs is that they strengthen the integrity of any foreseeable program of human lunar and Mars exploration. Until early 1993, planning for such a program evolved under the umbrella of the Space Exploration Initiative, also known as SEI, whose programmatic elements have been outlined in Stafford et al. (1991). While lunar/Mars program (LMP) planning in late 1993 has slowed to a crawl, it will undoubtedly revive at some point. The case for NEO missions that we develop here will apply equally well to future programs of human planetary exploration.

NEO missions buttress future LMPs by offering a compelling combination of benefits: gains in operational experience, a rich scientific return, an in-depth resource assessment, and reduction of technical risk in future interception or diversion scenarios. Asteroid missions lend themselves particularly to shakedown demonstrations of the major flight elements destined for use on later Mars expeditions.

Flight hardware intended for Mars transportation demands a particularly high degree of reliability, compared with that required for the few days of a round-trip to the lunar surface. Even after establishment of a lunar outpost, space transportation systems will not be capable of moving on immediately to Mars voyages lasting up to three years. We will need extensive experience in deep space operations before casting off for Mars, and the degree of confidence required in both machines and people may not be easy to acquire on brief flights within Earth-Moon space. Along with space station experience, prudent planning demands some type of demonstration of the Mars cruise hardware, involving less risk and mission complexity, before the first Mars expedition. While testing of surface systems can be accomplished on the Moon, elements such as the crew module, communications system, life support, schemes for radiation protection, propulsion, and mission operations will all need shakedown in a deep-space environment (Fig. 1).

Piloted NEO missions fulfill that need, and at a particularly opportune stage of an evolving LMP. A round trip of a year or so outside Earth-Moon space would extend the capability of human deep space-faring comfortably before the big leap to Mars. The one-year mission duration would parallel our experience aboard Mir and the future international space station, keeping such asteroid flights well within the base of human experience for microgravity exposure. One or more such "shakedown cruises" to asteroids will prove that the selected approach to Mars exploration is a sound one.

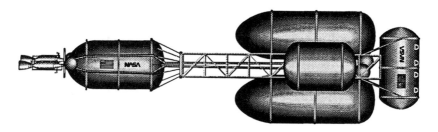

Figure 1.   Generic Mars cruise vehicle concept, with crew module at right shielded from nuclear propulsion system by propellant tanks (figure courtesy of the New Initiatives Office, NASA Johnson Space Center, Houston, Tex.).

Twenty years after the last Apollo scientific expedition to the Moon, it is worth remembering how we built operational confidence as we worked toward the first lunar landing. NASA mounted two lunar rehearsal missions, successively testing the operational hardware and software elements required for the first landing. Apollo 8, in December 1968, used the command and service modules (the interplanetary "cruise" vehicles) to conduct the first piloted lunar orbit operations. Apollo 10 saw the debut of the lunar module in lunar orbit, and this mission in May 1969 was a full dress rehearsal of the landing mission profile (Murray and Cox 1989).

Human NEO visits offer future LMPs a shakedown of all Mars hardware elements (except the lander) in a deep space environment. Yet the mission profile is much more benign than contemplated Mars scenarios. The flights can be conducted reasonably close to Earth-Moon space: some NEO mission profiles keep the spacecraft within a few tenths of one AU of Earth (Davis et al. 1993). Mission durations of about a year expose the crew to significantly less microgravity and radiation than a three-yr Mars landing would. Best of all from a crew standpoint, a Mars cruise vehicle configured for an NEO mission possesses a substantial abort capability. Figure 2 shows, for example, that for the 2009 profile to 1991 JW, the crew can abort to Earth just before the NEO rendezvous for just 6.0 km s^{-1}, returning to Earth four months later. For most of the mission, in fact, an abort is possible for much less than the 7 km s^{-1} ΔV capability required for completion of a Mars mission. NEO missions thus offer the opportunity to test Mars mission hardware via a gradual expansion of the flight envelope; such flights can serve the LMP well as dress rehearsals, much like the Apollo 8 and 10 precursors to the first lunar landing. Moreover, if nuclear propulsion is chosen for the Mars cruise vehicle, human NEO missions offer realistic flight tests, with the benefit of less demanding thrust profiles, before the commitment to operations on Mars.

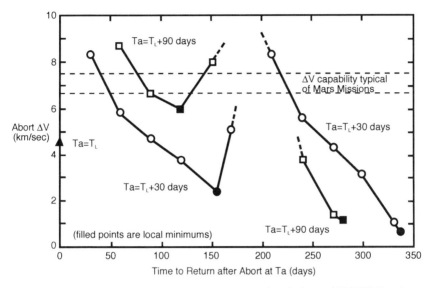

Figure 2.   Abort requirements for one-year round-trip mission to 1991 JW. T_a = time of abort, from launch; T_L = time of launch.

Future exploration of the Moon and Mars will place heavy emphasis on integration of scientific goals with the achievement of technical milestones. While NEO flights build deep space experience and reduce engineering risk, they also open up an entirely new "planetary" surface to *detailed* scientific exploration. We currently know NEOs only through our study of meteoritic fragments, and remotely glimpsed spectral analyses at optical, infrared, and radar wavelengths. Dramatic new insights into the nature of these objects will come only with returned physical samples (Swindle et al. 1992). A single human visit to a NEO would bring about an explosion in knowledge about small bodies, far beyond that realized by any practical program of robotic exploration. Astronauts would conduct detailed field studies, sample extensively and intelligently, and return a large, carefully documented collection of asteroidal materials to Earth. Together with robotic surveys of other asteroid types, a well-planned NEO visit by a human crew will yield far-reaching scientific results, and for a price that will increase only marginally the total cost of lunar/Mars exploration. This scientific return will be timely as well as revolutionary—coming well before the first Mars expedition, yet distinctly different in theme and content from lunar outpost science. NEO missions add a substantial bonus to the scientific payoff of an LMP exploration.

The same field work that astronauts conduct into the composition and history of NEOs will also produce a detailed assay of *in situ* resources. If resource potential is one of the criteria for target NEO selection, then the crew can verify the accessibility of desired materials, particularly water, volatiles, and free metals like iron and nickel (Lewis and Hutson 1993). Once sufficient quantities are confirmed, astronauts could demonstrate candidate sorting and extraction processes. A pilot plant, like those planned for the lunar outpost (Stafford et al. 1991), will prove the practical yield of such a process under field conditions, and pave the way for later emplacement of robotic extraction and delivery systems on NEOs. A human mission could even return a small quantity of refined asteroidal material (water, for example) to Earth—a down payment on later large-scale efforts for utilization of resources within an LMP.

No single NEO can supply a steady stream of resources to Earth-Moon space: launch windows are narrow, return opportunities are sporadic, and travel times far exceed those for lunar round trips. However, robotic extraction plants on a few dozen accessible asteroids could supply the quantities of water, other volatiles, bulk mass, and metals needed for a sustained interplanetary transportation system. A single NEO cannot serve as the "mother lode" for valuable *in situ* materials, but it can illustrate the resource potential of these near-Earth objects. A robotic demonstration plant, emplaced as part of a broader-based astronaut expedition, would be a solid start to the economic exploitation of the NEO population.

Shakedown of deep space hardware, scientific characterization of asteroids, chemical and physical surveys of asteroid resources—these benefits of human NEO flights synergistically deliver yet another: an early, robust ability to implement any number of asteroid interception and diversion strategies. Robotic surveys are unlikely to yield the depth of understanding necessary for such crucial future operations. The complex, multi-disciplinary investigation required, particularly of an NEOs interior structure, will be beyond the capability of even the next generation of remotely controlled spacecraft. A human visit to an NEO would not only maximize our understanding of that particular target, but, from several such expeditions, build broad practical experience in operations on and around this class of objects. While individual NEOs will have unique characteristics, they nevertheless fall into only a few classes with similar composition, physical evolution and collisional history. From previous human visits to a few representative NEOs, we would gain much-needed confidence in our ability to deal effectively with a similar body that may present an impact hazard. The knowledge and operational experience gained from human NEO flights would be cheap insurance, should we later need to mount robotic or piloted interception missions that must succeed.

Human NEO missions offer one final, important benefit to an LMP—a critical boost in programmatic momentum. Any successful lunar/Mars effort must be sustained over decades. In the last twenty years, national governments have found it difficult to sustain funding and public support for high technology programs, let alone those extending over the decadal time scales of an ambitious space exploration program. The goals of successive administrations often differ greatly, and economic conditions wax and wane on time scales shorter than the minimum period of steady funding needed to develop an LMP's hardware and operations concept. To sustain public interest and support, LMP planners must develop a series of challenging, timely milestones that demonstrate visible, meaningful progress. Human NEO flights can serve as ideal intermediate benchmarks, offering concrete operational and scientific progress while preparations continue for the first departure of humans to Mars.

III. ASTEROID MISSIONS IN THE CONTEXT OF LUNAR/MARS EXPLORATION

Human missions to NEOs, by paving the way for success on later, more ambitious Mars flights, can greatly enhance the scientific and programmatic strength of the entire lunar/Mars effort. NEO missions are hard to justify outside the larger programmatic context of an LMP, yet *their inclusion materially increases the chances for the programmatic and technical success of the larger effort.* The synergy between asteroid missions and lunar/Mars exploration is compelling. NEO flights are stepping stones to Mars in terms of operational experience. Missions to lunar outposts under consideration for the next decade will last about six weeks; durations of Mars expeditions that offer reduced microgravity exposure and long surface stays will last up to three years. The one-year durations of the human asteroid visits we propose represent a natural extension of the capability of human spacefaring beyond the Moon.

The growing NEO discovery rate has already yielded several attractive, round-trip mission profiles that include multi-week stay times and total flight durations of about a year. For targets larger than 0.5 km in diameter, opportunities for fast missions (<1 yr duration), requiring less ΔV than a round-trip lunar mission, occur on average about 9 to 10 times each year (Davis et al. 1993).

Very attractive asteroid targets for both robotic rendezvous, sample return, or human mission profiles are already in hand (see Davis et al. [1993] for a thorough discussion of the NEO mission trade space). Their Table I shows the range of launch dates, mission durations, and total ΔVs for the best NEO candidates. One NEO in particular, 1991 JW, offers particularly favorable mission opportunities.

Asteroid 1991 JW has the desired characteristics for easy access from Earth: low eccentricity ($e = 0.1183$), semimajor axis near 1 AU ($a = 1.0378$), and modest inclination ($i = 8.70539$). Twelve-month and six-month round trip missions are illustrated in Fig. 3, 4, and 5 for comparison with a modeled "ideal" asteroid, 1994"OK" (Davis et al. 1993). For NEO 1991 JW, the six-month mission is quite feasible, requiring a total ΔV of 10.9 km s^{-1} (lunar surface roundtrip requires roughly 9.4 km s^{-1}). A twelve-month round-trip to

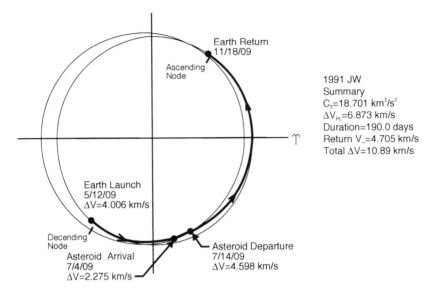

Earth Return
11/18/09

Ascending
Node

1991 JW
Summary
C_3=18.701 km^2/s^2
ΔV_{PL}=6.873 km/s
Duration=190.0 days
Return V_∞=4.705 km/s
Total ΔV=10.89 km/s

Earth Launch
5/12/09
ΔV=4.006 km/s

Decending
Node

Asteroid Arrival
7/4/09
ΔV=2.275 km/s

Asteroid Departure
7/14/09
ΔV=4.598 km/s

Figure 3.   Round-trip mission profiles to (A) 1991 JW in six months; (B) 1991 JW in one year; (C) the yet-to-be-discovered 1994"OK" in six months. The latter two missions require less ΔV than a one-way trip to the lunar surface.

this asteroid needs a total ΔV of only 5.9 km s^{-1}—less than the ΔV needed for minimum energy rendezvous for most NEOs. The May 2009 launch opportunity to 1991 JW offers a 3-month cruise to rendezvous, a stay time of 40 days, and a 7.5 month return leg, totaling 363 days.

A recent discovery has identified the first cometary body in the NEO population. Bowell (1992) showed that the NEO 1979 VA (4015) is the same object as comet Wilson-Harrington. This is the first object in the NEO population known to be volatile-rich, although it is highly probable that many of the C-type NEOs also contain significant amounts of volatiles. McAdams explored mission opportunities to 4015 1979 VA; he found that rendezvous missions could be accomplished for a total ΔV of 7.2 to 7.8 km s^{-1} and flight times of 4.4 to 5.1 yr, for three launch dates in 2000, 2004 and 2008. Gravitational encounters with Earth and Venus could reduce the ΔV requirements by about 1.2 km s^{-1} with an attendant increased flight time of over two more years. This ΔV is only about 25% larger than the 6 km s^{-1} that it takes to go from low Earth orbit to the lunar surface. We expect that 1979 VA is just the first of many such objects to be discovered, representing an assured source of volatiles among NEOs in accessible orbits. Orbits of other such water-rich objects may be even more attractive; the tantalizing prospect of rocket propellant derived from these naturally occurring "filling stations" makes them lucrative exploration targets.

Scientific payoff is not the primary rationale for solar system exploration by humans, but intellectual progress is integral to justifications of lunar/Mars exploration. NEO flights promise to add a totally new area of scientific return to an LMP's productivity. Missions to targets like 1991 JW, possible in the first decade of the next century, will not only expand our deep space experience base, but will also reap a scientific bonanza. The clues to early solar system history held by NEOs are highly complementary to the "evolved planet" studies that are the focus of lunar/Mars research. Given the dearth of hard knowledge about the composition and origin of NEOs, a human expedition to even one would produce ground-breaking results well in advance of the human-acquired Mars scientific return.

IV. ASTEROID FIELD EXPLORATION

Exploration of an NEO will be different from our previous experience with explorations of the lunar surface. The microgravity environment of such a small body is a key difference, but significant additional challenges will be encountered. The NEO surface will likely be covered with a thick blanket of regolith, pocked with impact craters, and, to the crew, will no doubt prove more geologically complex than expected. These challenges are tractable if we use both human crews and robots to exploit the limited exploration time available.

Davis et al. (1993) have calculated the available stay times in the object's vicinity on round-trip opportunities to NEOs during the period 1998 to 2010 (Table I). The stays range from as few as 10 to as long as 352 days, with an average for any given encounter of ~60 days. Although an NEO stay of 352 days is preferable to 10, one of the "laws" of field work in remote locations is that there is never enough time to do everything that is either planned or desired. This is particularly true when one considers the problems of the inevitable fatigue astronauts incur while working in a pressure suit, and the time demands of normal housekeeping operations aboard the Earth-NEO cruise vehicle. To overcome these obstacles, science operation plans will have to take advantage of the complementary talents of both human crews and robots. Robotic spacecraft are useful for routine, repetitive data-gathering, while humans perform best at compressing, integrating, and interpreting data, developing models, and testing them further. Smooth integration of these respective talents will be necessary for getting the maximum return from any exploration mission.

Although asteroids are small relative to planetary bodies like the Moon, they are still large in human terms. A human crew will be unable to explore every nook and cranny of an NEO during the limited stay time; therefore, the exploration plan must include robots to identify in real time those areas where human visits and sampling will be most productive. Activities appropriate for robots include: (1) producing multi-spectral maps of the surface materials of the asteroid; (2) surveying the spatial variations in geophysical parameters (e.g., gravity and magnetic anomalies); and (3) imaging the asteroidal surface from orbit for map-making and investigation planning. Analogous to photo-

TABLE I[a]

Round-Trip Opportunities to Near-Earth Asteroids (Launch Years 1998–2010)

Asteroid	Earth Launch Date	Launch Energy C3 (km s⁻¹)²	Midcourse Maneuver Date	Asteroid Arrival Date	Post-Launch ΔV (km s⁻¹)	Stay Time (days)	Asteroid Departure Date	Departure ΔV (km s⁻¹)	Earth Return Date	Earth Return V∞ (km s⁻¹)	Total ΔV (km s⁻¹)	Flight Time (days)
1982DB	01/07/98	30.88	—	01/14/00	1.155	222	08/23/00	0.083	02/08/02	–NA–	5.754	1492
	01/02/00	25.62	—	09/05/01	0.869	30	10/05/01	0.582	12/21/01	5.544	5.749	720
	01/10/00	25.93	—	10/17/01	0.701	70	12/26/01	0.565	02/07/04	–NA–	5.577	1489
	12/24/00	35.97	—	11/03/01	6.573	30	12/03/01	1.862	12/24/01	5.440	13.158	365
	12/21/01	32.95	—	01/30/02	0.955	30	03/01/02	0.668	02/11/04	6.354	6.224	782
	01/22/02	22.64	08/13/02	10/24/03	0.284	70	01/02/04	1.127	01/09/06	–NA–	5.584	1448
	01/26/04	17.50	08/19/04	08/21/05	0.894	53	10/13/05	1.495	01/06/08	–NA–	6.343	1440
	02/12/04	28.37	—	09/05/05	0.761	30	10/05/05	2.317	10/25/06	5.979	7.491	986
	01/29/06	12.04	08/21/06	06/20/07	1.647	43	08/02/07	1.817	01/04/10	–NA–	7.181	1436
	01/06/07	33.05	—	02/18/09	1.353	180	08/17/09	0.339	02/16/11	–NA–	6.297	1502
	02/03/08	5.96	08/15/08	04/11/09	2.663	60	06/10/09	2.139	01/31/12	–NA–	8.248	1458
	01/03/09	27.98	—	04/16/10	3.456	30	05/16/10	2.033	02/05/11	7.000	9.886	763
	01/08/09	28.36	—	12/02/10	0.928	45	01/16/11	0.336	02/07/13	–NA–	5.676	1490
	06/21/10	15.98	—	12/27/10	5.498	30	01/26/11	5.146	05/28/11	6.283	14.532	342
Anteros (1943)	05/28/99	24.72	04/28/00	09/03/00	1.842	233	04/24/01	0.492	05/27/02	–NA–	6.595	1094
	04/03/00	27.16	—	10/11/01	5.251	30	11/10/01	4.857	04/04/02	5.706	14.471	731
	12/04/01	32.50	—	05/08/02	4.591	14	05/22/02	4.871	12/04/02	5.389	14.045	365
	05/17/02	32.49	09/30/02	06/17/03	2.228	330	05/12/04	2.875	05/30/05	–NA–	9.685	1110

05/27/02	58.43	——	04/07/03	2.313	30	05/07/03	4.900	05/27/04	6.880	12.812	731
05/20/04	42.54	——	02/04/06	1.280	30	03/06/06	1.256	05/18/07	6.463	7.522	1093
05/26/04	30.52	06/08/05	10/11/05	1.155	142	03/02/06	1.076	05/26/07	-NA-	6.732	1096
05/30/06	17.91	03/14/07	07/14/07	2.873	30	08/13/07	0.406	05/27/09	-NA-	7.251	1093
05/08/07	28.39	——	08/09/08	3.969	30	09/08/08	4.316	05/08/09	5.873	12.699	731
05/24/09	35.71	——	07/25/10	0.595	237	03/19/11	1.765	05/25/12	-NA-	7.073	1096
05/31/09	38.68	——	01/05/10	2.645	30	02/04/10	5.974	06/01/11	6.880	13.451	731
Orpheus (3361) 01/25/98	30.40	——	06/09/98	2.125	352	05/27/99	1.444	04/04/01	-NA-	8.066	1164
01/13/02	36.60	——	05/11/02	2.250	30	06/10/02	2.827	04/28/03	-NA-	9.826	470
01/04/03	16.80	——	06/06/04	2.452	135	10/19/04	1.276	11/12/05	-NA-	7.652	1044
05/20/05	17.64	——	02/03/06	2.898	10	02/13/06	2.733	09/30/06	3.927	9.591	498
05/20/05	17.66	——	02/02/06	2.925	10	02/12/06	3.205	09/12/06	4.279	10.091	480
06/02/05	15.51	——	02/07/06	3.144	30	03/09/06	4.085	07/27/06	4.332	11.097	420
07/08/05	20.06	12/11/05	02/22/06	3.971	10	03/04/06	3.994	08/02/06	4.151	12.029	390
07/19/05	16.54	——	12/09/05	5.511	10	12/19/05	3.588	03/16/06	5.819	13.012	240
01/04/07	16.61	——	06/02/08	2.428	275	03/04/09	0.937	11/01/09	-NA-	7.281	1033
1989ML 12/29/08	16.30	——	06/17/09	3.024	10	06/27/09	3.459	01/13/10	5.455	10.427	380
01/03/09	16.79	——	06/13/09	3.142	10	06/23/09	3.409	01/03/10	5.875	10.516	365
01/22/09	26.06	——	06/04/09	3.766	10	06/14/09	3.502	12/18/09	6.679	11.622	330
02/09/09	47.44	——	05/26/09	4.833	10	06/05/09	3.886	12/06/09	7.444	13.922	300
01/14/09	21.79	——	06/01/09	3.807	10	06/11/09	7.519	10/11/09	9.474	15.502	270
07/07/02	13.83	10/31/04	06/02/03	0.685	167	11/16/03	1.103	07/07/05	4.574	6.126	1096

TABLE I[a] (cont.)

Round-Trip Opportunities to Near-Earth Asteroids (Launch Years 1998–2010)

Asteroid	Earth Launch Date	Launch Energy C3 (km s⁻¹)²	Midcourse Maneuver Date	Asteroid Arrival Date	Post-Launch ΔV (km s⁻¹)	Stay Time (days)	Asteroid Departure Date	Departure ΔV (km s⁻¹)	Earth Return Date	Earth Return V∞ (km s⁻¹)	Total ΔV (km s⁻¹)	Flight Time (days)
1989UQ	03/04/09	2.26	—	12/11/09	3.207	10	12/21/09	0.953	07/30/10	6.362	7.439	513
	04/02/09	6.21	—	12/16/09	3.119	10	12/26/09	0.989	07/26/10	6.458	7.566	480
	04/30/09	10.01	08/29/09	12/30/09	3.186	10	01/09/10	1.049	07/24/10	6.556	7.862	450
	05/28/09	24.09	08/29/09	01/01/10	2.809	10	01/11/10	1.081	07/22/10	6.671	8.124	420
	06/22/09	48.84	—	01/13/10	2.142	10	01/23/10	1.217	07/17/10	6.969	8.591	390
	07/10/09	69.13	—	01/27/10	1.865	10	02/06/10	1.612	07/10/10	7.636	9.473	365
	02/10/10	12.38	—	09/21/10	2.458	32	10/23/10	2.571	06/05/11	3.252	8.761	480
	02/26/10	13.11	—	09/14/10	2.655	47	11/01/10	2.794	05/22/11	—	9.213	450
	02/12/10	12.28	—	09/22/10	2.482	67	11/28/10	3.425	04/08/11	4.313	9.643	420
	03/09/10	13.84	—	09/13/10	2.845	80	12/02/10	3.605	04/03/11	4.522	10.245	390
	03/23/10	16.69	—	09/04/10	3.081	96	12/09/10	4.074	03/23/11	5.106	11.074	365
	03/06/10	15.73	—	08/02/10	3.765	10	08/12/10	2.698	11/06/10	6.088	10.341	345
	03/25/10	20.76	—	08/10/10	3.692	10	08/11/10	2.693	10/26/10	7.232	10.478	215
1994OK (IDEAL)	01/29/05	9.41	—	05/19/05	0.968	10	05/29/05	0.953	08/05/05	3.561	5.522	188
1991 JW	05/12/08	43.05	—	12/18/08	1.982	36	01/23/09	1.186	05/25/09	6.969	6.988	378
	05/16/09	21.16	—	08/23/09	1.790	40	10/02/09	0.628	05/15/10	5.208	5.900	363
	11/16/10	30.10	—	03/01/11	1.959	10	03/11/11	1.524	11/30/11	6.439	6.443	379

[a] Table from Davis et al. 1993. Orbital elements for near-Earth asteroids are in the Appendix of this book.

and reconnaissance-field mapping, and geophysical surveying carried out in terrestrial geologic investigations, these data sets will tell the crew the gross composition of the NEO, and the spatial variation of composition across the surface. They will also give the crew some insight into the internal make-up of the object, and, when correlated with surface maps, may permit development of limited three-dimensional models. All of these data sets will then enable the crew to plan further detailed field investigations to be conducted by both robotic and human explorers. This next phase of exploration will have two components: sampling of distinct multi-spectral units, and investigation of enigmatic areas. The latter includes those regions that are spectrally complex and thus unlikely to yield to simple sampling, or that are difficult to characterize from spectral data alone. Although available NEO spectra from terrestrial and space-based telescopes may suggest relative simplicity of composition, the complexity that will actually be found is likely to tax the planning and investigative capability of the human and robotic crew at any NEO.

There are a number of NEO characteristics that will guide field operations. The first of these is the microgravity environment, on the order of a few milli-g's. EVAs on an NEO will more closely resemble those conducted in low-Earth orbit (LEO) than on the lunar surface. The crew will need: (1) tethers and anchors to control astronauts and tools while collecting and documenting samples; (2) an equipment control strategy (to keep unattended tools and samples from drifting away); (3) remote manipulator systems, like the space shuttle's; and (4) manned maneuvering units for traversing the surface.

Field work on an asteroid will be more difficult than on the Moon or Mars, because NEOs lack sufficient gravity to enable the crew to "stand on the surface;" instead, they will be forced to hover gingerly above it. NEO surface explorers will require body restraints to reduce unwanted motion and concomitant consumable usage while working in microgravity. Studies of consumable usage during microgravity EVAs in LEO have shown that metabolic rates, when anchored only with a simple tether, are as much as 100% higher than when attempting the same task with foot restraints (J. Waligora, personal communication, 1993).

Translation and mobility aids for use in the milli-g gravity field of an NEO will likely involve a spectrum of approaches, ranging from simple tethers and foot restraints to complex, EVA-only vehicles, such as envisioned by Shields et al. (1988). At the low end of the spectrum, tethers, handrails, and slide wires have proven essential to weightless astronauts working in the space shuttle payload bay. These are simple restraints that ensure crewmembers can translate easily in spite of spurious trajectory inputs. For NEO field work, a network of anchors and slide wires webbing the asteroid's surface would permit tethered crewmembers to translate to particular work sites. A good analogy would be the roped routes used by expedition climbers to transport supplies over rough mountain terrain. At the next level of complexity, astronauts can use foot restraints attached to remote manipulator systems (RMS), like those currently employed on the Space Shuttle. Crewmembers use these

RMS-mounted foot restraints both to move to a work station outside the payload bay, and to provide a solid base to react against once at the work site. The ultimate in mobility aids would be the use of independent spacecraft, either manned maneuvering units (MMUs) that connect to life support backpacks, used during the 1980s for satellite rescue in the space shuttle program, or full-up spacecraft, such as the Manned Orbital Transfer Vehicle (MOTV) proposed by Shields et al. (1988) for use in geosynchronous satellite repair. The MMU option is relatively simple, in that it gives a suited astronaut the capability for personal translation without the need for tethers or restraints. The capacity for field work would be limited by the endurance of the backpack, its provisions for necessary tools, and the MMU's fuel capacity. The MOTV, in contrast, is a complete spacecraft, with pressurized crew compartments, a suite of RMS arms, hold-down arms for stabilizing the spacecraft, and a "man-in-the-can" hard "half-suit," entered from inside the vehicle. This partial suit is similar to hard diving suits for deep underwater work, and is just one example of the MOTV's deep-sea exploration ancestry. MOTV operations would encompass translation to the asteroid surface, anchoring there, and crew science activities using either the RMS arms, the hard "half-suit," or external EVA in a pressure suit.

Movement to and around the NEO, using either an MMU or an MOTV, will have the complexity of a satellite rendezvous. Orbital mechanics, rather than simple walking or driving, will govern movement around the body, and will make operations more complex than more familiar means of travel on a larger planetary body (Dupont et al. 1990). Translation across the surface will be more difficult than it first appears. A traverse to a specific location will be computed in an orbital framework, with the guidance system firing MMU or MOTV jets so as to minimize fuel expenditure, instead of the ballistic, point-to-point approach possible on a body such as the Moon.

A second driver in asteroid field work, harkening back to lunar field operations, is the dusty nature of the surface. Science investigations may take place on a body that has a uniform regolith, rather than outcrops and well-defined surface expressions of local geology (McKay et al. 1989; Asphaug and Nolan 1992). Contact between major rock units will likely be broad and diffuse, rather than sharp. As discussed above, multi-spectral "prospecting" will be required to define major spectral units prior to detailed study and sampling. If, as regolith estimates from the Martian satellites, Gaspra, and Ida suggest, rock outcrops are rare on NEO surfaces (Chapman et al. 1992, 1993), the best samples will come from large clasts found in impact ejecta. The regolith blanket may also affect the use of manned maneuvering units. Due to the microgravity environment, regolith stirred up by jet blast from an MMU or MOTV will not settle readily to the surface; it may reduce visibility enough to hamper the near-surface use of MMUs. Dupont et al. (1990) has likened the problem to that faced by divers working close to a silty lake bottom. In light of such concerns, dust control will be an important aspect of surface work, especially to control the introduction of dust into the working mechanisms of the EVA suit and the spacecraft interior. Although the properties of asteroid regolith are as yet unknown, the suspected similarity to lunar regolith suggests that asteroidal dust may be just as large a potential nuisance. Some mitigation strategies are outlined by Neal et al. (1988). The outer covering of the EVA suit should be as smooth and crease-free as possible, to limit the sticking and retention of dust. Mechanisms such as joints, bellows, cables and pulleys should be covered to prevent the introduction of dust, which could clog and jam these devices. One possible solution would be the use of an overgarment, donned and doffed outside the airlock at the beginning and end of each EVA period. The primary goal of any implementation should be to avoid, wherever possible, the introduction of dust-laden garmets and equipment into the spacecraft.

The third factor affecting NEO field work will be the complexity of maintaining constant voice and data communication. Because of the closeness of the horizon on a small body, astronauts or robots will not have to move far before losing sight of each other or the transfer vehicle. Dupont et al. (1990) determined that for a body the size of Phobos (28 km diameter), a $\sim$1.8 m (6 ft.) tall astronaut sees a horizon <270 m away. Maintaining constant voice and data contact between EVA crew, robots, and transfer vehicle will require some form of communications relay: a satellite, or a system of antenna towers and repeaters deployed at strategic locations on the NEO surface. The mechanics of orbits around such small bodies may make the communications satellite solution operationally complex, so simple towers and repeaters may provide a functional, if less elegant remedy.

The final challenge to NEO field exploration concerns the complexity of mission operations at a considerable distance from Earth. The likely light-time delay will make closely choreographed mission operations (characteristic of the space shuttle program) difficult at best. The crew must be capable of planning the best possible exploration program on-site, in real-time, based on the data that the robots and surface crew bring back. In a sense, mission control will operate *from* the transfer vehicle. Groundbased controllers will batch-process scientific data and provide advice on a day-to-day basis. Thus, the NEO mission will serve as a good rehearsal for Mars in terms of science planning. However, this approach demands a capable and well-trained crew, not only in mission operations, but in asteroid science as well. Only with the right mix of skills will the crew be able to extract the maximum science return from a limited NEO surface visit.

V. TECHNICAL CHALLENGES

Before astronauts can undertake detailed exploration of an NEO surface, they must confront and overcome nearly all of the same hazards facing a Mars expedition. These deep space challenges include prolonged exposure to microgravity, solar protons and galactic cosmic rays, highly reliable life support, crew health maintenance in the face of routine and space-peculiar ailments, and control of deep space operations despite the communications delays inherent in voyages beyond the Moon.

The most daunting physical hazard facing a crew on a typical one-year round-trip to an NEO is the debilitation caused by microgravity. Muscle atrophy, cardiac deconditioning, bone density changes, and immune system suppression are just a few of the physical effects of long-term microgravity exposure. Several astronauts aboard the Mir space station have endured a full year in microgravity; one of us (S. K.) spent 10 months on Mir on his longest visit. The Russian concensus is that missions up to 18 months in length can be conducted safely, with acceptable levels of muscle atrophy (reversible) and bone loss (permanent). The work capacity of the crew seems undiminished on these long-duration flights: successful EVAs have been conducted after 9 months in microgravity. Specific countermeasures on Mir include an average of two hours of daily exercise, dietary supplements, and daily wear of an elastic garment to apply loads to bones and skeletal muscles. Combinations of exercise and drugs to stem bone loss and physical deconditioning may eventually prove effective for longer missions, but more subtle effects than those confronted on Mir may still create problems. Few in the aerospace medical or astronaut communities are satisfied that current countermeasures are effective enough to support deep space missions over the one year threshold of human experience. Joint experiments over the next decade aboard Mir and the international space station will be necessary to find successful countermeasures for dealing with microgravity. Current plans call for joint research on at least three shuttle visits to the Mir station, and extended stays by U. S. astronauts aboard Mir, all beginning in 1995.

Long-duration missions beyond Earth's magnetosphere expose the crew to two radiation threats: prompt radiation effects from energetic solar flare particles, and a long-term increase in DNA damage (and cancer risk) from galactic cosmic rays (Letaw et al. 1987). The non-flare, background solar particle flux is not considered a serious hazard, but intense solar flares, occurring about once a decade, are particularly dangerous. For example, for each week that an unshielded astronaut crew spends outside the Earth's magnetosphere, they face one chance in 500 that they will receive a lethal radiation dose. Given timely onboard warning, a crew enroute to an NEO could wait out a days-long solar particle event by retreating to a small "storm shelter," shielded by at least 9 cm of aluminum, or its equivalent in water or propellant.

In contrast, the galactic cosmic ray (GCR) flux is hardly slowed by standard spacecraft shielding, and an increase in shield thickness against such energetic, heavy nuclei can result in a shower of damaging secondary particles within the spacecraft. Letaw et al. (1987) point out that for unshielded crews, the annual GCR dose equivalent ranges from 20 to 50 rem, much greater than the 0.5 rem limit for those in the general population. Truly effective shielding against the steady GCR flux would be prohibitively heavy, and the feasibility of magnetic shielding techniques to divert the particles is conjectural. The only practical solution at present is a combination of judicious shielding and limits on deep space exposure. For Mars missions, the recommended minimum amount of GCR shielding is 9 cm of aluminum (or equivalent) in all directions (Letaw et al. 1987). Even with such shielding, Mars crews must minimize their cruise phase in favor of a longer stay on the planet's surface, shielded by atmosphere. Unfortunately, a one-year NEO mission is practically all cruise, exposed to GCR. The bulk of an asteroid offers no additional protection during proximity operations. One bright spot is that typical NEO flight profiles last no longer than the cruise portion of "sprint" Mars missions, keeping NEO crews within the risk envelope contemplated for Mars expeditions.

The present corps of astronauts are classified as "radiation workers" by the National Council on Radiation Protection and Measurements (1989), bringing them under the umbrella of a key philosophy concerning exposure to ionizing radiation. This is the ALARA principle, meaning "as low as reasonably achievable." The ALARA principle is based on the concept that no exposure to ionizing radiation is considered either safe or routine; consequently, standards setting a minimum safe dosage are unacceptable, and all exposure to ionizing radiation must be actively managed through the use of dosimetry and establishment of dose limits. ALARA requires that radiation exposure always be kept to the lowest dosage reasonably achievable within the present limits of technology and economics. ALARA is the basis for nuclear industry regulations setting both the maximum acceptable lifetime dose of radiation, and the maximum acceptable dose from a single exposure. NASA uses ALARA to control the exposure of astronauts to ionizing radiation.

"Lifetime" dose limits define how much ionizing radiation that radiation workers can be exposed to during their careers; they are based on a presumed 3% increased risk of cancer mortality (Robbins 1990). Once the lifetime dose limit is reached, a radiation worker can no longer be required or allowed to work in an environment with the potential for further exposure. The "maximum single dose" limits specify the maximum exposure a radiation worker can sustain in one incident, regardless of the circumstances. Current standards allow workers to be exposed to radiation in excess of the maximum single dose limit only in extraordinary circumstances, and then only on a volunteer basis. An example of such a permissible excessive dose might be the rescue of a co-worker incapacitated in a radiation environment, like that in the vicinity of a nuclear mishap. However, it is important to note that receiving such an excessive dose does not increase a worker's lifetime dosage limits—a worker exposing himself to such an environment could significantly limit his career. In practical terms, a single NEO or Mars mission may be a "career-limiting" move for its astronaut crew, exposing them to enough radiation to ground them permanently.

The implication for astronaut crews is relatively simple; as their exposure to ionizing radiation accumulates throughout the course of a career, they may eventually reach their maximum lifetime dosage limit. Afterwards, they will not be allowed, under NCRP standards, to participate in flight activities that might involve additional exposure. At present, all shuttle flights take place within the confines of the Earth's magnetosphere. With the typical exposures encountered in LEO, it is unlikely that astronauts would be disqualified from future flights due to violations of a lifetime dosage limit. On flights outside the Earth's magnetosphere, however, the radiation environment makes it probable that a single, year-long asteroid mission could expose the crew to their lifetime dosage limit, thereby prohibiting them from future flights.

Whether crew members could voluntarily exceed their lifetime dosage limit is presently unclear. Radiation workers now are not allowed to do so, regardless of age or history of exposure; no opportunity has occurred to test this standard in the case of an astronaut. Note that ionizing radiation does not affect all humans equally—younger individuals are more susceptible to genetic damage, with adolescents at greatest risk. In addition to age-related differences, women up to age 40 have about 50% more sensitivity to exposure to ionizing radiation than men (Robbins 1990). The present career limit doses, recommended to NASA by the National Council on Radiation Protection and Measurements, range from 1 sev for a 25 year-old female astronaut to 4 sev for a 55 year-old, based on the formulae: career dose limit in sev (men) = 2 + 0.075(age 30); career dose limit in sev (women) = 2 + 0.075(age 38 [NCRP 1989]). These dose limits are based on exposure to blood forming organs, where developing leukemia is the primary risk. Other radiation limits stem from eye exposure and the subsequent risk of developing cataracts, and from skin exposure and the risk of developing skin cancer.

The implications to spacecraft and space suit designers are somewhat more complex. Since ALARA is specifically based on what is reasonably achievable, there are no set limits that can be used to specify the minimum spacecraft shielding required. Even so, the shortfalls in our ability to deal with radiation are many. The GCR flux in interplanetary space has been poorly observed, limited to some spotty Apollo and satellite data. Models of the free space (interplanetary) GCR flux match Earth-based measurements to within about 10%, giving some confidence that we understand potential cruise exposure levels. Energy transfer mechanisms between particles, shielding, and animal tissue are another source of uncertainty. Energy loss via ionization is well characterized, but the variety of nuclear interactions possible, coupled with limited nuclear cross section estimates, produces dose uncertainties of 20 to 30% (Wilson et al. 1991). On the other hand, we know that lighter elements possess the best shielding characteristics against cosmic rays, with hydrogen and water much more effective than metals (Letaw et al. 1987). The optimal shielding configuration is to surround the crew with their water stores, or liquid hydrogen or methane propellant. In short, before we can understand the magnitude of the risk and create effective countermeasures, we need more data and better predictive models on the nature of deep space radiation, particularly the GCR flux. All we can say now is that for a single NEO or Mars mission, radiation exposure may prove to be a tolerable hazard.

Exposure to nonionizing radiation has been recognized in recent years as an additional hazard that must be considered as part of a radiation risk assessment. Sources of nonionizing radiation include man made sources such as radio frequency (RF) exposure, from radars and high frequency radio transmissions, and natural sources such as ultraviolet and infrared radiation. Exposure to this environment will be highest during EVA, when the crew will be outside the shielded environment of the spacecraft. The space shuttle program controls exposure to radio frequency radiation by limiting use of high frequency transmitters during EVA; ultraviolet and infrared radiation are controlled by use of helmet visor coatings and insulating outer garments. Such measures are effective, but protection can be improved by implementing recommendations similar to those proposed by Neal et al. (1988), intended for use at a lunar outpost.

Any piloted deep space mission will require a proven, robust life support system, providing food, clean air and water, waste disposal, and a comfortable work environment. Such a system, reliable enough to function without re-supply over a 3-yr Mars mission profile, is beyond the current state of the art. An open-loop system (with no recycling of waste products) is prohibitively heavy, but systems that recycle water and solid waste, clearly necessary for such long flights, are still in the demonstration phase. The Mir station has a nearly closed water recycling loop, using a combination of approaches: excessive humidity is condensed and purified for drinking, water from urine is hydrolyzed to produce breathing oxygen, and shower and wash water is recycled for external use. About 15 to 20% of the water demand must still be supplied from the ground. The proposed international station will close only the water loop when it becomes permanently manned. Both that facility and Mir can serve as testbeds for development of cruise life support systems; for example, plants have been grown aboard Mir for atmospheric recycling and dietary variety. No clear technological approach has yet emerged as the answer to all life support needs and system design has not even begun. Fortunately, any LMP must confront and solve the life support problem, enabling parallel NEO missions. In turn, NEO flights will be used to test the life support systems over intervals approaching that of the Mars profile.

Keeping a crew physically and mentally healthy for a year or more in space will depend on much more than a good life support system. The crew size and gender mix will be important; those choices can be made after experience with operations of space stations. The crew skills needed for an NEO or Mars expedition dictate a larger group, but must be balanced against the greater amount of consumables required. Mir crews and controllers have avoided boredom on long-duration flights by having the ground provide a steady, but not overwhelming work schedule, with opportunities for optional activities if the core tasks are completed early. Mir crews also note the importance of regular opportunities for the crew to communicate informally with family and colleagues on Earth. A good mental outlook of the crew can best be maintained by planning for a busy cruise phase: outbound, concentrating on space-based astronomy, NEO remote sensing and preparations for surface operations; inbound to Earth, conducting sample and data analysis and further astronomical work.

Preventative health care will be mandatory on such long flights. A medical officer and a capable treatment facility can look after minor injuries, health monitoring, dental care, and unforeseen emergencies. Infection will be a special concern, as immune systems may be sluggish, wounds slow to heal, and latent viruses troublesome (Johnson and Brady 1992). Again, experience with space stations and lunar outposts will be key to overcoming these concerns. For example, Russian crews chosen for long-duration Mir flights undergo much more rigorous medical screening than astronauts slated for just a few weeks in orbit. Their pre-launch quarantine lasts three weeks, and sterilization routines are used to strip bacteria from the crew, their clothing, and Mir-bound equipment.

A last challenge to safe conduct of NEO or Mars missions will be developing a successful strategy for control of operations, despite one-way light travel times of up to one minute. Autonomous spacecraft systems and crew decision-making must still take advantage of ground advice in handling unexpected technical and scientific problems. Some experience will be gained from command and control of the space station, but the proximity to and reliance on mission control will reduce the parallels to NEO or Mars operations. Skylab experience illustrated the difficulty ground controllers may face in finding the right balance between offering timely advice and interfering with the crew's initiative. Over five years of Mir operations have resolved some of these issues: the crew and ground cooperate in prioritizing the work schedule, and the crew has a large measure of independence in executing research programs onboard. Mir spends a relatively large amount of time out of range of ground stations; this reality has accelerated the shift of autonomy toward the crew. On NEO flights, intelligent computer systems onboard would eliminate much of the need for ground monitoring of telemetry, moving the ground's role toward processing scientific data and providing strategic advice to the exploration crew. Testing these concepts, first on a space station, then on NEO missions, would resolve any shortcomings well in advance of the more demanding Mars profile.

None of the major hazards inherent in deep space travel is likely to prove intractable; a concerted effort to reach Mars will find ways to eliminate or minimize the risk to the crew. New technology will alleviate some risks, but an attractive approach is to minimize trip times, via nuclear propulsion or chemical sprint. Short interplanetary cruise periods will reduce exposure to microgravity and radiation hazards, and keep life support and crew morale concerns within the experience base of the space station.

VI. FIRST STEPS

Creating an opportunity for human exploration of the asteroids in parallel with an LMP requires that deliberate steps in that direction be taken now.

NEO missions will not be added at the last minute; LMP architecture will be determined long before development begins in earnest on the hardware. Similarly, if NEO flights are to benefit fully from the LMP concept, they must be built into the program from the start. Advocacy of human NEO missions must be constant, vocal, and technically well grounded if we are to turn potential into payoff.

Each new NEO discovery expands the range of human mission opportunities. As the discovery rate grows, so do the chances of identifying even more attractive NEO targets. Four of the seven best targets offering fast round trips were discovered only in the past three years. Already, there are about 22 known NEOs for which rendezvous is easier than placing a payload on the lunar surface (Davis et al. 1993). However, as fewer than 10% of the approximately 1700 NEOs larger than 1 km have been found (as of late 1992), identification of the best targets for robotic or piloted missions depends on a significant increase in the rate of NEO discovery. An augmented search for NEOs, as recommended by the NASA Asteroid Detection Workshop (Morrison 1992; see Chapter by Bowell and Muinonen), would detect perhaps 40 to 50% of the NEOs larger than 0.5 km in diameter, expanding the known set of asteroids with a round trip ΔV similar to that of 1991 JW.

Even with a boost in the discovery rate, target selection for NEO missions will be difficult without a follow-up program to observe and characterize the chemical and physical makeup of accessible NEOs . These observations should include photometry, radiometry, spectroscopy, and radar when possible. Under proper observing conditions, active radar imagery can overcome the size and distance handicaps limiting optical and infrared methods (Ostro 1989; see also his chapter).

Robotic visits to NEOs will be necessary in advance of human missions. The Gaspra and Ida encounters have already demonstrated the value of robotic asteroid studies with simple flybys—*in situ* robotic exploration will validate and expand upon these first close-up glimpses (Chapman et al. 1992, 1993). Robots can investigate regolith character (or lack thereof), composition, the degree of surface homogeneity, and some aspects of internal structure. Probes to several distinct NEO types would assess the range of compositions and surface morphologies within the NEO population; understanding the "end members" among that population would limit uncertainties in planning for eventual astronaut operations. We note that a robotic precursor mission to a particular NEO is not a prerequisite for a piloted mission, due to the long (several year) intervals between successive launch windows. A robotic flight to a similar type of NEO, and to a representative set of other examples, should suffice.

One encouraging trend in laying the groundwork for eventual human NEO missions is the growing international participation in the exploration effort. The NEO detection community already has strong overseas ties, and a series of robotic asteroid missions could be structured as a coordinated, multi-national program. The year 1994 will also mark the first joint Russian/American astronaut flight since 1975, to be followed by a series of shuttle flights to the Mir station, and long stays by U. S. astronauts aboard Mir from 1995 to 1997. Joint studies of long-duration space flight aboard both Mir and the international station are an essential first step to eventual human missions to NEOs and ultimately, Mars.

VII. CONCLUSIONS

A window for human exploration of asteroids will open when concerted efforts begin again to explore the Moon and Mars. The development of a Mars transportation capability (even a robust lunar capability) will inherently provide the means for safe and practical NEO exploration missions. However, NEO missions will not occur automatically with the building of Mars hardware, or as an afterthought once Mars expeditions begin. Rather, integration of NEO flights into LMP planning depends on early and sustained advocacy of the asteroid option. Currently, both the accessibility and scientific potential of NEOs are poorly understood by exploration planners. Human NEO missions not only bring solid operational and scientific benefits to an LMP, but strengthen the overall integrity of the program by broadening the science return, and sustaining momentum between the establishment of lunar outpost and the onset of Mars travel. For its part, the asteroid exploration community needs to raise its sights—exclusive focus on a few robotic explorers of limited capability is a self-fulfilling prophecy. Both NEO advocates and future LMP planners must move quickly to realize the synergies between human NEO missions and Mars exploration. The window to NEO exploration is opening—will human explorers be ready to jump through it?

REFERENCES

Asphaug, E., and Nolan, M. C. 1992. Analytical and numerical predictions for regolith production on asteroids. *Lunar Planet. Sci.* XXIII:43–44 (abstract).

Bell, J. F., Fanale, F., and Cruikshank, D. P. 1993. Chemical and physical properties of the Martian satellites. In *Resources of Near-Earth Space*, eds. J. S. Lewis, M. S. Matthews and M. L. Guerrieri (Tucson: Univ. of Arizona Press), pp. 887–901.

Bowell, E. 1992. 1979 VA = Comet Wilson-Harrington (1949 III). *IAU Circ.* Nos. 5585 and 5586.

Chapman, C. R., Davis, D. R., Neukum, G., Veverka, J., Belton, M. J. S., Johnson, T. V., Morrison, D., McEwen, A., and the Galileo Imaging Team. 1992. 951 Gaspra: preliminary Galileo SSI results on craters, collisions, and regolith. *Lunar Planet. Sci.* XXIII:219–220 (abstract).

Chapman, C. R., Belton, M. J. S., Veverka, J., and the Galileo Imaging Team. 1993. Galileo observations of the asteroid Ida. *Eos: Trans. AGU* 74:43, 384.

Davis, D. R., Friedlander, A. R., and Jones, T. D. 1993. Role of near-Earth asteroids in the space exploration initiative. In *Resources of Near-Earth Space*, eds. J. S. Lewis, M. S. Matthews and M. L. Guerrieri (Tucson: Univ. of Arizona Press), pp. 619–655.

Discovery. 1991. Near-Earth Asteroid Rendezvous (NEAR), Report of the Discovery Science Working Group, Executive Summary.

Dupont, A., Blackshear, J., Bailey, P., Ewan, P., and Kincade, R. 1990. Treatise On Spaceflight Amid A Two-Body Influence, Especially Related To The Mars-Phobos System And The Exploration Of Phobos. JSC Internal Report (Houston: NASA Johnson Space Center).

Gehrels, T. 1991. Scanning with charge-coupled devices. *Space Sci. Rev.* 58:347–375.

Gradie, J.C., Chapman, C. R., and Tedesco, E. F. 1989. Distribution of taxonomic classes and the compositional structure of the asteroid belt. In *Asteroids II*, eds. R. P. Binzel, T. Gehrels and M. S. Matthews (Tucson: Univ. of Arizona Press), pp. 316–335.

Hartmann, W. K., Tholen, D. J., and Cruikshank, D. P. 1987. The relationship of active comets, "extinct" comets, and dark asteroids. *Icarus* 69:33–50.

Helin, E. F., and Dunbar, R. S. 1990. Search techniques for near-Earth asteroids. *Vistas in Astron.* 33:21–37.

Helin, E. F., and Shoemaker, E. M. 1979. The Palomar planet-crossing asteroid survey, 1973-1978. *Icarus* 40:321–328.

Johnson, T. C., and Brady, J. N . 1992. A Scientific Role for Space Station Freedom: Research At The Cellular Level. AIAA Paper 92-1346 (Washington, D. C.: American Inst. of Aeronautics and Astronautics).

Jones, T. D., Lebofsky, L. A., Lewis, J. S., and Marley, M. S. 1990. The composition and origin of C, P, and D asteroids: Water as a tracer of thermal evolution in the outer belt. *Icarus* 88:172–192.

Langevin, Y., Bibring, J.-P., Gondet, B., and Cruikshank, D. P. 1991. ISM observations of the spectral characteristics of Phobos in the near-infrared. *Lunar Planet. Sci.* XXII:781–782 (abstract).

Letaw, J. R., Silberberg, R., and Tsao, C. H. 1987. Radiation hazards on space missions. *Nature* 330::709–710.

Lewis, J. S., and Hutson, M. L. 1993. Asteroidal resource properties suggested by meteorite data. In *Resources of Near-Earth Space*, eds. J. S. Lewis, M. S. Matthews and M. L. Guerrieri (Tucson: Univ. of Arizona Press), pp. 523–542.

Lewis, J. S., and Lewis, R. A. 1987. *Space Resources: Breaking The Bonds of Earth* (New York: Columbia University Press).

McFadden, L. A., Tholen, D. J., and Veeder, G. J. 1989. Physical properties of Aten, Apollo, and Amor asteroids. In *Asteroids II*, eds. R. P. Binzel, T. Gehrels and M. S. Matthews (Tucson: Univ. of Arizona Press), pp. 442–467.

McKay, D. S., Swindle, T. D., and Greenberg, R. 1989. Asteroidal regoliths: What we do not know. In *Asteroids II*, eds. R. P. Binzel, T. Gehrels and M. S. Matthews (Tucson: Univ. of Arizona Press), pp. 617–642.

Morrison, D., ed. 1992. *The Spaceguard Survey: Report of the NASA International Near-Earth Object Detection Workshop* (Pasadena: Jet Propulsion Laboratory).

Murray, C., and Cox, C. B. 1989. *Apollo: The Race to the Moon* (New York: Simon and Schuster).

National Committee on Radiation Protection and Measurements. 1989. Guidance on Radiation Received in Space Activities. Report 98. (Bethesda, Md.: Natl. Council on Radiation Protection and Measurement).

Neal, V., Shields, N., Jr., Carr, G., Pogue, W., Schmitt, H. H. and Schulze, A. E. 1988. Extravehicular Activity at a Lunar Base: Advanced Extravehicular Activity Systems Requirements Definition Study. NASA 9-17779.

NEAR (Near-Earth Asteroid Rendezvous) Science Working Group. 1986. JPL Report No. 86-7 (Pasadena: Jet Propulsion Laboratory).

Nelson, M. L., Britt, D. T., and Lebofsky, L. A. 1993. Review of asteroid compositions. In *Resources of Near-Earth Space*, eds. J. S. Lewis, M. S. Matthews and M. L. Guerrieri (Tucson: Univ. of Arizona Press), pp. 493–522.

Nichols, C. R. 1993. Volatile products from carbonaceous asteroids. In *Resources of Near-Earth Space*, eds. J. S. Lewis, M. S. Matthews and M. L. Guerrieri (Tucson: Univ. of Arizona Press), pp. 543–568.

Ostro, S. J. 1989. Radar observations of asteroids. In *Asteroids II*, eds. R. P. Binzel, T. Gehrels and M. S. Matthews (Tucson: Univ. of Arizona Press), pp. 192–212.

Robbins, D. 1990. In *The Biological Effects of Radiation on Humans: Proc. of the Tutorial On Radiation and the Space Exploration Initiative*, ed. N. A. Budden (Houston: NASA Johnson Space Center).

Shields, N., Jr., Schulze, A. E., Carr, G. P., and Pogue, W. P. 1988. Extravehicular Activity At Geosynchronous Earth Orbit: Advance Extravehicular Activity Systems Requirements Definition Study. NASA CR-NAS9-17779.

Shoemaker, E. M., Wolfe, R. F., and Shoemaker, C. S. 1990. Asteroid and comet flux in the neighborhood of Earth. In *Global Catastrophes in Earth History*, eds. V. L. Sharpton and P. D. Ward, Geological Soc. of America Special Paper 247 (Boulder: Geological Soc. of America), pp. 155–170.

Stafford, T. P., and the Synthesis Group. 1991. America at the Threshold: Report of the Synthesis Group on America's Space Exploration Initiative (Washington, D. C.: U. S. Government Printing Office).

Swindle, T. D., Lewis, J. S., and McFadden, L. A. 1992. Near-earth asteroids and the history of planetary formation. *Earth in Space* 4(6):11–14.

Tholen, D. J., and Barucci, M. A. 1989. Asteroid taxonomy. In *Asteroids II*, eds. R. P. Binzel, T. Gehrels and M. S. Matthews (Tucson: Univ. of Arizona Press), pp. 298–315.

Weissman, P. R., A'Hearn, M. F., and McFadden, L. A. 1989. Evolution of comets into asteroids. In *Asteroids II*, eds. R. P. Binzel, T. Gehrels and M. S. Matthews (Tucson: Univ. of Arizona Press), pp. 880–920.

Weissman, P. R., and Campins, H. 1993. Short-period comets. In *Resources of Near-Earth Space*, eds. J. S. Lewis, M. S. Matthews and M. L. Guerrieri (Tucson: Univ. of Arizona Press), pp. 569–617.

Wilson, J., Townsend, L. W., Schimmerling, W., Khandelwal, G. S., Khan, F., Nealy, J. E., Cucinotta, F. A., Simonsen, L. C., Shinn, L. A., and Norbury, J. W. 1991. Transport Methods And Interactions For Space Radiations. NASA RP-1257.

PART VI
Effects of NEO Impact

COMPUTER SIMULATION OF HYPERVELOCITY IMPACT AND ASTEROID EXPLOSION[a]

A. V. BUSHMAN
High Energy Density Research Center, Russian Academy of Sciences

A. M. VICKERY
University of Arizona

and

V. E. FORTOV, B. P. KRUKOV, I. V. LOMONOSOV, S. A. MEDIN,
A. L. NI, A. V. SHUTOV and O. Yu. VOROBIEV
High Energy Density Research Center, Russian Academy of Sciences

It has recently become apparent that large impacts of comets and asteroids with the Earth can have dramatic consequences for the origin and evolution of life on Earth. Such an impact was responsible for the extinction of the dinosaurs and a future impact of similar magnitude could cause the extinction of human beings. It is of vital interest, therefore, to be able to predict the results of large impacts on Earth and to explore various mechanisms, such as impacts and explosions, for deflecting Earth-bound asteroids. It is obviously impossible to investigate the consequences of such impacts and explosions at the proper size scale, and laboratory tests cannot be extrapolated directly to the relevant size scales. One powerful tool for studying these problems is numerical modeling, that is, computer simulations. These simulations require extensive knowledge of the physical and thermodynamic properties of relevant materials, that can be measured in the laboratory.

I. INTRODUCTION

In the last twenty years, the importance of large impacts for the origin and evolution of life on Earth has become increasingly apparent. It is now widely accepted that the impact of a 10-km diameter asteroid or comet caused the extinction of the dinosaurs and of a multitude of other life forms 65 Myr ago (Alvarez et al. 1980). This devastation of the dominant life forms in turn allowed the explosive evolution of mammals, culminating (so far) in Homo Sapiens. The realization that chance played such an important role in the evolution of terrestrial life is humbling; its corollary, that humankind could

---

[a] Editorial Note: A detailed description of the work in this chapter (see remarkable conclusion in Sec. V.) may be found in the book by Bushman et al. (1993).

be extinguished just as rapidly and thoroughly as the dinosaurs were, can be quite intimidating.

Part of the difficulty in realistically evaluating the dangers involved in these collisions as well as the optimum method for ameliorating the risk is that the relevant experiments cannot be carried out at the proper scale. The impactor that killed the dinosaurs had a mass of roughly 10^{15} kg and probably struck the Earth with a speed of about 20 km s^{-1}. In contrast, laboratory methods allow one to accelerate masses of about 1 to 10 g to speeds of 12 to 14 km s^{-1} (Al'tshuler 1965; Al'tshuler et al. 1981). There is a trade-off between mass and velocity in these instruments; basically, some fraction of the energy of the propellant can be transformed into kinetic energy of the projectile—the larger the projectile, the lower its velocity. Although collisions between mainbelt asteroids typically occur at 5 km s^{-1}, the collision between a comet and the Earth may occur at as much as 60 km s^{-1}. In any case, collisions that are likely to be a threat to life on Earth must involve impactors many of orders of magnitude larger than 10 g. Futhermore, the physical processes that dominate during the impact of small (laboratory-scale) objects may not be the same as those that dominate for the much larger asteroids or comets. Thus, many important aspects of large-scale hypervelocity impacts can only be studied by means of numerical simulations.

The validity of the numerical simulations depends not only on the quality of the code itself but also on the quality of the equation of state formulation and on that of the experimental data used to support it. We have developed new equation-of-state (EOS) formulations for iron and for silica (SiO$_2$), which were used in conjunction with our new numerical code. This new hydrocode uses such a mixed Lagrangian-Eulerian approach, with the problem subdivided into individual particles (Agureikin and Krukov 1986). A second code incorporates a modification of Godunov's method of movable grids, which allows accurate modeling of discontinuities such as shock waves, contact discontinuities between different materials, and free surfaces (Bushman et al. 1993). This code is also particularly useful for problems involving large deformations. We have tested these codes by modeling the impact of silicate and iron asteroids on the Earth. Because the codes gave the expected results, we then used them to model the impact of a spacecraft on an asteroid.

II. EQUATIONS OF STATE

An equation of state, broadly defined, is a description of the physical and thermodynamic properties of a material. These descriptions can be as simple as the familiar perfect gas EOS, which may be written as $P = (\gamma - 1)\rho E$ (where P is pressure, ρ is density, E is internal energy, and γ is a constant characteristic of the gas). For complicated materials, or for use over a wide variety of physical conditions, the EOS may be quite complex. This is particularly true for the study of impact cratering, which typically involves intrinsically complex materials such as rocks or ice, and for which pressures can range from hundreds of gigapascals (100 GPa = 1 Mbar) or more down to nearly zero. Because of this wide pressure range, the material may undergo several phase transitions, from a high-pressure phase (e.g., stishovite as opposed to quartz), through a melting, vaporization, and even ionization. Each phase change adds another degree of complexity to the EOS. For a basic introduction to equations of state and impact catering, see Melosh (1989); a classical description of the physics of shock waves is in Zel'dovich and Razier (1966); and a more modern, detailed survey is given by Bushman et al. (1993). Examples of papers on specific experimental techniques are by Jones et al. (1966), Holmes et al. (1989), Glushak et al. (1989), Vladimirov et al. (1984), Trainor et al. (1979), Bridgman (1949), Jayaraman (1983), Vohra et al. (1988), Gathers (1986), Neal (1979), and Brown and McQueen (1980).

For this study, two new EOS were formulated, one for iron and one for SiO_2. Because iron, like most metals, is a relatively simple substance, a detailed EOS based on free-energy potentials was developed. The semi-empirical model takes into account the cold lattice contribution, thermal vibrations of the atoms with anharmonic effects, and the contribution of thermally excited conducting electrons. The high-pressure phase is taken to be ϵ-iron, which transforms to the low-pressure phase, α-iron, at $P \cong 13$ GPa and room temperature, with a density decrease of approximately 8%. For this EOS, the α-iron phase is treated as ϵ-iron with porosity. Melting, vaporization, and ionization are all included. The SiO_2 EOS is of the simpler Mie-Gruneisen type because of the paucity of relevant thermodynamic data. This EOS is extended to the region of low density and generalized for the case of arbitrary energies. It is constrained to give the correct asymptotes to the behavior of an ideal gas and of hot dense matter. It includes the high-pressure phase transition of quartz to stishovite.

III. HYPERVELOCITY IMPACTS AND HYDROCODES

Hypervelocity impacts are those that occur at velocities greater than that of the speed of sound in the impacting bodies; because of this velocity contrast, information about the collision cannot be transmitted to the rest of the target (or impactor) as fast as the changes are occurring. The result is a large and discontinuous increase in the pressure, density, and energy of the material from ambient conditions; the locus of this discontinuous change is called a shock front. The requirements of conservation of mass, energy, and momentum across the shock front lead to the Rankine-Hugoniot equations that describe the change in pressure, density, and energy across the shock. Although in real materials the shock wave may have a finite rise-time, the mathematical description of a shock assumes an instantaneous change in pressure and the other material properties. In most cases, this assumption leads to predictions that are accurate to within the accuracy of actual measurements.

At sufficiently high pressures, familiar solid materials such as rocks behave as if they were fluids. Hydrocodes are computer programs that model fluid flow and which can therefore be used to study hypervelocity impact problems. The principle behind them is quite simple: they are based on conservation of mass, energy, and momentum and a description of physical and thermodynamic properties of the materials involved. In practice, however, designing a hydrocode is a difficult and arduous task.

The two most widely used approaches to designing hydrocodes are the Langrangian and Eulerian methods. In the Lagrangian method, the problem may be subdivided into cells defined by the position of their vertices. These vertices are allowed to move in response to pressure forces, and the cells thereby shrink or expand or change shape. The material within each cell becomes more or less dense and gains or loses energy because of the work done, and the pressure in each cell may increase or decrease. The cell vertices then move in response to the new pressure field, and the cycle continues. The problem with this formualtion for hypervelocity impact problems is that the compressions and distortions can cause the cells literally to turn inside out: the numerical uncertainty in the postion of the vertices can make it impossible to resolve a cell. In the Eulerian method, the problem is again divided into cells, but the cells do not move or change shape. Material is instead allowed to move from one cell to another, again in response to pressure forces. Although this approach obviates the problem of severe cell distortion, much care is needed to design an accurate way of moving the material—as well as the momentum and energy associated with it—from one cell to another and averaging the properties of the new components of the cell. This problem is particularly severe when more than one material is involved in the problem. Most modern hydrocodes use some mixture of the Langrangian and Eulerian techniques in order to take advantage of the strengths of each while minimizing the difficulties caused by their respective weaknesses.

The code used for this study is based on the method of individual Lagrangian particles (Harlow 1964; Marder 1975) instead of cells. Each particle is characterized by its mass, the position of its center, and by its velocity, density, and internal energy. Each particle may change shape, large particles may divide, and small particles (of the same material) may merge. There are two meshes involved: the Lagrangian mesh is defined by the positions of the cell centers (nodes); this mesh changes with time, and all parameters of the flow are defined for all nodes. The Eulerian mesh is constructed anew each time step in a rather arbitrary fashion. The pressure field determined in the Lagrangian mesh is interpolated onto the nodes of the Eulerian mesh, and the spatial derivatives of the flow field are calculated. These derivatives are interpolated back to the Lagrangian mesh, where they are used in conjunction with the equations for fluid flow to calculate the new density, velocity, and energy in each particle. The new velocities are then used to determine the new positions of the particles. Particles may be split or merge together in order to increase data regularity.

There are several advantages to this algorithm over more traditional particle-in-cell schemes. First the problems endemic to mixing of different materials are entirely avoided by requiring each particle to be compositionally homogeneous. (This is a problem for all codes that use, even partially, an Eulerian formulation.) Second, the calculation of density is done for each particle in the Lagrangian part of the computation which eliminates the possibility of powerful solution fluctuations, common in particle-in-cell methods where the density is calculated with respect to the Eulerian mesh. Third, because the density and other flow parameters change continuously, the only restriction on the minimum number of particles in a cell is the requirement of continuity. Thus the calculation may be performed with only one particle in a cell, which substantially reduces both the computing time and memory required. Fourth, computer time and memory are also minimized by the fact that only information about the particles, and not about the Eulerian mesh, needs to be saved from one time-step to another. Fifth, the algorithm is simple and homogeneous, which enables one to perform efficient parallelizations of computations on multiprocessor computers.

Furthermore, this general calculational method may be modified by changing the differencing scheme, the form of the Eulerian mesh, the method of particle splitting and merging, the method of interpolation between the meshes, and so on, in order to maximize performance for a given problem or computer. This method has been used successfully in the form of a dialogue program system on the computer BESM-6 (Agureikin and Krukov 1986), on a multiprocessor PS-2000 (Vilenkin et al. 1986), on an IBM PC-AT, and on a computing system based on T-800 transputers. The optimum performance to date has been achieved on computers with the SIMD (Single Instruction-Multiple Data) architecture.

IV. THE CALCULATION

A. Impact on Earth

Two calculations were done for the impact of a 10-km diameter asteroid on the Earth with a velocity of 20 km s^{-1}. Both were axisymmetric models of vertical impacts and neither included the effects of gravity. In the first calculation, the target consisted of a 5-km deep ocean, two crustal layers 3 and 8 km thick, respectively, underlain by a mantle (with initial density 3.3×10^3 kg m^{-3}) which extended to a depth of 34.6 km. The computational area measured 89.6 km along the z-axis and 40 km on the r-axis, with a steric resolution of 25 particles km^{-2}. The problem included a total of 89,600 particles. The second calculation was similar, except that it included an atmosphere. This atmosphere was modeled as an ideal gas with $\gamma = 1.4$ in a uniform layer with a density corresponding to that of air at a pressure of 50 kPa (0.5 bar) and a temperature of 300 K. The purpose of the first calculation was to test the ability of the code to perform such calculations, and the purpose of the second was to test the code's stability when materials of very different initial densities are involved. The first produced all the features expected for such an event, including the large deformations and displacements, and the formation of a large plume of water vapor. The second calculation clearly showed the formation of a shock wave in the atmosphere, with highly shocked gases trapped between the projectile and the shock wave. The very low-density wake behind the projectile was quickly filled in by the expanding hot gas. The atmospheric shock wave impinged on the ocean's surface obliquely and developed a distinctive three-part structure with a Mach stem and a reflected shock wave. Otherwise the results were similar to those of the first calculation, except that the amplitude of the shock wave penetrating the mantle was several percent smaller and the expansion of the water vapor plume is considerably impeded by the atmosphere.

The code that incorporates the Gudonov method of moving grids was used to model the impact of a 10 m iron asteroid on the Earth at 15 and 25 km s^{-1}. The front of the shock wave, the interface between the asteroid and the silicate target, and the free surface were all resolved accurately. Note that the problem involves a different time and space scale; for instance, the volume of the disturbed region at the end of the calculation is much greater than the initial volume of the asteroid. Use of the Lagrangian grid is prohibited by the large deformations involved, and the use of an Eulerian grid would require many more cells to ensure the same accuracy. In the case of the 15 km s^{-1} impact, the asteroid is melted and flows along the walls of the growing crater. In the case of the 25 km s^{-1} impact, the iron asteroid is vaporized. At the end of the calculation (25 ms after impact), the vapor is traveling upward with a velocity of approximately 4.3 km s^{-1}. The velocity is expected to increase as the highly shocked vapor expands and its internal energy is converted to kinetic energy (Vickery and Melosh 1990).

B. Impact of a Spacecraft on an Asteroid

One possible means of combatting an impending asteroidal collision with the Earth is to send a spacecraft loaded with explosives to intercept it. If the asteroid is sufficiently small, one could hope to break it apart and deflect its center of mass away from a collision course with Earth. Even if the collision were head-on, so that the center of mass was not deflected, the fragments might be small enough to be significantly slowed by the Earth's atmosphere and damage to the biosphere would by minimized. An asteroid too large to be broken apart might in principle be deflected, but the problem is beyond the scope of this chapter.

We model the spacecraft as a cylinder 6 m in diameter and 3 m high, which masses 1.455×10^5 kg and the silicate asteroid as a 100 m diameter sphere with a density $\rho_0 = 3.9 \times 10^3$ kg m^{-3}, corresponding to asteroid '1989c.' The relative collision velocity is 30 km s^{-1}. The primary energy source is not the spacecraft itself but the explosive with which it is packed. We assume that this energy is released instantaneously at the moment of impact, and we model the explosion by replacing the spacecraft with an equal volume of ideal gas with a polytropic constant $\gamma=3$, density $\delta_0 = 1.775 \times 10^3$ kg m^{-3}, and a specific internal energy $E_0 = 5.0 \times 10^9$ J kg^{-1}. The results show that the initial (maximum) pressure is 1.7×10^4 GPa. The pressure profiles have a sharp, detonation-like profile, and their amplitudes decline extremely fast; pressure at a depth of 15 m has declined to only 400 GPa. At a depth of several tens of meters, pressure in the shock wave drops below the pressure required for melting and vaporization. If these are required to demolish the asteroid, this goal may be unobtainable. We test the possibility of enhancing the effect of the explosion by following the first spacecraft with a second, similarly loaded with explosives, that impacts the asteroid in the center of the growing crater that results from the first impact. This does indeed extend the portion of the asteroid that experiences very high pressures but not enough to cause catastrophic disruption.

Another scenario that we tested was the explosion of a nuclear device at the bottom of a 75-m deep well drilled into a water-ice object 500 m in diameter. In this case, almost all the energy of the explosion is used to melt and vaporize the object. The outflux of highly shocked water vapor may impart sufficient impulse to change the orbit of the body, but this was not resolved in our calculations.

C. Discussion

Both of these scenarios for asteroid destruction and/or deflection have major practical problems. The first of these is is the extreme difficulty in ensuring the accurate rendevous of a spacecraft with a relatively small object in a not very well known heliocentric orbit. The farther the object is from Earth, the less deflection is required and the higher the relative velocity between spacecraft and asteroid. The latter has the advantage that the energy of the impact is greater for a given mass of the spacecraft; on the other hand, we have assumed that most of the energy delivered to the asteroid comes from the explosive carried by the spacecraft, so a change in the kinetic energy of the impact may not make a significant difference in the ultimate result. Furthermore, the greater the relative velocities, the more difficult to achieve a rendezvous at all. These problems are raised to some unknown but undoubtedly large power in the scenario that requires the impact of a second spacecraft in just the right place at just the right time.

The second scenario, involving a nuclear device detonated in a "well," has all these problems and more. In this case, the first spacecraft has to make a soft-landing and somehow excavate a hole big enough to contain a substantial nuclear device tens to hundreds of meters deep in the object. After this, either the original spacecraft or a second one must place the nuclear device at the bottom of the well without damaging it or causing the well to collapse. The device would be more effective if it were designed so that most of the energy of the explosion was directed downward (toward the center of the asteroid), but this adds difficulty that the device must be emplaced in a certain orientation.

V. CONCLUSIONS

We have developed new EOS formulations for iron (a possible impactor on Earth and a possible target in space) and for SiO_2, a common constituent of the Earth's crust and a more or less reasonable analog for silicate asteroids. We have used these EOS in conjunction with a newly developed hydrocode written to take advantage of parallel-processing computers. We tested the code (and the EOS) by simulating the impact of a 10-km diameter asteroid on Earth. Because this problem has previously been addressed by other workers using different codes, a favorable comparison of the results suggests that both codes are adequately modeling the phenomenon. We then calculated the results of the impact of an explosive-packed spacecraft with an asteroid and the detonation of a nuclear device in a hole drilled in an asteroid. We conclude that the destruction or significant deflection of a "killer asteroid" by these means presents an almost certainly insuperable technological problem.

REFERENCES

Agureikin, V. A., and Krukov, B. P. 1986. Individual particle method for calculation of multicomponent media flows with high strains. *Chislen. Metody Mekhaniki Sploshnoy Sredy.* 17(1):17 (in Russian).

Al'tshuler, L. V. 1965. Use of shockwaves in high-pressure physics. *Soviet Phys. Usp.* 8:52–91.

Al'tshuler, L. V., Bakanova, A. A., Dudoladov, I. P., Dynin, E. A., Trunin, R. F., and Chekin, B. S. 1981. Shock adiabats for metals. New data, statisical analysis and general regularities. *Soviet J. Appl. Mech. Tech. Phys.* 22:145.

Alvarez, L. W., Alvarez, W., Asaro, F., and Michel, H. V. 1980. Extra-terrestrial cause for the Cretaceous-Tertiary extinction. *Science* 208:1095–1108.

Bridgman, P. W. 1958. *The Physics of High Pressures* (London: G. Bell).

Brown, J. M., and McQueen, R. G. 1980. Melting of iron under core conditions. *Geophys. Res. Lett.* 7:533–536.

Bushman, A. V., Kanel', G. I., Ni, A. L., and Fortov, V. E. 1993. *Intense Dynamic Loading of Condensed Matter*, trans. J. W. Shaner (Washington, D. C.: Taylor and Francis).

Gathers, G. R. 1986. Dynamic methods for investigating thermophysical properties of matter at very high temperatures and pressure. *Rept. Prog. Phys.* 49:341–396.

Glushak, B. L., Zharkov, A. P., Zhernokletov, M. V. Ternovoy, V. Ya., Filimonov, A. S., and Fortov, V. E. 1989. Experimental investigation of the thermodynamics of dense plasmas formed from metals at high energy concentrations. *Zhurn. Eksper. Teoret. Fiziki* 96:1301–1318 (in Russian).

Harlow, F. H. 1964. The particle-in-cell computing method for fluid dynamics. In *Methods in Computational Physics*, vol. 1, eds. B. Alder, S. Fernbach and M. Rotenberg (New York: Academic Press), pp. 319–343.

Holmes, N. C., Moriarty, J. A., Gathers, G. R., and Nellis, W. J. 1989. The equation of state of platinum to 660 Gpa (6.6 Mbar). *J. Appl. Phys.* 66:2962–2967.

Jayaraman, A. 1983. Diamond anvil-cell and high-pressure physical investigations. *Rev. Modern Phys.* 55:65–108.

Jones, A. H., Isbell, W. H., and Maiden, C. J. 1966. Measurements of the very-high-pressure properties of materials using a light-gas gun. *J. Appl. Phys.* 66:3494–3499.

Marder, B. M. 1975. GAP–A PIC-type fluid code. *Math. Comput.* 29(130):434–446.

Melosh, H. J. 1989. *Impact Cratering: A Geologic Process* (New York: Oxford Univ. Press).

Neal, T. 1979. Determination of the Grüneisen γ for beryllium at 1.2 to 1.9 times standard density. In *High-Pressure Science and Technology*, vol. 1, eds. K. D. Timmerhaus and M. S. Barber (New York: Plenum Press), pp. 80–87.

Trainor, R. G., Shaner, J. W., Auerbach, J. M., and Holmes, N. C. 1979. Ultrahigh-pressure laser-driven shock-waves experiments in aluminum. *Phys. Rev. Lett.* 39:1154–1157.

Vickery, A. M., and Melosh, H. J. 1990. Atmospheric erosion and impactor retention in large impacts, with application to mass extinctions. In *Global Catastrophes in Earth History*, eds. V. L. Sharpton and P. D. Ward, Geological Soc. of America Special Paper 247 (Boulder: Geological Soc. of America), pp. 289–300.

Vilenkin, S. Ya., Krukov, B. P., Landin, A. A., Minin, V. F., and Sukhov, E. G. 1986. Development and realization of the method of numerical modeling of unsteady flows of multicomponent compressible media on the multiprocessor PS-2000. (Moskva: Inst. Priklad. Matem. Akd. Nauk, SSSR) (in Russian).

Vladimirov, A. S., Voloshin, N. P., Nogin, V. N., Petrovtzev, A. V., and Simonenko, V. A. 1984. Shock compressibility of aluminum at pa 1 Gbar. *Pis′ma v Zhurn. Eksper. Teoret. Fiziki* 39:69–72 (in Russian).

Vohra, Y. K., Duclos, S. J., Brister, K. E., and Ruoff, A. L. 1988. Static pressure of 255 GPa (2.55 Mbar) by X-ray diffraction: Comparison with extrapolation of the ruby pressure seal. *Phys. Rev. Lett.* 61:574–577.

Zel'dovich, Ya. B., and Raizer, Yu. P. 1967. *Physics of Shock Waves and High-Temperature Hydrodynamic Phenomena* (New York: Academic Press).

CONSEQUENCES OF IMPACTS OF COSMIC BODIES ON THE SURFACE OF THE EARTH

VITALY V. ADUSHKIN and IVAN V. NEMCHINOV
Institute for Dynamics of Geospheres, Russian Academy of Sciences

The impact of a very large asteroid can lead to extermination of humanity. Impacts of large objects can lead to global changes in the atmosphere. Medium-size and small objects can cause local and even regional catastrophes which may be considered by humanity as intolerable. A small cosmic body can devastate a large city, even if it explodes at some height above the surface. Now the cities cover only a small part of the Earth. But urbanization is going on all around the world. The average density of population is steadily increasing, as well as the price of life of a single person. At the same time nuclear power plants and chemical plants producing dangerous substances are being constructed. Humanity in some sense becomes more vulnerable. So we should try to estimate the threats from the medium-size and small cosmic bodies, especially keeping in mind that the frequency of such impacts is much higher than for the impacts of large bodies. Various factors have been taken into account, some of them presumably for the first time. The importance of some aspects of the atmospheric interactions has been reevaluated. The estimates presented in this chapter are based on the results of numerical simulations, modeling experiments, analysis of large yield nuclear tests, and scarce data of real impacts; but some estimates are very crude, and the physical models of numerical simulations are still incomplete. We have outlined the directions for further research.

I. GLOBAL, REGIONAL AND LOCAL EFFECTS

A. Introduction

The impacting cosmic objects (asteroids and comets) can be divided to several groups based on different considerations. The simplest way is to classify them according to their diameter. We are using the following 5 groups: very large objects (with diameter >10 km); large objects (with diameters in the range 2–10 km); medium size objects (0.2–2 km); small objects (30–200 m); very small objects (10–30 m and less).

Assuming that the typical density of the asteroids is 2.7 g cm^{-3} and the average velocity is 25 km s^{-1}, we obtain that kinetic energy E_k of very large objects is more than 10^8 Mt, for large objects the energy E_k lies in the range from 10^5 Mt to 10^8 Mt, for medium size objects from 10^3 to 10^5 Mt, for small objects from 3 to 10^3 Mt, and for very small objects E_k is less than 3 Mt. The classification based on kinetic energy of the impacting bodies seems to be appropriate to the problem being investigated. We shall use these energy groups in an attempt to facilitate description of consequences of impacts by different cosmic bodies.

We shall not discuss the impacts by very large objects or very small ones because the results are clear enough. In the first case the global catastrophe leads to extermination of humanity, in the latter case the result will be a local catastrophe, except for the case when the impactor directly hits a nuclear plant or some other dangerous object, that can have regional or even global consequences. Our main interest will be in the consequences of the impact by large, medium-sized, and small objects.

B. The Cretaceous/Tertiary Extinctions as an Example of Global Consequences

There have been at least two natural events that took place in the past giving us two experimental points when an impact led to global and local effects. Considering these events we can evaluate the consequences of future impacts and hazards to humanity. One of these natural events is believed to have been an impact of a large object (a 10 km asteroid) which took place 65 Myr ago (Alvarez et al. 1980). The mass of this body would be approximately 10^{15} kg. If we assume its velocity to be 20 km s^{-1}, then its kinetic energy was 6×10^7 Mt. Several sites seems to be associated with this event. The most probable is the Chicxulub structure with diameter 200 km or greater (see the Chapter by Smit). A huge mass of dust would have been lofted into the atmosphere due to ejection during the cratering process and formation of a huge mass of soot produced by giant fires ignited by thermal radiation (see the Chapter by Toon et al.). This would lead to effects similar to what was called nuclear winter effects (Crutzen and Birks 1982; Carrier et al. 1985) but on much larger scale. These problems will not be discussed in this chapter in detail. Here we only point out that the consequences to the biosphere were very severe, the impact caused mass extinctions of the living species, including the dinosaurs. However, humanity now has evolved beyond a crowd of dinosaurs and, being more vulnerable, is searching for possible threats from smaller cosmic objects.

C. The Tunguska Event as an Impact Causing Local Effects

A second natural event which definitely took place is the Tunguska event of 1908—probably a 60 m small object which exploded in the atmosphere above the surface of an unpopulated area in Siberia at a height of 8 km. This is close to the optimal height of a burst causing maximum damage to urban areas (Glasstone and Dolan 1977; Hills and Goda 1993), but happily there were no cities in this area at that time.

The field of directions of tree falls which was established by expeditions in 1958–1965 is presented in Fig. 1 (Zotkin and Tsikulin 1966; Fast et al. 1976). The peculiar feature of the contour of the devastated area, "the butterfly," was supposed to be the result of the interaction of ballistic shock wave caused by the flight of the body and the spherical shock wave caused by its explosion. This idea was confirmed by a simple modeling experiment, the explosion of a long detonation cord and an enhancement charge at its end above the model forest. This famous experiment (Zotkin and Tsikulin 1966) was conducted at the explosion branch of the Institute of Chemical Physics, which was later transformed into a branch of O. Yu. Schmidt Institute of Earth Physics and now constitutes the main part of the Institute for Dynamics of Geospheres. The destruction pattern obtained in this modeling experiment qualitatively resembles the actually observed damage and shows an important role of ballistic wave and inclination of the trajectory. On the basis of these experiments, several numerical gas-dynamic simulations have been fulfilled (Korobeinikov et al. 1990,1992). These simulations and the results obtained by analyzing barograms and seismograms (Turco et al. 1982) give the energy of this event from 10 to 30 Mt.

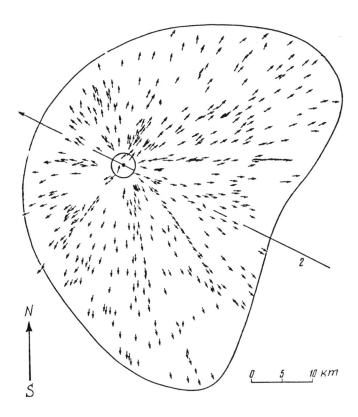

Figure 1.   The field of directions of the tree falls which was established as the result of 1958–1965 expeditions (Zotkin and Tsikulin 1966). Each arrow represents an average from 10 to 150 measurements. Point 1 is the proposed ground zero and the circle is the region of standing forest; line 2 is the proposed projection of trajectory.

D. Major Nuclear Exchange and the Impact of a Small Comet

The case of intermediate-energy release (approximately 6×10^3 Mt) has been studied in detail with respect to nuclear war (Crutzen and Birks 1982) only by theoretical means (happily for humanity). This value of energy is usually adopted as the energy that could be released in the course of major nuclear exchange if the START II disarmament treaty does not substantially reduce this quantity. The above-mentioned energy is equivalent to the kinetic energy of a small comet (200 m in diameter) with a velocity of 30 km s^{-1}. However, the results of the nuclear winter effect investigations cannot be used directly to evaluate the consequences of this impact because the nuclear explosions of a major nuclear exchange would be spread over a large territory (Carrier et al. 1985), and the comet or asteroid impact would lead to a more localized release of energy.

II. THE AIRBLAST CAUSED BY THE IMPACT

A. The Fireball and the Wake

The cosmic body moving through the atmosphere without fragmentation and disruption is seriously decelerated when the mass of the body is approximately equal to or less than the mass of the air in the column having the same cross section area as the body. So the approximate criterion for the negligible role of the deceleration is as follows:

$$\rho_a H (\sin \Theta)^{-1} = \rho_b R_b \tag{1}$$

where H is the characteristic scale of the atmosphere, Θ is the angle of inclination of the trajectory to the horizon, ρ_a is the atmospheric density, and ρ_b is the density of the body. During the flight, however, the body may change its shape and its area due to aerodynamic forces. The atmospheric breakup is negligible if radius R_b of the body is less than some definite critical value (Melosh 1989)

$$R_b^* = H (\rho_a/\rho_b)^{1/2} (\sin \Theta)^{-1}. \tag{2}$$

For the vertical impact ($\Theta = 90°$) of an icy body ($\rho_b = 1$ g cm^{-3}) and the atmosphere of the Earth ($\rho_a = 10^{-3}$ g cm^{-3}, $H = 8.5$ km), we obtain $R_b^* = 270$ m. If $R < R_b^*$ the fragmentation of the body and its dispersion can occur. Later we shall show than the relation (2) may overestimate the shielding effect of the atmosphere, and even the bodies with $R_b \approx 100$ m can reach the Earth's surface being only distorted by aerodynamic forces. So we may assume that medium-sized objects ($D > 200$ m) impact the surface almost without loss of mass and velocity.

After the impact of a cosmic body, its kinetic energy is transformed into the energy of the shock wave propagating downward and into the energy of the vapor plume arising from the surface. This plume begins to interact with the Earth's atmosphere and part of the energy is transformed into the energy of the airblast (O'Keefe and Ahrens 1982b; Ahrens and O'Keefe 1987; Roddy et al. 1987; Melosh 1989). The following simple relation gives us the criterion of deceleration of the plume by the atmosphere and transformation of the energy of the plume into the energy of the airblast:

$$M_v = \omega M_b = (4\pi\omega/3)\rho_b R_b^3 = 3\pi\rho_a^0 H^3 = M_a \tag{3}$$

or

$$R_b' = H(9\rho_a^0/4\omega\rho_b)^{1/3} \tag{4}$$

where M_v is the mass of the plume, M_b is the mass of the body, ρ_b is its density, R_b is the radius of the body, ρ_a^0 is the density of the atmosphere near the Earth's surface ($\rho_a \approx \rho_a^0$ for the altitudes $z \leq H$), M_a is the mass of air in the ejection cone with zenith angle 45°, ω is a nondimensional coefficient. For a stony body ($\rho_b = 2.7$ g cm^{-3}), in the atmosphere of the Earth ($\rho_a^0 = 10^{-3}$ g cm^{-3}, $H = 8.5$ km) we obtain $M_a = 5 \times 10^{12}$ kg and assuming $\omega = 1$ the critical radius $R_b' = 1.0$ km.

A characteristic dimension of a spherical airblast is defined as

$$R_a = [3E_a(\gamma - 1)/(4\pi\gamma p_a)]^{1/3} \tag{5}$$

where p_a is the initial pressure of the atmospheric gas, γ is the adiabatic exponent, E_a is the energy of the airblast, which is proportional to the kinetic energy of the body $E_k = M_b V^2/2$, where V is the velocity of the body: $E_a = \alpha E_k$. Thus, the radius of the fireball R_a is proportional to the radius of the body R_b:

$$R_a = \alpha^{1/3} R_b (\rho_b/\rho_a)^{1/3} (V/c_a)^{2/3} (3(\gamma - 1)/4\pi)^{1/3} \tag{6}$$

where c_a is the speed of sound and α is the coefficient of proportionality between E_a and E_k. The coefficients ω and α depend on the velocity of the projectile and the inclination of its trajectory, density, composition, strength and porosity of the projectile and the target (Ahrens et al. 1989; Melosh 1989). Moreover they are really not constants depending on this parameters as the energy budget and the mass of the ejecta feeding the blast wave are slowly varying with time. If we assume $\alpha = 0.3$ and $\gamma = 1.2$ then for $R_b = 100$ m, $\rho_b/\rho_a = 10^{-3}$ and $V = 30$ km s^{-1}, we obtain $R_a = 5$ km that is of the order of magnitude of the characteristic scale H of the atmosphere. So the two-dimensional effects are very important. The body with such a radius R_b may be fragmented but the fragments do not fall far from one another so a common fireball and a common crater are formed. If the average density of the impacting swarm is low enough, then the energy going into the ground is small ($\alpha \sim 1$) and the cratering effects become negligible.

The development of the airblast also is influenced by the existence of the wake formed by the body moving through the atmosphere. For simple estimates one can use the model of the wake as a cylindrical explosion. The transverse velocity V_t of a shock wave can be approximated by the simple equation

$$\frac{dR_s}{dt} = V_t = VR_b/R_s \tag{7}$$

where R_s is the radius of the shock wave and V is the velocity of the body. Integrating Eq. (7) one obtains

$$R_s^2 = 2(VtR_b) + R_b^2. \tag{8}$$

The characteristic time of the development of this rarefied hot channel before the impact is H/V, so if we assume $R_s >> R_b$, then the radius of the wake R_w before the impact at the height $z = H$ is

$$R_w = (2HR_b)^{1/2}. \tag{9}$$

For $R_b = 0.25$ km and $H = 8.5$ km we obtain $R_w = 2$ km. The wake is a rarefied channel through which some part of the energy can escape the dense layers of the atmosphere.

B. Overpressure and Dynamic Pressure

The destructive effects of the blast wave are usually related to values of the peak overpressure. At Nagasaki, dwellings collapsed at a distance up to 2 km from ground zero, where the peak overpressure Δp was estimated to be 3 psi (Glasston and Dolan 1977). The results of the nuclear tests show almost complete destruction of a one-story rambler-type house, and a two-story wood-frame house; an unreinforced brick house was destroyed at 4 to 5 psi, approximately equivalent to 0.3 bar or 0.03 MPa. The reinforced precast concrete house suffered only minor structural damage, but the rigid steel-frame house with aluminum panels collapsed at $\Delta p = 3$ psi.

Peak overpressure of the shock wave depends not only upon the distance to ground zero but also upon the height of burst. The value $\Delta p = 4$ psi was reached at a distance of 5 km for a 1 Mt nuclear explosion when it happened on the ground or at the distance of 9 km for the height of burst of 3.6 km (Glasstone and Dolan 1977). For a 30 Mt explosion simple estimates based on the hydrodynamic similarity give the radius $R = 25$ km for the explosion height of 10 km. So sizes of the rural areas which could be devastated by such impact are equivalent to the radius of a large city (e.g., Moscow).

There is another important quantity, dynamic pressure q, which defines the blast damage due to a drag force associated with strong winds accompanying a passage of the blast wave. For $\Delta p = 5$ psi the peak dynamic pressure q is 0.6 psi or 0.04 bar, and the maximum wind velocity is 160 miles per hour (equivalent to 260 km per hour). But at this preliminary stage of investigations of possible hazards we shall not go too far into such details. We shall mention only that a theoretically calculated zone of destruction of a forest array obtained by Korobeinikov et al. (1990) and given at Fig. 1 is contoured by the solid line.

Estimates of the devastation area due to the shock waves caused by the impacts of asteroids and comets had been given by Chapman and Morrison (1994). They took into account the overpressure in the shock wave front and dynamic pressure which caused the fall of trees. For calibration they used the Tunguska event (Zotkin and Tsikulin 1966) where the shock waves caused the fall of trees in the area of 2000 km^2, which is equivalent to an area of a circle with a radius $R_s = 25$ km. It was assumed that the area A_s of severe damage of the buildings is approximately equal to the area of the forest devastation. So the following expression can be used for estimations

$$A_s = \pi R_s^2 = 200 E_k^{2/3}, R_s = 8 E_k^{1/3} \qquad (10)$$

where R_s is in kilometers, A_s in square kilometers and yield E_k is in Mtons. The coefficients in Eq. (10) change with the height of the burst. The "optimum" height h (km) is $6.4 \times E^{1/3}$ (Glasstone and Dolan 1977; Hills and Goda 1993). If the height of the burst decreases to zero, the radius R_s decreases approximately 1.4 times. For $E = 30$ Mt one obtains $R_s = 18$ km.

C. The Atmosphere's Breakthrough

The scaling law (Eq. 10) is based on the theoretical investigations of a hydrodynamic problem of the shock wave propagation for point source energy release in the uniform atmosphere and on the numerous experimental investigations of blast waves produced by high-energy chemical explosives and nuclear devices for energies lower than ~ 10 Mt. But for high energies the radius R_s is of the same order of magnitude as the scale height H of the atmosphere or exceeds it (for $E = 10^3$ Mt we obtain $R_s = 80$ km).

The two-dimensional numerical calculations for the case of an explosion in a nonuniform atmosphere and various theoretical estimates of the atmosphere's breakthrough (Zel'dovitch and Raizer 1967) show that the shock wave, due to the decrease of the initial air density with the altitude above the Earth, moves upwards faster than in the radial direction. Numerical simulations (Jones and Sanford 1977; Jones and Kodis 1982) for a 500 Mt airblast at the ground (treated as a flat, free-slip, reflecting boundary) have shown that the dynamic pressure exceeds the threshold and the shock wave can knock down the trees at the distances less than 27.5 km instead of 45 km as follows from Eq. (10). Actually, not only the nonuniformity of atmosphere but the wake is also an important factor. This is confirmed by the numerical simulations of Nemchinov et al. (1993e) for a vertical impact of a 200 m spherical icy body with the density 1 g cm^{-3} striking the Earth's surface with the velocity of 50 km s^{-1}. The kinetic energy is 2 times larger than was considered by Jones and Sanford (1977). The density contours are presented at Fig. 2. In these calculations not only thermal radiation transfer was taken into account but in addition the wake formed behind the body during its flight through the atmosphere was considered as well. The wake expands not only during the flight of the body but also after the impact (expansion in one second to a radius of 2 km), and a great amount of mass and energy is injected into the wake. At two and a half seconds, the radius of the shock wave at the Earth's surface is 8 km and the diameter of the wake is approximately 4.5 km. So the high-pressure gas expands into the rarefied atmosphere through the wake.

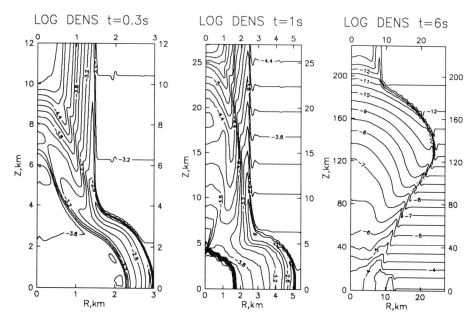

Figure 2.   The contours of constant density (g cm^{-3}) of the air for the vertical impact at the Earth of a 200 m icy body with velocity 50 km s^{-1} for the time 0.3, 1.0, and 6 s (a, b, and c, respectively).

D. The Mortality Due to Shock Waves

The mortality associated with the impact was estimated by Chapman and Morrison (1994) from the average world population density 10 persons per square kilometer. Using this density, we can obtain from Eq. (10) that area A_s embraces the number of people

$$N_s = 2 \times 10^3 E_k^{2/3}. \qquad (11)$$

We cannot simply assume that the area A_s, where the shock wave causes the complete devastation of the city, is at the same time the area where the shock wave leads to lethal consequences for all the inhabitants. This is only the number of persons at risk. For the Tunguska event with $E_k = 30$ Mt we obtain $N_s = 2 \times 10^4$. This number is very small in comparison with the casualties in Hiroshima and Nagasaki where the energy was ~20 Kt. The total number of killed was 68,000 (of the 250,000 total population) in Hiroshima on the area of 25 km^2 and 38,000 in Nagasaki (of the 170,000 population) on the area of 18 km^2. The number of injured was 76,000 in Hiroshima and 21,000 in Nagasaki. To obtain the mortality from shock waves we must exclude from the number of killed the casualties due to nuclear radiation, but it is approximately only 30% of all the casualties (Glasstone and Dolan 1977). We can also exclude the casualties due to burn injuries (they will be discussed later). Nevertheless the resulting number is much higher than obtained from Eq. (11). This is due to the fact that population densities in Hiroshima and Nagasaki were high enough (3000 and 2500 per km^2) and reached 10,000 per km^2 in areas close to ground zero (average population density of the five boroughs of New York City is about the same).

The actual number of personal casualties will be greatly dependent upon circumstances, but we can be sure that if a small cosmic body hits a very densely populated city, the number of dead and very seriously wounded could be higher by an order of magnitude or even more than is obtained from Eq. (11). The probability of such an event is nowdays much lower than the probability of the impact itself as the Earth is covered by cities over only about 10^{-4} of its total surface. However, the number of large cities is steadily increasing as well the average density of population.

As the mortality and the damage of the buildings can vary very seriously from one geographical region to another, one must think not only about the defense of the Earth as a whole (Ahrens and Harris 1992; see also the Chapters by Ahrens and Harris, and by Melosh et al.), but also about a local defense (by means of destruction of the cosmic body or deflection of its trajectory) to hit some unpopulated place on the Earth's surface. Here we can add only that the possibility of diverting the impacting body from a certain region, certain city, or from a certain country raises a new political and moral problem if such an alteration can be the cause of damage and mortality in another region or neighboring country. This problem needs a thorough investigation from different political and moral viewpoints, and special international law regulations should be adopted. Another method of reducing the casualties is to use evacuation, but the larger the area of possible devastation the more difficult it is to realize it.

III. THERMAL RADIATION AND FIRES

The area of forest fire ignition at the Tunguska was smaller than the area of devastation of the forest by shock waves. It can be seen from Fig. 3 where zones of radiant burning of trees in the region of the Tunguska fall (L'vov and Vasil'ev 1976) are presented. The area of radiant burning in this case is approximately 300 km^2 or equal to the area of the circular spot with the radius $R = 7$ km. In the estimates of Chapman and Morrison (1994) thermal radiation was not taken into account. However, in the case of an impact of larger cosmic bodies it was shown that the infrared radiation after ballistic reentry of ejecta (Melosh et al. 1990) and the visible and ultraviolet light

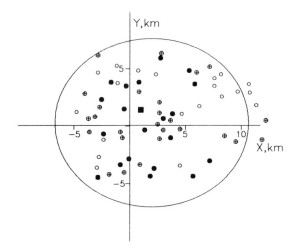

Figure 3.   Zones of radiant burning of trees in the region of the Tunguska fall
  burning: ○ = weak, ⊕ = moderate, ● = strong, ■ = Kulik's hut.

emitted by the air heated in the shock wave (Nemchinov and Svetsov 1991)
can ignite fires over huge territories.

Approximate radiant exposure for ignition of pine needles for $E = 20$ Mt is
80 J cm^{-2}, and slant range for this specific exposure for the nuclear explosion
with such energy is 12 miles, or 20 km, for a 12-mile visibility (Glasstone and
Dolan 1977). These estimates seem to contradict the experimental data of the
Tunguska fall but this contradiction can be easily overcome if we consider the
visibility and atmospheric conditions: very clear, 50 km; clear, 20 km; light
haze, 4 km; thin fog, 2 km; light to thick fog, 1 km or less. So we simply
assume that the weather conditions at this specific region and at this specific
time were not very good and light haze was covering the Earth's surface. The
attenuation of radiation in the atmosphere leads to the reduction of the radiant
exposure approximately 10 times, and we can obtain the observed area of fire
ignition.

A. The Role of the Wake Behind the Body

For intermediate size bodies a role of a thermal radiation until recently has not
been investigated. The results of a numerical simulation (Nemchinov et al.
1993e) of an impact of a small comet (200 m diameter icy body impacting the
Earth's surface at the velocity 50 km s^{-1}, mass $M = 4 \times 10^9$ kg, kinetic energy
$E = 1.2$ Gt) are presented at Fig. 4. In this figure the temperature contours
are given at several intervals of time. The simulations took into account the
decrease of the density of the atmosphere, and the wake which was formed
during the flight of the body through the atmosphere.

The velocity of a shock wave traveling upwards along the wake grows
with altitude and exceeds 40 km s^{-1} in one second after the impact. But

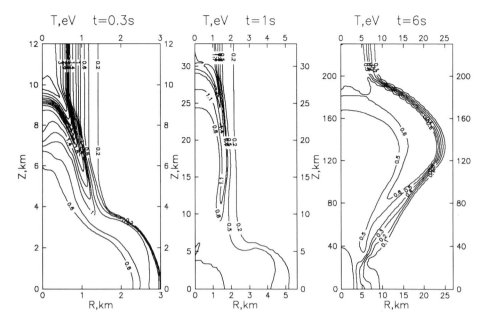

Figure 4. Plot of constant temperature contours in the atmosphere after a vertical impact of a 200-m icy body at a velocity of 50 km s^{-1}.

a gas in the wake in front of the shock wave also expands upwards with sufficiently high velocities, thus, decreasing the amplitude of the shock wave. The thermal radiation transfer also leads to decrease of the temperature. So maximum temperatures behind the shock wave are about 2 eV. This is greater than the maximum temperature of the plasma near the ground at the same time. The plasma at high altitudes become semi-transparent due to its expansion and reduction of the density of the ambient air. Hence, the radiation emission becomes more effective than in the lower atmosphere. The main new result is that the wake changes the shape of the fireball; the fireball becomes elongated rather than spherical. Thermal radiation flux on the Earth's surface illustrated in Fig. 5 was calculated ion the assumption that the atmosphere had perfect visibility. In one second the thermal radiation flux exceeds 100 W cm^{-2} at a distance of about 100 km. In two and a half seconds the flux exceeds 200 W cm^{-2}.

Taking into account the decrease of the density of the air with the height above the surface of the Earth, we obtain that the thermal radiation is mainly emitted from great heights, 30 to 60 km or even more, where the hot air becomes sufficiently transparent (Fig. 6). Therefore, the emitted light is transmitted to the Earth predominantly through the rarefied air without aerosols, moisture, and clouds, that is, through the layers where the visibility is close to perfect. This factor increases the flux of thermal radiation in comparison with these values that can be calculated from the simple scaling laws established

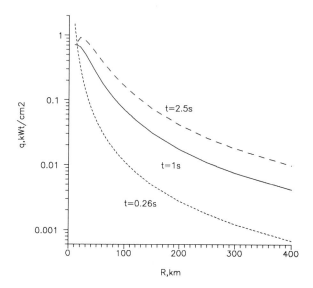

Figure 5. Radiation flux on the Earth's surface as a function of the distance from the point of impact of the 200 m icy body at a velocity of 50 km s^{-1}.

for the nuclear explosions (Glasstone and Dolan 1977; Crutzen and Birks 1982). For the nuclear explosions the approximate scaling law is

$$Q = \psi(E_a/4\pi r^2) \exp(-r/l) \qquad (12)$$

where Q is the energy delivered to the unit surface, E_a is the energy of the airblast proportional to the initial kinetic energy of the body, r is the radius, ψ is the coefficient of conversion of the airblast energy into the thermal radiation, and l is the visibility. We must keep in mind that the attenuation of the radiation is due not only to absorption but to scattering as well. The exponential law of attenuation assumed in Eq. (12) is an oversimplified one, but at this stage of investigation it is sufficient to use this expression. For the case of the thermal radiation emitted after the impact we can use another approximate relation

$$Q = \psi(E_a/4(\pi r^2 + z_0^2))\eta \qquad (13)$$

$$\eta = \exp(-H/l \sin\theta)$$

where z_o is the effective height of the emitting volume, H is the characteristic scale of the atmosphere, θ is the angle of ray inclination (with respect to the horizon), and η is the transmittance. Here we assume that the attenuation of radiation occurs only at the heights $z \leq H$, and the emitting volume is situated at $z_0 > H$.

If the visibility l is 12 km, $H = 8$ km, and $\theta = 45°$, then the transmittance $\eta = 10^{-1}$. Thus, we have to reduce the previous value of radiation flux

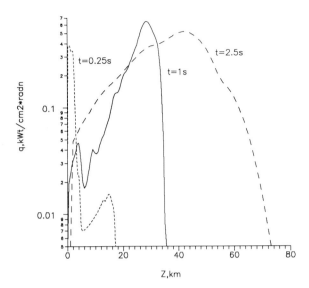

Figure 6. Radiation flux per unit angle which falls into an observer's view at a distance of 100 km from the point of impact as a function of altitude.

approximately 10 times. For $r \gg z_o$, assuming that visibility and the average angle of ray inclination are the same, one can obtain a simple equation

$$Q = (E_a/4\pi r^2)\eta\psi \tag{14}$$

and the effective radius of ignition of fires grows with the energy E_k of the body approximately as $R_f \sim E_k^{1/2}$. Therefore, for large bodies and large energies E_k, very large areas can be destroyed by fires, and the radius R_f increases with energy of impactor much faster than in the case of the shock waves.

B. The Scaling Law for Fire Ignition

The value of 100 J cm^{-2} can be adopted as the threshold of fire ignition (Glasstone and Dolan 1977). So for $E = 10^3$Mt, the radius R_f of the fire ignition is approximately 100 km, the area $A_f = 3 \times 10^4$km^2. If the airblast goes beyond the zone that catches on fire, the shock will tend to blow out the fire, as apparently happened at Tunguska. But for the large yield the radius R_f is so large that the shock wave deteriorates into an acoustical wave. The propagation time of a small-amplitude wave to a distance of 100 km is 300 s. Not only fires can be ignited by the high-velocity impact but sufficient energy after the ignition can be released by the combustion. Using the average density of Earth's population, we obtain the number of persons at risk $N_f = 3 \times 10^5$.

In Hiroshima and Nagasaki, 20 to 30% of all the total casualties were flash-burn injuries due to the direct action of thermal radiation impulse. Using the data obtained in Japan, it was expected (Glasstone and Dolan 1977) that the radiant exposures 20 J cm^{-2} would be the cause of the first-degree burns on unprotected skin (a reversible injury) for 82% of the population, but 15% would receive second-degree burns (which would heal in one or two weeks). Of course, the casualties by the direct action of thermal radiation impulse may be reduced by simple means of civil defense (shelters and other protection measures). We must mention also the eye injuries causing blindness and retinal burns, which could be especially great when impact occurs at night time, but they can be seriously reduced by means of adequate training provided there is warning of the possible impact.

As to the casualties by the fire ignition including suffocation from smoke, some measures of civil defense may be useful but they become more and more difficult as the area of fire ignition and the amount of heat and smoke release increase.

As we can easily see the thermal radiation can be more dangerous than the shock wave especially for large bodies and large kinetic energies

$$A_f = 30E_r, R_f = 3E_r^{1/2}, N_f = 3 \cdot 10^2 E_r \qquad (15)$$

where R_f is in km, A_f is in km^2, and E_r is the thermal radiation energy in megatons. For $E_r = 10^4$Mt one obtains, $R_f = 300$ km, $A_f = 3 \times 10^5$km^2, $N_f = 3 \times 10^6$. For $E_r = 10^5$Mt the previous parameters increase to $R = 1000$ km, $A_f = 3 \times 10^7$ km^2, $N_f = 3 \times 10^7$. The latest figure is of the order of magnitude of the losses in World War II. We suppose that humanity will assume such losses as prohibitive.

If we assume that the coefficient ψ of conversion of the airblast energy into thermal radiation energy is 40% and the ratio α of the airblast energy to the kinetic energy of the body is 50%, then the ratio of the thermal radiation energy to the kinetic energy of the body is 20%. So the threshold of the kinetic energy E_k of the asteroids or comets that will cause such great direct losses that definitely need protection is 10^5 to 10^6Mt; this corresponds to the energy of a comet with 1 km or 2 km diameter. This is a rather big size but we must keep in mind the rising average density of population of the Earth and rising cost of a human life, so this threshold may be decreased.

We must also keep in mind that the values of the coefficient ψ increase with the velocity of the body. This can be seen from the results of numerical simulations for various initial specific energies of the 100 Mt airblast presented in Fig. 7 (Nemchinov et al. 1993e). Also note, that the coefficient α is bigger for low-density comets than for stony asteroids.

C. Ignition of the Fires by Large-Body Impacts

For a larger comet ($M = 0.8 \times 10^{12}$kg and energy $E = 0.5 \times 10^6$Mt), the mean altitude of the plasma volume during the time of intensive radiation is $\sim$100 km (Nemchinov and Svetsov 1991); but for large radii we must take into account that the Earth is a sphere. Radiation impulse on the Earth's surface

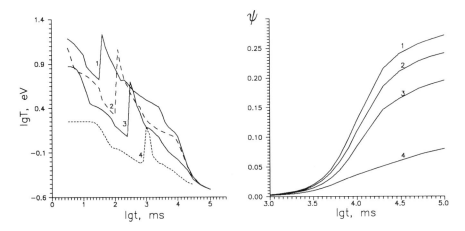

Figure 7.   A maximum temperature of the fire ball of a 100 Mt explosion near the Earth's surface and a portion of energy emitted versus time for a number of different initial specific energies: 2450 (1), 610 (2), 200 (3) and 50 kJ g^{-1} (4). Corresponding velocities of the impactor are 70 (1), 35 (2), 20(3) and 10 km s^{-1} (4).

obtained from numerical simulations was so great, that the energy absorbed by unit surface exceeds the value of 100 J cm^{-2} all over the area of direct vision; the plume direct vision radii proves to be > 1000 km. Thus, as a result of the radiation flux from the plasma plume after a high-speed impact, fires can arise on the areas with characteristic dimensions of the order of 1000 km. The quantity of smoke and soot aerosol ejected into the atmosphere has not yet been determined for such huge fires, but it can be essentially greater than what is proposed in the nuclear war scenario (Carrier et al. 1985).

D. Thermal-layer Effect: The Lifting of Soot and Aerosols by the Vortex and the Increase of Dynamic Pressure

A heated layer near the ground is formed due to combustion processes and absorption of the thermal radiation impulse. Even though this layer would be very thin, it could drastically change the flow field. High-speed jets moving along the surface would be formed augmenting the dynamic pressure effects, as mentioned in Sec. II.B.

This thermal-layer effect was discovered by Russian scientists as early as in the middle of the 1950s (Sadovsky and Adushkin 1988). Now it is being intensively studied by theoretical means, by estimates and numerical simulations, and by experimental laboratory modeling (Artem'ev et al. 1987,1988,1989; Bergel'son et al. 1987,1989). It also has been revealed and investigated by American scientists (Shreffler and Christian 1954; Glasstone and Dolan 1977; Mirels 1988; Reichenbach and Kuhl 1988).

We should also point out a very intriguing feature of a flow field in the case of the shock wave interacting with a thermal layer. A vortex-like structure is formed in the precursor (see Fig. 8, where only the lower part of

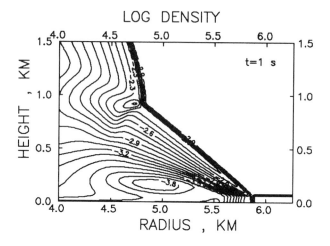

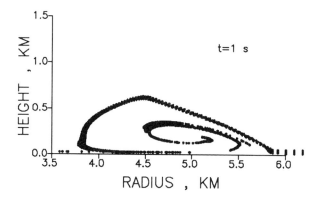

Figure 8.   (a) Plot of constant density contours in the atmosphere after a vertical
impact of a 200 m diameter icy body at a velocity 50 km s⁻¹ for the instant $t = 1$ s.
(b) Position of the passive markers initially contained in a thermal layer at the same
instant of time.

the fireball is plotted). The size of the precursor and the size of the vortex
are much larger than the thickness of the thermal layer and are of the order of
magnitude of the length of the thermal layer or of the distance of the shock
wave from the center of the explosion. The effects both of entraining soot
and aerosols by the vortex-like flow in the precursor formed after the impact
and by the dynamic pressure increase must be studied more thoroughly in the
future.

E. Thermal Radiation Emitted During the Flight of an Asteroid Through the Atmosphere

Let us consider some aspects of the flight of a large cosmic body through the atmosphere. Behind a bow shock wave which is formed in front of the body, the temperature of the air is high. The data on the thermodynamical properties of the heated air (Kuznetsov 1965) in the temperature range 0.7 to 12 eV can be approximated by a simple law

$$e = 31T^{1.5}\delta^{-1.2}, \delta = \rho/\rho_l \tag{16}$$

where e is the specific energy per unit mass in kJ g^{-1}, temperature T is in eV and δ is the relative density (normal density of the air $\rho_l = 1.29 \times 10^{-3}$ g cm^{-3}).

Using Eq. (16) it is easy to estimate the fundamental parameters of a strong shock wave from the Hugoniot relations (Nemchinov et al. 1989; Kiselev et al. 1991b; Nemchinov 1993)

$$T_s = 0.077U_s^{1.33}\delta_h^{0.08}$$

$$q_b = \sigma T_s^4 = 3.6 \times 10^{-6}U_s^{5.32}\delta_h^{0.32} \tag{17}$$

$$q_h = 0.071U_s^3\delta_h, \delta_h = \rho_h/\rho_l$$

where ρ_h is the initial density at the height h, U_s is the velocity of the gas behind the shock wave, T_s the temperature behind the shock wave, σ is the Stephan-Boltzman constant, q_b is the blackbody radiation flux, q_h is the hydrodynamic energy flux in MW cm^{-2}. So for velocities between 20 and 50 km s^{-1}, the temperature T_s is in the range 3 to 5 eV. The maximum of the blackbody spectrum is higher than the transparency boundary of the cold air ($\varepsilon^* \approx 6.5$ eV). A large part of the emitted radiation is absorbed before the shock wave. The heated air before the front begins to absorb photons with energies $\varepsilon < \varepsilon^*$; this screening effect may limit the radiation flux (Zel'dovich and Raiser 1967).

The role of the thermal radiation can be characterized by a parameter

$$\chi = q_b/q_h = 0.52 \cdot 10^{-4}U_s^{2.32}\delta_h^{-0.68}. \tag{18}$$

The Rosseland length l_r (in cm) in the temperature range 2 to 10 eV (Biberman 1970) can be approximated by the following expression

$$l_r = 0.1\delta^{-1.5}. \tag{19}$$

For the optically thick plasma behind the supercritical shock wave, radiation transfer can be described by the radiation heat-conduction approximation (Zel'dovitch and Raiser 1967), and the thickness of the heated layer x_t, in cm, is given by the following power law (Kiselev et al. 1991b)

$$x_t = 0.52 \cdot 10^{-5}U_s^{2.32}\delta_h^{-2.18}. \tag{20}$$

For example, one can estimate from Eqs. (18) to (20) parameters of the shock wave for the velocity $U_s = 50$ km s^{-1}. At the height $h = 25$ km the parameter $\chi = 4.1$, and, thus, the shock wave is supercritical, the thickness of the heated layer $x_t \sim 0.8$ m.

The expression (20) is valid for the case of a plane piston generating the shock wave, e.g., for $x_t < < D_b$, where D_b is the characteristic diameter of the body. If this unequality does not hold (large heights of flight, small bodies), the flow pattern becomes two- or even three-dimensional and the maximum temperature of the heated air becomes less than T_s. Results of such two-dimensional simulations for a 10 m cylindrical body moving with the velocity 68 km s^{-1} in the atmosphere at the height $h = 25$ km have shown that the size of the volume of the gas, having a temperature more than 1 eV, is 3 times more than the size of the body. The ablated material concentrates in a thin layer near the surface of the body and mixes with the air in the wake, but as the air plasma is opaque (see Eq. 19) the radiation emitted from the body's vaporized material can escape (and be registered) only from the regions far behind the body, where the luminous shock wave becomes transparent. This special regime was called the radiative hypersonic flow (Nemchinov and Popov 1983; Nemchinov 1993). The existence of such a mode of flight has been confirmed by modeling experiments using a hypersonic jet with velocities up to 50 km s^{-1} (Kiselev et al. 1991a).

For $D > > x_t$ the results of one-dimensional nonstationary simulations can be used to obtain the spectrum of radiation and the radiation flux density q_∞ at large distances from the shock wave. Numerical modeling of the structure and brightness of the shock wave for different velocities U_s have been carried out (Nemchinov et al. 1976,1978,1979). Special numerical technique used in this calculation (Nemchinov 1970; Kiselev et al. 1991b) allowed to take into account the very detailed spectrum and angular distribution of the radiation. A system of one-dimensional non-steady gas dynamic equations and spectral equations for radiation transfer was solved. Absorption and emission in lines were taken into account.

The shock waves that are under consideration are supercritical (in accordance to terminology of Zel'dovitch and Raizer 1967) but the hydrodynamic jump is not significantly screened by the heating layer and such shock waves appear to be quite intense sources of thermal emission (with an efficiency of the order of 12 to 30%). Some of the results for the case of relative density $\delta_h = 0.03$ and $\delta_h = 0.003$, corresponding to the height $h = 25$ km and $h = 40$ km, respectively, are presented in Table I, where V is the velocity of the shock wave, U_s is the velocity of the gas behind the front, q_h is the hydrodynamic energy flux, q_∞ is the flux of the emitted radiation, and ζ is the coefficient of conversion ($\zeta = q_\infty / q_h$). We see that there is the maximum of radiation efficiency when V is about 13 to 22 km s^{-1} (for $h = 25$ km). If we increase the air density up to $\delta_h = 0.1$ (it corresponds to the altitude 18 km) the energy flux of the emitted radiation for the velocity $V = 30$ km s^{-1} is as high as 4 MW cm^{-2}. So for this typical velocity of the impacting bodies the

TABLE I

$\delta_h = 0.03$, $h = 25$ km

V km s^{-1}	U_s km s^{-1}	q_∞ MW cm^{-2}	q_h MW cm^{-2}	ξ %
10.8	10	0.11	1.9	5.8
13.0	12	0.29	3.3	8.8
16.3	15	0.94	8.6	10.9
22.0	20	1.4	15.5	9.0
27.5	25	2.0	30.2	6.6
29.7	27	2.4	38.1	6.3
36.4	33	2.2	69.3	3.2
44.0	40	1.7	124	1.8
55.2	50	2.0	242	0.83
$\delta_h = 0.003$, $h = 40$ km				
22.8	20	0.4	1.7	29
32.9	30	1.1	5.8	22
43.5	40	1.8	13.7	12

flux q_∞ slightly increases when the altitude decreases.

Using the analogy between the hypersonic flight and the cylindrical explosion one can obtain the shape of the bow shock wave and estimate the energy losses due to thermal radiation

$$Q = \pi R_b^2 q_\infty^0 \varphi(D_{\min}, V) \qquad (21)$$

where q_∞^0 is the radiation energy flux at the blunt nose of the body, $D_{\min}$ is the minimum velocity where the radiation is still big enough (according to the data presented in the Table I we assume $D_{\min} = 10$ km s^{-1}).

The nondimensional coefficient at the right-hand side of Eq. (21) is approximately 5 for $V = 30$ km s^{-1}. Such an increase of energy losses in comparison with the losses from plane shock wave is due to the increase of the surface of the shock wave in the wake in comparison with the body's cross section.

The energy flux at the surface of the Earth q_e can be estimated from the following relation:

$$q_e = q_\infty^0 (R_b/h)^2 \varphi/4. \qquad (22)$$

For $V = 30$ km s^{-1}, $h = 25$ km we obtain $q_\infty^0 = 2.3$ MW cm^{-2}; for $R_b = 0.3$ km we obtain $q_e = 0.3$ kW cm^{-2} for the case of perfect visibility, and the radius of the area which is almost uniformly illuminated is approximately equal to h.

For the horizontal trajectory the value of q_e stays approximately the same during the characteristic time of irradiation $t_r = 2h/V$. In the previous example $t_r \approx 1.5$ s, so the radiant exposure (energy at the unit area of surface) $Q = q_e t_r = 0.45$ kJ cm^{-2}. Thus, even assuming that only 1/5 of the radiation emitted by the body reaches the surface of the Earth, we obtain that the radiant exposure exceeds the fire ignition threshold. So even for the grazing trajectories without direct collision with the Earth, the body can ignite fire in the long strip with a width approximately equal to the height of the flight, and the length of the strip corresponds to the length of the projection of the trajectory of the body moving approximately at the minimal height h_{min}. For the vertical impact the characteristic time of irradiation and the fluences are of the same order of magnitude at distances up to 25 to 30 km and the thermal layer (Sec. III.B) may be created at the surface at these or even large distances even before the impact.

IV. DUST CLOUDS

A. A Dust Cloud Caused By an Impact of a Large Asteroid

The impact on the Earth leads to the ejection of a large mass due to the cratering process. For a 10-km asteroid impact, numerical simulations (Roddy et al. 1987) show that the diameter of the transient crater reaches 80 km in 120 s. The mass of the ejecta (2×10^{17}kg) is 150 times larger than the mass of the asteroid; but 70% of the ejecta finally lie within 3 crater diameter (240 km). Most of the ejecta expands not higher than to the tropopause ($\sim$13 km altitude). Calculations show that 1.5×10^{17}kg were ejected to the level of tropopause, 1.6×10^{16}kg into the stratosphere, 9×10^{15}kg into the mesosphere (6 times larger than the body's mass).

B. Dust Clouds After an Impact of a Medium-size High-speed Comet

Let us decrease the diameter of the body down to 200 m ($\sim$50 times) and reduce the energy of the icy body to 1200 Mt. Even if we assume that the ejected mass is as low as 4×10^{11}kg in accordance to the result of nuclear tests (Glasstone and Dolan 1977), the huge cloud of dust is still formed.

The effect of contamination of the Earth's atmosphere due to impact of an icy comet was investigated by means of numerical simulations (Nemhinov et al. 1993b). The Monte Carlo method was used to simulate the movement of the dust cloud and 10,000 markers represented dust particles with the diameters in the range 0.01 to 10 mm. Each particle was carried along the air flow with the local velocity. In addition, it moved relative to the air due to the action of gravitational forces. The particles ejected by the soil erosion process were lifted by the air turbulence. To handle this phenomenon a semi-empirical diffusion coefficient was introduced. All particles that crossed the Earth's surface went into fallout and new particles were ejected from all points of the surface where the air velocity was large enough. The dust cloud influence on the air motion was assumed to be negligible.

The velocity field and the dust markers at 50 s are shown in Fig. 9. We can see that the dust from the zone with radii as large as 40 km is included in the motion, but the main part of the dust was raised from the region of 3 km. The top of the cloud reached 60 km. At 300 s, 5 min after the impact, the dust has lifted with the cloud to heights of more than 200 km. When it reaches

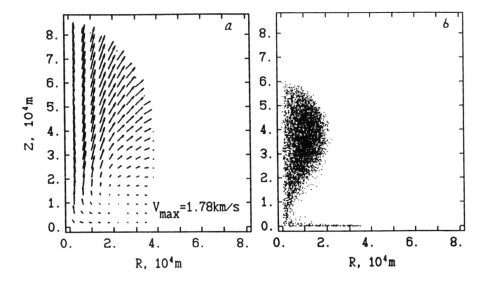

Figure 9.   (a) The velocity field at 50 s. (b) The dust markers at time 50 s.

300 km (Fig. 10), the ejected air and dust cloud lose their kinetic energy and begin to fall.

In Fig. 11 we see how the dust cloud after 11 min fell to the height of approximately 100 km and spread out to a diameter 400 km. This clouds contains small droplets due to condensation of vapor and very thin dust (from 10 μm to 10 mm) lofted from the surface of the Earth and ejected during the cratering process.

C. Dust Cloud After A Small Asteroid or Comet Impact

The cratering and ejection processes, and the formation and evolution of a dust cloud caused by an impact of a low-speed asteroid or high-speed comet can be investigated not only by theoretical means but also by using the experimental data. Heights of the cloud for high-yield Pacific tests done during operation Castle has been presented by Carrier et al. 1985. Stabilization altitude is reached by these clouds in about 6 to 7 min, with most of the rise occurring during the first 3 min. For $E = 16$ Mt the top of the cloud reaches 35 km and the bottom 17 km (in the tropics the tropopause usually occurs near 17 km at all seasons).

The estimates of the height of the cloud can be obtained from the solution of the problem of the motion of turbulent termic in the stratified atmosphere due to the Archimedes force (Gossard and Hook 1975):

$$h = 1.36 \left[\frac{\Pi_0}{v^2 N^2} \right]^{1/4}, N^2 > 0$$

(23)

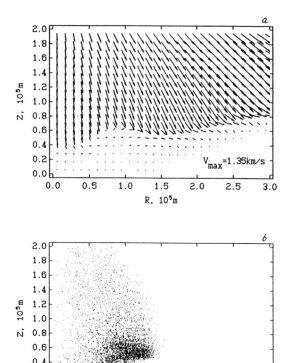

Figure 10.   (a) The velocity field at time 300 s. (b) The dust markers at time 300 s.

$$\Pi_0 = \int_0^\infty \int_0^\infty \left[\frac{\rho_0 - \rho}{\rho} \right] gr\,dr\,dz$$

where h is the height of the termic, N is the Vaisäla-Brunt coefficient, ν is the coefficient of turbulent diffusion, Π_o is the integral of buyoancy, r is radius, z is the height above the surface of the Earth, ρ is the density of the air and ρ_0 is the initial density, and g is the acceleration of gravity. Assuming that the cloud is purely gaseous, not containing any dust, and that the pressure is equal to the atmospheric pressure p_0, one obtains

$$\Pi_0 = \alpha E(\gamma - 1)/\gamma p_0 \tag{24}$$

where αE is the initial thermal energy of the cloud, and γ is adiabat exponent. Actually, the value of Π_0 is determined both by the density of the gas and the dust. Moreover the turbulence in the cloud is enhanced by the ejection processes. The amount of the dust ejected during the cratering process and entrained by the rising cloud and the coefficient of turbulence diffusion is

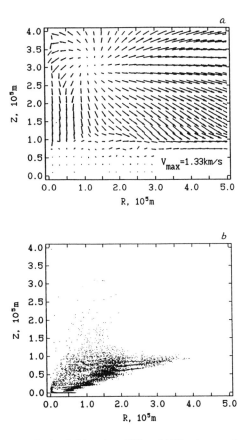

Figure 11. (a) The velocity field at time 680 s. (b) The dust markers at time 680 s.

difficult to determine purely by the theoretical methods. In the TNT explosion the process of dust formation, ejection and cratering are in many respects similar to that in the case of the impact of a cosmic body having low-velocity (10–15 km s^{-1}).

The snapshots of one such explosion with energy approximately 1 Kt TNT are presented in Fig. 12 (Adushkin et al. 1993). Using these snapshots one can easily imagine the immediate consequences of the impact of a cosmic body with much larger energy, e.g., 30 Mt (equivalent to the kinetic energy of a 100 m stony body with a velocity 12 km s^{-1}). The size of the fireball (and corresponding times) should be increased approximately 30 times. One of the main problems of the pollution of the atmosphere by such clouds is the problem of determining the maximum height. In this special case the temperature distribution above the surface of the Earth had peculiar features; due to special meteorological conditions an inversion layer was formed at the height of 0.8 to 1.5 km. Using the scaling laws of thermic dynamics (the equilibrium height h is proportional to $E^{1/4}$), one can use these experimental

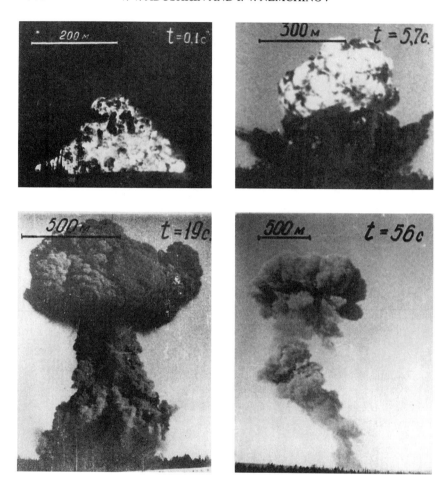

Figure 12.   The snapshots of a 1 Kt TNT explosion at t = 0.1, 5.7, 19, and 56 s.

results to predict the hydrodynamically similar evolution of the cloud caused by an impact on the surface of the Earth in the stratified atmosphere with the normal height of the temperature inversion layers (the height of the tropopause is 8 to 15 km). The height and the diameter of the cloud versus time are shown in Fig. 13. We can see that in spite of dust loading the cloud breaks through the tropopause and lifts a large mass of the dust to very high altitudes. One of the very important features of such experiments is the possibility to understand the influence of the real processes in a real geophysical medium; and we must mention the deformation and disruption of the cloud due to the winds blowing at this moment in the atmosphere.

The simple scaling law Eq. (23) for large yields can lead to large errors and numerical simulations should be used.   They can be checked by the experimental data obtained from nuclear tests.   The energy of the Tunguska

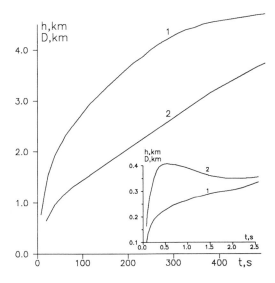

Figure 13. The height (1) and diameter (2) of cloud versus time.

event is only slightly less than the maximum energy of the nuclear devices exploded in the course of nuclear tests in the atmosphere in 1961 at Novaya Zemlya, namely 58 Mt. We see in Figs. 14 and 15 the results of the numerical simulation of gas dynamic processes in the atmosphere after that explosion at the height of 3.5 km (the numerical code is similar to the code used by Nemchinov et al. 1993c). The shock wave moves upwards, and accelerates, reaches 60 km in 2 min and 120 km in 3 min. A very interesting feature is that the air initially lifted behind the shock wave is falling back due to gravity (Fig. 15). The debris are concentrated in the vortex at the height of only 20 to 30 km. So for small bodies only the lower atmosphere is contaminated. For medium-sized bodies the upper atmosphere is also polluted.

D. Chemical Processes in the Atmosphere and Ionosphere Initiated by the Impact

The small droplets and the thin dust will stay for a long time in the atmosphere at great heights. For the impact of large bodies, the consequences of the global dust layer that would stay aloft in the stratosphere for several years was studied by Gerstl and Zardecki (1982). Solar radiative transfer calculations with the realistic optical parameters (scattering and absorption coefficients) were performed. A minimum mass of 10^{13} kg in the stratosphere was sufficient to reduce photosynthesis to 10^{-3} of normal.

The impact of a large body in the ocean would inject large quantities of water into stratosphere and lower mesosphere. The enhanced water vapor concentration would change the middle atmosphere (50–100 km) chemistry and heat budget (McKay and Thomas 1982). The increased mixing ratio of hydrogen decreases the ozone concentration above 60 km. The ozone

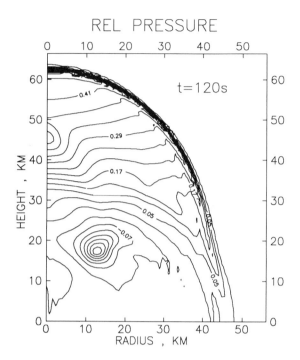

Figure 14.   The relative pressure contours at 2 min after a 58 Mt explosion at a height of 3.5 km. The values of $lg(\rho/\rho_a(z))$ are given at some of the curves; $\rho_a(z)$ is the initial atmospheric pressure at height z. The shock wave reaches a height of 60 m; the debris are located mainly at the height of 18 to 22 km.

reduction causes lowering of the average temperature. The creation of saturation conditions would have resulted in a layer of mesospheric ice noctilucent clouds.

An asteroid impact would produce large amounts of nitrogen oxides. Rapid production of NO_2 and subsequent rainout of HNO_3 and other strong acids can acidify surface waters (Lewis et al. 1982). These processes after impact of large bodies are discussed in detail in the Chapter by Toon et al. Similar processes for small- and medium-sized bodies are still to be studied. For example, the impact into water leads to the formation of a hollow water cylinder, raising very high into the upper atmosphere.

Numerical simulations of the impact of a 200 m water sphere (modeling a comet) with the velocity 50 km s^{-1} into the ocean with a depth of 4 km has been carried out (Nemchinov et al. 1993d). The hydrodynamic processes after the impact are similar to those by Ahrens and O'Keefe (1987) and Roddy et al. (1987). The maximum height of the water lip is 30 to 35 km in 30 to 40 s after the impact. The density of this upper part of this water splash is lower than is the normal density of the water; it is dispersed into small droplets. Whether they will evaporate and stay at these heights or will soon be rained

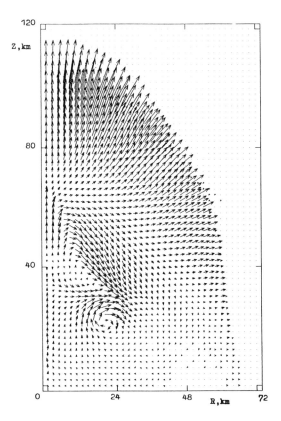

Figure 15.   The velocity field at the moment 3 min after the explosion. The shock
    wave reaches the height of 120 km, but the air behind it is falling back due to gravity.
    The debris are located mainly in the vortex at the height of about 20 to 25 km.

out is a key question to be solved.

The ocean water being evaporated releases natural salt (approximately 3%
by mass) and after dissociation, the chlorine which could severely diminish
the number density of the ozone.

Here we have briefly mentioned only the results of the release of natural
products. But different manmade products which are produced and stored
in the chemical plants and can be released after the destruction of tanks or
pipes by the shock waves or fires must also be taken into account. So we see
that very different chemical processes can be the result of the impact. The
long-term consequences caused by these injections are still to be studied.

V. WATER WAVES AND TSUNAMI

Huge tsunami waves would be generated by the impact of a cosmic body with
a sea or ocean surface; they would have long-distance propagation and cause
possible devastations.

Approximately 3/4 of the Earth's surface is covered by ocean with an average depth of 4 to 5 km. The impact into the ocean produces an intense compressive pulse in the water and a crater which is unstable and collapses. The compressive pulse and the excavation of the water crater produce a surge which is followed by subsequent waves generated by the crater slosh. Evaluating the dangerous consequences of the impacts, one must definitely take into account these deep water waves and tsunamis caused by the impact (Ahrens and O'Keefe 1987; Roddy et al. 1987; Hills and Goda 1993; see also the Chapter by Hills et al.).

A. Water Waves Caused by Impact

Numerical calculations (Ahrens and O'Keefe 1987; Roddy et. 1987) have given a very detailed picture of hydrodynamical processes caused by a vertical impact of a 10 km stony body with the velocity 20 km s^{-1} into the ocean with a depth of 5 km. The total energy of the impactor is equivalent to the energy of 6×10^7 Mt of TNT. In 120 s after the impact, the height of the rim crest of the ocean and crust is 35 km above the original water level. The height of the water wave is approximately 4 km. Smaller bodies have sizes less than the thickness of the ocean. Similar numerical simulations were carried out by Nemchinov et al. (1993d) for a medium-size comet, having the same velocity but a diameter of 2 km. The depth of the ocean was assumed to be 4 km, so it is twice as large as the diameter of the comet. At 37 s after the impact the maximum height of the water wave was 1.0 km and the location of this wave was 18 km from the center of the impact site.

The energy of the 150 m iron body moving at a velocity 20 km s^{-1} is 600 Mt. Such a body will go through the 600 m depth of sea water almost without deceleration and evaporates a large amount of water by the shock wave (Croft 1982). After striking the dense rocks at the sea bed, the body will explode and the resulting underwater explosion will somehow resemble the underwater explosion of a nuclear device.

B. Long-distance Propagation of the Water Waves

The propagation of water waves was modeled using the shallow water approximation. For cylindrical symmetry, the height of the wave is $\Delta \sim 1/r$, where r is the distance from the center. This law is in accordance with the empirical relation (Korobeinikov and Khristoforoff 1976) obtained as the result of analysis of the underwater TNT explosions. A decay of the water waves in the Bikini experiment also obeys this law (Glasstone and Dolan 1977). So we obtain the height of the wave $\Delta = 10$ m at a distance of 2000 km for a 2 km comet impacting the 4 km ocean, and at a distance of 300 km for a 150 m iron body impacting the sea which has a depth 600 m (this is the average depth of the Baltic sea). The chosen height of the water wave (10 m) corresponds to the amplitude of the most disastrous tsunamis recorded at the Kuril islands during this century (Shokin et al. 1989).

The interaction of the waves with the peculiarities of the bed topography has also been investigated by the two-dimensional numerical simulations in shallow water approximation. The results show that the height of the wave moving along a thin canyon or thin mountain ridge changes drastically. So, there is a possibility of large local and even regional increase of the disastrous consequences of the impacts at large distances.

C. Tsunami

As the water wave runs into shallows, its speed decreases and its front increases in sharpness. A huge tsunami hits the coast. The shores of sea and ocean usually are the regions of high density population and industry and this fact increases the hazards due to tsunami. Such low-lying areas as Holland, Denmark, etc., may be flooded, and large cities (like St-Petersburg) and even whole industrial regions situated there can be washed out. The problem of the evolution of the water wave into tsunami is far from completely understood. The influence of the real topography of the coastline is probably the most important factor which must be taken into account in future predictions of the hazards due to the tsunamis.

VI. CRATERING AND SEISMIC EFFECT

A. Craters and Ejecta

Let us estimate the minimal zone of destruction on the Earth's surface without resorting to intricate calculations. For this purpose we shall use data obtained by examining the Puchezh-Katunky crater which is 40 km in diameter (Ivanov 1992; Pevzner et al. 1992). It was probably created by a 2 km silicate asteroid moving with a velocity of 20 km s^{-1}. The ejecta from the crater falling on the surface and spreading out have destroyed the near-surface layers to a depth of 100 m at a distance of up to 40 km from the center of the impact. For large distances, the large boulders or even blocks and megablock falls, the blast wave and the seismic wave can also be a dangerous factor.

A more detailed picture of the cratering processes can be found elsewhere (Ivanov 1981; Melosh 1989). We shall only mention that for the oblique trajectories a simple rule of the thumb was proposed; an impact with the velocity V_Θ under the angle Θ to the horizon is equivalent to the vertical impact ($\Theta = 90°$) with the velocity $V_{90} = V_\Theta \sin \Theta$. The most probable angle of impact is 45° (Shoemaker 1962).

B. Amplitude of the Seismic Waves

For an estimate of the amplitude of the seismic waves we use the relations for maximal velocity of the displacement of the solid rocks obtained during underground nuclear explosions (Rodionov et al. 1971; Kostyuchenko et al. 1974).

$$U_m = 24(q^{1/3}/R)^{1.75} \tag{25}$$

where U_m is in cm s^{-1}, q is the energy in Kt TNT, and R is the distance in km. This formula gives the average values of horizontal velocity in body waves determined by measurements in rocks in the so-called near zone of explosion ($R < 100$ km), but with some restrictions we can use it for larger distances of the order of 1000 km.

We need to keep in mind that the explosion at the surface and the asteroid impact is less effective than the underground explosion. We must substitute into Eq. (25) a value $0.1q$ instead of q and thus we obtain a maximum velocity of seismic oscillations $U_m = 6$ cm s^{-1} for $E = 10^6$Mt at a distance of 1000 km. If we take into account the influence of the upper layers of the soft soil which usually are present in the populated areas, we need to increase this value of U_m approximately by a factor of 2. Therefore we obtain $U_m = 10$ cm s^{-1} at the distance of 1000 km from the impact site.

Following from an analysis of the destruction of usual-type buildings (Sadovsky and Kostyuchenko 1974), this value of U is a critical one; when the velocity is higher, considerable damage to buildings or even complete destruction occur.

Let us compare the seismic effect of an asteroid impact with the most destructive, catastrophic earthquakes. For an estimate of this effect we use the magnitude M in the Gutenberg-Richter scale (Sadovsky et al. 1987).

$$\lg E_s = 4.8 + (3/2)M \tag{26}$$

where E_s is the seismic energy in Joules. E_s is approximately 0.05 of the full energy of the explosion. For $E_k = 4 \times 10^{20}$ J, we obtain $M = 9$.

Similar values of M are obtained if we take the magnitude $M = 6.5$ for the most powerful underground explosion (Cannikin event at the Amchitka island in Aleutians) with energy $E = 5$ Mt (Rodean 1971) and then use the scaling law (Eq. 26). Earthquakes with larger amplitudes have not been registered during the last century. For the earthquake in China in 1920 which had a magnitude $M = 8.5$, more than 100,000 people were killed and the radius of the zone of devastation was as large as 600 km. This value is close to the value obtained by using the calculated maximal velocity. For the kinetic energy of 10^5 to 10^6Mt the area of devastation with $M = 9$ increases to 1000 km. In this region there may live as many as 3×10^7 people (for an average population density). Good steel-reinforced apartment houses and warning on possible impact can save lives.

C. Seismic Signal and Spectrum

All buildings, bridges, and other structures are able to vibrate to some extent without being damaged. Every structure and every element of a structure has many periods of vibration. For a majority of structures the most important is the longest one. The ground motion caused by an impact contains oscillations of many different frequencies. The short-period or high-frequency waves are absorbed by the ground more readily than are lower frequencies. The greater the distance from the explosion, the larger is the fraction of the seismic signal remaining in the low-frequency part of the spectrum.

The response of structures to seismic waves can be easily calculated only for idealized structures for which we know their size, shape, flexibility, natural vibration periods, dumping ratio of vibration and other structural details. However, we must take into consideration the complexity of the real structures and such processes as reflection, refraction, orientation of the structure, soil characteristics, local geological conditions. So there are different random factors.

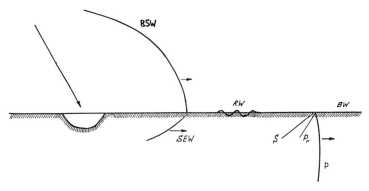

Figure 16. Wave picture. BSW = air blast shock wave; SEW = seismic wave in the ground; RW = Rayleigh wave; BW = body waves (P, reflected P_r and S).

To refine the consequences of such an event is very difficult. In Fig. 16 a simplified picture, illustrating the wave propagation in the air and in the ground and the interaction between these waves, is shown. In realty it is even more complicated. Very useful can be the results obtained during underground nuclear tests and large TNT explosions. A typical seismogram obtained after a 1 Kt TNT explosion is presented at Fig. 17.

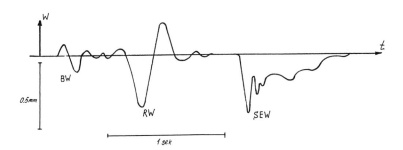

Figure 17. A typical seismogram of vertical vibrations $W(t)$ from a surface explosion of 1 Kt TNT at the distance 3 km.

If we go to much larger energies, several questions arise. One of the problems to be solved is the problem of evaluating the characteristics of the source of the seismic waves. We have to take into account the structure of the Earth to evaluate the spectrum of the seismic signal of such a mighty source and such a long trace of propagation of the seismic waves. During the nuclear tests it was established that considerable uncertainties exist due to factors that include large instabilities of the amplitude of seismic signals and the shape of the spectrum associated with propagation path effects, variations in upper mantle attenuation beneath different sites of energy release, and dependence of seismic coupling on source medium geology. Another problem to be solved is how to take into account the changes in the natural seismic regime of the region after the impact. The cratering process and the amplitude of the seismic waves or tsunamis depends very seriously upon whether a cosmic body hits the Earth's surface or explodes in the atmosphere.

VII. FRAGMENTATION AND DISRUPTION IN THE ATMOSPHERE

Large stony and iron meteoroids strike the Earth almost unchanged in shape and mass after the passage through the atmosphere. The absence of an impact crater from the Tunguska bolide shows that the atmosphere can eliminate cratering and diminish seismic effects for sufficiently small bodies. However, the famous 1 km Meteor Crater in Arizona which formed 50,000 yr ago, was made by the impact of an even smaller iron body of 30 m in diameter (Melosh 1989). Several smaller craters have been found on the Earth, e.g., the 100 m Kaali crater in Estonia formed approximately 3500 yr ago (Pirrus and Tiyurma 1987). Recently, a small crater (10 m in diameter and 3 m in depth) near Sterlitamak in Russia was formed by a very small iron body with a mass of about 325 kg (Ivanov and Petaev 1992). Though irons are constituting only 6 or 7% of the whole population of the impactors (Shoemaker 1983), we cannot absolutely exclude the possibility of the impact by such small bodies.

Relatively small objects behave in various ways during their flight, depending on their size, velocity, composition, shape and the inclination of the trajectory. The dominant physical process is disintegration due to aerodynamic forces (Passey and Melosh 1980; Melosh 1981,1989). Ablation, caused by thermal conductivity and radiation transfer from the shock-compressed air, can be neglected for bodies larger than 10 m (Nemchinov and Tsikulin 1962; Baldwin and Sheaffer 1971; Nemchinov et al. 1976; Biberman et al. 1980).

A. Simple Models of the Body's Deformation Due to Aerodynamic Forces

Heavily fragmented matter can be deformed easily and becomes similar to fluid (Grigoryan 1979). An assumption that the shape of a body is a cylinder, and under aerodynamic forces the ratio of its diameter to its height increases during the flight is often used (Melosh 1981; Ivanov et al. 1986; Chyba et al. 1993). The rate of flattening can be estimated from simple models. The pressure on the leading face of a blunt body is maximum at its critical point and decreases to the sides. The pressure gradient causes the motion of fluid particles (or quasi-fluid particles of the broken material). The velocity of the transverse motion V_t can be estimated from the following expression

$$V_t = (\rho_a/\rho_o)^{1/2}V \tag{27}$$

where V is the velocity of the body, ρ_o is its density, and ρ_a is the density of the atmosphere. For exponential atmosphere one can easily obtain the expression (2) determining critical radius. The critical diameter for aerodynamic dispersion of meteoroids in atmosphere of the Earth is 540 m, 330 m and 200 m, for icy, stony and iron bodies, respectively.

The equation of motion of the body, taking into account the increase of the effective cross section, has been calculated (Chyba et al. 1993). It was assumed that once an object spreads to, say, twice or 5 to 10 times its initial radius, its further spreading happens so quickly that it explodes. Chyba et al. (1993) predict the critical diameter to be 230 m, 90 m, and 20 m for the same objects at normal incidence. They came to the conclusion contrary to Korobeinikov et al. (1990,1992), that the Tunguska bolide was not a comet but a stony asteroid. This controversy has not been resolved as yet.

B. Numerical Modeling of Meteoroid Deformation and Disintegration

Straightforward numerical two-dimensional hydrodynamic simulation was carried out by the free-Lagrangian method (Hazins and Svetsov 1993; Ivanov et al. 1992). The Eulerean method with a special technique to track interfaces (Teterev et al. 1993) was also used. The icy cosmic body was considered as a fluid, with the equation of state of water, moving through the air of appropriate density. The calculations have revealed that the meteoroid is not significantly flattened. Instead, it gradually loses its mass due to the development of the Rayleigh-Taylor and Kelvin-Helmholtz instabilities and the blowing-off of the surface material by the air. A body tends to take a conical shape and to survive easily the passage through the atmosphere. For the velocity of 20 km s^{-1} a 200 m icy body dissipates before the impact less than 20% of its initial kinetic energy.

In Fig. 18 the density contours which approximately coincide with meteoroid boundaries are shown for different moments of time t. They were obtained by Eulerean method for a 200 m icy body with the initial velocity 50 km s^{-1}. We see the instability of the surface and the deformation of the body. The body gained a conical form and additionally a cavern at the blunt nose was formed. Before the impact a hollow channel in the body was formed. The elongated shell increased its diameter (approximately up to 300 m when the body was at the height of 2.5 km above the Earth). Later the body was almost totally disrupted into small fragments which reached distances up to 200 m from the axis (at $t = 1$ s, just before the impact). The remaining mass of these fragments is 80% of the initial mass and more than 70% of the initial energy. Inspite of the disruption of the body, a single bow shock wave comprised the fragments.

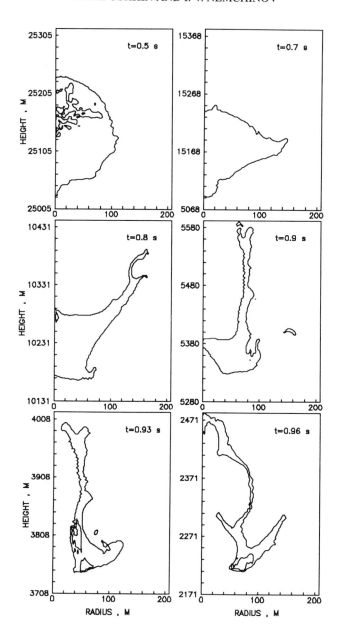

Figure 18.   Contours of dense parts of the body (with density of more than 0.8 g cm^{-3} obtained in the numerical simulation of the vertical flight of a 200 m diameter icy body with the initial velocity of 50 km s^{-1} through the atmosphere. Times are 0.5, 0.7, 0.8, 0.9, 0.93, and 0.96 s. At $t = 0$ the spherical body was at the height of 50 km above the Earth.

Similar numerical calculations were conducted for a large body with an initial diameter of 400 m. In this case the body was deformed but not disrupted and reached the surface of the Earth. Its total mass decreased only by 10%.

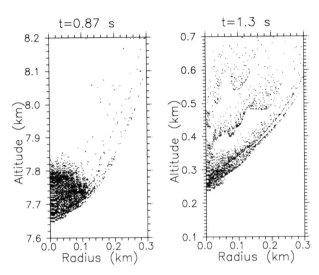

Figure 19.   Positions of representative particles in the sand bed model for two instances of the flight.   It was assumed that the meteoroid was instantaneously fragmented at the height of 25 km into 10^6 stony fragments, filling a 200 m diameter sphere, and having a velocity of 20 km s^{-1}. Fragments were divided into five groups of radii ranging from 0.1 to 10 m and average radii 1 m.

A sand bag model was also developed in which a meteoroid is represented as a conglomeration of independent particles moving through the atmosphere (Teterev and Nemchinov 1993). The particles transfer energy and impulse to the atmosphere and are enclosed by a single bow shock. This method also has shown that a heavily fragmented meteoroid takes a conical form (Fig. 19) and dissipates less energy than is predicted by simple theoretical models of disintegration. It was assumed that the meteoroid after the initial stage of fragmentation consists of 10^6 stony fragments, loosely packed in a 200 m diameter sphere with a velocity of 20 km s^{-1}. Fragments have average radii of 1 m but five groups with radii from 10 cm to 10 m were used in these simulations. Before the impact the diameter of the volume containing the main parts of the boulders increased approximately to 400 m. Due to the increase of volume of the body before the impact and the decrease of its average density, the mechanical impulse into the ground was less than for a more compact body and a large part of the kinetic energy of the body was transformed into the kinetic and thermal energy of the rising plume (O'Keefe and Ahrens 1982a).

These hydrodynamic calculations were accomplished only for meteoroides of a simple (spherical) initial shape, but many asteroids are binaries and each component can have a complicated shape. However, the simplified physical models were used (for instance, thermal radiation was not taken into account in these particular simulations), so the theory of meteoroid disintegration needs further development. Nevertheless, principal uncertainties hamper complete solution of the problem even for given mass and velocity of the body; these lie in the strength of the material, its composition and shape. Additional observations are needed to elucidate the properties of impacting cosmic bodies and to clarify the process of disintegration and fragmentation and thus make more certain the shielding effect of the atmosphere.

VIII. IONOSPHERIC OSCILLATIONS AND HEATING: ELECTRO-DYNAMIC AND MAGNETOHYDRODYNAMIC EFFECTS

The shock waves and seismic waves, the tsunami and the fires, the nuclear winter effects, etc., are now well known qualitatively (but still not quantitatively). In this chapter we do not dwell more on these effects but instead focus on new effects that have not yet been widely discussed.

A. The Explosion at Novaya Zemlya and Global Atmospheric Disturbances at Large Heights

The energy from the Tunguska event is only slightly less than the maximum energy of a nuclear device exploded in the course of nuclear tests in the atmosphere, in 1961 at Novaya Zemlya, namely 58 Mt. We see in Fig. 20 the results of the numerical simulation of gas dynamic processes in the atmosphere at 600 s after that explosion. One can see the density discontinuity located at the heights of about 120 to 180 km and radii less than 200 km. This is a reflected shock wave formed in the air previously ejected by the explosion and now falling back (V. V. Shuvalov, personal communication).

These layers of the atmosphere are ionized and, hence, the changes in the number density of electrons and in transmission of the electromagnetic waves also occur. Acoustic gravity waves are generated and they propagate a few times around the Earth. These ionospheric disturbances were registered at distances at least of 3000 km from the explosion. Of course, the same effects would be produced after the impact, but the amplitude of the ionospheric disturbances will be larger if the energy of the body is larger. We draw attention to this effect though it is not a direct threat to humanity. In this information age it can disrupt communications by radiowaves (Glasstone and Dolan 1977; Goodman and Aarons 1990). Such disruption (between different regions of one country or between different countries) may have dangerous consequences, when it occurs after catastrophic destruction of a large region.

If we increase the energy of the explosion, the velocity v of the air ejected upwards also increases. It can be estimated from a simple relation

$$v = (2\alpha E_k/(M_a + M_v))^{1/2} = V(\alpha M_b/(M_a + \omega M_b))^{1/2} \qquad (28)$$

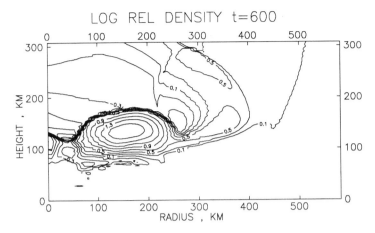

Figure 20. Plot of constant relative density contours at $t = 600$ s after a 58 Mt nuclear explosion. The values of $lg(\rho/\rho_a(z))$ are given on some of the curves; $\rho_a(z)$ is the initial air density at height z. Sharp changes in density coincide with the location of a shock wave.

where M_b is the mass of the body and M_a is the mass of the air above the point of the impact defined by Eq. (3). The maximum height Z_m of ejection can be estimated from the simple ballistic law

$$h = Z_m - Z_e = v^2/2g, \quad t_f = 4h/v \qquad (29)$$

where g is the acceleration of gravity and Z_e is the height of ejection. Having reached the maximum height the ejected atmosphere falls back, and its velocity is close to the ejection velocity (only the sign is changed). The time t_f is needed for the air parcel to rise to Z_m and then fall back. The atmosphere begins to oscillate and a shock wave is formed (Nemchinov et al. 1992; Nemchinov and Loseva 1994). The flow pattern resembles a fountain. At the height of ejection the radius of this fountain is $R_f = 2h$ and the cross section $S_f = 4\pi h^2$. The specific mass of the falling air $m_a = M_a/S_f$. For a 600 m body, assuming $V = 50$ km s^{-1}, and $\alpha = 0.3$, we obtain $v = 4$ km s^{-1}, $h = 800$ km, $R_f = 1600$ km, $t_f = 800$ s, and $m_a = 0.015$ g cm^{-2}. The air is cooled during the adiabatic phases of ejection and expansion and then is heated above the initial temperatures in the shock wave. The larger the energy of the body, the larger is the amplitude of oscillations, their area, and the temperature behind the reflected shock wave. It may reach several thousand degrees, so if $M_a >> M_b$ and the velocity of ejection is less than the escape velocity, the energy of the airblast and the evaporated masses of the body and of the surface layers of the Earth are retained in the upper atmosphere; and these layers have a low mass, so global changes of the chemistry and optical properties in the upper atmosphere may be the result. The long-term consequences of these effects are to be studied in the future.

B. Electrodynamic and Magnetohydrodynamic Processes Initiated by a High-speed Jet Moving Through the Atmosphere

For vertical impact the wake formed behind the body facilitates the ejection process. In the case of a medium-sized body (with typical size of 0.6 to 2 km) or larger bodies, the atmosphere does not decrease sufficiently the velocity of the plume. The impact with a velocity $V \sim 50$ km s^{-1} on the Earth's surface leads to the full vaporization of the body and of a significant part of the substance of the Earth's upper layers and even to the ionization of this vapor cloud and of the air behind the shock wave. As a result, a hypersonic jet of air and erosion plasma is formed. In Figs. 21 and 22 the temperature contours are shown for the case of a 2-km model SiO$_2$ projectile impacting the SiO$_2$ model Earth (these compositions are chosen for simplicity because the equation of state of SiO$_2$ is well known; more sophisticated analysis may be performed with more realistic compositions). The energy losses due to thermal radiation from a plume are not very high, so we can use the hydrodynamic similarity: all the moments of time and distances are proportional to the body size. For the initial velocity of the body 50 km s^{-1} the average velocity of the rising plume was 25 km s^{-1} (this corresponds with the calculations for different substances in the so-called "plane approximation" by Vickery [1986]). At $t = 0.6$ s the temperature in the SiO$_2$ plume is high enough—more than 1.8 eV. Due to the expansion of vapor its temperature is falling but at $t = 2$ s the temperature in the core of the plume is still high enough (approximately 1 eV), so it is ionized and still can interact with the magnetic field. Moreover in the case of the plume expanding into the atmosphere of the Earth, the shock wave in the air is formed with a temperature in the range of several eV. The ionized air behaves like a conducting piston and magnetohydrodynamic and electrodynamic effects are the consequences.

The kinetic energy E_j of the jet for a medium-sized body is far above the total energy of the geomagnetic field of the Earth (approximately equivalent to the energy of 200 Mt). The jet will propagate practically inertially with the constant mean velocity U. The interaction of this plasma jet with the Earth's magnetic field causes effects similar to those that are produced by cosmic nuclear explosions (Glasstone and Dolan 1977), but on a larger scale.

The electric field of polarization and electric current through the iono-sphere form a giant MHD-generator transforming the kinetic energy of the jet E_j into the thermal energy of the ionosphere Q (Nemchinov et al. 1993a). A coefficient of conversion $\zeta = Q/E_j$ is approximately equal to the following expression

$$\zeta = U^2 B^2 t^3 / (V_A M_j) \tag{30}$$

where B is the inductivety of the geomagnetic field near the Earth's surface and V_A is the effective Alfvén velocity in the ionosphere connected with the characteristic integral surface Alfvén conductivity of the ionosphere; $\sum_A = c^2/(4\pi V_A) \sim 1$ to 10 Ohm^{-1}; c is the velocity of light. Let us take for the characteristic time of interaction of the plasma jet with the magnetosphere

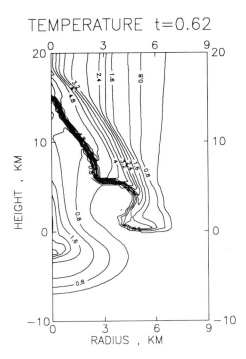

Figure 21.   Plot of constant temperature contours after a vertical impact of a 2-km sized SiO_2 body at a velocity 50 km s^{-1} on the Earth at $t = 0.62$ s. Sharp changes in the temperature coincide with the location of the interface between silicate vapor and the air. The shock wave in the air can also be seen above this interface. Maximum temperature in the vapor is 1.8 eV; in the air it is more than 4.8 eV.

$t = R_E/U \sim 600$ s, where R_E is the Earth's radius (by this time the interaction of the jet with the Earth's magnetic field is already global and the reduction of the geomagnetic field at the shock wave front takes place). Hence $\zeta \sim 10^{-3}$ to 10^{-2} and the total thermal energy release in ionosphere $Q \sim 10^{17}$ to 10^{18} J or 2×10^2 to 2×10^3 Mt of TNT. This energy release is greater then the thermal energy of the undisturbed ionosphere and the energy of the geomagnetic field at heights of more than 100 km.

The substantial ionospheric heating and the deformation of the magnetosphere can lead to different immediate consequences; for example, the destruction of the ozone layer, the disruption of the Van Allen radiation belts and precipitation of the trapped energetic particles from them, the large amplitude oscillations of the ionospheric and magnetospheric plasma, with the interruption in transmission of electromagnetic signals of different wavelengths (an important factor in our information age especially in the case of a catastrophic impact), the magnetic storms of a very high intensity to name a few. Long-term consequences, including biological, have yet to be evaluated.

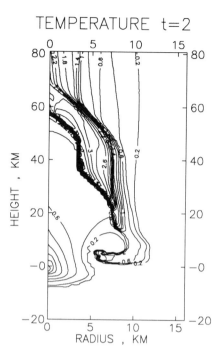

Figure 22.   Plot of constant temperature contours after a vertical impact of a 2-km silicate body at $t = 2$ s. The maximum temperature in the vapor is 1.1 eV and in the air it is more than 3.1 eV. The ionized jet interacts with the Earth's magnetic field.

C. The Role of the Wake for the Oblique Impact

The numerical simulations for the vertical impact have shown that the wake drastically increases the velocity of the plume. Although vertical impact is a convenient object for numerical simulations due to cylindrical symmetry, in real situations an impact is most likely to be oblique. Relatively simple three-dimensional hydrocode has been used to simulate the impact of the same asteroid falling with inclination angle 45° onto silicate Earth covered with exponential atmosphere (Artem'eva and Shuvalov 1994). The density contours are presented in Fig. 23. Due to the horizontal component of the body's momentum, the fireball has moved from the point of impact in the direction of the body's motion decreasing the mass and energy escaping through the wake. So for the oblique impacts with small angles of inclination the nonuniformity of atmosphere becomes the main factor determining the velocity of the atmospheric ejection and the estimates which do not take into account the wake are more appropriate. But a large program of further investigations is needed to obtain definite quantitative results for oblique impacts.

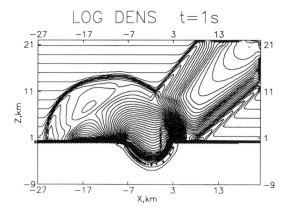

Figure 23.   Density contours at 1 s after the oblique (45°) impact of a 2-km sized body on the Earth.

IX. DANGEROUS OBJECTS ON THE EARTH'S SURFACE

After the 1920 disaster in China, especially after 1945, new problems arose. On the surface of the Earth hydroelectric dams, chemical plants, and nuclear power stations, and nuclear waste repositories are being built; related destruction of them may lead to severe consequences (*Proc. III Intl. Conf. on the Safety of Nuclear Power*; Melnikov et al. 1992).

A. Different Types of Dangerous Objects

Nuclear power plants, dangerous chemical plants producing poisonous substances or those which can pollute the environment if they are released, and radioactive wave repositories are some of the objects that must be protected from small asteroids or their fragments which could hit them. Either they must be protected by active means (e.g., by destroying asteroids with nuclear devices or ballistic weapons, or by altering the trajectories of such cosmic bodies), or by passive means (e.g., by reinforcing the structures of the objects, by changing their design, for instance, by placing them underground). Of course, these means are to be recommended only if there is a sufficiently large probability of such an accident and the hazards are large enough. We shall not describe here the usual structural damage from air blast (which can lead to deflection or even the full collapse of steel frames, brick walls, concrete and reinforced concrete-frame buildings, roofs, etc.). It may be excluded by placing the dangerous object underground. However, the directly coupled ground motion is a significant damage mechanism. Pipelines, tunnels, subways, underground structures all may collapse due to deformation and rupture.

B. Radioactive Waste Repository

The radioactive waste repository is a disposal system the objective of which is to isolate the radioactive substances it contains from the biosphere during the period when a significant hazard remains, within acceptable limits. The isolation capability of the repository depends on a system of barriers between the radioactive wastes and the biosphere. There are two groups of terrestrial repositories: near-surface disposal facilities and geologic disposal facilities. For disposal of low- and intermediate-level radioactive wastes, near-surface repositories have been operating for several decades, for example, in France, United Kingdom, the former U. S. S. R. and the United States.

There are national and international programs which have investigated the feasibility of a geological confinement system. Various geological media (e.g., salt, crystalline rocks such as granite, clays and tuff) and various designs have been investigated. Properly selected sites in aseismic regions would ensure that the wastes would not reach the biosphere for at least several thousand years. Several countries have already built the repositories (Belgium, Germany), in some others the repositories are under construction (Sweden, France). The U. S. A. is now trying to establish such a site in tuff (Yucca Mountains, Nevada) suitable as a repository. Other countries plan to build industrial repositories in the next 15 to 50 yr. We must underline the very long design time of these repositories.

C. Nuclear Power Plants

In 1990 there were 435 nuclear plants operating in the world in both industrialized and developing countries, supplying 17% of the world electricity needs. However, the Chernobyl accident clearly showed the risk associated with nuclear power and the international regime for nuclear safety is now being established. The probabilistic safety assessments identify the combination of events that, if they occur, will lead to a severe accident; it assesses the frequency of occurrence for such combinations and the consequences of those events. One must keep in mind that the design life of these plants has been set for a specific period, namely 40 to 50 yr, following past experience.

Let us evaluate the probability of an accident due to impacts. The surface of the Earth is $3 \times 10^8 \text{km}^2$. If the nuclear plants were distributed uniformly over the entire surface, the average surface area to each of these stations is $0.8 \times 10^6 \text{km}^2$ and the average radius of the area is 600 km. Of course, the number of these stations could be increased in the foreseeable future (a hundred years) by a factor of 10 reducing the radius to 200 km. If this radius is equal to the radius of a failure of the structures of the power plant by some of the factors, then each of the impacts will lead to the catastrophe which is of the order of magnitude of Chernobyl (100 millions Curie released) or greater.

If we take as a criterion of destruction of the construction of only the groundbased nuclear power station, the velocity of the soil 1 m s^{-1} leads not only to disruption of the joints and pipelines and leakage of radioactivity but even to the collapse of some walls and roofs. Then the distance for the impact of a 1 km body would be 200 km. As for the chemical plants constructed on the surface, the critical velocity is smaller ($10\text{--}15$ cm s^{-1}) and the distance is 1000 km.

D. Active and Passive Methods of Defense

If we change the mode of construction from groundbased plants to underground, then the distance would be less. For reinforced underground constructions the criterion is 10 m s^{-1} and the size of the destruction zone would shrink to 40 km. So we see that there are possible passive ways of diminishing the hazards of comets and asteroids, and not only active methods (destruction of asteroids and comets, alteration of their trajectories) could be used. The seismic waves are the only dangerous factors only in regions far from the shores of the seas and oceans or high enough above the water level. But nuclear power stations (and chemical plants) are often built near the seashore because it is easier to transport the heavy equipment to the construction sites, e.g., there are projects of construction of several underground nuclear power stations at Kolsky peninsula (Melnikov et al. 1992). We must keep in mind the hazards due to tsunami (see the Chapter by Hills et al.).

The main problem for the civil defense programs or for the construction of deeply buried chemical and nuclear power plants and for the reinforcement of dams and nuclear waste repositories is not only a scientific and technical one. It is also the problem of evaluating the real probability of such impacts by means of determination of a size-frequency distribution of cosmic bodies in space or tracking some of the Earth crossers with definite trajectories.

X. INVESTIGATION OF THE CURRENT IMPACTS OF METE-OROIDS ON THE SURFACE OF THE EARTH, THE MOON, MARS AND VENUS

There are at least two difficult problems arising when one tries to evaluate the real threat to humanity of the possible impacts of asteroids and comets on the surface of the Earth. One of the problems is to check the validity of the numerical simulations of physical, hydrodynamics and geophysical processes initiated by the impact, the simulations which have been already conducted to predict quantitatively the consequences of the impacts and are to be conducted in the future. One can use the geological data obtained in the course of the investigation of the natural events and data obtained in the course of nuclear tests. But the data related to atmospheric processes are almost completely absent; for instance, they are absent for the following processes: shock waves breaking through the nonuniform atmosphere, thermal radiation emitted during the flight of the body through the atmosphere and in the course of propagation of the shock wave to the high altitudes after the impact, spreading of the cloud of debris, etc. These processes have not been sufficiently investigated.

One can use the laboratory experimental simulations of these problems, and this is an effective tool of modeling. The description of these methods deserves a special chapter. We must mention for instance the modeling laboratory experiments on magnetohydrodynamic and electrodynamic effects which has recently begun (Nemchinov et al. 1993*a*); but it is difficult to satisfy the rules of the similarity for hydrodynamics processes, radiation heat transfer, and for other physical processes, keeping in mind the huge differences in scale.

Another problem is to determine the probabilities of such impacts for the cosmic bodies of different sizes, velocities, composition, and of other properties of these bodies. Such data are also necessary for planning measures to mitigate these hazards, and different measures of passive and active defense.

Several methods of obtaining necessary data are well known. First of all there is the study of the craters on the surface of the Earth and other planets, the Moon and other satellites. This has been summarized in the Chapters by Shoemaker et al., Neukum and Ivanov, and Rabinowitz et al.

A second method is the optical observations and tracking of these bodies by means of radars. These instruments now have necessary resolution for bodies with sizes ranging from 10 to 100 m (Gehrels 1991). As shown above, the impact of these small bodies can lead not only to local, but to regional catastrophes and in some cases may lead even to global consequences. The striking results of these observations are not only in the discovery of several near-Earth objects at very small distances from the Earth, some of which barely missed the Earth (Scotti et al. 1991; Gehrels 1993; Rabinowitz et al. 1993). The discovery rate of small Spacewatch objects having diameters less than 100 m is 10 to 100 times higher than that obtained via extrapolation of the mass-number distribution of larger objects. As has been emphasized by Gehrels (1993) there are 10^8 near-Earth asteroids with $D > 20$ m (see also the Chapter by Carusi et al.) and only 10^4 will be discovered and have their orbits precisely determined in 20 year effort.

Another method has been proposed recently (Melosh et al. 1993; Nemchinov et al. 1992), namely, to register the current impacts of such bodies on the surface of other planets and on the Moon or even to register the impact of smaller bodies (from 1 m to 10 m) and thus to obtain a size-frequency curve in the range of larger bodies by means of extrapolation and fill the gap between the data obtained in the course of optical observations of very small and very large cosmic bodies. Of course, this method does not exclude studying the craters formed after intrusion of large bodies into the atmosphere of Venus; as for the very dense atmosphere of Venus the threshold of ablation and aerodynamic breakup is much higher than for the Earth (Ivanov et al. 1992; Schultz 1992). Estimates for the near side of the Moon show that the objects 1 m in diameter strike the surface 3 times a year or once in 10 yr, depending on the assumed size-frequency distribution. This low rate makes the realization of such a method a very difficult problem—the duration of the observations must be very long. We hope that we have enough time before the following Tunguska fall occurs on the Earth and can use bigger targets (Mars, Venus and the Earth itself) to increase the rate of the impact and make the observations feasible. These method being realized in some form of special networks, including groundbased stations, and artificial satellites in a decade or two of operation can give valuable information about the impact rates of bodies with different velocities, size, shape, strength and composition, and on the hydrodynamic and physical processes which follow the entry of the body into the atmosphere, including fragmentation, disruption, radiation and formation of the wake, and the impact of the body, including cratering, ejection, airblast, radiation, dust cloud evolution, etc.

The proposed investigation by a special network during a decade or two can substantially increase our knowledge on the size-velocity distribution of the small cosmic objects in space, on their composition and at the same time our knowledge of the physical and hydrodynamical processes caused by the impacts, on the equation of state and on the optical properties of the substances of the cosmic bodies and of the substances of the upper layers of the Martian surface and of the lunar surface.

A. The Similarity of Hydrodynamical Processes for Impact on the Earth and Other Planets

The density of the Martian atmosphere ρ_a is 100 times less than of the air and from the relations (2) and (4) it follow that the critical radius $R_b^* = 30$ m. The process of the fragmentation and of the subsequent impact of the disrupted body on the Martian surface in some respects are similar to the process of fragmentation and of the impact on the Earth but for sizes 5 to 10 times smaller. Thus taking into account the numerical simulations described in Sec. VII.B, we conclude that the bodies with diameters 20 m or more are striking the Martian surface without being decelerated or seriously dispersed, and the bodies with the diameters 2 to 10 m probably are fragmented but the fragments would not fall far from one another and a common crater and a common fireball would be formed.

If we change the radius of the body R_b with the density of the atmosphere ρ_a in accordance with the relation (4) $(R_b \sim \rho_a^{1/3})$, then we obtain that the radius of the airblast is invariant but the radius of the wake near the body R_a and the ratio R_s/H do not remain the same and thus the strict similarity is violated; the ratio of the radius of the hole in the atmosphere to its characteristic scale at the moment of the impact in the case of Mars is smaller in comparison to that for the Earth.

A hot thermal layer at the surface of the planet may be formed due to the absorption of the light emitted by the fireball of the airblast. If we assume that the radius R_{HL} of this heated layer is determined by the energy of the thermal radiation per unit area of the surface exceeding some definite value Q_{HL} and assume that the thermal energy is proportional to the energy of the airblast, then in the case of perfect visibility we obtain $R_{HL} \sim \sqrt{E_a/Q_{HL}}$ or $R_{HL} \sim R_a^{3/2} \sim V^3 R_b^{3/2} \rho_b^{1/2}$. Changing the R_b proportionally to $\rho_a^{1/3}$ and retaining the same velocity V, one obtains $R_{HL} \sim \rho_a^{1/2}$. And the radius of the fireball remains the same. It is difficult to study the thermal layer effect by observing the development of the blast wave and of the fireball after the impacts of very small objects on the Martian surface, but processes of atmospheric breakthrough can be studied by such observations.

The numerical simulations (Nemchinov and Shuvalov 1992,1993; Nemchinov et al. 1993c) have been performed to check these considerations. The two-dimensional radiative hydrodynamic problem has been solved taking into account the formation and the expansion of the wake and the existence of the thermal layer at the surface.

In Fig. 24 the contours of constant pressure are shown for the impacts of the icy bodies with a velocity 50 km s^{-1} and the radii r_b = 10, 30, and 100 m and for times 1, 3, and 10 s. One can see that the larger is the radius of the body r_b, the larger is the influence of the stratification and of the wake, and the shape of the disturbed zone is farther from being spherical. Using the similarity rule $R_b \sim \rho_a^{1/3}$, one obtains that each of the above mentioned impacts on the surface of Mars is hydrodynamically similar to the impact on the surface of the Earth for the bodies with the radius R_b = 50, 150 and 500 m, respectively, and the same velocity. This was confirmed by the numerical calculations of a radiation hydrodynamic problem for the 200 m diameter body impacting the surface of the Earth with velocity 50 km s^{-1} (one can compare the shape of the shock wave in Fig. 24 with that in Fig. 2). As to the late-stage dissipation of the cloud, we must keep in mind that the atmospheric circulation processes in the middle atmosphere of Mars and the Earth are not identical (Leovy 1985; Zurek et al. 1992) but in both cases the initial disruption of the cloud is mainly due to the gradient of the wind velocities with the altitude.

B. The Similarity of the Hydrodynamical and Physical Processes for Impact on the Moon and Earth

As to the airless Moon, the gas dynamical and physical processes of the impacts of small bodies resembles very much the processes initiated by the impact of the very large cosmic bodies on the surface of the Earth. This is true at least at the initial stages of the impact. It is clear that if the speed of the body striking the Moon is high enough and the gravity has no serious influence on the expansion of the hot plume, the maximum velocity of the gas cloud does not decrease in the course of the expansion. The same holds for the large impacting bodies on the Earth, when atmospheric interaction processes are negligible. The simple geometric similarity rules can be applied except for the late stages of the excavation process. The necessity for these observations of the lunar impacts are caused by the uncertainties in the composition and the density of the impacting bodies. We must not only register the temperature of the vapor and the size of the vapor cloud but also the spectral characteristics of the emitted radiation.

C. Detection of the Impacts on the Moon and Mars

Discussing the problem of registration of the current impacts on the sur-

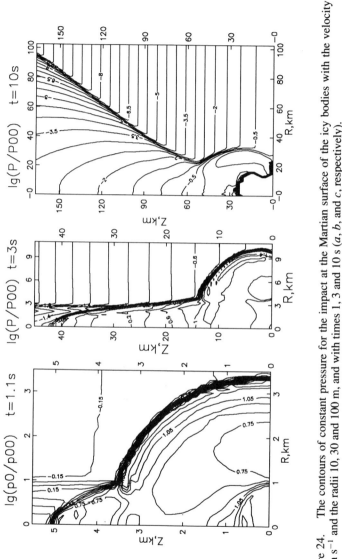

Figure 24. The contours of constant pressure for the impact at the Martian surface of the icy bodies with the velocity 50 km s^{-1} and the radii 10, 30 and 100 m, and with times 1, 3 and 10 s (a, b, and c, respectively).

face of the Earth and on the lunar surface, not only the obvious optical and spectroscopic methods through the telescopes from the Earth or special satellites can been proposed but also such methods as registration by radars of the ionospheric perturbations due to the long-range propagation of acoustic-gravitational waves in the Martian atmosphere (Nemchinov et al. 1993c). These latter methods do not demand the observation of the whole surface of the planet at once but give the possibility of registration by the satellite which is located far from the impact site at the moment of the impact.

As for impacts on the Moon, the resulting flash is brief; only a few milliseconds for a 1-m body. The radiative effects on the Moon are different from those on planets having atmospheres. In spite of high initial temperatures, the emitted radiative energy in lunar impacts is lower than the kinetic energy of the projectile due to the large optical thickness of the plasma plume. By the time that the plume is rarefied enough for the radiation to escape, most of the thermal energy is converted into kinetic energy. On planets with atmospheres the plume's kinetic energy is converted back into thermal energy of the atmospheric gases. However recording such flashes on the Moon is well within the capability of modest telescopes on the Earth's surface (Melosh et al. 1993). At the same time all these investigations can help to resolve the problem of the evolution of the atmosphere of the Earth and other planets (Ahrens et al. 1989).

D. Detection of Meteoroid Entry into the Earth's Atmosphere

The mass influx of the cosmic bodies onto Earth (Ceplecha 1992) probably has several maxima. The biggest is for the body's mass $\sim 10^{14}$kg. The second biggest reaches almost the same value and it belongs for bodies with masses of 1 Kt or radii approximately 10 m. These bodies are the least known in the solar system. For bodies less than approximately 0.1 m the statistics are large enough due to the observations of meteoritic entry into the atmosphere and there have been several attempts to evaluate the influx (McCrosky and Ceplecka 1969; McCrosky et al. 1978,1979). The mass of the body was obtained by means of the high-speed photography and the so-called physical theory of meteoroids (Bronsten 1983). One of the equations of this theory is the equation for the optical luminosity I of the meteor

$$I = \tau(-\frac{\mathrm{d}M}{\mathrm{d}t})V^2 \tag{31}$$

where τ is the dimensionless luminosity coefficient. This equation is written assuming that the luminosity is proportional to the kinetic energy of ablated mass and is defined by the rate of ablation rather than by the rate of energy deposition into the atmosphere.

The value of the coefficient τ usually used for determination of the so-called photometric mass of the meteoroid was obtained from the experiments with artificial meteors (Givens and Page 1971). But these experiments were carried with very small mass bodies and such values as well as the Eq. (31)

itself are unacceptable for large meteorites, where quite different radiation hydrodynamic regime take place: a main portion of thermal radiation is emitted by the air heated and compressed in the shock wave. There were attempts to compare observations of deceleration, ablation and luminosity for some bolides (ReVelle 1979; ReVelle and Rajan 1979; Nemchinov et al. 1989) using some data for meteors from the catalog (McCrosky et al. 1978,1979) and for meteors having been detected by a Prairie network and later found on the Earth's surface. Traditionally defined photometric mass of a body is strongly distinguished from its real mass (Nemchinov et al. 1989).

A number of satellites equipped with optical sensors that are now scanning or viewing the Earth are optimized for the detection of nuclear bursts in the atmosphere. But their instruments have a capability to detect and locate atmospheric entry above a certain luminosity threshold (Reynolds 1992; Jacobs and Spalding 1993; see also the Chapter by Tagliaferri et al.). On October 1, 1990 two geostationary satellites recorded a very intense flash over the western Pacific. The companion locator provided a very precise geographical position and altitude (30 km at peak signal). The data were obtained by non-imaging transient radiometers. The peak radiant intensity then corresponds to a source power of approximately 3.5×10^{11}W per steradian. Taking into account the above mentioned considerations, we shall not even try to obtain the mass of the body by using Eq. (31).

Assuming the emission of this fireball was isotropic the thermal energy radiated was approximately 2.2×10^{12}J. This energy is equivalent to 0.5 Kt of high explosives. If we assume 10% of the kinetic energy was converted into the energy of the emitted radiation and detected by the radiometer, the kinetic energy $E_k = 5$ Kt, and for a velocity of the body V from 20 to 30 km s^{-1}, one obtains that the mass M was 50 to 100t. If this was a stony body ($\rho_b = 2.7$ g cm^{-3}), its size R_b was approximately 2 m. There was a sharp increase in the intensity of the light during 0.4 s which corresponds to the distance along the trajectory 12 km for the velocity 30 km s^{-1}. So the body begins to radiate intensely at the height of approximately 40 km.

Numerical simulations assume the short time energy release (the explosions) at a height of 30 km were conducted by the method similar to that used by Nemchinov and Svetsov (1977) and by Nemchinov et al. (1993e). The energies of the airblast E_a were assumed 1 Kt and 10 Kt, and different initial specific energy ε have varied from 10 to 2500 kJ g^{-1} (they correspond to the velocities $V = \sqrt{2\varepsilon}$ from 5 to 70 km s^{-1}). Simulation shows that for velocities $V > 10$ km s^{-1} the total radiation emitted during 1 s is 20 to 30% of the total energy E_a. Approximately 1/2 of this energy is emitted in the spectral range 1 to 3 eV which was registered by the radiometer. And thus the energy $E_k = 3$ to 5 Kt is found, but the calculations show that for instant energy release, the duration of the maximum intensity radiation impulse is much shorter than was observed (<0.1 s). This is probably due to the fact that in reality the energy is released not instantly, but gradually during the flight.

For the 4 m diameter body the thickness of the layer of hot and compressed air is close to 0.2 m. The average optical thickness of such layer with a density 10^{-4} g cm^{-3} and temperature 2 to 5 eV is approximately 0.7 to 1.0. Its spectrum is close to the spectrum of a blackbody and we can use the results of numerical simulations presented in Table I (Sec. III.E). Such estimates give us the value of power source approximately equal to the observed for a 4 m body. They were confirmed by numerical simulations by Svetsov (1994).

In further investigations of such rare events of sufficiently large meteoroids, it is desirable to have additional information on the velocity of the body, inclination of the trajectory and the spectrum emitted. There is a specific type of hazards due to comets and asteroids in the world in which the number of nations possessing nuclear weapons is increasing—the chance that the effects of atmospheric entry and impact would be mistaken for an attack by a nation capable of nuclear response but not sophisticated enough to discern the natural event from a nuclear explosion (Gehrels 1991).

The process of radiation emission by cosmic body during its flight needs further investigation to obtain definite evidence as to whether it is a nuclear explosion or the entry of a meteoroid.

XI. CONCLUSIONS

We list the following conclusions.

1. The small cosmic body with kinetic energy E_k as low as 30 Mt (the Tunguska event), impacting surface or exploding in the atmosphere, can cause devastation of a large city($\sim$25 km in diameter) and the casualties may be as high as several million persons. Assuming that the impacting body has missed a densely populated city and the population in the region of the impact is the average population the casualties are approximately 2×10^4 persons.

2. The area A_s of devastation by the shock waves increases with the energy of the impactor more slowly than from the simple hydrodynamic similarity law ($A_s \sim E_k^{2/3}$). There are at least two factors decreasing the overpressure behind the shock wave: (i) the breakthrough effect due to air density decreases with the altitude; and (ii) the wake is formed behind the approaching body—the rarefied channel through which some part of the energy and the mass can escape the dense layers of the atmosphere.

3. The zone of fire ignition in the case of 30 Mt impact is lower than the zone of the destruction by the shock wave. For larger bodies, the role of the thermal radiation is probably underestimated due to the fact that the nonuniformity of the atmosphere and the wake drastically change the shape of the fireball—it becomes elongated rather than spherical. The thermal radiation is mainly emitted from large heights (30 to 60 km or even more) where the hot air becomes sufficiently transparent. The emitted light is transmitted through the layers of rarefied air where visibility is perfect enough. For the energy of the body 10^4Mt and 10^5Mt the area

of the fire ignition has a radius of approximately 400 km and 1000 km and the number of people who live in the region is 3×10^6 and 3×10^7, respectively (if the density population is equal to the average density population on the Earth). The latest number is of the order of magnitude of losses in the World War II. We suppose that these amounts must be assumed as prohibitive (the above mentioned energies correspond to the kinetic energy of a comet with diameters 0.5 and 1 km).

4. Dust clouds formed after the impact of cosmic bodies with kinetic energy larger than approximately 30 Mt are rising through the tropopause and the dust is lofted into the stratosphere (and even higher) where it would remain for a long time. The impact of a comet or an asteroid into the ocean or sea leads to the formation of a hollow water shell rising to a very large height (for a 200-m comet, the water lip reaches the altitudes of 30–35 km in 30–40 s). The injection of large amounts of water and especially salted water of seas and oceans may change the chemistry of the middle atmosphere.

5. Impact of asteroids and comets into the water causes big water waves and tsunamis when these waves reach shallows. The height of the wave is about 10 m at the distance of 2000 km from the site of impact of a 2-km diameter comet into the 4 km depth ocean and of 300 km for the 150 m iron body impacting the sea which depth is 600 m.

6. The earthquake in China in 1920 of magnitude 8.5 was the cause of 10^5 deaths in the area with the radius 600 km. Seismic waves caused by the impact of the body with the kinetic energy approximately 10^5 to 10^6 Mt causes an earthquake with a magnitude 9.0 and the zone of devastation is approximately 1000 km. In this area as much as 2×10^7 persons may live if the density population is equal to the average density on the Earth's surface.

7. The shielding effect of the atmosphere has probably previously been overestimated, and the medium-size objects (more than 200–400 m in diameter) strike the Earth almost without deceleration and loss of mass, although being deformed.

8. The ejection of the atmosphere by the large-scale airblasts caused by the impact leads to large-scale and global-wide oscillations of the atmosphere and ionosphere. This leads to heating of the upper atmosphere and to high intensity electromagnetic disturbances. High-speed plasma plumes interacting with the geomagnetic field initiate the magnetic disturbances propagating up to the radiation belts and increase the precipitation of the trapped particles.

9. Among the various consequences of the impacts of the cosmic bodies on the surface of the Earth new effects should be taken into account. Some of them are the following:

(a) Destruction of nuclear power stations, nuclear waste repositories, hydroelectric power stations and dams, and chemical plants by fires, shock waves, tsunamis, cratering processes and seismic effects.

(b) Pollution of the upper atmosphere by the man-made chemical products; released by the impact.

(c) Ionosphere response after the event; communication fall into disuse at radiowaves of different wavelengths which can lead to dangerous consequences in the information age.

(d) Magnetic storms of very high intensities.

10. The theoretical investigations should be supplemented by laboratory modeling experiments, especially of some of the new effects, and by observations of current impacts of small objects on the Moon, the Earth and other terrestrial planets. The results of the theoretical investigations could be refined by using more realistic physical models and more sophisticated codes. The preliminary results described above show that the main efforts in these investigations are to be exerted in the energy range between 10^3 and 10^5Mt or in the size range between 200 m and 1 km, although we must not exclude the energies as low as 30 Mt and as high as 10^6Mt.

11. There are a large number of factors making strict predictions very uncertain: the difference in the population density in different regions, in the types and structures of buildings, in the geology of different impact regions, in the configurations and bed topography of the oceans and seas, in weather conditions, the shape, strength, density and composition of the impactor, the inclination of its trajectory, etc. In future investigations some of these factors will be varied and then the result will be averaged, taking into account the probability of the impact into the ocean or onto the Earth's surface, and the probability of the velocity and the inclination of the trajectory, etc.

Further investigations of the consequences of the impacts on the Earth together with further observations and even direct investigations of the near-Earth-objects (their orbits, sizes, shapes, densities, composition, strength, etc.) can facilitate the achievement of such a stage of the problem under consideration. Then sufficiently well-founded recommendations for adopting some measures of active and/or passive defense from the hazards due to comets and asteroids could be outlined and adopted by the scientific community and by the governmental bodies of the United Nations.

Acknowledgments. We thank J. Hills, K. Zahnle and an anonymous referee for their helpful criticism, M. S.Matthews for significant improvement of language and style of the chapter, and T. Gehrels whose great efforts made possible the participation of Russian scientists in this book.

REFERENCES

Adushkin, V. V., Garnov, V., Divnov, I., Nemchinov, I., and Khristoforoff, B. 1993. Modeling of the evolution of the gas-dust clouds caused by asteroids and comets impacts. *Dokladi Rossiiskoi Akademii Nauk* 332(1):85–88 (in Russian).

Ahrens, T. J., and Harris, A. 1992. Deflection and fragmentation of near-Earth asteroids. *Nature* 360:429–433.

Ahrens, T. J., and O'Keefe, J. 1987. Impact on the Earth, ocean and atmosphere. *Intl. J. Impact Eng.* 5:13–32.

Ahrens, T. J., O'Keefe, J., and Lange, M. 1989. Formation of atmospheres during accretion of the terrestrial planets. In *Origin and Evolution of Planetary and Satellite Atmospheres*, eds. S. K. Atreya, J. B. Pollak and M. S. Matthews (Tucson: Univ. of Arizona Press), pp. 328–385.

Alvarez, L. W., Alvarez, W., Azaro, F., and Michel, H. 1980. Extraterrestrial cause for the Cretaceous-Tertiary extinction. *Science* 208:1095–1108.

Artem'eva, N. A., and Shuvalov, V. 1994. Oblique impact atmospheric effects. *Lunar Planet. Sci.* XXV:39–40 (abstract).

Artem'ev, V. I., Markovitch, I., Nemchinov, I., and Sulyaev, V. 1987. Two-dimensional, self-similar motion of a strong shock wave along a heated surface. *Soviet Phys. Dokl.* 32(4):245–246.

Artem'ev, V. I., Bergel'son, V., Kalmykov, A., Nemchinov, I., Orlova, T., Rybakov, V., Smirnov, V., and Khazins, V. 1988. Development of a forerunner in interaction of a shock wave with layer of reduced pressure. *Fluid Dyn.* 2:290–295.

Artem'ev, V. I., Bergel'son, V., Nemchinov, I., Orlova, T., Rybakov, V., Smirnov, V., and Khazins, V. 1989. Formation of new gas dynamic structures by the interaction of the thin channels of reduced density with the shock waves. *Mathematical Modeling* 1(8):1–11 (in Russian).

Baldwin, B., and Sheaffer, Y. 1971. Ablation and breakup of large meteoroids during atmospheric entry. *J. Geophys. Res.* 76:4653–4668.

Bergel'son, V. I., Nemchinov, I., Orlova, T., Smirnov, V., and Khazins, V. 1987. Self-similar development of a precursor in front of a shock wave interacting with a thermal layer. *Soviet Phys. Dokl.* 32(9):691–692.

Bergel'son, V. I., Nemchinov, I., Orlova, T., and Khazins, V. 1989. Self-similar flows upon instantaneous energy release in a gas containing channels of reduced density. *Soviet Phys. Dokl.* 34(4):350–352.

Biberman, L. M., ed. 1970. *Optical Properties of Hot Air* (Moscow: Nauka Press), (in Russian).

Biberman, L. M., Bronin, S., and Brykin, M. 1980. Moving of a blunt body through the dense atmosphere under conditions of severe aerodynamic heating and ablation. *Acta Astronautica* 7:53–65.

Bronsten, V. A. 1983. *Physics of Meteoritic Phenomena* (Dordrect: D. Reidel).

Carrier, G. F., Moran, W., Birks, J., Decker, R., Eadley, D., Friend, J., Jones, E., Katz, J., Keeny, S., Jr., Leovy, C., Longmire, C., McElroy, M., Press, W., Ruina, J., Shoemaker, E., Smith, L., Toon, O., and Turco, R. 1985. *The Effects on the Atmosphere of a Major Nuclear Exchange* (Washington D. C.: National Academy Press).

Ceplecha, Z. 1992. Influx of interplanetary bodies onto Earth. *Astron. Astrophys.* 263:361–366.

Chapman, C. R., and Morrison, D. 1994. Impacts on the Earth by asteroids and comets: Assessing the hazard. *Nature* 367:33–39.

Chyba, C. F., Thomas, P., and Zahnle, K. 1993. The 1908 Tunguska explosion: Atmospheric disruption of a stony asteroid. *Nature* 361:40–44.

Croft, S. K. 1982. A first-order estimate of shock heating and vaporization in oceanic impacts. In *Geological Implications of Impacts of Large Asteroids and Comets on the Earth*, eds. L. T. Silver and P. H. Schultz, Geological Soc. of America Special Paper 190 (Boulder: Geological Soc. of America), pp. 143–152.

Crutzen, P. J., and Birks, J. 1982. The atmosphere after a nuclear war: Twilight at noon. *Ambio* 11:114–125.

Fast, V. G., Barannik, A., and Rasin, S. 1976. Field of directions of tree fall in the vicinity of the Tunguska meteoroid fall. In *Questions of Meteoritics* (Tomsk: Izd. Tomsk Univ.), pp. 39–52.

Gehrels, T. 1991. Scanning with charge-coupled devices. *Space Sci. Rev.* 58:347–375.

Gehrels, T. 1993. The relative threat of small and large NEOs. Hazards Due to Comets and Asteroids, Jan. 4–9, Tucson, Ariz., p. 34, Abstract book.

Gerstl, S. A. W., and Zardecki, A. 1982. Reduction of photosynthetically active radiation under extreme stratospheric aerosol loads. In *Geological Implications of Large Asteroids and Comets on the Earth*, eds. L. T. Silver and P. H. Schultz, Geological Soc. of America Special Paper 190 (Boulder: Geological Soc. of America), pp. 201–210.

Givens, J. J., and Page, W. 1971. Ablation and luminosity of artificial meteors. *J. Geophys. Res.* 76:1039–1054.

Glasstone, S., and Dolan, P. 1977. *The Effects of Nuclear Weapons* (Washington, D. C.: U. S. Dept. of Defense and U. S. Dept. of Energy).

Goodman, J. M., and Aarons, J. 1990. Ionospheric effect on modern electronic systems. *Proc. IEEE* 78(3):512–528.

Gossard, E. A., and Hook, W. 1975. *Waves in the Atmosphere* (Amsterdam: Elsevier).

Grigoryan, S. S. 1979. Motion and disintegration of meteorites in planetary atmospheres. *Cosmic Res.* 17:724–740.

Hazins, V. M., and Svetsov, V. 1993. A conservative stable smoothness-enhancing free-Lagrangian method. *J. Comput. Phys.* 105:187–198.

Hills, J. C., and Goda, N. 1993. The fragmentation of small asteroids in the atmosphere. *Astron. J.* 105:1114–1144.

Ivanov, B. A. 1981. Mechanics of cratering. In *Itogi Nauki i Tekhniki. Mekhanika Devormiruemogo Tverdogo Tela*, vol. 14 (Moscow: VINITI), pp. 60–125.

Ivanov, B. A. 1992. Geomechanical models of impact cratering: Puchezh-Katunky structure. In *International Conf. on Large Meteorite Impacts and Planetary Evolution*, Sudbury, Canada, p. 40 (abstract).

Ivanov, B. A., Bazilevsky, A., Krynckov, V., and Chernaya, I. 1986. Impact craters on Venus: Analysis of Venera 15 and 16 data. *J. Geophys. Res.* 91:423–430.

Ivanov, B. A., Nemchinov, I., Svetsov, V., Provalov, A., Khazins, V., and Phillips, R. 1992. Impact cratering on Venus: Physical and mechanical models. *J. Geophys. Res.* 97:16167–16181.

Ivanov, B. A., and Petaev, M. 1992. Mass and impact velocity of the meteorite formed the Sterlitamak crater in 1990. *Lunar Planet. Sci.* XXIII:573–574 (abstract).

Jacobs, C., and Spalding, R. 1993. Fireball observation by satellite-based Earth-monitoring optical sensors. Hazards Due to Comets and Asteroids, Jan. 4–9, Tucson, Ariz., p. 45, Abstract book.

Jones, E. M., and Sanford, M., II, Jr. 1977. Numerical simulation of a very large explosion at the Earth's surface with possible application to tektites. In *Impact and Erosion Cratering*, ed. D. J. Roddy (New York: Pergamon Press), pp. 1009–1024.

Jones, E. M., and Kodis, J. 1982. Atmospheric effects of large body impacts. The first few minutes. In *Geological Implications of Impacts of Large Asteroids and Comets on the Earth*, eds. L. T. Silver and P. H. Schultz, Geological Soc. of America Special Paper 190 (Boulder: Geological Soc. of America), pp. 175–186.

Kiselev, Yu. N., Kosarev, I., Nemchinov, I., Rozhdestvenskii, V., Khristoforov, B., and Yur'ev, V. 1991*a*. Simulations of motions of large and fast meteoritic bodies. *Solar System Res.* 25(1):71–79.

Kiselev, Yu. N., Nemchinov, I., and Shuvalov, V. 1991*b*. Mathematical modeling of the propagation of intensely radiating shock waves. *Comput. Math. Math. Phys.* 31(6):87–101.

Korobeinikov, V. P., and Khristoforoff, B. 1976. Underwater explosion. In *Itogi Nauki i Tekhniki, Hydromekhanika*, vol. 9 (Moscow: VINITI), pp. 54–119 (in Russian).

Korobeinikov, V. P., Chushkin, P., and Shurshalov, L. 1990. Combined simulation of the flight and explosion of a meteoroid in the atmosphere. *Solar System Res.* 25:242–254.

Korobeinikov, V. P., Gusev, S., Chushkin, P., and Shurshalov, L. 1992. Flight and structure of the Tunguska cosmic body into the Earth's atmosphere. *Computers Fluids* 21:323–330.

Kostyuchenko, V. N., Rodionov V., and Sultanov D. 1974. Seismic waves of underground nuclear explosions. In *Peaceful Nuclear Explosions III* (Vienna: IAEA), pp. 447–461.

Kuznetsov, N. M. 1965. *Thermodynamic Functions and Shock Adiabats for Air at High Temperatures* (Moscow: Mashinostroyenie), (in Russian).

Lewis, J. S., Watkins, G., Hartman, H., and Prinn, R. 1982. Chemical consequences of major impact events on Earth. In *Geological Implications of Impacts of Large Asteroids and Comets on the Earth*, eds. L. T. Silver and P. H. Schultz, Geological Soc. of America Special Paper 190 (Boulder: Geological Soc. of America), pp. 215–221.

Leovy, C. B. 1985. The general circulation on Mars: Models and observations. In *Advances in Geophysics*, vol. 28, Part A, ed. S. Manabe (Orlando: Academic Press), pp. 327–346.

L'vov, Yn. A., and Vasil'ev, N. 1976. Radiant burning of trees in the vicinity of the Tunguska meteoroid fall. In *Questions of Meteoritics* (Tomsk: Izd. Tomsk. Univ.), pp. 53–57.

McCrosky, R. E., and Ceplecha, Z. 1969. Photographic networks for fireballs. In *Meteorite Research*, ed. P. M. Millman (Dordrecht: D. Reidel), pp. 600–612.

McCrosky, R. E., Shao, C., and Posen, A. 1978. Bolides of the Prairie Network I. General data and orbits. *Meteoritika* 37:44–59 (in Russian).

McCrosky, R. E., Shao, C., and Posen, A. 1979. Bolides of the Prairie Network II. Trajecteries and light intensities. *Meteoritika* 38:106–156 (in Russian).

McKay, P., and Thomas, G. 1982. Formation of noctilucent clouds by an extraterrestrial impact. In *Geological Implications of Impacts of Large Asteroids and Comets on the Earth*, eds. L. T. Silver and P. H. Schultz, Geological Soc. of America Special Paper 190 (Boulder: Geological Soc. of America), pp. 211–214.

Melnikov, N. N., Konuchin, V., and Naumov, V. 1992. *Underground Atomic Stations* (Apatity: Russian Academy of Sciences), (in Russian).

Melosh, H. J. 1981. Atmospheric breakup of terrestial impactors. In *Multi-Ring Basins*, eds. P. H. Schultz and R. B. Merrill (New York: Pergamon Press), pp. 29–35.

Melosh, H. J. 1989. *Impact Cratering: A Geological Process* (Oxford: Clarendon Press).

Melosh, H. J., Schneider, N., Zahnle, K., and Latham, D. 1990. Ignition of global wildfires at the Cretaceous/Tertiary boundary. *Nature* 343:251–254.

Melosh, H. J., Artemjeva, N., Golub, A., Nemchinov, I., Shuvalov, V., and Trubetskaya, I. 1993. Remote visual detection of impacts on the lunar surface. *Lunar Planet. Sci.* XXIV:975–976 (abstract).

Mirels, H. 1988. Interaction of moving shock with thin stationary thermal layer. In

Proc. of the 16th Intl. Symp. on Shock Tubes and Waves, ed. H. Grönig (Weinheim, N. Y.: VCH), pp. 177–183.

Nemchinov, I. V. 1970. On averaged radiation transfer equations and their applications to gas-dynamic problems. *Prikladnaya Matematika i Mekhanika* 34(4):706–721 (in Russian).

Nemchinov, I. V. 1993. Intensely radiating shock waves. *Khimitcheskaya Fizika* 12(3):320–333 (in Russian).

Nemchinov, I. V., and Loseva, T. 1994. Atmospheric oscillations initiated by the penetration of a comet or an asteroid into gaseous envelope of a planet. *Lunar Planet. Sci.* XXV:987–988 (abstract).

Nemchinov, I. V., and Popov, S. 1983. On thermal waves around meteoritic bodies during their hypersonic flight through the atmosphere. *Soviet Phys. Doklady* 269(3):578–580 (in Russian).

Nemchinov, I. V., and Shuvalov, V. 1992. The explosion in the atmosphere of Mars caused by a high-speed impact of cosmic bodies. *Lunar Planet. Sci.* XXIII:981–982 (abstract).

Nemchinov, I. V., and Shuvalov, V. 1993. Explosion dynamics in meteor impacts on the surface of Mars. *Solar System Res.* 26:333–343.

Nemchinov, I. V., and Svetsov, V. 1977. Calculation of development of a laser explosion in air with allowance for emission. *Zhurn, Prikladnoi Mekhaniki i Tekhnicheskoi Fisiki* 4:448–454.

Nemchinov, I. V., and Svetsov, V. 1991. Global consequences of radiation impulse caused by comet impact. *Adv. Space. Res.* 11(6):625–697.

Nemchinov, I. V., and Tsikulin, M. 1962. An estimation of thermal transfer by radiation in high-velocity meteorite motion in the Earth's atmosphere. *Geomagnetizm and Aeronomiya* 3:635–646 (in Russian).

Nemchinov, I. V., Orlova, T., Svetsov, V., and Shuvalov, V. 1976. On the role of radiation in high-velocity meteorite motion in the atmosphere. *Dokl. Akad. Nauk* 231(5):60–63 (in Russian).

Nemchinov, I. V., Svetsov, V., and Shuvalov, V. 1978. Structure of the heating layer ahead of the front of a strong intensely emitting shock wave. *Prikl. Mekh. Tekhnicheskaya Fiz.* 5:644–648.

Nemchinov, I. V., Svetsov, V., and Shuvalov, V. 1979. On the brightness of shock waves in low density air. *Zhurnal Prikl. Spectrosc.* 30(6):1086–1092 (in Russian).

Nemchinov, I. V., Novikova, V., and Popova, O. 1989. Analysis of observed motion and luminosity of high velocity meteorite bodies moving at high altitudes in the terrestial atmosphere. *Meteoritika* 48:124–137 (in Russian).

Nemchinov, I. V., Perelomova, A., and Shuvalov, V. 1992. Nonlinear response of the atmosphere to an impulsive disturbance. *Izvestiya, Russian Academy of Sci., Atmospheric and Oceanic Phys.* 28(3):178–182.

Nemchinov, I. V., Alexandrov, P., Artem'ev, V., Bergelson, V., and Rybakov, V. 1993*a*. On magnetohydrodynamic effects initiated by a high-speed impact of a cosmic body upon the Earth's surface. *Lunar Planet. Sci.* XIV:1063–1064 (abstract).

Nemchinov, I. V., Ivanov, B., Kozlov, I., Romanov, G., and Suvorov, A. 1993*b*. Contamination of Earth's atmosphere by dust particles ejected after the impact of small comets and asteroids. Hazards Due to Comets and Asteroids, Jan. 4–9, Tucson, Ariz., p. 59, Abstract book.

Nemchinov, I. V., Perelomova, A., and Shuvalov, V. 1993*c*. Determination of cosmic bodies size-velocity distribution by observation of current impacts on Mars. *Lunar Planet. Sci.* XXIV:1065–1066 (abstract).

Nemchinov, I. V., Popov, S., and Teterev, A. 1993*d*. Tsunamis caused by the impact of cosmic bodies in oceans or seas. Hazards Due to Comets and Asteroids, Jan. 4–9, Tucson, Ariz., p. 64, Abstract book.

Nemchinov, I. V., Popova, M., Shubadeeva, L., Shuvalov, V., and Svetsov, V. 1993*e*. Effects of hydrodynamics and thermal radiation in the atmosphere after comet impacts. *Lunar Planet. Sci.* XXIV:1067–1068 (abstract).

O'Keefe, J. D., and Ahrens, T. 1982*a*. Cometary and meteorite swarm impact on planetary surfaces. *J. Geophys. Res.* 87:6668–6680.

O'Keefe, J. D., and Ahrens, T. 1982*b*. The interaction of the Cretaceous/Tertiary extinction bolide with the atmosphere, ocean, and solid Earth. In *Geological Implications of Impacts of Large Asteroids and Comets on the Earth*, eds. L. T. Silver and P. H. Schultz, Geological Soc. of America Special Paper 190 (Boulder: Geological Soc. of America), pp. 103–120.

Passey, Q. R., and Melosh, H. 1980. Effects of atmospheric breakup on crater field formation. *Icarus* 42:211–233.

Pevzner, L. A., Kirjakov, A., Vorontsov, A., Masaitis, V., Maschak, V., and Ivanov, B. 1992. Vorotilovskay drillhole: First deep drilling in the central uplift of large terrestial impact crater. *Lunar Planet. Sci.* XXIII:1063–1064 (abstract).

Pirrus, E. A., and Tiyurma, P. 1987. Meteor craters of Estonia. In *All-Union Meteoritics Conference* (Moscow: Vernadsky Inst.), pp. 3–4 (abstract), (in Russian).

Proceedings III Intl. Conf. on the Safety of Nuclear Power (Sept. 1991) (Vienna: IAEA), 1992.

Rabinowitz, D. L. 1993. The size distribution of the Earth-approaching asteroids. *Astrophys. J.* 407:412–427.

Rabinowitz, D. L., Gehrels, T., Scotti, J. V., McMillan, R., Perry, M., Wisniewski, W., Larson, S., Howell, E., and Mueller, B. 1993. Evidence of a near-Earth asteroid belt. *Nature* 363:492–493.

Reichenbach, H., and Kuhl, A. 1988. Techniques for creating precursors in shock tubes. In *Proc. of the 16th Intl. Symp. on Shock Tubes and Waves*, ed. H. Grönig (Weinheim, N. Y.: VCH), pp. 847–853.

ReVelle, D. O. 1979. A quasi-simple ablation model for large meteorite entry: Theory vs observations. *J. Atmos. Terrestrial Phys.* 41:453–473.

ReVelle, D. O., and Rajan, R. 1979. On the luminous efficiency of meteoritic fireballs. *J. Geophys. Res.* 84:6255–6262.

Reynolds, D. A. 1992. Fireball observation via satellite. In *Proceedings of the Near-Earth Object Interception Workshop*, eds. G. H. Canavan, J. C. Solem and J. D. G. Rather (Los Alamos: Los Alamos National Lab), pp. 221–226.

Roddy, D. J., Shuster, S., Rosenblatt, M., Grant, L., Hassig, P., and Kreyenhagen, K. 1987. Computer simulations of large asteroid impacts into oceanic and continental sites-preliminary results on atmospheric, cratering and ejecta dynamics. *Intl. J. Impact Eng.* 5:123–135.

Rodean, H. C. 1971. *Nuclear-Explosion Seismology* (Livermore, Ca.: Univ. of California, U. S. AEC, Div. of Technical Information).

Rodionov, V. N., Adushkin, V., Kostychenko, V., Nikolayevsky, V., Romashov, A., and Tsvetkov, V. 1971. *Mechanical Effect of the Underground Explosions* (Moscow: Nedra), in Russian.

Sadovsky, M. A., and Adushkin, V. 1988. Effect of a heated wall layer on shock wave characteristics. *Trans. (Doklady) of the USSR Academy of Sciences. Earth Sci. Sects.* 300(3):12–15.

Sadovsky, M. A., and Kostyuchenko, V. 1974. On the seismic effect of the underground nuclear explosions. *Dokladi Akademii Nauk SSSR* 215(5):1097–1100 (in Russian).

Sadovsky, M. A., Bolkhovitinov, L., and Pisarenko, V. 1987. *Deformation of Geophysic Media and Seismic Process* (Moscow: Nauka Press), (in Russian).

Schultz, P. H. 1992. Atmospheric effects on ejecta emplacement and crater formation on Venus from Magellan. *J. Geophys. Res.* 97:16183–16248.

Scotti, J. V., Rabinowitz, D., and Marsden, B. 1991. Near miss of the Earth by a small asteroid. *Nature* 354:287–289.

Shreffler, R. G., and Christian, R. 1954. Boundary disturbances in high explosive shock tubes. *J. Applied Phys.* 25:324–331.

Shokin, Yu. I., Chubarov, L., Marchuk, A., and Simonov, K. 1989. *Numerical Experiment in the Problem of Tsunami* (Novosibirsk: Nauka Press), (in Russian).

Shoemaker, E. M. 1962. Interpretation of lunar craters. In *Physics and Astronomy of the Moon*, ed. Z. Kapal (Orlando Fla.: Academic press), pp. 283–359.

Shoemaker, E. M. 1983. Asteroid and comet bombardment of the Earth. *Ann. Rev. Earth Planet. Sci.* 11:461–494.

Svetsov, V. V. 1994. Radiation emitted during the flight: Application to assessment of bolide parameters from the satellite recorded light flashes. *Lunar Planet. Sci.* XXV:1365–1366 (abstract).

Teterev, A. V., and Nemchinov, I. 1993. The sand bag model of the dispersion of the cosmic body in the atmosphere. *Lunar Planet. Sci.* XXIV:1415–1416 (abstract).

Teterev, A. V., Misychenko, N., Rudak, L., Romanov, G., Smetannikov, A., and Nemchinov, I. 1993. Atmospheric breakup of a small comet in the earth's atmosphere. *Lunar Planet. Sci.* XXIV:1417–1418 (abstract).

Turco, R. P., Toon, O., Park, C., Whitten, R., Pollack, J., and Noerdlinger, P. 1982. An analysis of the physical, chemical, optical and historical impacts of the 1908 Tunguska meteor fall. *Icarus* 50:1–52.

Vickery, A. M. 1986. Effect of an impact-generated cloud on the acceleration of solid ejecta. *J. Geophys. Res.* 91:14139–14160.

Zel'dovitch, Y. B., and Raizer, Yu. 1967. *Physics of Shock Waves and High-Temperature Hydrodynamic Phenomena* (New York: Academic Press).

Zotkin, I. T., and Tsikulin, M. 1966. Simulation of the explosion of Tunguska meteorite. *Soviet Phys. Doklady* 11:183–186.

Zurek, R. W., Barnes, J., Haberle, R., Pollack, J., Tillman, J., and Leovy, C. 1992. Dynamics of the atmosphere of Mars. In *Mars*, eds. H. H. Kieffer, B. M. Jakosky, C. W. Snyder and M. S. Matthews (Tucson: Univ. of Arizona Press), pp. 835–933.

TSUNAMI GENERATED BY SMALL ASTEROID IMPACTS

J. G. HILLS
Los Alamos National Laboratory

I. V. NEMCHINOV and S. P. POPOV
Institute for Dynamics of Geospheres

and

A. V. TETEREV
Belorussian State University

The fragmentation of a small asteroid in the atmosphere greatly increases its cross section for aerodynamic braking and energy dissipation. This atmospheric protection produces a near threshold effect whereby the ground impact damage (craters, earthquakes, and tsunamis) produced by stony asteroids is nearly negligible if they are less than 200 m in diameter. Larger ones impact the ground at nearly the velocity they had at the top of the atmosphere to produce considerable impact damage. Water waves generated by an impactor are two-dimensional disturbances that fall off in height only inversely with distance from the impact point, so they have a long range. When these waves strike a continental shelf their speed decreases and their height increases. The average runup in height of a tsunami wave is more than an order of magnitude. Tsunamis are probably the most devastating form of damage produced by asteroids with diameters between 200 m and 1 km. An impact anywhere in the Atlantic Ocean by an asteroid more than 400 m in diameter would devastate the coasts on both sides of the ocean with tsunami wave runups of over 60 m high. The protection offered by the Earth's atmosphere produces an insidious situation whereby smaller, more frequent impactors such as Tunguska only produce air blast damage while objects 2.5 times larger than Tunguska, which hit every few thousand years, cause coherent destruction over many thousands of kilometers of coastal settlements. The smaller impactors give no qualitative warning of the enormous destruction wrought when an asteroid larger than the threshold diameter of 200 m hits an ocean. Studies of ocean sediments may be used to determine when coastal areas have been hit by tsunamis in the past. Tsunami debris has been found to be associated with the Cretaceous-Tertiary impact and should be detectable for smaller impacts.

I. INTRODUCTION

Waves produced in ocean impacts may be the most serious consequence of asteroid impacts short of the massive super killers such as the Cretaceous-Tertiary impactor. Just as on land, much of the kinetic energy of an asteroid that impacts the ocean goes into the formation of a crater, but the crater is not stable. A series of waves produced by the outward propagation of the crater rim and the refilling of the crater propagate radially away from the impact (Gault and Sonett 1982).

Asteroids with radii larger than the ocean depth produce tsunami-like waves with amplitudes comparable to the ocean depth at a short distance from the point of impact (cf., Ahrens and O'Keefe 1983,1987; Roddy et al. 1987). While these waves damp quickly, they are likely to produce catastrophic tsunamis all over the world. Changes in ocean bottom topography can scatter a considerable fraction of tsunami energies (Braddock 1970), which may cause these high-amplitude tsunamis to scatter around the continents to enter the other oceans beyond the one impacted. Because impacts by large asteroids have previously been studied in detail with regard to the Cretaceous-Tertiary impactor, they will not be discussed here.

In this chapter we are primarily concerned with the impacts of small (compared to the depth of the ocean) asteroids that produce waves with amplitudes that are much smaller than the depth of the ocean. Such tsunami waves in deep water do not dampen significantly until they run into shallows where they steepen into breakers and increase in height (Mader 1988). The average tsunami runup, the final height of the tsunami in units of the height of the wave in the deep water that produced it, is about 10 to 20 fold in the Hawaiian Islands (Mader 1991) although runups twice this value have been recorded. Similar runups may be expected elsewhere although the presence of a broad continental shelf can lead to the reflection of some of the tsunami energy (Nekrasov 1970).

The height of a tsunami in deep water only decreases inversely with the distance from its origin, so it can cause serious problems far from the impact point. This is a result of the wave being inherently two-dimensional. The intensity of an inherently three-dimensional disturbance such as an airburst or an earthquake falls off as the inverse square of the distance, so such a disturbance is far more localized than water waves. There are many anecdotal illustrations of the long-range nature of tsunamis, e.g., the earthquake in Chile in 1960 produced waves in deep water that traveled 150 degrees (over 17,000 km) around the Earth to produce tsunamis in Japan that were from 1 to 5 m high (average about 2 m) and killed at least 114 people with another 90 people missing (Takahasi 1961). [It is estimated that the full amplitude of the wave in deep water before hitting Japan was 40 cm, so the maximum height above normal sea level was 20 cm, and it had a period of 60 minutes (Iida and Ohita 1961). This requires an average tsunami runup of 10 fold and a maximum of 25 fold]. In the Hawaiian Islands, at 10,600 km from the epicenter, the major damage was in Hilo harbor where the maximum tsunami height was over 10 m and 61 people were killed (Cox 1961). We shall see that asteroid impacts can produce tsunamis vastly larger than these tsunamis and in areas, such as the Atlantic, where coastal settlements are poorly prepared for them.

II. IMPACTS INTO DEEP OCEANS

An analysis of experiments with underwater nuclear explosives shows that the full height of a deep-water wave at a distance r from the underwater detonation of energy Y is given by

$$h_W = 40,500 \text{ ft } \frac{(Y/\text{kton})^{0.54}}{r/\text{ft}} = 6.5 \text{ meters } \left(\frac{Y}{\text{gigaton}}\right)^{0.54} \left(\frac{1000 \text{ km}}{r}\right) \tag{1}$$

(Glasstone and Dolan 1977). This result is not sensitive to the depth at which the explosion occurs. The height h of the water wave above the ocean is half of the full height of the wave, so $h = 3.3$ m at 1000 km from a 1 gigaton explosion. A more recent analysis of Pacific test explosions in deep water with yields between 1 kiloton and 5 megatons and of modeled nuclear explosions of up to 100 megatons, shows a similar equation for the h above the ocean level:

$$h = \frac{1}{2}h_W = 4.5 \text{ meters } \left(\frac{Y}{\text{gigaton}}\right)^{1/2} \left(\frac{1000 \text{ km}}{r}\right). \tag{2}$$

The values given by this equation for $R > 100$ m are in satisfactory agreement with the values given by Eq. (1), considering the amount of extrapolation used.

Hills and Goda (1993) found the collision impact energies of comets, stony asteroids, and iron asteroids as a function of size and impact velocity taking into account the increase in their aerodynamic cross sections due to fragmentation. Figures 1 and 2 show the full height, H_W, (twice the height h above sea level) of a wave in deep water at a distance of 1000 km from the impact point for nickel-iron and stony meteorites, respectively, as a function of impactor radius for various impact velocities. The heights were found by putting the ground impact energies Y found by Hills and Goda (1993) into Eq. (1).

We note that for stony asteroids <100 m in radius the tsunami height is significantly less than it would be without air dissipation. There is a sharp increase in the tsunami wave produced by asteroids with radii above 100 m. This is also true of iron asteroids that are more than 40 m in radius.

For the larger asteroids, e.g., radii $R > 100$ m for stony asteroids, that suffer no significant energy dissipation in the atmosphere, the wave height in deep water ($h = h_w/2$) above mean sea level at distance r is given by

$$h = 7.8 \text{ meters } \left[\left(\frac{R}{203 \text{ meters}}\right)^3 \left(\frac{V}{20 \text{ km}}\right)^2 \left(\frac{\rho_M}{3 \text{ g cm}^{-3}}\right)\right]^{0.54} \left(\frac{1000 \text{ km}}{r}\right) \tag{3}$$

Here a stony asteroid with a radius of 203 m and a velocity of 20 km s^{-1} has an impact energy of 5 gigatons of TNT. An asteroid of this size or larger impacts Earth about every ten thousand years. Asteroids of sufficient size

Iron

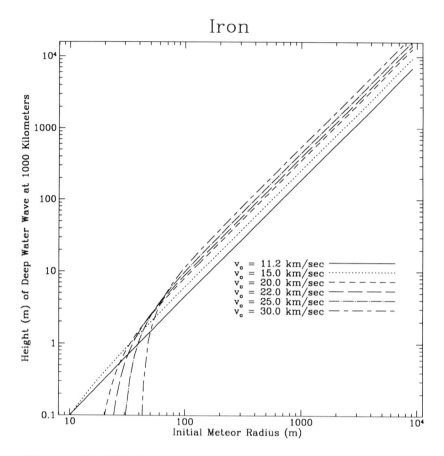

Figure 1.   The full height in meters of a wave in deep water at a distance of 1000 km
from the impact point in the ocean of a nickel-iron asteroid. This wave amplitude
is given as a function of impactor radius for various impact velocities. The height
of the wave above the mean ocean depth is half of the value shown. This wave runs
up in height by over an order of magnitude to produce tsunami when it runs into a
continental shelf.

produce craters that exceed the ocean depth. In this case, Eqs. (1) to (3) are
no longer valid. We discuss such impacts in the next section.

III. IMPACTS INTO A SHALLOW SEA

The average depth of the ocean is $d = 4$ to 5 km. If the depth of the crater
produced by the impactor is comparable to or larger than the local depth d
of the ocean or of a shallow sea, we can no longer use Eq. (1) to compute
the height of a wave in deep water far from the impact point. It is known
from nuclear weapon tests that an explosion in shallow water, e.g., Pacific test
Bikini Baker, deposits much less mechanical energy into the water than does

Soft Stone

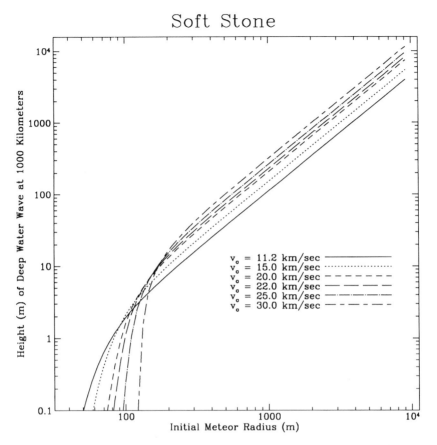

Figure 2.   The full height in meters of a wave in deep water at a ditance of 1000 km from the impact point in the ocean of a stony asteroid. This wave amplitude is given as a function of impactor radius for various impact velocities.

a similar explosion in deep water (Glasstone and Dolan 1977). Glasstone and Dolan find that the full height of the terminal wave at a distance r from an explosion in shallow water is given by

$$h_w = 1450 \text{ meters} \left(\frac{d}{r} \right) \left(\frac{Y}{\text{gigatons}} \right)^{0.25} \tag{4}$$

where d is the depth of the water and Y is the yield. We note that the terminal wave height is less sensitive to yield than the case for waves generated in deep water. There remains an inverse relationship between height and distance from the source. If we let $d = 5$ km, the average depth of the ocean, we find that Eqs. (1) and (4) give the same full height of $h_w = 8.1$ m at $r = 1000$ km for a yield of $Y = 1.5$ gigaton, which corresponds to a stony asteroid with a radius of 136 m or a diameter of 270 m if its impact velocity is 20 km s^{-1}. Schmidt and

Holsapple (1982) found that the depth of a crater in water is about 12 times the impactor diameter. This suggests that in shallow seas where the impactor diameter exceeds about 8% of the depth, it may be better to use Eq. (4) than Eq. (1) to determine the height of the wave in deep water far from the impact point. In the ocean, where $d = 5$ km, we should use Eq. (4) if the impactor diameter exceeds 400 m.

A Numerical Example

The three Europe-based authors of this review have done a computer simulation of the impact of a comet with a diameter of 2 km into an ocean with a depth of $d = 4$ km. This study provides a check on the validity of Eq. (4) for impacts of larger bodies into the ocean or of smaller bodies into shallow seas such as the Baltic. The model comet is spherical, has a density of 1 g cm^{-3}, and impacts the ocean at 20 km s^{-1}. Its impact energy is 600 gigatons of TNT.

They assumed the same equation of state for the comet as was used for the ocean water. This equation of state was calculated by G. S. Romanov and A. S. Smetannikov (Heat and Mass Transfer Institute, Minsk, Belorus) and provided in tabular form. They supposed that rocks of the oceanic bed are similar to granite and used Tillotson equation of state (Melosh 1989). The equation of state of air was borrowed from Kuznetsov (1965).

The hydrodynamic processes after the impact are qualitatively similar to those obtained by Ahrens and O'Keefe (1987), and Roddy et al. (1987) for an impactor 10 km across. Figure 3 a–d shows the interfaces between the atmosphere, water, and granite oceanic bed at 10, 20, 31, and 37 s after impact. The vertical and radial coordinates are expressed in km. In the last frame (37 s), the diameter of the crater is only 7 km, and the ratio of its diameter to that of the impactor is much smaller than that for the 10-km impactor. The depth of the crater is also smaller and the rim of the crater is below the original water surface layer, so the underwater crater in the solid material is much smaller than the case of the impact of the same size object on land. The maximum height of the hollow water cylinder of the comet debris is 48 km and for oceanic water splash it is close to 50 km. We have to keep in mind that the densities of these layers are lower than the normal density of water.

At 37 s the maximum height of the water wave moving out from the impact is $h = 1.3$ km. It is 18 km from the impact point and it is moving out at a speed of 0.5 km s^{-1}. Eq. (4) predicts that the height of the water wave above the surface $h = (h_w/2) = 0.5$ km, or about 0.4 that found in the numerical experiment, for $y = 600$ gigatons, $r = 18$ km, and $d = 2$ km. This discrepancy is not surprising. The height of the wave is still a significant fraction of the depth of the ocean, so it will exhibit some tsunami runup and will not have settled down to its final height in deep water. The calculation needs to be run over a longer period of time. Equation (1) predicts $h = 5.7$ km, which is clearly too high, as expected, because the crater diameter is much larger than

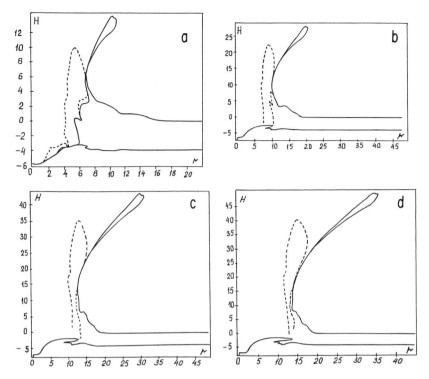

Figure 3.   The impact of a comet with a diameter of 2 km into an ocean with a depth of 4 km. The figures show the interfaces between the atmosphere, water, and granite oceanic bed at 10, 20, 31, and 37 s after impact.

the depth of the water.

There is a clear need to make further hydrodynamic calculations that will better determine the final wave heights as a function of impactor size, composition, and velocity in the critical regime between the limiting case values given by Eqs. (1) and (2). For impacts in the deep ocean this critical region may correspond to asteroids about 0.5 to 5 km in diameter. However, the above numerical illustration does show that it will probably be possible to estimate such terminal heights to within a factor of 2 or 3 even in a worst case, such as the numerical case discussed in this section, by taking a judicious average of the two limiting values.

IV. TSUNAMI RUNUPS

As the tsunami wave goes into a shoal, its speed decreases and its front increases in sharpness and amplitude until it breaks. We noted earlier that the average runup in height is about 10 to 20 fold in the Hawaiian Islands while the 1960 Chile tsunami produced runups in Japan that averaged 10 fold but reached about 25 fold in the Northern Islands.

An asteroid with a radius of 200 m (diameter 400 m) that drops anywhere in the mid Atlantic will produce waves in deep water that are at least $h = 3$ m high when they reach both the European and North American coasts [except for the dispersal of some of the wave energy at the largest distances due to differences in speed between the highest and lowest frequency components of the wave (C. Mader, personal communication)]. When it encounters land, this wave steepens and runs up into a tsunami with an average height of up to 60 m (if it follows the Hawaiian runups) that hits both sides of the Atlantic nearly simultaneously.

Tsunami Flood Plane

When the tsunami impacts the shore, the maximum distance X_{max} which the water moves inland from the sea depends on the maximum depth of the water at the shoreline, the runup height h_o, the slope of shore away from the coast, and on the roughness of the ground that the water moves across (cf., Mader 1991). If there is a flat coastal plane on which maximum water depth h is h_o, the depth at a distance X inland is given by

$$\frac{h}{h_o} = \left[1 + \left(\frac{X}{X_{max}}\right)\right]^{4/3} \tag{5}$$

where the maximum inward distance to which the water flows scales as

$$X_{max} = \frac{h_o^{4/3}}{n^2}A = Bh_o^{4/3} \tag{6}$$

where n is the Manning roughness number of the terrain over which the water is surging and A is a constant (Bretschneider and Wybro 1977). Here n varies from about 0.015 for very smooth terrain (e.g., mud flats and ice) to 0.070 for very rough coast areas (dense brush and trees and coarse lava formations). Developed areas typically have $n = 0.30$ to 0.035. For $n = 0.035$ and $h_o = 15$ m (50 ft), $X_{max} = 1.8$ km (5800 ft) (Bretschneider and Wybro 1977). Putting this scaling factor into Eq. (5), we find that

$$X_{max} = 1.0 \text{ km} \left(\frac{h_o}{10 \text{ meters}}\right)^{4/3}. \tag{7}$$

We note that a 100-m tsunami would travel inland about 22 km while 200-m ones would go 55 km. While there may be some difficulties in extrapolating Eq. (7) to these large values, it is clear that tsunami of these magnitudes would cause unprecedented damage to low-lying areas in North America such as Long Island and Delaware and it may totally submerge small, flat coastal countries in Europe such as Holland or Denmark.

A preliminary assessment (Mader and Hills, in preparation) has been made of the runups and coastal inundations produced by the impact of a large asteroid into the mid Atlantic. For numerical convenience the initial state of the disturbance was taken to be a crater 150 km in diameter with a depth equal to that of the ocean. This is smaller than the Cretaceous-Tertiary impact crater in the Yucatan and it may be produced by an object less than half the size of the K-T impactor, so this simulated water crater may correspond to a 10 Myr event rather than the 100 Myr event associated with the Cretaceous-Tertiary impact. When this wave ran up on the East Coast of the United States, it swept inland all the way to the foothills of the Appalachian mountains in the northern 2/3 of the country. Boston, New York, Washington, D. C. as well as all the other areas east of the Appalachians were covered with a considerable depth of water. Curiously, inland Florida was largely unaffected except in the Miami area, which was totally inundated. Florida was saved by the gradual slope of the ocean approaching its shore, which reflected a considerable fraction of the wave energy back into the Atlantic.

It is clear that detailed runup calculations like those discussed above need to be done for smaller impactors. However, such calculations require very much more finely detailed digital topographical maps of the ocean and the coast (which are not yet available) to resolve the features produced by waves propagating from smaller scale craters. Such work is planned for the future.

The physical damage caused by tsunami results principally from the impact of the debris carried by the moving water. There is much debris in developed areas. This debris acts as a battering ram that effectively scours away the area impacted by the tsunami flood. Because a disproportionate fraction of human resources are close to the coasts, tsunami are probably the most deadly manifestations of asteroid impacts apart from the very large Cretaceous-Tertiary superkillers.

V. EVIDENCE FOR TSUNAMI FROM IMPACTORS

Very large tsunamis have occurred. Deposits of unconsolidated corals hundreds of meters above sea level on the Hawaiian Islands of Lanai, Hawaii, Oahu, Molokai, and Maui provide evidence of giant tsunamis (Johnson and Kin 1993). On Lanai they are found as high as 326 m above sea level. A tsunami of similar height occurred in a fiord in Alaska in historical times. A tsunami at least 50 to 100 m in height has been found to be associated with the Cretaceous-Tertiary impactor (Bourgois et al. 1988).

Searches for tsunami in the geologic record have mostly been done in the past few years, so it is likely that new evidence for such tsunami will appear at an increasingly rapid rate. It may be especially profitable to search for tsunami waves produced by smaller impacts along the Atlantic coast, which is less prone to earthquake-induced tsunami. Geological (and perhaps archaeological) evidence for large tsunamis along the coasts of the major oceans (due to their large impact cross sections) may be the best counters for impacts of moderately large ($R = 100$–1000 m) asteroids. Impact tsunamis can be distinguished from tsunamis produced by earthquakes by the large-scale correlation of the tsunami damage along much of the coast rather than the more localized damage produced by earthquake tsunamis. It is also likely that large-scale tsunamis produced by impacts will have associated with them a world-wide (but usually thin) layer of platinum-iridium-rich dust which may be detected in the snow deposits of the Greenland and Antarctic ice caps (Hills and Goda 1993).

VI. CONCLUSIONS

The atmosphere is ineffective in preventing impact damage to the ground when the diameter of a stony asteroid exceeds 200 m. For iron meteorites that impact at greater than 20 km s^{-1}, the critical diameter is about 40 to 60 m. These properties cause a threshold effect whereby stony asteroids <200 m in diameter produce no significant ground (ocean) damage [but those larger than 60 meters in diameter can cause significant damage from airbursts (Hills and Goda 1993)], while those larger than this value can cause catastrophic tsunami.

The growth of the height of the tsunami wave in deep water with increasing impact energy slows considerably when the crater depth becomes comparable to the depth of the ocean. This occurs if the meteoroid diameter exceeds about 0.08 the depth of the water or at an impact energy of a few gigatons at a typical ocean depth. The probability is a few times 10^{-4} per year that an asteroid of suffecent size will impact an ocean on the Earth to produce tsunami with average height exceeding 60 m along the entire coast of the ocean.

A better physical understanding of the impact events and the coastal runups they produce will require more and improved hydrodynamic models. This is a challenging assignment. Computer simulations of water crater formation, such as the one discussed in this chapter, are ultimately limited by the fine detail associated with the front of the wave propagating out of the crater. The difficulty of modeling this detail has not yet allowed calculations to run long enough to reach the final wave heights associated with propagation in deep water. Runup calculations of particular harbors will require the availability of fine-structure (less than 100 m resolution) digital topographical maps of the ocean and coasts if they are to properly model the tsunami produced by small impactors. The entire research field of geologic assessments of tsunami produced by impactors is virtually nonexistent and needs to be initiated.

Acknowledgments. We would like to thank C. Mader, University of Hawaii, for many helpful comments that greatly increased the value of this review.

REFERENCES

Ahrens, T. J., and O'Keefe, J. D. 1983. Impact of an asteroid or comet in the Ocean and Extinction of terrestrial life. *J. Geophys. Res. Supp.* 88:799–806.

Ahrens, T. J., and O'Keefe, J. D. 1987. Impact on the Earth, ocean, and atmosphere. *Intl. J. Impact Eng.* 5:13–32.

Bourgeois, J., Hansen, T. A., Wiberg, P. L., and Kauffman, E. G. 1988. A tsunami deposit at the Cretaceous-Tertiary boundary in Texas. *Science* 241:567–570.

Braddock, R. D. 1970. Tsunami propagation over large distances. In *Tsunamis in the Pacific Ocean*, ed. W. M. Adams (Honolulu: East-West Center Press), pp. 285–303.

Bretschneider, C. L., and Wybro, P. G. 1977. Tsunami inundation prediction. In *Proc. of the Fifteenth Coastal Engineering Conference*, vol. 1, ed. C. L. Bretschneider (New York: American Soc. of Civil Engineers), pp. 1006–1024.

Cox, D. C. 1961. Effects of the May 1960 tsunami in Hawaii and other Polynesian islands. In *Proc. of the Tsunami Meetings Associated with the Tenth Pacific Science Congress*, ed. D. C. Cox (Honolulu: Univ. of Hawaii), pp. 87–95.

Gault, D. E., and Sonnett, C. P. 1982. Laboratory simulations of pelagic asteroidal impact: Atmospheric injection, benthic topography, and the surface wave radiation field. In *Geological Implications of Impacts of Large Asteroids and Comets on the Earth*, eds. L. T. Silver and P. H. Schultz, Geological Soc. of America Special Paper 190 (Boulder: Geological Soc. of America), pp. 69–92.

Glasstone, S., and Dolan, P. J. 1977. *The Effects of Nuclear Weapons*, 3rd ed. (Washington, D. C.: U. S. Government Printing Office).

Hills, J. G., and Goda, M. P. 1993. The fragmentation of small asteroids in the atmosphere. *Astronomical J.* 105:1114–1144.

Iida, K., and Ohta, Y. 1961. On the heights of tsunami associated with distant and near earthquakes. In *Proc. of the Tsunami Meetings Associated with the Tenth Pacific Science Congress*, ed. D. C. Cox (Honolulu: Univ. of Hawaii), pp. 105–123.

Johnson, C., and King, D. 1993. Can a Landslide Generate a 1000-ft Tsunami in Hawaii? Preprint.

Kuznetsov, N. M. 1965. *The Thermodynamic Properties and Shock Wave Adiabats of Air at High Temperatures* (Moscow: Mashinostroenie), (in Russian).

Mader, C. L. 1988. *Numerical Modeling of Water Waves* (Berkeley: Univ. of California Press).

Mader, C. L. 1991. Modeling Hilo, Hawaii tsunami inundations. *Sci. Tsunami Hazards* 9:85–94.

Melosh, H. J. 1989. *Impact Cratering: A Geological Process* (Cambridge: Oxford Univ. Press).

Nekrasov, A. V. 1970. Transformations of tsunamis on the continental shelf. In *Tsunamis in the Pacific Ocean*, ed. W. M. Adams (Honolulu: East-West Center Press), pp. 337–350.

Roddy, D. J., Shuster, S. H., Rosenblatt, M., Grant, L. B., Hassig, P. J., and Kreyenhagen, K. N. 1987. Computer simulations of large asteroid impacts into oceanic and continental sites-preliminary results on atmospheric, cratering, and ejecta dynamics. *Intl. J. Impact Eng.* 5:123–135.

Schmidt, R. M., and Holsapple, K. A. 1982. Estimates of crater size for large-body impacts: Gravitational scaling results. In *Geological Implications of Impacts of Large Asteroids and Comets on the Earth*, eds. L. T. Silver and P. H. Schultz, Geological Soc. of America Special Paper 190 (Boulder: Geological Soc. of America), pp. 93–101.

Takashasi, R. 1961. A summary report on the Chilean tsunami of May 24, 1960 as observed along the coast of Japan. In *Proc. of the Tsunami Meetings Associated with the Tenth Pacific Science Congress*, ed. D. C. Cox (Honolulu: Univ. of Hawaii), pp. 77–86.

ENVIRONMENTAL PERTURBATIONS CAUSED BY ASTEROID IMPACTS

OWEN B. TOON and KEVIN ZAHNLE
NASA Ames Research Center

RICHARD P. TURCO
University of California, Los Angeles

and

CURT COVEY
Lawrence Livermore Laboratory

We review the major mechanisms proposed to cause extinctions at the Cretaceous-Tertiary geologic boundary following an impact. We then consider how the proposed extinction mechanisms may relate to the impact of asteroids or comets in general. We discuss the limitations of these mechanisms in terms of the spatial scale that may be affected, and the time scale over which the effects may last. Our goal is to provide relatively simple prescriptions for evaluating the importance of colliding objects having a range of energies, and compositions. There are many uncertainties concerning the environmental effects of an impact that we seek to identify. For impactors with energies greater than about 10^6 megatons TNT equivalent, cooling and light loss due to dust lifted in the impact, as well as fires set by the lofted debris as it re-enters the atmosphere, appear to be the major short-term environmental hazards. Ultraviolet radiation enhancement due to ozone loss, and greenhouse warming due to stratospheric water injections may be important for several years following impacts larger than 10^5 megatons. Global acid rain is probably of secondary importance for energies below 10^8 megatons. However, ozone loss, water injections and acid rain effects are difficult to quantify because they depend on perturbations in chemistry and thermal structure that are likely to be large. For impactors with less than 10^5 megatons of energy, tidal waves may be the most destructive force let loose upon the global environment.

I. INTRODUCTION

The greatest natural disasters on Earth are caused by impacts of large asteroids and comets. Impacts of very large objects are rare compared to floods, earthquakes, and other more mundane hazards; indeed, they are so infrequent that they are normally disregarded on the time scale of human evolution. But should they be?

[791]

Modern debates about the damage that could be caused by asteroids and comets striking the Earth (see, e.g., Chapman and Morrison 1994; Ahrens and Harris 1992) have been provoked by the discovery of Alvarez et al. (1980) that the Cretaceous-Tertiary (K-T) mass extinction was caused by an impact. A favored extinction hypothesis postulates a blackout scenario in which dust raised by the impact prevents sunlight from reaching the surface for several months. Lack of sunlight terminates photosynthesis, prevents creatures from foraging for food, and leads to precipitous temperature declines (Alvarez et al. 1980; Toon et al. 1982). Obviously even much smaller impacts would have the potential to seriously damage human civilization, perhaps irreparably.

Detection of solar system objects on a collision course with Earth would clearly be of benefit if the threatening objects could be redirected (Ahrens and Harris 1992; Chapman and Morrison 1994). Modern astronomical networks can detect kilometer-sized bodies in Earth-crossing orbits, of which there are a thousand or so (Shoemaker et al. 1990). However, detecting and cataloging the much larger number of objects ten meters to a kilometer in diameter is a considerably more demanding project. To help assess the value of a search for smaller objects, we discuss in this chapter the mechanisms by which the K-T impactor caused a mass extinction, and how these mechanisms should scale with the magnitude (size and speed) of an impactor.

Fortunately, there are limited observational data upon which to base our discussion. Aside from controversial evidence for environmental changes which occurred at the K-T boundary, we have only the impact of the Tunguska meteorite in 1908, various large volcanic eruptions, and atmospheric nuclear weapons tests to compare with theories. Surprisingly, a rather consistent picture of the physical effects of impacts emerges, but with many unresolved questions. There are neither quantitative theories, nor clear observational data which relate measured physical changes in the environment to their effects on the Earth's biota or on human society. In the end the impact assessment of large impacts requires a subjective judgment of the importance of putative physical modification to the environment.

During the past decade there has been a creative burst in finding ways that impacts and explosions might damage the global environment. Table I outlines the primary mechanisms by which an impact could have caused the extinctions at the K-T boundary. In the following sections we discuss each of these mechanism in terms of the evidence for its importance, its robustness, and its likelihood of occurring after an impact. We also try to quantify, as simply as possible, the relations between impact energy and the magnitude of physical effects on the environment. Many of these mechanisms have a threshold for significant effects, which we identify. Finally, we estimate the smallest-sized objects that might produce important local, regional and global effects. One of our goals is to provide enough information to quantitatively compare different mechanisms and their implications.

TABLE I

Suggested Mechanisms for K-T Extinctions

Agent	Mechanism	Time Scale[a]	Geographic Scale (K-T)[b]
Dust loading	Cooling	Y	G
	Cessation of photosynthesis	M	
	Loss of vision	M	
Fires	Burning	M	G
	Soot cooling	M	
	Pyrotoxins	M	
	Acid rain	M	
NO_x generation	Ozone loss	Y	G
	Acid rain	M	R
	Cooling	Y	G
Shock wave	Mechanical pressure	I	R
Tidal wave	Drowning	I	R
Heavy metals, etc.	Poisoning	Y	G
Water/CO_2 injections	Warming	D	G
SO_2 injections	Cooling	Y	G
	Acid rain	Y	G

[a] I = instantly; M = months; Y = years; D = decades.
[b] R = Regional ($10^6 km^2$); G = global.

II. THE BLAST WAVE

A bolide generates shock waves as it propagates at hypersonic velocities through the atmosphere. If the atmospheric drag is great enough to stop the object, an airburst occurs (Chyba et al. 1993). Larger objects penetrate the atmosphere and crater the surface. In either case the object explosively surrenders its considerable kinetic energy. A prominent local effect, and the most important effect when the object disintegrates in the atmosphere, is the excitation of a powerful blast wave.

The chief factors determining the destructive potential of the blast wave are the energy released and the effective altitude of the explosion. The effective altitude of the explosion must be estimated by modeling the breakup of impactors traversing the atmosphere at hypersonic velocity. For most airbursts, the expected damage can be estimated from nuclear test data as well as from numerical models of such events. However, for very large events the local atmosphere can be blown away, with much of the shock energy dissipated to space rather than causing damage on the ground. In this case, the nuclear test data no longer apply and numerical modeling of blast waves is essential.

The blast wave consists of an abrupt pressure pulse (the shock) followed immediately by a substantial wind. The strength of this shock wave is usually characterized by the peak overpressure, which is defined as the difference between the ambient pressure and the pressure of the shock front. The peak overpressure is uniquely related to a maximum wind speed. For example, a 2 p.s.i. overpressure (atmospheric pressure is 14.7 p.s.i.) is accompanied by a wind speed of about 30 m s^{-1}, just below hurricane force, which can cause substantial damage to wood structures. A 4 p.s.i. overpressure corresponds to a wind of about 70 m s^{-1}, which is well above hurricane force. Even though humans can withstand considerably higher direct overpressures (severe injuries occur around 10 p.s.i.) flying debris even at 2 p.s.i. presents a severe hazard.

Nuclear weapons tests have provided an empirical relationship between the altitude of energy release and the distance over which a specific peak overpressure is reached. Based upon the information in Glasstone and Dolan (1977), Hills and Goda (1993) derived a convenient equation for the maximum distance at which an overpressure of 4 p.s.i. occurs:

$$r = ah - bh^2 E^{-1/3} + cE^{1/3}. \tag{1}$$

Here r is the maximum distance (km) of the overpressure contour from the point below a nuclear detonation at altitude h (km) and E is the energy of the explosion in megatons; a is 2.09, b is 0.449 and c is 5.08. Because only about half the energy in a nuclear weapon explosion goes into the shock waves (Glasstone and Dolan 1977), we will assume in using Eq. (1) that $E = 2\epsilon Y$, where Y is the kinetic energy of the impactor and ϵ is the fraction converted to shock energy.

As Hills and Goda (1993) point out, Eq. (1) implies an optimum burst height, $h_o = 2.3E^{1/3}$km. At the optimum burst height the area of destruction is approximately twice that for the same energy release occurring at the surface. Explosions that occur above 6.42 $E^{1/3}$km do not produce a 4 p.s.i. shock wave that reaches the surface.

Here we estimate airburst altitudes for various impactors incident at 45° (the most probable impact angle), following Chyba et al. (1993). Table II lists the smallest objects (radius r_s, with energy E_s) of a given type that according to Eq. (1) will generate a 4 p.s.i. shock wave at the surface. Also listed are the smallest objects (radius r_g with energy E_g) that reach the surface with at least half of their initial kinetic energy. These estimates depend upon the density ρ,

TABLE II

Properties of Objects that are Stopped by the Atmosphere

	Iron	Stone	Carbonaceous Chondrite	Short-period Comet	Long-period Comet
r_g, m	20	65	95	190	180
E_g, megaton	11	160	360	3×10^3	1×10^4
r_s, m	[a]	15	23	38	30
E_s, megaton		2	5	25	50
ρ, g cm^{-3}	7.8	3.5	2.5	1	1
v_i, km s^{-1}	15	15	15	25	50
S, dyne cm^{-2}	10^9	2×10^8	10^7	10^6	10^6
Q, erg g^{-1}	8×10^{10}	8×10^{10}	8×10^{10}	2.5×10^{10}	2.5×10^{10}

[a] At this strength and incidence angle, all iron objects that air burst will produce 4 p.s.i. shocks at the surface.

velocity v_i, strength S, and latent heat of ablation Q of the object (also given in Table II). Although object parameters are uncertain and the model is not without its deficiencies, most of the uncertainty implicit in Table II probably stems from unavoidable random variance in impact angle and impact velocity. For example, the 1σ variance in impact angle is 20°. This corresponds to an uncertainty of about a scale height in h, or to an order of magnitude in E_g. For example, E_g is 750 megatons for a stone incident at 65° (measured from the zenith), but only 75 megatons if incident at 25°. A noteworthy feature of Table II is the significant difference in the minimum energy required for various types of objects to reach the ground, or to produce a 4 p.s.i. shock wave that reaches the ground.

Figure 1 illustrates the area covered by the 4 p.s.i. contour for optimum height airbursts (Eq. 1) of various magnitudes. Also shown is the area covered by the 4 p.s.i. contour for different types of objects that are stopped in the atmosphere. Once the objects surpass the minimum sizes noted in Table II their 4 p.s.i. contours quickly approach those for an optimum height airburst.

The calculations in Fig. 1 labeled by the five types of objects are based on an explicit calculation of the expanding shock wave from a point explosion placed at the calculated airburst altitude (K. Zahnle, in preparation). Therefore these calculations are independent of the nuclear weapons data. These calculations allow us to treat fireballs that are large compared to a scale height. Such large explosions are relatively inefficient at producing shock wave damage at the ground because much of their initial energy is blown into space.

Based on comparison to atmospheric nuclear bomb tests of comparable damage potential, the Tunguska object, which struck Siberia in 1908, released an energy of 10 to 15 megatons and the explosion occurred near optimum burst height. The blast wave flattened some 2000 km^2 of unimproved

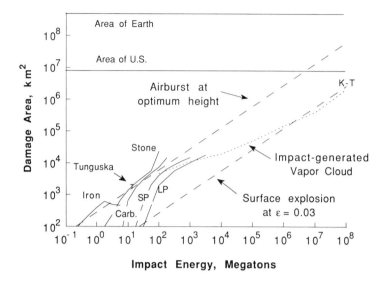

Figure 1. The area within the 4 p.s.i. overpressure contour as a function of the impact
    energy. The airburst expression assumes an optimum altitude for the explosion and
    that 100% of the impact energy goes into shock waves. These assumptions are both
    upper limits. The surface explosion curve is based on assuming that only 3% of the
    impact energy goes into shock waves as suggested by O'Keefe and Ahrens (1982)
    for the K-T event. The iron, stone, carbonaceous chondrite (Carb.), short-period
    comet (SP) and long-period comet (LP) curves are calculated as described in the
    text for impactors which deposit most of their energy in the atmosphere. Above
    a certain size (given in Table I) these bodies reach the ground, where their shock
    waves should not depend greatly on composition. The dotted curve is a calculation
    for the shock wave from the expanding vapor cloud from a stone which hits the
    ground. As for nuclear ground bursts, the stone hitting the ground produces a lesser
    shock than does a stone which explodes in the atmosphere. As the energy of the
    stone increases, it begins to blow away the atmosphere above the impact site which
    limits the amount of energy going into shock waves. For very high energies, the
    vapor cloud shock wave approaches the 3% efficiency suggested by O'Keefe and
    Ahrens (1982) for the K-T impactor.

coniferous forest. Tunguska is marked on Fig. 1. We note that there is no
direct determination of the kinetic energy of the Tunguska object. There are
some researchers who have suggested that its kinetic energy was much larger
than the 10 to 15 megatons that was released as shock energy (Turco et al.
1981,1982).

    We have also explicitly calculated the area within a 4 p.s.i. overpressure
contour for surface explosions. These calculations were done by modeling
the interaction of the expanding cloud of hot ejecta from the crater with the
atmosphere (Zahnle 1990; also see below). We find that blast waves from
low-energy surface explosions are comparable to airbursts (Fig. 1), as known
from nuclear weapons tests. However, for larger explosions the efficiency
of blast wave generation declines and approaches a few percent (O'Keefe

and Ahrens 1982). Effectively the coupling between the explosion and the atmosphere becomes inefficient when the characteristic radius reached by a strong shock, $R_s = 1.6Y^{1/3}$km (Y in megatons; note that a 4 p.s.i. shock is a weak shock), significantly exceeds the scale height of the atmosphere.

The K-T impact is also marked on Fig. 1. O'Keefe and Ahrens (1982) calculated that about 40% of the K-T impact energy (about 10^8 megatons) eventually reached the atmosphere, but that most of the energy went into long-term, low-grade heating and only 3% went to shocking air. In Fig. 1 we use Eq. (1) with $\epsilon = 3\%$ to estimate the area within the 4 p.s.i. contour for the K-T impactor. However, we note that this area estimate is hypothetical. There is no geologic evidence that defines the area over which the K-T impactor caused blast damage.

Nevertheless, Fig. 1 shows that blast waves from objects the size of the K-T impactor are not expected to devastate areas that are larger than a few percent of the Earth. Blast waves may cause much local damage, but are not a threat at the global scale.

III. TIDAL WAVES

The impact of a large object into the oceans is likely to generate an immense tidal wave. Bourgeois et al. (1988) present evidence for deposits from a tidal wave at the time of the K-T event whose amplitude they estimate to have been 50 to 100 m in Texas.

It is not practical to calculate the amplitude of tidal waves in detail because the amplitude depends upon the distance from the impact, the depth of the water into which the asteroid impacts, and the depth of the ocean between the impact site and the location of the wave. Moreover, the damage that might be caused by the wave depends upon its interaction with the coastal shelf where the wave comes ashore. Here we first estimate the amplitude of the waves in the open ocean generated by an impact. We do not expect tidal wave generation to be sensitive to the type of the impactor, except to the degree that various classes of objects are able to penetrate through the atmosphere (Table II). Next we determine how the wave will propagate in an idealized ocean of constant depth. Then we consider the effects of the tsunami along the coastal zone. Impacts which crater the ocean bottom, or the continents near the sea, may produce secondary tidal waves as the material ejected from the crater enters the water, or from the earthquakes the impact generates. We do not consider such tidal waves here. Rather we restrict ourselves to considering the waves generated by the impator itself hitting the ocean.

Laboratory experiments with high-velocity cm-sized projectiles led Gault and Sonett (1982) to conclude that the radius of the hole left in the water by an impacting body is approximately $R_w = 0.68\ Y^{1/4}$km (Y in megatons). Based on the discussion in Gault and Sonett (1982) we assume the amplitude of the wave at the edge of this hole D_0 would be the smaller of $0.66\ R_w$ or $0.3\ d_0$, where d_0 is the depth of the ocean. The ocean depth sets an upper limit on the amplitude of the wave produced by large impacts. In a 5 km deep ocean the transition occurs for an impact energy of about 125 megatons. Much of the energy of small oceanic impacts goes into the tsunami. For large impacts much of the energy of the impact initially may go into the waves, but this energy is largely dissipated at the ocean bottom. The fraction of the impactor energy going into the propagating wave drops in proportion to $(d_o/R_w)^2$. An impact of the size of the K-T object, with energy of about 10^8 megatons, into a 5 km deep ocean puts only about 0.1% of its energy into tidal waves according to these estimates.

We assume that the amplitude of the wave decays from a maximum value of D_o inversely as the distance from the edge of the hole. Then the area A contained within a particular amplitude contour D is given by

$$A(\text{depth} > D) = 0.29Y(1/D)^2 \text{ km}^2, Y < 0.2d_o^4$$

$$A(\text{depth} > D) = 0.13d_o^2 Y^{1/2}(1/D)^2 \text{ km}^2, Y > 0.2d_o^4 \tag{2}$$

where h and D are also in km.

Glasstone and Dolan (1977) present similar equations that were derived from underwater nuclear explosions in water of varying depth. Their equations suggest that

$$A(\text{depth} > D) = 0.05Y(1/D)^2 \text{ km}^2 \ Y < 28d_o^4$$

$$A(\text{depth} > D) = 0.2d_o^2 Y^{1/2}(1/D)^2 \text{ km}^2 \ Y > 10^3 d_o^4. \tag{3}$$

Equations (2) and (3) have similar forms. However, for small impacts Eq. (2) yields about 6 times more area, and for large impacts about 0.5 times as much area as Eq. (3) for the same energy. In addition, the cross-over point between the deep and shallow water equations occurs at significantly different energies, which is not clearly defined by Eq. (3). [Hills and Goda (1993) have used the Glasstone and Dolan (1977) deep water equations at all energies. This causes them to greatly overestimate the amplitude of the waves for larger impactors and to neglect the effects of ocean depth.]

The solid curves in Fig. 2 illustrate the area within which a 10 m or higher tidal wave would be present in a 4 km deep ocean, the mean depth of the Atlantic and Pacific Oceans, for various impactors and using Eq. (2). The dotted curve shows the extent of a 10 m wave in 2 km deep water, the average depth of the Gulf of Mexico, for a strong iron asteroid. The dashed curve, labeled "Glasstone & Dolan," shows the prediction of Eq. (3) in a 4 km ocean for a strong iron asteroid. The major difference between Eqs. (2) and (3) is at energies less than 10^3 megatons, where Gault and Sonnet's data (1982) suggest deeper waves. Whether this difference is real and represents the greater efficiency of an impact into the water compared to an explosion within the water, or is a problem in scaling the laboratory results to the real world, or is simply a representation of the uncertainty in the data and its application is unclear.

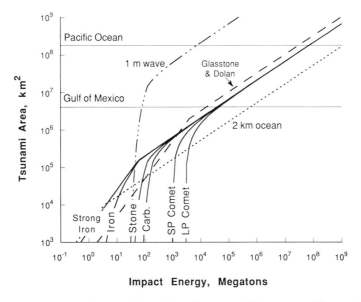

Figure 2. The area of ocean within which a deep water tidal wave would occur given an impact of specified energy. The 1-m wave curve is the area covered by waves 1-m in amplitude, generated by a stone hitting a 4-km deep ocean. The other curves are for the area within which 10-m amplitude waves occur for a 4-km deep ocean, except for the curve labeled "2-km ocean," which referes to a shallower ocean. All the calculations are based upon equations developed from the laboratory data of Gault and Sonett (1982), except for the curve labeled "Glasstone and Dolan," for which the equations came from the nuclear test data discussed by Glasstone and Dolan (1977).

A 10-m tidal wave on the open ocean is a very large wave. Even a 1-m wave could cause much damage, and for the same impact energy would reach 10 times farther and cover an area 100 times larger. The area covered by a 1-m wave is illustrated in Fig. 2 for the impact of a stony body in a 4 km deep ocean. A 10^4 megaton impact would create a one meter high, open-ocean tidal wave covering the entire Pacific Ocean. To put this in context, the 1960 Chilean tsunami, which killed 61 people in Hawaii and 199 in Japan, is calculated to have had an amplitude of 20 cm near Hawaii (Garcia and Butler 1977). Wave heights on the Hawaiian coast averaged about 3 m, but reached about 12 m in Hilo Bay, where most of the damage occurred (Eaton et al. 1961). Evidently special coastal configurations can strongly amplify the wave, causing localized damage. Van Dorn (1965) summarizes empirical studies of runup. He finds that in coastal regions with steep offshore slopes (1:40) the coastal waves are 10 to 20 times higher than the deep ocean waves. Broad offshore shelves or fringing reefs greatly reduce runup by a factor of 3 or so.

The literature on tsunamis provides further insight into the relative importance of impact generated waves. Van Dorn (1984) noted three tsunamis between 1952 and 1964 with wave energies exceeding 1 megaton. The 1960 tsunami, generated by a magnitude 8.6 Chilean earthquake, is thought to have been the largest of the century. The tsunami had an energy of about 5 megatons. Because impacts generate tsunamis efficiently, 5 megaton impacts in the Pacific would produce tsunamis comparable to the largest earthquakes. Oceanic impacts of this magnitude would be expected about once per century if the impactor penetrated the atmosphere; i.e., if it weren't for the atmosphere, impacts would produce tsunamis of the same frequency and magnitude as those from earthquakes. However, only iron impactors in this energy range actually do reach the surface (Fig. 2), and they are relatively infrequent. The threshold for a relatively common stony impactor to generate a significant tsunami is around 50 megatons, with a typical recurrence time of about a millennium. When such an event occurs, its tsunami is bigger than any that could be generated by an earthquake. We conclude, as did Hills and Goda (1993), that impacts into the ocean are likely to create tsunamis of a magnitude unprecedented in human history.

IV. THE EFFECTS OF DUST LOADING

The effects of dust loading on the climate, photosynthesis, and visibility—darkness at noon and impact winter—depend mostly on the quantity of submicron dust that reaches the stratosphere. First, we estimate the amount of dust of various sizes lofted by impactors with given compositions, energies and impact sites. Second, we address such factors as the global dispersal rate of the dust, and the time that the material remains in the atmosphere. Third, we calculate the effect of a given amount of dust on the Earth's radiation budget. Finally, we estimate the effect of low light levels on surface temperature, rainfall, and photosynthesis.

A. Large Dust Grains

The clay layer which marks the K-T boundary provides some direct information about the amount, properties, and distribution of dust produced by a large impact. The thickness of the clay layer is variable, and we are not aware of any statistical attempts to derive a global mean value. Often the average thickness is given as about 1 cm (Alvarez et al. 1980). However, after the impact carbonate deposition stopped for a time, but clay deposition did not, so that a portion of the K-T layer may represent the normal influx of clay to the stratigraphic column. Brooks et al. (1984) find that iridium is concentrated in the lowest 0.2 cm of the 0.8 cm K-T boundary clay from the location they studied. Smit and Romein (1985) describe a number of core samples and conclude that the thickness of the layer containing ejecta material (which they differentiate from the thicker clay layer) is less than 0.5 cm.

The ejecta layer contains large numbers of microtektite-like spherules. Smit and Klaver (1981) proposed that these spherules are the remains of the material melted by the impact. They may represent the bulk of the ejecta by mass, although their fractional contribution to the ejecta layer has not been quantified to our knowledge. We are not aware of published size distributions of the spherules, but Smit and Klaver (1981) state they are typically 0.5 to 1 mm in size, while Montanari et al. (1983) describe them as typically 0.2 to 0.5 mm in diameter with a maximum size of 1.3 mm. Hence the spherules are about the same size as small rain drops or drizzle drops. A 0.1 mm diameter sphere of density 2 g cm^{-3} will fall about 50 cm s^{-1}. The fall speed increases with the diameter and is independent of pressure except at high altitudes. Hence these spheres, if emplaced in the atmosphere at 100 km altitude, would remain airborne for less than two days. Although we show in Sec. V of the chapter that these particles can have a significant optical depth, and may be very important for starting global fires, they are of little importance for climate and photosynthesis due to their short lifetime.

One can quantify the production of large melt particles based partly on nuclear weapons tests (National Research Council 1985). Surface explosions create small particles mainly by vaporizing and melting rock, although there is also a small amount of sweep-up of surface materials by the blast waves. Therefore, the process by which nuclear weapons produce dust is similar to the melt droplet formation process during cratering discussed above for the mm-sized spherules. It is found that the dust mass in the stabilized plumes of large explosions is about 0.3 Tg/megaton (1 Tg = 10^{12}g), while the submicron fraction of the dust is about 8% (National Research Council 1985).

Zahnle (1990) shows (for a target and impactor of similar physical properties) that the mass of vapor and melt m_v generated by a high-velocity impactor of mass m_i cratering the surface is approximated by

$$m_v = m_i[(2/v)(4Qv/h)0.5]^{h-2} = 4Y v_i{}^{-h}[2(4Q_v/h)^{0.5}]^{h-2}. \quad (4)$$

Here v is the velocity of the impactor, Q_v is the latent heat of vaporization, and h is a constant with a value of about 0.6. If Q_v is 8×10^{10}erg g^{-1} and v_i is 25 km s^{-1}, then the vaporized and melted mass is about twice the mass of the impactor, or about 0.1 Tg/megaton. Hence the calculated melted mass per unit of energy from an impact is about the same as that observed to be lofted in the nuclear weapons tests.

O'Keefe and Ahrens (1982) and Melosh and Vickery (1991) have addressed theoretically the sizes of particles produced in impact-generated melts and vapors. Both groups concluded that the diameter of melt droplets should be on the order of microns at the (low) energies released in nuclear weapons tests, but hundreds of microns at the energy of the K-T impactor. In both studies the size of the droplets is proportional to the size of the ejecta plume which in turn is proportional to the size of the impactor. Hence the small sizes of the nuclear test particles, and the larger sizes of the clay layer spherules appear to be consistent with a melt/vapor origin.

Figure 3 illustrates the total mass of dust lifted by impacts for a range of energies using the expressions derived from nuclear surface tests, which

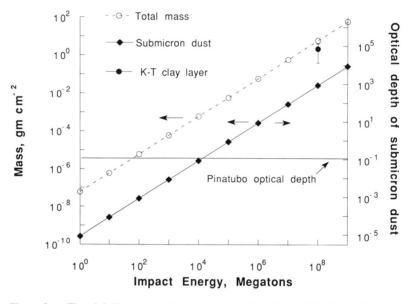

Figure 3.   The globally averaged mass concentration of dust lofted into the stratosphere and the optical depth of submicron dust, as a function of the yield of the
explosion. The total mass curve is derived from nuclear weapons explosions data,
and models of the amount of vapor and melt created by an impactor. The submicron
dust mass is determined from the amount of pulverized rock that is calculated to
be produced by an impact and from data on the mass fraction of submicron dust
from impacts. The K-T clay layer mass is based on typical estimates of the energy
of the impact, having an uncertainty of an order of magnitude (not illustrated), and
estimates of the thickness of the clay layer at the geologic boundary which vary
from 0.2 to 2 cm (a density of 2 g cm^{-3} was assumed). The optical depth for the
submicron dust, determined using a fixed extinction cross section discussed in the
text, can be determined from the right-hand scale.

as discussed above are within a factor of 3 of the melted and vaporized mass
determined from cratering studies. Most particles are too large to linger in
the atmosphere except for those originating from very small impacts. Also
shown in Fig. 3 is the amount of material found in the impact layer at the K-T
boundary. We have assumed that the density of the clay layer is 2 g cm^{-3},
and that the clay layer is between 0.2 and 2 cm thick. Remarkably, the clay
layer mass falls within the extrapolations of the nuclear weapons information.
A precise comparison cannot be made, because the energy of the K-T impact
is not known well. These results are consistent with the bulk of the K-T
layer being composed of large melt particles, which would not be optically
significant.

B. Submicron Dust

There remain several plausible sources of optically important submicron material from an impact. First, some fraction of the melted and vaporized ejecta

will lie in the submicron size range. Second, the melted and vaporized ejecta from a large impact will be partially blown into space. Re-entering debris with sizes larger than $100 \, \mu m$ should partially ablate (Hunten et al. 1980). The recondensed vapor could produce small particles. If the spherules are distributed ballistically (see below), late arriving spherules will collide at velocities of order 5 km s^{-1} with those spherules that had arrived earlier; these collision velocities are high enough to chip and fragment the particles. Finally, the impact not only vaporizes and melts debris, but the shock wave also pulverizes a significant mass of the target.

Bohor et al. (1987) have shown that the K-T ejecta layer contains grains of shocked quartz. These grains, which are in the hundred micron size range, are evidence of a component of the ejecta which has been shocked, but not melted. Using an argument analogous to that developed for vaporization by Zahnle (1990), we estimate that a mass $m_p = m_i (v_i / 2u_{He})^{2-h}$ is pulverized by impact. Here u_{He} is the Hugoniot elastic velocity (roughly 4×10^4cm s^{-1}). At $v_i = 25$ km s^{-1}, the mass of pulverized rock is about 125 times the mass of the impactor. O'Keefe and Ahrens (1982) performed numerical simulations of the K-T impact in which the mass of pulverized material hurled into the atmosphere was about 100 times the mass of the impactor. They also examined the size distributions of pulverized rock from nuclear tests and from laboratory impact studies and concluded that 0.1% was submicron in diameter. (By contrast, as noted earlier, 8% of the rock in stabilized nuclear clouds was submicron in size). Based on these arguments, we will assume 89% of the rock in stabilized nuclear clouds was submicron in size.) Based on these aruguments we will assume that a mass of material equal to 10% of the bolide mass (0.1% of the total ejecta) reaches the stratosphere as submicron dust.

Figure 3 shows the mass of submicron dust lofted by an impact. Here, we assume that the only source of submicron dust is 0.1% of the mass of the pulverized rock component, as discussed previously. Of course, most of the pulverized material is too large to have been globally distributed in the K-T layer. Comparing the total mass and submicron mass curves of Fig. 3 shows that the submicron dust is enriched by gravitational separation in the K-T clay layer relative to the pulverized rock component. We find from Fig. 3 that slightly less than 1% of the K-T boundary layer material may have been submicron in size. However, we emphasize that there are no data from the K-T layer supporting this mass fraction estimate. This estimate is uncertain by at least an order of magnitude. It may be that much of the pulverized material was trapped within larger rock fragments in the impact and was not globally distributed, suggesting a lower mass fraction of submicron material. Alternatively, there may have been other sources of submicron material we have not considered which would increase the mass fraction of submicron material. Toon et al. (1982) assumed that the entire mass of the clay layer represented submicron sized dust. Hence our current estimates of the submicron mass are a factor of 100 lower than those previously used in climate simulations.

The preceding discussion was largely oriented toward impacts occurring on the continental surfaces. Objects which do not reach the surface, such as the Tunguska meteorite, cannot contribute more than their own mass to atmospheric dust. It is likely that only a small fraction of the mass of a pulverized body actually lies in the submicron size range. Table II shows that only impactors with energies below 10^4 megatons will disintegrate in the atmosphere. As discussed below the amount of submicron mass injected by 10^4 megatons surface explosions (10% of the impactor mass) is not important climatologically. We conclude that airbursts will not contribute significantly to atmospheric dust loading.

Most asteroids and comets colliding with Earth strike the ocean. Gault and Sonett (1982) estimate that objects larger than 1/20 of the ocean depth will crater the ocean floor, while O'Keefe and Ahrens (1982) estimate those larger than 1/15 the ocean depth will begin to crater the ocean bottom. For example, a 10^4 megaton impact in a 5 km deep ocean would leave a crater. There is no reason to think that the amount of dust injected into the atmosphere from these oceanic cratering events will differ significantly from that injected by continental cratering events. Hence, we conclude that oceanic impactors with energies below 10^4 megatons can be ignored as contributors to atmospheric dust, while larger impacts can be treated using the same dust mass injections discussed above for continental impacts. We return to the fate and importance of the water injected by ocean impacts in Sec. VIII of this chapter.

C. Global Distribution of Dust

Before drawing conclusions about the environmental effects of impact generated dust we need to consider the mechanisms for spreading material over the Earth, and the lifetime of the dust in the atmosphere.

Large impacts may distribute ejecta ballistically over the planet (Zahnle 1990; Melosh et al. 1990), covering the planet nearly instantaneously. However, even impacts with energies on the order of 10 to 100 megatons are capable of lofting material into the stratosphere based on observations following nuclear tests and volcanic eruptions (Jones and Kodis 1982). Studies of similar situations in the context of nuclear winter imply global dispersion of the dust due to induced motions in the stratosphere (Malone et al. 1985,1986). Covey et al. (1990) performed three-dimensional simulations for a dust cloud containing micron-sized particles with a mass corresponding to an impact with an energy of 6×10^5 megatons. They found that this dust cloud spread to cover most of the Earth within a few weeks. Forced global spreading is not observed for volcanic clouds of moderate optical thickness. However, evidence for induced vertical motions in the tropics from the Pinatubo cloud suggests that such eruptions are on the verge of forcing significant changes in atmospheric circulation (Kinne et al. 1992). It would appear, therefore, that a dust cloud dense enough to alter significantly the radiation field regionally would also alter the dynamics of the stratosphere to ensure the rapid global distribution of the dust.

Toon et al. (1982) performed microphysical simulations of the evolution of a dust cloud in the stratosphere. They showed that the optical depth of the submicron dust from a large impact did not change significantly in the first month, but thereafter declined rapidly, so that little dust remained after about 6 months. These calculations considered only submicron dust, but larger particles if present would only briefly enhance the optical depth. Micron-sized dust typically falls out of the stratosphere in a few months, as is observed following large volcanic eruptions. For example, after the Pinatubo eruption a lingering layer of dust was observed by lidar and *in situ* sampling just above the tropopause (12–14-km) for about 6 months after the eruption, although no dust remained at altitudes of 18 km (Pueschel et al. 1994). (The longer lasting volcanic aerosol was composed of submicron sized sulfuric acid droplets created photochemically from the erupted sulfur dioxide.) Toon et al. (1982) showed that coagulation would occur rapidly in a dust cloud with even 1% of the mass of the K-T clay layer. Such coagulation would lead to large particle formation, and more rapid sedimentation removal than is indicated by the sizes of the original dust grains. Based on these results, we conclude that the optical effects of dust from even the largest impacts last no more than 6 months.

The removal of dust may occur by washout, as well as by sedimentation. A large impact into the ocean would not only hurl a cloud of dust into the atmosphere, but would also loft significant amounts of water vapor. Initially the dust particles and the water vapor may be segregated due to the large fall velocities of the dust particles at high altitudes. However, as discussed below, the water vapor would eventually condense to form ice crystals, which subsequently would fall out. In the lower atmosphere, ice clouds are observed to form preferentially on only a small fraction of the aerosols present in the atmosphere, although that fraction depends upon the circumstances of the ice cloud formation (Jensen and Toon 1994; Toon et al. 1989). Therefore, removal by direct incorporation of the dust into ice particles is likely to be inefficient. Falling snow flakes would collide with dust particles and scavenge them rather efficiently due to their high cross sectional area, but ice crystals would be less efficient at collisional removal of the dust. In the absence of any numerical simulations of these effects, it can only be noted that the ice clouds formed by oceanic impacts have the potential to sweep some or all of the dust from the sky. Unfortunately, the efficiency and time scale of this cleansing is not known at present. Of course, the ice crystals themselves may be radiatively significant as discussed in Sec. VIII of this chapter.

D. Radiative Effects of Dust

The optical depth is the fundamental parameter determining the influence of the dust on the radiation balance of the planet. It can be calculated given the mass and sizes of the dust particles and an assumed region of dispersal. The size distribution of submicron dust in low yield nuclear debris clouds is log-normal, with a mean radius of about 0.25 μm and a width (σ) of 2 (National Research Council 1985). The specific extinction coefficient (i.e., the extinction cross section per unit mass of dust) that results, $k_e = 3 \times 10^4 \text{cm}^2\text{g}^{-1}$, is near the highest value one would expect for dust. For larger particles the extinction would drop at least as rapidly as the inverse of the radius. The optical depth (Fig. 3) is $\tau = Mk_e$, where M is the submicron mass in the atmosphere per unit surface area. For reference, the optical depth of sulfuric acid aerosols resulting from volcanic eruptions such as that of El Chichon or Pinatubo is about 0.1 (Fig. 3). Such aerosols have optical properties similar to dust.

The fraction of incident sunlight reaching the surface through a layer of dust is the transmission of the layer. The transmission in a multiple scattering atmosphere dominated by dust or smoke is approximately an exponential function of the optical depth of the smoke or dust. The transmission $T = Ae^{-\tau/b}$, has $A = 0.9$, $b = 6.22$ for dust and $A = 0.8$, $b = 1.03$ for smoke (National Research Council 1985). These functions are plotted in Fig. 4. Soot is much more effective than dust in reducing the amount of sunlight reaching the surface. Dust and soot scatter sunlight back toward space, but soot also absorbs sunlight. In nuclear winter studies the estimated smoke absorption optical depth is about 2; hence the total smoke optical depth (including scattering) is about 3 (Turco et al. 1990). Figure 4 shows that a dust optical depth of about 20 would be required to have similar effects on sunlight transmission.

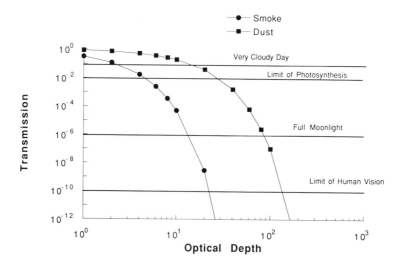

Figure 4.   The transmission of visible light through a given optical depth of smoke or dust. Smoke is less transmitting than dust because it absorbs much of the light. Also noted in the figure are the light levels which correspond to various natural phenomena.

Figure 5 combines the information in Figs. 3 and 4 yielding the effect of the submicron dust from a given impact on the transmission of sunlight. Also shown is the transmission which results if it is assumed that all of the dust produced from the vaporized portion of the ejecta material is submicron in size, which is equivalent to assuming that the entire K-T clay layer is submicron material. Toon et al. (1982), who assumed that the entire K-T clay layer is submicron material, concluded that impacts much smaller than the K-T event would produce significant loss of sunlight. However, the lower estimates of the submicron dust fraction developed here (Figs. 3 and 5) suggest that the K-T impact is near the boundary where significant radiative effects occur.

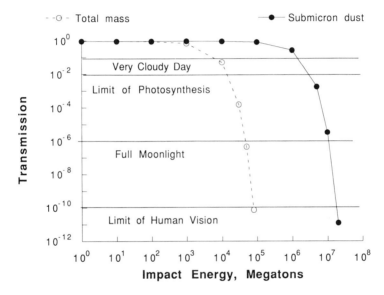

Figure 5.   The transmission of light as a function of the kinetic energy of the impacting body assuming either that all of the dust placed in the atmosphere is submicron (total mass), or that only 10% of the mass of the impactor (submicron dust) is injected as submicron dust. Also noted in the figure are the light levels that correspond to various natural phenomena.

E. Climatic Effects of Dust

We can now roughly assess the potential for impactors of various sizes to affect the global environment. However, it should be remembered that our estimate of the submicron mass fraction resulting from an impact is uncertain, and that the potential for the dust to be removed by falling ice crystals is also unknown. Therefore, there is a significant uncertainty—at least an order of magnitude— in the energies needed to produce various climate effects. Based on Fig. 3 we conclude that impacts with energies less than 10^4 megatons (which includes all impactors stopped by the atmosphere) do not inject enough submicron dust to have an optical depth that is larger than that of historic volcanic eruptions such as El Chichon (1982) or Pinatubo (1991). The radiative effects of dust and volcanic sulfuric acid clouds should be nearly the same, so that the observed volcanic optical depths and climate effects provide a direct means for scaling the climate effects of impacts. Hence, the climate effects of these impacts are within the noise level of year to year climate fluctuations.

Impacts with energies of 10^5 megatons should produce optical depths on the order of unity. Such optical depths may have occurred often during the past millennia following large volcanic eruptions such as that of Tambora (Stothers 1984), although no direct observations of such optical depths are available. Nevertheless, the resulting climate changes should be slightly larger than year to year variability.

The energy region between 10^5 and 10^6 megatons is a gray area spanning small effects and those that are obviously significant on a global scale. Covey et al. (1990) investigated a dust cloud with the global equivalent optical depth of about 3. According to Fig. 3 this optical depth could result from an impact with an energy of about 6×10^5 megatons. Using an atmospheric general circulation model (GCM) to simulate the climatic changes induced by the dust, Covey et al. (1990) found that the global average land temperature dropped by 8 K during the first two weeks after the simulated impact. Maximum cooling occurred in the Northern Hemisphere, where the impact and dust injection were assumed to occur, with large areas cooling from summertime norms down to 0 to 10°C. By 30 days after the impact, however, the dust had spread globally, diluting its effect enough to allow temperatures to recover to pre-impact values. This scenario, occurring at the height of the growing season, would have a severe impact on human agriculture, similar to and possibly considerably exceeding the "year without a summer" experienced in North America and Europe after the Tambora eruption (Stommel and Stommel 1983).

Impacts with energies above about 5×10^6 megatons should lower light levels below that required for photosynthesis, leading to the possibility of major effects on the biota. Impacts larger than 10^7 megatons should lower the light levels below the limit of human vision, making it difficult for animals without flashlights to forage for food. Such impacts approach the scale of the K-T event and would produce severe and longer lasting cooling.

Covey et al. (1990) used a general circulation model of the atmosphere (GCM) to consider a K-T event with an optical depth of 3000. In this case the global average land temperature dropped to about 0°C during the first 10 days after the impact and remained at this level until day 20, when the simulation ended. Examination of the geographical distribution of surface temperatures (Covey et al. 1990) shows that land areas near the oceans were up to 10 K above freezing while continental interiors were significantly below freezing. (Such models are not capable of predicting frost formation, but sporadic frost formation in the coastal regions is possible.) The warmer temperature along the coasts, a normal climatic feature of the wintertime Earth, is due to the immense heat capacity of the oceans. Even if no sunlight reached the surface, the oceans would cool only slightly during the first few weeks after an impact.

Covey et al. (1994) used a GCM to continue the simulations of Covey et al. (1990) for a year after impact. They found that the low land surface temperature found after 10 days were maintained for a year, despite removal of most of the dust after six months. This prolonged cooling occurred because the oceans, having been cooled during the first six months, take a long time to warm back to climatological norms. The thermal lag time of the climate system has been estimated at 10 to 100 yr in the context of anthropogenic global warming (Hoffert et al. 1980; Hansen 1985). Considering the relatively brief lifetime of the dust, the lower end of this time scale is an appropriate guess at how long the climate would take to recover after a K-T scale impact.

An additional climatic consequence of the dust injection may be worldwide drought. In the simulations of Covey et al. (1994), precipitation decreased by 95% for several months, recovering to half its pre-impact average at the end of one year of simulation. In these calculations, high-altitude heating of the air due to the dust's absorption of solar energy stabilized the atmosphere, suppressed convection and thereby confined precipitation to a thin layer immediately above the oceans. Such a result must be considered speculative, however, given the notorious difficulties involving the simulation of the hydrological cycle in GCMs.

V. FIRES

There is evidence at the K-T boundary for considerable soot production from burning forests (Wolbach et al. 1985,1990). The amount of soot, about 11 mg cm^{-2} or 5×10^{16} g for a worldwide layer, requires ignition of a large fraction of the Cretaceous world's ≈ 1 g cm^{-2} carbon in the above ground biomass (currently the world has about 0.2 g cm^{-2} of above ground biomass carbon) with very efficient conversion of the biomass into soot. It is unlikely that total combustion of the world's forest would occur from ignition near the impact site. However, for large impacts the debris from the crater explodes into space and re-enters the atmosphere over much of the globe (Argyle 1989; Zahnle 1990; Melosh et al. 1990). This re-entering material reaches a high temperature, and its downward thermal radiation can ignite fires over most of the Earth (Melosh et al. 1990).

The threshold energy for ballistic distribution can be estimated by comparing the mass of high-speed ejecta to the mass of air that must be thrust aside. Impact velocities on Earth are generally high enough that the impactor and a roughly comparable mass of target material are largely or partially vaporized or melted (Vickery and Melosh 1990; Melosh 1989). Hydrodynamic expansion of the hot vapor/droplet plume is the source of most high speed ejecta. The expansion velocity of the plume is determined approximately by the thermal energy of the vapor. The initial specific thermal energy of the vapor is of order $v_i^2/8$; thus the expansion velocity into a vacuum would be of order $v_i/2$ (Melosh and Vickery 1991). Global ballistic dispersion of ejecta demands ejection velocities of order 7 km s^{-1}, similar to that needed to attain a circular Earth orbit.

For simplicity, we treat the ejecta plume as a hemispherical cloud of hydrodynamically expanding rock (or water) vapor (Zahnle 1990). The vapor cloud is assumed to expand homologously and isotropically. Its interaction with the atmosphere is limited to imposing conservation of momentum so that along any trajectory the velocity at the leading edge of the cloud accounts for the inertia of the intercepted air. Thus if the vapor cloud is small it is stifled by the atmosphere; if it is large it breaks through. The smallest plumes to break through do so at the zenith, where the atmosphere is thinnest. Larger plumes open wider cones. As the impact energy increases, the cone widens, finally descending into a plane tangent to the surface, for which everything above the horizon is ejected (Fig. 6).

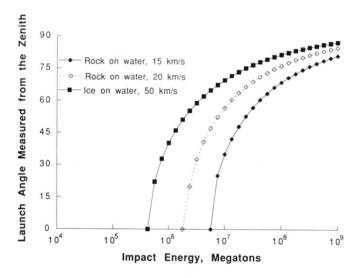

Figure 6.   The angle (measured from the zenith) above which ejecta escape the atmosphere with velocities >7.9 km s^{-1}, the velocity of a circular orbit skimming Earth's surface. This material can reach any place on Earth. The threshold for global ballistic dispersal is at roughly 10^6 megatons. The precise value of the threshold depends somewhat on impact velocity as well as on target and impactor composition. The figure shows results for impactors of several compositions hitting water targets. The results for rock targets differ only slightly from those shown for water targets.

Much of the ejecta that is lofted to space in the expanding vapor plume will condense and re-enter the atmosphere as sub-millimeter sized spherules. The re-entering ejecta stops mainly in the mesosphere, which heats up accordingly. The temperature of the mixed layer of particles and air can be estimated from conservation of energy, in which the kinetic energy of the particles is balanced against the specific heats of air and particles, the latent heats of melting and vaporization of the particles, and prompt radiative cooling of the layer (Zahnle 1990; Melosh et al. 1990). If the impact is large enough, the particulate optical depth is very large over the entire planet (Fig. 7), the temperature is high (Fig. 8), and downward thermal radiation a significant factor for life at the surface (Fig. 9).

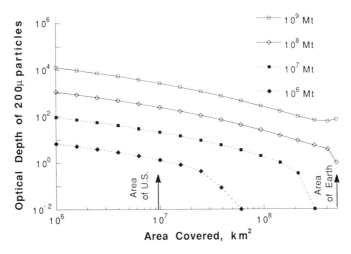

Figure 7.   The optical depth of the ballistic ejecta in the mesosphere as a function of the area of the Earth covered with material. In this example it is assumed that a rocky impactor has struck a rocky surface at 15 km s^{-1}, and that the ejecta that are entering the upper atmosphere have a typical size of 200 μm. For the same impact energy other types of impactors will cover a larger area of the Earth with a high optical depth of material, as can be determined from Fig. 6.

Figure 9 plots the total energy deposited (erg cm^{-2}) in the atmosphere by ballistic ejecta. Roughly half of this energy is radiated towards the surface. Perhaps half of the radiation directed toward the surface would penetrate the atmosphere and reach the ground. Because a wide range of orbits can reach a given place on Earth, this energy arrives over a period of tens of minutes. Over a comparable time, the radiation from an overhead Sun amounts to about 3×10^9erg cm^{-2}. Over much of the Earth, 10^7 megaton impact delivers an energy flux which is similar to solar radiation for tens of minutes. Larger impacts deliver proportionately more energy.

The energy deposited is compared to an estimated threshold for lighting fires in Fig. 9. For prolonged (>20 min) exposure, spontaneous ignition of both wet and dry wood begins at radiant fluxes exceeding 1.25×10^7erg s^{-1}cm^{-2} (Melosh et al. 1990); thus in Fig. 9 the approximate threshold for lighting fires is an energy input of 3×10^{10} to 10^{11}ergs cm^{-2}, with allowance for radiation to space and radiation absorbed by the atmosphere. For a K-T scale event ($\sim 10^8$ megaton) the amount of thermal radiation could have been

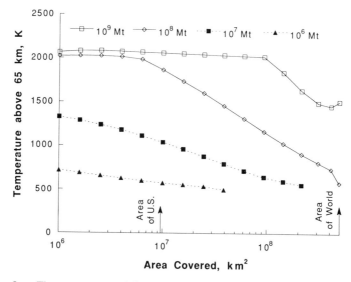

Figure 8.   The temperature of the re-entering ballistic ejecta in the mesosphere as a function of the area of the Earth covered. In this example it is assumed that a rocky impactor has struck a rocky surface at 15 km s^{-1}, and that the ejecta that are entering the upper atmosphere have a typical size of 200 μm. Other types of impactors will cover a larger area of the Earth with high-temperature material, as can be determined from Fig. 6. These temperatures generally are not high enough to melt rock, but they are high enough to radiate a significant amount of thermal energy.

high enough to ignite fires anywhere on the globe.

Not only are these fires likely to have killed many organisms, but the opacity of the smoke generated by the fires may have contributed to the opacity of the dust and augmented the surface cooling. Using the measured soot layer mass, and assuming a specific extinction coefficient for soot of $10^5 cm^2 g^{-1}$ yields a global soot optical depth of about 1000. Such a soot optical depth is capable of causing similar climatic effects to those discussed in the previous section for dust. The measured soot layer implies an enormous production of pyrotoxins. Crutzen (1987) estimates 10^{19}g of CO_2, 10^{18}g of CO, 10^{17}g of CH_4, and 10^{16}g of N_2O. This amount of CO_2, for example, is about three times the amount currently in the atmosphere. The enhanced CO_2 could lead to a temperature increase after the smoke cleared, but the increase would be limited by transfer of CO_2 into the oceans and back into the terrestrial biomass. At the time of the K-T event the ambient atmospheric CO_2 may already have been about 4 times higher than it is now so that this additional CO_2 from fires may have been of small consequence (Kasting et al. 1986).

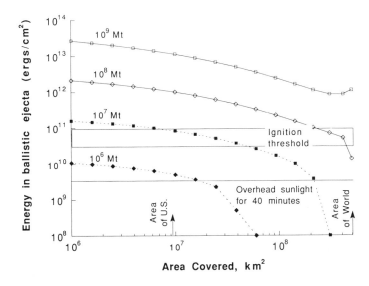

Figure 9.   The total energy deposited in the atmospere by the re-entering ballistic ejecta as a function of the area of the Earth covered. A large fraction of this energy, perhaps as much as 25%, would be re-radiated and reach the surface. In this example it is assumed that a rocky impactor has struck a rocky surface at 15 km s^{-1}, and that the ejecta that are entering the upper atmosphere have a typical size of 200 μm. Other types of impactors will cover a larger area of the Earth with high thermal radiation doses, as can be determined from Fig. 6.

VI. ACID RAIN

Acid rain has been a popular K-T extinction mechanism (Lewis et al. 1982; Prinn and Fegley 1987). Acid rain forms from the nitric oxide that is inevitably produced by strong shock waves in air. It is also possible to form acid rain from sulfur dioxide (or even from carbon dioxide) that is produced if the impact drives shock waves through a sulfate- (or carbonate-) rich target. Carbonate or sulfate production of acid rain is relatively unlikely, because only a small portion of the globe is covered by suitable sedimentary deposits. Because the K-T impactor appears to have struck just such a deposit, acid rain at the K-T boundary may have been much greater than usual, for an impact of that scale. Fires also produce nitrogen oxides and other chemicals that may lead to acid rain. Here we consider acid rain from NO produced in impact-generated shock waves.

There are four major processes for generating NO in an impact (Zahnle 1990):

1. NO is formed in shock waves when the meteoroid passes through (or explodes in) the atmosphere. We model the ablation, disruption, and deceleration of impacting bodies in the atmosphere following Chyba et al. (1993). This process is relatively efficient because the shocked gas

cools very quickly; this is the only process that can occur for smaller impacts. Typical yields of NO from shocked air are of order 10^9 moles per megaton of energy released in the atmosphere.

2. If the impactor is large enough to penetrate the atmosphere it craters the surface. NO is formed as the ejecta plume moves through the atmosphere, as long as the ejecta plume is moving faster than about 2 km s^{-1} (lower velocity shock waves in air do not get hot enough for N_2 to react with O_2 in order to produce NO). NO production is local to the impact site. A rough upper limit is reached when all the air above a plane tangent to the surface of the Earth is shocked, which yields no more than about 10^{15} moles of NO.

3. If the ejecta plume is massive enough and energetic enough, it punches through the atmosphere and enters near space. Most of the ejecta is launched on suborbital trajectories, but some escapes the gravity of Earth. This mechanism distributes debris ballistically over the entire globe (Argyle 1989; Zahnle 1990; Melosh et al. 1990). When the ejecta particles re-enter the atmosphere they can each produce a local shock wave. The NO formed in these shocks is distributed globally, but is largely confined to the mesosphere because typical ballistic particles are small enough ($100-500\mu m$, based on known terrestrial microtektite deposits) to be stopped at high altitudes. If all of the air above 50 km is processed in this way, the NO mixing ratio above 50 km can reach 1%, and the total NO yield can exceed 10^{15} moles.

4. If the impact is truly enormous, the energy content of the ballistically distributed ejecta is so large that its re-entry into the atmosphere will cause the entire atmosphere to be radiatively heated to $T > 1500$ K (Sleep et al. 1989). The atmosphere may become more than 0.2% NO, equivalent to 3×10^{17} moles of NO produced. Of course, in such an event acid rain mitigation would not be a priority.

The amount of NO produced by impact is shown as a function of initial kinetic energy in Fig. 10. Algorithms for NO production by the ejecta plume and by the atmospheric re-entry of ballistic ejecta closely follow those presented by Zahnle (1990) and Melosh et al. (1990).

According to Lewis et al. (1982), it takes some 3×10^{16} moles of HNO_3 to lower pH sufficiently to dissolve calcite in all the world's surface waters, assuming a mixed layer 75 m deep (this would correspond to a 600 micromolar solution of nitric acid). Calcareous plankton are considered to dissolve with the calcite. In Fig. 10 we assume that all impact-generated NO is converted to nitric acid, and that all the nitric acid quickly reaches the ocean, where it is confined to a surface mixed layer 75 m deep. We then estimate the fraction of Earth's surface waters that could be made acidic enough to dissolve calcite. We find, for example, that, if optimally distributed, 3×10^{14} moles of NO generated from a K-T sized impact could render as much as 1% of the world's surface waters corrosively acidic. It is clear from Fig. 10 that it is difficult

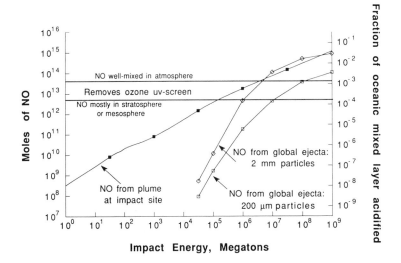

Figure 10. The number of moles of NO produced (1) by the entering body and its ejecta plume, if any, as it is blown skyward, and (2) by the re-entering ejecta assuming two different particle sizes. The consequences for these NO injections for ozone loss and for acidifying the ocean surface layers are also noted. Global ozone loss should occur for impacts of order 10^6 megaton, but acidification of the ocean surface layers by this mechanism is unlikely.

to generate enough acid rain by impact generation of NO to produce serious global effects.

There are two reasons why our estimates of NO generation are lower than those of Lewis et al. (1982) and Prinn and Fegley (1987). First, we find that much of the atmospheric energy deposition occurs at high altitudes where only a limited number of air molecules are available to be heated to high temperatures. The earlier studies assumed that the ejecta would deposit their energy at low altitudes where a greater mass of air could be heated. Second, we produce fewer molecules of NO per megaton of energy release. In the earlier studies it was assumed that the shocked air would cool quickly, as is the case for relatively small events like lightning bolts or nuclear weapons tests, and so the yield of NO would be fairly high. Here we allow the large volume of air shocked by the impact to cool more slowly, reducing the yield of NO.

Although the amount of nitric acid produced by a large impact is not enough to acidify the ocean surface layers, it is enough to produce a significant amount of acid rain. The global rainfall rate is about 5×10^{20} g yr^{-1}. We assume that 5 yr is required to remove all of the NO from the atmosphere, because much of the generated NO would be initially located at very high altitude. Then the typical concentration of nitric acid in rain with a production of 3×10^{14} moles of NO would be about 100 micromolar. Such concentrations of nitrate are already found in many regions experiencing high levels of acid rain. Near the impact site the rainfall following the impact may be acidified beyond current experience due to the generation of NO locally. However, globally the acidification of the rain may not be much different that currently is experienced in the eastern United States and in Europe.

VII. OZONE DEPLETION

Although conversion of impact-generated nitric oxide to acid rain is probably not a primary threat to most life on Earth, the ease with which nitrogen oxides can wipe out the ozone layer is a serious concern (Turco et al. 1982). According to Kasting and Ackerman (1985) it takes a volume NO mixing ratio of 2×10^{-7} to render the ozone ultraviolet screen ineffective. This is equivalent roughly to 4×10^{13} moles of NO, if uniformly mixed through the atmosphere. But, because almost all of the NO generated by an impact is put in the stratosphere and mesosphere, an impact generating some 5×10^{12} moles of NO would probably suffice to cleave the ozone shield.

After a major impact, the atmosphere would be highly perturbed chemically. Not only would nitrogen oxides be generated by the impactor, but also fires would pollute the atmosphere with a variety of compounds, and photolysis would be significantly reduced by low light levels. With low photolysis rates, current oxidation mechanisms would cease, allowing the build up of reduced compounds in the atmosphere. In addition, the normal wind patterns would be disturbed in unpredictable ways. The ozone layer could be depleted by the injection of large quantities of nitrogen oxides, by reactions occurring with dust and smoke particles, and heating of the ozone layer caused by re-entering debris, thermal radiation and enhanced solar absorption. The exact nature of these effects and their duration is impossible to predict accurately.

Because nitrogen dioxide is strongly absorbing in the near ultraviolet region, the NO_2 formed from the NO would greatly attenuate solar ultraviolet radiation (this, of course, applies to the period after the densest dust and smoke clouds had dissipated). Hence, in the early aftermath of a large impact, the danger of ultraviolet irradiation is probably negligible. (The NO_2 absorption would augment any surface cooling due, for example, to large dust loading.) However, after the NO_x had dissipated into the lower atmosphere, where removal is relatively rapid, and its mixing fraction reduced to one ppmv or less, the ozone layer could remain severely depleted.

The K-T event may have been a special case due to its impact on an evaporite deposit. After the clearance of the dust from the upper atmosphere in a few months, a long-lived dense aerosol layer may have been formed in the stratosphere as a result of sulfur injection (see below). The aerosols would have two important effects. First, the aerosols would speed the conversion of nitrogen oxides into nitric acid, which could then condense in the aerosols. While this would reduce the catalytic activity of NO, the aerosols could also activate other catalytic agents, notably chlorine. Second, the aerosols would scatter solar radiation and prevent large doses of ultraviolet radiation from reaching the surface (Vogelmann et al. 1992). The outcome, while uncertain, probably implies small ultraviolet radiation enhancements.

In general impacts with energies above about 10^5 megatons would significantly deplete the ozone layer. After the removal of the dust and NO_2, the latter by conversion to HNO_3, significantly increased doses of ultraviolet radiation would reach the surface for a period of perhaps several years. In the case of the K-T impact, however, the possible presence of a shielding layer of sulfate particles may have blocked some of the ultraviolet from reaching the surface.

VIII. WATER INJECTIONS

Emiliani et al. (1981) suggested that a massive injection of water vapor into the upper atmosphere may have occurred as a result of the impact at the K-T geologic boundary. The upper atmosphere currently has a very low humidity. The result of a substantial increase in upper atmosphere water vapor would be to increase the Earth's surface temperature due to the greenhouse effect (Manabe and Wetherald 1967).

In order to determine the potential for an impact to actually change the amount of water in the upper atmosphere, we must first determine the amount of water that could be lofted, and the region of the atmosphere where the water is injected. For reference the mass mixing ratio of water vapor currently is about 3 ppm above the tropopause yielding a water vapor mass above the tropopause of 2×10^{-4} to 6×10^{-4} g cm^{-2}.

Following the arguments in Zahnle (1990), and in parallel to the discussion presented above for vaporized rock, the amount of water vaporized by an impact (ignoring the density difference between the impactor and water) is given by $m_v = m_i \{2[0.5v_i(h/4Qv)^{0.5}]^{2-h} - 1\}$ if the radius of the impactor r_i is less than 0.75 d_o where d_o is the ocean depth. Otherwise, $m_v = m_i \{(0.75)d_o/r_i)[0.5v_i(h/4Qv)^{0.5}]^{2-h} - 1\}$. Here we assume that $Q_v = 2.5 \times 10^{10}$ erg g^{-1}. For an object substantially smaller than the ocean depth hitting Earth at 25 km s^{-1} this equation indicates that an amount of water about 11 times the mass of the impactor will be vaporized. The mass production rate is about 0.15 Tg of vaporized water per megaton of impact energy. For a 5 km radius body impacting in a 5-km deep ocean, an amount of water is vaporized which is about 3 times the mass of the impactor. Figure 11 illustrates this relationship as a function of impactor energy. Even impacts with energies as low as 10^4 megaton are capable of vaporizing an amount of water which is comparable to that above the tropopause.

In addition to vaporized water there will be water that is splashed to high altitude. Ahrens and O'Keefe (1983) investigated the amount of water lofted above the tropopause for an impact of a 5-km radius object into a 5-km deep ocean at 30 km s^{-1}. For a comet hitting the ocean they concluded that an amount of water equal to about 10 times the mass of the comet would be

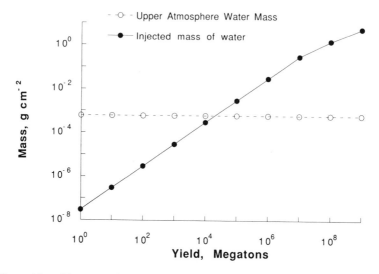

Figure 11.   The mass of water vaporized by a cometary impact of a given energy (at 25 km s^{-1} into a 5-km deep ocean) is compared with the mass of water that is present in the upper atmosphere at the current time.

lofted above the tropopause. A silicate impactor might loft 30 times its mass of water into the upper atmosphere. These numbers are consistent with those shown in Fig. 11, although somewhat larger because they include splashed as well as vaporized water.

Even for modest impacts with energies above about 10^2 megatons the water cloud will rise beyond 100 km (Jones and Kodis 1982), where it will form a great steam cloud. Portions of the cloud will be cold enough to form ice crystals which will fall downward and evaporate to humidify the lower atmosphere. However, the latent heat of water is significant so condensation will drive the cloud to adiabatically expand. Condensation after a 10^4 megaton impact may occur over several days, during which time the water will have been transported over great distances from the impact site. Part of the cloud will photolyze to form oxygen and hydrogen. The e-folding time for water vapor photolysis is about one day in free space so a considerable portion of the water vapor could be lost in this manner (Brasseur and Solomon 1986).

As noted by Emiliani et al. (1981) the atmosphere between about 100 mbar (16 km) and 2 mbar (44 km) can hold about 0.2 g cm^{-2} of precipitable water at ambient temperatures before condensation occurs.   At higher altitudes, injections of water in amounts that would lead to condensation would imply an atmosphere that is dominated by water vapor.   In order to satisfy the hydrostatic equation, not more that 2 g cm^{-2} of water could be placed above 2 mbar.   In reality both of these water injection amounts are upper limits because of self-limiting effects of water on the temperature profile. Water is a strong infrared radiator. Its presence in the upper atmosphere will lead to a temperature decline which will cause water to condense and precipitate out. No numerical simulations have been conducted to investigate this feedback, so the extent to which it limits the water injection is unknown. Although the radiative calculations are complicated by the potential change in the heat balance of the planet due to the dust clouds and due to the high-altitude ice clouds, it is likely that the water vapor cloud will cool to the skin temperature of the Earth, which is about 215 K, in about one week. The time scale is established by the time for a blackbody to dissipate the latent heat of 2 g of water condensing at 215 K. This temperature is also the ambient temperature of the stratosphere so we imagine a nearly isothermal upper atmosphere when the water vapor content is large. Calculations of the stratospheric temperature with higher than normal water vapor mixing ratios show that the upper atmosphere tends toward being isothermal at about the skin temperature as the water content increases (Manabe and Strickler 1964).

The water vapor content of the upper atmosphere at 215 K is about 0.1 g cm^{-2} assuming that the entire upper atmosphere is saturated with a vapor pressure of 13 dyne cm^{-2}. This temperature feedback indicates that the water remaining after a few weeks will not exceed that injected by 10^6 megaton impacts. Any additional water would be removed rapidly by condensation and subsequent sedimentation. Although this feedback limitation is only developed qualitatively here, and may vary substantially according to the actual temperature reached by the atmosphere above the tropopause, it suggests an upper limit to the amount of water in the upper atmosphere of about 250 times ambient. In this hypothetical water vapor rich atmosphere, the water vapor mass mixing ratio would vary from unity at about 80 km altitude to about 8×10^{-5} in the lower stratosphere. In atmospheres close to ambient, dynamical processes would cause the water vapor mixing ratio to become constant with altitude at the value found in the lower stratosphere over about a decade. However, a water-rich atmosphere is unstable with respect to vertical motions because any descending air parcels will have a water vapor partial pressure exceeding the vapor pressure, leading to rainout of the water, latent heat release and convective mixing. Once the mixing ratio becomes constant, the column amount of water would be about 1×10^{-2}g cm^{-2}, which is about 25 times ambient. It is likely other process are more important than the local vertical redistribution of water to achieve a constant mixing ratio. Probably the water in excess of ambient will be removed by horizontal transport to regions with temperatures below average. For example, the winter polar regions in the stratosphere and the summer mesopause currently have temperatures low enough to condense water at ambient concentrations. Such condensation is likely to return upper atmospheric water vapor concentrations to ambient levels over a period of a few years.

Given the possibility that large water vapor increases may occur for a few year period in the upper atmosphere, it is necessary to consider the implications for the surface temperature. Although an increase in water vapor mixing ratio will lead to a strong greenhouse effect, this will be countered to some extent by the increased albedo due to the formation of ice clouds. Whether such clouds will act to warm or cool the surface depends on their particle size (see, e.g., Lacis et al. 1992). If the cloud particles exceed several microns in sizes they will act to warm the surface by efficiently radiating infrared energy, but if they are smaller their reflection of sunlight will dominate and a cooling will occur. The sign of the impact also depends on the cloud optical depth. Of course, surface warming from such a greenhouse effect will not occur until the dust has cleared from the atmosphere, and sunlight has returned to warm the surface.

Even if the net effect is to produce a warming at the ground, the warming will be slow due to the large heat capacity of the atmosphere and oceans. It is important to distinguish between the response of the surface temperature to a small loss of sunlight or a large gain in infrared radiation, and the response of the surface temperature to a large loss of sunlight. In the case of a large loss of sunlight the land surface can cool quickly. Surface cooling cuts off convection so that the ground becomes decoupled from the atmosphere. The result is that surface air temperatures can decline over a few days (Turco et al 1983; Covey et al. 1990). The major factor which moderates this decline is the horizontal transport of heat from the oceans, which is not very effective far from the oceans. Small losses of sunlight or increases in energy at the surface causes changes over a much longer time scale. Small losses of sunlight do not cool the surface enough to prevent convective coupling of the surface to the atmosphere. Increases in energy received at the surface due to a greenhouse effect keep the surface coupled to the atmosphere due to convection. Therefore a large mass of air must be warmed. The typical time scale for such a perturbation is on the order of months (Manabe and Strickler 1964). With such a long response time of the atmosphere, the ocean surface layers become very effective at moderating the temperature response and the total response time rises to a decade or longer. Such slow responses are directly observed following volcanic eruptions. The volcanic aerosols, which themselves have a lifetime of about a year, produce a surface cooling of about 0.2 K which is about 25% of the cooling expected had they lasted for periods of a decade (Pollack et al. 1976).

We conclude that impacts with energies above about 10^4 megatons will produce a global scale water vapor cloud with a mass comparable to or larger than the water in the ambient upper atmosphere. The cloud would cool rapidly due to the high infrared opacity of water vapor—probably reaching temperatures of about 215 K—which would still allow upper atmospheric water vapor column concentrations to be about 250 times higher than at present for impacts with energies above 10^6 megatons. However, ambient levels of water vapor might be approached in just a few years as horizontal transport processes carry water rich air to colder parts of the atmosphere, such as the winter polar regions. Even the sign of the effect of such water vapor enhancements on the surface temperature is uncertain, because the ice clouds which form may have a significant albedo, but it is likely that the net effect will be to warm the surface. However, the warming effect will be slowed by the heat capacity of the oceans. Significant warming will only occur if the water is removed more slowly than the response time of the ocean surface layers which is about 10 yr.

IX. DISCUSSION

There are many uncertainties in evaluating the effects of impacts. The amount of injected water that remains in the upper atmosphere, and the effects of falling ice particles on dust removal have not been quantified. All aspects of atmospheric chemistry are highly uncertain, including the impact of nitrogen oxides on ozone. Dust injection estimates are hampered by lack of knowledge of the fraction of the debris that is submicron in size, and of the mechanisms that form submicron dust. There are also a number of processes that may have been important following the K-T impact that we have not discussed because they are specific to the location of that impact. For example, Sigurdsson et al. (1992) have pointed out that the impact site contains a thick evaporite deposit which is rich in sulfur. Hence, the impact could have released as much as 10^{17} moles of sulfur dioxide to the upper atmosphere. This material would subsequently have been converted to sulfuric acid aerosols which could have had a prolonged effect (either a warming or a cooling, depending on particle size and optical thickness) on the climate, and also could have created acid rain. The impact could also have released several times the present atmospheric abundance of carbon dioxide from the evaporite deposit, which together with the carbon dioxide from the fires and the injected water vapor could have led to a prolonged greehouse effect after the dust cleared (O'Keefe and Ahrens 1989). Unfortunately, the magnitude of the greenhouse effect is difficult to estimate partly because the removal of the water vapor and some of the carbon dioxide is likely to have been fairly rapid. Moreover, the Cretaceous is believed to have been a period of much higher ambient carbon dioxide levels than the present, so these injections may have only doubled the then ambient carbon dioxide.

Recognizing the fact that many of our estimates are only "order of magnitude" we summarize in Table III our best guesses for the impact energies at which various environmental effects become important.

Below about 10^4 megaton, only blast damage and tidal waves should be significant. For these energies blast damage will be local, meaning that it will not extend to areas larger than about 10^4 or 10^5 km^2. However, tidal waves could effect entire ocean basins. For instance, Fig. 2 shows that a 1 meter deep ocean wave, which rivals any tidal wave that has occurred during the past century due to earthquakes, could be generated over the entire Pacific Ocean at an impact energy of 10^4 megatons and over regions as large as the Gulf of Mexico at impact energies of 10^2 megatons.

In the energy range of 10^4 to 10^5 megatons water vapor injections grow to become significant on the global scale. If the submicron dust injection

TABLE III

Impact Magnitude for Effects to Occur

Energy Effect	10^4 Mt	10^5 Mt	10^6 Mt	10^7 Mt
Dust		Nuclear winter cooling	Photo- synthesis ceases	Vision ceases
Fires			Global ejecta, regional fires	Global fires
NO		Loss of O_3 shield		
Blast	Local			Regional
Tidal waves	Regional			
H_2O injections	Global			

fraction is higher than we have suggested, then dust injection could also be important in this energy range.

The energy range from 10^5 to 10^6 is transitional. The dust lifted in this energy range can produce optical depths from 1 to 10. Such optical depths equal or exceed those of historical volcanic eruptions and approach those having optical effects comparable to the optical effects of smoke generated by a nuclear war (Figs. 3 and 4). Cometary impactors in this energy range may throw material out across the globe (Fig. 6). The ejecta plumes of impacts between 10^5 and 10^6 megatons may produce enough NO to destroy the ozone shield (Fig. 10).

Between 10^6 and 10^7 megatons light levels drop below those which can support photosynthesis (Fig. 5). Ballistic ejecta re-entering the atmosphere will set fires over regional scales (above 10^6 km^2) and significant injections of water vapor into the upper atmosphere will occur.

At energies beyond 10^7 megaton light levels are too low to support vision, blast damage reaches the regional scale and fires are set globally.

Many of the mechanisms that have been discussed in this chapter have been investigated by only a single researcher, often in a preliminary fashion. Further studies will undoubtedly narrow many of the uncertainties that we have discussed. The complexity of the events that may follow a large impact are so great, however, that considerable uncertainty will remain after even the most exhaustive analysis. However, it is to be hoped that no large-scale terrestrial experiments occur to shed light on our theoretical oversights.

Acknowledgment. RT was partially supported by a grant by the Atmospheric Chemistry Program of the National Science Foundation, and by grants from the Upper Atmosphere Program of the National Aeronautics and Space Administration. Portions of this work were supported by NASA's Planetary Atmospheres Program managed by J. Bergstralh.

REFERENCES

Ahrens, T. J., and O'Keefe, J. D. 1983. Impact of an asteroid or comet in the ocean and extinction of terrestrial life. *Proc. Lunar Planet. Sci. Conf.* 13, *J. Geophys. Res. Suppl.* 88:799–806.

Ahrens, T. J., and Harris, A. W. 1992. Deflection and fragmentation of near-Earth asteroids. *Nature* 360:429–433.

Alvarez, L. W., Alvarez, W., Asaro, F., and Michel, H. V. 1980. Extra-terrestrial cause for the Cretaceous-Tertiary extinction. *Science* 208:1095–1108.

Argyle, E. 1989. The global fallout signature of the K-T bolide impact. *Icarus* 77:220–222.

Bohor, B., Modreski, P. J., and Foord, E. E. 1987. Shocked quartz in the Cretaceous-Tertiary boundary clays: Evidence for a global distribution. *Science* 236:705–709.

Brasseur, G., and Solomon, S. 1986. *Aeronomy of the Middle Atmosphere* (Dordrecht: D. Reidel).

Brooks, R. B., Reeves, R. D., Ryasn, D. E., Holzbecher, J., Collen, J. D., Neall, V. E., and Lee, J. 1984. Elemental anomalies at the Cretaceous-Tertiary boundary, Woodside Creek, New Zealand. *Science* 226:539–542.

Bourgeois, J., Hansen, T. A., Wiberg, P. L., and Kauffman, E. G. 1988. A Tsunami deposit at the Cretaceous-Tertiary Boundary in Texas. *Science* 214:567–570.

Chapman, C., and Morrison, D. 1994. Impacts on the Earth by asteroids and comets: Accessing the hazard. *Nature* 367:33–40.

Chyba, C. F., Thomas, P. J., and Zahnle, K. J. 1993. The 1908 Tunguska explosion: Atmospheric disruption of a stony asteroid. *Nature* 361:40–44.

Covey, C., Ghan, S. J., Walton, J. J., and Weissman, P. R. 1990. Global environmental effects of impact-generated aerosols; Results from a general circulation model. In *Global Catastrophes in Earth History*, eds. V. L. Sharpton and P. D. Ward, Geological Soc. of America Special Paper 247 (Boulder: Geological Soc. of America), pp. 263–270.

Covey, C., Thompson, S. L, MacCracken, M. C., and Weissman, P. R. 1994. Global climatic effects of atmospheric dust from an asteroid or comet impact on Earth. In preparation.

Crutzen, P. J. 1987. Acid rain at the K/T boundary. *Nature* 330:108–109.

Eaton, J. P., Richter, D. H., and Ault, W. U. 1961. The tsunami of May 23, 1960, on the island of Hawaii. *Bull. Seismolog. Soc. America* 51:135–157.

Emiliani, C., Kraus, E. B., and Shoemaker, E. M. 1981. Sudden death at the end of the Mesozoic. *Earth Planet. Sci.* 55:317–334.

Garcia, A. W., and Butler, H. L. 1977. Numerical simulation of Tsunamis originating in the Peru-Chile trench. In *Coastal Engineering: Proc. 15th Coastal Engineering Conference*, vol. 1 (New York: American Soc. of Civil Engineers), pp. 1025–1043.

Gault, D. E., and Sonett, C. P. 1982. Laboratory simulation of pelagic asteroidal impact: Atmospheric injection, benthic topography, and the surface wave radiation field. In *Geological Implications of Impacts of Large Asteroids and Comets on the Earth*, eds. L. T. Silver and P. H. Schultz, Geological Soc. of America Special Paper 190 (Boulder: Geological Soc. of America), pp. 69–72.

Glasstone, S., and Dolan, P. J. 1977. *The Effects of Nuclear Weapons*, 3rd ed. (Washington, D. C.: U. S. Government Printing Office).

Hansen, J., Russell, G., Lacis, A., Fung, I., Rind, D., and Stone, P. 1985. Climatic response times: Dependence on climate sensitivity and ocean mixing. *Science* 229:857–859.

Hills, J. G., and Goda, M. P. 1993. The fragmentation of small asteroids in the atmosphere. *Astron. J.* 105:1114–1144.

Hoffert, M. I., Callegari, A. J., and Hsieh, C. T. 1980. The role of deep sea heat storage in the secular response to climatic forcing. *J. Geophys. Res.* 85:6667–6679.

Hunten, D. M., Turco, R. P., and Toon, O. B. 1980. Smoke and dust particles of meteoric origin in the mesosphere and stratosphere. *J. Atmos. Sci.* 37:1342–1357.

Jensen, E., and Toon, O. B. 1994. Ice nucleation in the upper troposphere: Sensitivity to aerosol composition and size distribution, temperature, and cooling rate. *Geophys. Res. Lett.*, in press.

Jones, E. M., and Kodis, J. W. 1982. Atmospheric effects of large body impacts: The first few minutes. In *Geological Implications of Impacts of Large Asteroids and Comets on the Earth*, eds. L. T. Silver and P. H. Schultz, Geological Soc. of America Special Paper 190 (Boulder: Geological Soc. of America), pp. 175–186.

Kasting, J. F., and Ackerman, T. P. 1985. High atmospheric NOx levels and multiple photochemical steady states. *J. Atmos. Chem.* 3:321–340.

Kasting, J. F., Richardson, S. M., Pollack, J. B., and Toon, O. B. 1986. A hybrid model of the CO_2 geochemical cycle and its application to large impact events. *Amer. J. Sci.* 286:361–389.

Kinne, S., Toon, O. B., and Prather, M. J. 1992. Buffering of stratospheric circulation by changing amounts of tropical ozone: A Pinatubo case study. *Geophys. Res. Lett.* 19:1927–1930.

Lacis, A., Hansen, J., and Sato, M. 1992. Climate forcing by stratospheric aerosols. *Geophys. Res. Lett.* 19:1607–1610.

Lewis, J. S., Watkins, C. G. H., Hartman, H., and Prinn, R. G. 1982. Chemical consequences of major impact events on Earth. In *Geological Implications of Impacts of Large Asteroids and Comets on the Earth*, eds. L. T. Silver and P. H. Schultz, Geological Soc. of America Special Paper 190 (Boulder: Geological Soc. of America), pp. 215–221.

Malone, R. C., Auer, L. H., Glatzmaier, W. M. C., and Toon, O. B. 1985. The influence of solar heating and precipitation scavenging on the simulated lifetime of post-nuclear war smoke. *Science* 230:317–319.

Malone, R. C., Auer, L. H., Glatzmaier, W. M. C., and Toon, O. B. 1986. Nuclear winter: Three dimensional simulations including interactive transport, scavenging, and solar heating of smoke. *J. Geophys. Res.* 91:1039–1053.

Manabe, S., and Strickler, R. F. 1964. Thermal equilibrium of the atmosphere with a convective adjustment. *J. Atmos. Sci.* 21:361–385.

Manabe, S., and Wetherald, R. T. 1967. Thermal equilibrium of the atmosphere with a given distribution of relative humidity. *J. Atmos. Sci.* 24:241–259.

Melosh, H. J. 1989. *Impact Cratering: A Geologic Process* (New York: Oxford Univ. Press)

Melosh, H. J., and Vickery, A. M. 1991. Melt droplet formation in energetic impact events. *Nature* 350:494–497.

Melosh, H. J., Schneider, N. M., Zahnle, K., and Latham, D. 1990. Ignition of global wildfires at the Cretaceous/Tertiary boundary. *Nature* 343:251–254.

Montanari, A., Hay, R. L., Alvarez, W., Asaro, F., Michel, H. V., Alvarez, L. W., and Smit, J. 1983. Spheroids at the Cretaceous-Tertiary boundary are altered impact droplets of basaltic composition. *Geology* 11:668–671.

National Research Council. 1985. *The Effects on the Atmosphere of a Major Nuclear Exchange* (Washington, D. C.: National Academy Press).

O'Keefe, J. D., and Ahrens, T. J. 1982. Impact mechanisms of large bolides interacting with earth and their implication to extinction mechanisms. In *Geological Implications of Impacts of Large Asteroids and Comets on the Earth*, eds. L. T. Silver and P. H. Schultz, Geological Soc. of America Special Paper 190 (Boulder: Geological Soc. of America), pp. 103–120.

O'Keefe, J. D., and Ahrens, T. J. 1989. Impact production of CO_2 by the Cretaceous/Tertiary extinction bolide and the resultant heating of the Earth. *Nature* 338:247–249.

Pollack, J. B., Toon, O. B., Summers, A., Baldwin, B., Sagan, C., and van Camp, W. 1976. Stratospheric aerosols and climatic change. *Nature* 263:551–555.

Prinn, R., and Fegley, B. 1987. Bolide impacts, acid rain, and biospheric traumas at the Cretaceous-Tertiary boundary. *Earth Planet. Sci. Lett.* 83:1–15.

Pueschel, R. F., Ferry, G. V., Verma, S., and Howard, S. D. 1994. Northern polar vortex variability 17 Feb. 1992. *Geophys. Res. Lett.*, submitted.

Shoemaker, E. M., Wolfe, R. F., and Shoemaker, C. S. 1990. Asteroid and comet flux in the neighborhood of Earth. In *Global Catastrophes in Earth History*, eds. V. L. Sharpton and P. D. Ward, Geological Soc. of America Special Paper 247 (Boulder: Geological Soc. of America), pp. 155–170.

Sigurdsson, H., D'Hondt, S., and Carey, S. 1992. The impact of the Cretaceous/Tertiary bolide on evaporite terrane and generation of major sulfuric acid aerosol. *Earth Planet. Sci. Lett.* 109:543–559.

Sleep, N. H. Zahnle, K. J., Kasting, J. F., and Morowitz, H. J. 1989. Annihilation of ecosystems by large asteroidal impacts on the early earth. *Nature* 342:139–142.

Smit, J., and Klaver, G. 1981. Sanadine spherules at the Cretaceous-Tertiary boundary indicate a large impact event. *Nature* 292:47–49.

Smit, J., and Romein, A. J. T. 1985. A sequence of events across the Cretaceous-Tertiary boundary. *Earth Planet. Sci. Lett.* 74:155–170.

Stommel, H., and Stommel, E. 1983. *Volcano Weather* (Newport, R. I.: Seven Seas Pubs.).

Stothers, R. B. 1984. The great Tambora eruption in 1815 and its aftermath. *Science* 224:1191–1198.

Toon, O. B., Pollack, J. B., Ackerman, T. P., Turco, R. P., McKay, C. P., and Liu, M. S. 1982. Evolution of an impact-generated dust cloud and its effects on the atmosphere. In *Geological Implications of Impacts of Large Asteroids and Comets on the Earth*, eds. L. T. Silver and P. H. Schultz, Geological Soc. of America Special Paper 190 (Boulder: Geological Soc. of America), pp. 187–200.

Toon, O. B., Turco, R. P., Jordan, J., Goodman, J., and Ferry, G. 1989. Physical processes in polar stratospheric ice clouds. *J. Geophys. Res.* 94:11359-11380.

Turco, R. P., Toon, O. B., Park, C., Whitten, R. C., Pollack, J. B., and Noerdlinger, P.

1981. The Tunguska meteor fall of 1908: Effects on stratospheric ozone. *Science* 214:19–23.

Turco, R. P., Toon, O. B., Park, C., Whitten, R. C., Pollack, J. B., and Noerdlinger, P. 1982. An analysis of the physical, chemical, optical and historical impacts of the 1908 Tunguska meteor fall. *Icarus* 50:1–52.

Turco, R. P., Toon, O. B., Ackerman, T. P., Pollack, J. B., and Sagan, C. 1983. Nuclear winter: Global consequences of multiple nuclear explosions. *Science* 222:1283–1292.

Turco, R. P., Toon, O. B., Ackerman, T. P., Pollack, J. B., and Sagan, C. 1990. Climate and smoke: An appraisal of nuclear winter. *Science* 247:166–176.

Van Dorn, W. G. 1965. Tsunamis. *Adv. Hydrosci.* 2:1–48.

Van Dorn, W. G. 1984. Some Tsunami characteristics deducible from tide records. *J. Phys. Ocean.* 14:353–363.

Vickery, A. M., and Melosh, H. J. 1990. Atmospheric erosion and impactor retention in large impacts, with application to mass extinctions. In *Global Catastrophes in Earth History*, eds. V. L. Sharpton and P. D. Ward, Geological Soc. of America Special Paper 247 (Boulder: Geological Soc. of America), pp. 289–300.

Vogelmann, A. M., Ackerman, T. P., and Turco, R. P. 1992. Enhancements in biologically effective ultraviolet radiation following volcanic eruptions. *Nature* 359:47.

Wolbach, W. S., Lewis, R. S., and Anders, E. 1985. Cretaceous extinctions: Evidence for wildfires and search for meteoritic material. *Science* 230:167–170.

Wolbach, W. S., Gilmour, I., and Anders, E. 1990. Major wildfires at the Cretaceous/Tertiary boundary. In *Global Catastrophes in Earth History*, eds. V. L. Sharpton and P. D. Ward, Geological Soc. of America Special Paper 247 (Boulder: Geological Soc. of America), pp. 391–400.

Zahnle, K. 1990. Atmospheric chemistry by large impacts. In *Global Catastrophes in Earth History*, eds. V. L. Sharpton and P. D. Ward, Geological Soc. of America Special Paper 247 (Boulder: Geological Soc. of America), pp. 271–288.

EXTRATERRESTRIAL IMPACTS AND MASS EXTINCTIONS OF LIFE

MICHAEL R. RAMPINO
NASA, Goddard Institute for Space Studies

and

BRUCE M. HAGGERTY
New York University

Calculations of the probability of large-body impacts of various sizes on the Earth derived from the observed flux of Earth-crossing asteroids and comets, and of the estimated threshold impact size required to produce a global environmental disaster, support the hypothesis that large impact events may be a major causal factor in the record of mass extinctions of life. The impact hypothesis is further supported by the search for impact signatures at extinction boundaries in the geologic record, which has produced six cases of evidence considered diagnostic of large impacts (e.g., layers with high iridium, shocked minerals, microtektites), and at least six cases of elevated iridium of the amplitude that might be expected from collisions of relatively low Ir objects such as comets. Recent detailed studies of a number of extinction boundaries show evidence of similar biological, isotopic, and geochemical signatures of sudden environmental crisis and abrupt mass mortality consistent with the expected after effects of catastrophic impacts.

I. INTRODUCTION

Harold Urey (1973) published a prescient paper entitled "Cometary Collisions and Geological Periods," in which he hypothesized a general connection between large-body impacts and geologic boundaries based on the occurrence of tektites at a few boundaries, and suggested that similar glassy ejecta would be found to mark the boundaries of other geologic periods and epochs (Urey 1973). Back-of-the-envelope calculations of the energy involved ($\sim 10^{24}$J, or $\sim 10^8$Mt TNT equivalent) in the collision of a 10^{18}g object with the Earth suggested that "very great variation in climatic conditions covering the entire Earth should occur and very violent physical effects should occur over a substantial portion of the Earth's surface" (Urey 1973). At the time, the geological community, in the midst of the plate tectonic revolution, saw little need for a reconsideration of "catastrophist" views so universally rejected since the publication of Lyell's "Principles of Geology" in the early 1830s.

Surprisingly, seven years later, in 1980, L. Alvarez and colleagues produced hard evidence, in the form of a worldwide layer containing anomalous platinum-group metal abundances, for a large extraterrestrial impact at the time of the Mesozoic/Cenozoic Era boundary (the Cretaceous/Tertiary or K/T boundary, $\sim$65 Myr ago) (Alvarez et al. 1980). This major boundary in geologic time is marked by the abrupt mass extinction of some 70% of marine species (Raup 1991a) including 95% of marine calcareous plankton, the apparently sudden extinction of the dinosaurs (Sheehan et al. 1991; Chapter by Sheehan et al.), and other terrestrial animals and plants (Upchurch 1989; Sheehan and Fastovsky 1992). Shocked minerals (including stishovite), high-temperature microspherules, and most recently glassy microtektites, have since been found in the boundary layer (Chapter by Smit). A large (>200 km diameter) candidate crater, the Chicxulub impact structure in northern Yucatán (Hildebrand et al. 1991) dates from 65.2±0.4 Myr (Sharpton et al. 1992).

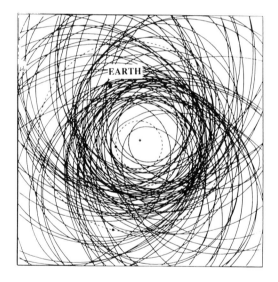

Figure 1.   Location of the Earth within a swarm of asteroids and comets shown by the orbits of the 100 largest known near-Earth asteroids (figure after Morrison 1992; courtesy of R. Binzel).

Astronomical observations show that the Earth is embedded in a swarm of Earth-crossing asteroids and comets (Fig. 1), and collision of these bodies with the planet must represent an on-going major geologic process (e.g., Morrison 1992; Shoemaker et al. 1990; Chapter by Rabinowitz et al.). Impact of objects greater than a few km in diameter (releasing energy $\geq 10^7$Mt TNT equivalent) have predictable far-field effects that should constitute a global environmental disaster (Morrison 1992; Chapter by Morrison et al.). The record of extinction of marine genera per geologic stage/substage over the last 540 Myr shows $\sim$24 well-defined peaks (5 major and $\sim$19 minor peaks) in the percentage of extinction of extant genera (Sepkoski 1982, 1992) (Fig. 2). These times represent extinction events in which it is estimated that ∼25% to >90% of extant marine species disappeared. From a review of the known flux of impactors of various sizes, the calculated environmental effects of large impacts, and recent detailed studies of a number of the extinction boundaries, we suggest that the temporal distribution, duration, and severity of mass extinction events may be explained within a working hypothesis of impact-generated disasters.

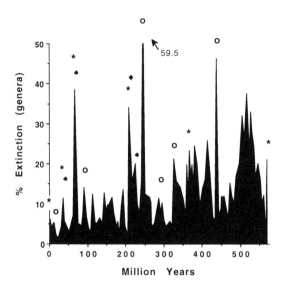

Figure 2.   Percent extinction of marine genera per geologic stage (or substage) during the Phanerozoic (data from Sepkoski 1992 and personal communication). Twenty-four local maxima are recognizable in Sepkoski's data; they are as follows (end of stage as dated by the DNAG time scale, or by Bowring et al. (1993) for the Cambrian, in Myr, may not be exactly coincident with the extinctions): 1. Tommotian (530); 2. Bottomian (520); 3. Dresbachian (515); 4. Trempealeauan (510); 5. Arenigian (478); 6. Caradocian (448); 7. Ashgillian (438); 8. Wenlockian (421); 9. Givetian (374); 10. Frasnian (367); 11. Visean (333); 12. Serpukhovian (320); 13. Stephanian (286); 14. Guadalupian (253); 15. Olenekian (245); 16. Carnian (225); 17. Norian (208); 18. Pliensbachian (193); 19. Tithonian (144); 20. Cenomanian (91); 21. Maastrichtian (66.4); 22. Late Eocene (36.6); 23. Middle Miocene (11.2); 24. Pliocene (1.64). Large (>70 km diameter) dated craters (diamonds), diagnostic stratigraphic evidence of impact (asterisks), possible stratigraphic evidence of impacts (open circles) (see text).

II. RATES OF ASTEROID AND COMET IMPACTS

The collision rates of asteroids and comets of various sizes with the Earth (Fig. 3), and the corresponding production of impact craters, can be calculated from the observed flux of Earth-crossing bodies of different types (Shoemaker et al. 1990; Chapter by Rabinowitz et al.). The diameter of craters produced by large-body impacts can be estimated by empirical formulae, and the estimated production rates of large craters from asteroid and comet collision on Earth during the last 100 Myr (with uncertainties of a factor of ~2) based on the recent flux are shown in Table I. Furthermore, population surges of asteroids ~25% above the mean level, due to the break-up of mainbelt asteroids of ~100 km diameter, may have occurred at average intervals of ~500 Myr, with duration above half-maximum of ~30 Myr (Shoemaker et al. 1990). Brief (a few Myr) increases in the comet flux up to 30 times the mean background, resulting from random (Shoemaker et al. 1990), or periodic (Rampino and Stothers 1984a; Davis et al. 1984) perturbations of the Oort Cloud are predicted at intervals of ~30 Myr, and the present flux may correspond to such a comet shower (Rampino and Stothers 1984a; Shoemaker et al. 1990).

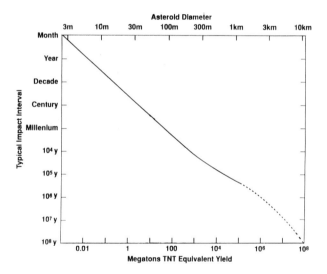

Figure 3.   Estimated frequency of impacts on the Earth from the present population of comets and asteroids, and evidence from impact craters. The megaton TNT equivalents of energy are as shown, as are possible (10^7Mt) and nearly certain (10^8Mt) thresholds of global catastrophe (figure after Morrison 1992).

Active comets are estimated to account for ≤50% of impact craters >20 km over the past 600 Myr, but some believe that the large craters that may have been associated with mass extinctions in the last 100 Myr were more likely formed by comets (Shoemaker et al. 1990). The production rate of craters larger than 20 km in diameter is estimated from astronomical data as $(4.9 \pm 2.9) \times 10^{-15}$km^{-2}yr^{-1} which is in excellent agreement with the rate of $(5.4 \pm 2.7) \times 10^{-15}$km^{-2}yr^{-1} estimated from the geologic record of impact craters over the last 120 Myr (Grieve 1991).

TABLE I

Estimate of Production of Craters by Asteroid and Comet Impact on Earth During the Last 100 Myr[a]

	Crater Diameters (km)				
	>30	>50	>60	>100	>150
Asteroids	58	8	3.2	0.3	0
Comets	24	8	5	1.6	1
Total	82	16	8	2	1

[a] From Shoemaker et al. 1990.

III. IMPACT-INDUCED CATASTROPHES: THRESHOLDS AND MECHANISMS

The large amount of fine ejecta and condensed vapor that was apparently distributed worldwide by the K/T boundary impact could have created a dense dust cloud while in the atmosphere, and Alvarez et al. (1980) suggested that the darkness and short-term surface cooling beneath such a cloud led to cessation of photosynthesis and to frigid conditions, resulting in the observed mass extinctions. The environmental effects of large impacts are summarized in the Chapter by Toon et al., and are outlined briefly here.

The energy released in collisions of asteroids with the Earth ($\sim$20 km s^{-1}) is estimated at $\sim$10^{5}J g^{-1}, whereas for comets ($\sim$50 km s^{-1}) it is $\sim$10^{6}J g^{-1}. The threshold impact energy required for global ballistic distribution of ejecta is estimated at $\sim$10^{23}J ($\sim$10^{7}Mt TNT equivalent) (Morrison 1992; Chapter by Morrison et al.); this amount of impact energy is also required to produce enough submicron dust ($\geq$10^{16}g) to bring global atmospheric transmission down to $<$10^{-2}, which is the limit of photosynthesis (Gerstl and Zardecki 1982). Events such as these are estimated from the population of Earth-crossing asteroids and comets to occur on time scales of about 2×10^7yr to $\sim$10^{8}yr for a 10^8 Mt event (Morrison 1992; Shoemaker et al. 1990) (Fig. 3). This would predict $\sim$5 to 6 major mass extinctions and between 18 and 27 minor mass extinctions in the last 540 Myr, in agreement with the actual number of 5 major and $\sim$19 minor observed extinction events.

Although the global dust cloud would most likely dissipate in 6 to 12 months (Pollack et al. 1983; Toon et al. 1982), calculations (Milne and McKay 1982) and experiments (Griffis and Chapman 1988) suggest that a few months of darkness would have been sufficient to produce the degree of extinction seen among pelagic phytoplankton at the K/T boundary. Analyses of carbon and nitrogen in the K/T boundary clay show that the amounts of marine and terrestrial biomass preserved at the boundary are approximately equivalent to the steady state global inventory (one generation) swept out of the ocean and buried by the ejecta, or burned in global wildfires (Gilmour and Anders 1989).

At times of heavy atmospheric dust loading, the Earth's surface no longer receives any solar radiation, has a net energy deficit, and drastic surface cooling is inevitable (Chapter by Toon et al.). Calculations with one-dimensional models of aerosol physics and climate found that light levels would have remained too low for visibility for up to 6 months, and too low for photosynthesis for up to one year after an impact providing 10^{16}g of dust, and continental temperatures in the models fell to below freezing for up to 2 years. Increased reflectivity from snow cover in continental interiors might provide a positive feedback process that could prolong cooling once the dust settled out of the atmosphere.

Three-dimensional global climate models (GCMs) predict a factor of two less surface cooling, but land-surface temperatures fell below freezing in less than a week (decreases of about 15°C), compared to about two months in the one-dimensional simulation (Covey et al. 1990; Chapter by Toon et al.). Such rapid and severe decreases in temperatures are predicted to have devastated global vegetation (Tinus and Roddy 1990).

In the case of an oceanic impact, water would probably comprise a significant fraction of the ejecta lofted to high altitudes. The water vapor should locally supersaturate the stratosphere, and thus much of the vapor might recondense and precipitate out of the atmosphere, perhaps within weeks to months. During this time, however, an enhanced greenhouse effect could result (Croft 1982; O'Keefe and Ahrens 1982; Emiliani et al. 1982). Condensation of the cloud of water vapor and meteoroid vapors could produce a combination of dust and ice particles in the upper atmosphere (McKay and Thomas 1982), leading to a decrease of ozone concentration above 60 km, a lowering of the average height of the mesopause, and atmospheric cooling. These conditions might give rise to a long-lived (possibly 10^5 to 10^6yr) global layer of mesospheric ice clouds.

The shock waves from the bolide and the transfer of energy from the fine ejecta could result in an immediate heat pulse with ≥ 15°C global temperature increase. The high temperature shock waves are predicted to create large amounts of NO (Chapter by Toon et al.), possibly producing nitric acid rain with a pH of ~ 1, and NO_x in the stratosphere would rapidly remove the ozone layer.

The impact fireball and the heat emitted globally by the re-entering impact ejecta are expected to ignite combustible material and create widespread wildfires (Melosh et al. 1990). Soot produced by the fires would add to the opacity of the atmosphere. Large amounts of such soot discovered at the K/T boundary support the burning of a significant fraction of the terrestrial biomass (Anders et al. 1986; Wolbach et al. 1988). Fires might also produce large amounts of CO_2, CO, CH_4, N_2O, and reactive hydrocarbons and oxides of nitrogen (NO + NO_2), leading to intense photochemical smog, and an increase in the greenhouse effect estimated to produce a heating of ~ 10°C (Wolbach et al. 1988). Noxious chemicals, such as polynuclear aromatic hydrocarbons (PAH) and CO from the wildfires (Venkatesan and Dahl 1989), trace metals released by the impactor, the fires, and/or by acid-enhanced weathering (Erickson and Dickson 1987), or possibly cyanide released in a cometary impact might contribute to the environmental disaster.

Impact into a carbonate-rich terrain (like that at Chicxulub) might lead to a release of CO_2 that could increase atmospheric carbon dioxide levels by factors of 2 to 10, causing a predicted rise in global temperatures of $\sim$2 to 10°C for 10^4 to 10^5yr after the impact (O'Keefe and Ahrens 1989). The loss of dimethyl-sulfide (DMS) production resulting from extinction of most of the calcareous nannoplankton at the K/T boundary should have led to a reduction in available cloud condensation nuclei. The resulting decrease of global cloud albedo is calculated to have caused a temperature increase of $\sim$8°C at the boundary (Rampino and Volk 1987).

IV. SPECIES KILL CURVES AND LARGE CRATERING EVENTS

Mass extinction events seem to represent a markedly different evolutionary regime from that which characterizes times of background extinction levels (Jablonski 1986). Compilation of the ranges of genera of marine organisms shows the extinction events as pulses of increased extinction rates usually within a single stage or substage (intervals of $\sim$2–6 Myr) (Sepkoski 1982,1992) (Fig. 2). However, compilations of species or genera over intervals that last several million years cannot resolve an event that occurred within a very narrow time horizon (McLaren 1985; Signor and Lipps 1982).

At major boundaries coinciding with mass extinctions, using high-resolution stratigraphic techniques, it is commonly possible to identify a global horizon or horizons of mass mortality at which time a large proportion of the biomass disappears over very brief intervals ($<$ a few times 10^4yr and possibly instantaneously) (McLaren and Goodfellow 1990). In addition to the many species that become extinct, tabulations of absolute abundances of species show that many of the surviving species had catastrophic decreases in populations. Study of the relative abundances of species (see, e.g., Keller 1989) may not reveal these catastrophic decreases in absolute abundances.

Many studies have sought to relate mass extinctions to purely terrestrial causes such as changes in climate (both warming and cooling) (Stanley 1984), sea-level fluctuations (Hallam 1989), and/or complex interactions between species, that might produce times of unusual stress on many different kinds of organisms (Buggisch 1991). In a test of the connection between extinction and changes in sea level and climate, Wise and Schopf (1981) examined the Pleistocene, an interval when climate and sea-level fluctuations were especially rapid and extreme, and concluded that marine extinctions were minor. However, the Pleistocene was a time of few shallow seas, and hence fewer fauna would be at risk during sea-level fall (Sheehan 1988).

Raup (1990,1991a, b) constructed a "species kill curve" from a fit to actual Phanerozoic fossil data, which represents a best fit to the mean waiting time between extinctions of various severity. Using impact crater data (Grieve 1991), he also computed the waiting time of terrestrial impact craters of

TABLE II

Large Impact Craters[a] and Correlative(?) Extinctions[b]

Name	~Diameter (km)	Age (Myr)	% Genera	% Species
Puzech-Katunki	80	220±10	20.1	43
Popigai	100	35±5[c]	11.4	25.5
Manicouagan	100	212±2	34.1	62
Chicxulub	~200	65.2±0.4	38.5	67

[a] Grieve 1991.
[b] Sepkoski 1992.
[c] Latest work indicates date of 36±1 Myr (J. Garvin 1993, personal communication).

various sizes, and a possible relationship between mass extinctions and impact cratering was then derived by eliminating the time variable (Fig. 4). We compared the theoretical curve with data representing specific large (>70 km) known impact craters with relatively well-defined ages (Grieve 1991) that overlap the ages of mass extinction boundaries (when true dating errors in both are taken into consideration) (Table II; Fig. 4). No other large, well-dated Phanerozoic craters are known to exist. The observed points agree with the predicted curve within the envelope of error estimated by Raup.

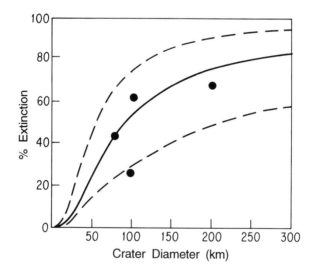

Figure 4.   Kill curve for Phanerozoic marine species, with estimated error bars plotted against estimated size of impact craters associated with extinctions of various kill magnitudes, assuming that the two are actually related (Raup 1991a, b). Estimated sizes of the largest craters with dates overlapping mass extinctions (Table II) are plotted against the species kill magnitude of those extinctions.

The shape of the species kill curve predicts that for craters smaller than ~50 to 60 km in diameter there will be no associated extinction pulse that stands above the ~20 to 25% background level of percent species extinction (Jansa et al. 1990; Jansa 1993), and indeed there are well-dated craters in this size range (Grieve 1991) (see, e.g., Kara, ~65 km; Montagnais, ~45 km) that apparently do not correlate with significant increases in extinction over background levels.

Using the known rate of impacts, and the hypothesized relationship between impacts and mass extinctions, Raup (1990) performed Monte Carlo simulations to determine the appearance of the extinction record supposing that impacts are sufficient to explain the entire record of the past 600 Myr. The results of these simulations compared well with the actual extinction record based on several criteria, including the severity of extinction events, and the general level of background extinction.

V. STRATIGRAPHIC EVIDENCE OF IMPACTS

Horizons containing shocked minerals, glassy microtektites/tektites, and/or other impact-related material, and ranging from regional to global in extent, are now documented at or near the Pliocene (~2.3 Myr) (Kyte 1988), Late Eocene/Early Oligocene (~36 Myr) (Alvarez et al. 1982; Ganapathy 1982), Cretaceous/Tertiary (65 Myr) (see, e.g., Alvarez et al. 1980), Triassic/Jurassic (~205 Myr) (Bice et al. 1992), and the Frasnian-Famennian (~367 Myr) (Claeys et al. 1992) extinction events. Large (>70 km diameter) impact craters are known with age determinations (Grieve 1991) that overlap the estimated ages of four of the extinction boundaries (Table II).

The search for stratigraphic iridium anomalies has been more problematic. The K/T iridium anomaly, averaging thousands of ppt, is globally well documented at more than 120 localities (Alvarez 1986; Chapter by Smit). Iridium anomalies have been searched for at other extinction boundaries, and, although elevated iridium levels (often 10 times or more above background values) have been found at a number of extinction boundaries, they are generally significantly weaker (100s of ppt) than the K/T iridium anomaly (Table III), and are commonly associated with nonchondritic element abundance patterns (Kyte 1988; Orth 1989; Orth et al. 1990; Wang et al. 1991). This has led to conclusions that the Ir peaks are probably unrelated to impact processes (Kyte 1988; Orth 1989; Orth et al. 1990), and to the belief that mass-extinction events other than the K/T are not marked by relative enrichments of platinum-group elements (Donovan 1989). However, preliminary studies suggest that such elevated iridium levels may be uncommon in the geologic record away from boundaries (Kyte and Wasson 1986; Alvarez et al. 1990), and three cases (the Eocene/Oligocene, Triassic/Jurassic, and Late Devonian) now exist where "small" iridium anomalies are accompanied by more diagnostic evidence of impacts (Fig. 2).

TABLE III
Elevated Iridium Reported at Extinction Boundaries

Locality	Ir (ppt)	Reference
	Pliocene	
Core EL 13-3, southeast Pacific	~5,000	Kyte 1988
	Middle Miocene	
DSDP Site 588B, South Pacific	152	Orth 1989
	Late Eocene	
Widespread	≤4,000	Various sources
	Cretaceous/Tertiary	
More than 100 sites, worldwide	1,000s ppt	Various sources
	Cenomanian/Turonian	
Western U.S., Colombia, S.A., Europe	≤560	Orth 1989; Orth et al. 1993
	Jurassic/Cretaceous	
Central Siberia	Average 7,800	Zhakarov et al. 1993
	Callovian/Oxfordian	
Spain and Poland	1,000–2,400?	Orth 1989
	Triassic/Jurassic	
Europe	≤400	McLaren and Goodfellow 1990; Olsen et al. 1990
	Permian/Triassic	
Changxing, China	8,000?	Orth 1989
Meishan, China	600±400	Dao-Yi et al. 1989
Meishan, China	2,000?	Dao-Yi and Zheng 1993
Nammal, Pakistan	366	Dao-Yi et al. 1989
Bolzano, Italy	230	Dao-Yi et al. 1989
San Antonio, Italy	3,000?	Brandner 1988
Tesero, Italy	135	Oddone and Vannucci 1988
Casera Federata, Italy	100–145	—
Butterloch, Austria	90–95	—
Carnic Alps, Austria	165	Holser et al. 1991
	230	—
Lalung, India	73	Bhandari et al. 1992
	114	—

TABLE III (cont.)

	Mississippian/Pennsylvanian	
Texas	380	Orth 1989
	Frasnian/Famennian	
China	230	Wang et al. 1991
Australia	300	Orth 1989
Europe	75–160	Orth 1989
	Ordovician/Silurian	
Anticosti Island, Canada	58	Orth 1989
Scotland	≤ 250	Orth 1989
China	≤ 230	Wang et al. 1993
	Precambrian/Cambrian	
China	2,900	Orth 1989;
	4,000	Dao-Yi et al. 1989

There are several reasons why impact-related Ir anomalies in the geologic record might show lower values than seen at the K/T boundary:

1. Meteorites differ in iridium content from $> 10^3$ ppb (some irons) to $\sim 10^{-2}$ ppb (eucrites and achondrites) (Palme 1982), and terrestrial impact melts range from > 30 ppb to background (0.01 ppb), thus impacts may produce distributed iridium anomalies of various concentrations. Comet impacts, although highly energetic, might produce only relatively minor iridium anomalies. For example, a 6.5 km diameter comet with a density of 0.5 g cm^{-3} would have a mass only $\sim 7\%$ of that of a 10 km asteroid with a density of 1.8 g cm^{-3}, but with an impact velocity up to 70 km s^{-1}, could produce an impact as energetic. Such an impact of a body of chondritic composition would produce ejecta with only 7% as much iridium as seen in K/T boundary sections. Because comets (e.g., 50% ice, 50% chondrite?) are poorer in Ir than chondritic meteorites, the amount of Ir in the ejecta could be even less. In support of a possible connection between relatively weak iridium peaks and impacts, we note that the widespread Australasian microtektites (0.76 Myr), clearly of impact origin, are associated with an iridium anomaly of only ~ 160 ppt (Koeberl 1993).

2. The primary carrier of the iridium, for example microspherules, may be inhomogeneously mixed in the boundary layer, leading to large variations in measured Ir content, even at the same geologic locality (Dao-Yi and Zheng 1993).

3. Recent theoretical studies (Vickery and Melosh 1990) suggest that very large impacts may produce relatively weak iridium anomalies because most of the vaporized impactor would be blown off of the Earth in the

energetic collision.

4. Many extinction events are represented primarily in shallow-water deposits, where reworking and diagenesis of sediments might be expected to dilute Ir concentrations, or could significantly change the elemental ratios of trace-metal anomalies (McLaren and Goodfellow 1990).

VI. EXTINCTIONS AND ENVIRONMENTAL PERTURBATIONS AT GEOLOGIC BOUNDARIES

A. Perturbations at Geologic Boundaries

Some confusion exists as to the definition of *abrupt* versus *gradual* events in the geological sciences (McLaren and Goodfellow 1990). Here, an event is considered abrupt if it occurred within a time span on the order of the limits of resolution in the geologic record imposed by sedimentation rates and sampling intervals. In practice, this means < a few times 10^4yr, or in some cases $\sim 10^5$yr. Of course, the event could still have been instantaneous. A series of such abrupt events might occur within a longer interval of time, producing a stepped record. However, high resolution stratigraphic techniques allow the global correlation of individual horizons (bedding planes) where major changes take place, and these horizons could represent essentially instantaneous and widespread events (McLaren and Goodfellow 1990).

An abrupt negative shift in δ^{13}C of several per mil in marine carbonates has been considered to be a good indicator of severe biomass loss (McLaren and Goodfellow 1990; Zachos and Arthur 1986). Such a shift may indicate a virtual cessation of global marine productivity, the so-called Strangelove ocean effect, accompanied by a homogenization of the δ^{13}C content of surface and deep waters (Hsu and McKenzie 1985,1990). Recently, it has been suggested that a negative shift in δ^{13}C in surface waters may also be the result of widespread combustion of terrestrial biomass, with transfer of isotopically light carbon to ocean surface waters (Ivany and Salawich 1993). The K/T boundary is marked by a worldwide negative anomaly in δ^{13}C of $\sim$3 per mil in the carbonate fine fraction ($<$63 μm, largely nannoplankton debris), and in individual planktonic foraminifers (Fig. 5) lasting for several hundred thousand years. The K/T boundary is also commonly marked by a drastic reduction in CaCO$_3$ deposition in the deep sea, reflecting primarily the mass extinction of calcareous nannoplankton and planktonic foraminifera (Fig. 5). Biogenic pelagic CaCO$_3$ deposition was depressed for at least 0.3 Myr to as long as 1 Myr in some areas. Benthic foraminifera were apparently little affected by the K/T boundary event (Widmark and Malmgren 1992), supporting the idea of a primarily surface-ocean perturbation.

In a number of the most complete K/T boundary sections (Smit 1990; also see his Chapter), a marked 2 per mil negative shift in δ^{18}O, beginning at the boundary and lasting at least several thousand years, has been interpreted as representing a rapid warming of $\sim$8°C (Fig. 5). A sharp positive sulfur

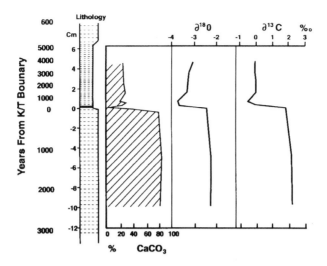

Figure 5.   Detailed K/T boundary section from Agost, Spain, showing negative oxygen- and carbon-isotope shifts at the K/T boundary (as marked by an iridium anomaly), and severe reduction in pelagic biogenic calcium carbonate deposition (figure after Smit 1990; see also his Chapter).

isotope excursion at the K/T boundary suggests a depletion of oxygen in ocean waters at the time (Kajiwara and Kaiho 1992).

If we use the K/T boundary as a model for impact-induced extinctions, we might expect other impact-related extinction boundaries to show (1) a globally synchronous or near-synchronous mass mortality level or levels; (2) a marked negative shift in $\delta^{13}C$ indicating a biomass loss and drop in productivity (the Strangelove ocean), proliferation of opportunistic species, followed by recovery and radiation of surviving biota; (3) a marked negative shift in $\delta^{18}O$ suggesting a brief global warming; (4) a positive shift in $\delta^{34}S$ suggesting anoxic ocean waters; (5) a reduction in marine biogenic $CaCO_3$; and (6) evidence of extraterrestrial impact (Ir and other characteristic trace elements, shocked minerals, microspherules, tektites, tsunami beds, etc.). However, extinctions resulting from other, non-impact causes might have some similar features.

McLaren and Goodfellow (1990) recognized such a common sequence among extinction horizons at the K/T, Triassic/Jurassic, Permian/Triassic, Frasnian/Famennian (Late Devonian), Ordovician/Silurian, and Precambrian/Cambrian boundaries. They found that the horizons of mass mortality are abrupt (at or below the time resolution of the geologic record), and commonly correlate closely with perturbations in the isotope systems of carbon, oxygen, and sulfur suggesting significant and sudden disturbances in ocean productivity, chemistry, circulation, and climate (McLaren 1985; Hsu and McKenzie 1985, 1990; Magaritz 1989).

Sudden and widespread ecological crises at a number of these boundaries are evidenced by the abrupt crash of planktonic communities, and by the proliferation of opportunistic and "disaster" forms such as stromatolites at the Permian/Triassic (Schubert and Bottjer 1992) and Frasnian/Famennian boundaries, fern-spore spikes at terrestrial K/T and Tr/J boundaries (Upchurch 1989; Olsen et al. 1990), and a fungal spore spike and evidence of gradual floral replacement at the Permian/Triassic boundary in Israel (Eshet 1990), Greenland (Balme 1979; Piasecki 1984), and the southern Alps (Visscher and Brugman 1986).

B. The Triassic-Jurassic Boundary

At the Triassic-Jurassic (Tr/J) boundary, iridium peaks of ∼200 to 400 ppt over a background of ∼30 ppt, and the occurrence of three closely spaced layers of shocked quartz suggest possible multiple impacts (McLaren and Goodfellow 1990; Bice et al. 1992). The vertebrate faunal and floral transitions at the end of the Triassic are quite marked in the Newark deposits of eastern North America, just prior to the eruption of the Newark Basalts (201±1 Myr, this may be the best date for the Triassic/Jurassic boundary). The pollen and spore transition, showing a fern spike followed by gradual replacement of flora, is estimated to have taken place in <20,000 yr, and the vertebrate extinction may have occurred over a period of at most 700,000 yr, but is poorly constrained (Olsen et al. 1990).

Weems (1992) recently argued that the Late Triassic vertebrate extinctions were gradual or perhaps stepwise, but this was based largely on tabulations of family diversity, with limited resolution of possible abrupt killing events in the record. Marine groups apparently also show an abrupt extinction, and the few calcareous nannoplankton then in existence exhibit an almost complete turnover at the boundary (Hallam 1990). Abrupt negative shifts in oxygen- and carbon-isotope ratios have been noted in some European marine boundary sites (McLaren and Goodfellow 1990).

C. The Permian-Triassic Boundary

The Permian-Triassic (P/Tr) boundary is marked by the disappearance of up to ∼96% of marine species (Raup 1979), 99% of reptile genera, and a major floral extinction, including the abrupt obliteration of the diverse *Glossopteris* flora in the Southern Hemisphere. Sudden disappearance of Late Permian pollen at an event horizon, and a global spike of fungal spores followed by the proliferation of acritarchs (e.g., in Israel, Europe, Greenland, Australia, and Pakistan) suggest a sudden global catastrophe, followed by a worldwide disaster ecology (Eshet 1990; Balme 1979; Visscher and Brugman 1986; Balme and Helby 1990). Planktonic and nektonic groups, and groups with planktonic growth stages also show a sudden crash at the boundary (Valentine 1986; Li et al. 1991), and biogenic carbonate decreases in many sections.

The end-Permian extinctions are commonly regarded as gradual in nature, and/or composed of a series of extinction episodes ranging from mid-Permian to early Triassic (Maxwell 1992; Erwin 1993). However, problems related to (1) the definition of the P/Tr boundary; (2) the Late Permian regression, which creates a scarcity of end-Permian shallow water sections, and likelihood of a Signor-Lipps effect; and (3) difficulties in correlation, especially between marine and nonmarine sections, have tended to obscure the evidence that the major extinctions and biomass loss apparently took place over an interval no longer than a few times 10,000 yr (McLaren and Goodfellow 1990).

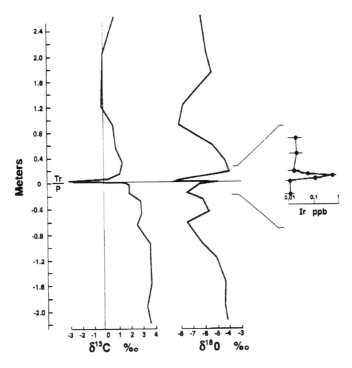

Figure 6.   Carbon- and oxygen-isotope values across the Permian-Triassic boundary, Meishan section (Site B), China, and abundance pattern of iridium at the boundary, which is marked by an abrupt change in ammonoids and bivalves. Ir reaches 600±400 ppt. High concentrations of microspherules are reported at the boundary (Dao-Yi et al. 1989). Recent studies suggest Ir as high as 2 ppb at Meishan, probably in microspherule-rich samples (Dao-Yi and Zheng 1993).

The P/Tr boundary is marked by significant shifts in carbon, oxygen, sulfur and strontium isotope ratios (Dao-Yi et al. 1989; Holser et al. 1991; Magaritz et al. 1988) (Fig. 6). A series of carbon-isotope excursions is protracted over a period of about ~1 to 3 Myr in the Carnic Alps (Holser et al. 1991) (Fig. 7), and the isotope shift has been described as gradual (Erwin 1993; Holser et al. 1991; Magaritz et al. 1988). However, the two major negative isotope shifts of ~1.5 and 2 per mil take place in two abrupt events, each of which occurs over <4 m of sediment (equivalent to <100,000 yr), with

the lower event near the base of the Tesero Horizon that defines the boundary
in the southern Alps (Holser et al. 1991). Both sharp negative excursions in
$\delta^{13}C$, separated by a partial recovery to more positive values, are marked by
Ir peaks (Fig. 6) (Holser et al. 1991).

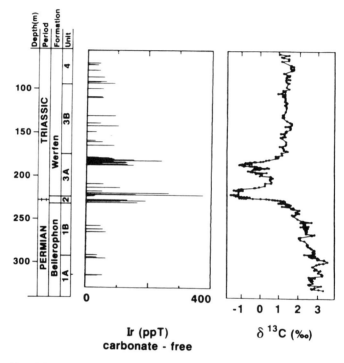

Ir (ppT)
carbonate - free

$\delta^{13}C$ (‰)

Figure 7.   Iridium measurements and carbon-isotope curve at the Permian-Triassic
boundary in Core GK-1 from the Carnic Alps, Austria. Abrupt decrease in $\delta^{13}C$ of
~2 per mil at the boundary with partial recovery and subsequent decreases of 0.5
to 1 per mil in the earliest Triassic, over about 50 m of section. The entire interval
may be as long as 2 Myr, but the most rapid negative shifts in $\delta^{13}C$ take place over
~4 m, equivalent to ~100,000 yr or less. Large negative shifts in $\delta^{18}O$ begin just
above boundary, and reach −5 per mil in amplitude. Ir measurements for the entire
core show elevated Ir (on a carbonate-free basis) at times of negative shifts in $\delta^{13}C$
(Attrep et al. 1991).

The major carbon-isotope anomaly is followed by abrupt negative shifts
in $\delta^{18}O$ in time intervals of 100,000 yr or less (Holser and Magaritz 1992;
Magaritz et al. 1988). In the southern Alps, a rapid drop in $\delta^{18}O$ of more
than −4 per mil follows the sudden negative $\delta^{13}C$ excursion of −1.5 per
mil. If the entire $\delta^{18}O$ signal is related to paleotemperature changes, then a
drastic warming is indicated. A marked and abrupt positive shift in $\delta^{34}S$ at
the boundary suggests the development of anoxic conditions in the oceans
(Dao-Yi et al. 1989).

Two or more iridium peaks in the hundreds of ppt range mark the boundary in Austria, Italy, China, and India (Erwin 1993; Dao-Yi et al. 1989; Holser and Magaritz 1992; Magaritz et al. 1988; Holser et al. 1991; Bhandari et al. 1992) (Table III). Larger Ir anomalies reported from Chinese—e.g., >2 ppb at Shangsi (Xu et al. 1985), and Italian sections ~4 ppb (Brandner 1988) have not been reproduced (Zhou and Kyte 1988), but the carrier phase for the Ir may be discontinuously distributed in the boundary layer(s), e.g., microspherules are reportedly common in Chinese sections (Dao-Yi et al. 1989). Recently, however, an Ir peak of 2 ppb was reported in a very thin band ~10 cm above the base of the Triassic in the Meishan section in China, coincident with a spherule-rich layer and a negative $\delta^{13}C$ shift of 6 per mil (Dao-Yi et al. 1993).

Large impacts are known or suspected to have occurred at or near the P/Tr boundary—the ~40 km wide Araguinha impact structure in Brazil has an $^{40}Ar/^{39}Ar$ age of 247±5.5 Myr (Engelhardt et al. 1992), and two very large (200 to 350 km wide) possible impact structures that may be Latest Permian age ($^{40}Ar/^{39}Ar$ ages of reset basement rocks are ~248 Myr) have been suggested for the Falkland Plateau (Rampino 1992).

D. The Frasnian-Famennian (Late Devonian) Boundary

Multiple Ir peaks of ~75 to 160 ppt, and up to ~300 ppt over a background of only ~20 to 30 ppt (Orth 1989; Orth et al. 1990; Wang et al. 1991) occur at or near the Frasnian-Famennian boundary, and within the Famennian (1 to 1.5 Myr after the F/F boundary) in widely separated areas (China, Australia). Although originally interpreted as nonimpact in origin, the subsequent discovery of glassy microtektites associated with the F/F boundary (Claeys et al. 1992) suggests a possible reinterpretation of the relatively weak Ir peaks as impact-related. Six impact craters (Grieve 1991) have radiometric ages that overlap the estimated age of the boundary (~367 Myr), although none of these seems to be large enough to be associated with a major mass extinction event. An impactor shower is a possibility (Rampino and Stothers 1984a; Hut et al. 1987), perhaps an asteroid surge that lasted ~30 Myr (Shoemaker et al. 1990), and includes increased rates of extinction from the Emsian through the Famennian (~390–360 Myr) (Sepkoski 1992).

Recent studies suggest that the F/F extinctions involved catastrophic collapse of plankton communities and ecosystems (McGhee 1988,1989), although the extinctions of different groups may have been in discrete steps (McLaren and Goodfellow 1990). An abrupt negative shift in $\delta^{13}C$ of 2 to 3 per mil occurs at the lower *triangularis* conodont zone boundary worldwide (e.g., Australia, Europe, China), indicating a sudden loss of ocean biomass (McLaren and Goodfellow 1990; Goodfellow et al. 1992). Several abrupt changes in conodonts apparently occurred within ~100,000 yr in the latest Frasnian. Sandberg et al. (1988), suggested that the major extinction event was very abrupt (<20,000 yr), with the demise of Frasnian reefs having taken place possibly ~1 Myr earlier. They also noted that debris flow and chaotic deposits, possibly related to tsunami, are common at the boundary in many places (e.g., Australia, China, Europe, Nevada).

In China, the large carbon-isotope shift of -3.5 per mil coincides with the Ir anomaly (Wang et al. 1991). The iridium peak in Australia, which occurs above the designated F/F boundary (based on conodonts), may have been preserved by the trapping of increased oceanic Ir by the "disaster" cyanophyte *Frutexites* in some sections after the sudden extinction and mass mortality marked by the negative carbon-isotope anomaly (McLaren 1985; McLaren and Goodfellow 1990). A large positive sulfur-isotope shift indicates reducing conditions in the oceans (Geldsetzer et al. 1987). $\delta^{18}O$ fluctuations of ~1 to 2 per mil may indicate climatic fluctuations of up to 8°C at the boundary (Li et al. 1991; Wang et al. 1991).

E. The Ordovician-Silurian Boundary

Negative carbon- and oxygen-isotope anomalies, and a positive sulfur-isotope excursion occur at or near the Ordovician-Silurian boundary (Long 1993; McLaren and Goodfellow 1990; Goodfellow et al. 1989). Although unresolved problems in correlation exist at the boundary, many of the major biotic changes apparently take place at a globally recognizable event horizon at the base of the *G. persculptus* graptolite zone, which, until stratigraphic redefinition in 1985, was the recognized Ordovician-Silurian boundary (Brenchley 1989). However, an initial extinction may have occurred about 2 Myr prior to the *persculptus* zone (Sheehan 1988; Brenchley 1989), and possibly, three separate episodes of extinction within <500,000 yr marked the latest Ordovician, with a pattern that suggests that trilobites, echinoderms and plankton living in temperate latitudes disappeared in the first event, followed by brachiopods and corals in shallow tropical seas in the later events. Phytoplankton and organisms with planktonic larval stages show an abrupt global extinction at or near the boundary, coincident with extinctions among benthic organisms (Colbath 1986).

Cooling and sea-level fall associated with glaciation may have been a factor in the Late Ordovician extinctions, as shallow, epicontinental seas populated by warm-water adapted species were drained (Sheehan 1988). There is some possible evidence for an impact or impacts at the time. An iridium peak (up to 230 ppt) and other trace-metal anomalies have been reported at or near the boundary in China (Wang et al. 1993), and Goodfellow et al. (1989) reported siderophile and chalcophile trace-metal anomalies at the Ordovician-Silurian boundary in the Northern Yukon Territory, coinciding with a mass mortality based on reduction in carbonate content, negative oxygen- and carbon-isotope excursions, and a positive sulfur isotope shift. They suggested that these anomalies may be impact related, and that an Ir peak could have been missed by the spotty sampling. Orth et al. (1986) found only a weak Ir peak (58 ppt) in a clay layer at the recognized O/S boundary at Anticosti Island, Quebec (no shocked quartz was found), but maximum values of 250 ppt Ir were reported from the Dobbs Linn, Scotland boundary site (Table III).

F. The Precambrian-Cambrian Boundary

Rocks spanning the suggested Precambrian-Cambrian boundary are marked by profound changes in global biota from nonskeletal organisms to those with mineralized skeletons (Magaritz 1989). Although inconsistent results make the events at the boundary difficult to sort out, the "China C marker," separating the Tommotian and Atdabanian Series (∼527 Myr), at the first appearance of trilobites, seems to be marked by a strong iridium anomaly (averaging ∼4,000 ppt), and increase in Os and other trace elements correlating with major negative carbon- and oxygen-isotope anomalies (Dao-Yi et al. 1989; Fan et al. 1984) (Fig. 8).

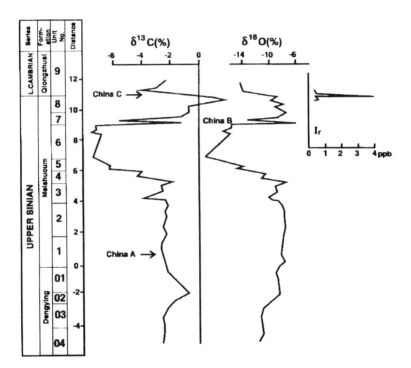

Figure 8.   Stable-isotope perturbations across the Precambrian-Cambrian boundary (defined by the China C marker) at the Meishucun section, China, and the position of iridium anomaly of ∼4 ppb (Dao-Yi et al. 1989). Extinction rates are highest at the China C marker (Dao-Yi et al. 1989).

The persistent carbon-isotope anomaly (e.g., in China, the Siberian Platform, Lesser Himalaya and Anti-Atlas) takes place just before the first appearance of larger skeletalized fossils, including archaeocyathids and trilobites (Dao-Yi et al. 1989). In Siberia and the Lesser Himalaya, the PC/C boundary zone is marked by a positive shift of ∼8 per mil followed by a negative shift of up to ∼7 per mil. These fluctuations have been interpreted as indicating marked changes in ocean fertility—a possible bloom of biomass in the latest Precambrian, followed by severe mass mortality and Strangelove ocean conditions in the Tommotian. Oxygen-isotope anomalies of up to ±4 per mil are reported in the same sequences, commonly with a pronounced negative shift at about the same level as the most negative carbon-isotope excursion (Dao-Yi et al. 1989; Aharon et al. 1987; Magaritz el al. 1986).

The large Lake Acraman impact structure (≥160 km in diameter) in Australia has an inferred age near the PC/C boundary; its ejecta layer in latest Precambrian sediments seems to be somewhat older than the China C marker (Gostin et al. 1989), but it might be coincident with the Vendian-Tommotian boundary, used by some to define the PC/C boundary.

G. The Eocene-Oligocene Boundary

The Late Eocene through Early Oligocene interval shows evidence of multiple impacts, both prior to and subsequent to the recognized E/O boundary. Iridium anomalies (commonly showing a doublet) range up to 4,000 ppt, but may be ≤200 ppt in some areas, and are associated with clinopyroxene microspherules. Tektites, microtektites, and shocked quartz are also associated with the transition from the Eocene to the Oligocene (Montanari 1990; Montanari et al. 1993). Biotic changes seem to have begun in the middle Eocene, but marine species apparently suffered a ~35% extinction from Late Eocene through Early Oligocene time (Sepkoski 1982, 1992). The times of microtektite falls may be associated with changes in microplankton communities that produce 1 to 2 Myr of stepwise events of accelerated faunal turnover, each characterized by generally less than 15% species extinction, but with a total turnover involving >60% of the plankton population (Keller 1986).

The Eocene-Oligocene boundary interval is apparently marked by a negative shift in the carbon-isotope ratio of benthic foraminifera of ~1.5 per mil (although disconformities complicate the interpretation) that correlates with an abrupt global increase in $\delta^{18}O$ in planktonic and benthic foraminifera, indicating cooling, ice-volume increases, or some combination of the two (Miller et al. 1991). Zachos et al. (1992) found the E/O transition to be marked by very rapid climatic changes, with a spike of ice-rafted sediment in the southern Indian Ocean (the onset of continental East Antarctic glaciation), and by rapid positive shifts in the carbon- and oxygen-isotope records of planktonic and benthic organisms at about 35.8 Myr. Although the positive oxygen-isotope shift was permanent, the pulse of ice rafting apparently lasted less than 100,000 yr, perhaps as briefly as 10,000 yr. The timing of the recorded impact events and the rapid climatic changes suggest a possible cause-and-effect, although perhaps delayed relationship, with positive feedbacks such as increased snow and ice cover.

A similar stepwise pattern of biotic change seems to occur at the Cenomanian-Turonian boundary (~92 Myr), where two closely spaced iridium peaks with values up to 560 ppt, coinciding with extinction steps, occur in western North America, Colombia, South America, and some places in Europe (Orth et al. 1993). Shocked quartz grains may be present at the boundary in Colorado (M. Rampino, unpublished data).

H. Extinctions and Evidence of Impacts: Correlations

Of the 24 Phanerozoic extinction peaks (local maxima) in the genus-level data of Sepkoski, 6 seem to be associated with significant stratigraphic evidence of impacts (large Ir anomalies, shocked quartz, and/or microtektites). At least 6 more are associated with "possible" evidence of impact consisting of lower concentrations of Ir, but still anomalous with respect to background values (Fig. 2). Several other boundaries (e.g., the Jurassic-Cretaceous, Callovian-Oxfordian, Devonian-Carboniferous) may show iridium enrichment, but data are scarce.

Although little work on iridium levels in sediments away from boundaries has been done, studies thus far suggest that iridium anomalies above background may be uncommon in the geologic record (Kyte and Wasson 1986; Alvarez et al. 1990). For example, Kyte and Wasson (1986) found no iridium peaks in a slowly deposited deep-sea core over an interval of ~30 Myr between the K/T boundary peak and slight Ir enrichments near the Eocene/Oligocene boundary. If this is the case, then a simple test of the significance of this correlation might be to consider the chances of hitting a target time series consisting of the 24 extinction events (with the age error bars in this case taken as the difference in ages given by the DNAG (Palmer 1983) and Harland-89 (Harland et al. 1990) time scales), with a time series consisting of the 6 times of diagnostic stratigraphic evidence of impacts. The chance of such a correlation calculated in this way is very low ($\sim 10^{-7}$). The accidental correlation of all 12 "diagnostic" and "possible" indicators of impact with the 24 mass extinctions is of even lower probability. These results suggest that, given the assumption that the two time series are related, the correlation of mass-extinction events and evidence of large impacts is extremely unlikely to be an accidental occurrence.

VII. POSSIBLE EXTINCTION PERIODICITY?

Approximately twenty four pulses of extinction at the generic level (using the metric of % extinction) have occurred during the the last 540 Myr, or a mean occurrence rate of one every ~23 Myr (Fig. 2). Raup and Sepkoski (1984) originally identified 12 extinction events at the family level by geologic stages over the last 250 Myr, and reported a statistically significant 26.4 Myr periodicity in the extinctions time series. Later studies found a similar periodicity in generic extinctions at the substage chronostratigraphic level (see Raup and Sepkoski 1986). Periods of ~26 to 31 Myr have been derived using various subsets of extinction events (family and genus levels), different geologic time scales, and various methods of time-series analysis (Rampino and Caldeira 1993). However, the regularity, statistical significance, and reality of the dominant periodicity have been subjects of debate (Stigler and Wagner 1987). The record of extinctions of nonmarine vertebrates may contain a similar ~30 Myr periodicity (Rampino and Stothers 1986). The issues of periodicity are discussed in detail in a book edited by Smoluchowski et al. (1986).

Part of the problem in acceptance of the periodicity as a real manifestation of the geologic record is the shortness of the record, and the reported apparent lack of evidence of a statistically significant periodicity in extinction events prior to 250 Myr ago. However, we find that when Fourier analysis is performed on an extended record of 22 extinctions (using the DNAG time scale) (Fig. 2) adding one Paleozoic extinction at a time from 250 Myr back to the Late Cambrian (~515 Myr), more than doubling the length of the series, a stable peak between 26.5 and 27.3 Myr remains the dominant feature of the Fourier spectrum (Fig. 9).

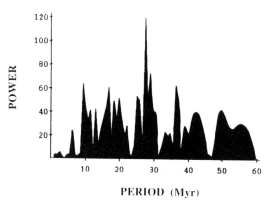

PERIOD (Myr)

Figure 9.   Fourier power spectrum for 22 extinction events from Fig. 2 (0–515 Myr), computed as described in Rampino and Caldeira (1993). The highest peak is at 27.3 Myr.

An indication as to the significance of the 27.3 Myr peak in the 0 to 515 Myr record was obtained by generating 1,000 pseudo-data sets, each containing the same number of randomly dated pseudo-events over the same time interval. Based on this analysis, the probability of generating higher spectral power at 27.3 Myr is ~2%, although further tests indicate that the probability of generating higher spectral power at any period between 10 and 65 Myr falls below the 95% confidence level. However, the robustness of the ~26 Myr periodicity in the extinction time series is attested to by the fact that the period is the dominant spectral feature in every extinction data set back to the Late Cambrian.

If the periodicity of ~26 to 30 Myr in the mass extinction pulses is real, it may be related to times of increased flux of long-period comets (comet showers) during the periodic passage of the solar system through the central plane of the Milky Way Galaxy (~26 to 36 Myr depending on galactic models), when gravitational perturbations of the Oort comet cloud by galactic tides or by massive objects may be more likely (Rampino and Stothers 1984*a*, *b*). Time-series analyses of terrestrial impact craters have provided some evidence of a possible 28 to 32 Myr periodicity in impacts (Rampino and Stothers 1984*a*, *b*; Alvarez and Muller 1984; Yabushita 1991), although these studies are plagued by small number statistics, and have been disputed (Grieve 1991). The recent impact-crater record has been interpreted as suggesting showers of objects at ~0, 35, 65, and 95 Myr (Hut et al. 1987).

VIII. CONCLUSIONS

The hypothesis concerning a connection between large-body impacts and mass extinctions is supported (1) by the discovery of the now well-documented large impact event at the K/T boundary by the Alvarez group and others; (2) by calculations relating to the catastrophic nature of the environmental effects in the aftermath of large impacts; (3) by the discovery of several additional layers of impact debris or possible impact material at, or close to, geologic boundary/extinction events; (4) by evidence that a number of extinctions were abrupt and perhaps catastrophic; and (5) by the accumulation of data on impact craters and astronomical data on comets and asteroids that provide estimates of collision rates of such large bodies with the Earth on long time scales.

Collision of asteroids and comets with the Earth is a normal geologic process, with frequency vs magnitude of events following an inverse power law relationship. The observed orbital elements and size-frequency distributions of Earth-crossing asteroids and comets predict that impactors greater than a few km in diameter should collide with the Earth on average every few tens of millions of years. These large impacts, releasing 10^7 to $> 10^8$ Mt TNT equivalent of explosive energy are calculated to produce global catastrophes involving possible loss of photosynthesis and "impact winter" conditions caused by global dust clouds of fine dispersed ejecta, surface incineration primarily by heat radiated from re-entering ballistic ejecta, nitric acid rain, loss of ozone screen, possible injection of large amounts of greenhouse gases (H_2O, CO_2) into the atmosphere, and other effects that would perturb climate and life. These environmental effects have been predicted to lead to mass mortality and subsequent extinction of a large fraction of extant species. Such large perturbations may also destabilize the climatic system, leading to longer term changes in the environment.

The ~540 Myr record of mass extinctions of ~25% to >90% of extant species may be largely explainable as a result of such impact-induced catastrophes and their aftermath, which perhaps exacerbate ongoing terrestrial environmental change. Extinction events show a generally common pattern of sharp negative shifts in carbon isotopes suggesting abrupt mass mortality at globally (or at least widespread) recognizable horizons, negative shifts in oxygen isotopes indicating sudden global warmings, and positive sulfur-isotope anomalies suggesting lowered oxygenation of ocean waters. Six of the ~24 pulses of extinction in the Phanerozoic (if one includes the Precambrian-

Cambrian boundary) seem to be associated with significant stratigraphic evidence of major impacts—layers containing high siderophile trace-element anomalies (especially iridium), shocked minerals, and/or glassy tektites and microtektites. An additional 6 extinction boundaries are associated with known layers of elevated iridium (and related trace-metal) concentrations above background that might be related to impacts of Ir depleted objects, possibly comets. Some evidence exists for an ~26 Myr periodicity in the Phanerozoic extinction record.

Acknowledgments. We thank C. Chapman and R. Binzel for use of figures, R. A. F. Grieve and P. M. Sheehan for critical reviews, and K. Caldeira, Y. Eshet, M. I. Hoffert, W. T. Holser, E. G. Kauffman, C. J. Orth, J. J. Sepkoski, Jr., E. M. Shoemaker, J. Smit, R. B. Stothers, E. Teller, and S. Yabushita for data, discussions, and criticism. P. Lewis helped with the manuscript. M. R. R. was supported by a NASA Grant.

REFERENCES

Aharon, P., Schidlowski, M., and Singh, I. B. 1987. Chronostratigraphic markers in the end-Precambrian carbon isotope record of the Lesser Himalaya. *Nature* 327:699–702.

Alvarez, L. W., Alvarez, W., Asaro, F., and Michel, H. V. 1980. Extra-terrestrial cause for the Cretaceous-Tertiary extinction. *Science* 208:1095–1108.

Alvarez, W. 1986. Toward a theory of impact crises. *Eos: Trans. AGU* 67:649–658.

Alvarez, W., and Muller, R. A. 1984. Evidence from crater ages for periodic impacts on the Earth. *Nature* 308:718–720.

Alvarez, W., Alvarez, L. W., Asaro, F., and Michel, H. V. 1982. Iridium anomaly approximately synchronous with terminal Eocene extinctions. *Science* 216:886–888.

Alvarez, W., Asaro, F., and Montanari, A. 1990. Iridium profile for 10 million years across the Cretaceous-Tertiary boundary at Gubbio (Italy). *Science* 250:1700–1702.

Anders, E., Wolbach, W. S., and Lewis, R. S. 1986. Cretaceous extinctions and wildfires. *Science* 234:261–264.

Attrep, M., Jr., Orth, C. J., and Quintana, L. R. 1991. The Permian-Triassic of the Gartnerkofel-1 Core (Carnic Alps, Austria): Geochemistry of common and trace elements II—INAA and RNAA. In *The Permian-Triassic Boundary in the Carnic Alps of Austria (Gartnerkofel Region)*, eds. W. T. Holser and H. P. Schönlaub, *Abhandlung Geol. Bundesanst.* 45:123–137.

Balme, B. E. 1979. Palynology of Permian-Triassic boundary beds of Kap Stosch, east Greenland. *Meddelelser Om Grönland* 200(6):1–35.

Balme, B. E., and Helby, R. J. 1990. Floral modifications at the Permian-Triassic boundary in Australia. In *The Permian and Triassic Systems and Their Mutual Boundary*, eds. A. Logan and L. V. Hills (Canadian Soc. Petrological Geology), pp. 433–444.

Bhandari, N., Shukla, P. N., and Azmi, R. J. 1992. Positive europium anomaly at the Permo-Triassic boundary, Spiti, India. *Geophys. Res. Lett.* 19:1531–1534.

Bice, D. M., Newton, C. R., McCauley, S., Reiners, P. W., and McRoberts, C. A. 1992. Shocked quartz at the Triassic-Jurassic boundary in Italy. *Science* 259:443–446.

Bowring, S. A., Grotzinger, J. P., Isachsen, C. E., Knoll, A. H., Pelechaty, S. M., and Kolosov, P. 1993. Calibrating rates of Early Cambrian evolution. *Science* 261:1293–1298.

Brandner, R. 1988. The Permian-Triassic boundary section in the Dolomites (Southern Alps, Italy), San Antonio section. *Ber. Geo. Bundesanst.* 15:49–56.

Brenchley, P. J. 1989. The late Ordovician extinction. In *Mass Extinctions, Processes and Evidence*, ed. S. K. Donovan (New York: Columbia Univ. Press), pp. 104–132.

Buggisch, W. 1991. The global Frasnian-Famennian-Kellwasser Event. *Geol. Rundschau* 80:49–72.

Claeys, P., Casier, J.-G., and Margolis, S. V. 1992. Microtektites and mass extinctions: Evidence for a Late Devonian asteroid impact. *Science* 257:1102–1104.

Colbath, G. K. 1986. Abrupt terminal Ordovician extinction in phytoplankton associations, southern Appalachians. *Geology* 14:943–946.

Covey, C., Ghan, S. J., Walton, J. J., and Weissman, P. R. 1990. Global environmental effects of impact-generated aerosols; Results from a general circulation model. In *Global Catastrophes in Earth History*, eds. V. L. Sharpton and P. D. Ward, Geological Soc. of America Special Paper 247 (Boulder: Geological Soc. of America), pp. 263–270.

Croft, S. K. 1982. A first-order estimate of shock heating and vaporization in oceanic impacts. In *Geological Implications of Impacts of Large Asteroids and Comets on the Earth*, eds. L. T. Silver and P. H. Schultz, Geological Soc. of America Special Paper 190 (Boulder: Geological Soc. of America), pp. 143–151.

Dao-Yi, X., and Zheng, Y. 1993. Carbon isotope and iridium event markers near the Permian/Triassic boundary in the Meishan section, Zhejiang Province, China. *Palaeoclimatol. Paleogeogr. Palaeoecol.* 104:171–176.

Dao-Yi, X., Zheng, Y., Yi-Yin, S., Jin-Wen, H., Qin-Wen, Z., and Zhi-Fang, C. 1989. *Astrogeological Events in China* (New York: Van Nostrand Reinhold).

Davis, M., Hut, P., and Muller, R. A. 1984. Extinction of species by periodic comet showers. *Nature* 308:715–717.

Donovan, S. K. 1989. Introduction. In *Mass Extinctions, Processes and Evidence*, ed. S. K. Donovan (New York: Columbia Univ. Press), pp. xi–xiv.

Emiliani, C., Kraus, E. B., and Shoemaker, E. M. 1982. Sudden death at the end of the Mesozoic. *Earth Planet. Sci. Lett.* 55:317–334.

Engelhardt, W. V., Matthai, S. K., and Walzebuck, J. 1992. Araguainha impact crater, Brazil. I: The interior part of the uplift. *Meteoritics* 27:442–457.

Erickson, D. J., and Dickson, S. M. 1987. Global trace-element biogeochemistry at the K/T boundary: Oceanic and biotic response to a hypothetical meteorite impact. *Geology* 15:1014–1017.

Erwin, D. H. 1993. *The Great Paleozoic Crisis* (New York: Columbia Univ. Press).

Eshet, Y. 1990. The palynostratigraphy of the Permian-Triassic boundary in Israel: Two approaches to biostratigraphy. *Israel J. Earth Sci.* 39:1–15.

Fan, D., Yang, R., and Huang, Z. 1984. The Lower Cambrian black shale series and the iridium anomaly in South China. In *Academia Sinica, Developments in Geoscience: Contribution to 27th Geological Congress, Moscow* (Beijing: Science Press), pp. 215–224.

Ganapathy, R. 1982. Evidence for a major meteorite impact on the Earth 34 million years ago: Implications for Eocene extinctions. *Science* 216:885–886.

Geldsetzer, H. H. J., Goodfellow, W. D., McLaren, D. J., and Orchard, M. J. 1987.

Sulfur-isotope anomaly associated with the Frasnian-Famennian extinction, Medicine Lake, Alberta, Canada. *Geology* 15:393–396.

Gerstl, S. A., and Zardecki, A. 1982. Reduction of photosynthetically active radiation under extreme stratospheric aerosol loads. In *Geological Implications of Impacts of Large Asteroids and Comets on the Earth*, eds. L. T. Silver and P. H. Schultz, Geological Soc. of America Special Paper 190 (Boulder: Geological Soc. of America), pp. 201–210.

Gilmour, I., and Anders, E. 1989. Cretaceous-Tertiary boundary event: Evidence for a short time scale. *Geochim. Cosmochim. Acta* 53:503–511.

Goodfellow, W. D., Geldsetzer, H. H. J., McLaren, D. J., Orchard, M. J., and Klapper, G. 1989. Geochemical and isotopic anomalies associated with the Frasnian-Famennian extinction. *Historical Biology* 2:51–72.

Goodfellow, W. D., Nowlan, G. S., McCracken, A. D., Lenz, A. C., and Gregoire, D. C. 1992. Geochemical anomalies near the Ordovician-Silurian boundary, Northern Yukon Territory, Canada. *Historical Biology* 6:1–23.

Gostin, V. A., Keays, R. R., and Wallace, M. W. 1989. Iridium anomaly from the Acraman impact ejecta horizon: impacts can produce sedimentary iridium peaks. *Nature* 340:542–544.

Grieve, R. A. F. 1991. Terrestrial impact: the record in the rocks. *Meteoritics* 26:175–194.

Griffis, K., and Chapman, D. J. 1988. Survival of phytoplankton under prolonged darkness: Implications for the Cretaceous-Tertiary boundary darkness hypothesis. *Palaeogeogr. Palaeoclimatol. Palaeoecol.* 67:305–314.

Hallam, A. 1989. The case for sea-level change as a dominant causal factor in mass extinction of marine invertebrates. *Phil. Trans. Roy. Soc. London* B325:437-455.

Hallam, A. 1990. The end-Triassic mass extinction event. In *Global Catastrophes in Earth History*, eds. V. L. Sharpton and P. D. Ward, Geological Soc. of America Special Paper 247 (Boulder: Geological Soc. of America), pp. 577–583.

Harland, W. B., Armstrong, R. L., Cox, A. V., Craig, L. E., Smith, A. G., and Smith, D. G. 1990. *A Geologic Time Scale* (Cambridge: Cambridge Univ. Press).

Hildebrand, A. R., Penfield, G. T., Kring, D. A., Pilkington, M., Camargo, Z. A., Jacobsen, S. B., and Boynton, W. V. 1991. Chicxulub Crater: A possible Cretaceous/Tertiary boundary impact crater on the Yucatán Peninsula, Mexico. *Geology* 19:867–871.

Holser, W. T., and Magaritz, M. 1992. Cretaceous/Tertiary and Permian/Triassic boundary events compared. *Geochim. Cosmochim. Acta* 56:3297–3309.

Holser, W. T., Schönlaub, H. P., Boeckelmann, K., and Magaritz, M. 1991. The Permian-Triassic of the Gartnerkofel-1 Core (Carnic Alps, Austria): Synthesis and conclusions. In *The Permian-Triassic Boundary in the Carnic Alps of Austria (Gartnerkofel Region)*, eds. W. T. Holser and H. P. Schönlaub, *Abhandlung Geol. Bundesanst.* 45:213–222.

Hsu, K. J., and McKenzie, J. A. 1985. A "Strangelove" ocean in earliest Tertiary. In *Carbon Cycle and Atmospheric CO_2, Natural Variations Archean to Present*, eds. E. T. Sundquist and W. S. Broecker, Geophysical Monograph 32 (Washington: D. C.: American Geophysical Union), pp. 487–492.

Hsu, K. J., and McKenzie, J. A. 1990. Carbon-isotope anomalies at era boundaries; Global catastrophes and their ultimate cause. In *Global Catastrophes in Earth History*, eds. V. L. Sharpton and P. D. Ward, Geological Soc. of America Special Paper 247 (Boulder: Geological Soc. of America), pp. 61–70.

Hsu, K. J., Oberhansli, H., Gao, J. Y., Shu, S., Haihong, C., and Krahenbuhl, U. 1985. "Strangelove ocean" before the Cambrian explosion. *Nature* 316:809–811.

Hut, P., Alvarez, W., Elder, W. P., Hansen, T., Kauffman, E. G., Keller, G., Shoemaker, E. M., and Weissman, P. R. 1987. Comet showers as a cause of mass extinctions.

Nature 329:118–126.

Ivany, L. C., and Salawitch, R. J. 1993. Carbon isotope evidence for biomass burning at the K-T boundary. *Geology* 21:487–490.

Jablonski, D. 1986. Evolutionary consequences of mass extinctions. In *Patterns and Processes in the History of Life*, eds. D. M. Raup and D. Jablonski (Berlin: Springer-Verlag), pp. 313–329.

Jansa, L. F. 1993. Cometary impacts into ocean: Their recognition and the threshold constraint for biological extinctioins. *Palaeogeogr. Palaeoclimatol. Palaeoecol.* 104:271–286.

Jansa, L. F., Aubry, M.-P., and Gradstein F. M. 1990. Comets and extinctions; cause and effect? In *Global Catastrophes in Earth History*, eds. V. L. Sharpton and P. D. Ward, Geological Soc. of America Special Paper 247 (Boulder: Geological Soc. of America), pp. 223–232.

Kajiwara, Y., and Kaiho, K. 1992. Oceanic anoxia at the Cretaceous/Tertiary boundary supported by the sulfur isotope record. *Palaeogeo. Palaeoclimatol. Palaeoecol.* 99:151–162.

Keller, G. 1986. Stepwise mass extinctions and impact events: Late Eocene to early Oligocene. *Marine Micropaleontol. 10:267–293.*

Keller, G. 1989. Extended period of extinctions across the Cretaceous/Tertiary boundary in planktonic foraminifera of continental shelf sections: Implications for impact and volcanism theories. *Geol. Soc. America Bull.* 101:1408–1419.

Koeberl, C. 1993. Extraterrestrial component associated with Australasian microtektites in a core from ODP Site 758B. *Earth Planet. Sci. Lett.* 119:453–458.

Kyte, F. T. 1988. The extraterrestrial component in marine sediments: Description and interpretation. *Paleoceanog.* 3:235–247.

Kyte, F. T., and Wasson, J. T. 1986. Accretion rate of extraterrestrial matter: Iridium in marine sediments deposited 33–67 Ma ago. *Science* 232:1225–1229.

Li, Z., Zhan, L., Yao, J., and Zhou, Y. 1991. On the Permian-Triassic Events in South China–Probe into the End Permian Abrupt Extinction and Its Possible Causes. Saito Ho-on Kai Special Pub. No. 3, pp. 371–385.

Long, D. 1993. Oxygen and carbon isotopes and event stratigraphy near the Ordovician-Silurian boundary, Anticosti Island Quebec. *Palaeogeogr. Palaeoclimatol. Palaeoecol.* 104:49–59.

Magaritz, M. 1989. ^{13}C Minima follow extinction events: A clue to faunal radiation. *Geology* 17:337–340.

Magaritz, M., Holser, W. T., and Kirschvink, J. L. 1986. Carbon-isotope events across the Precambrian/Cambrian boundary on the Siberian Platform. *Nature* 320:258–259.

Magaritz, M., Bar, R., Baud, A., and Holser, W. T. 1988. The carbon isotope shift at the Permian/Triassic boundary in the southern Alps is gradual. *Nature* 331:337–339.

Maxwell, W. D. 1992. Permian and Early Triassic extinction of non-marine tetrapods. *Palaeontology* 35:571–583.

McGhee, G. R., Jr. 1988. The Late Devonian extinction event: Evidence for abrupt ecosystem collapse. *Paleobiology* 14:250–257.

McGhee, G. R., Jr. 1989. The Frasnian-Famennian extinction event. In *Mass Extinctions, Processes and Evidence*, ed. S. K. Donovan (New York: Columbia Univ. Press), pp. 133–151.

McKay, C. P., and Thomas, G. E. 1982. Formation of noctilucent clouds by an extraterrestrial impact. In *Geological Implications of Impacts of Large Asteroids and Comets on the Earth*, eds. L. T. Silver and P. H. Schultz, Geological Soc. of America Special Paper 190 (Boulder: Geological Soc. of America), pp. 211–214.

McLaren, D. J. 1985. Mass extinction and iridium anomaly in the Upper Devonian of Western Australia: a commentary. *Geology* 13:170–172.

McLaren, D. J., and Goodfellow, W. D. 1990. Geological and biological consequences of giant impacts. *Ann. Rev. Earth Planet. Sci.* 18:123–171.

Melosh, H. J., Schneider, N. M., Zahnle, K. J., and Latham, D. 1990. Ignition of global wildfires at the Cretaceous/Tertiary boundary. *Nature* 343:251–254.

Miller, K. G., Berggren, W. A., Zhang, J., and Palmer-Julson, A. A. 1991. Biostratigraphy and isotope stratigraphy of Upper Eocene microtektites at Site 612: How many impacts? *Palaios* 6:17–38.

Milne, D. H., and McKay, C. P. 1982. Response of marine plankton communities to a global atmospheric darkening. In *Geological Implications of Impacts of Large Asteroids and Comets on the Earth*, eds. L. T. Silver and P. H. Schultz, Geological Soc. of America Special Paper 190 (Boulder: Geological Soc. of America), pp. 297–304.

Montanari, A. 1990. Geochronology of the terminal Eocene impacts; An update. In *Global Catastrophes in Earth History*, eds. V. L. Sharpton and P. D. Ward, Geological Soc. of America Special Paper 247 (Boulder: Geological Soc. of America), pp. 607–616.

Montanari, A., Asaro, F., Michel, H. V., and Kennett, J. P. 1993. Iridium anomalies of Late Eocene age at Massignano (Italy), and ODP Site 689B (Maud Rise, Antarctica). *Palaios* 8:420–438.

Morrison, D., ed. 1992. *The Spaceguard Survey: Report of the NASA International Near-Earth Object Detection Workshop* (Pasadena: Jet Propulsion Laboratory).

O'Keefe, J. D., and Ahrens, T. J. 1982. Impact mechanics of the Cretaceous-Tertiary impact bolide. *Nature* 298:123–127.

O'Keefe, J. D., and Ahrens, T. J. 1989. Impact production of CO_2 by the Cretaceous/Tertiary extinction bolide and the resultant heating of the Earth. *Nature* 338:247–249.

Oddone, W., and Vannucci, R. 1988. PGE and REE geochemistry at the B-W boundary in the Carnian and Dolomite Alps (Italy). *Mem. Soc. Geol. Italiana* 34:121–128.

Olsen, P. E., Fowell, S. J., and Cornet, B. 1990. The Triassic/Jurassic boundary in continental rocks of eastern North America; A progress report. In *Global Catastrophes in Earth History*, eds. V. L. Sharpton and P. D. Ward, Geological Soc. of America Special Paper 247 (Boulder: Geological Soc. of America), pp. 585–593.

Orth, C. J. 1989. Geochemistry of the bio-event horizons. In *Mass Extinctions: Processes and Evidence*, ed. S. K. Donovan (New York: Columbia Univ. Press), pp. 37–72.

Orth, C. J., Gilmore, J. S., Quintana, L. R., and Sheehan, P. M. 1986. Terminal Ordovician extinction: geochemical analysis of the Ordovician/Silurian boundary, Anticosti Island, Quebec. *Geology* 14:433–436.

Orth, C. J., Attrep, M., Jr., and Quintana, L. R. 1990. Iridium abundance patterns across bio-event horizons in the fossil record. In *Global Catastrophes in Earth History*, eds. V. L. Sharpton and P. D. Ward, Geological Soc. of America Special Paper 247 (Boulder: Geological Soc. of America), pp. 45–59.

Orth, C. J., Attrep, M., Jr., Quintana, L. R., Elder, W. P., Kauffman, E. G., Diner, R., and Villamil, T. 1993. Elemental abundance anomalies in the late Cenomanian extinction interval: A search for the source(s). *Earth Planet. Sci. Lett.* 117:189–204.

Palme, H. 1982. Identification of projectiles of large terrestrial impact craters and some implications for the interpretation of Ir-rich Cretaceous/Tertiary boundary layer. In *Geological Implications of Impacts of Large Asteroids and Comets on the Earth*, eds. L. T. Silver and P. H. Schultz, Geological Soc. of America Special Paper 190 (Boulder: Geological Soc. of America), pp. 223–233.

Palmer, A. R. 1983. The decade of North American geology 1983 geologic time scale.

Geology 11:503–504.

Piasecki, S. 1984. Preliminary palynostratigraphy of the Permian-Lower Triassic sediments in Jameson Land and Scoresby Land, East Greenland. *Bull. Geol. Soc. Denmark* 32:139–144.

Pollack, J. B., Toon, O. B., Ackerman, T. P., Mckay, C. P., and Turco, R. P. 1983. Environmental effects of an impact-generated dust cloud: Implications for the Cretaceous-Tertiary extinctions. *Science* 219:287–290.

Rampino, M. R. 1992. A large Late Permian impact structure from the Falkland Plateau. *Eos: Trans. AGU* 73:136.

Rampino, M. R., and Caldeira, K. 1993. Major episodes of geologic change: Correlations, time structure and possible causes. *Earth Planet. Sci. Lett.* 114:215–227.

Rampino, M. R., and Stothers, R. B. 1984*a*. Terrestrial mass extinctions, cometary impacts and the Sun's motion perpendicular to the galactic plane. *Nature* 308:709–712.

Rampino, M. R., and Stothers, R. B. 1984*b*. Geological rhythms and cometary impacts. *Science* 226:1427–1430.

Rampino, M. R., and Stothers, R. B. 1986. Geologic periodicities and the Galaxy. In The *Galaxy and the Solar System*, eds. R. Smoluchowski, J. N. Bahcall and M. S. Matthews (Tucson: Univ. of Arizona Press), pp. 241–259.

Rampino, M. R., and Volk, T. 1987. Mass extinctions, atmospheric sulphur and climatic warming at the K/T boundary. *Nature* 332:63–65.

Raup, D. M. 1979. Size of the Permo-Triassic bottleneck and its evolutionary implications. *Science* 206:217–218.

Raup, D. M. 1990. Impact as a general cause of extinction; a feasibility test. In *Global Catastrophes in Earth History*, eds. V. L. Sharpton and P. D. Ward, Geological Soc. of America Special Paper 247 (Boulder: Geological Soc. of America), pp. 27–32.

Raup, D. M. 1991*a*. A kill curve for Phanerozoic marine species. *Paleobiology* 17:37–48.

Raup, D. M. 1991*b*. *Extinction: Bad Genes or Bad Luck?* (New York: Norton).

Raup, D. M., and Sepkoski, J. J., Jr. 1984. Periodicity of extinctions in the geologic past. *Proc. Natl. Acad. Sci. U. S. A.* 81:801–805.

Raup, D. M., and Sepkoski, J. J., Jr. 1986. Periodic extinctions of families and genera. *Science* 231:833–836.

Sandberg, C. A., Ziegler, W., Dreesen, R., and Butler, J. L. 1988. Part 3: Late Frasnian mass extinction: Conodont event stratigraphy, global changes, and possible causes. *Cour. Forsch.-Inst. Senckenberg* 102:263–307.

Schubert, J. K., and Bottjer, D. J. 1992. Early Triassic stromatolites as post-mass extinction disaster forms. *Geology* 20:883–886.

Sepkoski, J. J., Jr. 1982. A Compendium of Fossil Marine Animal Families. Milwaukee Public Museum Contribution Biology and Geology No. 51.

Sepkoski, J. J., Jr. 1992. A Compendium of Fossil Marine Animal Families, 2nd ed. Milwaukee Public Museum Contribution Biology and Geology No. 83.

Sharpton, V. L., Dalrymple, G. B., Marin, L. E., Ryder, G., Schuraytz, B. C., and Urrutia-Fucugauchi, J. 1992. New links between the Chicxulub impact structure and the Cretaceous/Tertiary boundary. *Nature* 359:819–821.

Sheehan, P. M. 1988. Late Ordovician events and the terminal Ordovician extinction. *New Mexico Bureau of Mines & Mineral Resources Memoir* 44:405–415.

Sheehan, P. M., and Fastovsky, D. E. 1992. Major extinctions of land-dwelling vertebrates at the Cretaceous-Tertiary boundary, eastern Montana. *Geology* 20:556–560.

Sheehan, P. M., Fastovsky, D. E., Hoffmann, R. G., Berghaus, C. B., and Gabriel, D. L. 1991. Sudden extinction of the dinosaurs: Latest Cretaceous, Upper Great

Plains, U.S.A. *Science* 254:835–839.

Shoemaker, E. M., Wolfe, R. F., and Shoemaker, C. S. 1990. Asteroid and comet flux in the neighborhood of Earth. In *Global Catastrophes in Earth History*, eds. V. L. Sharpton and P. D. Ward, Geological Soc. of America Special Paper 247 (Boulder: Geological Soc. of America), pp. 155–170.

Signor, P. W., III, and Lipps, J. H. 1982. Sampling bias, gradual extinction patterns, and catastrophes in the fossil record. In *Geological Implications of Impacts of Large Asteroids and Comets on the Earth*, eds. L. T. Silver and P. H. Schultz, Geological Soc. of America Special Paper 190 (Boulder: Geological Soc. of America), pp. 291–296.

Smit, J. 1990. Meteorite impact, extinctions and the Cretaceous-Tertiary boundary. *Geologie en Mijnbouw* 69:187–204.

Smoluchowski, R., Bahcall, J. N., and Matthews, M. S., eds. 1986. *The Galaxy and the Solar System* (Tucson: Univ. of Arizona Press).

Stanley, S. M. 1984. Marine mass extinctions: A dominant role for temperature. In *Extinctions*, ed. M. H. Nitecki (Chicago: Univ. Chicago Press), pp. 69–117.

Stigler, S. M., and Wagner, M. J. 1987. A substantial bias in non-parametric tests for periodicity in geophysical data. *Science* 238:940–945.

Tinus, R. W., and Roddy, D. J. 1990. Effects of global atmospheric perturbations on forest ecosystems in the North Temperate Zone; Predictions of seasonal depressed-temperature kill mechanisms, biomass production, and wildfire soot emissions. In *Global Catastrophes in Earth History*, eds. V. L. Sharpton and P. D. Ward, Geological Soc. of America Special Paper 247 (Boulder: Geological Soc. of America), pp. 77–86.

Toon, O. B., Pollack, J. B., Ackerman, T. P., Turco, R. P., McKay, C. P., and Liu, M. S. 1982. Evolution of an impact-generated dust cloud and its effects on the atmosphere. In *Geological Implications of Impacts of Large Asteroids and Comets on the Earth*, eds. L. T. Silver and P. H. Schultz, Geological Soc. of America Special Paper 190 (Boulder: Geological Soc. of America), pp. 187–200.

Upchurch, G. R., Jr. 1989 Terrestrial environmental changes and extinction patterns at the Cretaceous-Tertiary boundary, North America. In *Mass Extinctions, Processes and Evidence*, ed. S. K. Donovan (New York: Columbia Univ. Press), pp. 195–216.

Urey, H. C. 1973. Cometary collisions and geological periods. *Nature* 242:32–33.

Valentine, J. W. 1986. The Permian-Triassic extinction event and invertebrate developmental modes. *Bull. Marine Sci.* 39:607–615.

Venkatesan, M. I., and Dahl, J. 1989. Organic geochemical evidence for fires at the Cretaceous/Tertiary boundary. *Nature* 338:57–60.

Vickery, A. M., and Melosh, H. J. 1990. Atmospheric erosion and impactor retention in large impacts, with application to mass extinctions. In *Global Catastrophes in Earth History*, eds. V. L. Sharpton and P. D. Ward, Geological Soc. of America Special Paper 247 (Boulder: Geological Soc. of America), pp. 289–300.

Visscher, H., and Brugman, W. A. 1986. The Permian-Triassic boundary in the Southern Alps: A palynological approach. *Mem. Soc. Geol. Italiana* 34:121–128.

Wang, K., Orth, C. J., Attrep, M., Jr., Chatterton, B. D. E., Hou, H., and Geldsetzer, H. H. J. 1991. Geochemical evidence for a catastrophic biotic event at the Frasnian/Famennian boundary in south China. *Geology* 19:776–779.

Wang, K., Orth, C. J., Attrep, M., Jr., Chatterton, B. D. E., Wang, X., and Li, J.-J. 1993. The great latest Ordovician extinction on the South China Plate: Chemostratigraphic studies of the Ordovician-Silurian boundary interval on the Yangtze Platform. *Palaeogeogr. Palaeoclimatol. Palaeoecol.* 104:61–79.

Weems, R. E. 1992. The "terminal Triassic catastrophic extinction event" in perspective: A review of Carboniferous through Early Jurassic terrestrial vertebrate extinction patterns. *Palaeogeogr. Palaeoclimatol. Palaeoecol.* 94:1–29.

Widmark, J. G. V., and Malmgren, B. 1992. Benthic foraminiferal change across the Cretaceous-Tertiary boundary in the deep sea; DSDP Sites 525, 527, and 465. *J. Foram. Res.* 22:81–113.

Wise, K. P., and Schopf, T. J. M. 1981. Was marine faunal diversity in the Pleistocene affected by changes in sea level? *Paleobiology* 7:394–399.

Wolbach, W. S., Gilmour, I., Anders, E., Orth, C. J., and Brooks, R. R. 1988. Global fire at the Cretaceous-Tertiary boundary. *Nature* 334:670–673.

Xu, D., Ma, S., Chai, Z, Mao, X., Sun, Y., Zhang, Q., and Yang, Z. 1985. Significance of a $\delta^{13}C$ anomaly near the Devonian-Carboniferous boundary at the Muhua Section, South China. *Nature* 314:154–156.

Yabushita, S. 1991. A statistical test for periodicity hypothesis in the crater formation rate. *Mon. Not. Roy. Astron. Soc.* 250:481–485.

Zachos, J. C., and Arthur, M. A. 1986. Paleoceanography of the Cretaceous/Tertiary boundary event: Inferences from stable isotopic and other data. *Paleoceanog.* 1:15–26.

Zachos, J. C., Breza, J. R., and Wise, S. W. 1992. Early Oligocene ice-sheet expansion on Antarctica: Stable isotope and sedimentological evidence from Kerguelen Plateau, southern Indian Ocean. *Geology* 20:569–573.

Zhakarov, V. A., Lapukhov, A. S., and Shenfil, O. V. 1993. Iridium anomaly at the Jurassic-Cretaceous boundary in northern Siberia. *Russian J. Geology Geophys.* 34:83–90.

Zhou, L., and Kyte, F. T. 1988. The Permian-Triassic boundary event: A geochemical study of three Chinese sections. *Earth Planet. Sci. Lett.* 90:411–421.

EXTINCTIONS AT THE CRETACEOUS-TERTIARY BOUNDARY: THE LINK TO THE CHICXULUB IMPACT

JAN SMIT

Vrije Universiteit

The Chicxulub impact on Yucatan is probably responsible for the Cretaceous/Tertiary K/T extinctions. The actual link between the Chicxulub impact and the extinctions is established through the abundant K/T outcrops with thick tsunami-generated sandstone layers around the Gulf of Mexico. The tsunami-generated layers are at the same stratigraphic position as the few mm fallout (fireball) layer in K/T sections far away from the impact. Both in the distal sections and in the sections close to Chicxulub in the Gulf of Mexico, the first sediments on top of the fallout layer are already strongly depleted in oceanic pelagic biota. The basal parts of the tsunami-generated layers contain melted and unmelted ejecta compatible in composition with the Chicxulub target rocks. The isotopic age of the Chicxulub melt is indistinguishable from the age of the impact glasses in the tsunami-generated layer. The K/T extinctions take place on two time scales. Initially, a mass mortality, reducing the standing crop of the oceans, immediately follows the impact. The actual extinctions probably extend for up to a few thousands of years after the impact, and are probably due to raised ocean-surface temperatures which lasted for several thousands of years. New opportunistic faunas show a rapid diversification and several consecutive blooms of different species occur. Many new species disappear quickly again. Terrestrial K/T extinctions are similar to the marine extinctions in that the rebound of life after the extinctions occurs along the same scenario. If raised atmospheric and ocean surface temperatures were necessary for inducing the actual extinctions, the Chicxulub impact may well have been the only one in the Phanerozoic to have caused a mass extinction, at least on the scale of the K/T extinctions.

I. INTRODUCTION

Mass extinctions occur frequently in the history of life. Some may take a few million years like the great Permian-Triassic extinctions, some happen literally overnight, like at the Cretaceous Tertiary boundary (Chapter by Rampino and Haggerty). Speculations on the whys and who-dun-its abound in the popular and scientific press, in particular on the extinction of the awe-inspiring dinosaurs (Chapter by Sheehan and Russell), because everyone likes to have a try at a possible scenario. Thus, when the impact extinction hypothesis finally gained respectability with the finding of the first hard (iridium anomaly) evidence (Alvarez et al. 1980; Smit and Hertogen 1980), the possibility had already been considered (de Laubenfels 1956; Urey 1973). But where and what is the evidence linking impact and extinction? Lines of evidence start with the geologic record, in particular the abundant record of marine planktic organisms, not with the dinosaurs. Surprisingly, the best evidence is not the extinction record. That will always be incomplete, or appear incomplete no matter how excellent the preserved geologic sections, because of the Signor-Lipps (1982) effect: A sudden truncation of species ranges will produce the illusion of a gradual record, because the chances of finding remains of the last surviving animal are minimal. Also, a sudden extinction can be explained away by assuming the rock record has a hiatus, which is very difficult to disprove. Perhaps the best evidence for a mass-extinction event is the rebound of life after the mass extinction (Chapter by Sheehan and Russell). The rapid worldwide adaptive radiation of surviving biota and the impressive speed by which evolution fills the empty niches with new species all are better proof of a worldwide ecological disaster than the extinction itself (Fig. 1) (Smit 1990).

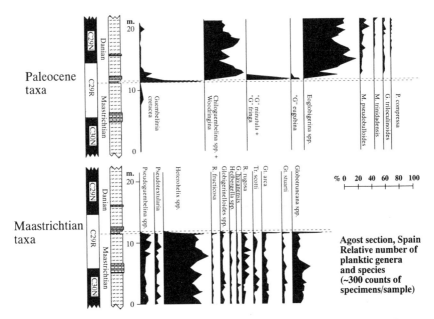

Figure 1.   Agost section, southern Spain. Plots of the relative abundance from 10 m (0.5 Myr) below to 10 m (1 Myr) above the K/T boundary. The relatively stable assemblages of the Maastrichtian are replaced by opportunist assemblages at the base of the Paleocene, showing large changes in relative abundance.

The prologue to the impact-extinction hypothesis began more than a century ago with the realization that vast terrestrial ecosystems dominated by dinosaurs had disappeared at the Cretaceous-Tertiary boundary. Imperfections in the geologic record were always blamed for the apparent sudden disappearance of the Cretaceous organisms, including dinosaurs and large numbers of marine organisms. In the early 1960s, geologists began to realize that deep-marine sequences, out of reach of erosional and other physical disturbing mechanisms, are in principle continuous, and that it might be possible to monitor evolutionary changes and extinctions continuously, i.e., in steps of < one thousand years. The Deep Sea Drilling Project vastly increased the number of continuous sections from the last 135 Myr, which is the age of the oldest existing ocean floor. Extinctions older than 135 Myr are therefore hard to investigate precisely because continuous records for those extinctions may not exist. That leaves the Cretaceous Tertiary (K/T) boundary as possibly the only one of the major mass-extinctions that can be monitored satisfactorily. Bramlette and Martini (1964), and Luterbacher and Premoli Silva (1964), studying continuous deepwater sections in France and in the Appenines of Italy, respectively, realized that planktonic unicellular organisms with calcareous skeletons (foraminifera and small algae [coccoliths]) disappeared suddenly and almost completely precisely at the K/T boundary and were replaced by a set of similar, but entirely new species. Because of their large numbers (up to 1 million cm^{-3} sediment) planktic foraminifers are ideal fossils to monitor the K/T extinctions. It soon became apparent that even in the most continuous rock-sequences, these foraminifers disappeared at a bedding plane—an almost zero instant in time—without any preceding changes leading towards their final demise. Reports to the contrary (Keller 1988; Keller et al. 1993) suffer from the Signor-Lipps (1982) effect, because the species inferred to disappear before the K/T boundary are rare. The plotting of relative species abundances instead of actual number of individuals of each species (Keller et al. 1988) also may create the illusion of gradual extinctions in which those at the K/T boundary are just one step of many (see Fig. 2 and Fig. 8, below). The sudden, unannounced extinctions ask for an explanation. Those explanations were sought in either Earth-bound, or extraterrestrial causes. The record of the planktic foraminifers suggests that the answers are not Earth-bound, because normal Earth-bound climate changes are gradual, not sudden. Planktic foraminiferal communities are far from uniform. Hemleben et al. (1989) demonstrated the wide ecological variety of planktonic foraminifers. Some live their entire life close to the ocean surface. Others live deeper, reproduce at ∼400 m depth, and are adapted to different temperatures and light. Some are carnivorous, some are grazers. Other groups carry symbionts. In other words, a foraminiferal assemblage is very sensitive to changes in ocean water chemistry, temperatures and salinity. If climate had changed due to terrestrial changes like increased volcanism, increase in seafloor spreading, sealevel changes or ice-ages, assemblages of foraminifers should have changed accordingly. This is not what happened at the K/T boundary; whatever their ecological niche, all planktic foraminifers suffered at the same geological instant. The discovery of the iridium anomaly gave a decisive boost to the extraterrestrial explanation, in particular the impact extinction hypothesis after the supernova explosion had to be abandoned because the K/T Os and Ir isotope ratios (Alvarez et al. 1980; Smit and Hertogen 1980) had a solar system signature. The impact extinction hypothesis is testable, and leads to testable predictions. In the years following the iridium discovery, a variety of additional evidence for impact was discovered. Microkrystites, possible condensates from the impact vapor cloud (Montanari et al. 1983; Smit et al. 1992), and shocked minerals with evidence for very high shock pressures (Bohor et al. 1984; Izett 1990) were found worldwide exactly at the K/T boundary. Finally, the ~210 km (possibly 300 km; Sharpton et al. 1993) Chicxulub K/T impact structure was found, surrounded by ejecta layers with impact glass (tektites, or their altered pseudomorphs) underlying thick, tsunami-generated sandstone layers. Because the ejecta are found precisely at the K/T boundary and because the $^{40}Ar/^{39}Ar$ 65.07±0.1 Myr age of tektites and the 64.98±0.08 Myr age of Chicxulub crater-melt (Swisher et al. 1992) are identical, the link of impact to the K/T extinctions is undeniable, but the exact mechanisms leading to extinctions are still not clear.

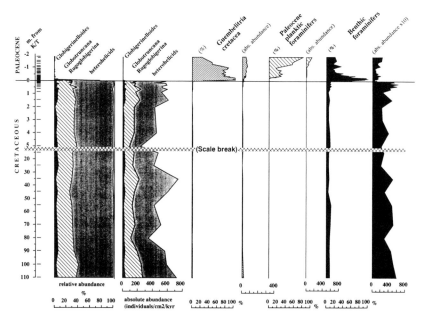

Figure 2.   El Kef, Tunisia section. Comparison of relative (absolute number of specimens/cm²/kyr) abundances (see also Fig. 8) of foraminifers. The absolute abundance emphasizes the mass mortality at K/T boundary. Although extinctions among benthic foraminifers were few, the absolute abundance shows that mass-mortality also has affected benthic populations. The relative abundance tends to mask this fact, suggesting a proliferation of benthic species instead.

II. DISTAL K/T SECTIONS

Undisturbed rock sequences containing a complete record of the events across the K/T boundary, where we can follow all events on time intervals of 500 yr or less are rare (Fig. 3). Representative sections are the Agost, Caravaca and Zumaya sections in Spain and the Kef section in Tunisia (Fig. 4). Time is monitored in the Agost section by magnetostratigraphy refined by analysis of the limestone-marl bedding rhythms, which are the preserved climate signals of the orbital precession/eccentricity variations (Kate and Sprenger 1992) (Fig. 5). The bedding rhythms suggest that no sediments are missing across the K/T boundary (Kate and Sprenger 1992), implying that changes in abundance and diversity of foraminifera can be followed in steps of less than one thousand years up to and across the impact ejecta layer.

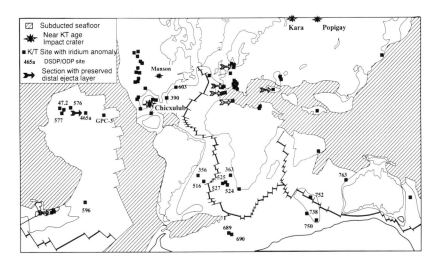

Figure 3.   K/T sites with an iridium anomaly plotted on a paleogeographic reconstruction of 65 Myr ago. Also indicated are those sections which have an undisturbed fallout layer (arrows).

A. Pre-Impact

Biotic changes before the deposition of the impact ejecta layer all appear related to background changes in climate. Some typical Cretaceous biota like the rudists and inoceramids (McLeod and Ward 1990) apparently disappear a few million years before the boundary, but assemblages of microplankton do not show significant changes (see Figs. 1 and 2; and 8 below) leading up to the K/T boundary ejecta layer. The claimed (Keller 1988) disappearance of species a few tens of centimeters below the K/T boundary is probably due to sampling bias, because a larger sample taken just below the K/T boundary contains all the "missing" species. Detailed iridium profiles measured below the K/T boundary (Alvarez et al. 1990) offer no evidence for enhanced Ir flux below the K/T ejecta layer. Some minor enhancement can be explained by diffusion or bioturbation.

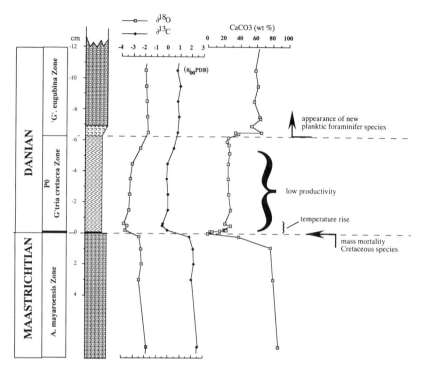

Figure 4.   Stratigraphy, detailed stable isotope (fine fraction, mainly coccoliths) and
CaCO₃ concentration (K/T boundary of the Agost, Spain, section). New planktic
species appear only after the ^{18}O values have returned to the same values as in the
upper Cretaceous.

B. Impact Ejecta Layer

The K/T boundary in distal sites (i.e., those >5000 km away from the Chicxulub crater) is marked by a few mm thick clay layer, the ejecta layer, which
separates sediments extremely rich in Cretaceous foraminifers from clay sediments poor in planktic biota (Fig. 6). The K/T boundary layer originates
from settling dust of the impact vapor cloud, which is dispersed worldwide.
In most K/T sections, the fallout layer is mixed with other sediments by bioturbation, and only in a few sections is the layer still intact (Fig. 3). The layer
has sharp upper and lower boundaries, as expected for a fallout layer. The
layer also contains anomalous concentrations of siderophile (platinum group)
metals. The maximum concentration is difficult to estimate, because iridium
appears somewhat mobile (Rocchia et al. 1992) and tends to concentrate in
the many pyrite and goethite concretions in the layer (Table I). Yet over 20
whole rock analyses of the ejecta layer in Spain yield an average of 25±5 ng/g
Ir, indicating a (CI chondrite) bolide/target ratio in the impact vapor cloud
of about 0.05. Abundant microspherules (>5000 cm^{-3}) occur almost worldwide (Fig. 7). Those spherules are usually altered but show an intricate texture

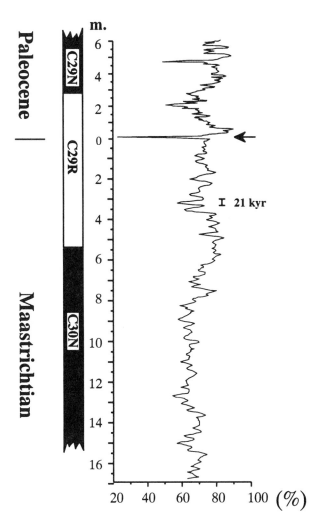

Figure 5.   Running average of CaCO₃ content (explained by orbital forcing according to the Milankovitch model). The dominant cycles are those of precession, on average 21 Kyr. Carboate and clay fluxes were calculated using average precession 21 Kyr increments.

of dendritic precursor minerals (mainly clinopyroxene, olivine, plagioclase) (Smit et al. 1992). Therefore these spherules classify as microkrystites, not tektites, and were probably a condensate from the impact vapor cloud. Some of the spherules have Ir concentrations of up to 500 ng/g and may have been ablated from the bolide itself (Robin 1992). Rare grains of quartz and other minerals with shock deformation planar features (PDF) occur as well (Bohor and Izett 1986). The grains are considerably smaller than the shocked grains

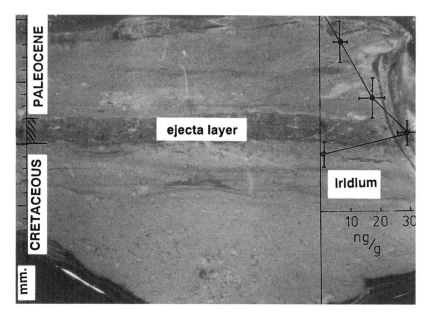

Figure 6.   Photograph of the fallout layer in the Agost section. The layer is undisturbed, and has sharp upper and lower boundaries. The dark sediment directly on top
is clay-rich as a result of the immediate decimation of oceanic carbonate producers
after the impact.

from the fallout layer in the U. S. Western Interior and Gulf of Mexico.

C. Post Impact

Organisms Which Produce Carbonate.   Already, the first millimeters of
sediment on top of the impact ejecta layer are strongly depleted in carbonate,
indicating that whatever mechanism removed ocean surface biota, had to
be an immediate result of the impact itself. This lends credit to proposed
mechanisms directly coupled to the impact such as the dust-cloud/darkness
scenario, flash heating of the atmosphere or wildfires.

Longer lasting mechanisms, in the order of a few hundred years, should
have left some signal in the sedimentary record. This is usually beyond strati-
graphical resolution, but in the Kef section, the most amplified across the K/T
boundary in the world, the first 2 cm of the boundary clay still contain a significant amount of planktic biota, gradually reducing over those 2 cm (Fig. 8).
Thus we observe that, although an initial mass-mortality immediately followed the deposition of the ejecta layer, an additional stress further reduced
the planktic biota. It is probably this additional stress—oxygen isotope analyses (Fig. 4) suggest a warming caused by a greenhouse atmosphere—which
has led to the final extinctions. Without the additional stress, the Cretaceous
biota might have returned from refugia of surviving populations, and extinctions on the scale as observed at the K/T boundary may never have taken place.
It seems therefore necessary to decouple the initial mass-mortality from the

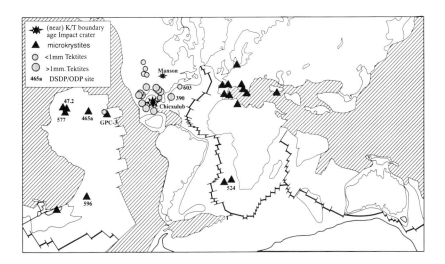

Figure 7.   Known distribution of tektites and microkrystites at the K/T boundary. The microkrystites are abundant outside the K/T tektite strewnfield, but extremely rare or absent within it.

mass extinctions. However, both work in concert and both were induced by the same impact. The boundary clay still has anomalous Ir concentrations at its base, exponentially decreasing upward. Diffusion from the ejecta layer and bioturbation may be responsible for the displacement of anomalous Ir, as shown in clearly bioturbated sequences (Michel et al. 1985). In other sections, close to continents and less bioturbated, like the Agost and Caravaca sections in Spain, surplus Ir may have been supplied by run-off from land. Stable carbon ^{13}C and oxygen isotopic ^{18}O ratios of the carbonate fraction in the basal part of the clay (Fig. 4), show large negative deviations from upper Cretaceous values (Hsu et al. 1982; Smit 1990; Zachos and Arthur 1986). These excursions are generally interpreted as resulting from the cessation of primary productivity (^{13}C), and the increase in ocean surface water temperatures (^{18}O) (Emiliani et al. 1981). The duration of the ^{18}O spike is a few thousand years, and that may have been the duration of the greenhouse atmosphere in the basal Paleocene. The new planktic faunas apparently could originate and proliferate only when the ^{18}O values returned to normal. The foraminiferal population in the clay is just a poor remnant of the Cretaceous populations. What is present is mostly reworked, but a few species are relatively abundant (*Guembelitria cretacea, Globigerinelloides sp.* and *Hedbergella monmouthensis*, heterohelicids) and therefore are most likely survivors. Those few survivors disappear or greatly decline in numbers (*G. cretacea*) when the new Paleocene planktic populations appear at the top of the boundary clay (Fig. 8). In the boundary clay, no new species appeared. The time of boundary clay deposition was an interesting time for the dinoflagellates. This group of phytoplankton did

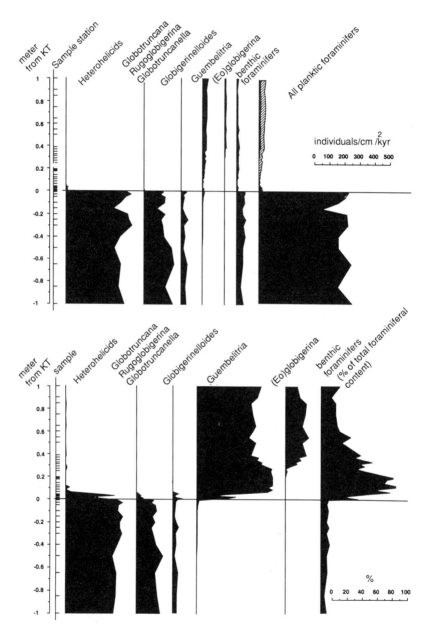

Figure 8.   Detailed comparison of relative and absolute foraminiferal abundance
changes across the K/T boundary in the El Kef, Tunisia, section. Same data set
as in Fig. 2. Although most Cretaceous species are still present at the base of the
boundary clay, their abundance is reduced.

not experience extinctions, probably because of their encystment capabilities. Yet within the boundary clay this group expanded and diversified. For instance, species adapted to brackish water were able to migrate to the open ocean. Apparently lack of competition enabled the dinoflagellates to move into formerly inaccesible niches. The initial recovery of the ocean surface began when new Paleocene foraminiferal populations appeared. At the same time, primary productivity of coccolithophorids also resumed, leading to renewed deposition of pelagic carbonate. However, sedimentation rates were still reduced in comparison with the upper Cretaceous. This agrees with the low ^{13}C values of basalmost Paleocene carbonate, indicating reduced primary productivity. Less easy to understand is why the sedimentation rate of the detrital clay fraction in basal Paleocene sediments also had been reduced worldwide (Fig. 9). Detrital clay is supplied by river discharge or by wind transport to the oceans. Reduced aridity (less evaporation), or increased plant cover may be factors controlling this clay flux. The K/T mass extinction had effectively removed large grazing herbivores for a few million years, allowing a lavish vegetation to develop. Also a change in climate to one more humid may have reduced soil erosion. In the U. S. Western Interior the pollen record shows moisture loving vegetation replacing seasonal vegetation just after the K/T boundary. It has been known for a long time that development of coal swamps initiated at the K/T boundary. For this reason the first extensive coal layer, the Z-coal, has long been considered as a proxy for the K/T boundary (Van Valen and Sloan 1977). The new planktic species appearing in the ocean were opportunists (e.g., the foraminifers *G. eugubina*, *G. fringa*, *G. minutula*). Those opportunists showed rapid evolutionary changes and changes in abundance. Well known are the blooms of "disaster species" *Thoracospheara* and *Braarudosphaera* (Percival and Fischer 1977). Successively, different species dominated over others and none of the species had a long lifespan (Fig. 1). The opportunistic assemblages disappeared after only a few tens of thousands of years. Similar developments occurred on the continent. It cannot directly be proven that dinosaurs disappeared as result of the impact, simply because of the lack of bones underlying the ejecta layer. However, ecologically it seems likely that they disappeared as the dominant form of life on land because of the response of life forms that were subordinate to the dinosaurs in the Cretaceous. It is common wisdom that the mammals flourished and diversified in the Paleocene because of the demise of the dinosaurs in a broad sense. But also on a finer scale this cause and effect relationship is evident (Fig. 10). The Bug Creek area in Montana is a classical region for continental K/T studies because localities with vertebrate faunas just above and below the boundary are abundant. Misinterpretations of the stratigraphic position of river channel deposits containing the vertebrate remains, and the misinterpretation of the position of the K/T boundary in the field (between 2 and 12 m too high) have led to the impression of gradual disappearance of dinosaurs and evolution of mammals (Sloan et al. 1986; Smit et al. 1987; Van Valen and Sloan 1977), still shown in many textbooks. The sediments in the first Paleocene stratigraphically superimposed river channels show a rapid diversification of eutherian (placental) mammal species within ~10^5 yr. For instance, ungulate mammal species increased from 1 to 8 (Rigby et al. 1987). One of the first primates also appeared during this first evolutionary pulse (Van Valen and Sloan 1977). Even more significant is the abundance of *Protungulatum* teeth (>1 million total) in the river channel that most likely was filled just after the K/T impact, the Bug Creek Anthills (Van Valen and Sloan 1977). It is the first deposit where the yield of dinosaur teeth is significantly less than in Cretaceous sediments (Rigby et al. 1987), and all those may have been reworked. The "bloom" of just one mammal species testifies to the ecological disappearance of the dinosaurs, no longer capable of restricting mammal populations. In the oceans stable planktic populations follow the opportunists after 5×10^4 to 10^5 yr. The species resemble planktic species still living today. These species last longer and their abundance patterns are not as erratic as those of the opportunists preceding them.

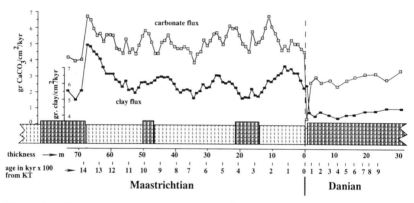

Figure 9. Zumaya section, Spain. flux (g/cm^2/kyr) of pelagic carbonate and clay across the K/T boundary is calculated using 21 kyr precession rhythms. The reduction in carbonate flux occurs precisely at the K/T boundary, the clay flux some 40 kyr later.

III. PROXIMAL K/T SECTIONS

Around the Chicxulub crater, K/T sequences are different from the distal sections (Fig. 11). In the Western Interior of North America the K/T boundary is marked by a dual-layered boundary clay (Izett 1990), and in marine sections around the Gulf of Mexico the K/T boundary is marked by thick, tsunami- or seiche-generated clastic deposits (Smit et al. 1992). The lower layer in continental deposits of the Western Interior is characterized by an abundance of spherules, dumbbells and other splashforms highly reminiscent of tektites, but altered to secondary minerals (goyazite and kaolinite in continental sites, smectite in marine sites). The lower layer contains moderately anomalous Ir (~0.32 ng/g) and rare shocked minerals. The thinner upper layer is often

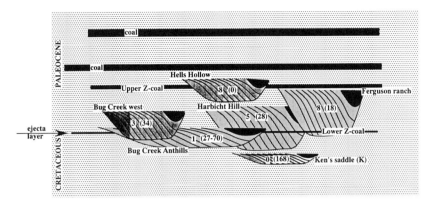

Figure 10.   Hell Creek/Bug Creek drainage, eastern Montana. Simplified reconstruction of the stratigraphic position of river channels with fossil-rich lag deposits,
with respect to the K/T boundary. Numbers in the channels indicate the number
of ungulate mammal species (Smit et al. 1987). In parentheses is the approximate
number of dinosaur teeth per ton of lag deposit sediment, sieved for teeth. Note
that channels not eroding into Cretaceous sediment do not contain dinosaur remains
(Sloan et al. 1986). The upper Z-coal has frequently been used to define the K/T
boundary in the field, but that level is between 2 and 11 m too high, leading to the
impression of significant mammal evolution below the K/T boundary.

sharply separated, but welded to the lower layer, and is characterized by highly
anomalous iridium concentrations (~14.6 ng/g; Izett 1990) and abundance of
shocked minerals and rare Ni-rich magnesioferrite spinels. One may hypoth-
esize for the origin of two layers that besides the Chicxulub impact, a second
(Manson?) impact has occurred, responsible for the upper layer, but that is
not necessarily so. Differential settling (coarse ejecta deposited first) and the
origin of each the two layers as ejecta from different phases of the impact
process (Bohor 1988; Smit et al. 1992) may explain the existence of two
layers, analogous to the sharply delineated superposition of the suevite over
the Bunte Breccia of the Ries crater ejecta (Horz 1982). The high concentrations of platinum group elements indicate that the upper layer, also termed
the fireball layer, has settled from the fine-grained impact vapor cloud, which
contains the vaporized remains of the bolide. Also, the ages of shocked zircon
grains in the fireball layer (Premo and Izett 1993) (300–575 Myr source rock
ages) appear to be too young for the Manson crater rocks (>1800 Myr old),
long considered as a possible K/T candidate impact crater, but now dated at
73 Myr (Izett et al. 1993).

   The thick clastic deposits at the K/T boundary around the Gulf of Mexico,
combine features of the distal K/T sections and the Western Interior K/T
sections, with the superimposed effects of tsunami waves generated by the
impact on Yucatan.   The depositional history of these clastic deposits is
therefore complex. However, within all clastic-deposit outcrops a consistent
stratigraphy can be distinguished, where each of the different units can be

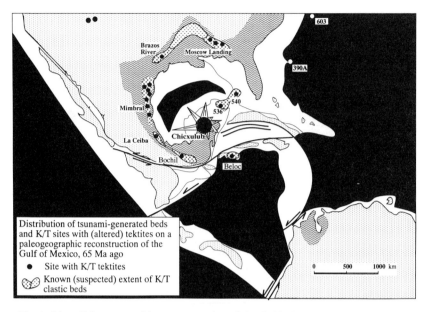

Figure 11.   Paleogeographic reconstruction of the Gulf of Mexico and Caribbean at
    K/T time. Areas where tsunami-generated beds were found are hatched, and the
    known sites with tektite remains are marked.

linked to consecutive phases of the Chicxulub impact.

Basically, the deep-water Gulf of Mexico sections are the same as the distal pelagic sections—the thick clastic deposits replacing the few mm thick ejecta layer. The boundary clay invariably occurs directly on top of the clastic layer.

The clastic deposits can be subdivided in three units (Figs. 12 and 13). The basal unit 1 is composed of coarse ejecta, reworked in channels with local seafloor rip-up material (oozes) by the first arriving tsunami waves. The ejecta are mostly altered impact glass fragments and tektites. In some sections rare unaltered relict impact glass fragments survived.

Also nonmolten, or partially molten clasts of shallow-water platform limestone of the same size as the tektites occur. These shallow-water limestone fragments are consistent with the stratigraphy of the Chicxulub area at the time of impact. In northern Yucatan, 2 to 3 km thick platform limestones alternating with evaporites form the upper part of the pre-impact Chicxulub stratigraphy, overlying a basement of early Paleozoic Pan-African terrain.

Sand and plant material, backwashed from coastal areas, arrives later in the deeper parts of the Gulf and overlie in thick (up to 9 m thick) channels, the ejecta-bearing channels (unit 2). Sedimentary features in unit 2 (Fig. 14) display several 180° paleocurrent direction changes in successive layers, showing a clear relation to up-surge and backwash stages of breaking mega-waves. The highest unit 3 consists of thin sandstone ripples alternating with fine silt. Like

Figure 12.   Arroyo de Mimbral, Mexico. Photograph of the clastic beds at the K/T boundary. Lowermost sediments are pelagic shales of the Cretaceous Mendez Formation. In the photograph is A. Montanari of Osservatorio Geologico Coldigoco. His hand is resting on a 50 cm thick layer with tektite remains. The laminated sediments below his hand are the unit 1 channels filled with ejecta and Mendez rip-up clasts. The laminated massive appearing middle part is the first channel of unit 2 with sands backwashed from coastal areas. The two ledges above the massive sandstone are the next two channels, still from unit 2. Top unit 3 displays three thin rippled layers, alternating with thin silts. About 15 cm of basal Paleocene Velasco shales are still visible below Quaternary conglomerate.

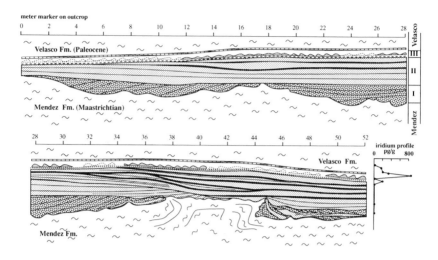

Figure 13.   Arroyo de Mimbral, Mexico. Simplified cross section through the outcrop, showing the different channellized units. Iridium abundances from Smit et al. (1991). As a result of differential loading by unit 1 channels, the soft Mendez oozes were squeezed away into diapiric structures between the channels. The top ripples (unit 3) are not confined in channels and occur continuously in every outcrop in eastern Mexico, even at those sites where the lower two units are not present, like at the Mesa de Llera, 30 km northwest of Mimbral. These ripples could have been made by "contour currents" generated by the developing seiche in the Gulf of Mexico.

the upper (fireball) layer in the continental sections, this upper unit contains anomalous iridium, in particular in the fine silt layers, shocked minerals and Ni-rich magnesioferrite spinels. Unit 3 reflects the waning activity of tsunami waves, allowing settling of clay and silt between arrival of the last tsunami or seiche waves. Thus, both in the Western Interior sections and in the marine Gulf coast sections coarse ejecta and fine-grained iridium-bearing ejecta are separated in an lower and upper layer. How much time is involved between deposition of the two different ejecta types is difficult to estimate. Regarding the amalgamated nature of the dual layers in the Western Interior, and the coherency of the tsunami-generated clastic layers, not more than a few days may have passed between deposition of the coarse molten and solid ejecta and the settling of the finer grained, iridium-rich impact vapor-cloud products.

Consistent with the distal sections in Spain, the planktic biota in the first pelagic sediments (the boundary clay) on top of the tsunami-generated layers are already strongly reduced, once again showing the quick response of the biosphere to the Chicxulub impact event.

IV. CONCLUSIONS

An impact with the size and characteristics of the Chicxulub impact may well be unique in the geologic record. The most amplified record of the extinctions

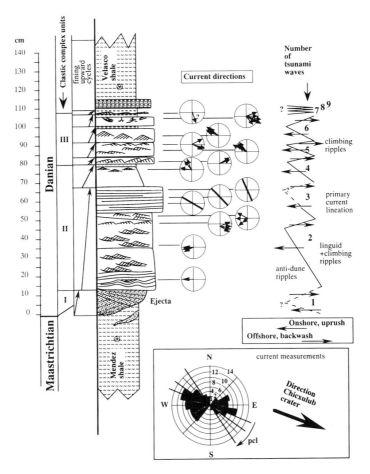

Figure 14. Tsunami generated clastic beds at the Cretaceous-Tertiary boundary at La Lajilla, Mexico, 40 km east of Ciudad Victoria. Measured paleocurrent directions display two dominant directions, different by 180°: one toward and one away from the Chicxulub crater. Those two directions alternate successively in stratigraphic sequence; they show up-surge and backwash stages of at least six individual tsunami waves.

shows a record on two time scales. A mass mortality immediately followed the impact, possibly within months. Ocean surface biota were strongly reduced, yet most species still may have been present. What exactly were the causes of the mass mortality can only be speculated upon. The actual extinctions may have been a matter of several hundred years, clearly also a result of the impact, but a longer lasting effect. The causes for the extinctions may be sought in the stable isotope record. ^{18}O changes of pelagic carbonate indicate a rise in ocean surface temperatures of about 10°C (Smit 1990). The high temperatures persist a few thousand years, and could be sustained by the combined effects of impact vaporization of limestone, releasing large quantities of CO_2 in the atmosphere, plus the released greenhouse gases of wildfires on the one hand and the reduced capability of storage of the CO_2 due to the removal of primary producers and destruction of forests on the other. It has been suggested that the Chicxulub crater is over 300 km in size, which would make it one of the largest young impact craters in the solar system (Sharpton et al. 1993). If, for producing extinctions of the magnitude at the K/T boundary not only an impact the size of Chicxulub would be necessary, but also an unusual target like the 2 to 3 km thick layer of limestone and evaporite cover in Yucatan (<2% of the Earth's surface is covered by such a thick limestone layer), then the K/T event may have been unique in the geologic record.

Acknowledgments. The author wishes to thank Th. B. Roep, W. Alvarez, A. Montanari, G. Ganssen, P. Smit and many others for invaluable help during this research project. M. Rampino and two anonymous reviewers helped to improve this manuscript. This research was supported by a fellowship of the Royal Dutch Academy of Sciences.

REFERENCES

Alvarez, L. W., Alvarez, W., Asaro, F., and Michel, H. V. 1980. Extraterrestrial cause for the Cretaceous-Tertiary extinction. *Science* 208:1095–1108.

Alvarez, W., Asaro, F., and Montanari, A. 1990. Iridium profile for 10 million years across the Cretaceous-Tertiary boundary at Gubbio (Italy). *Science* 250:1700–1702.

Bohor, B. F. 1988. Microspherules and the dualistic nature of the K-T boundary clay. *Lunar Planet. Sci.* XIX:333–334 (abstract).

Bohor, B. F., and Izett, G. A. 1986. Worldwide size distribution of shocked quartz at the K/T boundary: Evidence for a North American impact site. *Lunar Planet. Sci.* XVII:68–69 (abstract).

Bohor, B. F., Foord, E. E., Modreski, P. J., and Triplehorn, D. M. 1984. Mineralogic evidence for an impact event at the Cretaceous-Tertiary boundary. *Science* 224:867–869.

Bramlette, M. N., and Martini, E. 1964. The great change in calcareous nannoplankton fossils between the Maastrichtian and Danian. *Micropaleontology* 10:291–322.

de Lauben Fels, M. W. 1956. Dinosaur Extinction: One more hypothesis. *J. Paleontology* 30:207–218.

Emiliani, C., Kraus, E. B., and Shoemaker, E. M. 1981. Sudden death at the end of the Mesozoic. *Earth Planet. Sci. Lett* 55:317–334.

Hemleben, C., Spindler, M., and Anderson, O. R. 1989. *Modern Planktonic Foraminifers* (Berlin: Springer-Verlag).

Horz, F. 1982. Ejecta of the Ries Crater, Germany. In *Geological Implications of Impacts of Large Asteroids and Comets on the Earth*, eds. L. T. Silver and P. H. Schultz, Geological Soc. of America Special Paper 190 (Boulder: Geological Soc. of America), pp. 39–55.

Hsu, K. J., He, Q., McKenzie, J. A., Weissert, H., Perch-Nielsen, K., Oberhanstli, H., Kelts, K., LaBreque, J., Tauxe, L., Krahenbull, U., Percival, S. R., Wright, R., Karpoff, A. M., Petersen, N., Tucker, P., Poore, R. Z., Gombos, A. M., Pisciotto, K., Carman, M. F., and Schreiber, E. 1982. Mass mortality and its environmental and evolutionary consequences. *Science* 216:249–256.

Izett, G. A. 1990. *The Cretaceous/Tertiary Boundary Interval, Raton Basin, Colorado and New Mexico, and Its Content of Shock-Metamorphosed Minerals: Evidence Relevant to the K/T Boundary Impact-Extinction Theory*, Geological Soc. of America Special Paper 249 (Boulder: Geological Soc. of America).

Izett, G. A., Maurrasse, F. J.-M. R., Lichte, F. E., Meeker, G. P., and Bates, R. 1990. Tektites in Cretaceous-Tertiary Boundary Rocks on Haiti. U. S. Geological Survey Open-File Rept. 90-635, pp. 1–31.

Kate, W. G. T., and Sprenger, A. 1992. Rhythmicity In Deep Water Sediments, Documentation and Interpretation By Pattern and Spectral Analysis. Ph.D Thesis, Vrije Universiteit.

Keller, G. 1988. Extinction, survivorship and evolution of planktic Foraminifera across the Cretaceous/Tertiary boundary at El Kef, Tunisia. *Marine Micropaleontology* 13:239–263.

Keller, G., Barrera, E., Schmitz, B., and Mattson, E. 1993. Gradual mass extinction, species survivorship, and long-term environmental changes across the Cretaceous Tertiary boundary in high latitudes. *Geol. Soc. America Bull.* 105:979–997.

Luterbacher, H. P., and Premoli Silva, I. 1964. Biostratigrafia del limite cretaceo-terziario nell'Appennino centrale. *Rivista Italiana Paleontologia Stratigrafia* 70:67–128.

McLeod, K. G., and Ward, P. D. 1990. Extinction pattern of Inoceramus (Bivalvia) based on shell fragment biostratigraphy. In *Global Catastrophes in Earth History*, eds. V. L. Sharpton and P. D. Ward, Geological Soc. of America Special paper 247 (Boulder: Geological Soc. of America), pp. 509–518.

Michel, H. V., Asaro, F., Alvarez, W., and Alvarez, L. W. 1985. Elemental profile of iridium and other elements near the Cretaceous/Tertiary boundary in Hole 577B. In *Initial Reports of the Deep-Sea Drilling Project*, eds. G. R. Heath and L. H. Burckle (Washington, D. C.: U. S. Government Printing Office), pp. 533–538.

Montanari, A., Hay, R. L., Alvarez, W., Asaro, F., Michel, H. V., Alvarez, L. W., and Smit, J. 1983. Spheroids at the Cretaceous-Tertiary boundary are altered impact droplets of basaltic composition. *Geology* 11:668–671.

Percival, S. F., and Fischer, A. G. 1977. Changes in calcareous nannoplankton in the Cretaceus-Tertiary biotic crisis at Zumaya, Spain. *Evol. Theory* 2:1–35.

Premo, W. R., and Izett, G. A. 1993. U-Pb provenance ages of shocked zircons from K-T boundary, Raton Basin, Colorado. *Lunar Planet. Sci.* XXIV:1171–1172 (abstract).

Rigby, J., Newman, J. K., Smit, J., Kaars, S. V. d., and Sloan, R. E. 1987. Dinosaurs from the Paleocene part of the Hell Creek Formation, McCone County, Montana. *Palaios* 2:296–302.

Rocchia, R., Boclet, D., Bonte, P., Froget, L., Galbrun, B., Jehanno, C., and Robin, E. 1992. Iridium and other element distributions, Mineralogy, and magnetostratigraphy near the Cretaceous/Tertiary boundary in Hole 761C. In *Proc. Ocean Drilling Program, Scientific Results*, ed. U. von Rad and B. U. Haq (College Station, Tex.: Ocean Drilling Program), pp. 753–761.

Sharpton, V., Burjke, K., Camargo-Zanoguera, A., Hall, S. A., Lee, D. S., Marin, L. E., Reynoso, G. S., Muneton, J. M. Q., Spudis, P. D., and Fucugauchi, J. U. 1993. Chicxulub Multiring basin: Size and other characteristics derived from gravity analysis. *Science* 261:1564–1568.

Signor, P. W., III, and Lipps, J. H. 1982. Sampling bias, gradual extinction patterns, and catastrophes in the fossil record. In *Geological Implications of Impacts of Large Asteroids and Comets on the Earth*, eds. L. T. Silver and P. H. Schultz, Geological Soc. of America Special Paper 190 (Boulder: Geological Soc. of America), pp. 291–296.

Sloan, R. E., Rigby, J. K., Valen, L. M. V., and Gabriel, D. 1986. Gradual dinosaur extinction and simultaneous ungulate radiation in the Hell Creek Formation. *Science* 232:629–633.

Smit, J. 1990. Meteorite impact, extinctions and the Cretaceous-Tertiary boundary. *Geologie en Mijnbouw* 69:187–204.

Smit, J., and Hertogen, J. 1980. An extraterrestrial event at the Cretaceous-Tertiary boundary. *Nature* 285:198–200.

Smit, J., Kaars, W. A. van der, and Rigby, J. K. 1987. Stratigraphic aspects of the Cretaceous-Tertiary boundary in the Bug Creek area of eastern Montana, U. S. A. *Mem. Soc. Geol. France* 150:53–73.

Smit, J., Montanari, A., Swinburne, N. H. M., Alvarez, W., Hildebrand, A. R., Margolis, S. V., Claeys, P., Lowrie, W., and Asaro, F. 1992 Tektite-bearing, deep-water clastic unit at the Cretaceous-Tertiary boundary in northeastern Mexico. *Geology* 20:99–103.

Smit, J., Alvarez, W., Montanari, A., Swinburne, N., Kempen, T. M. v., Klaver, G. T., and Lustenhouwer, W. J. 1992. "Tektites" and microkrystites at the Cretaceous Tertiary boundary: Two strewnfields, one crater? *Lunar Planet. Sci. Conf.* 22:87–100.

Swisher, C. C., III, Nishimura, J. M. G., Montanari, A., Pardo, E. C., Margolis, S. V., Claeys, P., Alvarez, W., Smit, J., Renne, P., Maurasse, F. J. M. R., Curtiss, G., and McWilliams, M. 1992. Coeval ^{40}Ar/^{39}Ar ages of 65.0 million years ago from Chicxulub Crater melt-rock and Cretaceous-Tertiary boundary tektites. *Science* 257:954–958.

Urey, H. C. 1973. Cometary collisions and geological periods. *Nature* 242:32–33.

Van Valen, L., and Sloan, R. E. 1977. Ecology and the extinction of the dinosaurs. *Evol. Theory* 2:37–64.

Zachos, J. C., and Arthur, M. A. 1986. Paleoceanography of the Cretaceous/Tertiary boundary event: Inferences from stable isotopic and other data. *Paleoceanography* 1:5–26.

FAUNAL CHANGE FOLLOWING THE CRETACEOUS-TERTIARY IMPACT: USING PALEONTOLOGICAL DATA TO ASSESS THE HAZARDS OF IMPACTS

PETER M. SHEEHAN
Milwaukee Public Museum

and

DALE A. RUSSELL
Canadian Museum of Nature

The paradigm of catastrophic biotic change as a result of bolide impacts must be added to that of gradual evolutionary change through geologic time. The interaction between these two paradigms resulted in long periods of ecologic and taxonomic stasis broken by extinction events and followed by changes in dominant organisms. The pattern of change associated with the extinction of the dinosaurs is consistent with an impact-induced catastrophe, and the replacement of long-dominant dinosaurian herbivores by a radiation of mammalian herbivores derived from detritus-based food chains. Although modern civilization would be nearly destroyed by a K-T type impact, humans as a species would survive along with other organisms adapted to living on food reserves ("detritus"). The paleontological record should be carefully examined in order to reveal the effects of other, less energetic impact events.

I. INTRODUCTION

Mass extinctions provide a worse-case scenario for assessing the hazard of impacts to Earth's biota and human civilization. The impact event at the end of the Cretaceous Period (Alvarez et al. 1980) produced the most recent of only five major mass extinctions which occurred during the last 500 Myr (Raup and Sepkoski 1982; see also the Chapter by Rampino). These extinctions (Fig. 1) may have had varied causes, but all were characterized by such large loss of life that the ecosystems collapsed. Following each event completely new communities of organisms were formed as new groups of plants and animals radiated into the ecologic niches once occupied by the extinct organisms.

Estimates of the number of species that became extinct during mass extinctions range as high as 95% for the largest event at the Permian-Triassic boundary and from 60 to 80% for the other four events (Sepkoski 1989). Mass extinctions marked the end of long intervals of evolutionary stability during which natural selection was regulated by competition and predator-prey interactions (Fig. 1). These intervals of stability, named Ecologic Evolutionary

Units (EEUs) by Boucot (1983), dominated most of Phanerozoic time. During the EEUs plants and animals evolved in concert, and species seldom evolved radically new life styles because evolution was constrained ecologically.

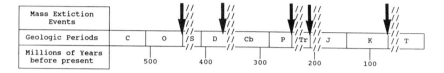

Figure 1.   Time-line showing the mass extinction events (arrows) and recovery intervals (hatching). Geologic periods are: C–Cambrian, O–Ordovician, S–Silurian, D–Devonian, Cb–Carboniferous, P–Permian, T–Triassic, J–Jurassic, K-Cretaceous, T–Tertiary.

Mass extinctions terminated EEUs, and, for intervals of a few million years following each event, the surviving organisms rapidly diversified, developing new life styles and radiating into ecologic niches vacated by the extinction (Sheehan 1991).   For example, after the Cretaceous extinction mammals ecologically replaced both herbivorous and carnivorous dinosaurs, as some mammals adapted to feeding on leafy vegetation, while others adapted to eating the newly evolved herbivorous mammals.

Of the five mass extinctions, the one at the Permian-Triassic boundary was by far the most severe.   Tectonic processes had united most of the tectonic plates into a supercontinent. The ocean basins deepened, sea level declined, anoxia was common, and the climate became less equitable, possibly triggering the mass extinction (Erwin 1991).

The remaining four extinctions, including the Cretaceous-Tertiary (K-T) event, were of similar magnitude. The Ordovician extinction coincided with a glacial interval during which the climate deteriorated and declining sea level drained extensive epicontinental seas (Sheehan 1988; Brenchley 1989). Search for evidence of an impact has been unsuccessful (Orth et al. 1986; Wilde et al. 1986). The Devonian extinction may have been associated with an impact (Claeys et al. 1992; Orth 1989), and evidence for an impact at the end of the Triassic Period has recently been documented (Bice et al. 1992). Of these events the K-T extinction is most clearly linked to an impact event (see the Chapter by Rampino and Haggerty.)

II. THE K-T EXTINCTION: A CHANGE IN PARADIGM

The K-T mass extinction is a contentious issue in paleontology (Alvarez et al. 1980; Alvarez 1986; Clemens et al. 1981; Russell 1982; Sloan et al. 1986; Archibald and Bryant 1990; Chapter by Smit).   The extinction coincided with the emplacement of one of the largest impact craters on Earth. At the extinction horizon is a thin clay layer derived from ejecta material. This clay layer contains numerous features linking it to the impact, including an iridium anomaly and presence of shock-metamorphosed quartz (Chapter by Smit). The clay layer is the only single geologic deposit known to be distributed worldwide.

The coincidence of this impact event with a mass extinction is incontrovertible, yet many paleontologists resist accepting a cause and effect relationship. This resistance may be in part because a new paradigm would be required to explain the evolutionary history of life on Earth. Paleontologists have operated under a long-accepted paradigm of gradual, evolutionary change dominated by competition between organisms. Prior to the discovery of the K-T impact, many had concluded that the K-T mass extinction was a gradual extinction during which newly evolving plants and animals replaced older ones through competitive interaction. If the impact rather than competition from mammals caused the extinction of the dinosaurs, the organisms may have been passive as a driving force in large-scale ecosystem change other than during adaptive radiations. The paradigm of evolutionary replacement needs to be revised.

A. The Extinction of Dinosaurs

In an attempt to refute the impact scenario, Sloan et al. (1986) assembled information on dinosaur occurrences in the western interior of North America through the last 10 Myr of the Cretaceous. They interpreted the data as showing a gradual decline in dinosaurian diversity. However, comparing the number of genera recorded by Sloan et al. (1986) between 2-Myr increments during this interval, Sheehan and Morse (1987) found no statistical evidence of decline in dinosaurian diversity.

It is noteworthy that no dinosaurs were recorded by Sloan et al. (1986) for the interval between 70 and 73 Myr ago during an advance of the sea. Although dinosaurs must have populated adjacent land areas during this time, no fossiliferous terrestrial strata are preserved for this interval within the region under consideration. Thus, the environment of sedimentation and the incompleteness of the sedimentary record, not simply the original abundance of dinosaurs, controls abundance in the fossil record.

Globally, only about 2100 articulated specimens identifiable to genus of dinosaurs have been collected (Dodson 1990), and these collections span an interval of 160 Myr. Russell (1994) estimates that no more than 28% of the total number of dinosaur genera in existence during the late Cretaceous (Campanian-Maastrichtian) have been identified, and this is the most thoroughly explored portion of the dinosaurian record. Obviously, the dinosaurian record is very incomplete, and short term trends in diversity (on the order of several Myr) are very difficult to assess.

To overcome problems of this kind, Sheehan et al. (1991) conducted a field study designed to test the gradualistic and catastrophic hypotheses. They examined a single stratigraphic sequence (the Hell Creek Formation in eastern Montana and western North Dakota) in which dinosaurian remains were preserved continuously and abundantly for the last 2 Myr of the Cretaceous.

They reasoned that if dinosaurs gradually declined, the pattern should be recognizable in a local area over a 2-Myr interval. They documented the distribution of skeletal parts in the lower (early), middle, and upper (late) thirds of the formation, with the assistance of large field parties. The thirds represented successive intervals, each about 730,000 yr long, with the top of the upper third located at the impact horizon. The relative proportions of dinosaurian families were found to have remained constant throughout the interval, and by implication the dinosaurian communities were also stable. No evidence of gradual extinction was found. The hypotheses of gradual decline failed the test, while the hypothesis for catastrophic extinction was consistent with the data.

Archibald and Bryant (1990) suggested that it was preferable to consider the entire vertebrate fauna, not just dinosaurs. Using data collected over several decades from the Hell Creek Formation, they found that 64% of vertebrate species survived the extinction. Archibald and Bryant (1990, p. 561) concluded that "...scenarios of catastrophic mass extinction are too severe in their estimates of potential faunal perturbation."

However, Sheehan and Fastovsky (1992) re-analyzed this body of data by dividing it into two subsets and found that 88% of land-dwelling species became extinct, while 90% of fresh-water-dwelling species survived. Extinction was attributed to participation in food chains directly linked to living plant material, and survival to food chains dependent on organic detritus left behind in lakes, streams, soil, and rotting logs (Sheehan and Fastovsky 1992). The food chain leading to dinosaurs had been essentially based on living plant tissue (Coe et al. 1987). The survival of mammals may have been related to habits which allowed them to feed in detritus-based food chains (Sheehan and Hansen 1986; Sheehan and Fastovsky 1992).

In the marine environment a similar pattern of extinction of species in food chains linked to phytoplankton and survival in food chains based on detritus feeding was found by Sheehan and Hansen (1986), Arthur et al. (1987), Gallagher (1991) and Rhodes and Thayer (1991).

Mammals appeared in the fossil record at about the same time as the earliest dinosaurs, but throughout Mesozoic time mammals remained physically small, and appear not to have been adapted to browsing green vegetation. Many were insectivores or omnivores, able to subsist on insects, other small arthropods, worms and larvae within the soil or rotting logs. The invertebrates in turn fed on saprophytic micro-organisms and dead material rather than on living plant tissues. Survivors of the impact were concentrated in these detritus-based food chains. Other organisms, including dinosaurs and many mammals, disappeared in the wake of the destruction of the standing crop of green plants.

During the first few million years of Tertiary time, mammals began a profound adaptive radiation and a general increase in body size as they began to imitate the feeding strategies of the extinct herbivorous and carnivorous dinosaurs. For the previous 140 Myr, this radiation apparently had been inhibited by the presence of dinosaurs. The fossil fields of Montana, the Dakotas, Alberta, and Saskatchewan constitute the only region where the terminal Cretaceous record has been studied in detail within a continental environment. This regional record is probably representative of the northern hemisphere. Land vertebrate assemblages in North America and Asia were broadly similar at that time, and there is no evidence of a late Cretaceous decline in dinosaurian diversity in central Asia (Jerzykiewicz and Russell 1991). All of the other large land areas, including Europe, belonged to another biotic realm, which was largely confined to the southern hemisphere (Bonaparte and Kielan-Jaworowska 1987; Russell 1982). Dinosaurian remains occur only sporadically throughout this vast region, and no well-sampled assemblages have been recovered (Russell 1982). A study of the plant record through the K-T boundary in New Zealand (Johnson 1991) is the only high resolution study from the southern hemisphere. There were very few extinctions of plants. A massive increase in the abundance of fern spores (indicating the presence of a disaster flora), which occurs across western North America and in Japan immediately above the impact clay, is absent in New Zealand. Nevertheless, dinosaurs had disappeared in the southern record by earliest Tertiary time, and severe extinctions probably took place among southern mammalian groups as well (cf., data in Gayet et al. 1991).

The pattern of extinction and survival among other groups of continental animals is probably significant. Pterosaurs, which may have been linked to marine ecosystems or possibly to scavaging dinosaurian carrion, became extinct. By Late Cretaceous time continental birds were of two major adaptive types: forest birds (enantiornithines) and shorebirds. Only the forest birds became extinct, and in spite of a very fragmentary record, their extinction may have been an important one (L. D. Martin, personal communication 1992).

The record of some groups of continental animals, such as insects and amphibians, shows little effect of the extinction (see LaBandeira 1992; Weems 1992). Insects and amphibians should show the effects of large thermal excursions (see also Archibald and Bryant 1990), and their survival suggests that thermal effects associated with a bolide impact were not capable, in themselves, of producing a global extinction. The survival of detritus-based ecosystems can also be taken as evidence that pollution through acid rain and the mobilization of heavy elements was not uniquely responsible for the extinctions. However, ecosystems within the semiarid interiors of continents, such as those of central Asia, where organic detritus was sparse, may have suffered much more severely than did coastal environments (Russell and Sues 1993). The lack of a strong extinction signal in vegetation in the southern hemisphere suggests that plant formations far removed from the impact site suffered relatively little damage. Nevertheless, the global atmospheric effects of the impact (evidently including darkness and drought) were apparently such that the production of plant food ceased for a period long enough to produce a collapse of dependent continental ecosystems through starvation.

B. A New Paradigm in the History of Life

The controversy over the extinction of dinosaurs is so contentious because we are in the midst of a paradigm change. From a traditional perspective, a previously unimportant group (mammals) acquired an adaptive advantage (for example, homeothermy or viviparity) and gradually expanded at the expense of the formerly dominant group (dinosaurs). From a catastrophic perspective, the rise to dominance of the survivors (mammals) is seen as a consequence of their "accidental" association with an ecosystem component that was not destroyed in the mass extinction.

After the impact scenario was proposed, some workers re-assessed their data and sought to document a gradual extinction. Titles of articles such as "Out with a whimper, not a bang" (Clemens et al. 1981) and "Gradual dinosaur extinction and simultaneous ungulate radiation in the Hell Creek Formation" (Sloan et al. 1986) illustrate traditional interpretations of the fossil vertebrate record in reaction to the proposal of an impact extinction scenario. A change in paradigm is at issue, and there is controversy because changes in paradigm are difficult to accept.

In our view, the unrelated coincidence of one of the largest known impact events with one of the greatest extinction events in the history of life is highly improbable. Accepting a catastrophic demise for the dinosaurs carries with it the implication that at least five times during the last 500 Myr an "external abiotic forcing factor" (Jablonski 1986, p. 132) interrupted long-term, evolutionary processes dominated by interoganismic competition and natural selection.

The innovative and relatively rapid diversification of mammals following the extinction of dinosaurs is a pattern that is being found after many mass extinctions. The pattern of recovery and the time needed for recovery of the marine fauna in the oldest mass extinction in the Ordovician is comparable to the youngest K-T mass extinction (Hansen and Sheehan 1989). At each major extinction, a long interval of stability in the major varieties of dominant organisms was terminated, and a rapid diversification and ecologic reorganization of survivors ushered in a new period of stability (Sheehan 1991).

During the long intervals of stability (EEUs), organisms evolve under a regime characterized by gradual change that is constrained by interactions with many associated species. To illustrate this continuity, the forms of most of the mammals belonging to a 12 Myr-old grassland community from Nebraska (Fig. 2) are recognizable because they were closely related to modern wolves, antelope, deer, horses, and elephants. They belonged to our own EEU.

However, 66 Myr ago (Fig. 3) the dinosaurs belonged to the previous EEU and were in anatomical continuity with their more remote dinosaurian predecessors. It is the "alien" appearance of another EEU that imparts much of the mystery and awe associated with dinosaurs.

The different appearances of dinosaurs and mammals is a measure of the importance of the impact event that occurred at the end of the Cretaceous.

Figure 2.   Reconstruction of a Nebraska scene about 12 Myr ago. (Figure courtesy of Milwaukee Public Museum.)

Figure 3. Reconstruction of a Montana scene about 66 Myr ago. (Figure courtesy of Milwaukee Public Museum.)

The physical environment of the major ecosystems remained the same, but their participants were completely reorganized. A morphologically diverse group (dinosaurs) was replaced by a poorly differentiated group (mammals; cf., Russell 1989, p. 212–213). Survival was evidently unrelated to how well organisms such as dinosaurs had adapted to their environments during previous times of normal selection (Jablonski 1986,1991; Sepkoski 1989). Survivors are likely to be small, generalized species. Organisms (such as some terminal Cretaceous mammals) which survive the still largely unexplored environmental pressures of a mass extinction are the ones which, over a few Myr, will restore the damaged ecosystems. Dinosaurs seem to embody an unexpected mosaic of adaptations (Coe et al. 1987; Russell 1989, p. 76; Russell and Russell 1993). However, it was the mammals, with their unique (and rather archaic) body plan, that produced adaptive mosaics which to a greater or lesser extent were unanticipated by their dinosaurian predecessors (cf., Janis and Damuth 1990). Although the combinations were different, many "dinosaurian" structures (horns, grinding teeth, long necks, hooves, tail clubs) were repeated in mammals because of their adaptive utility.

Mass extinctions only temporarily interrupt long-established trends toward increasing speed and complexity of behavioral responses (Vermeij 1987,1992). Evidence for the existence of these behavioral trends is preserved in skeletal structures reflecting increasing metabolic rates and greater brain body proportions in dinosaurs, which were continued in mammals and birds (Coe et al. 1987; Russell 1983; Wyles et al. 1983). During late Cretaceous time a variety of small, bipedal carnivorous dinosaurs is known to have possessed grasping hands and enlarged brains. They embodied a body plan that was more similar to that of humans than was that of the actual Cretaceous progenitors of humans. Extrapolating a curve which approximates maximum brain-body proportions in dinosaurs and later, in mammals through geologic time, it is reasonable to postulate a creature derived from a dinosaurian body plan would have attained human brain-body proportions in dinosaurs by now, if the dinosaurs had not become extinct (Russell and Seguin 1982; Russell 1987,1989). What then would have been the essential difference between this creature and the mammalian derivative that now dominates Earth? Mass extinctions may not fundamentally alter the long-term level of complexity attained by evolution, but rather the groups of organisms that actually attain these levels of organization (Russell 1989).

For the purpose of hazard assessment it is important to note that in the new paradigm, K-T type impacts pose a much more severe threat than under the paradigm of gradual change. Previously, impact events were thought to have little effect on the history of plants and animals. In the new paradigm K-T type impacts have resulted in drastic faunal turnover and ecologic reorganization.

III. THE PALEONTOLOGICAL RECORD AND ASSESSMENT OF IMPACT HAZARDS

A wide range of impact events has occurred throughout Earth history. Cretaceous-Tertiary sized events are rare, but constitute a worst-case scenario. Minor events such as the Tunguska event of 1908 produce very local effects that will seldom be recognized in the fossil record. As with others now interested in hazard assessment, paleontologists have only recently realized the significance of impact events. As the record of impacts of various sizes is explored by the paleontologic community over the next two decades, more attention will be focused on faunal changes associated with the smaller, more numerous events, in the expectation of providing important information of the assessment of high-probability impact hazards.

A. Threshold Events

Biotic effects may not scale linearly to the energy released in an impact. Dramatic increases of extinction may ensue as thresholds are breached and food webs broken. The elimination of one key species may cause the collapse of an entire community. Similar impactors may produce differing consequences because of target area effects. An oceanic impact will produce tsunamis, and an impact on a carbonate terrain may release enormous quantities of acids and greenhouse gases. Nor can the effects of an impact be predicted according to the size of the bolide, for its path, velocity and composition carry biological implications we are unable to assess (see various other chapters of this book).

Furthermore, biological systems are not homogenous in time and space. Sepkoski (1989) has noted several biological factors which may contribute to the vulnerability of Earth's biota to disruption. The degree to which faunas have differentiated zoogeographically may be important. The length of time since the last extinction may play a role in how well integrated communities are ecologically. The climatic regime on Earth has changed over time. At present the Earth has a steep climatic gradient from the equator to the poles, and organisms are adapted differently in each of the zones. During much of Earth history the latitudinal temperature gradient was less marked, and, as a result, equatorial and polar organisms were often less well differentiated. Climatic changes associated with impacts might have different global consequences today than in the past.

B. Extinctions Below the Scale of a Mass Extinction

During the past 500 Myr, the fossil record indicates that marine organisms underwent numerous extinctions and ecological reorganizations at levels smaller than the major extinction events. For example, during the mid-Paleozoic, reorganization of communities took place every few million years at what have been termed sequence boundaries (see, e.g., Brett et al. 1990). These reorganizations involved species extinction, but the new communities were derived from species that were common in similar communities prior to the reorganization. At these minor events taxa did not radiate into new life-styles, and the ecosystems did not collapse. Some of these minor events coincided with sea-level changes at sequence boundaries; but it is likely that others were linked to impacts. Methods must be found to distinguish changes caused by fluctuating climate and sea level from those caused by small impact events.

C. Possible Periodicity of Extinctions

At a slightly larger scale Sepkoski (1989) has recognized nine minor extinction events (in the range of 15% to 50% extinction of marine species) spaced at near 26-Myr intervals. One step in the cycle in the middle Jurassic lacks an extinction. Because extinction is an ongoing process (termed background extinction) that commonly results in 10 to 33% extinction of species during time intervals of this length, the actual rate of background extinction is increased only two or three times during the minor events. But the minor events are confined to a brief time interval compared to background extinction.

The large, impact-related K-T and Triassic-Jurassic mass extinctions are part of this periodicity. Of the smaller events, one in the Upper Eocene has some evidence for an associated impact, and the Cenomanian, Late Eocene, and Middle Miocene events may be associated with iridium anomalies (Sepkoski 1989; Orth 1989). The reality of the periodicity of these events has been much debated, but if it exists extraterrestrial forcing is indicated (the book edited by Smoluchowski et al. [1986] is mostly dedicated to this issue). At present we are in the middle of one of the cycles, which is good news in assessing current impact hazards.

IV. EFFECTS OF A MODERN K-T TYPE IMPACT EVENT

Obviously, the occurrence of a K-T type event would have devastating consequences for our civilization and the ecosystems in which we live. A detailed prediction of the consequences of a K-T sized impact is beyond our ability, but the types of changes can be inferred from changes associated with the K-T event.

Impact-induced mass extinctions have occurred in the past and will occur in the future. Ecosystems have always recovered, but during mass extinctions more than half of the species living on Earth became extinct. Species surviving the K-T event sustained enormous population losses, and many food chains were destroyed or severely damaged. The recovery lasted several million years, after which new groups of organisms dominated Earth.

Based on the changes at the K-T boundary, we have constructed one possible scenario of a modern K-T type impact. Our focus is at the beginning of the period of recovery from a modern K-T type impact, after easing of the immediate effects, such as months of darkness, continent-wide forest fires, acid rain, and possible tsunamis. Many species would have become extinct, and populations of surviving species would be at very low numbers. Ecologic collapse would be general. Some species would be in the beginning stages of rapidly evolving and adapting to new ways of life. Several millions of years would pass before the planetary ecosystems would stabilize.

All large mammals, including both domestic cattle and wild animals such as deer and wolves, would have starved in the decimated landscapes. Survival of small generalized species would be favored. Small, generalized carnivores or omnivores that could feed in detritus-based food chains including freshwater aquatic communities (such as racoons and otters) and insectivores in the soil fauna (such as shrews and moles) might have survived. The surviving mammals, including ourselves, would represent an average for brain-body proportions among pre-extinction mammals. Some species of insects, lizards, and some birds probably would survive also. It would be from among these survivors that the new herbivores and carnivores of the next EEU would be derived. After a few million years the landscape would include an array of animals as different from modern ones as were the dinosaurs of the Cretaceous (Fig. 3).

Immediately after the impact event stores of food including grain and canned food would allow humans to survive but in greatly reduced numbers. Vermin such as rodents, insects and fungi, which were previously adapted to feeding on human-produced food, would probably survive on the food stores also. These pests would significantly reduce the amount of remaining food available for human consumption.

Soils would recover from acid rain and heavy element pollution. Gardens could be cultivated because food crop species would almost certainly survive as seeds. Rapidly breeding small herbivores, such as insects and rodents would continue to be a problem. Newly evolving animals that presently do not even feed on plant matter, would pose a significant long-term threat to crops.

Crop species have been artificially selected for life under human care with the assistance of agricultural chemicals. Destruction of much of the basis of civilization would remove the agricultural infrastructure which provides fertilizers and pesticides. In the short term domestic cats might play a useful role in protecting food supplies. Wild plants would have strong advantages over the crop species, which, because of extensive artificial selection, will probably fair poorly in competition with the weeds in the absence of chemical herbicides and pesticides.

Human society would be vegetarian, because no livestock or wild game would be present. It would be many millennia before any herbivores evolved to body sizes large enough to replace cattle and other farm animals. Artificial selection would eventually allow domestication of some of the newly emergent plant and animal species. But initially food would be produced by individual subsistence farmers, not a sophisticated agricultural society.

On balance, it seems likely that humans would survive this horrendous catastrophe. However, the biosphere as a whole would be damaged on a scale approaching or exceeding the worst examples of human-induced environmental degradation in local areas.

V. CONCLUSIONS

The impact event at the K-T boundary represents a worse-case scenario for impact events. The K-T impact produced a massive restructuring of life on Earth. Changes of similar scale in the ecosystems occurred only five times during the last 500 Myr. The effect of an impact could be devastating to our civilization and all life on Earth.

Clearly, every attempt should be made to avert K-T type impacts. However, the probability of these impacts is very low, even in geologic time scales. As a result, developing a means of protecting Earth from K-T size impacts should be a long-term project. The immediate emphasis should be placed on smaller more frequent impact events.

The Spaceguard program will determine the nature of Earth-crossing objects, including their size-frequency distribution. As with any rapidly advancing field of science, peripheral fields such as paleontology, will be stimulated by newly available information. Over the next two decades the hazards of smaller, more frequent impacts will become more clearly understood. Accordingly, paleontologists should be encouraged to focus on the consequences of these smaller impacts as reflected in the fossil record, for they pose the most immediate threat.

REFERENCES

Alvarez, L. W., Alvarez, W., Asaro, F., and Michel, H. V. 1980. Extraterrestrial cause for the Cretaceous-Tertiary extinction. *Science* 208:1095–1108.

Alvarez, W. 1986. Towards a theory of impact crises. *EOS: Trans. AGU* 67:649–657.

Archibald, J. D., and Bryant, L. J. 1990. Differential Cretaceous/Tertiary extinctions of nonmarine vertebrates; evidence from northeastern Montana. In *Global Catastrophes in Earth History*, eds. V. L. Sharpton and P. D. Ward, Geological Society of America Special Paper 247 (Boulder: Geological Soc. of America), pp. 549–562.

Arthur, M. A., Zachos, J. C., and Jones, D. S. 1987. Primary productivity and the Cretaceous/Tertiary boundary event in the oceans. *Cretaceous Res.* 8:43–54.

Bice, D. M., Newton, C. R., McCauley, S., Reiners, P. W., and McRoberts, C. A. 1992. Shocked quartz at the Triassic-Jurassic boundary in Italy. *Science* 255:443–446.

Bonaparte, J., and Kielan-Jaworowska, Z. 1987. Late Cretaceous Dinosaur and Mammal Faunas of Laurasia and Gondwana. Occasional Paper of the Tyrrell Museum of Palaeontology No. 3 (Drumheller, Alberta: the Tyrrell Museum), pp. 24–29.

Boucot, A. J. 1983. Does evolution take place in an ecological vacuum? II. *J. Paleontology* 57:1–30.

Brenchley, P. 1989. The late Ordovician extinction. In *Mass Extinctions: Processes and Evidence*, ed. S. K. Donovan (New York: Columbia Univ. Press), pp. 104–132.

Brett, C. E., Miller, K. B., and Baird, G. C. 1990. A temporal hierarchy of pale-oecologic processes within a Middle Devonian epeiric sea. In *Paleocommunity Temporal Dynamics*, ed. W. Miller, III, Paleontological Soc. Special Publ. No. 5 (Knoxville, Tenn.: Paleontological Society), pp. 178–209.

Claeys, P., Casier, J.-G., and Margolis, S. F. 1992. Microtektites and mass extinctions: Evidence for a late Devonian asteroid impact. *Science* 257:1102–1104.

Clemens, W. A., Archibald, J. D., and Hickey, L. J. 1981. Out with a whimper not a bang. *Paleobiology* 7:293–298.

Coe, M. C., Dilcher, D. L., Farlow, J. O., Janzen, D. M., and Russell, D. A. 1987. Dinosaurs and land plants. In *The Origins of Angiosperms and Their Biological Consequences*, eds. E. M. Priis, W. G. Chaloner and P. R. Crane (Canibridge: Cambridge Univ. Press), pp. 225–258.

Dodson, P. 1990. Counting dinosaurs: How many kinds were there? *Proc. U. S. National Academy of Sci.* 87:7608–7612.

Erwin, D. H. 1991. The mother of mass extinctions. *Palaios* 6:517.

Gallagher, W. B. 1991. Selective extinction and survival across the Cretaceous/Tertiary boundary in the northern Atlantic costal plain. *Geology* 19:967–970.

Gayet, M., Marshall, L. G., and Sempere, T. 1991. The Mesozoic and Paleocene vertebrates of Bolivia and their stratigraphic context: A review. *Revista Tecnica de YPFB* 12:393–433 (Bolivia).

Hansen, T. A., and Sheehan, P. M. 1989. Rebounds from the Late Ordovician and Cretaceous extinctions: A comparison. *Geological Soc. of America Abstracts with Programs* 21:32.

Jablonski, D. 1986. Background and mass extinctions: The alternation of macroevolutionary regimes. *Science* 231:129–133.

Jablonski, D. 1991. Extinctions: A paleontological perspective. *Science* 253:754–757.

Janis, C. M., and Damuth, J. 1990. Mammals. In *Evolutionary Trends*, ed. K. J. McNamara (London: Belhaven Press), pp. 301–345.

Jerzykiewicz, T., and Russell, D. A. 1991. Late Mesozoic stratigraphy and vertebrates of the Gobi Basin. *Cretaceous Res.* 12:345–377.

Johnson, K. R. 1991. A southern hemisphere terrestrial Cretaceous/Tertiary boundary section: Macro- and microfloral record from the Paparoa Trough, South Island, New Zealand. *Geological Soc. of America Abstracts with Programs* 23(5):A358.

LaBandeira, C. C. 1992. Diversity, diets and disparity: Determining the effect of the terminal Cretaceous extinction on insect evolution. In *Fifth North American Paleontological Convention Abstracts*, eds. S. Lidgard and P. R. Crane, Paleontological Soc. Special Publ. No. 6 (Knoxville, Tenn.: Paleontological Society), pp. 174.

Orth, C. J. 1989. Geochemistry of the bio-event horizons. In *Mass Extinctions: Processes and Evidence*, ed. S. K. Donovan (New York: Columbia Univ. Press), pp. 37–72.

Orth, C. J., Gilmore, J. S., Quintana, L. R., and Sheehan, P. M. 1986. Terminal Ordovician extinction: geochemical analysis of the Ordovician/Silurian boundary, Anticosti Island, Quebec. *Geology* 14:433–436.

Raup, D. M., and Sepkoski, J. J. 1982. Mass extinctions in the marine fossil record. *Science* 215:1501–1503.

Rhodes, M. C., and Thayer, C. W. 1991. Mass extinctions: Ecological selectivity and primary production. *Geology* 19:877–880.

Russell, D. A. 1982. A paleontological consensus on the extinction of the dinosaurs? In *Geological Implications of Impacts of Large Asteroids and Comets on the Earth*, eds. L. T. Silver and P. H. Schultz, Geological Soc. of America Special Paper 190 (Boulder: Geological Soc. of America), pp. 401–405.

Russell, D. A. 1983. Exponential evolution: Implications for intelligent extraterrestrial life. *Adv. Space Res.* 3:95–103.

Russell, D. A. 1987. Models and paintings of North American dinosaurs. In *Dinosaurs Past and Present*, eds. S. J. Czerkas and E. C. Olson (Seattle: Univ. of Washington Press), pp. 115–131.

Russell, D. A. 1989. *An Odyssey in Time: The Dinosaurs of North America* (Toronto: Univ. of Toronto Press).

Russell, D. A. 1994. China and the lost worlds of the dinosaurian era. *Historical Biology*, in press.

Russell, D. A., and Russell, D. E. 1993. Mammal-dinosaur convergence. *National Geographic Res. Exploration* 9:70–79.

Russell, D. A., and Seguin, R. 1982. Reconstructions of the small Cretaceous theropod *Stenonychosaurus inequalis* and a hypothetical dinosauroid. *Syllogeus* 37:1–43.

Russell, D. A., and Sues, H. D. 1993. A comparison of terrestrial vertebrate extinctions during the Triassic-Jurassic and Cretaceous-Tertiary extinctions. *Geological Association of Canada Annual Meeting Abstracts* 18:A91.

Sepkoski, J. J. 1989. Periodicity in extinction and the problem of catastrophism in the history of life. *J. Geol. Soc.* 146:7–19.

Sheehan, P. M. 1988. Late Ordovician events and the terminal Ordovician extinction. *New Mexico Bureau of Mines & Mineral Resources Memoir* 44:405–415.

Sheehan, P. M. 1991. Patterns of synecology during the Phanerozoic. In *The Unity of Evolutionary Biology*, vol. 1, ed. E. C. Dudley (Portland, Or.: Dioscorides Press), pp. 103–118.

Sheehan, P. M., and Hansen, T. 1986. Detritus feeding as a buffer to extinction at the end of the Cretaceous. *Geology* 14:868–870.

Sheehan, P. M., and Fastovsky, D. E. 1992. Major extinctions of land-dwelling vertebrates at the Cretaceous-Tertiary boundary, eastern Montana. *Geology* 20:556–560.

Sheehan, P. M., and Morse, C. L. 1987. Cretaceous-Tertiary dinosaur extinction. *Science* 234:1171–1172.

Sheehan, P. M., Fastovsky, D. E., Hoffmann, R. G., Berghaus, C. B., and Gabriel, D. L. 1991. Sudden extinction of the dinosaurs: Latest Cretaceous, Upper Great Plains, U.S.A. *Science* 254:835–839.

Sloan, R. E., Rigby, J. K., Van Valen, L. M., and Gabriel, D. 1986. Gradual dinosaur extinction and simultaneous ungulate radiation in the Hell Creek Formation. *Science* 232:629–633.

Smoluchowski, R., Bahcall, J. N., and Matthews, M. S., eds. 1986. *The Galaxy and the Solar System* (Tucson: Univ. of Arizona Press).

Vermeij, G. J. 1987. *Evolution and Escalation* (Princeton, N. J.: Princeton Univ. Press).

Vermeij, G. J. 1992. Economics and evolution. In *Fifth North American Paleontological Convention Abstracts*, eds. S. Lidgard and P. R. Crane, Paleontological Soc. Special Publ. No. 6 (Knoxville, Tenn.: Paleontological Soc.), p. 298.

Weems, R. E. 1992. The "terminal Triassic catastrophic extinction event" in perspective: A review of Carboniferous through Early Jurassic terrestrial vertebrate extinction patterns. *Palaeogeogr. Palaeoclimatol. Palaeoecol.* 94:1–29.

Wilde, P., Berry, W. B. N., Quinby-Hunt, M. S., Orth, C. J., Quintana, L. R., and Gilmore, J. S. 1986 Iridium abundances across the Ordovician-Silurian stratotype. *Science* 233:339–441.

Wyles, J. S., Krunkel, J. G., and Wilson, A. C. 1983. Birds, behavior and anatomical evolution. *Proc. U. S. National Academy of Sci.* 80:4394–4397.

PART VII
Hazard Mitigation

DELFECTION AND FRAGMENTATION OF NEAR-EARTH ASTEROIDS

THOMAS J. AHRENS
California Institute of Technology

and

ALAN W. HARRIS
Jet Propulsion Laboratory

Collisions by near-Earth asteroids or the nuclei of comets pose varying levels of threat to man. A relatively small object, $\sim$100 m diameter, which might be found on an impact trajectory, could potentially be diverted from an Earth impacting trajectory by a rocket launched, 10^2 to 10^3 kg impactor, with a lead time of $\sim$10 yr. For larger bodies or shorter lead times, the use of kinetic energy impactors appears impractical because of the larger mass requirement. For any size object, nuclear explosions appear to be more efficient, using either the prompt blow-off from neutron radiation, the impulse from ejecta of a near-surface explosion for deflection, or, least efficiently, as a fragmenting charge.

I. INTRODUCTION

Several hundred asteroids and short-period comet nucleii with diameters $> 10^2$ m, have been discovered in Earth-crossing orbits. Upon extrapolating this known population of near-Earth objects (NEOs) to those not yet discovered, it is estimated that $\sim 2 \times 10^3$ objects $\gtrsim 1$ km in diameter are present in a transient population (Shoemaker et al. 1990; see Chapter by Rabinowitz et al.).

Comets are brought into the swarm of NEOs by gravitational perturbation mainly by Jupiter from their orbits in the Kuiper belt or Oort cloud (Weissman 1990). Some objects currently classed as near-Earth asteroids may be devolatilized comets. In the case of asteroids, the source of NEOs is largely from the main asteroid belt. Earth- or near-Earth-crossing objects are removed from this population either via collision with a planet or by gravitational perturbation which causes them to be ejected into hyperbolic orbits. The largest Earth-crossing asteroids have diameters approaching 10 km. It is unlikely that any objects larger than $\sim$5 km diameter remain undiscovered, but some 3 to 5 km objects are probably still to be found.

Scientific interest in NEOs is great because it appears that many of these objects are mainbelt asteroids which have been perturbed into terrestrial planet-crossing orbits, and thus give rise to a large fraction of the impact flux on terrestrial planet surfaces (Binzel et al. 1992). NEOs as small as 5 to 10 m in diameter can occasionally be telescopically observed (Scotti et al. 1991; see Chapter by Carusi et al.). Meteorites are fragments of NEOs that have survived passage through the Earth's atmosphere. Because the number distribution of different meteorite classes correlates poorly with asteroid type, as inferred from reflectance spectra of mainbelt asteroids, it may be that the present terrestrial meteorite collection is a poor sample of the asteroid population. To further study asteroids, one or more unmanned flyby or rendezvous missions to near-Earth asteroids (NEAs) are currently being planned by NASA (Veverka and Harris 1986; see Chapter by Cheng et al.). Finally, the composition of NEOs is of interest as these objects represent possible mineable resources which, in principle, could supply raw materials, including water, and hence, oxygen and hydrogen for extended space flights (see Chapter by Hartmann and Sokolov).

Earth-crossing asteroids have been recognized telescopically since 1932, when K. Reinmuth discovered 1862 Apollo. However, it was the American geologist, G. K. Gilbert whose work on Meteor Crater, Arizona, and many later workers, conclusively demonstrated that the impact of Earth-crossing asteroids and comets produce the ~120 known meteorite impact craters on the Earth and virtually all the craters on the Moon.

In 1980, a startling discovery was reported by Alvarez et al. (1980). They found a concentration of platinum group metals of 10^3 to 10^4 times normal crustal sedimentary abundances, and in approximately chondritic meteoritic proportions, at two sites in Europe, exactly at the Cretaceous-Tertiary (K/T) boundary. On this basis, they proposed that an asteroid or a comet, ~10 km in diameter, impacted the Earth 65 Myr ago and the prompt high speed projectile ejecta from this impact was lofted upward to the stratosphere and was dispersed worldwide. They further proposed that this dust layer gave rise to a temporary decrease in solar insolation and drastically, but temporarily, affected the world's climate, and hence the biotic food chain.

In the 13 yr since the publication of the Alvarez et al. hypothesis, probably the most significant discovery in paleontology in this century, the platinum-rich element layer has been uncovered at some 200 sites all over the Earth, in terrestrial sediments and shallow marine and deep marine sedimentary environments. The concentration of platinum-group elements in the impact ejecta varies such that it can be considered to consist of typically ~1% meteoritical material (Gubbio, Italy). In some localities it occurs up to ~10% concentration (Stevens Klint, Denmark). The platinum-rich element horizon is always found in sediments of exactly Cretaceous-Tertiary (65 Myr) age. In addition, the K/T ejecta layer contains mineral grains (quartz and feldspar) which bear distinctive shock-induced lamellae as well as spherules of impact-induced glass spheres. Such spherules have long been associated with impact craters both on the Earth and the Moon.

In 1992 good radiometric ages (Sharpton et al. 1992; Swisher et al. 1992) and stable isotope data associating the molten ejecta with a source terrane (Blum and Chamberlain 1992; Blum et al. 1993) have been obtained. The shocked mineral grains and shock-induced glasses are now believed to be part of ejecta from the ~250 km diameter Chicxulub Impact Crater located along the northwest coast of the Yucatan Peninsula in Mexico. The radiometric and stable isotope data reinforce the association of the Chicxulub Crater, which formed in shallow marine sediments, with the K/T boundary impact event. Previously Bourgeois et al. (1988) and Hildebrand et al. (1991) reported the occurrences of local, presumed tsunami-related, deposits directly underlying the platinum-rich element layer around the rim of the proto-Caribbean sea in Central America and in the Southeastern United States (see the Chapters by Smit and by Hills et al.). Such deposits were anticipated earlier, on a theoretical basis, by Ahrens and O'Keefe (1983).

The K/T boundary marks the end of the Mesozoic era and thus represents the demarcation between the age of reptiles (Mesozoic) and the age of mammals (Tertiary). Evidence that similar extinction-causing impacts occurred previously in Earth history is incomplete. The K/T event appears to be the most recent, and hence has the best preserved record, of five major extinction events which have been recognized by paleontologists to have occurred in the last 500 Myr of Earth history. Approximately 50% of all marine genera and probably 90% of all species became extinct at the K/T boundary. These included terrestrial, marine, micro- and macro-organisms, and the well-known groups of dinosaurs. An impact of such a large bolide on the Earth is believed to give rise to a large number of global physical (Gerstl and Zardecki 1982; O'Keefe and Ahrens 1982; Toon et al. 1982; Vickery and Melosh 1990) and chemical effects (O'Keefe and Ahrens 1989; Brett 1992; Sigurdsson et al. 1992) which are deleterious to life. However, the generation of environments which endanger the existence of a multitude of life forms and their detailed effect on the global environment are not yet thoroughly understood (Chapters by Sheehan and Russell and by Toon et al.).

As the Alvarez et al. hypothesis became more widely accepted in the 1990s scientists, and eventually governments, recognized that the Alvarez scenario bore many similarities to the post-nuclear winter environment (Covey et al. 1990; Crutzen and Birks 1982; Turco et al. 1990).

Sparked by public concern, the United States House of Representatives in 1991 requested the National Aeronautics and Space Administration to conduct studies of the asteroid-impact threat to the Earth's human population (Morrison 1992; Chapter by Morrison et al.) and possible measures which could be taken to prevent cosmically induced disasters with global consequences (Rather et al. 1992; Chapter by Canavan et al.). The recent Near-Earth Object Detection Workshop (Morrison 1992; Chapter by Morrison et al.) quantified the hazards to the world population from different sized Earth impactors based, in part, on the results of an earlier workshop (Shoemaker 1983) in 1981.

Using the estimated population of NEOs and their size distribution, objects with diameters of about 10 m impact the Earth almost annually, and although visible and audible for distances of 10^2 to 10^3 km, these objects largely break up and expend their typically 10 Kton (of TNT) energy in the atmosphere. Earth impact of NEOs of about 100 m diameter, e.g., the 1908 Tunguska event (energy $\sim$10 Mton) has a frequency of about once every $\sim$300 yr. Although the Tunguska bolide did not hit the Earth's surface, it nevertheless did great damage. These objects, although inducing local areas of devastation of $\sim$5 $\times$ 10^3 km^2, have an annual probability of leading to the death of a given individual of only $\sim$3 $\times$ 10^{-8}yr^{-1}. Although less frequent, once every 0.5 Myr, Earth impactors of the $\sim$1 km diameter size are inferred to be the minimum size which can induce global catastrophic effects ($\sim$25% human mortality). Thus the annual individual death probability from such an event is of the order of 5 $\times$ 10^{-7}. This is comparable to the annual worldwide probability of an individual succumbing in a commercial airplane accident. (Chapter by Morrison et al.)

When viewed in this way, it appears to us that a balanced response for society to deal with the NEO impact problem might be an expenditure perhaps up to some sizeable fraction of the amount of funding committed to air safety and control. We believe this would be in the range of 10^7 to 10^8 dollars per annum worldwide. As was concluded by the Near-Earth Object Detection Workshop, funding at this level would vastly improve our knowledge of the population and distribution of near-Earth objects using groundbased and possibly space-borne telescopes. We consider it premature to conduct detailed engineering studies or construct prototype systems for deflection of NEOs for several reasons. It is unlikely that such a system will ever be needed. If a body exists on a collision course, it is likely that it can be detected with enough lead time to construct a deflection system after the discovery and characterization of the body and its orbit. With today's technology, a deflection system is bound to be vastly more costly than a thorough groundbased survey. Thus for the present, in the absence of a discovered threatening body, it appears more useful to direct resources to improved searching rather than deflection technology.

In this chapter we examine the orbit perturbation requirements to deflect objects from the Earth, which might be found to have Earth-impacting trajectories. We then examine several physical means for both deflecting and explosively fragmenting such objects. We consider NEOs in three size ranges: 0.1, 1, and 10 km in diameter. Their collision fluxes, on the total area of the Earth are respectively, 10^{-3}, 10^{-5}, and 10^{-8} per year. Objects significantly smaller than 100 m pose little threat, because they do not penetrate the atmosphere intact. Short duration responses, which might be considered for new comets, have recently been described by Solem (1991,1992; Chapter by Solem and Snell). This study addresses the physical means of encountering NEOs with spacecraft-bearing energetic devices many years, or even decades, before projected Earth impact.

To quantify the present work especially with regard to nuclear explosive cratering in the low-gravity asteroid environment, we employ recent studies of cratering at varying gravities and atmospheric pressures (Housen et al. 1993; Schmidt et al. 1986) and impact ejecta scaling (Housen et al. 1983), which were not available to earlier studies (MIT Students 1968; Solem 1992).

II. NEAR EARTH ASTEROID ORBIT DEFLECTION CONSIDERA-TIONS

A. Short Time-Scale Deflection

On a time scale short compared to the orbit period P, the displacement δ achieved by a velocity change Δv is the same as for rectilinear motion:

$$\delta = \Delta v \cdot t \tag{1}$$

in either transverse or along the track of the orbital motion. For $t \lesssim P/2\pi$, to perturb a body $1\,R_\oplus$ in time t requires

$$\Delta v \sim \frac{R_\oplus}{t} \sim \frac{75 \text{ m s}^{-1}}{t, \text{ days}} \tag{2}$$

where $R_\oplus$ is an Earth radius.

B. Long Time-Scale Deflection

On a longer time scale, a transverse increment of velocity perpendicular to the orbital plane results in a change of inclination. If the Δv is applied as shown in Fig. 1a, the maximum displacement of the NEO, δ_{max}, occurs at 90° from the place in the orbit that the perturbation was applied. The semimajor axis remains unchanged. Then

$$\delta_{max} \cong \Delta v P/2\pi. \tag{3}$$

Thus, the perturbed NEO changes only its inclination.

If a perturbation Δv is applied radially as shown in Fig. 1b, the maximum radial displacement occurs 90° around and is the same as given by Eq. (3). However, as the body moves closer to the Sun, it moves faster, so that 180° around in its new orbit it is displaced along the orbit track by a distance

$$\delta_{max} \cong 2\Delta v P/\pi. \tag{4}$$

In the second half of the orbit, the displacement motion is reversed, so that the particle returns to the point of application of Δv at the same time as the unperturbed reference point, because the semimajor axis, and hence the orbit period was unchanged.

A velocity increment applied along the track of motion of an object produces a change of orbital semimajor axis, and hence orbital period, in

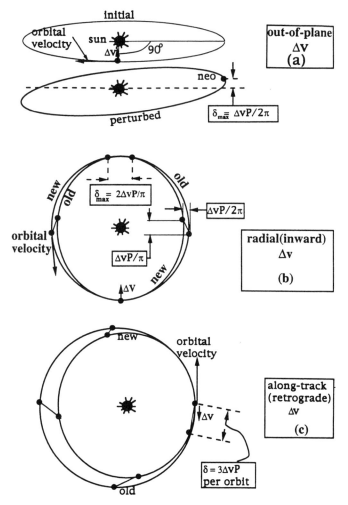

Figure 1. Sketch of the effect on the orbit of an NEO upon perturbing its velocity by
Δv: (a) perpendicular to its orbital plane; (b) radially in its orbital plane; and (c)
tangentially to its orbit, in its orbital plane.

addition to a change of eccentricity. Thus, the resultant displacement from
the original motion consists of an oscillatory component, but also a secular
drift, which continues to grow with successive orbits. The mean rate of drift
can be derived from the equation for the energy of orbital motion:

$$E = -\frac{GM_s}{2a} = -\frac{GM_s}{r} + \frac{1}{2}v^2 \tag{5}$$

where G is the universal constant of gravitation, M_s is the mass of the Sun, a
is the orbital semimajor axis, r is the distance from the Sun at the point in the

orbit where the velocity increment is to be applied, and v is the heliocentric velocity at that point in the orbit. After applying a velocity increment Δv along the time of motion, the energy of the new orbit is

$$E' = -\frac{GM_s}{2(a + \Delta a)} = -\frac{GM_s}{r} + \frac{1}{2}(v + \Delta v)^2 \qquad (6)$$

where Δa is the change in orbital semimajor axis. After expanding the above expression in terms of the small quantities, and differencing from the initial energy, we can obtain:

$$\frac{\Delta a}{a} \approx \frac{2\Delta v}{v_0}\sqrt{\frac{2a}{r} - 1} \qquad (7)$$

where $v_0 = \sqrt{GM_s/a}$ is the mean orbital velocity. From Kepler's third law, we have $P^2 \propto a^3$, and hence $\Delta a/a = (2/3)(\Delta P/P)$. The mean position in the new orbit diverges by a distance $-2\pi a \Delta P/P$ from the position in the original orbit, so the mean velocity of divergence is

$$\Delta v' = -\frac{2\pi a}{P}\frac{\Delta P}{P} = -v_0\frac{\Delta P}{P} \qquad (8)$$

and thus

$$\Delta v' = -3\Delta v\sqrt{\frac{2a}{r} - 1}. \qquad (9)$$

For a circular orbit, $a = r$, and the above expression reduces to

$$\Delta v' = -3\Delta v. \qquad (10)$$

After one orbit, the maximum deflection for a nearly circular orbit is (Fig. 1c)

$$\delta \simeq -3\Delta vP. \qquad (11)$$

Thus, over a time t long compared to P, an increment Δv applied parallel to v produces a deflection of

$$\delta \simeq 3\Delta vt. \qquad (12)$$

It can be seen that for an eccentric orbit, the greatest effect is achieved by applying Δv at the perihelion. For example, for an orbit of eccentricity 0.5, $r = 0.5a$ at perihelion and $\Delta v' = -3\sqrt{3}\Delta v \approx -5\Delta v$. At aphelion of the same orbit, $r = 1.5a$ and $\Delta v' = -\sqrt{3}\Delta v \approx -1.7\Delta v$. Thus for typical NEO orbits, the circular orbit value, $\Delta v' \sim 3\Delta v$ is accurate within a factor of 2 or so. In order to achieve a deflection distance of one Earth radius in a time $t(>P)$, the velocity increment required is:

$$\Delta v \approx \frac{R_\oplus}{3t} \approx \frac{0.07 \text{ m s}^{-1}}{t \text{ yr}}. \qquad (13)$$

Thus it appears that if a time of the order of a decade is available to achieve deflection, a velocity increment of $\sim$1 cm s^{-1} is sufficient.

III. IMPLEMENTATION OF ORBITAL DIVERSION

Several scenarios are considered, including deflection via kinetic energy impactor, mass driver systems, as well as nuclear explosive radiation and blow-off, and ejecta impulse from cratering explosions (see also the Chapter by Melosh et al.).

A. Direct Impact Deflection

It is feasible to deflect a small ($\sim$10^{2}m diameter) NEO via direct impact because:

1. The kinetic energy delivered for even a modest encounter velocity ($\sim$12 km s^{-1}) of an upper stage launched spacecraft is much more efficiently coupled (70 to 80%) to the asteroid (Smither and Ahrens 1992) than surface explosions. The energy density at 12 km s^{-1} is 70×10^{10}erg per g of impactor. This is much greater than typical chemical explosive energies (4×10^{10} erg g^{-1}), and as demonstrated below the ejecta throw-off from such an impact will suitably perturb the NEO.

2. Although cratering efficiency on a small (100 m diameter) NEO (escape velocity 5 cm s^{-1}) is unmeasured, extrapolating small-scale studies (at high and low gravities) suggests that the cratering efficiency may be $\sim$10^4 times (Holsapple 1993; Housen et al. 1993) the earthly value of 2.8 tons of rock per ton of equivalent explosive yield (Cooper 1977) For impact of an Earth-launched projectile into an NEO, two regimes of cratering mechanics need to be considered. At small scales (tens to perhaps hundreds of m), the strength of the target asteroid or comet needs to be considered as affecting, and perhaps even controlling impact cratering mechanics whereas for larger impact craters local gravity controls the cratering process.

Strength Regime. Housen et al. (1983) observed, on the basis of the experiments of Gault et al. (1963), that upon impact into basalt (here assumed to be typical of NEO materials), where the strength of the target controls the size of the crater, the cumulative mass of ejecta, $M(>v)$ traveling faster than v is given by

$$M(>v) = 0.05R_c^3\rho \left(\frac{Y}{\rho v^2} \right) \qquad (14)$$

where ρ and Y are the target density and material strength, respectively, and R_c is the radius of the crater produced. In the strength-controlled regime, the size of the crater is determined by the scale to which elastic deformation exceeds the yield strength of the target material, which in turn dictates the minimum speed with which broken fragments are ejected. We can derive the minimum speed v_{min} by equating the mass $M(>v_{min})$ with the total mass M_{ej}

evacuated from the crater. Assuming the crater is a half-oblate spheroid of radius R_c, and depth $\sim 0.4R_c$ (see, e.g., Pike (1980), we can relate the total mass excavated to the crater radius approximately:

$$R_c^3 \approx 1.2\frac{M_{ej}}{\rho}. \tag{15}$$

Inserting Eq. (15) into Eq. (14), and equation $M(> v_{min}) = M_{ej}$, we can rearrange the resulting expression to obtain the following for the minimum ejecta velocity:

$$v_{min} \approx \sqrt{0.06\frac{Y}{\rho}} \approx 0.24\sqrt{\frac{Y}{\rho}}. \tag{16}$$

Thus expression (14) is valid for $v \geq v_{min}$. Note that the elastic energy density per unit mass stored in a material when stressed to its yield point Y is $\sim YS/(2\rho)$, where S is the yield strain. Thus v_{min} is related dimensionally to the "rebound velocity" of the material, assuming a constant value of the yield strain for materials of various strengths. We note also that v_{min} is much greater than the escape velocity from a small asteroid, e.g., for a very weak material, $Y = 10^7$ dyne cm^{-2}, $v_{min} \approx 5$ m s^{-1}; for a 1 km diameter asteroid, the escape velocity is $v_{esc} \approx 0.5$ m s^{-1}.

Using the compilation of Holsapple and Schmidt (1982) of cratering experiments conducted in vacuum, for impacts into metals, the mass M_{ej} of material excavated as a function of impactor mass M_i, density ρ_i and impact speed v_i is given by

$$M_{ej} \approx 0.458 M_i \left(\frac{\rho}{\rho_i}\right)\left(\frac{\rho v_i^2}{Y}\right)^{0.709}. \tag{17}$$

The differential form of Eq. (14) can be written

$$dM = \frac{2v_{min}^2 M_{ej}}{v^3} dv \tag{18}$$

with v_{min} given by Eq. (16) and M_{ej} given by Eq. (17). The momentum impulse imparted to the NEO is thus

$$p = \int_{v_{min}}^\infty v\cos\theta dM = \int_{v_{min}}^\infty \frac{2v_{min}^2 M_{ej}\cos\theta}{v^2} dv = 2v_{min}M_{ej}\cos\theta \tag{19}$$

where θ is the angle from vertical of the ejecta spray, which we take to be $\sim 45°$. Thus

$$p \approx \sqrt{2}v_{min}M_{ej}. \tag{20}$$

By conservation of momentum, we obtain the recoil deflection velocity of an NEO of mass M_{NEO} which results from the impact:

$$\Delta v \approx \frac{M_i v_i + p}{M_{NEO} - M_{ej}} \approx \frac{M_i v_i + \sqrt{2}v_{min}M_{ej}}{M_{NEO} - M_{ej}}. \tag{21}$$

By substituting expression (17) for M_{ej} and (16) for v_{min}, and neglecting M_{ej} compared to M_{NEO}, we can rewrite the above to obtain an expression for the ratio M_i / M_{NEO}:

$$\frac{M_i}{M_{NEO}} \approx \frac{\Delta v}{v_i} \bigg/ \left[1 + 0.16 \left(\frac{\rho}{\rho_i} \right) \left(\frac{\rho v_i^2}{Y} \right)^{0.209} \right]. \qquad (22)$$

In the above expression, the numerator without the denominator gives the recoil which would result from an inelastic collision. The denominator is the "enhancement factor" which results from the added recoil from ejecta.

In Table I we list the masses required, impacting at 12 and 40 km s^{-1}, to produce a Δv of 1 cm s^{-1} in NEOs of diameters 0.1 and 1 km diameter, and yield strengths of 10^7 (soft rock or ice) and 10^9 (hard rock) erg cm^{-3}, respectively. We have taken $\rho = \rho_i = 2$ g cm^{-3} to compute these results.

TABLE I
NEO Deflection via Impact, Strength Regime for 1 cm s^{-1} Perturbation

v_i (km s^{-1})	Y (dyne cm^{-2})	M_i / M_{NEO}	M_i ($D = 100$ m)	M_i ($D = 1$ km)
12	10^7	2.0×10^{-7}	250 kg	250 T
40	10^7	4.5×10^{-8}	50 kg	50 T
12	10^9	3.1×10^{-7}	450 kg	450 T
40	10^9	7.6×10^{-8}	100 kg	100 T

From the above table, it is clear that the impactor mass required to deflect a 100 m NEO with a decade lead time is modest, less than 1/2 metric ton over a plausible range of strengths and impact velocities. However to deflect a 1 km NEO requires a very large impacting mass, generally hundreds of tons. As will be seen shortly, nuclear blasts offer a more economic solution for all but very small NEOs.

B. Gravity Regime

As an outgrowth of a program led by Robert Schmidt at Boeing Corporation (Seattle, WA) to study the scaling via variable gravity (centrifuge experiments) on gravity controlled impact and explosion craters, the cratering efficiency of these craters is better understood (see, e.g., Schmidt and Housen 1987; Holsapple and Schmidt 1980, 1982; Housen et al. 1983; and Holsapple 1993). They obtain the following empirical expression for the ejecta velocity distribution from impact craters where gravity dominates material strength:

$$M(>v) = 0.32 \rho R_c^3 \left(\frac{\sqrt{gR_c}}{v} \right)^{1.22}. \qquad (23)$$

As before, we can relate the total mass evacuated to the radius of the crater from Eq. (15). Equating $M(>v_{min})$ to M_{ej}, we can solve for v_{min} in the gravity regime:

$$v_{min} \approx 0.5 \sqrt{gR_c}. \qquad (24)$$

Note that this velocity is about that necessary to "hop" out of the crater from its center, and thus is in all cases less than the escape velocity from the asteroid. Because ejecta traveling at speeds less than the escape velocity will fall back on to the asteroid and not contribute to the momentum impulse, the integral analogous to Eq. (19) should be taken from a lower limit of v_{esc}, rather than v_{min}:

$$v_{esc} = \sqrt{\frac{2GM_{NEO}}{R_{NEO}}}.$$ (25)

Schmidt and his co-workers (see references above) also derive empirical relations for the total volume of material evacuated by impacts into various materials. For dry sand under gravity scaling, they obtain:

$$M_{ej} = 0.132 M_i \left(\frac{v_i^2}{R_i g}\right)^{0.51}.$$ (26)

By combining Eqs. (26) and (15) into Eq. (23), we can arrive at an approximate expression for $M(>v)$ as a function of impactor mass M_i and impactor velocity, v_i:

$$M(>v) \approx 0.047 \left(\frac{\rho_i}{\rho}\right)^{0.203} M_i \left(\frac{v_i}{v}\right)^{1.22}.$$ (27)

The differential form of this, neglecting the factor $(\rho_i/\rho)^{0.203}$, is:

$$dM = 0.057 M_i \frac{v_i^{1.22} dv}{v^{2.22}}.$$ (28)

The momentum impulse from the ejecta is thus:

$$p = \int_{v_{esc}}^{\infty} v \cos\theta \, dM = \int_{v_{esc}}^{\infty} 0.057 M_i \frac{v_i^{1.22}}{v^{1.22}} \cos\theta \, dv = 0.26 M_i \frac{v_i^{1.22}}{v_{esc}^{0.22}} \cos\theta.$$ (29)

Because the above integral involves velocities down to v_{esc}, the trajectories of ejecta at such low velocities are distorted, and the velocity after escape is less than that at the surface. We have numerically integrated the function with these corrections, and find for $\theta = 45°$,

$$p \approx 0.16 M_i \frac{v_i^{1.22}}{v_{esc}^{0.22}}.$$ (30)

By conservation of momentum, the recoil velocity of the asteroid is:

$$\Delta v \approx \frac{M_i v_i + p}{M_{NEO} - M_{ej}} \approx \frac{M_i v_i + 0.16 M_i \frac{v_i^{1.22}}{v_{esc}^{0.22}}}{M_{NEO} - M_{ej}}.$$ (31)

As before, we can neglect M_{ej} compared to M_{NEO}, and rearrange Eq. (31) to obtain the ratio of impactor mass to asteroid mass to achieve a deflection velocity Δv:

$$\frac{M_i}{M_{NEO}} \approx \frac{\Delta v}{v_i} \left/ \left[1 + 0.16 \left(\frac{v_i}{v_{esc}} \right)^{0.22} \right] \right. . \tag{32}$$

As with Eq. (22), the denominator is the "enhancement factor" due to recoil from ejecta.

In Table II we list the masses required, impacting at 12 and 40 km s^{-1}, to produce a Δv of 1 cm s^{-1} in NEOs of diameters 0.1 and 1 km diameter. We have taken $\rho = 2$ to compute these results.

TABLE II
NEO Deflection via Impact, Gravity Regime for 1 cm s^{-1} Perturbation

v_i (km s^{-1})	Diameter (km)	M_i/M_{NEO}	M_i
12	0.1	2.4×10^{-7}	240 kg
40	0.1	6.0×10^{-8}	60 kg
12	1.0	3.4×10^{-7}	340 T
40	1.0	8.6×10^{-8}	86 T

As with the strength regime, it is feasible to deflect a 100 m NEO by a few cm s^{-1} with a kinetic impactor, but the masses required to deflect a 1 km object are prohibitive, compared to deflection by nuclear explosion.

B. Mass Drivers for Deflection

As a long-term response, one might imagine employing a mass driver system which is in operation for many years. For a lead time of three decades prior to Earth encounter, and assuming a constant deflecting thrust over the entire time interval, a $\Delta v \sim 0.4$ cm s^{-1} would be required. (For a continuous acceleration, $\delta = 1/2 a t^2 = 1/2 (at)t = 1/2 \Delta v' t$. Hence from Eq. (12), $\delta \sim 3 \Delta v t/2$.) It might be technically feasible to deliver a reaction engine or "mass driver" to an asteroid which will launch ejecta mined from part of the asteroid. Such a device operating on a small asteroid over a decade time scale, provides the needed Δv. For an ejection velocity of ~ 0.3 km s^{-1}, the ejected mass necessary to produce a recoil of 0.4 cm s^{-1} is

$$\Delta m \sim \frac{0.4 \text{ cm s}^{-1}}{0.3 \text{ km s}^{-1}} m_a \tag{33}$$

where m_a is the NEO mass. For a 1 km diameter, 2 g cm^{-3} density NEO, the ejected mass is 14 Ktons. Although such a system might be technically feasible, it will become clear from what follows that nuclear energy offers a much less expensive solution.

C. Momentum Transfer from a Stand-Off Nuclear Blast

One means of coupling the energy of a nuclear explosion to an asteroid to deliver a momentum impulse to the asteroid is to use a neutron (and penetrative X-ray)-rich nuclear device in the stand-off mode (Hyde 1984). Specifically, we assume that neutrons from such a device penetrate a characteristic distance $t_0 \approx 20$ cm (for a mean atomic weight of 25 and an assumed NEO density of 2g cm^{-3}) along the line of travel in the process of being absorbed. This results from simply considering the absorption cross section of nuclei compared to their lattice spacing in typical mineral crystals. The absorption length is along the line of travel of the neutrons, thus for neutrons striking the surface at an oblique angle, the vertical depth of penetration is reduced by the cosine of the angle of incidence.

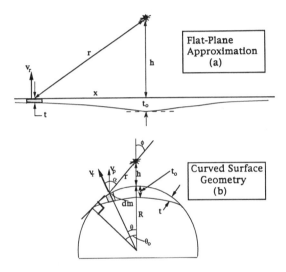

Figure 2.   Geometry of deposition of nuclear explosive radiation on (a) a half-space, and (b) a spherical NEO.

From Ahrens and Harris (1992), Eqs. (9) and (10), the vertical velocity of the layer spalled off the asteroid (Fig. 3) is

$$\Delta v_r = \frac{\gamma \Delta E}{c_p} \tag{34}$$

where γ is the thermodynamic Grüneisen ratio, assumed to be near unity, ΔE is the absorbed energy density, per mass, and c_p is the sound speed in the asteroid material, typically ~ 2 km s^{-1} for rocky regolith. To evaluate the integrated momentum impulse, we need to evaluate ΔE and the depth of penetration t as a function of distance away from the sub-blast point. Consider first the flat-plane approximation, where the ratio of stand-off distance h to

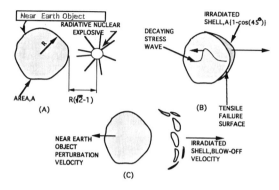

Figure 3.　Sketch of the use of nuclear explosive radiation to induce a (~ 1 cm s^{-1}) velocity perturbation in a NEO. (a) Nuclear explosive designed to provide a substantial fraction of its yield as energetic neutrons and gamma rays is detonated at an optimum height ($\sqrt{2} - 1)R$, above an asteroid (see Appendix), (b) Irradiated to a depth of ~ 20 cm, surface material subsequently expands and spalls away from the NEO, inducing a several kilobar stress wave in the NEO. (c) Blow-off of the irradiated shell induces a cm s^{-1} velocity perturbation in the NEO.

asteroid radius R is small, thus the surface of the asteroid can be regarded as a flat plane (see Fig. 2a).

In a ring of mass dm which lies a distance x from the sub-blast point, of width dx and thickness t, the characteristic depth of penetration of the neutrons is

$$t = t_0 \cos \phi \tag{35}$$

The angular area of the ring as seen from point of explosion is

$$dA = \frac{dx \cos \phi}{r} \cdot \frac{2\pi x}{r} \tag{36}$$

where the first term is the radial angular length and the second term is the circumferential angular length. The energy absorbed is thus

$$de = dA \cdot y/(4\pi) \tag{37}$$

where y is the neutron yield of the explosion. The mass contained in this ring is

$$dm = 2\pi \rho xt dx = 2\pi \rho xt_0 \cos \phi dx \tag{38}$$

where ρ is the density of the asteroid material, ~ 2 to 3 g cm^{-3}. The energy per unit mass in the annular volume is therefore

$$\Delta E = de/dm. \tag{39}$$

Hence,

$$\Delta E = \frac{y \cos \phi}{4\pi r^2 t \rho} = \frac{y}{4\pi r^2 t_0 \rho}. \tag{40}$$

Thus the ring of matter will be ejected upward from the asteroid surface at a velocity

$$\Delta v_r = \frac{\gamma y}{4\pi r^2 t_o \rho c_p} = \frac{\gamma y \cos^2 \phi}{4\pi h^2 t_o \rho c_p}. \tag{41}$$

The increment of momentum impulse to the asteroid is thus $dp = dm \cdot \Delta v_r$:

$$dp = \frac{\gamma y}{2h^2 c_p} x \cos^3 \phi dx. \tag{42}$$

We can substitute $\cos \phi = h/\sqrt{h^2 + x^2}$ in the above and integrate to obtain the total momentum transfer:

$$p = \frac{\gamma y h}{2c_p} \int_0^\infty \frac{x dx}{(h^2 + x^2)^{3/2}} = \frac{\gamma y}{2c_p}. \tag{43}$$

Interestingly, almost all of the physical parameters (h, ρ, t_o) disappear from the final result. One must be careful, however, to scrutinize the resulting absolute values of the energy density ΔE and the ejection velocity, Δv_r, implied by the chosen yield and stand-off distances. In particular, ΔE should not be so large that the surface material is melted or evaporated, nor should Δv_r be less than the escape velocity from the asteroid surface.

Now consider the more general case of a stand-off distance where $h/R \sim 1$, that is, where the curvature of the asteroid surface is important (see Fig. 2b). The depth of penetration of the neutrons becomes

$$t = t_o \cos \alpha \tag{44}$$

and the mass in a ring on the surface an angle θ away from the sub-blast point is

$$dm = 2\pi \rho t R \sin \theta R d\theta = 2\pi \rho t_o R^2 \cos \alpha \sin \theta d\theta. \tag{45}$$

The expressions for ΔE and Δv_r remain the same as above, but with r related to h, R, and ϕ as shown in Fig. 2b. In the spherical case, the mass increment is ejected radially, but only the component of velocity Δv_p parallel to the line from the center of the asteroid to the center of the explosion will contribute recoil momentum, thus $\Delta v_p = \Delta v_r \cos \theta$. The increment of momentum becomes

$$dp = dm \cdot \Delta v_p = \frac{\gamma y}{2c_p} \left(\frac{R}{r}\right)^2 \cos \alpha \sin \theta \cos \theta d\theta. \tag{46}$$

To obtain the total momentum, this must be integrated over the entire surface exposed to radiation. Because the integration is with respect to θ, the upper limit, where $\alpha = 90°$, is

$$\cos \theta_o = \frac{1}{1 + h/R}. \tag{47}$$

We can state the result in terms of the flat plane approximation, as follows:

$$p = A \frac{\gamma y}{2c_p} \tag{48}$$

where A is a constant that tends to 1 in the limit of $h/R \to 0$:

$$A = \left(\frac{R}{R+h}\right)^2 \int_0^{\theta_0} \frac{\sin^2 \alpha \cos \alpha \cos \theta}{\sin \theta} \, d\theta. \tag{49}$$

Note that we have made use of the sine law of triangles to eliminate r, i.e., $r/\sin\theta = (R+h)/\sin\alpha$. We have numerically evaluated the above integral to obtain the plot of A vs h/R in Fig. 4.

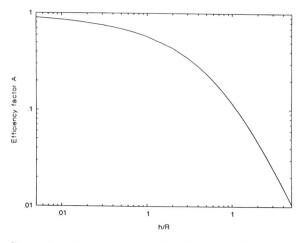

Figure 4.   Geometrical efficiency factor A (Eq. 49) vs normalized elevation h/R.

We now express the neutron yield y as

$$y = nW \tag{50}$$

where W is the total explosive yield. The momentum impulse p is related to the recoil velocity of the asteroid by dividing by the mass of the asteroid:

$$\delta v = \frac{p}{\frac{4}{3}\pi\rho R^3} \approx 0.1 \frac{nAW}{D^3}. \tag{51}$$

In the approximate expression, we have inserted nominal values of the constants as mentioned above, and converted units for convenience, to asteroid diameter in km, total explosive yield W in Kt, $(1 \text{ Kt} = 4 \times 10^{19}\text{erg})$, and δv in cm s^{-1}. For most cases of interest, the requirement that Δv_r be greater than the escape velocity from the surface is not an important limitation, nor is the requirement that the surface not be melted. Thus one can generally employ a stand-off distance of $h/R \sim 0.4$ as suggested by Ahrens and Harris (1992) (see also the Appendix at the end of this chapter), whence $A \sim 0.3$ (Fig. 4). Therefore, to obtain a δv of ~ 1 cm s^{-1} requires ~ 30 Kt of neutron energy yield for a 1 km diameter asteroid. If the efficiency of producing neutrons n lies between 0.03 and 0.3 as suggested by Ahrens and Harris (1992), then the required total explosive yield is 0.1 to 1 Kt, 100 Kt to 1 Mt and 100 Mt to 1 Gt, to deflect 0.1, 1, and 10 km asteroids by 1 cm s^{-1}.

D. Deflection by Surface Nuclear Explosive

Another approach to the use of nuclear explosives is to use a surface charge to induce cratering on the NEO. The thrown-off ejecta effectively induces a velocity change in the NEO and the ejecta is highly dispersed and is not expected to be a hazard when it is encountered by the Earth. This method suffers the disadvantage in that the NEO may be inadvertently broken into large fragments which may represent a hazard to the Earth. For 0.1, 1, and 10 km diameter, we examine the nuclear explosive surface charge required to perturb the NEO. Again we consider explosive cratering in the cases for strength and gravity controlled cratering.

Strength Regime. There is relatively little data for mass of the ejected volume of nuclear craters in the strength regime and there are virtually no calibration experiments. Schmidt et al. (1986, p. 190) give a value of ~ 10 kg mass of excavated "dry soft rock" per kg equivalent yield of high explosive. We take "soft rock" to imply a yield strength of $\sim 10^8$ dyne cm^{-2}. To estimate the recoil impulse of the ejecta from a surface nuclear explosion, we replace Eq. (17) by:

$$M_{ej} \approx 10\,W. \tag{52}$$

Equations (14), (16), (18), and (19) are all valid for this case, changing only the above expression for M_{ej}. In Eq. (16), we use $Y = 10^8$ dyne cm^{-2} to evaluate $v_{min} \approx 17$ m s^{-1}. Because for a nuclear explosive, the "impactor mass" is negligible, the momentum Eq. (21) consists only of the recoil momentum. Neglecting the mass of the ejecta compared to that of the NEO, we obtain

$$\Delta v \approx \frac{p}{M_{NEO}} \approx \frac{\sqrt{2} \cdot 10 \cdot v \cdot W}{M_{NEO}} \tag{53}$$

and thus,

$$W \approx 4 \times 10^{-5} \Delta v M_{NEO} \tag{54}$$

where Δv is in cm s^{-1}. Hence to deflect asteroids of 0.1, 1, or 10 km diameter by 1 cm s^{-1}, explosive yields of 40 t, 40 Kt, or 40 Mt are required, respectively.

Gravity Regime. Although a large number of large-scale surface explosion experiments have been conducted in air at Earth's gravity, the effects of both reduced (and increased) gravity and reduced (from atmospheric) and increased atmospheric pressure on cratering has been studied from surface

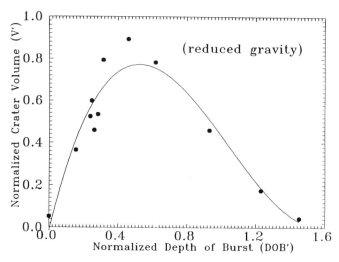

Figure 5. Normalized crater volume (Eq. [62]) vs normalized depth of burst (Eq. [60]) for 1.7 g cm^{-3} sand. Data taken by Johnson et al. (1969) at varying gravities 167, 373, 981 and 2453 cm s^{-2}. Polynomial is fit to data.

and buried explosive charges only in the laboratory. Johnson et al. (1969) first studied these effects using several gram charges and igniters at gravities varying from 167 to 2450 cm s^{-2} (Fig. 5).

Although the initial packing density of Johnson's et al.'s 1969 samples are reported as 1.7 g cm^{-3}, it appears possible that this density may have decreased substantially during the period that the test apparatus was subjected to reduced gravity. (These experiments were conducted within an aircraft flying a reduced gravity trajectory.) Moreover, the greater excavation efficiency of the propellant igniters vs detonators used in later experiments may account for the larger normalized crater volume vs depth of burst, as compared to later studies. Lower, but still highly variable, normalized crater volume cratering experiments are reported by Herr (1971) (Fig. 6). Herr conducted an extensive series of cratering tests on quartz sand and granulated carbonate material studied at variable atmospheric pressure and depth of burial.

The most extensive study of the effects of charge size, placement with respect to the free-surface, explosive type, soil density and atmospheric pressure on explosive (and impact cratering) have been conducted by Schmidt et al. (1986) and Housen et al. (1993). These data are also used to describe the effect of charge burial in the next section. On the basis of close modeling of nuclear explosion craters with small scale high explosive tests, we employ the formalism of explosive crater scaling of Holsapple and Schmidt (1980,1982) in which they define the cratering efficiency for gravity scaled cratering as:

$$\pi_v \equiv \rho V_e / W = a \pi_2{}^b \tag{55}$$

where a and b are constants which depend on charge type and placement, ρ

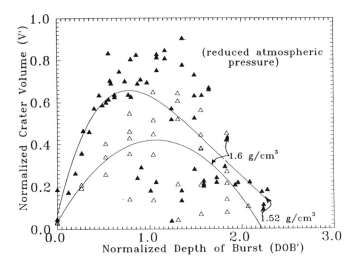

Figure 6.   Herr's (1971) data for experiments in 1.52 and 1.60 g cm^{-3} granular media for varying depths of burst and atmospheric pressures. Polynomial is fit to data. Coordinates are the same as in Fig. 5.

is the mass density of the media, V_e is the crater volume, and W is explosive equivalent mass, usually of TNT. The product

$$\rho V_e = M_{ej} \tag{56}$$

is the same quantity as in Eqs. 17 and 26. Here π_2 is the gravity scaling parameter and is defined as

$$\pi_2 \equiv \frac{g}{Q}\left(\frac{W}{\delta}\right)^{1/3} \tag{57}$$

where Q is the energy content per mass of TNT, 4.2×10^{10}erg g^{-1}, and δ is the mass density of the explosive charge.

For surface nuclear charges with high radiative outputs, the parameters a and b of Eq. 55 in the gravity regime are given by the curve of Schmidt et al. (1986, p. 192) as $a = 0.008$ and $b = -0.49$. Using Eq. (23) for the mass of ejecta traveling at velocities greater than v_{esc}, Eq. (15) and Eq. (55) in the integral analogous to that of Eq. (29) we find

$$p = \int_{v_{esc}}^{\infty} v\cos\theta \ dM = \cos\theta M_{ej}^{1.203}/(0.22 \ v_{esc}^{0.22}) \tag{58}$$

where M_{ej} is taken from Eq. 56.

Again in analogy to Eq. (31), we find

$$\Delta v = p/(M_{\text{NEO}} - M_{ej}). \tag{59}$$

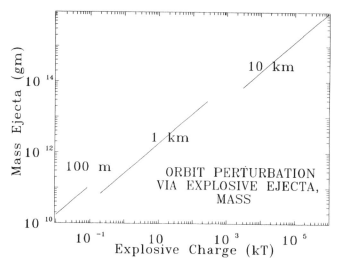

Figure 7. Mass ejecta accelerated to greater than escape velocity for cratering explosive charges on surface of 0.1, 1, and 10 km diameter NEO as a function of explosive yield.

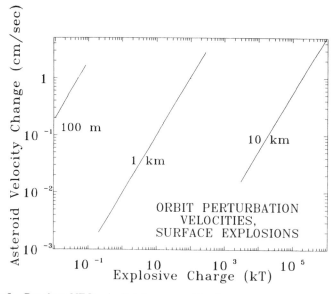

Figure 8. Resultant NEO velocity change resulting from momentum conservation vs surface charge for 0.1, 1 and 10 km objects.

The mass of ejecta and the velocity imparted to the NEO vs nuclear yield is plotted in Fig. 7 and Fig. 8. For a Δv of 1 cm s^{-1} for 0.1, 1, and 10 km NEOs require surface charges of 500 kg, 90 Kt and 0.2 Gt. These values are comparable to the yields calculated for 1 cm/sec velocity deflection via radiative stand-off explosions.

The extrapolation of cratering efficiency to the low gravity (0.003 cm s^{-2}) for a 0.1 km NEO should be regarded with caution. Moreover, whether cratering on asteroids is gravity or strength dominated may also depend on asteroid type.

Thus, surface explosions appear to be not substantially better than radiative stand-off explosions, in deflecting NEOs. However, it is clear that the knowledge requirements for surface cratering, especially with regard to material properties, for large NEO masses, are subject to greater uncertainty.

IV. DISPERSAL AND EXPLOSIVE FRAGMENTATION

Gravity Regime. By burial of a nuclear charge in a NEO, cratering efficiency can be drastically increased. There is good reason for desiring some nuclear charge burial, as surface exploded nuclear charges couple only a small fraction of their energy to rock (0.2 to 1.8%) for radiative and hydrodynamic coupling (Housen et al. 1993), whereas the large fraction of the energy of a deeply buried charge is coupled into rock. This approach can be employed with two objectives: the ejecta mass will increase down to some burial depth—however, its velocity will in general be lower and less momentum per unit of ejecta will be imparted to the NEO; secondly, the buried charge may be used to fragment and disperse the NEO.

In this section we use, as suggested by Housen et al. 1993, the normalized depth of burst, DOB' which is defined as

$$DOB' \equiv \frac{DOB}{a} \pi_2{}^{\alpha/3} \left\{ 1 + 0.9[P \pi_2{}^{\alpha/3}/(\rho g a)]^{0.7} \right\}^{\chi} \tag{60}$$

where DOB is the actual depth of burst, a is the radius of the equivalent sphere of TNT chemical explosive, P is the atmospheric pressure and the empirical parameter, $\alpha = 0.581$. Because nuclear explosives, in general, couple their yield less effectively than chemical explosives into Earth media we reduce nuclear yields by a factor of 1.6 when correlating to chemical explosive data. The exponent χ, is given as

$$\chi = \alpha/[0.7(3 - \alpha)]. \tag{61}$$

Similarly the normalized excavated volume V' is defined as

$$V' = \pi_v \pi_2{}^{\alpha} \left\{ 1 + 0.14[P \pi_2{}^{\alpha/3}/(\rho g a)]^{1/3} \right\}^{y} \tag{62}$$

where y is given as:

$$y = 9\alpha/(3 - \alpha). \tag{63}$$

Scaled results for small scale tests in quartz sand (1.8 g cm^{-3}), at normal and variable gravity up to 5×10^5 cm s^{-2} and variable atmospheric pressure, tests using a high density magnetite-bearing sand (3.08 g cm^{-3}), data from

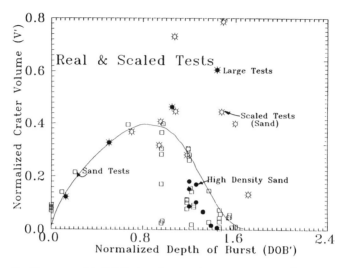

Figure 9. Housen et al.'s (1993) data for experiments in 1.8 g cm^{-3} sand, 3.1 g cm^{-3} sand, scaled laboratory (alluvium) and large scale nuclear and explosive tests. Coordinates are the same as in Figs. 5 and 6. Equation (64) is fit to these data.

large chemical (0.5 Kt-TNT) and nuclear (up to 100 Kt) explosions are shown in Fig. 9. These have been fit by Housen et al. to:

$$V' = 0.64 \log_{10} DOB' + 0.2884 - 9.7757 \times 10^{-3}/$$

$$(0.6959 - \log_{10} DOB')^{7.1805}. \qquad (64)$$

Figure 10 shows the equivalent radius of the mass of ejecta yielded, vs the DOB values, calculated from Eqs. (60) through (64). The peak values of each curve are those which result in "complete excavation," that is, the radius of the mass of the ejecta equals the radius of the NEO. The yield values required for an excavating charge are less by a factor of 3×10^3 to 5 in going from 0.1 to a 10 km NEO, than those calculated for fragmentation below. These charges are 800 kg for a 100 m NEO, 22 Kt for a 1 km NEO, and 0.6 Gt for a 10 km diameter NEO. The effect of gravity on the radius of excavated volumes is seen to be substantial. Notably, the optimum (largest radius of excavated volume) depth of charge decreases with increasing NEO size and surface gravity. Figure 10 also shows the radius of excavated volumes between craters on the Earth and a 10 km NEO differ by a factor of up to 5 in going from the gravity of a 10 km diameter object, 0.3 cm s^{-2} to that of the Earth (982 cm s^{-2}). Also shown is the curve of normalized charge depth vs radius of excavated volume when the slight effect of the Earth's atmosphere is taken into account.

Strength Regime. Small-scale fragmentation experiments on solid rocks demonstrate that the bulk of the fragments of a collisional disruption have velocities of ~ 10 m s^{-1}. However, the "core" or largest fragment has been

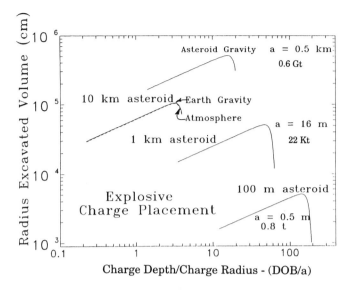

Figure 10.   Radius of excavated sphere of asteroidal material for 0.1, 1, and 10 km NEO vs normalized charge depth. Effect of nominal yield explosive for each size NEO is indicated. The effect of gravity is demonstrated by the curve labeled "Earth Gravity" which gives the excavated crater volume assuming terrestrial rather than asteroidal gravity for the 10 km asteroidal case, where a 0.6 Gt explosive charge yields a radius of excavated volume of crater corresponding to spheres of 5 and 1 normalized radius on the asteroid and Earth, respectively.   When the Earth's atmosphere is taken into account, the minor decrease in crater volume is indicated.

demonstrated to have a differential velocity of no more than ~ 1 m s^{-1} (see, e.g., Nakamura and Fujiwara 1991). From Eq. (2), if the body is fragmented ~ 75 days before Earth encounter then most of the $\gtrsim 10$ m fragments will still impact the Earth. For a small object (0.1 to 1 km), dispersal of the bulk of the fragments into the Earth's atmosphere may be sufficient, as long as no fragments $\gtrsim 10$ m are allowed. For a really large object (> 1 km) fragmentation would need to be conducted one or more orbits before intersection with the Earth to assure that most fragments miss the Earth. In general, the debris cloud would spread along the orbit according to Eq. (11) and in the transverse direction according to Eqs. (3) or (4). For a characteristic velocity of ejecta of 10 m s^{-1}, the debris cloud would be $\sim 10\,R_{\oplus}$ in radius (with some oscillation about the orbit) and grow in length by $\sim 200\,R_{\oplus}$ per orbit period. Thus, if the NEO were destroyed one orbit before encounter, the Earth might encounter as little as 0.1% of the debris. But more conservatively, if many large fragments with $\Delta v \lesssim 1$ m s^{-1} remained, as much as 10% of that mass might be intercepted. Thus fragmentation is likely to be a safe choice only for long lead-time response (decades) or for relatively small bodies where the fragments may be allowed to hit the Earth.

"Catastrophic disruption" is generally defined as fragmentation where the largest fragment is $\lesssim 1/2$ the total mass. The energy density to accomplish this decreases with increasing size of body, and becomes rather uncertain when extrapolated to 1 to 10 km size bodies (see, e.g., Housen and Holsapple 1990). However, for the present purpose, we are interested in the energy density necessary to break up a NEO so that all fragments are $\lesssim 10$ m in size. This is obviously a higher energy density than that required to just "break it in two," and we suggest that it should be of the order of the energy density needed to "break in two" a 10 m object E_{frac}, $\sim 10^7 erg\ g^{-1}$.

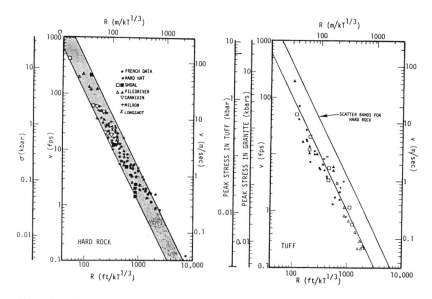

Figure 11. Particle velocity from contained explosion vs scaled radius. *Left*: hard rock; *Right*: tuff, representative of soft rock.

Because of the large energy requirements to fracture a well consolidated NEO, only nuclear explosives are considered. In order to relate the energy density as a function of radius r for a completely coupled (buried) nuclear charge of yield W, we employ the empirical relations of shock-induced particle velocity v vs energy scaled radius $r' \equiv r/(W)^{1/3}$ of Cooper (1977) (Fig. 11). For hard (mainly igneous) terrestrial rocks Cooper's compilation can be fit by

$$v\ (cm\ s^{-1}) = 5.72 \times 10^{10} r'^{-2}. \tag{65}$$

Similarly, for soft rocks, Cooper's compilation can be fit with

$$v\ (cm\ s^{-1}) = 2.90 \times 10^{10} r'^{-2}. \tag{66}$$

The shock wave internal energy per unit mass is equal to $v^2/2$:

$$E = v^2/2 \tag{67}$$

where v^2 can be specified by Eq. (65) or (66) and thus, $E = E_{frac} = 10^7 \text{erg g}^{-1}$. Upon substituting Eq. (66) into Eq. (67) for 1 Kt, we find $r = 36$ m. Thus, a 1 Kt explosive is expected to fragment a 36 m radius sphere of soft rock, if the explosive is placed well within the NEO. Also, a 1 Mt charge of explosive will fragment 360 m radii of soft rock and 1 Gt of explosive will fragment 3.6 km of soft rock. In contrast, for hard rock (Eq. [65]), which describes less attenuative rock, gives a radius of fracture of 70 m for 1 Kt explosion. From Eq. (66), to deliver 10^7erg g^{-1} to 0.1, 1, and 10 km diameter NEOs, similar to soft rock, requires 1 Kt, 1 Mt, and 1 Gt, centrally placed. For NEOs similar to hard rock 0.1, 1, and 10 km diameter require 3 Kt, 3 Mt and 3 Gt, of explosive, respectively.

The above discussion is based on the premise that the charge is buried to sufficient depth so as to obtain optimum fragmentation. Dispersal seems to require about the same energy as deflection, and also is benefited by charge burial. Hence, NEO deflection rather than destruction via fragmentation, appears to be the favorable choice.

V. MAXIMUM Δv ACHIEVABLE IN A SINGLE IMPULSE

Most of the deflection methods discussed above operate impulsively, by impacts or explosions. It is implicit in such methods that the impulse is applied unevenly throughout the NEO. Thus in the case of an incoherent NEO, various parts of the NEO will receive unequal increments of the velocity impulse; indeed the dispersion in Δv among various parts of the NEO is likely to be of the order of Δv itself. Thus the various parts of the NEO will have relative velocities with respect to the center of mass of the order of Δv. If Δv exceeds the NEO surface escape velocity, then the result of the impulse will be to disperse the various pieces of the NEO, rather than to move it coherently as a single body. If Δv is less than the surface escape velocity, then the separate pieces will not escape from one another, and deflection may occur coherently.

In the case of an initially coherent NEO, i.e., a solid rock, the shock wave propagating through the body as a result of a surface explosion or stand-off nuclear blast is likely to fracture the NEO at least into large pieces. Note that the energies calculated for comminution and dispersal of an asteroid by explosion are only about an order of magnitude greater than required to divert it by 1 cm s^{-1}. Hence even an impulsive Δv of only $\sim$10 cm s^{-1} would cause major fracturing of the NEO, so that in terms of its response to such an impulse, it should behave the same as an incoherent NEO, even if it were a single coherent rock before the impulse. We can appeal to terrestrial experience in this matter. Major earthquakes can deliver an impulsive Δv of the order of 1 m s^{-1} on a time scale of 10^1 to 10^2s, and such impulses do indeed cause 100 m objects (large buildings) to fracture and even fragment.

Thus it appears that for NEOs $\geq$100 m in diameter, the maximum single impulsive Δv that can be applied without danger of dispersing the NEO into large fragments is of the order of its surface escape velocity. This is $\sim$1 m s^{-1}

for a 1 km diameter NEO, and is directly proportional to diameter, i.e., ~ 10 cm s^{-1} for a 100 m NEO and ~ 10 m s^{-1} for a 10 km NEO. One can imagine that it would be desirable, indeed probably necessary, to apply several small velocity impulses to an object in order to divert it accurately. However there are limits to the number of impulses that could be economically employed, perhaps of order 10. It therefore appears reasonable to take $\sim 10 v_{esc}$ as a practical limit to the Δv which can be applied for diverting a NEO of a given size.

VI. CONCLUSIONS

We have examined the velocity criteria for perturbation of the orbits of Earth-crossing objects (asteroids and comets) so as to cause objects that have trajectories which intersect the Earth to be deflected. For objects discovered only as they approach on a collision course, the velocity perturbations required are tens to hundreds of m s^{-1}. Energy levels are prohibitive for larger bodies, and the required perturbation impulse would disrupt the body as discussed above.

We also note that perturbation of an object is most effective by applying a change in velocity Δv along its original orbit and thereby inducing a change in orbital period. An impulse Δv along the line of motion provides a larger deflection δ, after time t, of the order of $3\Delta vt$.

For a ~ 100 m diameter NEO, the kinetic energy of $\sim 10^2$ to $\sim 10^3$ kg impactors, intercepting at 12 km s^{-1} will provide enough energy to crater and launch ejecta in the low gravity environment of these objects to induce velocity perturbations of in the order of 1 cm s^{-1}. For larger diameter NEOs, deflection via this method appears impractical because of the large mass of impactors required. Mass drivers require launching $\sim 10^3$ to 10^4 tons of NEO material over an interval of 30 yr prior to encounter in order to deflect a 1 km NEO from Earth impact. Nuclear explosive irradiation may be used to blow-off a surface layer from one side of the asteroid, to produce a recoil deflection velocity. Charges of 0.1 to 1 Kt, 100 Kt to 1 Mt and 100 Mt to 1 Gt of nuclear explosives are required in order to blow off a shell sufficient to perturb the velocity of 0.1, 1, and 10 km NEOs by 1 cm s^{-1}. The ranges of explosives required correspond to a radiative efficiency range 0.3 to 0.03 assumed for the nuclear explosives. Surface charges of 500 Kg, 90 Kt and 200 Mt may be used to eject crater material to greater than local escape velocity, and hence, perturb 0.1, 1, and 10 km diameter NEOs by a velocity increment of ~ 1 cm s^{-1}. These estimates are based on using radiative nuclear charges and extreme extrapolation of the effect of gravity on gravity-dependent cratering. Burial of nuclear charges to induce fragmentation and dispersal requires *in-situ* drilling which is difficult on a low gravity object or technically challenging if dynamic penetration methods are to be employed. Optimally buried cratering charges required to excavate completely (working only against local gravity) 0.1, 1, and 10 km diameter NEOs require nuclear yields of 800 kg, 22 Kt and 0.6 Gt, respectively.

Upon examining the deflection or fragmentation options, deflection appears to be the most promising goal because charge burial is not required or desirable. For a small (100 m) NEO, the kinetic energy impact deflection method is both technically feasible and does not involve the politically complex issue of placing nuclear explosives on a spacecraft. For the 1 to 10 km diameter NEO, which includes the largest Earth-crossing objects, only the nuclear option is practical. For this task, deflection via nuclear explosive radiation appears to be the simplest method. This would appear to require less detailed knowledge of the physical characteristics of an Earth-crossing object, and the development of the charges required to deflect large Earth-crossing objects appear to be technically feasible.

Finally, we should note that while further study of the feasibility of diverting NEOs may be warranted, we do not believe it is appropriate now to conduct engineering designs of systems because of: (1) the low Earth impact probability of hazardous NEOs; (2) the high cost compared to low probability; and (3) the rapid changes in defense systems technology.

Acknowledgments. Research was supported at Caltech and JPL by NASA. We appreciate the helpful comments of the reviewers. TJA benefited from the technical discussions held at the LANL/NASA Workshop on Near Earth Object Interception, January 14–16, 1992, Los Alamos National Laboratory and we both benefited from attendance at the Tucson Conference, January 5–9, 1993, on the subject of this book, and the Erice Asteroid Hazard Workshop, 28 April–4 May, 1993. We thank R. Schmidt, K. Holsapple, K. Housen, and J. C. Solem for their preprints.

REFERENCES

Ahrens, T. J., and Harris, A. W. 1992. Deflection and fragmentation of near-Earth asteroids. *Nature* 360:429–433.

Ahrens, T. J., and O'Keefe, J. D. 1983. Impact of an asteroid or comet in the ocean and extinction of terrestrial life. *Proc. Lunar Planet. Sci. Conf.* 13, *J. Geophys. Res. Suppl.* 88:799–806.

Alvarez, L. W., Alvarez, W., Asaro, F., and Michel, H. V. 1980. Extra-terrestrial cause for the Cretaceous-Tertiary extinction. *Science* 208:1095–1108.

Binzel, R. P., Xu, S., Bus, S. J., and Bowell, E. 1992. Origins for the near-Earth asteroids. *Science* 257:779–782.

Blum, J. D., and Chamberlain, C. P. 1992. Oxygen isotope constraints on the origin of impact glasses from the Cretaceous-Tertiary boundary. *Science* 257:1104–1107.

Blum, J. D., Chamberlain, C. P., Hingston, M. P., Koeberl, C., Marin, L. E., Schuraytz, B. C., and Sharpton, V. L. 1993. Isotopic composition of K/T boundary impact glass with melt rock from Chicxulub and Manson impact structure. *Nature* 364:325–327.

Bourgeois, J., Hansen, T. A., Wiberg, P. L., and Kauffman, E. G. 1988. A tsunami deposit at the Cretaceous-Tertiary boundary in Texas. *Science* 241:561–570.

Brett, R. 1992. The Cretaceous-Tertiary extinction—A new fatal mechanism involving anhydrite target rocks. *Geochim. Cosmochim. Acta* 56:3603–3606.

Cooper, H. F., Jr. 1977. A summary of explosion cratering phenomena relevant to meteor impact events. In *Impact and Explosion Cratering*, eds. D. J. Roddy, R. O. Pepin and R. B. Merrill (New York: Pergamon Press), pp. 11–44.

Covey, C., Ghan, S. J., Walton, J. J., and Weissman, P. R. 1990. Global environmental effects of impact-generated aerosols; Results from a general circulation model. In *Global Catastrophes in Earth History*, eds. V. L. Sharpton and P. D. Ward, Geological Soc. of America Special Paper 247 (Boulder: Geological Soc. of America), pp. 263–270.

Crutzen, P. J., and Birks, J. W. 1982. The atmosphere after nuclear war: Twilight at noon. *Ambio* 11:115–125.

Gault, D. E., Shoemaker, E. M., and Moore, H. J. 1963. *Spray Ejected from the Lunar Surface by Meteoroid Impact*, NASA TN-D-1767.

Gerstl, S. A. W., and Zardecki, A. 1982. Reduction of photosynthetically active radiation under extreme stratospheric aerosol loading. In *Geological Implications of Impacts of Large Asteroids and Comets on the Earth*, eds. L. T. Silver and P. H. Schultz, Geological Soc. of America Special Paper 190 (Boulder: Geological Soc. of America), pp. 201–210.

Herr, R. W. 1971. *Atmospheric-Lithostatic Pressure Ratio on Explosive Craters in Dry Soil*, NASA TR-R-366.

Hildebrand, A. R., Penfield, G. T., Kring, D. A., Pilkington, M., Camargo, A. Z., Jacobsen, S. B., and Boynton, W. V. 1991. Chicxulub crater: A possible Cretaceous-Tertiary boundary impact crater on the Yucatan Peninsula, Mexico. *Geology* 19:867–871.

Holsapple, K. A. 1993. The scaling of impact processes in planetary sciences. *Ann. Rev. Earth Planet. Sci.* 21:333–373.

Holsapple, K. A., and Schmidt, R. M. 1980. On the scaling of crater dimensions, I: Explosive processes. *J. Geophys. Res.* 85:7247–7256.

Holsapple, K. A., and Schmidt, R. M. 1982. On the scaling of crater dimensions, II: Impact processes. *J. Geophys. Res.* 87:1849–1870.

Housen, K. R., and Holsapple, K. A. 1990 On the fragmentation of asteroids and planetary satellites. *Icarus* 84:226–253.

Housen, K. R., Schmidt, R. M., and Holsapple, K. A. 1983. Crater ejecta scaling laws: Fundamental forms based on dimensional analysis. *J. Geophys. Res.* 88:2485–2499.

Housen, K. R., Schmidt, R. M., Voss, M. E., and Watson, H. E. 1993. D. O. B. Scaling and Critical Depth of Burst for Cratering (Seattle: Boeing Corp.), DNA-TR-92-024.

Hyde, R. A. 1984. Cosmic Bombardment. Lawrence Livermore National Lab Internal Rept. UCID-20062.

Johnson, S. W., Smith, J. A., Franklin, E. G., Moracki, L. K., and Teal, D. J. 1969. Gravity and atmospheric pressure effects on crater formation in sand. *J. Geophys. Res.* 74:4838–4850.

MIT Students. 1968. *Project Icarus* (Cambridge, Mass.: The MIT Press).

Morrison, D., ed. 1992. *The Spaceguard Survey: Report of the NASA International Near-Earth-Object Detection Workshop* (Pasadena: Jet Propulsion Laboratory).

Nakamura, A., and Fujiwara, A. 1991. Velocity distribution of fragments formed in a simulated collisional disruption. *Icarus* 92:132–146.

O'Keefe, J. D., and Ahrens, T. J. 1982. Cometary and meteorite swarm impact on planetary surfaces. *J. Geophys. Res.* 87:6668–6680.

O'Keefe, J. D., and Ahrens, T. J. 1989. IMpact production of CO_2 by K-T extinction bolide and the resultant heating of the Earth. *Nauter* 338:247–249.

Pike, R. J. 1980. Control of crater morphology by gravity and target type: Mars, Earth, Moon. *Proc. Lunar Planet. Sci. Conf.* 11:2159–2189.

Rather, J. D. G., Rahe, J. H., and Canavan, G., eds. 1992. *Summary Report of the Near-Earth-Object Interception Workshop* (Washington, D. C.: NASA).

Schmidt, R. M., and Housen, K. R. 1987. Some recent advances in the scaling of impact and explosion cratering. *Intl. J. Impact Eng.* 5:543–560.

Schmidt, R. M., Holsapple, K. A., and Housen, K. R. 1986. Gravity effects in cratering. In *Pacific Enewetak Atoll Crater Exploration (PEACE) Program Enewetak Atoll, Republic of the Marshall Islands* (Seattle: Boeing Corp. and Defense Nuclear Agency), DNA-TR-86-182.

Scotti, J. V., Rabinowitz, D. L., and Marsden, B. G. 1991. Near miss of the Earth by a small asteroid. *Nature* 354:287–289.

Sharpton, V. L., Dalrymple, G. B., Marin, L. E., Ryder, G., Schuraytz, B. C., and Urrutia-Fucugauchi, J. 1992. New links between the Chicxulub impact structure and the Cretaceous-Tertiary boundary. *Nature* 359:819–821.

Shoemaker, E. M. 1983. Talk presented at Collision of Asteroids and Comets with the Earth: Physical and Human Consequences, July 13–16, Snowmass, Colo.

Shoemaker, E. M., Wolfe, R. F., and Shoemaker, C. S. 1990. Asteroid and comet flux in the neighborhood of Earth. In *Global Catastrophes in Earth History*, eds. V. L. Sharpton and P. D. Ward, Geological Soc. of America Special Paper 247 (Boulder: Geological Soc. of America), pp. 155–170.

Sigurdsson, H., D'Hondt, S., and Carey, S. 1992. The impact of the Cretaceous/Tertiary bolide on evaporite terrane and generation of major sulfuric acid aerosol. *Earth Planet. Sci. Lett.* 109:543–559.

Smither, C. L., and Ahrens, T. J. 1992. Melting, vaporization, and energy partitioning for impacts on asteroidal and planetary objects. In *Asteroids, Comets, and Meteors 1991*, eds. A. W. Harris and E. Bowell (Houston: Lunar and Planetary Inst.), pp. 561–564.

Solem, J. C. 1991. Nuclear Explosive Propelled Interceptor for Deflecting Comets and Asteroids on a Potentially Catastrophic Collision Course with Earth. Los Alamos National Lab Rept. LA-UR-91-3765.

Solem, J. C. 1992. Interception of Comets and Asteroids on Collision Course with Earth. Los Alamos National Lab Rept. LA-UR-231.

Swisher, C. C., III, Grajales-Nishimura, J. M., Montanari, A., Margolis, S. V., Claeys, P., Alvarez, W., Renne, P., Cedillo-Pardo, E., Maurrasse, F. J.-M. R., Curtis, G. H., Smit, J., and McWilliams, M. O. 1992. Coeval ^{40}Ar/^{39}Ar ages of 65.0 million years ago from Chicxulub crater melt rock and Cretaceous-Tertiary boundary tektites, *Science* 257:954–958.

Toon, O. B., Pollack, J. B., Ackerman, T. P., Turco, R. P., McKay, C. P., and Liu, M. S. 1982. Evolution of an impact-generated dust cloud and its effects on the atmosphere. In *Geological Implications of Impacts of Large Asteroids and Comets on the Earth*, eds. L. T. Silver and P. H. Schultz, Geological Soc. of America Special Paper 190 (Boulder: Geological Soc. of America), pp. 187–200.

Turco, R. P., Toon, O. B., Ackerman, T. P., Pollack, J. B., and Sagan, C. 1990. Climate and smoke: An appraisal of nuclear winter. *Science* 247:166–176.

Veverka, J., and Harris, A. W. 1986. Near Earth Asteroid Rendezvous (NEAR) Science Working Group Report. (Pasadena: Jet Propulsion Laboratory), JPL-86-7.

Vickery, A. M., and Melosh, H. J. 1990. Atmospheric erosion and impactor retention in large impacts, with application to mass extinctions. In *Global Catastrophes in Earth History*, eds. V. L. Sharpton and P. D. Ward, Geological Soc. of America Special Paper 247 (Boulder: Geological Soc. of America), pp. 289–300.

Weissman, P. R. 1990. The cometary impactor flux at the earth. In *Global Catastrophes in Earth History*, eds. V. L. Sharpton and P. D. Ward, Geological Soc. of America Special Paper 247 (Boulder: Geological Soc. of America), pp. 171–180.

APPENDIX

Analytic Derivation of "Optimum Stand-Off Distance" h

The solid angle subtended by a cone of apex half-angle θ is $2\pi(1 - \cos\theta)$, or in units of the total unit sphere, $(1 - \cos\theta)/2$ (Fig. 12). Thus the fraction of the NEO surface which is irradiated by a nuclear explosion at a stand-off distance h is

$$F = \frac{1}{2}(1 - \cos\theta). \tag{A1}$$

Similarly, the fraction of the radiative yield that intercepts the NEO surface is

$$f = \frac{1}{2}(1 - \cos\phi) = \frac{1}{2}(1 - \sin\theta). \tag{A2}$$

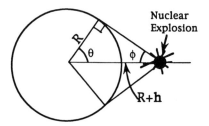

Figure 12. Geometry for calculating the optimum stand-off height h to deflect an NEO of radius R using the scheme shown in Fig. 3.

One criterion for an "optimum stand-off distance" h might be the distance at which the product $F \cdot f$ is maximum. Hence

$$\frac{d}{d\theta}(fF) = \frac{df}{d\theta}F + f\frac{dF}{d\theta} = 0. \tag{A3}$$

This reduces to the following equation:

$$\sin\theta - sin^2\theta = \cos\theta - \cos^2\theta. \tag{A4}$$

The solution of interest is $\theta = 45°$, when $f = F = (1 - \cos 45°)/2 = 0.146$ and $h = (\sqrt{2} - 1)R$, or $h/R = 0.414$. In Fig. 13, we show f, F, and the product $F \cdot f$ vs the stand-off distance ration h/R. It should be noted that the above criterion is not a rigorous condition. For example, to deflect a very small asteroid, it is likely that the explosive yield required would be small, but the Δv required may be close to the limit that would fracture and disperse the asteroid, thus it may be better to use a larger explosive yield at a larger stand-off distance in order to push as gently as possible on the largest possible

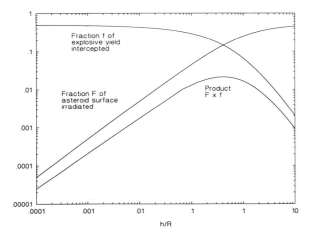

Figure 13.   Fraction f and F of the explosive yield intercepted by an NEO and of the NEO surface irradiated, vs normalized height h/R. Also shown is the product $f \times F$ vs h/R.

fraction of the asteroid's surface. Conversely, for a very large asteroid, it is more likely that the required Δv would be only a small fraction of that which would disrupt the body, but the total yield required would be large. Thus, in this case, it may be more efficient to use a lower stand-off distance to achieve a higher fraction of the yield intercepted by the asteroid.

DEFENDING THE EARTH AGAINST IMPACTS FROM LARGE COMETS AND ASTEROIDS

V. A. SIMONENKO, V. N. NOGIN, D. V. PETROV and
O. N. SHUBIN
Institute for Technical Physics, Chelyabinsk

and

JOHNDALE C. SOLEM
Los Alamos National Laboratory

We discuss the use of nuclear explosives for deflection, fracture, or fragmentation of massive comets and asteroids on potentially catastrophic collision course with Earth. We discuss the salient effects of a nuclear explosion near the surface of such an astral assailant, including: (1) an estimate of the radiation energy transfer to its surface; (2) ways to increase radiation energy transfer; (3) hydrodynamic motion and blow off from the surface, and the resulting momentum transfer; and (4) various ways the object can be disrupted by fracture or fragmentation. We estimate the effect of material properties on the mechanisms of disruption or deflection, as the assailant may be composed of dirty ice in the case of comets or rock or nickel-iron, in the case of asteroids. We define the scope of the technical problems associated with implementing such a defense and discuss some specific areas of research that must be addressed. We briefly discuss the international cooperation and stringent safeguards that must be implemented should such a defensive system be constructed.

I. INTRODUCTION

One of the numerous threats to the long-term survival of mankind is the collision of a large comet or asteroid with the Earth. There are a myriad of striking illustrations of such impacts on the Earth surface: meteorite craters, astroblemns and certain circular geologic structures (Khryanina 1987; Bjork 1961). Some of them have been thoroughly investigated and are understood (Bjork 1961; Masaitis et al. 1975; Khryanina and Zeilik 1980; also see the Chapter by Grieve and Shoemaker).

The consequences of such a collision could be a devastating climatic change, dwarfing the "nuclear winter" long feared as collateral effect of a global thermonuclear war. Evidence for severe climate alterations resulting from asteroid or comet impacts is reasonably well established. Perhaps the best known, and certainly the most extensively investigated, is the extinction of giant reptiles at the boundary of the Carboniferous and Paleogeneous

Ages (Alvarez et al. 1980). The frequency of such impacts is small but not insignificant—as evidenced by the numerous gigantic craters on the surface of the Earth. The most energetic event of this century occurred over the Tunguska Valley in 1908 (Krinov 1966). It was the result the explosive break-up of a chondritic asteroid at an altitude of several kilometers and it leveled about 2,000 square kilometers of forest. Its explosive yield has been estimated at 15 to 20 megatons (Mt) of TNT (Alekseev et al. 1990), although recent estimates have suggested that it may have been as high as 48 Mt (Hills and Goda 1993).

With our current technological and scientific understanding, what can mankind do to mitigate the threat from these astral assailants? It seems reasonably clear that we could set up defenses using rocket-delivered nuclear explosives. Indeed, among all the energy sources available to mankind, nuclear explosive devices possess the highest concentration of energy, and can be developed for yields of 100 Mt or more. For instance, a thermonuclear charge with a yield of 1 Mt has a mass of the order of 0.5 ton (Sikes and Davis 1987). It is interesting to note that to obtain an equivalent energy by impact of a body of the same mass, it must have a velocity of 4,000 km s^{-1}. Devices of this energy scale have been designed and tested, and we have extensive experience in the design and engineering of nuclear explosives. Under stringently regulated conditions, it appears possible to build a defensive device without large-scale testing.

We currently have a reasonably good understanding of the danger of such collisions with the Earth and possess the scientific and technological base for developing the means to protect mankind from such a global-scale catastrophe. This is why, during the last twenty years, the problem has attracted the interest of the broader scientific community and some governmental and industrial representatives in the developed countries. Here is a list of the problems that must be considered before development of a defensive system could be seriously undertaken.

1. Assessment of the number of dangerous cosmic bodies and their probability of collision with the Earth;
2. Assessment of the time required to detect and identify them as dangerous;
3. Understanding of the object's motion in the immediate vicinity of the Earth, penetration of the atmosphere, and detailed dynamics during collision with the Earth, with concomitant assessment of the local and global consequences of the collision;
4. Assessment of the required effect on the astral assailant, whether it be deflection or fragmentation;
5. Consideration of the needed nuclear devices, delivery means, and optimal regime of action;
6. Assessment of consequences of collision with fragments of an object that had been fractured by a nuclear explosive;

7. Consideration of ecological consequences for the Earth and space environment of nuclear explosions in the space.

TABLE I[a]

Yield Versus Mass for Nuclear Explosive Devices

Yield	Mass
1 Mt	0.5 ton
10 Mt	3 to 4 ton
100 Mt	20 to 25 ton

[a] Table from Sikes and Davis (1987).

For the moment, we will not concern ourselves with the political, sociological, economic, and management aspects of the problem. Turning our attention to the technical side of such an undertaking, we would like to answer a question of principle—is it possible to alter significantly the trajectory of such objects or to disrupt them by detonating nuclear explosions near their surface, and is it possible to calculate these effects? As mentioned above, a nuclear explosion possesses the greatest available concentration of energy, and the characteristic dimensions of the destruction zone, even for explosions of medium yield, are comparable to the dimensions of comets or asteroids that might pose a threat. For instance, explosion of a 100-kt device called "Sedan" produces a crater 370 m in diameter and 100 m in depth (Knowles and Brode 1977). During a nuclear explosion the assailant object is subject to various types of effects, the relative importance of which are determined by the nuclear explosion yield, distance from the object, the object dimensions, the materials of which it is composed, and the design of the nuclear device. Depending on the relationship between the specified parameters, the nuclear explosion effect on the object can result in its fragmentation, crushing, or deviation from its initial trajectory. During the fifty-year history of the development of nuclear explosive technology, the unique experience of design, fabrication, and assessment of effects of nuclear devices, a great deal of knowledge has been accrued, which can be used directly in the design of means of protection of the Earth from collision with a comet or asteroid. As shown in Table I, the yield available from tested nuclear devices is enough to deal with small objects, and the weight characteristics permit delivery by existing rockets. As necessary, the nuclear device can be increased by an order of magnitude and more, preserving specific characteristics, and, in this case, modification and elaboration of such devices would not require further testing. To arrange for maximum influence on asteroids of complex shape or consisting of several parts not strongly connected to one another, it is possible to apply several nuclear devices or a single device of optimum configuration. Special measures can be taken to decrease the danger of radiation contamination of the asteroid fragments. A key issue pursuant to application of nuclear explosive devices is a proper selection of the method of affecting the object, which in turn requires an adequate model of the evolution of the nuclear explosion near the object's surface.

An explosion in space is essentially different from an explosion under terrestrial conditions. The main differences are associated with absence of atmosphere, commensurability of the object's dimensions with linear scales of the phenomenon, the complex shape of the object, its relatively weak gravity, and the exotic set of materials of which the object can consist. The consequences of the influence of the nuclear explosion upon the astral assailant needs special study (see various other chapters in this book).

II. EFFECTS OF NUCLEAR EXPLOSION NEAR THE SURFACE OF THE ASSAILANT OBJECT

We will first consider a nuclear explosion with yield on the order of 1 Mt at the elevation of about 1 m above the surface of a comet or asteroid. Concerning the material of which the object is composed, besides alumino-silicates, which are the analogs of the terrestrial rocks, it is apparently reasonable to consider nickel-iron and dirty ices. For brevity, we will call all these materials the "ground." The evolution of a nuclear explosion near the surface of these objects is intrinsically different from the corresponding phenomena under terrestrial conditions, owing both to the absence of atmosphere and to the exotic set of materials that can compose the "ground," as mentioned above. The ground motion during the explosion is determined primarily by the fraction of the explosion energy transferred directly to the ground near the surface. The duration of energy release for powerful thermonuclear devices is on the order of one hundredth of a microsecond (Cooper et al. 1972). If the device is exploded above the surface, the main part of released energy will be radiated out of the nuclear device in the form of X-rays during a time period of several hundredths of a microsecond (Knowles and Brode 1977). The part of the X-radiation directed downwards illuminates the ground surface (see Fig. 1 below). The radiation flux incident upon different parts of the ground surface is determined by the physical characteristics of the nuclear device and its container and the geometrical divergence of the radiation. Though debris from the device and the ground can gain extremely high velocities (100–1000 km s^{-1}) as a result of radiation energy absorption, the time of the direct X-ray energy transfer into the ground is so small that the shifts of the surfaces are not important. The ground and the device can be regarded as motionless during the X-ray heating of the ground. As a result of this process a finite region of the ground is heated to a high temperature. It is lenticular in shape with a diameter of several meters and a thickness in the center of up to several tens of centimeters. The form and dimensions of this region depend on the explosion yield, elevation of the exploding device, irradiation characteristics of the device, and the ground characteristics. Due to the exceedingly high temperature of the heated zone, most of the incident energy reradiates from the ground and disperses into space. As a result, the share of the energy contained in the heated region is a small part of the full energy released during the explosion. Soon after completion of the radiation processes in the heated region the debris from the device, in the form of a hot plasma, impacts the object surface and transfers more energy to the ground. This stage continues for the first microsecond after the energy release. This process is described in detail by Cooper et al. (1972). The fraction of the total explosive energy delivered to the ground by radiative transfer is nearly 6%, and the fraction delivered by the device debris impact is nearly 2%. The main reason for the small contribution of the device debris, though its share is up to 25% the total balance of the released energy (Cooper et al. 1972), is the reradiation of the energy during the process of debris retardation by the surface of the object. Thus, the main part of the energy is transferred by means of absorption of direct radiation of the device incident on the object surface. For some special devices, a substantial fraction of the energy can be transferred by neutrons, and this energy will be deposited at greater depths in the ground. For our present considerations, we will ignore this contribution.

The ground heated by X-radiation and the device debris rapidly expands and some of the ground material is ejected upward. By conservation of momentum, a shock wave is formed propagating downwards from surface. Some general properties of radiation absorption by the ground and its subsequent conversion to kinetic energy can be followed on the basis of radiation transport solutions in diffusion approximation.

A. One-Dimensional Propagation of the Radiation Diffusion Front

We can acquire considerable insight by solving the planar one-dimensional problem of radiation propagation in a uniform semispace, without taking into account the fluid dynamic motion. The one-dimensional diffusion equation in slab symmetry is

$$\frac{\partial T}{\partial t} + b\frac{\partial T^4}{\partial t} = a\frac{\partial}{\partial x}T^n\frac{\partial T}{\partial t} \tag{1}$$

where T is temperature, t is time, and x is distance perpendicular to the surface. To simulate the heat flux from the nuclear explosion, we impose the boundary condition

$$\frac{lc\sigma}{6}\frac{\partial T^4}{\partial x} + \frac{c\sigma T^4}{4} = S_0 f(t/t_0) \tag{2}$$

where l is the Rosseland mean range in the diffusion limit, σ is the Stephan-Boltzmann constant, c is the speed of light, and the right-hand side is the radiative intensity. To simplify the source term, we let the dimensionless function $f(\tau)$ be normalized in the form

$$\int_0^\infty f(\tau)\,d\tau = 1. \tag{3}$$

To further simplify the problem, we let the Rosseland mean range be in the form

$$l = \frac{l_0 T^{n-3}}{\rho^m} = K T^{n-3} \tag{4}$$

i.e., we neglect density variations, and use perfect gas equations of state, so the specific internal energy is given by

$$\varepsilon = A_s T / (\gamma - 1) \tag{5}$$

where γ is the usual ratio of specific heats.

With these approximations, the coefficients in Eq. (1) are given by

$$a = \frac{4 l_0 c \sigma (\gamma - 1)}{3 A_s \rho^{m+1}} \quad \text{and} \quad b = \frac{\sigma (\gamma - 1)}{A_s \rho} \tag{6}$$

where ρ is the material density. It is convenient to rewrite the boundary conditions in the following form:

$$T^n \frac{\partial T}{\partial x} + P T^4 = F f(t/t_0) \tag{7}$$

where

$$P = \frac{3}{8K} \quad \text{and} \quad F = \frac{3 S_0}{2 K c \sigma}. \tag{8}$$

B. Solution Without Reradiation

It is difficult to obtain interesting analytic solutions of Eq. (1) using the coefficients given in Eq. (6), which give a complete description of the change in energy of both the matter and the radiation field as a function of time. For sufficiently low temperatures, the total energy is dominated by the energy of the matter as opposed to the radiation field. The final stages of energy absorption by irradiation is characterized by relatively low temperatures. Hence we are reasonably well justified in taking $b = 0$, rather than the expression in Eq. (6). Equation (1) then becomes

$$\frac{\partial T}{\partial t} = a \frac{\partial}{\partial x} T^n \frac{\partial T}{\partial t}. \tag{9}$$

To begin the analysis under the no-reradiation assumption, we let $f(t/t_0)$ be a constant. In this case there are only three dimensioned parameters in the problem:

$$[a] = \text{cm}^2 \cdot \text{keV}^{-n} \cdot \mu \text{s}^{-1}$$
$$[P] = \text{keV}^{n-3} \cdot \text{cm}^{-1} \tag{10}$$
$$[F] = \text{keV}^{n+1} \cdot \text{cm}^{-1}.$$

The possible derivative dimensioned combinations of Eq. (10) are of the form

$$B = a^\alpha P^\beta F^\delta. \tag{11}$$

To obtain a parameter with dimension $cm^{b_1} \cdot \mu s^{b_3} \cdot keV^{b_3}$, we must find the necessary values of α, β, γ as solutions to the simultaneous linear equations

$$\begin{cases} 2\alpha - \beta - \delta = b_1 \\ \alpha = b_2 \\ \beta(n-3) + \delta(n+1) - \alpha n = b_3. \end{cases} \tag{12}$$

In the general case, the discriminant of this system is not zero, so it has a single solution. Thus, there is a characteristic time, length, and temperature, which serve as the respective scales in the problem. The time scale is

$$t_* = \frac{4}{3} \frac{A_s l_0}{(\gamma - 1)\rho^{m-1}} \left(\frac{c\sigma}{4}\right)^{\frac{2-n}{4}} S_0^{\frac{n-6}{4}} \tag{13}$$

the length scale is

$$x_* = \frac{8}{3} \frac{l_0}{\rho^m} \left(\frac{c\sigma}{4}\right)^{\frac{3-n}{4}} S_0^{\frac{n-3}{4}} \tag{14}$$

and the temperature scale is

$$T_* = \left(\frac{4S_0}{c\sigma}\right)^{\frac{1}{4}}. \tag{15}$$

For an explosion having a yield of about 1 Mt, the flux falling onto the object's surface directly under the device is on the order of $10^{10} kj \, cm^{-2} \, \mu \, s^{-1}$. To simulate the process for a typical stony or chondritic asteroid, we will use the Rosseland mean range and equation of state for aluminum, corresponding to the following values: $\rho = 2.7 \, g \, cm^{-3}$, $l_0 = 0.49 \, cm^{1-3m} \, g^m keV^{n-3}$, $n = 7.3$, $m = 1.8$, $A/(\gamma - l) = 7.9 \times 10^4 \, kj \, g^{-1}$ (Avororin et al. 1990). Then for the characteristic parameters we obtain the following values:

$$t_* = 0.001 \, s \qquad x_* = 30.0 \, cm \qquad T_* = 3.14 \, keV \tag{16}$$

Because there is a characteristic time and dimension in the problem under consideration, there must be a similarity solution in time and space. The solution must have the following form:

$$T = T_* \vartheta \left(\frac{x}{x_*}, \frac{t}{t_*}\right) \equiv T_* \vartheta(\chi, \tau) \tag{17}$$

where the function ϑ is the solution to the equation

$$\frac{\partial \vartheta}{\partial \tau} = \frac{\partial}{\partial \chi} \left(\vartheta^n \frac{\partial \vartheta}{\partial \chi}\right) \tag{18}$$

with the boundary condition

$$\vartheta^n \frac{\partial \vartheta}{\partial \chi} + \vartheta^4 = 1. \tag{19}$$

To obtain the detailed time dependence of wavefront position and temperature, it is necessary to solve Eqs. (18) and (19) numerically. By analyzing the boundary conditions, however, it is possible to estimate the properties of the solution qualitatively. At early time, $\tau \ll 1$, the temperature at the boundary is small, $\vartheta \ll 1$, so we can neglect the ϑ^4 term in Eq. (19), which is responsible for reradiation of energy into vacuum. As a result, we obtain a self-similar problem with set flux at the boundary (Zel'dovich and Raizer 1966), whose solution is

$$r_f \sim \tau^{\frac{n+1}{n+2}} \quad \text{and} \quad \vartheta \sim \tau^{\frac{1}{n+2}}. \tag{20}$$

The magnitude of the temperature is limited by the largest value, $\vartheta = 1$. Consequently, the solution for $\tau \gg 1$ is close to solution of the problem with constant temperature at the boundary (Zel'dovich and Raizer 1966). Solutions for the heat wave front and the intensity are of the form $r_f \sim t^{\frac{1}{2}}$ and $S \sim t^{-\frac{1}{2}}$. Other characteristics of the solution are also functions of χ, τ and corresponding dimensional parameters. In particular, the time-integrated intensity (total energy fluence) through the boundary is

$$Q = Q_* \kappa(\tau) \tag{21}$$

where $\kappa(\tau)$ is a dimensionless function of dimensionless time, and

$$Q_* = \frac{4A_s l_0}{3(\gamma - 1)\rho^{m-1}} \left(\frac{c\sigma}{4}\right)^{\frac{2-n}{4}} S_0^{\frac{n-2}{4}} = t_* S_0. \tag{22}$$

For the example considered above $Q_* = 10^7 \text{ kj cm}^{-2}$.

C. Solution with Time-Dependent Intensity

If the function f is not constant, one more dimensioned parameter appears in the problem, namely t_0, the characteristic time of irradiation or the pulse length. In this case the solution is of the following form:

$$T = T_* \left(\frac{x}{x_*}, \frac{t}{t_*}, \frac{t_*}{t_0}, f\right) = T_* \vartheta(\chi, \tau, \lambda, f) \tag{23}$$

that is, it depends also upon dimensionless parameter $\lambda = t_*/t_0$ and the form of the function f. Analysis of such a solution is possible only for a specific shape of the function f. Although numerical calculations indicate that the function is essentially bell-shaped, the rise is sufficient to approximate it in a step-like form,

$$f(t/t_0) = \begin{cases} 1/t_0 & \text{if } 0 < t < t_0 \\ 0, & \text{otherwise.} \end{cases} \tag{24}$$

From numerical calculations we have learned that at time t slightly later than t_0, changes in T and Q become negligible. There is some period between the time when the X-radiation of the surface has ceased and the fluid dynamic motion has not yet begun, during which values of T and Q change slightly. Then the fraction of the energy absorbed is

$$\frac{Q_\infty}{Q_0} = \frac{t_* S_0 \kappa (1/\lambda)}{S_0 t_0} = \lambda \kappa (1/\lambda). \tag{25}$$

Thus, the dependence of Q_∞, T_∞ on parameters of the problem is reduced to dependence on the single parameter λ. Corresponding functions of λ can be tabulated using numerical calculations.

Note that if we compare two different magnitudes of the intensity (two explosions of different yield) with the same properties of the medium, the characteristic pulse length changes according to the relation

$$t_0 \sim t_* \sim S^{\frac{n-6}{4}} \tag{26}$$

and the value of λ remains constant, so the fraction of the energy absorbed by the surface remains the same. For the example chosen earlier

$$t_* \sim S_0^{0.325}. \tag{27}$$

So if the yield of the explosive is increased by an order of magnitude, the pulse length (irradiation time) must be increased a factor of 2 in order to retain the same coupling of energy to the ground. For the yields of interest, the pulse length scales appropriately with yield so that approximately the same fraction of energy is coupled to the ground, independent of the yield.

D. Fraction of the Radiation Energy Absorbed

To obtain a quantitative assessment of the fraction of the energy that is absorbed, we use one of the main properties of heat wave, namely the approximate constancy of the temperature in the heated region $T = T_f$. The energy in the heated region is

$$E \simeq \rho C_v x_f T_f = \rho C_v T_f(t) \int_0^t D_f(t) \, dt \tag{28}$$

where $D_f = dx_f/dt$ is the penetration velocity of the heat wave. To estimate the velocity of the heat diffusion front, we use an approximate solution to the problem of propagation of a planar heat wave with a set prescription for the temperature at the boundary. Assume that the speed of the heat wave is determined by the average temperature behind its front T_0 and the thickness of the heated region.

Again consider the one-dimensional problem in slab symmetry. Multiply both sides of the equation for nonlinear heat conductivity

$$\frac{\partial T}{\partial t} = a\frac{\partial}{\partial x}\left(T^n\frac{\partial T}{\partial x}\right) \tag{29}$$

by x^2 and integrate on the coordinate x, replacing temperature profile $T(x)$ for some average temperature T_0. The right-hand side of Eq. (29) becomes

$$\int_0^{x_f} x^2 a\frac{\partial}{\partial x}\left(T^n\frac{\partial T}{\partial x}\right)\,dx = \frac{2a}{n+1}\int_0^{x_f} T^{n+1}\,dx = \frac{2a}{n+1}T_0^{n+1}x_f \tag{30}$$

and the left-hand side of Eq. (29) becomes

$$\int_0^{x_f} x^2\frac{\partial T}{\partial t}\,dx = \frac{d}{dt}\int_0^{x_f} Tx^2\,dx = T_0x_f^2\frac{dx_f}{dt} \tag{31}$$

from which we obtain

$$x_f D_f = \frac{2a}{n+1}T_0^n(t) \tag{32}$$

and integrating Eq. (32), we obtain

$$x_f = \sqrt{\frac{4a}{n+1}\int_0^t T_0^n(t)\,dt}. \tag{33}$$

Now we can divide Eq. (32) by Eq. (33) to find the dependence of the front velocity upon the average temperature behind the front

$$D_f = \frac{T_0^n(t)}{\sqrt{\frac{n+1}{a}\int_0^t T_0^n(t)\,dt}}. \tag{34}$$

According to our assumption, this expression approximately describes dependence of the velocity of the planar diffusion front on time, not only in the case of a planar explosion, but for an arbitrary time dependence of temperature behind the front. It is applicable to the problem of propagation of a plane heat wave in a semispace with an arbitrary temperature dependence at the boundary. Numerical calculations show that this relation is of limited accuracy, but in the case of increasing (or not rapidly decreasing) temperatures, it is good to within about ten percent. This is good enough to carry out our analysis.

As a simplification, consider the case of a step function in intensity:

$$S(t) = \begin{cases} S_0 = Q_0/\tau, & \text{if } 0 < t < \tau; \\ 0, & \text{otherwise.} \end{cases} \tag{35}$$

Assuming that the temperature in the heated region changes only slightly, we can write Eq. (34) as

$$D_f \simeq \frac{T_f^{n/2}}{\sqrt{\frac{n+1}{a}t}}. \tag{36}$$

As has been shown above, for a constant flux S_0 during the characteristic period of time $t \simeq t_0$ the temperature settles to its highest value T_*. If the pulse length $t_0 >> t_*$, it is permissible to use the quasi-stationary approximation; the highest temperature "follows" the value of the incident flux

$$T_f \simeq \left[\frac{4S(t)}{c\sigma}\right]^{\frac{1}{4}}. \tag{37}$$

Using this approximation for the temperature, the thickness of the heated layer is $h_b \simeq x_f$, where x_f is given by Eq. (33). If the pulse length τ exceeds the characteristic settling time τ_b, then the final depth of the heating h_b occurs at $t = \tau_b$. For the opposite case, it occurs at $t = \tau$. We can determine τ_b by setting the diffusion front velocity equal to the isothermal sound velocity,

$$C_T = \sqrt{A_s T_f} \tag{38}$$

in the heated region. So from Eqs. (36) and (38), we obtain

$$\tau_b = \frac{a T_f^{n-1}}{(n+1)A_s}. \tag{39}$$

Thus, at values of the flux S_0 corresponding to $\tau_b > \tau$, the energy absorbed by the ground is given by

$$Q_a = h_b \rho C_v T_f = \frac{\rho A_s}{\gamma - 1} \left(\frac{4Q_0}{\tau c\sigma}\right)^{\frac{n+2}{8}} \sqrt{\frac{4a\tau}{n+1}} \tag{40}$$

where we have used T_f given by Eq. (37) with $S(t)$ as given by Eq. (35), and have evaluated $h_b \simeq x_f$ by substituting Eq. (37), with $S(t)$ as given by Eq. (35), into Eq. (33), and integrating. We have also used $\varepsilon = C_v T$ with Eq. (5). The fraction of the energy absorbed by the ground is then

$$\frac{Q_a}{Q_0} = \frac{\rho A_s}{\gamma - 1} \left(\frac{4}{c\sigma}\right)^{\frac{n+2}{8}} \left(\frac{4a}{n+1}\right)^{\frac{1}{2}} Q_0^{\frac{n-6}{8}} \tau^{-\frac{n-2}{8}} \tag{41}$$

where we have simply divided Eq. (40) by Q_0. Using the parameters for aluminum to represent a typical chondrite, we find the characteristic time $\tau_b = 1.3 \times 10^{-14} S_0^{1.58}$ and the fraction of absorbed energy $Q_a/Q_0 \simeq 0.02 Q_0^{0.163} \tau^{-0.66}$. In this case the time τ_b is several hundredths of a microsecond (the characteristic time of surface irradiation) at a flux of $S_0 \simeq 10^8$ kj cm^{-2} μ s^{-1}.

The above relation for the absorbed energy fraction is reasonably accurate for intensities greater than 10^8 kj cm^{-2} μ s^{-1}. At an intensity of 10^8 to 10^9 kj cm^{-2} μ s^{-1}, the absorbed energy fraction is $\sim$0.1. At higher intensities, the solution is inaccurate, because the settled temperatures are so high that radiation energy begins to make an important contribution to the full energy and Eq. (9) cannot be used in lieu of Eq. (1).

For heavier and more opaque materials such as iron, the solution will be applicable over the whole range of intensities of practical interest. For iron asteroids the characteristic time is $\tau_b = 7 \times 10^{-13} S_0^{1.63}$ and the fraction of absorbed energy is $Q_a/Q_0 \simeq 4 \times 10^{-4} Q_0^{0.19} \tau^{-0.69}$. For the intensities under consideration, the fraction of absorbed energy for an iron asteroid appears to be approximately 2/3 that of a stony asteroid. From the dependence of the absorbed energy fraction on Rosseland mean range and equation of state, we would conclude that the nuclear explosive will couple much more efficiently to icy comets than to chondritic or nickel-iron asteroids. We are somewhat over-estimating the absorbed energy fraction, however, because after completion of irradiation, the temperatures in the heated region are rather high, and reradiation from the free surface results in an appreciable decrease of the absorbed energy.

Three stages of an above-the-surface explosion are shown in Fig. 1. In Plot a, X-rays from the exploding device heat a lenticular region of the ground below it and some of the energy is lost by radiation from this region. In Plot b, the hot plasma from the ground expands and, by conservation of momentum, a shock wave is driven downward. Plasma from the expanding device debris collides with the plasma from the ground. Plot c depicts the state late in the explosion, showing the regions of the ground that are vaporized, melted, crushed, and fractured.

Numerical calculations lead to slightly smaller values of the absorbed energy than these analytic estimates, which are near to the results described by Cooper et al. (1972). The approximate solution we have obtained enables us to estimate the dependence of the absorbed energy on the parameters characterizing the ground properties. For our range of applicability, the absorbed energy fraction is

$$\frac{Q_a}{Q_0} \sim \sqrt{\frac{l_0 A_s}{(n+1)(\gamma - 1)\rho^{m-1}}}. \qquad (42)$$

III. EFFECTS OF NUCLEAR EXPLOSION BELOW THE SURFACE OF THE ASSAILANT OBJECT

The effect on an asteroid or comet could be substantially increased if the nuclear explosive could be buried in the ground. In this case, a considerably larger part of the explosive energy is transferred to the ground material

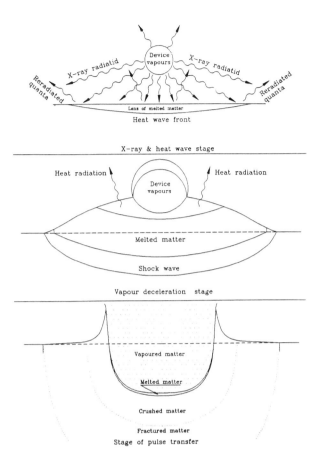

Figure 1.   Stages of a nuclear explosion detonated above the surface, but at fairly
low altitude. (a) X-rays from the exploding device heat a lenticular region of the
ground below it and some of the energy lost by radiation from this region. (b) The
hot plasma from the ground expands and, by conservation of momentum, a shock
wave is driven downward. Plasma from the expanding device debris collides with
the plasma from the ground. (c) Late in the explosion, showing the regions of the
ground that are vaporized, melted, crushed and fractured.

than for the explosion with some elevation above the surface. In the one-
dimensional case, when the nuclear device is surrounded by the ground, heat
wave propagation can be approximately described by the well-known self-
similar solution for an instantaneous point explosion (Zel'dovich and Raizer
1966). According to this solution, the heat wave "stops," when the fluid dynamic mechanism of energy transfer becomes dominant. For a one-megaton
device and ground similar to terrestrial rock, stopping would occur when the
heat wave is at a distance less than 2 m from the center of the explosion. The
stopping distance varies roughly as cube root of the yield. In the case of more
condensed and less transparent substances (iron, for instance), this distance

decreases considerably. If the ground layer thickness surrounding the nuclear charge exceeds the stopping distance, nearly all the energy of the explosion is transferred to the material of the comet or asteroid.

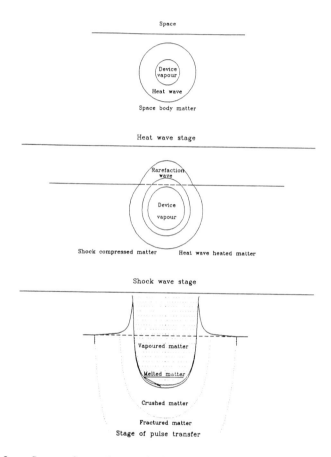

Figure 2.   Stages of a nuclear explosion detonated below the surface, but at a sufficiently shallow depth much of the surrounding material is ejected. (a) The heat wave stage, when thermal conductivity dominates the energy transport. (b) The detached shock wave, and the rarefaction wave above the surface. (c) A late stage in the explosions with various conditions of the ground material.

Three stages of a below-the-surface explosion are shown in Fig. 2. Plot a shows the heat-wave stage, when thermal conductivity dominates the energy transport. Plot b shows the detached shock wave, and the rarefaction wave above the surface. Plot c depicts a late stage in the explosions with various conditions of the ground material as in Fig. 1c.

If the thickness of surrounding layers is less than the stopping distance, then it is possible to estimate the energy transferred to an object by constructing a self similar solution for point explosion at the semispace boundary.

Consider the equation of nonlinear heat conductivity for the two-dimensional axisymmetric (cylindrical) problem:

$$\frac{\partial T}{\partial t} = \frac{1}{r}\frac{\partial}{\partial r}\left(raT^{n}\frac{\partial T}{\partial r}\right) + \frac{\partial}{\partial z}\left(aT^{n}\frac{\partial T}{\partial z}\right). \tag{43}$$

Let the energy be released instantaneously in some small region on the boundary of semispace $z = 0$. Let the temperature be zero $T(r, 0, t) = 0$ at the semispace surface. Multiplying Eq. (43) by z and integrating over the volume of the semispace we obtain

$$\int z\frac{\partial T}{\partial z}dV = \frac{d}{dt}\int zT\,dV = 0 \tag{44}$$

which implies that, during the heat wave propagation, the "temperature moment" is conserved; in other words,

$$\int zT\,dV = P = \text{const.} \tag{45}$$

It is more convenient to consider the problem in two-dimensional axisymmetric spherical coordinates:

$$\frac{\partial T}{\partial t} = \frac{1}{r^2}\frac{\partial}{\partial r}\left(r^2 aT^{n}\frac{\partial T}{\partial r}\right) + \frac{1}{r\sin\vartheta}\frac{\partial}{\partial\vartheta}\left(\frac{\sin\vartheta}{r}aT^{n}\frac{\partial T}{\partial\vartheta}\right). \tag{46}$$

There are only two-dimensioned parameters

$$[a] = \text{cm}^2\text{keV}^{-n}\mu s^{-1}$$
$$[P] = \text{keVcm}^4 \tag{47}$$

The single dimensionless combination, which can be composed of the radial coordinate r, the time t, and the parameters P and a, is

$$\xi = r\left(aP^{n}t\right)^{-\frac{1}{2(2n+1)}}. \tag{48}$$

The law of the diffusion front propagation appears at once:

$$r_f = \xi_0(\vartheta)\left(aP^{n}t\right)^{\frac{1}{2(2n+1)}}. \tag{49}$$

We now seek the solution, which should be in the form

$$T = At^{-\alpha}f(\xi, \vartheta) \tag{50}$$

where

$$\xi = \frac{r}{Bt^m} \qquad \text{and} \qquad m = \frac{1}{2(2n+1)}. \tag{51}$$

We obtain a relation between α and m by using Eq. (45)

$$\alpha = 4m = \frac{2}{2n+1}. \tag{52}$$

Setting $aA^n/B^2 = 1$, we obtain the equation for the function of representation

$$\alpha f + m\xi \frac{\partial f}{\partial \xi} + \frac{1}{\xi^2} \frac{\partial}{\partial \xi} \left(\xi^2 f^{\,n} \frac{\partial f}{\partial \xi} \right) + \frac{1}{\xi \sin \vartheta} \frac{\partial}{\partial \vartheta} \left(\frac{\sin \vartheta}{\xi} f^{\,n} \frac{\partial f}{\partial \vartheta} \right) = 0 \tag{53}$$

where

$$f(\frac{\pi}{2}, \xi) = 0, \qquad \text{and} \qquad f(\vartheta, \infty) = 0 \tag{54}$$

and we find expressions for A and B:

$$A = \left(\frac{P}{2\pi I_1 a^2} \right)^{\frac{1}{2n+1}} \tag{55}$$

$$B = \left(\frac{a^n P^n}{2\pi I_1} \right)^{\frac{n}{2(2n+1)}} \tag{56}$$

where

$$I_1 = \int \int f(\xi, \vartheta) \, \xi^{\,3} \cos \vartheta \, \sin \vartheta \, d\vartheta \, d\xi. \tag{57}$$

The ground energy in the region embraced by the heat wave is

$$E \sim \int T \, dV \sim t^{-\frac{n}{2(2n+1)}} \sim r_f^{-1}. \tag{58}$$

We find that the shock wave detaches from the heat wave. The spatial dependence the heat wave is given by $\xi_0(\vartheta)$. $f[\xi_0(\vartheta), \vartheta] = 0$. The location of the diffusion front is

$$r_f = \xi_0(\vartheta) Bt^m \tag{59}$$

and its velocity is

$$D_f = m\xi_0(\vartheta) Bt^{m-1}. \tag{60}$$

The approximate isothermal sound velocity in the heated region is

$$D_f = C_T = \sqrt{A_s \bar{T}} \tag{61}$$

where $\bar{T}$ is temperature averaged over the entire heated region

$$\bar{T} = At^{-4m} \bar{f}. \tag{62}$$

By equating the diffusion front velocity given in Eq. (60) to the sound velocity given in Eq. (61), we find the time at which the detachment occurs

$$t_T = \left\{ \left[\frac{\xi_0(\vartheta)}{2(2n+1)\sqrt{A_s \bar{f}}} \right]^{2(2n+1)} \frac{a^3 P^{n-1}}{(2\pi I_1)^{n-1}} \right\}^{\frac{1}{4n-1}} \tag{63}$$

and radius of detachment

$$r_T = \left[\frac{\xi_0^{4n}(\vartheta)}{2(2n+1)\sqrt{A_s \bar{f}}} \left(\frac{P}{2\pi I_1} \right)^{\frac{2n-1}{2}} a \right]^{\frac{1}{4n-1}}. \tag{64}$$

These results have some interesting qualitative consequences for deflecting or disrupting comets or asteroids with buried explosives. Let the center of an explosion having yield E_0 be at the depth $H_0 < r_{T0}$, where r_{T0} is the radius at which the heat wave stops. Then the temperature moment $P \sim T H_0^4 \sim E_0 H_0$ and

$$r_T \sim (E_0 H_0)^{\frac{2n^2-1}{(2n+1)(4n-1)}} \tag{65}$$

In general, for the ground composition, we will have $n \gg 1$, so

$$r_T \sim (E_0 H_0)^{1/4}. \tag{66}$$

We can also estimate the fraction of the energy transferred to the object from the energy remaining in the medium $E_* \sim r_T^{-1}$. This depends on the thickness of the layer surrounding the nuclear explosive, and can be approximated by using Eq. (66) to give

$$\frac{E_*}{E_0} \simeq \frac{H_0}{r_T} \sim \left(\frac{H_0^3}{E_0} \right)^{\frac{1}{4}}. \tag{67}$$

Of course, E_*/E_0 may not exceed 1, and $E_*/E_0 \simeq 1$ at $H_0 \simeq r_{T0}$.

IV. DEFLECTING OR FRAGMENTING THE ASTRAL ASSAILANT WITH A NUCLEAR EXPLOSIVE

We are interested in two effects: (1) changing of trajectory of the assailant; and (2) causing the assailant object to break up.

A. Size Limitations

We can obtain an estimate of maximum dimensions of an object that can be influenced by a nuclear explosion as follows. Let an explosion occur on the surface of an assailant object having mass m_a and radius R_a. Assume the mass m of the material ejected by the explosion is small compared with the object mass. The ejected mass has a distribution of velocities, but for simplicity, we will assume that all the ejected material has the same velocity, which represents an average $\bar{v}_m$ of the actual velocities. The ejection velocity, according to data published by Knowles and Brode (1977), is on the order of 100 m s^{-1}. This is a conservative estimate, and certainly varies somewhat with explosive yield and material properties. But for this analysis, we will assume it to be relatively constant.

Assuming $m << m_a$, conservation of momentum tells us that

$$m\bar{v}_\infty = m_a \Delta v_a \tag{68}$$

where $\bar{v}_\infty$ is the velocity of ejected material "at infinity" and Δv_a is velocity gained by the object as a result of the explosion. Similarly, conservation of energy tells us that

$$-G\frac{mm_a}{R_a} + \frac{m\bar{v}_m^2}{2} = \frac{m\bar{v}_\infty^2}{2} + \frac{m_a(\Delta v_a)^2}{2} \tag{69}$$

where G is gravitational constant. Combining Eqs. (68) and (69), we obtain

$$\bar{v}_\infty = \bar{v}_m \sqrt{1 - \frac{2Gm_a}{\bar{v}_m^2 R_a}}. \tag{70}$$

An obvious implication of Eq. (70) is that if our astral assailant is sufficiently massive, and we are limited in the velocity we can impart to the ejected material, the material will just fall back to the surface of the object (or at least remain in orbit around it) and, baring interaction with other objects, the assailant will resume its original trajectory. Let us define a critical radius for the object such that $\bar{v}_\infty = 0$. At this radius our efforts are futile. Then from Eq. (70)

$$R_{acrit} = \bar{v}_m \sqrt{\frac{3}{8\pi\rho_a G}} \tag{71}$$

where ρ_a is the average density of the object and we have used $m_a = 4\pi R_a^3 \rho_a / 3$. Similarly,

$$m_{acrit} = \frac{\bar{v}_m^3}{2G}\sqrt{\frac{3}{8\pi\rho_a G}}. \tag{72}$$

At $\rho_a \sim 5$ g cm^{-3}, $\bar{v}_m = 100$ m s^{-1} critical radius of the body $R_{acrit} \sim 60$ km and mass $m_{acrit} \sim 5 \times 10^{18}$ kg.

B. Orbital Deflection

We now estimate the change in the trajectory of the object as a result of a nuclear explosion. We assume that the body dimensions are small compared with the critical dimensions given above, in other words, $v_\infty \simeq v_m$. Assume the object is moving in the gravitational field of the Sun and that the orbital plane is unaltered by the explosion. The elliptical orbit of the object in polar coordinates is

$$r = \frac{p_0}{1 + \epsilon_0 \cos \phi}. \tag{73}$$

The eccentricity is given by

$$\epsilon_0 = \sqrt{1 + \frac{2E_a L_a^2}{\mu \alpha^2}} \tag{74}$$

where L_a is angular momentum relative to the system mass center, E_a is the energy of the assailant object, $\mu \simeq m_a$ is the reduced mass, $\alpha = Gm_a m_s$, $m_s = 2 \times 10^{30}$ kg is the Sun mass. The remaining coefficient in Eq. (73) is given by

$$p_0 = \frac{L_a^2}{\mu \alpha}. \tag{75}$$

Assume that the explosion occurs at the point with coordinates $\{r_0, \phi_0\}$. As a result of the explosion the object's trajectory

$$r = \frac{p_0}{1 + \epsilon_0 \cos(\phi_0 + \beta)} \tag{76}$$

changes to

$$r' = \frac{p_0 + \Delta p}{1 + (\epsilon_0 + \Delta \epsilon) \cos(\phi_0 + \Delta \phi + \beta)} \tag{77}$$

From Eqs. (76) and (77), we find that for small values of Δp and $\Delta \epsilon$,

$$\Delta \phi \simeq -\frac{\Delta p}{r_0 \epsilon_0 + \sin \phi_0} + \frac{\Delta \epsilon}{\epsilon_0} \tan^{-1} \phi_0. \tag{78}$$

Then the deviation of the trajectory at a distance from the Sun about the same as the radius of the Earth's orbit r_e is

$$\Delta r = r' - r = r_e \left[\frac{\Delta p}{p_0} \left(1 - \frac{r_e \sin(\phi_0 + \beta)}{r_0 \sin \phi_0} \right) + \frac{\Delta \epsilon r_e \sin \beta}{p_0 \sin \phi_0} \right]. \tag{79}$$

Differentiating Eq. (75) we obtain

$$\frac{\Delta p}{p_0} = 2 \frac{\Delta L_a}{L_a}. \tag{80}$$

Differentiating Eq. (74) and substituting Eq. (75) we obtain

$$\Delta \epsilon = \frac{p_0 E_a}{\alpha \epsilon_0} \left(\frac{\Delta E_a}{E_a} + 2 \frac{\Delta L_a}{L_a} \right). \tag{81}$$

And finally, substituting Eqs. (80) and (81) into Eq. (79) we obtain

$$\frac{\Delta r}{r_e} = \frac{2\Delta L_a}{L_a}\left(1 - \frac{r_e \sin(\phi_0 + \beta)}{r_0 \sin \phi_0}\right) + \frac{1}{\epsilon_0}\frac{E_a r_e}{\alpha}\left(\frac{\Delta E_a}{E_a} + 2\frac{\Delta L_a}{L_a}\right)\frac{\sin \beta}{\sin \phi_0}.$$

(82)

The energy of the object before the explosion was

$$E_a = \frac{m_a v_a^2}{2} - \frac{\alpha}{r_0}.$$

(83)

The change in its energy as the result of the explosion is

$$\Delta E_a = -m_a v_a \Delta v_a.$$

(84)

The object's initial angular momentum was

$$L_a \sim m_a v_a r_0.$$

(85)

and the change in angular momentum as the result of the explosion is

$$\Delta L_a \sim -m_a \Delta v_a r_0.$$

(86)

Thus the net deflection of the astral assailant is

$$\frac{\Delta r}{r_e} = -2\frac{\Delta v_a}{v_{ae}}(n - \cos \beta)$$

(87)

where $n = r_0/r_e$, and v_{ae} is velocity at the Earth's orbital radius. The parameter $\cos \beta$ can be expressed in terms of the orbit parameters

$$\cos \beta = \frac{1}{\epsilon^2}\left\{\left(\frac{k}{n} - 1\right)(k - 1) + \left[\epsilon^2 - \left(\frac{k}{n} - 1\right)\right]^2 \left[\epsilon^3 - (k - 1)^2\right]^{\frac{1}{2}}\right\}$$

(88)

where $k = v_{ae}^2/(GM_e/r_e)$.

As an example, we consider an assailant with $v_{ae} = 30$ km s^{-1}, $\epsilon_0 = 0.5$ (n can vary from 0.67 to 2.0). Figure 3 shows the dependence of $\Delta r/R_e$ on $n = r_0/r_e$. The largest value of the deflection Δr occurs at $n = 2.0$, when

$$\frac{\Delta r}{r_e} = -4\frac{\Delta v_a}{v_{ae}} = -4\frac{m}{m_a}\frac{v_a}{v_{ae}}$$

(89)

or

$$\frac{\Delta r}{R_e} = -4\frac{m}{m_a}\frac{v_a}{v_{ae}}\frac{r_e}{R_e}.$$

(90)

Assuming that the nuclear explosive has a yield of 1 Mt and is detonated on the surface, ejecting about a million tons of the ground material at a velocity of 100 m·s^{-1}, we are lead to the following conclusion; it is possible to perturb

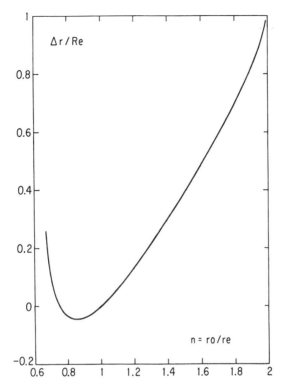

Figure 3.   Radial deflection of an astral assailant versus the radius at which it is intercepted.

the object's trajectory by a distance on the order of one Earth radius only if the object's mass is less than 3×10^{11} kg, i.e., it has a radius smaller than about 300 m.

Thus, using a 1-Mt nuclear explosive, the analysis shows it is possible to defend the Earth against 300-m astral assailants, even though such objects would have an impact energy equivalent to 10^4 Mt TNT. In this case, the detonation must occur when the object is quite far from the Earth, at a distance about equal to the Earth's orbital radius. Such a defense involves several formidable technical challenges: (1) detecting the objects; (2) identifying them as dangerous; and (3) intercepting them and delivering the nuclear explosives to their surface. We would like to point out that the problem of deflection is quite independent of the other problems, and the analysis we have described is rather conservative.

C. Fracture and Fragmentation

The zone of intensive crushing caused by a megaton explosion above the surface of an assailant object can be comparable with dimensions of the object itself. The explosion will not only change the object's energy and momentum, but can break the object into several large fragments, which are dangerous by themselves. Fragments blown from the surface could also inhibit the tactic of using multiple nuclear explosions to achieve the desired deflection, as these fragments will make it difficult for subsequent spacecraft to reach the surface. Clearly the extent to which fragmentation is an impediment to effective deflection is critically dependent on the composition of the object. If it is nickel-iron, it could be expected to be quite cohesive, and not break up easily. If it is a chondrite, the composite rocks may be very weakly bound. If there is danger of deleterious fragmentation, it will be preferable to blast the object to smithereens, each of which poses no particular threat as they would dissipate most of their energy penetrating the Earth's atmosphere.

The typical dimension of the crushed, or pulverized, zone for an underground explosion under terrestrial conditions is $r = 300$ m $Mt^{-1/3}$ for rocks and 600 to 900 m $Mt^{-1/3}$ for ice. Dimensions of the crushed zone obtained in the calculations for terrestrial conditions can be significantly less than for celestial bodies because: (1) the results of calculations for the explosion above a spherical surface of relatively small radius can strongly differ from calculations for an explosion above the surface of the semispace; and (2) the influence of low temperatures on strength of the assailant object material can also be important.

We can consider the possibility of crushing objects of considerably larger dimensions and mass (perhaps with radii of several kilometers) using a series of explosions on the surface. It would be necessary to determine their optimum number, yield, and relative location. For these purposes, nuclear devices with a special configuration can be used, but they will require making arrangements for landing on the surface of the object.

For fragmentation, the factors most crucial to the protection of the Earth are not the changes of the trajectory parameters, but the possibility of fracturing the astral assailant with nuclear explosions, the dimensions of fragments formed, the dispersal of the fragments in space up to that moment when they reach the Earth's atmosphere, the likelihood these fragments will burn out in the atmosphere, and the ecological consequences of such energy release in the atmosphere, or of the dispersed fall of the fragments on the surface of the Earth. It is important to recognize that the amount of the material that gets into the atmosphere from an object that is pulverized and dispersed is comparable with the amount of material thrown out during the eruption of a volcano of medium size, and the average density of particles flow is about 10^{-7} to 10^{-6} g cm^{-3}, if the explosion takes place at the distances of about 0.01 AU.

IV. CONCLUSIONS

Defending the Earth from the potentially catastrophic impact of comets or asteroids is within the scope of current science and technology. Nuclear explosive devices having the yield of more than 1 Mt can be used for fragmentation or deflection of dangerous objects. The effectiveness of an explosion depends on relative position of the nuclear device and the object. The easiest case to arrange is an explosion above the surface of the object. Explosion of a nuclear device very near the surface may require rendezvous, speed-matching and landing on the object. A subsurface explosion may require not only landing, but some sort of drill or penetrating apparatus to emplace the device. It might be possible to emplace the device with an impact penetrator, but such schemes are rather weighty and require detailed knowledge of the material composition of the object being penetrated. In all cases, the explosion effectiveness depends on the individual characteristics of the object, therefore it is desirable to have some preliminary knowledge of its shape, structure, and constituent materials. If time permits, this might be provided by a vanguard spacecraft.

A key issue for defending the planet is selection of method by which the threatening object will be neutralized. To deflect the object requires accurate determination of its orbital parameters. For example, at distance of 1 AU it is necessary to determine the speed within an accuracy of 10^{-4} to 10^{-5}. Furthermore, to deflect an object having a radius of 300 m at such a distance may require using a nuclear device having yield of ~ 1 Mt. An explosion of this yield can split the object into several pieces, each of which could be dangerous if it strikes the Earth. The large chunks will likely be surrounded by a cloud of small fragments, which will make it exceedingly difficult for a second spacecraft to get near enough to emplace another explosive. In view of these difficulties, it may be prudent to plan, from the very beginning, to shatter the object into small fragments that will be unable to penetrate the Earth's atmosphere. If a dangerous, friable, nonmetalic object of medium dimensions ($r = 300$ to 500 m) is first detected at small distances from the Earth (0.1 to 0.01 AU), shattering it into many fragments may be the only recourse. Under these circumstances it may be necessary to use a device of 1 to 100 Mt yield, depending on the space object's material composition and the nuclear explosive location relative to its surface. The consequences of a pulverized object entering the Earth's atmosphere can be compared with the consequences of ash and dust thrown out during the eruption of a volcano.

The complexity and cost of the of an astral assailant defense system make it unlikely that it will be built by one nation. It must be an international program of research with participation from all the nuclear-club countries as well as all other concerned nations of the world. The problem cannot be solved without using the latest information both in the sphere of nuclear explosive technology, and in the sphere of missile propulsion, guidance, and targeting technology. Most of this information is related to defense and is therefore likely to be classified or otherwise protected from general dissemination. Such protection could make collaborative enterprises very awkward. The way out of this situation could be to distribute the larger problems among separate countries and invoke international verification and control over the application of the resulting systems. We are keenly aware of the danger posed

by potential misuse of such a defensive system, and stringent safeguards must
be imposed.

This chapter has been concerned with the use of powerful explosions for
pulverizing, fragmenting, or deflecting large asteroids and comets on collision
course with the Earth. It is surely not an exhaustive analysis. In subsequent
works it is appropriate to consider in detail and describe quantitatively all the
mechanisms of energy transfer to such bodies, and the dependence of these
processes on properties of the materials of which the objects consist, on the
relative position of the device and the object surface, etc. With reasonable
completeness and wishing not to overload the text, we intended to specify
and illustrate the main mechanisms by which powerful nuclear explosive
devices can transfer energy to space objects, to point out the dependence of
these processes on the properties of the substances that are constituents of
such objects, and to characterize qualitatively the main processes influencing
the effectiveness with which such bodies could be deflected or destroyed.
But even with such cursory examination, we believe the following general
conclusion is convincing; by using nuclear explosions it is possible to alter
significantly the trajectories of such objects, or to change their state, which
will make it possible to prevent or substantially mitigate the consequences of
a collision with the Earth.

REFERENCES

Alekseev, A. S., Zeilik, B. S., and Voronin, Yu. A. 1990. *The Problems of Analyzing
of Circle Structures and the Problem of Space Protection of the Earth*, Siberian
Division of Academy of Science of the USSR, Computational Center, Preprint
No. 912 (Novosibirsk).

Alvarez, L. W., Alvarez, W., Asaro, F., and Michel, H. 1980. Extra-terrestrial cause
for the Cretaceous-Tertiary extinction. *Science* 208:1095–1108.

Avrorin, E. N., Vodolaga, B. K., Simonenko, V. A., and Fortov, V. E. 1990 *Powerful
Shocks and Extreme States of Matter* (Moscow: IVTAN Edition).

Bjork, R. L. 1961. Analysis of the formation of Meteor Crater, Arizona: A preliminary
report. *J. Geophys. Res.* 66:3379–3387.

Cooper, H. F., Jr., Brode, H. L., and Leigh, G. G. 1972. Some Fundamental Aspects of
Nuclear Weapons. Technical Report No. AWFL-TR-72-19 (Kirtland Air Force
Base, N. M.: Air Force Weapons Laboratory).

Hills, J. G., and Goda, M. P. 1993. The fragmentation of small asteroids in the
atmosphere. *Astron. J.* 105:1114–1144.

Khryanina, L. P. 1987. *Meteorite Craters on the Earth* (Moscow: Nedra).

Khryanina, L. P., and Zeilik, B. S. 1980. Geological structure of Shunak Crater (near
Balkhash Lake) and its evidences of meteorite impact. *Izvestia of AS USSR, Geol.
Ser.* 3:124–134.

Knowles, C. P., and Brode, H. L. 1977. The theory of cratering phenomena, an
overview. In *Impact and Explosion Cratering*, eds. D. J. Roddy, R. O. Pepin and

R. B. Merrill (New York: Pergamon Press), pp. 869–895.

Krinov, E. 1966. *Giant Meteorites* (New York: Pergamon Press).

Masaitis, V. L., Mikhailov, M. V., and Silivanovskaya, T. V. 1975. *Popigai Meteorite Crater* (Moscow).

Sikes, L. R., and Davis, D. M. 1987. The yields of Soviet strategic weapons. *Sci. Amer.*, pp. 29–37.

Morrison, D. 1990. Target Earth. *Sky & Tel.* 79:261–265.

Zeilik, B. S. 1991 *Shock-Blast Tectonics and Short Essay on Plate Tectonics* (Alma-Ata: Gylym).

Zel'dovich, Ya. B., and Raizer, Yu. P. 1966. *Physics of Shock Waves and High-Temperature Hydrodynamic Phenomena* (Moscow: Nauka).

THE COUPLING OF ENERGY TO ASTEROIDS AND COMETS

B. P. SHAFER, M. D. GARCIA, R. J. SCAMMON, C. M. SNELL and
R. F. STELLINGWERF
Los Alamos National Laboratory

J. L. REMO
Quantametrics Inc.

and

R. A. MANAGAN and C. E. ROSENKILDE
Lawrence Livermore National Laboratory

This chapter considers the coupling of energy to Earth-threatening asteroids or comets for the purposes of collision avoidance or collision damage mitigation through trajectory modification. The sources of energy considered are pulsed lasers, kinetic energy impactors, and nuclear (or conventional) explosives, detonated above or below the surface of the target object. Our study does *not* assess the feasibility of delivering a particular energy source to the target or address the multitude of system issues involved with delivery systems. Specifically, our study is limited to a determination of the impulses delivered to the target by the non-nuclear and nuclear energy sources listed above.

I. INTRODUCTION

The purpose of this chapter is to describe the coupling of energy to Earth-threatening asteroids or comets for the purposes of collision avoidance through trajectory modification, or failing that, collision damage mitigation. The sources of energy we will consider are listed in Table I. When the problem of diverting potentially hazardous near-Earth objects (NEOs) was considered initially, it was only natural that strategies that employ nuclear explosives be considered. "Peaceful" uses of nuclear explosives have been studied by the Nuclear Design Laboratories for decades. The reason for this is clearly illustrated in Table II, which lists specific energies (MJ kg^{-1}) of non-nuclear and nuclear energy sources. Nuclear explosives offer million-fold increases in energy content per unit mass. In the future, if it proves feasible to store and deliver sufficient quantities of antimatter to a NEO, specific energies could result that are hundreds or thousands of times greater than the specific energies delivered by conventional nuclear explosives.

TABLE I

Energy Sources

Pulsed lasers
Impact kinetic energy
Nuclear (or conventional) explosives in several delivery
or emplacement modes, i.e.,
Standoff (flyby)
Shallow burial (surface penetrator)
Deep burial (mining required)

TABLE II

Specific Energies

High explosive	6 MJ kg^{-1}
Kinetic energy (10 km s^{-1})	50 MJ kg^{-1}
Nuclear explosive	$4 \times 10^6 \text{ MJ kg}^{-1}$

In the following discussion, we attempt an even-handed treatment of non-nuclear and nuclear energy delivery systems. Clearly, non-nuclear diversion strategies have much less of an environmental and geopolitical downside than strategies that employ nuclear explosives. However, an objective consideration of energy coupling is a necessary input to any effort to design practical and effective strategies for the diversion of all classes of NEOs that put our planet at unacceptable risk.

This study is limited primarily to the impulses delivered to a NEO by the non-nuclear and nuclear energy sources listed in Table I. In particular, we will *not* assess the feasibility of delivering a particular energy source to the target. Clearly, the relative difficulties of energy delivery for each of the sources in Table I vary widely. For example, NEO diversion with a pulsed laser would likely use a groundbased laser, whereas a deeply buried nuclear explosive would require a rendezvous with the NEO and subsequent mining and emplacement. For the first case, the greatest challenge is the development of a laser with sufficient energy to propagate through the Earth's atmosphere with enough fluence remaining to divert an Earth-threatening NEO. In contrast, the energy source for the second case is well developed and by far the major challenge is the rendezvous and emplacement of the nuclear explosive. Clearly, any strategy for NEO diversion must consider the feasibility and relative difficulty of the delivery of a particular energy source to the target. However, for this chapter, we will put aside questions of feasibility and relative difficulty, and confine ourselves to studies of the impulse coupling efficiencies of the energy sources listed in Table I.

We now turn to the study of energy deposition and impulse production in NEOs. The five energy delivery mechanisms listed in Table I are depicted schematically in Fig. 1. The five mechanisms involve the deposition of photons, atomic particles, or solid projectiles in the target NEO. For photons and atomic particles, the ratio of particle energy to particle momentum is quite

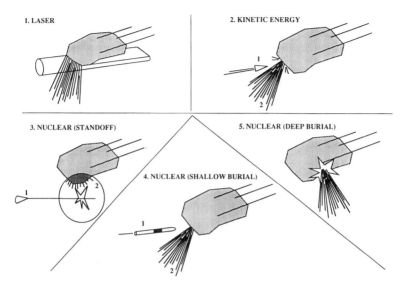

Figure 1.   Energy delivery mechanisms.

large and target impulse is primarily the result of energy deposition and sub-sequent target heating and expansion. For solid projectiles at low velocities, the ratio of projectile energy to projectile momentum is usually much lower and the impulse delivered to the target is the result of straightforward mo-mentum transfer. However, as the projectile velocity increases beyond a few km s^{-1}, energy deposition begins to dominate, as in the case of photons and atomic particles, and target impulse is primarily the result of target heating and expansion. In summary, for each of the five cases, with the exception of low-velocity, solid projectile impacts, energy deposition is followed by the expansion and blowoff of target material, resulting in the application of a recoil impulse to the fraction of the target remaining.

In summary, another matter that we must consider carefully is the relative danger to Earth from any proposed mitigation system. Depending upon the mass, composition, orbital parameters, and intercept distance of the target NEO, very large quantities of energy might be required for mitigation. In particular if a target NEO were sufficiently large and the warning time were suffi-ciently short, tens or even hundreds of megatons of TNT equivalents might be required. Such large energy stores may constitute a significant hazard in themselves, whether nuclear or non-nuclear. In particular, if a large, repet-itively pulsed laser were chosen for the mitigation mission, it would have a significant potential as a directed energy weapon against targets on or in the vicinity of the Earth. The hazard becomes particularly serious if the energy source for mitigation is a nuclear explosive.

In order to safeguard the general public from intentional or accidental misdirection of any mitigation system, we would propose the use of well-established and successful concepts and procedures employed for decades within the Department of Defense and the Department of Energy for the safety and security of nuclear and conventional weapon systems. A very careful risk/consequences analysis should be performed to determine the extent and cost of the safeguards system required for any mitigation concept deployed. An example here might be helpful. For the case of high-yield nuclear explosives, there are inherently robust functional safeguards that help to preclude unintentional full-yield nuclear detonations. In particular, the full-yield detonation of the nuclear fission trigger is required to enable the function of the high-yield thermonuclear stage. Accordingly, nuclear explosive safeguards focus on preventing unintentional, high-yield detonation of the nuclear fission triggers. Extensive effort is devoted to the design and testing of highly reliable systems containing redundant safety features to ensure that any nuclear detonation that occurs is intentional.

II. GROUND RULES, ASSUMPTIONS, AND CAVEATS

In the discussion that follows, we present estimates of the impulse coupling coefficients for each of the five energy delivery mechanisms depicted in Fig. 1. The impulse coupling coefficient C is defined as the ratio of the target recoil impulse (or impulse per unit area) to the energy (or energy per unit area) incident upon the target. We propose to define a nondimensional impulse coupling coefficient as shown in Fig. 2. Note that the coupling coefficient in common usage C and the newly defined coupling coefficient I^* have the same numerical values.

Common Usage (Mixed Units)

$$C = I/E \quad = \frac{dyne \cdot s/cm^2}{J/cm^2} \qquad \text{(Specific Energy)}$$

$$= \frac{dyne - s}{J} \qquad \text{(Total Energy)}$$

Proposed Change

$$C = I/E \quad = \frac{dyne \cdot s}{J} \quad = \frac{N \cdot s}{J} \times 10^{-5}$$

$$\text{Define } V_{REF} = 10^5 m/s \quad = 100 \text{ km/s}$$

Then define a nondimensional impulse

$$\boxed{I^* \equiv C \times V_{REF} \quad = \frac{I \times V_{REF}}{E}}$$

Numerical values of C (in old units) and I^* are identical.

Figure 2.   Impulse coupling coefficients.

In general, accurate estimates of impulse coupling coefficients require carefully controlled experiments or high-fidelity hydrocode simulations. Depending upon circumstances, either one-, two-, or three-dimensional hydrocode simulations may be required. Usually, however, simplified, empirically calibrated models are used to estimate the impulse coupling coefficients. This approach is usually adequate as long as the model is applied within the calibration range, or not far removed from it. Unfortunately, for some of the situations in this study, impulse coupling values are desired at energy levels that exceed our experimental knowledge base. Therefore, when we are required to extend our coupling estimates beyond their ranges of strict validity, we expect reduced confidence in our coupling estimates. Finally, we should also note that error estimates for the coupling coefficients are often not easily determined.

III. SIMPLIFIED ANALYSIS OF NEO DEFLECTION

Before examining each energy delivery mechanism in detail, we consider the general question of NEO deflection. A simple order-of-magnitude estimate of the specific impulse required to deflect an asteroid from its orbit around the Sun can be obtained from elementary dimensional considerations. The success of this approach depends upon the fact that the underlying dynamical phenomenon, namely planetary motion around the Sun, has an associated characteristic set of physical scales. The gravitational potential depends upon the solar mass $M_\odot$ ($M_\odot \equiv M_S$), the characteristic scale of distance R which is the mean radius of the Earth's orbit (1 AU), and these, together with the universal gravitational constant G determine a characteristic unit of time $\sqrt{R^3/GM_S}$, which is $1/2\pi$ of a year. (Strictly speaking in the two-body problem, the mass in this time scale should be $M_S + m$, but the mass m of the asteroid is negligible in comparison with the solar mass.)

Now, let us suppose that the asteroid of interest is following an elliptical or hyperbolic orbit about the Sun. At a certain instant in time, t_0, let an impulsive force $F_A\delta(t - t_0)$ be applied to the asteroid in some (as yet unspecified) direction $\hat{n}$. The magnitude of this impulse will be

$$I_A = \int_{-\infty}^{t>t_0} F_A\delta(t-t_0)\mathrm{d}t = F_A\sqrt{\frac{R^3}{GM_S}} = \Gamma F_G\sqrt{\frac{R^3}{GM_S}} = \Gamma m\sqrt{\frac{GM_S}{R}} \quad (1)$$

which has been rewritten in terms of the characteristic scales of the motion by expressing the impulsive force, $\Gamma = F_A/F_G$, nondimensionally in units of the gravitational force, $F_G = GM_S m/R^2$. Clearly, the ratio of these forces also represents, and is equivalent to, a nondimensional specific impulse, which is taken here to mean impulse per unit mass, of the form:

$$\Gamma = \frac{I_A/m}{\sqrt{GM_S/R}} \quad (2)$$

where the characteristic unit of specific impulse for planetary motion has the magnitude

$$\sqrt{GM_S/R} \approx 3 \times 10^6 \text{ dyne s g}^{-1} = 30 \text{ km s}^{-1}. \qquad (3)$$

It is instructive to digress briefly to elaborate on various definitions of the term "specific impulse." The nondimensional ratio involving an impulse per unit mass, which we have just introduced, appears naturally in our analysis of orbital deflection. However, we are aware that there are several other definitions in common use for the term, specific impulse. In the case of rocket thrust, this term is associated with:

$$\text{specific impulse} = \frac{\text{exhaust velocity}}{\text{local grav. accel.}} = \frac{[v]}{[g]} = \frac{[l/t]}{[l/t^2]} = [t] = \text{time} \qquad (4)$$

which has the dimensions of time. In the case of material ablation using a laser, one commonly compares the impulse achieved with the area exposed and identifies:

$$\text{specific impulse} = \frac{\text{impulse}}{\text{area}} = \frac{[Ft]}{l^2} = [Pt] = \text{pressure} \times \text{time} \qquad (5)$$

and its associated cgs-unit, 1 tap = 1 dyne s cm^{-2} = 1 g cm^{-1}s^{-1}. Alternatively, one may wish to compare this impulse achieved per unit area with the energy absorbed per unit area, whereupon one identifies:

$$\text{specific impulse} = \frac{\text{impulse/area}}{\text{fluence}} = \frac{[Ft/l^2]}{[E/l^2]} = \frac{[t]}{[l]} = \frac{1}{\text{velocity}} \qquad (6)$$

which has the dimensions of reciprocal velocity. Each of these measures of impulse has its utility within a specific technical community and application. On this account, it is useful to express any conclusions in several of these alternative ways in order to communicate the widest possible degree of understanding.

Now let us return to the original discussion. We assert that when the dynamical problem is expressed in these characteristic scales, the equations for the perturbed motion will depend, linearly, upon *only one small* nondimensional parameter, namely the ratio Γ. (Of course, the solution of the second-order ordinary differential equations of orbital motion also must depend upon two constants of the motion, the angular momentum l, and the total energy e. However, the magnitudes of these parameters, when expressed as nondimensional ratios in the appropriate characteristic units will be of order unity.) It immediately follows that any consequence of the perturbation must also depend upon this ratio and have its magnitude in these characteristic units. In particular as a result of the perturbation, we expect the asteroid's new orbit to cross the Earth's orbit at a slightly different point than it would have without the impulsive perturbation. This small linear displacement Δs, corresponds to a small angular displacement, $\Delta \varphi$, along the Earth's orbit, which being a nondimensional quantity derived from the motion, *must be of the order-of-magnitude of* Γ. Thus, we assert that apart from factors of order unity (which, of course, cannot be determined by these elementary considerations), the asteroid must be deflected through a distance on the order of the Earth's diameter, $2R_\oplus$, $2R_\oplus \equiv 2R_e$, (but, of course, henceforth we shall omit the factor of 2 in keeping with the spirit of this argument), which in nondimensional units means that

$$\frac{R_e}{R} \approx \frac{\Delta s}{R} \approx \Delta \varphi \approx \Gamma = \frac{I_A/m}{\sqrt{GM_S/R}}. \tag{7}$$

This implies that a specific impulse (i.e., impulse per unit mass) of the order-of-magnitude of

$$\frac{I_A}{m} \approx \frac{R_e}{R} \sqrt{\frac{GM_S}{R}} \approx 1 \text{ m s}^{-1} \tag{8}$$

would be required to achieve the desired spatial displacement.

Moreover, temporal displacements also must be governed by similar considerations. A change in the period of revolution ΔT and, therefore, the angular advance of the asteroid along its orbit in comparison with the advance of the Earth along its orbit, when expressed in terms of the time unit, $\sqrt{R^3/GM_S}$, also must be of the order of magnitude of Γ. This argument can be made a little more precise by estimating the change in the average orbital velocity, $\sqrt{GM_S/R}$ which then would be required to advance (or retard) the asteroid by an interval of time equivalent to that required to transit the diameter of the Earth. This time interval is

$$\frac{2R_e}{\sqrt{GM_S/R}} = \frac{2R_e}{R} \sqrt{\frac{R^3}{GM_S}} \approx 430 \text{ s.} \tag{9}$$

We shall estimate the change in velocity over one orbital period $T = 2\pi \sqrt{R^3/GM_S}$ by combining the differentials:

$$\Delta V \approx -\frac{1}{2} \frac{\Delta R}{\sqrt{R^3/GM_S}} \text{ and } \Delta T \approx 3\pi \frac{\Delta R}{\sqrt{GM_S/R}} \tag{10}$$

to get

$$\Delta V \approx -\frac{1}{6\pi} \frac{\sqrt{GM_S/R}}{\sqrt{R^3/GM_S}} \Delta T \approx -\frac{1}{3\pi} \frac{R_e}{R} \sqrt{\frac{GM_S}{R}} \approx -10 \text{ cm s}^{-1} \tag{11}$$

where we have taken for the change in the period of revolution the previous estimate for the interval of time to transit across the Earth's diameter. We have included additional numerical factors here to distinguish between short-term, evasive deflection on the scale of weeks or months and longer-term, more gentle deflection, which may be feasible with more advance warning on the scale of a year or a decade. Indeed, with advance warning of an impending disaster, by perhaps $n > 10$ orbital revolutions, a specific impulse, which has been reduced by this factor n and applied earlier in time, would achieve the same total angular displacement after that number of revolutions, because

$$\frac{n\Delta T}{\sqrt{R^3/GM_S}} \approx n\Gamma \approx \Delta\varphi \approx \frac{R_e}{R} \implies \frac{I_A}{m} \approx \frac{R_e}{nR}\sqrt{\frac{GM_s}{R}}. \qquad (12)$$

In this manner, the required specific impulse or velocity change could be reduced perhaps by yet another order of magnitude to something as low as 1 cm s^{-1}.

As indicated earlier, we have adopted a particular definition for the term, specific impulse, namely the impulse delivered per unit mass of the asteroid. However, once it has been established that the only small parameter in the analysis must have a magnitude of $\Gamma \approx R_e/R = 4 \times 10^{-5}$ (or lower), other measures of specific impulse also can be estimated. For example, within the rocket thrust community, one might want to compare the imparted velocity to the local gravitational acceleration of the asteroid:

$$\frac{\Delta V}{g_A} \approx \Gamma\sqrt{\frac{GM_S}{R}} \Big/ \frac{Gm}{R_A^2} \approx \frac{R_e}{R}\sqrt{\frac{GM_S}{R}} \Big/ G\rho_A R_A \approx 10^4 \text{ s} \qquad (13)$$

where we arbitrarily have chosen a 1-km diameter asteroid having a mean density of 3000 kg m^{-3}. (We also have omitted a divisor of $4\pi/3$ from the above estimate in keeping with the spirit of the dimensional argument.) In the case of irradiation by X rays or neutrons, one might wish to compare the imparted impulse to the projected area being exposed

$$\frac{I_A}{R_A^2} \approx \rho_A R_A \Gamma\sqrt{\frac{GM_S}{R}} \approx 15 \text{ Mtap} \qquad (14)$$

where Mtap = dyne·s cm^{-2} $\times 10^6$, or in terms of the impulse imparted per unit projected area to an energy of, say, $E_a \approx 1$ Mton $\approx 4 \times 10^{15}$ J absorbed over the same area:

$$\frac{I_A}{E_a} \approx \frac{\rho_A R_A^3 \Gamma}{E_a}\sqrt{\frac{GM_S}{R}} \approx \frac{\rho_A R_A^3}{E_a}\frac{R_e}{R}\sqrt{\frac{GM_S}{R}} \approx$$

$$10 \text{ dyne s J}^{-1} = 100 \text{ N s MJ}^{-1} \qquad (15)$$

(In order for this amount of energy to be absorbed by the asteroid, a much greater amount of energy must have been emitted by the source and then reduced by the square of the ratio of the source's standoff distance to the asteroid radius and by some effective absorption coefficient.) These elementary dimensional considerations yield order-of-magnitude estimates for the specific impulse required to deflect an asteroid from an impending collision with the Earth. The argument illustrates the power of utilizing characteristic physical scales for the phenomenon.

IV. APPROACH

In four of the five interaction mechanisms listed in Table I and shown in Fig. 1, energy is delivered as photons, atomic particles, or fission fragments. For a kinetic energy impactor (see Sec. V), an impact crater is produced in the target, followed by ejection of target material from the crater. For pulsed laser (energetic) photons or atomic particles, energy is deposited within the target, resulting in expansion and blowoff of the heated material near the surface.

As noted in Sec. II, detailed hydrocode simulations, or simpler, empirically calibrated models, or experimental measurements can be used to estimate impulse coupling coefficients. For the present study, we will rely primarily on the simpler models or experimental data for our coupling estimates. For a few selected cases, hydrocodes will be used to check the results of the simpler models. In the sequel, we will consider each interaction mechanism in detail. Discussions of the relevant issues and models or experimental measurements used to obtain the coupling coefficients are followed by the results obtained from the models, and discussions of the potential sources of error and ranges of applicability. The final section contains the summary and conclusions.

V. PULSED LASERS

Pulsed lasers offer an attractive non-nuclear option for NEO diversion if sufficiently high beam quality and total beam energy can be obtained. Given that these design issues can be overcome, a large groundbased laser offers a very short response time to a NEO threat. Although laser system design is beyond the scope and intent of this chapter, another option that should be considered is a space-based laser. Obtaining adequate laser power is clearly the driving issue for this option, but space basing may offer more optimal NEO engagement geometries than ground basing.

A simplified trajectory analysis revealed that for cases in which the NEO velocity exceeded the orbital velocity of the Earth, the optimal impulse direction is transverse to the NEO trajectory. However, for cases in which Earth velocity exceeded the NEO velocity, the optimal impulse direction is colinear with the NEO trajectory. Clearly, detailed systems studies are required to evaluate the relative merits of ground and space basing, in addition to developing optimal NEO engagement strategies for each basing concept.

The laser-target interaction occurs in two stages. Initially, only a fraction of the incident energy is absorbed by the target, and the remainder of the energy is reflected from the target. However, the absorbed energy fraction is sufficient to heat the target surface until a plasma is formed near the surface, which is, in general, much more absorbing than the initial target surface. As the pulse continues, more plasma forms and the plasma temperature and pressure continue to increase until the laser pulse terminates. The target impulse results primarily from the plasma pressure and ablation of the target surface caused by the intense thermal flux emitted by the plasma. If we increase the laser fluence for a fixed target material and constant laser wavelength and pulse duration, the impulse coupling coefficient increases rapidly, reaching a maximum value, then decreases gradually.

In general, laser-target interactions are very difficult to model, and complex radiation-coupled hydrocodes are required to obtain accurate predictions. Fortunately, a simplified model for predicting target impulses is available (Dingus and Goldman 1986). This model, termed the "plasma energy balance," enables fairly accurate estimation of target coupling coefficients, thereby avoiding the complexities of radiation-coupled hydrocode analyses. The plasma energy balance model has been validated by laser-target interaction experiments for a large number of target materials, laser wavelengths, and pulse durations.

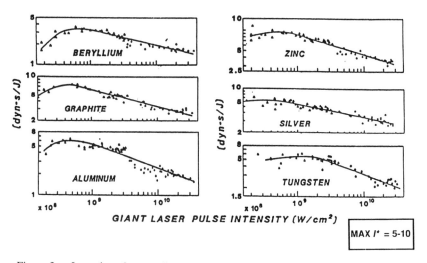

Figure 3.   Laser impulse coupling.

A crucial coupling issue is the sensitivity of target impulse coupling coefficients to variations in target material composition and laser wavelengths and pulse durations. Calculations performed with the plasma energy balance model have indicated that the maximum values of the impulse coupling coefficients exhibit weak sensitivity to target material and laser pulse duration. A slight increase in coupling coefficient may occur with decreasing laser wavelength. However, the impulse threshold fluence (fluence above which I^* increases rapidly) is a strong function of pulse duration for a fixed target material. This phenomenon is shown clearly in Fig. 3, in which experimental impulse coupling results for six elements (Be, C, Al, Zn, Ag, and W) are presented. The solid lines in each frame are least square fits to the experimental data. From the six data sets, it is apparent that MAX I^* does not vary much with atomic number, and lies in the range 5 to 10, with the exception of Be whose coupling value is approximately 3. To provide a reference point, we obtained a rough estimate of the laser output required to match the energy fluence delivered by a 1 Mton nuclear burst to a NEO. Neglecting beam losses, a laser output of 1 GJ s^{-1} for an uninterrupted period of 12 days would be required to match the nuclear burst. Beam losses would require either a more capable laser or longer irradiation times.

V. KINETIC ENERGY

Kinetic energy is another non-nuclear option for NEO diversion, but an interceptor rocket is required to deliver the kinetic energy payload. If warning times were too short, the long delivery times for chemical (or even nuclear) propulsion might preclude an intercept in time to deflect the NEO. We now consider the deposition of kinetic energy. The impact of a hypervelocity projectile produces a region of extremely high pressure in the impact zone and the projectile is deformed and heated during the initial phases of impact crater formation. As the crater formation proceeds, the projectile is deformed into a thin, hot, expanding shell which drives a strong shock into the target. Target impulse is produced when the diverging shock wave in the target is overtaken by the release wave from the free surface. Target material is then ejected as the pressure behind the diverging shock wave is released. The impulse delivered to the NEO is equal to the total momentum contained in the crater ejecta plus the initial momentum of the projectile. Depending upon the projectile impact velocity, the crater ejecta may consist of solid particles, molten droplets, vapor, or high-temperature plasma. For hypervelocity impacts ($\sim$6 km s^{-1}), the projectile mass is much smaller than the mass ejected from the impact crater. Therefore, for hypervelocities, the projectile behaves as a point source of kinetic energy delivered to the surface of the NEO and the detailed configuration of the projectile has little effect on the dynamics of crater formation.

To evaluate the impulse coupling coefficients, we must combine experimental results, hydrocode analyses, and velocity scaling relations to obtain I^* as a function of projectile velocity. By far, the majority of hypervelocity impact studies have employed metallic targets and impact velocities in the range 2 to 8 km s^{-1}. Typical data are presented in Eichelberger (1965), in which the dependencies of crater volume on projectile velocity and target strength are shown. Additional results are shown (see Eichelberger 1965) that relate target impulse to projectile velocity. However, for the NEO diversion problem, we must consider both metallic and nonmetallic targets and impact velocities potentially $>>$8 km s^{-1}. The lack of experimental data for nonmetallic targets at velocities >8 km s^{-1} makes the accurate estimation of impulse coupling coefficients for nonmetallic NEOs very difficult. In particular, the mechanism for late-stage crater formation in nonmetallic, stony NEOs is brittle fracture and pulverization, whereas the crater formation mechanism for ductile metallic NEOs is plastic flow and ductile void growth and coalescence. Therefore, in order to obtain good estimates for NEO impulse coupling coefficients, we must obtain reasonable models for the dependencies of impulse coupling on projectile velocity and target strength for stony as well as metallic NEOs.

By the use of dimensional analysis, we attempted to determine the variations of impact crater volume and target impulse as functions of projectile velocity. We were unable to determine a unique functional form for velocity scaling. However, using data for aluminum presented in Eichelberger (1965), we found that a quadratic velocity dependence produced a better fit to the data than a power law dependence. The methods we used for our dimensional analysis were taken directly from Langhaar (1962).

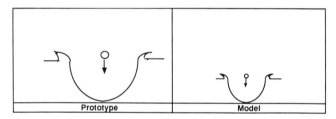

Figure 4.   Hyperveolcity impacts of prototype and model projectiles for similitude studies.

We have considered the very simple case of the impact of a spherical projectile on a half-space, i.e., semi-infinite target (Fig. 4). The projectile and target are composed of the same materials. For this simplified treatment, we have also chosen to neglect melting, vaporization, and ionization. We assume that a linear $u_S - u_p$ relation holds on the Hugoniot (where u_S and u_p are the shock velocity and particle velocity, respecitvely). Following Langhaar (1962), we specify the essential variables of the problem and determine the rank of the dimensional matrix r. Note that this specification may not be unique and some judgment is required. If we have n essential variables, then the number π (following Langhaar [1962], we note that the use of the symbol π to represent a dimensionless product is a convention and has no relation to the number 3.14159...) of nondimensional products is given by $\pi = n - r$. For the crater impact problem, we have specified the following variables as essential, where we have used the abbreviation P&T for projectile and target throughout: P&T density $= \rho_0$; P&T acoustic velocity $= c_0$; projectile velocity $= v_p$; P&T impact stagnation pressure $= p_S$; P&T flow stress $= Y$; projectile volume $= V_p$; crater volume $= V_c$; slope of $u_S - u_p$ curve on P&T Hugoniot $= s$.

From the jump conditions and linear $u_S - u_p$, we have the following:

$$u_S = c_o + s u_p;$$

$$p_S = \rho_o u_S u_p; \text{ and}$$

$$u_p = v_p/2 \text{ (symmetric impact)} \quad (16)$$

then

$$p_S = \rho_o(c_o + s v_p/2)(v_p/2)$$
$$= a v_p + v_p^2 \text{ (quadratic relation).} \quad (17)$$

We now proceed with the dimensional analysis. Our fundamental dimensions are M, L, T (mass, length, time). Following Langhaar (1962), our dimensional matrix is

	ρ_o	c_o	v_p	p_S	Y	V_p	V_c	s
M	1	0	0	1	1	0	0	0
L	-3	1	1	-1	-1	3	3	0
T	0	-1	-1	-2	-2	0	0	0.

$$(18)$$

The rank of the dimensional matrix is the size of the largest nonzero determinant of the matrix. Through trial and error, we found that colums 1, 2, and 6 produce a nonzero determinant, i.e.,

$$\begin{vmatrix} 1 & 0 & 0 \\ -3 & 1 & 3 \\ 0 & -1 & 0 \end{vmatrix} = (1) \begin{vmatrix} 1 & 3 \\ -1 & 0 \end{vmatrix} - (-3) \begin{vmatrix} 0 & 0 \\ -1 & 0 \end{vmatrix} = 3. \quad (19)$$

So $r = 3$ and $\pi = n - r = 8 - 3 = 5$. Therefore, we have five nondimensional products or Pi-Groups. The selection of the nondimensional products is somewhat arbitrary as long as they form a complete set. Considerable physical intuition may be required to choose the best set of nondimensional products. From Buckingham's theorem, it can be inferred that our impact process can be represented in the form $f(\pi_1, \pi_2, \ldots, \pi_5) = 0$. The first set of nondimensional products that we chose were:

$$\pi_1 = s$$
$$\pi_2 = V_c/V_p$$
$$\pi_3 = v_p/c_o$$
$$\pi_4 = \rho_o v_p^2/Y$$
$$\pi_5 = p_S/Y. \quad (20)$$

Next, we address the issue of similitude between a prototype system and a model of the prototype system. For kinetic energy impact, we have the situation shown in Fig. 4. Using * to represent model variables, complete similitude between model and prototype requires equality of Pi- Groups, i.e.,

$$\pi_i^* = \pi_i, \quad i = 1, 2, \ldots, 5$$
$$s^* = s$$
$$V_c^*/V_p^* = V_c/V_p$$
$$v_p^*/c_0^* = v_p/c_0 \tag{21}$$
$$\frac{p_0^* v_p^{2*}}{Y^*} = \frac{p_0 v_p^{2*}}{Y}$$
$$p_s^*/Y^* = p_s/Y.$$

We were hoping to use dimensional analysis to study velocity scaling. As noted in Eichelberger (1965), crater volume and target impulse exhibit the same velocity scaling, so we can infer target impulse velocity scaling from crater volume velocity scaling. For velocity scaling, the model and prototype have identical mechanical properties, i.e., $s^* = s$, $c_0^* = c_0$, $\rho_0^* = \rho_0$, and $Y^* = Y$.

When these conditions are inserted into Eq. 21, some very strong restrictions result.

$$V_c^*/V_p^* = V_c/V_p$$
$$v_p^* = v_p$$
$$v_p^{2*} = v_p^2 \tag{22}$$
$$p_s^* = p_s.$$

These equations indicate that only replica scaling is permissible, i.e., we are only allowed to vary projectile size from model to prototype, and of most importance is the restriction that model and prototype velocities must be equal. If we are to study velocity scaling, we must search for other nondimensional products.

Through trial and error, we arrived at another set of π's that helped answer some questions on velocity scaling. In the following, primes indicate the new π's and unprimed π's are the old π's.

$$\pi_1' = \pi_1 = s$$
$$\pi_2' = \pi_2/\pi_3 = \frac{V_c c_0}{V_p v_p}$$
$$\pi_3' = \pi_2/\pi_4 = \frac{V_c Y}{V_p \rho_0 v_p^2} \tag{23}$$
$$\pi_4' = \pi_2/\pi_5 = \frac{V_c Y}{V_p p_s}$$
$$\pi_5' = \pi_4/\pi_5 = \frac{\rho_0 v_p^2}{p_s}.$$

As before, similitude requires

$$\pi_i'^* = \pi_i', i = 1, 2, \ldots, 5$$

$$s^* = s$$

$$\frac{V_c^* c_0^*}{V_p^* v_p^*} = \frac{V_c c_0}{V_p v_p}$$

$$\frac{V_c^* Y^*}{V_p^* \rho_0^* v_p^{*2}} = \frac{V_c Y}{V_p \rho_0 v_p^2}$$

$$\frac{V_c^* Y^*}{V_p^* p_s^*} = \frac{V_c Y}{V_p p_s} \tag{24}$$

$$\frac{\rho_0^* v_p^{*2}}{p_s^*} = \frac{\rho_0 v_p^2}{p_s}.$$

To study velocity scaling, we must assume that the model and prototype material properties and projectile volumes are identical, i.e., $s^* = s$, $c_0^* = c_0$, $\rho_0^* = \rho_0$, $Y^* = Y$, and $V_p^* = V_p$.

Using these conditions in Eq. 24, we obtain

$$\frac{V_c^*}{v_p^*} = \frac{V_c}{v_p} \quad \text{(a)}$$

$$\frac{V_c^*}{v_p^{2*}} = \frac{V_c}{v_p^2} \quad \text{(b)}$$

$$\frac{V_c^*}{p_s^*} = \frac{V_c}{p_s} \quad \text{(c)}$$

$$\frac{v_p^{2*}}{p_s} = \frac{v_p^2}{p_s} \quad \text{(d)} \tag{25}$$

Equation 25 seems to indicate that crater volume scales with v_p, v_p^2, and p_s simultaneously, where $p_s = a v_P + b v_p^2$; however, this condition cannot hold in general. Note that the *only* way for Eq. 25 to hold rigorously is for $v_p^* = v_p$, in which case we cannot examine velocity scaling.

However, another way of looking at Eq. 25 is the following. Assume each of the four equations holds only for a particular range of velocities. In particular, Eq. 25a is assumed to hold only for very low velocity impacts, whereas Eqs. 25b and d are assumed to hold for very high velocities. However, Eq. 25c may be a key relationship in that it may hold for all velocity ranges. When the aluminum crater volume data were fit with a quadratic velocity function, the fit appeared to be better than the power function fit (Fig. 5). This outcome suggests that Eq. 25c is the key velocity scaling relation and that crater volume and target impulse scale with impact stagnation pressure. If p_s scaling is indeed the case, this is a fortunate situation for kinetic energy

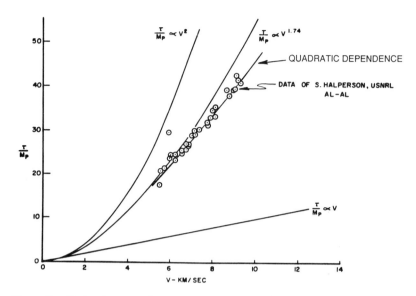

Figure 5. Ratio of crater volume to projectile mass as a function of impact velocity
for aluminum–aluminum impacts. Curves represent predictions obtained from
three theoretical approaches and are compared to experimental data of Halperson
(Eichelberger 1965). Figure adapted from a figure in *Behavior of Materials Under
Dynamic Loading*, 1965, and reprinted with permission from ASME.

coupling, because the impulse coupling will vary as v_p^2 for very high-velocity
impacts. This means that the coupling coefficient will level off, rather than
decrease with increasing v_p. If I^* approaches a constant as v_p becomes "very
large," we will have the total impulse increasing directly as v_p^2, i.e., the very
best coupling situation. We should note again, however, that these arguments
neglect melting, vaporization, or ionization of the target.

We now assess the dependencies of target impulse on impact velocity
and target strength. We consider velocity scaling first. The hydrocode results
presented in Eichelberger (1965) indicate that target impulse exhibits a power
law dependence on velocity, with a velocity exponent of 1.74. Strict kinetic
energy scaling would require an exponent of 2.00. An alternative quadratic
scaling law is derived above. However, the dimensional analysis was unable
to specify a unique form of the velocity scaling relation. As noted previously,
the quadratic form fits the crater volume data for aluminum as well or better
than the power law form.

Next, we consider the role of target strength. The dimensional analysis
indicates that the target impulse varies directly with target strength, provided
the simplifying assumption of constant flow stress holds. However, it remains
an open question of how well a constant flow (or crush) stress condition would
apply to crater formation and impulse production in a nonmetallic NEO.

We present estimates for I^* in Fig. 6, which were derived from several
models. These estimates are semi-quantitative at present. The lowest impulse

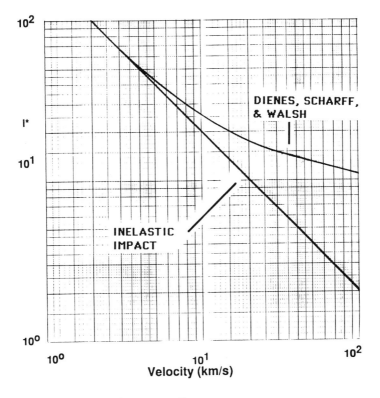

Figure 6.   Kinetic energy impuse coupling.

coupling coefficient results from inelastic impact, which is the downward sloping straight line in Fig. 6. This estimate for I^* is far too conservative for hypervelocity impacts. As the projectile velocity increases, either power law or quadratic scaling should apply. Power law scaling produces the upper curve in Fig. 6. Quadratic scaling would produce a curve that falls above the power law curve with a horizontal asymptote because I^* approaches a constant as the projectile velocity becomes very large. For the power law and quadratic models, the slopes of the asymptotes depend upon the velocity dependencies, but the specific values of I^* depend upon the value of flow stress assumed for the model. For a velocity of 20 km s^{-1}, I^* is estimated to lie in the range between 10 and 30. Much additional work that takes account of the characteristics of the target material characteristics will be required to sharpen these estimates.

VI. FLYBY NUCLEAR (OR CONVENTIONAL) EXPLOSIONS

The third NEO diversion mechanism shown in Fig. 1 is a standoff detonation of a nuclear (or conventional) explosive device. As noted in Table II, the specific energies of high explosives, hypervelocity impactors, and nuclear explosives are approximately 6, 50, and $4 \times 10^6 \text{MJ kg}^{-1}$, respectively. Because of their extremely high specific energies, nuclear explosives are a logical choice for the diversion of massive NEOs. This conclusion holds in spite of the obvious negative environmental and geopolitical consequences associated with the use of nuclear explosives. As in the case of a kinetic energy impactor, chemical or nuclear propulsion is required to deliver the nuclear explosive to the NEO, which may impose an unacceptable time delay if the warning time were too short. In principle, the detonation products from a high-explosive flyby could be used to divert a NEO. However, the million-fold reduction in specific energy requires a much larger chemical or nuclear propulsion system to deliver sufficient high explosives rapidly enough to be effective.

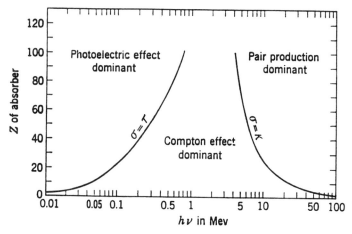

Figure 7.   Variations of photon interaction mechanisms with atomic number and photon energy. Figures 7–10 are adapted from *The Atomic Nucleus* (Evans 1955) and reprinted with permission from McGraw-Hill.

Detonations of chemical and nuclear explosives have some features in common, i.e., large amounts of energy are produced by relatively small amounts of material. For both detonations, the liberated energy is partitioned into the kinetic energy of atomic particles or molecules, and electromagnetic radiation. However, due to the million-fold greater energy density in nuclear explosives, the characteristics of the emitted particles and radiation differ widely for the high explosive and nuclear cases.

For high explosives, the chemical energy is converted into hot detonation products, which radiate roughly blackbody radiation at several thousand degrees Kelvin. For a nuclear explosive, fission and fusion energy sources emit energetic fission and fusion products, explosion debris, light ions, neutrons, electrons, and electromagnetic radiation, at temperatures of roughly tens of millions degrees Kelvin. In spite of a superficial similarity, standoff detonations of high explosives and nuclear explosives produce impulses by quite

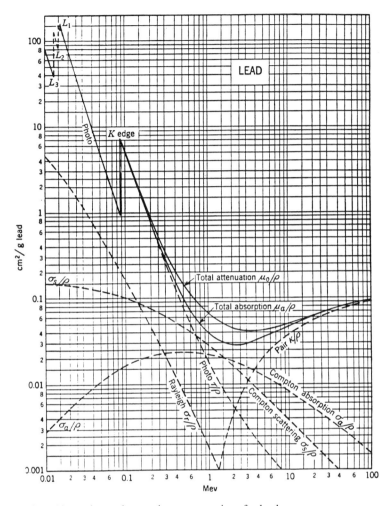

Figure 8.   Absorption and scattering cross sections for lead.

different mechanisms. High explosives produce an impulse by direct impingement of detonation products. In contrast, the major sources of impulse from nuclear explosives are neutrons and X rays, which penetrate deeply into the target, although explosion debris can also contribute significant impulse by direct impingement.

We now consider X rays and neutrons in detail, because these radiations are the principal sources of impulse from a standoff nuclear explosion. For both X ray and neutron deposition, a simplified, two-step process is used to calculate the target impulse. First, given the chemical composition of the target NEO, energy deposition profiles are calculated with one-dimensional radiation transport codes. From a deposition profile and a threshold energy for impulse production, the target impulse can be calculated using simplified models that ignore hydrodynamic details. Below we discuss the interactions of photons and neutrons in some detail.

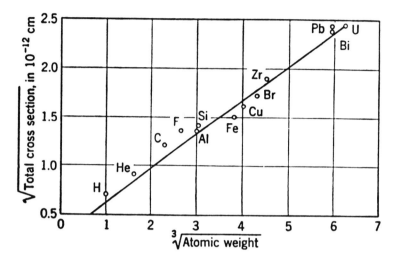

Figure 9.   Total cross sections for 19 MeV neutrons.

The intensity I of a photon beam of initial intensity I_0 after traversing a distance x into the target is given by

$$I = I_0 e^{-\mu x} \qquad (26)$$

where the linear attenuation coefficient μ is given by

$$\mu = Q_{ca} + Q_{cs} + \tau + \kappa \qquad (27)$$

with Q_{ca}, Q_{cs}, τ, and κ the linear Compton absorption, Compton scattering, photoelectric coefficient, and pair production attenuation coefficient, respectively (Evans 1955a). Figure 7 shows that the type of interaction depends on the photon energy (Evans 1955a). The linear attenuation coefficient measures the number of primary interacting photons and is to be distinguished from the absorption coefficient, which is always a smaller quantity and measures the energy absorbed by the medium. This is an important distinction for computing the effectiveness of momentum coupling to NEOs. The effects that photons produce in materials are almost exclusively due to secondary electrons with the energy absorption being the photon energy converted into kinetic energy of secondary electrons.

The dominance of the photoelectric cross section for photon energies <1 MeV is clearly shown in Fig. 8 (Evans 1955a). The contributions of Compton absorption and scattering as well as pair production dominate for higher photon energies.

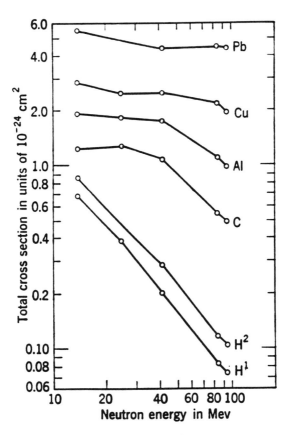

Figure 10.   Dependence of neutron cross sections on neutron energy and atomic mass.

The probability of a photoelectric cross section per atom Q_{PEa} cm²/atom is primarily related to interactions with the K-shell electrons and, therefore, yields only a partial cross section Q_{PEa} (K), which does not include the $L, M, \ldots$ shells. To approximately account for these interactions, the K shell cross sections are multiplied by a factor of 5/4. The total cross section is dependent on the atomic number Z and photon energy $h\nu$, and is approximated by

$$Q_{PEa} = \text{constant } Z^4/(h\nu)^3. \qquad (28)$$

The actual exponents of Z and $h\nu$ are both nonintegral functions of $h\nu$, and

$$Q_{PEa} = \text{constant } Z^n \qquad (29)$$

for fixed values of $h\nu$. The exponent n increases from 4.0 to 4.6 as $h\nu$ increases from 0.1 MeV to 3 MeV. Below 0.1 MeV, the photoelectric cross section includes absorption edges ($h\nu$ becomes smaller than the binding energy of some electrons, reducing the number ejected). NEOs with a high atomic

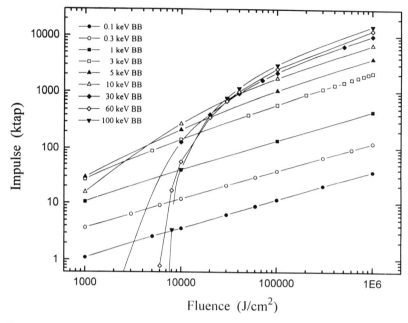

Figure 11.   X ray impulse as a function of fluence.

number and density (e.g., Fe, Ni, Mg, etc.) will absorb X rays over a shorter distance and undergo a specific impulse in a shorter time. Also, the X rays travel at the speed of light, while the neutrons travel at only a fraction of the speed of light.

The linear attenuation coefficient, Q_{PE} cm^{-1}, may be regarded as the effective energy absorption

$$Q_{PE} = Q_{PE_a}N \tag{30}$$

where N is the number of atoms per cm^3.

For any type of interaction, the mass attenuation coefficient is the linear coefficient divided by the density, ρ. The primary attenuation of photons is dependent on the sum of the cross sections presented by all the atoms in the mixture. Because chemical bonds are of the order of a few electron volts and, therefore, have little effect on the Compton, photo, or pair interactions, the overall mass attenuation coefficient is then

$$\mu/\rho = \sum_i \mu_i f_i / \rho_i \tag{31}$$

where f_i are the fractions by weight that make up the absorber. The collision probability for a photon in a distance dx is μdx, where $\mu dx \ll 1$. Therefore, the probability that a photon can travel a distance x without experiencing a

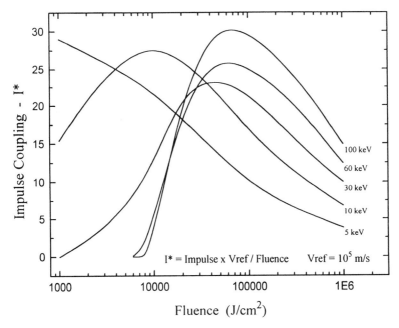

Figure 12. Nondimensional X ray impulse coupling as a function of fluence.

collision is $e^{-\mu x}$. The number N, of identical photons entering an absorber and traveling a distance x without collision, is $Ne^{-\mu x}$, and the number that have a collision between x and $x + dx$ is $N\mu e^{-\mu x}dx$. Summing over all possible path lengths from $x = 0$ to ∞ gives the mean free path $1/\mu$ before the first collision, i.e.,

$$\int_0^\infty e^{-\mu x}dx = 1/\mu. \tag{32}$$

Experimental determinations of the total neutron cross sections for a variety of elements have been made using mono-energetic neutrons. Generally, it was found that the nuclear cross section varies approximately with $A^{2/3}$, where A is the atomic mass number. This dependence is compatible with nuclear volume proportional to mass number, which implies that the nuclear radius can be written as $R = R_0 A^{1/3}$. The total cross section for high-energy neutrons is the sum of the absorption and scattering cross sections (Evans 1955b).

$$Q_{NT} = Q_{N\ abs} + Q_{N\ sc} = 2\pi(R + \lambda)^2 \tag{33}$$

where λ is the neutron wavelength and is related to the wave number K by

$$\lambda = 1/K = 4.55 \times 10^{-13}/[E(\text{MeV})]^{1/2} \text{ cm}. \tag{34}$$

The total cross section, Q_{NT}, for 19 MeV neutrons, is shown in Fig. 9 (Evans 1955b). The straight line is the function

$$(Q_{NT})^{1/2} = (2\pi)^{1/2}(R_0 A^{1/3} + \lambda) \tag{35}$$

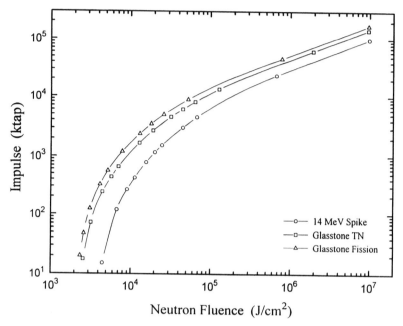

Figure 13.   Neutron impulse as a function of fluence.

where $R_0 = 1.4 \times 10^{-13}$ cm and $\lambda = 1.04 \times 10^{-13}$ cm.

The above describes how the absorption of fast neutrons, such as those derived from a nuclear explosion, will depend on the atomic mass number. The lower density NEO materials (primarily H, C, N, and O) will allow neutrons to penetrate further, allowing a larger mean free path over which the momentum can couple. This is a critical parameter in determining the effectiveness of NEO orbit modification by nuclear radiation.

At higher neutron energies, Fig. 10 (Evans 1955b) suggests that not all the neutrons that strike the target area, $\pi(R + \lambda)^2$, are initially captured and that as the neutron energy is increased, low atomic mass NEO target material becomes more transparent, with the mean free path of the neutron (before absorption) increasing. More energetic neutrons change the effectiveness of the energy coupling and the statistics of momentum transfer to the NEO.

The most accurate techniques for analyzing and computing the dynamics of the interactions of the various nuclear radiation components with solid targets are the multi-dimensional radiation transport codes and hydrocodes that utilize an equation of state, radiation transport, and detailed material characteristics and interactions. However, the basic physics of the interaction of nuclear radiation can be understood without going into the complex details of the many processes relevant to the main issue of NEO hazard mitigation, which is velocity change through ablation-based impulse generation. A simple model has been developed (Hammerling and Remo 1993) that uses (target) mate-

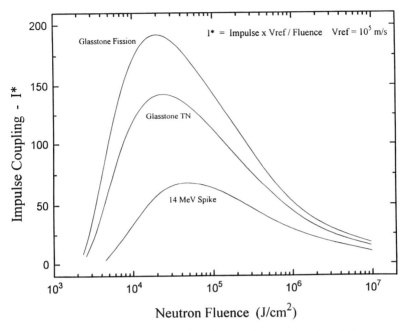

Figure 14.   Nondimensional neutron impulse coupling as a function of fluence.

rial and nuclear explosion radiation properties to compute one-dimensional radiation-induced materials removal and impulse generation. This model is not meant to take the place of more detailed computations that involve the use of elaborate hydrocodes and long run times. However, for situations where extreme accuracy is not needed, this model can rapidly compute the momentum transfer and associated velocity change from the fractional distribution of the various nuclear explosive energy sources. The energy spectrum of the principal radiative contributions interacting with the NEO target materials will determine the appropriate mass absorption or opacity coefficients. However, incorporation of a single spectral value into a simple X ray and neutron ablation model of an asteroid or comet established a bias based on the use of a dominant (absorption) line rather than a Planckian radiation spectrum, which overemphasizes the importance of elements having strong absorption for that particular line. For this reason, realistic atomic abundances with all the significant absorption lines integrated over the blackbody spectrum should be used for the model.

The specific impulse I (momentum per unit area), by X and gamma ray ablation and neutron absorption, is determined by the differential relationship

$$dI = \rho V dx \tag{36}$$

where

$$V^2/2 = e - e_v \tag{37}$$

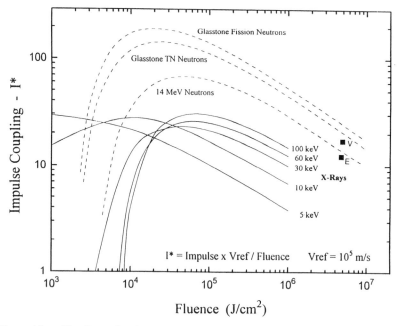

Figure 15.   Nondimensional X ray and neutron impulse coupling as a function of fluence.

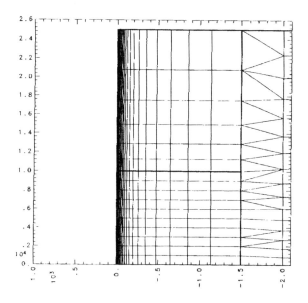

Figure 16.   Initial hydrocode mesh.

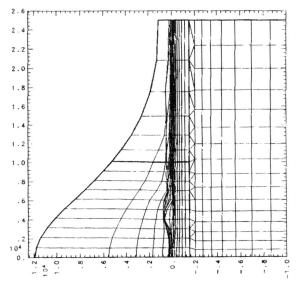

Figure 17. Deformed hydrocode mesh.

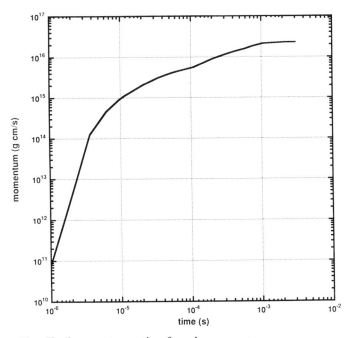

Figure 18. Total momentum vs time for volume source.

and ρ is the mass density, V is the velocity of the differential mass, e and e_v are the incident energy per unit mass and the vaporization energy per unit mass, respectively, and dx is the radial differential to the NEO surface; e

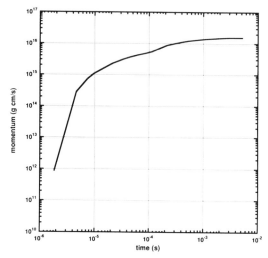

Figure 19.   Total momentum vs time for edge source.

is related to the absorbed fluence F, and the mass attenuation coefficient $\hat{\mu}$ (where $\hat{\mu} = \mu/\rho$ by

$$e = \hat{\mu} F. \tag{38}$$

The specific impulse is then given by

$$dI = \sqrt{2}(\hat{\mu} F - e_v)^{1/2} \rho dx. \tag{39}$$

It is assumed that the fluence varies with the penetration depth according to

$$F = F_0 \exp(-\rho \mu x / \cos \Theta) \tag{40}$$

and for radiation normal to the surface, $\Theta = 0$. From this equation, it is clear that the atomic composition (opacity) will dominate the fluence and, as will be shown below, the impulse and velocity change.

The fluence contribution from each fraction of yield n_i of the total yield Y of the nuclear device at burst distance L from the NEO is

$$F_0 = n_i Y / 4\pi L^2. \tag{41}$$

The scaled variables are introduced: $I = (\sqrt{e_v}/\hat{\mu})I^*$; $F_0^* = F_0/F_{min} > 1$; and $F_{min} = e_v/\mu$. Combining and using the scaled variables and integrating, the scaled impulse is then found,

$$I^* = 2\sqrt{2} \left\{ (F_0^* - 1)^{1/2} - \tan^{-1}(F_0^* - 1)^{1/2} \right\}. \tag{42}$$

The associated velocity change DV of a NEO of mass m per unit area and opacity $\hat{\mu}$ is

$$DV \cong 2 \left[\sqrt{2e_v}/m\hat{\mu} \right] \left\{ (F_0^* - 1)^{1/2} - \tan^{-1}(F_0^* - 1)^{1/2} \right\}. \tag{43}$$

Each component of radiation flux contributes to the total orbital velocity change. Depending on the stand-off nuclear explosion design, the yield will provide a specific spectrum of nuclear radiation and debris. The interaction of the radiation with the NEO will depend on the NEO composition. Values for asteroid material properties, examples of nuclear fractional energy distributions for both fission and fusion devices, and data on gamma ray, X ray, and neutron mass absorption coefficients can be found in Evans (1955a, b), Hammerling and Remo (1993), Remo (1993) and Glasstone and Dolan (1977a). We have used this procedure to calculate the target coupling coefficients for a sequence of blackbody X ray spectra and three neutron spectra. The blackbody temperatures studied were 0.1, 0.3, 1, 5, 10, 30, 60, and 100 keV. The neutron spectra consist of a 14 MeV spike and fission and thermonuclear neutron spectra taken from Glasstone and Dolan (1977b). The results from the X ray and neutron calculations follow.

Impulses for a pure silicon dioxide target as a function of incident fluence and blackbody temperature are presented in Fig. 11. These results were used to compute I^* as a function of incident fluence and blackbody temperature shown in Fig. 12. MAX (I^*) for blackbody X ray spectra is seen to fall between 22 and 30. Likewise, the impulses produced by the three neutron spectra are shown in Fig. 13, followed by plots of I^* as a function of fluence for the three spectra in Fig. 14. MAX (I^*) for neutrons is seen to fall between 60 and 180, a significantly greater range than for X rays (Fig. 15). An independent check of an impulse coupling value for 14 MeV neutrons was made using a two-dimensional hydrocode. Results were obtained for two different energy source terms, the first a volume source at the burst point and the second an edge source at the surface of the asteroid. Plots of the initial and distorted meshes are presented in Figs. 16 and 17. Plots of total momentum vs time for volume and edge sources are presented in Figs. 18 and 19. Finally, hydrocode results are compared with the previous impulse results in Fig. 15. Although not perfect, the agreement is good (see points labeled V and E).

The effect of target porosity is addressed below in Appendix I. This effect is examined because it is probable that many NEOs are covered by a porous surface layer (regolith) that is the result of collisions with other bodies and accretion of dust and small particles on the surface of the NEO. A simplified model of target blowoff from porous targets is developed and the impulse for a porous target is compared with the impulse for a full-density target. Typically, a target porosity of 50% results in a target impulse reduction of 20% or less. Finally, highly simplified models for deposition, ablation, and spall have been used to estimate the velocities imparted to various sizes and shapes of asteroids. These results are presented in Appendix II.

VII. SHALLOW OR DEEPLY BURIED NUCLEAR (OR CONVENTIONAL) EXPLOSIONS

In this section, we present detailed calculations of impulse coupling for buried nuclear and conventional explosive charges in several media. As noted previously, the buried explosive strategy presents a difficult challenge for charge emplacement, especially for high-strength materials. However, it is potentially attractive because the momentum coupling efficiency and the coupling coefficient I^* are 1 to 2 orders of magnitude higher than for any of the other methods. This inherent high efficiency results from the fact that energy is hydrodynamically and mechanically coupled to the surrounding material throughout the vaporization or detonation phase and the expansion of the explosive cavity. Losses occur only through inelastic material response, propagation of the stress wave beyond the elastic radius at ranges far from the explosive, and eventual venting of the cavity gases. For detonations near or below optimum burial depth in weak materials, up to 30% or more of the total explosive energy may be coupled into kinetic energy of the ejecta mound. This far exceeds the efficiencies achievable with radiative coupling or device debris expansion from stand off detonations.

The dynamic phenomenology of buried cratering detonations has been exhaustively discussed by other authors (Burton et al. 1975; Terhune et al. 1970; Kreyenhagen and Schuster 1977; Westine 1970; Nordyke 1961,1977; Knowles and Brode 1977; Melosh 1980), and cratering has also been investigated with scaling and dimensional analysis methods (Baker et al. 1973; Chabai 1965,1977; Holsapple and Schmidt 1980; Schmidt and Housen 1987; Killian and Germain 1977; White 1971,1973; Housen et al. 1983; Cooper and Sauer 1977). These topics will not be reviewed here. However, it is essential to note that there is one fundamental difference between cratering on Earth and the asteroid deflection case. The dynamics of cratering events on Earth or any large body are strongly influenced by the gravitational field. First, the ejecta are decelerated by gravity, and some of the ejecta material will fall back to partially refill the initial or true crater. As the depth of burial increases, the average ejecta velocities decrease and more of the material falls back into the crater. It is for this reason that the apparent crater volume increases up to a certain scaled depth of burial, known as "optimum cratering depth," and then progressively decreases for deeper burial as the ejecta refill the crater. Optimum scaled burial depth for cratering events in typical Earth materials is about 40 to 60 m kt$^{-1/3}$. The crater volume rapidly decreases for greater depths, and eventually a mound or "retarc" is formed by the bulked ejecta material. The situation is completely different in the microgravity of a NEO, because the escape velocity is on the order of a few cm/s or less. Velocities of material in the ejecta mound are ~ 10 to 500 m s^{-1}; hence, virtually all of the material far exceeds the escape velocity and there will be little or no fallback. Rather than displaying an optimum depth of burial, momentum coupling for a NEO will increase monotonically with increasing depth, up to the point at which the material response becomes elastic and the explosion is fully contained. Cratering will not occur at depths greater than this and no momentum would be coupled to the body. The approach to maximize momentum coupling is thus to bury the detonation as deeply as possible while still assuring effective cratering of the body. The cratering process in a gravitational field is also influenced by overburden stresses and atmospheric pressure. These effects are likewise negligible and can be ignored for the NEO case. Because gravitational effects are not significant, simple cube-root energy scaling will apply for material dynamics and momentum coupling (as long as material properties are not rate-dependent). Thus, calculations can be conveniently performed for an arbitrary energy yield and scaled to any other yield of interest.

A. Numerical Modeling Approach

We selected five materials of widely varying properties for consideration in this study: (1) Bearpaw clay shale, a weak, saturated rock with high total porosity and water content; (2) granite, a high-strength silicate rock with negligible porosity and water content; (3) nickel-iron, a dense, very high-strength material with negligible porosity; (4) water ice (no porosity); and (5) snow with 43% porosity. The basic physical properties assumed for these materials are summarized in Table III. The Bearpaw shale is not similar to any asteroid or comet material, but serves as a useful reference case because it is a very weak, fully saturated material that will give efficient, almost-hydrodynamic coupling of momentum. This represents coupling efficiency close to the upper limit that would be expected for typical crustal rocks on Earth. The other four materials are analogous (although not identical) to materials of interest in NEO deflection applications. (For a more detailed discussion of NEO material properties see the Chapter by Remo.) We have performed numerical simulations for 1-kiloton nuclear and high explosive events at five to seven burial depths in each material. The depth range considered was from 11 to 88 m for Bearpaw shale and granite. Shallower depths were used for the dense, very high-strength nickel-iron material, and greater depths for the lower density ice and snow. The depth sequences were terminated at the deep end because the threshold between cratering and containment was being approached for most of the materials. The nickel-iron events failed to crater at depths beyond 33 to 35 m $kt^{-1/3}$, slightly deeper than the range employed here. No calculations were made for very shallow burial because accurate treatment of the highly distorted flow for near-surface bursts requires an arbitrary Lagrange-Euler (ALE) numerical method or a similar automatic rezone technique, which was not available in the codes used for this study. The calculations did not include gravity, atmospheric pressure, or rate-dependence effects. Therefore, cube-root scaling of lengths with energy yield $Y^{1/3}$ is appropriate and all results will be expressed in scaled form.

The approach used to develop accurate constitutive models for the materials is to simulate planar plate impact data, spherical divergent wave experiments, and large-scale field tests (if available). The spherical divergent wave experiments are crucial because they expose the material to a broad range of peak stresses (including stresses below a few GPa that are predominant for cratering phenomenology), and to divergent stress-strain loading and unloading paths similar to those experienced in cratering events. Dynamic data were

TABLE III

Summary of Basic Physical Properties Assumed for Materials
Used in the Cratering Calculations

	Bearpaw Shale	Hard Granite	Nickel-Iron[a]	Ice	Snow
Bulk density (g cm^{-3})	2.217	2.65	7.856	0.917	0.520
Sonic velocity (m s^{-1})	1960	6530	5940	3860	126
Bulk modulus (GPa)	8.5	60.0	165.9	8.85	0.0053
Poisson's ratio	0.48	0.23	0.284	0.32	0.32
Water content by weight	0.142	0.0	0.0	1.0	1.0
Unconfined strength (MPa)	3.0	10.4	600.0	15.0	0.1
Ultimate strength (MPa)	5.0	374.0	600.0	17.5	0.1
Tensile strength (MPa)	0.2	2.0	400.0	3.8	0.01
Void porosity	0.0	0.0	0.0	0.0	0.433
Scaled depth of burial	11.1−	11.1−	3.7−	22.1−	22.1−
Range (m kt$^{-1/3}$)	88.4	88.4	29.8	138.4	169.0

[a] Normal strength. Properties of the "low-strength" nickel-iron considered
in this study are identical for the "normal-strength" nickel-iron, with the
exception that the material strength was decreased to match that of granite
(see discussion in the text).

not available for nickel-iron, so an approximate model had to be synthesized
based on other information. The experiments were simulated with the SOC
one-dimensional Lagrangian finite-difference code (see Seidl 1965; Cherry
and Petersen 1970; Schatz 1973,1974; Snell 1994). Physical models provided in the code include: (1) nuclear energy source (Butkovich 1967); (2)
JWL (Jones, Wilkins, Lee) high-explosive burn and equation of state (Lee et
al. 1968,1973; Dobratz 1981); (3) high-pressure, high-temperature gas equation of state for geologic materials (Butkovich 1967; Butkovich et al. 1981);
(4) multiphase water vaporization and adiabatic release (Schatz 1974); (5)
dynamic quartz-stishovite phase transformation between 10 and 40 GPa for
silicate rocks (Wackerle 1962; McQueen et al. 1967; Kieffer 1977); (6) ductile yield and brittle fracture (Schatz 1974); (7) damage-dependent strength
degradation of failed material (Budiansky and O'Connell 1976; Taylor et al.
1986; Schatz 1974); (8) irreversible hysteretic compaction of voids (Crowley 1973; Schatz 1974); and (9) anisotropic tensile fracture and directional
crack opening (Maenchen and Sack 1964; Burton and Schatz 1975; Burton
et al. 1977; Bryan et al. 1977). In addition to the standard models, special
treatments had to be developed for modeling the water ice (Anderson 1968;
Bakanova et al. 1975; Gaffney 1985,1979,1980; Gaffney and Matson 1980;
Larson 1978,1984; Larson et al. 1973), and snow (Gaffney et al. 1985; J.
B. Johnson 1991, personal communication; Johnson et al. 1992,1993; M.
Furnish 1992, personal communication). Ice displays extremely complex behavior involving dynamic yielding, partial melting, and nonequilibrium phase
transformations in the moderate stress regime from 0.15 to 2.2 GPa (1.5 to

22 kbar). Snow undergoes strong porous compaction at low stresses below 25 MPa, and shows dilatant pore recovery on unloading at very low stresses less than 10 MPa. Models for these materials were developed by matching detailed dynamic data in the regimes of interest.

To perform the cratering calculations, the constitutive models developed and validated in SOC were implemented in the Sandia dynamic finite-element code PRONTO (Taylor and Flanagan 1987,1989). All calculations presented here utilized two-dimensional, cylindrically axisymmetric geometry. The energy source was located at the appropriate burial depth on the axis of rotational symmetry. The upper surface of the body was treated as a free surface and was assumed to be flat (i.e., curvature was ignored, a good approximation provided the radius of the body is at least ten or more burial depths). The side and bottom boundaries of the calculational domain were sufficiently far removed from the energy source so that no artificial reflections from the boundaries returned to the crater region within the time period of interest. The material was assumed to be uniform in all cases. The energy sources were spherical. The nuclear calculations were initiated by depositing all the energy in the vaporized device canister at zero time (pill source approximation). The high explosive was detonated from the center outward. The calculations were run for at least five to six stress wave transit times from the explosive to the free surface. By this time, essentially all of the material in the ejecta mound was in ballistic free fall and the momentum had reached its final asymptotic value. Because of the extremely low wave velocities in the porous snow, it proved unfeasible to run the deeply buried calculations for the full five transit times. However, close examination of the results showed that there were no significant stress gradients in the mound and the material was in free fall, demonstrating that the momentum coupling was complete when the runs were terminated. The momentum recorded in these calculations is the vertical component, i.e., the momentum component perpendicular to the free surface. In axisymmetric geometry, the horizontal momentum components around the axis of symmetry cancel out exactly. The vertical component is the quantity of interest for determining the delta-velocity and the deflection of the body. Here, we have chosen to integrate the momentum over a true crater volume somewhat arbitrarily defined as an inverted cone with its apex at the center of the energy source. This approach will slightly underestimate the total vertical momentum in most cases, because some material initially located below the energy source receives a positive vertical velocity and is ejected at late time due to spallation and rebound effects. The error will be small, because this deep material is ejected with much lower velocities than the shallower overlying material. An alternative method would have been to integrate the vertical momentum over all failed material in upward-directed ballistic trajectories at the end of the calculation. This approach would risk overestimating the momentum, because the dynamics of the deep, low-velocity material below the source is somewhat sensitive to the material properties and especially to the poorly known tensile strength. We considered it preferable to use a conservative assumption that was consistent for all materials and calculations, hence, the approach of integrating over a conical true crater volume. With a few exceptions discussed later, all of the material within the true crater volume was fully failed, had positive (upward-directed) vertical velocities, and was in ballistic free fall when the calculation terminated. Thus, the vertical momenta derived here are expected to be accurate representations of the true, asymptotic late-time momentum coupled to the ejecta mound. In a few cases, we made rough estimates of the error introduced by ignoring material outside the conical volume. It was found that the momentum was underestimated by about 10% for a deeply buried event and 20% for a shallow event. Attempting to apply corrections of this magnitude is not worthwhile, because uncertainties introduced by assumptions about the material properties and other factors are probably larger than this. These calculations are intended as semi-quantitative scoping studies, to provide basic information for certain classes of materials that are similar to some asteroid and comet materials. Accurate, detailed calculations for an actual deflection event would, of course, require much more knowledge about the composition, material properties, and geometry of the target body.

B. Results and Discussion

The calculational results are presented as curves of scaled momentum (kg-m/s-kt) vs scaled depth of burial (m kt$^{-1/3}$). All energies are referenced to the standard value of 1 kt (1 kt = 4.184 × 10^{12}Joule). Values read from the calculated curves are easily scaled to any other energy yield Y(kt); to obtain the true depth of burial, multiply the scaled depth by $Y^{-1/3}$; to obtain the true momentum, multiply the scaled momentum by Y. Given the momentum coupled to the NEO, the change in velocity and the resultant deflection can be directly computed, as described in the Chapter by Solem and Snell. The examples presented in that chapter employ the first two sets of calculations discussed below.

Scaled momentum coupling curves for nuclear and high explosive events in Bearpaw shale and granite are displayed in Fig. 20. As expected, the weak, saturated Bearpaw shale produces much more efficient momentum coupling than the high-strength granite. The high explosive coupling efficiencies are higher than the nuclear efficiencies by a factor of 1.3 to 1.4 for both materials. Interestingly, the curves for both nuclear and high explosive coupling are almost linear over the scaled depth range examined here. This fact provides a convenient approximation that can be employed in analytic treatments of momentum coupling and deflection (see the Chapter by Solem and Snell). Unfortunately, the curves for the other materials do not follow a linear behavior and this simple approximation cannot be applied.

The scaled momentum coupling curves for nickel-iron are displayed in Fig. 21. The effective bulk strength of nickel-iron asteroids is uncertain and may, in fact, vary over a rather broad range (J. L. Remo 1993, personal communication). Accordingly, we have performed two sets of calculations,

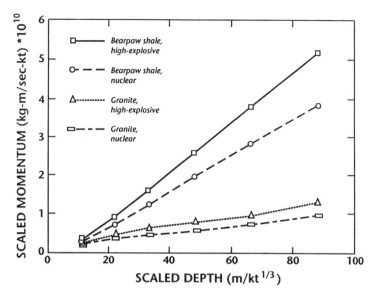

Figure 20.   Scaled momentum coupling vs scaled depth of burial, for nuclear and high explosive events in Bearpaw shale (fully saturated, very weak rock) and in granite (dense, high-strength rock).

the first for a normal or realistic strength representative of competent material (lower pair of curves), and the other for a greatly reduced strength equal to that of granite (upper pair of curves). The low-strength case might be appropriate for a severely flawed material with many large-scale fractures caused by previous collisions with other bodies. The coupling efficiencies are greater for the low-strength material, as expected. The influence of strength was further explored by performing several calculations for a range of assumed strengths at a fixed scaled burial depth of 23 m $kt^{-1/3}$. As shown in Fig. 22, the momentum coupling varies only slightly for strengths of 0.3 and 0.6 GPa. As the strength is increased to 0.8 GPa, the coupling becomes more sensitive and decreases steeply. At a strength of 1.2 GPa, the material did not fracture all the way to the free surface and the event failed to crater. No material would be ejected in this case and no momentum would be imparted to the object. Strength is a key parameter that must be understood and correctly modeled to obtain accurate predictions.

Another illuminating result emerged from the investigation of strength effects. In empirical studies of cratering on Earth, the energy yield Y is often taken as the principal scaling variable. Length dimensions are scaled as Y^{-n} where the exponent n is between 1/3 and 1/4, and volumes are scaled as Y^{-3n}. This simple approach is sometimes adequate because the densities of crustal silicate rocks typically fall within a narrow range between 1.6 and 2.7 g cm^{-3}, and other physical properties are likewise roughly comparable. It

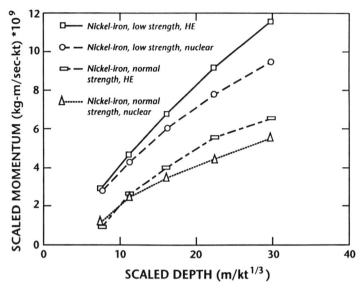

Figure 21.   Scaled momentum coupling vs scaled depth of burial for nuclear and high
explosive events in nickel-iron. Two different strength descriptions were assumed, a
realistic strength for the competent material ("normal strength," lower two curves),
and a reduced strength equal to that used for granite ("low strength," upper two
curves).

has been demonstrated that data for a much wider range of material properties
and physical conditions can be collapsed by using more general scaling laws
(Schmidt and Housen 1987). A simple example is the incorporation of the
density variable ρ into the scaling law, with lengths scaling as $Y^{-1/3}/\rho$. (To
achieve rigorous similitude for this type of scaling, the energy source characteristics and other physical properties would also have to be modified when
ρ is changed (Schmidt and Housen 1987).) We have performed calculations
for nickel-iron and granite with identical strength models, but with densities
that differ by a sizable factor (7.8 g cm^{-3} vs 2.65 g cm^{-3}). This provides an
opportunity to examine the validity of a density scaling law. It is necessary
to note that the Hugoniots and other physical properties of nickel-iron and
granite are quite different, so we would not anticipate exact agreement of
the density scaling. Figure 23 compares the momentum coupling curves vs
the density-scaled depth of burial. Despite the aforementioned differences in
material properties, the density-scaled coupling curves for the two materials
are in reasonably good agreement. Incorporating density into the scaling law
and using equal strengths for the two materials were sufficient to collapse the
coupling curves over most of the scaled depth range. The nickel-iron curves
are somewhat lower at the shallow depth end of the range, a fact which is undoubtedly attributable to incorrect similitude for some of the other parameters
of the system.

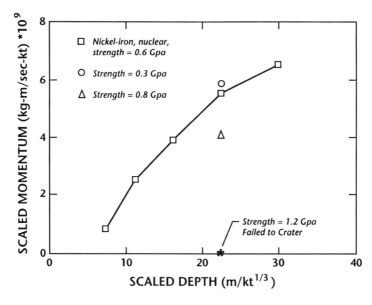

Figure 22.   Illustrations of the influence of strength variation on momentum coupling (nuclear sources in nickel-iron). The highest strength case failed to crater, i.e., the event was fully contained.

Momentum coupling curves for ice and for 43% porous snow are presented in Fig. 24. The efficiencies are dramatically lower for snow due to dissipation by irreversible porous compaction at low stress levels. Almost all of the void porosity is removed at a stress of 25 MPa. This effect is obviously much more important for coupling applications than the solid phase transitions that occur in ice at higher pressures. The porosity of 43% was chosen as a typical value for consolidated snow and because experimental data were available at this density. Unconsolidated or "fluffy" snow of higher porosity would give even greater dissipation and less efficient coupling than this material. The porosity variable has a major influence, and information about the porosity and its depth dependence would be needed to make accurate predictions of coupling for the porous material near the surface of a comet.

The complex phase transition behavior of ice in the 0.2 to 2.2 GPa stress range can potentially influence the accuracy of the calculational models. The calculations discussed above assume that ice transformed to the high-density phase above 0.6 GPa and unloaded metastably, i.e., that no reversions to the low-density phases occur. This is the most pessimistic assumption, because unloading along the steep high-density phase path gives the greatest dissipation and the least efficient momentum coupling. To examine the influence of different hypothetical unloading behaviors, we have performed additional calculations assuming that extreme phase reversal takes place during unloading (Fig. 25; compare dashed and dotted curves). The behavior shown here

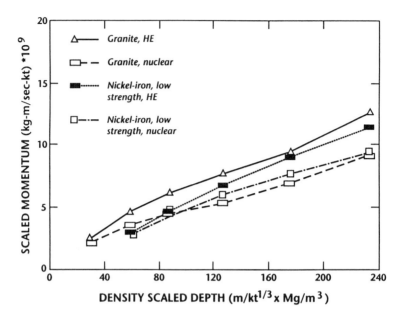

Figure 23.   Scaled momentum coupling vs density scaled depth of burial for nuclear
events in granite and in nickel-iron with reduced strength equal to that of granite.
The coupling curves for the two different materials lie close together when equal
strengths are assumed and the burial depth is scaled with density.

is probably quite unrealistic, because the experimental data indicate that no
significant phase reversal occurs at pressures above 0.2 GPa (0.002 Mbar),
at least on the microsecond time scale of the dynamic experiments (Gaffney
1985; Larson 1984). However, this extreme assumption will give an estimate
of the largest possible effect of phase reversion. Momentum coupling curves
calculated with no phase reversion and with the extreme phase reversion are
compared in Fig. 26. Although the close-in waveforms were somewhat modified by the phase reversion model, the difference in momentum coupling is
negligible. Clearly, the occurrence of phase reversion on unloading does not
have a large influence on momentum coupling, and detailed modeling of this
behavior is not critical for the present application.

The coupling curves for ice and snow are compared with the silicate rock
materials in Figs. 27 and 28, respectively. The ice curves lie approximately
midway between Bearpaw shale and granite (Fig. 27). The snow curves are
lower and are more nearly comparable to the granite (Fig. 28; note expanded
scale). Again, we note that the low coupling efficiency for snow is due to
the irreversible porous compaction at low stresses. These results demonstrate
that void porosity can be comparable to strength in importance and must be
properly taken into account in the calculational models.

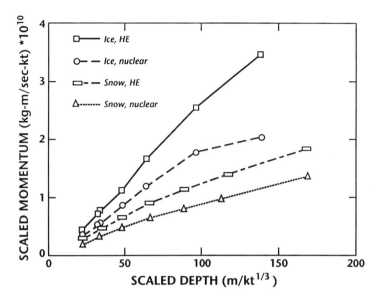

Figure 24.   Scaled momentum coupling vs scaled depth of burial for nuclear and high explosive events in ice and in snow.

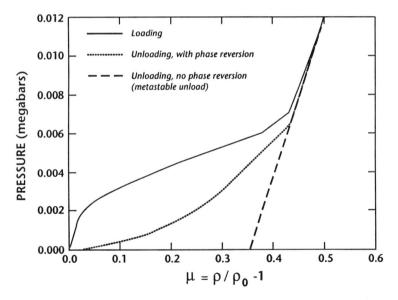

Figure 25.   Pressure vs excess compression behavior of ice at low to moderate pressures. Two possible unloading curves are shown: with no phase reversion (dashed), and with extreme phase reversion (dotted).

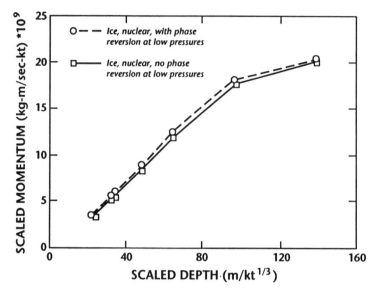

Figure 26.   Influence of phase reversion on momentum coupling for nuclear events
in ice.

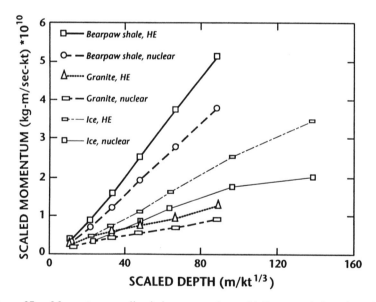

Figure 27.   Momentum coupling in ice: comparison with Bearpaw shale and granite.

C. Summary of Impulse Coupling for Buried Events

A summary of the impulse coupling coefficients for buried nuclear and high
explosive events is given in Table IV. At typical cratering depths, the best

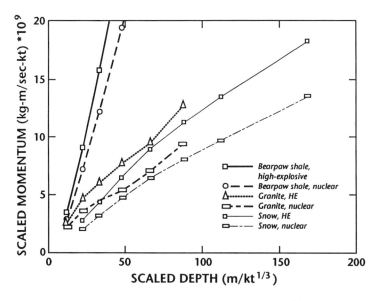

Figure 28.   Momentum coupling in snow: comparison with Bearpaw shale and granite.

		I^*
Pulsed Lasers		2 - 10
Kinetic Energy	10 km / s	25 - 50
	100 km / s	10 - 20
Standoff X - Rays	5 keV	4 - 28
	10 keV	7 - 27
	30 keV	10 - 23
	60 keV	12 - 26
	100 keV	15 - 30
Standoff Neutrons	14 MeV	10 - 70
	Thermonuclear	14 - 140
	Fission	16 - 190
Deep Burial	Bearpaw Shale	60 - 900
	Hard Granite	40 - 220
	Nickel - Iron	20 - 130
	Ice	80 - 480
	Snow	50 - 320

Figure 29.   Specific results for impulse coupling.

coupling is achieved for relatively weak and nonporous materials (saturated Bearpaw shale, ice). The coupling efficiency tends to progressively decrease for highly porous materials (snow) and for increasing material strength (granite, nickel-iron). Note that deeply buried explosives produce the highest values of I^* for any of the diversion schemes addressed in this study.

TABLE IV
Impulse Coupling Coefficients (I^*) for Minimum and
Maximum Depths of Burial (DOB)[a]

	Bearpaw Shale	Hard Granite	Nickel-Iron[a]	Ice	Snow
Min DOB, High explosive	90	50	30	110	70
Min DOB, Nuclear	60	40	20	80	50
Max DOB, High explosive	1220	300	160	830	440
Max DOB, Nuclear	900	220	130	480	320

[a] See Table III.

$$I^* = \frac{I \times V_{REF}}{E}$$

FOR SURFACE LOADING WITH PHOTONS

AND NEUTRONS, AND $I^* < $ MAX I^*

$I^* \sim 1/\sqrt{E}$

$I \sim \sqrt{E}$ (sublinear dependence)

FOR KINETIC ENERGY LOADING

$I^* \sim E^{-a}$ ($a = 0 - 0.13$)

$I \sim E^b$ ($b = 0.87 - 1.0$)

Figure 30.   General results for impulse coupling.

IX. SUMMARY AND CONCLUSIONS

A concise summary of the results obtained for this chapter is presented in Figs. 29 and 30. These results are intended to provide reasonable estimates of target impulse coupling coefficients for scientists and engineers interested in the problem of NEO diversion. For future studies of asteroid and comet diversion, the impulse measure derived in Sec. III may prove to be more useful than the measure I^* derived in Fig. 2. If a serious quest were undertaken to design and build a NEO diversion system, many more detailed simulations and experiments would be required.

Acknowledgments. The authors would like to acknowledge the assistance of E. C. Schwegler, Jr., of the Los Alamos National Laboratory, in performing the neutron impulse calculations, and T. J. Lawrence, United States Air Force Academy, for reviewing this chapter and providing many useful suggestions.

REFERENCES

Anderson, G. D. 1968. The Equations of State of Ice and Composite Frozen Soil Materials. U. S. Army Cold Regions Research and Engineering Laboratory (Hanover, N. H.), Report 257.

Bakanova, A. A., Zubarev, V. N., Sutulov, Yu. N., and Trunin, R. F. 1975. Thermodynamic properties of water at high pressures and temperatures. *Soviet Physics-JETP* 41(3):544–548.

Baker, W. E., Westine, P. S., and Dodge, F. T. 1973. *Similarity Methods in Engineering Dynamics: Theory and Practice of Scale Modeling* (Rochelle Park, N. J.: Hayden Book) p. 252.

Bryan, J. B., Snell, C. M., Heusinkveld, M., Burton, D. E., Bruce, A. L., Lettis, L. A., and Butkovich, T. R. 1977. Controlled blasting calculations and experiments. In *Proceedings of American Nuclear Society Topical Meeting on Energy and Mineral Resource Recovery Research*.

Budiansky, B., and O'Connell, R. J. 1976. Elastic moduli of a cracked solid. *Intl. J. Solids Structures* 12:81–97.

Burton, D. E., and Schatz, J. F. 1975. Rock Modeling in TENSOR74, A Two-Dimensional Lagrangian Shock Propagation Code. Lawrence Livermore National Laboratory Report UCID-16719 (Livermore, Calif.: Lawrence Livermore National Lab).

Burton, D. E., Snell, C. M., and Bryan, J. B. 1975. Computer design of high-explosive experiments to simulate subsurface nuclear detonations. *Nuclear Tech.* 26:65–87.

Burton, D. E., Lettis, L. A., Bryan, J. B., Butkovich, T. R., and Bruce, A. L. 1977. Anisotropic Creation and Closure of Tension Induced Fractures. Lawrence Livermore National Laboratory Report UCRL-79578 (Livermore, Calif.: Lawrence Livermore Lab).

Butkovich, T. R. 1967. The Gas Equation of State for Natural Materials. Lawrence Livermore National Laboratory Report UCRL-14729 (Livermore, Calif.: Lawrence Livermore Lab).

Butkovich, T. R., Moran, B., and Burton, D. E. 1981. A Model for Calculating Shock Loading and Release Paths for Multicomponent Geologic Media. Lawrence Livermore National Laboratory Report UCRL-53178 (Livermore, Calif.: Lawrence Livermore National Lab).

Chabai, A. J. 1965. On scaling dimensions of craters produced by buried explosives. *J. Geophys. Res.* 70:5075–5098.

Chabai, A. J. 1977. Influence of gravitational fields and atmospheric pressures on scaling of explosion craters. In *Impact and Explosion Cratering*, eds. D. J. Roddy, R. O. Pepin and R. B. Merrill (New York: Pergamon Press), pp. 1191–1214.

Cherry, J. T., and Petersen, F. L. 1970. Numerical simulation of stress wave propagation from underground nuclear explosions. In *American Nuclear Society*

Symposium on Engineering with Nuclear Explosives, vol. 1, CONF-700101, Jan. 14–16, Las Vegas, Nev., pp. 142–220.

Cooper, H. F., Jr., and Sauer, F. M.. 1977. Crater-related ground motions and implications for crater scaling. In *Impact and Explosion Cratering*, eds. D. J. Roddy, R. O. Pepin and R. B. Merrill (New York: Pergamon Press), pp. 1133–1163.

Crowley, B. K. 1973. Effects of porosity and saturation on shock-wave response of tuffs. *Intl. J. Rock Mech. Mining Sci.* 10:437.

Dingus, R. S., and Goldman, S. R. 1986. Plasma energy balance model for optical-laser induced impulse in vacuo. In *Proceedings of the International Conferences on Lasers '86*, ed. R. W. McMillan, Nov. 3–7, Orlando, Fla., pp. 111–122.

Dobratz, B. M.. 1981. LLNL Explosive Handbook, Properties of Chemical Explosives and Explosive Simulants. Lawrence Livermore National Laboratory Report UCLR-52997 (Livermore: Calif.: Lawrence Livermore National Lab).

Eichelberger, R. J. 1965. Hypervelocity impact. In *Behavior of Materials Under Dynamic Loading*, ed. N. J. Huffington (New York: American Soc. of Mechanical Engineers), pp. 175–187.

Evans, R. D. 1955*a*. *The Atomic Nucleus* (New York: McGraw Hill), pp. 711–716.

Evans, R. D. 1955*b*. *The Atomic Nucleus* (New York: McGraw Hill), pp. 456–458.

Gaffney, E. S. 1979. Equations of state of ice and frozen soils. *Proc. Lunar Planet. Sci. Conf.* 10:416–418.

Gaffney, E. S. 1980. The identification of ice VI on the Hugoniot of ice Ih. *Geophys. Res. Lett.* 7:407–409.

Gaffney, E. S. 1985. Hugoniot of water ice. In *Ices in the Solar System*, eds. J. Klinger, D. Benest, A. Dollfus and R. Smoluchowski (Dordrecht: D. Reidel), pp. 119–148.

Gaffney, E. S., and Matson, D. L. 1980. Water ice polymorphs and their significance on planetary surfaces. *Icarus* 44:511–519.

Gaffney, E. S., Blaisdell, G. L., and Brown, J. A. 1985. Shock compression of water ice and snow. Symposium on Planetary Ice, American Geophysical Union Fall Meeting, Dec. 9–13, San Francisco, Calif.

Glasstone, S., and Dolan, P. J. 1977*a*. *The Effects of Nuclear Weapons*, 3rd ed. (Washington, D. C.: U. S. Dept. of Defense and the Energy Research and Development Administration), pp. 7–12.

Glasstone, S., and Dolan, P. J. 1977*b*. *The Effects of Nuclear Weapons*, 3rd ed. (Washington, D. C.: U. S. Dept. of Defense and the Energy Research and Development Administration), pp. 364–365.

Hammerling, P., and Remo, J. L. 1993 NEO interaction with X-ray and neutron radiation. In *Proceedings of the Near-Earth Object Interception Workshop*, eds. G. H. Canavan, J. C. Solem and J. D. G. Rather (Los Alamos: Los Alamos National Lab), pp. 186–193.

Holsapple, K. A., and Schmidt, R. M. 1980. On scaling of crater dimensions. I. Explosive processes. *J. Geophys. Res.* 85:7247–7256.

Housen, K. R., Schmidt, R. M., and Holsapple, K. A. 1983. Crater ejecta scaling laws: Fundamental forms based on dimensional analysis. *J. Geophys. Res.* 88:2485–2499.

Johnson, J. B., Brown, J. A., Gaffney, E. S., Blaisdell, G. L., and Solie, D. J. 1992. Shock Response of Snow—Analysis of Experimental Methods and Constitutive Model Development. Cold Regions Research and Engineering Laboratory Report 92-12 (Hanover, N. H.: Cold Regions Research and Engineering Lab).

Johnson, J. B., Solie, D. J., Brown, J. A., and Gaffney, E. S. 1993. Shock response of snow. *J. Applied Physics*, 73:4852–4861.

Kieffer, S. W. 1977. Impact conditions required for formation of melt by jetting in silicates. In *Impact and Explosion Cratering*, eds. D. J. Roddy, R. O. Pepin and R. B. Merrill (New York: Pergamon Press), pp. 751–769.

Killian, B. G., and Germain, L. S. 1977. Scaling of cratering experiments—an analytical and heuristic approach to the phenomenology. In *Impact and Explosion Cratering*, eds. D. J. Roddy, R. O. Pepin and R. B. Merrill (New York: Pergamon Press), pp. 1165–1190.

Knowles, C. P., and Brode, H. L. 1977 The theory of cratering phenomena, an overview. In *Impact and Explosion Cratering*, eds. D. J. Roddy, R. O. Pepin and R. B. Merrill (New York: Pergamon Press), pp. 869–896.

Kreyenhagen, K., and Schuster, S. 1977. Review and comparison of hypervelocity impact and explosion cratering calculations. In *Impact and Explosion Cratering*, eds. D. J. Roddy, R. O. Pepin and R. B. Merrill (New York: Pergamon Press), pp. 983–1002.

Langhaar, H. L. 1962. *Dimensional Analysis and Theory of Models* (New York: J. Wiley & Sons), pp. 18–43.

Larson, D. B. 1978. Explosive energy coupling in ice and frozen soils. In *Proceedings of the Third International Conference on Permafrost*, vol. I, July 10–13, Edmonton, Alberta, Canada, pp. 806–812.

Larson, D. B. 1984. Shock wave studies of ice under uniaxial strain conditions. *J. Glaciology* 30(105):235–240.

Larson, D. B., Bearson, G. D., and Taylor, J. R. 1973. Shock wave studies of ice and two frozen soils. In *Second International Conference on Permafrost*, Yakutsk, U. S. S. R., 13–28 July, pp. 318–325.

Lee, E. L., Hornig, H. C., and Kury, J. W. 1968. Adiabatic Expansion of High Explosive Detonation Products. Lawrence Livermore National Laboratory Report UCRL-50422 (Livermore, Calif.: Lawrence Livermore National Lab).

Lee, E. L., Finger, M., and Collins, W. 1973. JWL Equation of State Coefficients for High Explosives. Lawrence Livermore National Laboratory Report UCID-16189 (Livermore, Calif.: Lawrence Livermore National Lab).

Maenchen, G., and Sack, S. 1964. The TENSOR Code. In *Methods in Computational Physics*, vol. 3, eds. B. Alder, S. Fernbach and M. Rotenberg (New York: Academic Press), pp. 181–210.

McQueen, R. G., Marsh, S. P., and Fritz, J. N. 1967. Hugoniot equation of state of twelve rocks. *J. Geophys. Res.* 72:4999–5086.

Melosh, H. J. 1980. Cratering mechanisms—observational, experimental, and theoretical. *Ann. Rev. Earth Planet. Sci.* 8:65–93.

Nakamura, A., and Fujiwara, A. 1991. Velocity distribution of fragments formed in a simulated collisional disruption. *Icarus* 92:132–146.

Nordyke, M. D. 1961. Nuclear craters and preliminary theory of mechanics of explosive crater formation. *J. Geophys. Res.* 66:3439–3459.

Nordyke, M. D. 1977. Nuclear cratering experiments: United States and Soviet Union. In *Impact and Explosion Cratering*, eds. D. J. Roddy, R. O. Pepin and R. B. Merrill (New York: Pergamon Press), pp. 103–124.

Remo, J. L. 1993. Asteroid/meteorite analogs and material properties. In *Proceedings of the Near-Earth Object Interception Workshop*, eds. G. H. Canavan, J. C. Solem and J. D. G. Rather (Los Alamos: Los Alamos National Lab), p. 168.

Schatz, J. F. 1973. The Physics of SOC and TENSOR. Lawrence Livermore National Laboratory Report UCRL-51352 (Livermore, Calif.: Lawrence Livermore National Lab).

Schatz, J. F. 1974. SOC73, A One-Dimensional Wave Propagation Code for Rock Media. Lawrence Livermore National Laboratory Report UCRL-51689 (Livermore, Calif.: Lawrence Livermore National Lab).

Schmidt, R. M., and Housen, K. R. 1987. Some recent advances in the scaling of impact and explosion cratering. *Intl. J. Impact Eng.* 5:543–560.

Seidl, F. C. P. 1965. SOC, A Numerical Model for the Behavior of Materials Exposed to Intense Impulsive Stresses. Lawrence Livermore National Laboratory Report UCID-5033 (Livermore, Calif.: Lawrence Livermore National Lab).

Snell, C. M. 1994. User's Manual for SOC Solid Dynamics Code. LAMS Report in Publication (Los Alamos: Los Alamos National Lab).

Taylor, L. M., and Flanagan, D. P. 1987. PRONTO 2D, A Two-Dimensional Transient Solid Dynamics Program. Sandia National Laboratories Report SAND-86-0594 (Albuquerque, N. M.: Sandia National Lab).

Taylor, L. M., and Flanagan, D. P. 1989. PRONTO 3D, A Three-Dimensional Transient Solid Dynamics Program. Sandia National Laboratories Report SAND-87-1912 (Albuquerque, N. M.: Sandia National Lab).

Taylor, L. M., Chen, E. P., and Kuszmaul, J. S. 1986. Microcrack induced damage accumulation in brittle rock under dynamic loading. *Computer Methods in Applied Mech. Eng.* 55:301–320,

Terhune, R. W, Stubbs, T. F, and Cherry, J. T. 1970. Nuclear cratering on a digital computer. In *Proceedings of Symposium on Engineering with Nuclear Explosives*, vol. 1, American Nuclear Society CONF-700101, Jan. 14–16, Las Vegas, Nev., p. 334.

Wackerle, J. 1962. Shock wave compression of quartz. *J. Applied Phys.* 33:922–937.

Westine, P. S. 1970. Explosive cratering. *J. Terra-mechanics* 7:9.

White, J. W. 1971. Examination of cratering formulas and scaling methods. *J. Geophys. Res.* 76:8599–8603.

White, J. W. 1973. An empirically derived cratering formula. *J. Geophys. Res.* 78:8623–8633.

APPENDIX I

Impulse Coupling in Porous Media

We developed a simplified model for the impulses applied to a rigid wall by full-density (a) and porous (b) hot slabs containing the same amounts of deposited energy. For this study, the porosity factor is f. The conditions in both slabs for $t = 0$ and $t > 0$ are shown in Fig. 31. Our goal is to obtain a rough comparison of the impulses produced by cases (a) and (b). To do this, we calculate, for each case, the impulse applied to each wall from $t = 0$ until the arrival of the release wave at the wall. Although this model does not account for the impulse that results from the tail of the release wave, this model is sufficiently accurate to obtain the ratio of the impulse of case (b) to the impulse of case (a).

From Fig. 31, the initial pressures from cases (a) and (b) are p_a and p_b, respectively. The times for release wave arrival are t_0^a and t_0^b, respectively. Assuming a polytropic expansion, the calculation proceeds as follows:

$$pV^\gamma = \text{constant} = C \tag{A.1}$$

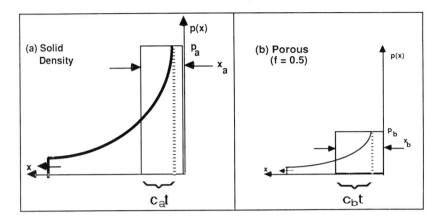

Figure 31.   Comparison of pressure profiles in full density and porous hot slabs.

then

$$p = C\rho^{\gamma}. \tag{A.2}$$

Now, estimate the impulses for cases (a) and (b).

$$I_a \doteq p_a t_0^a = \frac{p_a x_a}{c_0^a}$$

$$I_b \doteq p_b t_0^b = \frac{p_b x_b}{c_0^b} \tag{A.3}$$

where

$$c_0^2 = \left(\frac{\partial p}{\partial \rho}\right)_s = C\gamma\rho^{\gamma-1} \tag{A.4}$$

$$c_0 = \sqrt{\gamma C}\rho^{\frac{\gamma-1}{2}}. \tag{A.5}$$

The impulse ratio $I \equiv I_b/I_a$ is given by

$$I = \frac{p_b x_b}{c_0^b} \Big/ \frac{p_a x_a}{c_0^a} = \left(\frac{p_b}{p_a}\right)\left(\frac{x_b}{x_a}\right) \Big/ \left(\frac{c_0^b}{c_0^a}\right)$$

$$= \left(\frac{C\rho_b^{\gamma}}{C\rho_a^{\gamma}}\right)\left(\frac{x_b}{x_a}\right) \Big/ \left(\frac{\sqrt{\gamma C}\rho_b^{\frac{\gamma-1}{2}}}{\sqrt{\gamma C}\rho_a^{\frac{\gamma-1}{2}}}\right)$$

$$I = \left(\frac{\rho_b}{\rho_a}\right)^{\gamma}\left(\frac{x_b}{x_a}\right)\left(\frac{\rho_b}{\rho_a}\right)^{-\frac{\gamma-1}{2}}$$

$$\rho_b = (1-f)\rho_a \tag{A.6}$$

$$x_b = x_a/(1-f) = x_a(1-f)^{-1}.$$

Therefore

$$\frac{\rho_b}{\rho_a} = 1 - f, \quad \frac{x_b}{x_a} = (1 - f)^{-1} \tag{A.7}$$

then

$$I = (1 - f)^{\gamma}(1 - f)^{-1}(1 - f)^{-\frac{\gamma-1}{2}} \tag{A.8}$$
$$I = (1 - f)^{\frac{\gamma-1}{2}}.$$

In general, for gaseous molecules (or atoms) $\gamma = (\xi + 2)/\xi$, where ξ is the number of quadratic terms in the Hamiltonian of the molecules (or atoms) in the heated slab. Next, evaluate the exponent of $(1 - f)$.

$$\frac{\gamma - 1}{2} = \frac{\frac{\xi+2}{\xi} - 1}{2} = \frac{1}{\xi}. \tag{A.9}$$

Then our general expression becomes

$$I = (1 - f)^{\frac{1}{\xi}}. \tag{A.10}$$

We consider two specific cases, the first for a monatomic gas ($\xi = 3$) and the second for a diatomic gas ($\xi = 5$). If $f = 0.5$, we have

$$I = (0.5)^{1/3} = 0.794 \quad \text{(monatomic)}$$
$$I = (0.5)^{1/5} = 0.871 \quad \text{(diatomic)}. \tag{A.11}$$

Even for porosities as high as 0.5, target impulse reductions of only 10 to 20% result.

APPENDIX II

Deflection of Asteroids by Surface Energy Deposition

This appendix considers asteroid deflection using energy deposition on the asteroid's surface from an external source. Highly simplified models for deposition, ablation, and spall have been used to estimate the velocities imparted to various sizes and shapes of asteroids. The case considered is that of either neutron or X-ray deposition from a nuclear device, but the model applies with minor modifications to laser deposition or other energy sources. The main issues considered here are the minimum energy requirements for ablation and spall deflection, taking into account the many geometrical factors. We find that velocity changes on the order of 1 m s^{-1} should be possible with a 1 Mton device for asteroids of about 1/2 km diameter. Various scaling laws and comparisons with simple formulae are considered. One major outstanding problem is the probability of fragmentation with such a technique. The details of the analysis follow.

Evaluation of the standoff burst scenario requires an understanding of the efficiency of the process, the details of the deposition mechanism, and the mechanical reaction of the main body. This appendix deals mainly with the first of these three items, with emphasis on the geometric penalties and the energy thresholds to be overcome. This takes us one step beyond the simple scaling laws considered in previous papers. The other two issues will be approached with detailed hydrodynamic models; research is currently underway.

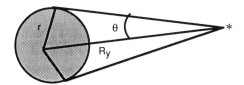

Figure 32.   Basic geometry of a standoff burst irradiating a spherical object.

Although no asteroids are perfect spheres, the spherical geometry is useful for obtaining an understanding of the scaling involved in irradiating a finite-sized body. The basic geometry considered is shown in Fig. 32, where the following quantities are defined: r = radius of asteroid; R_y = standoff distance of burst from center of asteroid; and θ = half angle intercepted by asteroid. Once the problem has been analyzed for a spherical asteroid, we generalize to the case of a sphere (or other shape) with a flat surface facing the burst.

Geometry plays an important role in the case of a standoff burst for several reasons. First, the standoff distance and object shape determine how much of the yield of the device actually is deposited on the object. Second, they determine the area of the object that will be ablated. Third, they determine the local angle of deposition, which, in turn, determines the depth of the ablated layer. Finally, the shape of the deposition area will affect the direction of the ablated velocity and the efficiency of the deflection process. Combining these geometric effects with some simple energy considerations results in a simple, but accurate, model of the burst process.

By far the most severe penalty for standoff bursts of spherically symmetrical devices is the fraction of energy radiated away into space. For the spherical asteroid shown in Fig. 32, the energy interception angle is given by

$$\sin(\theta) = \frac{r}{R_y}. \tag{A.12}$$

If we define E_y = total radiated yield of the device and E_d = total energy deposited on the asteroid, then θ determines the fraction of energy deposited by considering the solid angle subtended:

$$E_d = E_y \frac{(1 - \cos(\theta))}{2} \tag{A.13}$$

or combining Eqs. A.12 and A.13:

$$E_d = \frac{E_y}{2}\left[1 - \sqrt{1 - \left(\frac{r}{R_y}\right)^2}\right]. \tag{A.14}$$

At $r = R_y$, exactly 1/2 of the energy falls on the object, while at large R_y, the fraction deposited approaches $(1/4)(r/R_y)^2$.

The sphere result may be contrasted to that of a disk of radius r. We imagine that this flat area is on the surface of an object also of radius r (this makes the scales comparable to the sphere case). In this case, Eq. (A.12) becomes:

$$\tan(\theta) = \frac{r}{R_y - r}. \tag{A.15}$$

Combining Eqs. A.13 and A.15 produces:

$$E_d = \frac{E_y}{2}\left[1 - \frac{1}{\sqrt{1 + \left(\frac{r}{R_y-r}\right)^2}}\right]. \tag{A.16}$$

In both limits, Eqs. (A.14) and (A.16) agree. Figure 33 shows these two results, where E_d is in units of E_y and R_y is in units of r. At a distance of 5 times the radius, the deposited energy has fallen a factor of 100 below the device yield, while at a standoff equal to the radius the disk intercepts about a factor of 2 more energy than the sphere. Just above the surface, the spherical deposition drops much more rapidly than the disk case. This means that if a flat area can be found on an otherwise spherical asteroid, this will be the optimal spot for a near surface burst. Because such flat areas are often associated with craters on observed asteroids, we refer to this scenario as the "crater burst." Another method to avoid this penalty would be to somehow direct the device energy toward the asteroid. Laser irradiation would fall into this category, allowing all of the device energy to be deposited. We now continue with the blowoff impulse calculations by estimating the blowoff mass and velocity.

We introduce the depth of penetration of the irradiation on the asteroid surface: d_O = depth of penetration at normal incidence. Two cases will be considered: uniform deposition to depth d_O and exponential deposition with e-folding length d_O. These two cases should provide approximate bounds for the deposition profiles encountered in practice. In a rough sense, the first case is an approximation to the deposition profile of energetic particles, such as neutrons; the second is appropriate to high-energy radiation, such as X-rays. For a sphere, the angle of incidence produces an actual depth that is d_O at the center of the radiated region and drops to zero at the edge. The area averaged mean depth can be shown to be approximately $(2/\pi)^2 d_O = 0.41 d_O$ in this case. For a flat surface, the depth falls to zero at infinite radius and we

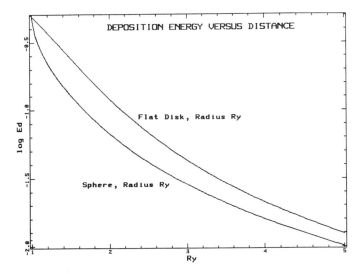

Figure 33.   Deposition fraction of energy falling on a sphere and a disk as a function of standoff distance.

ignore this correction. Because we are concerned with the amount of ablated material, we define D as the mean depth of vaporized layer. For the uniform deposition case:

$$D = d_o \quad \text{(flat surface)} \tag{A.17}$$

$$D = 0.41 d_o \quad \text{(sphere)} \tag{A.18}$$

provided that the layer is heated above the vaporization temperature. Introduce the following: ρ = density of the material, g cc^{-1}; H_m = heat of melt, erg g^{-1}; H_v = heat of vaporization, erg g^{-1}; C_v = specific heat of the solid/liquid phase, erg g-K^{-1}; T_v = temperature of vaporization, K.

The threshold specific energy required to vaporize the material (from an initial temperature of zero K) is

$$e_{th} = C_v T_v + H_m + H_v. \tag{A.19}$$

The mass of the deposition layer is

$$\text{mass} = A\rho D. \tag{A.20}$$

The specific energy of a uniformly heated layer will be

$$e_n = \frac{E_d}{\text{mass}}. \tag{A.21}$$

So Eqs. A.17 and A.18 apply only if $e_n > e_{th}$, otherwise $D = 0$.

For the exponential deposition case, the situation is different. Here the specific energy is a function of the depth x measured from the surface of the material:

$$e = e_n \left[\exp \left(-\frac{x}{d_o} \right) \right] \qquad (A.22)$$

where e_n is given by Eqs. (A.20) and (A.21) with $D = d_O$, because the integral over x of Eq. (A.22) is just $e_n d_O$, as in the uniform deposition case. This deposition profile has two consequences. First, there will be a tail of deep deposition that will not be sufficient to vaporize the material and this energy will be lost to the ablation process. It is easily shown that this lost energy is exactly equal to e_{th} and that the depth of the ablated layer is

$$D = d_o ln \left(\frac{e_n}{e_{th}} \right) \quad \text{(exponential deposition)} \qquad (A.23)$$

and, as in the uniform case, an additional factor of 0.41 must be included for spherical geometry. The consequence of Eq. (A.23) is that the ablated layer is thicker and cooler than the uniform case for heating well above threshold. On the other hand, the energy of the ablation will be

$$e_{abl} = e_n - e_{th} \quad \text{(uniform deposition)} \qquad (A.24)$$

$$e_{abl} = e_n - 2e_{th} \quad \text{(exponential deposition).} \qquad (A.25)$$

This means that less energy is available in the exponential case and thus a larger source of energy will be required to initially vaporize the material if the deposition is exponential.

Once the mass and energy of the ablated material is are known, it is a simple matter to compute the ablation velocities. First, we estimate γ, ratio of specific heats of the ablated material. We will take $\gamma = 5/3$ in the computations below, but keep in mind that the ablated material will probably be dissociating and ionizing and these processes could lower γ to values as low as 1.1. The ablation energy then implies an ablation velocity of

$$v_{abl} = \sqrt{\gamma(\gamma - 1)e_{abl}} \qquad (A.26)$$

where the gamma dependent term is the ideal gas value. For $\gamma = 5/3$, almost exactly 1/2 of the ablation energy is converted to kinetic energy of the expanding material; the other half remains as thermal. For lower gamma, more of the energy is locked into internal modes and v_{abl} is reduced.

One final geometric correction is needed for the spherical asteroid; the ablation velocities are not all directed toward the burst point, but are normal to the asteroid surface. Averaging over angle gives the projection correction:

$$\text{proj} = 1 \quad \text{(flat surface)} \qquad (A.27)$$

$$\text{proj} \approx \frac{1}{2}\left(1 + \frac{r}{R}\right) \quad \text{(sphere)}. \tag{A.28}$$

If we define M as the total asteroid mass, then conservation of momentum produces V, the reaction velocity (differential change in velocity in the direction away from the burst):

$$V = (\text{proj} \times \text{mass})\frac{v_{abl}}{M} \tag{A.29}$$

For the sake of comparison, we adopt as a standard model a spherically symmetric burst with a yield of 1 Mton in either neutrons or x-radiation. In Fig. 34, we show the reaction velocity increment as a function of standoff distance for neutrons (uniform deposition to a depth of 10 cm) onto stony meteors of radius 100, 200, and 300 m. Note that the scale extends from the center of each asteroid to 600 m, with the left-most zero of each curve occurring at each asteroid surface. The right-most zero value indicates that beyond this distance the energy density is not sufficient to vaporize any of the asteroid material. The peak reaction velocity is attained for each case at a standoff distance of about 20 to 30% of the radius, with peak change in velocity being 260 cm s^{-1} for the 100 m case, 65 cm s^{-1} for the 200 m case, and 28 cm s^{-1} for the 300 m case. These values would be sufficient to deflect a dangerous asteroid if detected early enough. For reference, at $r = 500$ m, the peak V is 5.3 cm s^{-1}; at $r = 750$ m, V-peak 3.3 cm s^{-1}; and at $r = 1$ km, V-peak $= 1.5$ cm s^{-1}. Figure 35 shows the same plot, but for the case of 1 MT of X-rays (assuming a deposition depth of 0.01 cm and an exponential deposition profile) onto a stony asteroid. Here the peak is a little closer to the asteroid surface, the peak velocities are down by about a factor of 3, and the threshold distance is much larger in this case (60 km, in fact, for the $r = 100$ m case), because of the much smaller deposition depth. Figure 36 shows the same plot, but for the case of 1 Mton of neutrons onto an iron-nickel asteroid. The peak velocities are about 60% of the stony/neutron case shown in Fig. 34 and the higher density for this material has narrowed the range of effective standoff distances relative to the stony case. Figure 37 shows the case of 1 Mton of X-rays onto an Fe/Ni asteroid. Again, the shape has changed as in Fig. 35 and the maximum velocity is down by about a factor of 3 from Fig. 36.

As noted in the previous section, the pressures generated by the ablation process can be substantial and the possibility of fragmenting, fracturing, or spalling layers from the asteroid needs to be considered. The details of the energy pulse length and shape are important in determining the peak pressure, as is the internal geometry of the object in determining the location and magnitude of tensile waves within the asteroid. In view of the large pressures predicted here at time zero, it is likely that for all cases near the "optimum" standoff distance, the spall strength of the material (which is on the order of a few tens of kbar or less) will be exceeded. For nearly instantaneous deposition, the unloading wave at the front surface is likely to produce "front surface spall," which will detach a layer of material in which the pressure

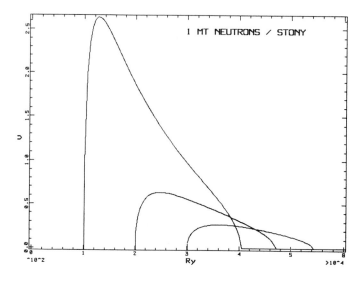

Figure 34.   Deflection velocity (cm s^{-1}) as a function of distance (cm) from the
center of the asteroid for the standard case with 1 Mton of neutron deposition onto
stony asteroids with radii of 100, 200 and 300 m.

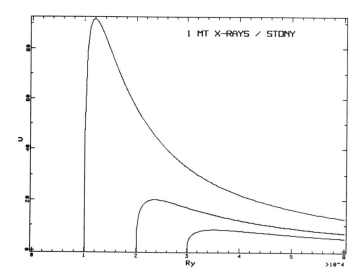

Figure 35.   Deflection velocity (cm s^{-1}) as a function of distance (cm) from the
center of the asteroid for the standard case with 1 Mton of X ray deposition onto
stony asteroids with radii of 100, 200 and 300 m.

exceeds the spall strength. Because this layer can be much thicker than
the vaporized material, this is a possible acceleration mechanism for the
asteroid. We estimate its effectiveness by modifying the model derived above

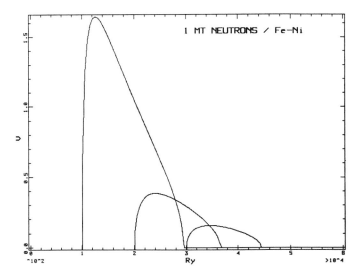

Figure 36.   Deflection velocity (cm s^{-1}) as a function of distance (cm) from the center of the asteroid for the standard case with 1 Mton of neutron deposition onto Fe/Ni asteroids with radii of 100, 200 and 300 m.

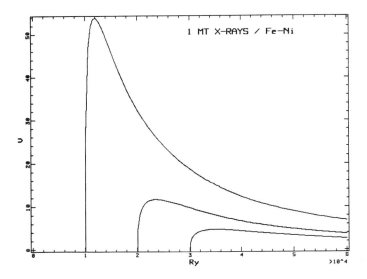

Figure 37.   Deflection velocity (cm s^{-1}) as a function of distance (cm) from the center of the asteroid for the standard case with 1 Mton of X ray deposition onto Fe/Ni asteroids with radii of 100, 200 and 300 m.

as follows; we derive a new energy threshold based on the fracture strength of the material from the Grüneisen pressure relation

$$e_{th} \ (\text{spall}) = \frac{P_{\text{spall}}}{\Gamma \rho} \tag{A.30}$$

where Γ is the Grüneisen coefficient, taken to be unity here, and P_{spall} is the spall threshold pressure. For very rapid loading processes (high strain rates), this quantity is known to be much larger than the static spall strength and we take its value to be 100 kbar for the present estimate. The resulting threshold energy is 3.3×10^{10} erg g^{-1} for the stony case, roughly a factor of 3 smaller than the vaporization threshold energy. For the Fe/Ni case, the threshold is 1.33×10^{10} erg g^{-1}. In addition, the velocity of the spalled material will be approximately the speed of sound in the solid, which we take to be 2 km s^{-1}.

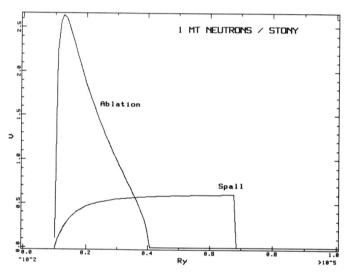

Figure 38. Front surface spall reaction velocity compared to the standard case for neutron deposition.

Figure 38 shows a comparison of the reaction velocities due to ablation vs those obtained by the front surface spall process with the above assumptions, for the radius = 100 m case from Fig. 34. The lower ablation velocity in the spall case dominates the result, with reaction velocities down by a factor of 5 relative to the standard vaporization model. The lower threshold energy allows a larger standoff, however, and the curve peaks for this particular case at the cutoff distance of about 7 times the asteroid radius. Note that there is a range from 4 to 7 r in which the spall process is effective, but no material is vaporized. This may be a parameter region of reduced fragmentation danger. The Fe/Ni case is similar, but the maximum spall distance extends to 9.5 asteroid radii.

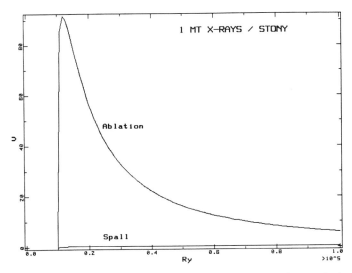

Figure 39.   Front surface spall reaction velocity compared to the standard case for X ray deposition.

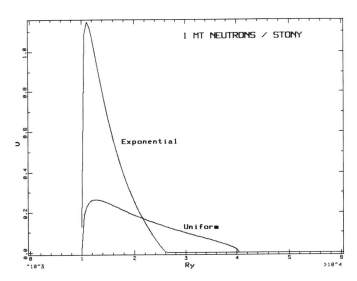

Figure 40.   Effect of changing neutron deposition to an exponential deposition profile in the standard model.

Figure 39 shows the same comparison, but for X ray deposition. Here the extreme thinness of the spalled layer (even though thicker than the vaporized layer) plus the fixed ablation velocity results in a negligible reaction velocity (less than 1 cm s^{-1}).

Additional issues with the spall mechanism include the need for a better estimate of the thickness and speed of the spalled layer and an estimate of when the effects of spall at other surfaces (especially the rear surface) will occur, canceling the desired effect. Experiments by Nakamura and Fujiwara (1991) with impacts of small spheres on larger spheres suggest that one or more layers will tend to be ejected from the surface of a targeted asteroid, leaving a large core with little or no velocity perturbation. Detailed hydrodynamic models are being constructed to address these questions and will be reported separately.

The above models use exponential deposition in small depths to model X rays and uniform deposition in larger depths to model neutrons. Neither model is the entire truth, but the actual case should lie between these two extremes. For reference, we show here the effect of changing the deposition profile alone. Figure 40 shows the standard $r = 100$ m neutrons/stony case from Fig. 34, together with the result of changing to an exponential profile with a 10 cm e-folding depth. As expected, the thicker ablated layer produces a much larger velocity near the peak, while the larger required threshold energy produces a narrower range of possible burst radii. The peak reaction velocity attained here is over 10 m s^{-1}.

TERMINAL INTERCEPT FOR LESS THAN ONE ORBITAL PERIOD WARNING

JOHNDALE C. SOLEM and CHARLES M. SNELL
Los Alamos National Laboratory

We derive a series of expressions to delineate the utility, performance, and range of applicability of rocket interceptors designed to deflect or pulverize comets or asteroids on collision course with Earth. The interaction is calculated for both kinetic-energy deflection and nuclear-explosive deflection and uses a fairly general relationship between the energy deposited and the blow-off mass. In the nuclear-explosive case, we calculate the interceptor mass and cratering effect for detonations above the surface and below the surface as well as directly on the surface of the assailant. Because different assailants could possess a wide range of densities and material properties, the principal value of this work is to show the relationships among the salient parameters. However, using typical values for the various physical properties, we make the following observations: (1) kinetic-energy deflection with chemically fueled rockets is very limited in its range of effectiveness. It can effect ocean diversion of rocky asteroids with diameter smaller than about 70 m, if the interceptor is launched when the distance to the assailant is more than 1/30 AU. At shorter range, interceptors become impractically massive, and the probability of fracture of the assailant increases dramatically. Because 70-m chondrites dissipate almost all their energy traversing the upper atmosphere, it is unlikely that kinetic-energy deflection will have much use; (2) nuclear explosive deflection is imperative for assailants greater than about 100 m detected closer than 1/30 AU because of interceptor size. Deflection of nuclear surface bursts offers a three-to-four order of magnitude reduction in interceptor mass over kinetic-energy deflection. The advantage of nuclear explosive deflection decreases slightly with specific impulse and decreases dramatically with assailant velocity. Fragmentation is a problem for nuclear explosive intercepts launched closer than about 1/3 AU; (3) nuclear penetrators offer no advantage for deflection, but are better for pulverization; (4) nuclear stand-off deflection greatly reduces fragmentation probability, but with a substantial increase in interceptor mass.

I. INTRODUCTION

The problem of preventing a collision with a comet or asteroid, which we call "astral assailants" at the risk of creating a pathetic fallacy, can be considered in two domains: (1) actions to be taken if the collision can be predicted several orbital periods in advance, and can be averted by imparting a small change in velocity (most effectively at perihelion); and (2) actions to be taken when the object is less than an astronomical unit (AU) away, collision is imminent, and deflection or disruption must be accomplished as the object closes on Earth.

[1013]

We call the first domain of actions, "remote interdiction," and the second domain of actions, "terminal interception."

If all of the Earth-threatening asteroids were known, the orbits could be calculated and the process of deflection could be carried out in a leisurely manner. Remote interdiction would be the option of choice. But 99% have not yet been discovered (Morrison 1991; Chapters by Bowell and Muinonen and by Rabinowitz et al.). Furthermore, there are an enormous number of unknown long-period comets for which a thorough search is completely impractical (Chapter by Shoemaker et al.).

Asteroids in the 100-m size range are exceedingly difficult to detect unless they are very close. Comets in this size range are more conspicuous owing to their coma, but they move a lot faster and can be in retrograde orbits or out of the plane of the ecliptic. In either case, it seems likely that we will have little time to respond to a potential collision.

In this chapter, we consider the dynamics of the terminal intercept problem. We explore the possibility of using kinetic-energy deflection as well as nuclear explosives. Nuclear explosives can be employed in three different modes depending on their location at detonation: (1) buried below the assailant's surface by penetrating vehicle or other means; (2) detonated below but very near the assailant's surface; or (3) detonated some distance above the surface. See also related Chapters by Ahrens and Harris and by Hills et al.

II. INTERCEPTION AND DEFLECTION SCENARIO

Figure 1 shows the interception scenario. In plot 1a, the asteroid or comet is headed toward Earth at a velocity v. The interceptor traveling at an oppositely directed velocity V is about to engage the assailant object. The assailant has a mass M_a, and the interceptor, because it has long since exhausted its fuel, has its final mass M_f. We cannot hope to deflect the assailant like a billiard ball because $M_a >> M_f$. So the interceptor must supply energy to blow off a portion of the assailant's surface M_e, as shown in plot 1b. The blow-off material is very massive compared to the interceptor, $M_a >> M_e >> M_f$. One might think that a conventional high explosive would suffice, but the energy it would supply would be relatively insignificant. Standard high explosive releases 10^3 calories $= 4.184 \times 10^{10}$ erg g^{-1}. An asteroid moving at 25 km·s^{-1} has a specific energy of 3.125×10^{12} erg g^{-1}—about 75 times the specific energy of high explosive. If the interceptor is moving at the same speed in the opposite direction ($V = v = 25$ km·s^{-1}), the interceptor would impact with a specific energy 300 times that of high explosive. There is enormous kinetic energy available; a chemical energy release would be in the noise. However, even this tremendous kinetic energy would be completely swamped by a nuclear explosive. The yield-to-weight ratio of nuclear explosives is generally measured in kilotons per kilogram, that is, tons per gram. A typical specific energy is a million times that of chemical high explosive, or about four orders of magnitude higher that the kinetic energy of the interceptor collision.

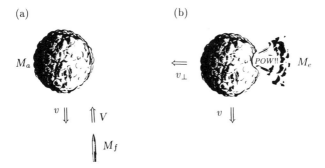

Figure 1.   Interception scenario (a) interceptor about to engage the assailant. (b) Interceptor supplies energy to blow off a portion of the assailant's surface and imparts a transverse velocity.

III. KINETIC-ENERGY DEFLECTION

The final velocity of an interceptor missile relative to the Earth, or the orbit in which it is stationed, is given by the rocket equation,

$$V = g I_{sp} \ln \frac{M_i}{M_f} \tag{1}$$

where M_i and M_f are the initial and final mass of the interceptor and I_{sp} is the specific impulse of the rocket fuel. In general, the time required to reach this relative velocity will be short compared to the total flight time. The time elapsed from launch to intercept is

$$\Delta t = \frac{R_l}{v + V} \tag{2}$$

where R_l is the range when the interceptor is launched and v is the speed at which the assailant is closing on the Earth. So the range at which the assailant is intercepted will be given by

$$R_i = R_l \left(1 - \frac{v}{v + V} \right). \tag{3}$$

If the impact gives the assailant a transverse velocity component $v_\perp$ then the threatening assailant will miss its target point by a distance

$$\varepsilon = R_l \frac{v_\perp}{v} \left(\frac{V}{v + V} \right) \tag{4}$$

where we have neglected the effect of the Earth's gravitational focusing and used a linear approximation to Keplerian motion. To obtain the transverse velocity component, we would use the kinetic energy of the interceptor to blast a crater on the side of the assailant. The momentum of the ejecta would be balanced by the transverse momentum imparted to the assailant. From Glasstone's empirical fits (Glasstone 1962), the mass of material in the crater produced by a large explosion is

$$M_e = \alpha^2 E^\beta \tag{5}$$

where α and β depend on the location of the explosion, the soil composition and density, gravity, and a myriad of other parameters. Clearly the crater constant α and the crater exponent β will vary depending on whether we are considering an assailant composed of nickel-iron, stony-nickel-iron, stone, chondrite, ice, or dirty snow. For almost every situation, however, we find $\beta \simeq 0.9$.

The kinetic energy available when the interceptor collides with the astral assailant is

$$E = \frac{1}{2} M_f (V + v)^2 \tag{6}$$

Only a fraction of the interceptor's kinetic energy is converted to kinetic energy of the ejected or blow-off material. Let this fraction be equal to $\frac{1}{2}\delta^2$, or

$$\delta = \sqrt{2 \frac{\text{ejecta kinetic energy}}{\text{interceptor kinetic energy}}}. \tag{7}$$

The reason for this strange definition is that it greatly simplifies the algebra. We will call the parameter δ the energy fraction. Then the transverse velocity imparted to the assailant is

$$v_\perp = \delta \frac{\sqrt{M_e E}}{M_a} = \frac{\delta}{M_a} \sqrt{\frac{M_e M_f (V + v)^2}{2}} = \frac{\alpha \delta}{M_a} \left(\frac{M_f (V + v)^2}{2} \right)^{\frac{\beta+1}{2}} \tag{8}$$

where M_a is the mass of the comet or asteroid. We can combine Eqs. (4), (5), and (8) to obtain

$$\varepsilon = \alpha \delta R_l \frac{V (V + v)^\beta}{M_a v} \left(\frac{M_f}{2} \right)^{\frac{\beta+1}{2}} \tag{9}$$

for the miss distance. Equation (9) reveals the importance of the intercept velocity V, which is proportional to specific impulse I_{sp}. If $V \ll v$, the deflection is proportional to V, and if $V \gg v$, the deflection is proportional to $V^{\beta+1} \sim V^2$.

A. Optimum Mass Ratio for Kinetic Energy Deflection

The energy on impact is proportional to the final mass of the interceptor and the square of its relative velocity, as given in Eq. (6). The smaller its final mass, the higher its relative velocity, so there is some optimum mass ratio that produces the greatest deflection for a given initial mass. This would be the optimal interceptor design—the most bang for the buck.

Substituting Eq. (1) into Eq. (9), setting

$$\frac{d\varepsilon}{d(M_i/M_f)} = 0. \tag{10}$$

and solving, we find the mass ratio that produces the largest value of ε,

$$\frac{M_i}{M_f} = e^Q \tag{11}$$

where

$$Q = 1 - \frac{v}{2gI_{sp}} + \sqrt{1 + \frac{1-\beta}{1+\beta}\frac{v}{gI_{sp}} + \left(\frac{v}{2gI_{sp}}\right)^2}. \tag{12}$$

We note that this optimal mass ratio depends only on the velocity of the assailant relative to Earth v and the interceptor's specific impulse I_{sp}. The value of β is a constant of the assailant's soil composition and is very close to 0.9, and $g \simeq 980$ cm·s^{-2}. In the limit of very high specific impulse, the optimum mass ratio is

$$\frac{M_i}{M_f} = e^2. \tag{13}$$

This limit can be approached, but is not realistic owing to v/gI_{sp} limitations. The maximum displacement of the impact location on Earth is then given by

$$\varepsilon = \frac{\alpha\delta v^\beta R_l}{M_a}\left(\frac{M_i e^{-Q}}{2}\right)^{\frac{\beta+1}{2}}\frac{gI_{sp}Q}{v}\left(1 + \frac{gI_{sp}Q}{v}\right)^\beta. \tag{14}$$

Remarkably, when Eq. (12) is put into Eq. (14), the resulting exceedingly complex expression can be put in dimensionless form. Figure 2 plots the dimensionless parameter $\varepsilon M_a/\alpha\delta v^\beta R_l M_i^{\frac{1}{2}(\beta+1)}$ versus the dimensionless parameter gI_{sp}/v for $\beta = 0.8$, 0.9, and 1.0. It shows the increasing advantage to higher specific impulse derived from Eq. (14).

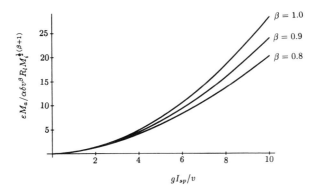

Figure 2.   Dimensionless plot of kinetic-energy deflection.

A great deal of physical insight can be obtained just by studying the axis labels of the dimensionless plot. From the ordinate, we see that for the same value of gI_{sp}/v, which is more or less fixed by interceptor design, the asteroid deflection ε is:

1. Proportional to the range of the assailant at launch (R_l);
2. Inversely proportional to the mass of the assailant (M_a);
3. Nearly proportional to the velocity of the assailant relative to Earth ($v^\beta \simeq v^{0.9}$);
4. Nearly proportional to the initial mass of the interceptor ($M_i^{\frac{1}{2}(\beta+1)} \simeq M_i^{0.95}$);
5. Proportional to the crater constant (α);
6. Proportional to the square root of the fraction of interceptor kinetic energy converted to blow off kinetic energy ($\frac{1}{2}\delta^2$).

Equation (14) can be rearranged to give the required initial mass or mass in orbit of the interceptor,

$$ M_i = 2e^Q \left[\frac{M_a v \varepsilon}{\alpha \delta R_l g I_{sp} Q} \left(\frac{1}{v + g I_{sp} Q} \right)^\beta \right]^{\frac{2}{\beta+1}}. \tag{15} $$

The mass given by Eq. (15) will generally be the largest single factor in the cost of a defensive system of this sort. To appreciate the magnitude of the problem, it is now necessary to put in a few numbers. The best chemical fuels might have a specific impulse as high as 500 s, which we will use to make the point. The density of potential astral assailants varies greatly, from less than 1 g·cm^{-3} for a snow-ball comet to a little over 1 g·cm^{-3} for a dirty-ice comet to about 3 g·cm^{-3} for a chondrite to about 8 g·cm^{-3} for a nickel-iron asteroid. An agreeable average is 3.4 g·cm^{-3}. The velocity of the assailant relative to Earth could range from 5 km·s^{-1} for an asteroid in nearly coincident orbit with Earth to 70 km·s^{-1} for a long-period comet in retrograde orbit near the plane of the ecliptic. We will take 25 km·s^{-1} for this example.

Because the material properties of asteroids and comets vary so widely, an estimate of the crater constant and crater exponent is somewhat arbitrary. Here we will make an estimate for impact cratering using medium hard rock. Glasstone uses $\beta \simeq 0.9$ and $\alpha \simeq 8.4 \times 10^{-4}$ g$^{\frac{1}{2}(1-\beta)}$·cm$^{-\beta}$·s$^\beta$ for an explosive buried at the optimal depth for maximum ejection of dry soil. For a surface burst, Glasstone takes $\alpha \simeq 1.6 \times 10^{-4}g^{\frac{1}{2}(1-\beta)}$·cm$^{-\beta}$·s$^\beta$. The correct value of α for the impact crater is somewhere between a surface burst and an optimally buried explosion. For the purpose of the estimating the crater size for kinetic-energy deflection, we will take $\alpha \simeq 2 \times 10^{-4}$ g$^{\frac{1}{2}(1-\beta)}$·cm$^{-\beta}$·s$^\beta$. Kreyenhagen and Schuster (1977) have noted that impacts in the 20 km·s^{-1} range couple 50 to 80% of their energy to the ground, while surface bursts couple only 1 to 10%. We will assume about 60% coupling and about half of that goes to the blow off. Thus about 30% of the interceptor's kinetic energy is converted to

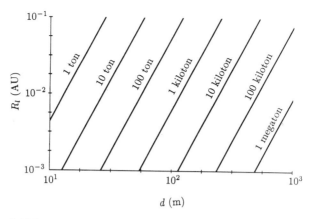

Figure 3.   Initial masses of optimally designed interceptors using kinetic-energy deflection for ocean diversion (1 Mm).

kinetic energy of the blow off, corresponding to $\delta = 0.775$.

Figure 3 shows the initial mass of the interceptor required to deflect the astral assailant by 1 Mm, as a function of the assailant's diameter d and its range R_l when the interceptor is launched. We have assumed an assailant density of $\rho = 3.4$ g·cm^{-3}, an assailant velocity of $v = 25$ km·s^{-1}, a crater exponent of $\beta = 0.9$, a crater constant of $\alpha = 2 \times 10^{-4}$ g$^{\frac{1}{2}(1-\beta)}$·cm$^{-\beta}$·s$^{\beta}$ and an energy fraction of $\delta \simeq 0.775$. A one-megameter deflection is typical of the course change required to divert the assailant from impact in a populated area to a nearby ocean. To interpret the figure for a ten-megameter deflection, which would be conservative for missing the planet entirely ($R_{\oplus} = 6.378$ Mm), multiply the masses by about a factor of ten ($M_i \propto \varepsilon^{\frac{2}{\beta+1}}$, so a factor of 10 in ε corresponds to a factor of 11.3 in M_i).

An ocean impact is not without damage (Ahrens and O'Keefe 1987), but, in general, the damage will be far less than if the impact were in a populated area. Roughly speaking, the height of the wave and the distance from impact are jointly proportional to the square root of the energy deposited in the water (Mader, personal communication). A 100-m diameter chondrite might typically impact with 100 megatons of energy. Taking about 5% of that energy (Van Dorn 1961) as coupled to the water, a water wave of about 3 m in height would be encountered 100 km from the impact point.

Figure 3 makes a clear statement about the applicability of kinetic-energy deflection. Kinetic-energy deflection is practical only for assailants considerably less than 100 m in diameter. To handle a 100-m assailant would require a 1000 ton interceptor even if launched when the assailant was still 0.1 AU away. The mass would go to 10,000 tons if the assailant were deflected to miss the planet entirely rather than diverted to an ocean. Thus dealing with assailants larger than 100 m requires another technology.

IV. NUCLEAR EXPLOSIVE DEFLECTION

Much more deflection can be obtained if a nuclear explosive is used to provide the cratering energy. In this scenario, most of the weight after the rocket fuel is expended would be the nuclear explosive, which produces a yield of

$$E = \varphi M_f \tag{16}$$

where φ is the yield-to-weight ratio. Again, $\delta^2/2$ of this energy goes into the dirt ejected from the crater, so the transverse velocity imparted to the assailant is

$$v_\perp = \frac{\delta}{M_a}\sqrt{\varphi M_f M_e} = \frac{\alpha\delta}{M_a}\left(\varphi M_f\right)^{\frac{\beta+1}{2}}. \tag{17}$$

We can combine Eqs. (4), (5), and (17) to obtain

$$\varepsilon = \frac{\alpha\delta R_l}{M_a v}\frac{V\left(\varphi M_f\right)^{\frac{\beta+1}{2}}}{V + v} \tag{18}$$

for the displacement.

A. Optimum Mass Ratio for Nuclear Explosive Deflection

Substituting Eq. (1) into Eq. (18) and solving Eq. (10), we find the logarithm of the mass ratio that produces the largest value of ε,

$$Q = -\frac{v}{2gI_{sp}} + \frac{1}{2}\sqrt{\frac{8v}{(1+\beta)gI_{sp}} + \left(\frac{v}{gI_{sp}}\right)^2}. \tag{19}$$

In the limit of very high specific impulse, the optimum mass ratio is

$$\frac{M_i}{M_f} = 1. \tag{20}$$

In the limit of very low specific impulse, the optimum mass ratio is

$$\frac{M_i}{M_f} = \exp\left(\frac{2}{1+\beta}\right). \tag{21}$$

The maximum displacement of the impact location on Earth is then given by

$$\varepsilon = \frac{\alpha\delta R_l}{M_a v}\frac{gI_{sp}Q(\varphi M_i e^{-Q})^{\frac{\beta+1}{2}}}{gI_{sp}Q + v}. \tag{22}$$

For a surface burst, Glasstone uses $\beta = 0.9$, but takes $\alpha \simeq 1.6 \times 10^{-4}\mathrm{g}^{\frac{1}{2}(1-\beta)} \cdot \mathrm{cm}^{-\beta} \cdot \mathrm{s}^\beta$. He describes the material as dry soil. Medium strength rock would be more consistent with $\alpha \simeq 10^{-4}\mathrm{g}^{\frac{1}{2}(1-\beta)} \cdot \mathrm{cm}^{-\beta} \cdot \mathrm{s}^\beta$, and, in the 20-kt range,

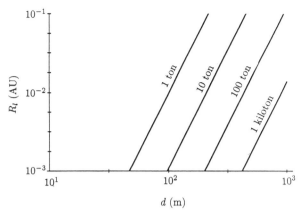

Figure 4.   Initial masses of optimally designed interceptors using nuclear-explosive deflection for ocean diversion (1 Mm).

would roughly agree with Cooper (1976). If about 5% of the nuclear explosive energy goes into kinetic energy of the blow off, then $\delta = 1/\sqrt{10} \simeq 0.316$.

Equation (22) can be rearranged to give the required initial mass of the interceptor,

$$M_i = \frac{e^Q}{\varphi} \left[\frac{M_a v \varepsilon}{\alpha \delta R_l} \left(1 + \frac{v}{g I_{sp} Q} \right) \right]^{\frac{2}{\beta+1}} \qquad (23)$$

where now Q is given by Eq. (19).

It is generally known that the yield of nuclear warheads can be a few kilotons per kilogram if they weigh more than about a hundred kilograms. For the purpose of these estimates, we will take a conservative value of $\varphi = 1$ kiloton·kilogram^{-1}. Figure 4 is analogous to Fig. 3, using the values of α and δ given above. Ocean deflection of 1 Mm is sought and the following values are used: $\rho = 3.4$ g·cm^{-3}, $v = 25$ km·s^{-1}, $\beta = 0.9$, $\alpha = 10^{-4}$ g$^{\frac{1}{2}(1-\beta)}$·cm$^{-\beta}$·s$^{\beta}$ and $\delta \simeq 0.316$.

A good way to compare kinetic-energy deflection with nuclear-explosive deflection is to look at the ratio of the initial masses of the interceptors. If we divide Eq. (23) by Eq. (15), we see that all variables drop out except specific impulse I_{sp}, the assailant's velocity v, the energy fraction δ, and the cratering constant α. For a comparison of the techniques, we would keep the same values of I_{sp} and v. We define the ratio

$$R_m = \frac{M_i \text{ given by Eq. (23)}}{M_i \text{ given by Eq. (15)}}. \qquad (24)$$

The appropriate dimensionless ratio for the comparison is

$$R_m \frac{\alpha_n \delta_n}{\alpha_k \delta_k} \qquad (25)$$

where the subscripts n refer to the parameters for nuclear-explosive deflection and the subscripts k refer to the parameters for kinetic-energy deflection. This is the actual ratio of initial interceptor weights for kinetic-energy versus nuclear-explosive deflection. Figure 5a shows this ratio as a function of assailant velocity v for specific impulse I_{sp} = 500 s. Figure 5b shows the same ratio as a function of specific impulse I_{sp} for assailant velocity v = 25 km·s^{-1}. Figure 5c shows the same ratio as a function of both specific impulse and assailant velocity. For the numerical examples we have chosen, we have

$$\frac{\alpha_n \delta_n}{\alpha_k \delta_k} = \frac{10^{-4} \times 0.316}{2 \times 10^{-4} \times 0.775} = 0.204. \tag{26}$$

So for our particular selection of parameters, we can read the mass ratios in Figs. 5a, 5b, and 5c by multiplying the number on the vertical axis by 0.204.

From Figs. 5a, 5b, and 5c, we learn the following qualitative features:

1. The interceptor weight is about three orders of magnitude less for nuclear-explosive deflection than for kinetic-energy deflection;
2. The advantage of nuclear-explosive deflection decreases significantly with assailant velocity;
3. The advantage of nuclear-explosive deflection decreases slightly with specific impulse.

B. Penetrators

The biggest crater is not produced by a surface burst, but by an explosive buried some distance below the surface. Clearly, if it is buried too deeply, it will produce no crater at all. It will either fracture the assailant, or all the energy will be absorbed and a cavity will be formed below the surface. The optimum depth for cratering is a function of all the usual parameters describing material properties, but quite importantly, gravity, which, to a large extent, can be ignored for comets and asteroids.

1. Numerical Calculations of Momentum Coupling for Buried Explosions. The expressions cited above for ejecta mass and energy coupling fractions are empirical and are based upon cratering data obtained in the natural gravitational field of Earth. It is well known that the crater formation process can be strongly influenced by gravity and by local material properties. To obtain a more quantitative picture of momentum coupling from buried explosions, we have performed two-dimensional finite difference calculations for events in two geologic materials with dramatically different properties. It has been shown that such calculations can simulate the details of energy coupling, velocity and stress waveforms, and all aspects of the dynamic crater formation process (Burton et al. 1975). For accurate results, it is essential that the high-pressure equation of state and the low-pressure solid constitutive response of the material be well known. This requirement is not fulfilled for most asteroid and comet materials, and obtaining better data is an important priority. For purposes of this study, we have selected two Earth materials

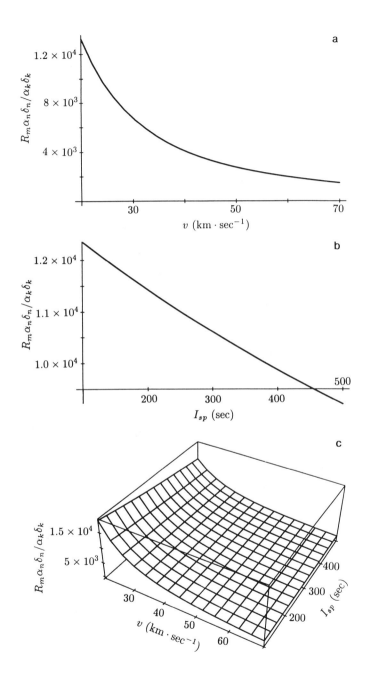

Figure 5. Ratio of kinetic-energy interceptor mass to nuclear-explosive interceptor mass. (a) Ratio vs v for $I_{sp} = 500$ s. (b) Ratio vs I_{sp} for $v = 25$ km·s^{-1}. (c) Ratio vs I_{sp} and v.

that do have reliable equations of state and constitutive models: (1) Bearpaw clay shale, a very weak, fully saturated rock with high water content; and (2) granite, a strong, dense silicate rock with very low porosity and water content. The basic physical properties are summarized in Table I. The strength of the granite is somewhat lower than that of small intact samples, but is representative of the bulk or *in situ* strength that is governed by flaws, cracks, joints, and other large-scale heterogeneities. These two materials were deliberately chosen to represent rather extreme cases. The weak Bearpaw shale will give efficient, almost hydrodynamic coupling, whereas the high-strength granite will produce inefficient coupling. Other typical Earth media, such as porous rock or rocks of intermediate strength, would be expected to fall between these two extremes. However, asteroid materials could lie outside this range. For example, a dense very high strength iron or nickel-iron composition could give significantly lower coupling than even the granite. A low-density material such as ice could give coupling comparable to or better than the Bearpaw shale. The depth-of-burial scaling employed here does not take account of the initial material density. This is not a major factor for cratering on Earth, because virtually all experiments have been conducted in materials within a narrow density range of about 1.6 to 2.6 $g \cdot cm^{-3}$. Including the density in the scaling law can improve the correlation for a wider range of materials.

We performed numerical simulations for 1-kiloton events at six depths of burial ranging from 11 to 88 m. Gravity was turned off in all cases, because overburden stresses and gravitational deceleration exert a negligible effect on the cratering dynamics for a small asteroid. For example, the gravitational acceleration at the surface of a 1-km-diameter asteroid with a mean density of $2.7 g \cdot cm^{-3}$ is only 3.8×10^{-5} of that on Earth, or 3.7×10^{-2} $cm \cdot s^{-2}$. If we assume that vertical velocities within the ejecta mound are $\sim 10^4$ $cm \cdot s^{-1}$, it is evident that the material will loose much less than 1% its velocity over a characteristic distance of 1 km. In fact, virtually all of the ejecta material will far exceed the escape velocity of the body and will suffer only negligible loss due to gravitational deceleration. Likewise, the dynamic stresses in the cratered material are typically 10 to 10^5 Pa. The overburden stress at a burial depth of 100 m in the 1-km body is only $\sim 10^{-4}$ Mpa, or at least 5 orders of magnitude less than the dynamic stresses. Overburden will thus have no measurable influence on the material behavior, either in the hydrodynamic- or strength-dominated regimes. Gravitational and overburden effects may begin to become significant and should be taken into account for bodies that are tens to hundreds of km in diameter. Gravity will tend to decrease the momentum coupling efficiency for very large bodies, but can be ignored for most applications under consideration here.

In the absence of gravity, it is an excellent approximation to scale the depth of burial inversely as the cube root of explosive yield. Two different energy source types were considered, high explosive and nuclear. High explosives are not feasible for asteroid deflection because of the very low energy-to-weight ratio. However, for a given energy, high explosives couple

TABLE I
Summary of Basic Physical Properties for Materials Used
in the Cratering Calculations[a]

	Bearpaw Shale[b]	Hard Granite[c]
Density (g·cm^{-3})	2.217	2.65
Sonic velocity (m·s^{-1})	1960	6530
Bulk modulus (GPa)	8.5	60.0
Poisson's ratio	0.48	0.23
Water content by weight	0.142	0.0
Unconfined strength (MPa)	3.0	10.4
Ultimate strength (MPa)	5.0	374.0
Tensile strength (MPa)	0.2	2.0
High Explosive		
a (g·cm·s^{-1}·kiloton^{-1})	-3.597×10^{14}	9.371×10^{13}
b (g·s^{-1}·kiloton^{-1})	6.233×10^{15}	1.342×10^{15}
Nuclear Explosive		
a (g·cm·s^{-1}·kiloton^{-1})	-2.343×10^{14}	1.041×10^{14}
b (g·s^{-1}·kiloton^{-1}	4.564×10^{15}	9.399×10^{14}

[a] These two materials were chosen to represent extreme cases. The weak Bearpaw shale will give efficient, almost-hydrodynamic coupling, whereas the high-strength granite will produce inefficient coupling. Other typical media, such as porous rock, ice, or snow, would be expected to fall between these two extremes. Also shown are the coefficients for obtaining the perpendicular momentum of the ejecta as given by Eq. 27 for the fitted lines in Fig. 6.

[b] Bearpaw clay shale is a very weak, fully saturated rock with high water content.

[c] Granite is a strong, dense silicate rock with very low porosity and water content.

more efficiently than nuclear, because material near the source is shocked to lower pressures and less waste heat is deposited. The high-explosive results are presented here as a matter of interest. The energy deposited in the case of kinetic-energy deflection is characteristically somewhere between high explosive and nuclear explosive. The calculations were run for at least five stress wave transit times from the explosive to the free surface. By this time, the material in the ejecta mound was in ballistic free fall and the momentum had reached its final asymptotic value. Only the normal component of momentum (perpendicular to the asteroid surface) was recorded, because the horizontal momentum components cancel out in axisymmetric geometry. Results of all the calculations are displayed in Fig. 6, which shows the scaled perpendicular momentum as a function of scaled depth of burial. It is noteworthy that relationship is nearly linear for all cases examined here. Although the approximation is not perfect, the largest error is only about 8%. The linear relationship provides a convenient description of momentum coupling for use in parameter studies. Accordingly, we have adopted an equation of the following form to approximate the average perpendicular momentum:

$$p_\perp = aE + bDE^{\frac{2}{3}} \qquad (27)$$

where D is the depth of burial. The coefficients a and b are dependent on the material and energy source type and have been evaluated for the range

$$1.1 \times 10^3 \text{ cm} \cdot \text{kiloton}^{-\frac{1}{3}} \le DE^{-\frac{1}{3}} \le 8.8 \times 10^3 \text{ cm} \cdot \text{kiloton}^{-\frac{1}{3}}.$$

Coefficients for the fits shown in Fig. 6 are listed in Table I. The linear approximation is valid only over the depth range examined in the calculations and will give meaningless results outside this range.

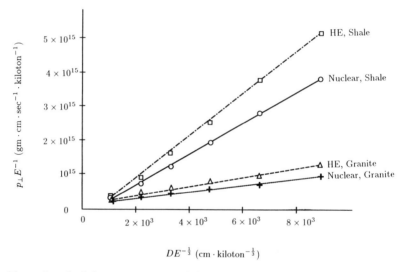

Figure 6.   Scaled momentum vs scaled depth of burial from numerical simulations at six different depths with linear fits.

Because there is no gravity in these calculations, the perpendicular component of the ejecta momentum, for a given material and scaled depth of burial, is simply proportional to the energy yield of the explosive.

2. *Comparison with Empirical Formulas.* It is of interest to compare the results of these calculations with the Glasstone empirical treatment. The mass of crater ejecta is given by Eq. (5), and the kinetic energy of the ejecta is

$$E_e = \frac{\delta^2}{2}E. \qquad (28)$$

Assuming the ejecta can be characterized by an average velocity V_e,

$$\frac{1}{2}M_e V_e^2 = E_e \qquad (29)$$

and

$$V_e = \sqrt{\frac{2E_e}{M_e}} = \delta \sqrt{\frac{E}{M_e}}. \tag{30}$$

Using Eq. (5), the average ejecta momentum can then be obtained,

$$p_\perp = M_e V_e = \alpha \delta E^{\frac{\beta+1}{2}}. \tag{31}$$

If we adopt the Glasstone value of the crater constant $\alpha = 8.4 \times 10^{-4} \mathrm{g}^{\frac{1}{2}(1-\beta)} \cdot \mathrm{cm}^{-\beta} \cdot \mathrm{s}^\beta$ at an optimum depth of burial $D_{opt} \simeq 4.6 \times 10^3 E^{-\frac{1}{3}} \mathrm{cm} \cdot$ kiloton$^{-\frac{1}{3}}$, and assume $\delta = 0.3$ and the nuclear explosive energy is $E = 1$ kiloton $= 4.186 \times 10^{19}$ erg, we obtain the following values for the momentum coupling:

$$p_\perp = (\mathrm{g} \cdot \mathrm{cm} \cdot \mathrm{s}^{-1}) \begin{cases} 1.05 \times 10^{16}, & \beta = 1.0; \\ 1.10 \times 10^{15}, & \beta = 0.9; \\ 1.15 \times 10^{14}, & \beta = 0.8. \end{cases} \tag{32}$$

These empirical values are compared with the numerical calculations in Fig. 6. A logarithmic scale is used for the vertical axis to encompass the entire range of the data. The point for the nominal cratering exponent $\beta = 0.9$ falls almost in the middle of the calculations. Thus, Glasstone's empirical fits are at least qualitatively reasonable and consistent with the calculational analysis. Of course, there is clearly a substantial range of variability in momentum coupling, and quantitative predictions for a specific material and depth of burial are best obtained from detailed calculations. But from Fig. 7, it is clear that the values we have taken as approximations to α, β, and δ for both the kinetic-energy and nuclear-surface-burst cases are reasonably well justified. For the kinetic-energy case, we took a crater exponent of $\beta = 0.9$, a crater constant of $\alpha = 2 \times 10^{-4} \mathrm{g}^{\frac{1}{2}(1-\beta)} \cdot \mathrm{cm}^{-\beta} \cdot \mathrm{s}^\beta$ and an energy fraction of $\delta \simeq 0.775$, which correspond to a scaled momentum of $6.78 \times 10^{14} \mathrm{g \cdot cm \cdot s}^{-1} \cdot \mathrm{kiloton}^{-1}$. We said this case would correspond to an explosion between optimum depth of burial and a surface burst. From Fig. 7, it corresponds to a scaled depth of burial between 2×10^3 and 4×10^3 cm·kiloton$^{-\frac{1}{3}}$. For the nuclear-surface-burst case, we took a crater exponent of $\beta = 0.9$, a crater constant of $\alpha = 10^{-4} \mathrm{g}^{\frac{1}{2}(1-\beta)} \cdot \mathrm{cm}^{-\beta} \cdot \mathrm{s}^\beta$ and an energy fraction of $\delta \simeq 0.316$, which correspond to a scaled momentum of $1.38 \times 10^{14} \mathrm{g \cdot s}^{-1} \cdot \mathrm{kiloton}^{-1}$. From Fig. 7, it appears that this would correspond to a rather shallow scaled depth of burial—roughly 5×10^2 cm·kiloton$^{-\frac{1}{3}}$, which is a near-surface burst and is somewhat shallower than the range of depth examined in these calculations.

The momentum coupling obtained from the cratering calculations can be directly applied to compute the transverse velocity and deflection of the assailant. In this case, the equations take a particularly simple form. By conservation of momentum, the transverse velocity of the assailant is

$$v_\perp = \frac{p_\perp}{M_a}. \tag{33}$$

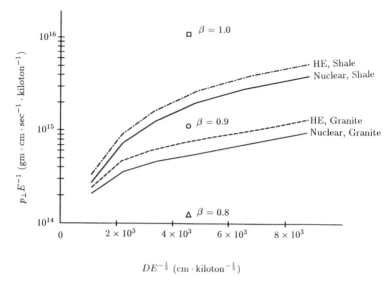

Figure 7.   Linear fits to scaled momentum vs scaled depth of burial with points from empirical formulas for various values of the cratering exponent.

From Eq. (4), the deflection is

$$\varepsilon = R_l \frac{p_\perp}{M_a v}\left(\frac{V}{v + V}\right).$$ (34)

3. The Possibility of Fracture. All of these calculations assume that the asteroid material is uniform, i.e., large-scale inclusions or heterogeneities are ignored, and depth-dependent variations of material properties such as porous near-surface regolith layers are not considered.   The interface above the detonation is treated as a planar free surface, which is equivalent to assuming that the burial depth is small relative to the asteroid radius. If the depth of burial becomes comparable to the radius, the geometry is more complicated because reflections will return from the sides and opposite face of the body within the time interval important for dynamic cratering. There is clearly a possibility of fracturing much or all of the entire asteroid for configurations of this type. Also, material spalled from the opposite face of the body will tend to partially cancel the effect of material ejected from the near-face crater, thus reducing the efficiency of momentum coupling. Based on the behavior of a wide variety of geologic materials on Earth, the opposite face should be at a range of $\sim\!2 \times 10^4$ to 10^5 cm·kiloton$^{-\frac{1}{3}}$ from the detonation point to assure fully elastic response of the material and eliminate significant fracturing or spallation. Even larger separations might be needed for extremely weak materials or for double and multiple bodies with interfaces of near-zero tensile strength.   These considerations depend on the specific situation and are a major reason why prior survey of the body might be required for design of a successful cratering event. If the assailant were fractured rather than deflected, the problem may be exacerbated rather than eliminated. Several chunks in close proximity will penetrate the atmosphere as if they were a single body. Furthermore, they will present a most disagreeable multiple target, making a second shot difficult or impossible.

4. *Optimal Configuration for a Penetrator.* The relative velocity will provide adequate kinetic energy to bury the nuclear explosive at significant depths. In order to penetrate into the assailant, the nuclear explosive must be fitted with a weighty billet: a cylinder of metal (probably tungsten) that will erode during penetration. The billet will add weight to the package that must be delivered. We would like to optimize the system to minimize the mass in orbit, M_i. Thus we need to decide on the optimum trade-off between mass invested in the nuclear explosive and mass invested in the billet. Say the nuclear explosive is a cylinder with length equal to its diameter, a reasonably realistic assumption. The mass of the nuclear explosive is

$$M_x = 2\pi r^3 \rho_x \tag{35}$$

where r is the cylinder radius and ρ_x is the average density of the explosive assembly. The billet in front of the explosive is also a cylinder, and must provide the material to erode during penetration. In general, the billet will penetrate a ρr of the assailant approximately the same as its own ρr, so

$$\rho_b l_b \simeq \rho_a D \tag{36}$$

where ρ_b and l_b are the density and length of the billet, and ρ_a is the density of the assailant. In this approximation, the mass of the billet is

$$M_p = \pi r^2 \rho_a D. \tag{37}$$

By combining Eqs. (27), (35), and (37), we find

$$\frac{p_\perp}{M_f} = \frac{a\varphi\xi}{1+\xi} + \left(\frac{4\varphi^2 \rho_x^2}{\pi}\right)^{\frac{1}{3}} \frac{b}{\rho_a(1+\xi)} \tag{38}$$

where $M_f = M_x + M_p$ and $\xi = M_x/M_p$. To optimize the system, we wish to obtain the largest possible value of $p_\perp/M_f$. If

$$\frac{a\rho_a}{b}\left(\frac{\pi\varphi}{4\rho_x^2}\right)^{\frac{1}{3}} < 1 \tag{39}$$

which would generally be the case, $p_\perp/M_f$ increases monotonically with ξ. In other words, the smaller the billet, the better. A shallow depth of burial— close to the equivalent of a surface burst—will always perform better in an integrated system of this type than a deep burial.

The conclusion is that a penetrator has no value in enhancing deflection, but may be of great value if we choose to pulverize the astral assailant.

B. Stand-off Deflection

The fracture problem can be much mitigated by detonating the nuclear explosive some distance from the astral assailant. Figure 8 illustrates the difference between a surface burst and a stand-off detonation. Rather than forming a crater, the neutrons, X-rays, γ-rays, and some highly ionized debris from the nuclear explosion will blow off a thin layer of the assailant's surface. This will spread the impulse over a larger area and lessen the shear stress to which the assailant is subjected. Of these four energy transfer mechanisms, by far the most effective (at reasonable heights of burst) is neutron energy deposition, suggesting that primarily fusion explosives would be most effective.

The problem of calculating the momentum transferred from a stand-off detonation is discussed in other chapters. A complete description requires computer simulations. However, some general statements can be made. At an optimal height of burst, about 2 to 8% of the explosive's energy is coupled to the assailant's surface, again depending on the assailant's actual composition and the neutron spectrum and total neutron energy output of the explosive. This corresponds to an energy fraction δ of 0.2 to 0.4. Most of the energy is deposited within 10 cm of the surface. The cratering constants can still be used as in Eq. (5), but for this surface blow off, $\beta \simeq 1$ and α ranges from 10^{-6} to 2×10^{-6} cm^{-1}·s. If we select an assailant for which $\delta = 0.3$ and $\alpha = 1.5 \times 10^{-6}$ cm^{-1}·s, we find that the blow-off fraction will be about a factor of 35 times smaller than the surface burst. The blow-off fraction given would be in the range of 1% for $R_l = 0.1$ AU and in the range of 1/3% for $R_l = 1/3$ AU. Similarly, from Eq. (23) we find that the initial mass of the interceptor would have to be about 40 times as large. So in Fig. 4 the mass would be multiplied by 40, i.e., ranging from about 40 tons to about 40 kilotons. The latter would not be very practical.

V. SUMMARY AND COMMENTS

The problem of terminal interception of comets and asteroids on collision course with Earth has two components: (1) detection of these relatively small assailants; and (2) smashing or deflecting them should they be on an endangering path. In this chapter, we have addressed the latter issue. The relationships we have derived should guide thinking on how to counter such assailants. The main value is to show the functional relationship among the parameters. This chapter is not intended to be an exhaustive study, and much research will be required to evaluate the constants in the equations we have derived. But the following observations are compelling.

1. Kinetic energy deflection with chemically fueled rockets is very limited in its range of effectiveness. It can effect ocean diversion of rocky asteroids

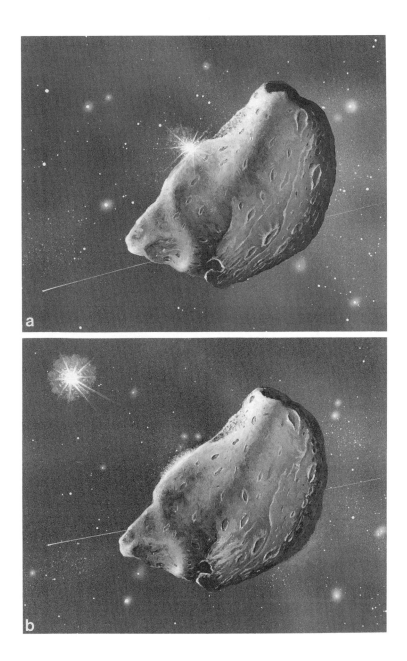

Figure 8.   Illustration of nuclear-explosive deflection techniques. (a) Surface burst. (b) Stand-off detonation.

with diameter smaller than about 70 m, if the interceptor is launched when the distance to the assailant is more than 1/30 AU. At shorter range, interceptors become impractically massive, and the probability of fracture increases dramatically. Because 70-m chondrites dissipate almost all their energy traversing the upper atmosphere, it is unlikely that kinetic-energy deflection will be very useful (Hills and Goda 1993).

2. Nuclear-explosive deflection is imperative for assailants greater than about 100 m detected closer than 1/30 AU because of the enormous mass of the interceptor required for kinetic-energy diversion.

3. Nuclear-surface-burst deflection offers a three-to-four order of magnitude reduction in interceptor mass. The advantage decreases slightly with specific impulse; the advantage decreases dramatically with assailant velocity; and fragmentation is a problem for intercepts closer than about 1/30 AU.

4. Nuclear penetrators offer no advantage for deflection, but are better for pulverization.

5. Nuclear stand-off deflection reduces fragmentation probability, but involves a substantial increase in interceptor mass

The assailant object depicted in Fig. 1 is roughly spherical in shape. In fact, comets or asteroids are generally quite aspherical as illustrated in Fig. 8, and many may be loose associations of two objects—binaries. [The two near-Earth objects for which we have resolved images, Gaspra and Toutatis, may well be binaries.] All the deflection techniques except the stand-off nuclear burst make a crater that is small compared to the characteristic dimension of the assailant. The linear momentum impulse will be imparted along a line connecting that crater and the center of mass, with corrections for local geology and topography. An aspheric object will also receive some angular momentum, depending on the location of the crater and the object's inertial tensor. The size of the impulse will depend on material properties, geology, and topography.

Thus, it will be necessary to characterize the geology and mechanical properties of the assailant when using the cratering deflection techniques. Such characterization could be accomplished by a vanguard spacecraft. Stand-off deflection is much less sensitive to these details. In general, linear momentum will be imparted along the line connecting the detonation point with the center of mass—a large lever arm. Little angular momentum will be imparted, and this will depend on relative projected areas of various assailant topographic features compared with components of the inertial tensor. Thus, besides its inherent fracture-mitigation virtues, the stand-off deflector demands substantially less information about the object it is deflecting.

REFERENCES

Ahrens, T. J., and O'Keefe, J. D. 1987. Impact on the Earth, ocean, and atmosphere. *Intl. J. Impact Eng.* 5:13–32.

Burton, D. E., Snell, C. M., and Bryan, J. B. 1975. Computer design of high-explosive experiments to simulate subsurface nuclear detonations. *Nuclear Tech.* 26:65–87.

Cooper, H. F., Jr. 1976. Estimates of Crater Dimensions for Near-Surface Explosions of Nuclear and High Explosive Sources. Univ. of California/Lawrence Livermore National Lab. Rept. UCRL 13875.

Glasstone, S., ed. 1962. *The Effects of Nuclear Weapons*, ref. ed., (Washington, D. C.: U. S. Atomic Energy Commission), pp. 289–296.

Hills, J. G., and Goda, M. P. 1993. The fragmentation of small asteroids in the atmosphere. *Astron. J.* 105:1114–1144.

Kreyenhagen, K., and Schuster, S. 1977. Review and comparison of hypervelocity impact and explosion cratering calculations. In *Impact and Explosion Cratering*, eds. J. D. Roddy, R. O. Pepin and R. B. Merrill (New York: Pergamon Press), pp. 983–1002.

Morrison, D. 1991. Public Forum. Presented to the *International Conference on Near-Earth Asteroids*, San Juan Capistrano Research Inst., San Juan Capistrano, Ca., June 30–July 3.

Van Dorn, W. G. 1961. Some characteristics of surface gravity waves in the sea produced by nuclear explosives. *J. Geophys. Res.* 66:3845–3852.

VEHICLE SYSTEMS FOR MISSIONS TO PROTECT THE EARTH AGAINST NEO IMPACTS

JOSEPH G. GURLEY
Los Angeles, California

WILLIAM J. DIXON
Manhattan Beach, California

and

HANS F. MEISSINGER
Microcosm, Inc.

An active defense against impacts by massive NEOs, involving either deflection of the object, or its fragmentation, will require a space vehicle to carry a weapon or other device to the vicinity of the threat object. Vehicle systems to mitigate impact threats are discussed in terms of requirements, mission types, and key system parameters. For most threats, the necessary technology is available or can be developed with low risk. Emphasis is placed on systems based on near-term technology in the critical areas of launch systems, propulsion, communications and control, and guidance. The most attractive missions are early interceptions of asteroids or short-period comets, discovered many decades prior to impact. For such targets, there is only a small additional risk, and large cost savings, if major development costs are delayed until an actual threat is detected. This will permit the incorporation of advanced technologies which might achieve some combination of reduced cost, improved performance, and enhanced reliability.

I. VEHICLE SYSTEMS

This chapter treats vehicles to be used in a system of active defenses against threatened impacts by massive near-Earth objects (NEOs), building on a recent examination of this problem by a NASA workshop (Rather et al. 1992; Chapter by Canavan et al.). Its first objective is to demonstrate the maturity of the spaceflight technology needed to respond to a threatened impact, in the event that such an impact should be predicted by a near-term surveillance and tracking system. Other objectives are to characterize practical vehicle systems, to quantify some of their most important parameters, and to consider how future technological developments may modify these preliminary results. The approach taken is to describe conceptual vehicle systems based on high-confidence, near-term extrapolations from current technology. As the feasibility of an effective defense against an impacting body is contingent on an effective surveillance and tracking system, providing adequate warning time, this study emphasizes the urgent necessity of developing and deploying such a surveillance system.

Missions for defense against impacts by NEOs can be characterized by the attributes of the threat object and by the technological means used to counter that threat. For purposes of exposition, we can distinguish two basic classes of threats, one consisting of asteroids and short- or medium-period comets (period $\lesssim 200$ yr), and the other of long-period comets. And, for characterizing vehicle systems, we can distinguish two basic classes of interceptions, those involving rendezvous and/or docking, and those involving high approach velocity. Approach velocities for the latter class of trajectories are generally comparable to the velocities with which the objects approach the Earth—roughly 10 to 30 km s^{-1} for asteroids, and 15 to 70 km s^{-1} for long-period comets.

These two binary partitions of the mission space result in four possible combinations. One of these combinations, a rendezvous mission to a long-period comet, is impractical; it requires either a very long lead time, incompatible with current or foreseeable detection technology, or a very great propulsive impulse (ΔV), incompatible with current or foreseeable propulsion technology. The remaining three combinations, rendezvous and high-velocity missions to asteroids, and high-velocity missions to long-period comets, are discussed below (Sec. III), following a discussion of general requirements applicable to all (Sec. II). The chapter concludes with an analysis of the implications of advanced technology (Sec. IV).

II. REQUIREMENTS

A. Functional and Performance Requirements

The mission concepts and vehicle systems described here represent engineering solutions to address the functional and performance requirements needed to mitigate the risk of a destructive impact. The principal functional requirements are (1) delivery of a payload (a weapon or other threat mitigation device) to the threat object or its vicinity; (2) providing communications, command, and control support for the payload and vehicle; and (3) protection of the payload and vehicle from the various environmental stresses associated with space flight. Performance requirements, specifying quantitative performance parameters, are logically derived from these functional requirements (Table I).

With today's technology, spaceflight capabilities are essentially propulsion limited. Therefore the most important performance requirements, in terms of impact on system costs, are those pertaining to the payload delivery function, especially the mass of the payload to be delivered, the timeliness of the delivery, and the total propulsive velocity increment ΔV required. The concept descriptions given below will discuss factors affecting these parameters and other critical system attributes.

TABLE I

Requirements Summary

Functional Requirements	ΔV (km s⁻¹)	Min. Mass Concerns	Performance Requirements[a]			Engineering & Technology Concerns & Issues
			Nav, Guidance, Control	Communications	Automation	
1.0 Payload delivery						
1.1 Launch to LEO	c. 10	I_{sp} 450 s	Inertial		Autonomous	Launch vehicle selection
1.2 Inject to Sun orbit	6–9(R); 3.3 (HV)		GPS/Inertial	TDRSS(?)	Autonomous	High-thrust cryogenic upper stage(s)
1.3 Cruise	c. 0.1		DSN	DSN	Ground control	Storable liquids; multiple restarts
1.4 Arrive at target	.1–3(R); 0.3(HV)	I_{sp} 300 s	DSN/Optical	DSN	Autonomous homing	Acquisition range; ephemeris error
1.5 Orbit asteroid	c. 0.1(R)		DSN/Optical (R)	DSN (R)	Semi-autonomous	Orbit stability
2.0 Support P/L Ops						
2.1 Initial recon			DSN/Optical	DSN	Semi-autonomous	Sensor capabilities
2.2 Activate weapon		Re-entry protection		DSN	Ground-controlled arming	Safety & reliability
2.3 Verify effects			DSN	Semi-autonomous	Platform & sensor selection	
3.0 Environment						
3.1 Noise & vibration		Structure 20%				Primary launch & injection phases
3.2 Thermal		"				Range of solar distances 0.3–4 AU
3.3 Particles & Rad		"				Natural radiation + nuclear effects

[a] Entries marked (R) or (HV) apply exclusively to rendezvous or high-velocity approach missions, respectively.

B. States of Readiness

The reaction time for a program to protect the Earth from the threat of an impact must be compatible with the warning times associated with the class of objects it is designed to deal with. Once an asteroid or comet has been discovered and its orbit determined, and it is judged to be a threat, the response time available before a vehicle must be launched to intercept it may range from a few weeks to many years. Most Earth-crossing asteroids with a size large enough to be a global threat will probably be identified several decades before the projected impact, but long-period comets may be first discovered only months, or possibly weeks, before impact. Therefore, launch delays of several years may suffice for asteroid missions, but a response within a few weeks may be required for comet missions.

The implications for vehicle system design are illustrated by considering the analogous situation with regard to other space programs. At one extreme, the Galileo Project, intended to put a scientific spacecraft into orbit about Jupiter, was launched in 1989, 15 years after the start of mission definition studies, and 12 years after congressional approval in 1977. (Galileo will not arrive at Jupiter until 1995, 18 years after project approval.) At the other extreme is the ICBM program, which achieved the ability to launch on about 10 minutes' notice, from any of a number of launch sites, to strike any one of many possible targets. The response times for these two programs are, therefore, 12 to 15 years for Galileo, and 10 minutes for the ICBM program.

Figure 1 shows how program costs and defensive capability vary as a function of the system's "state of readiness," which in this context may be defined as the system response time, from the recognition of a threat to the launch of an interceptor to mitigate that threat. The defensive capability is represented by a "benefits" curve giving a rough estimate of the expenditure which would be acceptable, to reduce the probable property damage and casualties to zero. The shape of this curve reflects the bimodal population of threat objects; for the majority of threats—those due to Earth-crossing asteroids and short-period comets—warning times are on the order of decades, while for others (primarily long-period comets) they are only a few months. The absolute dollar magnitude of the benefits, estimated here as on the order of $100 M per year, is subject to large uncertainties, which might be reduced by further analysis of data concerning perceived risks and risk-reduction expenditures for disasters of other types (see Chapter by Canavan).

The "cost" curve, shown as a double line on Fig. 1, starts low, assuming that money for vehicle systems need be spent only after a threat is identified and an appropriate defense can be tailored to the specific orbital, mass, and material characteristics of that particular threat. Because of the infrequent need (about once every 1000 yr, to intercept objects of 100 m diameter and up) and assuming a relatively modest unit cost on the order of $1 B to $3 B, the average annual expenditure is only $1 M to $3 M, in addition to the annual

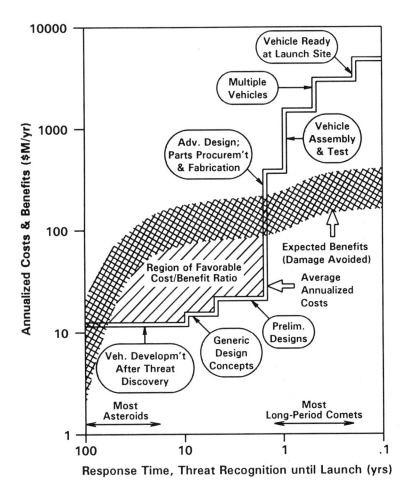

Figure 1. Cost/benefit analysis of NEO defense. (Balloon captions indicate the level of pre-discovery development activities.)

expense of the detection and tracking facilities of about $10 M (Morrison 1992).

The cost curve turns steeply upward at the point where the necessary response time is shorter than the minimum time (about 1 to 3 yr) required to design, build, and deploy an interceptor vehicle. To defend against the hazard of a long-period comet discovered only weeks or even months before the projected impact, it is necessary to have developed in advance the designs, the hardware and software for the launch vehicle and the intercepting space vehicle, and the payload, to deflect or otherwise mitigate the threat. Because the size and shape of the cometary body would be unknown in advance, it might be necessary to have several alternative payloads available; and because the target could approach from any celestial direction, we would need the ability to launch from any of several launch sites, in order to match the inclination of the space vehicle's departing launch asymptote to the demands of the target. As in the case of the ICBM program, all of the development money would have to be spent in advance of the discovery of the threat, and an arsenal of payloads, space vehicles, and launch vehicles would have to be maintained in a continuous state of readiness to support a launch on a few weeks notice. The program costs are very high, because of the necessity of deploying vehicles even when an actual threat does not occur, and because of the high unit cost of the fast-response, high-ΔV systems required.

From Fig. 1 we conclude that there is a broad range of system response times for which benefits exceed program costs, with a margin sufficient to make the results insensitive to reasonable changes in the assumptions. The most economically attractive strategy, which minimizes the cost/benefits ratio, is to defend against asteroids and short-period comets only, with vehicle development delayed until a specific threat has been identified. With several decades of warning time available for most impact events, this system will require little initial expenditure above that needed for the discovery and warning system, and will achieve a major reduction in the risk of a catastrophic impact. The much greater cost of a program with a response time short enough to defend against the threat of impacts by long-period comets, is not economically feasible, at least in the immediate future. Such a program, requiring a vehicle system that is much more expensive, maintained in a continuous high state of readiness, and supported by a much more complex detection system, would be orders of magnitude more costly, while providing a relatively small increment in damage reduction.

C. Effects of Target Size and Orbital Properties

The wide diversity of sizes, shapes, and material properties of potential impacting objects, and the wide range of orbit parameters, challenge the systems engineer to devise system concepts which are either extremely versatile, or tailored to match the requirements of an individual threat.

Target size is one of the most critical parameters of the threat object, because the payload mass which must be delivered is roughly proportional to the mass of the object, and therefore to the cube of a linear dimension. This makes it very important to identify the range of sizes presenting the greatest threat. Significant risk of property damage and loss of life may result from impacts of objects in the size range (variously defined as the maximum dimension or the mean diameter) from about 30 m to 3 km. Here we adopt the widespread (but not universal) opinion that target sizes from 1 to 2 km pose the greatest threat. The lower bound is large enough that serious global damage to the environment may result, and the upper bound is high enough that there is small probability of near-term discovery of an unknown Earth crosser of larger size.

Objects in the size range from 100 to 1000 m also present a significant threat of local or regional damage, and are sufficiently bright for optical detection during passages near Earth. The interception of objects in this size range may be required, even if the primary objective is protection against kilometer-sized objects.

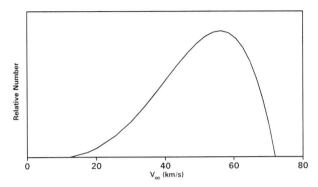

Figure 2.   Distribution of long-period comet velocities relative to Earth.

Orbital properties of asteroids can be deduced from the extensive data base of discovered Earth-orbit crossers (Shoemaker et al. 1990; see also the Chapter by Rabinowitz et al.). For long-period comets on Earth-crossing orbits, we can draw certain inferences from the limited number which have been observed. While inside the orbit of Jupiter, they follow nearly parabolic trajectories, with a nearly isotropic distribution of orbit planes. Impacts are equally divided between the day hemisphere (impact after perihelion) and the night hemisphere (impact before perihelion). Comet velocities relative to the Earth vary from 12 to 72 km s^{-1}, with a distribution which is biased towards the higher values (Fig. 2). For a population model in which the number with a perihelion distance less than q is proportional to $q^{3/2}$, the median perihelion distance for Earth-orbit-crossing comets is 0.63 AU, a value which is used here as representative of the class.

D. Deflection Velocity versus Orbit and Lead Time

The mean time interval between impacts by objects larger than 100 m is on the order of 1000 yr, which is very long compared to the time, about 25 yr (Morrison 1992; Chapter by Bowell and Muinonen), required for a newly deployed detection system to survey the entire current population of Earth crossers. Therefore it is highly probable that the warning time available, when an asteroid threat is discovered, will be many decades, and not just a few months or even a few years.

A long warning time is favorable for a system which deflects the asteroid trajectory by a small velocity increment (typically on the order of 1 cm s^{-1}) at a time before impact of 10 years or more. The velocity impulse needed, for an object whose orbit is similar to that of a specific asteroid, can be calculated as follows:

1. Orbital elements.   The orbital elements are propagated forward to an

epoch when the heliocentric distance at the nodal crossing equals 1 AU.

2. Orbital displacement. Using the equations of Keplerian motion, a spherical distribution of unit velocity impulses at a particular deflection time is mapped into an ellipsoidal distribution of position displacements at the nominal impact time.

3. Aimpoints. The displacement ellipsoid is projected onto a plane normal to the direction from which the asteroid approaches the Earth, generating a two-dimensional distribution of aimpoint vectors. The semimajor axis of the aimpoint ellipse represents the "aimpoint sensitivity," the approach trajectory offset due to unit velocity impulse in the optimum direction.

4. Required deflection. Division of the effective Earth radius (taking into account the increase due to gravitational focusing), by the aimpoint sensitivity gives the required velocity impulse for successful deflection.

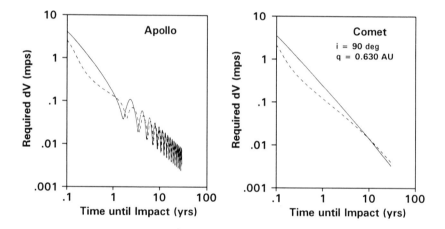

Figure 3.   Required deflection velocity, as a function of time before impact, for an asteroid and for a typical long-period comet. Solid line: impact before perihelion. Dashed line: impact after perihelion.

Approximations valid for very short times, and others for very long times, show a required velocity impulse which is inversely proportional to the time until impact. Results from a general solution, for the asteroid Apollo and for a fictitious comet, covering a broad range of times, are presented in Fig. 3. Apollo was chosen as a representative asteroid threat because its orbital elements are typical of those of potential impactors, and its calculated impact velocity (20.3 km s^{-1}) is near the median for known Earth-crossing asteroids. The fictitious comet was assigned orbital elements representing median values for Earth crossers, an inclination of 90 degrees and a perihelion distance of 0.630 AU.

III. MISSION AND VEHICLE CONCEPTS USING NEAR-TERM TECHNOLOGY

A. Mission Types

Proposed space missions to counter the threat of a destructive impact generally fall into one of two basic categories, divert missions or fragmentation missions. The objective of a divert mission is to deflect the trajectory of the threat object so that it will miss the Earth. A fragmentation mission's objective is to shatter the threat object into pieces too small to penetrate the Earth's atmosphere and cause serious damage. The divert option is generally preferable, because an attempted fragmentation may leave a few fragments large enough to cause severe damage. However, it may not be feasible, using a single impulse, to impart a velocity impulse $\gtrsim 1$ cm s^{-1} without fracturing the target. This effectively restricts the use of deflection to those asteroids and short-period comets which can be discovered several decades before the time of the threatened impact. For a long-period comet, fragmentation could be the only option which is technologically feasible.

Divert missions may be further partitioned into rendezvous missions and high-velocity-approach missions. For "slow" asteroids (those which would impact Earth with velocities of 20 km s^{-1} or less) a rendezvous mission may be preferred, because it allows ample time to study the target object at close range, and plan the divert maneuver at a time of favorable orbital phase (near perihelion) and favorable rotational attitude (to minimize the chances of fragmentation). For "fast" asteroids, with greater impact velocities, a high-velocity-approach mission, similar to a scientific flyby, may be a necessity because of the excessive mass of propulsive fuel needed for a rendezvous. These considerations lead us to propose three types of missions for early consideration: a rendezvous and divert mission to a slow asteroid; a high-velocity approach and divert mission to a fast asteroid; and a fragmentation mission to a long-period comet.

A typical NEO interception mission consists of the following phases:

1. Launch and low Earth orbit. A standard launch vehicle places the spacecraft in a low Earth orbit (LEO), with an altitude of about 200 km. Launch time of day and azimuth are chosen so that the plane of the orbit contains the desired asymptotic velocity vector on departure from Earth. A coast phase of fractional orbit duration is used to control the phase angle of that asymptotic velocity.

2. Injection. A short-duration, high-thrust propulsion stage accelerates the vehicle to an Earth-escape trajectory that results in a heliocentric transfer orbit to the vicinity of the target object. State of the art propulsion systems for this maneuver use cryogenic propellants with an I_{sp} of about 450 s.

3. Heliocentric transfer. For an asteroid mission, the heliocentric transfer requires about one to three years. Most of this time is spent in an unaccelerated or coasting mode, but several small propulsive maneuvers are required. During the first 30 days, one or more guidance adjustments are made, to correct for errors in the injection maneuver. There may also be a small orbital maneuver about midway through the transfer orbit, primarily to reconcile the orbital planes used for Earth departure and target arrival. Then, during the final 30 days, there may be one or more guidance adjustments to correct for errors observed during the course of the Earth-based radiometric tracking conducted up to that point. During this and later phases of the mission, storable propellants with I_{sp} on the order of 300 s are used. For a mission to a long-period comet, a shorter flyout time and an interception much closer to Earth are required because of limitations in near-term detection capabilities and current propulsion technology.

4. *Approach.* Onboard sensors, such as an optical imager, detect the target object and provide line-of-sight data for homing guidance. For a rendezvous mission, a major propulsive maneuver reduces the closing velocity to a much smaller value, and subsequent corrective maneuvers approximate an exponential deceleration profile. The rendezvous phase terminates with the spacecraft hovering near the target, or in a loose orbit about it. For a high-velocity-approach deflection mission, the aimpoint is offset from the target to cause a flyby on the side from which the impulse is desired, for a standoff burst. For a fragmentation mission, the spacecraft is targeted to penetrate to a point deeply buried within the target object.

5. *Orbital phase.* For a rendezvous mission, the orbital phase may involve a wait of several months, until the target object reaches a favorable point in its orbit (generally the perihelion point). Orbits of mean radii between about 1.5 and 100 times the radius of the target are predicted to exhibit reasonable stability in the presence of perturbing forces—primarily those due to the irregular shape of the target, at close range, or due to solar gravitational forces and radiation pressure at large range. The orbital phase provides opportunity for close-range observation of the size, shape, mass, and rotational parameters of the target object, and perhaps measurements or tests bearing on its composition and mechanical properties, and for positioning the payload device (assumed to be a chemical or nuclear explosive device) near or on the surface, as required for an effective divert impulse.

6. *Divert or fragmentation phase.* For a deflection mission, the payload, a chemical or nuclear explosive device, is detonated at a point on or near the surface of the target object. For fragmentation, detonation should occur only after penetration to a point well below the surface. For asteroids larger than about 100 m, a nuclear device is necessary. For smaller objects, the use of chemical explosives, placed on or below the surface, may be feasible, but would be far more costly in terms of payload mass.

7. *Assessment.* An assessment phase is needed in order to verify that the mission has been successful, with the target object's velocity altered sufficiently to ensure that it will miss the Earth. Alternatives to consider are (a) Earth-based observations; (b) instrumented subsatellites or probes, deployed from the primary vehicle during the orbital phase, and placed where the target object itself will shield them from the effects of the weapon; or (c)

an independent observation vehicle, separately launched from Earth.

Table II lists estimates of some key parameters for each of the three types of NEO interception missions identified above. Numerical estimates for the vehicle mass were arrived at by starting at the bottom of the table, and entering values of the achievable deflection velocity or of the fragment size needed to achieve successful negation of the threat. The mass of the negation device was estimated as 1000 kg for a divert velocity of 1 cm s^{-1} (Ahrens and Harris 1992; Solem 1993; see also the Chapter by Ahrens and Harris), or 5000 kg for fragmentation, assuming a threat object of diameter about 1000 m. The larger mass assumed for fragmentation allows for a penetration device and for the more powerful detonation needed to ensure thorough fragmentation. Spacecraft mass on arrival at the target was calculated by adding an allowance of 250 kg for sensors and spacecraft systems (including guidance, navigation, communications, and control), plus 20% for structural and thermal systems. Mass during the cruise phase includes fuel for the indicated ΔV for approach guidance and rendezvous, assuming an I_{sp} of 300 s. Mass in LEO includes the injection fuel, assumed to have an I_{sp} of 450 s.

Most of the functions required of the NEO interception spacecraft are similar to those performed by the current generation of Earth-orbital systems and planetary probes, but several items are peculiar to this mission. The first is the need for protective packaging to ensure the intact survival and recovery of the payload (assuming that it is a nuclear device), in the event of a malfunction during the boost or injection phases of the mission. The technical problems are similar to those encountered in the use of radioisotope thermoelectric generators (RTGs) in missions to the outer planets, but are more severe because of the greater mass of nuclear material involved, and because of the potential for a high-level nuclear detonation. There are also political problems to be addressed, concerning the use of nuclear explosives for missions of this type. A second issue is the design of the terminal guidance required for rendezvous with the target or to hit a very precisely defined aimpoint near it or within it. One difficulty is that the target is a very dim object with an ephemeris error which, for an asteroid, is generally about 50 km, and for a comet may be much greater. A third issue is that of the mass of the propulsion system, relative to the limited lift capabilities of existing launch vehicles.

B. Missions for Deflection of Asteroids and Short-Period Comets

In order to quantify the required velocity impulse, or ΔV, needed for interception missions to asteroids, we have examined two-impulse transfers from LEO to asteroid rendezvous, and single-impulse high-velocity-approach trajectories, to several asteroids. For each asteroid, there are two distinct Earth-impact geometries; an approach from the day hemisphere, or an approach from the night hemisphere. Table III shows the results for the dayside impact cases, for each of four asteroids which span a broad range of possible Earth-impact velocities. Results for nightside impacts are essentially identical. For a rendezvous, the ΔV required increases as the impact velocity increases, but for

TABLE II
Key Parameters of Typical NEO Intercept Missions

Threat Object	Asteroid or SPC[a]	Asteroid or SPC	LP Comet
Earth Impact Velocity (km s^{-1})	13–20	20–35	15–70
Warning time (yr BI[a])	30–100+	30–100+	1–5
Launch time (yr BI)	11–13	16–18	0.5–2
Mass in LEO (1000 kg)	6.5–28	3.5	80–250
Injection ΔV (km s^{-1})	6–9	3.3	10–15
Mass in heliocentric cruise (kg)	1700–3800	1700	8800
NEO encounter time (yr BI)	10	15	0.1–1
Approach velocity (km s^{-1})	0.1–2.5	10–30	40–70
Approach guidance accuracy (m)	1000	10	5
Approach ΔV (km s^{-1})	0.3–2.8	0.3	1.0
Mass on arrival (kg)	1500	1500	6300
Time in vicinity of NEO	5–15 months	10 s	5 s
Mass of negation device (kg) (assumed)	1000	1000	5000
Negation type	Deflection	Deflection	Fragmentation
Deflection ΔV (cm s^{-1})	1.0	1.0	NA
Largest fragment dimension (m)	NA	NA	25–50

[a] SPC = Short Period Comet.

a high-velocity approach the ΔV is essentially independent of impact velocity. For low-energy impactors, such as the asteroids Anza and Geographos, the optimal rendezvous requires an injection from LEO into an orbit nearly matching that of the asteroid, and then the rendezvous is completed with very little additional fuel consumption. For more energetic objects, like Apollo or Icarus, the optimum mission profile involves injection onto a flyout trajectory to a point near the asteroid's aphelion point, where a much larger fraction of the mission ΔV is used for the final rendezvous maneuver. Figure 4 shows, for the asteroid Apollo, the launch time/arrival time combinations which result in low values of mission ΔV, for both rendezvous and high-approach-velocity missions.

TABLE III
Mission ΔV for Asteroid Intercept Missions

Asteroid Name	Earth Impact V	Rendezvous Missions			High-Velocity Approach
	km s^{-1}	Injection km s^{-1}	Rendezvous km s^{-1}	Total ΔV km s^{-1}	Total ΔV km s^{-1}
Anza	14.20	6.27	0.01	6.28	3.27
Geographos	16.63	8.70	0.06	8.76	3.28
Apollo	20.26	6.57	2.45	9.02	3.27
Icarus	30.59	9.04	7.46	16.50	3.36

An alternative mission plan, to achieve rendezvous with a very fast asteroid while keeping the mission ΔV low, is to use gravity-assist flybys of one or more of the principal planets. A trajectory using one or two flybys of Venus, followed by one or two flybys of Earth, can reduce the velocity impulse for a rendezvous with the asteroid Apollo from 9 km s^{-1} to about 5.5. A preliminary study of mission opportunities shows that such trajectories are generally available, but that they typically require a mission duration of 6 to 10 yr, compared to the 1 to 3 yr required for the direct transit.

For objects with Earth impact velocities exceeding 20 km s^{-1}, the rendezvous and deflection mission may require a propulsive ΔV of impractical magnitude. The high-velocity approach and deflection mission is a promising alternative for such objects (see last column of Table III and Fig. 4), because of the much smaller ΔV. An onboard sensor, a guidance computer, and a propulsion system for terminal maneuvers are required in order to achieve the hypervelocity interception capability needed for this mission. The terminal guidance technology required is similar to that developed by the Defense Department, under the SDIO (Space Defense Initiative Office), for the exoatmospheric interception of ballistic missiles. Some of these guidance problems can be alleviated by deploying a separate reconnaissance spacecraft, a vehicle with only a lightweight payload of sensors and a beacon, to achieve a rendezvous and perform short-range observations, before the main interceptor arrives.

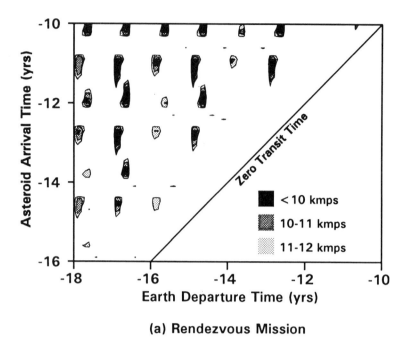

(a) Rendezvous Mission

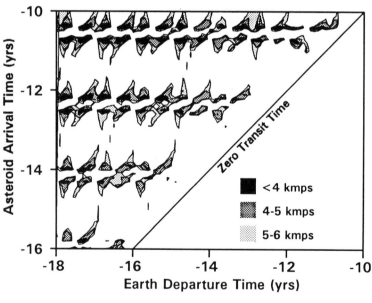

(b) High-Velocity-Approach Mission

Figure 4.   Mission opportunity maps for interception missions to the asteroid Apollo.
(Time is measured relative to the predicted impact event.)

For low- to moderate-impact velocities (20 km s^{-1} or less), a rendezvous mission may be preferred even though the high-velocity approach requires less ΔV. Negative aspects of the high-velocity interception mission include a requirement to deliver a larger divert velocity impulse to the target, because the interception generally occurs near the Earth's orbit, at a solar distance of about 1 AU, rather than at the perihelion point. In addition, the inflexible timing of the detonation could prove troublesome in the case of an irregularly shaped target object, because it removes the capability to choose a favorable target rotational attitude. This may increase the risk of target fragmentation, possibly leaving one or more large fragments on a collision course with Earth.

C. Missions to Long-Period Comets

Missions to long-period comets are characterized by short warning times, high values of propulsive ΔV, and large payload mass. Discovery will usually occur when the comet is within 5 AU of the Sun, because of its remoteness and low visibility while distant from the Sun. Additional delays, before sufficient observations are available for an accurate determination of the orbit, are likely to reduce the warning time, from the first recognition of the impact danger until the projected impact, to two years or less.

The mission ΔV for a rendezvous is unacceptably large, but periodic opportunities exist for high-velocity approach missions (Fig. 5). However, the ΔV is large nevertheless, because of the short warning time, the high likelihood of a substantial out-of-the-ecliptic motion, and the comet's high value of heliocentric velocity (essentially equal, at a solar distance of 1 AU, to the solar escape velocity, 42.1 km s^{-1}). Note that, for comets with perihelion distances between 0.5 and 1 AU, the night hemisphere or before-perihelion impactors are more difficult to reach than the day hemisphere or after-perihelion ones; the out-of-ecliptic distance is much greater in the former case.

Most comets are thought to be structurally weak, composed of loosely compacted ices and dust, and could not support the mechanical shock accompanying an attempted deflection impulse of the large magnitude needed. Therefore the feasibility of a deflection mission to a long-period comet is doubtful, and it is preferable to attempt a fragmentation mission, whose objective is to shatter the target into many small pieces. For fragmentation to be fully successful, the largest remaining fragment should be too small to survive passage through the Earth's atmosphere. Estimates of the efficacy of protective shielding by the atmosphere suggest a value on the order of 30 m. The global damage inflicted would be substantially reduced by even coarser fragmentation.

For effective fragmentation, the payload, a rather large nuclear device (probably several megatons TNT equivalent), must impact and perhaps penetrate the nucleus of the comet before detonation, rather than explode above the surface. Various techniques have been suggested for burying the nuclear charge. Drilling a well and statically emplacing the charge, the technique used for underground testing on Earth, is impractical because of the power

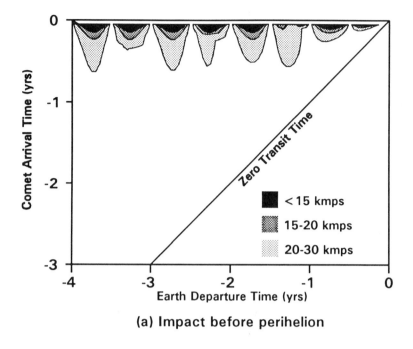

(a) Impact before perihelion

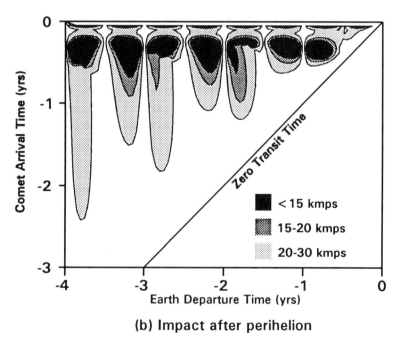

(b) Impact after perihelion

Figure 5.   Mission opportunity maps for a high-velocity approach to a long-period
comet in a steeply inclined orbit.

required for drilling and the implied necessity of a rendezvous with the very rapidly moving body. More practical alternatives are kinetic penetrators, and precursor explosions by either chemical or nuclear devices. If the prevailing theories concerning the weak strength of cometary material hold up after more detailed explorations, then the passive kinetic penetrator is the preferred solution, especially in view of the very high approach velocity which characterizes Earth-to-comet trajectories. If the fragmentation technique is to be used against bodies of greater structural strength, then a precursor explosion, analogous to dual-charge rounds used with anti-tank weapons, might be required.

Whether the payload is to penetrate the nucleus of the target, or explode on or above the surface, the terminal guidance requirements of the preceding section will apply. However, for the comet mission, they will be exacerbated by the higher approach velocity, and by the presence of the cometary coma and tail which will interfere with the ability of optical sensors to locate the nucleus accurately.

Various factors, including the extremely limited warning time and the high likelihood of a substantial out-of-the-ecliptic motion, lead to these conclusions about the prospects for near-term defenses against impacts by long-period comets:

1. Readiness. Providing defense against impacts by long-period comets requires the highest state of program readiness, and therefore is the least attractive from an economic perspective, of the several missions examined.
2. Vehicle size. Interception of a long-period comet requires a larger payload and a higher mission ΔV, and therefore a larger, more expensive space vehicle, than an asteroid mission.
3. Uncertainties. Finally, at the time the spacecraft with its payload is launched, the size and mass of the target will not yet be accurately known, and its shape can only be guessed at. These uncertainties will lead to the selection of a payload (undoubtedly a nuclear explosive device) which is oversized compared to the mean estimate of the requirements, provided the launch vehicle will permit. This will further increase the mission cost, and further compromise its economic feasibility. Uncertainties in the mechanism of penetration and in the mechanics of fragmentation will introduce engineering concerns and design penalties, and further reduce the probability of achieving mission success.

IV. IMPLICATIONS OF ADVANCED TECHNOLOGY

A. Key Technologies

Advances in spacecraft and space mission technology expected in the near future will significantly enhance the ability of averting the threat of asteroid or comet impact. Of particular interest are advances in propulsion technology, miniaturization of spacecraft subsystems and components, refinement of terminal guidance, and enhanced automation and robotics capabilities. The resulting benefits include major reductions of initial gross spacecraft mass and, hence, reduced launch vehicle size and cost; the ability to intercept an incoming NEO earlier and at greater distance from Earth, thereby reducing the required deflection impulse; and enhanced response capability to NEO impact threats with a short reaction time due to late detection.

It would be difficult and risky to project advanced technology developments, qualitatively and quantitatively, over more than a few decades, and to assess development costs and risk. Instead, the benefits of greater threat mitigation capabilities based on likely technology advances in the areas mentioned will be discussed in general, and trends of anticipated system performance gains illustrated by examples.

B. Propulsion Technology

Extensive efforts to increase the specific impulse of space propulsion systems have been in progress in laboratory and test facilities over several decades, both in chemical and electric propulsion. After reaching a mature state, this advanced technology will be applicable to future NEO threat mitigation missions and will greatly enhance mission performance.

Technology advances of particular concern here are those in nuclear-thermal rocket engines with specific impulses up to about 1000 s (Jones 1992a, b; see also the Chapters by Venetoklis et al. and by Willoughby et al.) and electric propulsion systems such as ion and pulsed plasma thrusters with specific impulses up to about 3000 to 5000 s (Jones 1992a, b; Pollard et al. 1993). Compared with current high-performance cryogenic chemical rockets, such systems achieve very large reductions in propellant mass ratios for high-ΔV missions (Table IV). The significant ratio of the required propellant mass M_p to the final spacecraft mass M_f is

$$M_p/M_f = M_i/M_f - 1 = \exp(\Delta V/g I_{sp}) - 1 \qquad (1)$$

where M_i is the initial spacecraft mass, and ΔV is the velocity impulse (3 km s^{-1} in the example used in the table).

Thrust magnitudes given in Table IV are representative of the types of propulsion systems considered. Because of the low thrust acceleration of typical electric propulsion units, as low as 0.05 N per kW of applied power, the thrust duration can become extremely long.

The required propulsion system power P is proportional to the product of thrust force F and the specific impulse I_{sp}, and inversely proportional to the propulsive efficiency η of the system, in accordance with

$$P = k F I_{sp}/\eta \qquad (2)$$

where $k = 0.490$ kW N^{-1} s^{-1}, with P in kW, F in N, I_{sp} in s, and η in %. In the above example of a 5000 kg spacecraft with $I_{sp} = 3000$ s, the required power

TABLE IV

System Characteristics with Conventional and Advanced Propulsion

(Assumed spacecraft dry mass M_f = 5000 kg; total ΔV = 3 km s^{-1})

Propulsion Type	High-Performance Cryogenic Chemical	Nuclear Thermal Rocket	Electric Thruster (ion or pulsed plasma)	
			Ion / Pulsed Plasma	Ion / Pulsed Plasma
Specific Impulse I_{sp} (sec)	450	900	3000	5000
Propellant ratio, M_p/M_f	0.973	0.405	0.107	0.063
Propellant mass M_p (kg)	4865	2025	535	315
Typical thrust level F (N)	10,000	10,000	1.0 / 5.0	1.0 / 5.0
Average thrust acceleration (cm s^{-2})	150.7	171.2	.019 / .095	.0194 / .097
Thrust phase duration (days/min)	0/33	0/29	183/0 / 37/0	179/0 / 36/0
Electric thruster power P (kw)	NA	NA	22.6 / 113	37.7 / 188.5
Comments	Single propulsion stage assumed; ΔV much greater than 3 km s^{-1} requires multiple stages. M_P nearly doubles for storable biprop (I_{sp}, 300 s).	Single propulsion stage is more realistic than for chemical rockets.	Pulsed plasma thruster with higher thrust level gives lower thrust phase duration. Calls for nuclear electric rather than solar electric power source. Single propulsion stage is sufficient, even for ΔV much larger than the assumed 3 km s^{-1}.	

level is 113 kW, assuming 65% propulsion system efficiency (see Table IV). To provide this large amount of power a space nuclear power generator such as the SP 100 system, which has been under development for a number of years, might be considered. Solar-electric power generation would be suitable only for lower thrust levels, e.g., for applications requiring a moderate total ΔV and lower spacecraft mass, and for missions restricted to solar distances not much greater than 1 AU.

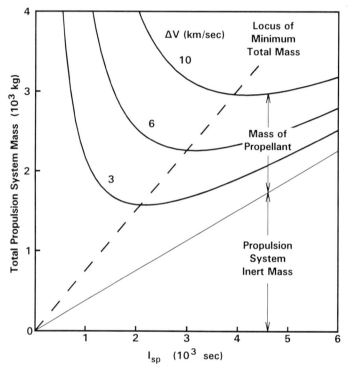

Figure 6. Propulsion system mass vs I_{sp} for several values of ΔV.

The optimal I_{sp} value at which an electric propulsion system should be designed to operate depends on the required total impulse and its specific mass in kg/kW. A wide I_{sp} range can be used with only slight variations of thruster efficiency. Figure 6 illustrates the trade-off between the mass of the thrust system and power source, which is proportional to I_{sp}, and the propellant mass which is inversely proportional to I_{sp}. The minimum total mass and the optimum I_{sp} increase with the total velocity impulse for which the system is designed.

With the greatly reduced propellant mass ratios achievable through electric propulsion, transfers to NEO intercept can be accomplished with a correspondingly reduced initial spacecraft mass and a smaller-size launch vehicle. The technique of planetary gravity assist that has been proposed to reduce ΔV requirements for asteroid intercept missions using chemical propulsion would become unnecessary, and the longer trip time and greater mission profile complexity associated with it are avoided. Also, the need for multiple propulsion stages is circumvented because the $\Delta V/g I_{sp}$ ratio generally is much smaller than 1.0. The same stage that provides the initial ΔV increment to start the transfer phase can also be used to provide the retro-velocity for rendezvous with the target or for orbit insertion, as called for in the scenario.

The foregoing discussion does not reflect the inherent inefficiency of extremely low acceleration during geocentric departure. Conventional high-thrust propulsion retains the advantage of achieving the required heliocentric departure velocity V_{∞}, starting from the initial Earth-orbital velocity V_o, by applying the comparatively small increment

$$\Delta V_i = \sqrt{2V_o{}^2 + V_\infty{}^2} - V_o \tag{3}$$

With extremely low-thrust propulsion, the required velocity increment is approximately equal to the sum $V_o + V_\infty$, where the first term represents the velocity increment for a spiral escape from Earth orbit and the second term is the heliocentric departure velocity. This sum is always greater than the ΔV_i requirement in the high-thrust case. For departure from a 500-km circular Earth orbit, with $V_\infty = 6$ km s^{-1}, the two values are 13.61 and 4.71 km s^{-1}, respectively.

C. Miniaturization of Spacecraft Components

Major reductions of spacecraft subsystem mass have been achieved in the last two decades, notably in structures, power generation and control, and electronic subsystems, including guidance and control components, sensors, transmitters, receivers and antennas, computers, and data handling and storage. In Earth-orbiting spacecraft these advances have led to major mass and size reductions, making possible light satellite ("LightSat") designs with a gross mass as low as one half to one quarter the mass of earlier-generation spacecraft, 10 to 20 yr ago. Much of this progress in miniaturization derives from military space system evolution, driven by the need for deploying numerous small satellites at reduced launch cost.

For purposes of mitigating asteroid and comet threats, the benefits of subsystem miniaturization are reflected in a reduction of the initial and final mass. However, in a mission with a large nuclear explosive device as payload, this subsystem mass reduction becomes less significant in the overall mass budget. Moreover, with the advent of high-I_{sp} propulsion systems the M_i/M_f ratio will be drastically reduced. Together, these considerations tend to reduce the importance of subsystem miniaturization.

As an example, consider an interceptor spacecraft with a total mission velocity of 6 km s^{-1}, in which the subsystem mass is reduced by two thirds but the payload mass is fixed. Mass characteristics with and without miniaturization are listed in Table V for both a chemical propulsion system with $I_{sp} = 450$ s and an advanced (nuclear-electric) propulsion system with $I_{sp} = 3000$ s.

TABLE V

Mass Reduction Through Subsystem Miniaturization for Systems with Low and High Specific Impulse

(Assumed total $\Delta V = 6$ km s^{-1})

	450		3000	
Specific impulse I_{sp} (s)				
Mass ratio, M_i/M_f	3.79		1.22	
Propellant ratio, M_p/M_f	2.79		0.22	
Subsystem miniaturization	No	Yes	No	Yes
Mass breakdown (kg)				
Structures & subsystems	600	200	600	200
Propulsion	800	800	1100^a	1100^a
Payload (nuclear device)	1000	1000	1000	1000
Final mass, M_f	2400	2000	2700	2300
Propellant mass, M_p	6696	5580	594	506
Inital mass, M_i	9096	7580	3294	2806
M_i reduction due to miniaturization		1516		443

^{a} Based on 113 kW propulsive power and a specific mass of 10 kg/kW.

Only for chemical propulsion are the M_i savings large enough to permit the use of a smaller, lower-cost launch vehicle through subsystem mass reduction. Thus, any weight benefits from subsystem miniaturization will have to be assessed relative to advances in other aspects of space system technology.

D. Terminal Guidance

The high precision required for delivering the NEO threat mitigation device at exactly the intended location on or above the surface of the small target object demands an extremely high terminal guidance accuracy.

In past and current planetary flyby and orbit missions, terminal guidance accuracies of several kilometers in directions tangential and normal to the approach trajectory have been considered adequate and have been achieved consistently by correction maneuvers performed days or weeks before arrival at the target. Closest approach distances in such missions typically range from several hundreds to thousands of km. Terminal guidance corrections are performed with the aid of Earth-based and/or spacecraft-based error detection techniques. Optical sensing by an instrument on the spacecraft combined with inertial navigation is used to project the closest approach point, and hence, the residual guidance errors that must be corrected.

In the asteroid or comet intercept mission terminal guidance must be refined to reduce the terminal approach errors by several orders of magnitude, to a range of hundreds or even tens of meters, depending on the target size and the type of intercept intended.

Table VI lists some threat mitigation and other mission modes and summarizes terminal guidance requirements and issues of concern. Modes (a) and (b) achieve target deflection by kinetic energy, chemical high-explosive, or nuclear explosive energy transfer (see the Chapter by Ahrens and Harris). Mode (c) may be used to initiate surface operations, such as implanting an explosive or propulsive energy transfer device, or to explore physical target characteristics, e.g., in a precursor mission conducted prior to threat mitigation. Modes (d) and (e) may serve various options, such as close observation, reconnaissance, or deferred-time detonation. All of these scenerios demand the development of advanced guidance techniques to meet the unprecedented requirements of high terminal accuracy.

In a close intercept mission, optical detection and tracking of a small, 100 m class target approaching at a relative velocity of 20 to 30 km s^{-1} will have to be performed by an onboard homing sensor. A detection range of at least several thousand km is required to allow enough response time for guidance corrections. Even at such ranges the time-to-go before encounter is of the order of only 100 s. If half of this time is lost due to a maneuver execution delay, the maneuver velocity requirement is doubled. Target detection and tracking at these large ranges could be facilitated if a radio beacon were placed on the target by a precursor mission, if enough advance warning time is available.

Guidance accuracy requirements critically depend on the selected en-

TABLE VI

Target Acquisition and Terminal Guidance Requirements

Intercept Mode	Mission Type	Approach Velocity (km s^{-1})	Required Acquisition Range[a] (km)	Required Terminal Accuracy[a] (m)	Issues and Concerns
(a) Direct impact at high velocity	Close intercept: Kinetic energy deflection / Nuclear explosive deflection or fragmentation	15–30	3000	50	Early acquisition, prompt lateral guidance maneuver / Critical guidance accuracy with advanced sensor technology
(b) Tangential impact at high velocity	Close intercept: Kinetic Energy Deflection / Nuclear explosive deflection or fragmentation	15–30	3000	20	Early acquisition, prompt lateral guidance maneuver / Critical guidance accuracy with advanced sensor technology
(c) Soft landing, following retro maneuver	Distant intercept: Implant energy transfer device or mass driver / Implant nuclear device	2–5	300	100	Early retro maneuver, several days before acquisition[b] / High terminal accuracy / Soft landing adaptable to uncertainty in gravity
(d) Injection into near-circular orbit, following retro maneuver	Distant intercept: Close observation / Deferred detonation, standoff or surface	2–5	300	200	Early retro maneuver, several days before acquisition[b] / Moderate terminal accuracy / Orbit insertion adaptable to uncertainty in gravity
(e) Zero-velocity rendezvous/formation flying	Distant intercept: Close observation / Deferred detonation, standoff or surface	2–5	300	500	Early retro maneuver, several days before acquisition[b] / Moderate terminal accuracy / Requires periodic altitude correction

[a] Rough estimates; 100 to 300 m class target object assumed. Less terminal accuracy required with larger target.
[b] Earthbased relative trajectory information and maneuver commands can be used.

counter mode. A direct, frontal impact, mode (a), or a nearly tangential impact, mode (b), are likely options for deflecting the target object away from Earth impact in a close-intercept scenario (see Chapter by Ahrens and Harris). In either mode, the impactor miss at Earth is roughly equal to the product of the imparted velocity impulse and the time to Earth encounter. Frontal impact requires a lower terminal guidance accuracy than tangential impact and depends less critically on early guidance error detection by the homing sensor, and therefore appears preferable.

In the distant intercept scenario, the optimum orientation of the deflection impulse is generally parallel or anti-parallel to the heliocentric target velocity. If the diversion mission can be performed years in advance of the projected Earth impact, intercepting the target at its perihelion provides the highest impulse effectiveness per unit impact deflection distance at Earth. In this scenario modes (c), (d), or (e) are likely candidates for executing the target deflection, at lower guidance accuracy requirements compared with modes (a) or (b). The initial relative approach speed before rendezvous at the distant intercept point is of the order of one or several km s^{-1}. On approaching the zero-range-rate and zero-range-error condition, by intermittent retro-thrust application, the sensitivity to thrust and coast duration increases, along with rapid improvement of the error detection accuracy.

The problems outlined above must be resolved in developing sufficiently accurate, timely and dependable terminal guidance capabilities. In the close-intercept scenario, the large target detection range and the extreme angular resolution requirements are critical for mission feasibility, and place the greatest demands on sensor technology advances. Earth-based remote guidance is not sufficiently accurate for the terminal approach phase because of the small target size, and the large (20 to 40 s) communication delays are not consistent with instant guidance error correction at the approach speeds involved. Distant intercept missions, with an order of magnitude lower approach speed, require much lower target detection range and guidance error correction capabilities.

Hypervelocity target intercept techniques have been under development by the U. S. military for purposes of ballistic missile defense (see the Chapter by Nozette et al.). Such techniques, once they become available without security restrictions, promise to provide critically needed advances toward solving the difficult autonomous terminal guidance problems, especially in the close-intercept mission class.

E. Automation and Robotics

Advances in automation and robotics technology will be essential to several phases, or elements, of NEO intercept and threat mitigation missions: assembly in low Earth orbit of separately launched segments of very large interceptors; autonomy of the final phase of intercept guidance; and target surface activities such as implanting and operating high-energy propulsion systems or mass drivers, performing subsurface placement of a nuclear explosive, or collecting and processing surface material for propulsive purposes.

The automation and robotics needs depend on the mission objectives and scenario details. In the U. S., major advances in automation and robotics technology for space operations have been achieved during the last decade, in connection with on-orbit assembly of the Space Station, and for projected lunar and planetary surface operations. In addition to full autonomy, there have been new developments in remote control and supervised autonomy ("telepresence"). Continuing development in these areas are essential for the feasibility of future threat mitigation missions.

F. Trades between Technology Development and Mission/System Design Requirements

Mission and system engineering methodology used in designing conventional space missions is also applicable for selecting the best design approach for NEO threat defense missions. Figure 7 illustrates this methodology schematically as applied to NEO threat defense and indicates the flow of trade studies between system design and technology requirements. A first step in performing such trades is to identify relevant figures of merit and to use them to assess the different conceptual approaches. The goal is to achieve a proper balance between overly complex and costly system implementations based on currently available technology, and overly demanding technological developments aimed at reducing complexity and cost. In addition to the cost, other criteria such as development time, development and schedule risk, and probability of success, are of critical importance in these trade studies.

With the time available for mission and system evolution as well as for technology advancement anticipated to stretch over several decades, rather than a few years as in more conventional missions, these trade studies will need repeated updates and iterations, as new advances or breakthroughs in capability are achieved. With the mission cost exceeding that of conventional missions by at least an order of magnitude, the incentives to find the most cost-effective approach are correspondingly greater. Conducting exploratory precursor missions may not only help to demonstrate feasibility, but also contribute to validating the results of initial system and technology trades.

G. Summary of Technical Issues

Technology requirements of NEO defense missions are dictated by the nature of the threat, i.e., the object type and size, its orbit characteristics and arrival velocity, and the time remaining before the predicted collision. Such missions place unprecedented demands on the technology to be used for their implementation, particularly the launch, terminal guidance and landing phases, and impulse generation at the target. Advanced propulsion technology, and robotics and automation will play key roles. Available study results favor nuclear detonation over propulsive impulse generation, or kinetic energy exchange by direct impact, for targets much larger than 100 m, with a launch mass reduction of several orders of magnitude.

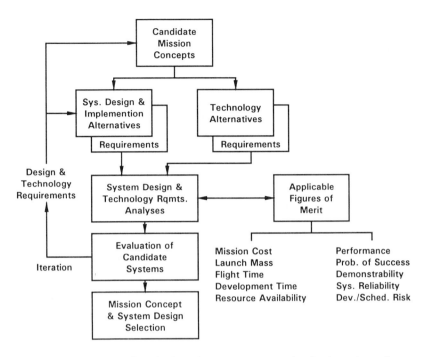

Figure 7.   Methodology for selection of system concepts and technology alternatives.

Space mission and system engineering methods used today in more conventional missions provide excellent tools for system evaluation, comparison, and selection of the best concept from a feasibility, risk and cost standpoint. Assessment of technology requirements is a principal step in this selection process, to exploit promising innovations and to eliminate impractical, potentially risky or unacceptably costly implementation concepts. This is a process that will require repeated iteration. Table VII summarizes technology drivers, limiting factors, and development priorities associated with various mission phases, reflecting the concerns discussed earlier in this section, as well as other related issues. The high cost of developing such technology (although not covered in this chapter) is a principal limiting factor concerning almost all items listed.

TABLE VII

Technology Drivers and Limiting Factors

Activity or Mission Phase	Technology Drivers	Limiting Factors or Constraints	Development Priority
1. Test and demonstration	Test facility development and testing	Realism and validity	High
	Precursor missions to NEO	Enough time available	High
2. Launch (*see also* Propulsion)	Launch vehicle capability	Unprecedented vehicle size	Very high
	Launch readiness	Long standby periods	High
	Unprecedented scope	Coordination	Medium
	On-orbit assembly	Feasibility and risk	High
3. Interplanetary transfer (*see also* Propulsion)	Complex mission planning & execution, possibly with gravity assist	Launch windows	Medium
4. Guidance, navigation & control	Target detection, terminal guidance sensing	Sensor limitations	Very high
	Pinpoint terminal accuracy		
	Autonomous rendezvous & landing	Risk, safety	High
5. Propulsion (Phases 2, 3 & 4)	High specific impulse	Safety	High
	Endurance & reliability	Development cost & schedule	Medium
6. Nuclear detonation	Control, safety, & high yield	Adequate test program	Very high
		Long development time	
7. Communications & tracking	Comm time delay compensation	Autonomy requirements	High
	Continuous coverage	Possible Sun interference	(NA)
8. Automation & robotics	On-orbit assembly	Human assistance essential	Very high
	Remote autonomous control near or at target	Backups burden design, increase launch weight	High
	Safety & reliability		

REFERENCES

Ahrens, T. J., and Harris, A. W. 1992. Deflection and fragmentation of near-Earth asteroids. *Nature* 360:429–433.

Jones, L. W. 1992*a*. Electric propulsion. *Aerospace America* Dec. 1992, p. 42.

Jones, L. W. 1992*b*. Nuclear thermal propulsion. *Aerospace America* Dec. 1992, p. 28.

Morrison, D., ed. 1992. *The Spaceguard Survey: Report of the NASA International Near-Earth-Object Detection Workshop* (Pasadena: Jet Propulsion Laboratory).

Pollard, J. E., Marvin, D. C., Janson, S. W., Jackson, D. E., and Jenkin, A. B. 1993. *Electric Propulsion Flight Experience and Technology Readiness*, Report No. ATR-93(8344)-2 (El Segundo, Ca.: The Aerospace Corporation).

Rather, J. D. G., Rahe, J. H., and Canavan, G., eds. 1992. *Summary Report of the Near-Earth-Object Interception Workshop* (Washington, D. C.: NASA).

Shoemaker, E. M., Wolfe, R. F., and Shoemaker, C. S. 1990. Asteroid and comet flux in the neighborhood of Earth. In *Global Catastrophes in Earth History*, eds. V. L. Sharpton and P. D. Ward, Geological Soc. of America Special Paper 247 (Boulder: Geological Soc. of America), pp. 155–170.

SPACE LAUNCH VEHICLES

P. L. RUSTAN
The Ballistic Missile Defense Organization

There are a variety of space launch vehicles available worldwide with lift capabilities ranging from small 373 kg payloads to low-Earth orbit (LEO) to those capable of lifting large 10,444 kg payloads to Earth escape trajectories. The opening of the former Soviet Union space program to western nations has further expanded the availability of launch vehicles worldwide. The Proton booster is expected to be particularly cost efficient in lifting large payloads, at an expected price of 50 to 70 million U. S. dollars for a lift capability of 16,039 kg to LEO at an orbital inclination of 51.6°. These launch vehicles provide the capability to send a variety of payloads on interplanetary trajectories which can intercept an asteroid or comet in order to destroy it or deflect its path away from the Earth.

A critical decision that will have to made when dealing with a potential threat of a comet or asteroid to planet Earth will be the choice of a space launch vehicle. The launch vehicle is the mechanism used to take a system to space. Once several basic questions such as the size of the comet or asteroid, the warning time, and the comet or asteroid position at the time of encounter have been answered, the launch vehicle to take a given payload to the target will have to be selected. Two types of launch vehicles are being used in the U. S. today, the reusable Space Shuttle and the expendable vehicles which have been developed for commercial and military applications. This chapter summarizes the characteristics of the best known launch vehicles being produced by the United States and other industrialized countries around the world, and provides the analysis framework for choosing the optimal upper stage propulsion technology.

I. LAUNCH VEHICLE SUMMARY

The design of a launch vehicle and the execution of its mission is closely related to the propulsion technology being used. A propulsion system is needed to accomplish four basic functions (Sackheim et al. 1991): (1) to lift the launch vehicle and its payload from the launch pad and place it into LEO; (2) to transfer the payload from a holding LEO to a planetary encounter; (3) to provide retropropulsion on approaching the target in some cases; and (4) to perform attitude control and orbit corrections. Ascent propulsion capability and the physical limitations imposed by celestial mechanics establish the constraints on payload mass, volume, and configuration for each launch vehicle. The propulsion capability of the launch vehicle is determined by the kind of propellant being used, either solid or liquid. Liquid propellants have higher specific impulse and a unique restart capability but pose more difficult handling issues than solid propellants. These constraints have led to the development of a wide range of launch vehicles in terms of fuel, lift capability, and launch site.

A summary of available launch vehicles is provided describing the range of vehicles from the smaller launch vehicles capable of placing a payload in LEO to the largest vehicles such as the U. S. Space Shuttle and the Russian Energia (Isakowitz 1991). The five key parameters in the selection of the launch vehicle are: (1) the lift capability to LEO; (2) the state of development and success history; (3) the location of the launch sites; (4) the number of stages; and (5) the cost of the system. The following tables and figures summarize the capabilities of the launch vehicles. Figure 1 summarizes the performance capability for selected small orbital launch vehicles (Isakowitz 1991). The Scout launch vehicle is no longer in production, and an upgraded Scout, the Scout II, is under development. The rest of the vehicles are all new designs, with the exception of the Minuteman II/III derivative which is built of salvaged Minuteman stages. For each vehicle, the vehicle manufacturer, LEO lift capability, Geostationary Transfer Orbit (GTO) lift capability, and thrust at lift off are shown. Figure 2 summarizes the performance capability for selected large orbital launch vehicles. The Ariane 4 will soon be surpassed by the Ariane 5 which is currently under development. This figure is not all inclusive, but is intended to represent the range of heavy lift capabilities worldwide. For each vehicle, the vehicle manufacturer, LEO lift capability, GTO lift capability, and thrust at lift off are shown.

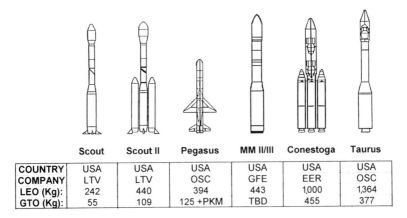

	Scout	Scout II	Pegasus	MM II/III	Conestoga	Taurus
COUNTRY	USA	USA	USA	USA	USA	USA
COMPANY	LTV	LTV	OSC	GFE	EER	OSC
LEO (Kg):	242	440	394	443	1,000	1,364
GTO (Kg):	55	109	125 +PKM	TBD	455	377

Figure 1.   Small orbital launch vehicles capability summary.

Table I lists the costs, inventory and lead time of a variety of launch vehicles. The Peacekeeper/Minuteman entries are blank because this con-

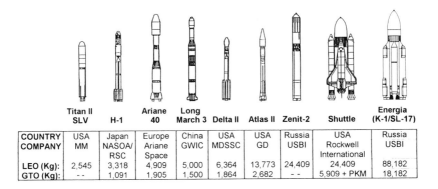

	Titan II SLV	H-1	Ariane 40	Long March 3	Delta II	Atlas II	Zenit-2	Shuttle	Energia (K-1/SL-17)
COUNTRY	USA	Japan	Europe	China	USA	USA	Russia	USA	Russia
COMPANY	MM	NASOA/ RSC	Ariane Space	GWIC	MDSSC	GD	USBI	Rockwell International	USBI
LEO (Kg):	2,545	3,318	4,909	5,000	6,364	13,773	24,409	24,409	88,182
GTO (Kg):	- -	1,091	1,905	1,500	1,864	2,682	- -	5,909 + PKM	18,182

Figure 2.   Large orbital launch vehicles capability summary.

figuration is still in the conceptual stage. The Proton and Energia launch vehicles do not have published lead times due to the recent restructuring of the Russian space program. The launch costs are in 1991 dollars and the inventory columns reflects the status of the program. A hot line reference for inventory indicates that the production facilities are ready to produce a launch vehicle to order.

The Proton vehicle produced by the Russian Krunichev Enterprise will be marketed internationally under a joint venture agreement with an American firm, Lockheed Corporation. Under this agreement Lockheed will provide $5 million to continue the operation of the Krunichev plant in Moscow that produces the Proton (Lawler 1993). The Proton will be marketed for civilian and commercial satellites worldwide, in direct competition with Arianespace. The Ballistic Missile Defense Organization of the U. S. Department of Defense (DoD) and NASA have encouraged these types of agreements and are aggressively pursuing additional work to use Russian launch vehicles and space related technologies in DoD programs.

Table II summarizes the capabilities of various launch vehicles to LEO (185 km circular), as well as Earth escape trajectories. The vehicle name followed by the launch inclination in parentheses is shown in the first column. For the Pegasus vehicle the launch inclination is independent of the launch site because it is an air launched vehicle, and the inclination shown is only an example. The main three column headings use a standard notation to express the energy of a hyperbolic escape trajectory. $C_3 = 0$ refers to an escape trajectory which results in an energy (and velocity) relative to the Earth of zero when the payload leaves the Earth's sphere of influence. $C_3 = 5$ refers to an Earth escape trajectory which has an excess departure energy (5 km^2 s^{-2}) over a $C_3 = 0$ trajectory, and is normalized with respect to mass. The $C_3 = 0$ trajectory represents the minimum energy needed to deliver a payload on an

TABLE I

Launch Vehicle Costs and Availability[a]

Launch Vehicle	Typical Cost ($M)	Inventory	Lead Time (Months)
Scout	10 to 12	Three (gov't owned)	36
Pegasus	11.5	Hot line	18 to 20
Scout (enhanced)	15	In development	First launch TBD
Pegasus XL-C	14	In development	18 to 20
MM II/III STAR 37	≈7	Conversion in development	12 to 36
Conestoga	10 to 20	In development	First launch 1993
Taurus	18	In development	18
Peacekeeper/MM	—	—	—
Peacekeeper (enhanced)	—	—	—
Titan II/SLV	43	41	12
H-2	100 to 120	In development	—
Ariane 40	60	Hot line	36
Long March	25	Hot line	12 to 36
Delta II	51 to 71	In production	36
Atlas II	70 to 80	In production	36
Zenit-2	80	Hot line	≈36
Titan III	120 to 130	Operational	24 to 36
Titan IV	170	In production	36
Proton	50 to 70	In production	—
Shuttle	245	Operational	36
Energia	110+	Operational	—

[a] Status of 1993.

interplanetary trajectory, while the $C_3 = 5$ km^2 s^{-2} trajectory provides a faster trajectory than a $C_3 = 0$ trajectory to a given point in the solar system.

The first column of Table II summarizes the LEO payload capability. The next two columns summarize the Earth escape trajectory capability assuming a LEO parking orbit is used before injecting the payload on an escape trajectory. For each escape trajectory two columns are shown; one for a solid upper stage and one for a liquid upper stage. The $C_3 = 5$ columns show the performance assuming a LEO parking orbit and then injection into an escape trajectory with an energy of 5 km^2 s^{-2}. The final three columns show the performance assuming a direct ascent into a GTO, and an Earth escape trajectory from GTO.

Figures 3 and 4 summarize the data presented in Table II. Figure 3 shows the payload capability of the various launch vehicles to LEO and to $C_3 = 0$ from LEO using a solid propellant upper stage. Figure 4 shows the payload capabilities to GTO, which is almost the same as $C_3 = 0$ from LEO. The launch vehicle capabilities span three orders of magnitude, and thus there is a launch vehicle for almost any lift requirement.

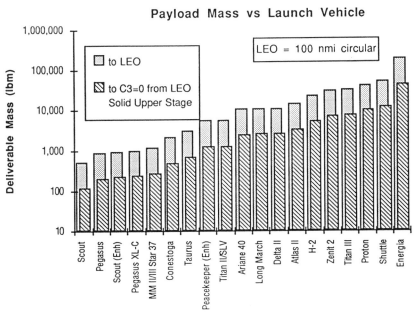

Figure 3.   Launch vehicle performance to LEO and Earth escape from LEO.

As shown previously, the useful payload mass is closely related to the launch vehicle. In order to size the payload, the required momentum change to divert the asteroid and the impact point must be known. For large momentum changes at relatively nearby impact points, only nuclear explosives can deliver enough momentum change to the asteroid's orbit to ensure a sufficient orbit

TABLE II
Launch Vehicle Performace Summary[a]

Launch Vehicle (inclination)	To LEO Payload (kg)	C₃=0 from LEO Useful Payload		C₃=5 from LEO Useful Payload		To GTO Payload (kg)	C₃=0 from GTO Useful Payload	
		Solid (kg)	Liquid (kg)	Solid (kg)	Liquid (kg)		Solid (kg)	Liquid (kg)
Scout (37.7°)	241	55	48	49	43	54	39	38
Pegasus (75°)	393	90	78	80	69	125	88	88
Scout (enh., 37°)	439	101	88	90	77	109	77	76
Pegasus XL-C (75°)	465	107	93	95	82	147	104	103
MM II/III Star 37	540	124	108	111	95			
Conestoga	998	230	199	205	176	454	320	318
Taurus (30°)	1,361	313	271	279	240	376	266	264
Peacekeeper (enh.)	2,540	584	505	521	447	431	304	297
Titan II/SL V (90°)	2,540	584	505	521	447	431	304	297
Ariane 40 (5.2°)	4,899	1,127	975	1,004	862	1,901	1,342	1,330
Long March CZ-3 (31°)	4,990	1,148	993	1,023	878	1,497	1,057	1,048
Delta II (28°)	4,990	1,148	993	1,023	878	1,819	1,284	1,273
Atlas II (28°)	6,350	1,461	1,264	1,302	1,118	2,676	1,889	1,873
H-2 (30°)	10,433	2,398	2,078	2,138	1,830	3,992	2,811	2,795
Zenit 2 (51.6°)	13,835	3,161	2,735	2,818	2,419	4,082	2,875	2,859
Titan III (28°)	14,742	3,338	2,889	2,976	2,555	5,670	3,523	3,493
Proton (51.6°)	19,505	4,484	3,901	3,997	3,422	3,629	2,556	2,541
Shuttle (28°)	24,358	5,602	4,847	4,994	4,287	5,897	4,163	4,128
Energia (51.6°)	87,998	20,240	17,512	18,040	15,488	18,144	12,810	12,701

[a] Table from Launch Vehicle Payload Weights, Vehicles & Payloads Session, Near-Earth-Object Detection Workshop (Rather et al. 1992). See text for explanation of the column headings.

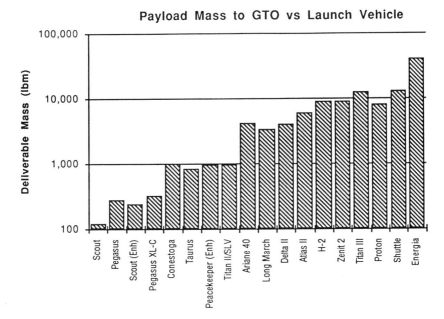

Figure 4.   Launch vehicle performance to GTO.

change.   Figure 5 shows the nuclear weapons weight required to deliver a specific amount of explosive power.   The yield curve ranges from the equivalent of small tactical nuclear fission warheads to large nuclear fusion bombs.

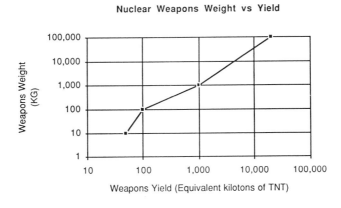

Figure 5.   Nuclear weapons yield comparison with weapons weight.

Once the warhead size is known, and the point of impact in the solar system has been selected, an analysis can be performed to select the propulsion system to deliver the warhead to the asteroid.   This propulsion system can either be a chemical system such as cryogenic oxygen/hydrogen, or a nuclear electric system such as a nuclear power system coupled with a suite of ion thrusters, or a combination of both.

II. CONCLUSION

A wide range of launch vehicles are available that can place an interceptor into an interplanetary transfer trajectory. Vehicles such as the Titan II can launch a roughly 500 kg interceptor towards an asteroid or comet. This interceptor could carry a warhead with a few hundred kiloton nuclear explosive to divert the comet or asteroid. The largest launcher, the Energia, can launch a 20,000 kg interceptor carrying a nuclear warhead with almost ten megatons of explosive power. The Energia launch vehicle, as well as the cost efficient Proton, are now available due to the recent opening of the former Soviet Union space program to western nations. The worldwide supply of launch vehicles provides the ability to carry the smallest as well as the largest conceivable warheads to protect the Earth from threatening asteroids or comets.

REFERENCES

Isakowitz, S. J. 1991. *International Reference Guide to Space Launch Systems*, (Washington, D. C.: American Inst. of Aeronautics and Astronautics).
Rather, J. D. G., Rahe, J. H., and Canavan, G., eds. 1992 *Summary Report of the Near-Earth-Object Interception Workshop* (Washington, D. C.: NASA).
Lawler, A. 1993. Lockheed, Krunichev Forge Team. *Space News* Jan. 4–10.
Sackheim, R. L., Wolf, R. S., and Zafran, S. 1991. Space propulsion systems. In *Space Mission Analysis and Design*, eds. J. R. Wertz and W. J. Larson (Boston: Kluwer), pp. 579–583.

THE ROLE OF NUCLEAR THERMAL PROPULSION IN MITIGATING EARTH-THREATENING ASTEROIDS

ALAN J. WILLOUGHBY and MELISSA L. McGUIRE
Analex Corporation

STANLEY K. BOROWSKI
NASA Lewis Research Center

and

STEVEN D. HOWE
Los Alamos National Laboratory

The feasibility of using nuclear thermal propulsion systems is examined for solving local and catastrophic threats to Earth from collision-bound asteroids. The advantages and issues of nuclear propulsion for deflecting and controlling spinning near-Earth objects (NEO) are identified, as are the propellant requirements, options and logistic strategies. Operations and issues are examined in the context of a prudent and predictable mitigation process with little chance of increasing the risk to Earth. Not within the scope of this chapter are terminal intercepts of NEOs or long-period comets. Nuclear thermal rocket engines being considered for human planetary travel, and with significant development history, are found to be very suitable for deflecting NEOs of all sizes. Extraterrestrial volatiles, particularly water, are the recommended propellant. To deflect very large global-threat NEOs, attention should be given to extending engine burn life and emplacement of a propellant tank infrastructure. Rotational accelerations during deflection are found to be very significant. Two approaches to rotational control are defined, and a rotational "breakup by spin-up" scheme is also suggested.

I. THE NEO THREAT MITIGATION PROCESS

A. Earth Protection Objectives

Given the discovery of an Earth-threatening asteroid, humanity's most important objective will be to prevent the collision with Earth and its consequences, in the most reliable and predictably effective way possible. We must also assure that the problem is not aggravated by hasty actions nor by physically unpredictable countermeasures. At the same time, during the years preceding detection of Earth-threatening objects, our solution on standby should not be more risky than the threat.

B. Mitigation Overview

Deflecting an asteroid is a momentum task, not an energy task. Countermeasures (such as nuclear weapons) which deposit excess energy into an asteroid or comet in an effort to change its momentum, do so at the risk of breaking the object. Each new fragment must be evaluated as a potential threat, thus compounding the problem. If the excess energy is delivered as sharp pulses, this risk is aggravated. By contrast, propulsive methods can achieve safe, controllable deflections by imparting momentum with a minimum of excess energy, and by imparting it gradually.

Nuclear propulsion's characteristics suit it uniquely to solving this most grave of problems confronting humanity and all life on Earth. High specific impulse gives even near term nuclear rockets speed and payload advantages, compared to any chemical rockets (Gilland and Oleson 1992). Given short warning of a collision, nuclear rockets can better perform intercepts at a maximum range from Earth. Or, given a long warning, nuclear rockets can more quickly deliver more people and/or equipment to the threat object.

Any rocket or mass driver can provide linear momentum to the asteroid. But nuclear rockets are more propellant versatile and fuel efficient than chemical rockets, and simpler than many mass drivers. A nuclear thermal rocket simply heats propellant in a reactor then expands it through a nozzle to produce thrust. It is ideal for taking advantage of extraterrestrial materials to solve the hazardous NEO problem.

Nuclear thermal rocket (NTR) programs, conceived for human transport to Mars, had the wrong emphasis for NEO deflection missions. To maximize specific impulse, Earth-based hydrogen propellant has been selected, as opposed to more easily attained extraterrestrial propellants, such as water. Short burns of expendable rockets are the norm for a Mars mission, rather than long-life reusable systems. Little analytic attention has been given to high-inclination and high-eccentricity destinations. Despite these apparent mismatches, we show that even the near-term nuclear thermal rockets now envisioned are adequate for most cases of controlled NEO deflection. Larger NEOs would require longer rocket life, and multiple rockets. If nuclear rocket technology were optimized for the NEO threat problem as its most critical mission, its routine missions of deep planetary science and safe human space transport would benefit immensely.

The question of human vs robotic on-site NEO operations remains open, and will depend on the progress in robotics during the next few decades. As of now, humans seem to be needed. Nuclear rockets would provide excellent leverage for economical, safe human round trips, particularly to difficult destinations. If totally robotic operations do become feasible for such missions, the relative simplicity of nuclear rockets would help facilitate this change to robotics.

Controlling the rotation of the object being deflected is a major design consideration for achieving assured, predictable deflection. In general, NEOs can be spun or despun more readily than they can be pushed. Approaches are defined which deal with this issue during linear translation. A serendipitous countermeasure for many threatening NEOs also results from their easy spinnability. NEOs that have low tensile strength (due either to inherent material properties or to prior fractures) could be deliberately broken apart by centrifugal force when they are spun up to faster rotation rates.

Nuclear thermal rockets will probably be the solution of choice when timely warning of Earth-threatening NEOs is achieved. They appear to be cheaper, simpler, more dependable and more predictable than chemical rockets and/or nuclear weapons. To provide humanity secure asteroid protection, we will need to supplement our warning efforts and science missions with operational practice at spinning, despinning, pushing and breaking NEOs of all types.

C. Options for NEO Mitigation

Completion of detection of Earth-threatening objects is an accomplishment we are only likely to achieve in a decade or two (Chapter by Bowell and Muinonen). Threats destined to hit us any sooner may injure Earth without warning. Because warning is essential, it makes sense to consider equally all negation options that can be available at that future time. Solid core nuclear thermal propulsion is one of the options.

Given warning, our time available to respond is either long or short. This depends partly upon how many tasks we have left undone. Our available response time will be severely limited if we are underprepared, are dealing with long-period comets, or are dealing with relatively small objects which have escaped our long term search.

In time-limited crisis situations, rendezvous is not possible. But ready nuclear thermal rockets would achieve more distant intercepts than would conventional rockets. Intercepts are addressed in the Chapter by Venetoklis et al. We would need nuclear warheads to attempt deflection. There would be little room for error and a likely lack of knowledge of the specific threat NEO. Warheads could compound our crisis instead of solving it. The advantage of nuclear rockets may not be sufficient. By contrast, given the more probable case of long warning, we can deflect our threat object with deliberate assurance and with repeat chances.

The event and decision diagram shown in Fig. 1 presumes adequate warning. The first step is a flyby mission to characterize the specific threat object. We wish to determine its specific size, composition, structure and rotation. If we have been prudent enough to have performed prior exploration of all types of asteroids and comets, each threat object should then compare closely to some well-understood NEOs. Nuclear rockets could improve this flyby mission if it were urgent to explore the threat object while it is still at a difficult location, or could upgrade the flyby to a rendezvous. But in general, a successful flyby would not demand a nuclear rocket.

Understanding the specific physical nature of the threat object will help us select one of three options in our logical flow of events. It may be relatively small and hence easily accelerated. It may fracture easily. Or, it may be both difficult to accelerate (large) and difficult to fracture or disassemble.

Our options are: (1) carefully push it with our rocket; (2) break it apart and disperse it in a predictable fashion; or (3) make propellants, and then push it with rocket(s). Propellant production may take place on the threat object itself (3a), or on other asteroids (3b). Option (3b) will be shown to be much more practical than it may seem to be at first blush. Nuclear rockets can greatly facilitate all of these options, as will be discussed.

Working missions are indicated for many of these operations. It is premature to conclude whether these working missions must involve humans on site, or whether purely robotic missions would be feasible and reliable. This answer will depend on the state-of-the-art of robotics at the time. Virtual reality and other technologies may make future robotic missions a viable choice. But as of now, human missions are envisioned (Chapter by Jones et al.) (Fig. 1).

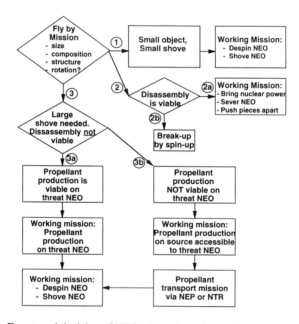

Figure 1. Events and decision of NEO mitigation, given adequate warning.

II. NUCLEAR THERMAL PROPULSION SYSTEMS

A. Nuclear Engine Candidates

Four technologies are currently candidates for solid core nuclear thermal rockets (Clark et al. 1991; Borowski et al. 1993) We will merely summarize their strengths and limitations as they relate to the NEO deflection mission.

Of the American technologies, the NERVA has the greatest heritage and therefore the lowest development risk. Detailed designs exist and numerous tests were completed in the 1960s and early 1970s. A NERVA-derived reactor (NDR) has been studied extensively for possible human space travel. (NDR is sometimes called the "Enabler.") With improvements including composite nuclear fuels, coatings and state-of-the-art rocket technology, the NDR would have a nominal specific impulse (I_{sp}) of 925 s. The NDR may evolve to higher temperatures and specific impulses if binary or ternary carbide nuclear fuels were incorporated.

The engine burn life for NDR depends upon the nuclear fuel type and temperature, and the propellant type. With hydrogen propellant and composite nuclear fuel, the NDR has 1.5 hr nominal burn life at 2750 K. Dependent upon the chamber pressure and nozzle expansion ratio, this rocket could operate at an I_{sp} of 925 s. Life will increase to 4.5 hr if we reduce the temperature to 2600 K. The corresponding loss of I_{sp} would be 20 to 30 s. About 10 hr of burn life could be achieved if the temperature was lowered to 2400 K, but I_{sp} would drop to 825 to 850 s.

The particle bed reactor (PBR) is our second NTR technology option. It may achieve higher power density than the NERVA concept due to the large heat transfer area of its very small coated fuel spheres. If proof-of-concept tests prove successful, the PBR should have a better specific impulse of about 1000 s. The issues of burn life and non-hydrogen propellants seem about the same as for the NDR.

The fast neutron ceramic-metal (CERMET) reactor concept is our third current candidate for the nuclear thermal rocket. Claims of more rugged construction would need to be verified. Specific impulse would be about the same as for NDR. The CERMET has refractory metal interfaces with the propellant. Its fuel elements should be much less chemically reactive with the propellant, leading to longer burn life. As a fast reactor, it will be less sensitive to injected reactivity from various propellants than will the thermal reactors.

Our fourth nuclear thermal rocket concept already addresses the burn-life problem, pending system level verification tests. The CIS reactor, named for the new Commonwealth of Independent States, was developed during fuel element and reactor subsystems tests in the former Soviet Union. Although it was begun four years after the NERVA program, it has continued and advanced well beyond NERVA with the development of ternary carbide nuclear fuel.

The CIS concept utilizes a heterogeneous design that separates the fuel and moderator permitting much higher fuel temperatures. Specific impulse is enhanced by the heterogeneous design because power density and materials selection are tailored for increasing temperature operation along the path of the propellant. Because of the high temperature capability of ternary carbide, an I_{sp} of 960 s and good life are expected.

Even longer burn life is achievable by backing off on maximum temperature. The CIS reactor has demonstrated life (Clark et al. 1993) of 1 hr at a gas exit temperature of 3100 K, 4.5 hr at 2900 K, and 6000 hr at 2000 K. At NERVA exit temperatures of about 2750 K, the CIS elements could give 25 hr of life, instead of 1.5 hr.

B. Propellant Options

Whatever our deflection scheme and our energy source, reaction mass must be ejected to change the momentum of the threat NEO. This reaction mass must be (1) brought from Earth, (2) taken from the subject NEO, or (3) brought from elsewhere in the solar system. For relatively massive NEOs we will need a fairly large ejection (propellant) mass, even to impart a small velocity change. It is very expensive and difficult to bring large masses of anything out of Earth's deep gravity well. Nuclear rocketry gives us better options.

In principle, a nuclear thermal rocket (NTR) can use any fluid it can pump through its core as a propellant. The fluid can contain impurities, or can be a mixture of fluids. Dirty, greasy comet water; asteroid water with dissolved silicates, etc., methane, ammonia, or carbon dioxide are good examples of easily produced extraterrestrial propellants. Quality control, we expect, could be fairly lenient.

By contrast, chemical rockets require separate fuels and oxidizers of very meticulously controlled quality. They could be very difficult to produce in space. Impure or improper mixtures could lead to combustion instabilities and explosive destruction of our rocket engines.

Water is addressed in this analysis as the primary candidate for propellant of extraterrestrial origin. It is abundant in asteroid and comet bodies and it is relatively easy to extract. If pure hydrogen were demanded, water could be further electrolyzed into hydrogen and oxygen. But this seems quite unnecessary, even counterproductive.

Current NTR concepts are being designed with only hydrogen propellant in mind. If water or other oxygen bearing propellants were used, adequate nuclear fuel life in the oxygen environment would need to be assured. The operating temperature could be held low, or an oxide based fuel form could be developed. Low temperature would limit the specific impulse. But this would not be a critical issue if extraterrestrial propellants were abundantly available. An evolutionary NDR might not have such limitations.

For current carbon-based NDR and PBR concepts, the projected nuclear fuel types are not resistive to oxygen attack at high temperatures; nor is burn life longer than a few hours a design goal, even when using pure hydrogen. When burn life becomes a constraint for NEO deflection (as it does for the large NEOs), then there are a few solutions: use more engines, evolve to better nuclear fuels, or operate at lower temperatures while sacrificing some specific impulse. At lower temperatures propellant dissociation is low, thus burn life is longer and is less sensitive to specific propellant composition.

The benefits of extraterrestrial propellants and reusable rockets are very great. More serious investigation of nuclear rocket burn life for nonhydrogen propellants would be valuable for reasons of economical routine space flight, not just for an occasional NEO deflection mission.

Zubrin (1990) has looked extensively at extraterrestrial propellant candidates in nuclear thermal rockets. Based on successful tests, he regards 2800 K as a safe temperature for carbide, uranium-thorium oxide and CERMET fuels. The ideal specific impulse of water would be 370 s. Some tests and other data suggest that carbide or CERMET fuels can attain temperatures approaching 3200 K. The ideal I_{sp} of water at this temperature would rise to 418 s.

Recent studies of Mars missions using NTR propulsion have considered engine sizes of 15, 25, 50 and 75 klb$_f$ (Borowski et al. 1993). The 25 klb$_f$ engine would have a mass of 3.4 t (inclusive of the internal shield) and a thermal power output of over 500 MW. The NDR's jet efficiency (jet power/reactor power) is about 98.5%. These engines scale well. Double or triple the thrust (which is double or triple the power output) corresponds to masses of only 4.7 t and 6.4 t, respectively. These engines can be used singly, or clustered for redundancy. Although NTR engines could be made in a broader range of sizes, it appears unnecessary to construct unique engines for an occasional deflection mission.

We can compute the mass flow rate of propellant (dm_p/dt) for these engines by noting that thrust T is related to the propellant exhaust velocity v_e by

$$T = (dm_p/dt)v_e \qquad (1)$$

which is further related to specific impulse through the gravitational acceleration constant g_o by

$$v_e = g_o I_{sp} \qquad (2)$$

Table I summarizes the propellant flow rates for the three larger nuclear engines mentioned above. Their metric thrust equivalents are 111.2 kN, 222.4 kN and 333.6 kN, respectively. The mass flow rates (in metric units of kilograms per second and tonnes per hr) are valid for any engine at the specified thrust. These rates are calculated for water across a broad temperature range, and for hydrogen propellant across a likely range of I_{sp}. This table does not imply that the exact same engines would use either hydrogen or water, but rather that they could be the same thrust. The propellant's behavior as a neutron moderator will also dictate, in part, specific engine designs.

III. NEO DEFLECTION MISSION OPERATIONS

A. Imparting Momentum to NEOs

The velocity changes (dv) needed to avoid NEO collisions with Earth are on the order of 1 cm s^{-1} (if applied near perihelion, more than a full revolution prior to collision). For cases where a small multiple of this velocity increment is needed, the unitary nature of this Δv allows for easy recalculation.

Because the velocity change is very small compared to rocket exhaust, the relative mass change of the NEO plus rocket and propellant is also very small. The exponential rocket equation is unnecessary to the analysis. Simple

TABLE I
Nuclear Thermal Rocket Mass Flow Rates

	Engine Thrust (English/metric)					
	25 klb$_f$		50 klb$_f$		75 klb$_f$	
	111.2 kN		222.4 kN		333.6 kN	
			Mass Flow Rates (metric/metric)			
Specific Impulse	(kg s^{-1})	(t hr^{-1})	(kg s^{-1})	(t hr^{-1})	(kg s^{-1})	(t hr^{-1})
H$_2$O propellant						
I_{sp} = 102 s	11.2	400.3	222.3	800.3	333.5	1200.6
I_{sp} = 255 s	44.5	160.2	88.9	320.0	133.4	480.2
I_{sp} = 408 s	27.8	100.1	55.6	200.2	83.4	300.2
H$_2$ propellant						
I_{sp} = 875 s	13.3	46.8	26.7	93.2	40.0	140.0
I_{sp} = 925 s	12.3	44.3	24.5	88.2	36.8	132.1
I_{sp} = 1000 s	11.3	40.7	22.7	81.7	34.0	122.4

conservation of momentum (Newton's Third Law) closely defines our mass of propellant m_p at exhaust velocity v_e needed for our NEO of mass M_{NEO}:

$$M_{NEO}\,(dv) = m_p v_e. \tag{3}$$

How wide of a range of specific impulse should we examine? We should include very low specific impulse. Zuppero and Landis (1991) have studied extracting and delivering comet water to Earth vicinity markets nuclear steam rockets. The delivered quantity (and profits) are maximized with very low I_{sp}. Our challenge differs, however, in that unexploited excess energy would be acceptable for our mission.

Very high specific impulses are important too. They are possible with gas core nuclear rockets (5000 s) or with ion propulsion (2000 to 10,000 s).

The broad range of reaction masses vs NEO size and I_{sp} is displayed in Fig. 2. This logarithmic graph emphasizes that propellant requirements for each nuclear rocket concept are essentially the same, and depend weakly on the choice of hydrogen versus water. An NTR with water has an I_{sp} similar to a chemical rocket burning hydrogen and oxygen. However, the NTR can more readily obtain its propellant in space.

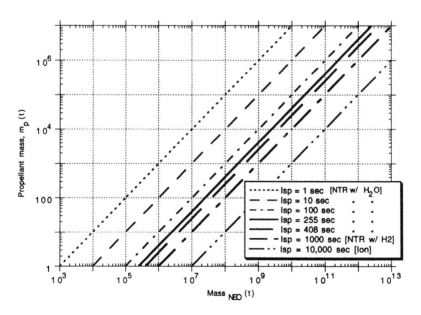

Figure 2.   Propellant mass for NEO deflection of $\Delta v = 1$ cm s^{-1}.

This same graphical information may be more useful in tabular form. Table II shows our propellant requirements in metric tonnes as a function of NEO mass and I_{sp}. Our odd, but prudently selected, specific impulse numbers previously shown in Table I, lead to tonnages in round numbers for easy interpretation in Table II. Approximate diameters of typical NEOs are included for easier visualization.

Asteroid sizes in Table II are labeled with the thresholds of the catastrophes we are trying to prevent. See the Chapter by Morrison et al. for a more detailed summary of these thresholds and their uncertainties. Stony asteroids with diameters of 50 m provided the threshold of destructive airbursts at lower altitudes. (Smaller objects break up safely at higher altitude [see the Chapter by Tagliaferri et al].) At diameters of about 100 m (perhaps a little larger for stones), asteroids begin to penetrate the atmosphere and cause craters. These diameters correspond to about 2×10^5 and 10^6 tonnes. We use 2.7 g cm^{-3} as a typical asteroid density (Millis and Dunham 1989). Hills et al. (in their Chapter) have identified 200 m diameter (10^7t) as a threshold size for the generation of tsunami (tidal waves).

Large craters and intense regional damage would occur from asteroids between 10^6 and 10^{10} tonnes (<2 km diameter stony objects). Global effects would result from NEOs of 10^{10} to 10^{12} tonnes (2 to 10 km diameter stony objects). Mass extinctions would be expected from objects of 10^{12} tonnes or greater (Chapter by Toon et al.). The more rare metallic objects reach these thresholds at even smaller sizes. Greater collision velocities also reduce these threshold sizes. Thresholds are not sharp; they are uncertain by about an order of magnitude with respect to object mass.

Which deflection cases fall under our option (1), a small object requiring a small shove? Realize that the object being shoved is massive compared to our nuclear rocket (dry mass a few tens of tonnes); but, NEOs of up to a few times 10^7t (up to 300 m diameter) can be adequately pushed by an NTR with an ordinary load of propellant. If we can find these small objects, it is relatively simple to protect ourselves against airbursts, small craters and small tsunami.

Can nuclear electric propulsion (NEP) do better because of its higher specific impulse? Galecki and Patterson (1987) have examined Mars cargo vehicles in our range of interest. One such vehicle, a 3 MW argon ion system at an I_{sp} of 9591 s, could haul an 80 t payload from the Moon to Mars using about 25 t of argon propellant. This suggests that ion systems could deflect NEOs up to a few times 10^8 tonnes; but ion propulsion systems lack high mass flow rates. Over 300 days of thrusting were needed in this case. This would not be as effective as a shorter time for changing the NEO orbit. Higher power systems could operate more quickly, but this size NEO (10^8 t) seems about the limit of the NEP solution.

As the NEOs become larger, up toward the 10^9 t range, the mass of propellant needed rises well beyond 1000 t. Because we would desire far better than a close miss, the propellant amounts in Table II should be at least doubled. Delivery of sufficient propellant from Earth to the NEO would be very expensive in these more challenging cases. Given the human consequences

TABLE II
Propellant Mass Needed for NEO Deflection for $\Delta v = 1$ cm s^{-1}

Approx. NEO Diameter	NEO Mass (t)	Threshold of:	Isp (sec)						
			1.02	10.2	102	255	408	1019	10190
			Steam		H2O ~1000 °K	H2O ~2000 °K	H2O ~3100 °K	H2 ~3100 °K	(ions)
10 m	1E+03		1 t	100 kg	10 kg	4 kg	2.5 kg	1 kg	0.1 kg
20 m	1E+04		10 t	1 t	100 kg	40 kg	25 kg	10 kg	1 kg
40 m	1E+05	Airbursts	100 t	10 t	1 t	400 kg	250 kg	100 kg	10 kg
100 m	1E+06	Craters	1000 t	100 t	10 t	4 t	2.5 t	1 t	100 kg
200 m	1E+07	Tsunamis	10000 t	1000 t	100 t	40 t	25 t	10 t	1 t
400 m	1E+08		1E+05 t	10000 t	1000 t	400 t	250 t	100 t	10 t
1 km	1E+09	Global effects	1E+06 t	1E+05 t	10000 t	4000 t	2500 t	1000 t	100 t
2 km	1E+10		1E+07 t	1E+06 t	1E+05 t	40000 t	25000 t	10000 t	1000 t
4 km	1E+11	Mass extinction	1E+08 t	1E+07 t	1E+06 t	4E+05 t	2.5E+05 t	1E+05 t	10000 t
10 km	1E+12		1E+09 t	1E+08 t	1E+07 t	4E+06 t	2.5E+06 t	1E+06 t	1E+05 t
20 km	1E+13		1E+10 t	1E+09 t	1E+08 t	4E+07 t	2.5E+07 t	1E+07 t	1E+06 t

Small shove. Use propellants from anywhere.

Use extraterrestrial volatiles as NTR propellant.

NTR engine life should be long.

Not a reasonable solution.

of massive regional destruction and huge tidal waves, it would certainly be worth doing.

A more economical solution does exist for deflecting these large NEOs: options (3) in Fig. 1. We can haul our relatively light propellant production unit to NEO itself or to a "nearby" asteroid, then extract the water or volatiles we need. Becuase each 333.6 kN engine (or cluster) can flow 480 t hr^{-1} of propellant for several hours, we would need only a few engines to exhaust the necessary propellant mass.

At the threshold of global effects, NEOs of 10^{10}t, engine burn life becomes an important issue. A few hundred hours of burn time may be needed. Engine life should be extended to the nuclear fuel burn-up limit. If a conventional limit of 1% burnup of ^{235}U is assumed, then ideally less than five engines of the 333.6 kN size would be needed. About thirteen such engines are needed if we account for margin and contingency.

Given the serious nature of the NEO threat and the remoteness of operations, the engines should be used to their full capability. Our reactors can go as far as perhaps 5% ^{235}U burnup before ceasing to function. This would reduce our engine count by a factor of 5, of course, and make the deflection of very large NEOs even more tractable. It also seems prudent, in such a grave situation, to build much larger engines and to design for higher burn-up levels.

For most NEOs of global consequence ($>10^{11}$t), the challenge of delivering 10^5 to 10^7 tonnes of propellant from Earth does not even seem like an option. The poor mass fraction of rocket delivery from the Earth's gravity well makes the task as formidable as the consequences. But producing this amount of extraterrestrial volatiles is not unreasonable.

B. Manipulating Rotating NEOs

Will any push on a rotating NEO achieve what we want? Must we remove its spin first? How much will its spin change as we push it? To address these questions, let us examine a common shape, the contact binary. Let us compare the spin imparted if we accelerate by 1 cm s^{-1} worth of push, but without regard for spin control.

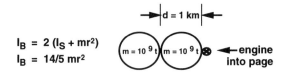

$$I_B = 2(I_S + mr^2)$$
$$I_B = 14/5 \; mr^2$$

Figure 3.   Rotational inertia of contact binary asteroid.

Figure 3 shows an idealized contact binary consisting of two rigidly joined spheres of 10^9 tonnes each. The moment of inertia of a sphere (I_{isp}) about an axis through its center is 2/5 mr^2. The moment of inertia of our binary (I_B)

through its center is found to be $14/5 \, mr^2$ from the transfer theorem

$$\mathbf{I}_{x'} = \mathbf{I}_x + mr_{x'-x}^2. \tag{4}$$

A 111.2 kN engine is mounted at one end as shown. The torque of this small engine at 1 km radius will impart our binary asteroid with an angular acceleration of 1.6×10^{-7} rad s^{-2}. A burn of 8000 t of propellant would be needed to impart a velocity change of 1 cm s^{-1}. The mass flow would be 44.5 kg s^{-1} for 1.8×10^5 s. Integrating over this time, the increase in spin would be 2460 revolutions day^{-1}. Clearly from this example, our asteroids can be spun (or despun) much more readily than they can be pushed the requisite amount.

The spinning problem is compounded by the fact that real asteroids may be irregularly shaped. Their inertial tensors $\mathbf{I}$ have nonzero cross products. This means that a spin imparted about one axis will cross couple into the others. The exact nature of $\mathbf{I}$ will be difficult to predict upon first observation, as it will be spinning slowly about its primary axis.

Two control rotational approaches are illustrated in Fig. 4. In the first case, the NEO is despun in a trial and iteration process. The main pusher rockets are then mounted to push nearly through the center of gravity with minimum moment arms. Several rockets (only one is shown) should be placed to give balanced torque in each direction about the approximated center of mass. Despite minimal torques, we can anticipate that the object will begin to rotate slightly about each axis as shown. Sensors placed about the NEO can detect these rotational accelerations. A logic network then controls the relative thrust of each engine, just as if the NEO were a spacecraft. Or if desired, small vernier rockets of any type can be placed with long moment arms to keep the NEO properly aimed. Their thrust should also contribute to the desired velocity change.

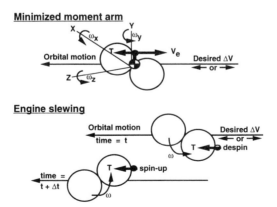

Figure 4.   Rotational control approaches during asteroid deflection.

An engine slewing technique is the second case shown in Fig. 4. Here we allow the NEO to continue spinning. By thrust vectoring our rocket engines, we keep the impulse directed along the orbital path to either accelerate or decelerate the NEO. Either direction is desirable for long warning deflections, but only acceleration is shown in the figure. During one phase of the thrusting the spin will be increased; while during another phase it is decreased. Several engines would probably be used, because complex rotations will sometimes put any given engine in an unworkable thrusting position. Sensors and logic networks are deemed necessary in this method also.

Realizing how easily NEOs can be spun, and how strong the internal tensile stresses must be for large spinning objects, we can see that spinning might be a solution as well as a problem. Controlled break-up by spin-up could send our threat NEO flying into fragments, with better dispersal than fracturing a nonspinning object. It could, given sufficient time to reassess the orbits of each fragment, solve our problem completely or reduce it to only one smaller threatening fragment.

The keys to a controlled breakup will be knowledge and timing. Detailed survey of the object should identify its materials, fractures and homogeneity. We expect fractures to be the norm, because it is prior collisions in the main belt which have caused asteroids to become Earth crossers. NEOs of cometary origin may not be fractured, but should have low tensile strength inherently. Timing of the breakup can be aided by well-placed high explosives, or by partial deliberate severing.

C. Propellant Production Strategies

How difficult would it be to produce enough water in space to fill our propellant needs? A large percentage of the near-Earth asteroids are carbonaceous and contain 5% to 20% water (Lewis and Lewis 1987). They suggest it can be extracted by heating to 250 to 300°C. Nichols (1993) recommends 400°C. It might be done by solar power, but in the age of space nuclear power, it could be done more easily using reactors. For example, to obtain 5000 t of water to move a 10^9t NEO twice our unit velocity change, we would need to process 50,000 t of material at 10% water content. At a typical density of 2 g cm^{-3}, this would be an area 100 m by 100 m$\times$2.5 m deep. Such a task seems far simpler than delivering 5000 t of propellant from Earth.

However, we must protect our Earth against any threat object, not merely ones which lend themselves readily to propellant production. What about threat NEOs that prove to be dry stone or metal? How could we wring propellant from them? Must we shift our propulsion strategy to some other form of mass driver, one that ejects raw material or iron slugs, for example?

These solutions generally presume that our solar system is quite empty. The truth, however, is that a huge population of objects will have been catalogued during the Spaceguard Survey (Morrison 1992). The NASA International Near-Earth-Object Detection Workshop showed that after 25 yr, Earth-crossing asteroids greater than 100 m diameter would be 12% to 55% completely surveyed. Only 12% of this population of about 320,000 would identify 38,400 objects. If the expected half are C class, we would have almost 20,000 objects to choose from which contain at least 10^5t of water in each. More recently reported simulations (see the Chapter by Bowell and Muinonen) suggest an even more complete survey. Perhaps it could reach 100,000 such sources. If we have warning, it would be practically impossible not to have known volatile sources convenient to and nearly coplanar to the threat NEO. We may choose several of many good sources.

Delivery of propellant to the threat NEO benefits from an efficient space cargo system. Ion propulsion systems can be recommended for their excellent payload mass fraction. Or, nuclear thermal systems can be recommended for good mass fraction and commonality with the pusher system.

It is evident that propellant tanks will be a more pressing need than nuclear engines for deflecting large NEOs. If we conduct normal space activities with long-range wisdom, this tankage infrastructure will already be in space by the time our warning system alerts us to specific dangers. Each launch from Earth should leave its propellant tanks to useful locations in space. It is wasteful to let tanks re-enter or drift away after expending most of the energy to deliver them to useful destinations. If saved, the tanks will become reusable assets for a space economy built upon space resources. An NEO deflection mission will call upon these space assets, just as our terrestrial emergency systems call upon our terrestrial assets and technology.

IV. RECOMMENDED PRECURSOR EXPERIMENTS

The space activities which ought to precede the deflection and spin-up operations described here are not hard to deduce. They fall into scientific and operational categories. Scientific missions should provide us an intimate knowledge of every type of NEO, both physical and chemical. These missions can also be engineering precursors for the propulsion and other technologies we will need later. Sample returns should be done to define the right processes for propellant production from NEOs. Operational missions should include pilot propellant production. They should be augmented by a series of experiments on nonthreatening, small NEOs. These experiments should despin, spin and push NEOs. Controlled fracturing of spinning NEOs should be perfected also, to complete our arsenal of safety countermeasures.

This recommended series of precursor experiments are exactly the ones that would be needed to implement an asteroidal resource program (see the Chapter by Hartmann and Sokolov). In fact, resource utilization and asteroid defense are entirely synergistic. Eventually, routine space resource operations will provide the infrastructure to implement an occasional defensive deflection mission, simply and reliably.

V. CONCLUSIONS

Nuclear thermal rockets appear very well suited to deflecting any size NEO from a collision with Earth in a controlled manner, given adequate warning.

For the large NEOs (over 10^8t), extraterrestrial volatiles will be our preferred propellant.

For NEOs over 10^{10}t, nuclear engine life becomes important. Nuclear engine concepts need to be selected that use durable nuclear fuels, such as ternary carbide which the CIS has already developed. A propellant tank infrastructure also becomes important for cases in which propellant is produced at a time or place separate from the NEO deflection burn. A tank infrastructure should be established in advance in space by abandoning short-sighted current practices of dropping and discarding tanks.

REFERENCES

Borowski, S. K., Corban, R. R., McGuire, M. L., and Beke, E. G. 1993. Nuclear Thermal Rocket/Vehicle Design Options for Future NASA Missions to the Moon and Mars. AIAA-93-4170 (Huntsville, Al: AIAA).

Clark, J. S., McDaniel, P., Howe, S. D., and Stanley, M. 1991. Nuclear Thermal Propulsion Technology: Summary of FY 1991 Interagency Panel Planning, by NASA-LeRC, AFPL, LANL, and INEL. AIAA 91-3631 (Cleveland, Oh: AIAA).

Clark, J. S., McIlwain, M. C., Smetanikov, V., D'yakov, E. K., and Pavshook, V. A. 1993. U.S./CIS eye joint nuclear rocket venture. *Aerospace America* 31(7):28–35.

Galecki, D. L., and Patterson, M. J. 1987. *Nuclear Powered Mars Cargo Transport Mission Utilizing Advanced Ion Propulsion*, NASA TM-100109.

Gilland, J. H., and Oleson, S. R. 1992. *Combined High and Low Thrust Propulsion for Fast Piloted Mars Missions*, NASA CR-190788.

Lewis, J. S., and Lewis, R. A. 1987. *Space Resources: Breaking the Bonds of Earth* (New York: Columbia Univ. Press) pp. 247–248.

Millis, R. L., and Dunham, D. W. 1989. Precise measurement of asteroid sizes and shapes from occultations, In *Asteroids II*, eds. R. P. Binzel, T. Gehrels and M. S. Matthews (Tucson: Univ. of Arizona Press), pp. 167–168.

Morrison, D., ed. 1992. *The Spaceguard Survey: Report of the NASA International Near-Earth-Object Detection Workshop* (Pasadena: Jet Propulsion Laboratory).

Nichols, C. R. 1993. Volatile products from carbonaceous asteroids. In *Resources of Near-Earth Space*, eds. J. S. Lewis, M. S. Matthews and M. L. Guerrieri (Tucson: Univ. of Arizona Press), pp. 543–568.

Zubrin, R. M. 1990. Missions to Mars and the Moons of Jupiter and Saturn Utilizing Nuclear Thermal Rockets with Indigenous Propellants. AIAA 90-0002 (Reno, Nev.: AIAA).

Zuppero, A., and Landis, G. A. 1991. Optimum rocket propulsion for energy-limited transfer. Resources of Near-Earth Space, Jan. 7–10, Tucson, Ariz., Abstract book, p. 25.

APPLICATION OF NUCLEAR PROPULSION TO NEO INTERCEPTORS

PETER VENETOKLIS and ERIC GUSTAFSON
Northrop Grumman Corporation

and

GEORGE MAISE and JAMES POWELL
Brookhaven National Laboratory

This chapter discusses nuclear thermal propulsion (NTP), both in general and as applied to a near-Earth object interceptor. The fundamentals of NTP are presented, followed by a historical perspective, summary descriptions of 14 different NTP systems, and more detailed descriptions of four "near-term" designs. Application to a vehicle is illustrated via a design exercise, wherein NTP-related issues such as decay heat management, propellant nuclear heating, and vehicle configuration are addressed. It is demonstrated that a compact NTP system such as that based on the particle bed reactor enables a dramatic increase in the achievable final velocity of a NEO interceptor with respect to modern-day chemical propulsion systems. The chapter provides the reader with analytical tools and methodologies sufficient for first-order performance prediction of propulsive stages equipped with NTP systems, and should provide the reader with an understanding of the basics of the application of NTP.

I. INTRODUCTION

The potentially catastrophic effect of a large comet or asteroid striking the Earth has been the subject of much public discussion and speculation for quite some time. Modern theorists believe that such impacts have occurred repeatedly in Earth's history, causing dramatic climate changes and drastically altering Earth's ecosystem. Many experts believe that the extinction of the dinosaurs was caused by an asteroid impact near Mexico's Yucatan peninsula, and warn that the probability of such an event occurring again is not negligible.

In parallel with this topic there has been, in recent years, a resurgence of interest in nuclear thermal propulsion (NTP) systems for space exploration and other missions. NTP systems provide a dramatic improvement in rocket propulsion efficiency over the best modern rocket engines, which rely on chemical combustion. Dramatic performance improvements over current systems appear directly relevant to a NEO interceptor mission, as the mission may be extremely demanding of propulsive performance.

This chapter was written to provide the reader with familiarity of both the underlying principles of NTP and the application of NTP to design of a NEO interceptor. The fundamentals of nuclear propulsion and why NTP provides better performance than chemical rockets are reviewed in the first part of the chapter. The remainder of the chapter is devoted to a design exercise illustrating how NTP can be applied to a NEO interceptor vehicle. Because NTP systems offer substantial performance improvements over chemical systems, the focus of the design exercise is on a situation where high performance is more critical, i.e., one where response time after detection is short, making successful intercept time-critical.

In the case of the design exercise, a successful intercept is defined as one where the interceptor arrives at the target NEO in time to permit alteration of the NEO's trajectory away from Earth impact. If a fixed divert impulse is assumed, e.g., that generated by a 100 megaton nuclear device, then the minimum range from Earth that the diversion must occur to avoid the Earth can be defined in terms of the NEO's mass and orbital characteristics. This minimum range dictates the maximum time after detection the interceptor vehicle has to reach the NEO. This time interval translates to interceptor velocity, and increasing interceptor velocity increases the range at which the intercept can occur. Thus, higher interceptor velocity permits successful intercept of larger NEOs.

The design exercise describes some of the issues involved in analysis of NTP-equipped systems. Although NTP systems are theoretically more than twice as efficient as chemical systems, certain traits of NTP systems conspire to detract from that performance advantage. These traits typically cause NTP-equipped vehicles to be less mass-efficient than their chemical counterparts, partially offsetting the efficiency advantage. The main purposes of the design exercise are to demonstrate the effect of these NTP-unique issues on mission and configuration analyses, familiarize the reader with the potential pitfalls of applying NTP technology to space vehicles, and to show that, even with these pitfalls included, NTP still offers better performance to chemical propulsion for NEO missions.

II. NUCLEAR ROCKET CONCEPTS

As was mentioned in the introduction, nuclear rockets have a clear advantage over chemical rockets for terminal intercept of Earth-threatening NEOs. In this section we explain the basis for the advantage, provide a brief historical sketch of nuclear rocketry, and discuss the main types of nuclear rockets that have been studied. It is not the purpose of this section to present an extensive discourse on nuclear rocketry. Excellent books (Bussard and DeLauer 1958) and conference proceedings (El-Genk and Hoover 1993) already exist that provide in-depth coverage of the field. To avoid confusion, it should be pointed out at the very beginning there are two main classes of nuclear rockets: nuclear thermal propulsion (NTP) and nuclear electrical propulsion

(NEP). In a NEP device, the fission power is first converted to electrical power which is then used to accelerate charged particles to very high velocities. The NEP typically has a very high specific impulse but a very low thrust. Although both types are being considered for planetary exploration, the long acceleration time associated with an NEP system more than negates the terminal velocity advantage enabled by its I_{sp} for the problem at hand, i.e., terminal intercept of NEO. Because only the NTP is applicable, our attention will be confined to this type of rocket.

A. Advantages of Nuclear Over Chemical Rockets

A solid core nuclear thermal rocket is illustrated schematically in Fig. 1. Cold, high-pressure hydrogen is ducted into the core from a cryogenic propellant tank. It is then heated while flowing through the core. The hydrogen is finally expanded through a supersonic nozzle, producing thrust.

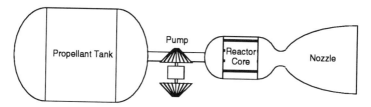

Figure 1.   Schematic representation of a nuclear thermal propulsion system.

The most important advantage of a solid core nuclear rocket over chemical rocket is that we are able to use a low molecular-weight gas (hydrogen) as the propellant. This performance advantage is readily seen by comparing the specific impulse (I_{sp}) of rockets with different propellants. Specific impulse of a rocket engine (Roy 1965) is defined as:

$$I_{\text{sp}} = \frac{F}{g_0 \dot{m}} \qquad (1)$$

where F is the thrust produced, $\dot{m}$ is the rate of consumption of the propellant and V_0 is the acceleration due to gravity at sea level. (The term g_0 is needed only because I_{sp} is normally represented in seconds.) Applying the momentum equation to the rocket, one can readily show that the specific impulse is proportional to the velocity of the propellant exiting the nozzle,

$$I_{\text{sp}} = \frac{1}{g_0}(V_{\text{EXIT}}). \qquad (2)$$

Thus, the most direct way one can improve the performance of a rocket is to increase the propellant exhaust velocity.

For an ideal rocket (no friction losses in the nozzle and expansion to vacuum) the exhaust velocity (Shapiro 1963) is expressed by:

$$V_{EXIT} = \sqrt{\frac{2g_0 kTR}{(k-1)m} T_0}$$

(3)

and the I_{sp} by

$$I_{sp} = \frac{1}{g_0}\sqrt{\frac{2g_0 k}{(k-1)}\frac{R}{m} T_0}$$

(4)

where k is the ratio of specific heats of the propellant, R is the universal gas constant, m is the molecular weight of propellant gas, and T_0 is the propellant temperature in nozzle chamber.

If we assume that the maximum operating temperature T_0 of the rocket is fixed by the temperature limit of the solid materials that come in contact with the hot gases, then the remaining variables that can influence I_{sp} are shown in the proportionality:

$$I_{sp} \propto \sqrt{\frac{k}{(k-1)m}} \cdot$$

(5)

For diatomic hydrogen H_2 ($k = 1.4$ and $m = 2$) we get:

$$\sqrt{\frac{k}{(k-1)m}} = 1.3228.$$

(6)

For comparison, a chemical rocket burning hydrogen and oxygen generates steam (H_2O) as the exhaust gas, with $k = 1.33$ and $m = 18$. For this case we find:

$$\sqrt{\frac{k}{(k-1)m}} = 0.471.$$

(7)

Thus, we see that operating at the same nozzle chamber temperature, a rocket with H_2 exhaust will produce a specific impulse 2.8 times higher than a comparable H_2O exhaust. We have oversimplified the comparison somewhat because the specific heats do not remain constant during expansion and the exhaust of a typical hydrogen-oxygen chemical rocket is not pure steam. However, the basic conclusion remains unaltered: the specific impulse of a solid-core nuclear rocket is more than twice as good as the best chemical rocket.

To illustrate the advantage of a high specific impulse rocket in effecting a terminal intercept, we need only examine what is commonly called the "ideal rocket equation" (derivation from Newton's second law is straightforward and left to the reader):

$$\Delta V = I_{sp} g_0 \ln\left(\frac{M_0}{M_f}\right)$$

(8)

Here ΔV is the velocity increase of a spacecraft, M_0 is its total initial mass of the vehicle (which includes the propellant mass) and M_f is the final mass after the propellant has been consumed. Clearly, if we have two interceptors of the same initial mass with the same amount of fuel the one with an I_{sp} 2.8 times higher will achieve a final velocity which is 2.8 times larger and will, consequently, reach the threatening NEO in a much shorter time.

So far we have only discussed the solid-core NTR and the advantages in using hydrogen as the propellant. As shown by Eq. (4) the specific impulse can also be increased by increasing T_0. To take full advantage of T_0 we must consider liquid-core and gaseous core nuclear rockets. These designations refer to the thermodynamic phase of the uranium fuel while the reactor is operating. The advantage of taking on the additional complications of containing a liquid or gaseous core is a major increase in performance. The specific impulse is proportional to the square root of the nozzle chamber temperature T_0, and the rocket performance can be enhanced substantially by increasing this temperature. Clearly, in a solid-core reactor the maximum temperature that the gas can attain is the melting temperature of the solid fuel (~ 3000 K). Thus, the I_{sp} is limited to about 1000 s. In a liquid core rocket we intentionally let the fuel melt so we can exceed the ~ 3000 K limit and operate in the 5000 to 6000 K range. Besides the T_0 gain, we note that there is an additional advantage for operating at these temperatures; the hydrogen is dissociated and the molecular weight drops from 2 to 1. The I_{sp} of a liquid-core rockets is about 1500 to 2000 s. With gaseous core rockets there is no upper limit for the fuel itself, and the limit is set by the ability to cool the enclosing structure. With a gaseous core reactor specific impulse values of 6000 s appear to be attainable.

B. Brief Historical Sketch

Historically, the nuclear powered rocket is not a new concept. Almost as soon as controlled fission was first achieved, suggestions appeared for harnessing this energy to propel various vehicles including spacecraft (Murray 1954). While for other modes of transportation (nuclear airplanes, submarines) the nuclear fuel provided an almost limitless source of power, for a space rocket the advantages were less obvious. We still had to expend a propellant to provide the thrust and this propellant had to be carried along. The advantage, as pointed out earlier, was that we could choose the propellant with a very low molecular weight, i.e., hydrogen. In the United States, the first period of intense research and development in nuclear rocketry spanned from the period form 1955 to 1973. It was called Project NERVA, an acronym for Nuclear Engine for Rocket Vehicle Applications. A total of 20 reactors were designed and tested at a cost of $1.4 billion (Clark et al. 1993). In spite of many successful ground firings the program was canceled (in 1973) before any flight tests could be conducted. A revival of interest in nuclear rocketry started with the Strategic Defense Initiative (SDI) in 1983 and was subsequently reinforced by the Space Exploration Initiative (SEI) in 1987. In

1985 the SDI Organization funded a program to develop a nuclear powered rocket based on the particle bed reactor (PBR) design. (The sponsorship of the program was taken over by the U. S. Air Force in 1991 and became the Space Nuclear Thermal Propulsion program. As of the preparation of this chapter, the SNTP program was being terminated due to funding constraints.) In response to the SEI needs, the effort, spearheaded by NASA/Lewis Research Center, has consisted mostly of design studies. The final down-selection of the rocket design has not yet taken place and consequently serious funding has not been made available. As part of the process of selecting a nuclear rocket design for SEI missions, which includes a manned mission to Mars before 2019, NASA/Lewis Research Center hosted a workshop where 16 different nuclear rocket concepts were presented (Clark et al. 1993).

Since the breakup of the Soviet Union we have learned that the Soviet Union has had a very active nuclear rocket program spanning several years. Many details of their program were revealed at the Semipalatinsk-21 Conference held in the Republic of Kazakhstan, September 22–26, 1992. Notable among their achievements are high temperature fuels and structural materials. The Russians have indicated strong interest in collaborating with the United States in any future development of nuclear rockets.

C. Current Status of Nuclear Rocketry

A large variety of nuclear rocket engines have been proposed and studied since the inception of the idea of using fission energy to propel a rocket vehicle. An excellent up-to-date compilation of various rocket types can be found in the 1991 NASA NTP Workshop Proceedings (Clark et al. 1993). The purpose of this Workshop was to review the broad spectrum of NTP concepts as candidates for the SEI program, specifically, for manned flight to Mars prior to 2019. Industry, universities, and national laboratories were asked to propose NTP concepts for consideration. A total of 14 different concepts were proposed. These are summarized in Table I. Some of these have received very extensive development—others are basically "paper studies."

The 14 concepts were evaluated by the Technology Review Panel and preliminary ratings were determined. Various evaluation criteria were used; however, technology readiness level (TRL) was weighted rather highly. The reasons for this are quite obvious, because a flight-qualified engine is needed before 2016. Based on this evaluation the three concepts selected for further evaluation were ENABLER, CERMET, and PBR. In addition to these U.S. designs, NASA has included a fourth, which is based on the Russian NTP technology. This is now identified as the Confederation of Independent States (CIS) concept. Each of these four concepts is discussed below.

Before we proceed, a note of caution is in order. The four designs selected for further study for the manned Mars mission (before 2019) are not necessarily the best for terminal interception of an incoming NEO. The design criteria for these two classes of missions are very different. For example, whereas high terminal velocity of the vehicle is merely one desirable attribute

TABLE I

NTP Concepts Considered at NASA NTP Workshop

Concept	TRL	I_{sp} (s)	Fuel	Temperature (K)
Solid Core				
NERVA Baseline	6	825–850	Duplex	2270
ENABLER	4–5	925–1080	UC-ZrC-C	2700–3000
CERMET	4–5	832	UO_2-W	—
Wire Core	2	930	UN-W	3030
Adv. Dumbo	3–4	—	UC-ZrC	2700–3300
Pellet Bed	3	998	UC/TaC	3100
Particle Bed	4	1000–1200	UC-ZrC	3000–3500
Low Pressure	1–2	1050–1210	UC-ZrC	3000–3800
Foil Reactor	1–2	990	UO_2	2700–3400
Liquid Core				
Liquid Annulus	1	1600–2000	?	3000–5000
Droplet Core	1–2	1600–3000	?	5000–7000
Gaseous Core				
Open Cycle	1–2	5200	U Plasma	?
Vapor Core	1–2	1280	UF_4-HfC	6000–8000
Light Bulb	1–2	1870	?	7200

for the manned Mars mission, for the NEO intercept it is the most important requirement. Also, the payload masses will undoubtedly be very different, as will the flight profiles. In addition there is the political question as to when (if ever) we choose to develop this terminal intercept capability. As the mission requirements for NEO intercepts crystallize, a detailed tradeoff of different concepts should be performed. In the absence of such information, the four concepts discussed below are simply those that could probably be ready for deployment within the next 25 yr.

1. ENABLER. The ENABLER nuclear rocket is based upon upgraded NERVA technology. As mentioned previously, the NERVA system had several successful ground firings before the program was terminated in 1973. The basic reactor core geometry of the ENABLER is very similar to the NERVA design. As illustrated in Fig. 2, the core is made up of a bundle of hexagonal fuel elements. Each of these fuel elements, 2.5 cm across the flats, is penetrated by 19 axial coolant passages. For every two fuel elements, there is one hexagonal moderator/support element. arranged as shown in Fig. 2. The moderator is also cooled by hydrogen, although much less heat is produced there than in the fuel elements. The bundle of fuel and support elements are enclosed in a high-strength steel reactor vessel which also serves as the rocket motor case. The one important difference between the ENABLER and NERVA is in the makeup of the nuclear fuel. Whereas NERVA used UC fuel beads randomly distributed in a graphite matrix, the ENABLER uses a composite matrix shown at the bottom of Fig. 2. With this improved fuel the reactor can operate at a higher chamber temperature (2000°C or higher) and the release of fission products into the exhaust is significantly reduced.

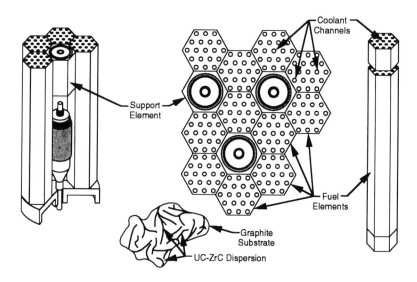

Figure 2.   ENABLER Reactor configuration.

2. CERMET. The name CERMET derives from the fuel form of this particular reactor design. The CERMET fuel consists of UO_2 (CERamic) fuel particles in a tungsten (METal) matrix. The 60/40 fuel/matrix mixture is formed into prismatic fuel elements similar in shape to those of the ENABLER design, but somewhat larger (4.9 cm) across the flats. Each fuel element contains 331 cooling passages. After fabrication, the fuel elements are clad with tungsten/rhenium, which contains fission products. Unlike the other reactor designs discussed in this chapter, the CERMET is a "fast" reactor. This means that the neutrons are not "thermalized" (reduced to thermal velocities) prior to inducing fission. Therefore, the CERMET design does not incorporate a moderator and the uranium fuel is fully enriched. Considerable research and development was applied to the CERMET design in the 1960s. Numerous tests were performed at elevated temperatures and at very rapid heat-up rates, including in-core tests. The fuel performed very well with minimal release of fission products. The test campaign was terminated before actual ground firings of a CERMET-based rocket could be performed.

3. Particle Bed Reactor. The PBR design seeks to maximize the heat transfer efficiency from fuel to coolant/propellant by maximizing the surface area-to-volume ratio of the fuel. In its simplest form, the PBR consists of a packed bed of small fuel particles, through which coolant is pumped.

Figure 3 depicts a representation of a PBR "fuel element." The fuel element consists of the particle bed and the structure required to support it and to permit its incorporation into an engine system. Simply put, the fuel element accepts cold hydrogen and ejects hot hydrogen. Two concentric, porous, cylindrical "frits" control coolant flow and retain the particles between them. Coolant flows radially inward, first through the outer, or cold, frit, through the particle bed itself, and finally through the inner, or hot, frit. The cold frit, so named because it encounters the coolant at cryogenic temperatures, is designed to control the axial flow distribution of the coolant. This maximizes heat transfer efficiency by matching flow to the bed's power output, which varies axially. The heat released by nuclear fission within the fuel particles transfers into the propellant as the propellant passes through the $\sim$1 to 2 cm thickness of the bed. The heated propellant then passes through the inner, or hot, frit, which acts as a screen to retain the particles within the fuel element. The heated propellant then flows out the lower end of the fuel element. The frit assembly sits within a hexagonal block, which serves as a neutron moderator and permits assembly of multiple elements into a core.

A number of fuel elements are assembled into a reactor core. A cylindrical pressure vessel contains the core. Flow enters the core from one end, passes through the moderator blocks to cool them, and then enters the inlet annulus to the cold frit. Heated coolant ejects out the other end of each of the fuel elements into a plenum. Here the flow from all the elements combines and passes through a sonic throat. A supersonic nozzle expands the flow to produce thrust. The core also contains a number of control and safety devices.

4. CIS Concept. This design (Nuclear Thermal Propulsion Technical In-

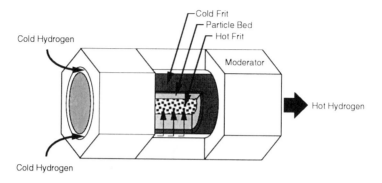

Figure 3.   PBR fuel element.

terchange Meeting 1992) is based on the nuclear rocket technology developed
in the former Soviet Union (now the Commonwealth of Independent States,
or CIS). The basic fuel form, depicted in Fig. 4, can best be described as a
"twisted ribbon." These ribbons, each 10 cm long, are collected into 4.5 cm
diameter bundles. Ten of these bundles, placed end-to-end, comprise one fuel
element. The hydrogen coolant flows axially through the passages between
the twisted ribbons. The convoluted path substantially enhances mixing and
heat transfer rate.

A reactor core consists of 102 fuel elements arranged in a hexagonal
pattern, surrounded by a zirconium hydride moderator. The fuel is a mixed
carbide consisting of uranium, niobium, zirconium and carbon ($(U, Nb, Zr)C$).
This fuel has a high melting point and is resistant to erosion from hot, high
pressure hydrogen.

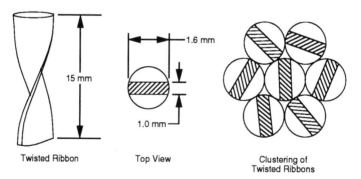

Figure 4.   CIS Reactor concept configuration.

5. *Performance Summary.* Published performance quotes for the four
"near-term" designs described in detail in Secs. II.C.1–4 are identified in Table II. Note that engine system thrust-to-weight values may be "shielded" or
"unshielded." Some designs incorporate a radiation shield inside the pressure
vessel for the purpose of protecting control components. Because most NTP

applications will require some form of radiation shield, one must be careful when comparing different designs to allow for the presence or absence of an internal shield. Determining the sufficiency of this shield requires consideration of the unique aspects of the mission that the NTP system is being used for. Radiation shielding is discussed further in Sec. III.B.

Note also that some performance parameters are expressed as ranges. These represent published values, and, in the case of thrust level, indicate scalability of the design. All these concepts are scalable to some degree, limited at the lower end by the need to achieve reactor criticality. Thrust-to-weight ratios will vary with size (i.e., thrust level).

TABLE II
"Near-Term" NTP Concept Performance Characteristics

Concept	Thrust (kN)	I_{sp} (s)	T/We (unshielded)	T/We (shielded)
ENABLER	334	925	4.0	2.3
CERMET	445	800–900	5.0	
PBR	89–356	1000	30.0	
CIS Concept	334	959		4.6

In the illustrative mission studies shown below we have used the characteristics of the PBR-type nuclear rocket, because the authors have the greatest familiarity with this concept (one of the authors (J. Powell) is the inventor of this concept).

III. NTP-EQUIPPED VEHICLE SIZING AND CONFIGURATIONS

The application of NTP to a vehicle designed to intercept NEOs is illustrated most clearly by example. First-order vehicle configurations suitable for predicting approximate performance and for comparing NTP and chemical systems are easily developed utilizing the rocket equation (Eq. 8), a few mass estimates, and some geometry.

A. Assumptions for a Design Exercise

An interceptor vehicle using a NTP system will typically begin its journey from low Earth orbit (LEO). United Nations guidelines (United Nations 1992) direct that space activities involving nuclear power or propulsion start from a long-life nuclear-safe orbit (NSO) of 800 km altitude or greater. Although no U. N. guideline exists for nuclear propulsion applications, if this consideration were held for NTP, it would handicap NTP systems in comparison to conventional chemical systems. This is due to the fact that a launch vehicle's payload delivery capability to 800 km altitude is less than its capability to LEO (typically ∼185 km). Normally, comparison of NTP and chemical systems needs to account for this difference in "starting mass." However, the NEO intercept problem provides two reasons to ignore this difference.

First, because our planned interceptor is delivering a payload that essentially consists of a large nuclear bomb loaded with plutonium, the environmental hazard from the payload exceeds that of the NTP system, thus saddling the chemical interceptor with the same environmental considerations as the NTP interceptor. Second, diversion of a "dinosaur killer" Earth-impacting body would probably be of critical global importance, far outweighing the risks of a LEO start of a nuclear engine. For these reasons, we chose to use the same starting mass for our NTP and chemical interceptors.

The interceptor vehicle will consist of several components. A payload sits atop a propulsive stage consisting of a propellant tank, engine system, avionics, and ancillary structure. "Sizing" a vehicle involves determining the masses of each of these stage components, as well as the payload.

Let us begin with the rocket equation. The ideal velocity increase of a space vehicle, known as ΔV, is determined by the relationship:

$$\Delta V = I_{sp} g_0 \ln \left(\frac{M_0}{M_f} \right). \tag{9}$$

Our goal in developing a NEO interceptor is to determine a vehicle's ΔV for a given set of initial conditions, and use that information to determine time to intercept. In this design exercise, we are assuming an I_{sp} of 1000 s, for reasons detailed in Sec. II. This leaves only the mass ratio of the vehicle to be computed.

Mass ratio is often determined last in vehicle configuration exercises, because ΔV is usually a fixed parameter based on trajectory requirements. Because initial mass m_0 is a function of the launch vehicle used and its payload delivery capability to the target orbit, final mass is usually the dependent variable, and is varied by varying payload. However, we chose to approach vehicle sizing from a different perspective.

A likely candidate payload for an interceptor is a large nuclear warhead, as the energy of its detonation is sufficient to apply a nontrivial impulse to a large NEO. If we fix the yield from this device at, say, 100 megatons, we can estimate the interceptor payload mass to be approximately 20 metric tonnes (t) (Canavan et al. 1992). This payload alone is near the upper limit of current United States launch vehicle capability (Isakowitz 1991), and addition of a high-performance propulsive stage increases the total interceptor mass well beyond any current or planned United States launch vehicle. This fact is not dependent on the propulsion technology used for the interceptor, and delivery of a large warhead to a NEO necessitates use of a large launch vehicle or multiple smaller launch vehicles. With this in mind, an initial mass greater than current U. S. capability can rationally be assumed for the purposes of this design exercise. A somewhat arbitrary value of 100 t was chosen for initial mass (chosen because a payload fraction of 20% of the initial mass is a reasonable starting assumption). This initial mass falls well below capabilities required for SEI, and is achievable with current technology. In fact, the CIS

Energia booster can be configured to deliver payloads well in excess of 100 t to low Earth orbit (Isakowitz 1991).

The fixed initial and payload masses provide to us a stage mass of 80 t. Before we can break this figure into component masses, we must make a few more assumptions. We need to determine propellant mass, propellant tank mass, engine mass, and mass of other components (avionics, RCS, and ancillary structure). Propellant tank mass is a function of propellant mass, while the other masses can be estimated independently of propellant quantity. Typically, engine system mass, which includes the reactor, pressure vessel/nozzle, propellant feed system, turbopump, and thrust vector controllers, is expressed as a function of engine thrust. Table II identifies thrust-to-weight (T/We) ratios for the near-term NTP designs. For reasons discussed later, we have selected a thrust level of 300 kN for our PBR engine, and its 30:1 T/We results in a system mass of 1000 kg.

B. Mitigation of Nuclear Radiation Effects

The very compact design of the PBR core causes the radiation leakage from the engine to be much higher than the leakage from a typical large power plant reactor. Approximately 10 to 15% of the neutrons generated in a PBR will leak out, compared to much less than 1% for a commercial power reactor. This is true for most or all NTP systems to some degree. And as a result radiation shielding is needed to protect vehicle and engine components, propellant and payload from the effects of the nuclear radiation (neutrons and gamma rays).

The most significant vehicle component requiring shielding is the propellant. The neutrons and gamma rays which are incident upon the propellant tank are scattered and absorbed within the propellant. The energy absorbed is converted to heat, increasing the temperature of the liquid hydrogen. In order to prevent boiling of propellant within the propellant tank, this heating must be reduced by shielding the propellant tank from the reactor. The radiation shielding required is extremely dependent on the vehicle design and parameters such as the propellant tank diameter and the distance between the reactor and the tank. Materials suitable for radiation shielding include graphite, tungsten, lithium hydride and boron. The optimal shield composition is also dependent on the vehicle design and the degree to which the nuclear heating rate must be decreased. For this mission, a heating rate reduction of about 70% is sufficient to preclude boiling. A shield which consists primarily of graphite is near optimal in minimizing mass. We have estimated the shield mass for adequately mitigating propellant heating to be 250 kg for our design exercise. This shield is placed directly above the reactor, both for maximum effectiveness in covering the field of view and to permit easy cooling with propellant.

Another area of concern is the effect of the radiation leakage on the operation of the engine control electronics, which are located near the engine. Electronic component performance is adversely affected by the radiation emitted by the engine. This leads to a requirement that shielding for electronics must reduce the radiation intensity by several orders of magnitude, as opposed to less than one order of magnitude for the propellant. Because the electronic components have a much smaller exposed area, localized shielding is much more mass-effective than the full-area shield applied for the propellant tank. We have estimated that 50 kg of shielding is sufficient to protect the engine control equipment, making our total shielding allocation 300 kg.

Although the combined shielding effect of the radiation shield, the propellant tank, and the distance from the engine to the front of the stage provides substantial mitigation of the radiation impinging on the payload and avionics (located at the top of the stage), the effects of radiation must be assessed on these components. In particular, because the payload is a nuclear explosive device, radiation impingement may be of serious concern, and additional shielding may be warranted.

C. Engine Thrust Level Selection

Determining the proper thrust level for the interceptor engine involves a trade between the mass of the engine and the ΔV losses due to nonimpulsive energy addition (a phenomenon commonly referred to as finite-burn or "gravity" losses). A vehicle generating a finite thrust level will change velocity over a finite period of time, rather than instantaneously, as assumed in the ideal rocket equation. Adding energy in this manner is less efficient than adding energy impulsively, with the result that the ΔV generated is less than that predicted by the rocket equation. Reducing the burn time (by increasing thrust) reduces the finite-burn loss penalty, but penalizes the vehicle's mass ratio. Normally, maximizing the ΔV addition requires optimizing the vehicle's thrust level, but this optimization involves analyses beyond the scope and purpose of the design exercise, and interested readers are referred to any of the myriad texts (Griffin and French 1991; Wertz and Larson 1991) on space vehicle design for further details. A good rule of thumb for on-orbit departures suggests that the optimum vehicle thrust-to-weight ratio at ignition should be 0.25 to 0.4, depending on generated velocity and specific impulse. This fact is reflected in the 300 kN thrust level we selected, which results in an ignition thrust-to-weight ratio of 0.3. Once the ideal ΔV generated is known, finite-burn losses can be determined.

To the 1300 kg mass identified for a shielded PBR engine we add 2500 kg for avionics, ancillary structure, and other fixed mass. This leaves us with 76.2 t mass for the propellant and propellant tanks. Propellant tank mass is often expressed as a percentage of propellant mass, and a typical value used for hydrogen tanks in SEI missions is 16% (Wickenheiser et al. 1991). But, because the missions these tanks are used for involve storage of cryogenic hydrogen propellant for months, these tanks are heavily insulated. Our interceptor requires storage of hydrogen in the tank for hours or days, rather than months, and we can use much lighter insulation than that required by SEI missions. We translate the reduced insulation requirement to a tank fraction of 12%:

$$M_t = 0.12M_p \tag{10}$$

where M_t is the propellant tank empty mass in kg and M_p is the total propellant mass in kg. This results in a mass of 8.2 t for the empty propellant tank and 68.0 t for the propellant.

D. Management of Reactor Decay Heat

Although we now have sufficient information for determining the vehicle mass ratio, we must consider another unique aspect of NTR systems before proceeding. Nuclear fission occurs spontaneously in radioactive elements, releasing energy in the process. Some of this energy manifests itself as localized heating. In the case of uranium, the period of time of decay is long enough to make the heating effect negligible. However, some of the fission products generated during operation of the PBR have much shorter decay times, with a correspondingly larger heating effect. This is not a problem during engine operation, because the propellant removes the generated heat from the engine. After engine shut-down, however, the heat generated by spontaneous decay of the fission product inventory within the reactor is sufficient to melt the internal engine structure unless cooling is provided.

For the NEO interceptor application, it can be argued that what happens to the stage is irrelevant after it drives the payload to its final velocity. This argument has merits, both because the stage ends up in a heliocentric orbit with little or no likelihood of Earth re-encounter and because the nature of the NEO threat places an additional premium on performance. However, because the purpose of this chapter is as much to educate the reader as it is to demonstrate the advantages of NTP, decay heat management is included as a mission ground rule. In addition, the flight tests that will almost certainly be conducted if a NEO interceptor is built will operate under more stringent requirements, and operational systems usually mirror flight test systems.

One cooling strategy involves flowing propellant through the deactivated reactor until the short-life fission products have decayed and the heat deposition rate has decreased sufficiently. This cooling requirement imposes a penalty on NTP-equipped vehicles, because the cooling propellant must be accelerated along with the rest of the vehicle. Although impulse is generated during the cool-down process, it adds to the vehicle's energy inefficiently. Determining cool-down propellant requires consideration of the internal structure of the engine, the engine power level, and the duration of full power operation. An analysis performed for the SNTP program was used to generate Fig. 5, which depicts cool-down propellant (expressed as a fraction of ΔV propellant) vs the duration of full-power operation. This figure was generated for a PBR-type reactor configuration at a power level of 1600 MW_t, although it should provide a close approximation of cool-down fraction for any configuration at a similar power level. To negate the fact that heating in the reactor during fission product decay is distributed differently than during full-power operation, the data in the figure represents a mean reactor outlet temperature of 1500 K, rather than the normal >3000 K operating temperature.

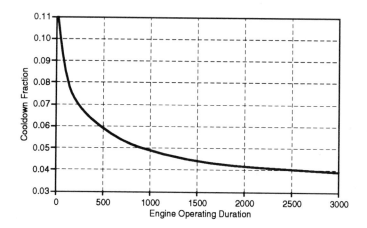

Figure 5.   Cool-down fraction vs engine full-power operating duration.

E. Computation of Interceptor Propellant Loading

Although iteration is required to determine the cool-down propellant fraction for a fixed total propellant mass, an approximation is sufficient for this design exercise. For the purposes of computing vehicle mass ratio, cool-down propellant is considered part of the vehicle final mass. Our propellant capacity of 68.0 t allows estimation of engine run time at ~2200 s, as propellant mass flow rate = thrust÷I_{sp}. Figure 6 suggests that an assumed cool-down fraction of 4.5% is appropriate and somewhat conservative. This permits determination of the propellant mass available for generating ΔV:

$$M_p = 1.045 M_{p\Delta V} \tag{11}$$

where M_p is the total propellant mass in kg and $M_{p\Delta V}$ is the impulse propellant mass in kg.

This relationship allows us to compute $M_{p\Delta V}$, M_f, and the vehicle mass ratio:

$$M_{p\Delta V} = M_p \div 1.045 = 65.1 \text{ t} \tag{12}$$

$$M_f = M_0 - M_{p\Delta V} = 34.9 \text{ t} \tag{13}$$

$$M_0/M_f = 2.8653 \tag{14}$$

F. Interceptor Performance Prediction

Applying Eq. (8) (the ideal rocket equation) with the mass ratio in Eq. (14) yields a ΔV of 10.3 km s^{-1}. We estimated an I_{sp} of 500 s for the decay heat removal phase of operation, based on reactor outlet temperature and analytic models. This I_{sp} applies to the 2.9 t of cooldown propellant which, using Eq. (8) again, generates 0.4 km s^{-1} of additional ΔV during cooldown. Our total ideal ΔV is 10.7 km s^{-1}. As noted earlier, we now have enough information to estimate finite-burn losses. A numerical integration was performed to estimate a loss of 9% for this velocity I_{sp} and ignition thrust-to-weight. This loss factor indicates that the actual ΔV generated is 9.7 km s^{-1}.

Because we are interested in interceptor performance in heliocentric space, we can convert this ΔV to excess hyperbolic velocity (V_{hp}). Assuming a circular starting orbit of 185 km altitude, we can compute V_{hp} with the following relationship (valid for circular starting orbits only) (Bate et al. 1971):

$$V_{hp} = \sqrt{\left[\sqrt{\frac{\mu}{6378 + H}} + \Delta V\right]^2 - \frac{2\mu}{6378 + H}} \qquad (15)$$

where H is the circular parking orbit altitude in km and μ is the Earth gravitational parameter (398,600 km^{3}s^{-2}).

V_{hp} for our PBR-based interceptor is 13.6 km s^{-1}. V_{hp} in this case represents the interceptor's "final" velocity with respect to the Earth as it leaves Earth's gravitational sphere of influence and enters the Sun's. Because this application of NTP is being considered for terminal intercept, V_{hp} is an appropriate figure of merit for comparing propulsion technologies. Translation of V_{hp} to intercept time requires characterization of the threat trajectory and some orbital mechanics calculations, and is outside the scope of this chapter.

G. Equivalent Chemical Vehicle Sizing

The steps used for sizing the PBR-equipped interceptor can also be used for sizing an equivalent chemical stage, assuming certain changes in the ground rules. As noted in the previous section, consideration of a NSO start in this case is driven by the payload, not the propulsion system, so the same starting orbit was used for the chemical interceptor. Chemical stages are more mass-efficient than NTP stages due to higher average propellant density. For this exercise, we utilized a tank fraction of 5% to reflect this fact. The vehicle sizing parameters are listed in Table III.

TABLE III
Chemical Interceptor Sizing Inputs

Engine thrust	300	kN
Engine Mass	400	kg
Engine I_{sp}	450	s
Propellant tank fraction	5.00%	
Ancillary mass	2,500	kg
Payload mass	20,000	kg
Initial mass in low Earth orbit	100,000	kg

Because no radiation shielding or cool-down is required, and we are assuming the same initial and payload masses, we can proceed directly to computation of the vehicle mass ratio. Simple arithmetic provides us with a propellant mass of 73.4 t, and a mass ratio of 3.7594. With this mass ratio and the chemical system I_{sp}, the rocket equation predicts an ideal ΔV of 5.8 km s^{-1}. Finite-burn losses are lower for lower ΔVs and shorter run times (due to the lower I_{sp}), and are estimated at 4% in this case. Therefore, the effective ΔV for the chemical equivalent of our interceptor is 5.6 km s^{-1}. V_{hp} for the chemical interceptor is 7.6 km s^{-1}, indicating that the PBR-equipped interceptor is $\sim$80% faster than the chemical interceptor. Our NTP-equipped interceptor requires about 21 hr to reach the effective end of Earth's gravitational sphere of influence ($\sim$10^6 km), while the chemical equivalent requires $\sim$36 hr to reach the same distance.

A final note on finite-burn losses. The 9% loss quoted for the nuclear interceptor is substantial, and is a result of the long run time associated with the higher I_{sp} of nuclear engines. Vehicle sizing with NTP systems often involves optimizing engine thrust in order to maximize the actual V produced. This optimization trades the losses associated with long run times caused by small engines against the lower ideal V associated with lower mass fractions caused by large engines.

H. NTP Interceptor Configuration

Table IV summarizes the basic mass breakdown developed for the PBR-based interceptor. Note that the mass analysis has been simplified for this exercise, and more detailed mass allocations would be appropriate for an actual interceptor design study. Note also that, for simplicity, no trapped or reserve propellant has been allocated.

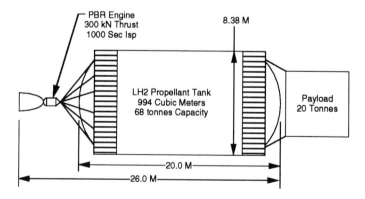

Figure 6.   Interceptor vehicle configuration.

The propellant mass allocated for the NTP interceptor was used to determine tank dimensions. A liquid hydrogen density of 70.5 kg m^{-3} and an ullage allowance of 3% were used to determine a propellant tank volume of

994 m^3. If we assume our tank diameter to be the same as the Space Shuttle external tank (8.38 m) (Isakowitz 1991) and we assume the same 0.7:1 elliptical end domes, our tank length is 20.0 m. Our engine is estimated to be 4 m long, and we have allocated 2 m for separation between the tank and the engine (this last value is another optimization variable, as it has a direct impact on radiation shielding requirements). Our interceptor stage is therefore 26 m long, and is depicted in Fig. 6.

TABLE IV
Interceptor Mass Breakdowns (kg)

	NTP Interceptor		Chemical Interceptor
Payload		20,000	20,000
Stage Empty Mass		12,000	6,600
Engine system	1,000		400
Radiation shield	300		—
Ancillary mass	2,500		2,500
Propellant tank	8,200		3,700
Propellant		68,000	73,400
Impulse propellant	65,100		73,400
Cooldown propellant	2,900		—
Initial mass in LEO		100,000	100,000
Stage propellant fraction		0.85	0.92

IV. SAFETY ISSUES

Of all the issues related to the development and implementation of nuclear rocket systems, both technical and programmatic, safety stands out as having received the most attention. The topics of nuclear energy and radiation generally produce strong emotional reactions in people, no matter what their positions are. Therefore, no discussion of nuclear propulsion would be complete without a review of the basic safety issues and principles.

As noted previously, the United Nations has established guidelines for space nuclear power systems. These guidelines do not apply to space nuclear propulsion systems, but their intent is relevant. The purpose of the U. N. guideline with regard to a minimum orbital altitude for space nuclear power systems is to ensure that, in the event of a failure or other mission terminating event, no radioactive material re-enter the Earth's biosphere. The minimum specified orbital altitude of 800 km ensures that the space nuclear reactor will not de-orbit for hundreds or thousands of years, long enough to ensure that radiation from the decay of fission products decreases to a nonhazardous level.

The same principle can be applied to NTP systems. In fact, this principle has been applied in numerous safety analyses performed for the various nuclear propulsion studies conducted over the last few years. NTP systems have several major advantages over power systems with regard to safety. First, the missions for which they have been considered typically involve departure from Earth orbit to some other target. This takes the reactor away from the Earth, where the possibility of re-entry drops to nearly zero. Second, the short operating time relative to power systems dramatically reduces the production of long-life fission products, with the result that decay radiation output drops to a safe level much more quickly. "Safe" radiation levels are dictated by the National Council on Radiation Protection, and are periodically revised.

Safety concerns with NTP systems are greatly alleviated by the fact that the NTP systems discussed in this chapter all use uranium for fuel. Uranium's extremely low level of radioactivity means that, prior to start-up, the NTP reactor can be approached without any protection and without any concern over radiation. Redundant safety systems installed in the reactor positively prevent it from "turning on" inadvertently, even in the worst conceivable scenarios. This ensures that the reactor can be delivered to space with no hazard to the environment in the event of a launch vehicle failure. Observing NSO considerations once the NTP-equipped vehicle is in space completes the safety story, indicating that the issue of safety with respect to the reactor can be properly addressed, and that solutions are available that satisfy all public requirements.

V. CONCLUSIONS

The ideal rocket equation predicts that, all other things being equal, doubling specific impulse doubles final velocity. Because all other things in this case are not equal, we performed the design exercise to assess the impact of the unique considerations associated with NTP on performance. The assumptions made were biased against NTP to some degree, in order that the lower maturity of the technology be accounted for, but it was demonstrated that the $\sim$2:1 improvement in final velocity is valid. This conclusion is predicated upon use of a high thrust-to-weight system like the PBR. Low thrust-to-weight systems will suffer in comparison, because their inert mass will be a larger percentage of their total mass. The reader can prove this by repeating the design exercise with one of the heavier propulsion systems.

The improvement in final velocity suggests an appreciable benefit in terminal intercept of a threatening NEO. Twice the final velocity suggests that an NEO can be intercepted at twice the range from the Earth. With a fixed available divert impulse, we can successfully divert a NEO double the mass divertable by a chemical interceptor.

The benefits of NTP can be utilized in different manners, as well. The improved performance permits delivery of a much larger payload (the reader can verify that this payload mass is approximately 43 t) in the same time.

Conversely, the greater performance can be used to deliver the same payload in the same time span, but with a much lower initial mass (<50 t). This last option may well be the most attractive aspect of NTP in this era of tight budgets.

This design exercise was deliberately simple in its treatment of the non-NTP aspects of vehicle design. Many considerations neglected here will alter vehicle performance, and design options exist that offer improved final velocities. It is hoped that this treatment of the application of NTP has provided the reader with the tools and insight to incorporate NTP into conceptual designs for whatever mission is desired. The issues considered and techniques applied have broad applicability, and NTP's performance offers enormous benefits to a broad range of space missions.

REFERENCES

Bate, R., Mueller, D., and White, J. 1971. *Fundamentals of Astrodynamics* (New York: Dover).

Bussard, R. W., and DeLauer, R. D. 1958. *Nuclear Rocket Propulsion* (New York: McGraw-Hill).

Canavan, G. H., Solem, J. C., and Rather, J. D. G., eds. 1992. *Proceedings of the Near-Earth Object Interception Workshop* (Los Alamos: Los Alamos National Lab).

Clark, J. S., McDaniel, P., Howe, S., Helms, I., and Stanley, M. 1993. *Nuclear Thermal Propulsion Technology: Results of an Interagency Panel in FY 1991*, NASA TM-105711.

El-Genk, M. S., and Hoover, M. D., eds. 1993. *Transactions of the Tenth Symposium on Space Nuclear Power Systems* (Albuquerque, N. M.: Univ. of New Mexico, Inst. for Space Nuclear Power Studies).

Griffin, M., and French, J. 1991. *Space Vehicle Design* (Washington, D. C.: American Inst. of Aeronautics and Astronautics).

Isakowitz, S. ed. 1991. *International Reference Guide to Space Launch Systems* (Washington, D. C.: American Inst. of Aeronautics and Astronautics).

Murray, R. L. 1954. *Introduction to Nuclear Engineering* (Englewood Cliffs, N. J.: Prentice Hall).

Nuclear Thermal Propulsion Technical Interchange Meeting, vol. 1. 1992. Proceedings from Conference held 20–23 Oct. at NASA Lewis Research Center (Cleveland, Oh.).

Roy, A. E. 1965. *The Foundations of Astrodynamics* (New York: McMillan).

Shapiro, A. H. 1953. *The Dynamics and Thermodynamics of Compressible Fluid Flow*, vol. 1 (New York: Ronald Press).

United Nations. 1992. *Principles Relevant to the Use of Nuclear Power Sources in Outer Space*, Resolution to the General Assembly, A/SPC/47/L6, 28 Oct.

Wickenheiser, T. J., Gessner, K. S., and Alexander, S. W. 1991. *Performance Impact on NTR Propulsion of Piloted Mars Missions With Short Transit Times*, AIAA 91-3401 (Washington, D. C.: American Inst. of Aeronautics and Astronautics).

Wertz, J., and Larson, W., eds. 1991. *Space Mission Analysis and Design* (Dordrecht: Kluwer).

APPENDIX

List of Acronyms and Symbols

CERMET	Ceramic-metallic
CIS	Commonwealth of Independent States
C_p	Propellant specific heat
ΔV	Change in vehicle velocity
F	Engine thrust
g_0	Gravitational acceleration constant
H	Orbital altitude
I_{sp}	Specific impulse
k	Ratio of specific heats
LEO	Low Earth orbit
m	Propellant molecular weight, also meters
$\dot{m}$	Propellant mass flow rate
M_0	Vehicle initial mass
M_f	Vehicle final mass
M_p	Total vehicle propellant mass
$M_{p\Delta V}$	Impulse propellant mass
M_t	Vehicle propellant tank empty mass
μ	Earth gravitational parameter
NEO	Near-Earth object
NERVA	Nuclear engine for rocket vehicle applications
NSO	Nuclear-safe orbit
NTP	Nuclear thermal propulsion
PBR	Particle bed reactor
SDI	Strategic defense initiative
SEI	Space exploration initiative
SNTP	Space nuclear thermal propulsion
T_0	Engine chamber temperature
TRL	Technology readiness level
T/We	Engine thrust-to-weight ratio
V_{EXIT}	Engine exhaust velocity
V_{hp}	Excess hyperbolic velocity

NON-NUCLEAR STRATEGIES FOR DEFLECTING COMETS AND ASTEROIDS

H. J. MELOSH
University of Arizona

and

I. V. NEMCHINOV and Yu. I. ZETZER
Russian Academy of Sciences

A number of authors have recently suggested that the only plausible defense against large Earth-threatening comets or asteroids is the use of very large (Gigaton) nuclear weapons. However, it can be plausibly argued that the mere existence of an arsenal of such weapons constitutes a danger to humanity far greater than the threat they are intended to mitigate. In this chapter we explore other means of deflecting threatening asteroids. Beginning from the fundamental physical requirements for asteroid deflection, we examine the capabilities of kinetic energy deflection (i.e., a high-speed collision with another, smaller object), mass drivers, solar sails, a new solar collector strategy that we discovered in preparing this study, and the more highly speculative uses of beamed energy, such as lasers and microwaves. It will be seen that even kinetic energy deflection, if augmented by the "billiard shot" strategy of first deflecting a smaller asteroid to collide with a larger one, can provide a plausible alternative to nuclear weapons. Mass drivers offer good performance at the expense of technical complexity, solar sails probably suffer too drastically from low thrust to be plausible, while the solar collector approach offers a very favorable performance, although it suffers from a number of problems whose solutions are not yet well defined. Beamed energy deflection is probably far in the future, but we nevertheless examine some physical constraints on such systems.

I. INTRODUCTION

If one accepts the premise that comets and asteroids with diameters greater than about 1 km constitute a serious threat to civilization, then it is natural to seek a defense against them. Ahrens and Harris (1992) and Canavan et al. (1992) argue that very large nuclear weapons are the only plausible defense against such objects. However, the mere existence of huge, deployable nuclear weapons itself constitutes a terrifying threat to civilization that may overshadow the danger posed by comets and asteroids. It thus seems imperative to examine carefully non-nuclear strategies for deflecting such objects.

Deflection of a comet or asteroid on a collision course with the Earth requires only one thing; that its velocity be changed. The magnitude of the necessary velocity change depends on the length of time before impact, with longer advance times favoring small velocity changes, and the mode of application of the impulse. Ahrens and Harris (1992) show that if thrust is applied along the direction of motion of a body in a near-circular orbit, then the displacement δ depends on the velocity change ΔV and time t before impact t as $\delta = 3\Delta vt$. On the other hand, if thrust is applied in a perpendicular direction, then the deflection reduces to the linear estimate $\delta = \Delta vt$. It is thus clearly desirable to apply thrust in the direction of motion. To deflect a comet or asteroid by a distance equal to the Earth's radius ten years ahead of a projected impact requires a velocity impulse on the order of 1 cm s^{-1} (Ahrens and Harris 1992):

$$\Delta v = \frac{7 \text{ cm s}^{-1}}{t \text{ (yr)}}. \tag{1}$$

This estimate applies to impulsive velocity changes, such as would be applied by the explosion of a nuclear weapon. Melosh and Nemchinov (1993) show that the same factor of 3 also applies to steady thrust applied along the direction of orbital motion, so that the displacement due to a constant acceleration a is given by $\delta \approx 3/2at^2$. The acceleration necessary to deflect an object by the radius of the Earth is:

$$a \approx \frac{4.7 \times 10^{-7} \text{ cm s}^{-2}}{t(\text{yr})^2}. \tag{2}$$

Whether the velocity change is applied impulsively or steadily, the only means of achieving it is through Newton's third law of action and reaction. Momentum must be delivered to the comet or asteroid. Deflection of these objects is difficult because they are so massive. Although the requisite velocity changes or accelerations are small, the sheer mass of the objects to be deflected makes this a challenging operation. Most deflection schemes envisage imparting a high velocity to a small amount of mass. This mass is usually part of the comet or asteroid itself. Schemes that involve landing propellant on the asteroid to provide working mass are extremely wasteful (every kilogram of propellant landed on the comet or asteroid requires thousands of times more mass to loft it from the Earth and deliver it to the object), and will not be considered here. The most favored nuclear scenario (Ahrens and Harris 1992) involves detonation of a large neutron bomb near the comet or asteroid and imparts an impulse in reaction to the heat-induced spallation of a surface layer.

Because the working mass comes from the comet or asteroid itself, the major limitation on deflection schemes is the amount of energy delivered, not mass. This contrasts sharply with rockets that carry their own propellant, for which mass itself is the limiting factor. In the case of rockets, designers strive to achieve the highest exhaust velocity possible, so that the limited mass of propellant may impart the maximum momentum impulse. However, when reaction mass is abundant, the maximum momentum of the ejecta, $p = mv_e$, where m is the ejected mass and v_e is its velocity, for a fixed energy, $E = 1/2mv_e^2$, invested in ejecting it, is achieved by the *lowest* ejection velocity, thereby maximizing the momentum per unit of energy, $P/E = 2/v_e$. Of course, ejection velocities cannot be arbitrarily low, as this may make the requisite mass too large to be practical, but it is important to keep in mind that low ejection velocities are generally preferable. The optimum (and probably not achievable) application of fixed energy is if the comet or asteroid is split into two equal parts, each of which is given an equal and opposite velocity impulse.

Although the principle is simple, the different means by which mass can be ejected from a comet or asteroid are diverse. Nuclear weapons can certainly deliver a large amount of concentrated energy, but this energy is not easily coupled to the principal task of ejecting mass. In the standoff scenario (Ahrens and Harris 1992) only 1 to 10% of the neutron bomb's energy is coupled to the spalled layer. Other scenarios employ the kinetic energy of an impacting object to create a crater and eject material, mass drivers landed on the comet or asteroid's surface to quarry and eject material, solar sails to provide a steady, weak thrust without mass ejection, a solar collector to heat surface material to the point that it spontaneously boils off, and various types of beamed energy to heat and thus eject surface materials. We will not discuss the use of chemical explosives, because the energy density in projectiles striking at even modest velocities far exceeds that available from chemical reactions (the energy per unit mass of TNT is equivalent to that of a projectile traveling at only 3 km s^{-1}). The abilities of these different schemes to deflect comets and asteroids will be analyzed below. The merit of each scheme will be compared using plots of a type advocated by Canavan (Canavan et al. 1992) in which the vertical axis is the diameter of a comet or asteroid that can be deflected from a collision with the Earth and the horizontal axis is time before impact at which the action is commenced. Note that in the following pages the cumbersome phrase "comet or asteroid" will be replaced simply by "asteroid," with the understanding that either one is meant. As it is believed that asteroids dominate the current cratering flux on the Earth for craters smaller than 50 km diameter (Shoemaker et al. 1990), this shorthand is appropriate.

II. KINETIC ENERGY DEFLECTION

One of the conceptually simplest means of deflecting a threatening asteroid is simply to strike it with a massive projectile, whether that be another asteroid, a rocket, or a rocket-propelled mass. One of the major problems with this scheme is the danger of fragmenting the asteroid and thus increasing the damage upon collision with the Earth. However, recent images of the asteroids Gaspara (ca. 19 km long) (Belton et al. 1992) and Ida (ca. 52 km long) (Kerr 1993) shows that they have been repeatedly struck by large objects in the past that created large craters on their surfaces without apparently fragmenting them, so it may be worthwhile to consider this possibility.

If a projectile with mass m_p and diameter L strikes an asteroid at velocity v_i at normal incidence (natural impacts take place over a variety of angles; however, the maximum linear momentum transfer occurs at normal incidence, and we assume this configuration in all the following), it delivers a direct momentum impulse of $m_p v_i$. This direct impulse, however, is augmented by the reaction to ejecta thrown out of the crater. As we will show, the momentum of this ejecta far exceeds the direct impulse. The momentum of the projectile itself can thus be neglected in most cases. The kinetic energy of the projectile is thus used to power the ejection of a large quantity of relatively low-velocity crater ejecta.

The size of the crater produced by a given impact can be estimated by a variety of crater scaling laws (Melosh 1989). We use a scaling law originally proposed by Holsapple and Schmidt (1982) and recently updated (Schmidt and Housen 1987).

$$D = 1.16 \left(\frac{v_i^2}{gL} \right)^{0.22} L \qquad (3)$$

where D is the transient crater diameter, g is the surface acceleration of gravity and the constant and fractional exponent in the equation are appropriate for competent rock. The dimensionless combination in the parentheses is the Froude number.

This scaling law assumes that gravity, not strength, limits the growth of the crater. Although some arguments have been made in favor of strength scaling in the past, a recent numerical investigation (Asphaug and Melosh 1993) of the formation of the crater Stickney on the 27 km long Martian moon Phobos has shown that for impacts on such small bodies the shock wave fragments the rock surrounding the impact site long before the crater opens, so that even if the asteroid were strong to begin with, the impact itself converts the surrounding material to rubble before the crater forms.

Although Eq. (3) tells us how large a crater is produced by an impact, it does not tell us how fast the ejecta is moving. An extension of the crater scaling laws (Housen et al. 1983), however, gives us the mass m_{ej} ejected at a velocity equal to or greater than v

$$m_{ej} = \pi \rho R^3 \alpha \left(\frac{\sqrt{gR}}{v} \right)^{1.7} \qquad (4)$$

where $R = D/2$ is the crater radius, ρ is the asteroid density and α is the crater depth-diameter ratio, $\alpha = H/D$, here taken to be 0.2. Differentiating this equation with respect to v gives the mass dm_{ej} ejected at velocities between v and $v + dv$. Multiplying this mass by $v \sin \phi$, where ϕ is the angle of ejection (typically 45°), and integrating from the escape velocity of the asteroid up to the impact velocity yields the total momentum impulse delivered from the reaction to the ejecta, ΔP

$$\Delta P = \int_{v_{\text{esc}}}^{v_i} v \sin \phi \, dm_{ej}. \qquad (5)$$

Substituting Eq. (4) and evaluating numerical factors gives

$$\Delta P = 0.075 \rho D^{3.85} g^{0.85} v_{\text{esc}}^{-0.7} \qquad (6)$$

where the v_i term has been neglected because the low-velocity term dominates the equation; as expected, the more abundant low-velocity ejecta makes the largest contribution to the momentum impulse. Further substituting for the crater diameter D from Eq. (3), and using the fact that on a spherical asteroid of radius r_a, the acceleration of gravity $g = 4/3\pi G \rho r_a$ and escape velocity $v_{\text{esc}} = \sqrt{8/3\pi G \rho r_a}$ where G is Newton's gravitational constant, the expression for the momentum impulse delivered by the impact becomes

$$\Delta P = 0.064 \rho^{0.65} L^3 v_i^{1.69} G^{-0.35} r_a^{-0.70}. \qquad (7)$$

It is easy to show that this impulse is much larger than the projectile's own momentum, which we neglect in the following discussion. Equation (7) gives us the basis for constructing a plot of asteroid diameter vs deflection time, because the velocity impulse imparted to the asteroid of mass m is $\Delta v = \Delta P/m$. Using Eq. (1) for impulsive deflection, assuming that $\rho = 3000$ kg m^{-3} for both target and projectile, the diameter d_a of an asteroid that can be deflected by an impact of a projectile with diameter L is given by

$$d_a \, (\text{km}) = 0.14 L(m)^{0.81} t \, (\text{yr})^{0.27} v_i \, (\text{km s}^{-1})^{0.46}. \qquad (8)$$

Figure 1 shows the predictions of this equation for projectile diameters of 1, 10 and 100 m and an impact velocity of 10 km s^{-1}. Deflection of asteroids in the 1 to 5 km diameter range thus requires projectiles on the order of 10 m diameter that weigh on the order of 1500 tons—something that could conceivably be launched, but which would require on the order of 50 Space Shuttle flights, not to mention the fuel to get it from low Earth orbit to the asteroid. This mass is to be compared to the ~20 tons mass (Canavan et al. 1992) of a 0.1 to 1.0 GTon nuclear warhead (Ahrens and Harris 1992) needed to do the same job. Under these circumstances kinetic energy deflection looks relatively unfavorable, although not impossible.

However, another possibility that may make kinetic energy deflection more attractive is the indirect "billiards shot" scenario (Canavan et al. 1992) in which a small, favorably situated asteroid is itself deflected by some means into collision with a larger threatening asteroid. In this case, 100 m diameter projectiles might not be out of the question.

To complete this discussion we examine the conditions under which an impact disrupts the asteroid, rather than merely forms a crater upon it. We

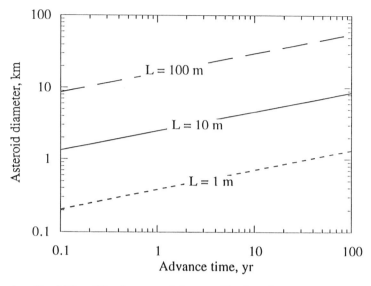

Figure 1.   Capability of kinetic energy deflectors. The three lines represent impacts
   by projectiles of 100, 10 and 1 m in diameter, and show how large an asteroid may
   be deflected from a collision with the Earth as a function of the time elapsed between
   the impact on the asteroid and the predicted collision with Earth.

assume that the asteroid is a strengthless collection of rubble bound together
only by its self gravity. This model is supported by numerical computations
that suggest that the impacts visible on the surfaces of known asteroids were
sufficient to fracture them (Asphaug and Melosh 1993). In addition, the
observed tidal breakup of comet Shoemaker-Levy 9 after a close pass by
Jupiter in 1992 indicates that it had a strength of at most 10^{-4} bar (Scotti
and Melosh 1993), and was thus essentially strengthless. Even if strength
does play a role in inhibiting the catastrophic disruption of asteroids, the
zero-strength criterion derived below will give a lower limit on the asteroid
diameter at disruption.

   In the older literature, a disruption threshold was established by equating
the kinetic energy of the projectile to the gravitational binding energy of
the asteroid (Melosh 1989). However, it is now clear that this criterion is
much too conservative. One of the principal characteristics of impacts is that
the projectile's energy is initially concentrated in a region of the target of
roughly the same size as the projectile (the "isobaric core") and then falls
off rapidly with distance from this region. The impact energy in the target
is thus very heterogeneously distributed and is thus not very effective at
disrupting it. A better, although rough, criterion is to use the Holsapple-
Schmidt scaling law, Eq. (3), and suppose an asteroid is disrupted when the
predicted crater diameter equals the diameter of the asteroid itself. Although
crude, this criterion actually matches several computations of the impact

breakup of Mercury quite well (Tonks and Melosh 1992). A slightly different criterion that gives nearly the same answer begins with an estimate of the particle velocity at the antipode of the impact site by presuming that outside the isobaric core the particle velocity falls as the inverse square power of distance. If the particle velocity at the antipode exceeds the escape velocity, then the asteroid is considered to be disrupted. This criterion has the simple functional form

$$r_{disrupt} = 0.35 \left[\frac{L^2 v_i}{\sqrt{G\rho}} \right]^{1/3}. \tag{9}$$

For projectiles with diameters of 1, 10 and 100 m striking at 10 km s^{-1}, as illustrated in Fig. 1, the asteroids struck must be more than 0.2, 0.92, and 4.3 km in diameter, respectively, to escape disruption. These diameters fall to the left of the plot in Fig. 1 (for 0.2 km diameter it falls on the ordinate itself), so that disruption does not constitute a serious problem for kinetic energy deflection schemes unless the advance time is very short and an extremely large projectile must be used.

III. MASS DRIVERS

Mass drivers were originally developed in support of space settlement projects in the mid 1970s (O'Neill 1977). They were designed to accelerate large quantities of loose surface material from the Moon's surface to space construction sites. However, the reaction from this acceleration can also be used to change the velocity of a small asteroid and thus deflect it from collision with the Earth. Although the details of mass drivers vary, most use electromagnetic forces acting through either sliding metal or plasma contacts to accelerate buckets containing surface material.

To deflect an asteroid, a mass driver would be flown to the asteroid and firmly anchored to its surface. Reaction mass would be quarried from the asteroid, placed into the buckets, then flung into space in the right direction to apply a consistent acceleration along the asteroid's orbital path. Thus, on a rotating asteroid it could not operate continuously, but must eject material only when it is pointed in the right direction (although perhaps some sort of flexible steering system could be developed). Ejection of material would take place over a long period of time, so the effectiveness of mass drivers is governed by Eq. (2) for continuous accelerations. Aside from the very considerable physical plant that must be landed on the asteroid, and the need for low-gravity mining techniques, the main limitation on the mass driver is energy. The thrust developed by a mass driver is simply the mass flow rate dm/dt times the ejection velocity, v_e. The acceleration achieved is this thrust divided by the asteroid's mass m. If we presume a high efficiency (near 90%) conversion of electrical energy into kinetic energy of the ejecta, then the power needed to sustain the mass flow rate is simply $W = 1/2 dm/dt v_e^2$. This power could be supplied either from solar or nuclear sources. As we are trying to avoid nuclear solutions, even the use of nuclear reactors in space, we presume that the power to run the mass driver is obtained from solar energy. The efficiency e of conversion of solar energy to electricity is near 10% for either thermomechanical (solar boilers, turbines and generators) or probable near-term photovoltaic systems (Stone 1993). For comparison with other deflection schemes, we rate the effectiveness of mass drivers in terms of the area of a solar collector needed to intercept the power to run the mass driver. This power is thus $W = eAS$, where A is the area of the solar collector ($A = \pi/6d^2$ for a circular solar collector of diameter d) and S is the solar constant, about 1 kW/m^2 at 1 AU from the Sun.

Substituting for all the terms described above, using Eq. (2) for the required acceleration and solving for the diameter of the deflected asteroid d_a, we get

$$d_a \text{ (km)} = 2.8 \frac{t \text{ (yr)}^{2/3} d \text{ (km)}^{2/3}}{v_e \text{ (km s}^{-1})^{1/3}}. \tag{10}$$

Note the very different time dependence in Eqs. (10) and (8). This is mainly because kinetic energy deflection is achieved by a single impulse whereas the mass driver produces a continuous acceleration. Long advance times thus favor steady acceleration schemes such as mass drivers. This is shown graphically in Fig. 2, where the capabilities of mass drivers for 1 and 10 km diameter solar collectors are compared. In this figure we assume an ejection velocity v_e of 300 m s^{-1}. Higher ejection velocities require less mass to be processed, but do not use the available energy as efficiently. For the same energy, the 8 km s^{-1} mass driver suggested by Canavan et al. (1992) can deflect asteroids only 1/3 the size of those shown in Fig. 2.

Because the mass driver uses material from the asteroid itself, one point to keep in mind is that it should not use up the entire asteroid. Although low ejection velocities use the available energy more efficiently, they also require more mass. The ejection velocity should not be so low that the entire asteroid must be processed into the mass driver. The total mass ejected during operation of the mass driver is simply $(dm/dt)t$. It is easy to use the same arguments leading to Eq. (10) to show that the diameter of an asteroid which is just consumed by its mass driver is

$$d_{\text{consumed}} \text{ (km)} = 0.146 \frac{t \text{ (yr)}^{1/3} d \text{ (km)}^{2/3}}{v_e \text{ (km s}^{-1})^{2/3}}. \tag{11}$$

For the conditions shown in Fig. 2 d_{consumed} is always much smaller than the asteroid itself. In spite of the technical difficulties associated with the landing and operation of a mass driver plant, Fig. 2 shows that mass driver systems with modest ($\sim$1 km diameter) solar collector areas are capable of deflecting asteroids in the 1 to 10 km diameter range with a few years advance warning. Continuing study of the other costs associated with this deflection system may thus be warranted.

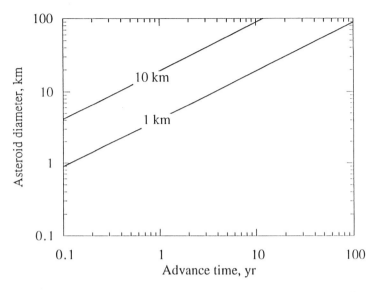

Figure 2. Capability of mass drivers. The mass driver is categorized by the diameter of a solar collector (at 1 AU) needed to supply operating power at 10% overall efficiency. The lines for 1 and 10 km diameter circular collectors show that modest-size systems may be capable of diverting asteroids in the 1 to 10 km range.

IV. SOLAR SAILS

Whereas mass drivers use material from the asteroid itself, solar sails use no ponderable matter at all. They depend on solar radiation pressure to generate a small but steady thrust that can potentially be harnessed to deflect a threatening asteroid. The pressure exerted by solar radiation is given by $2S/c$ for perfect reflection, where S is the solar constant and c is the speed of light. Although S is more than 1 kW m^{-2} at 1 AU from the Sun, the speed of light is so large that the overall thrust is very low—only about 6×10^{-6}N m^{-2} of reflector. Nevertheless, solar sails have been proposed as a practical, fuelless interplanetary propulsion system (Friedman 1988; Wright 1992), and it is worth while investigating them as potential tools for deflecting asteroids. Using Eq. (2) for a steady acceleration and equating it to the acceleration a_s applied by a circular solar sail of diameter d to an asteroid of mass m, $a_S = (\pi/2d^2S/c)/m$, and solving for the diameter d_a of the asteroid that can be deflected,

$$d_a \text{ (km)} = 0.086d \text{ (km)}^{2/3}t \text{ (yr)}^{2/3}. \tag{12}$$

The predictions of this equation are shown in Fig. 3 for sail diameters of 10, 100 and 1000 km. It is clear that truly enormous structures are necessary to deflect asteroids in the 1 to 10 km diameter range. Although 1 km diameter solar sails are well within the reach of current technology (Friedman 1988; Wright 1992), launching and assembling 100 km diameter sails presents a

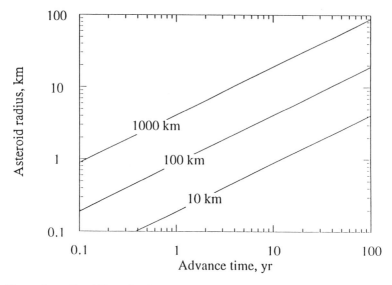

Figure 3.   Capability of solar sails. The three lines are for different solar sail diameters. Even small asteroids require enormous solar sails (100 to 1000 km in diameter) which, along with the technical difficulty of tethering them to the asteroid, makes such a deflection system look very unfavorable.

challenge that is beyond current abilities.

Another disadvantage of solar sails is the difficulty of attaching a tether to a rotating, perhaps precessing, asteroid. Although methods of solving this problem exist, they are mechanically complex, involving a cinch to hold the asteroid itself, a system of gimbals through which the thrust is transmitted, and a long tether to the solar sail. Although this system of asteroid deflection seems environmentally cleaner than the others, it requires space constructions on a scale that is far ahead of current technology.

V. SOLAR COLLECTOR

Solar sails, described above, suffer from very low thrust and mechanical complexity. In the course of studying solar sails, however, we discovered an approach that is arguably better than any other previously proposed. This scheme uses the solar sail as a light collector, focusing sunlight onto the surface of the asteroid and generating thrust as the asteroid's surface layers vaporize. In effect, the system is a solar-powered mass driver, but it does not require any equipment to be landed and operated on the asteroid's surface. This approach imitates the known effect of sunlight in generating the gas jets that perturb cometary orbits (Crifo 1987; Peale 1989; Whipple 1950). Although this system is highly speculative, and involves technology somewhat beyond present capabilities, we will analyze it in detail simply because it is a new concept that has not been considered by other authors. Furthermore, the initial analysis is highly favorable, although a number of serious problems with this concept remain to be solved.

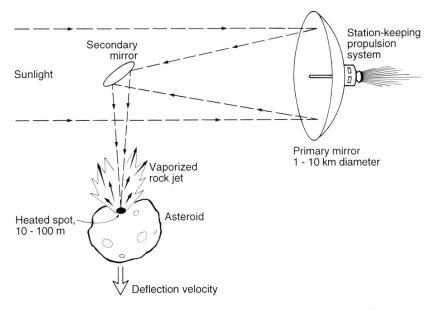

Figure 4.   Schematic illustration of a solar collector deflection system. Sunlight collected by a large, light primary mirror is focused on the surface of an asteroid or comet and evaporates material from its surface. The reaction to this vapor plume changes the asteroid's orbit. The primary mirror requires a modest station-keeping propulsion system, such as a high specific impulse solar-electric system.

The irradiation of an asteroid or comet by concentrated solar radiation is represented schematically in Fig. 4. For definiteness we assume that a several km diameter solar collector with 1 km focal length concentrates solar radiation on the asteroid's surface. Assuming that the divergence of the radiation is 10 milliradian (the same as the divergence of light from the Sun's disc at 1 AU), we obtain an illuminated spot diameter of about 10 m. The solar constant at 1 AU is about 1 kW m^{-2}. A well-metallized film reflects 90 to 95% of the incident radiation, so the light intensity q at the surface of the asteroid may be about $q = 10$ MW m^{-2}. This is sufficient to evaporate stony or icy bodies having low thermal conductivity (but perhaps not solid iron asteroids, which fortunately are rare).

The thrust developed by evaporated surface material is very similar to that of a mass driver $F = \beta dm/dt\, v_e$ where β is a numerical factor (taken here to be 0.5) that accounts for the hemispherical rather than linear expansion of the gas, dm/dt is the mass flow rate, and v_e is the ejection velocity from the asteroid. The mass flow rate is essentially the rate of evaporation of surface material, given by the incident power divided by the heat of evaporation of the material H_{vap}, plus whatever heat is absorbed in raising the temperature of the ejected gas. We take $H_{vap} = 15$ MJ kg^{-1} for silicates and 3 MJ kg^{-1} for ice, and assume an ejection velocity close to the sound speed, $v_e = 1$ km s^{-1}. This ejection velocity is also close to the average molecular speed of typical silicate vapor molecules (O_2, MgO, SiO) at the vaporization temperature (Hashimoto 1990) of 1500 to 2000°C (depending on pressure). In the case of water ice, another 1 MJ kg^{-1} must be added to raise the average molecular speed to 1 km s^{-1}. For silicates the resulting thrust is about 3.3×10^{-5} N W^{-1} of collected energy, or 0.03 N m^{-2} of solar collector at 1 AU—a factor of about 3000 better than a pure solar sail, even for silicate asteroids (icy objects enjoy a factor of 4 further advantage).

The process of evaporation has been simulated in laboratory experiments by means of a high-power laser. Although the laser beam is both more collimated and monochromatic than sunlight, we used it merely to simulate the effect of rapid optical energy deposition on a solid surface, and we believe that it provides an adequate simulation to this extent. We first studied the irradiation time needed to bring the surface of the asteroid to evaporation temperatures. The rotation of an asteroid with typical surface velocity about 1 m s^{-1} limits the heating time to 10 s for a 10 m spot. A pulse of laser radiation with duration 0.5 msec was focused on a specimen of basalt that simulates a silicate asteroid. Figure 5 is a photo of the emission of the basalt vapor jet generated by this irradiation. These experiments and theoretical estimates show that at a laser light intensity about $q = 1$ GW m^{-2}, evaporation begins after 0.1 msec. The delay is longer for lower intensities: it scales roughly as q^{-2}. Thus for $q = 10$ MW m^{-2} evaporation should begin on a time scale of about 1 s—short compared to the time it takes surface material to rotate through the irradiated spot on the asteroid. The illuminated spot thus does not need to track the surface; as the asteroid rotates under the spot the evaporation continues steadily, the rotation serving to bring fresh material into the heated area. Note that the thrust may also lead to the change of the rotation velocity.

We note that asteroids and especially comets are not believed to be homogeneous (Dodd 1981), so that it is unlikely that the material evaporated off the surface will be pure silicate gas. Silicate meteorites may contain volatiles, including organic compounds, immersed in a stony matrix. Using typical values of the thermal diffusivity of 10^{-5} to 10^{-6} m^2 s^{-1}, the thickness of the layer heated by a 10 s irradiation may reach several mm. This value is comparable to the size of typical inclusions, such as chondrules or breccia fragments (Dodd 1981). The sudden evaporation and expansion of volatile-rich inclusions may cause a drastic increase in local pressure and consequent fracture of the matrix. The gas expends some of its force accelerating these particles and the clast-ladened gas is thus ejected with relatively low velocity. This effect is equivalent to a decrease in the heat of evaporation of the asteroid material. This enhanced mass ejection may cause a substantial increase of the thrust because the ratio of the momentum P to the energy E is inversely proportional to the velocity of ejection v_e : $P/E \propto 1/v_e$. It is still more

Figure 5.   Laser irradiation of a basalt test "asteroid" in the laboratory. The base of
the luminous plume is 1 cm in diameter and the laser energy density is 1 GW m^{-2}.

difficult to analyze the irradiation of the comets because little is known about their composition and structure. A conventional viewpoint is that comets are comprised of ice with silicate inclusions. To estimate the efficiency of the irradiation of such inhomogeneous systems, a second set of experiments with high-power lasers was performed.

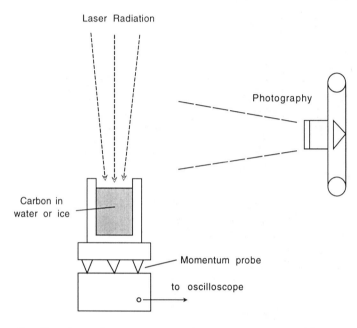

Figure 6.   Experimental setup to measure the momentum generated by evaporation of test materials, such as water or ice with included carbon black, or basalt as it is evaporated by laser irradiation.

Our experimental setup is schematically represented in Fig. 6. In these experiments laser pulses with durations of 0.4 msec and 50 nsec were focused on water or ice containing carbon powder in different concentrations and measurements of the momentum P were performed. The ratio P/E of the momentum P to the laser energy E for the 0.4 msec laser pulse as a function of the weight percent of carbon and the laser energy density E/A, where A is the area of the irradiated spot, is shown in Fig. 7. Within the range of the carbon concentrations under study there is a single dependence of P/E vs E/A. Note that the value of P/E for the water with the carbon powder is about ten times greater than that for basalt. Irradiation of ice with carbon powder, all other things being the same, results in a two-fold decrease of P/E; i.e., an increase of cohesion leads to a decrease in the ejected mass. In this case, the entire layer of ice 3-cm thick was transformed into a mixture of water and small fragments of ice. Although the energy in the laser pulse was nominally sufficient to melt only 5 to 10% of the ice, it proved to be enough to disintegrate it. A similar processes may take place on comets irradiated by focused solar radiation. In this case the irradiation time necessary for deflection may be much less than that for silicate asteroids. More detailed experimental study is required on the irradiation of inhomogeneous media or porous, dirty ice (Kömle et al. 1991) similar to the "fluffy stuff" of the real comets.

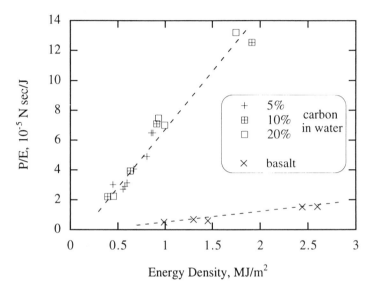

Figure 7. The ratio P/E of momentum to laser energy as a function of the laser energy density q for the material combinations listed in the legend. Note that the evaporation of water is not sensitive to the amount of carbon black in the ranges listed. The results for ice (not shown) give about half the momentum as water.

Bringing together the equations described above, we estimate the diameter d_a of an asteroid that can be deflected by a circular solar collector of diameter d a time t before impact on Earth

$$d_a \text{ (km)} = 3.4d \text{ (km)}^{2/3} t \text{ (yr)}^{2/3}. \tag{13}$$

Thus, a 1 km diameter solar collector operating for a year can deflect asteroids up to 3.4 km in diameter. Estimating a mass of 5 g m^{-2} for an aluminized kapton collector, which is well within current technology (Friedman 1988), such a collector would weigh about 4 tons and could easily be lofted by the Space Shuttle. This same collector, functioning for a decade, could deflect a 10 km diameter asteroid. In contrast, the nuclear standoff scenario requires a 0.2 to 2.0 GTon nuclear weapon weighing more than 20 tons (Canavan et al. 1992), detonated ten years in advance (Ahrens and Harris 1992) (note that this estimate corrects a number of obvious arithmetic errors in this reference).

Figure 8 illustrates the deflection capability for 1 and 10 km diameter solar collectors, and for both silicate asteroids and icy comets. The plot also

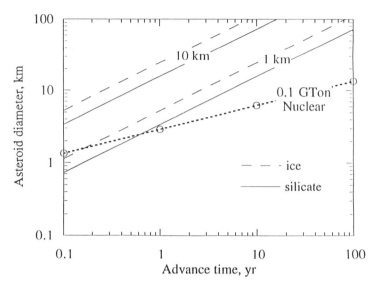

Figure 8.   Asteroid deflection capabilities of solar collectors vs nuclear weapons.
This plot shows the diameter of the asteroid (or comet) that can be deflected as
a function of the time before impact. The pairs of solid and dashed lines are for
silicate and icy bodies, respectively, that can be deflected by either 1 km or 10 km
diameter solar collectors. The heavy dotted curve with representative points is
for the nuclear stand-off scenario employing a 0.1 GTon neutron bomb with an
(optimistic) assumed conversion of 0.3 into neutron energy.

illustrates the deflection capability of a 0.1 GTon nuclear weapon, using the
most optimistic scenario of Ahrens and Harris (1992). Note the different
time dependencies of the two scenarios. This is because nuclear deflection
applies a single impulse, whereas solar evaporation affords a steady push.
Long advance times thus favor the solar collector approach.

The advantages of the solar collector approach are that no tethers or
special surface preparation on the asteroid are needed, the method provides
a steady, gentle push that minimizes the danger of disrupting the asteroid,
or even disrupting possible gravitationally bound binary objects, and the
system is reusable. Furthermore (and this could be very important), the solar
collector is big, slow and fragile, so it is not easily misused as a weapon, in
strong contrast to the nuclear deflection scenario.

Two disadvantages of this approach are that the solar collector must be
delivered to the asteroid, although this may be accomplished by using the
collector initially as a true solar sail to reach its target, and a modest station-
keeping propulsion system is required to keep the solar collector properly
oriented and on target (a solar radiation pressure of about 10 N must be countered for a 1 km diameter solar collector). Such a station-keeping system
might use a high specific impulse solar-electric propulsion system to reduce
the required propellant mass. Such an ion propulsion system would also func-

tion with a minimum of vibration that might adversely affect the collector's performance.

The major problem faced by the solar collector is probably degradation of the collector surface by hot gas and dust ejected from the heated spot on the asteroid. More study of this problem is required, but it seems that it could be greatly ameliorated by the use of small, rugged secondary or tertiary mirrors (necessary anyway to keep the heated spot in the right location relative to the asteroid's orbit) in the vicinity of the asteroid, while the large and fragile primary collector could stand off some distance by virtue of a long focal length (C. Steffens, 1993 personal communication). The secondary mirror surface might be protected by bleeding a small quantity of gas over it, thus deflecting expanding gases and small dust particles. Even with secondary mirrors, however, the thermodynamics of radiation transport requires that the last mirror in the optical system subtend a solid angle of at least $\pi(T_S/T_{sun})^4$ where T_{sun} is the Sun's surface temperature (5770 K) and T_S is the temperature of the heated spot. For a spot temperature of 2000 K, this requires a subtended angle of about 0.05 steradians, so the required mirror will have to be relatively close to the heated spot and may intercept a significant amount of ejecta. Clearly, more study of this problem is needed.

Although it is not our desire to go into engineering details of this system, preferring instead to point out the high theoretical potential of the solar collector concept, a few points need to be made. First, the surface of the mirror does not need to be accurately figured to the standards of fine optical instruments such as telescopes, nor does it need to be optically aligned with high accuracy. The fractional-wavelength accuracy required of telescope mirrors is necessary to attain a diffraction-limited image at the focus. Instead, all we require is that the 1/2-degree diameter disc of the Sun be crudely imaged on the surface of an asteroid. Indeed, there is no need for an image at all—a nonimaging collector such as those devised by Winston (Walford and Winston 1989) would perform its task just as well. The collector itself would probably achieve its shape by a combination of rotation about its axis to provide rigidity and a system of cables operated from an axial "mast" pulling on stiff splines to warp the collector into the desired shape. A Cassegrain-type arrangement might be desirable, in which the concentrated beam of sunlight is reflected by a secondary mirror through a hole in the center of the collector. In this arrangement the surface of the collector faces directly away from the expanding gas cloud and the integrity of the reflecting surface can be maintained more easily. The tertiary mirror that is subjected to the brunt of the expanding gasses may require its own propulsion system to keep it on station, or perhaps a series of such mirrors might be anchored to the asteroid and used alternately as they rotated into a favorable configuration with the main collector. Other questions focus on the reliability of a space-based system that is required to function over a period of years. In any event, a variety of solutions are available to the problems faced by this system, and only future study will show whether this possibility is really viable.

VI. BEAMED ENERGY: LASERS AND MICROWAVES

An obvious solution to the problem of solar mirror degradation by ejecta is to transmit the energy not simply by collection of natural sunlight, but by first converting the sunlight into a small-divergence laser or microwave beam. Such a system would allow the solar collector to stand off at large distances from the asteroid, thus keeping it safe from evaporated material. The disadvantage of such systems are their greatly increased technical complexity and the inefficiency inherent in the conversion of solar radiation to laser light or microwaves. Although systems of this type clearly require space-based technology that is far in the future, it is still possible to outline the fundamental physical requirements needed to create them.

Some experiments with laser irradiation are reported in the preceding section. However, those experiments were designed to simulate the effect of concentrated solar radiation. There are several new features to consider for an asteroid deflection system specifically designed to use lasers. As in all the other systems, of course, the main goal is to ablate material from the asteroid's surface and deflect the asteroid by the resulting momentum impulse. The principal difference between laser radiation and concentrated sunlight is the laser beam's high intensity and small divergence, which gives it the ability to act on asteroids at large distances. Using a hypothetical 1 km diameter gas lens (Michaelis et al. 1991; Phipps 1992) a divergence of 5×10^{-10} radian can probably be obtained. In this case the maximum standoff distance may be as great as 0.3 AU. A 10^9 to 10^{11}W laser operating for 30 to 50 days is capable of deflecting a 1 km radius asteroid from a collision course with the Earth. The energy necessary to drive the laser may be derived from a solar collector. In this case a collector diameter between ~ 3 and ~ 10 km will be needed, assuming a somewhat optimistic 10% conversion efficiency of solar radiation to laser light. A separate plot of asteroid diameter vs advance time for different solar energy collector diameters is not needed for this case: Because the assumed efficiencies are the same, the reader should refer to the plot (Fig. 2) for mass drivers. Microwaves may also be focused into beams of high intensity and small divergence by the use of large phased antenna arrays which, although fragile, might easily be constructed in space. There are two modes of microwave irradiation of an asteroid, depending upon the incident energy flux. The first mode, initiated at energy fluxes greater than 10^8W m^{-2}, generates a plasma sheath over the target which then heats and vaporizes the surface, deflecting the asteroid in reaction to the momentum of the expanding vapor plume. Although the impulse efficiency of this mode is low ($P/E < 10^{-5}$N-sec J^{-1}), it may be the only solution possible for metallic objects or asteroids with high electrical conductivities. In the second mode, appropriate for dielectrics with lower conductivity, microwave energy is deposited beneath the surface of the asteroid. The characteristic depth of penetration of microwave energy beneath the asteroid's surface is given by (Landau and Lifshitz 1960, p. 265)

$$l = \frac{\lambda}{2\pi \sqrt{\epsilon'} \tan \delta} \tag{14}$$

where λ is the microwave's wavelength in free space, ϵ' is the real part of the complex permittivity $\epsilon = \epsilon' + i\epsilon''$, and the loss tangent is $\tan \delta = \epsilon''/\epsilon'$. This equation is valid only for $\epsilon'' << \epsilon'$; it does not hold for metals. When energy is input at a constant rate the temperature rises until conductive losses balance the heating by microwave dissipation. The deeper the microwaves penetrate, the longer the time required. The characteristic heating time τ_0 is the thermal time constant time of this layer, $\tau_0 = l^2/\kappa$, where κ is the asteroid's thermal diffusivity. For example, adopting $\lambda = 3$ cm, $\epsilon' = 4$, $\kappa = 10^{-5}$ to 10^{-6} m^2 s^{-1} and a relatively absorptive dielectric ($\tan \delta = 0.1$), $l = 2.4$ cm and $\tau_0 = 60$ to 600 s, which is the length of time irradiation must occur for the temperature to build up to its maximum. In this case an average microwave flux intensity $q = 3 \times 10^6$ W m^{-2} would yield temperatures in the vicinity of 10^3 K after about 25 s of heating. For a more transparent dielectric ($\tan \delta = 10^{-4}$), $l = 24$ m and $\tau_0 > 10^8$ s (comparable to the total time available to deflect the asteroid, so the maximum temperature is never reached). Nevertheless, a temperature of 10^3 K would be achieved in about 2.4×10^4 s (about 7 hr of heating) throughout the entire thickness of the surface layer.

Experiments on microwave heating were carried out at microwave flux densities of up to 10^{11} W m^{-2} and wavelengths of 3 and 12 cm in both pulsed and continuous irradiation modes. We found that Plexiglas targets crack and spall around the region of maximum energy release, which is some distance below the surface. A cavity sometimes develops in this region under the action of both thermoelastic stresses and vapor pressure. The formation of a cavity depends on the energy flux density; at 3×10^6 W m^{-2} it requires several times 10^7 J m^{-2} to open one. Cavity formation and rupture, in addition to the direct spallation, may eject large amounts of mass at velocities of tens of m s^{-1} and thus provide a relatively large impulse for the energy absorbed. Values of P/E for this process approach 10^{-4} to 10^{-3} N-sec/J, higher than that measured for lasers and perhaps also higher than for a solar collector. The success of this method requires that asteroids are relatively transparent to microwaves. Of the principal classes of meteorites, iron, stony-iron, and stony, the greatest depth of penetration occurs for the last. Some experimental work was also performed on the L-chondrite Tzarev. The measured microwave reflection coefficient was in the range 0.55 to 0.35 in the frequency range of 10 to 20 GHz, but a complete measurement of the complex permittivity was not possible. Heating rates of 5 ± 3 K s^{-1} were observed at flux densities of 1.5 to 3×10^5 W m^{-2}.

Phased array antennas can be used to focus microwave beams. However, a fundamental limitation on the beam is that its angular divergence (in radians) is of order of the wavelength λ divided by the array aperture a. To focus microwaves on a spot 100 m in radius at a distance of only 0.003 AU requires a phased array 160 km in diameter. The total radiated power must exceed about 10 GW for energy fluxes on the asteroid to reach $10^6 W\, m^{-2}$.

The creation of such huge laser or microwave systems with small beam divergences is difficult even to imagine in the near future. But we must keep in mind the possible progress in laser or microwave techniques. This approach has at least one major advantage—the system can be built beforehand at the Earth or at the Moon or at a special space construction site. A laser or microwave deflection system based near the Earth or Moon is well suited to the deflection of small bodies (100–200 m in diameter) which are more difficult to detect at large distances from the Earth but whose impacts may still be very damaging. Deflection of larger objects requires mobile laser or microwave generators that can be transported to the near vicinity of the asteroid. Such a system thus requires both a propulsion system and a source of energy, perhaps a solar-electric power system. A system of this type would seem to have few advantages over the solar collector approach.

VII. CONCLUSIONS

A number of viable approaches exist for deflecting Earth-threatening asteroids in the 1 to 10 km diameter range without the use of large nuclear weapons. Although all of these approaches probably require more development and technical skill than the use of a nuclear explosive, the danger inherent in the mere existence of an arsenal of large nuclear weapons makes the nuclear approach very questionable. Of the different non-nuclear deflection schemes, the solar collector approach promises to be the easiest to realize in the near term, if the problem of secondary mirror degradation by ejecta can be solved satisfactorily. This approach requires little in the way of new technology, although construction of such a system would still be a formidable challenge to our current space capabilities. A small (40 m diameter) solar collector has already been deployed in space as part of an experimental program aboard the Russian Mir space station on the night of 3–4 February 1993, and the Russian plans eventually envision collectors up to 600 m in diameter. Large laser systems or mass drivers are also capable of deflecting asteroids with high efficiency, although such systems require still more technical development; perhaps in the more distant future systems of this kind will be found superior. Kinetic energy deflectors might be made effective through the "billiard shot" strategy, but they will always suffer from the danger of fragmenting the asteroid rather than just altering its path. Direct experience with collisions on asteroids is probably needed before this danger can be evaluated. Solar sails seem to have too many problems to consider seriously. Microwaves are attractive from the point of view of efficiency, but because of their long wavelengths compared to laser light require vast space constructions to focus them even over modest solar system distances.

It is thus clear that nuclear weapons are not the only possible solution to the problem of deflecting "killer" asteroids. Inasmuch as nuclear weapon deflection systems are likely to offer more hazard to the human race than the asteroids they are designed to deflect, it seems logical to bend our energies to extending and refining the current work on non-nuclear deflection systems.

Acknowledgments. The authors wish to thank A. G. Petschek and R. H. Frisbee who made a number of important comments on the solar collector approach and to two anonymous referees who did not like the solar collector at all. We hope that we have adequately addressed their concerns. We are grateful to V. A. Rybakov and S. A. Medvedyk for conducting the laser experiments and I. B. Monastyrsky and E. V. Ratnikov for their participation in the microwave experiments.

REFERENCES

Ahrens, T. J., and Harris, A. W. 1992. Deflection and fragmentation of near-Earth asteroids. *Nature* 360:429–433.

Asphaug, E., and Melosh, H. J. 1993. The Stickney impact of Phobos: A dynamical model. *Icarus* 101:144–164.

Belton, M. J. S., Veverka, J., Thomas, P., Helfenstein, P., Simonelli, D., Chapman, C., Davies, M. E., Greeley, R., Greenberg, R., Head, J., Murchie, S., Klaasen, K., Johnson, T. V., McEwen, A., Morrison, D., Neukum, G., Fanale, F., Anger, C., Carr, M., and Pilcher, C. 1992. Galileo encounter with 951 Gaspra: First pictures of an asteroid. *Science* 257:1647–1652.

Canavan, G. H., Solem, J. C., and Rather, J. D. G., eds. 1992. *Proceedings of the Near-Earth-Object Interception Workshop* (Los Alamos: Los Alamos National Lab).

Crifo, J. F. 1987. Improved gas-kinetic treatment of cometary water sublimation and recondensation: Application to comet P/Halley. *Astron. Astrophys.* 187:438–450.

Dodd, R. T. 1981. *Meteorites* (Cambridge: Cambridge Univ. Press).

Friedman, L. 1988. *Starsailing* (New York: J. Wiley & Sons).

Hashimoto, A. 1990. Evaporation kinetics of forsterite and implications for the early solar nebula. *Nature* 347:53–55.

Holsapple, K. A., and Schmidt, R. M. 1982. On the scaling of crater dimensions—2. Impact processes. *J. Geophys. Res.* 87:1849–1870.

Housen, K. R., Schmidt, R. M., and Holsapple, K. A. 1983. Crater ejecta scaling laws: Fundamental forms based on dimensional analysis. *J. Geophys. Res.* 88:2485–2499.

Kerr, R. A. 1993. Galileo reveals a badly battered Ida. *Science* 262:33.

Kömle, N. I., Steiner, G., Dankert, C., Dettleff, G., Hellmann, H. Kochan, H., Baguhl, M., Kohl, H., Kölzer, G., and Thiel, K. 1991. Ice sublimation below artificial crusts: Results from comet simulation experiments. *Planet. Space Sci.* 39:515–524.

Landau, L. D., and Lifshitz, E. M. 1960. *Electrodynamics of Continuous Media* (New York: Pergamon Press).

Melosh, H. J. 1989. *Impact Cratering: A Geologic Process* (New York: Oxford Univ. Press).

Melosh, H. J., and Nemchinov, I. V. 1993. Solar asteroid diversion. *Nature* 366:21–22.

Michaelis, M. M., Dempers, C. A., Kosh, M., Prause, A., Notcutt, M., Cunningham, P. F., and Waltham, J. A. 1991. A gas-lens telescope. *Nature* 353:547–548.

O'Neill, G. K. 1977. *The High Frontier: Human Colonies in Space* (New York: Morrow).

Peale, S. J. 1989. On the density of Halley's comet. *Icarus* 82:36–49.

Phipps, C. 1992. Laser deflection of NEO's. In *Proceedings of the Near-Earth-Object Interception Workshop*, eds. G. H. Canavan, J. C. Solem and J. D. G. Rather (Los Alamos: Los Alamos National Lab), pp. 256–260.

Schmidt, R. M., and Housen, K. R. 1987. Some recent advances in the scaling of impact and explosion cratering. *Intl. J. Impact Eng.* 5:543–560.

Scotti, J., and Melosh, H. J. 1993. Estimate of the size of comet Shoemaker-Levy 9 from a tidal breakup model. *Nature* 365:733–735.

Shoemaker, E. M., Wolfe, R. F., and Shoemaker, C. S. 1990. Asteroid and comet flux in the neighborhood of Earth. In *Global Catastrophes in Earth History*, eds. V. L. Sharpton and P. D. Ward, Geological Soc. of America Special Paper 247 (Boulder: Geological Soc. of America), pp. 155–170.

Stone, J. L. 1993. Photovoltaics: Unlimited electrical energy from the sun. *Physics Today* 46:22–29.

Tonks, W. B., and Melosh, H. J. 1992. Core formation by giant impacts. *Icarus* 100:326–346.

Walford, W. T., and Winston, R. 1989. *High Collection Nonimaging Optics* (San Diego: Academic Press).

Whipple, F. L. 1950. A comet model. I. The acceleration of Comet Encke. *Astrophys. J.* 111:411–418.

Wright, J. L. 1992. *Space Sailing* (Philadelphia: Gordon and Breach).

PART VIII
Considerations for Future Work

THE IMPACT HAZARD: ISSUES FOR THE FUTURE

DAVID MORRISON
NASA Ames Research Center

and

EDWARD TELLER
The Hoover Institution

The reality of cosmic impacts and their hazard is now widely accepted. In this chapter we discuss some of the public policy implications of this knowledge, including issues associated with the development and possible testing of defensive technologies. Such issues should be the subject of vigorous public debate, carried out in an open manner with wide international participation.

I. REALITY OF THE HAZARD

Recognition that impacts from asteroids and comets constitute a credible and significant hazard to life and property is a recent phenomenon. Indeed, the concept of hypervelocity impacts is itself a rather new idea, and it was not until the second half of the present century that the impact origin of the lunar craters was generally accepted in the scientific community. Today, however, the fact of rare cosmic impacts on the Earth and other planets is unquestioned, and the interpretation that such impacts constitute a danger to both individuals and society as a whole is gaining acceptance (Chapman and Morrison 1994; Teller 1993). We have now reached the third stage, that of seriously discussing ways to protect ourselves from cosmic impacts (Ahrens and Harris 1992; Chapters by Canavan et al. and by Ahrens and Harris).

The impact hazard is qualitatively different from other natural hazards in three ways:

1. Destructive impacts are exceedingly rare, with the intervals between even the smallest such events amounting to many human generations (Shoemaker 1983; Chapman and Morrison 1994). No one alive today, therefore, has ever witnessed such an event, and indeed there are no credible historical records of human casualties from impacts in the past millennium. Consequently, it is easy to dismiss the hazard as negligible or to ridicule those who suggest that it be treated seriously (Chapter by Morrison et al.).

2. Cosmic impacts are capable of producing destruction and casualties on a scale that far exceeds any other natural disasters; the results of impact by an object the size of a small mountain exceed the imagined holocaust of a full-scale nuclear war (Chapter by Toon et al.). Even the worst storms or floods or earthquakes inflict only local damage, while a large enough impact could have global consequences and place all of society at risk (Chapter by Morrison et al.). Impacts are, at once, the least likely but the most dreadful of known natural catastrophes.

3. It is possible to avoid many (perhaps most) impacts by the application of current technology (Chapter by Canavan et al.). We cannot tame a typhoon or suppress an earthquake, but we do have the means to detect most incoming projectiles and either deflect or destroy them (Ahrens and Harris 1992; see also their Chapter). Therefore, investigation of the impact hazard can and should naturally lead to a practical program to mitigate these hazards.

Although no destructive impact has been documented in historical times, the credibility of the impact hazard is based on several concrete events or discoveries. The example of the 1908 Tunguska event with its explosive yield in excess of 10 megatons of TNT is sobering (Krinov 1963); had this impact taken place in a heavily populated region instead of the wilderness of Siberia, the consequences would have been grim indeed. In addition, the discovery that the mass extinction of the dinosaurs and many other species at the end of the Cretaceous was caused by an impact of the order of 10^8 megatons (Alvarez et al. 1980) has sensitized the scientific community to the destructive potential of large impacts and aroused widespread public attention. Finally, the increasing rate of discovery of NEOs (near-Earth objects), and especially of small Earth-crossing asteroids, permits quantitative predictions of contemporary impact rates (Shoemaker 1983). All of these factors contribute to the current public concern over impact risk and the attention that official scientific and governmental bodies have directed to this issue (Chapter by Park et al.).

Congressional perception of these issues was succinctly stated for the public record by the House Committee on Science and Technology in the NASA Multiyear Authorization Act of 1990: "The Committee believes that it is imperative that the detection rate of Earth-orbit-crossing asteroids must be increased substantially, and that the means to destroy or alter the orbits of asteroids when they do threaten collisions should be defined and agreed upon internationally. The chances of the Earth being struck by a large asteroid are extremely small, but because the consequences of such a collision are extremely large, the Committee believes it is only prudent to assess the nature of the threat and prepare to deal with it."

II. PUBLIC POLICY ISSUES

The nature of NEOs and the hazard they pose are legitimate and interesting scientific subjects. They also raise issues of public policy, especially as we consider ways of preventing impacts. At the current stage, we cannot answer most of the policy questions, but it is appropriate to raise them for discussion.

Today we recognize that cosmic impacts have played an important (and possibly dominant) role in the evolution of life on Earth (see the Chapters by Smit and by Sheehan and Russell). Impacts by NEOs are a part of the natural environment of the planet. We believe that it is appropriate to advocate the application of technology to eliminate terrestrial impacts. Individuals can formulate their own answers to this question, but we cannot imagine society generally retreating from the opportunity to protect itself from such natural hazards. The development of technology in the past few centuries has been toward increasing understanding and control of natural forces in an effort to improve human life. Protecting ourselves against impacts is a natural extension of those trends, comparable to efforts to develop new drugs and treatments for disease.

We may also ask if our current understanding of the impact hazard warrants taking action, and in particular the expenditure of public funds. Can we justify raising an alarm about a danger that most of the public is unaware of and many consider to be ridiculous? Most of those who study NEOs are confident that current impact rates are known to within a factor of a few for the size range of interest (roughly 50 m to 20 km diameter), and therefore that we can calculate impact probabilities to a similar level of accuracy (see the Chapter by Rabinowitz et al.). It also seems clear that, while we can carry out such statistical calculations, we have not yet made a serious effort to survey and inventory the potentially dangerous Earth-crossing asteroids (see the Chapter by Bowell and Muinonen). Our current catalogs include fewer than 10% of the population that concerns us. Therefore it seems entirely appropriate to point out that the hazard is real and to urge that steps be taken to better quantify the risk, leading toward an international survey of the inner solar system for any NEOs that might constitute a present danger. Many of us are also willing to argue that expenditure of tens of millions of dollars per year in public funds for a program of NEO discovery and research is justified, relative to expenditure on other government programs in support of both science and risk reduction (see Chapters by Morrison et al., Harris et al., Canavan, and Weissman).

Improvement in our general understanding of the NEO population will come from current efforts to upgrade observing programs and introduce CCD technology on several existing telescopes (Chapter by Carusi et al.). However, a nearly complete inventory aimed at identifying specific threats requires the development of the Spaceguard Survey or its equivalent (Morrison 1992; Chapter by Bowell and Muinonen). Only if we actually look will we know if the statistical hazard we are currently discussing corresponds to a concrete threat to ourselves or our children.

Most people would agree that if an NEO of any size greater than a few tens of meters were found on a probable impact trajectory, the presence of this threat would stimulate an interest in countermeasures. A specific threat does wonders in focusing our thinking, and society often seems at its best when reacting to a perceived emergency. However, no such specific threat has yet been identified, and therefore we may ask how much effort and expenditure of funds should be applied to an abstract threat that may be—indeed is most likely to be—very remote in time.

Even if we concede that it is appropriate to expend funds on development of countermeasures, we must consider issues of priority. Do we know enough about NEOs to determine how they might be deflected or destroyed? Or should we first undertake a scientific program to understand them better, including direct exploration of NEOs by spacecraft? We must consider trade-offs between our desire to develop a robust defensive capability and the requirements for better understanding of the targets in advance of such engineering work.

A program of investigation of NEOs might include an accelerated survey to locate suitable targets, astronomical observations of comets and asteroids supported by laboratory investigations, and direct examination of NEOs by spacecraft. In 1986 an international flotilla of spacecraft investigated Comet Halley, and NASA plans a near-Earth asteroid rendezvous (NEAR) mission to asteroid Eros in 1996 (Chapter by Cheng et al.). However, there are at present no plans for landing on either a comet or asteroid or returning samples from such objects to Earth, so it is unlikely that we will obtain the engineering information we need within the next decade or more.

Many of these policy issues were discussed at an international meeting held in the spring of 1993 at Erice, Sicily. At the conclusion of the meeting, the participants issued the following summary statement:

We the undersigned participated in the Erice International Seminars on Planetary Emergencies: 17th Workshop: The Collision of an Asteroid or Comet with the Earth, meeting in Erice, Sicily from 28 April to 4 May, 1993. Following presentations and discussions on the threat to humanity from cosmic impact, the undersigned concur as to the following points: (1) Cosmic impact is an environmentally significant phenomenon which has played a major role in the evolution of life on Earth. (2) In any year there is a very low probability that a large cosmic impact may occur which would destroy human civilization or even a significant fraction of life on Earth. However, the threat is real and requires further internationally coordinated public education efforts. (3) A significant near-term cosmic impact threat identified at the Workshop is a naturally produced atmospheric explosion of a small NEO being mistaken for a nuclear explosion, at a time and place of international tension. These events have been observed. Such an event could be misinterpreted as a nuclear attack and trigger an unfortunate reaction. (4) The gathering of additional physical knowledge of NEOs and their effect on the Earth is a scientifically and socially important endeavor. These multi-disciplinary efforts should be conducted in an open coordinated international manner. Dedicated international astronomical facilities similar to the proposed Spaceguard System should be developed. The defense-related assets and technologies of the former cold war combatants can contribute to the gathering of these data, through ground and robotic space observation. These skills and technologies, necessary for any large, complex investigation, should be well utilized now that the threat of global thermonuclear war has been reduced. (5) The study of potential mitigation systems should be continued. Many of us believe that unless a specific and imminent threat becomes obvious, actual construction and testing of systems that might have the potential to deflect or mitigate a threat may be deferred because technology systems will improve.

III. PROPOSALS CONCERNING EXPERIMENTATION

One of the more divisive issues raised in this debate is that of testing a defensive system. Great divergence of opinion is to be expected when we discuss development of technology for the deflection of NEOs. In the Erice statement quoted above this is the one issue upon which not all the participants could agree. Development and testing were also strongly criticized by Sagan (1992).

If one accepts the need for experimentation to develop the technical means to deflect an asteroid or comet, there remain questions concerning the proper approach to such experimentation. In the absence of a specific threat, we would not wish to undertake active experiments until we obtained additional information on the physical properties of comets and asteroids, presumably through space missions to representative bodies (see the Chapters by Chapman et al., Rahe et al., Cheng et al., and Nozette et al.). We would also require careful investigation to ensure that testing of deflection technologies be done in ways that minimize the possibility of the object or its fragments ever impacting the Earth. In general, this means selecting targets whose orbits do not bring them close to our planet.

Several technical approaches to deflection have been considered (see Chapters by Canavan et al., Ahrens and Harris, and Melosh et al.). These include (a) deflection by steady ejection of material by heating or a mass driver; (b) deflection by high velocity impact; (c) deflection by surface or subsurface detonation of a chemical explosive; and (c) deflection (or destruction) by a nuclear explosive.

The first of these—the continuous application of low thrust—is a technique that lies beyond current technology. While such an approach may be important in the future for the utilization of asteroidal material (see Chapters by Hartmann and Sokolov, and Melosh et al.), it is not practical as a near-term deflection technique. The second (kinetic energy) and third (conventional explosives) are also discussed by Melosh et al. in their Chapter. The advantage of kinetic energy deflection is that the energy deposited into the target could exceed up to a thousandfold the energy of an equal weight of high explosive. Both approaches require greater knowledge of the physical properties of NEOs, and for purposes of our discussion can be discussed with the fourth option, that of nuclear explosives. The use of nuclear explosives is obviously the most controversial option. Any use of nuclear explosives in space, either for testing or for actual deployment against a threatening NEO, will require international consensus and modification or exception to current international agreements. However, nuclear explosives provide the only means to deliver the required deflection impulse to a large (>1 km) asteroid or comet with current launch vehicles (Ahrens and Harris 1992).

If one is to conduct a test program, where should it be done? Because the greater hazard is associated with the larger (>1 km) objects (Morrison 1993; Chapman and Morrison 1994), experimentation with NEOs of this size scale should have priority. If the testing were restricted to cis-lunar space, it would be necessary on average to wait more than a millennium for one of the currently known NEOs larger than 1 km. Even after a Spaceguard Survey has discovered all such NEOs, the average waiting time would be of order a century (Chapter by Bowell and Muinonen). On the other hand, if we do not restrict ourselves to near-Earth space, missions with existing launch vehicles to reach currently known large NEOs are possible nearly every year (Morrison and Niehoff 1979; Veverka et al. 1989; Chapter by Cheng et al.), and when the Spaceguard Survey is complete an order of magnitude more potential targets will be available. Thus for reasons of access alone, experiments should not be restricted to the vicinity of the Earth. In addition, it is simpler to ensure that no debris from an experiment will impact our planet if the experiments are conducted on distant NEOs, perhaps Amors which are not currently in Earth-crossing orbits.

Experiments involving surface or subsurface explosives or hypervelocity impacts will produce ejecta. Indeed, it is the momentum carried by these ejecta that provide the means to alter the orbit of the NEO itself. Most such ejecta will be too small to constitute a danger even if the fragments were to impact the Earth. However, any experimental program must guard against the possibility of large fragments that could themselves produce a hazard. A stand-off nuclear explosion is less likely to generate large fragments (Ahrens and Harris 1992). One of the objectives of an experimental program would be to determine the efficacy of stand-off nuclear explosions as a means of altering NEO orbits.

We presume that any testing program would involve several impulses of differing magnitude applied to the target object. The immediate effects of the explosions would be observed from a safe distance by several appropriately instrumented rendezvous spacecraft, and between explosions the NEO could be observed at close hand and any changes in its orbit precisely determined before the next phase of the program. It would also be of interest to determine the intrinsic strength and interior structure of any target NEO, and to carry out such experiments on a variety of NEOs (e.g., a metallic asteroid, a stony asteroid, and an inactive or extinct comet nucleus). Considerable study would be required before a recommendation should be made concerning such experiments in NEO deflection. In the end, any such measures must be taken by an appropriate international body. It is our expectation that it would require many years—perhaps decades—to gain international approval to implement a testing program. The political problems associated with this activity are a greater challenge than the technical ones. Discussion of a possible testing program is itself an unprecedented experiment in international politics, and we expect that the experience gained in the process of discussion of proposals to carry out a venture of this magnitude for the common protection of our planet would be as valuable as results of the experiments themselves. Conversely, it is as risky to defer international collaboration until a specific threat is identified as it would be to defer experimentation until an emergency is upon us.

IV. THE DECISION-MAKING PROCESS

One of us (E. T.) had two major experiences relevant to the question of what kind of action scientists should take in connection with a policy problem such as the impact hazard. One experience was the dramatic development of theoretical physics in central Europe at the beginning of this century. It had two major characteristics: It produced great and surprising additions to our knowledge in the form of relativity and quantum mechanics; and it cost practically nothing. The second experience was connected with the coming of Hitler and of the Second World War together with military-technological innovations including the atomic bomb. These developments involved the expenditure of great amounts of public money. These two developments have greatly influenced our world in this century. One of the unfortunate consequences was secrecy in government and science and reliance on a government-science-military elite for important decisions.

During the Second World War, this secrecy was on the whole justified. After the Second World War, the justification for such secrecy declined, but in practice the apparatus for classifying results of scientific research and confining debate on important issues to members of a government-science-military elite continued. Fortunately, developments in world politics during the last couple of years have eliminated most justifications for official secrecy and strengthened the arguments for openness.

This history is relevant to the present discussion because in a regime of secrecy, where information is available to relatively few individuals, recommendations by small groups of people can have very great weight. For many decades, we have seen the consequences of decision-making by relatively small groups. We should be deeply grateful that the outcome of international developments during the past decade have been favorable to a pluralistic, rather than restricted, discussion of issues of broad international significance, such as the impact hazard.

In the present situation, there can be little doubt that the important decisions connected with the danger of asteroid impact should be made by open democratic means. Under these circumstances, scientists and engineers should limit themselves to the finding and publication of relevant facts. Of course, interpretations and value judgments are useful and important as well. But the resolution of points of disagreement and the formulation of policies to deal with the impact issue must reside with the population at large through their legitimate representatives. As experts, we must participate in this process, but we cannot and should not dictate the outcome of the public discussions that have already begun on this issue. It is the job of the scientist to find and explain the facts. It is the job of the citizen to apply value judgments and decide on strategy and funding levels.

In this spirit, we conclude this short chapter by recommending a principle for action in the present period. Our results should be widely publicized and explained, with secrecy restrictions abolished as completely and rapidly as possible. This principle holds particularly for issues associated with the use of nuclear energy. The present book and conference are an example of such open discussion and debate. Further, both the decisions and implementation of any programs to deal with the impact hazard should be shared by the international community. All part of the world are equally at risk from impacts, and we all share a common interest in our self-protection from such cosmic catastrophes. One of us (E. T.) urges that experimentation should not be delayed except for strong reasons, since procedures for protection need to be decided on the basis of data on comets and asteroids, part of which can be obtained only through experimentation.

REFERENCES

Ahrens, T. J., and Harris, W. A. 1992. Deflection and fragmentation of near-Earth asteroids. *Nature* 360:429–443.

Alvarez, L. W., Alvarez, W., Asaro, F., and Michel, H. V. 1980. Extra-terrestrial cause for the Cretaceous-Tertiary extinction. *Science* 208:1095–1108.

Chapman, C. R., and Morrison, D. 1994. Impacts on the Earth by asteroids and comets: Assessing the hazards. *Nature* 367:33–40.

Krinov, E. E. 1963. The Tunguska and Sikhote-Alin meteorites. In *The Moon, Meteorites, and Comets*, eds. B. M. Middlehurst and G. P. Kuiper (Chicago: Univ. of Chicago Press), pp. 208–234.

Morrison, D., ed. 1992. *The Spaceguard Survey: Report of the NASA International Near-Earth-Object Detection Workshop* (Pasadena: Jet Propulsion Laboratory).

Morrison, D. 1993. An international program to protect the Earth from impact catastrophe: Initial steps. *Acta Astron.* 30:11–16.

Morrison, D., and Niehoff, J. 1979. Future exploration of asteroids. In *Asteroids*, ed. T. Gehrels (Tucson: Univ. of Arizona Press), pp. 227–250.

Sagan, C. 1992. Between enemies. *Bull. Atomic Sci.* 48:24–26.

Shoemaker, E. M. 1983. Asteroid and comet bombardment of the Earth. *Annual Rev. Earth Planet. Sci.* 11:461–494.

Teller, E. 1993. Why now? In *Proceedings of the Near-Earth-Object Interception Workshop*, eds. G. H. Canavan, J. C. Solem and J. D. G. Rather (Los Alamos: Los Alamos National Lab), pp. 175–179.

Veverka, J., Langevin, Y., Farquhar, R., and Fulchignoni, M. 1989. Spacecraft exploration of asteroids: The 1988 perspective. In *Asteroids II*, eds. R. P. Binzel, T. Gehrels and M. S. Matthews (Tucson: Univ. of Arizona Press), pp. 970–993.

THE DEFLECTION DILEMMA: USE VERSUS MISUSE OF TECHNOLOGIES FOR AVOIDING INTERPLANETARY COLLISION HAZARDS

ALAN W. HARRIS
Jet Propulsion Laboratory

GREGORY H. CANAVAN
Los Alamos National Laboratory

CARL SAGAN
Cornell University

and

STEVEN J. OSTRO
Jet Propulsion Laboratory

A system capable of deflecting a near-Earth object (NEO, an asteroid or comet in Earth-approaching orbit) out of an Earth-impacting trajectory could also be used to deflect a nonmenacing NEO so it impacts the Earth. We calculate the expected frequency of opportunities to misuse a deflection system as a function of NEO diameter, the capability of the putative deflection system, and the fraction of the full near-Earth asteroid (NEA) population that is known. Our principal result, which is nearly independent of other assumptions, can be simply stated: the frequency of opportunities to misuse a deflection system, for NEAs of a given size, is $\sim 100(\Delta v)^2$, or $\sim 1/t_r^2$, times the natural impact frequency with the Earth of NEAs of the same size. Here Δv is the deflection velocity in meters per second that a hypothetical system is capable of achieving; equivalently, t_r is the time in years that the given system needs to deflect an object by one Earth radius, i.e., the response time required in legitimate use. For a system that would be effective against objects discovered only days or weeks before impact, opportunities for misuse might be so frequent as to be continuously present. For a less capable system, the frequency of opportunities for misuse may be only once a century or less, but still more frequent than the need to use it. Unwillingness or inability to develop a deflection capability in advance of need leaves us vulnerable to that fraction of NEO impacts by bodies (mainly long-period comets) that may not be discovered with enough lead time to construct a defensive system. But the potential for misuse of a system built in advance of an explicit need may in the long run expose us to a greater risk than the added protection it offers. This is the deflection dilemma.

I. INTRODUCTION

In its annual motion about the Sun, the Earth moves through a cloud of asteroids and comets in orbits which cross the Earth's, the near-Earth objects (NEOs). Occasional collisions with members of this population are inevitable. Once we recognize that collisions with NEOs larger than a few hundred meters in diameter could threaten the global civilization, means for mitigating this threat seem clearly worth considering. Deflection methods which have been discussed (cf., Ahrens and Harris 1992; Chapters by Ahrens and Harris and by Canavan) include mass driver engines propelling reaction mass into space, high speed collisions, and sub-surface or standoff thermonuclear explosions. These same capabilities can in principle be used to alter the orbit of an object on a nonintercepting trajectory so that it does impact the Earth. Some have warned (Sagan 1992) that through negligence, fanaticism, or madness, the technology to deflect asteroids and comets might be used to generate a global catastrophe on a time scale much shorter than the waiting time for the natural catastrophe that this technology is intended to circumvent. Those who take seriously a probability of 10^{-3} in a century of a catastrophic asteroidal impact must surely take seriously, say, a probability of order unity in a century that an opportunity will exist to misuse deflection technology to cause such an impact. The cure, it is suggested, may be worse than the disease.

In this chapter, we calculate the expected frequency of opportunities to misuse a deflection system as a function of NEO diameter, the capability of the putative deflection system, and the fraction of the NEA population that is known—that is, the number of nearby bodies available for misuse. We find that opportunities to misuse a deflection system are much more frequent than are occasions to use it for its intended purpose. That result is very robust in that it depends on very few auxiliary assumptions. It is expressly not our purpose here to discuss the plausibility that some nation or group might seize an opportunity to deflect an NEO toward the Earth. Nor do we address in any detail the technical feasibility or cost of developing a deflection system. Such systems are physically quite possible in terms of energy and momentum considerations, so the potential capability to develop and misuse such systems must be taken seriously.

II. OFFENSIVE USE OF A DEFLECTION SYSTEM

Consider a system capable of deflecting an asteroid by a velocity increment Δv. According to Ahrens and Harris (1992; Chapter by Ahrens and Harris), in a time $t < P/2\pi$ (where P is the period of the asteroid) the displacement is $\sim(\Delta v)t$. For $t > P/2\pi$, the displacement along-track is $\sim 3(\Delta v)t$, and across-track, $\sim(\Delta v P/2\pi)$. Thus the requirement that such a system be capable of deflecting an incoming asteroid by a distance of the order of the Earth's radius,

R_E, requires a response time t_r for applying Δv of:

$$t_r \approx \frac{R_E}{\Delta v} \tag{1}$$

where $t_r < P/2\pi$ and

$$t_r \approx \frac{R_E}{3\Delta v} \tag{2}$$

where $t_r > P/2\pi$. In the case of $t_r > P/2\pi$, we have assumed a displacement along-track because it is most efficient. Note that the above relations between t_r and Δv are very conservative. Because of inevitable uncertainties in the exact trajectory of a threatening object, one would no doubt want to deflect it by a comfortable margin, perhaps several Earth radii; thus, a realistic value of t_r for a given value of Δv, or vice-versa, might be several times larger than given by the above expressions.

Now consider the possibility of offensive misuse of the same system. In a time T, by how far might the same system displace an available asteroid toward the Earth? This case differs in one subtle way from the above, because the option to displace along-track instead of across-track is not a free choice— one must displace the asteroid in a prescribed direction to hit the Earth. Or to consider the matter in reverse, the areal phase space from which an asteroid could be diverted from its natural course so it hits the Earth has an across-track dimension

$$\delta_\perp \approx \Delta v \left(\frac{P}{2\pi} \right) \tag{3}$$

and an along-track dimension

$$\delta_\parallel \approx 3(\Delta v)T. \tag{4}$$

In this case, we have assumed that $T > P/2\pi$. Thus the area of the phase space from which one might choose potential deflectable asteroids is

$$A \sim \pi \delta_\perp \cdot \delta_\parallel \approx \frac{3}{2}(\Delta v)^2 PT. \tag{5}$$

The collisional cross section of the Earth, allowing for the gravitational focusing of bodies approaching at about 10 km s^{-1}, is $A_E \approx 2\pi R_E^2$. The ratio A/A_E is equal to the frequency with which an asteroid passes within a divertable range of the Earth (for a system capable of diverting by Δv) to the natural frequency of collisions with the Earth of objects of the same size. In other words, this is the ratio of the chance that a system *could* be misused to the chance that it would be needed for the legitimate task of diverting an asteroid. One further factor might be taken into account. The collision frequency of asteroids hitting the Earth does not depend on whether or not we know about them. But only asteroids discovered in advance are available for deflection toward the Earth. Thus the ratio of "chance to misuse" to "need to deflect" is $\Omega = fA/A_E$, where f is the fraction of the asteroid population in question which has been discovered and tracked:

$$\Omega = \frac{\text{chance to misuse}}{\text{need to deflect}} \approx \frac{fA}{A_E} \approx \frac{3f(\Delta v)^2 PT}{4\pi R_E^2}. \tag{6}$$

For purposes of discussion, we evaluate the above expression, using a typical NEA orbit period, $P \sim 4$ yr, and misuse deflection time, $T \sim 4$ yr. An upper limit to T is about 10 yr, a time in which other nations could detect a deflection and have time to develop their own countermeasures. Thus we obtain:

$$\Omega \approx 100 f \Delta v^2 \tag{7}$$

where ΔV is in m s^{-1}. We can express the above ratio in terms of the required response time, t_r, for legitimate use of the hypothesized system, by substituting Eq. (1) or (2) into Eq. (7):

$$\Omega \approx \frac{(0.4 \text{ to } 4)f}{t_r^2} \tag{8}$$

where t_r is in years. Within the parentheses, the coefficient 0.4 applies for short response time ($t_r < P/2\pi \sim 0.5$ yr), and 4 for longer response times, with a smooth transition in between.

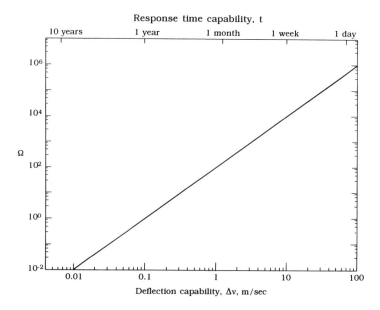

Figure 1.   The ratio Ω of the frequency of opportunities to misuse a deflection system to the frequency of occasions requiring the use of such a system, as a function of the capability of the system in terms of the Δv it is capable of delivering to an asteroid (bottom scale) or, equivalently, the time t_r required to move an asteroid by one Earth radius at that Δv (top scale).

Figure 1 is a plot of Ω vs Δv from expression (7), for $f = 1$. We will discuss other values of f in Sec. IV, below. The equivalent scale in units of t_r across the top of the figure is derived from Eqs. (1) and (2) blended smoothly between short and long t_r. It is noteworthy that for a defense system with a response time t_r of the order of 1 yr, *the frequency of opportunities for misuse is about the same as the frequency of situations requiring the use of such a system.* This result is independent of almost all assumptions, such as the collision frequency itself. For even shorter response times (that is, systems capable of larger Δv), the frequency of opportunities for misuse can greatly exceed the frequency of need to use such a system.

III. DEFLECTION VELOCITY

What is a plausible range of possible deflection velocities that may be achievable by a deflection system? Ahrens and Harris (1992; Chapter by Ahrens and Harris) estimate that even employing the technologically easiest method, stand-off nuclear explosions, a 1 km asteroid can be diverted ~ 1 m s^{-1} with an explosive energy of about 1 to 10 MT. The deflection velocity Δv is proportional to the explosive energy and inversely proportional to the asteroid mass, thus:

$$\Delta v \approx \frac{E}{3D^3} \qquad (9)$$

where E is the explosive energy in megatons and D is the asteroid diameter in km.

It is not possible to apply a single impulse Δv greater than about the object's surface escape velocity without disrupting the body rather than deflecting it in one piece. The surface escape velocity from a sphere of diameter D is

$$v_e = \left(\frac{2\pi G\rho}{3} \right)^{1/2} D \approx (0.65 \text{ m s}^{-1})(D \text{ in km}). \qquad (10)$$

One can imagine using multiple impulses. Indeed, for accurately "herding" an asteroid toward the Earth, this would be a necessity. Thus the maximum deflection velocity achievable is

$$\Delta v \approx n v_e \approx n(0.65 \text{ m s}^{-1})(D \text{ in km}). \qquad (11)$$

where n is the number of impulses applied. At least a few impulses would be required just to achieve the needed accuracy, and more than a few hundred might become impractical.

In Fig. 2, we have plotted the limits on Δv derived from Eqs. (9) and (11), as a function of asteroid diameter, for $E = 10$, 100 and 1000 MT total explosive energy, and for $n = 5$, 50 and 500 impulses. In the following discussion, we will take $E = 100$ MT and $n = 50$ to define a nominal limit on Δv vs diameter (solid line in Fig. 2), but results for other assumptions can be easily derived from the figures. We note that sizable values of Δv (>1 m s^{-1}) can be

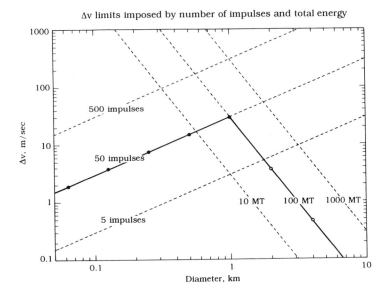

Figure 2.   Limits on deflection velocity Δv imposed by the number of impulses n required (such that each individual impulse is less than the surface escape velocity of the object), and the total explosive energy E required to achieve the deflection. In the discussion, we assume nominal limits of $n = 50$ impulses and $E = 100$ MT (solid lines). Values for each of the size objects considered in Figs. 3 and 4 are indicated by dots.

obtained for large asteroids ($D>1$ km) with only a few 10 MT weapons, such as were once the mainstays of the U. S. and Soviet nuclear arsenals.

IV. POPULATION OF NEAs AVAILABLE FOR MISUSE

In order to apply a system to deflect an asteroid toward the Earth, one must discover and track enough bodies to have an available divertable asteroid in a reasonable amount of time. How many such bodies might one expect to be available? Using estimates of the total population of NEAs vs size (Morrison 1992; Chapter by Rabinowitz et al.), we have derived the differential populations in factor-of-two size bins for all NEAs larger than 50 m in diameter. Taking the Spaceguard survey (Morrison 1992; Chapter by Bowell and Muinonen) as a representative example of a possible search program, we have computed the number of NEAs in each size bin that would be discovered as a function of time. From the graphs and tables given in the above references, we find that the fraction f, or completeness of the survey, can be well represented as an exponential function:

$$f = 1 - e^{-t/t_0} \tag{12}$$

where t is time and t_0 is the characteristic time scale of discovery. Table I lists the results of these analyses. The collision frequencies listed are estimated as (total number in size bin)$\times 4.2 \times 10^{-9}$ yr^{-1} (cf., Chapter by Rabinowitz et al.). The final column is the fraction f, which appears in Eqs. (6), (7) and (8).

Note the flatness of the discovery spectrum. Over a range of one order of magnitude in diameter, from ~ 0.1 to ~ 1.0 km, the population of NEAs varies by nearly 3 orders of magnitude; yet the number of NEAs discovered after 10 yr varies by only a factor of 5. A fair fraction of this decrease with increasing size is due to the asymptotic approach to completeness in the larger size bins.

From the results in Table I, we can estimate the frequency with which objects over a range of sizes might be divertable toward the Earth, as a function of the capability of a putative deflection system. Figure 3 is a plot of that frequency, using the assumed results of a ten-year Spaceguard survey as an illustration. Note that over the range of size from ~ 0.1 to ~ 1.0 km, the relative frequency of possible misuse at any given value of Δv varies by only about one order of magnitude. For each size object considered, we have indicated the limiting value of Δv from Fig. 2 (solid dots for $n \leq 50$, open circles for $E \leq 100$ MT).

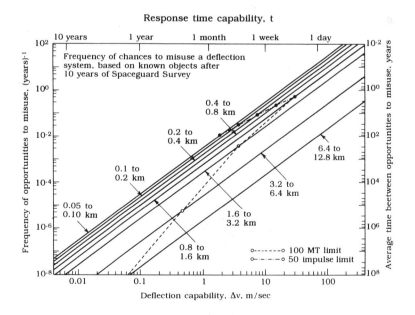

Figure 3.   The frequency of opportunities to misuse a deflection system for various NEA diameters vs the deflection capability Δv or, equivalently, the deflection response time t_r. We indicate, from Fig. 2, the maximum deflection velocity for each size object that can be achieved by 50 impulses (filled circles), or by 100 MT total explosive impulse (open circles). This plot is based on the fraction of the NEA population which could be discovered in ten years by the Spaceguard survey.

TABLE I
Number of NEAs and Number Discovered in 10 Years by Spaceguard

Diameter Range (km)	Total Number of NEAs	Collision Frequency (yr^{-1})	Characteristic Time Scale of Discovery t_0 (yr)	Number Discovered after 10 yr	Fraction f Discovered after 10 yr
0.05–0.10	1,700,000	7.1×10^{-3}	2400	7100	0.0042
0.1–0.2	250,000	1.1×10^{-3}	470	5400	0.021
0.2–0.4	50,000	2.1×10^{-4}	130	3600	0.073
0.4–0.8	11,000	4.8×10^{-5}	43	2400	0.207
0.8–1.6	2,600	1.1×10^{-5}	14	1400	0.523
1.6–3.2	700	2.9×10^{-6}	4.5	620	0.89
3.2–6.4	60	2.5×10^{-7}	1.5	60	1.0
6.4–12.8	5	2.1×10^{-8}	0	5	1.0

Finally, one can estimate the frequency of opportunities to misuse a deflection system as a function of survey completeness. Using Eq. (12) and the data in Table I, we have computed this frequency as a function of duration of a Spaceguard-level survey, for each size bin, assuming the limiting values of Δv shown in Fig. 3. One should not take the time scale literally, because time intervals toward the right side of the plot are long compared to the expected rate of advance of technology. At the left margin, we see the frequency of opportunities to misuse a deflection system with our present level of survey completeness and at the right margin the frequency given complete knowledge of the NEA population in each size bin. Figure 3 can be thought of a "snapshot" cutting across Fig. 4 (cf., the dots in the two figures), to show the dependence on Δv at a given time. We note, for example, that the opportunity to deflect a 1 km NEA into an Earth-impact trajectory presents itself today only about once a century (or 10^{-2} yr^{-1}), while after a decade of a Spaceguard-level survey opportunities present themselves about once a year.

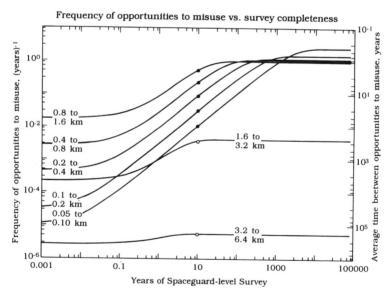

Figure 4. The frequency of opportunities to misuse a deflection system for various NEA diameters vs the completeness of discovery of the NEA population, parameterized in terms of the Spaceguard survey estimated performance. For each size bin, we have taken the maximum Δv as given in Figs. 2 and 3. The left margin corresponds to present-day knowledge of the population, and the right margin to complete knowledge.

It may be instructive to consider a couple of cases based on currently known NEAs. Yeomans and Chodas (see their Chapter) list all known Earth approaches by comets or asteroids to within 10 lunar distances for the interval 2001 to 2200. In addition to the closest approach, they list the minimum separation of the orbits at the time of encounter, necessary for estimating $\Delta v_\perp$ (the cross-track deflection velocity required) which is generally the larger component. The asteroid 4179 Toutatis will pass within about 0.01 AU from the Earth in 2004. The two orbits miss each other by only 0.006 AU, requiring (from Eqs. 3 and 4) $\Delta v_\perp \approx 45$ m s^{-1}. This could be applied as little as a year in advance. The along-track adjustment required to cause a collision is only $\delta_\parallel \approx 0.008$ AU, so from Eq. (4), $\Delta v \approx 1$ m s^{-1} if applied 10 yr in advance, but is still only a fraction of $\Delta v_\perp$ even if applied only 1 year in advance. Toutatis is ~ 4 km in diameter. Thus, by Eq. (9), the total explosive energy required to deflect it by ~ 50 m s^{-1} is $\sim 10^4$ MT. This is an aggregate yield about equal to the present global stockpile of nuclear arms. Referring to Fig. 4, note that for bodies as large as 3.2 to 6.4 km in diameter, our present knowledge of the population is nearly complete, so further surveying will not change the statistics much. From Fig. 3, note that the 3.2 to 6.4 km diameter line at a Δv of 50 m s^{-1} indicates a frequency of possible misuse of about once per 10 yr. Hence the upcoming close pass of Toutatis by the Earth represents about an expected level of opportunity for misuse of a deflection system on a very large NEA.

Another example is 1991 OA, an asteroid ~ 1 km in diameter, which will pass within 0.015 AU of the Earth in 2070. What makes this close approach unusual is that it is the closest to intersection of the Earth's orbit with that of any other known object in the next century, 0.003 AU, and hence has the minimum $\Delta v_\perp$, ~ 23 m s^{-1}, needed to cause a collision. The impulse required to deflect 1991 OA into an impact trajectory can be delivered by only ~ 60 MT. From Fig. 4, the frequency of opportunities to deflect an object this size toward the Earth with a 100 MT delivered impulse should occur about every 50 yr, with our present knowledge of the population. This is again consistent with the fact that the "best" deflection opportunity now known occurs in 2070. However, note that with complete knowledge of the population, an opportunity to deflect a 1 km asteroid toward the Earth with 100 MT total impulse would occur every few years.

V. CONCLUSIONS

The possibility of misusing a deflection system depends strongly, and almost solely, on the capability of such a system (i.e., on the deflection Δv it is able to achieve or, equivalently, on the minimum response time, t_r, it requires). Several conclusions can be drawn.

1. A system of very low capability ($\Delta v \leq 0.1$ m s^{-1}), such as might suffice to deflect NEAs discovered long in advance of a collision event, poses minimal threat of being misused to deflect asteroids toward the Earth. On the other hand, the response time t_r required to move an asteroid away from the Earth with such a limited system is > 1 yr, thus calling into question the need to build such a system in advance of a discovery of an

object on a collision trajectory.

2. A system of moderate capability, $\Delta v \sim 1$ m s^{-1}, would have potential application for protecting against long-period comets, where the response time is about one year or less. The probability that such a system could be misused is small, but is about 100 times greater than the probability that it would need to be used.

3. A highly capable system, able to deflect an object with only a few days' warning, coupled with a Spaceguard-level search for NEAs, presents a virtual continuum of opportunities for misuse. Such a high-capability system is not required for deflection of large long-period comets (see conclusion 2 above). Its only legitimate application would be for very fast response to approaching small asteroids. Because such small objects constitute only a very small fraction of the NEO collision hazard, and because a deflection system effective for such objects has significant potential for misuse, it appears imprudent to build such a system—at least at this time.

Beyond protecting the Earth against impacting NEOs, there are other benign motivations for developing an asteroid orbital engineering capability (see, e.g., Lewis et al. 1993). Some authors (see, e.g., O'Leary 1977; Gaffey and McCord 1977) have proposed doing so to utilize mineral resources in asteroids, and Herrick (1979) suggested a scenario for crashing a part of the asteroid 1620 Geographos into Central America, to excavate a new Atlantic-Pacific canal. We must caution that any such orbital engineering systems present the same or greater risk for misuse or accidental mishap as a defensive deflection system.

Acknowledgements. We are grateful to J. Burns, D. Morrison, B. Murray and E. Shoemaker for helpful comments. The work at JPL was supported under contract from NASA. C. S. was supported by a NASA Grant.

REFERENCES

Ahrens, T. J., and Harris, A. W. 1992. Deflection and fragmentation of near-Earth asteroids. *Nature* 360:429–433.

Gaffey, M. J., and McCord, T. B. 1977. Mining outer space. *Technology Rev.* 79:3–11.

Herrick, S. 1979. Exploration and 1994 exploitation of Geographos. In *Asteroids*, ed. T. Gehrels (Tucson: Univ. of Arizona Press), pp. 222–226.

Lewis, J. S., Matthews, M. S., and Guerrieri, M. L., eds. 1993. *Resources of Near-Earth Space* (Tucson: Univ. of Arizona Press).

Morrison, D., ed. 1992. *The Spaceguard Survey: Report of the NASA International Near-Earth-Object Detection Workshop* (Pasadena: Jet Propulsion Laboratory).

O'Leary, B. 1977. Mining the Apollo and Amor asteroids. *Science* 197:363–366.

Sagan, C. 1990. Between enemies. *Bull. Atomic Sci.* 48:24–26.

COST AND BENEFIT OF NEAR-EARTH OBJECT DETECTION AND INTERCEPTION

GREGORY H. CANAVAN
Los Alamos National Laboratory

The value of defenses against near-Earth object (NEO) impacts is bounded by the losses expected in their absence. For current estimates of impact frequencies, damage durations of a few decades, and economic costing, small, intermediate, and large NEOs all make contributions to expected losses large enough to justify their detection. Estimates of detection and interception costs determine the lowest cost NEO defenses and the proper mix between detection and defense. Coupling this result with estimates of the benefits of defense determines the maximum NEO diameters for which defenses are cost effective. Extending the detection many orbits before impact determines optimal detection ranges, costs, and technologies close to Spaceguard's. Extension of existing technology for detection and defense are appropriate for most NEOs.

I. INTRODUCTION

This chapter discusses the value of defenses against NEOs. It does so in three steps. The first is determining the value of detection and defense, which this chapter does by estimating the expected losses in their absence, starting from the treatment in Canavan (1993a). The second is determining the cost of optimal mixes of detection and defense, which it bases on the marginal costs estimated in the first step. The third step is estimating the value of detection and defense, which it does by calculating the reduction in the expected losses when optimal detection or defenses are deployed.

The first step starts with a review of the data on the impact frequency by NEOs of various diameters, continues with the development of improved models for estimating the losses they are expected to produce, and concludes with the calculation of the value of defenses by estimating the losses expected in their absence, which it does by summing the product of their impact frequency and expected losses over NEO diameters to give improved estimates of expected losses. Its goal is not to estimate losses precisely, but to bound the contributions to these losses from NEOs of different sizes, show the sensitivity of expected losses to uncertainties in the data and physical processes, and suggest areas for further theory and experiments.

The second step determines the optimal combination of NEO detection and interception. It gives simple estimates of the costs of each and equates their marginal costs to produce the lowest cost combinations for a range of

NEO diameters. That provides a concrete basis for the discussion of the proper mix between detection and defense, which has been a contentious subject in recent NEO meetings. It then discusses the properties of the resulting optima and their sensitivities to NEO and defense parameters, which illuminate the main areas of the analysis needing additional work. The calculations are simplest for long-period comets (LPCs), which are treated explicitly. These calculations show how to optimize the detection and deflection components of affordable defensive systems around the parameters of the LPCs and interceptors. The model is then coupled with analytic estimates of the benefits of defense to determine the maximum NEO diameter for which defenses are cost effective, which is reasonable, relevant, and only weakly dependent on model parameters. Defenses for LPCs could arguably be developed over a period of a few decades for expenditures that are modest compared to expected losses.

The third step extends the analysis, with less precision, to advanced interceptors, whose more efficient deflection decreases the optimal detection time and cost for optimal defense; smaller NEOs, whose smaller diameters drive detection to shorter times and ranges; and intercept on final approach. Defense costs would then be dominated by fixed costs for nuclear interceptors launched on demand, but the costs for kinetic energy intercepts are such that distant nuclear intercepts could be appropriate for the deflection of large NEOs, and short-range nonnuclear intercepts appropriate for small NEOs. In the final step, the analysis is extended to estimate the value of detection of NEOs many orbits before impact, which with marginal costs of detection calculated above imply an optimal detection range, cost, and performance roughly consistent with that of Spaceguard. The technology and cost for detection and defense appear to be consistent with current developments. They could produce economically viable defenses for a range of NEO diameters.

II. EXPECTED LOSSES FROM NEO IMPACT

The Chapter by Morrison et al. summarizes NEO impact frequency data. An earlier version was used to define the NEO threat for the Interception Workshop. However, a preliminary study of the Spacewatch optical search data showed interesting discrepancies—particularly for NEOs less than a few hundred meters across. Canavan (1993a) gave a quick assessment of the impact of these new data on estimated losses from NEO impacts, but found the error bars to be about as big as the mean values. At the subsequent NEO meeting at the University of Arizona, new Spacewatch data showed even more significant discrepancies (Rabinowitz 1993). That led to a refinement of the earlier estimates of losses. Those estimates were presented and discussed at the Erice Meeting on NEOs, which made it clear that the initial interpretation of the Spacewatch data overestimated the number of NEOs with diameters greater than a few tens of diameters, which necessitated a still further revision of the loss estimates. The resulting analysis and estimates are summarized in this section, which describes the analysis, estimates the losses from NEOs of all sizes, and estimates the uncertainties in those estimates. It uses the damage model developed in the earlier version of this chapter to estimate the area of destruction from an NEO of given mass and energy, and that to estimate the expected loss for a given area of destruction.

The analysis identifies a natural division into three diameter ranges. The smallest NEOs, which have diameters below about 200 m, make an enhanced contribution, particularly if the Spacewatch impact frequency data are used. Small metallic NEOs could make a significant contribution to expected losses, although these losses are sensitive to uncertainties in impact frequency, composition, and atmospheric interaction. Small stony NEOs make a comparable contribution. Land impacts of intermediate NEOs, which have diameters from 200 m to 2 km, make a contribution comparable to that from all small NEOs. The losses due to tsunamis generated by their ocean impact could be an order of magnitude larger. Large NEOs, which have diameters greater than about 2 km, could produce global damage, which could be bounded by warning and preparation to $\approx\$400$ M/yr, relatively insensitive to uncertainties and model parameters.

A. Impact Frequency Data

Figure 1 is a reproduction of the first figure of Chapman and Morrison (1993), which is a fit to Shoemaker's lunar crater data (Shoemaker 1983). For "small" NEOs, with diameter $D<200$ m or energy $Y\leq10^3$megaton (MT$\approx4.2\times10^{15}$Joules—the yield of a million tons of high explosives or one ton of nuclear explosives), the impact frequency scales roughly as $1/Y^{8/9}$, or $1/(D^3)^{8/9} = D^{-8/3}$. For "intermediate" NEOs with 200 m$<D<2$ km, or 10^3MT$<Y<10^7$MT, the impact frequency scales roughly as $1/Y^{2/3}$, or $(1/D^3)^{2/3} = D^{-2}$. For "large" NEOs with $D>2$ km or $Y>10^7$MT, the impact frequency falls roughly as $1/Y$, or D^{-3}. The last differs from the scaling of Fig. 2 of that paper, which falls more slowly than $Y^{-2/3}$ for $Y>10^6$MT. This change is presumably made to accommodate an assessment that an NEO the size of the K/T impactor hits about once every 5×10^7yr rather than approximately every 10^7yr.

The additional line joining the solid circles at 5 and 50 m represents the Spacewatch data. The earlier version of those data indicated much higher impact frequencies below about 300 m (Canavan 1993a). The Spacewatch data have subsequently been revised downward into agreement for diameters greater than about 50 m (Rabinowitz 1993). The Spacewatch impact frequency of about one per month at $D = 5$ m is about a factor of 30 higher than "hazard of impacts." Such a collision frequency is not inconsistent with other optical (Spalding 1993) and infrared (Tagliaferri 1993) measurements at small diameters, but the conversion of Spaceguard's measured brightnesses into diameters is indirect and the conversion of the optical and infrared signatures into incident kinetic energies is incomplete. The impact of this discrepancy is assessed parametrically below.

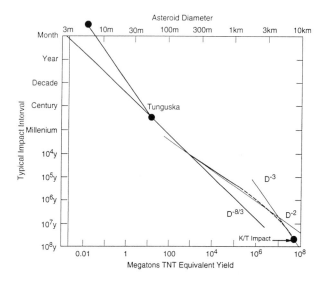

Figure 1.   Typical NEO impact intervals as a function of diameter from crater and
Spacewatch data.

B. Impact Frequency Probability Density Function

The cumulative impact frequency at a given diameter D discussed in the previous section is the integral over the collision frequencies of NEOs that "diameter and larger" (Morrison 1993). In each of the ranges of small (<200 m),
intermediate (200 m to 2 km), and large (>2 km) diameter NEOs discussed
above, the cumulative impact frequency can be represented by a power law
approximation

$$N(D) = KD^{-\alpha} \tag{1}$$

in terms of which the cumulative collision frequency is the integral over the
NEO impact frequency probability density function

$$n(D) = \alpha KD^{-(\alpha+1)} \tag{2}$$

where K and α are constants that can be evaluated from the measured impact
rates in each range. For small NEOs, Fig. 1 gives a cumulative impact
frequency that scales as $D^{-8/3}$, which gives $\alpha \approx 8/3$. The impact frequency at
$D \approx 200$ m is $\approx 10^{-4} \mathrm{yr}^{-1}$ which determines K to be

$$K = D^{\alpha}N(D) \approx (200\mathrm{m})^{8/3} \times 10^{-4}\mathrm{yr}^{-1} \approx 140\mathrm{m}^{8/3}\mathrm{yr}^{-1}. \tag{3}$$

$D = 200$ m is the break point between the scaling for small and intermediate
NEOs, so it can also be used to evaluate the parameters for intermediate
NEOs, for which $\alpha \approx 2$ and

$$K = (200 \mathrm{\ m})^2 \times 10^{-4}\mathrm{yr}^{-1} \approx 4 \mathrm{\ m}^2\mathrm{yr}^{-1}. \tag{4}$$

For large NEOs, $\alpha \approx 3$, and

$$K \approx (2000 \text{ m})^3 \times 10^{-5} \text{yr}^{-1} \approx 8 \times 10^4 \text{m}^3 \text{yr}^{-1}. \tag{5}$$

C. NEO Damage Radius, Area, and Fraction

Hills and Goda (1993) can be used to infer the damage from stony NEOs tens to hundreds of meters in diameter, which may break up in the atmosphere during entry, but can still cause damage on the ground by depositing kinetic energy in the atmosphere as they decelerate. The shock so produced gives a radius of destruction of approximately

$$R \approx bY^{1/3} \approx b \left(\frac{\rho V^2}{4} \right)^{1/3} D \tag{6}$$

where $Y = mV^2/2$ is the kinetic energy of an NEO of velocity v, density $\rho \approx 3$ kg m^{-3}, and mass $m = 4\pi\rho(D/2)^3/3$. The empirical parameter $b \approx 0.047$ (m s^2/kg)$^{1/3}$ corresponds to an over-pressure of ≈ 2 psi, which would destroy most buildings (Glasstone 1962). In this approximation, R is proportional to D and depends only on the NEO energy. Canavan (1993a) shows that the damage radius R scales roughly linearly on the initial NEO diameter, which produces this level of destruction over a fraction of the Earth's surface of about

$$f = \frac{\pi R^2}{4\pi R_e^2} \approx b^2 \left(\frac{\rho V^2}{32R_e^3} \right)^{2/3} D^2 = CD^2 \tag{7}$$

where R_e is the Earth's radius and $C \approx 6 \times 10^{-10}$ m^{-2} for a typical NEO V of 20 km s^{-1}. Hills and Goda estimate that only stony NEOs with diameters greater than about 50 m or metallic NEOs with diameters greater than 6 m penetrate the atmosphere far enough to cause damage on the ground.

D. Expected Loss

The U. S. Gross National Product is about \$5T yr^{-1}, which is about a quarter of the Earth's total gross product, $G \approx \$20T$ yr^{-1}. That value is used below as a surrogate for the total loss, because given adequate warning, evacuation and proper preparation could reduce loss of life and limit damage to the loss of production from the area devastated for the period of time required for recovery. For an impact that renders a fraction f of the Earth's surface unproductive for a time T, a phase-space estimate of the expected loss is fGT. The calculations below assume a recovery time of $T = 20$ yr for all NEO diameters; the impact of varying T with f is explored in the earlier version of this chapter (Canavan 1993a). This recovery time $T = 20$ yr is also the reciprocal of the real interest rate of 5% yr^{-1}, which is appropriate to discount or capitalize losses of comparable uncertainty. The loss fGT is a function of NEO diameter through f. The expected differential loss dL from NEOs with diameters between D and $D+dD$ is given by

$$dL = GT \, n(D)f(D)dD \, . \tag{8}$$

The expected integral loss from a range of diameters is calculated by integrating dL to arrive at

$$L = GT \int dx \, n(x)f(x) \tag{9}$$

where the limits of the integral cover the range of diameters of interest. The following sections evaluate these differential and integral losses for small, intermediate, and large NEOs.

1. Expected Loss from Small NEOs. The differential loss from stony NEOs with $D < 200$ m is from Eqs. (2), (7), and (8)

$$dL/dD = GT \alpha K D^{-(\alpha+1)}CD^2 \tag{10}$$

which is shown in Fig. 2. dL/dD has a peak value of about $100M yr^{-1}m at $D = 1$ m, but for the $\alpha = 8/3$ of Fig. 1, it falls as $D^{1-\alpha} = D^{-5/3}$ to about $0.15M yr^{-1}m by $S = 50$ m, which is the minimum size stony NEO that can cause ground damage in the Hills-Goda analysis. The integrated loss from small NEOs with diameters up to $D > S$ is

$$L = GT \int_S^D dx \, n(x)f(x) \tag{11}$$

which reduces to

$$L = GT \int_S^D dx \, \alpha K x^{-(\alpha+1)}Cx^2$$
$$= GT C \alpha K \left[(D)^{2-\alpha} - (S)^{2-\alpha} \right] / (2 - \alpha). \tag{12}$$

For $D >> S$, the integrated losses rapidly rise by about 200 m to the limiting value

$$L \approx GT \alpha K C S^{2-\alpha} / (\alpha - 2) \tag{13}$$

which depends on $S^{2-\alpha} = 1/S^{2/3}$. Figure 3 shows L as a function of S for $D = 200$ m. For $S < 50$ m, L scales approximately as $1/S^{2/3}$; for larger S, it falls more rapidly. For Hills and Goda's $S = 50$ m, the expected integral loss is about $6M yr^{-1}, which would only justify a defensive program of about that magnitude. Hills and Goda's prediction of the threshold for small NEO breakup have been confirmed roughly by others, but is probably only accurate to within about a factor of 2. A value of $S = 25$ m would justify a program of about $13M yr^{-1}; a value of 100 m would only justify $2M yr^{-1}. Because the Spacewatch data departs from that used above only for smaller stony NEOs than could penetrate and cause damage, the discrepancy in impact

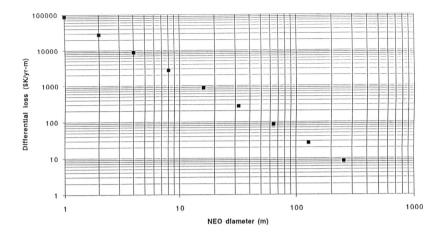

Figure 2.   Differential losses from small NEOs as a function of the minimum penetrating diameter.

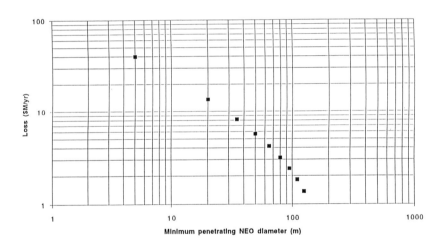

Figure 3.   Integral losses from small, stony NEOs as a function of the minimum penetrating diameter.

frequency of Fig. 1 does not impact these assessments of the expected damage from small stony NEOs. These \$2 to 13M yr^{-1} losses would only justify a defense program of that order, but it is not clear that a significant effort could be executed for that amount. However, the detection of small NEOs could apparently be justified on the basis of the reduction in the false alarm rates in missile warning systems (Tagliaferri 1993).

Because of their greater strength, smaller metallic NEOs can penetrate the atmosphere deeply enough to produce damage. That is offset by their small percentage, about 3% of the total impacts, which multiplies the losses calculated above. Losses from small metallic NEOs are most sensitive to the diameter of the smallest penetrating NEOs, which is $\approx$6 m for metallic NEOs impacting with relative velocities $\leq$15 km s^{-1}. In Fig. 1, this sensitivity emphasizes the discrepancy between the lunar crater and Spacewatch data. To fall from a few impacts per month at 5 m to 0.3 per century at 30 m, the Spacewatch data would have to fall roughly as D^{-4}, which would make losses sensitive to minimum diameter.

The uncertainties in the Spacewatch data depend on the conversion of brightnesses into diameters. That suggests a straightforward way of assessing the losses from metallic NEOs. If the impact frequency is evaluated at 50 m, where the two data sets agree, varying α from 8/3 to 4 spans the range of values of diameter that could ultimately result from the reduction of the Spacewatch brightnesses to diameters. That variation is shown in Fig. 4. For $\alpha = 8/3$, the integral loss is about \$1M yr^{-1}. The expected loss reaches a value of about \$7M yr^{-1} by $\alpha = 4$.

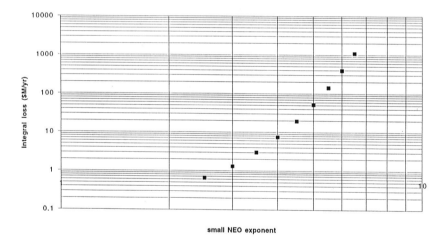

Figure 4.   Integral losses from small, metallic NEOs as a function of the power law of their cumulative distribution function.

This loss is, however, quite uncertain because of the strong $D^{2-\alpha} = 1/D^2$ scaling on the smallest penetrating diameter. If that diameter was increased by a factor of 2, which is well within the uncertainty of current models, the loss would be decreased by a factor of 4 to $\approx$\$2M yr^{-1}, which is insignificant. If the smallest penetrating diameter was decreased by a factor of 2, which is barely credible, the expected losses from small metal NEOs would be almost 4 times those from small, stony NEOs. For plausible variations, metallic NEOs could make a small but interesting contribution to the total losses from NEOs;

thus, it would appear worthwhile to resolve the uncertainties in the reduction of the Spacewatch data, and to improve the rough estimates of the strength and penetration of the small metallic NEOs and of their damage radii. While small NEOs make a large contribution to the total number of impacts, except for extreme assumptions about collision frequencies and distributions, they do not make a significant contribution to the total losses from all NEOs.

2. *Expected Loss from Intermediate NEOs.* The impact frequency of NEOs with intermediate diameters of ≈200 m to ≈2 km is characterized by a power law with $\alpha = 2$. NEOs of these diameters survive to impact the ground, so their damage mechanisms shift to a combination of conventional hypervelocity impact, which have been studied extensively (Shelkov 1991), and novel mechanisms such as tsunamis and other phenomena, which have not (Hills and Goda 1993). If the damage radius continues to scale as Eq. (7) for NEO diameters up to ≈2 km, an assumption implicit in other analyses (Morrison 1993), the expected loss from intermediate size NEOs is

$$L = \int_{200}^{D} \mathrm{d}x \, n(x)f(x)GT = 2KCGT \ln(D/200 \text{ m}) \tag{14}$$

which is shown by the bottom curve in Fig. 5. The loss increases slowly to a limiting value, which is ≈2 × 4 m² yr⁻¹ 6 × 10⁻¹⁰m⁻² $20T yr⁻¹ 20 yr ln (2000/200)≈$5M yr⁻¹. This loss would only justify a premium on the order of that estimated above for small stony NEOs. Moreover, it is quite uncertain; the actual losses could be smaller or larger, because the damage mechanisms in this intermediate range have been studied less than the direct damage mechanisms for small NEOs or the global ones for large NEOs.

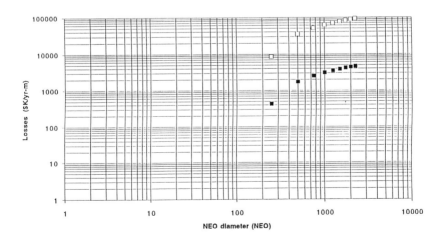

Figure 5. Integral losses from land (bottom curve) and ocean (top curve) impacts of intermediate NEOs.

One phenomenon with the potential to greatly increase the losses from intermediate NEOs is tsunamis generated by their impact in the oceans, which could cause destruction in the densely populated areas within tens of kilometers of shore all the way around the ocean. At the Arizona meeting this phenomenon was discussed by Hills et al. (1993), who give a simple model for evaluating the potential impact of tsunamis. It cites Glasstone for the deep-water wave height h_W as a function of the NEO energy released, and for the observation that "this result is not sensitive to the depth at which the explosion occurs"—and hence to the details of energy deposition. For nominal NEO densities and velocities, the deep-water wave height is

$$h_W \approx 15 \text{ m}(D/400 \text{ m})^{3/2}(1000 \text{ km}/r) \tag{15}$$

where D is the NEO diameter and r is the distance from impact. The quantity of interest is the height of the wave when it runs up on the shore, which can be written as $h_0 \approx Z h_W$, where $Z \approx 10$ to 40 is an amplification factor to account for concentration at the shoreline. A nominal value of $Z = 20$ is used below. Hills et al. argues that the tsunami could cause damage inland to a maximum distance of

$$
\begin{aligned}
X\text{max} &= 1 \text{ km}(h_0/10 \text{ m})^{4/3} \\
&\approx 1 \text{ km}(1.5Z)^{4/3}(1000\text{km}/r)^{4/3}(D/400 \text{ m})^2.
\end{aligned} \tag{16}
$$

The area damaged is $\approx 2\pi R_O X_{\max}$, where R_O is the effective radius of the ocean basin the NEO hits. The Earth has about a half-dozen basins, which cover about four-fifths of its surface; thus,

$$R_O \approx (4/5 \times 4\pi R_e^2/6)^{1/2} \approx 0.7R_e \approx 4700 \text{ km}. \tag{17}$$

If a 400 m NEO hit near the middle of such a basin, the damage would extend inland $X_{\max} \approx 1 \text{ km} (1.5 \times 20)^{4/3} (1000 \text{ km}/4700 \text{ km})^{4/3} \approx 12 \text{ km}$. That would give a total area of destruction of about $2\pi R_O X_{\max} \approx 2\pi \ 4700 \text{ km} \ 12 \text{ km} \approx 3.5 \times 10^5 \text{ km}^2$, or a fraction of about 7×10^{-4} of the surface of the Earth. Because $X_{\max}$ scales as $1/r^{4/3}$, the damage could be greater for impacts closer to shore or at a steep angle of incidence. Value is concentrated near shorelines, so the fraction of value destroyed might be larger by a factor of W than the fraction of the surface destroyed. That concentration is not explicitly estimated in the treatment below, which treats W as a parameter. Equation (7) infers a fractional destruction of value for land impacts of $f = CD^2$. For tsunamis, the fraction of surface area and value also scale on D^2 through $X_{\max}$. Thus, the ratio of the damage done by tsunamis to that by land impacts of NEOs in the range of 200 m to 2 km from Eq. (14) is thus about

$$\frac{W2\pi R_O X_{\max}}{CD^2 4\pi R_e^2} \approx \frac{W\pi \ 0.7R_e \ 12 \text{ km} \ (D/400 \text{ m})^2}{CD^2 4\pi \ R_e^2} \approx 7W. \tag{18}$$

If the concentration of value near the oceans is a factor of $W \approx 3$, the tsunami damage from an ocean impact would be greater than that from the land impact of an intermediate diameter NEO by a factor of ≈ 20. Because the contribution from Eq. (14) is about \$5M yr^{-1}, that would bring the total losses from intermediate NEOs up to about \$100M yr^{-1}, as shown by the top curve in Fig. 8 below, which is about equal to the contributions calculated for small metallic NEOs above and large NEOs below. Thus, with tsunamis included, intermediate NEOs have the potential of generating losses large enough to justify modest defenses.

The loss from intermediate NEOs is equally sensitive to all NEO diameters—each octave in diameter contributes about equally. Thus, NEOs of all intermediate sizes would have to be detected and intercepted for maximum benefit. If defenses were developed that could only detect NEOs larger than 200 m or intercept NEOs smaller than 2 km in diameter, the expected loss and resulting value of the defenses would be reduced accordingly.

The amount of warning likely varies with NEO diameter. For NEOs 2 km across, it could be from years to centuries, if the NEO was detected several orbits before impact. In that case, warning alone could be adequate to provide for the evacuation that would be needed to prevent great loss of life. Defenses would be an option if they could be executed for a cost of about \$100M yr^{-1}. Even if a 2 km NEO was only detected on final approach at 1 AU, that would still leave several months for evacuation. However, for a 200 m NEO, the detection range might be reduced to ≈ 0.1 AU and the response time to a few days, which might be too short for evacuation, in which case defenses would be required to prevent great loss of life and achieve the full benefit.

3. Expected Loss from Large NEOs. In general, the expected loss from the largest NEOs is treated as the product of an infinite loss and a small (possibly zero) probability (Morrison 1993). The economic model used here assumes, instead, that with warning and evacuation, loss of life could be limited from even large NEO impacts. Then the damage could be bounded by the economic value of the facilities destroyed, the supplies needed to survive until recovery, and the global production lost during the time it took the Earth's atmosphere and civilization to recover. That cost is given by the product of the Earth's product G and the recovery time of about $T = 20$ yr assumed. In that case, the loss from large NEOs is

$$L = \int_{2\,\text{km}}^{\infty} dx\, n(x) f(x) GT = GT \int_{2\,\text{km}}^{\infty} dx\, n(x) 1 = GT N(2\,\text{km}). \quad (19)$$

Figure 1 gives a collision frequency of about one impact every million years for NEOs of 2 km or greater diameter, or $N(2\,\text{km}) \approx 10^{-6}\,\text{yr}^{-1}$. Thus, the expected loss from large NEOs is

$$L \approx \$20T\ \text{yr}^{-1} \times 20\ \text{yr}\ 10^{-6}\ \text{yr}^{-1} \approx \$400M\ \text{yr}^{-1} \quad (20)$$

which would justify defenses of about that magnitude. This is about 100 times greater than the $\approx \$5M\ \text{yr}^{-1}$ losses from land impact of intermediate NEOs, and about 4 times the $\approx$\$100M yr^{-1} rough losses from tsunamis from ocean impacts of intermediate NEOs discussed in the previous section.

For large NEOs, reducing losses to these levels is contingent on adequate warning and preparation, without which the expected losses could be global, catastrophic, and unbounded. In order to reduce the expected losses to the $\approx$\$400M yr^{-1} of Eq. (20), it would be necessary to have at least detection, so that evacuation of the area of destruction and storage of supplies could be completed in time. These measures would ultimately have to be supplemented by defenses for NEOs large enough to literally fracture the Earth, but that requires diameters orders of magnitude greater than the few kilometers associated with global climate impact. Equation (7) is intended to be used for $f << 1$, but taking f to be unity gives a rough estimate of the NEO diameter required for catastrophic destruction of $D \approx 1/\sqrt{C} \approx 40$ km, which is consistent with other estimates (Fig. 1 in Morrison 1993). For NEOs of that diameter and larger, defenses would be required.

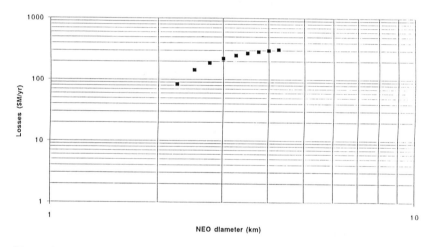

Figure 6.　Avoidable integral losses as a function of the largest NEO diameter that can be addressed.

Equation (20) assumes that it is possible to prepare for NEOs of all sizes. It becomes increasingly difficult to deflect NEOs of very large diameter. If above some diameter it became impossible to deflect them, the benefits of defenses would be decreased. That reduction can be estimated by taking the upper limit of the integral in Eq. (19) to be the diameter of the largest NEO that can be negated, which produces the available loss from large NEOs, which is

$$L = \int_{2\,\mathrm{km}}^{D} \mathrm{d}x\, n(x)f(x)GT = GT \int_{2\,\mathrm{km}}^{D} \mathrm{d}x\, n(x)1$$
$$= GT\,K\left[(2\,\mathrm{km})^{-\alpha} - D^{-\alpha}\right] \qquad (21)$$

which is shown in Fig. 6. There is a rapid increase in the loss as the diameter addressed is increased from 2 to 4 km. Beyond that, the losses converge rapidly, because the impacts of larger NEOs are so infrequent. Figure 1 shows that for $D > 3$ km, α falls off even faster than the $\alpha = 2$ used in constructing Fig. 6, so the bulk of the benefits of detection can actually be realized for diameters of ≈ 4 km. Defenses are an additional option, if cost effective, i.e., if implementable for costs of $\approx\$400$M yr^{-1}.

E. Summary of Loss Estimates

This section bounds the value of defenses against NEO impacts by calculating the expected losses in their absence by summing the product of their expected impact frequency and expected damage loss over all NEO diameters from a standard economic model of expected loss to produce bounded estimates for NEOs of all sizes. The results are summarized in the differential and integral loss rates of Figs. 7 and 8, which identify three diameter ranges of interest. The smallest NEOs with diameters below about 200 m have collision frequencies that fall rapidly with diameter. There is an enhanced contribution from small metallic NEOs, which have a maximum integral loss of $\approx\$3$M yr^{-1}, sensitive to poorly known physical parameters and processes. Small stony NEOs have an integral loss of $\approx\$6$M yr^{-1}, which would not justify significant defenses. Land impacts of intermediate sized NEOs with diameters from 200 m to 2 km produce integral losses of $\approx\$5$M yr^{-1}. Losses from tsunamis generated by their ocean impacts are estimated to be about $\approx\$100$M yr^{-1}, which would justify defenses costing about that much. Large NEOs with diameters over about 2 km contribute expected losses of $\approx\$400$M yr^{-1}, which are relatively insensitive to uncertainties and model parameters. The losses from these infrequent, large NEOs dominate the total.

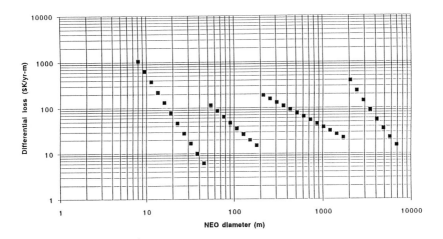

Figure 7.   Differential losses for NEOs of all diameters.

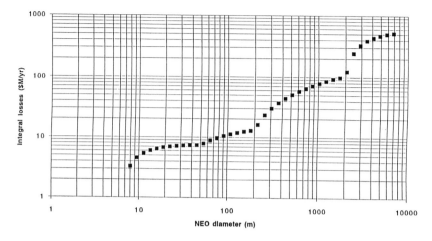

Figure 8.   Integral losses for NEOs of all diameters.

There is an important distinction between the losses from the different ranges of NEOs. Even without preparation, the losses from intermediate NEOs would be large, but tolerable. For large NEOs, losses are contingent on adequate warning and preparation, without which expected losses would be unbounded. To reduce the expected losses from large NEOs to the $\approx\$400M$ yr^{-1} value of Eq. (20), it is necessary to have at least detection, so that evacuation and preparation could be completed in time. That done, the losses of Eq. (20) would remain as a residual, which could be further reduced by effective defenses. For intermediate NEOs, expected losses are bounded even without detection. Detection alone could permit evacuation, which would reduce loss of life from the largest of them, although the loss of productive capability would still remain at the indicated levels.

III. COST OF NEO DEFENSES

In this section cost of optimized NEO defenses is estimated. It first provides a simple model for the cost of detection as a function of NEO range and diameter. It then derives a model for NEO deflection that depends on the same parameters. It then finds the proper combinations of detection and deflection that minimize cost, which also determines the optimal detection sensor range. The result is a combination that properly equates marginal costs for detection and deflection.

A. Detection

Canavan (1993*b*) estimates roughly the costs of an optimized visible telescopic search for NEOs. Phipps performs an independent and somewhat broader estimate of the requirements and costs for search (Phipps 1993).

Each concludes that for searches that are largely in opposition to the Sun, the cost for search out to a distance R for objects of diameter $\geq D$ is given roughly by

$$C_S \approx A(R/D)^2(1+R)^2. \tag{22}$$

An object in opposition at a range R (AU) from the Earth is a distance $1+R$ from the Sun, so it scatters a flux proportional to $D^2/(1+R)^2$, of which a fraction $\approx A_{opt}/R^2$ would be captured by a sensor of area A_{opt}. Thus, the signal is proportional to $A_{opt}D^2/R^2(1+R)^2$. To produce a given signal, the area (and hence cost) of the sensor required would scale as $A_{opt} \propto R^2(1+R)^2/D^2$, giving Eq. (22). Equation (22) is inaccurate for $R>>1$ due to brightness limitation and the large numbers of objects detected, but the optimal detection range determined below is within its range of validity.

The search cost C_S includes the cost for both constructing and operating the detection system, which are comparable for most military defensive systems. If R is measured in AU and D is measured in km, the constant coefficient is estimated in Canavan (1993b) as $A \approx \$50\text{M-km}^2/\text{AU}^4$ for a ground-based telescope system and $\approx \$1\text{B-km}^2/\text{AU}^4$ for a space-based system. The calculations below assume that this scaling can be used for larger ranges than are normally considered, although the optimal intercept times and ranges are such that the approximate form of Eq. (22) is fairly accurate. This ignores a number of practical background issues that are addressed at the end of the chapter.

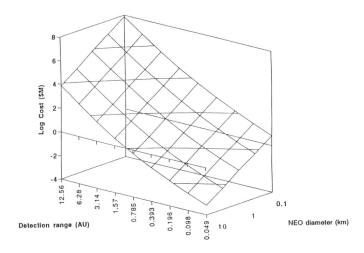

Figure 9.   Detection cost as a function of NEO diameter and range.

Figure 9 shows the detection costs from Eq. (22) as functions of R and D. At $R \approx 1$ AU the detection costs for $D \approx 1$ km are $\approx \$200\text{M}$ in rough accord with the Spaceguard study (pp. 49–61 in Morrison 1993). That system and cost would also detect 10 km NEOs at $\approx$3 AU and 0.1 km NEOs at 0.2 AU. Detection costs for 1 km NEOs would increase to $\approx$\$20B by $R\approx$3 AU. Detection costs vary as $1/D^2$, so searching to 1 AU for 0.1 km NEOs would cost a like amount. While these detection costs become large compared to those for Spaceguard, they remain commensurate with the few \$100M yr^{-1} expected losses from large NEOs estimated above, which discount to a total loss of a few \$B, justifying detection costs of that order of magnitude.

B. Deflection

Canavan et al. (see their Chapter) consolidate various estimates of the interceptor masses required to deflect or destroy NEOs of various sizes as a function of the reaction time available. There is a correlation between the cost and mass of space payloads, so these masses can be roughly converted into the costs. The costs vary for different defenses—mass drivers, kinetic energy, nuclear explosives, etc. This chapter parameterizes their costs and provides example calculations for nuclear and kinetic energy deflection.

The interceptor delivers a final mass M_f with a specific energy density ϕ, which depends on the intercept concept. For nuclear explosives, $\phi \approx 2$ megaton (MT)/tonne$\approx 9 \times 10^{12}$ Joule/kg. For kinetic energy deflection by head-on impacts, the specific energy is about that of the NEO, which is $\approx(30$ km s$^{-1})^2/2$. This energy release $M_f\phi$ ejects a mass M_e at a velocity v_e, whose recoil deflects the NEO. Conservation of energy assuming complete transfer gives the energy imparted to the NEO by subsurface but non-optimally buried bursts as approximately

$$M_f\phi \approx M_e v_e^2. \tag{23}$$

Conservation of momentum gives the velocity increment imparted to the NEO as

$$m\Delta v \approx M_e v_e \tag{24}$$

where m is the mass of the NEO. With Eq. (23), this gives

$$\Delta v \approx \frac{M_e v_e}{m} \approx \frac{M_f\phi}{m v_e} \tag{25}$$

although in order to avoid fracturing the NEO, it might be necessary to deliver this impulse through a sequence of a few tens of smaller explosions with that total mass.

C. Displacement

The displacement a given deflection produces later on depends on where and when it is applied in the NEO's trajectory. The displacement can be written as

$$\delta \approx k\Delta vt \tag{26}$$

where t is the time between deflection and impact, ΔV is the velocity increment from Eq. (25), and k is a numerical parameter. Ahrens and Harris (1992) show that $k \approx 3$ to 5 for deflection many orbits prior to impact. However, interceptor release can only occur after detection. For LPCs, that is likely to take place on the first appearance, in which case, most of the response time available must be used for interceptor fly out. Then, $k \approx 0.1$ is a more appropriate value (Chapter by Canavan et al.) which is the value used below. For defense of the whole Earth, it is necessary to generate a displacement $\delta \approx R_e$, which with Eqs. (25) and (26) gives

$$M_f \approx \frac{R_e m v_e}{\phi k t}. \tag{27}$$

In space, mass is directly related to cost, so this interceptor mass can be converted into a rough interceptor cost. An interceptor final mass of $M_f \approx 10$ tonnes would require about 30 tonnes into deep space and perhaps 100 tonnes into low-Earth orbit (LEO), which is about the limit of what a fully integrated international effort could now produce. The booster to reach LEO, such as a Russian Energia, could cost on the order of $100M. The upper stage and controls needed for rendezvous could cost another $\approx$ $100M. The nuclear explosive could add $\approx$$100M more. If life-cycle operational costs were roughly equal to the total cost of the booster, payload, and controls; the total cost for the interceptor might be about $500M, or $50M/tonne of interceptor final mass. Assuming that these values could be scaled continuously to other masses gives a deflection cost of

$$C_D \approx B M_f \tag{28}$$

where M_f is given by Eq. (27) and $B \approx$ $50M/tonne. Adding a fixed cost to Eq. (8) to account for small payloads for small NEOs would increase detection and total costs, but would not affect the optimizations below. Figure 10 shows the deflection cost C_D for times to deflect of 0.01 to 2 yr. C_D scales on m/t. With 1 yr to deflect, the variable scaling costs for nuclear deflection are $\approx$$1M for $D = 1$ km and $\approx$$1B for $D = 10$ km. According to Eq. (27), $M_f \propto 1/\phi$, so the costs for kinetic energy deflection would be higher by about a factor of 10^4, or $\approx$$10B for a $D = 1$ km, and a year's warning.

D. Cost of Defense

The total cost of defense is the sum of the costs of detection and deflection, which is

$$
\begin{aligned}
C = C_S + C_D &\approx A \left(\frac{R}{D}\right)^2 (1 + R)^2 + B M_f \\
&\approx A \left(\frac{vt}{D}\right)^2 (1 + vt)^2 + \frac{B R_e m v_e}{\phi k t}
\end{aligned}
\tag{29}
$$

where the detection range R is replaced by the detection time t, through the NEO velocity using $R \approx vt$. Figure 11 shows the total cost C for nuclear

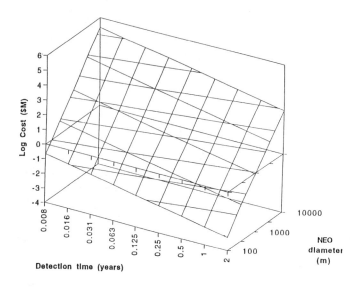

Figure 10.   Deflection cost as a function of NEO diameter and range.

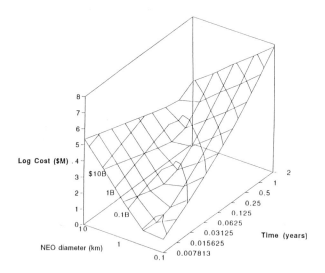

Figure 11.   Detection plus deflection cost as a function of NEO diameter and range
for nuclear deflection.

deflection. For short times and small D the total costs are $\approx$\$10M; for large
diameters they rise to $\approx$\$100B. For long times, the total cost is dominated by
the cost of detection, which increases as t^4, and is largest for small NEOs.

There is a progression in minimum-cost systems that increases from a

≈$10M system that could detect and deflect 0.1 to 0.3 km NEOs with 0.01 yr warning, through a ≈$100M system that could detect ≈1 km NEOs with 0.1 yr warning, to a ≈$1B system that could detect and deflect ≈3 km NEOs with 0.5 yr warning. Of interest here is that for a given D, the costs for deflection dominate for short times, while those for detection dominate for long times, so that the total cost exhibits a minimum. For D = 3 km, at t = 0.01 yr, C≈$10B. C then falls to ≈$300M at ≈0.25 yr before rising again to ≈$10B at 1 yr. For D = 1 km, the minimum is ≈$50M at t = 0.06 yr; for D = 10 km it is ≈$5B at ≈1 yr.

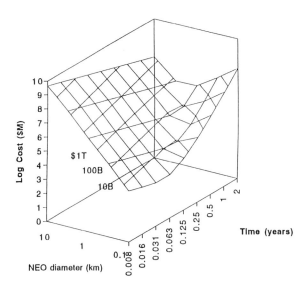

Figure 12.   Detection plus deflection cost as a function of NEO diameter and range for kinetic energy.

Figure 12 shows the costs for kinetic energy deflection by hard collisions. For a given time and diameter, detection costs are the same as nuclear, but deflection costs are higher by a factor of ≈10^4. Compared to Fig. 11, the detection costs on the left of Fig. 12 are raised by ≈10^4, so that the minima are generally higher and at larger times. The total costs for kinetic energy are comparable to those for nuclear deflection at small D and long t, where detection costs dominate. There is again a progression in minimum-cost systems, but for kinetic energy, it increases from a ≈$10B system for 0.1 to 0.3 km NEOs with 0.1 yr warning to a ≈$100B system for ≈1 km NEOs with 1 yr warning. Kinetic energy deflection also shows a minimum cost for each D, which is at ≈0.5 yr for 1 km and increases with D. For reaction times from years to decades, there is a cost penalty of about a factor of 100 for the use of kinetic rather than nuclear deflection.

E. Optimal Detection Times

In the total cost of Eq. (29), deflection costs dominate at small t, but decrease as $1/t$, and detection costs increase as t^2 to t^4, becoming dominant at large t. Thus, there is a minimum in the total cost of defense at an optimal detection time. It is not possible to determine that minimum analytically for the general case, but for $R \ll 1$ AU, i.e., $t \ll 1/6$ yr, Eq. (29) can be approximated by $C \approx A(vt/D)^2 + BR_e mv_e/\phi kt$, which is minimized by the choice

$$t_{opt} \approx \left(\frac{BR_e mv_e D^2}{2\phi kAv^2} \right)^{1/3}. \tag{30}$$

For large R, $C \approx AR^4/D^2 + BM_f \approx A(vt)^4/D^2 + BR_e mv_e/fkt$, which is minimized by the choice

$$t_{opt} \approx \left(\frac{BR_e mv_e D^2}{4\phi kAv^4} \right)^{1/5}. \tag{31}$$

Figure 13 shows a composite of these two limits as a function of NEO diameter and the specific energy of the deflection mechanism, joined at $t_{opt} = 1$ AU/$v \approx 1/6$ yr. For the $\phi \approx 10^{13}$ J kg^{-1} of nuclear deflection, the optimal time of deflection for $D = 100$ m is $\approx 10^{-3}$ yr; by 300 m it increases to ≈ 0.01 yr, in accord with Fig. 11. It increases to ≈ 0.3 yr by $D = 3$ km, which at an NEO velocity of $v = 30$ km s^{-1} corresponds to a detection range of ≈ 2 AU, which, with advanced detection concepts, appears possible for objects of that size. For larger D, t_{opt} increases according to Eq. (31). The optimal detection time increases as ϕ decreases. For the $\phi \approx 10^9$ J kg^{-1} of hard kinetic deflections, it is $\approx 1/30$ yr at $D = 100$ m, and $\approx 1/2$ yr at $D = 1$ km, in accord with Fig. 12.

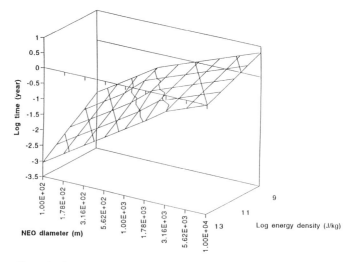

Figure 13.   Optimal detection time as a function of NEO diameter and deflection mechanism specific energy.

When t_{opt} is small, it scales on NEO diameter as $(mD^2)^{1/3} \propto (D^3 D^2)^{1/3} \propto D^{5/3}$, which is much stronger than its scaling on other parameters. t_{opt} scales on A and B only through $(B/A)^{1/3}$, which is a weak dependence that suppresses the impact of uncertainties in detection and deflection costs on optimal detection times and ranges. If space sensors were required, they might cost about an order of magnitude more than the groundbased sensors assumed in Fig. 13, which would reduce t_{opt} by a factor of $\approx 1/10^{1/3} \approx 0.5$. t_{opt} scales as $1/v^{2/3}$, so it decreases for higher speed NEOs, although the optimal range for detection, $vt_{opt} \alpha v^{1/3}$, increases for faster LPCs.

When t_{opt} is large, it scales on NEO diameter as $(mD^2)^{1/5} \propto (D^3 D^2)^{1/5} \propto D$, as seen from 3 to 10 km in Fig. 13. This linear increase of t_{opt} with D also holds for larger D for both nuclear and kinetic deflection. While t_{opt} is relatively insensitive to most other parameters, one exception should be noted. Advanced interceptors with much higher fly-out velocities could intercept much sooner after detection, which increases k from ≈ 0.1 to ≈ 3, an increase of ≈ 30. According to Eq. (31), the impact of such an increase would be to reduce the optimal detection time by about a factor of $30^{1/5} \approx 2$. For a 10 km NEO, that would reduce the optimal detection time and range from 1 yr and 6 AU to about 1/2 yr and 3 AU.

F. Optimal Defense Costs

When the optimal detection time is substituted back into Eq. (29), it gives the optimized (minimum) total cost, C_{opt}, which for t_{opt} small, is

$$C_{opt} \approx 3A^{1/3} \left(\frac{BR_e mv_e v}{2Dfk} \right)^{2/3} \tag{32}$$

and for t_{opt} large, is

$$C_{opt} \approx 5(A/D^2)^{1/5} \left(\frac{BR_e mv_e v}{4\phi k} \right)^{4/5}. \tag{33}$$

Figure 14 shows the composite C_{opt} as a function of D and ϕ, joined at 1 AU. The near curve for nuclear and the far curve for kinetic energy differ because of their specific energies by about a factor of $\approx (10^4)^{2/3} \approx 500$. For t_{opt} small, C_{opt} scales on D as $(m^2/D^2)^{1/3} \propto (D^6/D^2)^{1/3} = D^{4/3}$, which is slightly softer than the $D^{5/3}$ scaling of t_{opt}. C_{opt} scales on the cost of detection only as $A^{1/3}$, which is weak; it scales as $B^{2/3}$ on the cost of deflection, which is slightly stronger. Note that $C_{opt} \propto v_e^{2/3}$ is not a weak scaling. That reflects the sensitivity of total costs to the coupling achieved, which is somewhat dependent on the penetration possible at the very high closing velocities encountered in deflecting LPCs on final approach (Canavan et al. 1994). In this optimal defense, 1/3 of the total cost C_{opt} is devoted to detection; 2/3 to interception. That allocation of costs between detection and defense is

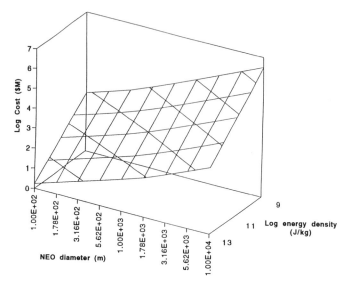

Figure 14.   Optimal detection plus deflection cost as a function of NEO diameter and deflection mechanism specific energy.

determined by the form of Eq. (29) and is independent of the specific cost and performance parameters assumed.

For t_{opt} large, C_{opt} scales on D as $D^{-2/5}m^{4/5} \propto D^{-2/5}(D^3)^{4/5} \propto D^2$, which is stronger than the linear scaling of t_{opt}. It is also noticeably stronger than the $D^{4/3}$ scaling for smaller D. C_{opt} depends weakly on detection costs through $A^{1/5}$; more strongly on deflection costs through $B^{4/5}$; 1/5 of the total cost is devoted to detection; 4/5 to deflection. Total costs are almost linearly sensitive to NEO and coupling parameters. Based on the expected losses of a few \$B given above, it would appear that the costs for nuclear deflection are acceptable for LPC diameters up to $\approx$ 10 km, which covers the bulk of the threat. Kinetic energy deflection would only appear affordable to diameters up to about 100 m. Note that C_{opt} scales as $1/k^{4/5}$; thus, the 30-fold increase in k possible with advanced interceptors could reduce the total cost of intercept by a factor of $30^{4/5} \approx 15$. For a 10 km NEO that would mean a reduction from \$1B to $\approx$\$100M.

It is argued that defenses should not be started until a threatening NEO is actually detected. Delaying preparations for defense would impose a delay of some time T between detection of the LPC and the launch of the interceptor. With no preparations whatsoever, it could take decades to develop and execute needed precursor experiments on coupling efficiencies, incorporate them into appropriate interceptors, and launch them at the LPC. Delays that long are clearly inappropriate for the optimal time scales deduced above. If coupling experiments were done in advance, the time for integration of space vehicles might be reduced to 2 to 3 yr. If the experiments were performed and the intercept vehicles developed, but not integrated with their boosters and explosives, the delay might be reduced to on the order of a year. If all elements were developed, but not integrated, the delay could be made controllable in the range of months to weeks.

Such delays would not impact the cost of detection, but would reduce the interceptor fly-out time to $t - T$, which would increase the cost for deflection and through it the total cost for defense. With delays, the optima shift to a few times T, so that the optimal costs with delays will be roughly the costs without delays at a few times T. For $t >> t_{opt}$, the total costs increase roughly as t^2 for t_{opt} small, and as t^4 for t_{opt} large; thus, significant delays rapidly increase total costs. For this reason, delays between detection and launch for integration or any other reason should be kept small compared to optimal detection times in order to keep from adversely impacting the effectiveness of defenses. The calculations above, which indicate the optimal time for the detection of objects of any diameter, can be used to determine the delays that can be introduced without compromising the basis for optimization.

IV. EFFECTIVENESS OF OPTIMIZED DEFENSES

Section II estimates the expected losses from NEOs of various diameters. Section III estimates the cost of optimized detection and deflection of NEOs of a given diameter. This section combines those results to determine the NEO diameters for which detection and deflection are economically viable. It does so by equating their marginal benefits to their marginal costs, and solving for the diameter at which they are equal. As above, the Earth's gross product $G \approx \$2 \times 10^{14}$ yr^{-1} is used as a surrogate for total losses. Thus, if disruption was global and lasted for about $T \approx 20$ yr, the expected loss would be about $GT \approx \$4 \times 10^{15}$. It is shown that cost-effective defenses could eliminate most of that loss.

A. Optimized LPC Defenses

LPCs are a stressing threat that would only be detected on final approach, as assumed in the above analysis. Large LPCs could cause global damage. LPCs with the diameters greater than the few kilometers needed for global effects have an integrated impact frequency of $f \approx 10^{-7}$yr^{-1}, which would lead to an expected loss of about $L = fGT \approx \$400$M yr^{-1}. Discounted at a real interest rate of $i = 5\%$ yr^{-1}, which is appropriate for such risks, the present value of the expected loss is $L/i = fGT/i \approx \$8$B, the value suggested above. However, in balancing benefits and costs, it is appropriate to do so at the margin.

The impact frequency's probability density function can be approximated by a power law $n(D) \approx \alpha K D^{-(\alpha+1)}$, where K and α are constants. The impact frequency of NEOs with diameters $>D$ is

$$f(D) = \int_D^\infty dD\, n(D) = K D^{-\alpha} \tag{34}$$

whose variation with D can be used to evaluate the decrease of the loss $L = fGT$ of the previous paragraph as the NEO diameters addressed is increased, which gives the benefit gained from defending against them. The resulting marginal benefit is

$$-dL/dD = -GT\,df/dD = GT\alpha K D^{-(\alpha+1)} \tag{35}$$

which is equated to the differential cost of the additional coverage to determine the maximum NEO diameter that should be defended against. That differential cost is available from Eqs. (32) and (33), and can be written as

$$C = HD^\beta \tag{36}$$

where H and β are constants determined by Eqs. (32) and (35). Multiplying this expression by the discount rate i to convert the total cost of defenses into an annual expenditure and differentiating it with respect to D gives

$$i\,dC/dD = i\beta H D^{\beta-1}. \tag{37}$$

Equating this cost with the benefits of Eq. (35) produces

$$D_{max} = \left(\frac{GT\alpha K}{i\beta H}\right)^{1/(\beta+\alpha)} \tag{38}$$

as the maximum diameter D_{max} for which defenses are cost effective. While the exponent α is not known with precision for large NEOs, Fig. 1 gives a rough value of $\alpha \approx 3$ for $D > 3$ km, where Fig. 14 gives $\beta \approx 2$, which reduces Eq. (38) to

$$D_{max} \approx \left(\frac{3GT K}{2iH}\right)^{1/5} \tag{39}$$

which scales only weakly on loss and cost parameters. The benefit parameters G and T are estimated above. K can be determined from the integral impact frequency of $\approx 10^{-7}$ yr^{-1} for diameters greater than ≈ 4 km to be $K \approx 10^{-7}$yr$^{-1} \times (4$ km$)^3 \approx 6.4 \times 10^{-6}$ km^{3}yr^{-1}. The cost coefficient H can be determined from Fig. 14, for the parameters above, for $D > 3$ km to be $H \approx \$20$M km^{-2}. This combination of parameters gives $D_{max} \approx 8$ km, which is about the size of the K/T impactor that ended the age of the dinosaurs. It would take a factor of ≈ 30 change in any of the parameters to alter D_{max} by a factor of 2. For this D_{max}, Fig. 14 gives a defense cost of $C \approx \$1$B, and Fig. 13 gives a detection time of ≈ 1 yr and range of ≈ 6 AU, which should be possible for an object of this size. Defenses of this magnitude could arguably be developed over a period of a few decades for the implied expenditures of $\approx iC \approx 0.05$ yr$^{-1}\$1B\approx \$50$M yr^{-1}. Note that this allowed expenditure is far below the expected losses of $fGT \approx \$400$M yr^{-1} estimated above, because

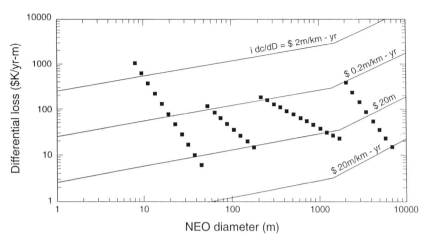

Figure 15.   Cost and benefit of optimized defense against NEOs of various sizes.

the defenses are determined by the equality of marginal (rather than total) costs and benefits.

It is also possible to solve for this value of D_{max} graphically, which provides a convenient vehicle for assessing sensitivities to uncertain parameters. Figure 15 adds the differential costs calculated in this section to the differential benefits shown in Fig. 7. The bottom solid line shows $i\,dC/dD \equiv iC'$ from Eq. (37) for nominal cost parameters, i.e., $\beta \approx 2$ and $H \approx \$20M\ km^{-2}$ for $D > 3$ km, and $\beta \approx 4/3$ and $H \approx \$40M/km^{4/3}$ for $D < 1$ km. The line for idC/dD intersects the dL/dD line for large NEOs at $D \approx 7$ km, which agrees with the value from Eq. (39) to within the accuracy of the parameters used. The added curves above are for costs 10, 100, and 1000 times the nominal parameters—the last also being approximately the cost for kinetic energy deflection. For nominal detection and deflection costs, defenses would be cost effective for NEOs up to ≈ 8 km in diameter. For 10 times nominal costs, defenses would be cost effective for NEOs up to ≈ 3.5 km in diameter. For costs greater than 100 times nominal, defences would not be cost effective for large NEOs. Figure 15 also shows that for nominal costs, defenses would be cost effective for smaller NEOs as well, which is addressed further in the next section.

B. Small and Intermediate NEO Diameters

The analysis above can be extended to smaller NEOs, although a few caveats are in order. The first is that the detection of smaller NEOs is a difficult task; it is not clear that the model for scaling detection performance and cost used above, which is grounded in optical search, is appropriate for them. Advanced sensor technologies might be needed and available. The second is that the expected losses from smaller NEOs are less well understood; hence, the benefits that would be generated by defending against them are not as well quantified. A full effectiveness treatment must, therefore, await the resolution of the outstanding issues in these two areas.

That said, a few observations are in order. Advanced detection technologies should still scale somewhat like Eq. (22) for detection, i.e., for any concept, costs should increase with range and decrease with NEO diameter in some fashion. Thus, smaller diameters are likely to drive detection to shorter ranges. It is difficult to secure the orbits of small objects from prior observations, which means that small NEOs are likely to be detected on final approach. In that case, their intercept would be much like that of the LPCs treated above, so that the value $k = 0.1$, and the figures based on it, are also appropriate for smaller NEOs.

From Fig. 15 it appears that expected differential losses from small and intermediate NEOs are comparable to those from large NEOs. For nominal differential losses and costs, it indicates that defenses would be cost effective for at least NEOs of intermediate and large diameter—if the costs of detection could be held to the nominal costs of Fig. 9 and the costs of interception could be scaled to the smaller interceptor payloads required.

If costs were increased by a factor of 10, defenses would be cost effective for large NEOs with diameters <4 km, most intermediate and small stony NEOs, and metallic NEOs smaller than $\approx$30 m in diameter. Thus, apart from a few gaps, they would be effective for essentially all NEOs. If costs were increased by a factor of 100 from their nominal values, defenses would not be effective for large, intermediate, or small stony NEOs. They would still be effective for small metallic NEOs with diameters less than $\approx$20 m, although that result is sensitive to their impact frequency, composition, and strength, which are not well known.

If costs were increased by a factor of 1000, defenses would appear to be effective only for metallic NEOs with diameters under $\approx$10 m, although that result is sensitive to the assumption that the detection costs assumed above could be met for objects that small. Small metallic NEOs are a special case. Because of their strong scaling on diameter, even for 1000 times nominal costs, defenses against NEOs bigger than about 15 m would be justified, although they would have to be accomplished for the cost of Fig. 14, which is only $\approx$0.05 $\times$ \$1B yr$^{-1}\approx$50M yr^{-1}. It is not clear that the required defense could be provided for that amount. Moreover, Fig. 5 shows that the optimum detection for such NEOs is at about 0.001 yr, or about 8 hr, at about 0.01 AU. At such ranges, radars could be used; their costs should be comparable to those assumed in Eq. (22).

The resulting cost for defense is so small that it would be lost in the fixed costs for nuclear interceptors that were launched on demand. However, for 10 m NEOs, the costs for kinetic energy intercepts are reduced enough to suggest an alternative approach. Small nonnuclear satellites with adequate mass and thrust could be placed into orbit before detection, and then maneuvered into the path of the NEO during the few hours between detection and impact. When they were overrun, the NEO's own kinetic energy would fragment it into many pieces, each of which could be too small to survive atmospheric reentry. This is the analog of using "brilliant pebbles" to destroy missiles launched upwards from the ground (Canavan and Teller 1990). It appears that the latter's sensors and propulsion concepts could also be used to maneuver interceptors in front of small NEOs. If so, intermediate and distant nuclear intercepts could be used for deflection of large NEOs, and short-range nonnuclear intercepts could be used for small and intermediate NEOs.

C. Value of Detection Alone

Equation (39) and Fig. 15 indicate that for nominal parameters, defenses would only be cost effective for NEO diameters under about 8 km. For larger diameters, detection alone could be effective and affordable. The derivative with respect to D of the annualized cost for the detection of NEOs with diameters greater than D is

$$i\,dC_S/dD \approx -2iAR^2(1+R)^2/D^3 \tag{40}$$

Figure 16 adds $i\,dC_S/dD \equiv iC_S'$ from Eq. (40) for three values of A. The center curve is for the nominal $A = \$50\text{M-km}^2/\text{AU}^4$. The top curve is for $A = \$500\text{M-km}^2/\text{AU}^4$, and the bottom curve is for $A = \$5\text{M-km}^2/\text{AU}^4$, which respectively shift the costs up or down by a factor of 10 or left or right by a factor of ≈ 2. All three curves lie below the differential losses L' for large NEOs, which means that detection would be cost effective for any of the three costs shown. The center curve for nominal costs intersects L' for intermediate NEOs at about $D = 2$ km, which means that it would be cost effective to detect them, too, because evacuating people and goods from the coastal areas to be destroyed would reduce expected losses by an amount equal to the marginal costs of the defenses. The intersection for the higher cost curve is shifted to $D \approx 4$ km. Conversely, the intersection for the lower cost curve is shifted to $D \approx 400$ m, which means that detection of most intermediate NEOs would be effective if possible at these costs.

The benefits of detection come in two stages. In the absence of any detection capability, the impact of an NEO greater than a few kilometers across could cause climatic disruptions for a period of decades. That could lead to global starvation in addition to the direct damage, which would be catastrophic. With detection radii on the order of a few AU, detection should be given sufficiently far before impact to make it possible to assemble adequate supplies to survive this transient period of starvation/production shortfall and to evacuate the area destroyed directly. In that case the losses could be limited to the global production disrupted by the impact, which is of the order of the fGT used above. The case of detection many orbits before impact is treated explicitly below.

D. Detection Many Orbits Before Impact

Current defensive concepts propose searching the region containing most of

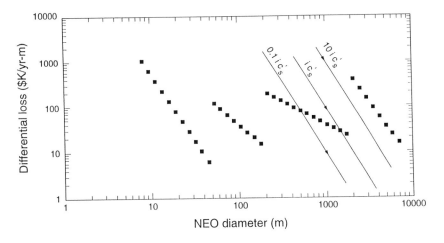

Figure 16.   Cost and benefit of detection only of NEOs of various sizes.

the large NEOs for several decades to detect those on threatening trajectories, which could then be deflected many orbits before impact. Evaluating the value of detection many orbits before impact is complex, but the main features can be illustrated by an extension of the model above, which calculates the probability of impact without detection, determines how it varies with the performance and cost of the search system, and equates its derivative with respect to detection radius to the derivative of the cost of detection to determine the optimum detection radius.

It is assumed that the detection radius searches out to a radius R_d, which means that in each repetition, the search covers a volume proportional to R_d^2. If the NEOs are distributed over a volume of radius R, a phase space estimate of the probability of detection per revolution is proportional to $q = (R_d/R)^2$. The Earth has a radius R_e, so there is also a probability per pass that it will sweep over and hit an NEO that is proportional to $p = (R_e/R)^2$. The model ignores the NEOs' true orbits and correlations between revolutions, assuming that successive revolutions represent independent samplings of the NEO distribution, and associates search repetitions with Earth orbits or years.

The total number of NEOs that could cause catastrophic damage is N; they are assumed to be spread uniformly throughout a sphere of radius R, so their average density is $n \approx N/(4\pi R^3/3)$. R can be determined from the estimate that there are about 2000 NEOs 1 km or larger in diameter, which produce a collision with the Earth about once every million years (Morrison 1993). If the Earth simply swept over them, the geometric collision frequency would be $f \approx nV_e \pi R_e^2$, where $V_e \approx 30$ km s^{-1} is the Earth's velocity in its orbit, which gives $n \approx f/V_e \pi R_e^2$, or $R \approx (3NV_e R_e^2/4f)^{1/3}$. For the parameters above, this gives an estimate of $R \approx 5 \times 10^8$ km ≈ 3.4 AU. That is a bit small to enclose all of the NEOs that could threaten the Earth, but it gives an estimate of the density of NEOs throughout space, including the positions of the undiscovered ones.

The probability that a given NEO will impact the Earth on the first revolution is $p \approx (10^4 \text{ km}/3.4 \text{ AU} \times 1.5 \times 10^8 \text{ km/ AU})^2 \approx 4 \times 10^{-10}$. The probability that it will miss on the first and hit on the second pass is $(1-p)p$. The probability that it will hit within the first t periods is

$$
\begin{aligned}
P &= p + (1-p)p + (1-p)^2 p + \ldots + (1-p)^{t-1} p \\
&= p[1 - (1-p)^t]/[1 - (1-p)] \\
&= 1 - (1-p)^t
\end{aligned} \tag{41}
$$

which rises linearly to ≈ 0.03 in 10^8yr. There are $N \approx 2000$ large NEOs. The probability that none of them will impact in time t is $(1-P)^N$, so the probability that there will be at least one impact is $U = 1 - (1-P)^N$, which reaches 0.5 in about 10^6yr, as expected from the total collision frequency used in constructing the model. The probability that a given NEO will not be detected until it impacts is

$$
\begin{aligned}
Q &= p + (1-q)(1-p)p + \ldots + (1-q)^{t-1}(1-p)^{t-1}p \\
&= p[1 - (1-q)^t (1-p)^t]/[1 - (1-q)(1-p)].
\end{aligned} \tag{42}
$$

The probability that at least one of the N large NEOs will impact without detection is $V = 1 - (1-Q)^N$, which is shown in Fig. 17. As R_d and q grow, the NEO volume is swept out after smaller numbers of orbits. For $R_d = 2$ AU the probability of impact without detection is $\approx 5 \times 10^{-7}$ at $t = 1$; it saturates at $\approx 2 \times 10^{-6}$ after ≈ 5 periods. For $R_d = 2$ AU, $q \approx 0.35$, so after 5 periods, $(1-q)^t \approx 0.65^5 \approx 0.1$, after which Q and V become independent of t. For $R_d = 1$ AU, V saturates at $\approx 10^{-5}$ in ≈ 30 yr; for 0.5 AU at $\approx 3 \times 10^{-5}$; for $R_d = 0.25$ at $\approx 10^{-4}$; and for $R_d = 0.125$ at $\approx 3 \times 10^{-4}$. For $R_d = 0$, V reduces to U and saturates at ≈ 0.5 in 10^6yr.

For impact without warning, the expected loss is about GT; thus, after t searches with detection radius R_d, the expected loss is about $GT V(t; R_d)$, which is proportional to V of Fig. 17. At any time t, the variation of the expected loss with detection radius, i.e., the marginal loss, is $GT \, dV/dR_d$. That can be determined numerically and equated to the marginal cost of detection range, which from Eq. (22) is

$$
dCs/dR = 2A[R(1+R)^2 + R^2(1+R)]/D^2. \tag{43}
$$

Figure 18 shows the marginal benefits minus costs, $-GT \, dV/dR_d - dCs/dR$, as a function of t and R_d for an $A = \$50\text{M-km}^2/\text{AU}^4$ sensor system and $D = 1$ km. For $R_d > 2$ AU, detection costs dominate and the marginal benefits less costs are negative for all times. For short times, i.e., very fast searches, marginal benefits less costs are negative for all R_d. For modest R_d and times of a few decades, benefits rise sharply as the risk of undetected

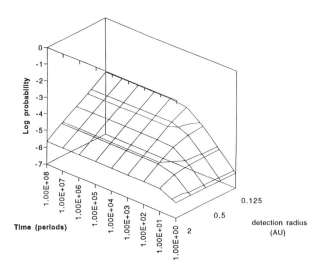

Figure 17.   Probability of large NEO impact without detection as a function of the detection radius of the search system.

impact compounds. For such R_d and times, the zero contour, on which $-GT\,dV/dR_d = dCs/dR$, moves from $R_d = 0$ at $t = 3$ yr out to $R_d \approx 1.3$ AU by $t = 10$ yr, and ≈ 1.5 AU by $t = 30$ yr. It remains at about that value for larger times. This contour indicates the R_d for which the marginal benefits of detection are just offset by their marginal costs, which is the optimal design R_d for the sensor system.

Because these costs and benefits are to be discounted back to the present with a rate of $\approx 5\%$ yr^{-1}, which implies a time horizon of ≈ 20 yr, the optimal value for R_d is about 1.3 to 1.5 AU. Because the marginal search costs, dCs/dR, depend strongly on R_d, modest changes in parameters would not affect this result greatly. It is interesting that for nominal parameters this model gives a value of R_d close to the ≈ 200 million kilometers ≈ 1.33 AU of Spaceguard. For $R_d = 1.5$ AU and $D = 1$ km, Fig. 9 gives a cost of $\approx\$200$M, which would correspond to annual expenditures of about 5% yr$^{-1} \times \$200M\approx\$10$M yr^{-1}, which is about the amount requested for Spaceguard. Thus, detection many orbits prior to impact appears to be cost effective with about the current level of technology proposed. It is possible to extend this calculation to an optimal combination of detection and defense, although that is not done here. Note, however, that the short time horizons implied by typical discount rates implies that the bulk of the contribution would be from relatively short times, in which case the analysis should reduce to a result like that of the LPC detection and defense studied above.

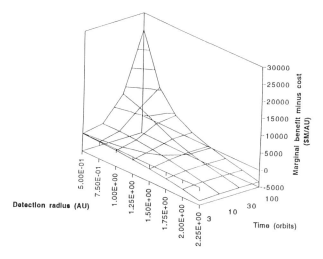

Figure 18.   Benefit minus cost for detection system as a function of the duration of the search and detection radius of the search system.

V. SUMMARY AND CONCLUSIONS

This chapter determines the value of detection and defense, presents simple estimates of the marginal costs of NEO detection and interception, derives optimal combinations of them that give the lowest-cost defenses for a range of NEO diameters and reaction times, and determines the detection range for which their marginal benefits equal the marginal cost of their construction and operation. In doing so, it determines the optimal mixes of detection and deflection for LPCs and NEOs of all sizes. It bounds the benefits of defenses by estimating the expected losses in their absence. It derives simple scaling models for the performance and cost of detection and deflection systems and algebraic models of detection of NEO.

The optimization is accomplished in two steps. In the first, the marginal costs of detection and deflection are calculated and equated to determine optimal detection times, ranges, and costs, which scale strongly on NEO diameter, but weakly on other performance and cost parameters. In the second they are equated to the marginal benefits, i.e., the reductions in expected losses from the first section, to explicitly determine the detection range and deflection characteristics of affordable defensive systems for LPCs, small NEOs, and large NEOs detected either on final approach or many orbits before impact.

The maximum LPC diameter for which defenses are cost effective is ≈8 km. For this diameter, optimal defenses would cost ≈$1B and would require roughly the detection capability of the proposed Spaceguard system. Defenses of this magnitude could arguably be developed over a period of a few decades for expenditures of ≈$50M yr^{-1}, which is commensurate with expected losses. Thus, in the case of LPCs, balanced defenses appear both attainable and attractive. Delays longer than the optimal detection times could increase these costs significantly. Advanced interceptors could decrease optimal detection times and ranges by factors of ≈ 2 and the cost for optimal defense by a factor of ≈ 10 for large NEOs. For large NEOs, detection and warning alone would also be cost effective.

The analysis can be extended with less precision to smaller NEOs, although there are concerns about the models for detection, cost, and expected losses. Smaller diameters drive detection to shorter ranges, where radars and other sensors could be used. The cost for nuclear defenses could be larger than those for kinetic energy interceptors, which could be placed into orbit before detection, maneuvered into the path of the NEO, and used to convert its own kinetic energy into the means to fragment them into pieces too small to survive re-entry.

An extension of this model produces an estimate of the value of detection many orbits before impact. When the marginal costs for early detection are equated to the marginal expected losses for search times of a few decades, the optimal detection range is ≈ 1.5 AU, which is about that of the proposed Spaceguard system. The required sensor system could apparently be produced with current technology for roughly the costs implied by the marginal analysis. Thus, for nominal performance and costs, detection appears both affordable and desirable for the full range of NEO hazards now projected. Defense through deflection is also appropriate for most catastrophic threats. Such defenses could be developed through programs that would be modest compared to current space defense expenditures.

REFERENCES

Ahrens, T. J., and Harris, A. W. 1992. Deflection and fragmentation of near-Earth asteroids. *Nature* 360:429–433.

Canavan, G. 1993*a*. Value of space defenses. In *Proceedings of the Near-Earth Object Interception Workshop*, eds. G. H. Canavan, J. C. Solem and J. D. G. Rather (Los Alamos: Los Alamos National Lab), pp. 261–274.

Canavan, G. 1993*b*. Acquisition and track of near-Earth objects. In *Proceedings of the Near-Earth Object Interception Workshop*, eds. G. H. Canavan, J. C. Solem and J. D. G. Rather (Los Alamos: Los Alamos National Lab), pp. 212–220.

Canavan, G., and Teller, E. 1990. Strategic defence for the 1990s. *Nature* 344:699–704.

Chapman, C., and Morrison, D. 1994. Impacts on the Earth by asteroids and comets: Assessing the hazard. *Nature* 367:33–40.

Glasstone, S. 1962. *The Effects of Nuclear Weapons* (Washington, D. C.: U. S. Atomic Energy Commission).

Hills, J. G., Bourgeois, J., and Sandford, M. T. 1993. Deepwater waves and tsunami generated by small asteroid impact. Hazards Due to Comets and Asteroids, Jan.

4–9, 1993, Tucson, Ariz., Abstract book, p. 40.

Hills, J. G., and Goda, M. P. 1993. The fragmentation of small asteroids in the atmosphere. *Astron. J.* 105:1114–1144.

Morrison, D. 1993. The impact hazard. In *Proceedings of the Near-Earth Object Interception Workshop*, eds. G. H. Canavan, J. C. Solem and J. D. G. Rather (Los Alamos: Los Alamos National Lab), pp. 51–53.

Phipps, C. 1993. Dynamics of NEO interception. In *Proceedings of the Near-Earth Object Interception Workshop*, eds. G. H. Canavan, J. C. Solem and J. D. G. Rather (Los Alamos: Los Alamos National Lab), pp. 111–115.

Rabinowitz, D. L. 1991. Detection of Earth-approaching asteroids in near real-time. *Astron. J.* 101:1518–1529.

Shelkov, E. 1991. Space patrol—Earth's protection from space objects. In *Proceedings, Conference on Planetary Emergencies*, ed. A. Zichichi (River Edge, N. J.: World Scientific).

Shoemaker, E. 1983. Asteroid and comet bombardment of the Earth. *Ann. Rev. Earth Planet. Sci.* 11:461–494.

Spalding, R. 1993. Satellite monitoring of fireball events. Paper presented at Erice International Seminar on Planetary Emergencies on Planetary Emergencies, 17th Workshop: The Collision of an Asteroid or Comet with the Earth, April 18–May 4, Erice (Italy).

Tagliaferri, E. 1993. Asteroid detection by space based sensors. Paper presented at Erice International Seminar on Planetary Emergencies, 17th Workshop: The Collision of an Asteroid or Comet with the Earth, April 18–May 4, Erice (Italy).

THE COMET AND ASTEROID IMPACT HAZARD IN PERSPECTIVE

P. R. WEISSMAN
Jet Propulsion Laboratory

The potential hazard from comet and asteroid impacts is one of a number of serious natural and man-made calamities facing modern society. However, only three of these currently have the potential to wipe out a significant fraction of human life on this planet: impacts, nuclear war, and the AIDS epidemic. How urgent then is it that action be taken at this time to mitigate the possibly catastrophic effects of impacts, and what fraction of available resources should be given over to this problem? By at least two measures commonly used to estimate the seriousness of potential threats, frequency of occurrence and annual fatality rate, impacts do not demand priority attention at this time. In addition, the credibility of the impact hazard with the public and with government decision makers is likely too poor at this time to support a drive for major expenditures on defensive systems. A program of public education and ongoing research is recommended to gain public support, to better establish the nature of the impact hazard, and to provide a database for eventual mitigation development. Waiting a decade or more to begin work on a defensive system subjects the Earth's population to minimal risk, while allowing emerging technologies to develop which may sharply reduce the cost and/or complexity of such a defense.

I. INTRODUCTION

To present-day astronomers, the idea of comets and asteroids striking the Earth with catastrophic consequences, is obvious. The surfaces of the terrestrial planets and the Moon, as well as the satellites of the giant planets, bear testimony to the violent early bombardment history of the solar system. Age dating of craters on the Earth shows that this is an ongoing process, and the presence of comets and asteroids in Earth-crossing orbits allow calculation of the probability of random impacts by this population of objects (Weissman 1990; Shoemaker et al. 1990).

To the general public, however, the threat of cataclysmic meteorite impacts seems a remote one. There is no known incident of a major crater-forming impact in recorded human history. The only large, documented airburst event which resulted in substantial surface damage is the 1908 Tunguska explosion, and even that event is much less widely known to the general public than similar terrestrial catastrophes such as the 1883 Krakatau volcanic eruption, which was of comparable magnitude. Lesser events such as the Sikhote-Alin meteorite fall in 1947 (Krinov 1963) or the October 1992 fall of a 27 pound meteorite in Peekskill, New York, which damaged a Chevy Malibu sedan (di Cicco 1993), are looked upon as curiosities, but not generally as harbingers of much worse catastrophes.

There are thus two important aspects to the comet and asteroid hazard problem. First, how can astronomers best describe the impact hazard problem to the general public and to decision makers in government, so as to establish the genuine nature of the hazard and maintain credibility, while not creating undue alarm? Second, what weight should be given to the problem and what resources diverted to meet it, in the light of numerous other hazards encountered by society in an already imperfect world?

These questions deal with subjects which go far beyond the normal scientific questions that we as astronomers are asked to consider. They involve economic, social, political, and moral issues, as well as scientific and technological ones. This researcher can not claim to be an expert in all those areas. However, the organizers of the meeting have asked the author to provide such a discussion, with particular attention to the question of whether a technology program to develop an active impactor defense should be initiated in the near future.

The discussion that follows should thus be looked upon as one by an informed citizen and generalist, and not a specialist and expert in each of the many disciplines involved. The discussion will be technical, but will also involve opinions and judgements which are solely those of this researcher; wherever possible, those opinions and judgements will be clearly identified as such. In bringing the comet and asteroid hazard problem to the fore, it is our responsibility as scientists to provide the best possible information in an unbiased manner, so that a full public discussion of the issues and informed decision making can take place.

In attempting to place the impact hazard problem in perspective, there are several major ways in which the problem can be considered. First, what is the **urgency** of the problem? How important is it that some action be taken now? What are the consequences of not taking action? Second, what is the **uncertainty** in our knowledge of the potential impactors? Do we know enough about their physical structure, population, and orbital dynamics to properly evaluate the hazard and to design mitigation technologies? Third, what is the **priority** of the problem? Are there other more immediate problems demanding attention and resources? Should the impact hazard take precedence over other hazards confronting society and the environment? Fourth, what is the current **credibility** of the impact hazard problem in the eyes of the public, of government officials, and of the media? Will requests for funding for increased telescopic searches, spacecraft missions to near-Earth objects, and study of mitigation technologies fall on deaf ears, or worse yet, be greeted with derisive laughter?

In addition, is deflection of asteroids and comets **plausible** at this time with currently available or soon-to-be available technologies? Can deflection technologies be developed and tested **legally** at this time, or would they require renegotiation of existing international treaties? Can such technologies be tested **safely** so that they do not pose a risk to international stability or result in costly accidents, and so that they do not somehow, inadvertently, increase the impact hazard to the Earth? These questions will be discussed below.

II. URGENCY OF THE IMPACT HAZARD PROBLEM

The question of urgency is one that lends itself most readily to a scientific analysis. The current knowledge of Earth-crossing and Earth-approaching comet and asteroid orbits is sufficient to estimate impact probabilities and energies with a fairly high degree of accuracy. Estimates of the number of objects and their mass distributions, while still somewhat uncertain, particularly with regard to comets, is in rough agreement with the frequency of impacts observed from counted craters on dated surfaces (Grieve 1987). Thus, one can readily estimate the risks to the Earth's population of waiting to take action on this problem.

For the sake of this discussion, assume that a rudimentary asteroid and comet deflection capability can be developed in a time period of ten years. Also, assume that the applicable technology will be nuclear warheads on conventional chemical rockets; this is the basic deflection system that has currently been proposed (Ahrens and Harris 1993); non-nuclear deflection is also possible, but likely would require far more development effort and time. All of the major areas of technology required for a nuclear weapons-based mitigation defense are currently in hand. They include: large launch vehicles (e.g., Titan IV, Proton, Shuttle, Energia), nuclear warheads up to 10 megatons (or more?) in explosive yield, near-Earth object detection and tracking telescopes, and technologies for navigating a warhead to, and homing in on a specified target. There are many questions that still need to be answered in integrating the various technological elements into a functioning defensive system, and many questions about how it should be used. Assume that the ten-year development program will answer all of those questions.

Ten years does not seem unreasonably long or short for such a program. Consider that the Manhattan Project produced not one, but two different, functional nuclear weapons within five years of the start of the program. The Apollo Program accomplished a manned landing on the Moon (and successful return) in just over eight years from the speech by President Kennedy initiating the program in 1961. In each of these cases, much of the required technology was not in hand at the start of the program, and numerous major design problems needed to be solved in the course of development. In comparison, integration of existing technologies into a rudimentary asteroid/comet defense seems a relatively easily obtainable, though likely expensive, goal.

Of course, it is possible that the greater complexity, bureaucracy, and fear of failure that has plagued major defense and space projects in recent years would increase the development time. As a hedge against this possibility, the equations below are written in a form so that the reader can insert his or her own values for any of the critical parameters.

What then is the risk to which the Earth's population is exposed by not immediately developing a deflection defense against asteroids and comets at this time? Currently, no known object is on an Earth-impacting trajectory for the predictable future. What is the probability that existing searches will discover such a body next year? That probability is given by

$$p = 1 - (1 - p_i)^{ND} \approx p_i ND \qquad (1)$$

where p_i is the mean probability of impact for an Earth-crossing asteroid or comet, N is the number of years between discovery and impact to be considered, and D is the number of objects discovered per year.

Consider the Earth-crossing asteroids first. The mean probability of an impact per object is 4.2×10^{-9} yr^{-1} (Shoemaker et al. 1990). Discovery rates ranged from 15 to 50 yr^{-1} between 1988 and 1992 (Morrison 1992); as a rough mean, a value of 35 yr^{-1} will be used here. Then, the probability of discovering an object next year that will impact within 10 yr is

$$p = 4.2 \times 10^{-9} \times 10 \times 35 = 1.5 \times 10^{-6} \qquad (2)$$

a fairly small number.

One can further extend the calculation by asking what is the probability that an Earth-crossing asteroid will be discovered at any time in the next N years, with an impact occurring within those N yr? Assuming the same numbers as above, but allowing discovery rates to grow 20% per year, and summing the probability over each of the next 10 years, the resulting value is $p = 1.5 \times 10^{-5}$, still a rather small number.

Suppose then that all of the estimated 1030 Earth-crossing asteroids (ECAs) with diameters >1 km are discovered (Shoemaker et al. 1990; other estimates put the number of ECAs greater than 1 km diameter as high as 2100; see the Chapter by Rabinowitz et al.). What is the probability that one of them will impact within the next 10 years?

$$p_{ECA} = 4.2 \times 10^{-9} \times 10 \times 1030 = 4.3 \times 10^{-5} \qquad (3)$$

still a fairly low probability event. Note that this is the probability for an event anywhere on the Earth's surface, much of which is either ocean or sparsely populated land areas.

One can perform a similar calculation for smaller, Tunguska-type events. For these locally destructive events, one can restrict the hazard estimate to airburst events over heavily populated areas. The annual probability of a Tunguska-type event anywhere on the Earth is estimated to be $\sim 4 \times 10^{-3}$ (Morrison 1992). Only 29% of the Earth's surface is land area, and much of that is uninhabited (Kurian 1989). In the United States, urban areas and transportation account for 2.9% of land use; the world average is closer to 2% (Ehrlich and Ehrlich 1989). Ninety-two percent of the land in the United States is classified as pasture, rangeland, crops, ungrazed forest, desert, swamp, tundra, national parks, wildlife refuges, surface mining, and transmission lines (Ehrlich et al. 1977). Farm buildings account for 1.2% and military bases 1.3%. To be conservative, assume that worldwide one would be concerned about airbursts over a land area twice that of the urban fraction in the United States, or 5.8%. Then the annual probability of a Tunguska-type event over a populated area is 6.7×10^{-5}. The probability of a Tunguska event over a populated area occurring in the next 10 years is 6.7×10^{-4}.

Similar calculations can be performed for long and short-period comets. Weissman (1982) found a mean impact probability for long-period comets (LPC) of 2.2×10^{-9} per perihelion passage (see also the Chapter by Marsden and Steel) Weissman (1990) estimated that an average of 10.1 long-period comets brighter than absolute magnitude $H_{10} = 11$ (radius >1.2 km) pass perihelion inside the Earth's orbit per year, based on Everhart's (1967) flux of long-period comets, corrected for observational selection effects. Using the cometary mass distribution found by Weissman (1990), this implies 57.2 Earth-crossing comets >1 km in diameter per year. Thus the probability of a long-period comet impact sometime in the next 10 years is

$$p_{\mathrm{LPC}} = 2.2 \times 10^{-9} \times 10 \times 57.2 = 1.3 \times 10^{-6}. \qquad (4)$$

This is less than the hazard posed by the Earth-crossing asteroids, even when one considers the higher mean impact velocities of the long-period comets. Substantially higher cometary fluxes are expected during "cometary showers" (Hills 1981) when large numbers of long-period comets are perturbed out of the Oort cloud by a close stellar passage or an encounter with a giant molecular cloud (Weissman 1990). However, current evidence is that we are not experiencing a cometary shower, and the onset time for such a shower is $\sim 10^4$ to 10^5 yr, so they do not pose an immediate threat.

There are 26 short-period comets (SPC) in Earth-crossing orbits listed in the most recent comet catalog (Marsden and Williams 1993). Of these, four are lost (their orbits are not well determined and they have not been observed on recent returns), one disintegrated in 1853 (P/Biela), and one is no longer Earth-crossing (P/Lexell). Taking the 26 orbits as representative, the mean impact probability is 7.3×10^{-10} yr^{-1} (Chapter by Shoemaker et al.). The estimated total number of active short-period comets larger than 1 km in diameter is ~ 40 to 100 (Weissman 1990; see also the Chapter by Shoemaker et al.). Thus the probability of an impact in the next 10 years is given by

$$p_{\mathrm{SPC}} = 7.3 \times 10^{-10} \times 10 \times 100 = 7.3 \times 10^{-7}. \qquad (5)$$

One possible additional source of impactors not considered here is extinct short-period comets. It has been suggested that some fraction of the known Earth-crossing asteroids (possibly as much as 50%) are short-period comet nuclei that have evolved to dormant, inactive states (Wetherill 1988; Weissman et al. 1989). These objects are presumably already included in the estimated Earth-crossing population of 1030 to 2100 objects >1 km in diameter. However, there may additionally be a substantial number of extinct short-period comets in more typical Jupiter-crossing orbits, analogous to the active Jupiter and Halley family comets. Shoemaker et al. (see their Chapter) argue that such objects may account for as much as an additional 25% in the total impact rate. That estimate is very uncertain because of the complex observational and dynamical selection effects involved, and because of assumptions as to whether the source of the short-period comets is the Oort cloud or the Kuiper belt. For the purposes of the discussion here, those objects will be ignored; although they may raise the total impact rate by $\sim$25%, that is not significant in the context presented herein.

The total impact hazard for all of the possible sources discussed above can be approximated by summing the individual probabilities. The result is a probability of 7.2×10^{-4} for an impact sometime in the next ten years. That total is dominated, as one would expect, by the smallest, most frequent impactors, the locally destructive Tunguska airburst events. If one considers only impacts by objects >1 km in diameter, then the probability is 0.5×10^{-4} of such an impact in the next ten years. That rate is dominated by near-Earth asteroids, and is in good agreement with estimates by Shoemaker et al. (1990).

The question then is whether these probabilities are sufficient motivation to justify action by government agencies, in particular development of a deflection capability. Governments regularly construct and/or promote defensive systems against natural disasters such as earthquakes, floods or storms, through mechanisms such as building codes, levees, warning and evacuation plans, etc. These defensive systems are typically scaled to deal with expected events that occur, within a factor of two, once in 100 years. For the impact hazard, the frequency with which events might occur is less than once in 10^4yr for Tunguska-like airbursts (over populated areas), and less than once in 10^5yr for major impacts (anywhere on Earth). Thus, major destructive impacts are not frequent enough to fit within the context of normal disaster planning.

In addition, pragmatic and/or parochial disaster planners will note that the land area of any single nation is only a small fraction of the total target area of the Earth. The United States occupies 6.4% of the land area of the Earth, or only 1.9% of the total area. For Russia, the corresponding numbers are 11.5% and 3.3%; for the member nations of the European Space Agency those numbers are 1.7% and 0.5%. Because only currently space-faring nations are likely to be able to do anything about the impact hazard, will they commit resources to defending what will likely be someone else's territory against these low probability events?

At some point, impacts become a global hazard because of climatic effects, and thus a major impact anywhere on the Earth will likely affect all nations. The threshold for globally catastrophic events is highly uncertain.

Toon et al. (see their Chapter) put the threshold at an impact energy of 10^5 to 10^6 megatons, or an impactor diameter of 1 to 2.2 km (assuming an asteroid with a density of 3.5 g cm^{-3} and an impact velocity of 20 km s^{-1}). Based on the estimates above, the probability of such an impact is 0.5×10^{-5}yr^{-1} for the lower limit of the energy range, or about five times less for the upper limit.

This, of course, is the conundrum of the impact hazard. The probability of major impacts occurring on the Earth, or Tunguska-type airbursts over populated areas, is typically much lower than most natural or man-made disasters, but the possibly lethal results may be very much greater. How does one properly allocate resources to such rare but devastating events?

III. UNCERTAINTY OF THE IMPACT HAZARD

As already noted, impact frequencies are fairly reliably known for both comets and asteroids, with the major uncertainty coming from estimates of the sizes of each population, and the mass distributions of the individual objects. Estimates are generally much better for asteroids than for comets. However, the cratering rate on the Earth and Moon obtained by counting craters on dated surfaces (Grieve 1987), serves as a direct check on those estimates and shows that they are correct to within a factor of 2 to 3.

Considerably less is known about the physical nature of the individual impactors. Without such a database, it is difficult to develop precise models for how to deflect these objects. Of particular interest are the internal structure, bulk density and material strengths. Some of this information can be inferred from meteorite samples recovered on the Earth, but the meteorites are a biased sample, both dynamically and compositionally, and likely do not contain any samples of cometary materials.

Current models of the internal structure of both asteroids and comets have tended to focus on what are known as rubble pile or fractal models (Davis et al. 1989; Weissman 1986; Donn 1991). In the case of the asteroids, it is suspected that many are reassembled fragments of larger objects, bound only by self-gravity, while for the comets the nuclei are believed to be weakly bonded, primordial agglomerations of small icy planetesimals. Support for these models have come from radar observations of two near-Earth asteroids, 4769 Castalia and 4179 Toutatis, both of which appear to show bimodal structure (Ostro et al. 1990, 1993), and from observations of random and tidal disruption of cometary nuclei, most recently comet Shoemaker-Levy 9 which apparently disrupted during a pass within the Roche limit of Jupiter in 1992 (Marsden 1993). In addition, theoretical modeling of the effects of impacts on asteroidal bodies (Nolan et al. 1992) has suggested that the internal structure of many small asteroids could be highly fractured, even if they appear to be single, unified bodies.

Additional uncertainty exists with regard to asteroid and comet regoliths. It was expected that there would be very little regolith on small asteroids because of the inability of their weak gravitational fields to retain even low-velocity ejecta. However, Galileo spacecraft visual and infrared imaging of asteroids 951 Gaspra (Belton et al. 1992; Weissman et al. 1992) and 243 Ida (Belton et al. 1993; Carlson et al. 1993) has implied a substantial regolith on each of the asteroids' surfaces. Little is known about cometary regoliths. It is expected that cometary nuclei cover themselves with a nonvolatile lag deposit of large grains, but it is not known if this material is simply a loose agglomeration or a welded surface layer (Rickman 1991).

The uncertainty about the internal structure of comets and asteroids, and the existence of regoliths on small asteroids, both create problems for deflecting these bodies by means of large explosions on their surfaces or nearby in space. Because of the regolith, coupling of the blast to the underlying "bedrock" (if there is a single, unit structure) may be highly inefficient. If the entire body is fragmented, in essence a continuous regolith, then coupling of the blast to the object may be extremely poor. A deflecting blast may instead result in fragmentation of the asteroid or comet nucleus, with little or no change in orbital parameters. Thus, instead of a single large object on an Earth-impacting trajectory, one may produce a "shotgun blast" of smaller impacts. It is entirely possible that the cumulative effect of those numerous smaller impacts may be much greater than a single impact by an equal mass object.

The Galileo flybys of Gaspra and Ida were not able to measure the asteroids' masses, and hence their bulk densities. Indirect estimates of asteroid masses of a few of the largest mainbelt objects have been made based on perturbations of other mainbelt asteroids during close approaches (Schubart and Matson 1979). These values have tended to confirm expectations of density based on spectroscopic type and meteorite analogs. However, the errors in such estimates are typically ~ 10 to 50%, and no measurements have been made of the density of small asteroids, similar to the Earth-crossing asteroids.

Estimates of comet nuclei densities have also been performed indirectly, based on fits to observed nongravitational forces (from jetting of surface volatiles) in the orbital motion of comets. Estimates for comet Halley range from 0.2 to 0.5 to 1.2 g cm^{-3} (Rickman 1989; Sagdeev et al. 1987; Peale 1989), with error bars extending over the entire range of possibilities. Thus, for the moment, cometary bulk densities are essentially unknown.

In the case of near-Earth asteroids, UBV photometry and visual and infrared spectroscopy has allowed the identification of the surface compositions of some of the near-Earth asteroids (NEAs), and these have been matched to meteorite analogs, fragments of the NEAs recovered on the Earth's surface. However, the bulk composition of the individual asteroids is still unknown, though the Galileo measurements did show evidence for some compositional heterogeneity on Gaspra (Granahan et al. 1992). Although considerable data was obtained on cometary composition from the Halley flyby missions in 1986 (Krankowsky 1991), there is still a great deal more that needs to be studied, in particular about the hydrocarbon component of the nucleus. The discovery that a substantial fraction of the nucleus mass was contained in pure hydrocarbon, or "CHON" (for carbon-hydrogen-oxygen-nitrogen) particles was one of the major surprises of the Halley spacecraft missions. Also, there is evidence for chemical heterogeneity among the individual nucleus fragments (Mumma et al. 1993). These current unknowns concerning composition will introduce additional uncertainty in estimating the coupling between the deflecting blast and the comet or asteroid.

To remove these uncertainties, a series of spacecraft missions are required to study the composition and physical structure of Earth-approaching comets and asteroids. These must be rendezvous missions so as to allow precise determinations of the mass and bulk density of the objects, as well as higher gravity harmonics which would be a clue to internal structure. The spacecraft must carry science instruments which will provide the elemental, molecular, and mineralogic compositions of each object. Internal structure should be probed using either microwave sounding techniques (likely possible for comets) or through direct seismic experiments. Rendezvous missions to multiple objects are required so as to examine compositional and structural diversity among these populations, and thus establish the range of parameters that could be expected in defending against an impact by any random object.

IV. PRIORITY OF THE IMPACT HAZARD

The hazard posed by impacts of comets and asteroids is not the only problem facing society. Currently identified ecological problems include overpopulation, global warming, global climate change (from volcanic aerosols), ozone depletion, and deforestation. Furthermore, there are human problems such as malnutrition, disease (in particular, but not only, AIDS), and pollution, and political problems such as nuclear proliferation and ethnic conflicts. Additionally, some areas of technical investigation, such as earthquake prediction, have the potential for preventing substantial loss of life and/or economic damage. These lists are not meant to be all-inclusive, but rather provide a sample of the global questions facing modern society.

All of these hazards place demands on governments for solutions, and for the resources to achieve those solutions. Many of the hazards are interrelated, in both positive and negative ways. For example, deforestation provides land for growing food and for accommodating population growth. On the other hand, malnutrition and disease serve as a check on overpopulation, though certainly not a very humane one.

What priority then should be given to the impact hazard problem? Is it more important than all of these other hazards? Potentially, very large impacts, comparable to the late Cretaceous event, could result in massive global fatalities. But such events have a mean frequency of once every 50 Myr. Smaller impacts may still result in sufficient climatic change to cause global crop failure and famines. Using the estimate of Toon et al. (see their

Chapter), the threshold for global effects occurs for impacts of objects 1 to 2.2 km in diameter, or with frequencies of about once every 2×10^5 to 10^6yr.

Among the hazards listed above, only two likely have the potential for massive, near-term loss of life on a global scale: nuclear war and AIDS. The threat of nuclear annihilation has decreased substantially in recent years as a result of the end of Cold War. However, many nations still possess nuclear weapons and others are attempting to obtain them. Some of the present or potential nuclear-capable nations are in what would be considered trouble spots, e.g., the Middle East, the former Soviet Union, and so there is a heightened potential for nuclear incidents, with unknown consequences.

The AIDS epidemic has now spread worldwide; an estimated 10 million people are infected with the AIDS virus including 2 million in the United States (Karplus 1992). AIDS-related deaths in the United States averaged 15,700 yr^{-1} from 1987 to 1989 (Wright 1991). Intensive medical research efforts to develop a cure and/or a vaccine have so far produced only limited results. It is entirely possible that a solution may appear at any time, but at present the disease continues to spread at an alarming rate.

Each of these two hazards clearly demand immediate and substantial attention and resources. Each has received substantial resources, both in the United States and in other developed countries. Given the immediate nature of these threats, it is entirely logical that they have priority over the impact hazard.

The other hazards listed above fall into two groups: immediate problems that continue to result in high death rates, and long-range problems whose effects are small now but have the potential to become major calamities in the foreseeable future. Examples of the first type of hazard are malnutrition and disease; examples of the latter are global warming and overpopulation. Note that for these problems, the phrase "foreseeable future" refers to the next 50 to 100 years. This is a relatively short time span as compared with the frequencies derived earlier for impact events.

It is worthwhile to discuss one of the lesser hazards listed above in somewhat more detail. During this author's preparation for the Hazards Due to Comets and Asteroids meeting, an article appeared in the *Los Angeles Times* (Roark 1992) describing the United Nations' efforts at dealing with common childhood diseases in underdeveloped countries (Fig. 1). The article reported that approximately 2.1 million children would die in the coming year due to preventable childhood diseases, because of a lack of vaccination programs in the underdeveloped nations. An additional 6.6 million children will die of curable diseases such as pneumonia and diarrheal diseases, which are treatable with common antibiotics. The estimate is necessarily statistical and based on past experience. However, the uncertainties in the estimate are likely relatively small, on the order of perhaps 10 to 20%. Thus, it is not a question of whether or not these children *may* die, but rather only the precise number that *will* die.

Annual death rates provide one basis for comparing the relative impor-

U.N. Calls Many Child Deaths Preventable

By ANNE C. ROARK
TIMES STAFF WRITER

Pneumonia is now the biggest killer of children in the world, resulting in 3.6 million deaths annually, but in most cases the cure is a five-day course of antibiotics that costs only 25 cents, according to a United Nations report released today.

The means of stopping pneumonia and dozens of other childhood diseases are now "available and affordable," the report said, but countries are not making the necessary investments in basic medical care, sanitation and education.

"The present neglect," the report said, "is a scandal of which the public is largely unaware."

Each week, the report found, a quarter of a million children die of malnutrition and diseases that are either curable or preventable.

"No famine, no flood, no earthquake, no war has ever claimed the lives of 250,000 children in a single week," said James P. Grant, executive director of the United Nations Children's Fund. "Yet malnutrition and disease claim that number of child victims *every week.*"

Child Deaths

More than 65% of the 12.9 million child deaths in the world each year are caused by pneumonia, diarrheal diseases, vaccine-preventable diseases or some combination of the three. Here are the main causes of deaths under age 5 in developing countries in 1990.

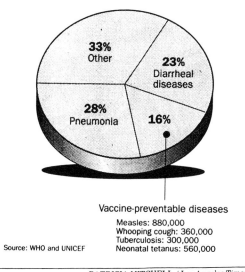

Source: WHO and UNICEF

Vaccine-preventable diseases
Measles: 880,000
Whooping cough: 360,000
Tuberculosis: 300,000
Neonatal tetanus: 560,000

PATRICIA MITCHELL / Los Angeles Times

Figure 1.   Front page article from the *Los Angeles Times*, December 17, 1992, describing statistics of child fatalities in underdeveloped countries due to a lack of basic medical treatments and vaccination programs (copyright 1992, The Los Angeles Times. Reprinted with permission.).

tance of individual hazards. Chapman and Morrison (1994) estimated that the nominal threshold impactor that will cause sufficient global climatic disruption to result in starvation of 10^9 people, a 1.5 km diameter impactor, similar to the event discussed in Sec. II, results in an average annual worldwide fatality rate of ~ 3000 yr^{-1}. The actual deaths likely all occur within one to two years after the impact, but the low expected frequency of that event results in the modest annual fatality rate. Thus, on the basis of annual fatalities, the impact hazard is orders of magnitude less lethal than the current lack of minimal medical programs for children in underdeveloped nations.

These figures can be compared with annual deaths in the United States from various other causes (data for 1987) (Wright 1991): fires and burns, 5000; drowning, 5000; falls, 12,000; motor vehicle accidents, 46,000; homicide, 19,000; suicide, 30,000; cancer, 473,000.

Thus, there are a wide range of hazards, either natural or man-made, which are either comparable to or greatly exceed the annual death rate expected from impacts. For each of these hazards there are oftentimes simple technological fixes which would contribute to greatly decreasing the death rates, e.g., mandatory seat belt laws for motorists; life jacket laws for boaters; smoke detectors in homes and workplaces. For others of these hazards, continued research can likely also reduce death rates, e.g., cancer. In yet other cases, there are legal remedies which could have potentially large effects, e.g., gun control to decrease homicide and suicide rates.

The point is that based on normal methods for evaluating risks such as annual fatality rates or frequency of occurrence, the impact hazard does not appear to have higher priority than many other problems with which society currently deals. Again, the problem here is how to respond to a very low probability event with a very high potential damage level.

What priority do non-scientists give the impact hazard? In 1992, the Democratic vice-presidential candidate (and eventual winner), Senator Albert Gore, Jr., published a book on environmental problems facing society, *Earth in the Balance* (1992). The book described a wide range of ecological problems, including overpopulation, ozone depletion, global warming, climate fluctuation (from volcanic dust input), and deforestation. No mention was made of the impact hazard, despite the fact that Senator Gore was chairman of one of the subcommittees that oversaw the NASA budget. The Senator (now Vice-President) either judged that the impact hazard was not significant enough to include, or that it was not a sufficiently credible threat.

Similarly, a recent book by W. J. Karplus with the intriguing title *The Heavens Are Falling: The Scientific Prediction of Catastrophes in Our Time* (1992) discusses eight major hazards facing society, including ozone depletion, climate change, overpopulation, AIDS and earthquakes. The book does not mention the asteroid and comet hazard.

Thus, the impact hazard is just one of a large number of problems currently facing society. If resources are spent to mitigate impacts on the Earth by large asteroids or comets, then they will likely come from resources currently spent on other, more immediate problems, at a likely cost of human lives. Deciding on what priority to give each individual hazard will continue to be a difficult process.

V. CREDIBILITY OF THE IMPACT HAZARD

The issue of credibility is a very important one for the impact hazard. As often noted, this topic has a high "giggle factor" and it is often not taken very seriously by the public or press. If an effective defense is to be developed against asteroid and comet impacts, then there must be widespread public understanding and support of the efforts to deal with the hazard. As noted in the introduction, public experience is that comets and asteroids do not strike the Earth; there is no known incident of a major cratering event in human history. As reporting of random meteorite falls like the 1992 Peekskill, New

York event becomes more widespread, public opinion may begin to accept the idea of larger impacts being possible.

One problem for those advocating an impact hazard defense and/or detection system is that their recommendations often appear to be self-serving. Astronomers who study small bodies have advocated an observing program that emphasizes searching for large (>1 km) Earth-crossing asteroids and comets (Morrison 1992). These are, in general, the same objects that those astronomers are currently discovering with their existing search programs. Thus, their recommendations can be viewed as motivated by a desire to obtain additional funding and instrumentation for their ongoing work. Conversely, recommendations by scientists and technologists involved with nuclear weapons programs have been to emphasize the danger of the smaller, Tunguska-like airburst events (Canavan et al. 1993). These smaller impactors are far more frequent, and thus more likely to create alarm. At the same time, they are likely small enough to be destroyed or deflected with currently existing technologies and warhead yields, similar in many ways to concepts studied for ballistic missile defense.

At first glance, the two groups can not both be right about which size range of objects poses the greatest hazard. A cynical observer would conclude that one or both groups is perhaps biasing its conclusions to fit its own needs. This perception of self-serving conclusions is further re-enforced by the declining funding situation at present for both planetary science and defense-related technology efforts in the United States and elsewhere. An example of this type of skeptical reaction by one very influential newspaper is shown in Fig. 2, an article, from *The Wall Street Journal*, entitled "Never Mind the Peace Dividend, the Killer Asteroids are Coming" (Davis 1992).

However, a more careful examination is clearly required. Although the recommendations may appear self-serving, that does not stop them from being correct. A medical doctor will benefit financially from the illness of his patients, but that does not mean that he or she will deliberately make them ill, or that he or she will find illnesses to be treated when none in fact exist (sadly, there are exceptions where doctors have been accused and convicted of precisely such crimes). In addition, legitimate scientific analyses often disagree when there is a lack of definitive data, or when the parties to the debate come from different communities of scholars, with different training and different philosophical outlooks. The scientists and technologists best able to advise the public on the impact hazard and possible deflection techniques are necessarily those who are already expert in these fields. As knowledge of this problem matures, it is likely that agreement will be reached on more and more of the relevant issues.

The problem then is how to create an atmosphere of credibility for the impact hazard efforts. The answer is to approach the problem slowly, and to conduct a patient campaign of public education in this area. This will clearly take time, but if demands for resources are rushed before the public acceptance of impacts is firm, then it is more likely that resistance will form

Never Mind the Peace Dividend, The Killer Asteroids Are Coming!

By Bob Davis

Staff Reporter of THE WALL STREET JOURNAL

WASHINGTON—Talk about Star Wars. The National Aeronautics and Space Administration is about to recommend that the Earth start planning to assemble an arsenal of nuclear missiles to head off an attack by asteroids. NASA astronomers figure that big asteroids smack into the Earth only once every 500,000 years, but say it's never too early to prepare.

"If you did find a [big asteroid], you'd have a danger to the Earth—something capable of killing one billion people," warns David Morrison, a NASA astronomer. Dr. Morrison persuaded Congress to fund two studies of the asteroid peril, which the agency plans to release in a few weeks.

At a meeting in January at Los Alamos National Laboratory in New Mexico, the home of The Bomb, NASA convened a group of weapons scientists to figure out ways to fight asteroids. Their conclusion: Build an armada of 10 ground-based missiles, each equipped with a 100-megaton warhead—bigger than any nuke ever exploded on Earth—and keep them ready for asteroid attack. Then, if astronomers spot the Big One, they can launch nukes and explode them in front of the incoming asteroid millions of miles in space.

The detonations would blow off a few inches of rock and dirt from the asteroid—enough to change its course slightly and save the Earth. "You can impart a more gentle push to the thing" than trying to blast it out of the sky, says Johndale Solem, a physicist at Los Alamos.

Dr. Solem says the Los Alamos group rejected a lot of fanciful ideas before settling on nukes to hunt asteroids. One scientist proposed launching 20,000 spears at a rogue asteroid. Another suggested nudging small asteroids into Earth orbit, and using them to attack bigger asteroids. That plan was called Brilliant Mountains—a big brother to the controversial Star Wars proposal called Brilliant Pebbles.

All this talk distresses Dr. Morrison, who simply wants Earthlings to spend $50 million to catalog all the one-half-mile wide or larger asteroids and comets that pass near Earth. If one of these asteroids is bound for Earth, it should make a few passes before impact. "You'd have decades of warning," he says.

But that isn't enough for Dr. Solem. He wants to hunt down football-field-sized asteroids, which could level New York City. To combat that threat, you'd need a nuclear armada, a prospect that moved a weapons scientist at Los Alamos to shout: "Nukes forever."

Figure 2. *Wall Street Journal* article on the impact hazard, March 25, 1992. The fact that a respected and influential publication like the *Wall Street Journal* treats some of the suggested responses to the impact hazard with such skepticism, is clear evidence of a credibility problem with the media and public. Reprinted by permission of The Wall Street Journal, copyright 1992, Dow Jones & Company, Inc. All rights reserved worldwide.

and will subsequently be difficult to overcome. Consider how society has reacted to other new environmental hazards as they have been discovered. The ozone depletion problem due to chlorofluorocarbons was discovered by Molina and Rowland (1974). The idea initially received a great deal of attention and public debate, and some of the responses to it were openly hostile. Because the problem involved what was then an $8 billion industry in the United States, manufacturers of chlorofluorocarbons were particularly anxious to refute Molina and Rowland's conclusions (Ehrlich and Ehrlich 1989). At one point, a leading industrial journal went so far as to charge that the ozone problem was a KGB plot to destabilize Western industry.

Support for the ozone problem oscillated back and forth for over a decade until the discovery of the Antarctic ozone hole in 1985. At that point the weight of evidence became great enough to force worldwide action on the problem, though even then there still was much resistance and foot-dragging because of the economic issues involved (Ehrlich and Ehrlich 1989).

Similar conclusions can be drawn from looking at initial public reaction to questions raised over pesticides, acid rain, global warming, etc. New scientific or technological problems are greeted with curiosity and interest, but resistance often builds because of vested interests that may be threatened by the solution. If the evidence is not overwhelming then it will be debated and a range of conclusions will be drawn, which usually extends from "no problem" to "immediate crisis." Additional data must be accumulated to resolve these questions. At the same time, the public, including the press and government decision makers, must be educated as to the nature of each problem.

The impact hazard problem is still a very immature one, in that the public education process has only just begun. Before public opinion will support the use of substantial government resources, i.e., tax dollars, to developing an asteroid and comet defense, it must first be convinced that the threat is genuine. That convincing will take time.

VI. OTHER ISSUES: PLAUSIBILITY, LEGALITY, AND SAFETY

Plausibility. Current expectations about the technology necessary to deflect asteroids and comets have centered around the use of nuclear weapons, launched on conventional rockets. Clearly, some form of rudimentary defensive system could be constructed against potential impactors, particularly the smaller objects, using this technology. But will such a defensive system work? As argued above, the current state of knowledge of comet and asteroid internal structure and regoliths is not sufficient to predict accurately the effects of a deflecting blast. In particular, the likely rubble pile nature of asteroids and cometary nuclei raises the threat that much of the energy will be expended disrupting the object rather than deflecting it. The consequences of disruption could well be to increase the lethal effects of the impactor(s).

The studies of possible deflection technologies to date (see, e.g., Canavan et al. 1993) are still fairly modest, and have not yet considered the problem in the detail necessary to know how well such a system would work. Before one goes forward with an actual hardware program, these studies need to be performed at a far more detailed level, and covering the full range of possible asteroid and comet parameters (unless better data becomes available). In addition, all aspects of a defensive system must be considered, not just the deflecting rockets, but also the detection system that will find and track potential impactors. The current Spaceguard proposal (Morrison 1992) outlines plans for a system that will detect the >1 km objects and some fraction of the smaller objects. If the decision is taken to try and defend against smaller objects, then a far more comprehensive detection system must be developed, and the costs and relative merits of such a system must be weighed very carefully.

One should also consider the possible role of emerging technologies in developing an effective impactor defense. Just because one can build a defensive system with the technology now in hand, does not mean that a defense should be built. It is highly possible that new technologies that are currently not anticipated, will provide breakthroughs that make development of an impactor defense far easier and cheaper. Predicting what those technologies will be is probably a futile exercise, but there is certainly potential for very large advances in the near future in superconductivity, artificial intelligence, and miniaturization of electronics. The point is that one cannot say what new and useful technologies will be available in 20 or 50 years, yet there certainly are going to be some, and that is a very short time to wait relative to the time scale of the impact hazard.

Legality. The 1967 Outer Space Treaty bans the use of nuclear weapons in space (Florini 1985). Specifically, the treaty says,

"States Parties to the Treaty undertake not to place in orbit around the earth any objects carrying nuclear weapons or any other kinds of weapons of mass destruction, install such weapons on celestial bodies, or station such weapons in outer space in any other manner."

Use or testing of nuclear weapons in space is also banned by the 1963 Treaty Banning Nuclear Weapons Test in the Atmosphere, in Outer Space, and Under Water.

Development of an asteroid/comet defense, as currently envisioned, would violate these treaties because of the use of nuclear weapons to provide the deflection impulse. It should, of course, be possible to negotiate exceptions to the treaties so as to make a planetary defense system legal under international law. However, such negotiations should not be entered into lightly, because the treaties provide valuable safeguards which make nuclear war less likely.

Any weapon that could be used to deflect asteroids could obviously also be used directly against nations on the Earth. As the world powers continue to "build down" nuclear weapons inventories, individual nations will require assurances of their own security before they agree to give up their weapons. The development of any nuclear defensive capability against impactors will have to be accomplished under international scrutiny, and will have to provide sufficient safeguards against malicious use, including against possible terrorist actions.

A potentially even worse scenario is the malicious use of the deflection capability to target an asteroid passing close to the Earth onto an impacting trajectory (Sagan 1992). Although such misuse may seem farfetched, its potential for creating a global catastrophe makes the development of an impactor defensive capability itself a risky venture.

Safety. Ultimately, one of the important goals of any impact hazard defense must be that it is safe to use, and that it does not pose a greater threat than the impact hazard itself. Given that the possible global damage from the larger impacts is so great, the potential threat from the defensive system must be much, much less.

The section above has mentioned some political concerns associated with the safety of a standby defensive system. Other concerns would be the safe and successful launch of the large warheads (a problem presumably already solved by nuclear weapons designers), and the subsequent orbital evolution of a deflected or disrupted comet or asteroid. If a threatening impactor is disrupted then, as its debris will all still be in Earth-approaching orbits, the larger fragments will need to be tracked and have their orbits predicted well into the future.

If a defensive system is developed, there will naturally be a desire to test it. Such tests must be performed in a dynamically safe area of the solar system, well away from the Earth. If the test is done on actual near-Earth objects, then there is a potential for accidentally deflecting an object onto an impacting trajectory, or for fragmenting the object, should the test go awry. Tests should be performed on analog objects in the main belt, and well away from the dynamical and secular resonances that deliver objects into Earth-crossing orbits.

VII. DISCUSSION

There is no doubt that impacts of comets and asteroids pose a genuine threat to human life on the Earth, with possibly extremely lethal consequences. There is also no doubt that impacts are only one of a large number of environmental threats that are currently recognized and which society must consider. None of the problems can be ignored, and none of the problems can consume all of the resources available to deal with them.

The story of "Chicken-Little" is firmly entrenched in the public consciousness. That story unfortunately creates a negative reaction to claims of impending disaster as a result of comet and asteroid impacts. In addition, common experience says that such events are not very likely, if they occur at all.

As scientists we know that impacts do occur, and that they have occurred on Earth with devastating consequences. Scientific investigations of past impact events, such as the late Cretaceous extinction, have tended to focus on the largest and most destructive events. This creates an unconscious bias within the scientific community that any future events will have similar consequences. That is certainly true for some, very rare events. The question we can not presently answer is at what level do smaller, more frequent impact events still have globally catastrophic consequences?

In addition, catastrophic events that occur frequently on the astronomical or geological time scales with which we are familiar, have a very low probability on human time scales. Again, this difference in perception heightens the concern of the scientists relative to that experienced by the general public.

Furthermore, there is a very different perception among the public with regard to natural and man-made disasters. On the assumption that human beings have some control over their technological creations, systems such as air travel, nuclear power, and food production are compelled to have very high safety standards. This is particularly true in instances where humans must surrender their control to others; deaths in air travel are investigated intensely, while a much higher annual death rate from motor vehicle accidents is accepted with far less questioning. In contrast, humans regularly decide to accept the risk of living in earthquake zones, on flood plains, or in areas frequented by tornados or hurricanes, either because the frequency of such catastrophes is not high enough to evoke concern, or because of overriding economic and social requirements, or because they are viewed as natural disasters, and thus, "nobody's fault."

As a natural disaster, and a very infrequent one, the impact of comets and asteroids is less likely to evoke concern than ozone depletion or global warming, because the latter are looked upon as man-made disasters, and the likely time scale for those problems to become serious is far shorter. This ignores an important fact about the impact hazard, that it is random and could occur at any time, but that too is part of the nature of the human response to hazards.

Lastly, a significant problem is what might be termed the "gender gap" of the impact hazard. During the Tucson meeting, a woman astronomer attending some of the sessions commented to friends on what she perceived as the "little boys and their bombs" aspect of the ongoing studies. It is a fact that the vast majority of scientists involved with this problem are male: the Morrison (1992) report, for example, was authored by 22 men and 2 women, and reviewed by 4 men; the senior authors of the Canavan et al. (1993) report were all men, as are the senior authors of all the chapters in this volume. This is out of proportion to the ratio of men to women in planetary astronomy and in defense technology. It is entirely possible that one-half of the Earth's population may regard the attention given the impact hazard as an unnecessary and wasted exercise.

This chapter has attempted to show that rapid action in developing an asteroid and comet defense is neither necessary, nor prudent. Expected frequencies of impacts are low compared to other natural disasters and it is unlikely that waiting a modest period of time to start a technological program will substantially endanger human beings on this planet. The current state of knowledge of the physical nature of asteroids and comets is not sufficient to design a deflection defense at this time. There exist a large number of other problems facing society which are more immediate in nature, and which threaten far higher annual fatality rates. The current public acceptance of the reality of the impact hazard is poor and there exists a genuine need for a program of public education before governments can be convinced to devote substantial resources to this problem. Lastly, a variety of technological, legal, and safety issues must be solved before development of an asteroid and comet defense can go forward.

At the present time, the best advice that I can offer to my scientific and technological colleagues is, "GO SLOW." Premature or overly ambitious attempts to divert substantial resources to dealing with the impact hazard are likely to be counter-productive. The correct course is to prepare carefully the public for the problem by a program of public education, along with ongoing observational studies to find potential impactors and to improve our knowledge of asteroids and comets. Because many of the important questions can only be answered by direct, *in situ* measurements, a program of spacecraft rendezvous missions is also called for, to as large and diverse a number of these objects as possible.

A technological program to develop a defensive system at this time would be premature, in large part because of the lack of sufficient knowledge of the targets to design a successful deflection or destruction capability, and also because of the likelihood of being overtaken by technological progress and breakthroughs long before such a capability is actually needed.

This chapter has likely raised more questions than it has answered. In general, these are political, social and moral issues which need to be debated openly by all concerned parties. One wishes for the wisdom of Solomon to provide the answers. The decisions that need to be made will not be easy, nor will the patience that must be exercised in educating the public while building the case for action on the impact hazard problem. As scientists, our task will be to increase our understanding of the hazard, inform the public and government as to what we have learned, and recommend prudent courses of action. It is a heavy responsibility, and one that no scientist involved in this area should take lightly.

Acknowledgments. The author is deeply grateful to A. Harris for his advice and assistance during the writing and review of this chapter. The author also thanks S. Ostro, C. Sagan, D. Morrison, C. Chapman, and two anonymous referees for useful discussions and comments. This work was supported by the Planetary Geology and Geophysics Program and was performed at the Jet Propulsion Laboratory under contract with the National Aeronautics and Space Administration.

Note Added in Proof

The recent impacts of the fragments of comet Shoemaker-Levy 9 on Jupiter have greatly heightened public awareness and interest in the impact hazard. This is a fortuitous event; it should greatly aid the education process advocated in this chapter. However, public interest in such entertaining science is often fleeting, and thus the need still exists for a long-term program of public education and discussion. Comet Shoemaker-Levy 9 has helped provide a very valuable step in the right direction, but it will still require a conscientious effort on the part of this community to turn that into the reality of an observational program to search for potential Earth impactors, and missions to investigate them with interplanetary spacecraft. At present, given the preliminary reports on the impacts by observers, there is no evidence that would argue for starting a weapons program to provide a defensive capability at this time, and the conclusions of this chapter on that subject remain the same.

REFERENCES

Ahrens, T. J., and Harris, A. W. 1992. Deflection and fragmentation of near-Earth asteroids. *Nature* 360:429–433.

Belton, M. J. S., Veverka, J., Thomas, P., Helfenstein, P., Simonelli, D., Chapman, C., Davies, M. E., Greeley, R., Greenberg, R., Head, J., Murchie, S., Klaasen, K., Johnson, T. V., McEwen, A., Morrison, D., Neukum, G., Fanale, F., Anger, C., Carr, M., and Pilcher, C. 1992. Galileo encounter with 951 Gaspra: First pictures of an asteroid. *Science* 257:1647–1652.

Belton, M. J. S., Chapman, C. R., Davis, D., Veverka, J., Gierasch, P., Thomas, P., Helfenstein, P., Klaasen, K. P., Harch, A., Johnson, T. V., Davies, M. E., Anger, C., Carr, M. H., McEwen, A., Fanale, F., Greeley, R., Greenberg, R., Head, J. W., Ingersoll, A., Neukum, G., and Pilcher, C. 1993. Galileo/SSI observations of 243 Ida. *Bull. Amer. Astron. Soc.* 25:1136 (abstract).

Canavan, G. H., Solem, J. C., and Rather, J. D. G., eds. 1993. *Proceedings of the Near-Earth Object Interception Workshop* (Los Alamos: Los Alamos National Lab).

Carlson, R., Weissman, P. R., Smythe, W. D., Segura, M., Matson, D. L., Johnson, T. V., Fanale, F. P., Granahan, J., McCord, T. B., Soderblom, L. A., Kieffer, H. H., Danielson, G. E., Encrenaz, T., and Taylor, F. W. 1993. Ida: Preliminary results from the Galileo Near Infrared Mapping Spectrometer. *Bull. Amer. Astron. Soc.* 25:1137 (abstract).

Chapman, C., and Morrison, D. 1994. Impacts on the Earth by asteroids and comets: Accessing the hazard. *Nature* 367:33–40.

Davis, B. 1992. Never mind the peace dividend, the killer asteroids are coming! *Wall Street Journal* March 25, p. B1.

Davis, D. R., Weidenschilling, S. J., Farinella, P., Paolicchi, P., and Binzel, R. P. 1989. Asteroid collisional history: Effects on sizes and spins. In *Asteroids II*, eds. R. P. Binzel, T. Gehrels and M. S. Matthews (Tucson: Univ. of Arizona Press), pp. 805–826.

di Cicco, D. 1993. New York's cosmic car conker. *Sky & Telescope* 85:46.

Donn, B. 1991. The accumulation and structure of comets. In *Comets in the Post-Halley Era* eds. R. L. Newburn, Jr., M. Neugebauer and J. Rahe (Dordrecht: Kluwer), pp. 335–359.

Ehrlich, P. R., and Ehrlich, A. H. 1989. *Healing the Planet* (New York: Addison-Wesley).

Ehrlich, P. R., Ehrlich, A. H., and Holdren, J. P. 1977. *Ecoscience* (San Francisco: W. H. Freeman).

Everhart, E. 1967. Intrinsic distributions of cometary perihelia and magnitudes. *Astron. J.* 72:1002–1011.

Florini, A. 1985. *Developing the Final Frontier: International Cooperation in the Peaceful Uses of Outer Space* (New York: United Nations Association U. S. A.).

Gore, A. 1992. *Earth in the Balance* (New York: Houghton-Mifflin).

Granahan, J. C., Fanale, F. P., Robinson, M., Carlson, R. W., Kamp, L. W., Klaasen, K. P., Belton, M. J. S., Cook, D., McEwen, A. S., Soderblom, L. A., Carcich, B. T., Helfenstein, P., Simonelli, D., Thomas, P. C., and Veverka, J. 1992. Galileo's multispectral synergistic view of 951 Gaspra. *Eos: Trans. AGU* 74:197 (abstract).

Grieve, R. A. F. 1987. Terrestrial impact structures. *Ann. Rev. Earth Planet. Sci.* 17:245–270.

Hills, J. G. 1981. Comet showers and the steady-state infall of comets from the Oort cloud. *Astron. J.* 86:1730–1740.

Karplus, W. J. 1992. *The Heavens Are Falling: The Scientific Prediction of Catastrophes in Our Time* (New York: Plenum).

Krankowsky, D. 1991. The composition of comets. In *Comets in the Post-Halley Era*, eds. R. L. Newburn, Jr., M. Neugebauer and J. Rahe (Dordrecht: Kluwer), pp. 855–877.

Krinov, E. L. 1963. The Tunguska and Sikhote-Alin meteorites. In *The Moon, Meteorites and Comets*, eds. B. M. Middlehurst and G. P. Kuiper (Chicago: Univ. of Chicago Press), pp. 208–234.

Kurian, G. T., ed. 1989. *Geo-Data: The World Geographical Encyclopedia* (Detroit: Gale Research).

Marsden, B. G. 1993. Comet Shoemaker-Levy (1993e). *IAU Circ.* 5744.

Marsden, B. G., and Williams, G. V. 1993. *Catalogue of Cometary Orbits 1993*, 8th ed. (Cambridge, Mass.: Smithsonian Astrophysical Observatory).

Molina, M., and Rowland, F. S. 1974. Stratospheric sink for chlorofluoromethanes: Chlorine atom catalysed destruction of ozone. *Nature* 249:810–814.

Morrison, D., ed. 1992. *The Spaceguard Survey: Report of the NASA International Near-Earth Object Detection Workshop* (Pasadena: Jet Propulsion Laboratory).

Mumma, M. J., Stern, S. A., and Weissman, P. R. 1993. The origin of comets: Reading the Rosetta stone. In *Protostars and Planets III*, eds. E. H. Levy and J. I. Lunine (Tucson: Univ. of Arizona Press), pp. 1177–1252.

Nolan, M. C., Asphaug, E., and Greenberg, R. 1992. Numerical simulation of impacts on small asteroids. *Bull. Amer. Astron. Soc.* 24:959 (abstract).

Ostro, S. J., Chandler, J. F., Hine, A. A., Rosema, K. D., Shapiro, I. I., and Yeomans, D. K. 1990. Radar images of asteroid 1989 PB. *Science* 248:1523–1528.

Ostro, S. J., Jurgens, R. F., Rosema, K. D., Winkler, R., Howard, D., Rose, R., Slade, M. A., Yeomans, D. K., Campbell, D. B., Perillat, P., Chandler, J. F., Shapiro, I. I., Hudson, R. S., Palmer, P., and de Pater, I. 1993. Radar imaging of asteroid 4179 Toutatis. *Bull. Amer. Astron. Soc.* 25:1126 (abstract).

Peale, S. J. 1989. On the density of Halley's comet. *Icarus* 82:36–49.

Rickman, H. 1989. The nucleus of comet Halley: Surface structure, mean density, gas and dust production. *Adv. Space Res.* 9(3):59–71.

Rickman, H. 1991. The thermal history and structure of cometary nuclei. In *Comets in the Post-Halley Era*, eds. R. L. Newburn, Jr., M. Neugebauer and J. Rahe (Dordrecht: Kluwer), pp. 733–760.

Roark, A. C. 1992. U. N. calls many child deaths preventable. *Los Angeles Times*, Dec. 17, p. A1.

Sagan, C. 1992. Between enemies. *Bull. Atomic Sci.* 48:24–26.

Sagdeev, R. Z., Elyasberg, P. E., and Moroz, V. I. 1987. Is the nucleus of comet Halley a low density body? *Nature* 331:240–242.

Schubart, J., and Matson, D. L. 1979. Masses and densities of asteroids. In *Asteroids*, ed. T. Gehrels (Tucson: Univ. of Arizona Press), pp. 84–97.

Shoemaker, E. M., Wolfe, R. F., and Shoemaker, C. S. 1990. Asteroid and comet flux in the neighborhood of Earth. In *Global Catastrophes in Earth History*, eds. V. L. Sharpton and P. D. Ward, Geological Soc. of America Special Paper 247 (Boulder: Geological Soc. of America), pp. 155–170.

Weissman, P. R. 1982. Terrestrial impact rates for long and short-period comets. In *Geological Implications of Impacts of Large Asteroids and Comets on the Earth*, eds. L. T. Silver and P. H. Schultz, Geological Soc. of America Special Paper 190 (Boulder: Geological Soc. of America), pp. 15–24.

Weissman, P. R. 1986. Are cometary nuclei primordial rubble piles? *Nature* 320:242–244.

Weissman, P. R. 1990. The cometary impactor flux at the earth. In *Global Catastrophes in Earth History*, eds. V. L. Sharpton and P. D. Ward, Geological Soc. of America Special Paper 247 (Boulder: Geological Soc. of America), pp. 171–180.

Weissman, P. R., A'Hearn, M. F., McFadden, L. A., and Rickman, H. 1989. Evolution of comets into asteroids. In *Asteroids II*, eds. R. P. Binzel, T. Gehrels and M. S. Matthews (Tucson: Univ. of Arizona Press), pp. 880–920.

Weissman, P. R., Carlson, R. W., Smythe, W. D., Byrne, L. C., Ocampo, A. C., Kieffer, H. H., Soderblom, L. A., Fanale, F. P., Granahan, J. C., and McCord, T. B. 1992. Thermal modeling of asteroid 951 Gaspra. *Bull. Amer. Astron. Soc.* 24:933 (abstract).

Wetherill, G. W. 1988. Where do the Apollo objects come from? *Icarus* 76:1–18.

Wright, J. W., ed. 1991. *The Universal Almanac 1991* (New York: Andrews and McNeel).

EVALUATING SPACE RESOURCES IN THE CONTEXT OF EARTH IMPACT HAZARDS: ASTEROID THREAT OR ASTEROID OPPORTUNITY?

WILLIAM K. HARTMANN
Planetary Science Institute

and

ANDREI SOKOLOV
Moscow, Russia

Current investigations indicate the potential for enormous economic value in the materials of Earth-approaching asteroids. A quantitative example is given for a 2-km asteroid, identified as metallic by radar techniques. The response to the asteroid threat should thus be considered only in coordination with the response to what we call the asteroid opportunity. Comparison of the multi-decade time scale of development of our space capabilities with the multi-century time scale of a global threat indicates that it is prudent to concentrate what efforts we can afford on a proactive feasibility study of asteroid resource exploitation, rather than a reactive program solely designed to destroy asteroids or alter their orbits. The program we advocate would be a more positive investment in the future than expending the same funds only to develop a defense. It would aim at cataloging, visiting, and processing materials of NEOs. A byproduct of such a program would be development of the ability to modify asteroid orbits, thus abolishing the asteroid threat. Given the complementary nature of the American and Russian space capabilities, an international partnership would be a much more effective approach to the problem, from the overall perspective of the human species, than unilateral national programs. Multinational collaboration on an upgraded Mir space station or Mir-2 station, rather than building the redundant, downsized American station, could release American funding for a broad NEO program and other projects, and provide a proving ground for developing long-term human capabilities in space.

I. ON THE VALUE OF ASTEROIDS

Our approach to the problem of the threat posed by near Earth objects (NEOs, including asteroids, extinct comet nuclei, and active comet nuclei) is essentially a minority report in this book. To us, it seems extraordinary that a major international conference on the asteroid threat could occur with virtually no discussion of the positive opportunity offered by asteroids. In short, the asteroid threat should be considered only in the context of the asteroid opportunity.

Technical studies of the asteroid opportunity are not new. Gaffey and McCord (1977) pointed out that modest size asteroids may contain materials worth more than $1 trillion that could supply Earth's needs of certain metals and other materials for decades. Herrick (1979) and Morrison and Niehoff (1979) touched on these ideas in a Space Science Series volume (the Herrick manuscript was written in 1971). O'Leary et al. (1979) published an analysis of asteroid prospecting and retrieval. These ideas have been further developed at a popular level (see, e.g., Hartmann 1982), a semi-technical level (see, e.g., Hartmann et al. 1984), and a technical level (see, e.g., Lewis and Lewis 1987).

To the objection that the asteroid opportunity is too subjective or visionary to consider in a volume of this sort, we offer two answers: first, the question of economic value is quantifiable; second, any reaction to the asteroid threat requires a broad, visionary discussion of the future, because we are asked to consider threats that may be a million years in materializing (i.e., reaching a probability of occurrence greater than one-half in the time considered) at the level of global disaster (Chapman and Morrison 1994), and a century or more at the level of localized, urban-scale disasters. The asteroid opportunity, moreover, is being examined in technical detail today; for example, the Space Science Series volume on space resources (Lewis et al. 1993) is an essential companion to this volume. To take a specific example, Lewis and his colleagues have pursued a multi-year multi-million dollar effort at the University of Arizona to identify processing techniques for space resources.

Hartmann et al. (1984) reviewed projections of resource needs in the next century and noted that most serious studies have predicted either catastrophic shortages by the end of the 21st century or radical changes in global sociological/political structures and lifestyles. In most such studies, the projected 21st century lifestyles were generally "poorer," based on definitions involving criteria such as per capita energy availability or per capita materials availability.

Hartmann et al. noted that virtually all of these economic projections assumed a *closed, finite, terrestrial environment*, and that virtually none considered the addition of resources that might be derived from our larger, cosmic environment. Input of *extraterrestrial* metals and solar energy resources would radically alter the equation.

For example, our current policy of exploration for *terrestrial* ores and fossil fuels has led to processing of progressively lower-grade ores and fuels, and begun the exhaustion of the most accessible resources. The processing of lower-grade ores and fuels, together with increased throughput volumes per year, has created environmental, economic, and health problems, because the products are dumped into the terrestrial biosphere. The consequences of continued terrestrial exploitation by developing third world countries is indicated not only by environmental problems in the United States, but also by the serious environmental problems that resulted from the shorter time scale industrialization of the Soviet Union.

Recent studies of interplanetary bodies, including Earth-approaching asteroids, indicate that the possibility of finding and processing extraterrestrial resources in space is credible, and that these resources could alleviate at least some of the problems that will result if we attempt continued economic development in a closed terrestrial system.

The meteorites in our museums—primitive chondritic stones (with dispersed metals), basalt-like achondrites, water-rich carbonaceous chondrites, stony irons, and pure irons—are a sampling of the materials that are to be found in larger bodies in space.

The pure metal asteroids might seem the most promising, but Lewis and Lewis (1987) emphasized the economic value of metals dispersed in chondritic asteroids. Considering a 1-km diameter chondritic stone asteroid as a 1-km ore body which could be mined over a period of years, they compute an economic value of $150 billion (1987 market prices) for the 7500 tons of platinum group metals alone, which they propose could be returned to Earth in the equivalent of 300 Space Shuttle loads; "thus the platinum-group byproduct would by itself more than pay for all the shuttle flights that could be flown over a 25 year period. This is also enough platinum to meet Earth's total use rate for several decades" (Lewis and Lewis 1987, p. 262). They assume that the iron, nickel, and cobalt resources in this metal could be utilized in space to help build space processing stations—an investment in Earth's future. If Lewis and Lewis are correct, economic value of Earth-approachers is not restricted to metallic asteroids.

In their summary, Lewis and Lewis remark also on iron asteroids of 1-km diameter, "Each one-kilometer-sized metallic asteroid will provide a billion tons of iron, 200 million tons of nickel, 10 million tons of cobalt, and 20,000 tons of platinum metals: net market value, about $1 trillion."

These considerations establish a two-fold need to evaluate the asteroid opportunity in parallel with evaluating our response to the asteroid threat: (1) the asteroids contain materials that are likely to be needed by human society over the next century or so; (2) the value of these materials offers, in principle, the chance for the asteroid program to be put on a self-paying basis.

Developments in asteroid research since Lewis and Lewis published in 1987 appear to create even more interesting and credible examples. For example, Ostro et al. (1991) concluded from a radar return signature that they had obtained the first direct observational evidence of a metallic Earth-approacher analogous to an iron meteorite. Meteorite statistics and asteroid surveys have long given evidence that a few Earth-approaching asteroids must be metal-rich, although chondritic rocky materials (either carbonaceous or ordinary) constitute the majority of the Earth-approaching asteroids (McFadden 1983; Hartmann et al. 1987; McFadden et al. 1989).

About 4% of observed falls are irons or stony-irons (Wasson 1974). The percentage of metallic Earth-approaching asteroids is probably comparable. It could be smaller if there is a large component of interplanetary carbonaceous objects that break up in the atmosphere; however, spectral studies of kilometer-scale Earth approaching asteroids do not suggest a large component of carbonaceous NEOs at that scale. For example, the summary by Hartmann et al. (1987) indicates only one C, P, or D low-albedo class object out of 13 Earth approachers, with eight S-class objects and no M-class objects. In a sample of 18 Atens and Apollos, McFadden et al. (1989) list about 4 Cs, 10 Ss, and 1 M (allowing for some ambiguous classifications). Metallic asteroids are difficult to detect spectrally because metals have few diagnostic absorption features, although the asteroids of M spectral class are widely regarded as being irons or stony irons. A few Earth-approaching Ms are now known. A November 1992 data summary includes 1986 DA and 3554 Amun as Ms; both were discovered in 1986, and both have diameters near 2 km.

Because the M classification still has some compositional ambiguity, radar studies of these objects are especially important. Having studied the radar return signal of the object temporarily designated as 1986 DA, Ostro et al. concluded that it "is a piece of NiFe metal," not a stony-iron, and that it has a diameter of 1.7 to 2.9 km.

Let us follow the techniques of Lewis and Lewis (1987) and others to illustrate the economic value of materials in such an object, or similar objects on closer orbits. It is difficult to be certain from the result of Ostro et al. that 1986 DA is 100% nickel-iron, although they argue against a typical stony-iron composition. To be conservative, assume that 90% of its mass is metal equivalent to iron meteorites. The metal mass would be about 3.0×10^{13}kg. From this we can calculate the approximate values shown in Table I, although there is some range in the figures for different types of meteoritic metal and different purities of ores.

TABLE I
Economic Analysis of 2-km M-Class Metal Rich Asteroids,
Such as 1986 DA

Component	Fraction of Metal by Mass	Mass (kg)	Estimated Value $/(kg)	Total Estimated Current Market Value
Iron	0.89	2.7×10^{13}	0.1	$3 trillion
Nickel	0.10	3.0×10^{12}	3	$9 trillion
Cobalt	0.005	1.5×10^{11}	25	$4 trillion
Platinum-group metals	15 ppm	4.5×10^{8}	20,000	$9 trillion
			Total Value	$25 trillion[a]

[a] trillion = 10^{12} (U. S. definition).

The platinum group metals are perhaps the most interesting because their high value/kg make them economically attractive to return to Earth (Lewis and Lewis 1987). (Other materials might be used initially to construct the necessary infrastructure in space.) Note that the large supply of platinum-group metals, though supplied over a period of years, would probably depress the world market price for those materials. An evaluation of this effect is underway (Kargel, in preparation). In any case, it is clear from the numbers that the economic value of the platinum-group metals necessary to depress the world market value seriously would be enormous, and therefore attractive to pursue in any case.

Of course, the values cited above must be compared against the costs of extracting the materials. These cannot easily be quoted because the techniques are still under development (Lewis et al. 1993). Two important points, however, are that: (1) the values are very large compared to space development budgets over a decade (such as the cost of the Apollo program [$30 billion] or the projected space station cost [$10–50 billion]); and (2) the true costs of raw materials on Earth (including environmental impact mitigation) are increasing, while costs of operating in space are clearly decreasing, as indicated over recent decades, as space infrastructure development continues. (This can be expected regardless of the motivation for space activity, whether scientific, political, military, or commercial.) Thus, over the time scales which must be considered in the context of the asteroid threat, we project that acquisition of asteroid resources will eventually become cost effective.

II. SOCIO-POLITICAL IMPLICATIONS

Examination of the geopolitical causes of warfare in this century, with its attendant death, destruction, and misery, indicates that national efforts to gain political control over strategically valuable resources ranks very high as a motivation for conflict. Comparison of the American responses in the cases of the Persian Gulf war and the Yugoslavian warfare supports this point. A "Persian Gulf" mentality has dominated international strategic politics in this century; national planners assume that it is worth a high risk to control regions that contain abundant fossil fuels and resources. Therefore, identification of an abundant cosmic resource base in space could, if handled with adequate political acumen, reduce a major cause of human conflict and encourage a more stable, peaceful terrestrial society.

As an example, Lewis and Lewis (1987) emphasize the political history of South Africa, which contains 96% of the world's platinum metal reserves, and the dominant supplies of additional strategic materials such as chromium. These materials, required by industry, are relatively deficient in the northern hemisphere, a circumstance that lent a certain attraction to European colonization and control.

III. THE NEED FOR A NEW, INTERNATIONAL APPROACH

The possibility that we can find numerous bodies in space that contain many of these resources raises a dramatic chance, but not a guarantee, that humans can alter geopolitical strategies to achieve a lower-risk situation. On the one hand, an unscrupulous race by national or corporate entities to control the resources in space could lead to a repetition in the 21st century of the conflicts of the 18th to 20th centuries. One might even say "pointless conflicts," in terms of long-term geopolitical outcome; note the current preeminent position of countries defeated in World War II, after only three or four decades. On the other hand, we could move in a new direction. A mandate to develop new international socio-political instruments to exploit asteroid materials and distribute the benefits to terrestrial society would allow our global societies to escape the "Persian Gulf Syndrome." The challenges in this regard are socio-political as much as scientific and technical. The asteroid opportunity offers a chance to make experimental responses to those challenges.

From certain political perspectives, it could be argued that a country, corporation, or other entity making the investment should reap all the monetary reward for itself. This view in fact has been a driver for much of the world's material progress, but at the same time, we see that in its most rapacious form it will encourage discord in the future. If developed countries exploit asteroid materials solely for themselves, in order to increase their per capita consumption rates of metals and other commodities, differences in living standards will continue to diverge. Whether this is right or wrong, it will exacerbate perceived inequities. A future, in which undeveloped countries' populations perceive that they are frozen out of the benefits of material resources and increased living standards, is likely to be less stable politically; this will lead to further conflicts and destruction, regardless of questions of justification. In the interests of political stability that allow a pursuit of happiness for many populations, the evolution of new geopolitical instruments is needed to allow global and mutually beneficial investment in, and return from, the asteroid opportunity.

In addition to these relatively practical considerations, there is a more purely political issue of perception. If the United States is perceived, rightly or wrongly, as keeping its space weapons program alive, thinly disguised as an asteroid protection program, this will increase political instability in the world. On the contrary, an international program to develop new asteroidal resources for global commerce and benefit could build a more stable world community.

IV. SHAPING AN INEXPENSIVE NEO PROGRAM

Making such an effort would involve several steps: first, telescopic reconnaissance and cataloging of NEO properties. This is already underway, although it would be prudent to speed its course. (Fortunately, this would also be the first step in any reactive NEO defense program, which gives us more time to debate the two approaches. Indeed, this is the philosophy behind the Spaceguard NEO Survey [Chapter by Morrison et al.]. We note, however, that the very name "Spaceguard" betrays that the survey is conceived of, or is being marketed as, a defense against a perceived threat, not as an opportunity to build a new future for humanity. We regard this mindset as a mistake.)

A second step would be investment in probes to visit NEOs. This should be expanded in order to correlate spectral classification (made by groundbased telescopes) with actual composition and with meteorite type, so that we know more precisely what composition to expect in each asteroid target. Rendezvous missions, rather than flyby missions, are the most efficient robotic missions because they can utilize gamma-ray and X-ray spectrometers and other devices to sample composition. Space probes to NEO are among the cheapest investments in spacecraft missions as shown by recent NASA-supported studies of Discovery class missions (see, e.g., Davis et al. 1990). Studies of ΔV requirements to reach Earth-approaching asteroids, by Davis et al. and others, show that there are many NEOs *easier to reach and return from* than Mars, and a few easier to reach and return from than the moon. Examples are listed in Table II. This table illustrates that many different spectral classes of asteroids are readily accessible. These include class S (chondritic or other stony meteoritic material?), class V (basaltic material), and class C (carbonaceous material?). The table also illustrates that some classes typical of more remote outer solar system material, such as D (reddish carbonaceous material, typical of comets and Trojan asteroids) are also known and accessible, but at higher ΔV cost in the known sample. Interestingly, the table includes one "asteroid" (4015) now known to be a dormant comet nucleus (see TABLE II, footnote *e*).

A third step in shaping the NEO program would be the investment in human flights to NEOs. This would allow realistic experiments to test practical operations in asteroid environments. These include sample collection, tests of ore separation schemes ranging from magnetic rakes to sort iron-rich metallic grains from regoliths, to "black box" chemical processing experiments that could be developed by different commercial interests. Steps 2 and 3 would allow assessment of the opportunities and costs for asteroid resource acquisition. For example, if regoliths exist on asteroids in the size ranges in question, "magnetic rake" sorting techniques might allow inexpensive separation of metallic components, even on rocky objects. Similarly, recent observations of Toutatis by Ostro (see his Chapter) and colleagues indicate that some of the Earth-approaching NEOs are compound bodies or rubble pile structures; this means that they would be easy to disaggregate for the purpose of processing ores. On the negative side, as pointed out to us by R. Binzel, such physiognomy might make them "structurally unstable," meaning that they could disaggregate during orbit modification attempts (a danger if one is trying to nudge them out of Earth-approaching orbit), and that impulse or vibration could cause them to seek a new equilibrium configuration. Experiments in resource acquisition and processing could be terminated at any time if the potential rewards looked unpromising. Moreover, even complete termination of the effort would leave humanity with a valuable capability to operate in space and continue scientific studies of other bodies—a capability that would not necessarily arise if all the emphasis toward an NEO is on defense against a perceived NEO threat.

A fourth step would be experiments in modifying the orbits of NEOs. This would have two motivations: (a) it would be attractive from the viewpoint

TABLE II

ΔV s[a] to Reach Selected NEOs in Order of
Ease of Round Trip

	q	Q	Spectral Class[b]	$\Delta V_{outbound}$[c]	$\Delta V_{roundtrip}$[d]
Low ΔV Asteroids					
1943 Anteros (1973EC)	1.06	1.80	S	5.27	5.5–13
4660 Nereus (1982DB)	0.95	2.02	EMP or C	4.45	5.5–13
1989ML	1.10	1.42	?	4.25	6.3–12
3361 Orpheus (1982HR)	0.82	1.60	?	5.29	7.2–13
3757 (1982XB)	1.02	2.66	S	5.30	10.9
3908 (1980PA)	1.04	2.81	V	5.61	11.3
1980AA	1.05	2.73	?	5.40	11.5
3288 Seleucus (1982DV)	1.10	2.96	S	5.91	12.2
3352 McCauliff (1981CW)	1.18	2.57	?	5.90	12.4
Additional Objects of Interest:					
Moon	—	—	(Satellite)	5.7	11.4
3551 (1983RD)	1.07	3.12	V	6.48	12.9
4015 (1979VA)[e]	1.00	4.28	CF	6.79	13.1
2100 Ra-Shalom (1978RA)	0.47	1.19	C	7.95	14.9
2201 Oljato (1947XC)	0.63	3.72	Unique	8.04	15.8
Mars	—	—	(Planet)	4.9–11.7[f]	18.8
3552 Don Quixote (1983SA)	1.21	7.26	D	11.81	23.5

[a] $\Delta V_{outbound}$ = km s^{-1} to go from low Earth orbit (LEO) to soft landing on asteroid surface. $\Delta V_{roundtrip} = \Delta V_{outbound}$ plus km s^{-1} to launch from asteroid to LEO, for characteristic, not optimum, situation. (See also footnote d.)

[b] Data courtesy of D. J. Tholen, November 1993.

[c] Adapted from 1992 data table courtesy of J. S. Lewis, and data in Davis et al. (1990).

[d] Double figures with dashes give typical values in time interval 1990–2010 from LEO to landing on object and back to Earth, depending on launch date and assuming no aerobraking on return to Earth (from Davis et al. 1993). Single figures are derived from the Shoemaker figure of merit as outlined in Shoemaker and Helin (1978), and give typical round-trip ΔV s (see footnote a). The round-trip figure for Mars is calculated in an equivalent way, with no aerobraking at Mars or Earth.

[e] Recently identified with comet Wilson-Harrington, observed in 1949.

[f] Low figure assumes complete aerobraking of hyperbolic approach speed in Martian atmosphere; high figure assumes rocket braking during approach and during landing.

of bringing economically valuable objects into orbits requiring lower ΔV energy expenditures to reach from Earth; and (b) it would directly reduce or end the threat of NEO impacts. Thus, addressing the asteroid threat would be a natural eventual *byproduct* of any proactive program to exploit the asteroid opportunity.

Programs motivated mainly by the asteroid threat and programs motivated mainly by the asteroid opportunity have many parallel and identical aspects, particularly the first step, an intensive NEO survey program. However, because the predicted time scale for the NEO threat to global civilization to reach a value $\gtrsim 1/2$ is of the order of 10^4 to 10^6 yr (Chapman and Morrison 1994), it appears prudent to organize our program to capitalize primarily on the asteroid opportunity. If a greater-than-expected NEO impact threat materializes from the survey, the goals can be modified; if not, an aggressive program is likely to achieve the ability to modify orbits, and thus *end the threat* within one or two centuries, and at the same time provide humanity with an enormous, productive capability in space. A purely defensive program would achieve few of these benefits.

Some elements of such a program are nearly in place today. For example, the United States and Russia maintain strikingly complementary space capabilities, although little has been done to exploit the complementarity. For example, the United States has a major investment in the Space Shuttle transportation system, and is known for the high technology and miniaturization of its space hardware. Russia has the production capacity for bigger rocket boosters than are being built elsewhere, and also maintains a working space station, permanently inhabited for a number of years, in which individual cosmonauts have lived as much as a year at a time (Fig. 1).

The space station is a particularly interesting case in point. The American Space Station program, as announced initially by President Reagan, is a carry-over from the cold war. It was conceived in large part as an effort by the U. S. to build a station bigger and better than the Soviet stations. Although it would be a valuable advance in developing human capabilities to operate in space, it is widely seen as crippling scientific research in the American space program because of its impact on the NASA budget. In the contemporary era, with the budgetary hardships faced in both countries, it seems difficult to justify two independent space station programs. On the contrary, a partnership to upgrade and refurbish the Mir station, or move on jointly to so-called Mir 2 concepts, would allow both countries to save money. Objectives that could be met include: (1) development of a base for future piloted asteroid flights and sample return (interplanetary vessels could be designed to operate in space only, from the station; human return and sample return to Earth could use separate vehicles already in existence); (2) financial support for the Mir program that would be consistent with American policies of support for democratic changes in Russia; (3) reduction of American Space Station expenses; and (4) freeing of U. S. funds for the remaining technical program of NEO reconnaissance, planetary exploration, and a more productive human flight program aimed at asteroid exploration. To make such a move realistic, we would need a sociopolitical/legal mechanism to redirect many of the present American space station jobs (rather than cancelling them) to the Mir upgrade and asteroid exploration hardware programs, as well as to other technical programs such as solar panel development for spaceflight and commercial use.

Figure 1.   The existing Russian Mir space station, in its 1991–1993 configuration. This station has demonstrated many of the objectives proposed for the down sized American station, as well as some of the capabilities needed for long duration flights and human exploration of asteroids. Russian-American partnership on development and refurbishment of this station could save money now committed to the redundant, downsized U. S. Station, and release U. S. funds for an expanded NEO exploration program (painting by A. Sokolov).

Since the above paragraph was included in the original submitted version of this chapter, the U. S. and Russia announced cooperation on the design of Mir 2 and a down-scaled U. S. station. We applaud this move. Although the stations' missions do not directly involve staging for interplanetary flight, these programs will increase humanity's capability to operate in space, and this in turn will increase our capability to address the asteroid threat, and the asteroid opportunity, in the ways proposed in this chapter.

V. SUMMARY

In short, if we focus our response to the asteroid threat on the asteroid opportunity, we will make a much more productive investment in the future than if we mount a purely reactive, defensive program. In either case, a comprehensive NEO survey is imperative.

Is an NEO survey merely a boondoggle for asteroid-minded astronomers? In answer, it is highly appropriate in this book to quote a pioneer of astronomy at the University of Arizona, A. E. Douglass, who founded the University's Steward Observatory seven decades ago, and also serendipitously discovered the science of dendrochronology, or tree ring dating, which allows precise, absolute dating of southwestern archaeological ruins. Douglass thus had good credentials to point out that the practical consequences of seemingly academic knowledge are hard to predict. At the dedication of the Observatory in 1922, he made the following remark:

> Scientific research is business foresight on a large scale. It is knowledge obtained before it is needed. Knowledge is power, but we cannot tell which fact in the domain of knowledge is the one which is going to give the power, and we therefore develop the idea of knowledge for its own sake, confident that some one fact or training will pay for all the effort.

In that spirit we urge a positive program designed not only to catalog NEO discoveries and data, but to reconnoiter them at close hand, with both robotic and human exploration, which is well within our power, and to keep our eyes open not just to their threat of impact on Earth, but also to the possible benefits of their materials to terrestrial society.

Acknowledgments. We thank R. Binzel, colleagues at PSI, and an anonymous reviewer for helpful critiques. We also thank J. Lewis and D. Davis for assistance with data for Table II. The work was partially supported by a contract from NASA's Planetary Geology and Geophysics Program. Parts of this chapter are adopted from a book to be published in Russian by the same authors.

REFERENCES

Davis, D. R., Hartmann, W., Friedlander, A., Collins, J., Niehoff, J., and Jones, T. 1990. The Role of Near-Earth Asteroids in the Space Exploration Initiative (Tucson: Planetary Science Institute).

Davis, D. R., Friedlander, A., and Jones, T. 1993. Role of near-Earth asteroids in the Space Exploration Initiative. In *Resources of Near-Earth Space*, eds. J. S. Lewis, M. S. Matthews and M. L. Guerrieri (Tucson: Univ. of Arizona Press), pp. 619–655.

Gaffey, M., and McCord, T. 1977. Mining the asteroids. *Mercury* 6(6):1.

Hartmann, W. K. 1982. Mines in the sky are not so wild a dream. *Smithsonian* (September), p. 70.

Hartmann, W. K., Miller, R., and Lee, P. 1984. *Out of the Cradle* (New York: Workman Publishing Co.).

Hartmann, W. K., Tholen, D. J., and Cruikshank, D. P. 1987. Relationship of active comets, "extinct comets," and dark asteroids. *Icarus* 69:33–50.

Herrick, S. 1979. Exploration and 1994 exploitation of Geographos. In *Asteroids*, eds. T. Gehrels and M. S. Matthews (Tucson: Univ. of Arizona Press), pp. 222–226.

Kargel, J. S. 1993. Economic potential of asteroidal sources of precious metals. *Geol. Soc. of America Bull.*, submitted.

Lewis, J. S., and Lewis, R. A. 1987. *Space Resources* (New York: Columbia Univ. Press).

Lewis, J. S., Matthews, M. S., and Guerrieri, M. L., eds. 1993. *Resources of Near-Earth Space* (Tucson: Univ. Arizona Press).

McFadden, L. A. 1983. Spectral Reflectance of Near-Earth Asteroids: Implications for Composition, Origin, and Evolution. Ph.D. Thesis, Univ. of Hawaii.

McFadden, L. A., Tholen, D., and Veeder, G. 1989. Physical properties of Aten, Apollo, and Amor asteroids. In *Asteroids II*, eds. R. P. Binzel, T. Gehrels and M. S. Matthews (Tucson: Univ. of Arizona Press), pp. 442–467

Morrison, D., and Niehoff, J. 1979. Future exploration of the asteroids. In *Asteroids*, eds. T. Gehrels and M. S. Matthews (Tucson: Univ. of Arizona Press), pp. 227–250

O'Leary, B., Gaffey, M., Ross, D., and Salkeld, R. 1979. Retrieval of asteroidal materials. In *Space Resources and Space Settlements*, eds. J. Dillingham, W. Galbreath and B. O'Leary, NASA SP-428.

Ostro, S. J., Campbell, D., Chandler, J., Hine, A., Hudson, R., Rosema, K., and Shapiro, I. 1991. Asteroid 1986DA: Radar evidence for a metallic composition. *Science* 252:1339.

Shoemaker, E. M., and Helin, E. 1978. Earth-approaching asteroids as targets for exploration. In *Asteroids: An Exploration Assessment*, eds. D. Morrison and W. C. Wells, NASA CP-2053, pp. 245–256.

Wasson, J. 1974. *Meteorites* (New York: Springer-Verlag).

THE LESSON OF GRAND FORKS: CAN A DEFENSE AGAINST ASTEROIDS BE SUSTAINED?

ROBERT L. PARK
The American Physical Society

LORI B. GARVER
National Space Society

and

TERRY DAWSON
U. S. House of Representatives

A standing defense against large asteroid and comet impacts is rendered impractical by the long interval between events. Governments, which are under constant pressure to respond to immediate crises, are unlikely to sustain a defense against an infrequent and unpredictable threat. Nor can it be argued that such short-term priorities are misplaced. Indeed, civilization will do well to survive long enough to be threatened by a major asteroid impact. The emphasis should be on early detection, thus allowing sufficient time to mount a response to a specific threat.

I. INTRODUCTION

Our understanding of the history of Earth and its inhabitants is undergoing a radical change. The gradual processes of geologic change and evolution, it is now clear, are punctuated by natural catastrophes on a colossal scale—catastrophes resulting from collisions of large asteroids and comets with Earth. It is, to use the popular term, a "paradigm shift."

This "new catastrophism," is not unlike the revolutions brought about by the heliocentric solar system of Copernicus, or Darwinian evolution, or the big bang. In retrospect, such revolutionary ideas always seem obvious. On reading the *Origin of Species*, Thomas Huxley remarked simply: "Why didn't I think of that." Now, looking at the Moon, we find ourselves wondering why it took so long to ask whether the process that cratered its surface is still going on.

Scientists have an obligation to share this latest epiphany. To do so it will be necessary to translate geologic times and astronomical distances for people who scale time by a life span and distance by the circumference of the Earth. On the scale of a human lifetime, the relics of early civilization seem very ancient, but compared to the frequency of major asteroid or comet impacts, the entire history of homo sapiens is a barely resolvable interval.

II. ATTENTION SPAN

The long time scale between major impacts has implications for public policy. Governments do not function on geologic time. On the North Dakota prairie near the town of Grand Forks, lie the abandoned ruins of America's ballistic missile defense system. The roots of weeds are already beginning to force their way into invisible cracks in the concrete and pry the silos apart.

Built in accordance with the ABM treaty, the Grand Forks facility was meant to defend our retaliatory capacity. It was declared operational in 1975— and decommissioned the same year (Lakoff and York 1989). National leaders had been persuaded by some scientists that the Grand Forks facility would meet the threat to our intercontinental ballistic missile fleet, even though other scientists warned that the system was dangerous and ineffective. It was closed because the money to operate it was needed for other projects that were deemed to be more urgent.

The lesson of Grand Forks is as old as human history: societies will not sustain indefinitely a defense against an infrequent and unpredictable threat. Governments often respond quickly to a crisis, but are less well suited to remaining prepared for extended periods. Even on the brief scale of human lifetimes, resources are eventually diverted to more immediate problems, or defenses are allowed to decay into a state of unreadiness. According to news accounts, in the great flood of 1993, the U. S. Corp of Engineers prepared to close the massive iron gates in the vast complex of levees on the Mississippi and its tributaries only to discover that some of the gates had been removed and sold for scrap. Periodic inspections had been suspended to save money. Nor can it be argued that such short-term priorities are necessarily misplaced. Indeed, civilization will do well to survive long enough to be threatened by a major asteroid impact; our own destructive impulses or the unanticipated consequences of our technologies seem likely to do us in first. It is unrealistic to expect governments to sustain a commitment to protection against a rare occurrence when they are constantly under pressure to respond to some perceived immediate crisis.

Particularly now, with nuclear weapons being dismantled by the major powers, any talk of a nuclear defense against such an unlikely hazard as cosmic collisions will be seen as an effort by the weapons community to sustain itself. The risk of diversion of any mitigation system to military uses must be regarded as a more immediate hazard. Indeed, we are concerned that the entire discussion of mitigation is premature and serves largely to divert attention from the primary task, which is to define the asteroid hazard.

Given the frequency of past collisions, major impact is unlikely to occur in the next century. On the other hand, all of modern technology is squeezed into the present century, and the pace of technological advance is accelerating.

It would be presumptuous to suppose that defenses devised today will be of more than historical interest to our scientific heirs a century from now, or a millennium, or a thousand millennia, when the rock finally comes.

Discussion of mitigation may serve one public purpose. It is important that devastation not be accepted as inevitable, otherwise society might prefer not to know when it is coming. An asteroid interception workshop hosted by NASA in 1992 (Canavan et al. 1993) concluded that available technology can deal effectively with a threatening asteroid, given warning time on the order of several years. That conclusion validates the view that current efforts should concentrate on detection and orbit determination.

The challenge of science is to identify objects that threaten Earth and work out the timetables for their arrival. Here the challenge is straightforward and technical. What technology or technologies provide the most cost effective way of identifying candidates for significant collisions? If there are no candidates in the next several decades, is it necessary to further refine the orbits or devise mitigation schemes? If we are threatened on the time scale of a few decades, would it be desirable to employ planetary radars to refine the orbits? These are, in any case, the sort of technical issues scientists are trained to resolve.

III. MISTAKEN IDENTITY

The emphasis has properly been on impacts that would be expected to have global consequences. Even for objects too small to produce more than local effects, however, it has been pointed out that an impact might be misidentified as a nuclear explosion (Canavan and Solem 1993). Misidentification would be most likely among nations that have recently joined the ranks of "nuclear powers" and would therefore be expected to have less sophisticated means of verification.

It is more than a hypothetical concern. We recall that the 1978 South Indian Ocean anomaly, detected by a Vela satellite, was suspected at the time of being a South African-Israeli nuclear test. In spite of the failure to find any confirming evidence from intelligence sources or atmospheric monitoring, it created international tensions that lasted for years. At the time, there were suggestions that it might have been an artifact produced by a micrometeorite impact on the Vela satellite itself, but little serious consideration seems to have been given to the idea that the satellite had observed the fireball from an asteroid impact in the atmosphere. A 1990 satellite observation of an apparent asteroid impact fireball over the Western Pacific has been described by Reynolds (1993). The danger of misidentification, which grows as weapons proliferate among less sophisticated nations, is meliorated in part by publicizing the possibility. The only sure means of avoiding an unfortunate response, however, would be for everyone to know the impact is coming. Which again places the emphasis on detection.

IV. INVOLVING CONGRESS

Efforts to persuade governments to invest significant resources in evaluation of the hazard of asteroid impacts must overcome what has been called "the giggle factor." Clearly, elected officials in Washington are not being inundated with mail from constitutents complaining that a member of their family has just been killed or their property destroyed by a marauding asteroid. Indeed, the prevailing view among government officials who hear about this issue for the first time is that the epoch of large asteroid strikes on Earth ended millions or billions of years ago.

Congressional involvement has been confined to the Committee on Science, Space and Technology of the U. S. House of Representatives, whose current chair, George Brown of California, has maintained an interest in the asteroid issue for several years. The Committee directed NASA to conduct two international workshops on the asteroid threat (House Committee on Science, Space and Technology 1990). The objective of the first was to determine the extent to which the threat is "real," and to define a program for significantly increasing the detection rate of large asteroids in Earth-crossing orbits. The second dealt with the feasibility of preventing large asteroids from striking Earth (see the Chapter by Canavan et al.).

In March of 1993, the Space Subcommittee held a formal hearing to examine the results of the two workshops. Some members remain skeptical that the threat is real. But even among those who recognize that it is only a question of when a major impact will occur, there was no sense of urgency. Given the severe constraints imposed by the current budget situation, therefore, it seems unlikely that Congress would agree to devote more than a few million dollars per year to asteroid detection and research. If prudently spent, however, even that modest level of resources should significantly speed up the process of cataloging Earth-crossing asteroids. Perhaps the major impact of the workshops has been in NASA itself. The Agency now seems persuaded that near-Earth asteroids are deserving of scientific attention, and that efforts should be made to increase the rate at which such objects are identified.

V. INVOLVING THE DEPARTMENT OF DEFENSE

In addition to cataloging the orbits of large near-Earth objects, the primary focus of the astronomical community for the foreseeable future will be to study their origin and composition, and to determine the size distribution of objects striking Earth. The frequency of impacts of objects of various sizes is known only to limited precisions (Chapter by Rabinowitz et al.). In particular, objects up to several meters in diameter explode in the atmosphere without reaching the surface. Although the energy released in these explosions may be many times greater that released by the Hiroshima bomb, they most frequently occur over the ocean or sparsely inhabited regions of Earth and go unreported.

The system of military surveillance satellites, however, which exists to detect nuclear detonations or missile launches, are well suited to detection and evaluation of small asteroid impacts. Indeed, some useful data on such impacts may already exist on archived computer tapes, covering the past twenty years. In any case, if "tasked" to do so, the military satellites could provide a rich source of information on the size distribution of Earth-impacting asteroids (see the Chapter by Tagliaferri et al.). Representative George Brown (1993) sent a letter to Secretary of Defense Les Aspin requesting that the Defense Department provide active support to the astronomical community in collecting and disseminating scientifically useful information concerning asteroid strikes that do not reach the surface.

VI. A STRATEGY FOR BEING TAKEN SERIOUSLY

Congress is unlikely to take any action in the absence of public pressure. Once the public understands that Earth and the life on it have been shaped by cosmic collisions (and the process is continuing), they will be more likely to support the science needed to evaluate the threat. The scientific community must, therefore, concentrate on public education.

Media coverage of the asteroid issue has thus far been intense. A report issued by the American Institute of Aeronautics and Astronautics, "Dealing with the threat of an asteroid striking the Earth," spawned 200 times the number of press clippings as does their usual release; and in addition to the press, the movie *Jurassic Park*, which attributes the extinction of dinosaurs to the Cretaceous-Tertiary impact 65 million years ago, may turn out to be the most-watched movie of all time. In a public relations sense, the battle is largely won.

All of this creates a dilemma. While it is important to inform the public, it is dangerous to encourage fear mongering. A few more Swift-Tuttles (see Glossary) could undermine the credibility of the scientific community on this issue. Scientists would do well, for example, to avoid such terms as "near miss." The public understands "near-miss" as the draft of wind from a truck that passes as you step off the curb—not a truck that went by six hours earlier.

Sensational accounts of the asteroid and cometary traffic are not, unfortunately, confined to checkout-counter tabloids. Even in such staid newspapers as the *New York Times* and *Washington Post*, articles may include a well-reasoned discussion of relative risk, but the headline writers find "doomsday rock," "space bullets" and "killer comet" irresistible. These headlines exploit the excessive fear engendered by events people feel powerless to control. The image of an indifferent mountain of stone and metal guided by the immutable laws of physics toward an inevitable rendezvous with Earth, is the stuff of nightmares. Remarkably, however, Nature has apparently provided a non-threatening demonstration. The impact of comet Shoemaker-Levy 9 on the back side of Jupiter in July of 1994 provides an historic opportunity to educate the public without terrorizing anyone.

Shoemaker-Levy 9, in its last pass by the Jupiter, broke into a string of 21 major pieces. The energy released by the impacts of the full string will be equivalent to about a billion megatons of TNT. Although the pieces will impact on the side of Jupiter away from Earth, millions of amateur astronomers will be watching to see the flashes reflected from Jupiter's moons. A few hours later, the rotation of Jupiter will bring the impact region into view. There is great disagreement about what will be seen, but no one suggests that it will not be spectacular.

The asteroid-comet community needs only to insure that everything is fully and accurately explained; the message will take care of itself: (1) the energy deposited by cosmic impacts is enormous; (2) this is a process that is still going on.

REFERENCES

Brown, G. E., Jr. 1993. Letter to Secretary of Defense Aspin, dated 22 April 1993.
Canavan, G. H., and Solem, J. C. 1993. Summary of Workshop. In *Proceedings of the Near-Earth-Object Interception Workshop*, eds. G. H. Canavan, J. C. Solem and J. D. G. Rather (Los Alamos: Los Alamos National Lab), pp. 20–48.
Canavan, G. H., Solem, J. C., and Rather, J. D. G., eds. 1993. *Proceedings of the Near-Earth-Object Interception Workshop* (Los Alamos: Los Alamos National Lab).
House Committee on Science, Space and Technology. 1990. In *House Report 101-763*.
House Subcommittee on Space. 1993. *Hearing on the Results of the Near-Earth-Object Workshops, 24 March 1993*.
Lakoff, S., and York, H. F. 1989. *A Shield in Space?: Technology, Politics, and the Strategic Defense Initiative* (Berkeley: Univ. of California Press).
Reynolds, D. A. 1993. Fireball observation via satellite. In *Proceedings of the Near-Earth-Object Interception Workshop*, eds. G. H. Canavan, J. C. Solem and J. D. G. Rather (Los Alamos: Los Alamos National Lab), pp. 221–226.

APPENDIX

EARTH-CROSSING ASTEROIDS

B. G. MARSDEN and G. V. WILLIAMS
Harvard-Smithsonian Center for Astrophysics

The following tabulation lists 103 earth-crossing (or potentially earth-crossing) asteroids. The rationale for selection is that at least one of the two (generally) M and two N criteria (Marsden 1993) is less than 0.075 AU, and the objects are listed in increasing order of this minimum value, all values of M and N up to this value being shown. On the assumption that the Earth's orbit is a circle of radius 1 AU in the plane of the ecliptic, M represents the distances from the ecliptic when the object is at the Earth's distance from the Sun (or at perihelion if this is not possible), and N represents the distances of the object's nodes on the ecliptic from the Earth's orbit. Although other recent listings (see, e.g., Shoemaker 1993) may utilize alternative procedures to establish whether an asteroid could be Earth-crossing (even, perhaps, after tens of thousand of years), the M and N criteria have the benefits of simplicity and immediate applicability, and they should identify all known potential hazards for a century or two into the future, even when orbits are quite imprecise (also see the Chapter by Rabinowitz et al.).

The present list is complete through the data given in the August 1993 batch of *Minor Planet Circulars*, and for objects observed at multiple oppositions (or for long arcs at a single opposition) the elements used are the J2000.0 osculating values at this time.

Some of the short-arc objects are shown with letters preceding the provisional designation. R indicates that radar observations are available and substantially improve the determination of the orbit, L means that the object is lost, and V means that it is "very" lost. Following the M and N values is the actual minimum distance and date during 1900-2200, provided that this distance, taken in some instances from the Chapter by Yeomans and Chodas, is less than 0.025 AU (and for the L and V cases, only the discovery apparitions are considered).

Understandably, the great majority of these values are early in the table, and the rather large value of M remaining for 1991 OA will undoubtedly be diminished due to rather large planetary perturbations between now and the date shown for a close encounter.

Next come the perihelion and aphelion distance q and Q (in AU), the large value of Q for 1991 OA confirming the possibility of substantial perturbations by Jupiter; this object and the eight with even larger Q could be candidates for cometary origin, and among the eight is (4015) Wilson-Harrington, which seems actually to have exhibited cometary behavior in 1949. Of the objects of small Q, 1991 VG also has a very small orbital inclination i to the ecliptic, shown (in degrees) in the next column, and it is widely thought to be a man-made object launched in the 1970s; N tends to be more significant than M as a criterion for close approach only for the high-i objects.

The absolute visual magnitude H ranges from 14 to 29, and although albedos are unknown, these figures suggest a size range from a few kilometers down to the same number of meters. Of the largest objects, the most potentially dangerous to the Earth during the foreseeable future would seem to be (4179) Toutatis, which makes essentially quadrennial approaches, and the 3:1 resonance also with Jupiter renders the orbit quite chaotic (Whipple 1993).

The final columns list the discovery data (date, place and observers). This is the information as contained for the relevant provisional designations in the files of the Minor Planet Center. In several instances independent discoveries were later reported by other observers.

REFERENCES

Marsden, B. G. 1993. To hit or not to hit. In *Proceeding of the Near-Earth-Object Interception Workshop*, eds. J. D. G. Rather, J. H. Rahe and G. H. Canavan (Los Alamos: Los Alamos National Lab), pp. 67–71.

Shoemaker, E. M. 1993. Asteroid tables. In *The Spaceguard Survey: Report of the NASA International Near-Earth-Object Detection Workshop*, ed. D. Morrison (Pasadena: Jet Propulsion Laboratory), pp. A.1–A.4.

Whipple, A. L. 1993. In "A Tiny Visitor Departs," by A. M. *Sky & Tel.* 86(2):78.

APPENDIX

Object	Desig.	M	N	Actual (year)	q	Q	i	H	Discovered (Y M D)	Discovery	Discoverer
(2201) Oljato	1947 XC	0.000 0.009	0.002	—	0.629	3.722	2.5	15.2	1947 12 12	Flagstaff	Giclas
	V 1993 HD	0.000 0.056	0.002	—	0.485	2.405	5.7	23.0	1993 04 20	Kitt Peak	Spacewatch
(1981) Midas	1973 EA	0.001 0.051	0.001	—	0.622	2.930	39.8	15.0	1973 03 06	Palomar	Kowal
	V 1993 KA$_2$	0.001 0.016	0.008	0.001 (1993)	0.502	3.953	3.2	29.0	1993 05 21	Kitt Peak	Spacewatch
	V 1991 BA	0.001 0.020	0.021	0.001 (1991)	0.713	3.772	2.0	28.5	1991 01 18	Kitt Peak	Spacewatch
	1989 UP	0.002 0.040	0.003	0.016 (2129)	0.982	2.745	3.9	20.5	1989 10 27	Kitt Peak	Spacewatch
Hermes	V 1937 UB	0.003 0.005	0.015	0.005 (1937)	0.619	2.673	6.1	18.0	1937 10 28	Heidelberg	Reinmuth
	L 1993 BD$_3$	0.003	0.027	—	1.022	2.249	0.9	26.0	1993 01 26	Kitt Peak	Spacewatch
(4581) Asclepius	1989 FC	0.004 0.046	0.017	0.005 (1989)	0.657	1.388	4.9	20.5	1989 03 31	Palomar	Holt & Thomas
	L 1991 VA	0.004	0.008	0.012 (1991)	0.926	1.932	6.5	26.5	1991 11 01	Kitt Peak	Spacewatch
	L 1991 TU	0.005 0.023	0.006	0.005 (1991)	0.945	1.887	7.7	28.5	1991 10 07	Kitt Peak	Spacewatch
	1988 XB	0.005 0.021	0.041	—	0.761	2.175	3.1	17.5	1988 12 05	Gekko	Oshima
(4660) Nereus	1982 DB	0.005 0.007	0.028	—	0.953	2.026	1.4	18.3	1982 02 28	Palomar	Helin
(4179) Toutatis	1989 AC	0.005	—	0.008 (2060)	0.903	4.110	0.5	14.0	1989 01 04	Caussols	Pollas
(2340) Hathor	1976 UA	0.007 0.041	0.029	0.010 (2004)	0.464	1.224	5.8	20.3	1976 10 22	Palomar	Kowal
	1993 EA	0.007 0.027	0.055	0.006 (2086)	0.527	2.017	5.1	17.0	1993 03 03	Kitt Peak	Spacewatch
	1988 TA	0.007 0.037	0.060	—	0.803	2.278	2.5	21.0	1988 10 05	Palomar	Mueller & Phinney
	R 1986 JK	0.007	0.062	0.010 (1988)	0.894	4.700	2.1	19.0	1986 05 05	Palomar	Shoemaker
	V 1992 YD$_3$	0.051	0.007	0.021 (1958)	1.006	1.325	27.1	26.0	1992 12 27	Kitt Peak	Spacewatch
	1993 KH	0.008	0.009	—	0.850	1.615	12.8	19.0	1993 05 24	Siding Spring	McNaught
(2101) Adonis	R 1990 OS	0.009 0.018	—	0.020 (2053)	0.903	2.440	1.1	20.0	1990 07 21	Palomar	Helin
	1990 HA	0.010 0.062	0.060	—	0.792	4.366	3.9	17.0	1990 04 17	Kleť	Mrkos
	L 1990 UA	0.012 0.017	—	—	0.770	2.672	1.0	19.5	1990 10 16	Palomar	Helin
	1936 CA	0.012 0.021	—	0.007 (2177)	0.442	3.309	1.4	18.7	1936 02 12	Uccle	Delporte
	V 1993 HP$_1$	0.012	0.014	—	0.973	2.993	8.0	27.0	1993 04 27	Kitt Peak	Spacewatch

APPENDIX (cont.)

Object		M	N	Actual		q	Q	i	H				Discovery	
(3361) Orpheus	1982 HR	0.013	0.034	0.073	0.017	2194	0.819	1.599	2.7	19.0	1982	04	24	Cerro El Roble	Torres
	1991 VH	0.047		0.013			0.973	1.300	13.9	17.0	1991	11	09	Siding Spring	McNaught
(2135) Aristaeus	1989 UQ	0.015	0.021		0.016	2120	0.673	1.158	1.3	19.0	1989	10	26	Caussols	Pollas
	1977 HA	0.015		0.015			0.795	2.405	23.0	17.9	1977	04	17	Palomar	Helin & Bus
	1991 VG	0.015	0.025	0.020	0.072	0.003	1991	0.976	1.077	1.4	28.8	1991	11	06	Kitt Peak	Spacewatch
	1980 AA	0.015		0.059			1.052	2.733	4.2	19.0	1980	01	13	Kleť	Mrkos
L	1993 KA	0.033		0.016			1.007	1.503	6.0	26.0	1993	05	17	Kitt Peak	Spacewatch
R	1990 MF	0.018	0.032				0.950	2.543	1.9	18.5	1990	06	26	Palomar	Helin
(3362) Khufu	1984 QA	0.018	0.057		0.016	1917	0.526	1.453	9.9	18.1	1984	08	30	Palomar	Dunbar & Barucci
L	1990 SM	0.019					0.485	3.829	11.6	16.5	1990	09	22	Siding Spring	McNaught & McKenzie
L	1989 VB	0.019		0.051			1.005	2.724	2.1	20.0	1989	11	01	Siding Spring	Parker
R	1991 AQ	0.020	0.035	0.030	0.010	2165	0.499	3.818	3.2	17.5	1991	01	14	Palomar	Helin
(3757)	1982 XB	0.020		0.030			1.016	2.654	3.9	19.0	1982	12	14	Palomar	Helin
L	1990 UN	0.021	0.035				0.807	2.610	3.7	23.5	1990	10	22	Kitt Peak	Spacewatch
L	1991 GO	0.022	0.030				0.663	3.257	9.7	19.0	1991	04	11	Kitami	Endate & Watanabe
L	1983 LC	0.023	0.025				0.764	4.495	1.5	19.0	1983	06	13	Palomar	Helin & Dunbar
(4034)	1986 PA	0.023		0.061			0.589	1.531	11.2	18.1	1986	08	02	Palomar	Helin
(4769) Castalia	1989 PB	0.023					0.549	1.577	8.9	16.9	1989	08	09	Palomar	Helin
L	1991 BN	0.023	0.040		0.043		0.869	2.017	3.4	20.0	1991	01	19	Kitt Peak	Spacewatch
	1992 NA	0.023		0.052			1.049	3.731	9.8	16.5	1992	07	01	Siding Spring	McNaught
L	1993 DA	0.023		0.023			0.848	1.023	12.4	26.0	1993	02	17	Kitt Peak	Spacewatch
	1988 EG	0.024	0.034				0.636	1.903	3.5	19.0	1988	03	12	Palomar	Alu
L	1993 FA1	0.024		0.024			1.014	1.839	20.5	25.0	1993	03	28	Kitt Peak	Spacewatch
L	1992 JD	0.025		0.025			1.002	1.067	13.5	25.0	1992	05	03	Kitt Peak	Spacewatch
(5011) Ptah	6743 P-L	0.026			0.019	2170	0.818	2.452	7.4	17.0	1960	09	24	Palomar	PLS
(3200) Phaethon	1983 TB	0.026			0.020	2099	0.140	2.403	22.1	14.6	1983	10	11	Earth orbit	IRAS
(2061) Anza	1960 UA	0.026					1.051	3.484	3.8	16.6	1960	10	22	Flagstaff	Giclas
L	1978 CA	0.059		0.026			0.883	1.366	26.1	18.0	1978	02	08	ESO	Schuster
(4450) Pan	1987 SY	0.027	0.043				0.596	2.287	5.5	17.1	1987	09	25	Palomar	Shoemaker

Name	Designation						q	a	i	H	Year	Mo	Dy	Observatory	Discoverer
(1862) Apollo	1932 HA		0.028	0.032			0.647	2.295	6.4	16.2	1932	04	24	Heidelberg	Reinmuth
	6344 P-L	V	0.028				0.940	4.297	4.6	21.6	1960	09	24	Palomar	PLS
	1989 QF		0.028	0.041			0.676	1.627	3.9	17.0	1989	08	31	Palomar	Shoemaker
(2102) Tantalus	1975 YA			0.029			0.905	1.675	64.0	15.3	1975	12	27	Palomar	Kowal
	1992 UY4		0.030				1.009	4.300	2.8	17.5	1992	10	25	Palomar	Shoemaker
	1992 DU	L		0.030			0.957	1.363	25.1	25.0	1992	02	26	Kitt Peak	Spacewatch
(3908)	1980 PA		0.031				1.041	2.807	2.2	17.4	1980	08	06	ESO	Schuster
	1989 UR		0.031	0.069			0.695	1.465	10.3	18.0	1989	10	28	Palomar	Mueller & Mendenhall
	1991 CS			0.032	0.023	2040	0.938	1.308	37.1	17.5	1991	02	13	Siding Spring	McNaught
	1991 TT	L		0.032			1.002	1.385	14.8	26.0	1991	10	06	Kitt Peak	Spacewatch
	1991 EE		0.033				0.844	3.648	9.8	17.5	1991	03	13	Kitt Peak	Spacewatch
(3671) Dionysus	1984 KD			0.034	0.022	2123	1.003	3.387	13.6	16.3	1984	05	27	Palomar	Shoemaker
	1991 JX		0.036				1.011	4.029	2.3	18.5	1991	05	10	Palomar	Helin
	1991 JW		0.052	0.036			0.915	1.161	8.7	19.5	1991	05	08	Palomar	Lawrence & Helin
(3551)	1983 RD		0.037				1.071	3.112	9.5	16.8	1983	09	12	Palomar	Dunbar
(5604)	1992 FE		0.037	0.055			0.551	1.303	4.8	17.0	1992	03	26	Siding Spring	McNaught
(4183) Cuno	1959 LM		0.038				0.719	3.243	6.8	14.5	1959	06	05	Boyden	Hoffmeister
	1954 XA	V	0.038	0.068			0.509	1.046	3.9	19.0	1954	12	05	Palomar	Abell
(1566) Icarus	1949 MA		0.040				0.187	1.969	22.9	16.4	1949	06	27	Palomar	Baade
	1979 XB	V	0.040	0.057			0.649	3.876	24.9	19.0	1979	12	11	Siding Spring	Russell
(4953)	1990 MU		0.041				0.556	2.687	24.4	14.3	1990	06	23	Siding Spring	McNaught
	1989 JA		0.041	0.043			0.913	2.627	15.2	17.0	1989	05	01	Palomar	Helin
	1987 SF3		0.042				1.047	3.457	3.3	19.0	1987	09	26	Palomar	Shoemaker
(5189)	1990 UQ		0.044	0.062			0.810	2.292	3.6	17.5	1990	10	20	Siding Spring	McNaught
(4486) Mithra	1987 SB		0.045	0.052			0.742	3.655	3.0	15.4	1987	09	22	Smolyan	Elst, Shkodrov & Ivanova
	1993 BX3		0.046				1.003	1.787	2.8	20.5	1993	01	31	Siding Spring	McNaught

Object		M	N	Actual	q	Q	i	H	Year	Mo	Dy	Discovery	
(1620) Geographos	1951 RA	0.046	0.064		0.827	1.662	13.3	15.6	1951	09	14	Palomar	Wilson & Minkowski
	1993 DQ₁	0.047	0.050		1.038	3.042	10.0	17.0	1993	02	26	Kitt Peak	Spacewatch
	1991 RB	0.047	0.055		0.749	2.151	19.5	19.0	1991	09	04	Siding Spring	McNaught
(4015) Wilson-Harrington	1979 VA	0.048	0.048		0.997	4.285	2.8	16.0	1979	11	15	Palomar	Helin
	1989 DA	0.048			0.987	3.337	6.4	18.0	1989	02	27	Palomar	Phinney
	1991 DG	0.048	0.064		0.909	1.945	11.2	19.0	1991	02	20	Siding Spring	McNaught
	L 1991 XA	0.049			0.979	3.563	5.3	24.0	1991	12	03	Kitt Peak	Spacewatch
	L 1950 DA	0.053	0.074		0.838	2.530	12.1	15.9	1950	02	23	Lick	Wirtanen
	1991 VK	0.055	0.072		0.911	2.776	5.4	17.0	1991	11	01	Palomar	Helin & Lawrence
(1943) Anteros	1973 EC	0.056			1.064	1.796	8.7	15.8	1973	03	13	El Leoncito	Gibson
	1989 AZ	0.057			0.876	2.413	11.7	19.5	1989	01	08	Palomar	Shoemaker
(3122) Florence	1981 ET₃		0.057		1.021	2.516	22.2	14.2	1981	03	02	Siding Spring	UCAS
	1991 VE	0.058			0.299	1.482	7.2	19.0	1991	11	03	Palomar	Helin & Lawrence
(1685) Toro	1948 OA	0.063			0.771	1.963	9.4	14.2	1948	07	17	Lick	Wirtanen
	L 1991 JR		0.063		1.039	1.770	10.1	22.5	1991	05	08	Kitt Peak	Spacewatch
	1992 BF	0.064			0.662	1.154	7.3	19.0	1992	01	30	Palomar	Lawrence & Helin
	1992 SK	0.070	0.064		0.843	1.654	15.3	17.5	1992	09	24	Palomar	Helin & Alu
(5143) Heracles	1991 OA	0.065		0.015 2170	1.035	3.980	5.5	17.5	1991	07	16	Palomar	Holt
	1991 VL	0.065			0.419	3.249	9.2	13.9	1991	11	07	Palomar	Shoemaker
(3752) Camillo	1985 PA		0.067		0.986	1.841	55.5	15.5	1985	08	15	Caussols	Helin, Barucci & Heudier
(2063) Bacchus	1977 HB	0.072			0.701	1.454	9.4	16.4	1977	04	24	Palomar	Kowal
	1984 KB	0.074			0.522	3.910	4.8	15.5	1984	05	27	Palomar	Shoemaker
(1917) Cuyo	1968 AA		0.074		1.063	3.235	24.0	13.9	1968	01	01	El Leoncito	Cesco & Samuel

GLOSSARY

GLOSSARY*

a	*see* semimajor axis.
ablation	removal of material by attrition, e.g., by passage through the atmosphere.
ABM	anti-ballistic missile.
absolute magnitude	the magnitude of an asteroid at zero phase angle and at unit heliocentric and geocentric distances.
accretion	process by which matter is assembled to form larger bodies such as stars, planets, and satellites.
achondrite	meteorite of nonsolar composition, also known as differentiated stony meteorite.
acritarchs	a diverse group of microscopic organisms that produce fossils composed of hollow, organic-walled structures.
agglutinate	small objects consisting of glass and fragments of minerals or rocks, welded together into an aggregate, produced by micro-meteorite impact into fine-grained unconsolidated regolith.
albedo, Bond	fraction of the total incident light reflected by a spherical body. Bolometric Bond albedo refers to reflectivity over all wavelengths.
albedo, geometric	ratio of planet brightness at zero phase angle to the brightness of a perfectly diffusing disk with the same position and apparent size as the planet. Bolometric geometric albedo refers to reflectivity over all wavelengths.
Amor asteroids	asteroids having perihelion distance $1.0167 \text{ AU} < q \leq 1.3 \text{AU}$.

* We have used some definitions from *Glossary of Astronomy and Astrophysics* by J. Hopkins (by permission of the University of Chicago Press, copyright 1980 by the University of Chicago), from *Astrophysical Quantities* by C. W. Allen (London: Athlone Press, 1973), and from *The Planetary System* by David Morrison and Tobias Owen (Reading, Mass.: Addison-Wesley Publishing Co., 1988). We also acknowledge definitions and helpful comments from various chapter authors.

anomalous tail
also called antitail and type III tail of a comet; a tail with apparent direction in the sky considerably divergent from the projected antisolar direction, in particular a tail that appears to point toward the Sun. Such a tail is essentially a thin sheet of solid particles confined to the orbital plane of the comet, outside the orbit and well behind the Sun-comet direction.

anorthosite
an igneous rock made up almost entirely of plagioclase feldspar.

anoxia
a environment where available oxygen is less than that required by most organisms.

aphelion
in the orbit of a solar system body, the most distant point from the Sun.

Apollo asteroids
asteroids having semimajor axis $a \geq 1.0$ AU, and perihelion distance $q \leq 1.0167$ AU.

archaeocyathids
an extinct group of cup-shaped, reef-forming organisms known only from the Cambrian Period of geologic time.

arming
the changing from a safe condition to that ready for firing, for example, a fuse. This may be done before flight or by some mechanism(s) during flight.

aspect
angle between the rotation axis of the body and the radius vector of the Earth.

astenosphere
the mobile region of a planet just below the lithosphere.

asteroid
one of a multitude of objects ranging in size from sub-km to about 1000 km, most of which lie between the orbits of Mars and Jupiter.

asteroid belt
a region of space lying between the orbits of Mars and Jupiter, where majority of the asteroids are found. *See also* mainbelt asteroids.

asteroids, nomen-
clature
see permanent designation *and* preliminary designation of asteroids.

astrobleme
an impact structure.

astrometry
precision measurement of position and/or velocity of an astronomical object.

astronautics
the science of locomotion outside the Earth's atmosphere, involving the problems of artificial Earth satellites and of interplanetary travel.

Aten asteroids
asteroids having semimajor axis $a < 1.0$ AU, and aphelion distance $Q > 0.9833$ AU.

AU
astronomical unit. The mean distance of the Earth from the Sun, equal to about 1.496×10^8 km.

$B(a,0)$	mean opposition magnitude; in the B band defined by $B(a,0) = B(1,0) + 5\log a(a - 1)$, where $B(1,0)$ is the old-style absolute magnitude and a is the semimajor axis in AU.
$B(1,0)$	old-style B-band absolute magnitude, now superseded by H. $H \approx B(1,0) - 1.0$ mag.
backscatter	scattering of radiation (or particles) through angles greater than $90°$ with respect to the original direction of motion.
ballistic missile	a missile which is not controlled or guided for the major part of its flight, thus resembling the flight of a shell from a gun.
barycenter	the center of mass of a system.
basalt	a dark, fine-grained, mafic igneous rock composed primarily of plagioclase and pyroxene.
baud	transmission rate of one bit per second.
B class	a subclass of the C asteroids, distinguished by higher albedos than the average C type.
benthic	life forms living on the bottom of the sea.
binary carbide nuclear fuel	a solid solution of uranium and zirconium ceramic carbides having an operating temperature capability in excess of 2900 K.
biomass	the total mass of all the organisms of a given type or in a given area; global biomass represents the total mass of terrestrial organisms.
bioturbation	disturbance of sedimentary strata by burrowing, plowing, digging, and eating organisms like worms, echinoids.
bi-propellant rocket	a rocket whose basic components are (1) combustion chamber/nozzle assembly including a propellant injector, a combustion chamber and an exhaust nozzle; (2) propellant control valves used for starting and stopping the propellant flow or varying the flow through valves to give a variable thrust; (3) propellant feed system consisting of storage tanks and a pressurizing system for two separate propellants.
blackbody	an idealized body which absorbs all radiation of all wavelengths incident on it. The radiation emitted by a blackbody is a function of temperature only. Because it is a perfect absorber, it is also a perfect emitter.
bolometric	including radiation over all wavelengths.
breccia	a clastic rock composed of angular, broken rock fragments that are embedded into a finer-grained matrix.
brecciation	breakage of a rock into smaller fragments.

brightness temperature
temperature a blackbody would have to have in order to emit radiation of the observed intensity at a given wavelength.

bulk modulus
a material property; an elastic modulus is applied to a body having uniform stress distributed over the whole of its surface and defined by the ratio of the intensity of stress to the fractional change in the volume of the body.

burn life
useful operational lifetime of a nuclear reactor determined by material erosion or by constraints of uranium-fuel consumption.

c
speed of light defined to be exactly 2.99792458×10^8 m s^{-1}.

calcareous
material containing or composed largely of calcium carbonate ($CaCO_3$).

carbonaceous chondrite
a chondritic meteorite, containing more than 0.2%, by weight, of carbon. Most chondrites are highly oxidized and have nearly solar composition for all but the most volatile elements. It is the most primitive (least processed) type of meteorite.

catastrophic disruption
term applied to collisional breakup when the mass of the largest post-impact fragment is $\leq 50\%$ of the original target mass.

CCD
charge-coupled device. Solid-state detector for low-light level imaging (*see* pixel).

C class
common asteroid type in the outer part of the main belt; typically have flat spectra longward of 0.4 μm and are presumably similar in surface composition to some carbonaceous chondrites. The relative strength of a ultraviolet absorption feature may be correlated with the presence of water of hydration. B, F and G are subclasses of the C class.

CERMET
a fast neutron reactor concept which uses a ceramic/refractory metal fuel element design.

chalcophile
applied to elements with a strong affinity for sulfur, which concentrate in sulfides in the Earth's mantle rather than the core.

chaotic orbit
orbit characterized by at least one Lyapunov characteristic exponent being strictly positive.

Charpy impact specimen
rectangular (metallic) prism, 55 mm by 10 mm square; center notch is 5 mm deep.

Charpy impact technique
impact with high strain.

chondrite
originally defined as a meteorite that contained chondrules; now also implies a chemical composition similar to that of the Sun, for all but the most volatile elements.

chondrule | approximately spherical assemblages, characteristic of most chondrites, that existed independently prior to incorporation in the meteorite. Chondrules show evidence for partial or complete melting.

clast | a rock fragment produced by mechanical weathering of a larger rock and included in another rock.

clastic | sediments consisting of grains.

clathrate | a structure formed by the systematic inclusion of certain molecules in cavities within a crystal lattice.

clathrate hydrate | a type of solid molecular compound in which one component, such as CH_4, is trapped by van der Waals forces in cavities in the lattice structure of another compound, such as H_2O ice. Such a structure has been suggested for the icy material in comets to explain the nearly simultaneous appearance in the spectra of emission bands of such radicals as CN, CO, and OH as comets approach the Sun.

color index | the difference in magnitudes between any two spectral regions. Color index is always defined as the short-wavelength magnitude minus the long-wavelength magnitude. In the UBV system, the color index for an A0 star is defined as $B - V = U - B = 0$; it is negative for hotter stars and positive for cooler ones.

coma | the usually spherical region of diffuse gas, $\sim$150,000 km in diameter (variations can be very large), which surrounds the nucleus of a comet. Together, the coma and the nucleus form the comet's head.

comet | a diffuse body of gas and solid particles (such as CN, C_2, NH_3, and OH), which orbits the Sun. The orbit is highly elliptical or even parabolic (average perihelion distance less than 1 AU; average aphelion distance, roughly 10^4 AU). Comets are unstable bodies with masses on the order of 10^{18} g whose average lifetime is about 100 perihelion passages. Periodic comets comprise 20% of the known comets.

comet nucleus | the solid macroscopic structure which is the ultimate source of the material observed in the cometary coma, tails and trail. The physical nature of the typical nucleus is yet unclear; models include the icy-conglomerate ("dirty snowball"), the "rubble pile" variant, the "fluffy particle aggregation," and the "muddy iceball" concept.

comets, nomenclature	when a newly discovered comet is confirmed, the IAU Central Bureau for Astronomical Telegrams assigns a provisional designation consisting of the year of discovery followed by a lowercase letter in order of discovery for that year. The discoverer's name is attached to the designation. If a reliable orbit is later established, the comet is given a permanent designation consisting of the year of perihelion passage followed by a Roman numeral in order of perihelion passage. Comets with orbital periods > 200 yr are called "long-period"; those with periods < 200 yr are indicated by P/ preceding the name of the comet; $P < 20$ yr called "short period" (Jupiter family) and those of $20 < P < 200$ yr "intermediate period" (Halley family).
comets, tail	type I = ion tail; type II = a tail composed of small solid particles of dust or ice; type III = anomalous tail (*see* separate entry for definition).
commensurate orbits	a term applied to two bodies orbiting around a common barycenter when periods are in low integer ratio.
Commonwealth of Independent States (CIS) reactor	a thermal neutron reactor design developed in the former Soviet Union which uses a low-temperature moderator and high-temperature uranium carbide fuel.
composite nuclear fuel	uranium carbide and zirconium carbide veins in graphite having an operating temperature capability up to 2800 K.
conjunction	*see* elongation.
conodont	tiny, phosphatic, tooth-shaped fossils that occur in Cambrian to Triassic sediments, and are useful for correlation of rock layers.
continuous wave radar	a radar system in which a transmitter sends out a continuous flow of radio frequency energy to the target, which reflects a small fraction to a receiving antenna. The reflected wave is typically distinguishable from the transmitted signal by a slight change in frequency. The C.W. method enables moving targets to be distinguished against a stationary reflecting background and is more conservative of bandwidth than pulse radar.
contour current	currents in the ocean which follow contour lines along the continental rise.
Coordinated Universal Time (UTC)	the time scale available from broadcast time signals; UTC is maintained within +0.90 second of UT (*see* Universal Time) by the introduction of one-second steps.
cosmic rays	atomic nuclei (mostly protons) that are observed to strike the Earth's atmosphere with high energies.

counterglow (Gegenschein)	a very faint glow (about one degree across) which can occasionally be seen in a part of the sky opposite the Sun. Its optical spectrum indicates that it is sunlight reflected by dust.
crust	the outermost, highly differentiated, solid layer of a planet or satellite, mostly consisting of crystalline rock or ice.
cyanophyte	cyanobacteria; members of a group of primitive one-celled organisms that carry out photosynthesis in the presence of light, with concomitant production of oxygen.
D class	a rare asteroid type in the main belt, becoming increasingly dominant beyond the 2:1 Jovian resonance. Its spectrum is neutral to slightly reddish shortward of 0.5 μm, very red longward of 0.55 μm, and for some objects the spectrum tends to flatten longward of 0.95 μm. Coloring may be due to kerogen-like materials.
declination	angular distance north or south of the celestial equator.
Deep Space Network (DSN)	NASA radiotelescopes in Australia, Spain and the U. S. for communication with spacecraft.
degassing	the release of volatiles from the interior of a planet, most commonly through volcanic activity.
delivery system	a mode of weapon delivery to an intended target, for example by a rocket.
ΔV	change in vehicle velocity.
deposition	the act of depositing something, as mass, momentum, or energy, into a body or material.
detonate	the action by which an explosive device is fused and explosive yield is created.
detritus feeding	feeding on dead organic matter rather than on living plants.
digenesis	the changes which take place in sediments at low temperatures and pressures after deposition.
diapir	structure in which lower strata pierce through upper layers.
differentiation (in a planet)	a process whereby the primordial substances are separated. Generally, metal sinks to the center to form a core, displacing the lighter silicates which form the crust plus mantle.
distal	an area far away from the source of sediment particles.
dm_p/dt	mass flow rate of propellant.
Doppler spectrum	spectrum of radar echo Doppler shifts due primarily to planet rotation.
drift scanning	CCD scanning.
dunite	an ultramafic rock composed of at least 90% olivine.

dynamic pressure	the kinetic energy per unit volume of the fluid, equal to $1/2\,pq^2$, where p is the density and q the resultant velocity.
e	*see* eccentricity.
Earth-crossing asteroids (and comets) (ECA)	objects whose orbits can intersect the capture cross section of the Earth as a result of long-range perturbations by the planets.
E class	a rare asteroid type with featureless 0.3 to 1.1 μm spectra (identical to M and P classes) but distinguished by high albedo. Surface composition may be similar to enstatite chondrites.
eccentricity of an elliptical orbit	the amount by which the orbit deviates from a circle centered at the attracting body: $e = (Q - q)/(Q + q)$, where Q is the aphelion distance and q the perihelion distance.
ecliptic	plane of the Earth's orbit.
ejecta	materials ejected from a crater either by an impact or by the action of volcanism.
ejecta blanket	the deposit surrounding an impact crater composed of material ejected from the crater during its formation.
electric propulsion	propulsion concepts which use electrical energy to accelerate propellant particles to high-exhaust velocities via thermal, electrostatic or electromagnetic means.
elongation	the angle planet-Earth-Sun. Eastern elongations appear east of the Sun in the evening; western elongations, west of the Sun in the morning. An elongation of 0° is called conjunction; 180° is called opposition; and 90° is called quadrature.
emissivity	ratio of the radiation emitted by a body to that emitted by a blackbody at the same temperature.
enabler	NERVA Derived Reactor (NDR).
endogenic	originating within a planetary or planetesimal object.
energy coupling	the physical process by which incoming energy interacts with target material to create a resultant useful response in the material, e.g., material blow-off, momentum deposition, or target fragmentation.
enstatite chondrite	collective name for the EH and EL classes of chondritic meteorite, highly reduced chondrites with Mg/Si ratio around 0.83.
eon	10^9 yr; *see* Gyr.
EOS	the equation-of-state of a substance describes the dependence of its thermodynamic properties, such as pressure and energy, upon density and temperature.

ephemeris (*pl.*, ephemerides)	list of computed positions occupied by a celestial body at specified times.
Ephemerides of Minor Planets (EMP)	published yearly by the Russian Academy of Sciences, Institute of Theoretical Astronomy, St. Petersburg, Russia.
epicontinental sea	shallow seaway covering a continent, common at many times in the geologic past, but the only modern example is Hudson Bay.
escape velocity	the velocity required to escape entirely from the gravitational field of an object; also the minimum impact velocity for any body arriving from a very great distance.
eucrite	class of achondritic meteorite consisting of Ca-pyroxene and plagioclase; the class is believed to have originated on the asteroid Vesta.
eV	electron volt = 1.602×10^{-12} erg.
EVA	extra vehicular activity, when a person is outside the spacecraft.
evaporites	sediments formed by precipitation of salts from natural brines during evaporation in isolated arms of the sea.
exogenic	originating externally to a planetary or planetesimal object.
explosive yield	a measure of an explosive's output, usually in terms of blast overpressure.
fall	a meteorite that was observed while falling; usually recovered soon after the fall and relatively free of terrestrial contamination and weathering effects.
family	statistically significant cluster of asteroids in proper element space which may share a common origin, perhaps by the collisional disruption of a larger parent body (*see* Hirayama family).
F class	subclass of C asteroids, distinguished by a weak to nonexistent ultraviolet absorption feature.
feldspars	common aluminous silicate minerals in meteorites and other rocks. Plagioclase feldspars are members of a solid solution series which varies continuously from sodium-rich to calcium-rich compositions.
find	meteorite that was not seen to fall, but was found and recognized subsequently.
fireball	*see* meteor.
fluence	the flux of mass, momentum, or energy impinging upon a body or material.
flyby	a space mission with only a brief transit in the vicinity of the target celestial body.

foraminifers   unicellular protists, producing a skeleton consisting of a row of chambers, arranged in a variety of ways; the skeleton is usually calcareous.

Fourier analysis   a method whereby any periodic function is broken down into a convergent trigonometric series of sine and cosine functions.

fuel (nuclear)   fissionable compounds used to generate heat in nuclear reactors and rockets, and which remain in the rocket. Distinct from "rocket fuel," which combines with oxidizers to release energy, and exhausts out of the rocket as propellant.

G V-band slope parameter in the H,G magnitude system. $G \approx 0$ pertains to steep phase curves, such as those of low-albedo asteroids; $G \approx 1$ to shallow phase curves, such as those of icy satellites.

Galileo   spacecraft with a mission to observe the Galilean satellites, the four large satellites of Jupiter.

$g_\circ$ gravitational acceleration constant = 9.80665 m s^{-2}.

gardening   reworking and overturning of a regolith, principally by micrometeoroid bombardment.

G class   a subclass of the C asteroids, distinguished by a strong ultraviolet absorption feature.

Gegenschein   *see* counterglow.

genus (pl., genera)   a taxonomic unit of organisms, referring to groups of related species.

geocentric   Earth-centered.

geosynchronous orbit (GEO)   orbital period of exactly one sidereal day, or geostationary orbit (low-inclination geosynchronous orbit).

Giotto   spacecraft sent by European Space Agency to flyby P/Halley in 1986 and P/Grigg-Skjellerup in 1992.

Global Positioning System (GPS)   a system of many navigational satellites in 12-hr orbits about the Earth operated by the U. S. Air Force.

goyazite   Ba/Sr bearing alumino-phosphate mineral.

granite   an igneous rock associated primarily with the Earth's continental crust, composed chiefly of quartz and alkali feldspar.

graptolite   a class of extinct (Cambrian-Carboniferous) colonial marine organisms that secreted a chitinous exoskeleton with characteristic growth bands and lines.

gravitational constant, G the constant of proportionality in the attraction between two unit masses a unit distance apart.

gravity scaling	scaling of explosion or impact craters where only gravity limits the size of the crater.
greenhouse effect	the effect of heat retention in the lower atmosphere as a result of absorption and re-radiation of long wave ($>4\,\mu$m) terrestrial radiation by clouds and gases (e.g., water vapor, carbon dioxide, methane).
GTO	geosynchronous transfer orbit.
Gyr	gigayear = 10^9 yr.
H	absolute magnitude in the H,G magnitude system. H pertains to the V band unless subscripted otherwise (e.g., H_B). It is the time-averaged magnitude of an asteroid, calculated at zero phase angle and unit heliocentric and geocentric distances.
H(α)	reduced V band magnitude of an asteroid at phase angle α as calculated using the H,G magnitude system, where H is the absolute magnitude, G is a slope parameter.
half-life	the time required for disintegration of half of the radioactive atoms in a sample.
heat transfer	the mechanism of transferring heat from one medium to another by conduction or by other means, such as radiation and convection.
heavy bombard-ment	the period of time, beginning during planetary formation and apparently lasting until 3.8 Gyr ago, when the cratering rate was high throughout at least the inner solar system (*see* late-heavy bombardment).
height-of-burst (HOB)	the height above the ground where a weapon is fused, thereby creating explosive yield.
heliocentric	Sun-centered.
heterogeneous reactor design	a concept which employs separated and different materials for the moderator and fuel, each optimized for maximum performance.
Hill sphere	the approximately spherical region within which a planet, rather than the Sun, dominates the motion of particles.
Hirayama family	refers specifically to one of the clusters of asteroids first noted by K. Hirayama in the early 20th century.
homeothermy	metabolic regulation of an organism's body temperature.
Hugoniot stress	stress derived from an isentropic adiabatic curve which is a function of two parameters, the initial volume and the initial pressure.

homing (guidance)	passive homing is a system in which a missile homes onto a source of energy radiated by the target. If the missile receives reflections back from the target of its own transmission, it is called active homing.
hydrazine	a colorless liquid, N_2H_4, propellant with a high freezing point of $+2°C$. It is spontaneously ignitable with nitric acid and concentrated hydrogen peroxide; can also be used alone as a propellant gas.
hypergolic	propellants that ignite spontaneously upon contact with an oxidant.
I	moment of inertia, a tensor the subscripts of which refer to an axis or to the shape of the body (e.g., S for sphere, B for binary); not to be confused with specific impulse, I_{sp}.
I_{sp}	*see* specific impluse.
i	inclination of an orbit. The angle between an asteroid's orbit and the plane of the ecliptic (or between a satellite's orbit and the planet's equatorial plane).
IAU	International Astronomical Union.
IAUC	International Astronomical Union Circulars.
igneous	melting and subsequent solidification of a rock.
impact horizon	a level in a rock stratigraphic column with evidence (e.g., iridium spike, shock metamorphosed quartz) that the sediments were associated with an impact.
impact melt	target material that was melted by the heat generated by an impact.
impact strength	energy density (specific energy times target density) required to produce a barely catastrophic outcome.
inclusions	aggregates of mineral grains that existed independently prior to incorporation in the meteorite.
inertial guidance system	control of a missile by a dead-reckoning mechanism, which calculates the actual flight path with the aid of specially sensitive gyroscopes and/or accelerometers (i.e., an inertial measurement unit) and makes the required corrections to the flight path by the movement of trajectory controls.
inertial space	a fixed reference point in space used in dynamics or trajectory models. Usually the Earth or Sun is used.

infrared	that part of the electromagnetic spectrum that lies beyond the red, having wavelengths from about 0.75 nm to a few millimeters (about 10^{11} to 10^{14} Hz). Infrared radiation can be produced by atomic transitions, or by vibrational (near-infrared) and rotational (far-infrared) transitions in molecules.
inoceramids	large clams, like rudists, restricted to the Cretaceous and Jurassic Periods.
intercept	the process of "hitting" the target in the terminal stage of missile flight.
interceptor	a missile which homes in on a target either to impact it or to fly within a given distance of it.
International Ultraviolet Explorer (IUE)	an Earth-orbiting observatory.
interstellar grains	small solid particles that exist in interstellar space.
ion rocket propulsion	a form of propulsion in which ionized gas is accelerated to high velocities in an electric field.
IRAS	Infrared Astronomical Satellite.
iron meteorite	a meteorite composed primarily of metallic iron and nickel and thought to represent material from the core of a differentiated parent body.
isomer	one of a number of molecules that all have the same elemental composition but which differ from each other in structure.
isotope	any of two or more forms of the same elements whose atoms all have the same number of protons but different numbers of neutrons.
Jacobi ellipsoid	a triaxial figure assumed by a rapidly rotating body of low strength with its specific angular momentum exceeding a critical value. Its shape is determined by self-gravity and centrifugal force, and depends only on the body's density and rotation rate.
K	Kelvin, temperature unit; zero point at about $-273°$ Centigrade (Celsius, C).
Kepler velocity	orbital velocity of a gravitationally bound object around the central object, i.e., the velocity that leads to a centrifugal force exactly balancing the gravitational attraction between the two objects.

Kepler's laws	(1) A planetary orbit is an ellipse with the Sun at one focus. (2) (Law of Areas) The radius vector of a planetary orbit sweeps out equal areas in equal times. (3) (Harmonic Law) The square of the period is proportional to the cube of the distance.
kerogen	bituminous material.
kiloton (Kt)	energy equivalent to 1000 tons of TNT (4.185×10^{12} Joules).
kinetic energy	a state variable of a body equal to 1/2 times its mass times its velocity squared.
Kirkwood gaps	regions in the asteroid zone which have been cleared of asteroids by the perturbing effects of Jupiter; named for the American astronomer, Daniel Kirkwood, who first noted them in 1866.
klbf	a kilopound force, 1000 pounds of force.
km	kilometer = 10^3 m = 10^5 cm.
kN	a kilonewton, 1000 Newtons of force.
K/T event	the major break in the history of life on Earth (a mass extinction) that occurred 65 Myr ago, between the Cretaceous and Tertiary periods, apparently due to the impact of an asteroidal object. The K was chosen by the International Commission of Stratigraphy because in the hierarchy of the periods in a stratigraphic time scale, three C's existed as initials: Cambrian, Carboniferous and Cretaceous. The K is from *Kreide* , German for chalk. Also, the word Cretaceous comes from the island of Crete, *Kreta* in German. T is for *Tertiar* –Tertiary.
Kuiper belt	a ring of 10^8 to 10^{10} remnant icy planetesimals beyond the orbit of Neptune.
lag-deposit	coarse-grained deposit in which the fine components are winnowed out by currents.
Lagrangian points, L	the five equilibrium points in the restricted three-body problem. Two of the Lagrange points (L_4 preceding, and L_5 following Jupiter in its orbit) are located at the vertices of equilateral triangles formed by the two primaries (e.g., Sun and Jupiter) and are stable; the other three are unstable and lie on the line connecting the two primaries (L_3 opposite Jupiter, L_2 behind, and L_1 in front of Jupiter as seen from the Sun) (*see* Trojans).
laser	light amplification by stimulated emission of radiation. A device capable of generating high-intensity coherent electromagnetic radiation.
late-heavy bombardment	a period of time from about 4.2 to 3.8×10^9 yr ago when most of the basins and other craters were formed on the Moon and terrestrial planets.

libration	a small oscillation around an equilibrium configuration, such as the angular change in the face that a synchronously rotating satellite presents towards the focus of its orbit.
lightcurve	brightness values plotted as a function of time. Note that this plot does not necessarily have to show variability.
lightcurve amplitude	peak-to-peak value in magnitudes of a lightcurve showing variability.
lithosphere	the stiff upper layer of a planetary body, including the crust and part of the upper mantle, lying above the weaker asthenosphere.
long-period comet	a comet with an almost parabolic orbit and a period of revolution round the Sun exceeding 200 yr. Some have orbital periods of millions of years.
low Earth orbit (LEO)	satellite orbit of altitude about 200 km, just high enough to avoid imminent orbital decay and re-entry due to atmospheric drag.
low-Z material	a material of low atomic number; a material characterization parameter used to qualitatively characterize its response to nuclear irradiation.
Lyapunov characteristic exponents	numbers which indicate how fast nearby orbits diverge and thus the degree of unpredictability of such orbits.
magnitude	an arbitrary number, measured on a logarithmic scale, used to indicate the brightness of an object. Two stars differing by 2.5 magnitudes differ in luminosity by a factor of 10 according to $m_1 - m_2 = 2.5 \log (l_2/l_1)$. One magnitude difference is the fifth root of 100, or a factor of 2.512. The brighter the star, the lower the numerical value of the magnitude and very bright objects have negative magnitudes. The star Vega (α Lyrae) is defined to be magnitude zero in the UBV system, which is identified with specific ultraviolet, blue and visual light filters.
mainbelt asteroids	occupy the main asteroid belt between the orbits of Mars and Jupiter.
mantle	interior zone of a planet or satellite below the crust and above the core.
mare (pl. maria)	an area on the Moon or Mars that appears darker and smoother than its surroundings. Lunar maria are scattered basaltic flows.
mass driver	a device used for propulsion whereby material is thrust in one direction with a useful force gained in the opposite direction.
matrix	the fine-grained material that occupies the space in a rock, such as a meteorite, between the larger, well-characterized components such as chondrules or inclusions.

M class	common asteroid type in the main belt with featureless 0.3 to 1.1 μm spectra identical to E and P classes, but distinguished by moderate albedos. Presumed to have metallic (Ni-Fe) compositions, but with varying metal contents.
mean motion	average daily motion of an orbiting body, 360/P deg/day where P is the orbital period in days.
megaton (Mt)	energy equivalent of 10^6 tons of TNT (4.185×10^{15} Joules).
mesopause	the temperature inversion at about 80 km height in the terrestrial atmosphere, which separates the mesosphere below from the thermosphere above; commonly marked by clouds composed of ice crystals or meteoric dust.
metamorphic rock	any rock produced by the physical and chemical alteration (without melting) of another rock that has been subjected to high temperature and pressure.
metamorphism	solid-state modification of a rock, e.g., recrystallization, caused by elevated temperature (and possibly pressure).
meteor	the light and ionization produced by a solid particle from space experiencing frictional heating when entering a planetary atmosphere; also used for the glowing phenomenon itself. If particularly bright, it is described as a fireball.
meteorite	a natural object of extraterrestrial origin that survives passage through the atmosphere.
meteoroid	a natural solid object moving in interplanetary space, of a size smaller than about 0.1 km and much larger than a molecule.
meteor shower	many meteors appearing to radiate from a common point in the sky.
microkrystites	term to distinguish clinopyroxene crystal-bearing impact glass droplets from entirely glassy microtektites.
microwave	an electromagnetic wave (in the radio region just beyond the infrared) with a wavelength from about 1 mm to 30 cm ($\sim 10^9$ to 10^{11} Hz).
microspherule/ microtektite	microscopic spheres (also teardrops, dumb-bells), which are commonly attributed to meteorite impacts, found distributed in sediments. Microspherule refers to spherical objects of various compositions; microtektite refers to glassy spheres that represent impact-melt material that was thrown into space and later landed and solidified.
missile	a system of weapon delivery involving the use of rocket motors or powerful engines; these could be either guided or unguided high-flying ballistic missiles.

mitigation	the making of comets and asteroids less hostile.
moderator	a low atomic number material (e.g., hydrogen, lithium or water) used in reactor design to collisionally reduce neutron energy to levels which improve the fission probability of the uranium fuel atoms.
momentum	a state variable of a body equal to the product of the mass of the body and its velocity.
m_p	mass of propellant; in an NTR, the propellant is used both as the reactor coolant and the rocket exhaust mass.
MPC	*Minor Planet Circulars*
msec	millisecond = 10^{-3}s.
Myr	10^6 yr.
nanoplankton	plankton that are too small to be caught by standard nets. Commonly refers to *Coccolithophores* , a family of one-celled, marine protists that are covered by calcareous plates (coccoliths).
nektonic	an adjective applied to free-swimming organisms in aquatic environments.
NEO	near-Earth object; an asteroid or comet on a trajectory which passes close to Earth.
NERVA (Nuclear Engine for Rocket Vehicle Application)	a homogeneous, thermal neutron NTR concept employing uranium fuel in a graphite moderator. NERVA was developed and tested during the 1960s and 1970s by NASA and the Atomic Energy Commission.
NERVA-Derived Reactor (NDR)	a state-of-the-art embodiment of the NERVA concept utilizing composite fuel, improved materials and improved rocket components. Also called an enabler.
nm	nanometer = 10^{-9} m.
nodes	the points at which an object's orbit crosses the plane of the ecliptic. The longitude of the ascending node is one of the six orbital elements and measures the angle between the ascending node and vernal equinox, measured in the plane of the ecliptic.
nongravitational effects	systematic effects remaining in the motion of comets after all gravitational perturbations by attractions from the principal planets have been allowed for. Such effects are generally associated with a jet effect on the comet's motion caused by asymmetric loss of volatiles and dust into the coma as the rotating nucleus is heated by sunlight.

nuclear device or weapon	a collection of parts and components capable of giving a nuclear yield, usually consisting of a nuclear explosives package, arming, fusing, and firing components and electronics, in a case to hold it all together.
nuclear electric propulsion (NEP)	an electric propulsion system using a nuclear reactor power source for generating electrical energy, which is then used to accelerate a propellant.
nuclear explosive yields	1 Kton = 4 x 10^{19} erg.
nuclear fuel	the fissionable material that remains in a nuclear rocket's reactor to generate power; it is different from "rocket fuel," which combines with an oxidizer and exhausts from chemical rockets.
nuclear propulsion	a rocket system capable of high thrust based on a working fluid exhausting from a high-temperature nuclear reactor.
(NTR) nuclear thermal rocket	a rocket concept which uses a nuclear fission reactor as a direct source of thermal power for propellant heating and thrust generation.
nucleus (of a comet)	a small body composed of volatile ices (primarily water), silicate dust, and hydrocarbons, orbiting the Sun. When it approaches the Sun, ices near the nucleus surface sublimate and the evolving gases carry entrained dust particles with them, producing an extended atmosphere, or "coma."
obliquity	the angle between a planet's axis of rotation and the pole of its orbit.
Ω	longitude of ascending node; the angle between some line in the reference plane (usually the direction to the vernal equinox) and the point where the body crosses the reference plane moving south to north.
ω	argument of perihelion or periapse for a planet or satellite. Angular distance (measured in the plane of a body's orbit) in the direction of motion from the ascending node to the perihelion point.
$\tilde{\omega}$	longitude of perihelion or periapse for a planet or satellite = $\Omega + \omega$.
Oort cloud	a cloud of comets having semimajor axes $\geq$20,000 AU inferred by J. H. Oort in his empirical study of the orbits of long-period comets. Comets in this shell can be sufficiently perturbed by passing stars or giant molecular clouds so that a fraction of them acquire orbits that take them within the orbits of Saturn and Jupiter.
opposition	*see* elongation.

opposition effect	an enhancement in the brightness of an object when observed at phase angles $<7°$, in excess of that predicted by a linear extrapolation of the brightness-phase relation from larger phase angles.
orbital elements	six quantities that fully describe an orbit; along with time, they specify the position of an orbiting body along its path. a, e, i, Ω, ω, and T. See seperate entry for each element.
ordinary chondrite	collective name for the most common variety of chondritic meteorite, subdivided into H, L and LL groups on the basis of Fe content and distribution.
osculating orbit	the path that an asteroid would follow if it suddenly became subject only to the inverse-square attraction of the Sun or other central body. In practice, secondary bodies such as Jupiter produce perturbations, thus osculating orbital elements are subject to variations over time.
P/	*see* comets, nomenclature.
p	geometric albedo. The ratio of the brightness of an asteroid to that of a perfectly scattering screen of the same cross-sectional area and in the same place, both being illuminated and viewed normally.
paired falls	meteorite specimens originally recovered some distance apart and hence given separate names, but later recognized as fragments of a single parent mass, on the basis of classification, cosmic ray or gas-retention age, texture, or other diagnostic features.
pallasite	class of stony-iron meteorites in which the Fe-Ni metal forms a continuous framework enclosing nodules of the silicate olivine.
particle bed reactor	a compact, high-power density, thermal neutron NTR concept with a high heat transfer capability achieved by directly flowing coolant over a bed of small fuel spheres.
payload	a scientific package or a warhead and fuse contained at the front end of a missile; the working package delivered to some target or location.
payload mass fraction	the percentage of a spacecraft's original pre-departure mass allotted to useful payload.
P class	a common asteroid type in the outer main belt with a heliocentric distribution that peaks near the 3:2 Jovian resonance. Their spectra are featureless from 0.3 to 1.1 μm (identical to E and M classes), but the class is distinguishable by low albedos.
Peierls stress	spatial derivative of the lattice friction force on a dislocation divided by the length of the slip.

pelagic biota	life forms which float on ocean surfaces.
perihelion	point of least heliocentric separation for a body in an eccentric orbit.
perihelion distance	in the orbit of a solar system body, the closest point to the Sun.
permanent designation	the numbers and names, beginning with 1 Ceres, given to asteroids for which orbits are accurately determined.
perturbation	a force on an object other than the inverse square gravitational attraction of the central object about which it is orbiting.
Phanerozoic Era	the period in history of the Earth, from 500 Myr ago to the present, for which there is a well-developed record of shelly fossils.
phase angle	α, the solar phase angle; the angle subtended at the center of the planet by the vector directions to the Sun and observer.
phase curve	a plot showing the brightness (reduced to common heliocentric and observer distances) of a planet, satellite, or asteroid vs phase angle.
phase integral	the relationship between the geometric and Bond albedos of a body; $A = pq$, where A is the bolometric Bond albedo, p is the bolometric geometric albedo and q is the phase integral.
photometry	the measurement of light intensities.
phytoplankton	plankton that carry out photosynthesis and are the basis of aquatic food chains.
pixel	picture element. Electronic images are composed of pixels arranged in rows and columns. The read-out of a CCD is from row to row to the read-out end-register.
plagioclase	a mineral group, formula $(Na,Ca)Al(Si,Al)Si_2O_8$; a solid solution series from $NaAlSi_3O_8$ (albite) to $CaAl_2Si_2O_8$ (anorthite), triclinic. It is one of the most common rock-forming minerals.
planetesimal	small rocky or icy body formed from the primordial solar nebula, ranging in size up to 10 km or more, out of which all larger solar system members are presumed to have accumulated.
planetocentric	centered on a planet. A satellite is in a planetocentric (as opposed to heliocentric) orbit. A planetocentric coordinate system is subtended at the planet's center.
plankton	minute aquatic organisms that drift with water movements; having in general no locomotive organs.

plasma	the completely ionized gas, the so-called fourth state of matter, in which the temperature is too high for atoms, as such, to exist and which consists of free electrons and free atomic nuclei.
PLS	Palomar-Leiden Survey of faint asteroids.
polarization	the process affecting radiation, especially light, such that the vibrations assume definite form. Polarization is defined as negative if the light reflected from a boundary is greater in the plane given by the scattering plane (source-boundary-observer) than in the perpendicular plane. If the light intensity is greater in the perpendicular direction, the polarization is called positive, and if it is the same in both perpendicular and parallel directions, the light is unpolarized.
porosity	the property of a material which allows a gas or liquid to flow through it, i.e., it has internally distributed material voids.
Poynting-Robertson effect	A dissipative force due to the anisotropic loss of momentum by a particle through re-radiation of solar energy. This causes aphelion collapse such that a circular orbit is soon attained; thereafter the particle spirals slowly in toward the Sun. Small particles (below 1 cm) are most severely affected because the force varies as the reciprocal of its size.
precession	a slow, periodic conical motion of the rotation axis of a spinning body.
preliminary designation of asteroids	the system for designating asteroids upon discovery and before their orbits are well-enough determined that they could be given a permanent number and name. The designations are supplied by the Minor Planet Center and consist of the year of discovery, and an uppercase letter to indicate the halfmonth in that year (A = Jan. 1–15, B = Jan. 16–31, ..., Y = Dec. 16–31, I being omitted).
primitive	in planetary science and meteoritics, a type of object or rock that is little changed chemically since its formation—hence representative of the conditions in the solar nebula at the time of formation of the solar system.
prograde motion	motion in the same direction as the prevailing direction of motion. As viewed from the north, prograde motion is counterclockwise, or west to east.
propellant	a fuel for a rocket engine, either liquid, solid, gaseous or any combination of these.
proper elements	orbital elements, a, e, i, from which the effects of planetary perturbations have been removed.
protoplanet	a precursor body from which a planet develops.

proximal   an area close to the source of sediment particles.

pterosaur   flying reptile.

pulse radar   a radar system in which rhythmic radio frequency pulses are directed toward a target by a transmitter, typically followed by a pause for a receiving antenna to record the reflected energy off the target.

pyroxenes   a group of common rock-forming silicates which have ratios of metal oxides (MgO, FeO or CaO) to SiO_2 of 1:1. These are called metasilicates. Pure members of this group are: $MgSiO_3$ (enstatite), $FeSiO_3$ (ferrosilite). Pure $CaSiO_3$ does not crystallize with the pyroxene structure. Ca does substitute for up to 50% of the Mg and Fe in the pyroxene structure.

q *see* perihelion distance.

Q *see* aphelion.

Q class   a rare asteroid classification denoted by moderate albedos and spectra with a strong absorption feature shortward of 0.7 μm and a modest absorption feature centered near 1 μm. The spectra are interpreted as being similar to those of ordinary chondrites. At present this asteroid type has only been identified for 1862 Apollo and a few other near-Earth asteroids.

quadrature   *see* elongation.

radar   the use of reflected or automatically re-transmitted radio waves to gain information about a distant object (*see* continuous wave radar and pulse radar).

radar cross section effective projected area of a radar target calculated on the assumption of a perfect, isotropic reflector. It is often expressed as a dimensionless quantity normalized by the true projected area of the target (planet or asteroid).

radioactivity   the spontaneous disintegration of certain heavy elements accompanied by the emission of (a) α particles which are helium atoms divested of their outer electrons; (b) β particles which are electrons; and (c) γ rays which are electromagnetic radiations like radio waves and X-rays.

Radio-Isotope Thermoelectric Generator (RTG)   typically providing spacecraft primary power in the 0.1 to 1.0 kW range. Not used in near-Earth missions because of potential re-entry and contamination hazard from the radioactive core element.

R class   a rare asteroid classification exemplified by 349 Dembowska and denoted by moderately high albedos and spectra with a strong absorption feature shortward of 0.7 μm and a fairly strong absorption feature centered near 1 μm.

reactor	the core of an atomic power plant.
refractory	term describing the high-temperature stability of an element or phase; the opposite of volatile.
regolith	layer of fragmental incoherent rocky debris as surface terrain; it is produced by meteoritic impact on the surfaces of the planets, satellites or asteroids.
regolith breccia	fragmental breccias containing some identifiable regolith component such as solar wind gas.
resonance	the enhanced response of any oscillating system to an external stimulus that has the same driving frequency as the natural frequency of the system; higher-order resonances occur when these frequencies are commensurable.
restricted three-body problem	two bodies assumed to be point masses and called primaries revolve around their center of mass under the influence of their mutual attraction. The problem is to determine the motion of a third body attracted by the previous two but not influencing their motion.
retrograde motion	the opposite of prograde motion.
right ascension	angular distance east of the vernal equinox, as measured on the celestial equator.
rms	root mean square; the square root of the mean square value of a set of numbers.
Roche limit	the minimum distance at which a fluid satellite influenced by its own gravitation and that of a central mass can be in mechanical equilibrium. For a satellite of zero tensile strength, and the same mean density as its primary, in a circular orbit around its primary, this critical distance is 2.44 times the radius of the primary.
rocket payload	the useful load in the final stage of a rocket.
Rosseland mean opacities	a coefficient of opacity which is a weighted inverse mean of the opacity over all frequencies. It is applied when the optical depth is very large and the radiative transport reduces to a diffusion process.
rubble pile	an asteroid that has experienced an impact with enough energy to shatter it, but not enough to disperse the fragments, which remain held together by their own gravity.
rudists	clams, pelecypods, related to the oysters, known exclusively from the Cretaceous and Jurassic Periods; capable of forming true reefs at the expense of corals.
saprophytic	living on dead or decaying organic matter.

Schmidt telescope — a type of telescope (more accurately, a large camera) which compensates for the coma produced by a spherical concave mirror through the use of a thin correcting lens placed at the opening of the telescope tube.

S class — a common asteroid class in the inner main belt with moderate albedos and reddish spectra shortward of 0.7 μm and moderate to nonexistent absorption features in the near-infrared; may be similar to stony-iron meteorites, but their meteoritical interpretation is uncertain.

secular perturbations — averaged perturbations experienced by planets and asteroids when the effects that depend upon the actual positions of the objects in their orbits are eliminated.

secular resonance — a situation in which the rate of the precession of the proper longitudes of the nodes or perihelion of an asteroid's orbit (called nodal or apsidal frequencies) is equal to one of the nodal or apsidal frequencies associated with the mutual secular perturbations of the major planets.

seiche — a standing wave, usually produced in enclosed basins such as deep lakes.

semimajor axis — in the orbit of a solar system body, its the mean distance from the Sun.

shock wave — the waves formed by a body or impactor when its speed through the air or other solid, gaseous, or liquid medium exceeds the speed of sound in that material.

short-period comet — a comet with an orbital period of <200 yr.

sidereal period — the time it takes for a planet or satellite to make one complete rotation or revolution relative to the stars.

siderophile — applied to elements with a weak affinity for oxygen and sulfur, and soluble in molten iron, which are found in iron meteorites and probably are concentrated in the Earth's core.

Signor-Lipps effect — the tendency of abrupt extinctions to appear gradual as a result of sampling effects, and the imperfection of the geologic record. Because the last occurrence of a species can occur before its extinction, but not after, the last recorded occurrence of a fossil species in the geologic record will most likely predate its extinction. This gives the impression that a series of extinctions took place over a period of time, when in fact they may have been simultaneous.

silicate — any of a wide range of rocks and minerals composed in part of silica (silicon and oxygen).

SNC meteorite	uncommon, but apparently genetically related meteorite types which are highly differentiated (Shergottites, Nahklites, and Chassignites). They may originate from Mars.
SNR	signal-to-noise ratio.
solar nebula	the gas-dust disk that surrounded the proto-sun. The term protoplanetary cloud is sometimes used as a synonym for the solar nebula.
solar sail	a large-area solar-light absorbing panel which converts a photon fluence into useful momentum as a means of propulsion in space.
solar wind	the energetic charged particles that flow radially outward from the solar corona, carrying mass and angular momentum away from the Sun.
Spaceguard	proposed international discovery program of NEOs.
Spacewatch	an ongoing CCD search for near-Earth asteroids and a radial velocity search for planets around other stars are conducted at the 0.9-m Newtonian and 1.8-m alt-az Spacewatch telescopes of the University of Arizona on Kitt Peak.
species	fundamental taxonomic unit of organisms. In living organisms, species are groups of individuals that look alike and can interbreed. In paleontology, species are defined according to morphological similarities.
specific energy	kinetic energy per unit mass.
specific impulse	I_{sp}, a performance parameter of a rocket propellant, equal to the thrust force divided by the mass flow rate.
standard thermal model	a simplistic thermal model for asteroids and other airless bodies that assumes the ideal situation of a nonrotating spherical body in instantaneous equilibrium with insolation. It also assumes that the subsolar and sub-Earth points on the body coincide. The thermal emission is a function only of subsolar distance.
stishovite	a high-density form of SiO_2, produced only at extremely high pressures and commonly associated with impact events.
stony-iron meteorite	a rare differentiated meteorite, composed of a mixture of silicates with metallic iron-nickel, thought to have originated near the core-mantle boundary of a differentiated parent body.
stratigraphy	the branch of the geological sciences concerned with the study of stratified rocks in time and space.
strength scaling	scaling of explosion or impact crater where only material strength limits size of craters.
stromatolite	a laminated, mounded structure built up over long periods of time by sediment.

Swift-Tuttle   comet P/Swift-Tuttle was once believed possibly to impact the Earth some time in the foreseeable future.

symbionts   life forms which live in symbiosis with a host organism, to the benefit of both.

synchronous   rotation of a body so that it always keeps the same face toward
rotation   another object; the situation where the periods of rotation and revolution of an orbiting body are equal.

synodic period   the time it takes for a planet or the Moon to return to the same position relative to the Sun, as seen from the Earth (or more generally for a planet or satellite as seen from some other planet).

tails of comets   *see* comets, tail.

taxonomic clas-   a system for categorizing similar observed properties of aster-
sifications of   oids, such as color or spectral properties and albedo.
asteroids

T thrust, the propulsive force imparted to a space object by ejecting propellant mass rearward from its propulsion system.

T class   low albedo asteroids having spectra with a moderate absorption feature shortward of $0.85\,\mu$m and generally flat in the near-infrared.

terminal gui-   guidance applied to a missile between the termination of the
dance   mid-course guidance and impact with, or detonation in close proximity of, the target.

ternary carbide   a solid solution of uranium, zirconium and niobium ceramic
nuclear fuel   carbides having an operating temperature capability in excess of 3000 K.

terrane   area with a coherent set or group of rocks.

terrestrial planets Mercury, Venus, Earth and Mars.

thermal conduc-   the proportionality constant that gives the amount of heat con-
tivity   ducted through a unit cross section in unit time under the influence of unit heat gradient.

thermal emission   the emission of electromagnetic radiation from a body due to its temperature and emissivity.

thermal inertia   a material parameter which indicates the rate at which a body's temperature responds to changing heat input. It is proportional to the square root of the product of thermal conductivity and volume heat capacity.

thermo-nuclear   atomic reaction involving the fusion of atoms; e.g., hydrogen atoms joining to form helium atoms.

thermophysical model	a thermal model for asteroids and other airless bodies that lies between the Standard Thermal and Isothermal Latitude Models. It takes into account the thermophysical properties of the body and may also include spin axis and direction as well as shape.
thruster	a device used for propulsion; working material is thrust outward in one direction with a useful force gained in the opposite direction.
time-delay integration (TDI)	CCD scanning.
Tisserand invariant	a pseudo-constant of the motion in the restricted three-body problem based on the Jacobi integral, used to identify returning short-period comets, even though their orbits may have been perturbed by a close Jupiter encounter.
Titius-Bode law	proposed by Titus in 1766 and advanced by Bode in 1772 for remembering the distances of the planets from the Sun. Take the series 0, 3, 6, 12, . . .; add 4 to each member of the series, and divide by 10. The resulting sequence 0.4, 0.7, 1.0, 1.6, . . . gives the approximate distance from the Sun (in AU) of Mercury, Venus, Earth, Mars, . . ., out to Uranus. The law fails for Neptune and beyond. Its value at 2.8 spurred a search for a "missing" planet between Mars and Jupiter, leading to discovery of the largest asteroids, i.e., Ceres and Juno.
Tracking and Data Relay Satellite System (TDRSS)	NASA's geosynchronous satellites in two orbital positions to provide almost complete global coverage of LEO satellites.
trajectory	the flight path of a projectile, missile, rocket or satellite.
trilobites	an extinct group of arthropods that somewhat resembled modern horseshoe crabs, except that their bodies were separated into three lobes front-to-back. Trilobites were very abundant in the Paleozoic Era of geologic time from the Cambrian Period to the end of the Permian Period.
Trojans	asteroids in orbits librating around two of the Lagrangian points, namely preceding and following Jupiter in its orbit, equidistant from the Sun and Jupiter (*see* Lagrangian points).
tsunami	a large sea wave.
Universal Time (UT)	a measure of time that conforms, in close approximation, to the mean diurnal motion of the Sun and serves as the basis of all civil timekeeping.

ultraviolet	the part of the electromagnetic spectrum that lies at wavelengths shorter than 0.35 nm. Ultraviolet absorption features in asteroid spectra result from charge transfer mechanisms.
use control	the action of controlling the use of a bomb through certain discriminating hardware and procedures which preclude unauthorized use of a weapon, while allowing authorized use.
v_e	propellant exhaust velocity.
Väisälä orbits	orbits computed on the assumption that the heliocentric radial velocity is zero, i.e., the object is taken to be at perihelion or aphelion (or in a circular orbit as a special case). Series of Väisälä orbits with different eccentricities can be derived very simply from only two observations; often useful in identifying further observations.
V class	a rare asteroid classification exemplified by 4 Vesta. Spectra are very red shortward of $0.5\,\mu$m, moderately red from 0.5 to $0.7\,\mu$m, and show a strong near-infrared absorption feature centered around $0.95\,\mu$m. Surface composition may be similar to basaltic achondrites.
vernal equinox	the intersection of ecliptic and celestial equator where the Sun is moving from south to north.
vivipary	birth of young rather than production of an external egg.
volatile	an element that condenses from a gas or evaporates from a solid at a relatively low temperature.
X-rays	very high frequency, short wavelength electromagnetic waves given off by high-temperature sources or materials at certain discrete frequencies under controlled conditions.
zodiac	a belt around the sky that is 18° wide and centered on the ecliptic, within which are found the Moon and planets, in addition to the twelve zodiacal constellations the signs of the zodiac.
zodiacal light	a faint glow that extends away from the Sun in the ecliptic plane of the sky, visible to the naked eye in the western sky shortly after sunset or in the eastern sky shortly before sunrise. Its spectrum indicates it to be sunlight scattered by interplanetary dust. The zodiacal light contributes about a third of the total light in the sky on a moonless night.

ACKNOWLEDGMENTS

ACKNOWLEDGMENTS

The editors thank J. E. Frecker, who volunteered as one of the proofreaders of this book and M. Schuchardt and K. Swarthout for their invaluable assistance with graphics and photograph reductions. They also would also like to acknowledge those individuals who helped greatly with the production of this book: M. F. A'Hearn; C. Alcock; W. Arens-Fischer; J. K. Beatty; M. J. S. Belton; J. Benson; J. Bourgeois; L. Brookshaw; J. C. Browne; J. A. Burns; J. R. Clark; S. F. Cox; E. Crawley; S. C. Crow; J. D. Drummond; J. Erickson; P. Farinella; D. B. Fastovsky; T. Faÿ; U. Fink; M. J. Gaffey; F. Hörz; K. Housen; H. Igleseder; S. Isobe; W. Jian-min; Z. Jia-xiang; Ľ. Kresák; R. J. Lawrence; T. Lawrence; L. A. Lebofsky; C. Levasseur-Regourd; D. H. Levy; J. S. Lewis; C. Mader; A. J. Maury; M. C. McCracken; R. S. McMillan; J. Pike; D. A. Reynolds; E. Roemer; H. H. Schmitt; J. V. Scotti; P. M. Sforza; I. I. Shapiro; W. Sichao; A. Sokolsky; V. S. Solomatov; M. V. Sykes; C. B. Tarter; W. Tedeschi; J. Telles; S. Thompson; K.-V. Tran; W. Z. Wisniewski; S. Wyckoff; B. H. Zellner; and A. Zuppero.

The following authors wish to acknowledge specific funds involved in supporting the preparation of their chapters.

Bottke, W. F., Jr.: NASA Planetary Geology and Geophysics Program Grant NAGW-1029
Bowell, E.: NASA Grant NAGW-1470 *and* NAGW-3397
Chapman, C. R.: PSI Contribution No. 318
Grieve, R. A. F.: Geological Survey of Canada Grant 43592
Hartmann, W. K.: NASA Contract NASW-4726: PSI Contribution No. 309
Rampino, M. R.: NASA Grant NAGW-1697
Sagan, C.: NASA Grants NAGW-1870
Toon, O. B.: NASA Grants NAGW-2183 & NAG1-7726, *and* NSF Grant ATM-89-11836

INDEX

INDEX